21st International Conference on Few-Body Problems in Physics

AA004171

EPJ Web of Conferences
Volume 113 (2016)

Chicago, Illinois, USA
18 – 22 May 2015

Editors:

C. Elster
D. R. Phillips
C. D. Roberts

ISBN: 978-1-5108-2207-8

Printed from e-media with permission by:

Curran Associates, Inc.
57 Morehouse Lane
Red Hook, NY 12571

Some format issues inherent in the e-media version may also appear in this print version.

Printed by Curran Associates, Inc. (2016)

For additional information, please contact EDP Sciences – Web of Conferences at the address below.

EDP Sciences – Web of Conferences
17, Avenue du Hoggar
Parc d'Activité de Courtabœuf
BP 112
F-91944 Les Ulis Cedex A
France

Phone: +33 (0) 1 69 18 75 75
Fax: +33 (0) 1 69 28 84 91

contact-edps@webofconferences.org

Additional copies of this publication are available from:

Curran Associates, Inc.
57 Morehouse Lane
Red Hook, NY 12571 USA
Phone: 845-758-0400
Fax: 845-758-2633
Email: curran@proceedings.com
Web: www.proceedings.com

TABLE OF CONTENTS

EPJ Web of Conferences **113**, 00001 (2016)
DOI: 10.1051/epjconf/201611300001

21st International Conference on Few-Body Problems in Physics
MAY 18-22 2015 | CHICAGO

Preface

Few-Body 21 (FB21), the Twenty-first International Conference on Few-Body Problems in Physics, was held from 18-22 May 2015 in the West Loop of downtown Chicago, Illinois, USA. It was jointly organized by Ohio University and Argonne National Laboratory.

The study of few-body systems has a long and rich history. Traditionally the three-body system was a testing ground for two-body potentials. From the 1960s through to the 1990s few-body physics emerged as a field where precision comparisons between theory and experiment were carried out in order to tease out subtle details of the inter-particle dynamics.

Few-body systems are a key milepost on the road from two-body to many-body dynamics. They are the first place that genuine three-body forces can be observed, and their dynamics displays unique features, e.g. the Efimov effect and associated universal physics. The advent of tunable interactions in cold atomic gases has opened up the possibility to explore all these aspects of the few-body problem in a new realm.

Few-body physics now stretches from this domain of neV energies to investigations of hadron properties at GeV energies – and encompasses significant swaths of contemporary nuclear physics in between. The meetings in this series have marked our progress to this point, the first being held in London in 1959, with subsequent meetings in Brela (1967), Birmingham (1969), Budapest (1971), Los Angeles (1972), Quebec (1974), Dehli (1976), Graz (1978), Eugene (1980), Karlsruhe (1983), Sendai (1986), Vancouver (1989), Adelaide (1992), Williamsburg (1994), Groningen (1997), Taipei (2000), Durham (2003), Santos (2006), Bonn (2009), and Fukuoka (2012).

FB21 presented progress since the Fukuoka meeting on problems that can be understood in terms of a few effective degrees of freedom. This included experimental and theoretical investigations pertaining to atomic and molecular physics, the role of clustering in the structure and reactions of light nuclei, few-nucleon systems and their interactions, the physics of hadrons, and hypernuclei.

Over 240 participants from 29 countries around the world attended the conference. FB21 featured 31 invited plenary talks, 32 invited lead parallel session talks and 127 contributed talks, resulting in a total of 190 oral presentations. In addition, there were more than 30 poster presentations in a separate poster session on Tuesday afternoon. The week in Chicago provided a snapshot of our vigorous and diverse field.

A unique highlight of this Chicago meeting was the conference banquet, held in Bruce Wayne's Penthouse, aka the *Penthouse* at Wyndham Grand Chicago Riverfront Hotel. Fickle May weather saw 25°C at the conference opening turn into 5°C on banquet evening; but this did not deter 112 conference participants from enjoying sweeping views of the Chicago River and skyline from the open-air patio, followed by a three-course meal inside the Penthouse, whose floor-to-ceiling windows offered 360° views of Chicago at dusk, sunset, and lighting-up of the Windy City. Many of the diners went on afterwards to complete the evening with visits to some of Chicago's numerous music bars and clubs, featuring blues, jazz, rock and reggae.

The meeting was only possible because of generous sponsorship from Ohio University's Institute of Nuclear and Particle Physics, Argonne National Laboratory's Theory Group, Argonne National Laboratory's Physics Division, The Department of Physics and Astronomy at Ohio University, The National Superconducting Cyclotron Laboratory at Michigan State University, The Triangle Universities Nuclear Laboratory, Jefferson Laboratory, and the International Union of Pure and Applied Physics. The APS Topical Group on Few-body Systems sponsored a reception in conjunction with the Tuesday poster session.

The meeting was conducted in accordance with the principles passed by IUPAP's General Assembly in 2008. In particular, no bona fide scientist was excluded from participation on the grounds of national origin, nationality, or political considerations unrelated to science.

That the meeting went smoothly is attributable to the excellent work of Ms. Debra Beres, who managed myriad administrative details with calm efficiency. Ms. Beres was ably assisted by Ms. Tracey Stamcik. At Ohio University, Mr. Wayne Chiasson and Ms. Julie Goettge worked out a number of financial aspects, and assisted us with the construction of the conference packets. Ohio University graduate students Nick Compton, Linda Hlophe, Cody Parker, Andrea Richard, and Arbin Thapaliya acted as conference helpers throughout the week and did a very professional job.

The publication of these proceedings was enabled by the meeting's sponsors and additional financial support from Ohio University's Office of Research. We also thank the many scientists who refereed the contributions, and the Scientific Committee who helped co-ordinate that process. The contributions in this volume reflect the impressive breadth and excellent quality of the contributions to the meeting. That – combined with the strong presence of younger scientists in Chicago – augurs well for the next conference in this series, which will take place in Caen, France in July 2018.

Scientific Organizing Committee

Ani Aprahamian	University of Notre Dame
John Arrington	Argonne National Laboratory
Doerte Blume	Washington State University
Carl Brune	Ohio University
Cheng Chin	University of Chicago
Gail Dodge	Old Dominion University
Charlotte Elster (co-chair)	Ohio University
Benjamin F. Gibson	Los Alamos National Laboratory
Peter Mueller	Argonne National Laboratory
Daniel Phillips (co-chair)	Ohio University
Lucas Platter	University of Tennessee
Craig Roberts (co-chair)	Argonne National Laboratory
Michael Thoennessen	NSCL, Michigan State University
Robert Wiringa	Argonne National Laboratory

International Advisory Committee

Hans-Jürgen Arends (Germany)
Nir Barnea (Israel)
Vladimir Belyaev (Russia)
Doerte Blume (USA)
Jaume Carbonell (France)
Joe Carlson (USA)
Xiaolong Cui (China)
Jens Dilling (Canada)
Robert Edwards (USA)
Francesca Ferlaino (Austria)
Victor Flambaum (Australia)
Haiyan Gao (USA)
Benjamin F. Gibson (USA)
Ronald Gilman (USA)
Chris Greene (USA)
Bo Höistad (Sweden)
Calvin R. Howell (USA)
Randall Hulet (USA)
Aksel S. Jensen (Denmark)
Nasser Kalantar-Nayestanaki (Netherlands)
Cynthia Keppel (USA)
Alejandro Kievsky (Italy)
Gastão Krein (Brazil)
Terry Mart (Indonesia)
Indranil Mazumdar (India)
Judith McGovern (United Kingdom)
Ulf-G. Meißner (Germany)
Richard G. Milner (USA)
Yongseok Oh (South Korea)
Makato Oka (Japan)
Allena Opper (USA)
Giuseppina Orlandini (Italy)
Assumpta Parreño (Spain)
Teresa Peña (Portugal)
Michael Pennington (USA)
Sofia Quaglioni (USA)
Stephanie Reimann (Sweden)
Kenshi Sagara (Japan)
Hideyuki Sakai (Japan)
Martin Savage (USA)
Achim Schwenk (Germany)
Kimiko Sekeguchi (Japan)
Sofianos A. Sofianos (South Africa)
Michael Thoennessen (USA)
Peter Tandy (USA)
Ulrike Thoma (Germany)
Anthony W. Thomas (Australia)
Henryk Witała (Poland)
Hushan Xu (China)
Yupeng Yan (Thailand)
Bing-Song Zou (China)

FB21

21st International Conference
ON FEW BODY PROBLEMS IN PHYSICS

MAY 18 - 22 2015

Main Topics

- *Atomic and molecular physics including ultracold gases*
- *Cluster approaches to structure and reactions of light nuclei*
- *Few-nucleon systems and their interactions*
- *Hadron physics*
- *Strange and exotic matter including hypernuclei*

PLENARY SPEAKERS

Aldo Antognini
ETH ZÜRICH, SWITZERLAND

Sinya Aoki
UNIVERSITY OF KYOTO, JAPAN

Konstanze Boretzky
GSI DARMSTADT, GERMANY

Eric Braaten
THE OHIO STATE UNIVERSITY, USA

Ian Cloet
ARGONNE NATIONAL LAB, USA

Will Detmold
MASSACHUSETTS INSTITUTE OF TECHNOLOGY, USA

Jo Dudek
OLD DOMINION UNIVERSITY, USA

Nadia Fomin
UNIVERSITY OF TENNESSEE, USA

Martin Freer
UNIVERSITY OF BIRMINGHAM, UK

Johann Frenje
MASSACHUSETTS INSTITUTE OF TECHNOLOGY, USA

Haiyan Gao
DUKE UNIVERSITY, USA

Mario Gattobigio
UNIVERSITY OF NICE, FRANCE

Rudi Grimm
UNIVERSITY OF INNSBRUCK, AUSTRIA

Hans-Werner Hammer
TU DARMSTADT, GERMANY

Emiko Hiyama
RIKEN, JAPAN

Selim Jochim
UNIVERSITY OF HEIDELBERG, GERMANY

Kate Jones
UNIVERSITY OF TENNESSEE, USA

Daekyung Kang
LOS ALAMOS NATIONAL LAB, USA

Dean Lee
NORTH CAROLINA STATE UNIVERSITY, USA

Alessandro Lovato
ARGONNE NATIONAL LAB, USA

Victor Mokheev
JEFFERSON LAB, USA

Wilfred Nörtershäuser
TU DARMSTADT, GERMANY

Saori Pastore
UNIVERSITY OF SOUTH CAROLINA, USA

Mike Pennington
JEFFERSON LAB, USA

Sofia Quaglioni
LAWRENCE LIVERMORE NATIONAL LAB, USA

Cheng-Ping Shen
IHEP BEIJING, CHINA

Patricia Solvignon-Slifer
UNIVERSITY OF NEW HAMPSHIRE, USA

Artemis Spyrou
MICHIGAN STATE UNIVERSITY, USA

Shih-Kuang Tung
UNIVERSITY OF CHICAGO, USA

Hui Zhai
TSINGHUA UNIVERSITY, CHINA

Nikolai Zinner
UNIVERSITY OF AARHUS, DENMARK

SCIENTIFIC ORGANIZING COMMITTEE

Craig Roberts, Organizer
ARGONNE NATIONAL LABORATORY, USA

Charlotte Elster, Organizer
OHIO UNIVERSITY, USA

Daniel Phillips, Organizer
OHIO UNIVERSITY, USA

Ani Aprahamian
UNIVERSITY OF NOTRE DAME, USA

John Arrington
ARGONNE NATIONAL LABORATORY, USA

Doerte Blume
WASHINGTON STATE UNIVERSITY, USA

Carl Brune
OHIO UNIVERSITY, USA

Cheng Chin
UNIVERSITY OF CHICAGO, USA

Gail Dodge
OLD DOMINION UNIVERSITY, USA

Peter Mueller
ARGONNE NATIONAL LABORATORY, USA

Lucas Platter
UNIVERSITY OF TENNESSEE, USA

Michael Thoennessen
NSCL, MICHIGAN STATE UNIVERSITY, USA

Robert Wiringa
ARGONNE NATIONAL LABORATORY, USA

To find out more about FB21 VISIT: www.fewbody21.us OR E-MAIL: fewbody21@gmail.com

Argonne NATIONAL LABORATORY · NSCL · TUNL · OHIO UNIVERSITY · INSTITUTE OF NUCLEAR & PARTICLE PHYSICS INPP · Jefferson Lab Thomas Jefferson National Accelerator Facility

Attendee List for FEW BODY 21

May 18, 2015 Chicago, Illinois
Participants: 241

Bijaya Acharya
University of Tennessee
1413 Circle Drive
Knoxville, TN 37996
United States
acharyab@phy.ohiou.edu

Patrick Achenbach
Mainz University
J.J.-Becherweg 45
Mainz, 55099
Germany
patrick@kph.uni-mainz.de

Mahdi Ahmadian Shalchi
Instituto de Fısica Teorica/UNESP
R. Dr. Bento Teobaldo Ferraz, 271
Bairro: Barra-Funda 01.140-070
Sao Paulo
Brazil
mashalchi@gmail.com

Mohammad Ahmed
NCCU/TUNL
PO Box 90305
Durham, NC 27708
United States
mohammad.ahmed@duke.edu

Shunichi Ando
Sunmoon University
School of Mechanical & ICT
Convergence Engineering
Asan
Korea, Republic of
sando@sunmoon.ac.kr

Aldo Antognini
ETH
Otto-Stern-Weg 5
Zurich
Switzerland
aldo@phys.ethz.ch

Sinya Aoki
YITP, Kyoto University
Kitashirakawa Oiwakecho
Sakyo-ku
Kyoto, Kyoto 606-8502
Japan
saoki@yukawa.kyoto-u.ac.jp

Ani Aprahamian
University of Notre Dame
936 N Notre Dame Ave
South Bend, IN 46617
United States
aapraham@nd.edu

John Arrington
Argonne National Laboratory
9700 S. Cass Ave, Bldg #203
Argonne, IL 60439
United States
johna@anl.gov

Marlène Assié
IPN
15 rue Georges Clémenceau
Orsay
France
assie@ipnorsay.in2p3.fr

James Babb
ITAMP, Harvard-Smithsonian
60 Garden St., MS 14
Cambridge, MA 02138
United States
jbabb@cfa.harvard.edu

Sonia Bacca
TRIUMF
4004 Wesbrook Mall
Vancouver, BC V6T 2A3
Canada
bacca@triumf.ca

Andrew Bacher
Indiana University Bloomington
610 S Highland Ave
Bloomington, IN 47401
United States
bacher@indiana.edu

Zara Bagdasarian
Forschungszentrum Jülich
Wilhelm-Johnen-Straße
Jülich, 52428
Germany
z.bagdasarian@fz-juelich.de

Sam Bailey
University of Birmingham
C9 Kenilworth Court
Hagley Road
Birmingham, B16 9NT
United Kingdom
s.c.bailey@pgr.bham.ac.uk

Nir Barnea
The Hebrew University
Givaat-Ram
Jerusalem, 91904
Israel
nir@phys.huji.ac.il

Vadim Baru
University of Bochum
Ruhr-Universität Bochum
Bochum, 44780
Germany
vbaruru@gmail.com

Adnan Bashir
Universidad Michoacana
Circuito Presea Generalisimo Morelos 486
Fraccionamiento Morelia 450
Morelia, Michoacan 58080
Mexico
adnan@ifm.umich.mx

Yajnavalkya Bhattacharya
York University
55 Westway Crescent
Vaughan, ON L4K 5M1
Canada
yajnaval@gmail.com

Elmar Biernat
CFTP, Inst. Superior Técnico,
Universidade de Lisboa
Avenida Rovisco Pais 1
Lisboa, Lisboa 1049-001
Portugal
elmar.biernat@tecnico.ulisboa.pt

Daniele Binosi
ECT*
Strada delle Tabarelle 286
Villazzano
Trento, 38123
Italy
binosi@ectstar.eu

Michael Birse
University of Manchester
School of Physics and Astronomy
University of Manchester
Manchester, UK M13 9PL
United Kingdom
mike.birse@manchester.ac.uk

Doerte Blume
Washington State University
Dept. of Physics and Astronomy
Pullman, WA 99164
United States
doerte9999@gmail.com

Eric Braaten
Ohio State University
191 W Woodruff Ave
Columbus, OH 43210
United States
braaten@mps.ohio-state.edu

Fabio Braghin
Federal University of Goias - Brazil
Institute of Physics - Federal University of Goias
Goiania, Goias/Brazil 74001970
Brazil
fbraghin.ufg@gmail.com

Carl Brune
Ohio University
Department of Physics and Astronomy
Ohio University
Athens, OH 45701
United States
brune@ohio.edu

Tongtong Cao
University of South Carolina
1800 Senate St
APT 206
Columbia, SC 29201-3833
United States
caot@email.sc.edu

Lei Chang
University of Adelaide
Department of Physics
Adelaide, SA 5005
Australia
lei.chiong@gmail.com

Chen Chen
USTC
Department of Modern Physics
University of Science and Technology of China
Hefei, 230026
China
seracron@ustc.edu.cn

Cheng Chin
University of Chicago
929 E. 57th St
Chicago
United States
cchin@uchicago.edu

Ho-Meoyng Choi
Kyungpook National University
80 Daehakro, Bukgu
Daegu
Korea, Republic of
homyoung@knu.ac.kr

Izabela Ciepał
Institute of Nuclear Physics PAS
Radzikowskiego 152
Kraków
Poland
izabela.ciepal@ifj.edu.pl

Ian Cloet
Argonne National Laboratory
9700 S Cass Ave
Argonne, IL 60439
United States
icloet@anl.gov

Nicholas Compton
Ohio University
Edwards Accelerator Laboratory
Ohio University
Athens, OH
United States
nc543611@ohio.edu

Sidney Coon
University of Arizona
Department of Physics
1118 E. Fourth Street
Tucson, AZ 85721
United States
coon@physics.arizona.edu

Benjamin Crowe III
North Carolina Central University
1801 Fayetteville Street
Durham, NC 27707
United States
bcrowe@nccu.edu

Kevin Daily
Purdue University
525 Northwestern Ave
West Lafayette, IN 47906
United States
daily5@purdue.edu

Jurij Darewych
York University Toronto
4700 Keele St. N
Toronto, ON M3J 1P3
Canada
darewych@yorku.ca

Barry Davids
TRIUMF
4004 Wesbrook Mall
Vancouver, BC V6T2H4
Canada
davids@triumf.ca

Sergio Deflorian
University of Trento
via Sommarive, 14
Trento
Italy
s.deflorian.2@unitn.it

Arnoldas Deltuva
Vilnius University
A. Gostauto 12
Vilnius
Lithuania
arnoldas.deltuva@tfai.vu.lt

William Detmold
Massachusetts Institute of Technology
77 Massachusetts Ave, 6-303
Cambridge, MA 02139
United States
wdetmold@mit.edu

Jose D'Incao
JILA, University of Colorado
440 UCB
Boulder, CO 80309-0440
United States
jpdincao@jila.colorado.edu

Minghui Ding
Peking University
No.5 Yiheyuan Road, Haidian District
Beijing
China
mhding@pku.edu.cn

Michael Distler
University Mainz
J.J.Becher-Weg 45
Mainz, 55128
Germany
distler@uni-mainz.de

Reinhard Doerner
Goethe-University
Max-von-Laue.Str. 1
Frankfurt a.M., 60438
Germany
doerner@atom.uni-frankfurt.de

Jérémy Dohet-Eraly
TRIUMF
4004 Wesbrook Mall
Vancouver, BC V6T 2A3
Canada
jdoheter@triumf.ca

Michael Doring
George Washington University
Dept. of Physics, 725 21st St, NW
Washington, DC 20052
United States
doring@gwu.edu

Evangeline Downie
George Washington University
725 21st Street NW
Washington, DC 20052
United States
edownie@gwu.edu

Gordon Drake
University of Windsor
Department of Physics
401 Sunset Street
Windsor, ON N9E 3M7
Canada
gdrake@uwindsor.ca

Jozef Dudek
Jefferson Lab/Old Dominion University
12000 Jefferson Ave
Newport News, VA 23606
United States
dudek@jlab.org

Gernot Eichmann
University of Giessen
Heinrich-Buff-Ring 16
Giessen, 35392
Germany
gernot.eichmann@theo.physik.uni-giessen.de

Bruno El-Bennich
U. Cruzeiro do Sul & IFT State University of São Paulo
Laboratório de Física Teórica e Computacional, Universidade
Rua Galvão Bueno 868, Bloco B
São Paulo, São Paulo 01415-000
Brazil
elbennich@me.com

Charlotte Elster
Ohio University
Dept. of Physics and Astronomy
Athens, OH 45701
United States
elster@ohio.edu

Shimpei Endo
LKB, ENS
24 rue Lhomond
Paris, 75231
France
shimpei.endo@lkb.ens.fr

Evgeny Epelbaum
Ruhr-Universitaet Bochum
Universitaetsstr. 150
Bochum, Germany 44780
Germany
evgeny.epelbaum@rub.de

Vasily Eremenko
Ohio University
251B Clippinger Labs
Ohio University
Athens, OH 45701
United States
eremenko@ohio.edu

Lei Feng
The University of Chicago
5470 S Harper Ave
APT 1C
Chicago, IL 60615
United States
l.ray.feng@gmail.com

Fabrizio Ferrari Ruffino
Università di Trento
via Sommarive 14
Trento
Italy
f.ferrariruffino@unitn.it

Igor Filikhin
North Carolina Central University
1801 Fayetteville St.
Durham, NC 27707
United States
ifilikhin@nccu.edu

Daniel Fischer
Missouri S&T
1315 N Pine St
Rolla, MO 65409
United States
fischerda@mst.edu

Nadia Fomin
University of Tennessee
1408 Circle Drive
Knoxville, TN 37996
United States
nfomin@utk.edu

Tobias Frederico
Instituto Tecnológico de Aeronáutica
Pr. Mal. Eduardo Gomes 50
São José dos Campos, São Paulo
12242200
Brazil
tobias@ita.br

Martin Freer
University of Birmingham
School of Physics and Astronomy
University of Birmingham
Birmingham, B15 2TT
United Kingdom
M.Freer@bham.ac.uk

Johan Frenje
Massachusetts Institute of Technology
175 Albany Street, NW17-235
Cambridge, MA 02139
United States
jfrenje@psfc.mit.edu

Forrest Friesen
Duke/TUNL
BOX 90319
Durham, NC 27708
United States
fqf@tunl.duke.edu

Christopher Fritzsch
WWU Münster
Wilhelm-Klemm-Straße 9
Institut für Kernphysik
Münster, NRW 48149
Germany
c.fritzsch@uni-muenster.de

Yasuro Funaki
RIKEN
Hirosawa 2-1, Wako 351-0198
Saitama
Japan
funaki@riken.jp

Fei Gao
Peking University
Department of Physics
Peking University
Beijing
China
hiei@pku.edu.cn

Haiyan Gao
Duke University & Duke Kunshan University
Dept of Physics
Duke University
Durham, NC 27707
United States
gao@phy.duke.edu

Eduardo Garrido
Instituto de Estructura de la Materia
Serrano 121
CIF: Q2818002D
Madrid, E-28006
Spain
e.garrido@csic.es

Ashot Gasparyan
Ruhr University of Bochum
Institut für Theoretische Physik II
Fakultät für Physik und Astronomie
Ruhr-Universitä
Bochum
Germany
ashot.gasparyan@rub.de

Mario Gattobigio
INLN - University of Nice
1361 Route des Lucioles
Valbonne, 06560
France
mario.gattobigio@inln.cnrs.fr

Alexandros Gezerlis
University of Guelph
Department of Physics
University of Guelph
Guelph, ON N1G 2W1
Canada
gezerlis@uoguelph.ca

Benjamin Gibson
LANL & Physical Review
T-02, MS-B283
Los Alamos National Lab
Los Alamos, NM 87545
United States
bfgibson@lanl.gov

Luca Girlanda
University of Salento
Via per Arnesano
Lecce, Italy 73020
Italy
girlanda@le.infn.it

Jacek Golak
Jagiellonian University
Institute of Physics
Reymonta 4
Cracow, Malopolska PL-30384
Poland
jacek.golak@uj.edu.pl

Chris Greene
Purdue University
300 Forest Hill Dr
West Lafayette, IN 47906
United States
chgreene@purdue.edu

Harald Griesshammer
Institute for Nuclear Studies
Department of Physics
George Washington University
Washington, DC 20052
United States
hgrie@gwu.edu

Katherine Grzywacz Jones
University of Tennessee
1408 Circle Drive
Knoxville, TN 37919
United States
kgrzywac@utk.edu

Qingze Guan
Washington State University
Department of Physics and Astronomy
Pullman, WA 99163
United States
gqz0001@gmail.com

Mohammadreza Hadizadeh
Ohio University
Department of Physics & Astronomy
Clippinger Labs 240
Athens, OH 45701
United States
hadizadm@ohio.edu

Gerald Hale
Los Alamos National Laboratory
T-2, MS B214
Los Alamos, NM 87545
United States
ghale@lanl.gov

Hans Hammer
TU Darmstadt
Institut fuer Kernphysik --
Theoriezentrum
Schlossgartenstr. 2
Darmstadt, 64289
Germany
Hans-Werner.Hammer@physik.
tu-darmstadt.de

Zhonglin Han
Duke/TUNL
BOX 90319
Durham, NC 27708
United States
mpaul@tunl.duke.edu

Nathan Harshman
American University
4400 Mass. Ave. NW
Washington, DC 20016
United States
harshman@american.edu

Marc Herrmann
University of Iowa
382 Westgate St Apt 4
Iowa City, IA 52246
United States
marc-herrmann@uiowa.edu

Emiko Hiyama
RIKEN
2-11 Hrosawa
Wako, Saitama 351-1098
Japan
hiyama@riken.jp

Linda Hlophe
Ohio University
251B Clippinger Laboratories
Athens, OH 45701
United States
lh421709@ohio.edu

Roy Holt
Argonne National Laboratory
Physics Division
Argonne, IL 60439
United States
holt@anl.gov

Yusuke Horinouchi
University of Tokyo
7-3-1
Hongo, Bunkyo-ku, Tokyo
Japan
yusukehorinouchi@gmail.com

Wataru Horiuchi
Hokkaido University
Kita 10 Nishi 8, Kita-ku
Sapporo, 0600810
Japan
whoriuchi@nucl.sci.hokudai.ac.jp

Calvin Howell
Duke University/TUNL
Box 90308
Duke University
Durham, NC 27708
United States
howell@tunl.duke.edu

Wansoo Huh
Soongsil University
Dongjakku Sangdodong
511
Seoul, Seoul 156-743
Korea, Republic of
wshuh@ssu.ac.kr

Nils Hüsken
WWU Münster
Institut für Kernphysik
Wilhelm-Klemm-Straße 9
Münster, 48149
Germany
nils.huesken@googlemail.com

Jongwon Hwang
Seoul National University
1 Gwanak-ro, Gwanak-gu
Seoul
Korea, Republic of
hjw8707@snu.ac.kr

Yordanka Ilieva
University of South Carolina
712 Main St
Columbia, SC 29208
United States
ilieva@sc.edu

Souichi Ishikawa
Hosei University
2-17-1 Fujimi
Chiyoda, Tokyo 102-8160
Japan
ishikawa@hosei.ac.jp

Chen Ji
TRIUMF
4004 Wesbrook Mall
Vancouver, BC V6T2A3
Canada
jichen@triumf.ca

Selim Jochim
Heidelberg University
Im Neuenheimer Feld 226
Heidelberg, 69120
Germany
jochim@uni-heidelberg.de

Jacob Johansen
The University of Chicago
929 E 57th St.
GCIS ESB11
Chicago, IL 60637
United States
jacobjjohansen@uchicago.edu

Philip Johnson
American University
Department of Physics
Washington, DC 20016
United States
pjohnson@american.edu

Michael Jones
Michigan State University/NSCL
640 S. Shaw Lane
East Lansing, MI 48824
United States
jonesm@nscl.msu.edu

Nasser Kalantar-Nayestanaki
KVI-CART/University of Groningen
Zernikelaan 25
Groningen, 9747AA
Netherlands
nasser@kvi.nl

Hiroyuki Kamada
Kyushu Institute of Technology
1-1 Sensuicho, Tobata
Kitakyushu, 804-8550
Japan
kamada@mns.kyutech.ac.jp

Daekyoung Kang
Los Alamos National Laboratory
Theoretical Division, MS B283, LANL
Los Alamos, NM 87545
United States
4charm@gmail.com

Vladimir Karmanov
Lebedev Physical Institute
Leninsky prospect 53
Moscow, 119991
Russian Federation
karmanov@sci.lebedev.ru

Cynthia Keppel
Thomas Jefferson National Accelerator Facility
Suite 4 / Room C114
12000 Jefferson Avenue
Newport News, VA 23606
United States
keppel@jlab.org

Roman Kezerashvili
New York City College of Tech., City Univ. of New York
300 Jay Street
Brooklyn, NY 11201
United States
rkezerashvili@citytech.cuny.edu

Lev Khaykovich
Bar Ilan University
Physics Department
Ramat Gan
Israel
lev.khaykovich@biu.ac.il

Alejandro Kievsky
INFN-Pisa
Physics Department
Largo B. Pontecorvo 3
Pisa, 56127
Italy
kievsky@pi.infn.it

Sunji Kim
Seoul National University
23-416, Seoul National University, 1 Gwanak-ro,
Gwanak-gu, Seoul,151-747, South Korea
Seoul
Korea, Republic of
sjkim06@snu.ac.kr

Tom Kirchner
York University
Department of Physics and Astronomy
4700 Keele Street
Toronto, ON M3J 1P3
Canada
tomk@yorku.ca

Johannes Kirscher
The Hebrew University
Givat-Ram
Jerusalem, 91904
Israel
j.kirscher@mail.huji.ac.il

Keiichi Kisamori
CNS, The University of Tokyo
2-1 Hirosawa
Wako, Saitama 351-0198
Japan
kisamori@cns.s.u-tokyo.ac.jp

Barbara Klos
University of Silesia, Katowice
Uniwersytecka 4
40-0007 Katowice
Poland
barbara.klos@us.edu.pl

Sebastian Koenig
The Ohio State University
Department of Physics
191 W Woodruff Ave
Columbus, OH 43212
United States
koenig.389@osu.edu

Adam Kozela
Institute of Nuclear Physics PAN
ul. Radzikowskiego 152
31-342 Kraków, Poland
Kraków
Poland
adam.kozela@ifj.edu.pl

Gastao Krein
Instituto de Fisica Teorica - UNESP
Rua Dr. Bento Teobaldo Ferraz, 271 - Bloco II
Sao Paulo, SP 01140-070
Brazil
gkrein@ift.unesp.br

Eva Kuhnle
Heidelberg University
Im Neuenheimer Feld 226
Heidelberg, 69120
Germany
kuhnle@physi.uni-heidelberg.de

Anna Kwiatkowski
TRIUMF
4004 Wesbrook Mall
Vancouver, BC V6T2A3
Canada
aniak@triumf.ca

Georgios Laskaris
Duke University
912 Lancaster Street
Durham, NC 27701
United States
georgios.ilias.laskaris@gmail.com

Rimantas Lazauskas
IPHC Strasbourg
23 rue du Loess
Strasbourg
France
rimantas.lazauskas@iphc.cnrs.fr

Dean Lee
North Carolina State University
Box 8202
Department of Physics
Raleigh, NC 27695
United States
dean_lee@ncsu.edu

Tsung-Shung Lee
Argonne National Laboratory
9700 S. Cass Ave
Argonne, IL 60439
United States
lee@phy.anl.gov

Winfried Leidemann
University of Trento
Dipartimento di Fisica, Universita` di Trento
Via Sommarive 14
Trento, Italy 38123
Italy
leideman@science.unitn.it

Sofia Leitão
Centre for Theoretical Particle Physics, University of Lisbon
CFTP, Departamento de Física
Avenida Rovisco Pais,1
Lisbon, 1049-001
Portugal
sofia.leitao@tecnico.ulisboa.pt

Mantile Lekala
University of South Africa
Department of Physics
Florida Science Campus
Johannesburg
South Africa
lekalml@unisa.ac.za

Jesper Levinsen
Monash University
Melbourne, Victoria 3800
Australia
jesper.levinsen@monash.edu

Richard Lombardini
St. Mary's University, San Antonio
One Camino Santa Maria
San Antonio, TX 78228
United States
rlombardini@stmarytx.edu

Diego Lonardoni
Argonne National Laboratory
9700 South Cass Avenue
Building 203, Room B246
Argonne, IL 60439
United States
lonardoni@anl.gov

Alessandro Lovato
Argonne National Laboratory
9700 South Cass Avenue
Argonne, IL 60439
United States
lovato@alcf.anl.gov

Amy Lovell
Michigan State University/NSCL
640 S. Shaw Lane
East Lansing, MI 48824
United States
lovell@nscl.msu.edu

Joel Lynn
Los Alamos National Laboratory
1985 Elderberry St
Los Alamos, NM 87544
United States
joel.lynn@lanl.gov

Yukie Maeda
University of Miyazaki
1-1 Kibanadai-Nishi, Miyazaki-city
Miyazaki, 889-2192
Japan
yukie@cc.miyazaki-u.ac.jp

Maxim Mai
University Bonn, HISKP
Endenicher Allee 33
Bonn
Germany
vladimirjuni@googlemail.com

Ronald Malone
Duke/TUNL
BOX 90319
Durham, NC 27708
United States
rm216@phy.duke.edu

Arman Margaryan
Duke University
Physics Bldg., Science Dr.
Box 90305
Durham, NC 27708
United States
amargaryan90@gmail.com

Pieter Maris
Iowa State University
Dept. of Physics and Astronomy
Iowa State University
Ames, IA 50011
United States
pmaris@iastate.edu

Johann Marton
Stefan Meyer Institute, Austrian Academy of Sciences
Boltzmanngasse 3
Vienna, 1090
Austria
johann.marton@oeaw.ac.at

Hiroshi Masui
Kitami Institute of Technology
165, Koen-cho
Kitami, Hokkaido 090-8507
Japan
hgmasui@mail.kitami-it.ac.jp

Judith McGovern
University of Manchester
School of Physics and Astronomy
Schuster Building, Brusnwick St
Manchester, UK M13 9PL
United Kingdom
judith.mcgovern@manchester.ac.uk

Mirko Miorelli
TRIUMF/UBC
4004 Wesbrook Mall
Vancouver, BC V6T2A3
Canada
mmiorelli@triumf.ca

Victor Mokeev
Jefferson Lab
12000 Jefferson Ave Ste #5
Newport News, VA 23606
United States
mokeev@jlab.org

Sergej Moroz
CU Boulder
Department of Physics
Boulder, CO 80305
United States
sergej.moroz@colorado.edu

Herve Moutarde
Irfu, CEA-Saclay
CEA, Centre de Saclay
Gif-sur-Yvette, France 91191
France
herve.moutarde@cea.fr

Peter Mueller
Argonne National Laboratory
9700 S Cass Ave
Argonne, IL 60439
United States
pmueller@anl.gov

Luke Myers
Jefferson Lab
12000 Jefferson Avenue
Bldg 12_1/C102
Newport News, VA 23606
United States
lmyers@jlab.org

Fred Myhrer
University of South Carolina
Dept. Physics and Astronomy
Columbia, SC 29208
United States
myhrer@physics.sc.edu

Pascal Naidon
RIKEN
2-1 Hirosawa
Wako, Saitama 351-0198
Japan
pascal@riken.jp

Rodrigo Navarro Perez
Universidad de Granada
Av Fuentenueva S/N
Granada, Granada 18071
Spain
rnavarrop@ugr.es

Amy Nicholson
UC Berkeley
366 LeConte Hall #7300
Berkeley, CA 94720
United States
anicholson@berkeley.edu

Hisako Niko
Springer
van Godewijck straat 30
Dordrecht, 3311GX
Netherlands
Hisako.Niko@springer.com

Wilfried Noertershaeuser
TU Darmstadt
Institut fuer Kernphysik
Schlossgartenstr. 9
Darmstadt
Germany
wnoertershaeuser@ikp.tu-darmstadt.de

Ann Orel
University of California, Davis
4001 9th St. N Apt. 1806
Arlington, VA 22203
United States
aeorel@ucdavis.edu

Giuseppina Orlandini
University of Trento
Department of Physics
Via Sommarive 14
Trento, 38123
Italy
giuseppina.orlandini@unitn.it

Nigel Orr
LPC-Caen
LPC-ENSICAEN
Bd Marechal Juin
Caen, 14050
France
orr@lpccaen.in2p3.fr

Shinsho Oryu
Tokyo University of Science
2641 Yamazaki
Noda, Chiba 278-8510
Japan
oryu@rs.noda.tus.ac.jp

Michael Papenbrock
Uppsala University
Box 516
Uppsala, 75120
Sweden
michael.papenbrock@physics.uu.se

Cody Parker
Ohio University
Dept. of Physics & Astronomy
251B Clippinger Laboratories
Athens, OH 45701
United States
cp260509@ohio.edu

Colin Parker
University of Chicago
929 E 57th St
Chicago, IL 60637
United States
cvparker@uchicago.edu

Saori Pastore
University of South Carolina
712 Main St
Columbia, SC 29205
United States
saori.pastore@gmail.com

Chao Peng
Duke University
24 Cedarwood Way
Apt. I
Newport News, VA 23608
United States
cp121@duke.edu

Michael Pennington
Jefferson Lab
12000 Jefferson Avenue (Suite 1)
Newport News, VA 23188
United States
michaelp@jlab.org

Jesus Perez-Rios
Purdue University
Department of Physics and Astronomy
525 Northwestern Ave.
West Lafayette, IN 47907
United States
jperezri@purdue.edu

Stefan Petschauer
TU München
James-Franck-Straße 1
Garching, 85747
Germany
stefan.petschauer@ph.tum.de

Daniel Phillips
Ohio University
242C Clippinger Laboratory
Athens, OH 45701
United States
phillips@phy.ohiou.edu

Maria Piarulli
Old Dominion University
Department of Physics
Norfolk, VA 23529
United States
mpiar001@odu.edu

Steven Pieper
Argonne National Laboratory
Physics, Bldg 203
Argonne, IL 60439
United States
spieper@anl.gov

Lucas Platter
University of Tennessee, Knoxville
Department of Physics and Astronomy
1408 Circle Drive
Knoxville, TN 37996
United States
lplatter@utk.edu

Wayne Polyzou
University of Iowa
Department of Physics and Astronomy
Iowa City, IA 52242
United States
polyzou@uiowa.edu

Sixue Qin
Argonne National Laboratory
9700 S Cass Ave
B-230 Bldg 203
Argonne, IL 60439
United States
sqin@anl.gov

Jianwei Qiu
Brookhaven National Laboratory
Physics Department
Brookhaven National Lab
Upton, NY 11973
United States
jqiu@bnl.gov

Sofia Quaglioni
Lawrence Livermore National Laboratory
P.O. Box 808
L-4141
Livermore, CA 94551
United States
quaglioni1@llnl.gov

Francesco Raimondi
TRIUMF
4004 Wesbrook Mall
Vancouver, BC
Canada
fraimondi@triumf.ca

George Rawitscher
University of Connecticut
Department of Physics
2152 HIlside Rd U-3046
Storrs, CT 06269-3046
United States
george.rawitscher@uconn.edu

Khepani Raya
University of Michoacan
Francisco J Mugica
Morelia
Mexico
khepani@gmail.com

Stephanie Reimann-Wacker
Lund University
Sölvegatan 14A
(Box 118)
22100 Lund, Sweden 22100
Sweden
reimann@matfys.lth.se

Paul Reimer
Argonne National Laboratory
Physics Division Bldg 203
9700 S. Cass Ave
Argonne, IL 60439
United States
reimer@anl.gov

Zhongzhou Ren
Nanjing University
Department of Physics
Nanjing
China
zren@nju.edu.cn

Andrea Richard
Ohio University
Edwards Accelerator Lab
1 Ohio University
Athens, OH 45701
United States
ar286106@ohio.edu

Matteo Rinaldi
Perugia University & INFN
Via Alessandro Pascoli
Perugia
Italy
matteo.rinaldi@pg.infn.it

Craig Roberts
Argonne National Laboratory
Physics Division
9700 South Cass Avenue
Argonne, IL 60439
United States
cdroberts@anl.gov

Julie Roche
Ohio University
Edwards Accelerator Lab
1 Ohio University
Athens, OH 45701
United States
rochej@ohio.edu

José Rodríguez Quintero
University of Huelva & CAFPE
Dpto. Fisica Aplicada, Fac. CCEE
Universidad de Huelva
Huelva, Huelva 21071
Spain
jose.rodriguez@dfaie.uhu.es

Eduardo Rojas
Universidad de Antioquia
P.O. Box 1226 Medellin, Columbia
Medellin, Antioquia
Colombia
rojas4000@yahoo.com

Carolina Romero Redondo
Lawrence Livermore National Laboratory
PO Box 808. L-414
Livermore, CA 94551
United States
c.romeroredond@gmail.com

Dagmara Rozpedzik
Jagiellonian University
Lojasiewicza 11
Krakow
Poland
dagmara.rozpedzik@uj.edu.pl

Olga Rubtsova
Moscow State University
Skobeltsyn Institute of Nuclear Physics
1(2) Leninskie gory, GSP-1
Moscow, 119991
Russian Federation
rubtsova-olga@yandex.ru

Yuta Sada
RCNP, Osaka University
10-1 Mihogaoka, Ibaraki
Osaka, Osaka 567-0047
Japan
sada@rcnp.osaka-u.ac.jp

Mark Saffman
University of Wisconsin Madison
Department of Physics
1150 University Avenue
Madison, WI 53706
United States
msaffman@wisc.edu

Kenshi Sagara
Kyushu University
Yakuin 4-12-13-501, Chuo-ku
Fukuoka, Fukuoka-ken 810-0022
Japan
sagara@phys.kyushu-u.ac.jp

Daris Samart
School of Physics, Suranaree University of Technology
111 University Avenue, Nakhon Ratchasima 30000
Nakhon Ratchasima
Thailand
jod_daris@yahoo.com

Helios Sanchis Alepuz
Justus Liebig University Giessen
Heinrich-Buff-Ring 16
Giessen, Hessen 35390
Germany
hsanchisalepuz@gmail.com

Daniel Sayre
Lawrence Livermore National Laboratory
9000 East Ave.
Livermore, CA 94550
United States
sayre4@llnl.gov

Jean Antoine Scarpaci
CSNSM/CNRS/UPS
bat 104
Orsay, France 91405
France
scarpaci@csnsm.in2p3.fr

Matthias Schindler
University of South Carolina
Department of Physics and Astronomy
712 Main St
Columbia, SC 29208
United States
mschindl@mailbox.sc.edu

Soeren Schlimme
University Mainz
J.J.Becher-Weg 45
Mainz, 55128
Germany
schlimme@kph.uni-mainz.de

Eva Schulhoff
University of Windsor
401 Sunset Ave
Windsor, ON
Canada
schulh1@uwindsor.ca

Michael Schulz
Missouri S&T
1315 N Pine St.
Rolla, MO 65409
United States
schulz@mst.edu

Wolfgang Schweiger
University of Graz
Universitaetsplatz 5
Graz, 8010
Austria
wolfgang.schweiger@uni-graz.at

Jorge Segovia
University of Salamanca
Plaza de la Merced s/n
Salamanca, Salamanca 37008
Spain
segonza@usal.es

Kimiko Sekiguchi
Tohoku University
6-3 Aoba, Aramaki, Aoba-ku
Sendai
Japan
kimiko@lambda.phys.tohoku.ac.jp

Rie Sekine
Hokkaido University
N10W8
Sapporo
Japan
sekine@nucl.sci.hokudai.ac.jp

Oleksandr Shebeko
Institute for Theoretical Physics
Academicheskaya Str.1, Kharkov 61108
Kharkov
Ukraine
shebeko@kipt.kharkov.ua

Chengping Shen
Beihang University
School of Physics, Beihang University
Beijing
China
shencp@ihep.ac.cn

Ik Jae Shin
Institute for Basic Science
70, Yuseong-daero 1689-gil,
Yuseong-gu
Daejeon
Korea, Republic of
geniean@ibs.re.kr

Haik Simon
GSI Helmholtzzentrum für
Schwerionenforschung GmbH
Planckstr. 1
Darmstadt, 64291
Germany
h.simon@gsi.de

Roman Skibinski
Jagiellonian University
Institute of Physics
ul. Lojasiewicza 11
Krakow, PL-30348
Poland
roman.skibinski@uj.edu.pl

Dane Smith
Ohio State University
191 W. Woodruff Ave.
Columbus, OH 43210
United States
smith.7991@osu.edu

Artemis Spyrou
Michigan State University
640 S. Shaw Ln
East Lansing, MI 48823
United States
spyrou@nscl.msu.edu

Rajesh Srivastava
Indian Institute of Technology Roorkee
Department of Physics, IIT Roorkee
Roorkee, 247667
India
rajsrfph@gmail.com

Vincent Sulkosky
University of Virginia
1004 Cedar Grove Rd
Ruckersville, VA 22968
United States
vasulk@jlab.org

Renat Sultanov
Saint Cloud State University
720 4th Ave. So.
Centennial Hall 367 B
Saint Cloud, MN 56301
United States
rasultanov@stcloudstate.edu

Vladimir Suslov
North Carolina Central University
1801 Fayetteville St. Crest
Durham, NC 27707
United States
vsuslov@nccu.edu

Arbin Thapaliya
Ohio University
Clippinger Labs 251B
Athens, OH 45701
United States
at311509@ohio.edu

Michael Thoennessen
Michigan State University
640 South Shaw Lane
East Lansing, MI 48824
United States
thoennessen@nscl.msu.edu

Ye Tian
University of South Carolina
248 South Marion
Columbia, SC 29205
United States
tianye8001@gmail.com

Werner Tornow
Duke University & TUNL
Department of Physics
Duke University
Durham, NC 27708
United States
tornow@tunl.duke.edu

Shalva Tsiklauri
The City University of New York - BMCC
199 Chambers St.
New York, NY 10007
United States
stsiklauri@bmcc.cuny.edu

Shih-Kuang Tung
Northwestern University
1930 Ridge Ave Apt. B106
Evanston, IL 60201
United States
zekest@gmail.com

Yiharn Tzeng
Academia Sinica
128 Academia Road, Section 2,
Nankang, Taipei 1152, 11529
Taiwan
phtzengyh@sinica.edu.tw

Masahito Ueda
The University of Tokyo
Department of Physics
University of Tokyo, 7-3-1 Hongo,
Bunkyo-ku
Tokyo, 113-0033
Japan
ueda@phys.s.u-tokyo.ac.jp

Yury Uzikov
Joint Institute for Nuclear Researches
Jolioitt-Curie,6
Dubna, Moscow region 141980
Russian Federation
uzikov@jinr.ru

Jared Vanasse
Duke University
Physics Bldg., Science Dr. Box 90305
Durham, NC 27708
United States
jjv9@phy.duke.edu

Artem Volosniev
Aarhus University/TU Darmstadt
Ny Munkegade 120
Aarhus
Denmark
artem@phys.au.dk

Joseph Walshe
University of Birmingham
Edgbaston
Birmingham, West Midlands B15 2TT
United Kingdom
j.walshe@pgr.bham.ac.uk

Shin Watanabe
Kyushu University
Department of Physics, Kyushu
University 6-10-1
Hakozaki, Higashi-ku, Fukuoka
812-8581
Fukuoka
Japan
s-watanabe@phys.kyushu-u.ac.jp

Kyle Wendt
University of Tennessee
522 Farragut Commons Drive
Knoxville, TN 37934
United States
kylewendt@gmail.com

Richard Williams
University of Giessen
Institut fuer Theoretische Physik I
Heinrich-Buff-Ring 16
Giessen, Hesse 35394
Germany
richard.williams@theo.physik.
uni-giessen.de

Robert Wiringa
Argonne National Laboratory
9700 S. Cass Ave.
Argonne, IL 60439
United States
wiringa@anl.gov

Rachel Wooten
Purdue University
3355 Peppermill Dr
West Lafayette, IN 47906
United States
wootenr@purdue.edu

Yangqian Yan
Washington State University
350SE Spring Street
Pullman, WA 99163
United States
yanyangqian@gmail.com

Dylan Yost
Colorado State University
1875 Campus Delivery
Colorado State University
Fort Collins, CO 80523
United States
dylan.yost@colostate.edu

Zhenhua Yu
Tsinghua Unversity
Institute for Advanced Study
Room 302
Beijing, 100084
China
huazhenyu2000@gmail.com

Nicholas Zachariou
University of South Carolina
712 Main Street
Columbia, SC 29208
United States
nicholas@jlab.org

Hui Zhai
Tsinghua University
Institute for Advanced Study
Tsinghua University
Beijing, Beijing 100084
China
huizhai.physics@gmail.com

Xilin Zhang
University of Washington
3910 15th Ave NE
Physics and Astronomy Building (PAB)
Seattle, WA 98195
United States
zhangxilin83@gmail.com

Nikolaj Zinner
Aarhus University
Department of Physics and Astronomy
Ny Munkegade 120
Aarhus C, DK-8000
Denmark
zinner@phys.au.dk

Bing-Song Zou
ITP/CAS Beijing
Zhong Guan Cun East Street 55 #
Beijing
China
zoubs@itp.ac.cn

Alex Zylstra
Massachusetts Institute of Technology
77 Mass Ave, NW17-256
Cambridge, MA 02139
United States
azylstra@psfc.mit.edu

21st International Conference on Few-body Problems in Physics

May 18-22, 2015

Crowne Plaza Metro, Chicago, IL

Sunday, May 17

4 pm-7 pm Registration open

Monday, May 18

8-10 am Registration open

8:50 am *Welcome*

C. Roberts (Argonne)

R. Janssens (Argonne)

9-10:30 am *Lasers for Few-body Physics*

Chair: D. Phillips (Ohio)

S. Jochim (Heidelberg) "Few-body physics with ultracold atoms"

J. Frenje (MIT) "High Energy Density Plasmas for studies of Basic Nuclear Science and Nuclear Astrophysics"

W. Noertershaeuser (TU, Darmstadt) "Laser Spectroscopy of Few-Electron Systems at the Extremes: Structure of Light Nuclei and QED in Strong Fields"

10:30-11 am Coffee break

11 am-12:30 *From Mesons to Hypernuclei*

Chair: C. Roberts (Argonne)

J. Dudek (Old Dominion/JLab) "Hadron resonances from QCD"

W. Detmold (MIT) "Properties of light nuclei from lattice QCD"

E. Hiyama (RIKEN) "Recent progress of hypernuclear physics"

12:30-2 pm Lunch break

2-3:50 pm Parallel sessions:

Salon A: Continuum QCD (Chair: Roberts)

L. Chang (Adelaide) "Pion electromagnetic properties in continuum QCD"

D. Binosi "From continuum QCD to hadron observables"

B. El-Bennich "Flavorful Hadron Physics"

E. Rojas "Insights into the quark-gluon vertex from lattice QCD and meson spectroscopy"

R. Williams "Confronting hadron phenomenology with the quark-gluon vertex"

Salon B: Extracting three-nucleon forces (Chair: Epelbaum)

K. Sekiguchi (RIKEN) "Recent results on deuteron–proton scattering"

R. Skibinski "Recent developments in the study of 3N systems within the new chiral forces."

I. Ciepal "Experimental Studies of Coulomb-Force Effects in dp breakup at Medium Energy"

B. Klos "Experimental study of Three-Nucleon Dynamics in the dp breakup reaction"

Y. Maeda "Study of 3NF effects at the intermediate energies by pd breakup measurements"

Salon C: Few-body universality from atoms to nuclei (Chair: Hammer)
M. Ueda (Tokyo) "Relationship between the Onset of the Limit Cycle and the Universal Three-Body Parameter in Efimov Physics"
L. Khaykovich "Universal Loss Dynamics in Unitary Bose Gas"
P. Naidon "Scattering of universal fermionic clusters"
A. Kievsky "Weakly bound state with spin-isospin symmetry"
N. Barnea "The nuclear contact and the photoabsorption cross section"

Salon D: Light nuclei in astrophysics (Chair: Brune)
B. Davids (TRIUMF), "Astrophysical reactions of light and intermediate mass nuclei"
X. Zhang "How well do we understand 7Be + p → 8B + γ? An EFT perspective"
J. Dohet-Eraly "Towards an ab initio description of light-nuclei radiative captures"
H. Masui "Deuteron-like correlation of valence nucleons for the T = 0 channel in ^{18}F"
S. Ishikawa "Low-lying ^{12}C continuum states in three-α model"

3:50-4:20 Coffee break

4:20-6:10 Parallel sessions:
Salon A: Pions and Resonances (Chair: Mokeev)
M. Doering (George Washington) "Baryon spectroscopy: new results and perspectives"
E. Biernat "Quark model with chiral-symmetry breaking and confinement in the CST"
B. Liu "Few-body methods and results for hadrons in medium"
W. Schweiger "The Pion-Cloud Contribution to the Electromagnetic Nucleon Form Factors"
Y. Tian "Exclusive π–Electroproduction off the Neutron in Deuterium in the Resonance..."

Salon B: Two-photon physics (Chair: McGovern)
M. Ahmed (North Carolina Central) "Photon-Induced Reaction Studies at the High Intensity Gamma Ray Source (HIGS) for A = 3 to 16"
L. Myers "Recent Deuteron Compton Scattering Results and ... Neutron Polarizabilities"
H. Griesshammer "News on Deuteron and ^3He Compton Scattering in Chiral EFT..."
M. Distler "Inclusive electron scattering on light nuclei: constraining the two-photon..."
M. Miorelli "Electric dipole polarizability: from few- to many-body systems"

Salon C: Unequal masses and the Super-Efimov effect (Chair: Ueda)
E. Kuhnle (Heidelberg) "Efimov physics in an ultracold Bose-Fermi mixture of ^{133}Cs and ^6Li"
J. Johansen "Efimov Physics in a ^6Li-^{133}Cs Atomic Mixture"
S. Endo "Third virial coefficient of a two-component unitary Fermi gas across..."
S. Moroz "Super Efimov effect and its extension to mass imbalanced systems"
Z. Yu "Revealing the origin of super-Efimov states in the hyper-spherical formalism"

Salon D: Fusion reactions and Inertial Confinement Fusion plasmas (Chair: Frenje)
D. Sayre (Livermore) "Measurement and Analysis of the T+T Neutron Spectrum from ICF"
G. Hale "Spectra for the A=6 reactions calculated from a three-body resonance model"
A. Bacher "T(t,2n)4He Reaction: energy dependence of its mechanism"
C. Romero-Redondo "Advances in the ab initio description of nuclear three-cluster systems"
C. Parker "The 3H(d, γ) Reaction at Ec.m. ≤ 300 keV"

Tuesday, May 19

9-10:30 am *Recent progress in Efimov physics*
Chair: D. Blume (Washington State)
S.-K.Tung (Chicago) "Geometric scaling of Efimov states in ultracold ^{6}Li-^{133}Cs mixtures"
D. Kang (Los Alamos) "How to extract two- and three-body contacts"
N. Zinner (Aarhus) "Exploring the few- to many-body crossover using cold atoms in 1D"

10:30-11 am Coffee break

11 am-12:30 *Nuclei as few-body systems*
Chair: E. Garrido (Madrid)
M. Freer (Birmingham) "Clustering in light nuclei"
H.-W. Hammer (TU, Darmstadt) "Few-body universality in halo nuclei"
S. Quaglioni (Livermore), "Ab initio calculations of reactions with light nuclei"

12:30-2 pm Lunch break

2-3:50 pm Parallel sessions:

Salon A: Distribution functions (Chair: Arrington)
J. Roche (Ohio) "Precision studies of the DVCS process at JLab"
H. Moutarde "Overlap Representation of GPDs in the Dyson-Schwinger Formalism"
J. Rodriguez "A DSE-inspired approach for the computation of the pion's …"
H.-M. Choi "Light-Front Quark Model Phenomenology Consistent with Chiral…"
E. Pace "The ^{3}He spectral function in light-front dynamics"

Salon B: Three-nucleon systems at low energy (Chair: Sekiguchi)
C. Howell (Duke),"Few-nucleon research at TUNL: probing 2N and 3N forces with neutrons"
L. Girlanda "Tuning the 3N force from 3N scattering data"
R. Malone "Neutron-neutron quasifree scattering in nd breakup at 10 MeV"
K. Sagara "Systematic study on Space-Star Anomaly in pd breakup"
S. Koenig "The proton-deuteron scattering length in pionless effective field theory"

Salon C: Electronic dynamics in few-body systems (Chair: Greene)
A. Orel (UC-Davis) "Studies of HeH: Dissociative Excitation and Mutual Neutralization"
M. Schulz "Experimental Test of Theoretical Analysis of Coherence Effects in Atomic…"
T. Kirchner "A theoretical analysis of recent ion-lithium collision experiments…"
R. Sultanov "Low temperature HD + H_2 inelastic scattering of astrophysical interest"
D. Fischer "Collision dynamics studied with a polarized MOT target"

Salon D: Clusters and halos (Chair: Platter)
E. Garrido (Madrid) "Rotational character of the ^{12}C spectrum investigated through inelastic cross sections via photon emission"
Y. Funaki "Hoyle band and α condensation in ^{12}C"
J. Walshe "Exotic cluster structures: the search for nuclear water"
M.D. Jones "Further insights into the reaction ^{14}Be(CH2,X)^{10}He"
S. Watanabe "4B and 3B channel-coupling effects on ^{6}Li elastic scattering in CDCC"

3:50 pm POSTER SESSION BEGINS

3:50-4:20 pm Coffee break

4:20-6:10 Parallel sessions:

Salon A: Precision Spectroscopy and the Proton Radius Puzzle (Chair: Mueller)
D. Yost (Colorado State) "Frequency comb spectroscopy of the 1S/3S-3D transitions in H"
C. Ji "Understanding The Proton Radius Puzzle: Nuclear Structure Effects in Light Muonic.."
C. Peng "Proton Charge Radius (PRad) Experiment at Jefferson Lab"
G. Drake "Electron Shake-up and Shake-off Following 6He β Decay"
J. Marton "Strong interaction studies with kaonic atom"

Salon B: Four nucleons (Chair: Barnea)
A. Deltuva (Lisbon) "Reactions in the four-nucleon system above breakup threshold"
R. Lazauskas "Description of four-nucleon collisions including the breakup"
K. Kisamori "Study of tetra-neutron system via...reaction ^{4}He(^{8}He,^{8}Be)4n"
W. Leidemann "The resolution of resonances with the LIT method"
P. Maris "^{4}He and ^{6}Li with consistent chiral interactions"

Salon C: Pairing in medium-mass nuclei (Chair: Mazumdar)
M. Assie (IPN, Orsay) "Study of pairing and quartetting in medium mass nuclei"
Y. Tzeng, "Contributions of chiral two- and three-nucleon interactions to closed shell nuclei"
A. Kwiatkowski "High-precision atomic mass measurements for ... exotic nuclei"
S. Kim "Spectroscopy of 17C via one-neutron knockout reaction"
W. Horiuchi "5B calculation with Pauli constraint for shell- and cluster-structure in ^{16}O"

Salon D: Few-hadron hypernuclei (Chair: Hiyama)
P. Achenbach (Mainz) "Experimental investigations of the hypernucleus H4Lambda"
S. Petschauer "Baryonic forces in SU(3) chiral effective field theory"
N. Zachariou "Determination of Polarization Observables ... for FSIs in $\gamma d \to K^+\Lambda n$"
B. Gibson "Analysis of a possible nnΛ resonance"
H. Kamada "A Λnn three-body resonance"

5:30-7:50 pm Reception sponsored by the Few-body Topical Group of the APS

7:50 pm POSTER SESSION ENDS

Wednesday, May 20

9-10:30 am *The electroweak structure of few-nucleon systems*
Chair: G. Orlandini (Trento)
A. Antognini (PSI) "Spectroscopy of muonic atoms: atomic to nuclear & particle physics"
N. Fomin (Tennessee) "Hadronic weak interaction studies at the SNS"
S. Pastore (South Carolina) "Electromagnetic structure of light nuclei"

10:30-11 am Coffee break

11 am-1 pm *Nuclear forces twenty years after AV18*
Chair: R. Wiringa (Argonne)
S. Aoki (Kyoto) "Hadron interactions from lattice QCD"
A. Lovato (Argonne) "Electromagnetic and neutral-weak response functions of ^{4}He & ^{12}C"
J. Arrington (Argonne) "Recent results on short-range correlations in light nuclei"
D. Lee (North Carolina State) "Nuclear scattering from ab initio lattice simulations"

1 pm-... Lunch and free afternoon

6:30-9:30 pm Conference Banquet at the Penthouse at Wyndham Grand Chicago Riverfront Hotel

Thursday, May 21

8:30-10:20 am Parallel sessions:

Salon A: Large-x physics (Chair: Cloet)

R. Holt (Argonne) "Quarks in Few Body Systems"

P. Reimer "Nuclear structure and the flavor dependence of the EMC effect"

A. Bashir "From continuum QCD to hadron form factors"

C. Chen "Higher-order effective fermion interactions and emergent symmetries"

F. Gao "Leading twist Distribution Amplitudes of heavy-quark vector mesons"

Salon B: Strange *and exotic matter* (Chair: Gibson)

Y. Sada (RCNP-Osaka) "Exclusive Analysis of the in-flight 3He(K−, Λp)n missing reaction to search for the K−pp bound state"

R. Kezerashvilii "K- pp and K- K- p Clusters"

D. Lonardoni "Strangeness in nuclei and neutron stars: a challenging puzzle"

S. Ando "$^{6}_{\Lambda\Lambda}$He in cluster effective field theory"

I. Filikhin "Hyperon energy in ^{6}He and ^{7}He"

Salon C: Efimov effect and collisions (Chair: Khaykovich)

R. Doerner (Frankfurt) "Observation of the Efimov state of the Helium trimer"

Y. Yan "Incorporating exact two-body propagators for zero-range interactions in N-body..."

J. d'Incao "Three-body physics in strongly correlated spinor condensates"

J. Perez-Rios "Reactivity in ion-neutral high density media"

S. Oryu "Effective Screened Coulomb Potential from e−-e−to 208Pb-208Pb Systems"

Salon D: Halos around A=20 (Chair: Aprahamian)

N. Orr (Caen) "Exploring the structure of the most neutron rich boron and carbon isotopes"

T. Frederico "Momentum distributions in light halo nuclei and structure constraints "

B. Acharya "Properties of ^{22}C at Leading Order in Halo Effective Field Theory"

I. Mazumdar "Finite-range & zero-range interactions to study halo nuclei: a comparative..."

Z. Ren "Investigation of ^{23}N in a three-body model"

10:20-10:50 am Coffee break

10:50-12:40 pm Parallel sessions:

Salon A: Novel structure and states (Chair: Qin)

B. Zou "Some Recent Progress on Hyperon Resonances"

V. Baru "Light-quark mass dependence of the X(3872) as a molecular state"

G. Krein "Mesonic decay of charm hypernuclei λc"

M. Mai "Recoil corrections in antikaon-deuteron scattering"

F. Braghin "Higher-order effective fermion interactions and emergent symmetries"

Salon B: Electromagnetic probes of light nuclei (Chair: Leidemann)

S. Bacca (TRIUMF) "Theory of electromagnetic reactions on light nuclei"

B. Schlimme "Deuteron form factor measurements at low momentum transfer"

D. Rozpedzik "Electron and photon scattering on deuteron with the improved chiral NN..."

G. Laskaris "Photodisintegration of ^{3}He with double polarizations"

V. Sulkosky "Results on Double-polarization Asymmetries in the Quasielastic ^{3}He(e,e'd)..."

Salon C: Connecting few- and many-body physics (Chair: Schulz)
S. Riemann (Lund) "Confined Bose-Einstein condensates under rotation: From the few-body to the thermodynamic limit"
D. Hudson-Smith "Inducing Resonant Interactions in Ultracold Atoms with an Oscillating…"
P. Johnson "Nonuniversal effective interactions for harmonically trapped ultracold atoms "
K. Daily "Few-body systems in the adiabatic hyperspherical representation"
R. Wooten "Few-body treatment of the quantum Hall system"

Salon D: Relativity (Chair: Frederico)
W. Polyzou (Iowa) "Relativistic Few-Body Methods"
M. Hadizadeh "The Relativistic Three-Nucleon Bound State in Three-Dimensions"
V. Karmanov "Direct Bethe-Salpeter solutions in Minkowski space"
S. Leitao "Quarkonia and heavy-light mesons in a covariant quark model"
A. Shebeko "Clothed Particles in QED and QCD"

12:40-2:20 pm Lunch break

2:20-3:50 pm *The nucleon as a few-body problem*
Chair: M. Pennington (JLab)
I. Cloet (Argonne) "Perspectives on hadron structure from continuum QCD"
H. Gao (Duke) "Experimental progress and status on nucleon form factors"
V. Mokeev (JLab) "Studies of excited nucleon state structure with CLAS and CLAS12"

3:50-4:20 pm Coffee break

4:20-5:50 pm *Nuclear reactions as a few-body problem*
Chair: C. Elster (Ohio)
K. Jones (Tennessee) TBA
H. Simon (GSI) "Exploring the proton and neutron-rich continuum"
A. Spyrou (Michigan State) "Neutron-neutron correlations in the decay of light neutron-unbound nuclei"

Friday, May 22

8:30-10:20 am Parallel sessions:

Salon A: Novel facilities and methods (Chair: El-Bennich)

J. Qiu (Iowa State) "Electron-ion collider: taking us to the next QCD frontier"
T. Keppel "Exploring Pion and Nucleon Structure with Tagged Structure Functions"
G. Eichmann "Two-photon physics from the quark level"
Y.-X. Liu "Few-body methods and results for hadrons in medium"
H. Sanchis-Alepuz "Glueballs in the Dyson-Schwinger framework: Asymptotics & ..."

Salon B: Nucleon-nucleon scattering (Chair: Howell)

E. Epelbaum (Ruhr U., Bochum) "Chiral nuclear forces: recent developments"
R. Navarro Perez "The Falsification of Nuclear Forces"
M. Piarulli "Minimally non-local nucleon-nucleon potentials in χEFT at order Q4"
Z. Bagdasarian "Nucleon-Nucleon Scattering at Small Angles, measured at ANKE-COSY"
A. Gasparyan "Nucleon-nucleon scattering in the $^1S_0$ partial wave in the modified..."

Salon C: Advances in the three-body problem (Chair: Kievsky)

O. Rubtsova (Moscow State) "Quantum Scattering Theory in a Discrete Representation"
R. Sekine "Solving time-dependent few-body Schrödinger equation with basis expansion..."
V. Eremenko "About Faddeev-AGS equations for (d, p) reactions on heavy nuclei"
V. Suslov "The general configuration-space Faddeev formalism for studying pd scattering"
A. Lovell "Two neutron decay of ^{16}Be"

Salon D: Efimov physics in nuclei? (Chair: Freer)

A. Nicholson (Berkeley) "Universal noise and Efimov physics"
J. Kirscher "On the pion-mass dependence of light nuclei"
G. Orlandini "Is the 0+ resonance of the α-particle a 'breathing mode'?"
M. Shalchi "Scaling functions of 2n separation energies of ^{20}C with finite-range potentials"
J. Vanasse "Pionless EFT analysis of three-nucleon scattering at low energy"

10:20-10:50 Coffee

10:50-12:40 Parallel sessions:

Salon A: Hadronic reactions and structure (Chair: Krein)

E. Downie (George Washington) "The A2 nucleon polarizability program at MAMI"
C. Fritzsch "Studies on η meson production in dpcollisions at the ANKE spectrometer"
N. Huesken "Studies on η meson production and...with WASA-at-COSY"
S.-X. Qin "Practical corollaries of transverse Ward-Green-Takahashi identities "
J. Segovia "Strong diquark correlations inside the proton"

Salon B: Weak interactions and symmetry violation (Chair: Bacca)

M. Schindler (S.Carolina) "Hadronic parity violations in a few-body systems"
D. Samart "Parity-violating nucleon-nucleon potential in the 1/Nc expansion"
Y. Uzikov "Proton-Deuteron Scattering and Test of Time-Reversal Invariance"
F. Myhrer "Neutrino pion production off deuteron"
J. Golak "Break-up Channels in Muon Capture on 3He"

Salon C: Low-dimensional systems (Chair: Riemann)

M. Saffmann (Wisconsin) "Coherence and entanglement of few-atom ensembles"

J. Levinsen "Strong-coupling ansatz for the 1D Fermi gas in a harmonic potential"
D. Blume "Static and dynamic properties of one-dimensional ultracold few-atom systems "
N. Harshman "Few-Body Symmetries and Algebraic Solvability in One-Dimensional Traps"
A. Volosniev "Simulating Effective Spin Chains with Strongly Interacting Atomic Gases"

Salon D: Into the continuum with ab initio methods (Chair: Lee)
A. Gezerlis (Guelph) "Chiral 2N and 3N interactions and quantum Monte Carlo applications"
I. Shin "^{6}Li in ab initio approach: no core full configuration and Gamow shell model"
K. Wendt "Infrared extrapolations for the no-core shell model"
F. Raimondi "Ab initio many-body calculations of single-nucleon transfer reactions with ...."
J. Hwang "Study of ^{19}C by one-neutron knockout"

12:40-2:20 pm Lunch break

2:20-3:50 pm *Meson spectroscopy*
Chair: T. Keppel (JLab)
M. Pennington (JLab) "Precision focus on the meson spectrum opens new window on QCD confinement"
C.-P. Shen (IHEP-Beijing) "Recent empirical developments in the study and understanding of XYZ states"
E. Braaten (Ohio State) "Towards a Theoretical Understanding of the XYZ Mesons from QCD"

3:50-4:05 pm Coffee break
4:05-5:35 pm *Few-body physics in cold atomic gases: beyond the Efimov effect*
Chair: C. Chin (Chicago)
M. Gattobigio (Nice) "Universality in Efimov physics: exploration in the N-body sector"
H. Zhai (Tsinghua) "Few-body problem in spin-orbit coupled quantum gases"
R. Grimm (Innsbruck) TBA

5:35-5:45 pm Closing

edp sciences

EPJ Web of Conferences

Statement of Peer review

In submitting conference proceedings to EPJ Web of Conferences, I certify to the Publisher that I adhere to the **Policy on Publishing Integrity** of the journal in order to safeguard good scientific practice in publishing.

1. All articles have been subjected to peer review administered by the proceedings editors
2. Reviews have been conducted by expert referees, who have been requested to provide unbiased and constructive comments aimed, whenever possible, at improving the work.
3. Proceedings editors have taken all reasonable steps to ensure the quality of the materials they publish and their decision to accept or reject a paper for publication has been based only on the merits of the work and the relevance to the journal.

Title, date and place of the conference

21st International Conference on Few-body Problems in Physics
May 18-22, 2015
Chicago, IL, USA

Proceedings editor(s)

Ch. Elster

D R Phillips

C D Roberts

Date and editor's signature

February 10, 2016

17 avenue du Hoggar - PA de Courtabœuf - BP 112 - 91944 Les Ulis Cedex A (France)
Tel: 33 (0)1 69 18 75 75 Fax : 33(0)1 69 07 45 17 – www.edpsciences.org

EPJ Web of Conferences **113**, 01001 (2016)
DOI: 10.1051/epjconf/201611301001

Hadron Resonances from QCD

Jozef J. Dudek[1,2,a] for the Hadron Spectrum Collaboration

[1] *Department of Physics, Old Dominion University, Norfolk, VA 23529, USA*
[2] *Theory Center, Jefferson Lab, 12000 Jefferson Avenue, Newport News, VA 23606, USA.*

Abstract. I describe how hadron-hadron scattering amplitudes are related to the eigenstates of QCD in a finite cubic volume. The discrete spectrum of such eigenstates can be determined from correlation functions computed using lattice QCD, and the corresponding scattering amplitudes extracted. I review results from the **Hadron Spectrum Collaboration** who have used these finite volume methods to study $\pi\pi$ elastic scattering, including the ρ resonance, as well as coupled-channel πK, ηK scattering. The very recent extension to the case where an external current acts is also presented, considering the reaction $\pi\gamma^\star \to \pi\pi$, from which the unstable $\rho \to \pi\gamma$ transition form factor is extracted. Ongoing calculations are advertised and the outlook for finite volume approaches is presented.

1 Introduction

Quantum Chromodynamics presents us with a unique challenge within the Standard Model – the observed spectrum of strongly interacting particles, hadrons, does not appear to trivially follow from the basic quark and gluon fields whose interactions are described by the QCD lagrangian. Indeed the observed experimental spectrum shows a degree of regularity yet to be explained from first principles, and, to date, seems to lack states one might expect within QCD where the gluonic field plays an active role: glueballs and hybrid mesons.

The excited spectroscopy of QCD is the study of *resonances* – excited hadrons are not asymptotic states, they decay rapidly to lighter stable hadrons, and appear in experiments as enhancements in the scattering of these stable hadrons. It follows that to rigorously study the excited spectrum of QCD we should calculate *scattering amplitudes* and examine their resonant content.

Lattice QCD is a first-principles approach to the theory which considers the quark and gluon fields on a finite discretized grid of points, computing correlation functions as the average over a set of Monte-Carlo generated field configurations. The spectrum of eigenstates for example can be determined by considering the Euclidean time dependence of two-point correlation functions using operators having the quantum numbers of hadrons, constructed from quark and gluon fields. Lattice QCD is thus a controlled approximation in which the artificially introduced scales, the lattice spacing and the lattice volume, can be varied, and ultimately the true theory reached in the appropriate limits. In practice many calculations also opt to work with values of the light-quark mass parameter larger than that found experimentally, thus reducing the computational cost.

[a] e-mail: dudek@jlab.org

By necessity Lattice QCD calculations are done in a finite Euclidean volume, and in such a box, where asymptotic states cannot be prepared, there is no scattering in the usual sense, so it is not immediately obvious how one is to go about computing scattering amplitudes. Fortunately, it is the case that the *discrete spectrum of states* in a finite volume is determined by the scattering amplitudes, and by computing the spectrum in one or more volumes, we may hope to determine these scattering amplitudes. Then by examining the energy dependence of these amplitudes, we may find their resonant content and make statements about the excited spectrum of QCD.

2 Scattering amplitudes & the spectrum in a finite volume

The effect of a finite periodic boundary on the spectrum of eigenstates of a quantum system can be illustrated by considering non-relativistic quantum mechanics in one space dimension. Free particles of definite momentum are described by plane-wave wavefunctions, e^{ipx}, and application of a periodic boundary condition at $x = L$ indicates that free-particle momentum must be quantized in units of $\frac{2\pi}{L}$. For two identical particles interacting via an arbitrary finite-range potential, applying the boundary condition to a scattering solution outside the range of the potential, $\cos(p|x| + \delta(p))$, leads to a condition $p = \frac{2\pi}{L} - \frac{2}{L}\delta(p)$, which determines the allowed discrete values of momentum (and hence energy) in terms of the elastic scattering phase shift, δ. It follows that the discrete spectrum at any given volume, L, contains information about the scattering amplitude in infinite volume.

The more complicated case of elastic scattering in a quantum field theory in $3 + 1$ dimensions, having a cubic periodic boundary, was worked out by Lüscher for the rest frame, and extensions to moving frames appeared later [1–5]. Conceptually the result is similar to that above, with some slight complications arising because the cubic symmetry of the boundary is not commensurate with the angular-momentum basis we use to characterize partial waves. These complications can be handled by considering the reduced symmetry group of the boundary, at rest or boosted, which has a finite number of irreducible representations.

Going beyond elastic scattering, the formalism for *coupled-channel* scattering has been derived independently in a number of differing approaches to field theory [6–9], all leading to the same condition relating the scattering t-matrix to the discrete spectrum of states in a finite volume:

$$\det\left[\mathbf{t}^{-1}(E) + i\rho(E) - \mathbf{M}(E, L)\right] = 0. \tag{1}$$

The determinant is over the space of open scattering channels[1], and $\rho(E)$ is the phase space matrix for these channels, with $\mathbf{M}(E, L)$ being a matrix of known, essentially kinematic, functions. The discrete values of $E(L)$ which solve this equation provide the spectrum of QCD in this finite $L \times L \times L$ volume. It follows that a first-principles calculation of this spectrum in one or more volumes could yield information about $\mathbf{t}(E)$, and we turn our attention to lattice QCD as the tool to determine the spectrum.

3 Determining the spectrum using lattice QCD

Two-point Euclidean correlation functions have a spectral representation,

$$\langle 0|\mathcal{O}_i(t)\mathcal{O}_j^\dagger(0)|0\rangle = \sum_n e^{-E_n t}\langle 0|\mathcal{O}_i|n\rangle\langle 0|\mathcal{O}_j|n\rangle^*, \tag{2}$$

[1] and over the partial waves subduced into the particular irreducible representation under consideration

where the discrete spectrum of eigenstates of QCD having the quantum numbers of the operators $O_{i,j}$ appears. These operators should be constructed from the basic discretized quark and gluon fields of lattice QCD, and ideally to determine the spectrum efficiently we would like to make the values of $\langle 0 | O_{i,j} | \mathfrak{n} \rangle$ as large as possible for the low-lying states we are interested in. This can be achieved by developing operators which 'closely resemble' the eigenstates.

A set of operators which have been shown to efficiently interpolate 'single-meson' states [10–13] take the form $\bar{\psi}\Gamma \overleftrightarrow{D} \dots \overleftrightarrow{D} \psi$, i.e. a (smeared) quark bilinear featuring Dirac gamma structure and some number of gauge-covariant derivatives. Operators resembling a pair of mesons can be constructed as products of the bilinears presented above, e.g. $\sum_{\vec{p}_1 \vec{p}_2} C(\vec{p}_1, \vec{p}_2)\, \pi(\vec{p}_1)\, \pi(\vec{p}_2)$ where $\pi(\vec{p}) \sim \sum_{\vec{x}} e^{i\vec{p}\cdot\vec{x}} \bar{\psi} \mathbf{\Gamma} \psi(\vec{x})$ where $\mathbf{\Gamma}$ is such that the bilinear has pseudoscalar quantum numbers and overlaps strongly with the ground state pion. The generalized Clebsch-Gordan coefficients, $C(\vec{p}_1, \vec{p}_2)$, are chosen to ensure that the operator transforms in the desired irreducible representation of the reduced rotational group.

Calculation of correlation functions using these operators can be done in an efficient manner using the *distillation* technology first presented in Ref. [14]. A spectrum of excited levels can be extracted from a *matrix* of correlation functions, computed using a basis of operators of the type described above. The variational best estimate for the spectrum is obtained by solving the generalized eigen-problem [15–17],

$$\mathbf{C}(t)\, v^{\mathfrak{n}} = \lambda_{\mathfrak{n}}(t)\, \mathbf{C}(t_0)\, v^{\mathfrak{n}}, \tag{3}$$

where the eigenvalue $\lambda_{\mathfrak{n}}(t) \to e^{-E_{\mathfrak{n}}(t-t_0)}$ supplies us with the energy of the $\mathfrak{n}^{\text{th}}$ state.

4 Elastic $\pi\pi$ scattering and the ρ resonance

One of the simplest scattering problems in hadron physics is $\pi\pi$ in isospin-2, where experimentally there are found to be no low-lying resonances. The minimal quark content of this channel is $uu\bar{d}\bar{d}$ and a suitable basis of operators resembles pairs of operators capable of interpolating pions: $\sum_{\vec{p}_1 \vec{p}_2} C(\vec{p}_1, \vec{p}_2)\, \pi(\vec{p}_1)\, \pi(\vec{p}_2)$. The **Hadron Spectrum Collaboration** computed [13] finite volume energy spectra in a calculation where the light-quark masses took a value such that $m_\pi = 391\,\text{MeV}$. The spectra were obtained on three different volumes and in several different moving frames. Using the elastic Lüscher formalism and its moving frame extensions, phase shift values spanning the entire elastic region (below 4π threshold) for S and D waves were obtained as presented in Figure 1. The results show the same qualitative features as experiment, that of a weakly repulsive interaction which is strongest in S-wave.

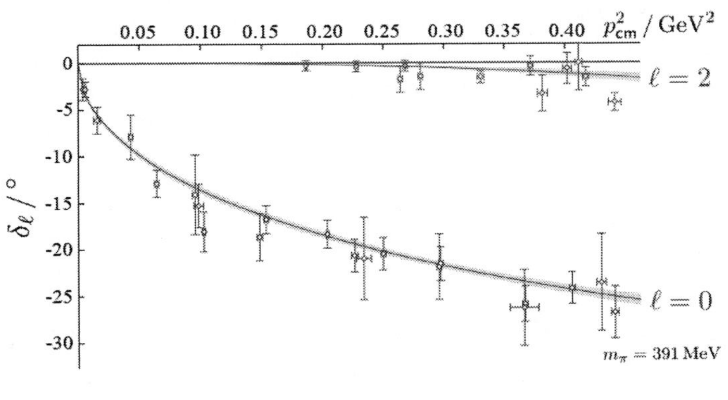

Figure 1. Elastic scattering phase shift in degrees for isospin-2 $\pi\pi$ in S-wave and D-wave plotted as a function of cm-frame momentum. Scattering length parameterizations describe the energy dependence. Computed with quark masses corresponding to a pion mass of 391 MeV. Figure taken from [13].

A slightly less simple scattering problem is $\pi\pi$ in isospin-1, where on the basis of experiment, we expect a vector resonance, the ρ. A suitable operator basis for the computation of correlation functions supplements the "$\pi\pi$-like" operators presented previously with $\bar{\psi}\mathbf{\Gamma}\psi$ operators likely to overlap well with a tightly-bound $q\bar{q}$ state (as we expect the ρ to dominantly be). Computation [18] on the same set of lattices as described above leads to the elastic scattering phase shift presented in Figure 2. The energy dependence of the scattering amplitude can be described well by a relativistic Breit-Wigner form corresponding to a single isolated resonant state. The small width $\Gamma \sim 12\,\mathrm{MeV}$ of the extracted resonance is a result of the small phase space for decay of an 855 MeV resonance into two 391 MeV pions. If we define a $\rho \rightarrow \pi\pi$ coupling via $\Gamma = \frac{g_R^2}{6\pi}\frac{p^3}{m_R^2}$, the extracted coupling has value $g_R = 5.70(11)$, which is compatible with the experimental value [19] of $\sim 6.0(1)$, in line with the expectation that it is almost quark mass independent [20], and only the phase space is changing as the pion mass decreases to its physical value.

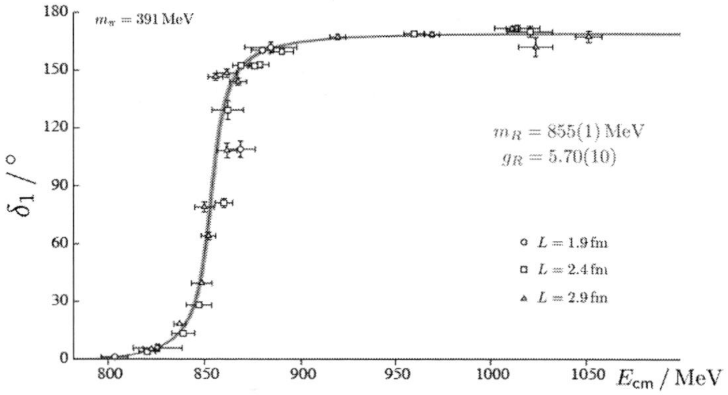

Figure 2. Elastic scattering phase shift in degrees for isospin-1 $\pi\pi$ in P-wave plotted as a function of cm-frame energy. A relativistic Breit-Wigner parameterization describes the energy dependence. Computed with quark masses corresponding to a pion mass of 391 MeV. Figure taken from [18].

A recent calculation [21][2] performed with lighter quark masses such that $m_\pi = 236\,\mathrm{MeV}$ is presented in Figure 3 where we observe the expected increase in width, decrease in ρ mass, and negligible change in g_R.

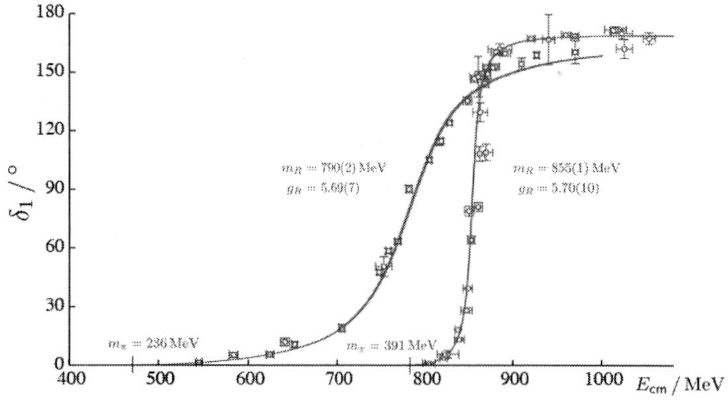

Figure 3. Elastic scattering phase shifts in degrees for isospin-1 $\pi\pi$ in P-wave plotted as a function of cm-frame energy. Comparison of calculations performed with quark masses corresponding to pion masses of 391 MeV and 236 MeV. Figure taken from [21].

[2] this calculation also considered coupled channel amplitudes into $K\overline{K}$, finding essentially no coupling at low energies.

5 Coupled-channel πK, ηK scattering

Refs [22, 23] from the **Hadron Spectrum Collaboration** present the first lattice QCD calculation to extract *coupled-channel* scattering amplitudes, considering the strangeness-1, isospin-1/2 channel, which at low energies features πK and ηK scattering. Spectra were extracted from correlation functions computed with operators resembling πK $(\sum_{\vec{p}_1\vec{p}_2} C(\vec{p}_1,\vec{p}_2)\,\pi(\vec{p}_1)\,K(\vec{p}_2))$, ηK $(\sum_{\vec{p}_1\vec{p}_2} C(\vec{p}_1,\vec{p}_2)\,\eta(\vec{p}_1)\,K(\vec{p}_2))$ and $q\bar{q}$-like constructions $(\bar{\psi}\Gamma\psi)$. Some example spectra are shown in Figure 4 where we see that a large number of energy levels can extracted, even in cases of approximate degeneracy, with good statistical precision.

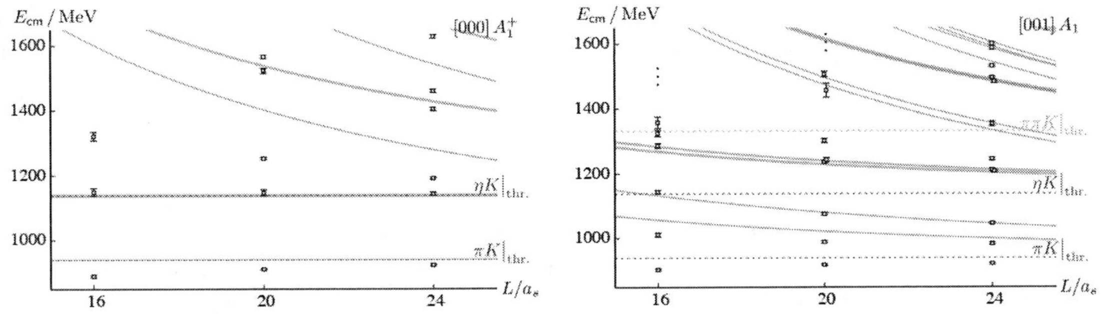

Figure 4. Spectrum of QCD eigenstates with strangeness-1 and isospin-1/2 extracted from lattice QCD calculations on three volumes (16^3, 20^3, 24^3) with quark masses such that $m_\pi = 391$ MeV. (a) Rest frame spectrum in A_1^+ irrep (dominated by $J^P = 0^+$). (b) Spectrum in a frame moving with one unit of lattice momentum in the irrep A_1. Red, green lines show the positions of non-interacting πK, ηK energy-levels respectively. Figure taken from [22, 23].

Extraction of the coupled-channel t-matrix from these levels is challenging. At any given energy the unitarity preserving, time-reversal invariant t-matrix is described by $\frac{1}{2}N(N+1)$ real numbers. This means that Equation 1, which relates the finite volume spectrum to the t-matrix features $\frac{1}{2}N(N+1)$ unknowns for each energy level and this clearly cannot be solved on a level-by-level basis.

An approach to get around this problem is to parameterize the energy dependence of the t-matrix and to attempt to describe the entire spectrum of states, even in multiple different volumes, by varying parameters until the best match to the lattice QCD spectrum is obtained. This was done in [22, 23], trying a varied set of parameterizations based upon the K-matrix, and consistent results for the energy dependence of the amplitudes were obtained. An example is presented in Figure 5 for S-wave, P-wave and D-wave scattering, where the amplitudes are expressed in terms of phase shifts and an inelasticity parameter defined below:

$$t_{\pi K,\pi K} = \frac{1}{2i\rho_{\pi K}}\left(\eta\, e^{2i\delta_{\pi K}} - 1\right),\; t_{\eta K,\eta K} = \frac{1}{2i\rho_{\eta K}}\left(\eta\, e^{2i\delta_{\eta K}} - 1\right),\; t_{\pi K,\eta K} = \frac{1}{2\sqrt{\rho_{\eta K}\rho_{\pi K}}}\left(\sqrt{1-\eta^2}\, e^{i(\delta_{\pi K}+\delta_{\eta K})}\right) \quad (4)$$

The πK P-wave amplitude is found to feature a bound state only slightly below threshold – that this $K^\star$ does not appear as a resonance as it does in experiment is a consequence of the quark mass used, for any smaller light quark mass, the state would appear above threshold with a decay width into πK.

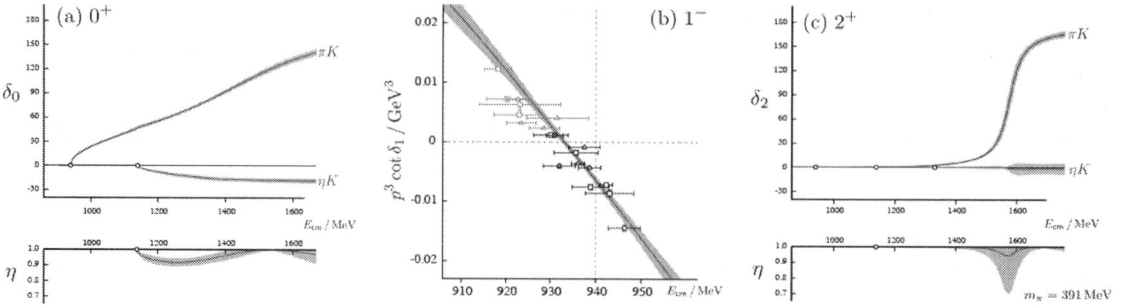

Figure 5. Coupled channel πK, ηK amplitudes determined at $m_\pi = 391$ MeV. (a) Phase shifts and inelasticity in $J^P = 0^+$ partial wave. (b) πK phase shift for $J^P = 1^-$ in threshold region – zero crossing of $p^3 \cot \delta_1$ corresponds to a vector bound state. (c) Phase shifts and inelasticity in $J^P = 2^+$ partial wave – rapid increase in πK phase indicative of a narrow tensor resonance, essentially decoupled from the ηK channel. Figure taken from [22, 23].

The S-wave t-matrix shows only very weak coupling between the πK and ηK channels. The gradual increase in the πK phase shift around 1400 MeV corresponds to a pole singularity far into the complex plane which may be identified with a broad scalar resonance. The experimental $K_0^\star(1430)$, which decays to πK and not ηK may be related to this state. Another S-wave singularity is also present causing the relatively rapid rise in phase shift at πK threshold. This pole lies on the real axis, below the threshold on the unphysical sheet and is thus considered a *virtual* bound state. Suggestions from unitarized chiral perturbation theory are that the experimental resonance singularity known as the κ may become just such a virtual bound state as the mass of the pion is increased [20].

The determination of the $J^P = 2^+$ amplitude is somewhat less rigorous owing to the neglect of the $\pi\pi K$ channel in this first calculation. Nevertheless a consistent description of the finite volume spectrum can be obtained considering only πK and ηK scattering, and the extracted amplitude shows decoupling of the two channels and a clear narrow tensor resonance decaying to πK. This state is likely related to the experimental $K_2^\star(1430)$.

6 $\pi\gamma^\star \to \pi\pi$ and the unstable ρ transition form factor

An extension of the finite volume formalism to deal with the case where external currents are present (building upon the ideas outlined in [24]) has recently appeared [25, 26], suggesting that the finite volume matrix elements of the current which can be extracted from three-point correlation functions can be related to their infinite volume counterparts.

The development of the distillation technology needed to efficiently calculate the required three-point correlation functions in the case of the vector current relevant to electromagnetic amplitudes was presented in [27], and the first calculation of an electromagnetic scattering amplitude, $\pi\gamma^\star \to \pi\pi$, has very recently appeared [28]. In this paper, the **Hadron Spectrum Collaboration** determined matrix elements of the type $\langle \pi | \bar{\psi}\gamma_\mu\psi | \pi\pi \rangle$ on a lattice of volume $\sim (2.4\,\mathrm{fm})^3$ with $m_\pi = 391$ MeV and from these extract the scattering amplitude illustrated in Figure 6. The ρ resonance which couples to both $\gamma\pi$ and $\pi\pi$ clearly dominates the amplitude, and the transition form factor for unstable $\rho \to \pi\gamma^\star$ was extracted as a function of photon virtuality, Q^2, in a rigorous manner from the residue of the ρ pole singularity in the complex energy plane.

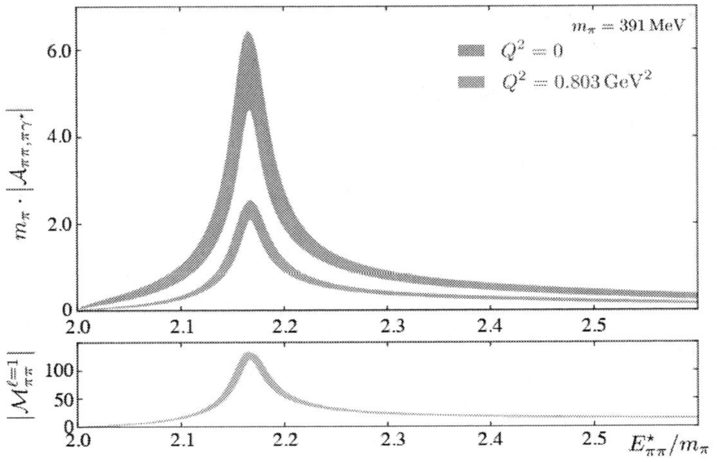

Figure 6. Upper panel: Amplitude for $\pi\gamma^\star \to \pi\pi$ in P-wave as a function of cm-frame energy at two values of photon virtuality – the bump due to the ρ resonance. Lower panel: the corresponding amplitude for elastic scattering $\pi\pi \to \pi\pi$. Both computed on a 20^3 lattice with $m_\pi = 391$ MeV. Figure taken from [28].

7 Outlook

Combining carefully constructed operators with efficient correlation construction technology and variational analysis techniques, it has proven possible to extract large numbers of discrete excited states from lattice QCD calculations. Novel approaches have been applied to extracting scattering amplitudes from these spectra, particularly in the coupled-channel case where unitarity preserving parameterizations are required. Non-resonant and resonant cases have been considered, and recently we have seen the important extension to the case of coupling to external currents, opening up the possibility of considering photo/electro-production amplitudes, weak meson decays and other more exotic situations.

Ongoing calculations within the **Hadron Spectrum Collaboration** include meson-meson scattering systems featuring charmed mesons [29, 30], as well as other coupled-channel systems in the light quark sector such as $\pi\eta$, $K\bar{K}$ in which the a_0 resonance appears. Obvious future extensions of these calculations include systems with baryons, e.g. $N^\star$ resonances in πN scattering, as well as other radiative transitions, including those between charmonium resonances [31, 32] which are well measured experimentally. A formalism to deal with three-hadron scattering amplitudes will be required to study higher-lying resonances, and progress is being made in developing the appropriate theory [33–36].

Acknowledgements

I am indebted to my colleagues within the Hadron Spectrum Collaboration, in particular Raul Briceno, Robert Edwards, Christian Shultz, Christopher Thomas and David Wilson, without whose hard work and original ideas the results presented above could not have been obtained. I am supported by the US Department of Energy through Early Career award contract DE-SC0006765 and contract DE-AC05-06OR23177, under which Jefferson Science Associates, LLC, manages and operates Jefferson Laboratory.

References

[1] M. Luscher, Nucl.Phys. **B354**, 531 (1991)
[2] K. Rummukainen, S.A. Gottlieb, Nucl.Phys. **B450**, 397 (1995), hep-lat/9503028
[3] C. Kim, C. Sachrajda, S.R. Sharpe, Nucl.Phys. **B727**, 218 (2005), hep-lat/0507006

[4] Z. Fu, Phys.Rev. **D85**, 014506 (2012), 1110.0319
[5] L. Leskovec, S. Prelovsek, Phys.Rev. **D85**, 114507 (2012), 1202.2145
[6] S. He, X. Feng, C. Liu, JHEP **0507**, 011 (2005), hep-lat/0504019
[7] M.T. Hansen, S.R. Sharpe, Phys.Rev. **D86**, 016007 (2012), 1204.0826
[8] R.A. Briceno, Z. Davoudi, Phys.Rev. **D88**, 094507 (2013), 1204.1110
[9] P. Guo, J. Dudek, R. Edwards, A.P. Szczepaniak, Phys.Rev. **D88**, 014501 (2013), 1211.0929
[10] J.J. Dudek, R.G. Edwards, M.J. Peardon, D.G. Richards, C.E. Thomas, Phys.Rev.Lett. **103**, 262001 (2009), 0909.0200
[11] J.J. Dudek, R.G. Edwards, M.J. Peardon, D.G. Richards, C.E. Thomas, Phys.Rev. **D82**, 034508 (2010), 1004.4930
[12] C.E. Thomas, R.G. Edwards, J.J. Dudek, Phys.Rev. **D85**, 014507 (2012), 1107.1930
[13] J.J. Dudek, R.G. Edwards, C.E. Thomas, Phys.Rev. **D86**, 034031 (2012), 1203.6041
[14] M. Peardon et al. (Hadron Spectrum), Phys.Rev. **D80**, 054506 (2009), 0905.2160
[15] C. Michael, Nucl.Phys. **B259**, 58 (1985)
[16] M. Luscher, U. Wolff, Nucl.Phys. **B339**, 222 (1990)
[17] B. Blossier, M. Della Morte, G. von Hippel, T. Mendes, R. Sommer, JHEP **0904**, 094 (2009), 0902.1265
[18] J.J. Dudek, R.G. Edwards, C.E. Thomas (Hadron Spectrum), Phys.Rev. **D87**, 034505 (2013), 1212.0830
[19] K. Olive et al. (Particle Data Group), Chin.Phys. **C38**, 090001 (2014)
[20] J. Nebreda, J. Pelaez., Phys.Rev. **D81**, 054035 (2010), 1001.5237
[21] D.J. Wilson, R.A. Briceno, J.J. Dudek, R.G. Edwards, C.E. Thomas (2015), 1507.02599
[22] J.J. Dudek, R.G. Edwards, C.E. Thomas, D.J. Wilson (Hadron Spectrum), Phys.Rev.Lett. **113**, 182001 (2014), 1406.4158
[23] D.J. Wilson, J.J. Dudek, R.G. Edwards, C.E. Thomas, Phys.Rev. **D91**, 054008 (2015), 1411.2004
[24] L. Lellouch, M. Luscher, Commun. Math. Phys. **219**, 31 (2001), hep-lat/0003023
[25] R.A. Briceno, M.T. Hansen (2015), 1502.04314
[26] R.A. Briceno, M.T. Hansen, A. Walker-Loud, Phys.Rev. **D91**, 034501 (2015), 1406.5965
[27] C.J. Shultz, J.J. Dudek, R.G. Edwards, Phys. Rev. **D91**, 114501 (2015), 1501.07457
[28] R.A. Briceno, J.J. Dudek, R.G. Edwards, C.J. Shultz, C.E. Thomas, D.J. Wilson (2015), 1507.06622
[29] G. Moir, M. Peardon, S.M. Ryan, C.E. Thomas, L. Liu, JHEP **1305**, 021 (2013), 1301.7670
[30] L. Liu, G. Moir, M. Peardon, S.M. Ryan, C.E. Thomas, P. Vilaseca, J.J. Dudek, R.G. Edwards, B. Joo, D.G. Richards (Hadron Spectrum), JHEP **07**, 126 (2012), 1204.5425
[31] J.J. Dudek, R.G. Edwards, D.G. Richards, Phys. Rev. **D73**, 074507 (2006), hep-ph/0601137
[32] J.J. Dudek, R. Edwards, C.E. Thomas, Phys. Rev. **D79**, 094504 (2009), 0902.2241
[33] K. Polejaeva, A. Rusetsky, Eur.Phys.J. **A48**, 67 (2012), 1203.1241
[34] R.A. Briceno, Z. Davoudi, Phys.Rev. **D87**, 094507 (2012), 1212.3398
[35] M.T. Hansen, S.R. Sharpe (2015), 1504.04248
[36] M.T. Hansen, S.R. Sharpe, Phys.Rev. **D90**, 116003 (2014), 1408.5933

EPJ Web of Conferences **113**, 01002 (2016)
DOI: 10.1051/epjconf/201611301002

Exploring the few- to many-body crossover using cold atoms in one dimension

Nikolaj Thomas Zinner[1,a]

[1] *Department of Physics and Astronomy, Aarhus University, Ny Munkegade 120, DK-8000 Aarhus C, Denmark*

Abstract. Cold atomic gases have provided us with a great number of opportunities for studying various physical systems under controlled conditions that are seldom offered in other fields. We are thus at the point where one can truly do quantum simulation of models that are relevant for instance in condensed-matter or high-energy physics, i.e. we are on the verge of a 'cool' quantum simulator as envisioned by Feynman. One of the avenues under exploration is the physics of one-dimensional systems. Until recently this was mostly in the many-body limit but now experiments can be performed with controllable particle numbers all the way down to the few-body regime. After a brief introduction to some of the relevant experiments, I will review recent theoretical work on one-dimensional quantum systems containing bosons, fermions, or mixtures of the two, with a particular emphasis on the case where the particles are held by an external trap.

1 Introduction

Atomic gases cooled to extremely low temperatures are a very rich and expanding field of physics where one may explore many different features of quantum few- and many-body systems in a highly controlled manner [1–4]. In these systems one has accurate control over both geometry and interactions, and a hotly pursued feature has been the realization of systems in one dimension (1D) [5–12]. The interactions are also controllable in 1D [13] and one may even get to the hard-core regime and study the Tonks-Girardeau limit [14, 15] in experiments [7, 8]. In the latest developments it has now become possible to use 1D setups to reach the limit of small particle number [16] and few-body fermion systems have been realized in several experiments in the Jochim group in Heidelberg [17–21]. A lot of theoretical research has been devoted to explore few-body physics in 1D and also to try to extrapolate into the many-body regime. Here I will review some of these developments with an excusable bias toward some of the work performed in our collaboration. Unless otherwise stated I will focus exclusively on 1D systems with short-range interactions that are implemented by a Dirac delta function. The basic Hamiltonian is

$$H = \sum_i \left[\frac{p_i^2}{2m} + V_e(x_i)\right] + \sum_{i<j} V(x_i - x_j), \tag{1}$$

[a] e-mail: zinner@phys.au.dk

where the sums run over the number of particles, $i = 1, \ldots, N$, x_i is the coordinate and p_i the momentum of the ith particle. The external trapping potential $V_e(x)$ is assumed to be the same for all particles. The mass m is assumed to be the same for all particles in most of the discussion below, except for the section on mixtures where I will explicitly state when different masses are considered. The two-body interactions are given by terms of the form $V(x_i - x_j) = g\delta(x_i - x_j)$, where g is the coupling strength. When I talk about strong interactions it is the limit where $1/g \to 0$. For fermions the interactions are typically only between pairs of opposite spin projection. However, due to the Pauli principle enforcing the antisymmetry of pairs with identical spin projection, the interaction terms in the Hamiltonian above will have no consequence for identical fermion pairs. For bosons with more than one internal component, one has to consider the possibility of interactions between the pairs with the same internal components (intraspecies) and pairs with different internal components (interspecies). Intra- and interspecies interactions do not have to be the same in general. They will all be assumed to be given by Dirac delta functions. In the discussions below I will make it clear whenever this is important.

2 Bosonic systems

A particular focal point of attention in the work on few-boson systems has been the transition from weak to strong interactions. In the presence of strong short-range interactions, a pair of bosons will have a wave function that has to vanish when the two particles overlap. As noticed in the famous paper of Girardeau [15], this means that the ground state may now be described by a totally antisymmetric fermionic wave function if one makes appropriate changes of signs to ensure bosonic symmetry. For instance, if you take N identical bosons with no internal degrees of freedom and put them into some trapping potential, then the ground state for strong interactions is obtained by taking the N lowest single-particle eigenfunctions of the single-particle Schrödinger equation and construction their totally antisymmetry (Slater determinant) N-body wave function, and finally taking the norm of that to make sure that the wave function does not change sign under exchange of any two particles.

A number of studies have explored the physics of few-boson systems as the interactions are changed from weak to strong using the multi-configurational method [22–24] for double-wells and exact diagonalization [25–28] or diffusion Monte Carlo [29, 30] in a harmonic trap. In the latter case of a harmonic trap, Girardeau has given a number of exact solutions for strong interactions [31] that also apply for bosons with multiple internal states (spinor bosons). The fact that the exact solution of the two-body problem in a harmonic trap is available [32] has been used to propose analytical N-body states [33]. This is very similar to using Jastrow factors to incorporate correlations [34, 35].

In recent years, we have explored the physics of two-component Bose systems in harmonic traps, i.e. similar mass bosons with two internal degree of freedom that we denote A and B. In the limit where the AA and BB interactions are negligible compared to AB interactions, we found that one may solve the three-body problem exactly for strong interactions [36]. The ground state turns out to have an energy that is half an odd integer times the harmonic trap quantum, and its wave function (which is doubly degenerate) displays the precursor of ferromagnetism, i.e. the spatial ordering of the particles is either AAB or BAA, but *never* ABA. On the other hand, one can find excited state in the system that *only* display the ABA spatial ordering. In our latest work [37], we confirm that this kind of fractional energy behavior persists for larger systems and we present analytical results for the four-body state (two A and two B atoms) for strong interactions. Indeed, already for balanced ten particle systems with strong AB interactions one can see the perfect ferromagnetic ordering emerge in the harmonic trap with a profile that approximates that expected in the many-body limit [37]. Another very recent study focusing on the strongly interacting limit also finds ferromagnetic behavior in this parameter regime [38].

3 Fermionic systems

The solution of the two-component Fermi system in 1D in known in the homogeneuos case from the nested Bethe ansatz of Gaudin and Yang [39]. However, in the case of a non-trivial confining potential no exact solution to the N-body problem is known for general interaction strength. Here I will focus on harmonically trapped fermions but stress that many of the results and methods can be applied to other external confining geometries as well. An approach under exploration is to use Bethe ansatz in conjunction with a variational principle [40] while also the multi-configurational Hartree method has been applied [41] However, at this point it is safe to say that many studies rely in one way or another on exact diagonalization [42–49]. In the strongly interacting limit, some exact solutions are in fact known [50–52], although we caution that using the spin formalism of for instance Ref. [50] can lead to errors already at the level of four particles [43]. In our recent work on few-fermion physics, we have considered three-body problems both for two- [53] three-component systems [54]. A new variational approach was introduced in Ref. [53] which aims to find an accurate energy and wave function for the three-body problem using knowledge of the extreme limits of vanishing and infinite interaction strength. This provides a very accurate approximation that even extends to mass-imbalanced systems. Most recently, we have have considered dynamics of strongly interacting few-body systems in the case where one has time-dependent external traps. This may lead to dynamical engineering of magnetic quantum states [55].

The interesting question of how to get from the few- to the many-body limit has received much attention lately, in particular for strongly interacting particles. Again, Girardeau has written down some exact solutions to the two-component Fermi system in a trap with strong interactions [56] that are simple and beautiful, and have captured much attention. While they are of great theoretical use, they have little experimental relevance as they are superpositions of states that are adiabatically connected to experimentally relevant states for large but finite interaction strength. This was predicted in Ref. [57] and has recently been confirmed experimentally in Ref. [21]. Studies of the few- to many-body limit have been done using diffusion Monte Carlo [58], exact diagonalization [59, 60, 62], and most recently also coupled-cluster methods from quantum chemistry [63]. A particularly interesting feature is the fact that one may map the strongly interacting problem onto a spin model of the Heisenberg type [60–62], but with nearest-neighbor interactions that now depend on the external confinement and are not constant in general. It is interesting to note that the ansatz solution of the strongly interacting problem presented in Ref. [62] allows one to obtain a very accurate expression for the Heisenberg Hamiltonian in the case of harmonic confinement.

In our group we have pioneered the use of the Lee-Suzuki effective interaction method combined with exact diagonalization for 1D systems with short-range interactions [37, 59]. It provides a large reduction in computational time over traditional exact diagonalization with the bare Dirac delta function interaction. This has allowed us to consider for instance the polaron problem, i.e. a Fermi gas of one component (spin up) and a single (impurity) particle of the other component (spin down), for up to ten particles. Here one see a very characteristic phase separation with the impurity going to the middle of the trap causing a deep indentation in the majority particle density. This is also seen in the recent study of Ref. [62] where the many-body limit could be explored analytically using a very accurate ansatz wave function.

4 Mixed systems

Last but certainly not the least interesting are the mixed systems. This should be understood as those systems which are mixtures of fermions and bosons that may be either of equal or unequal mass. As in all the previous examples, Girardeau has once again been one of the first to consider such systems [64, 65]. In Ref. [64] a very interesting Bose-Fermi mapping was introduced. However, as we have pointed out in Ref. [57], these mappings are not useful for systems under non-trivial external confinement, and have to be applied with great care. While for three-body problems with equal mass Bose-Bose (spinor) mixtures one can indeed obtain the correct results for strong interactions [66], this does not extend to larger systems using Bose-Fermi mappings [57]. In Ref. [67] a four-body Bose-Fermi mixture with equal mass is considered in the strongly interacting regime, and the results are compared to a numerical study of the trapped system using the density-matrix renormalization group (DMRG) method. The Bose-Fermi mapping predicts a totally fermionized state in the strongly interacting limit that is identical to that obtained in a four-body system of identical (spinless or spin-polarized) fermions. This latter state has a characteristic four-hump structure in the harmonic trap. The DMRG calculations in Ref. [67] seem to confirm that this is indeed the strongly interacting ground state. However, applying the formalism of Ref. [57] one may confirm that this is not the case, and that the density akin to four spinless fermions is only found in an excited state. This leads one to suspect that the DMRG (built on a variational principle) may have gotten stuck in an excited state. This is plausible due to the (quasi)-degenerate manifold of state found in the strongly interacting limit [57]. This highlights the need for (semi)-analytical insight into the strongly interacting regime in spite of the success of the DMRG method over the past decades.

The problem of mixtures of Bose and Fermi atoms has already an intricate structure at the level of three atoms [54, 68, 69]. In particular, the strongly interacting limit and the (quasi)-degenerate manifold of state for large but finite interaction depend delicately on the composition, i.e. whether one considers two bosons and a fermion (all of equal mass) or vice versa [68, 69], or one can consider taking three altogether different components [54]. In light of this it is a very interesting question how can we determine which final state for strong interaction will be reach given an initial state for vanishing interaction strength? More precisely, if we adiabatically increase the interacting strength from zero to a very large value (effectively infinite), which state is reached? This question was explored in great detail in Ref. [70], and further elaborated by the same author in Ref. [71]. Using the symmetries of the problem and general mathematical arguments, Ref. [71] argues that for the case of four particles or less one may solve this issue by using the symmetries of the system. On the other hand, when the particle number is larger than four, one can no longer expect this and must resort to other means. This is an interesting question that certainly deserves further exploration.

Another important mixed system is the unequal mass case [72–76]. Some recent results have demonstrated a phase separation in the light component for Fermi-Fermi mixtures [75], which can also be seen for a single heavy impurity in a light Bose system [74]. In fact, the case of a single impurity in a Bose gas is often called the 'Bose polaron' problem and is much studied at present in the few-body limit [73, 74, 76]. In our recent work we have explored the dynamics when the boson-impurity interaction is suddenly switched on. While the formalism used there works for general dimensions, we focused on the 1D case in Ref. [73] where this kind of quench protocol leads to oscillatory behavior in the weakly interacting regime. In Ref. [74] we present our results on the static configuration of the Bose polaron using a new formalism that is a hybrid of the hyperspherical adiabatic approach. What is particularly interesting about this latter work is the fact that one can extrapolate to large particle numbers with miniscule computational effort. In fact, as we argue in Ref. [74] we expect the approximations we make to work better for larger systems. By comparison to numerically exact results [37] for systems with ten particles or less we confirm that the (semi)-analytical method put forth in Ref. [74] is accurate within a few percent already for eight bosons and an impurity. In addition, this new formalism can handle any trapping potential that can be different for the bosons and the impurity, it can tackle any mass ratios, and the traps may even be displaced from one another.

5 Outlook

As I have argued above, the experimental situation is very promising for studying few-body physics in 1D, as well as for gaining knowledge about how one extrapolates to the many-body regime. I like to call this few-body physics in a many-body world. A considerable body of theoretical work has been put forth in this area and more is certainly to come. Many new avenues are being explored and I believe that new aspects of 1D physics will be found. I think that an important outstanding question concerns the notion of strong in real life. What does 'strong interaction' really mean? Above I have used the mathematical definition which means that you take a coupling constant to infinity. Some recent experimental results [21] strongly suggest that we understand that limit [57]. This was done with relatively few particles and I would really like to see how things change as we increase the number of particles, preferably one at a time. I am also very interested in exploring dynamics in 1D systems with strong interactions such as quantum state transfer [61] and dynamical quantum magnetic effects [55]. An intriguing question that comes to mind in this respect is that of spin-charge separation. It is know from 1D Luttinger liquid formalism that the low-energy excitations of a two-component Fermi gas in 1D split into charge density and spin density waves, and furthermore that these waves have different velocities, hence we get spin-charge separation. As far as I know, at this point it has not been unambiguously observed in electronic systems such as nanowires. What I would really like to see is how this effect emerges from a few-body perspective.

Acknowledgements

I thank my collaborators Artem Volosniev, Amin Dehkharghani, Oleksandr Marchukov, Dmitri Fedorov, Aksel Jensen, Manuel Valiente, David Petrosyan, Jonathan Lindgren, Christian Forssén, Jimmy Rotureau, and Hans-Werner Hammer. The experimental group of Selim Jochim in Heidelberg is gratefully acknowledged for discussions and for making their data available to us. This work is supported in part by the Danish Council for Independent Research DFF and the DFF Sapere Aude program.

References

[1] I. Bloch, J. Dalibard, and W. Zwerger, Rev. Mod. Phys. **80**, 885 (2008).
[2] M. Lewenstein *et al.*, Adv. Phys. **56**, 243 (2007).
[3] T. Esslinger, Ann. Rev. Cond. Mat. Phys. **1**, 129 (2010).
[4] N. T. Zinner and A. S. Jensen, J. Phys. G:Nucl. Part. Phys. **40**, 053101 (2013).
[5] H. Moritz, T. Stöferle, M. Köhl, and T. Esslinger, Phys. Rev. Lett. **91**, 250402 (2003).
[6] T. Stöferle, H. Moritz, C. Schori, M. Köhl, and T. Esslinger, Phys. Rev. Lett. **92**, 130403 (2004).
[7] T. Kinoshita, T. Wenger, and D. S. Weiss, Science **305**, 1125 (2004).
[8] B. Paredes *et al.*, Nature **429**, 277 (2004).
[9] T. Kinoshita, T. Wenger, and D. S. Weiss, Nature **440**, 900 (2005).
[10] E. Haller *et al.*, Science **325**, 1224 (2009).
[11] E. Haller *et al.*, Nature **466**, 597 (2010).
[12] G. Pagano *et al.*, Nature Phys. **10**, 198 (2014).
[13] M. Olshanii, Phys. Rev. Lett. **81**, 938 (1998).
[14] L. W. Tonks, Phys. Rev. **50**, 955 (1936).
[15] M. D. Girardeau, J. Math. Phys. **1**, 516 (1960).
[16] F. Serwane *et al.*, Science **332**, 336 (2011).

[17] G. Zürn *et al.*, Phys. Rev. Lett. **108**, 075303 (2012).
[18] G. Zürn *et al.*, Phys. Rev. Lett. **111**, 175302 (2013).
[19] A. Wenz *et al.*, Science **342**, 457 (2013).
[20] S. Murmann *et al.*, Phys. Rev. Lett. **114**, 080402 (2015).
[21] S. Murmann *et al.*, arXiv:1507.01117 (2015).
[22] S. Zöllner, H.-D. Meyer, and P Schmelcher, Phys. Rev. A **74**, 063611 (2006); Phys. Rev. A **75**, 043608 (2007); Phys. Rev. Lett. **100**, 040401 (2008).
[23] E. Tempfli, S. Zöllner, and P. Schmelcher, New J. Phys. **11**, 073015 (2009).
[24] S. Zöllner, H.-D. Meyer, and P Schmelcher, Phys. Rev. A **78**, 013629 (2008);
[25] F. Deuretzbacher, K. Bongs, K. Sengstock, and D. Pfannkuche, Phys. Rev. A **75**, 013614 (2007).
[26] M. A. Garcia-March and Th. Busch, Phys. Rev. A **87**, 063633 (2013).
[27] S. Campbell, M. A. Garcia-March, T. Fogarty, and Th. Busch, Phys. Rev. A **90**, 013617 (2014).
[28] M. A. Garcia-March *et al.*, Phys. Rev. A **90**, 063605 (2014).
[29] M. A. Garcia-March *et al.*, Phys. Rev. A **88**, 063604 (2013).
[30] M. A. Garcia-March *et al.*, New J. Phys. **16**, 103004 (2014).
[31] M. D. Girardeau, Phys. Rev. A **83**, 011601(R) (2011).
[32] T. Busch, B.-G. Englert, K. Rzążewski, and M. Wilkens, Found. Phys. **28**, 549 (1998).
[33] I. Brouzos and P. Schmelcher, Phys. Rev. Lett. **108**, 045301 (2012).
[34] I. Brouzos and A. Förster, Phys. Rev. A **89**, 053632 (2014).
[35] B. Wilson, A. Förster, C. C. N. Kuhn, I. Roditi, and D. Rubeni, Phys. Lett. A **378**, 1065 (2014).
[36] N. T. Zinner *et al.*, Europhys. Lett. **107**, 60003 (2014).
[37] A. S. Dehkharghani *et al.*, Scientific Reports **5**, 10675 (2015).
[38] P. Massignan, J. Levinsen, and M. M. Parish, arXiv:1507.02814 (2015).
[39] M. Gaudin, Phys. Lett., Sect. A **24**, 55 (1967); C. N. Yang, Phys. Rev. Lett. **19**, 1312 (1967).
[40] D. Rubeni, A. Förster, and I. Roditi, Phys. Rev. A **86**, 043619 (2012).
[41] I. Brouzos and P. Schmelcher, Phys. Rev. A **87**, 023605 (2013).
[42] P. O. Bugnion and G. J. Conduit, Phys. Rev. A **87**, 060502(R) (2013).
[43] S. E. Gharashi and D. Blume, Phys. Rev. Lett. **111**, 045302 (2013).
[44] T. Sowiński, T. Graß, O. Dutta, and M. Lewenstein, Phys. Rev. A **88**, 033607 (2013).
[45] P. D'Amico and M. Rontani, J. Phys. B **47**, 065303 (2014); Phys. Rev. A **91**, 043610 (2015).
[46] S. E. Gharashi, X. Y. Yin, and D. Blume, Phys. Rev. A **89**, 023603 (2014).
[47] R. Lundmark, C. Forssén, and J. Rotureau, Phys. Rev. A **91**, 041601(R) (2015).
[48] S. E. Gharashi, X. Y. Yin, and D. Blume, Phys. Rev. A **91**, 013620 (2015).
[49] T. Sowiński, M. Gajda, and K. Rzążewski, Europhys. Lett. **109**, 26005 (2015).
[50] L. Guan, S. Chen, Y. Wang, and Z.-Q. Ma, Phys. Rev. Lett. **102**, 160402 (2009).
[51] C. N. Yang, Chin. Phys. Lett. **26**, 120504 (2009).
[52] X. Cui and T.-L. Ho, Phys. Rev. A **89**, 023611 (2014).
[53] N. J. S. Loft *et al.*, Eur. Phys. J. D **69**, 65 (2015).
[54] A. G. Volosniev *et al.*, Few-Body Syst. **55**, 839 (2014).
[55] A. G. Volosniev, H.-W. Hammer, and N. T. Zinner, arXiv:1507.00186 (2015).
[56] M. D. Girardeau, Phys. Rev. A **82**, 011607(R) (2010).
[57] A. G. Volosniev *et al.*, Nature Commun. **5**, 5300 (2014).
[58] G. E. Astrakharchik and I. Brouzos, Phys. Rev. A **88**, 021602(R) (2013).
[59] E. J. Lindgren *et al.*, New J. Phys. **16**, 063003 (2014).

[60] F. Deuretzbacher *et al.*, Phys. Rev. A **90**, 013611 (2014).
[61] A. G. Volosniev *et al.*, Phys. Rev. A **91**, 023620 (2015).
[62] J. Levinsen, P. Massignan, G. M. Bruun, and M. M. Parish, Science Adv. **1**, e1500197 (2015).
[63] T. Grining *et al.*, arXiv:1507.03174 (2015); arXiv:1508.02378 (2015).
[64] M. D. Girardeau and M. Olshanii, Phys. Rev. A **70**, 023608 (2004).
[65] M. D. Girardeau and A. Minguzzi, Phys. Rev. Lett. **99**, 230402 (2007).
[66] F. Deuretzbacher *et al.*, Phys. Rev. Lett. **100**, 160405 (2008).
[67] B. Fang, P. Vignolo, M. Gattobigio, C. Miniatura, and A. Minguzzi, Phys. Rev. A **84**, 023626 (2011).
[68] N. L. Harshman, Phys. Rev. A **86**, 052122 (2012).
[69] R. E. Barfknecht, I. Brouzos, A. Förster, Phys. Rev. A **91**, 043640 (2015).
[70] N. L. Harshman, Phys. Rev. A **89**, 033633 (2014).
[71] N. L. Harshman, arXiv:1501.00215 (2015); arXiv:1505.00659 (2015).
[72] N. P. Mehta, Phys. Rev. A **89**, 052706 (2014).
[73] A. G. Volosniev, H.-W. Hammer, and N. T. Zinner, Phys. Rev. A **92**, 023623 (2015).
[74] A. S. Dehkharghani, A. G. Volosniev, and N. T. Zinner, arXiv:1503.03725 (2015).
[75] D. Pęcak, M. Gajda, and T. Sowiński, arXiv:1506.03592 (2015).
[76] N. P. Mehta, arXiv:1508.00156 (2015).

EPJ Web of Conferences **113**, 01003 (2016)
DOI: 10.1051/epjconf/201611301003

Cluster Symmetries and Dynamics

Martin Freer[1,a]

[1] *School of Physics and Astronomy, University of Birmingham, Birmingham, B15 2TT, UK*

Abstract. Many light nuclei display behaviour that indicates that rather than behaving as an A-body systems, the protons and neutrons condense into clusters. The α-particle is the most obvious example of such clustering. This contribution examines the role of such α-clustering on the structure, symmetries and dynamics of the nuclei ^{8}Be, ^{12}C and ^{16}O, recent experimental measurements and future perspectives.

1 Introduction

The behaviour of nuclei as a function of their internal energy is rich and varied. Collective degree of freedom result of transferal to rotational modes, above decay threshold particle emission is possible and if those decay thresholds are connected with the decay of composite particles, e.g. α-particles, then cluster emission can occur. Particularly in light nuclei, the nature of the intrinsic structure prior to such cluster decays is found to be closely related to the decay partition. This has been encapsulated in the Ikeda diagram [1] (Fig. 1). Here as the internal energy of the nucleus is increased to the point an α-decay threshold is encountered then the nucleus is able to dissipate the internal energy into the mass of the clusters. For example, at an excitation energy of 7.27 MeV the 3α decay threshold in ^{12}C is encountered. The Ikeda picture would suggest that at this point that one possible structural mode for ^{12}C is that of 3 α-clusters. Interestingly, the 3α-decay threshold is lower than in energy than the ^{8}Be+α two-body decay threshold (Fig. 2). The ^{8}Be nucleus is, however, itself unbound to decay to 2α-particles by 93 keV. So by the same arguments ^{8}Be would contain a 2α-cluster structure.

Excited states located just above such decay thresholds thus have the maximum structural overlap with such clusters, with increasing excitation energy allowing additional degrees of freedom to mix. Famously, the 7.65 MeV Hoyle state (Fig. 2) has been linked to 3α-cluster structure (as outlined in Ref. [2] and references therein), which has been linked, in turn, to the triple α-process [3, 4] responsible for the synthesis of carbon-12. Here, first two α-particles fuse to form ^{8}Be whose α-decay results in an equilibrium concentration of ^{8}Be. The second step is the capture by ^{8}Be to form ^{12}C at an excitation above the ^{8}Be+α threshold followed by electromagnetic decay to the ^{12}C ground state (see Fig. 2). Hoyle recognised the need for a $J^{\pi} = 0^{+}$ state close to this energy in order to account for the absolute abundance of ^{12}C and the relative abundance of ^{12}C and ^{16}O [5]. The presence of the Hoyle-state at 7.65 MeV resonantly boosts the capture process by a factor of close to 10-100 million. Hoyle predicted the existence of a state at 7.68 MeV [5], and whilst visiting Caltech, convinced the local group to search for the state. They measured the ^{14}N(d,α)^{12}C reaction using a high resolution spectrometer at which point a state at 7.68 MeV was observed [6]. Subsequent measurements refined

[a]e-mail: M.Freer@bham.ac.uk

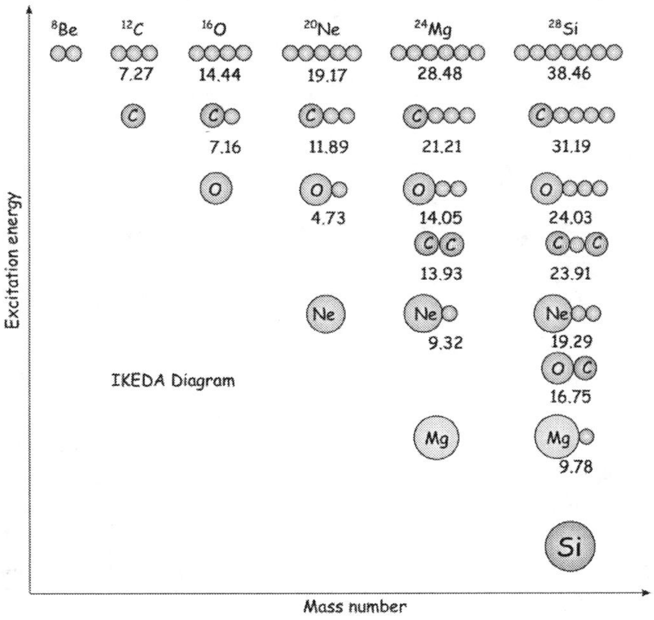

Figure 1. The Ikeda diagram [1]. This approach reveals that cluster structures, rather than appearing in the ground-state, should appear close to the cluster decay thresholds. At this point there is the possibility to convert the internal excitation energy into the mass of the cluster constituents.

the energy of the state to 7.653±0.008 MeV and indicated the most probable spin and parity to be 0^+ [7]. The connection between the existence of organic life, and ultimately human-kind, has led to the interpretation [8] that the prediction of the existence of 7.65 MeV state by Hoyle was an example of the anthropic principle, an idea introduced by Carter in 1973 [9]. The principle relies on the fact that intelligent life exists, to assert certain properties of the universe must exist, i.e. we exist therefore so must the 7.65 MeV state in ^{12}C. The question as to if Hoyle deployed the anthropic principle or not has been the subject of some debate, reviewed by Kragh [10].

The existence of 0^+ cluster states close to the α-decay thresholds of ^{8}Be and ^{12}C is mirrored in other light nuclei, which are α-conjugate systems. In ^{16}O the α-decay threshold is 7.16 MeV and the first excited 0^+ state lies at 6.05 MeV. This latter state has been identified with a ^{12}C+α cluster structure [11]. In this instance the Ikeda picture is fulfilled. It is interesting to ponder why this might be the case: why is it that nature arranges for such cluster states to be close to the decay thresholds, recognising the mass partition? Weakly bound nuclei close to decay thresholds can be thought of as open quantum systems where the properties of the unbound states influence, or couple to, the bound states below the threshold. Okołowicz, Nazarewicz and Płoszajczak have recently explored the link between the appearance of cluster states at threshold and the role of the continuum [12]. This is also discussed in Ref. [13].

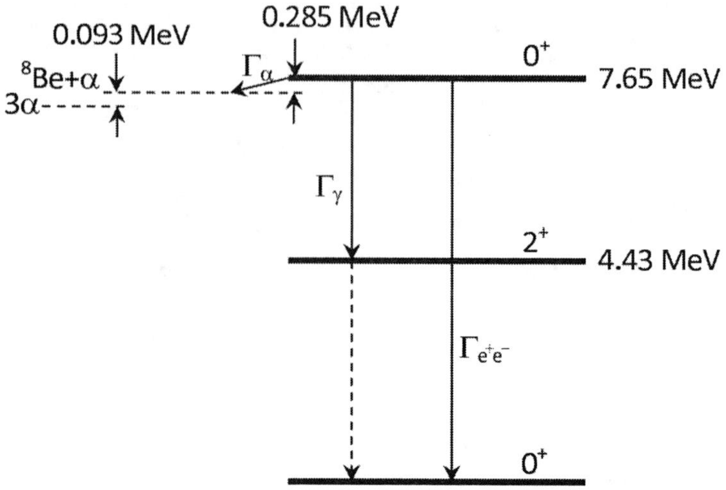

Figure 2. The energy levels of ^{12}C relevant for the triple-α process. The synthesis of ^{12}C takes place in two stages (i) the formation of ^{8}Be and then (ii) ^{8}Be+$\alpha \rightarrow ^{12}$C. The 7.65 MeV Hoyle state promotes the reaction rate associated with this second step by close to eight orders of magnitude. The presence of the Hoyle-state close to the 3α decay threshold, would suggest it possesses a 3α cluster structure.

The transition from a shell-model ground-state, where the degrees of freedom are those of a strongly correlated single-particle system, to a clustered excited state, mirror the transitions that take place in the complex many-body system of water. Water is a many body system where the interplay between hydrogen bonds and van der Waals interactions not only describe the bulk fluid but the dynamics of the formation of water clusters [14, 15]. Such clusters are challenging to observe from the experimental perspective. As heat is added to water, the excess energy results in the formation of steam or water vapour. Here these droplets can be viewed as liberated water clusters. In the nuclear analogue, as energy is added, α-particles can be evaporated - when the decay threshold is reached. If clusters can exist in water, can they also form within the fermionic neutron-proton nuclear fluid?

Fig. 3 shows a modified version of the Ikeda diagram which shows the water-line linked with the explicit precipitation of clusters above, and the shell-model like structures below. The question as to the structure of nuclei above and below this line can at least be answered from the theoretical perspective. The two densities shown in Fig. 3 correspond to Antisymmetrized Molecular Dynamics (AMD) calculation for ^{12}C [16]. The densities show the resulting intrinsic structures for the ground-state, 0_1^+, and second 0^+ state, 0_2^+. Both reveal the three α-structure, with it being more explicit in the case of the excited state, which is linked to the ^{12}C Hoyle-state. However, it is clear that the triangular symmetry is present also in the ground-state. The energy levels of ^{12}C calculated using the AMD approach are shown in Fig. 4. There is seen to be excellent agreement with experiment, a feature which is not found in the case of mean-field inspired models such as the no core shell model (NCSM) [19].

An experimental verification of the structure of ^{12}C linked to these two intrinsic configurations has been the subject of intensive experimental work.

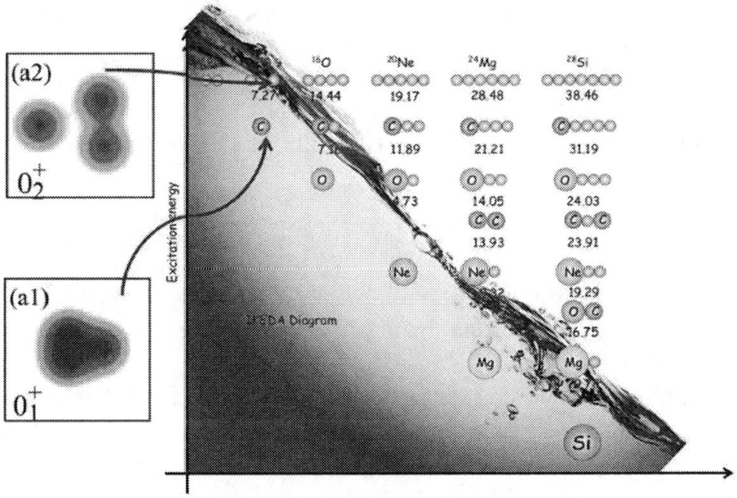

Figure 3. An adapted version of the Ikeda diagram, Fig. 1, delineating the change in structure that occurs above and below the α-decay threshold. The AMD calculations [16] for ^{12}C show that above the threshold (0_2^+) the 3α cluster structure is evident. For the ground state, 0_1^+, the cluster structure is still apparent, but the clusters are in a more compact configuration.

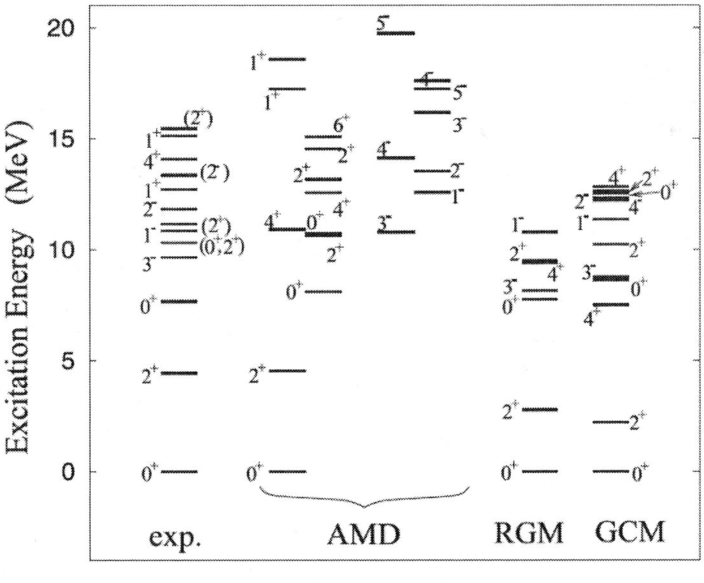

Figure 4. Energy levels for ^{12}C predicted by the AMD approach [16], resonanting group method (RGM) [17] and generator coordinate method (GCM) [18], from [16].

2 Experimental Signatures for the 3α System

In 1938 Hafstad and Teller [20] identified for α-particle systems that there should be a set of dynamical symmetries. In the case of ^{12}C the dynamical symmetries of 3α-system correspond to a spinning top with a triangular point symmetry (D_{3h}). The rotational properties of these states are given by

$$E_{J,K} = \frac{\hbar^2 J(J+1)}{2\mathcal{I}_{Be}} - \frac{\hbar^2 K^2}{4\mathcal{I}_{Be}} \tag{1}$$

where $\mathcal{I}_{Be}$ is the moment of inertia corresponding to two touching α-particles, which can be determined from the ^{8}Be ground-state rotational band [20]. K is the projection of the angular momentum onto the symmetry axis of the 3α system.

One would expect that based on this structure there should be a number of rotational bands with different values of K. For K^π=0$^+$, the rotations will be around an axis which lies in the plane of the three α-particles, generating a series of states 0$^+$, 2$^+$, 4$^+$ .... These correspond to the rotation of a ^{8}Be nucleus - the rotation axis passing through the centre of the third α-particle. The next set of rotations corresponds to the rotation around an axis perpendicular to the plane of the triangle, with each α-particle having one unit on angular momentum - giving $L = 3 \times 1\hbar$; K^π=3$^-$. Rotations around this axis and that parallel to the plane combine to give a series of states 3$^-$, 4$^-$, 5$^-$.... The next set of collective states then correspond to each α-particle having $L = 2$; K^π=6$^+$... Such an arrangement possesses a D_{3h} point group symmetry.

2.1 Excited States Linked to the Hoyle-state

If the structure of the excited Hoyle-state was close to that of the ground state, then the coupling of rotational modes would then produce a corresponding 2$^+$ state at 4.4 MeV (the first excited state in ^{12}C is at 4.4 MeV) above 7.65 MeV, i.e. 12.05 MeV. There is no known 2$^+$ state at this energy, this itself points to a more complex structure. The closest state which has been reported with J^π = 2$^+$ is at 11.16 MeV [21]. This state was originally observed in the ^{11}B(^{3}He,d)^{12}C reaction, however has not been observed in measurements subsequently. A re-measurement of this reaction using the K600 spectrometer at iThemba in South Africa demonstrates that the earlier observation of a state at 11.16 MeV was almost certainly an experimental artifact and no such state exists [22].

If the Hoyle-state is more extended than the ground-state (as suggested in Fig. 3), and the system behaves in a rotational fashion, then the 2$^+$ state would be lower in energy. Alternatively, the Hoyle-state could possess no collective excitations. Studies of the ^{12}C(α,α') and ^{12}C(p,p') [23] reactions indicate the presence of a 2$^+$ state close to 9.6-9.7 MeV with a width of 0.5 to 1 MeV. The state is only weakly populated in these reactions, presumably due to its underlying cluster structure, and is broad. Consequently, its distinction from other broad-states and dominant collective excitations (e.g. the 9.6 MeV, 3$^-$ state) made its unambiguous identification in these measurements challenging. Further, and perhaps definitive, evidence for such an excitation comes from measurements of the ^{12}C(γ,3α) reaction performed at the HIGS facility, TUNL [24]. Here a measurable cross section for this process was observed in the same region of 9-10 MeV which cannot be attributed to other known states in this region. Furthermore, the angular distributions of the α-particles are consistent with an L=2 pattern, demonstrating a dominant 2$^+$ component.

Based on a rather simple description of this state in terms of three α-particles with radii given by the experimental charge radius, it is possible to use the 2 MeV separation between the Hoyle-state and the proposed 2$^+$ excitation to draw some conclusions as to the arrangements of the clusters [23]. A linear arrangement of the 3α-particles, in which the separation is close to that of the ^{8}Be ground state, would give a separation of between the Hoyle-state and the 2^+ state of close to 1 MeV - as opposed to the 2 MeV that is observed experimentally. The data would then indicate that rather than a linear arrangement of the three clusters, a more appropriate description would be a loose arrangement of the α-particles in something approaching a triangular structure.

A natural extension of the rotational model is that there should also be a collective 4^+ state. Using the simple $J(J+1)$ scaling, a 4^+ excitation close to $E_x(^{12}C) = 14$ MeV would be expected. Measurements of the two reactions $^9Be(\alpha,3\alpha)n$ and $^{12}C(\alpha,3\alpha)^4He$ have been performed [25]. These measurements indicate a candidate state close to 13.3 MeV with a width estimated to be 1.7 MeV. It is believed that this is not a contaminant and is observed with similar properties in all spectra. Angular correlation measurements made using the ^{12}C target are not definitive, but indicate a 4^+ assignment.

2.2 Dynamical Excitations of the Ground State

As indicated in Fig. 3 the triangular structure seen in the AMD calculations for the Hoyle-state is also found in the ground-state. Here the α-particles exist in a more compact configuration and at such separations the internal fermionic degrees of freedom become important and the Pauli principle plays a role in perturbing the cluster structure. It is well-known that the ground-state has a significant overlap with shell-model like structures, whereas approaches the NCSM [19] fail to reproduce the properties of the Hoyle-state.

Nevertheless, the triangular symmetry that is observed in the calculations may also be found in the experimental spectroscopic fingerprint. The rotations corresponding to $K^\pi=0^+$ have long been known to correspond to the 0^+ ground state, 2^+ 4.4 MeV and 4^+ 14.1 MeV excited states. The next collective rotation occurs when each α-particle is provided with one unit of angular momentum around the 3-fold symmetry axis that passes through the centre of the triangle. These three units of angular momenta are linked to $K^\pi=3^-$. The collective excitations are then 3^-, 4^-, 5^-..... A precision measurement of the width of the 9.64 MeV, 3^-, state has recently been published [26] indicating that α-clustering may play a non-negligible role in the structure of this state. A collective 4^- excitation of this state has been suggested at 13.35 MeV [27, 28], with a 5^- member at 22.4 MeV [29]. The rotational behaviour of the ground-state rotational band and the $K^\pi=3^-$ band (associated with the 9.64 MeV state) are indicative of a D_{3h} dynamical symmetry reflecting the underlying triangular 3α-structure of ^{12}C [29].

3 The 4α system ^{16}O

The work by Hafstad and Teller [20] indicates the collective properties of the 4α system should be described by the tetrahedral symmetry group; T_d. Here the properties are those of a spherical top, with equal moments of inertia. If one assumes the separation of the α-particles is that which is associated with the ^8Be ground state, $\mathcal{I}_{Be}$, then the rotational energies are given by

$$E_J = \hbar^2 \frac{J(J+1)}{4\mathcal{I}_{Be}} \tag{2}$$

The rotation of the tetrahedral structure corresponds to the equivalent rotation of two ^8Be nuclei around their symmetry axis and hence the $4\mathcal{I}_{Be}$ in the denominator. The symmetry then dictates that all values of J are permitted except J=1, 2 and 5; states with J= 0, 4 and 8 have even parity and J=3, 7 and 11 have negative parity. A key feature of this structure would be degenerate 6^+ and 6^- states. A similar conclusion can be found in the recent work of Bijker and Iachello [30]. The experimentally observed states at 6.130 MeV, 3^-; 10.356, 4^+ and 21.052 MeV 6^+ have been linked in this latter work to the collective excitations of the tetrahedral structure. These same calculations predicted states at

6.132, 10.220 and 21.462 MeV and electromagnetic transition strengths $B(E3)$ and $B(E4)$ of 181 and 338 e^2fm^{2L} compared with experimental values of 205(10) and 378(133) e^2fm^{2L}. The comparison between experiment and theory is compelling.

An alternative theoretical approach is provided by the Alpha Cluster Model (ACM) calculations of Bauhoff, Shultheis and Shultheis [11]. These calculations identify a number of cluster structures, including a tetrahedral arrangement of the four α-particles in the ground-state. In addition, a planar arrangement of α-particles is found for the first excited 0^+ state. These structures gives rise to rotational bands. The main difference between the ACM and Algebraic Cluster Model (ACM') of Ref. [30] is evident in the assignment of the 10.356 MeV 4^+ state. The ACM assigns it to the planar rotational structure, whereas as noted above the ACM' links it to the tetrahedral ground-state. What is clear from measurements of the α-decay branching ratios for decay to the ^{12}C ground state and first excited states is that the states in the ACM planar band, above the alpha-decay threshold, all have very similar decay properties - they predominantly decay to the ground state [31]. This similar structure conflicts with the tetrahedral interpretation and indicates a collective excitation built around a ^{12}C+α cluster structure where the total angular momentum of the state is generated by the orbital motion of the α-particle around the ^{12}C core.

To arrive at a better understanding of the cluster symmetries of ^{16}O further electromagnetic transition strengths need to be determined. These include states above the α-decay threshold, where small branching ratios ($<10^{-5}$) make such studies very challenging.

4 Summary

One of the original predictions of the structure of light nuclei [20] suggested the ground state of nuclei such as ^{8}Be, ^{12}C and ^{16}O are composed of geometric arrangements of α-particles. Though this idea has largely fallen from favour, it is now clear that the two nuclei ^{8}Be and ^{12}C have a spectroscopic finger print that reveals this underlying cluster structure. In these systems the cluster symmetries remain, even if the clusters themselves are not completely intact. The 4α system ^{16}O presents the next challenge to our understanding. Within a model which exploits the dynamical symmetries of the tetrahedral system, T_d, there appears to be a reasonable basis for the description of the experimental states. However, there remain inconsistencies from the experimental perspective, which mean that there remains experimental challenges to provide data that can definitively constrain the symmetries and structure.

Acknowledgements

The author would like to acknowledge the contribution from his colleagues at Birmingham who have contributed significantly through the experimental programme that underpins the present contribution.

References

[1] K. Ikeda, *et al.*, Prog. Theor. Phys. Suppl., Extra Numbers, 464 (1968)
[2] M. Freer and H.O.U. Fynbo, Progress in Particle and Nuclear Physics **78**, 1 (2014)
[3] E. Öpik, Proceedings of the Royal Irish Academy **A 54**, 49 (1951)
[4] E.E. Salpeter, Astrophysical Journal **115**, 326 (1952)
[5] F. Hoyle, Astrophysical Journal, Supplement Series **1**, 12 (1954)
[6] D.N.F. Dunbar, R.E. Pixley, W.A. Wenzel and W. Whaling, Phys. Rev. **92**, 649 (1953)

[7] C.W. Cook, W.A. Fowler, C.C. Lauritsen and T. Lane, Phys. Rev. **107**, 508 (1957)

[8] J.D. Barrow and F.J. Tipler, The Anthropic Cosmological Principle, Cambridge University Press (1986)

[9] B. Carter, Confrontation of Cosmological Theories with Observational Data, Dordecht: Reidel 291 (1974)

[10] H. Kragh, Archive for History of Exact Sciences **64**, 721 (2010)

[11] W. Bauhoff, H. Schultheis, and R. Schultheis, Phys. Rev. C **29**, 1046 (1984)

[12] J. Okołowicz, W. Nazarewicz and M. Płoszajczak, Fortschritte der Physik **61**, 66 (2013)

[13] M. Freer, Nature **487**, 309 (2012)

[14] R. Ludwig, Angew. Chem. Int. Ed., **40** 1808 (2001)

[15] F.N. Keutsch and R. J. Saykally, PNAS **98**, 10533 (2001)

[16] Y. Kanada En'yo Prog. Theor. Phys. **117**, 655 (2007)

[17] T. Marumori, J. Phys. Soc. Jpn. **44**, 225 (1978); M. Kamimura, Nucl. Phys. **A351**, 456 (1981)

[18] E. Uegaki, S. Okabe, Y. Abe and H. Tanaka, Prog. Theor. Phys. **57**, 1262 (1977)

[19] P. Navrátil, *et al.*, Phys. Rev. Lett. **84**, 5728 (2000)

[20] L. R. Hafstad, E. Teller, Phys. Rev. **54**, 681 (1938)

[21] G. M. Reynolds, D. E. Rundquist and R. M. Poichar, Phys. Rev. C **3**, 442 (1971)

[22] F.D. Smit, *et al.*, Phys. Rev. C **86**, 037301 (2012)

[23] M. Itoh *et al.*; Nucl. Phys. **A738**, 268 (2004); Phys. Rev. C **84**, 054308 (2011); M. Freer *et al.*; Phys. Rev. C **80**, 041303(R) (2009); W.R. Zimmerman *et al.*; Phys. Rev. C **84**, 027304(BR) (2011).

[24] W.R. Zimmerman *et al.*; Phys. Rev. Lett. **110**, 152502 (2013).

[25] M. Freer *et al.*; Phys. Rev. C **83**, 034314 (2011).

[26] Tz. Kokalova, M. Freer, Z. Buthelezi, J. Carter, R. W. Fearick, S. V. Förtsch, H. Fujita, R. Neveling, P. Papka, F. D. Smit, J. A. Swartz, and I. Usman Phys. Rev. C **87**, 057307 (2013)

[27] M. Freer *et al.*, Phys. Rev. C **76**, 034320 (2007)

[28] O.S. Kirsebom *et al.*, Phys. Rev. C **81**, 064313 (2010)

[29] D.J. Marín-Lambarri, R. Bijker, M. Freer, M. Gai, Tz. Kokalova, D.J. Parker, C. Wheldon, Phys. Rev. Lett. **113**, 012502 (2014)

[30] R. Bijker and F. Iachello, Phys. Rev. Lett. **112**, 152501 (2014)

[31] C. Wheldon, *et al.*, Phys. Rev. C **83**, 064324 (2011)

EPJ Web of Conferences **113**, 01004 (2016)
DOI: 10.1051/epjconf/201611301004

Few-Body Universality in Halo Nuclei

H.-W. Hammer[1,2,a]

[1] *Institut für Kernphysik, Technische Universität Darmstadt, 64289 Darmstadt, Germany*
[2] *ExtreMe Matter Institute EMMI, GSI Helmholtzzentrum für Schwerionenforschung GmbH, 64291 Darmstadt, Germany*

Abstract. Few-body systems with resonant S-wave interactions show universal properties which are independent of the interaction at short distances. These properties include a geometric spectrum of three- and higher-body bound states and universal correlations between few-body observables. They can be observed on a wide range of scales from hadrons and nuclei to ultracold atoms. In this contribution, we focus on few-body universality in halo nuclei which can be considered as effective few-body systems consisting of halo nucleons and a core. This concept provides a unifying framework for halo nuclei with calculable corrections. Recent progress in this field with an emphasis on the possibility of finding Efimov states in halo nuclei is discussed.

1 Introduction

Particles with resonant interactions have universal properties that are independent of their interaction at short distances. The scattering of particles with short-range interactions at sufficiently low energy is determined by their S-wave scattering length a. Let us consider equal mass bosons of mass m for simplicity. If a is much larger than the range of the interaction R, the system shows universal properties [1]. The simplest example is the existence of a shallow two-body bound state with binding energy

$$B_2 = \frac{1}{ma^2} + O(R/a)\,, \tag{1}$$

and mean-square separation $a^2/2$ if a is large and positive.[1]

If a third particle is added, a three-body parameter κ_* is required to fully characterize the universal properties. For fixed scattering length, this implies universal correlations between different three-body observables such as the Phillips line [2]. Moreover, the Efimov effect [3] generates a universal spectrum of three-body bound states with binding energy

$$B_3 = -\frac{1}{ma^2} + \left(e^{-2\pi/s_0}\right)^{n-n_*} \exp\left[\Delta(\xi)/s_0\right] \frac{\kappa_*^2}{m} + O(R/a)\,, \tag{2}$$

where the angle ξ is defined by $\tan\xi = -(mB_3)^{1/2}\, a$, the index n labels the three-body states, $\Delta(\xi)$ is a universal function with $\Delta(-\pi/2) = 0$, and $s_0 = 1.00624...$ is a transcendental number (see Ref. [4] for more details). This spectrum is illustrated in Fig. 1 in the two-dimensional plane spanned by the

[a] e-mail: Hans-Werner.Hammer@physik.tu-darmstadt.de
[1] We use natural units with $\hbar = c = 1$.

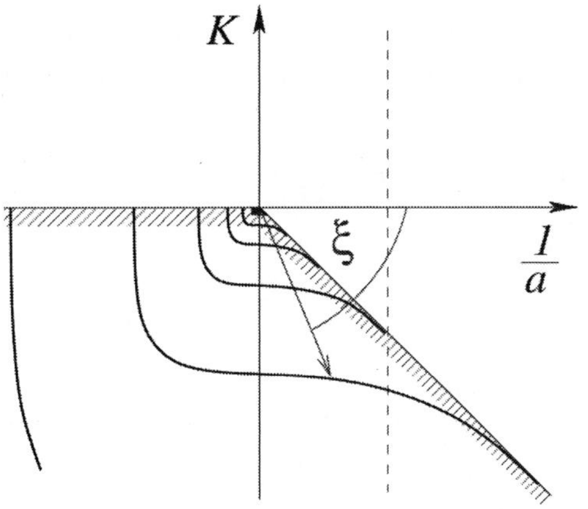

Figure 1. Illustration of the Efimov spectrum: The energy variable $K = \mathrm{sgn}(E)\sqrt{m|E|}$ is shown as a function of the inverse scattering length $1/a$. The solid lines indicate the Efimov states while the hashed areas give the scattering thresholds. The dashed vertical line illustrates an exemplary system with fixed scattering length.

energy variable $K = \mathrm{sgn}(E)\sqrt{m|E|}$ and the inverse scattering length $1/a$. The solid lines indicate the Efimov states while the hashed areas give the scattering thresholds below which the bound states can exist. The dashed vertical line illustrates an exemplary system with fixed scattering length.

In the unitary limit $1/a = 0$, Eq. (2) reduces to the geometric spectrum

$$B_3 = \left(e^{-2\pi/s_0}\right)^{n-n_*} \frac{\kappa_*^2}{m} + O(R/a). \tag{3}$$

It is now evident that the three-body parameter κ_* is chosen as the binding momentum of the state with label n_*, but other definitions are possible.

The spectrum shown in Fig. 1 is invariant under discrete scaling transformations with the preferred scaling factor $\lambda_0 = e^{\pi/s_0}$:

$$\kappa_* \longrightarrow \kappa_*, \qquad a \longrightarrow \lambda_0^n a, \qquad B_3 \longrightarrow \lambda_0^{-2n} B_3, \tag{4}$$

where n is any integer. This discrete scale invariance holds for all few-body observables and is a clear signature of a renormalization group limit cycle in the three-body system [5]. Its manifestation in observables is often referred to as *Efimov physics*.

If more particles are added, no new parameters are needed for renormalization at leading order [6]. As a consequence, all four-body observables in the universal regime are governed by the same limit cycle and can be characterized by a and κ_*. A similar behavior is expected for higher-body observables. In ultracold atoms, these properties have now been verified for up to five particles [7, 8]. For a review of the status in nuclear and particle physics, see Ref. [9].

2 Halo EFT

Here, we use this universality as the starting point for an effective field theory (EFT) expansion for strongly-interacting quantum systems with short-range interactions. This universal EFT is applicable to any system with resonant interactions, i.e., any system with short-range interactions and large scattering lengths. Examples include nuclear halo states, ultracold atoms close to a Feshbach resonance, and hadronic molecules in particle physics. The breakdown scale M_{high} of this theory is set by the lowest energy degree-of-freedom not explicitly included in the theory. In nuclear and particle physics, this is typically given by one-pion exchange such that $M_{\text{high}} \sim m_\pi$. In ultracold atoms, M_{high} is determined by the scale set by the van der Waals interaction. The typical momentum scale of the theory is $M_{\text{low}} \sim 1/a \sim k$. For momenta k of the order of the breakdown scale M_{high} or above, the omitted short-range physics is resolved and has to be treated explicitly. This framework exploits the appearance of a large scattering length $a \gg 1/M_{\text{high}}$, independent of the mechanism generating it.

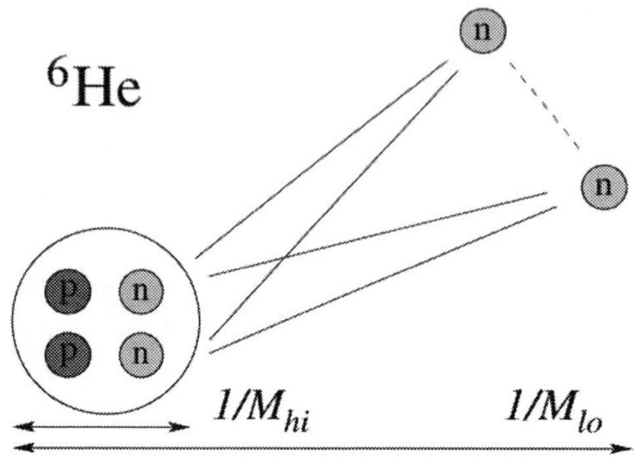

Figure 2. Illustration of the antisymmetrization of neutrons in the halo nucleus ^{6}He treated as $\alpha + n + n$ in the framework of halo EFT. The exchange illustrated by the dashed line involves active degrees of freedom and is antisymmetric, while the exchanges indicated by the solid lines are not included.

The application of these ideas to nuclear physics has led to the development of the so-called pionless EFT, which is applicable for typical momenta below the pion mass. (See Refs. [10–12] for reviews and references to earlier work.) If one also introduces field operators for certain tightly-bound clusters of nucleons, such as α particles, the pionless EFT is usually referred to as halo EFT [13, 14]. This approach can be viewed as a generalization of nuclear cluster models. One might worry about the missing antisymmetrization of nucleons in the core, which are not active degrees of freedom in halo EFT, and the halo nucleons (see Fig. 2 for an illustration in the case of ^{6}He treated as $\alpha + n + n$.). However, the effect of exchanging nucleons in the core with halo nucleons on observables is governed by the same expansion in $M_{\text{low}}/M_{\text{high}}$ as the EFT itself. Therefore, these effects are controlled and remain small in its applicability domain.

Halo EFT complements ab initio approaches to halo nuclei by providing universal relations between different few-body observables. These relations can be combined with input from ab initio theory or experiment to predict halo properties. Halo EFT has recently been applied to describe the structure and reactions of many known and predicted one- and two-neutron halo systems. For one-neutron halos it essentially reproduces the effective range expansion, but their electromagnetic structure and reactions can be predicted [15–20]. Moreover, the natural accuracy limits of cluster models can be estimated from the order at which local gauge invariant couplings to the currents appear.

The Efimov effect provides a natural binding mechanism for two-neutron halos, but the contribution of higher partial waves and partial wave mixing complicate the situation compared to ultracold atoms. While halo EFT naturally accommodates resonant interactions in higher partial waves [13, 14], there is no Efimov effect in this case [21, 22]. The possibility of observing excited Efimov states in two-neutron halo nuclei was investigated in Refs. [23–25]. The two-neutron halo nucleus ^{6}He is particularly interesting due to the dominance of P-wave neutron-α interactions. Its structure and renormalization properties were recently investigated in Refs. [26, 27]. Finally, the electromagnetic structure and breakup of two-neutron halo nuclei has also been calculated in halo EFT [28–30]. See Ref. [31] for a discussion of the Coulomb dissociation of ^{11}Li at this conference.

Proton halo systems are less common due to the delicate interplay between attraction from the strong interaction and the repulsion from the Coulomb interaction. The presence of the Coulomb barrier introduces the Coulomb momentum as a new scale into the problem, which is given by the inverse of the Bohr radius of the system. The additional Coulomb scale complicates the power counting since it is independent of the hadronic scales. Higa et al. [32], e.g., treated the Coulomb momentum as a high-momentum scale in their study of $\alpha - \alpha$ scattering. However, this treatment is not always appropriate. The correct scaling of the Coulomb momentum depends on the system considered. This is one of the issues investigated in recent studies of proton halo nuclei and their electromagnetic structure and reactions [33–38]. For a discussion of radiative proton capture on ^{7}Be at this conference, see Ref. [39]. Finally, halo nuclei containing strange baryons are also amenable to halo EFT [40–43], but the scarcity of data in this sector presents a particular challenge.

In the remainder of this proceedings contribution, we review some of the calculations discussed above in more detail. A special emphasis is placed on the possibility of finding Efimov states in halo nuclei.

3 Excited Efimov states in halo nuclei

As discussed above, the Efimov effect provides a natural binding mechanism for two-neutron halo nuclei. Thus the universal EFT, which is based on an expansion around the Efimov limit, becomes a powerful tool to describe their properties. Since the strength of the interaction between the neutrons and the core is fixed, however, the identification of Efimov physics is more delicate than for ultracold atoms. In particular, the log-periodic dependence of observables on the scattering length can not be used to identify Efimov physics. Instead one looks for excited states which (approximately) satisfy the universal scaling relation for bound states encapsulated in Eq. (2).

Halo nuclei have been discussed as possible candidates for Efimov states since more than 30 years [44]. It is particularly instructive to plot the candidate nuclei in a two-dimensional plane characterized by the neutron-core energy E_{nn} and neutron-neutron energy E_{nn} in units of the three-body ground state energy E_{gs} introduced by Amorim et al. [45]. If a given nucleus lies within a certain boundary curve that weakly depends on the mass number A of the core, it should display an excited Efimov state if the ground state is bound by the Efimov effect. The status for ^{11}Li, ^{12}Be, ^{14}Be, ^{18}C, and ^{20}C according to a recent EFT analysis [23] assuming that these nuclei can be treated as S-wave two-neutron halo nuclei at leading order is summarized in Fig. 3. Note that the point for ^{20}C has shifted as indicated by the arrow from using the neutron separation energies of the recent AME2012 atomic mass evaluation [46] instead of the older values used in Ref. [23]. As a consequence, none of the depicted nuclei are expected to exhibit an excited Efimov state.

The most promising system known so far is the ^{22}C halo nucleus which was found to display an extremely large matter radius [47] and has a significant S-wave component in the n-^{20}C system [48]. Acharya et al. [24] used halo EFT to explore the correlation between the n-^{20}C energy and the two-neutron separation energy of ^{22}C. They showed that an excited Efimov state in ^{22}C is excluded by combining this correlation with the matter radius measurement of Tanaka et al. [47]. See Ref. [31] for a more detailed discussion of the work at this conference.

Whether heavier two-neutron halos exist is still an open question, although there is some experimental evidence that the ground states of ^{31}Ne and ^{37}Mg have a low one-neutron separation energy and are deformed P-wave halos [49, 50]. This makes it worthwhile to investigate the possibility for Efimov states in heavier nuclei.

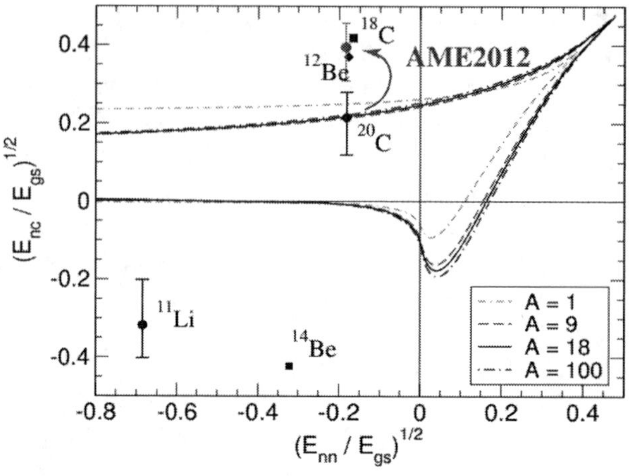

Figure 3. Boundary curves for the existence of an excited Efimov state as function of the neutron-core energy E_{nn} and neutron-neutron energy E_{nn} in units of the three-body ground state energy E_{gs}. The boundary curves depend weakly on the core mass A. The shift of ^{20}C from using the values of the newer AME2012 atomic mass evaluation [46] is indicated by the arrow.

There has been much interest recently, both experimentally and theoretically, in determining precise values for masses, understanding shell evolution and the location of the dripline in the neutron rich calcium isotopes [51–56]. Coupled-cluster calculations of neutron rich calcium isotopes that included coupling to the scattering continuum and schematic three-nucleon forces, suggested that there is an inversion of the *gds* shell-model orbitals in ^{53,55,61}Ca. In particular it was suggested that a large S-wave scattering length might occur in ^{61}Ca with interesting implications for ^{62}Ca.

Hagen and collaborators [25] used the coupled-cluster method combined with modern chiral effective theory interactions and follow the method outlined in Ref. [57] to compute the elastic scattering of neutrons on ^{60}Ca. They analyzed the resulting phase shift data to obtain quantitative estimates for the scattering length and the effective range and showed that a large scattering length can be expected in this system. The results obtained from the ab initio calculations were then used as input for halo EFT. They analyzed the implications of the coupled cluster results for the ^{60}Ca-n-n system. Specifically, they focused on the signals of Efimov physics that are a consequence of the large scattering length in the ^{60}Ca-n and n-n systems.

This is illustrated in Fig. 4, where the universal correlation between the ^{61}Ca-n scattering length and the two-neutron separation energy of ^{62}Ca is shown. For ^{62}Ca, the discrete scaling factor governing the energy spectrum is approximately $16^2 = 256$ [1], which is slightly more favorable than in the case of three identical bosons. The asymptotic scaling ratio applies for deep states and in the unitary limit of infinite scattering length. For levels near threshold, however, the ratio of energies can be significantly smaller (see Fig. 1 and the corresponding discussion in Ref. [1]). In the case of ^{62}Ca, the whole energy region between $S_n \approx 5 - 8$ keV and the breakdown scale $S_{\text{deep}} \approx 500$ keV is available for Efimov states. In the shaded region of Fig. 4 halo EFT breaks down. Around a two-neutron separation energy of about 230 keV indicated by the dashed vertical line, the ^{60}Ca-n scattering length jumps from plus to minus infinity and an excited Efimov state appears. It is thus conceivable that ^{62}Ca would display an excited Efimov state and unlikely that it would not display any Efimov states at all.

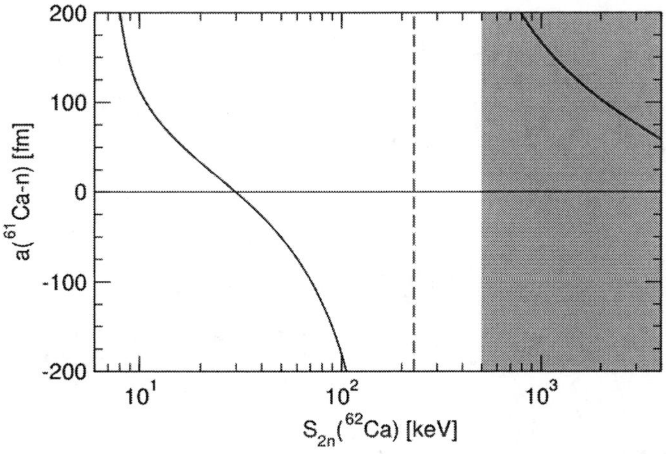

Figure 4. The correlation between the ^{61}Ca-n scattering length and the two-neutron separation energy S_{2n} of ^{62}Ca. The emergence of an excited Efimov state around S_{2n} = 230 keV is indicated by the vertical dashed line. In the shaded region, halo EFT breaks down.

4 Conclusion

In this proceedings contribution, we have reviewed the manifestations of few-body universality in halo nuclei with a special focus on the possibility of observing Efimov states. While the ground state of many S-wave halo nuclei is close to the Efimov limit, there is currently no observed halo nucleus that displays an excited Efimov state or is likely to display such a state. There is some theoretical evidence that the situation might be different for ^{62}Ca. The results of Hagen et al. [25] imply that ^{62}Ca is possibly the largest and heaviest halo nucleus in the chart of nuclei. They showed that as a result a large number of observables would display characteristic features of Efimov physics. Measurements of these observables clearly pose a significant challenge for experiment. For example, ^{58}Ca is the heaviest Calcium isotope that has been observed experimentally [58]. However, future radioactive beam facilities might provide access to calcium isotopes as heavy as ^{68}Ca and thereby facilitate a test of this hypothesis.

Another important aspect from few-body universality in halo nuclei beyond Efimov physics is the possibility to formulate an effective field theory exploiting the universal aspects of the resonant interactions involved. This effective field theory is called halo EFT. It is complementary to ab initio approaches to halo nuclei by providing universal relations between different few-body observables. These relations can be combined with input from ab initio theory or experiment to predict halo properties. Moreover, they allow to test the consistency of different approaches and/or experiments. A particular strength of this approach lies in the possibility to describe the electroweak structure and reactions of halo nuclei in a model independent way with controlled error estimates. A brief overview of the activities in this exciting area with a large future potential was given in Section 2 and will not be repeated here.

Acknowledgements

This work was supported by the BMBF under contract 05P12PDFTE and by the Helmholtz Association under contract HA216/EMMI. Collaborations and discussions on this topic with B. Acharya, D. Canham, C. Forssén, G. Hagen, P. Hagen, D. Phillips, L. Platter, and E. Ryberg are gratefully acknowledged.

References

[1] E. Braaten and H.-W. Hammer, Phys. Rept. **428**, 259 (2006) [cond-mat/0410417].

[2] A. C. Phillips, Nucl. Phys. A **107**, 209 (1968).

[3] V. Efimov, Phys. Lett. B **33**, 563 (1970).

[4] E. Braaten, H.-W. Hammer, and M. Kusunoki, Phys. Rev. A **67**, 022505 (2003) [cond-mat/0201281].

[5] P.F. Bedaque, H.-W. Hammer, and U. van Kolck, Phys. Rev. Lett. **82**, 463 (1999) [nucl-th/9809025].

[6] L. Platter, H.-W. Hammer, and U. G. Meißner, Phys. Rev. A **70**, 052101 (2004) [cond-mat/0404313].

[7] F. Ferlaino, R. Grimm, Physics **3**, 9 (2010).

[8] A. Zenesini et al., New J. Phys. **15**, 043040 (2013) [arXiv:1205.1921].

[9] H.-W. Hammer and L. Platter, Ann. Rev. Nucl. Part. Sci. **60**, 207 (2010) [arXiv:1001.1981 [nucl-th]].

[10] S. R. Beane, P. F. Bedaque, W. C. Haxton, D. R. Phillips, and M. J. Savage, *At the frontier of particle physics, Vol. 1, M. Shifman (ed.)* (World Scientific, Singapore, 2001) 133 [nucl-th/0008064].

[11] P. F. Bedaque and U. van Kolck, Ann. Rev. Nucl. Part. Sci. **52**, 339 (2002) [nucl-th/0203055].

[12] E. Epelbaum, H.-W. Hammer, and U.-G. Meißner, Rev. Mod. Phys. **81**, 1773 (2009) [arXiv:0811.1338 [nucl-th]].

[13] C. A. Bertulani, H.-W. Hammer, and U. van Kolck, Nucl. Phys. A **712**, 37 (2002) [nucl-th/0205063].

[14] P. F. Bedaque, H.-W. Hammer, and U. van Kolck, Phys. Lett. B **569**, 159 (2003) [nucl-th/0304007].

[15] H.-W. Hammer and D. R. Phillips, Nucl. Phys. A **865**, 17 (2011) [arXiv:1103.1087 [nucl-th]].

[16] G. Rupak and R. Higa, Phys. Rev. Lett. **106**, 222501 (2011) [arXiv:1101.0207 [nucl-th]].

[17] L. Fernando, R. Higa, and G. Rupak, Eur. Phys. J. A **48**, 24 (2012) [arXiv:1109.1876 [nucl-th]].

[18] G. Rupak, L. Fernando, and A. Vaghani, Phys. Rev. C **86**, 044608 (2012) [arXiv:1204.4408 [nucl-th]].

[19] B. Acharya and D. R. Phillips, Nucl. Phys. A **913**, 103 (2013) [arXiv:1302.4762 [nucl-th]].

[20] X. Zhang, K. M. Nollett, and D. R. Phillips, Phys. Rev. C **89**, 024613 (2014) [arXiv:1311.6822 [nucl-th]].

[21] Y. Nishida, Phys. Rev. A **86**, 012710 (2012) [arXiv:1111.6961 [cond-mat.quant-gas]].

[22] E. Braaten, P. Hagen, H.-W. Hammer, and L. Platter, Phys. Rev. A **86**, 012711 (2012) [arXiv:1110.6829 [cond-mat.quant-gas]].

[23] D. L. Canham and H.-W. Hammer, Eur. Phys. J. A **37**, 367 (2008) [arXiv:0807.3258 [nucl-th]].

[24] B. Acharya, C. Ji, and D. R. Phillips, Phys. Lett. B **723**, 196 (2013) [arXiv:1303.6720 [nucl-th]].

[25] G. Hagen, P. Hagen, H.-W. Hammer, and L. Platter, Phys. Rev. Lett. **111**, 132501 (2013) [arXiv:1306.3661 [nucl-th]].

[26] J. Rotureau and U. van Kolck, Few Body Syst. **54**, 725 (2013) [arXiv:1201.3351 [nucl-th]].

[27] C. Ji, C. Elster, and D. R. Phillips, Phys. Rev. C **90**, 044004 (2014) [arXiv:1405.2394 [nucl-th]].

[28] P. Hagen, H.-W. Hammer, and L. Platter, Eur. Phys. J. A **49**, 118 (2013) [arXiv:1304.6516 [nucl-th]].

[29] P. Hagen, *Effective Field Theory for Halo Nuclei*, Dissertation, University of Bonn (2014), http://hss.ulb.uni-bonn.de/2014/3526/3526.htm .

[30] B. Acharya, *Properties of One- and Two-Nucleon Halo Nulcei in Effective Field Theory*, PhD thesis, Ohio University (2015), http://rave.ohiolink.edu/etdc/view?acc_num=ohiou1423835581 .
[31] B. Acharya and D. Phillips, these proceedings [arXiv:1508.02697].
[32] R. Higa, H.-W. Hammer, and U. van Kolck, Nucl. Phys. A **809**, 171 (2008) [arXiv:0802.3426 [nucl-th]].
[33] R. Higa, EPJ Web Conf. **3**, 06001 (2010) [arXiv:1001.0540 [nucl-th]].
[34] E. Ryberg, C. Forssén, H.-W. Hammer, and L. Platter, Phys. Rev. C **89**, 014325 (2014) [arXiv:1308.5975 [nucl-th]].
[35] X. Zhang, K. M. Nollett, and D. R. Phillips, Phys. Rev. C **89**, 051602 (2014) [arXiv:1401.4482 [nucl-th]].
[36] E. Ryberg, C. Forssén, H.-W. Hammer, and L. Platter, Eur. Phys. J. A **50**, 170 (2014) [arXiv:1406.6908 [nucl-th]].
[37] X. Zhang, K. M. Nollett, and D. R. Phillips, arXiv:1507.07239 [nucl-th].
[38] E. Ryberg, C. Forssén, H.-W. Hammer, and L. Platter, arXiv:1507.08675 [nucl-th].
[39] X. Zhang, K.M. Nollet, D.R. Phillips, these proceedings [arXiv:1508.06935].
[40] H.-W. Hammer, Nucl. Phys. A **705**, 173 (2002) [nucl-th/0110031].
[41] S. I. Ando, G. S. Yang, and Y. Oh, Phys. Rev. C **89**, 014318 (2014) [arXiv:1310.1432 [nucl-th]].
[42] S. I. Ando and Y. Oh, Phys. Rev. C **90**, 037301 (2014) [arXiv:1407.1608 [nucl-th]].
[43] S. I. Ando, U. Raha, and Y. Oh, Phys. Rev. C **92**, 024325 (2015) [arXiv:1507.01260 [nucl-th]].
[44] D.V. Fedorov, A.S. Jensen, and K. Riisager, Phys. Rev. Lett. **73**, 2817 (1994).
[45] A. E. A. Amorim, T. Frederico, and L. Tomio, Phys. Rev. C **56**, R2378 (1997) [nucl-th/9708023].
[46] M. Wang et al., Chin. Phys. C **36**, 1603 (2012).
[47] K. Tanaka et al., Phys. Rev. Lett. **104**, 062701 (2010).
[48] W. Horiuchi and Y. Suzuki, Phys. Rev. C **74**, 034311 (2006).
[49] T. Nakamura et al., Phys. Rev. Lett. **112**, 142501 (2014).
[50] N. Kobayashi et al., Phys. Rev. Lett. **112**, 242501 (2014).
[51] A. Lapierre et al., Phys. Rev. C **85**, 024317 (2012).
[52] A.T. Gallant et al., Phys. Rev. Lett. **109**, 032506 (2012).
[53] J. D. Holt and A. Schwenk, J. Phys. G **39**, 085111 (2012).
[54] G. Hagen, M. Hjorth-Jensen, G.R. Jansen, R. Machleidt, and T. Papenbrock, Phys. Rev. Lett. **109**, 032502 (2012).
[55] W. Nazarewicz et al., Phys. Rev. C **53**, 740 (1996).
[56] J. Erler et al., Nature **486**, 509 (2012).
[57] G. Hagen and N. Michel, Phys. Rev. C **86**, 021602(R) (2012).
[58] O. B. Tarasov et al., Phys. Rev. Lett. **102**, 142501 (2009).

EPJ Web of Conferences **113**, 01005 (2016)
DOI: 10.1051/epjconf/201611301005

Ab initio calculations of reactions with light nuclei

Sofia Quaglioni[1,a], Guillaume Hupin[2,1,b], Angelo Calci[3,c], Petr Navrátil[3], and Robert Roth[4]

[1] *Lawrence Livermore National Laboratory, P.O. Box 808, L-414, Livermore, California 94551, USA*
[2] *Institut de Physique Nucléaire, Universitée Paris-Sud, IN2P3/CNRS, F-91406 Orsay Cedex, France*
[3] *TRIUMF, 4004 Wesbrook Mall, Vancouver, British Columbia, V6T 2A3, Canada*
[4] *Institut für Kernphysik, Technische Universität Darmstadt, D-64289 Darmstadt, Germany*

Abstract. An *ab initio* (i.e., from first principles) theoretical framework capable of providing a unified description of the structure and low-energy reaction properties of light nuclei is desirable to further our understanding of the fundamental interactions among nucleons, and provide accurate predictions of crucial reaction rates for nuclear astrophysics, fusion-energy research, and other applications. In this contribution we review *ab initio* calculations for nucleon and deuterium scattering on light nuclei starting from chiral two- and three-body Hamiltonians, obtained within the framework of the *ab initio* no-core shell model with continuum. This is a unified approach to nuclear bound and scattering states, in which square-integrable energy eigenstates of the A-nucleon system are coupled to $(A-a)+a$ target-plus-projectile wave functions in the spirit of the resonating group method to obtain an efficient description of the many-body nuclear dynamics both at short and medium distances and at long ranges.

1 Introduction

Understanding the structure and the dynamics of nuclei as many-body systems of protons and neutrons interacting through the strong force is one of the central goals of nuclear physics. As systems that are amenable for accurate first-principle or *ab initio* calculations, light nuclei represent an excellent testing ground in such an effort. Until about eight years ago, with the exception of studies involving $A = 4$ nucleons or less, most of the progress in this direction has been achieved through *ab initio* calculations of well-bound nuclear states and their properties. However, efforts to generalize *ab initio* many-body techniques to also encompass $A > 4$ nuclei that are weakly bound, or that only exist as low-energy resonances, as well as to explain light-nucleus scattering and low-energy reactions have been growing, driven by new ideas, the rising availability of data on exotic nuclei, and increased access to high-performance computing. At the same time, effective field theory (EFT) and renormalization group transformations are providing nuclear theorists with a systematic approach to, and a much-improved understanding of the nuclear forces, and are yielding increasingly precise interactions and theoretical uncertainties, which are the basis for accurate predictions. The combination of these developments is opening the exciting prospect of a predictive and quantified theory of

[a] e-mail: quaglioni1@llnl.gov
[b] e-mail: hupin@ipnorsay.in2p3.fr
[c] e-mail: calci@triumf.ca

light-nucleus reactions important for applications, e.g.: scattering processes widely used in material science to probe thin films, or analyze the surface of artworks and archeological artifacts; thermonuclear fusion reactions crucial for big-bang and stellar nucleosyntesis; or those that power research facilities directed toward developing fusion energy. For the success of such a program it is essential to properly understand three key elements and their interplay: the role of nucleon-nucleon (NN) and three-nucleon (3N) forces; the influence of open channels and continuum degrees of freedom; and the propagation of the uncertainties in interaction models into error bars on computed observables. In this contribution we will review recent progress in addressing the first two of these elements by means of *ab initio* calculations of light-nucleus structure and dynamics carried out within the no-core shell model with continuum (NCSMC) approach. In particular, in Sec. 2 we will briefly outline the main aspects of the NCSMC formalism, while results for nucleon- and deuterium-nucleus dynamics employing chiral EFT NN+3N forces will be discussed in Sec. 3. Concluding remarks and future prospects will be presented in Sec. 4.

2 Formalism

To achieve a unified *ab initio* description of structural and reaction properties of light nuclei, we work within the theoretical framework of the NCSMC, introduced in Refs. [1] and [2]. In this approach, the translational-invariant ansatz for the many-body wave function in each partial wave of total angular momentum J, parity π and isospin T, is given by a superposition of square-integrable energy eigenstates of the compound A-nucleon system and continuous cluster states according to:

$$|\Psi_A^{J^\pi T}\rangle = \sum_\alpha c_\alpha^{J^\pi T}|A\alpha J^\pi T\rangle + \sum_\nu \int dr\, r^2 \frac{\gamma_\nu^{J^\pi T}(r)}{r} \mathcal{A}_\nu |\Phi_{\nu r}^{J^\pi T}\rangle\,. \tag{1}$$

Here $|A\alpha J^\pi T\rangle$ represent antisymmetric, translation-invariant discrete eigenstates of the A-nucleon Hamiltonian, labelled by the energy index α. They are obtained ahead of time by means of the *ab initio* no-core shell model (NCSM) [3] through the diagonalization of the A-nucleon microscopic Hamiltonian within a model space spanned by many-body harmonic oscillator (HO) wave functions with up to $N_{\rm max}$ HO excitations above the unperturbed configuration and frequency $\hbar\Omega$. Further

$$|\Phi_{\nu r}^{J^\pi T}\rangle = \Big[\Big(|A-a\ \alpha_1 I_1^{\pi_1} T_1\rangle |a\ \alpha_2 I_2^{\pi_2} T_2\rangle\Big)^{(sT)} Y_\ell(\hat r_{A-a,a})\Big]^{(J^\pi T)} \frac{\delta(r - r_{A-a,a})}{r r_{A-a,a}}\,. \tag{2}$$

are continuous resonating-group method (RGM) [4, 5] basis states, representing an $(A-a)$-nucleon target and a-nucleon projectile (with $a < A$), whose centers of mass are separated by the relative coordinate $\vec r_{A-a,a}$, traveling in a ${}^{2s+1}\ell_J$ wave of relative motion (with s the channel spin, and ℓ the relative momentum of the system). The eigenstates $|A-a\ \alpha_1 I_1^{\pi_1} T_1\rangle$ and $|a\ \alpha_2 I_2^{\pi_2} T_2\rangle$ of each cluster of nucleons are obtained analogously to, and consistently with those of the A-nucleon system and are characterized by total angular momentum $I_{1,2}$, parity $\pi_{1,2}$, isospin $T_{1,2}$, and energy index $\alpha_{1,2}$. The index $\nu = \{A{-}a\,\alpha_1 I_1^{\pi_1} T_1;\ a\,\alpha_2 I_2^{\pi_2} T_2;\ s\ell\}$ collects all quantum numbers associated with this continuous basis, and its full antisymmetrization is recovered by introducing an appropriate inter-cluster antisymmetrizer $\mathcal{A}_\nu$, which accounts for exchanges of nucleons between target and projectile. The unknown discrete $c_\alpha^{J^\pi T}$ and continuous $\gamma_\nu^{J^\pi T}(r) = (\mathcal{N}^{-1/2}\chi)_\nu(r)$ variational amplitudes, are the solutions of the orthogonalized coupled-channel equations

$$\begin{pmatrix} E_\alpha\, \delta_{\alpha\alpha'} & (h\mathcal{N}^{-\frac12})_{\alpha\nu'}(r') \\ (h\mathcal{N}^{-\frac12})_{\alpha'\nu}(r) & (\mathcal{N}^{-\frac12}\mathcal{H}\mathcal{N}^{-\frac12})_{\nu\nu'}(r,r') \end{pmatrix} \begin{pmatrix} c_{\alpha'} \\ \frac{\chi_{\nu'}(r')}{r'} \end{pmatrix} = E \begin{pmatrix} \delta_{\alpha\alpha'} & (g\mathcal{N}^{-\frac12})_{\alpha\nu'}(r') \\ (g\mathcal{N}^{-\frac12})_{\alpha'\nu}(r) & \frac{\delta_{\nu\nu'}\delta(r-r')}{rr'} \end{pmatrix} \begin{pmatrix} c_{\alpha'} \\ \frac{\chi_{\nu'}(r')}{r'} \end{pmatrix}, \tag{3}$$

which follow from the many-body Schrödinger equation when using the ansatz (1) for the many-body wave function and projecting over the model space spanned by the $|A\alpha J^{\pi}T\rangle$ and $|\Phi^{J^{\pi}T}_{\nu r}\rangle$ basis states. In Eq. (3), E denotes the total energy of the system and the two by two block-matrices on the left- and right-hand side of the equation represent, respectively, the NCSMC Hamiltonian and norm (or overlap) kernels. The upper diagonal blocks are given by the Hamiltonian (overlap) matrix elements over the square-integrable part of the basis. In particular, as the basis states are NCSM eigenstates of the A-nucleon Hamiltonian, these are trivially given by the diagonal matrix of the E_{α} eigenergies (the identity matrix). Similarly, those over the orthonormalized continuous portion of the basis appear in the lower diagonal block and are obtained from the norm (or overlap) RGM kernel $\mathcal{N}_{\nu\nu'}(r,r') = \langle\Phi^{J^{\pi}T}_{\nu r}|\mathcal{A}_{\nu}\mathcal{A}_{\nu'}|\Phi^{J^{\pi}T}_{\nu' r'}\rangle$ and Hamiltonian kernel $\mathcal{H}_{\nu\nu'}(r,r') = \langle\Phi^{J^{\pi}T}_{\nu r}|\mathcal{A}_{\nu}H\mathcal{A}_{\nu'}|\Phi^{J^{\pi}T}_{\nu' r'}\rangle$ [6, 7]. The off diagonal blocks contain the couplings between the two sectors of the basis, with $g_{\alpha\nu}(r) = \langle A\,\alpha J^{\pi}T|\mathcal{A}_{\nu}|\Phi^{J^{\pi}T}_{\nu r}\rangle$ and $h_{\alpha\nu}(r) = \langle A\,\lambda J^{\pi}T|H\mathcal{A}_{\nu}|\Phi^{J^{\pi}T}_{\nu r}\rangle$. The scattering matrix (and from it any scattering observable) follows from matching the solutions of Eq. (3) with the known asymptotic behavior of the wave function at large distances by means of the microscopic R-matrix method [8, 9].

3 Unified description of structure and reaction observables with chiral two- and three-nucleon forces

The *ab initio* NCSMC was initially developed to compute nucleon-nucleus collisions starting from a two-body Hamiltonian and applied to the description of the unbound ^{7}He nucleus [1, 2]. The approach was later extended to include explicit 3N forces, which are necessary to obtain a truly accurate and quantitative description of light-nucleus scattering process [10, 11]. In the following we review applications of such an extended approach, using an Hamiltonian based on the chiral N^{3}LO NN interaction of Ref. [12] and N^{2}LO 3N force of Ref. [13], constrained to provide an accurate description of the $A = 2$ and 3 [14] systems and unitarily softened via the similarity-renormalization-group (SRG) method [15–19] to minimize the influence of momenta higher than a low-momentum resolution scale of $\Lambda_{\rm SRG} = 2.0$ fm^{-1}. In particular, Sec. 3.1 discusses predictions of elastic scattering and recoil of protons off ^{4}He [20], as well as a study of continuum and $3N$-force effects on the energy levels of ^{9}Be [21]. In Sec. 3.2 we review the first application to describe more challenging deuterium-nucleus dynamics [22], and discuss our ongoing effort to describe deuteron-induced nucleon transfer reactions.

3.1 Nucleon-nucleus scattering

The recent NCSMC study of low-energy cross sections for elastic scattering and recoil of protons from ^{4}He of Ref. [20] is an excellent example of how the combination of efficient *ab initio* approaches, accurate nuclear interactions and high-performance computing capabilities can lead to a predictive theory of light-nucleus reactions important for applications. In this case, accurate ^{4}He$(p,p)^4$He and ^{1}H$(\alpha,p)^4$He angular differential cross sections for a variety of proton/^{4}He incident energies and detection angles are key to the feasibility and quality of ion-beam analysis applications aimed at determining the concentrations and depth profiles of helium and hydrogen, respectively, at the surface of materials or in thin films [23–25]. Starting from SRG-evolved chiral NN+3N forces with $\Lambda_{\rm SRG} = 2.0$ fm^{-1}, accurate results for p–^{4}He scattering were obtained by employing an $N_{\rm max} = 13$, $\hbar\Omega = 20$ MeV HO model space, proton-^{4}He binary-cluster basis states including up to the first seven ($I_1^{\pi_1}T_1 = 0^+_1 0, 0^+_2 0, 0^- 0, 2^- 0, 2^- 1, 1^- 1$ and $1^- 0$) ^{4}He eigenstates, and the first fourteen (of which three $3/2^-$ and two $1/2^-$) square-integrable eigenstates of the ^{5}Li compound nucleus. As can be seen from Fig. 1 the computed NCSMC angular differential cross sections agree very well with data both for ^{4}He$(p,p)^4$He and ^{1}H$(\alpha,p)^4$He processes. For comparison, in panels (a) and (b) we also show as black dashed lines

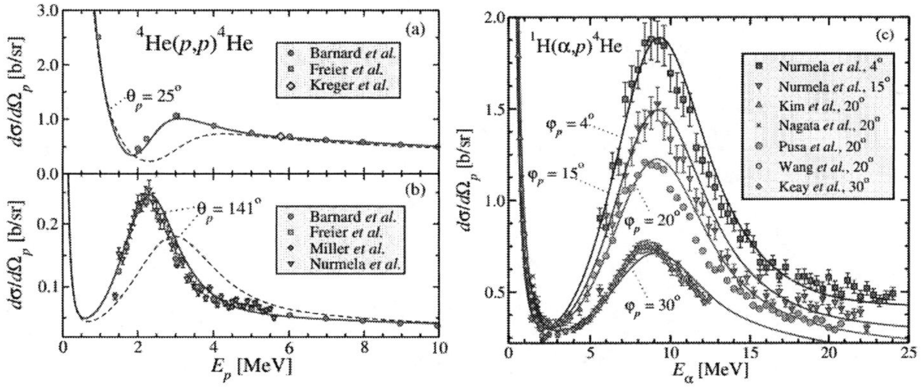

Figure 1. Computed (lines) ^{4}He$(p, p)^4$H angular differential cross section at forward scattering angle $\theta_p = 25°$ (a) and backscattering angle $\theta_p = 141°$ (b) as a function of the proton incident energy compared with measurements (symbols) from Refs. [26–30], and ^{1}H$(\alpha, p)^4$He angular differential cross section at the proton recoil angles $\varphi_p = 4°, 16°, 20°$, and $30°$(c) as a function of the incident ^{4}He energy compared with data (symbols) from Refs. [23, 30–37]. The solid lines corresponds to the present NCSMC calculations. Also shown (black dashed lines) are the results of Ref. [38], i.e. without ^{5}Li square intregrable eigenstates.

the results [38] obtained within a model space spanned by p–^{4}He cluster states alone, that is by retaining only the second term of Eq. (1) and solving $\mathcal{N}^{-\frac{1}{2}}\mathcal{H}\mathcal{N}^{-\frac{1}{2}}\chi = E\chi$. While the high-energy tail of the cross section can already be described well working within such more limited model space, the inclusion of square-integrable eigenstates of the compound ^{5}Li system is essential at lower energy, where it efficiently compensates for missing higher excitations of the ^{4}He core.

At the same time, a proper treatment of continuum degrees of freedom is also indispensable to draw reliable conclusions on the influence of 3N forces in the low-lying spectra of loosely bound systems, such as ^{9}Be, for which all excited states lie above the n+^{8}Be threshold. The positive parity resonances of this nucleus are in general found too high in energy compared to experiment, in *ab initio* calculations that treat them as bound states such in the NCSM study of Ref. [39]. In the same setting, counter to expectations, the splitting between the lowest $5/2^-$ and $1/2^-$ resonances tends to be overestimated when 3N effects are included. A recent NCSMC study of ^{9}Be as a linear combination of 9-body square-integrable eigenstates and ^{8}Be+n binary-cluster states with the ^{8}Be in its ground and 2^+ states [21] helped shed some light on the interplay of continuum degrees of freedom and 3N force effects in this nucleus. As can be seen in Fig. 2, for all energy levels the inclusion of the continuum significantly improves the agreement with experiment. In particular, the splitting between the $5/2^-$ and $1/2^-$ levels is substantially reduced when the continuum is included due to a shift towards lower energies of the $^2P_{1/2}$ resonance. However, the most dramatic continuum effects were found in the positive-parity resonances, shown in Fig. 2(b). The $1/2^+$ and the $3/2^+$ S-wave resonances are several MeV lower in the NCSMC, close to their experimental value.

3.2 Deuterium-induced reactions

One of the most challenging and notable applications of the NCSMC accomplished so far is the simultaneous study of the ^{6}Li ground state and d–^{4}He elastic scattering using NN+3N forces from chiral effective field theory (softened – as in the previous section – by means of an SRG transformation

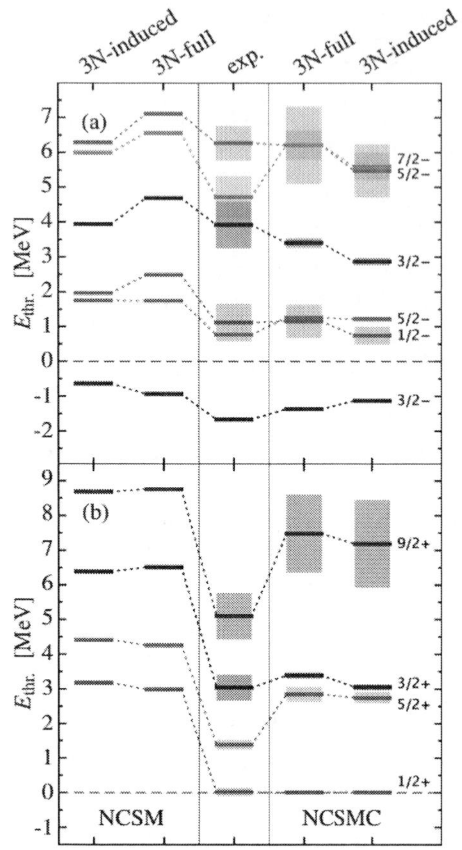

Figure 2. Computed low-lying energy spectrum of ^{9}Be (a) negative- and (b) positive-parity states compared to experiment. The chiral $NN + 3N$ Hamiltonian of Ref. [40] with evolution parameter $\lambda = 2$ fm^{-1} was used. The negative(positive)-parity NCSMC calculation, coupling ^{9}Be square integrable eigenstates with ^{8}Be$(0^+, 2^+)$ binary cluster states, was performed in a $N_{\rm max} = 12(11)$, $\hbar\Omega = 20$ MeV HO model space. Here $E_{\rm kin}$ indicates the center-of-mass energy with respect to the ^{8}Be+n threshold.

with $\Lambda_{\rm SRG} = 2.0$ fm$^-1$) [22]. Unless otherwise stated, all calculations were carried out using the ansatz of Eq. (1) with fifteen discrete eigenstates of the ^{6}Li system and continuous d-^{4}He(g.s.) binary-cluster states with up to seven deuteron pseudostates in the $^3S_1$–$^3D_1$, $^3D_2$ and $^3D_3$–$^3G_3$ channels. Convergence in the HO expansions was approached at at $N_{\rm max} = 11$, using a frequency of 20 MeV, around which the ^{6}Li g.s. energy calculated within the square-integrable basis of the NCSM becomes nearly insensitive to $\hbar\Omega$ [41].

Similar to the proton-^{4}He scattering discussed earlier, also in this case the square-integrable eigenstates of the compound (here ^{6}Li) system play a crucial role in achieving an accurate description of the dynamic process. Besides helping to account for the polarization of the ^{4}He, of which we can (computationally) afford to include only the g.s., here they also contribute to the description of the projectile distortion. This is exemplified by the fairly rapid convergence, shown in the left panel of Fig. 3, of the d–^{4}He scattering phase shifts with respect to the number of deuterium pseudostates. The high quality of the results obtained with the chiral NN+3N forces is corroborated by the good agreement with experiment of Fig. 3 (right panel), presenting ^{4}He$(d,d)^4$He angular distributions in the $2.93 \leq E_d \leq 12.0$ MeV interval of incident energies.

At the same time, the inclusion of continuum degrees of freedom is essential to reach a true understanding of a nucleus with low breakup threshold such as ^{6}Li, the ground state of which lies only 1.47 MeV (compared to its absolute energy of nearly 32 MeV) below the ^{4}He+d separation energy. As shown in Fig. 4, combined with the inclusion of the $3N$ force, the NCSMC yields a rather good agreement with the observed spectrum. As one should expect, for the ground state and the narrow

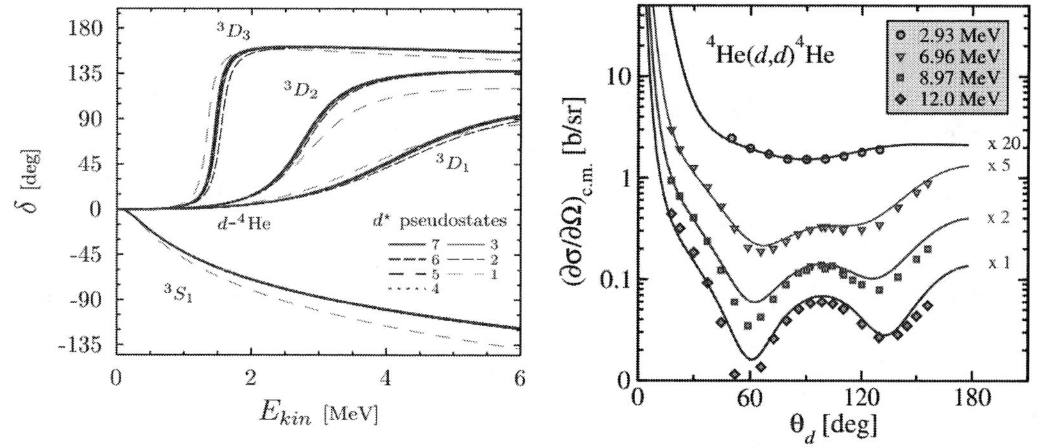

Figure 3. Panel (a): Computed d-^{4}He S- and D-wave phase shifts at $N_{\max} = 9$, obtained with fifteen square-integrable ^{6}Li eigenstates, as a function of the number of ^{2}H pseudostates (up to seven) in each of the $^3S_1$–$^3D_1$, $^3D_2$ and $^3D_3$–$^3G_3$ channels. The two-body part of the SRG-evolved N^{3}LO NN potential (NN-only) with $\Lambda = 2.0$ fm^{-1} was used. Panel (b): Computed ^{2}H$(\alpha, d)^4$He using the NN+$3N$ Hamiltonian (lines) and measured (symbols) center-of-mass angular distributions at $E_d = 2.93, 6.96, 8.97$ [45], and 12 MeV [46], scaled by a factor of 20, 5, 2, and 1, respectively. All positive- and negative-parity partial waves up to $J = 3$ were included in the calculations.

3^+ resonance the NCSMC energy levels are also in good agreement with those obtained from an extrapolation to $N_{\max} = \infty$ of a traditional NCSM calculation. This is a strong indication that the slight overestimation of the first excited state is likely due to remaining deficiencies in the adopted 3N force model, particularly concerning the strength of the spin-orbit interaction. However, only within the NCSMC do the computed wave functions present the correct asymptotic, which for the g.s. are Whittaker functions. This is essential for the extraction of the asymptotic normalization constants. The present calculation reproduces the empirical binding energy of ^{6}Li, yielding an asymptotic D- to S-state ratio of the ^{6}Li wave function in $d + \alpha$ configuration of -0.027 in agreement with a determination from ^{6}Li-^{4}He elastic scattering [42] and the value previously obtained by Nollett et al. using variational Monte Carlo [43]. Contrary to the lighter nuclei, the empirical value of this ratio was still uncertain for ^{6}Li, with different determinations disagreeing even as to its sign [44].

The above accurate studies of nucleon- and deuterium-^{4}He elastic scattering with chiral NN+3N interactions set the stage for the most advanced *ab initio* calculation of the ^{3}H$(d, n)^4$He fusion, currently under way. This study is being carried out within an over-complete NCSMC model space including n+^{4}He(g.s.) and d+^{3}H(g.s.) continuous basis states, as well as square-integrable discrete eigenstates of the compound ^{5}He nucleus. Figure 5 shows preliminary results for the n–^{4}He scattering phase shifts from zero to 24 MeV in the center-of-mass energy. Despite the fairly small size of the harmonic oscillator basis, the calculation is in close agreement with experiment. In particular, besides a slight shift of the P-wave resonances to lower energies, the inclusion of d+^{3}H channels leads to the appearance of a resonance in the $^2D_{3/2}$ partial wave, just above the d+^{3}H threshold. This is the exit channel of the deuterium-tritium fusion.

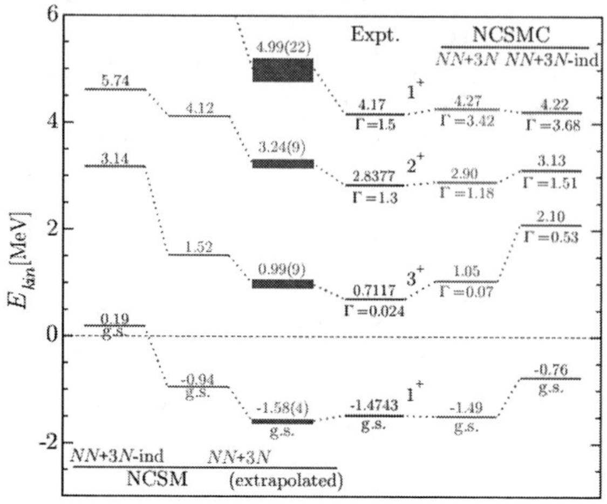

Figure 4. Computed NCSMC low-lying spectrum of positive-parity states (and their widths Γ) for the ^{6}Li nucleus with (red lines) and without (blue lines) inclusion of the initial chiral $3N$ force compared to experiment (black lines). Also shown on the left-hand side are the best ($N_{\max} = 12$) and extrapolated ($N_{\max} = \infty$) NCSM energy levels. The zero energy is set to the respective computed (experimental) d+^{4}He breakup thresholds.

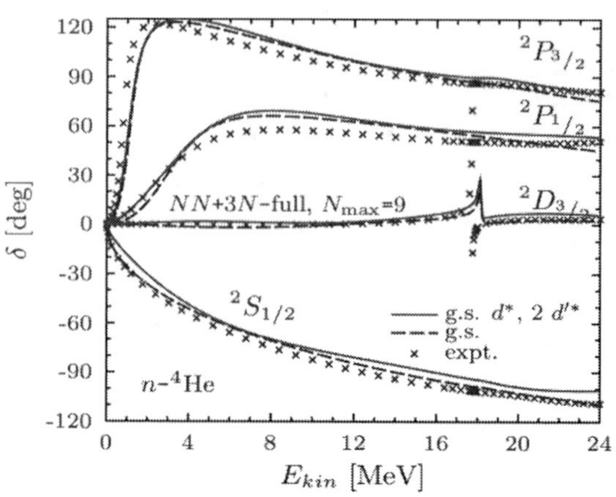

Figure 5. Preliminary results (lines) for the n–^{4}He scattering phase shifts from zero to 24 MeV in the center-of-mass energy compared to experiment (crosses). All calculations were performed at $N_{\max} = 9$ within an over-complete NCSMC model space including n+^{4}He(g.s.) and d+^{3}H(g.s.) continuous basis states with up to two ^{2}H pseudostates in the $^3S_1$–$^3D_1$ (d^*) and $^3D_2$ (d'^*) channels, as well as square-integrable discrete eigenstates of the compound ^{5}He nucleus.

4 Conclusions and future prospects

A unified *ab initio* description of light-nucleus structure and reaction properties starting from accurate NN+3N forces is now becoming possible, thanks in part to the NCSMC approach. This is an *ab initio* theory including the continuum which combines the efficient description of short- and medium-range correlations of the NCSM with the ability of the RGM of describing the scattering physics of a system. In this contribution, we reviewed recent applications of the approach to the description of elastic scattering and recoil of protons from ^{4}He, continuum and 3N-force effects on the energy levels of the ^{9}Be, and ^{6}Li structure and d–^{4}He dynamics using chiral NN+3N forces. Building on this work, we are now pursuing several new applications to light-nucleus structure and reaction properties with chiral NN+3N forces. The obtained accurate nucleon-^{4}He scattering wave functions are currently being used as input for the study of the ^{4}He$(p, p'\gamma)^4$He and ^{4}He$(n, n'\gamma)^4$He bremsstrahlung, the radiative process by which a photon is emitted as a result of the nuclear collision between a nucleon and a ^{4}He nucleus. The former is one of the few measured light-nucleus bremsstrahlung cross sections, while the latter is a necessary preliminary step for the study of the more complicated ^{3}H$(d, n\gamma)^4$He bremsstrahlung cross section, which could be used to diagnose plasmas in fusion experiments but is not known well enough. The general NCSMC framework applicable to targets heavier than ^{4}He is now being applied to study continuum and 3N force effects on the low-lying spectrum of the ^{11}Be one-neutron halo nucleus and its photodissociation into n+^{10}Be. The obtained accurate ^{6}Li ground state and d–^{4}He scattering wave functions set the stage for the first *ab initio* study of the ^{2}H$(\alpha, \gamma)^6$Li radiative capture, responsible for the big-bang nucleosynthesis of ^{6}Li, as well as the calculation of ^{6}Li g.s. properties including the effect of the continuum degrees of freedom. Finally, we are in the process of carrying out the first *ab initio* calculation of the ^{3}H$(d, n)^4$He fusion using chiral NN+3N forces.

Acknowledgements

Prepared in part by LLNL under Contract DE-AC52-07NA27344. This material is based upon work supported by the U.S. Department of Energy, Office of Science, Office of Nuclear Physics, under Work Proposal Number SCW1158, the NSERC Grant Number 401945-2011, the Deutsche Forschungsgemeinschaft through Contract SFB 634, the Helmholtz International Center for FAIR within the framework of the LOEWE program launched by the State of Hesse, and the BMBF through Contract No. 06DA7074I. TRIUMF receives funding via a contribution through the Canadian National Research Council. Computing support for this work came from the LLNL institutional Computing Grand Challenge program, the National Energy Research Scientific Computing Center (edison) supported by the Office of Science of the U.S. Department of Energy under Contract No. DE-AC02-05CH11231, the LOEWE-CSC Frankfurt, the computing center of the TU Darmstadt (lichtenberg), and from an INCITE Award on the Titan supercomputer of the Oak Ridge Leadership Computing Facility (OLCF) at ORNL.

References

[1] S. Baroni, P. Navrátil, S. Quaglioni, Phys. Rev. Lett. **110**, 022505 (2013)
[2] S. Baroni, P. Navrátil, S. Quaglioni, Phys. Rev. C **87**, 034326 (2013)
[3] P. Navrátil, J.P. Vary, B.R. Barrett, Phys. Rev. Lett. **84**, 5728 (2000)
[4] Y.C. Tang, M. LeMere, D.R. Thompsom, Phys. Rep. **47**, 167 (1978)
[5] K. Langanke, H. Friedrich, *Advances in Nuclear Physics* (Plenum, New York, 1987)
[6] S. Quaglioni, P. Navrátil, Phys. Rev. Lett. **101**, 092501 (2008)
[7] S. Quaglioni, P. Navrátil, Phys. Rev. C **79**, 044606 (2009)
[8] M. Hesse, J.M. Sparenberg, F. Van Raemdonck, D. Baye, Nucl. Phys. A **640**, 37 (1998)
[9] M. Hesse, J. Roland, D. Baye, Nucl. Phys. A **709**, 184 (2002)
[10] K.M. Nollett, S.C. Pieper, R.B. Wiringa, J. Carlson, G.M. Hale, Phys. Rev. Lett. **99**, 022502 (2007)
[11] M. Viviani, L. Girlanda, A. Kievsky, L.E. Marcucci, Few-Body Syst. **54**, 885 (2013)
[12] D.R. Entem, R. Machleidt, Phys. Rev. C **68**, 041001 (2003)
[13] P. Navrátil, Few-Body Syst. **41**, 117 (2007)
[14] D. Gazit, S. Quaglioni, P. Navrátil, Phys. Rev. Lett. **103**, 102502 (2009)
[15] S.D. Głazek, K.G. Wilson, Phys. Rev. D **48**, 5863 (1993)
[16] F. Wegner, Ann. Phys. **506**, 77 (1994)
[17] S.K. Bogner, R.J. Furnstahl, R.J. Perry, Phys. Rev. C **75**, 061001 (2007)
[18] H. Hergert, R. Roth, Phys. Rev. C **75**, 051001 (2007)
[19] E.D. Jurgenson, P. Navrátil, R.J. Furnstahl, Phys. Rev. Lett. **103**, 082501 (2009)
[20] G. Hupin, S. Quaglioni, P. Navrátil, Phys. Rev. C **90**, 061601 (2014)
[21] J. Langhammer, P. Navrátil, S. Quaglioni, G. Hupin, A. Calci, R. Roth, Phys. Rev. C **91**, 021301 (2015)
[22] G. Hupin, S. Quaglioni, P. Navrátil, Phys. Rev. Lett. **114**, 212502 (2015)
[23] P. Pusa, E. Rauhala, A. Gurbich, A. Nurmela, Nucl. Instr. and Meth. B **222**, 686 (2004)
[24] F. Pászti, Nucl. Instr. and Meth. B **66**, 83 (1992)
[25] Y. Avrahami, E. Zolotoyabko, Nucl. Instr. and Meth. B **120**, 84 (1996)
[26] G. Freier, E. Lampi, W. Sleator, J.H. Williams, Phys. Rev. **75**, 1345 (1949)
[27] W.E. Kreger, W. Jentschke, P.G. Kruger, Phys. Rev. **93**, 837 (1954)
[28] P. Miller, G.C. Phillips, Phys. Rev. **112**, 2043 (1958)

[29] A.C.L. Barnard, C.M. Jones, J.L. Weil, Nucl. Phys. **50**, 604 (1964)
[30] A. Nurmela, E. Rauhala, J. Räisänen, J. Appl. Phys. **82**, 1983 (1997)
[31] S. Nagata, S. Yamaguchi, Y. Fujino, Y. Hori, N. Sugiyama, K. Kamada, Nucl. Instr. and Meth. B **6**, 533 (1985)
[32] J.E.E. Baglin, A.J. Kellock, M.A. Crockett, A.H. Shih, Nucl. Instr. and Meth. B **64**, 469 (1992)
[33] Y. Wang, J. Chen, F. Huang, Nucl. Instr. and Meth. B **17**, 11 (1986)
[34] I. Bogdanović Radović, O. Benka, Nucl. Instr. and Meth. B **174**, 25 (2001)
[35] S.K. Kim, H.D. Choi, Nucl. Instr. and Meth. B **174**, 33 (2001)
[36] J.C. Keay, D.C. Ingram, Nucl. Instr. and Meth. B **211**, 305 (2003)
[37] J.F. Browning, J.C. Banks, W.R. Wampler, B.L. Doyle, Nucl. Instr. and Meth. B **219-220**, 317 (2004)
[38] G. Hupin, J. Langhammer, P. Navrátil, S. Quaglioni, A. Calci, R. Roth, Phys. Rev. C **88**, 054622 (2013)
[39] C. Forssén, P. Navrátil, W.E. Ormand, E. Caurier, Phys. Rev. C **71**, 044312 (2005)
[40] R. Roth, S. Binder, K. Vobig, A. Calci, J. Langhammer, P. Navrátil, Phys. Rev. Lett. **109**, 052501 (2012)
[41] E.D. Jurgenson, P. Navrátil, R.J. Furnstahl, Phys. Rev. C **83**, 034301 (2011)
[42] E.A. George, L.D. Knutson, Phys. Rev. C **59**, 598 (1999)
[43] K. Nollett, R.B. Wiringa, R. Schiavilla, Phys. Rev. C **63**, 024003 (2001)
[44] D. Tilley, C. Cheves, J. Godwin, G. Hale, H. Hofmann, J. Kelley, C. Sheu, H. Weller, Nucl. Phys. A **708**, 3 (2002)
[45] J.H. Jett, J.L. Detch, N. Jarmie, Phys. Rev. C **3**, 1769 (1971)
[46] L.S. Senhouse, T.A. Tombrello, Nucl. Phys. **57**, 624 (1964)

EPJ Web of Conferences **113**, 01006 (2016)
DOI: 10.1051/epjconf/201611301006

Experiments towards resolving the proton charge radius puzzle

A. Antognini[1,2,a], K. Schuhmann[1,2], F. D. Amaro[3], P. Amaro[4], M. Abdou-Ahmed[5], F. Biraben[6], T.-L. Chen[7], D. S. Covita[8], A. J. Dax[2], M. Diepold[9], L. M. P. Fernandes[3], B. Franke[9], S. Galtier[6], A. L. Gouvea[3], J. Götzfried[9], T. Graf[5], T. W. Hänsch[9,b], M. Hildebrandt[2], P. Indelicato[6], L. Julien[6], K. Kirch[1,2], A. Knecht[2], F. Kottmann[1], J. J. Krauth[9], Y.-W. Liu[7], J. Machado[4,6], C. M. B. Monteiro[3], F. Mulhauser[9], F. Nez[6], J. P. Santos[4], J. M. F. dos Santos[3], C. I. Szabo[6], D. Taqqu[1], J. F. C. A. Veloso[8], A. Voss[5], B. Weichelt[5], and R. Pohl[9]

[1] *Institute for Particle Physics, ETH, 8093 Zurich, Switzerland*
[2] *Paul Scherrer Institute, 5232 Villigen-PSI, Switzerland*
[3] *LIBPhys, Physics Department, Universidade de Coimbra, 3004-516 Coimbra, Portugal*
[4] *Laboratório de Instrumentação, Engenharia Biomédica e Física da Radiação (LIBPhys-UNL) e Departamento de Física da Faculdade de Ciências e Tecnologia da Universidade Nova de Lisboa, Monte da Caparica, 2892-516 Caparica, Portugal*
[5] *Institut für Strahlwerkzeuge, Universität Stuttgart, 70569 Stuttgart, Germany*
[6] *Laboratoire Kastler Brossel, UPMC and CNRS, 75005 Paris, France*
[7] *Physics Department, National Tsing Hua University, Hsincho 300, Taiwan*
[8] *i3N, Universidade de Aveiro, Campus de Santiago, 3810-193 Aveiro, Portugal*
[9] *Max Planck Institute of Quantum Optics, 85748 Garching, Germany*

Abstract. We review the status of the proton charge radius puzzle. Emphasis is given to the various experiments initiated to resolve the conflict between the muonic hydrogen results and the results from scattering and regular hydrogen spectroscopy.

1 The proton charge radius puzzle

The historical route to the proton charge radius (r_p) is from elastic electron-proton scattering. In a completely complementary fashion, it has been obtained also from "high-precision" laser spectroscopy of hydrogen (H). Since a few years, "high-sensitivity" laser spectroscopy of muonic hydrogen (μp) offers a third way. The value extracted from μp with a relative accuracy of 5×10^{-4} is an order of magnitude more accurate than obtained from the other methods. Yet the value is 4% smaller than derived from electron-proton scattering and H spectroscopy with a disagreement at the 7σ level [1–5].

In the last five years as summarized in [6, 7] various cross checks and refinements of bound-state QED calculations needed for the extraction of r_p from μp have been performed, together with investigations of the proton structure. Several suggestions in the field of "beyond standard model" BSM physics have been articulated, re-analysis of scattering data have been carried out and new experiments have been initiated. Despite this, presently the discrepancy still persists and the resolution

[a] e-mail: aldo@phys.ethz.ch
[b] Also at: Ludwig-Maximilians-Universität, 80539 Munich, Germany.

of the proton radius puzzle remains unknown. In this article, we summarize mainly the ongoing experimental activities which hold the potential to unravel the proton radius puzzle.

As the atomic energy level are slightly modified by the nuclear finite size, it is possible to deduce the nuclear charge radius by performing spectroscopy of the atomic energy levels. In leading order, the energy shift caused by the nuclear finite size is

$$\Delta E_{\text{finite size}} = \frac{2\pi\alpha}{3}|\phi^2(0)|^2 R_E^2 = \frac{2m_r^3\alpha^4}{3n^3} R_E^2 \tag{1}$$

where $\phi(0)$ is the wavefunction at the origin in coordinate space, m_r the reduced mass of the atomic system, α the fine structure constant and n the principal quantum number. R_E is the charge radius of the nucleus defined in a covariant way as the slope of the electric form factor (G_E) at zero momentum exchange Q^2

$$R_E = -6\frac{dG_E}{dQ^2}\bigg|_{Q^2=0}. \tag{2}$$

Non-relativistically, R_E is the second moment of the electric charge distribution ρ_E of the nucleus $R_E^2 \approx \int d\vec{r}\rho_E(\vec{r})r^2$.

The m_r^3 dependence of Eq. (1) reveals the advantages related with muonic atoms. As the muon mass is 200 times larger than the electron mass, the muonic wavefunction strongly overlaps with the nucleus ensuing a large shift of the energy levels due to the nuclear finite size. Thus, the muonic bound-states represent ideal systems for the precise determination of nuclear charge radii.

The proton form factors can be obtained from unpolarized differential cross section measurements of electron scattering off protons. In the one-photon approximation the elastic cross section is

$$\frac{d\sigma}{d\Omega} = \frac{d\sigma}{d\Omega}\bigg|_{Mott} \times \frac{1}{1+\tau}\left(G_E^2(Q^2) + \frac{\tau}{\epsilon}G_M^2(Q^2)\right) \tag{3}$$

where the Mott cross section applies for point-like particles, G_E and G_M are the electric and magnetic form factors of the proton, $\tau = Q^2/4M^2$ and $\epsilon^{-1} = 1 + 2(1+\tau)\tan^2(\theta/2)$ are kinematical variables. Commonly, the Rosenbluth separation is applied to disentangle the charge from the magnetic contributions by using the angle-dependence at fixed Q^2. Therefore, by measuring the differential cross section at various Q^2 and angles θ, one obtains $G_E(Q^2)$ and $G_M(Q^2)$ and via Eq. (2) the radius.

2 Muonic hydrogen and possible new physics explanations

The CREMA collaboration has measured two transition in μp: from the triplet ($2S_{1/2}^{F=1} - 2P_{3/2}^{F=2}$) [1] and the singlet ($2S_{1/2}^{F=0} - 2P_{3/2}^{F=1}$) [2] 2S-states yielding a radius of $r_p = 0.84087(39)$ fm. More specifically the two measured energy splittings, from the triplet $h\nu_t$ and from the singlet $h\nu_s$ states, can be combined to obtain both the 2S Lamb shift $E_L = \Delta E(2S - 2P_{1/2})$ and the hyperfine splitting E_{HFS}:

$$E_L = \frac{1}{4}h\nu_s + \frac{3}{4}h\nu_t - 8.8123(2)\text{ meV}, \tag{4}$$

$$E_{\text{HFS}} = h\nu_s - h\nu_t + 3.2480(2)\text{ meV}, \tag{5}$$

where the numerical values follow from reliable proton-independent corrections of the 2P states. Experimentally [2]

$$E_L^{\text{exp}} = 202.3706(23)\text{ meV} \tag{6}$$

equivalent to a frequency 48932.99(55) GHz, limited by statistics while the systematics effects is at the 300 MHz level. From theory [8]

$$E_L^{\mathrm{th}} = 206.0336(15)\,[\mathrm{meV}] - 5.2275(10)\,\Big[\frac{\mathrm{meV}}{\mathrm{fm}^2}\Big] r_p^2 + 0.0332(20)\,[\mathrm{meV}]. \tag{7}$$

The first term accounts for QED contributions, the second for finite size effects, and the third for the two-photon exchange (TPE) contribution. From the Lamb shift an improved r_p value free from uncertainties related with the HFS splitting has been determined, and from the HFS the Zemach radius, albeit not with the same accuracy as the charge radius.

The consistency of the two μp measurements, represents an important cross check of the muonic results. The typical systematic effects affecting the atomic energy levels are substantially suppressed in μp due to the stronger binding. The internal fields and the level separation of the muonic atoms are greatly enhanced compared to regular atoms making them insensitive to external fields (AC and DC Stark, Zeeman, black-body radiation and pressure shifts). Thus μp turns out to be very sensitive to the proton charge radius (m_r^3-dependence) and insensitive to systematics which typically scales as $\sim 1/m_r$. The possible involvement of weakly bound three-body systems in the muonic hydrogen spectroscopy experiment [9] has been ruled out by three-body calculations [10], and by the experimental non-observation of sizable additional line broadening, line splitting and event rate decrease.

The bound-state QED corrections which give rise to Eq. (7) have been computed by several groups [11, 12] as summarized in [8] and updated recently in [13, 14]. Particular attention has been devoted to the TPE contribution which has been computed in two frameworks: one making use of dispersion relations and measured inelastic structure functions of the proton [15], the second from chiral perturbation theories [16]. Both ways provide consistent predictions. In the dispersion-based approach a subtraction term in form of an integral from zero to infinite Q^2 is necessary which can not be fully constrained by data. At intermediate Q^2 indeed a modeling of the proton is necessary [17]. The large majority of the community, agrees with a value for this subtraction term which is two order of magnitude smaller than the measured discrepancy of 0.3 meV [18–20]. Even if improbable, in principle still a very un-smooth and physically unmotivated proton structure could be constructed which could shift the μp transition to explain the measured discrepancy [21]. However, the published subtraction functions proposed to solve the proton radius puzzle would affect through the Cottingham formula the proton and neutron masses by 600 MeV [22] which is quite implausible when compared with the measured and computed neutron-proton mass difference of 1.29 MeV [23].

Explaining the discrepancy by this subtraction term can be interpreted as an exotic hadronic effect. Other effects occurring in the vicinity of the nucleus, as the breakdown of the perturbative approach in the electron-proton interaction [24, 25], the interaction with sea $\mu^+\mu^-$ and e^+e^- pairs [26] etc., have been suggested and need to be further investigated but at moment are not yet conclusive. Several BSM extensions have been proposed but the vast majority of them have difficulties to resolve the measured discrepancy without conflicting with other low energy constraints. Still some BSM theories can be formulated but they require fine-tuning (e.g. cancellation between axial and vector components), targeted coupling (e.g. preferentially to the muon or to muon proton) and are problematic to be merged in a gauge invariant way into the standard model [27–29].

3 Hydrogen experiments

In a simplified way, the hydrogen S-state energy levels can be described by

$$E(nS) = \frac{R_\infty}{n^2} + \frac{L_{1S}}{n^3} \tag{8}$$

where $R_\infty = 3.289\,841\,960\,355(19) \times 10^{15}$ Hz is the Rydberg constant and approximately

$$L_{1S} \simeq 8171.636(4)\,[\mathrm{MHz}] + 1.5645\left[\frac{\mathrm{MHz}}{\mathrm{fm}^2}\right] r_p^2 \tag{9}$$

the Lamb shift of the ground state given by bound-state QED calculations. The different n-dependence of the two terms in Eq. (8) permits the determination of both R_∞ and r_p, from at least two transition frequencies in H, assuming Eq. (9).

Being the most precisely measured transition ($u_r = 4\times10^{-15}$) [30, 31] and because it is showing the largest sensitivity to the Lamb shift contributions usually the 1S-2S transition is used. By combining it with a second transition measurement, it is possible to determine the proton charge radius. When taken individually, the various r_p values extracted from H spectroscopy by combining two frequency measurements (2S-4S, 2S-12D, 2S-6S, 2S-6D, 2S-8S, 1S-3S as "second" transition) are statistically compatible with the value from μp. Only the value extracted by pairing the 1S-2S and the 2S-8D transitions is showing a 3σ deviation while all the others differ only by $\lesssim 1.5\sigma$.

So the 4.4σ discrepancy between the proton charge radius from μp and H spectroscopy emerges only after an averaging process (mean square adjustments of all measured transitions) of the various "individual" determinations and consequently is less startling than it looks at first glance. A small systematic effect common to the H measurements could be sufficient to explain the deviation between μp and H results. This fact becomes even more evident if we consider the frequency shifts (absolute and normalized to the linewidth) necessary to match the r_p values from μp and H as summarized for selected transitions in Table 1. Obviously the discrepancy can not be solved by slightly tuning

Table 1. Hypothetical shift of the transition frequency value needed to match the r_p from H and μp, expressed also relative to the stated experimental accuracy σ, and to the transition effective linewidths Γ_{eff}.

Transition	Shift rel. to uncertainty	Absolute shift	Shift rel. to effective linewidth
μp(2S-2P)	$100\,\sigma$	75 GHz	$4\,\Gamma_{\mathrm{eff}}$
H(1S-2S)	$4'000\,\sigma$	40 kHz	$40\,\Gamma_{\mathrm{eff}}$
H(2S-4P)	$1.5\,\sigma$	9 kHz	$7\times10^{-4}\,\Gamma_{\mathrm{eff}}$
H(2S-2P)	$1.5\,\sigma$	5 kHz	$7\times10^{-4}\,\Gamma_{\mathrm{eff}}$
H(2S-8D)	$3\,\sigma$	20 kHz	$2\times10^{-2}\,\Gamma_{\mathrm{eff}}$
H(2S-12D)	$1\,\sigma$	8 kHz	$5\times10^{-3}\,\Gamma_{\mathrm{eff}}$
H(1S-3S)	$1\,\sigma$	13 kHz	$5\times10^{-3}\,\Gamma_{\mathrm{eff}}$

(shifting) the measured values of the 1S-2S transition in H and the 2S-2P transitions in μp because it would require displacements corresponding to 4000σ and 100σ, respectively. Expressing the required frequency shift relative to the linewidth as in the last column allows to better recognize some aspects of the experimental challenges. For example a shift of only $7\times10^{-4}\,\Gamma$ of the $2S-4P$ transition would be sufficient to explain the discrepancy. A control of the systematics which could distort and shift the line shape on this level of accuracy is far from being a trivial task. Well investigated are the large line broadenings owing to inhomogeneous light shifts which results in profiles with effective experimental widths much larger than the natural linewidths [3].

Another exemplary correction relevant in this context, named quantum interference, has been brought recently back to attention [32] and has been applied also to muonic atoms [33]. An atomic transition can be shifted by the presence of a neighboring line, and this energy shift δE, as a rule of thumb, maximally amounts to [32] $\frac{\delta E}{\Gamma} \approx \frac{\Gamma}{D}$ where D is the energy difference between the two resonances and Γ the transition linewidth. Thus, if a transition frequency is aimed with an absolute accuracy of Γ/x, then the influence of the neighboring lines with $D \leq x\Gamma$ has to be considered.

The precise evaluation of these quantum interference effects are challenging because they require solving numerous differential equations describing the amplitude of the total excitation and detection processes from initial to final state distributions and because it depends on experimental parameters such as the angular position and polarization sensitivity of the detectors, the laser intensity, direction and polarization, the initial population distribution among states, etc.

Generally speaking, transition frequencies involving states with large n are more sensitive to systematical effects caused by external fields. Emblematic is for example the n^7-dependence of the DC Stark effect. Motivated by the possibility that minor effects in H could be responsible for the observed discrepancy, various activities have been initiated in this field:

- *$2S - 4P_{1/2}$ and $2S - 4P_{3/2}$ at MPQ Garching [34]:* They are aiming at improving previous measurements by a factor of 5 down to an accuracy of few kHz which would yield a r_p with less than 2% accuracy when paired with the 1S-2S transition. Preparation of the 2S state by means of optical excitation and an almost 4π Lyman-alpha detection system are key elements to control the line pulling due to quantum interference on the $1 \times 10^{-4}\Gamma$ level of accuracy required.
- *$1S - 3S$ transition at LKB and MPQ [35, 36]:* The 2010 results from the LKB group delivered the second most precise transition frequency measurement in H with a total uncertainty of 13 kHz corresponding to a relative accuracy of 4.5×10^{-12}. The error budget was dominated by statistics (12 kHz) and uncertainties in the velocity distribution of the atomic beam (3 kHz). A 1% accuracy of the proton radius will require a measurement of the 1S-3S transition with accuracy of about 2 kHz. To reach this goal, the Paris group is presently pursuing the measurement of the 1S-3S transition at 205 nm wavelength using cw spectroscopy along the same line of investigations as in previous experiment. Special emphasis is devoted to the velocity dependent systematic effects. Oppositely, the MPQ group, to circumvent the difficulties related with the generation of the 205 nm light, has devised an experiment which uses pico-second frequency comb pulses.
- *$2S - 2P$ classical Lamb shift in Toronto [37]:* The measurement of the 2S-2P energy splitting alone can lead to r_p. Indeed, as this transition does not depend on the Rydberg structure there is no need to combine it with a second transition frequency measurement [1]. Microwave spectroscopy based on the Ramsey method of separated oscillatory field is used for this purpose. A factor of 5 improvement is anticipated, which implies a determination of r_p to the 0.6% level. To reach this ambitious goal the position of the line has to be determined with 1 part in 10^4.

The "second" (beside the 1S-2S transition) transition frequency measurement in H can be interpreted as a measurement of the Rydberg constant. An alternative way to an independent determination of the Rydberg constant, is to perform optical spectroscopy of H-like ions between circular Rydberg states where the nuclear size corrections are basically absent, the QED contributions small, and the linewidths narrow [38], or via spectroscopy of positronium and muonium [39].

4 Scattering experiments

The Mainz A1 collaboration at MAMI has measured in 2010 1422 precise relative e-p cross sections in the low-Q^2 regime (0.0038 GeV^2 to 0.98 GeV^2) and a wide range of beam energy and scattering angles [5]. Two spectrometers were moved with overlapping angle settings while a third spectrometer was kept fixed and used as a luminosity monitor.

[1] Note that all atomic transition frequencies expressed in Hz depend on R_∞, thus indirectly via the SI units system also the 2S-2P splitting depends on R_∞. Yet the to date accuracy of the 2S-2P measurements are several order of magnitude worse that the R_∞ accuracy.

As data are available only down to a minimal Q^2, extrapolation to $Q^2 = 0$ is required to determine the proton charge radius. Numerous works have been concerned with the issues related with this extrapolation procedure. The Mainz group found a satisfactory goodness of fit ($\chi^2 = 1.14$ for 1422 points) through the use of flexible fitting functions (splines and polynomials) and the resulting radius reads [5] $r_p = 0.879(8)$ fm in agreement with the CODATA06 value of 0.8768(69) fm based mostly on atomic measurements. It is important that the fit function be flexible enough to adequately reproduce the data, without being so flexible that over-fitting occurs. One solution to this problem is given in [4, 41] where the low-Q^2 behavior of the form factor is constrained by using large-r assumption of the charge distribution. Reanalysis of the world data using these constraints yields $r_p = 0.879(11)$ fm [42]. Another approach to address issues of over- or under-fitting data is the use of bounded polynomial z expansion (after conformal mapping) and constraining the expansion coefficients to decrease "perturbatively" with increasing order [40].

Conformal fits to the form factors were performed in [43] yielding $r_p = 0.870(23)(12)$ fm in agreement with the CODATA value. Another group fitting the 1422 Mainz data points found $r_p = 0.840(15)$ fm [44] in agreement with the PSI result but with a $\chi^2 = 1.4$ (using more flexible functions the same group found $\chi^2 = 1.1$). Noteworthy, the value of $r_p \approx 0.84$ fm from analysis of scattering data using dispersion relations and vector-mesons dominance models was obtained by the same group prior to the publication of the muonic result. The use of a form factor model (for all Q^2-range) based on dispersion relation and vector-meson dominance introduce rigidity in the model which results in the larger χ^2. So tension exists between the use of a physically motivated model giving a poorer fit or very flexible fit functions without physical constraints yielding better χ^2. This tension could arise by inappropriateness of the theoretical model, insufficient treatment of experimental details, or due to an underestimation of the scattering cross section uncertainties.

A wide-ranging study of possible systematic effects of the 2010 Mainz data has been recently reported in [40]. Special attention was devoted to the extrapolation procedure, to normalization factors needed to smoothly combine the various spectrometer settings and to refinement of the radiative corrections. This reanalysis yields $r_p = 0.895(20)$ fm. When applied to the world data (excluding Mainz 2010) $r_p = 0.918(24)$ fm is found [40]. Even though some inconsistencies between data sets were found, which have led to the increased uncertainty of the r_p extracted from scattering, it remains difficult to reconcile the scattering results with the muonic results. The only sure conclusion is that analysis of low-Q^2 scattering data is not simple and remains a matter of discussion. Because data at still lower Q^2 would be beneficial, two electron-proton experiments have been initiated:

- *PRad experiment at Hall B in JLAB [45]:* This experiment planned to operate at Q^2 down to 2×10^{-4} GeV^2 aims to obtain r_p with sub-percent accuracy. The experiment is based on a windowless target and a downstream calorimeter which allows to extend the cross sections measurements to smaller scattering angles. The need to measure relative cross sections at about 0.2% level requires knowledge of the angle to 10 μrad accuracy which makes this experiment very challenging.
- *Initial state radiation at MAMI, A1 collaboration [46]:* Making use of the initial state radiation techniques, where the initial electron momentum is degraded by photon emission, the momentum transfer to the proton is reduced. The scattered electron is measured with the usual spectrometers but no information on the photon is observed. However, by comparing measurements with Monte Carlo simulations accurate form factors can be determined down to Q^2 of 2×10^{-4} GeV^2.

Other scattering experiments can provide very important information:

- *MUSE: muon scattering experiment at PSI [47]:* The MUSE experiment plans to measure $\mu^+ - p$ and $\mu^- - p$ as well as $e^+ - p$ and $e^- - p$ scattering down to 0.002 GeV^2. By comparing negative with positive charges they will have an handle on the insidious TPE contribution. Comparison between electron and muon cross sections allows the elimination of common systematical effects, including some extrapolation uncertainties. Thus, this experiment has not only the potential to measure the proton charge radius absolute value to 2% accuracy, but also a possible difference between radii extracted from electron and muon scattering down to a relative accuracy of 1%. In this way possible muon-specific interactions can be disclosed.

- *Deuteron scattering at Mainz [48]:* New e-d scattering data have been collected in Mainz, with the aims to extract a new value of the deuteron charge radius and break-up information which may be used to compute the deuteron polarizability contribution in muonic deuterium.

Of relevance in this context is also the planned measurement at JLAB [49] aiming at the electric form factor of the mirror nuclei ^{3}He and ^{3}H to extract their charge radii difference, and the TREK program at J-PARC that scrutinizes K-decays to search for BSM physics motivated by the r_p puzzle.

5 Muonic deuterium and muonic helium ions

In 2009 the CREMA collaboration measured two 2S-2P transitions in muonic deuterium (μd). The 2S-2P energy splitting in μd was determined with about 1 GHz accuracy, which corresponds to a relative accuracy of 20 ppm and 5% of the linewidth.

Evaluation of the most challenging systematic effect, the quantum interference effect has been recently completed [33]. Due to the proximity of two 2P states (4Γ apart), the quantum interference effects might be considerable. However, a quantitative evaluation of this effect, when accounting for the used excitation and detection schemes, the detector geometry, the laser direction and polarization etc., yields a line shift $\delta E \leq 0.001\Gamma$, thus far below the statistical accuracy of our experiment.

Moreover, the theory in muonic deuterium has only recently converged [52] to a state which allows a precise determination of the deuteron charge radius from the Lamb shift measured in μd. An impressive progress as been achieved in recent years both on the "purely" QED sector [12, 50, 51], as well as in the computation of the TPE contribution [53–55], yielding [52]:

$$\Delta E_{\mu\mathrm{d}}(2S-2P_{1/2}) = 228.7766(10)\,[\mathrm{meV}] - 6.1102(3)\left[\frac{\mathrm{meV}}{\mathrm{fm}^2}\right] r_d^2 + 1.7091(200)\,[\mathrm{meV}] \qquad (10)$$

where the first term represent basically the "pure" bound-state QED contributions, the second term with r_d in fm the leading finite size contribution (including mixed radiative-finite-size corrections) and the third the TPE contribution.

Combining our measurements with the prediction of Eq. (10) we will obtain r_d with a relative accuracy of $u_r = 4 \times 10^{-4}$ limited by the TPE contribution. A second route to a precise r_d value is to combine the r_p extracted from μp with the H-D isotopic measurements of the 1S-2S transition [30]. A comparison of these two numbers will check the consistency of the muonic results and will give new constraints to BSM theories, e.g. if and how the "new force carrier" can couple to the neutron.

In 2013 and 2014 we have measured for the first time 2 transitions frequencies in $\mu^4\mathrm{He}^+$ and 3 in $\mu^3\mathrm{He}^+$ with relative accuracy of about 40 ppm. The uncertainty of the transition frequency measurement is entirely given by statistical uncertainty, since systematics effects or uncertainty related with the laser frequency calibration are < 2 ppm. To extract the nuclear charge radii from these measurements the corresponding theoretical predictions have to be known. Preliminary values read

$$\Delta E_{^4\mathrm{He}}(2S-2P_{1/2}) = 1668.669(20)\,[\mathrm{meV}] - 106.340\left[\frac{\mathrm{meV}}{\mathrm{fm}^2}\right] r_{^4\mathrm{He}}^2 + 9.52(30)\,[\mathrm{meV}] \qquad (11)$$

$$\Delta E_{^3\mathrm{He}}(2S-2P_{1/2}) = 1644.658(20)\,[\mathrm{meV}] - 103.508\left[\frac{\mathrm{meV}}{\mathrm{fm}^2}\right] r_{^3\mathrm{He}}^2 + 14.66(40)\,[\mathrm{meV}]. \qquad (12)$$

Final numbers can not be given at this stage because it requires sorting out of the several contributions calculated by various groups using different frameworks [12, 50, 51, 56]. Similar to μd, in recent years there has been an impressive progression of the TPE predictions using state-of-the-art nuclear potentials, rendering the inelastic nuclear contribution with 5% accuracy [57]. Such an accuracy opens the way to alpha-particle and helion radii determination with a relative accuracy better than 1×10^{-3} which will be compared with the very precise value available from scattering [58]. Still, an important contribution of the TPE in μ^4He^+ and in μ^3He^+, related with the intrinsic nucleon polarizabilities, has not yet been addressed by the community. "Simple" scaling as used in μd [55] probably does not apply in this situation because of the smaller separation between nuclear and nucleon energies.

Besides providing insights into the r_p puzzle these nuclear radii represent benchmarks to check few-nucleon ab-initio calculations, or vice versa to fix low-energy coefficients (e.g. c_D or c_E of the three-nuclei-interaction) describing the nuclear potential in effective field theories [59]. Moreover they can be used as anchor point for the ^{6}He-^{4}He and ^{8}He-^{4}He isotopic shift measurements [60]. The radii extracted from μHe^+ measurements will be used to disentangle the 4σ discrepancy between two ^{3}He-^{4}He isotopic shifts measurements [61, 62], and their knowledge opens the way to enhanced bound-state QED tests for one- and two-electrons systems in "regular" He^+ [63] and He [64].

6 Conclusions

Various attempts have been made to find a solution of the proton charge radius puzzle which has been exposed by the Lamb shift measurement in muonic hydrogen. A plethora of theoretical works has been devoted in refining and rechecking the underlying theory necessary to extract the charge radius from the muonic measurements, in proposing theories beyond the standard model, in reanalyzing scattering data, and investigating proton structure. After all, the puzzle still persist. As a next step, new experimental inputs are required to provide guidance. Understanding of nuclear effects will be of primary importance for the interpretation of the next muonic measurements.

Acknowledgements

This work is supported by the Swiss National Science Foundation Projects No. 200021L_138175 and No. 200020_159755, the DFG_GR_3172/9-1, the ERC StG. #279765, by the FCT, Portugal under Contract No. SFRH/BPD/76842/2011 SFRH/BPD/74775/2010, through the projects No. PEstOE/FIS/UI0303/2011, PTDC/FIS/117606/2010, SFRH/BPD/92329/2013 and SFRH/BD/52332/2013. We acknowledge fruitful discussions with J. Arrington, N. Barnea, S. Bacca, M. Birse, E. Borie, C. Carslon, M. Distler, M. Gorchtein, C. Ji, S. Karshenboim, J. McGovern, K. Pachucki and S. Schlimme.

References

[1] R. Pohl et al., Nature **466**, 213 (2010).
[2] A. Antognini et al., Science **339**, 417 (2013).
[3] B. de Beauvoir et al., Eur. Phys. J. D **12**, 61 (2000).
[4] I. Sick, Few-Body Systems **55**, 903 (2014).
[5] J. C. Bernauer et al., Phys. Rev. C **90**, 015206 (2014).
[6] R. Pohl, R. Gilman, G. A. Miller and K. Pachucki, Ann. Rev. Nucl. Part. Sci. **63**, 175 (2013).
[7] C. E. Carlson, Progress in Particle and Nuclear Physics **82**, 59 (2015).
[8] A. Antognini et al., Annals Phys. **331**, 127 (2013).
[9] U. D. Jentschura, Ann. Phys. 326, **500** and **516** (2011).

[10] J.-P. Karr and L. Hilico, Phys. Rev. Lett. **109**, 103401 (2012).
[11] K. Pachucki, Phys. Rev. A **60**, 3593 (1999).
[12] E. Borie, arXiv:1103.1772 v7 (2011).
[13] C. Peset and A. Pineda, arXiv:1508.01948 (2015).
[14] S. G. Karshenboim et al., Journal of Physical and Chemical Reference Data **44**, 031202 (2015);
[15] C. E. Carlson and M. Vanderhaeghen, Phys. Rev. A **84**, 020102 (2011).
[16] C. Peset and A. Pineda, Nuclear Physics B, **887**, 69 (2014).
[17] R. J. Hill and G. Paz, Phys. Rev. Lett. **107**, 160402 (2011).
[18] M. C. Birse and J. A. McGovern, Eur. Phys. J. A **48**, 120 (2012).
[19] M. Gorchtein, F. J. Llanes-Estrada and A. P. Szczepaniak, Phys. Rev. A **87**, 05250 (2013).
[20] J. M. Alarcón, V. Lensky, V. Pascalutsa, Eur. Phys. J. C **74**, 2852 (2014).
[21] G. A. Miller, Phys. Lett. B **718**, 1078 (2013).
[22] A. Walker-Loud, C. E. Carlson, and G. A. Miller, Phys. Rev. Lett. **108**, 232301 (2012).
[23] Sz. Borsanyi et al., Science **347**, 1452 (2015).
[24] U. D. Jentschura, Phys. Rev. A, **92**, 012123 (2015).
[25] K. Pachucki and K. A. Meissner, arXiv:1405.6582 (2014).
[26] U. D. Jentschura, Phys. Rev. A **88**, 062514 (2013).
[27] S. G. Karshenboim, D. McKeen, M. Pospelov, Phys. Rev. D **90**, 073004 (2014).
[28] C. E. Carlson and M. Freid, arXiv:1506.06631 (2015).
[29] P. Brax and C. Burrage, Phys. Rev. D **91**, 043515 (2015).
[30] C. G. Parthey et al., Phys. Rev. Lett. **107**, 203001 (2011).
[31] A. Matveev et al., Phys. Rev. Lett. **110**, 230801 (2013).
[32] M. Horbatsch and E. A. Hessels, Phys. Rev. A **82**, 052519 (2010).
[33] P. Amaro et al., Phys. Rev A **92** 022514 (2015).
[34] A. Beyer et al., Ann. Phys. **525**, 671 (2013).
[35] S. Galtier et al., Optics Communications **324**, 34 (2014).
[36] E. Peters et al., Ann. Phys. **525**, 29 (2013).
[37] A. Vutha et al., BAPS.2012.DAMOP.D1.138, (2012).
[38] J. N. Tan, S. M Brewer, N. D Guise, Phys. Scr. T144:014009 (2011).
[39] D. A. Cooke et al., Hyperfine Interact. **233** 67 (2015).
[40] G. Lee, J. R. Arrington and R. J. Hill, arXiv:1505.01489 (2015).
[41] I. Sick and D. Trautmann, Phys. Rev. C **89**, 012201(R) (2014).
[42] J. Arrington and I. Sick, arXiv:1505.02680 (2015).
[43] R. J. Hill and G. Paz, Phys. Rev. D **82**, 113005 (2010).
[44] I. Lorenz, U. G. Meißner, H. W. Hammer, Y. B. Dong, Phys. Rev. D **91**, 014023 (2015).
[45] A. Gasparian et al., EPJ Web Conf. **73**, 07006 (2014).
[46] M. Mihovilovic, H. Merkel and A. Weber, PoS Bormio2014, 051 (2014).
[47] M. Kohl et al., Eur. Phys. J. Web Conf. **66**, 06010 (2014).
[48] M. Distler and S. Schlimme private communications (2015).
[49] L. S. Myers, D. W. Higinbotham and J. R. Arrington, arXiv:1408.5283 (2014).
[50] E. Y. Korzinin, V. G. Ivanov, S. G. Karshenboim Phys. Rev D, **88**, 125019 (2013).
[51] B. J. Wundt and U. D. Jentschura, Phys. Rev. J. D **65**, 357 (2012).
[52] J. J. Krauth et al., ArXiv:1506.01298 (2015).
[53] O. J. Hernadez et al., Phys. Lett. B **736**, 344 (2014).

[54] K. Pachucki and A. Wienczek, Phys. Rev. A **91**, 040503(R) (2015).
[55] C. E. Carlson, M. Gorchtein, and M. Vanderhaeghen, Phys. Rev. A **89**, 022504 (2014).
[56] A. P. Martynenko, arXiv:hep-ph/0612298v4 (2014).
[57] C. Ji, N. Nevo Dinur, S. Bacca and N. Barnea, Phys. Rev. Lett. **111**, 143402 (2013).
[58] I. Sick, arXiv:1505.06924 (2015).
[59] M. Piarulli et al., Phys. Rev. C **87**, 014006 (2013).
[60] Z. T. Lu et al., Rev. Mod. Phys. **85**, 1383 (2013).
[61] R. van Rooij et al., Science **333**, 196 (2011).
[62] C. Pastor et al., Phys. Rev. Lett. **108**, 143001 (2012).
[63] M. Herrmann et al., Phys. Rev. A **79**, 052505 (2009).
[64] D. Z. Kandula et al., Phys. Rev. A **84**, 062512 (2011).

EPJ Web of Conferences **113**, 01007 (2016)
DOI: 10.1051/epjconf/201611301007

Hadronic Weak Interaction Studies at the SNS

Nadia Fomin[1,a]

[1] *University of Tennessee, Knoxville, TN 37996*

Abstract.
Neutrons have been a useful probe in many fields of science, as well as an important physical system for study in themselves. Modern neutron sources provide extraordinary opportunities to study a wide variety of physics topics. Among them is a detailed study of the weak interaction. An overview of studies of the hadronic weak (quark-quark) as well as semi-leptonic (quark-lepton) interactions at the Spallation Neutron Source (SNS) is presented. These measurements, done in few-nucleon systems, are finally letting us gain knowledge of the hadronic weak interaction without the contributions from nuclear effects. Forthcoming results from the NPDGamma experiment will, due to the simplicity of the neutron, provide an unambiguous measurement of the long range pion-nucleon weak coupling (often referred to as h_π), which will finally test the theoretical predictions. Results from NPDGamma and future results from the $n +^3 He$ experiment will need to be complemented by additional measurements to completely describe the hadronic weak interaction.

1 Introduction

The hadronic weak interaction has been an object of experimental study for many decades, but it still holds many mysteries. While parity violation has been observed many times in nuclear systems, corrections for nuclear effects are required in order to isolate any NN components. To avoid this, it is desirable to use the simplest systems available to conduct these studies. The hadronic weak interaction program at the SNS uses two of the smallest and well understood nuclei for this purpose.

The Fundamental Neutron Physics Beamline (FnPB) [1] at the SNS will be home to a series of experiments that aim to answer fundamental questions in nuclear physics, including studies of the hadronic weak interaction between nucleons. The NPDGamma and $n +^3 He$ experiments are set to test the strength of the hadronic weak interaction between nucleons. This interaction was first parametrized in terms of meson exchange in the DDH model [2] over thirty years ago. However, the meson exchange couplings are still largely unknown today. Modern theoretical frameworks include Effective Field Theory [3] as well as Lattice QCD [4]. EFTs are model independent and are able to provide systematic theoretical error estimates, as well as a connection to QCD via symmetry principles. Additionally, EFTs are useful in interpreting experimental data and calculations of parity-violating observables in systems of up to five nucleons are currently possible. Lattice QCD is on its way to becoming a reliable tool for hadronic parity violation calculations, and future experiments present exciting opportunities to test precise predictions.

[a] e-mail: nfomin@utk.edu

The first SNS experiment to take data is NPDGamma, which aims to measure a parity violating asymmetry in the direction of gamma rays emitted from cold neutron capture on parahydrogen [5, 6]. This asymmetry is directly related to the long range pion-nucleon weak coupling, h_π^1 in the DDH framework of meson exchange [2]. In the DDH model, physical observables can be written in terms of several total meson couplings, with several experiments necessary to isolate them all. In case of NPDGamma, the experimental observable is a parity violating asymmetry between the spin of the neutron and the direction of the outgoing gamma in cold neutron capture on hydrogen. In the DDH model, this asymmetry is given by:

$$A_\gamma = -0.11h_\pi^1 - 0.001h_\rho^1 + 0.004h_\omega^1 \tag{1}$$

where the subscripts reference the meson, and the superscript indicate the isospin change. The weak couplings (h's) are expected to be of the same order, meaning that NPDGamma is primarily sensitive to the h_π^1 coupling, and isolates the ΔI=1 part of the hadronic weak interaction, unencumbered by nuclear effects. The DDH model predicts h_π^1 to be ~4.5x10^{-7}, with a "reasonable range" of -1 to 11 (x10^{-7}). The high flux at the FnPB has allowed for a measurement with 1.3×10^{-8} precision.

A more modern approach to the problem is to use effective field theories [3] which are model independent and provide systematic theoretical error estimates. In effective field theory, the NPDGamma observable is given by [7]:

$$A_\gamma = \frac{4}{3}\sqrt{\frac{2}{\pi}}\frac{M^{3/2}}{\kappa_1(1-\gamma a^{^1S_0})}g^{^3S_1-^3P_1}, \tag{2}$$

where all parameters are known experimentally, and a measurement of A_γ will isolate $g^{^3S_1-^3P_1}$ low energy coupling constant. Finally, a first calculation of h_π^1 has been done on the lattice [4], with a result of $1.099 \pm 0.505^{+0.058}_{-0.064}[x10^{-7}]$.

The n+^{3}He experiment to search for a parity-odd asymmetry in the $\vec{n} +^3 He \rightarrow p + T + 765keV$ reaction is running at the Spallation Neutron Source. This is a natural follow-on to NPDGamma and is sensitive to $\Delta I = 0, 1$, and 2 parts of the hadronic weak interaction. The projected error on the asymmetry is 1.6×10^{-8} and the experiment is expected to finish taking data in 2015. Unlike NPDGamma, this asymmetry probes $\Delta I = 0, 1$, and 2 components of the NN weak interaction and needs additional DDH couplings to be described [8]:

$$A_p = -0.185h_\pi^1 - 0.038h_\rho^0 + 0.023h_\rho^1 - 0.001h_\rho^2 - 0.050h_\omega^0 - 0.023h_\omega^1. \tag{3}$$

The $\vec{n} +^3 He \rightarrow p + t + 765keV$ process has also been calculated in n chiral EFT [9].

2 $\vec{n} + p \rightarrow d + \gamma$: The NPDGamma experiment

The NPDGamma experiment isolates the ΔI=1 part of the weak interaction by measuring the correlation between the neutron spin and the direction of the outgoing gamma in the capture of cold neutrons on protons. The deuteron is the most loosely bound of all nuclei, meaning the weak NN interaction is dominated by the long-range component. The simple nature of the interaction also eliminates the need for any nuclear physics corrections, making the results easy to interpret.

The experimental apparatus is shown in Fig. 1. NPDGamma ran on the FnPB [10] at the SNS from 2011 until June, 2014. Cold neutrons were polarized via a supermirror located ~ 15 m downstream of the moderator face. The neutron beam is pulsed at 60 Hz, meaning that the neutron velocities and energies can be inferred from their times of flight. This allows for the spins of the neutrons to be reversed on a pulse-by-pulse basis through the use of an RF spin rotator [11]. The data were collected

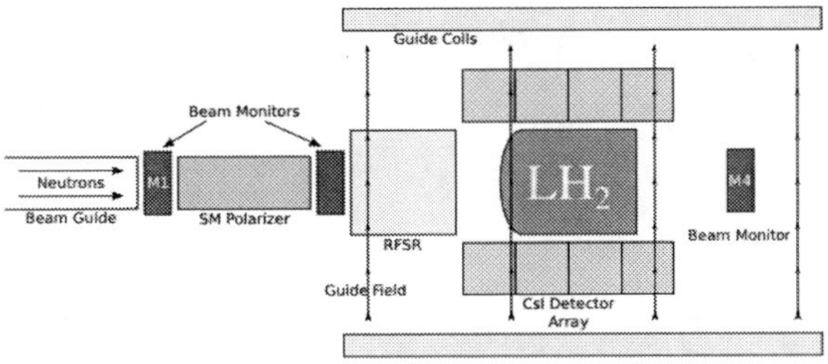

Figure 1. Layout of the NPDgamma apparatus. See text for details.

in spin octets (↑↓↓↑↓↑↑↓ and ↓↑↑↓↑↓↓↑) in order to cancel any time-dependent systematic effects to the second order. The neutrons are then captured on a 16 L target made of parahydrogen at ~ 16 K. The gammas from the neutron capture on protons are detected via a 3π acceptance detector array surrounding the hydrogen target. There are 48 detectors, arranged in 4 rings of 12 around the target. Each detector consists of 2 CsI crystals (15.2x15.2x15.2 cm^3), viewed by a vacuum photodiode [12]. The neutron flux makes a counting experiment impractical, so the detectors are operated in current-mode. Finally, the experimental apparatus is held inside a ~10 Gauss magnetic field to maintain the neutron polarization on the way to the hydrogen target.

A raw asymmetry is formed for octets (8 accelerator pulses worth) of data, 4 of which will have had neutrons with spins aligned with the magnetic field and 4 anti-aligned. Asymmetries are formed between pairs of detectors at the same angle with respect to the magnetic field in order to cancel pulse-to-pulse beam fluctuations in the following manner:

$$A_{raw} = \frac{(Y_i^{\uparrow} + Y_{i+6}^{\downarrow}) - (Y_i^{\downarrow} + Y_{i+6}^{\uparrow})}{Y_i^{\uparrow} + Y_{i+6}^{\downarrow} + Y_i^{\downarrow} + Y_{i+6}^{\uparrow}}, \tag{4}$$

where the Y's are the yields in pairs of detectors.

Each detector has a sensitivity to the up-down and left-right asymmetries given by the so-called "geometric factors". This sensitivity depends on the position of the detector with respect to the point of neutron capture. The geometrical factors are average energy-weighted functions that are a measure of the emission direction of a photon from the target that deposits energy in a given detector. They were obtained by a combination of Monte Carlo techniques and measurements with a known ^{137}Cs source.

To extract the physics asymmetries (A_{UD} and A_{LR}), several corrections need to be made. Multiplicative corrections for the polarization of the incoming neutron beam, depolarization in the hydrogen target, the efficiency of the RF spin rotator are applied. In addition to capture on the parahydrogen, neutrons are also captured by the aluminum in the cryogenic target vessel. To account for this contribution to our detector signals, data were taken on a specially constructed aluminum target, and the up-down and left-asymmetries also determined. To extract the proton physics asymmetries, we fit the data to the following expression:

$$A_{raw}[i] = (F_H \cdot A_{UD}^H[i] \cdot G_{UD}^H[i] + F_H \cdot A_{LR}^H[i] \cdot G_{LR}^H[i] + F_{AL} \cdot A_{UD}^{AL}[i] \cdot G_{UD}^{AL}[i] + F_{AL} \cdot A_{LR}^{AL}[i] \cdot G_{LR}^{AL}[i]) \cdot C_{eff}$$

where F_H and F_{AL} are the prompt signal captures of neutron capture on hydrogen and aluminum for detector element i,G_{XY} are the geometric factors for detector element i for the hydrogen and aluminum

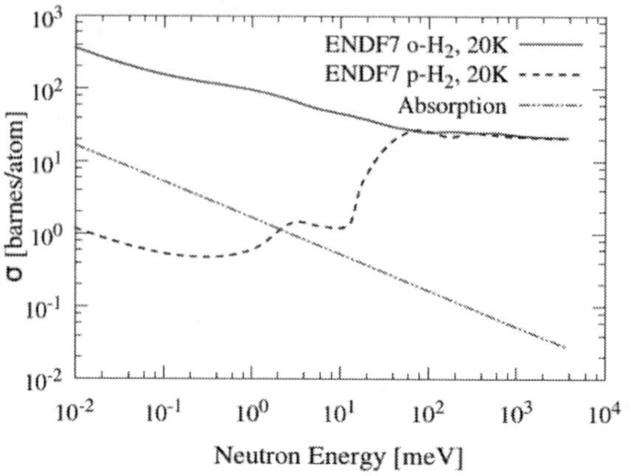

Figure 2. Neutron hydrogen scattering cross sections for ortho-hydrogen, para-hydrogen as well as absorption.

targets, and C_{eff} is the correction factor that's a product of the neutron beam polarization, RFSF efficiency, and the depolarization factor.

Neutron beam polarization and RFSF efficiency were measured in a separate experiment using transmission measurements through a polarized ^{3}He cell of both neutron spin states. Depolarization due to spin-flip scattering on hydrogen was calculated using Monte Carlo techniques and known scattering cross-sections. Fig. 2 shows the scattering cross sections on the two molecular states of hydrogen: ortho (proton spins aligned) and para (proton spins anti-aligned). Polarized neutrons that scatter on ortho-hydrogen depolarize before capture, which is the reason that the hydrogen target was operated at < 20 K, where the admixture of orthohydrogen is very small at thermodynamic equilibrium. However, as hydrogen is circulated through the target vessel, some part of the volume may not be in thermodynamic equilibrium, and the maximum orthohydrogen concentration can be inferred from a cross section measurement recently performed [13] using the NPDGamma apparatus. Depolarization values were calculated using thermodynamic equilibrium and maximum orthohydrogen contamination scenarios. An average of the two values will be used as a correction in the final analysis.

The NPDGamma experiment completed its production data taking in June of 2014. There are three independent analysis of the hydrogen data being performed. They are in good agreement with each other as can be seen from raw detector asymmetries in Fig.3. While the parity violating asymmetry result is not yet available, the uncertainty on the measurement will be $< 1.4\times10^{-8}$.

3 $\vec{n} +^3 He \rightarrow p + T + 765keV$: The n+^{3}He experiment

The study of the hadronic weak interaction continues at the FnPB at the SNS with the n+^{3}He experiment, which aims to measure the directional asymmetry in the capture of cold neutrons on ^{3}He. This asymmetry is sensitive to a combination of weak couplings, and will complement other measurements of the kind. The goal is achieve a combined statistical and systematic uncertainty of $<1.6\times10^{-8}$.

The experimental concept is similar to that of NPDGamma. The same supermirror polarizer is used to produce a polarized neutron beam. A new spin rotator has been constructed to allow the neutron spin to be rotated to point either along or against the direction of momentum. The spin flipper is a double cosine theta coil design providing a uniform interior transverse RF magnetic field oscillating at $\sim$ kHz. The neutrons are captured and detector in a target-detector wire chamber filled

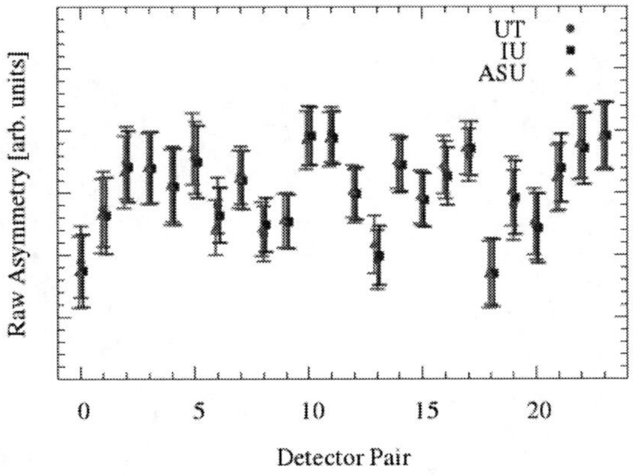

Figure 3. Raw hydrogen data asymmetries (arbitrary units) from three independent analyses.

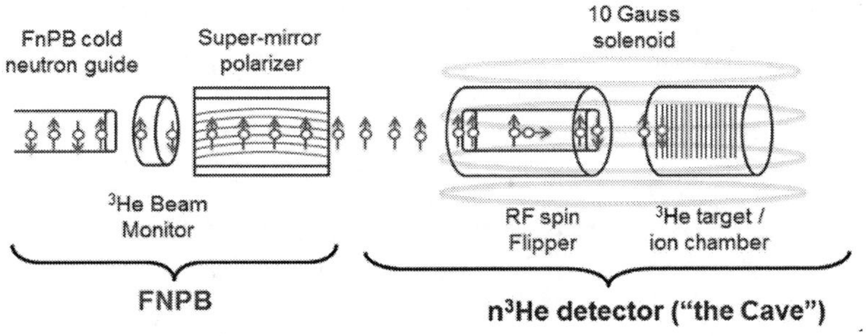

Figure 4. Experimental layout of the n+3He apparatus. Details in text.

with ^{3}He gas so as to be completely absorbing. Neutrons that capture on ^{3}He produce an excited ^{4}He nucleus which then decays into a proton and a triton. Both the proton and triton then ionize the gas as they move through the gas volume and the resulting electrons are carried away by the anode wires. There are 17 high voltage frames with 8 wires each and 16 signal frames with 9 wires each, for a total of 144 signals to be read out. The range of the tritons (~1cm) is much smaller than that of the protons (4 cm), and most of the energy is carried away by the protons. As with NPDGamma, the detector is operated in current mode.

The ion chamber was commissioned in late 2014. The pulse integrals of the individual detector signal wire signals are shown in Fig. 5. As can be seen, most of the neutron beam is stopped in the first third of the wire chamber.

The n+^{3}He experiment is to take production data until the end of 2015 calendar year to achieve its goal of $<1.6\times10^{-8}$ total uncertainty.

4 Summary

The FnPB at the SNS has been home to two important measurements of the hadronic weak interaction. NPDGamma's result will be the first measurement of the long-range $\Delta I=1$ component of the weak NN interaction that's free from nuclear uncertainties. The n+^{3}He experiment continues this study as part

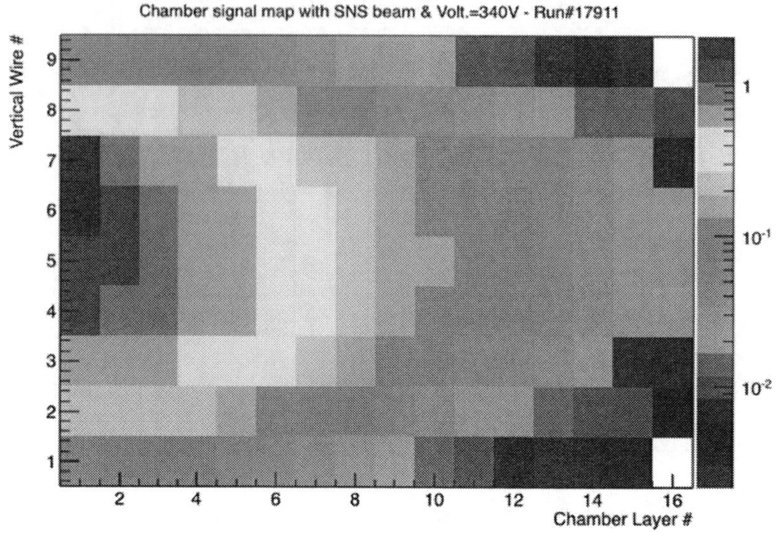

Figure 5. Integrated wire signals in the ion chamber.

of a larger effort in the field to finally pin down the weak couplings between nucleons after decades of measurements.

References

[1] N. Fomin and others, Nucl. Instr. Meth. A773 (2015) 45.

[2] B. Desplanques *et al*, Annals Phys. 124 (1980) 449.

[3] Shu-Lin Zhu *et al*, Nuc. Phys. A478 (2005) 435.

[4] J. Wasem, Phys. Rev. C **85**, 022501 (2012).

[5] N. Fomin, AIP Conf. Proc. 1560 (2013) 145.

[6] M. Gericke *et al*, Phys. Rev. C **83**, 055505 (2011).

[7] M. R. Schindler and R. P. Springer, Prog. Part. Nucl. Phys. **72**, 1 (2013) [arXiv:1305.4190 [nucl-th]].

[8] M. Viviani, R. Schiavilla, L. Girlanda, A. Kievsky and L. E. Marcucci, Phys. Rev. C **82**, 044001 (2010)

[9] M. Viviani, A. Baroni, L. Girlanda, A. Kievsky, L. E. Marcucci and R. Schiavilla, Phys. Rev. C **89**, 064004 (2014) doi:10.1103/PhysRevC.89.064004 [arXiv:1403.2267 [nucl-th]].

[10] N. Fomin *et al.*, Nucl. Instrum. Meth. A **773**, 45 (2014)

[11] P.-N. Seo *et al.*, Phys. Rev. ST Accel. Beams **11**, 084701 (2008) [arXiv:0710.2871 [nucl-ex]]

[12] M. T. Gericke *et al.*, Nucl. Instrum. Meth. A **540**, 328 (2005)

[13] K. B. Grammer *et al.*, Phys. Rev. B **91**, no. 18, 180301 (2015)

EPJ Web of Conferences **113**, 01008 (2016)
DOI: 10.1051/epjconf/201611301008

Electromagnetic structure of light nuclei

Saori Pastore[1,2,a]

[1] *Department of Physics and Astronomy, University of South Carolina, Columbia, SC 29208, USA*
[2] *Theoretical Division, Los Alamos National Laboratory, Los Alamos, NM 87545, USA*

Abstract. The present understanding of nuclear electromagnetic properties including electromagnetic moments, form factors and transitions in nuclei with A $\leq$ 10 is reviewed. Emphasis is on calculations based on nuclear Hamiltonians that include two- and three-nucleon realistic potentials, along with one- and two-body electromagnetic currents derived from a chiral effective field theory with pions and nucleons.

1 Introduction

A major goal in nuclear physics is to understand nuclear structure and dynamics in terms of underlying interactions occurring between individual nucleons. Studies grounded on this basic picture of the nucleus are referred to as *ab initio*. An exceptionally powerful tool to asses the validity of our theoretical models is to investigate nuclear electromagnetic (e.m.) observables, such as ground state properties, *e.g.*, e.m. moments and form factors, as well as e.m. reactions, *e.g.*, photon- and electro-induced reactions. In these processes, external e.m. probes interact with the nuclear charge and current distributions with a strength characterized by the fine-structure constant $\alpha \sim 1/137$. The small value of the fine-structure constant allows for a perturbative treatment of the e.m. interaction, while non-perturbative physics pertains only to the nuclear target. For light nuclei, terms that go beyond the leading order contribution in the $Z\alpha$-expansion (where Z is the number of protons) can be safely disregarded, leaving us with relatively simple reaction mechanisms and manageable formal expressions. For example, at leading order, the cross section associated with the inclusive electro-nucleus scattering processes is factorized into the leptonic tensor, which is completely specified by the measured electron's kinematic variables, and the hadronic one associated with the nuclear target, and proportional to matrix elements squared of the nuclear e.m. charge and current operators. A clear connection between measured quantities, *i.e.*, cross sections, and calculated matrix elements is then realized. Experimental data of e.m. observables are, in most cases, known with great accuracy providing us with viable and strong constraints on our models. Likewise, for light nuclei, theoretical calculations are affected by relatively small numerical errors because for these systems the many-body problem can be solved exactly or within controlled approximations. This allows for solid comparisons between experimental data and theoretical predictions.

In Fig. 1, a cartoon picture of the double differential cross section for electron scattering off nuclei is represented. Different values of energy ω transferred to the system, correspond to different excitation energies of the nucleus. By varying ω, we can access the ground state (elastic peak), low-lying

[a] e-mail: saori@lanl.gov

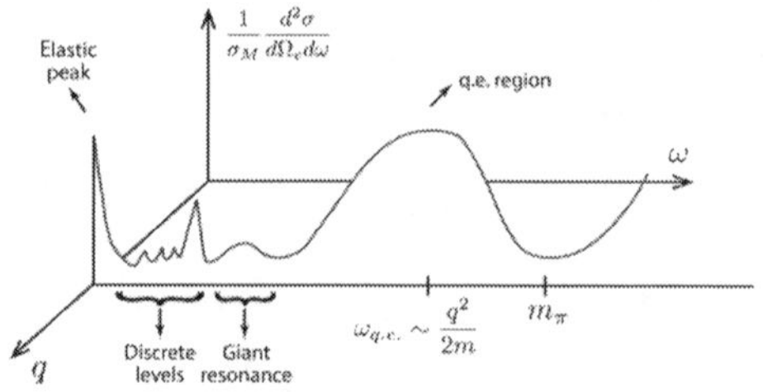

Figure 1. Schematic representation of the double differential cross section for electron scattering off nuclei at fixed value of momentum transfer.

(discrete) nuclear excited states, giant resonance modes, and the quasi-elastic energy region up to the pion-production threshold. For each value of excitation energy ω, one can study the matrix elements' behavior as a function of the momentum $|\mathbf{q}|$ transferred to the nucleus. In particular, by varying $|\mathbf{q}|$ one can explore the e.m. charge and current distributions with a spatial resolution $\propto 1/|\mathbf{q}|$. In this talk, I will focus on *ab initio* calculations of ground state nuclear e.m. properties, that is e.m. moments and elastic form factors, as well as widths of e.m. transitions occurring between low-lying nuclear states. These studies have been recently reported in a topical review on e.m. reactions on light nuclei [1], where more details and references to original articles can be found. Recent developments on theoretical *ab initio* investigations on other very interesting e.m. processes in light nuclei, such as photo-absorption and radiative capture reactions, Compton scattering, sum rules ..., are well represented in this conference, see, *e.g.*, contributions by X. Zhang, J. Dohet-Eraly, H. Griesshammer, M. Miorelli, S. Bacca, N. Barnea, D. Rozpedzik, and A. Lovato in these proceedings.

A theoretical understanding and control of nuclear e.m. structure and dynamics is a necessary prerequisite for studies on weak induced reactions, such as neutrino-nucleus interactions. The experimental data acquisition for this kind of processes is comparatively more involved owing to the tinier cross sections and to the fact that neutrinos are chargeless particles and, thus, they are hard to collimate and detect. An important advance in this direction has recently been carried out by Lovato and collaborators [2], and for a status report on *ab initio* calculations of weak response functions in ^{4}He and ^{12}C I refer to the plenary talk of A. Lovato (the associated contribution can be found in these proceedings). Moreover, a theoretical understanding of the structure and dynamics of light nuclei is a necessary prerequisite for research projects aimed at studying larger nuclear systems. For these reasons, it is imperative to first validate our theoretical understanding of e.m. reactions on light nuclei.

2 Nuclear Hamiltonians and electromagnetic currents

In the *ab initio* framework, the nucleus is described as a system made of A non-relativistic point-like nucleons interacting among each other via many-body forces and its energy is approximated by the following Hamiltonian:

$$H = \sum_i K_i + \sum_{i<j} v_{ij} + \sum_{i<j<k} V_{ijk} \,, \tag{1}$$

where K_i is the non-relativistic single-nucleon kinetic energy, while v_{ij} and V_{ijk} are two-nucleon (NN) and three-nucleon (3N) potentials, respectively. Implicit in the equation above is the assumption that four-nucleon forces and higher terms in the many–body expansion are suppressed. The NN and 3N potentials are phenomenological in nature in that they involve a number of parameters—subsuming

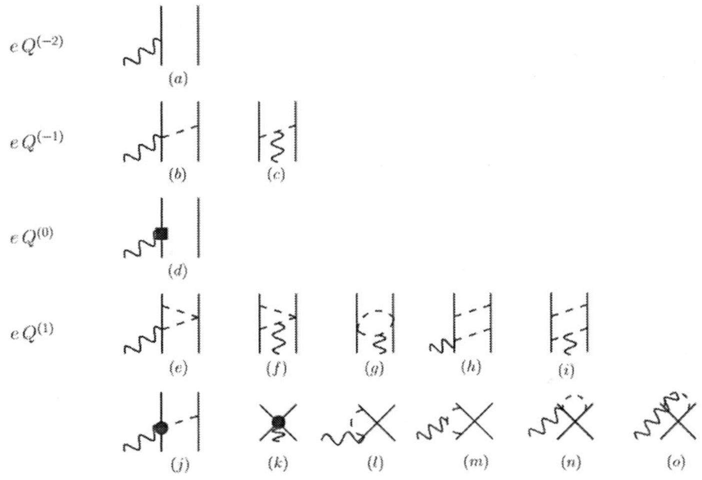

Figure 2. Diagrams illustrating one- and two-body EM currents entering at LO ($e\,Q^{-2}$), NLO ($e\,Q^{-1}$), N2LO ($e\,Q^{0}$), and N3LO ($e\,Q^{1}$). Nucleons, pions, and photons are denoted by solid, dashed, and wavy lines, respectively.

underlying Quantum Chromodynamics (QCD) effects—that are fixed by fitting experimental data. For example, NN potentials are constrained to reproduce a large number of NN scattering data, along with the deuteron binding energy. Nuclear forces belonging to this class of highly accurate nuclear potentials are referred to as 'realistic'. Most realistic potentials describe the long range ($\propto 1/m_\pi$ where m_π is the pion mass) part of the nuclear interaction in terms of one-pion-exchange interaction mechanisms. Different dynamical schemes are implemented to account for intermediate and short range effects, among which multiple-pion-exchange, contact interactions, heavy-meson-exchange, or excitations of nucleons into virtual Δ-isobars. Here, the realistic potentials utilized to solve the Schrödinger equation $H|\Psi\rangle = E|\Psi\rangle$ (where $|\Psi\rangle$ is the nuclear wave function) are the Argonne $v18$ [3] (AV18) NN potential in combination with either the Urbana IX [4] or Illinois-7 [5] 3N potentials, as well as combinations of NN and 3N potentials derived from chiral effective field theory (χEFT)[6–9].

Nuclear charge (ρ) and current ($\mathbf{j}$) operators describe the interactions of nuclei with external e.m. probes. They are also expanded in a series of many-body operators as

$$\begin{aligned} \rho &= \sum_i \rho_i(\mathbf{q}) + \sum_{i<j} \rho_{ij}(\mathbf{q}) + \ldots\,, \\ \mathbf{j} &= \sum_i \mathbf{j}_i(\mathbf{q}) + \sum_{i<j} \mathbf{j}_{ij}(\mathbf{q}) + \ldots\,, \end{aligned} \tag{2}$$

where $\mathbf{q}$ is the momentum transferred to the nucleus. In Impulse Approximation (IA), that is retaining only one-body operators in the equations above, nuclear e.m. charge and current distributions are simply the sums of those associated with individual protons and neutrons. The non-relativistic charge operator for point-like nucleons is simply the proton charge, while the nucleon current consists of a convection term associated with the current generated by moving protons and a spin-magnetization term associated with the spins of both protons and neutrons. The IA picture of the nucleus is, however, incomplete as it fails to explain, *e.g.*, the measured magnetic moments of light nuclei. Corrections that account for processes in which external e.m. probes couple to pairs of interacting nucleons, described by two-body current operators, need to be incorporated in the theoretical *ab initio* description. Meson-exchange currents (MECs)—postulated in the '40s by Villars [10] and Miyazawa [11]—follow naturally once meson-exchange mechanism are invoked to describe interactions between individual nucleons. They account for processes in which the external e.m. probe couples with mesons being exchanged between nucleons. The first evidence of meson-exchange effects in light nuclei can be traced back to the 1972 work by Riska and Brown [12], in which MECs were found to provide the missing 10% correction to the IA value necessary to reach agreement between the calculated and the measured cross sections for the radiative capture of proton on neutron at thermal neutron energies. Since then, MECs have evolved into highly sophisticated and accurate currents. In their most recent formulation [13, 14], in order to assure consistency between nuclear forces and e.m. currents, MECs are constructed from realistic NN and 3N potentials so as to satisfy the continuity equation. Addition of these MECs corrections to the IA picture successfully explains a wide number of e.m. nuclear observables in light nuclei [15, 16].

Recent years have witnessed the tremendous development and success of χEFT [17–19] that reinforces and grounds the achievements of conventional theoretical approaches. The relevant degrees of freedom of nuclear physics are bounds states of QCD, *i.e.*, pions, nucleons, and Δ's, ... On this basis, their dynamics is completely determined by that associated with the underlying degrees of freedom of quarks and gluons, that is QCD. However, at low energies, QCD does not have a simple solution because the strong coupling constant becomes too large and perturbative techniques cannot be applied to solve it. χEFT is a low-energy approximation of QCD valid in the energy regime where the typical momenta involved, generically indicated by Q, are such that $Q \ll \Lambda_\chi \sim 1$ GeV, where Λ_χ is the chiral-symmetry breaking scale. χEFT provides us with effective Lagrangians describing the interactions between pions, nucleons, and Δ's that preserve all the symmetries, in particular chiral symmetry, exhibited by the underlying theory of QCD at low-energy. These effective interactions, and the transition amplitudes derived from them, can be expanded in powers of the small expansion parameter Q/Λ_χ, restoring, in practice, the possibility of applying perturbative techniques also in the low-energy regime. The unknown coefficients of this expansion in small momenta—referred to as low-energy constants (LECs)—while being tied to QCD effects and therefore attainable from QCD calculations, are, in practice, fixed by comparison with the experimental data. Due to the chiral expansion, it is then possible to evaluate nuclear observables to any degree ν of desired accuracy, with an associated theoretical error roughly given by $(Q/\Lambda_\chi)^{(\nu+1)}$. This calculational scheme has been widely utilized to study both nuclear forces and nuclear electroweak currents. The many-body operators emerging from direct evaluations of the transitions amplitudes with interactions provided by χEFT Lagrangians involve multiple-pion exchange operators, as well as contact-like interaction terms. Nuclear two– and three–body interactions were first investigated in the late '90s by Ordòñez, Ray, and van Kolck using a χEFT with pions and nucleons [20–22]. Currently, chiral NN (3N) potentials commonly used in *ab initio* calculations include up to next-to-next-to-next-to leading order or N3LO (next-to-next-to leading order or N2LO) corrections in the chiral expansion [6, 7, 23].

Vector e.m. currents have been first derived from a χEFT with pions and nucleons by Park, Min, and Rho in Ref. [24]. The resulting operators account for two-pion exchange terms entering at N3LO in the chiral expansion. These currents have been utilized in a number of so called 'hybrid' calculations[1] of nuclear e.m. observables, including magnetic moments and M1 properties of A = 2–3 nuclei and radiative capture cross sections in A = 2–4 systems [25–27]. More recently, χEFT e.m. currents and charge have been derived up to one loop contributions included within two different implementations of time ordered perturbation theory: one is by the JLab-Pisa group (see Refs. [28–31]) and the other one is by the Bochum-Bonn group (see Refs. [32, 33]). In this talk, I focus on results obtained using chiral e.m. currents, and compare, where possible, different theoretical evaluations

[1] In this kind of calculations, matrix elements of the chiral e.m. charge and current operators are evaluated in between wave functions obtained from conventional realistic potentials, as opposed to potentials derived consistently from χEFT.

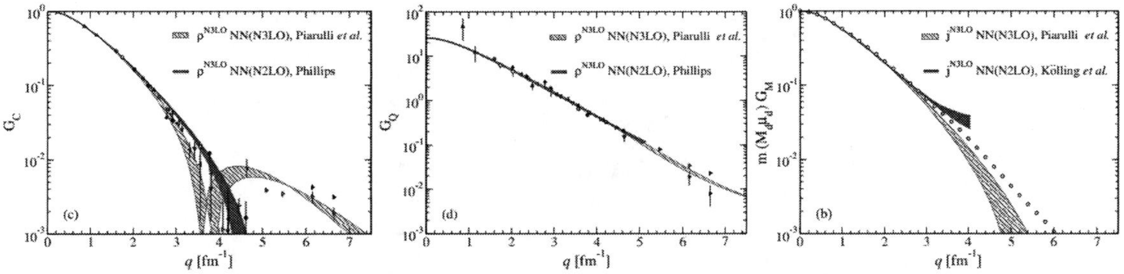

Figure 3. Left: Deuteron charge form factor from Ref. [31] (magenta hatched band) and Ref. [36] (purple solid bands) compared with experimental data (black points). **Middle**: Same as left panel but for deuteron quadrupole form factor. **Right**: Deuteron magnetic form factor from Ref. [31] (magenta hatched band) and Ref. [37] (purple solid band) compared with experimental data (open circles).

against the experimental data. For results based on conventional e.m. MEC currents I refer the reader to the review articles of Refs. [1, 15] and references therein.

Before I proceed presenting applications to e.m. observables, I will briefly describe the e.m. operators as they emerge from a χEFT with pions end nucleons. I will start off with the vector e.m. current which is diagrammatically represented in Fig. 2. The leading order (LO) contribution to the e.m. current illustrated in panel (a) is simply given by the non-relativistic one-body current used in IA calculations, while the N2LO one-body operator of panel (d) is a relativistic correction to the LO IA current. Currents of one- and two-pion range, describing long and intermediate range dynamics, enter at NLO and N3LO—panels (b), (c), (e)—(j). Short-range dynamics is encoded by the contact currents of panel (k). Unknown LECs enter the tree-level diagram of panel (e) and contact currents of panel (k). LECs entering the contact terms are of two kinds, namely minimal and non-minimal. The former enter also the NN chiral potential at NLO, and can then be constrained to NN scattering data; the latter need to be fixed from e.m. experimental data. A common procedure implemented to reduce the number of unknown non-minimal LECs (there are 5 of them) is to impose that the two LECs entering the isovector part of the tree-level current illustrated in panel (e) are in fact saturated by the Δ-couplings entering the Δ transition e.m. current [29, 31]. The remaining three LECs are commonly fixed so as to reproduce the magnetic moments of the deuteron, triton, and ^{3}He [31].

Early investigations on the e.m. charge operator in χEFT have been carried out in Refs. [34–36], and more recently loop corrections have been derived in Ref. [32], and subsequently Ref. [30]. In closing, we note that the structure of the charge operator is quite different from that of the vector e.m. current. Two-body corrections, in this case, are expected to be relatively small. In fact, leading two-body operators of one-pion range are suppressed as they enter at N3LO (as opposed to NLO as seen in the case of the vector currents), while there are no free LECs entering the charge operator up to N4LO [30].

3 Deuteron, ^{3}He and ^{3}H electromagnetic form factors

For A = 2-3 nuclei, theoretical calculations performed by different groups are available, which makes it possible to compare them not only with the experimental data but also between themselves to test the solidity of the *ab initio* prescription. The left and middle panels of Fig. 3, show the deuteron charge and quadrupole form factors, respectively, calculated by Piarulli and collaborators [31] (magenta hatched bands) and Phillips [35, 36] (purple bands). Both calculations are based on chiral NN potentials. In particular, Piarulli *et al.* use wave functions from the chiral NN potential at N3LO [38], while Phillips those from the NN interaction at N2LO [39]. The thickness of the bands represents

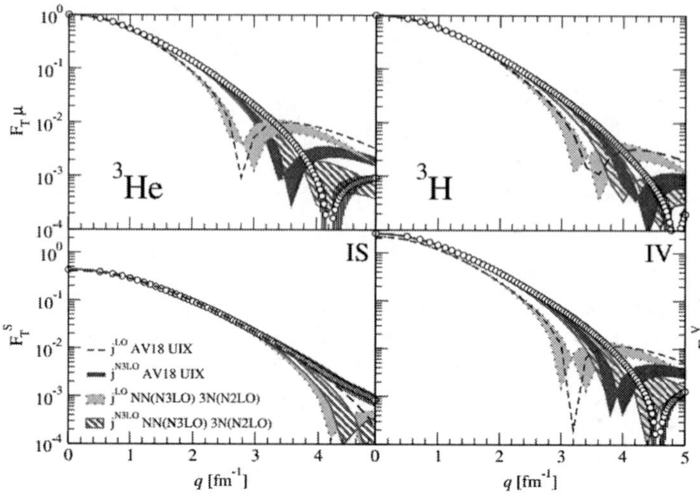

Figure 4. The ^{3}He and ^{3}H magnetic form factors, and their isoscalar and isovector combinations from Ref. [31]. Blue dashed lines (cyan bands) and red (magenta hatched) bands represent calculations based on the AV18+UIX [NN(N3LO)+3N(N2LO)] nuclear Hamiltonians with cumulative χEFT e.m. currents at LO and N3LO, respectively.

the sensitivity of the results to different cutoffs utilized to regularize the divergent behavior at high momenta of the chiral operators' matrix elements [31, 35, 36]. The two calculations very nicely agree with the experimental data for low-value of momentum transferred ($q \simeq 3$ fm^{-1}) and exhibit a similar (small) cutoff dependence. In the case of the charge form factor, as q increases, both theoretical calculations exhibit a more pronounced cutoff dependence, and differ between each other, an indication that this observable is sensitive to the nuclear wave functions utilized in the calculations. In the case of the quadrupole form factor, the agreement with the experimental data is seen up to $q \simeq 6$ fm^{-1}, well beyond the expected regime of validity of the χEFT framework. In the right panel of Fig. 3, we compare the results for the deuteron magnetic form factor obtained by Piarulli *et al.* [31] (hatched magenta band) based on chiral N3LO potential [38] and chiral e.m. currents at N3LO [29], with the fully consistent χEFT calculations by Kölling *et al.* [37] (solid purple band) based on the chiral NN potential at N2LO [39] and chiral e.m. currents at N3LO [32, 33]. The theoretical results are in very good agreement with each other and with the experimental data for values of momentum transferred $q \simeq 3$ fm^{-1}, and present a comparable cutoff dependence.

Form factor calculations of $A = 3$ nuclei have been reported in Ref. [31]. Here, we show results for the trinucleon magnetic form factors obtained utilizing the chiral e.m. currents at N3LO of Refs. [28, 29] and two sets of nuclear Hamiltonians, namely the AV18 [3] NN plus UIX [4] 3N potentials, and the N3LO [38] NN and N2LO [9] 3N potentials. Calculations in IA are given in light blue (based on chiral interactions) and blue (based on conventional interactions), while full calculations that include the complete e.m. current up to N3LO are given in magenta (based on chiral interactions) and red (conventional interactions). In the figure, the top panels show the ^{3}He and ^{3}H magnetic form factors, while the bottom ones show their isoscalar (F_T^S) and isovector (F_T^V) combinations [31]. As is well know from studies based on the conventional approach (see Ref. [15]), two-body e.m. currents are crucial to improve the agreement between the observed positions of the zeros and the predicted ones at LO (or IA). Despite the excellent agreement between theory and experiment for $q \lesssim 2$ fm^{-1}, the theory underpredicts the data at higher momentum transfers, while the zeros are found at lower values of q than observed. The theoretical description of the first diffraction region is still incomplete.

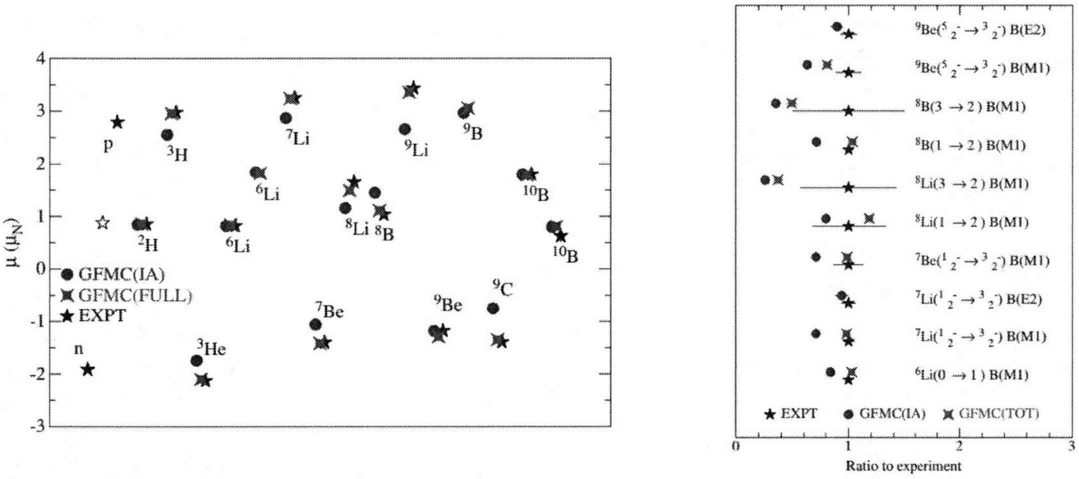

Figure 5. Left: Magnetic moments in nuclear magnetons for $A \leq 10$ nuclei from Ref. [40]. Black stars indicate the experimental values, while blue dots (red diamonds) represent GFMC calculations which include the LO or IA (N3LO) e.m. currents from χEFT. **Right**: Transition widths from Ref. [40] normalized to the experimental values for $A = 6$–9 nuclei. The notation is as in left panel.

4 Magnetic moments and electromagnetic transitions in $A \leq 10$ nuclei

Moving on to larger nuclear systems, we find a number of Greens's Function Monte Carlo calculations [41] based on the AV18 [3] plus IL7 [5] nuclear Hamiltonian that use the chiral e.m. currents up to N3LO from Refs. [28, 29, 31]. Magnetic moments of light nuclei [40] are summarized in the left panel of Fig. 5, where IA results are given by blue dots, while calculations that include the full chiral e.m. current operator are indicated by red diamonds to be compared with the experimental data represented by black stars. First, we note that corrections from two-body currents are found to be small where the IA picture is satisfactory (see, *e.g.*, ^{6}Li, ^{9}Be, ^{10}B), and large where the IA picture is incomplete (see, *e.g.*, ^{7}Li, ^{7}Be, ^{9}C, ^{9}Li). Corrections from two-body components can be as large as 40%, as seen in the case of ^{9}C, and are crucial to reach (or improve) the agreement with the experimental data. It is also interesting to note that two-body effects, while being significant for the ^{9}C and ^{9}Li magnetic moments, are found to be negligible for those of ^{9}Be and ^{9}B. This behavior can be explained considering the dominant spatial symmetries of the nuclear wave functions for these $A = 9$ systems. For example, the dominant spatial symmetry of ^{9}Be (^{9}B) corresponds to an $[\alpha, \alpha, n(p)]$ structure [42]. Therefore, the unpaired nucleon outside the α clusters does not interact with other nucleons. As a consequence, two-body currents, that describe the coupling of external e.m. probes with pair of interacting nucleons, produce a negligible correction. On the other hand, the dominant spatial symmetry of ^{9}C (^{9}Li) corresponds to an [α, ^{3}He (^{3}H), pp (nn)] structure, and two-body correlations contribute in both the trinucleon clusters and in between the trinucleon clusters and the valence pp (nn) pair, resulting in a large two-body current contribution.

GFMC calculations of selected E2 and M1 transitions in low-lying nuclear states [40] are summarized in the right panel of Fig. 5. Predictions in IA are represented by blue dots, while those obtained with the full chiral e.m. current operator are represented by red diamonds. Calculations for E2 transitions implicitly include the effect of two-body currents via the Siegert theorem, where the charge density is used in IA. Also for these observables the effect of two-body e.m. currents can be large, and for cases in which the experimental errors are relatively small, *e.g.*, ^{7}Li($1/2^- \rightarrow 3/2^-$), ^{8}B($1^+ \rightarrow 2^+$), it is found that their inclusion leads to agreement with the experimental data. This scheme has been most recently utilized to study e.m. (both E2 and M1) transitions occurring in ^{8}Be [43, 44]. It is found that the agreement between the calculated and the experimental M1 widths is not satisfactory. Nevertheless, chiral two-body e.m. currents provide corrections at the 20%-30% level, which, in all but one case, improves the IA values. It is possible that the systematic underprediction of these observables is due to a poor knowledge of the small components entering the calculated GFMC nuclear wave functions [44].

5 Summary and outlook

In this talk, I presented an overview on the present status of *ab initio* calculations of e.m. observables, including e.m. moments and form factors, as well as e.m. transitions in light nuclei. The emphasis was on calculations that account for NN and 3N potentials in the nuclear Hamiltonians utilized to generate the wave functions, and one- and two-body e.m. current operators. I focused on results that account for two-body operators that have been derived from a χEFT formulation with pions and nucleons, including up to corrections of two-pion range. The *ab initio* prescription is extremely successful in explaining the experimental data, provided that many-body effects in both the e.m. currents and nuclear Hamiltonians are accounted for. χEFT based calculations of $A = 2$ and 3 nuclei e.m. form factors [31] nicely agree with the experimental data in the low-energy regime of applicability of χEFTs, with two-body corrections in the e.m. currents playing an important role in improving the agreement between the calculated and the experimental values of the trinucleon magnetic form factors. E.m. two-body current operators provide a 40% correction in the calculated magnetic moment of ^{9}C [40], and corrections at the 20%-30% level in M1 transitions occurring in ^{8}Be [44]. There are many interesting e.m. observables that can be accessed within this formalism. For example, few (or no) *ab initio* calculations of e.m. (charge and magnetic) form factors in $A > 4$ nuclei currently exist [1], and it would be interesting to perform them to have a deeper insight on nuclear e.m. structures. A complete microscopic profile of nuclei includes also studies of e.m. reactions such as radiative captures and photonuclear reactions. From the theoretical point of view, *i)* the construction of chiral potentials compatible with Quantum Monte Carlo computational calculations [45], opens up the possibility of performing consistent Quantum Monte Carlo calculations that use chiral potentials and chiral currents; *ii)* the construction of the chiral NN potential with the explicit inclusion of Δ-excitation [46], allows study of the effects of Δ-isobars in chiral two-body e.m. current operators [28]. Weak processes are also being vigorously studied within the χEFT formulation. Among these studies are the derivation of the axial two-body current operator up to one-loop [47], as well as the construction of two-body operators entering pion-production reactions induced by neutrino scattering off nuclei (see contributions by F. Myhrer in this proceedings).

Acknowledgements

This work was supported by the National Science Foundation, grant No. PHY-1068305. I am grateful and owe sincere thanks to R. Schiavilla, R.B. Wiringa, Steven Pieper, L. Girlanda, M. Piarulli, M. Viviani, L.E. Marcucci, A. Kievsky, and S. Bacca.

References

[1] S. Bacca, S. Pastore, J. Phys. **G41**, 123002 (2014), 1407.3490

[2] A. Lovato, S. Gandolfi, J. Carlson, S.C. Pieper, R. Schiavilla, Phys. Rev. Lett. **112**, 182502 (2014)

[3] R.B. Wiringa, V.G.J. Stoks, R. Schiavilla, Phys. Rev. C **51**, 38 (1995)
[4] B. Pudliner, V. Pandharipande, J. Carlson, R.B. Wiringa, Phys. Rev. Lett. **74**, 4396 (1995)
[5] S.C. Pieper, AIP Conference Proceedings **1011**, 143 (2008)
[6] R. Machleidt, D. Entem, Physics Reports **503**, 1 (2011), and references therein.
[7] E. Epelbaum, U.G. Meißner, Ann. Rev. Nucl. Part. Sci. **62**, 159 (2012), and references therein.
[8] D. Gazit, S. Quaglioni, P. Navrátil, Phys. Rev. Lett. **103**, 102502 (2009)
[9] P. Navrátil, Few-Body Systems **41**, 117 (2007)
[10] F. Villars, Phys. Rev. **72**, 256 (1947)
[11] H. Miyazawa, Progress of Theoretical Physics **6**, 801 (1951)
[12] D. Riska, G. Brown, Phys. Lett. B **38**, 193 (1972)
[13] L.E. Marcucci, M. Viviani, R. Schiavilla, A. Kievsky, S. Rosati, Phys. Rev. C **72**, 014001 (2005)
[14] L.E. Marcucci, M. Pervin, S.C. Pieper, R. Schiavilla, R.B. Wiringa, Phys. Rev. C **78**, 065501 (2008)
[15] J. Carlson, R. Schiavilla, Rev. Mod. Phys. **70**, 743 (1998), and references therein.
[16] A. Lovato, S. Gandolfi, R. Butler, J. Carlson, E. Lusk et al., Phys. Rev. Lett. **111**, 092501 (2013)
[17] S. Weinberg, Phys. Lett. B **251**, 288 (1990)
[18] S. Weinberg, Nucl. Phys. B **363**, 3 (1991)
[19] S. Weinberg, Phys. Lett. B **295**, 114 (1992)
[20] U. van Kolck, Phys. Rev. C **49**, 2932 (1994)
[21] C. Ordóñez, U. van Kolck, Phys. Lett. B **291**, 459 (1992)
[22] C. Ordóñez, L. Ray, U. van Kolck, Phys. Rev. C **53**, 2086 (1996)
[23] E. Epelbaum, H.W. Hammer, U.G. Meißner, Rev. Mod. Phys. **81**, 1773 (2009), and references therein.
[24] T.S. Park, D.P. Min, M. Rho, Nucl. Phys. A **596**, 515 (1996)
[25] Y.H. Song, R. Lazauskas, T.S. Park, D.P. Min, Phys. Lett. B **656**, 174 (2007)
[26] Y.H. Song, R. Lazauskas, T.S. Park, Phys. Rev. C **79**, 064002 (2009)
[27] R. Lazauskas, Y.H. Song, T.S. Park, Phys. Rev. **C83**, 034006 (2011)
[28] S. Pastore, R. Schiavilla, J.L. Goity, Phys. Rev. C **78**, 064002 (2008)
[29] S. Pastore, L. Girlanda, R. Schiavilla, M. Viviani, R.B. Wiringa, Phys. Rev. C **80**, 034004 (2009)
[30] S. Pastore, L. Girlanda, R. Schiavilla, M. Viviani, Phys. Rev. C **84**, 024001 (2011)
[31] M. Piarulli, L. Girlanda, L. Marcucci, S. Pastore, R. Schiavilla et al., Phys. Rev. **C87**, 014006 (2013)
[32] S. Kölling, E. Epelbaum, H. Krebs, U.G. Meißner, Phys. Rev. C **80**, 045502 (2009)
[33] S. Kölling, E. Epelbaum, H. Krebs, U.G. Meißner, Phys. Rev. C **84**, 054008 (2011)
[34] M. Walzl, U.G. Meißner, Phys. Lett. B **513**, 37 (2001)
[35] D.R. Phillips, Phys. Lett. B **567**, 12 (2003)
[36] D.R. Phillips, J. Phys. **G34**, 365 (2007)
[37] S. Kolling, E. Epelbaum, D. Phillips, Phys. Rev. **C86**, 047001 (2012)
[38] D.R. Entem, R. Machleidt, Phys. Rev. C **68**, 041001 (2003)
[39] E. Epelbaum, W. Glöckle, U.G. Meißner, Nucl. Phys. A **747**, 362 (2005)
[40] S. Pastore, S.C. Pieper, R. Schiavilla, R. Wiringa, Phys. Rev. **C87**, 035503 (2013)
[41] J. Carlson, S. Gandolfi, F. Pederiva, S.C. Pieper, R. Schiavilla, K.E. Schmidt, R.B. Wiringa (2014), 1412.3081
[42] R.B. Wiringa, Phys. Rev. **C73**, 034317 (2006)

[43] V. Datar, D. Chakrabarty, S. Kumar, V. Nanal, S. Pastore et al., Phys. Rev. Lett. **111**, 062502 (2013)

[44] S. Pastore, R.B. Wiringa, S.C. Pieper, R. Schiavilla, Phys. Rev. C **90**, 024321 (2014)

[45] J.E. Lynn, J. Carlson, E. Epelbaum, S. Gandolfi, A. Gezerlis, A. Schwenk, Phys. Rev. Lett. **113**, 192501 (2014), 1406.2787

[46] M. Piarulli, L. Girlanda, R. Schiavilla, R. Navarro Pérez, J.E. Amaro, E. Ruiz Arriola, Phys. Rev. **C91**, 024003 (2015), 1412.6446

[47] A. Baroni, L. Girlanda, S. Pastore, R. Schiavilla, M. Viviani (2015), arXiv:1509.07039

EPJ Web of Conferences **113**, 01009 (2016)
DOI: 10.1051/epjconf/201611301009

Hadron Interactions from lattice QCD

Sinya Aoki[1,2,a]

[1] *Yukawa Institute for Theoretical Physics, Kyoto University, Kyoto 606-8502, Japan*
[2] *Center for Computational Sciences, University of Tsukuba. Ibaraki 305-8571, Japan*

Abstract. We review our strategy to study hadron interactions from lattice QCD using newly proposed potential method. We first explain our strategy in the case of nuclear potentials and its application to nuclear physics. We then discuss the origin of the repulsive core, by adding strange quarks to the system. We also explore a possibility for H-dibaryon to exist in flavor SU(3) limit of lattice QCD. We conclude the paper with an application of our strategy to investigate the maximum mass of neutron stars.

1 Introduction

The current understanding of the structure of matters is as follows. At some microscopic level, the building blocks of matters are atoms, which are composed of heavy and compact nuclei in the center surrounded by electrons. A nucleus is made of protons and neutrons, which are also composed of more fundamental objects, quarks. A word hadron is a generic name for composite particles made of quarks such as proton and neutron.

In this talk, we consider interactions among such hadrons. For example, protons and neutrons interact with each other to form nuclei, bound states of them. Such interactions is called the nuclear force, the origin of which has been explained by the exchange of new particles, called pion[1]. We now know not only proton and neutron, collectively called nucleons, but also pions are composite particles of quarks, together with gluons, which mediate the strong force between quarks. Fig. 1 shows some phenomenological estimates of the nuclear potential as a function of a distance between two nucleons, whose derivative with respect to the distance gives the force between two nucleons, the nuclear force. As can be seen from the figure, the potential shows a rather complicated behavior as a function of the distance due to the compositeness of nucleons.

The dynamics of quarks and gluons is known to be governed by the theory called Quantum Chromodynamics (QCD). Therefore, in principle, hadron interactions such as the nuclear force can be derived from QCD. Interactions of QCD becomes so strong at long distance that no isolated quarks can be observed. This phenomenon is called quark confinement. The perturbative expansion relied on the weak coupling does not work for QCD since its coupling constant is large as the quark confinement indicates. Lattice field theories have been proposed to investigate strongly interacting theories such as QCD, by defining quantum field theories on a discrete space-time (lattice) instead of the continuous one. QCD defined on N^4 lattice (lattice QCD) is equivalent to well-defined statistical system with a

[a] e-mail: saoki@yukawa.kyoto-u.ac.jp

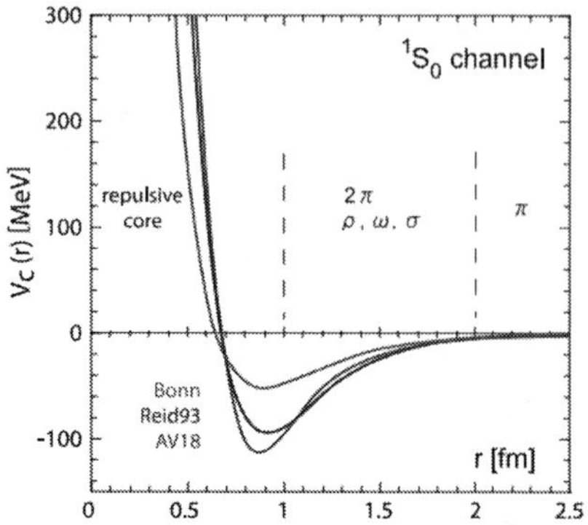

Figure 1. Three examples of the phenomenological nuclear potential, Bonn[2], Reid93[3] and Argonne v_{18}[4]. Taken from Ref. [5].

finite lattice spacing a and the finite extension $L = Na$, which is manifestly gauge invariant and non-perturbatively defined[6]. Because of these properties, numerical simulations based on Monte-Carlo method can be successfully applied to lattice QCD[7].

Fundamental degrees of freedom in QCD are quarks and gluons, from which stable hadrons such as nucleon and pions emerge as physical degrees of freedom by QCD interactions. Interactions among these stable particles are also described by the same theory, QCD. A main purpose of this talk is to review our attempts to understand hadron interactions based on QCD by using lattice QCD techniques.

2 Strategy and nuclear potentials

2.1 Our strategy

We first briefly explain our strategy to investigate hadron interactions in lattice QCD[5, 8], which consists of three main steps.

In the 1st step, we calculate potentials between two hadrons by numerical simulations in lattice QCD. There are many subtleties and discussions on how to define potentials in QCD, but we do not discuss them in this talk. Please see Ref. [9] for more details.

In the 2nd step, we calculate physical observables such as binding energy and scattering phase shift of the two hadrons, using potentials obtained in the 1st step.

In the 3rd step, we employ the same 2-body potentials to investigate few hadron systems or structures of may body systems such as heavy nuclei and nuclear matters.

The 1st step in the above has been first applied to the nuclear potential [5], which has shown that the proposal seems working well, and Ref. [5] has received general recognition[10]. After this success, a research group named Hadrons to Atomic nuclei from Lattice QCD (HAL QCD) Collaboration has bend formed to investigate various aspects of hadronic interactions by this method, called the HAL QCD method.[1]

2.2 Example: Nuclear potentials

Let us consider the nuclear potential, as an example how the strategy in Sect. 2.1 works.

In the 1st step, we calculate the nuclear potential in lattice QCD. Fig. 2 shows the nuclear potential for the spin-singlet ($S = 0$) sector, calculated in lattice QCD at the lattice spacing (space

[1]The binding energy of two nucleon systems has been calculated directly in lattice QCD (for example, see [11, 12]), but results disagree with ours. Recently, however, it turns out that plateaux of the binding energy in the finite box from two different source operators, wall and smeared ones, differ significantly [13], so that the claimed strong binding of the two nucleon systems at heavy pion masses [11, 12] should be taken with a grain of salt. We therefore do not consider the direct method in this report.

resolution) $a \simeq 0.1$ fm with $m_u = m_d < m_s$, where m_u, m_d, m_s are up, down, strange quark masses, respectively[14]. In this calculation, pion mass m_π, which is a measure of quark mass in the simulation, is about 700 MeV, while $m_\pi \simeq 140$ MeV in Nature. Lighter the quark mass, harder the simulation in lattice QCD. This is a reason why we employed $m_\pi \simeq 700$ MeV for this study. As can be seen from Fig. 2, qualitative features of nuclear potential in Fig. 1 are reproduced in our lattice QCD simulations. Namely, the potential in Fig. 2 shows attractions at medium and long distances, while it gives repulsions at short distance, called a repulsive core.

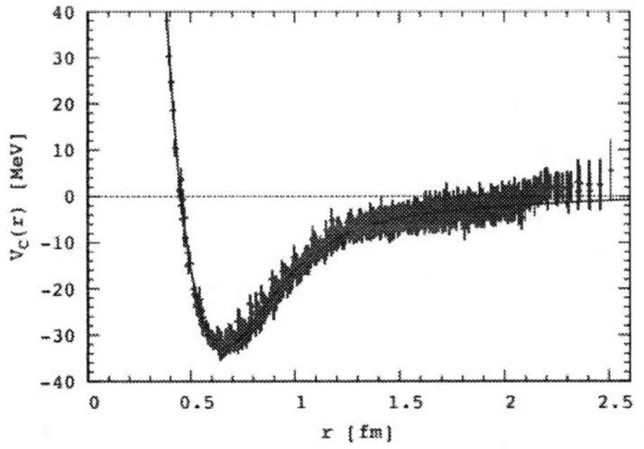

Figure 2. Spin-singlet nuclear potential calculated in lattice QCD, together with the multi-Gaussian fit by the solid line. Taken from Ref. [14].

In the 2nd step, we employ the potential obtained in lattice QCD to calculate the scattering phase shift of two nucleons. In practice, the potential in Fig. 2 is fitted with some functional form. For example, the multi-Gaussian function is used in Fig. 2, shown by the solid line. We then solve the Schrödinger equation to determine the phase shift at a given energy. Fig. 3 shows the scattering phase shift in the spin-singlet channel as a function of the laboratory energy $E_{\rm lab}$ with the rest energy of two nucleons subtracted[14]. The scattering phase shift obtained form lattice QCD has a reasonable shape. Its strength, however, is weaker than the experimental one (solid line), probably due to the heavier quark mass in lattice QCD simulations: The scattering length, calculated as $\lim_{k\to 0} \tan\delta(k)/k$, is 1.6 ± 1.1 fm in this lattice QCD simulation at $m_\pi \simeq 700$ MeV, while it should be about 20 fm at the physical pion mass, $m_\pi \simeq 140$ MeV. Lattice QCD simulations at the physical pion mass are definitely required to reproduce the correct scattering phase shift and scattering length in lattice QCD. Such attempts are currently ongoing by using the K computer in Kobe, Japan, whose peak speed is 10 PetaFlops[16].

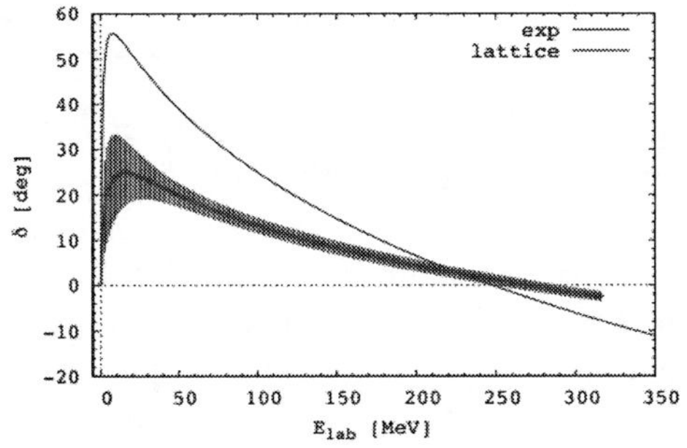

Figure 3. The scattering phase in the spin-singlet channel in the laboratory frame obtained from the lattice nuclear potential in Fig. 2, together with experimental data by the solid line[15]. Taken from Ref. [14].

In the 3rd step, using the potential obtained in lattice QCD, we estimate the binding energy of some nuclei. For example, Fig. 4 shows the binding energy of a 4 nucleon ground state in $(L,S)J^P = (0,0)0^+$, configuration corresponding to the ^{4}He nucleus, as a function of a number of bases in the stochastic variational method[17]. The potentials used in this calculation, different from the one in

Fig. 2, are obtained in the flavor SU(3) limit that $m_u = m_d = m_s$ at $a \simeq 0.12$ fm and $L \simeq 4$ fm with $M_{\rm PS} \simeq 470$ MeV, where $M_{\rm PS}$ is the mass of pseudo-scalar meson (π and K). In the calculation of the ground state energy for ^{4}He, the Wigner type force, where the odd part of potentials is equal to the even part, as well as the Serber type force, where the odd part is set to zero, are employed for a comparison. As seen in Fig. 4, results form two different forces agree, and the binding energy of ^{4}He is about 5.1 MeV. Again, a smaller binding energy than Nature is partly due to the heaver quark masses and to the flavor SU(3) limit, in addition to absence of 3- and 4- nucleon forces. See Ref. [17] for more details. Recent applications of lattice QCD potentials to medium-heavy nuclei can be found in Ref. [18].

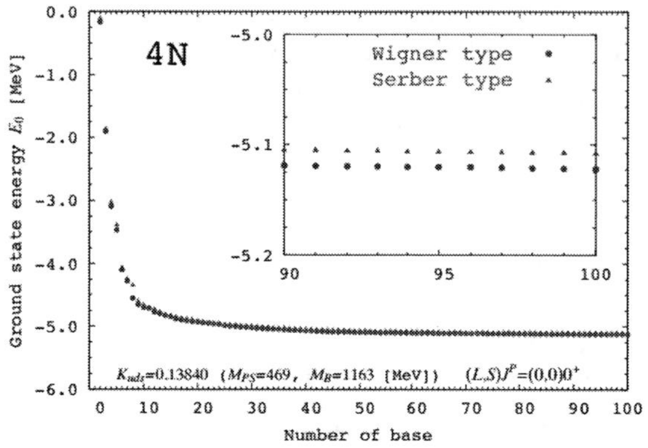

Figure 4. Biding energy of 4 nucleon ground state in lattice QCD with the flavor SU(3) limit, as a function of a number of bases. Two types of 2-body potentials are employed for a comparison. Taken from Ref. [17].

3 Repulsive core

The short distance repulsion in the nuclear force, first introduced by Jastrow[19] and called the repulsive core, plays several important roles in nuclear physics and astrophysics. For example, it explains stability of matters. A nucleus would collapse to a very dense object if the nuclear force had only attractions without repulsive core. The repulsive core sustains neutron stars against their gravitational collapses and ignites the Type II supernova explosions as the gravitational collapse bounds by the repulsion[20].

3.1 The origin of the repulsive core

It is interesting and important to ask what is the origin of the repulsive core. Since quarks are fermions, two can not occupy the same position with same quantum numbers due to the Pauli exclusion principle. However, quarks have 3 colors (red, blue and green), 2 spins (↑ and ↓) and 2 flavors (up and down), so that six quarks can occupy the same position. Since this means that two nucleons occupy the same position, the repulsive core can not be explained simply by the Pauli exclusion principle alone. However, allowed color combinations are restricted once spin and flavor quantum numbers are fixed, as an example for two protons with opposite spins is shown in the top of Fig. 5. In this example, up quarks with ↑ in each proton can not have the same color. Similarly, up quarks with ↓ also can not. It has been pointed out that this restriction for color combinations plus one gluon-exchange interactions between quarks may be an origin of the repulsive core[21, 22].

In order to check whether the origin of the repulsive core is related to the partial restriction of color combinations, it is useful to consider what happen if an extra degree of freedom, a strange quark, is added to the system. In this system, we have the $\Lambda(uds)$ baryon, which is composed of up, down

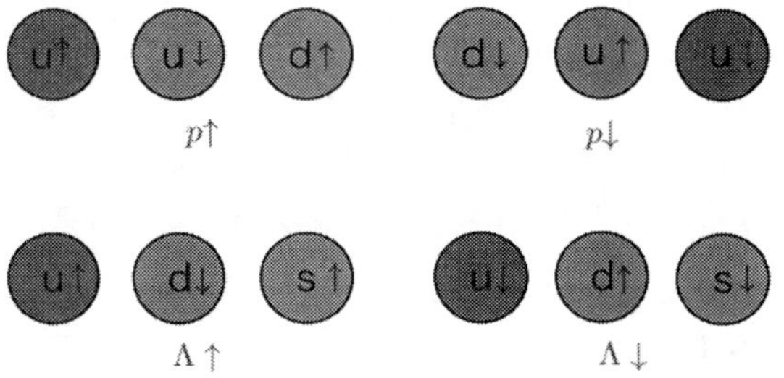

Figure 5. (Top) An example of quark configurations for two protons with opposite spins. (Bottom) An example of quark configurations for two Λ's with opposite spins.

and strange quarks. If we consider two Λ's with opposite spins, all color combinations are allowed as drawn in the bottom of Fig. 5, where two quarks with the same flavor (up, down or strange) in each Λ have same color. If the origin of the repulsive core is related to the partial restriction of color combinations, the potential between two Λ's in the spin singlet channel has no repulsive core.

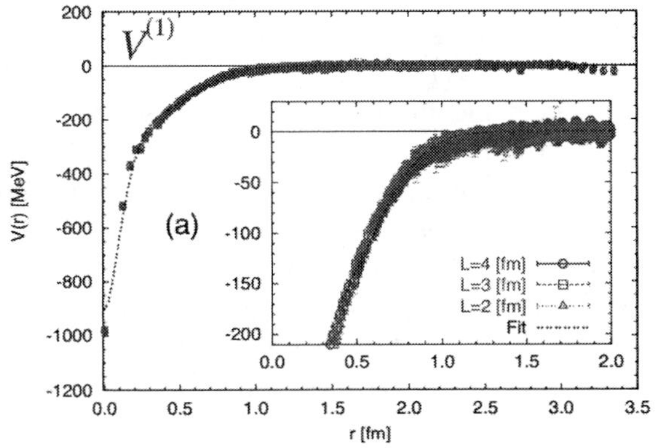

Figure 6. Flavor-singlet potential $V_C^{(1)}(r)$ for the box size $L \simeq 2, 3, 4$ fm at $M_{\rm PS} = 1015$ MeV. Taken from Ref. [24].

To confirm this, we have performed 3-flavor lattice QCD simulations in the flavor SU(3) limit ($m_u = m_d = m_s$)[23, 24]. In the flavor SU(3) limit, two octet baryons are classified by the irreducible representations of the SU(3) group as $\mathbf{8} \otimes \mathbf{8} = (\mathbf{27} \oplus \mathbf{8}_s \oplus \mathbf{1})_{S=0} \oplus (\mathbf{10}^* \oplus \mathbf{10} \oplus \mathbf{8}_a)_{S=1}$, where S is the total spin. The system of two Λ's with opposite spins we are interested in belongs to the flavor singlet representation **1**, whose potential is denoted as $V_C^{(1)}$. More precisely the single state is denoted as

$$|\text{singlet}\rangle = -\sqrt{\frac{1}{8}}|\Lambda\Lambda\rangle + \sqrt{\frac{3}{8}}|\Sigma\Sigma\rangle + \sqrt{\frac{4}{8}}|N\Xi\rangle. \qquad (1)$$

Fig. 6 shows the flavor-singlet potential $V_C^{(1)}(r)$ as a function of r on volumes $L \simeq 2$, 3 and 4 fm at $a \simeq 0.12$ fm and $M_{\rm PS} \simeq 1015$ MeV in the SU(3) limit[24]. As can be clearly seen in the figure, the attractive instead of repulsive core indeed appears in this setup. This result strongly suggests that partial restriction of color combinations due to the Pauli principle among quarks is important for the repulsive core.

3.2 H dibaryon

Since the force in the flavor-singlet channel is attractive at all distances, an existence of bound states may be expected. In this channel, a six quark bound state made of *uuddss*, name H-dibaryon, has been predicted[25], but not observed yet experimentally. We have investigated whether the H dibaryon appears for the flavor-singlet channel in the SU(3) limit, by using the singlet-potential $V_C^{(1)}(r)$.

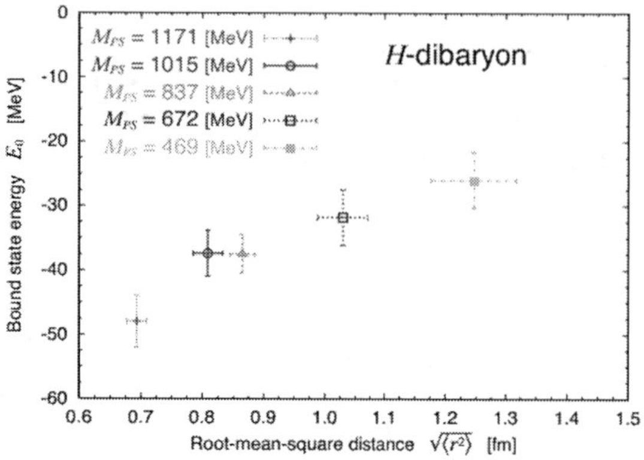

Figure 7. The binding energy $B_H = -E_0$ and the root-mean-square distance $\sqrt{\langle r^2 \rangle}$ of the bound state in the flavor singlet channel at each $M_{\rm PS}$. Taken from Ref. [17].

For this purpose we fit the potential in Fig. 6 at $L \simeq 4$ fm by the analytic function composed of an attractive Gaussian core plus a long range (Yukawa)^2 attraction as $V_C^{(1)}(r) = b_1 e^{-b_2 r^2} + b_3(1 - e^{-b_4 r^2})(e^{-b_5 r}/r)^2$. An example of the fitted result with $\chi^2/\text{dog} \simeq 1$ is shown by the dashed line in Fig. 6. By solving the Schrödinger equation with the fitted potential in the infinite volume, we have found one bound state, the H-dibaryon, in the flavor SU(3) limit, whose binding energy and wave function are obtained. Fig. 7 shows the binding energy and the root-mean-square(rms) distance of the H-dibaryon obtained from our potential at several values of $M_{\rm PS}$. The rms distance $\sqrt{\langle r^2 \rangle}$ is a measure of the "size" of the H-dibaryon, which is compared to that of the deuteron, 3.8 fm, in Nature. Although the current result is obtained in the flavor SU(3) limit with heavier up and down quark masses than Nature, this comparison suggests that H-dibaryon is much more compact than the deuteron. As $M_{\rm PS}$ decrease, the binding energy decreases and the rms distance increases. Note that, despite that the attractive potential becomes stronger as $M_{\rm PS}$ decreases[17, 24], the binding energy B_H decreases du to the fact that the increase of the attraction toward the lighter $M_{\rm PS}$ is compensated by the increase of the kinetic energy for the lighter baryon mass. As a result, the size of H-dibaryon also increases.

We close this section by concluding that the H-dibaryon exists in the flavor SU(3) limit, whose binding energy shows a mild quark mass dependence and is about 20 $\sim$ 50 MeV at this range $M_{\rm PS}$. (See [26] for the direct calculation of the binding energy for the H-dibaryon in 2+1 flavor QCD.)

4 Summary and more

The potential method, called the HAL QCD method, is new but very useful to investigate not only the nuclear force but also general baryonic interaction in (lattice) QCD[27–29]. A comparison of the nuclear potential with the flavor-singlet potential in the flavor SU(3) limit leads to some understanding of the repulsive core, that the Pauli exclusion principle plays an important role for its existence. The HAL QCD method can be easily extended also to meson-baryon and meson-meson interactions[30].

Our strategy based on the HAL QCD method to investigate nuclear physics and astrophysics is as follows. We first extract potentials form lattice QCD using the HAL QCD method. Using the sophisticated many-body methods combined with these potentials obtained in QCD, we then investigate structures of nuclei and nuclear matter, based on which we finally study phenomena in astrophysics such as properties of neutron stars and supernova explosions.

As a demonstration how the above strategy indeed works, we show results in Ref. [31], which give quark mass dependences of the nuclear matter, the neutron matter and the maximum mass of neutron stars, using nuclear potentials obtained in the flavor SU(3) limit[17].

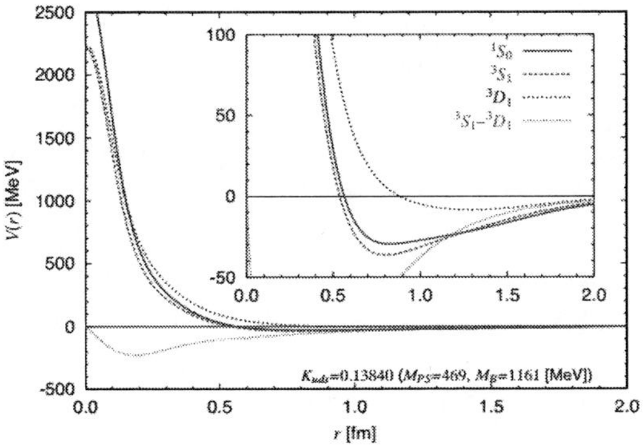

Figure 8. The lines represent the χ^2 fits of the nuclear potentials for S and D wave extracted from lattice QCD at $M_{\rm PS} \simeq 470$ MeV in the flavor SU(3) limit. Taken from Ref. [31].

Fig. 8 shows the nuclear potentials in S and D waves obtained from fits to the lattice QCD data in the flavor SU(3) limit at $M_{\rm PS} \simeq 470$ MeV[17]. Using these potentials, the equation of state (EOS)

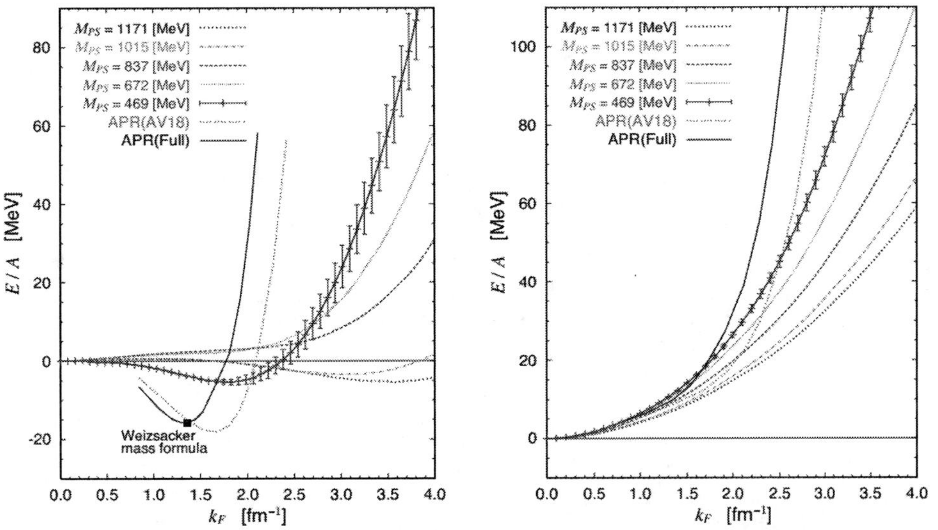

Figure 9. (Left) Ground state energy per nucleon (E/A) for the symmetric nuclear matter as a function of the Fermi momentum k_F, obtained by the BHF theory with the nuclear potentials in lattice QCD. Filled square indicates the empirical saturation point, together with the curves labeled as APR from Ref. [32]. Errors for the result at $M_{\rm PS} \simeq 470$ MeV represent statistical uncertainties. (Right) Ground state energy per nucleon for the pure neutron matter as a function of the Fermi momentum k_F. Details of the calculations are the same as the symmetric nuclear matter. Both are taken from Ref. [31].

for the nuclear matter and the neutron matter has been investigated by the Brueckner-Hatree-Fock (BHF) theory[31].[2] Fig. 9 (Left) shows the ground state energy per nucleon (E/A) for the symmetric nuclear matter ($Z = N = A/2$) with the proton number Z, the neutron number N and the mass number $A = N + Z$, as a function of the Fermi momentum k_F for different $M_{\rm PS}$. The most important feature of the symmetric nuclear matter in Nature is its saturation property that E/A has a minimum at normal

[2] Currently, more sophisticated many-body calculations are attempted in collaboration with experts in this area.

nuclear matter density ρ_0. The empirical saturation point from the Weizsäcker mass formula gives $(k_F, E/A) \simeq$ (1.36 fm^{-1}, -15.7 MeV), denoted by the filled square in Figure 9 (Left), where we also show the results of Ref. [32], obtained by the variational method using AV18 nuclear potentials[4] with and without phenomenological three-nucleon force. Our lattice data show the saturation not only at $M_{\rm PS} \simeq 470$ MeV (the lightest quark mass) but also at $M_{\rm PS} \simeq 1020, 1170$ MeV (the heaviest two quark masses), which indicates that the saturation originates from a suitable balance between the repulsive core and the intermediate attraction of the nuclear force. An appearance of the saturations at more than $10\rho_0$ at $M_{\rm PS} \simeq 1020, 1170$ MeV, however, might be an lattice artifact, and needs to be checked.

Fig. 9 (Right) shows E/A for the pure neutron matter ($N = A$) as a function of k_F. The neutron matter is not self-bound due to large Fermi energy. The corresponding pressure is calculated as $P = \rho^2 \frac{\partial(E/A)}{\partial\rho}$ with the density $\rho = \frac{k_F^3}{3\pi^2}$, so that P increases rapidly as $M_{\rm PS}$ decreases.

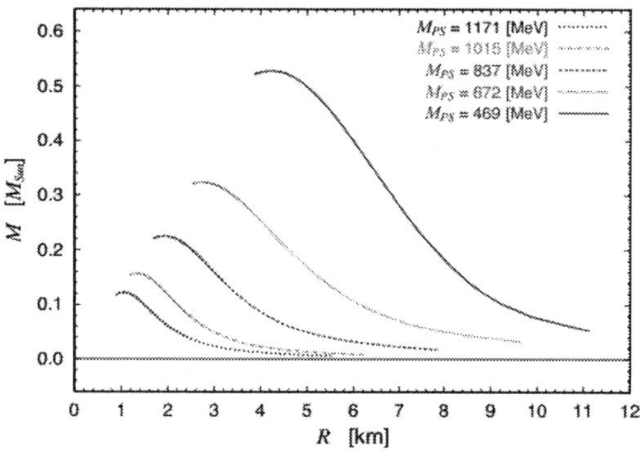

Figure 10. Mass-radius relation of the neutron star. Neutron-star matter consists of n, p, e^- and μ^- with charge neutrality and chemical equilibrium. The EOS for the nucleons is obtained by an interpolation between the symmetric nuclear matter and the pure neutron matter under the parabolic approximation. Taken from Ref. [31].

To obtain the mass (M) and radius (R) relation of the neutron star, the Tolman-Oppenheimer-Volkoff (TOV) equation is solved under the charge neutrality and beta-equilibrium for neutron, proton electron and muon, with the EOS of asymmetric nuclear matter approximated as

$$\frac{E}{A}(\rho, x) = \frac{E_{Z=N}}{A}(\rho) + (1-2x)^2 \epsilon_{\rm sym}(\rho), \quad \epsilon_{\rm sym}(\rho) = \frac{E_{Z=0}(\rho)}{A} - \frac{E_{Z=N}(\rho)}{A} \tag{2}$$

for the proton fraction $x = \rho_x/\rho$.

Fig. 10 shows the M-R relation of the neutron star at different $M_{\rm PS}$. As $M_{\rm PS}$ decreases, the M-R curve shifts to the upper right direction, suggesting the stiffening of the EOS. The maximum mass of the neutron star ($M_{\rm max}$) is found to be 0.53 times the solar mass $M_\odot$ at our lightest $M_{\rm PS} \simeq 470$ MeV, which is too small to account for the observed neutron stars, mainly due to the larger $M_{\rm PS}$. A naive extrapolation of $M_{\rm max}$ and the corresponding radius to $M_{\rm PS} \simeq 140$ MeV, with a function $f(M_{\rm PS}) = a/(M_{\rm PS} + b) + c$, gives $M_{\rm max} = 2.2 M_\odot$ and $R = 12$ km. To confirm these predictions, we will have to repeat the procedure explained here with the nuclear potentials at $m_\pi \simeq 140$ MeV.

Acknowledgements

The author would like to thank other members of the HAL QCD Collaboration for discussions. This work is supported in part by the Grant-in-Aid of the Japanese Ministry of Education (No. 25287046), MEXT SPIRE and JICFuS.

References

[1] H. Yukawa, Proc. Math. Phys. Soc. Japan **17**, 48 (1935).
[2] R. Machleidt, Phys. Rev. **C63**, 024001 (2001).
[3] V. G. J. Stoks, R. A. M. Klomp, C. P. F. Terheggen and J. J. de Swart, Phy. Rev. **C49**, 2950 (1994).
[4] R. B. Wiringa, V. G. J. Stoks and R. Schiavilla, Phy. Rev. **C51**, 38 (1995).
[5] N. Ishii, S. Aoki and T. Hatsuda, Phys. Rev. Lett. **99**, 022001 (200).
[6] K. G. Wilson, Phys. Rev. D **10**, 2445 (1974).
[7] M. Creutz, Phys. Rev. D **21**, 2308 (1980).
[8] S. Aoki, T. Hatsuda and N. Ishii, Prog. Theor. Phys. **123**, 89(2010).
[9] S. Aoki *et al.* [HAL QCD Collaboration], Prog. Theor. Exp. Phys. **2012**, 01A106 (2012).
[10] In *Research Highlights 2007*, Nature **450**, 1130 (2007).
[11] T. Yamazaki, K. i. Ishikawa, Y. Kuramashi and A. Ukawa, Phys. Rev. D **86** (2012) 074514 [arXiv:1207.4277 [hep-lat]]; Phys. Rev. D **92** (2015) 1, 014501 [arXiv:1502.04182 [hep-lat]].
[12] S. R. Beane *et al.* [NPLQCD Collaboration], Phys. Rev. D **85** (2012) 054511 [arXiv:1109.2889 [hep-lat]]; Phys. Rev. D **87** (2013) 3, 034506 [arXiv:1206.5219 [hep-lat]]; Phys. Rev. C **88** (2013) 2, 024003 [arXiv:1301.5790 [hep-lat]].
[13] T. Iritani for the HAL QCD Collaboration, *Lattice QCD studies of baryon interactions from HAL QCD method and Lüsher's finite volume method*, in the proceedings of *LATTICE 2015* [arXiv:1511.05246 [hep-lat]].
[14] N. Ishii *et al.* [HAL QCD Collaboration], Phys. Lett. **B712** , 437 (2012).
[15] http://www.nn-online.org/
[16] http://www.aics.riken.jp/en/k-computer/about/
[17] T. Inoue *et al.* [HAL QCD Collaboration], Nucl. Phys. A**881**, 28 (2012).
[18] T. Inoue *et al.* [HAL QCD Collaboration], Phys. Rev. C **91**, no. 1, 011001 (2015).
[19] R. Jastrow, Phys. Rev. **81**, 165 (1951).
[20] R. Tamagaki et al., Prog. Theor. Phys. Suppl. **112**, 1 (1993).
H. Heiselberg and V. Pandharipande, Ann. Rev. Nucl. Part. Sci. **50**, 481 (2000).
J. M. Lattimer and M. Prakash, PhysRept. **333**, 121 (2000).
[21] M. Oka, K. Shimizu and K. Yazaki, Phys. Lett. B **130**, 365 (1983).
[22] M. Oka, K. Shimizu and K. Yazaki, Nucl. Phys. A **464**, 700 (1987).
[23] T. Inoue *et al.* [HAL QCD collaboration], Prog. Theor. Phys. **124**, 591 (2010).
[24] T. Inoue *et al.* [HAL QCD Collaboration], Phys. Rev. Lett. **106**, 162002(2011).
[25] R. L. Jaffe, Phys. Rev. Lett. **38**, 195 (1977) [Erratum-ibid. **38**, 617 (1977)].
Reviewed in, T. Sakai, K. Shimizu and K. Yazaki, Prog. Theor. Phys. Suppl. **137**, 121 (2000).
[26] S. R. Beane *et al.* [NPLQCD Collaboration], Phys. Rev. Lett. **106** (2011) 162001 [arXiv:1012.3812 [hep-lat]].
[27] H. Nemura, N. Ishii, S. Aoki and T. Hatsuda, Phys. Lett. B **673**, 136 (2009) .
[28] F. Etminan *et al.* [HAL QCD Collaboration], Nucl. Phys. A **928**, 89 (2014).
[29] M. Yamada *et al.* [HAL QCD Collaboration], PTEP **2015**, no. 7, 071B01 (2015).
[30] Y. Ikeda *et al.*, Phys. Lett. B **729**, 85 (2014).
[31] T. Inoue *et al.* [HAL QCD Collaboration], Phys. Rev. Lett. **111**, no. 11, 112503 (2013).
[32] A. Akmal, V. R. Pandharipande and D. G. Ravenhall, Phys. Rev. C **58**, 1804 (1998).

EPJ Web of Conferences **113**, 01010 (2016)
DOI: 10.1051/epjconf/201611301010

Ab initio calculation of the electromagnetic and neutral-weak response functions of ^{4}He and ^{12}C

A. Lovato[1,a], O. Benhar[2,b], J. Carlson[3], S. Gandolfi[3], Steven C. Pieper[1], N. Rocco[2], and R. Schiavilla[4,5]

[1] *Physics Division, Argonne National Laboratory, Argonne, IL 60439*
[2] *INFN and Department of Physics, "Sapienza" University, I-00185 Roma, Italy*
[3] *Theoretical Division, Los Alamos National Laboratory, Los Alamos, NM 87545*
[4] *Theory Center, Jefferson Lab, Newport News, VA 23606*
[5] *Department of Physics, Old Dominion University, Norfolk, VA 23529*

Abstract. Precise measurement of neutrino oscillations, and hence the determination of their masses demands a quantitative understanding of neutrino-nucleus interactions. To this aim, two-body meson-exchange currents have to be accounted for along within realistic models of nuclear dynamics. We summarize our progresses towards the construction of a consistent framework, based on quantum Monte Carlo methods and on the spectral function approach, that can be exploited to accurately describe neutrino interactions with atomic nuclei over the broad kinematical region covered by neutrino experiments.

1 Introduction

The description of neutrino interactions with nuclei, besides being interesting in its own right, provides an essential input for neutrino-oscillation experiments, such as DUNE [1], which plan to measure neutrino oscillation parameters and the neutrino mass hierarchy and determine the charge-conjugation and parity (CP) violating phase. Such experiments make use of nuclear targets as detectors that, while allowing for a substantial increase in the event rate, demand a quantitative understanding of neutrino-nucleus interactions.

The MiniBooNE collaboration has reported a measurement of the charged-current quasielastic (CCQE) neutrino-carbon inclusive double differential cross section, exhibiting a large excess with respect to the predictions of the relativistic Fermi gas model [2]. It has been suggested that the discrepancy might be due to the occurrence of events with two particle-two hole final states [3, 4] that are not taken into account by the relativistic Fermi gas. Within a realistic model of nuclear dynamics, their occurrence arises naturally owing to two-body meson-exchange currents and/or correlations induced by the nuclear interaction.

Because neutrino beams are always produced as secondary decay products, their energy is not sharply defined, but broadly distributed. Therefore, the observed cross section for a given energy and angle of the outgoing lepton includes contributions from energy- and momentum-transfer regions

[a] e-mail: lovato@anl.gov
[b] e-mail: benhar@roma1.infn.it

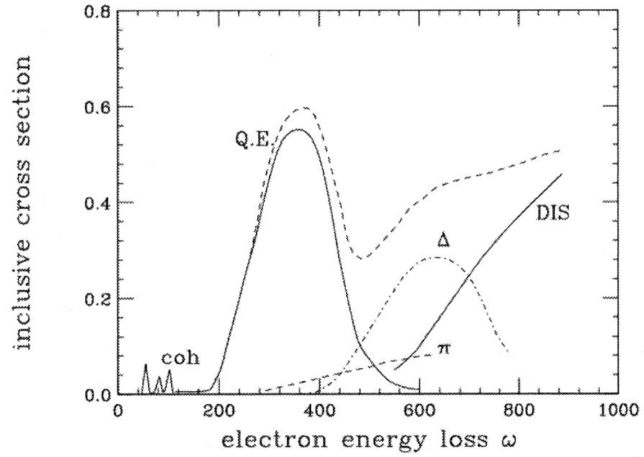

Figure 1. Schematic representation of the inclusive electron-nucleus cross section at beam energy around 1 GeV, as a function of energy loss, from Ref. [5].

where different mechanisms are at play. They can be best identified in electron-scattering experiments, in which the energy of the incoming electron is precisely known. The typical behavior of the double differential inclusive cross section for the process

$$e + A \to e' + X\,, \tag{1}$$

with a beam energy around 1 GeV where only the outgoing electron is detected, is shown in Figure 1, taken from Ref. [5]. The target nucleus in its ground state and the undetected hadronic final state are denoted by A and X, respectively. At small energy transfer the structure of the low-lying energy spectrum and collective effects are important. In the quasielastic peak region, the cross section is dominated by scattering off individual nucleons although a significant contribution from nucleon pairs is also present. The Δ-resonance region is characterized by one or more pions in the final state, while at large energy transfer the deep inelastic scattering contribution is the largest.

The ground state of the nucleus does not depend upon the momentum transfer of the process and can be accurately described within nonrelativistic many-body theory (NMBT). On the other hand, the final state does depend on the momentum transfer and the nonrelativistic approximation can be safely applied only in the region of low and moderate momentum transfer, corresponding to $|\mathbf{q}| \lesssim 500$ MeV. The treatment of the region of high momentum transfer requires a theoretical approach in which the accurate description of the nuclear ground state provided by NMBT is combined with a relativistically consistent description of both the final state and the nuclear current.

Setting up a framework in which two-body currents and nuclear correlations induced by realistic Hamiltonians are consistently accounted for is the main topic of this proceeding.

2 Lepton-Nucleus cross section

The double differential inclusive cross-section of the electromagnetic (EM) and neutral-current (NC) lepton-nucleus transition $\ell + A \to \ell' + X$ is proportional to the product of the leptonic and the hadronic tensors,

$$\frac{d\sigma}{d\Omega_\ell dE_\ell} \propto L_{\mu\nu} W^{\mu\nu} \tag{2}$$

The leptonic tensor $L_{\mu\nu}$ is completely determined by lepton kinematics, whereas the hadronic tensor,

$$W_{\mu\nu} = \sum_X \langle 0|J_\mu{}^\dagger|X\rangle \langle X|J_\nu|0\rangle \, \delta^{(4)}(p_0 + q - p_X) \, , \tag{3}$$

containing all the information on strong interaction dynamics, describes the nuclear response of the target to the transition currents J_μ from the initial state $|0\rangle$, with four momenta p_0, to the final state $|X\rangle$, with four momenta p_X.

In general, the nuclear current consists of one- and multi-nucleon contributions; the latter arise from processes in which the interaction with the beam particle involves single or multiple mesons exchanged between target nucleons, and are usually referred to as meson-exchange currents (MEC)

$$J^\mu = J_1^\mu + J_2^\mu + J_3^\mu = \sum_i j_i^\mu + \sum_{j>i} j_{ij}^\mu + \sum_{k>j>i} j_{ijk}^\mu \, . \tag{4}$$

The relation between the MEC and the nuclear hamiltonian will be discussed in the following section.

3 Moderate momentum transfer regime

In the moderate momentum transfer regime, both the initial and final states appearing in Eq. (3) are eigenstates of the nonrelativistic nuclear Hamiltonian H

$$H|0\rangle = E_0|0\rangle \quad , \quad H|X\rangle = E_X|X\rangle \, . \tag{5}$$

The nuclear Hamiltonian describing the dynamics of the point like nonrelativistic nucleons is given by

$$H = \sum_{i=1}^{A} \frac{\mathbf{p}_i^2}{2m} + \sum_{j>i=1}^{A} v_{ij} + \sum_{k>j>i=1}^{A} V_{ijk} \, . \tag{6}$$

In the above equation, $\mathbf{p}_i$ is the momentum of the i-th nucleon, while v_{ij} and V_{ijk} are the two- and three-nucleon potentials, respectively.

For nuclei as large as ^{12}C, the ground state wave function can be obtained via the Green's Function Monte Carlo (GFMC) approach [6, 7]. It has to be remarked that, within the limits of applicability of NMBT, the GFMC is truly an ab initio approach, allowing *exact* calculations of a number of nuclear properties.

A key feature of the description of neutrino-nucleus interactions at low and moderate momentum transfer is the possibility of employing a set of electroweak charge and current operators consistent with the Hamiltonian of Eq. (6). The nuclear electromagnetic current, $J^\mu_{\rm em} \equiv (J^0_{\rm em}, \mathbf{J}_{\rm em})$, trivially related to the vector component of the weak current, is constrained by H through the continuity equation [8]

$$\boldsymbol{\nabla} \cdot \mathbf{J}_{\rm em} + i[H, J^0_{\rm em}] = 0 \, . \tag{7}$$

Since the two- and three-nucleon potentials v_{ij} and V_{ijk} do not commute with the charge operator $J^0_{\rm em}$, the above equation implies that $J^\mu_{\rm em}$ involves two- and three-nucleon contributions, as shown in Eq. (4).

The one-body electroweak operator is obtained from a nonrelativistic expansion of the covariant single-nucleon currents, while the two-body charge and current operators that we employ in our calculations are derived within the conventional meson-exchange formalism [9, 10]. Since the contribution

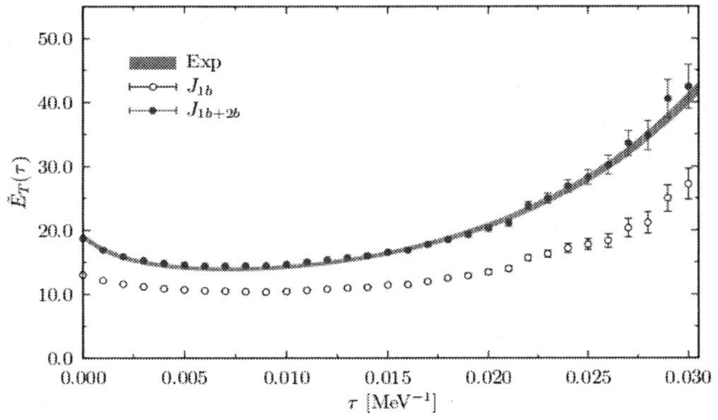

Figure 2. Euclidean electromagnetic transverse response function of ^{12}C at $q = 570$ MeV (adapted from Ref. [18]). The theoretical predictions obtained including only one-body and both one- and two-body transition operators are represented by open circles and solid circles, respectively. The shaded band refers to the Euclidean response extracted from Ref. [19].

of three-body currents associated with the three-body nuclear potential have been found to be generally negligible, we have not included these terms in our calculations. Nonrelativistic MEC have been used in analyses of a variety of electromagnetic moments and electroweak transitions of s- and p-shell nuclei at low and intermediate values of energy and momentum transfers [11–17], improving on the description of the experimental data with respect to the one-body approximation.

In the moderate momentum-transfer regime, the nuclear cross section can be rewritten in terms of the response functions $R_{\mu\nu}(q,\omega)$, obtained from Eq. (3) replacing the components of the current operator with their expressions obtained in the nonrelativistic limit. We have used the GFMC method to calculate the imaginary-time response function–the so-called Euclidean response function–to the electromagnetic and neutral weak currents of ^{4}He and ^{12}C. The Euclidean response function is defined as the Laplace transform of the response,

$$E_{\alpha\beta}(q,\tau) = C_{\alpha\beta}(q) \int_{\omega_{\rm th}}^{\infty} d\omega\, e^{-\tau\omega} R_{\alpha\beta}(q,\omega)\,, \tag{8}$$

where $\omega_{\rm th}$ is the inelastic threshold and the $C_{\alpha\beta}$ are q-dependent normalization factors. In $R_{\alpha\beta}(q,\omega)$ the ω-dependence enters via the energy-conserving δ-function and the dependence on the four-momentum transfer $Q^2 = q^2 - \omega^2$ of the electroweak form factors of the nucleon and N-to-Δ transition in the currents. Once the latter dependence has been removed, as described in Ref. [18], the Euclidean response can be expressed as a ground-state expectation value,

$$\frac{E_{\alpha\beta}(q,\tau)}{C_{\alpha\beta}(q)} = \frac{\langle 0|O_\alpha^\dagger(\mathbf{q})e^{-(H-E_0)\tau}O_\beta(\mathbf{q})|0\rangle}{\langle 0|e^{-(H-E_0)\tau}|0\rangle}\,. \tag{9}$$

In the above equation H is the nuclear Hamiltonian of Eq. (6) (here, the AV18+IL7 model), τ is the imaginary-time, and E_0 is a trial energy to control the normalization. The calculation of the above matrix element is carried out with GFMC methods [20] similar to those used in projecting out the exact ground state of H from a trial state.

It has to be noted that evaluating the matrix element of Eq. (9) for a nucleus as large as ^{12}C is computationally nontrivial, as it requires computer capabilities and resources available on only forefront computers.

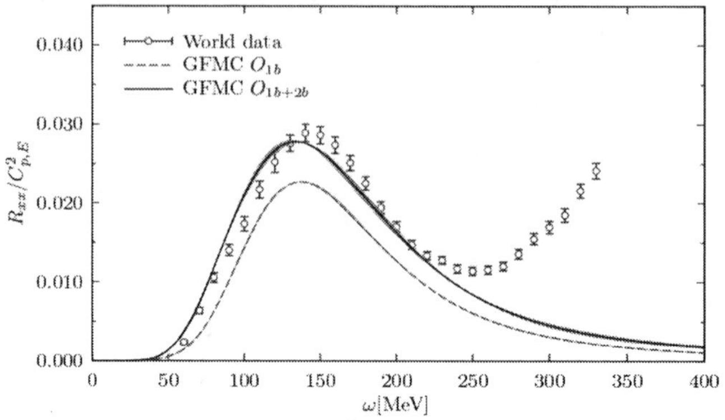

Figure 3. Electromagnetic transverse response functions of ^{4}He at q = 500 MeV. Experimental data are from Ref. [23].

In figure 2, taken from Ref. [18], the electromagnetic transverse Euclidean response function of ^{12}C, is compared to the one obtained from the analysis of the world data carried out by Jourdan [19] represented by the shaded band. Two-body MEC contributions substantially increase the one-body electromagnetic transverse response function. This enhancement is effective over the whole imaginary-time region we have considered, with the implication that excess transverse strength is generated by two-body currents not only at energies larger than the one corresponding to the quasi-elastic peak, but also in the quasi-elastic and threshold regions. The full predictions for the transverse Euclidean response functions is in excellent agreement with the experimental data.

The inversion of a Laplace transform subject to statistical Monte Carlo errors, needed to retrieve the energy dependence of the responses, is long known to involve severe difficulties. However, there are techniques developed in condensed matter theory and other contexts, that seem to have successfully overcome the inherent ill-posed nature of the problem. One of these is known as the *maximum entropy technique* [21, 22]; we have recently used it to perform stable inversions of the ^{4}He electromagnetic Euclidean response.

The ^{4}He electromagnetic transverse response function (at q = 500 MeV), obtained from inversion of the corresponding Euclidean response is shown in figure 3. The inversions are, to a very large degree, insensitive to the choice of default model response [18] needed by the maximum entropy method. Results obtained with one-body only (dashed line) and (one+two)-body (solid line) currents are compared with an analysis of the experimental world data [23] (empty circles). In the quasi-elastic region there is excellent agreement between the full theory and experiment. Two-body currents significantly enhance the transverse response function, not only in the dip region, but also in the quasi-elastic peak and threshold regions, providing the missing strength that is needed to reproduce the experimental results.

4 Relativistic regime

The dynamical model discussed in section 3 can be also employed to describe the nuclear response in the kinematical region of large momentum transfer, in which relativistic effects play a major role. In this regime, the final state includes at least one particle carrying a large momentum $\sim \mathbf{q}$, and fully relativistic expressions need to be retained for the transition currents.

The impulse approximation (IA) scheme and the spectral function formalism allow one to circumvent the difficulties associated with the relativistic treatment of the nuclear final state and current operator, while at the same time preserving essential features (such as correlations) inherent to the realistic description of nuclear dynamics described in the previous section. The IA scheme is based on the fact that in the kinematical region of large momentum transfer, in which $|\mathbf{q}|^{-1} \ll d$, with d being the average nucleon-nucleon separation distance in the target, the nuclear scattering can be reasonably assumed to reduce to the incoherent sum of elementary scattering processes involving individual nucleons.

The ground state of the target nucleus, which does not depend on $\mathbf{q}$, is still an eigenstate of the nuclear Hamiltonian of Eq. (6). Within the IA the final state factorizes

$$|X\rangle = |\mathbf{p}\rangle \otimes |n_{A-1}, \mathbf{p}_n\rangle \,. \tag{10}$$

In the above equation, the state $|\mathbf{p}\rangle$ is a plane wave describing a noninteracting nucleon, while $|n_{A-1}, \mathbf{p}_n\rangle$ is an eigenstate of the $(A-1)$ nuclear Hamiltonian of Eq. (6) describing the $(A-1)$-system recoiling with momentum $\mathbf{p}_n$. Using the above equation, it is possible to rewrite the nuclear transition matrix element of the one-body current between free nucleon states – which can be computed exactly, retaining the fully relativistic expressions of the currents – and the nuclear amplitude involving the target ground state and the state of the spectator $(A-1)$-system. By using Eq. (10) in Eq. (3), the nuclear cross section can be written in the form [5]

$$d\sigma_{IA} = \sum_i \int d^3k\, dE\, P_i(\mathbf{k}, E) d\sigma_i \,, \tag{11}$$

where $d\sigma_i$ is the cross-section describing scattering on the individual i-th nucleon, the momentum and removal energy of which are distributed according to the spectral function $P_i(\mathbf{k}, E)$. Highly accurate theoretical calculations of the spectral function can be carried out for uniform nuclear matter, exploiting the simplifications arising from translation invariance [24]. The results of these calculations have been combined with the information obtained from coincidence $(e, e'p)$ experiments at moderate energy and momentum transfer, to obtain spectral functions of a variety of nuclei within the local density approximation (LDA) [25].

Note that, because of the strong spatial-spin-isospin correlations that are present among nucleons in the nucleus, the recoiling $(A-1)$-nucleon system is not necessarily left in a bound, one hole, state. In fact, two hole-one particle states, in which one of the spectator nucleons is excited to the continuum, typically contribute $15-20\%$ of the spectral function normalization, the corresponding strength being located at large momenta and energies.

Neglecting the contributions of final states involving more than two nucleons in the continuum, the cross section can be written as

$$d\sigma = d\sigma_{1p1h} + d\sigma_{2p2h} \propto L^{\mu\nu}(W_{\mu\nu}^{1p1h} + W_{\mu\nu}^{2p2h}) \,. \tag{12}$$

Carrying out calculations of nuclear amplitudes combining fully relativistic MEC currents and a description of nuclear dynamics taking into account short range correlations requires a generalization of the factorization ansatz. The authors of Refs. [26, 27] have extended the factorisation ansatz of Eq. (10) to allow for a consistent treatment of the amplitudes involving one- and two-nucleon currents. The resulting expression is

$$|X\rangle = |\mathbf{p}\mathbf{p}'\rangle \otimes |m_{A-2}, \mathbf{p}_m\rangle \,, \tag{13}$$

where the states $|\mathbf{p}\mathbf{p}'\rangle$ and $|m_{A-2}, \mathbf{p}_m\rangle$ describe two non-interacting nucleons of momenta $\mathbf{p}$ and $\mathbf{p}'$ and the $(A-2)$-particle residual system, respectively. The two-particle two-hole contribution to the

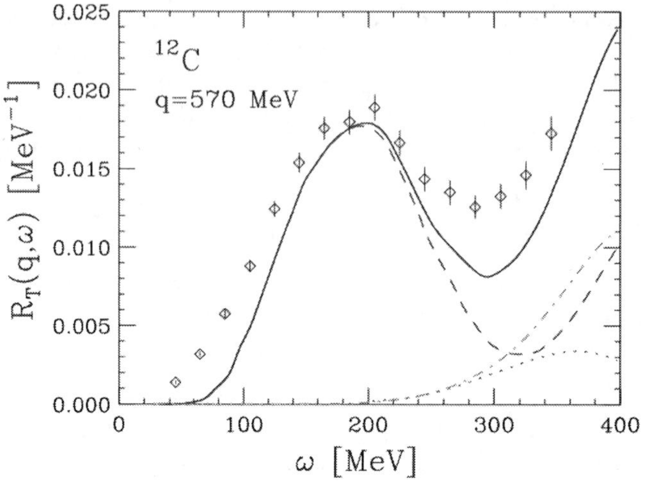

Figure 4. Electromagnetic transverse response of ^{12}C at momentum transfer $|\mathbf{q}| = 570$ MeV, taken from Ref. [27]. The solid line represents the results of the full calculation, whereas the dashed line has been obtained including only amplitudes involving the one-body current. The contributions arising from the two-nucleon current are illustrated by the dot-dashed and dotted lines, corresponding to the pure two-body current transition probability and the interference term, respectively. The experimental data are taken from Ref. [19].

hadronic tensor obtained from the extended factorisation ansatz is a sum of three terms,

$$W^{2p2h}_{\mu\nu} = W^{2p2h}_{\mu\nu\,11} + W^{2p2h}_{\mu\nu\,12} + W^{2p2h}_{\mu\nu\,22}, \tag{14}$$

where $W^{2p2h}_{\mu\nu\,11}$ and $W^{2p2h}_{\mu\nu\,22}$ involve the squared amplitudes of the matrix elements of one- and two-nucleon currents, while $W^{2p2h}_{\mu\nu\,12}$ describes the interference between the amplitudes involving one- and two- body currents [27].

Figure 4, taken from Ref. [27], shows the transverse electromagnetic response of ^{12}C at $|\mathbf{q}| =$ 570 MeV computed using the carbon spectral function of Ref. [25] and approximating the two-hole spectral function of carbon with that of uniform nuclear matter at density corresponding to Fermi momentum $k_F = 221$ MeV. The fully relativistic expression of the MEC described in Refs. [28, 29] was used, with the same form factors and Δ-width. The solid line represents the results of the full calculation, whereas the dashed line has been obtained including only the amplitudes involving the one-body current. The contributions arising from the MEC are illustrated by the dash-dot and dotted lines, corresponding to the pure two-body current transition probability and the interference term, respectively. The latter turns out to be sizable, its contribution being comparable to the total two-body current response for $\omega < 350$ MeV. It has to be noted that these results still need to be improved, as they do not include the corrections taking into account the effects of final state interactions. The data resulting from the analysis of [19] are also included for comparison.

A comparison between the results of figure 4 and the GFMC results of figure 3 shows distinctive discrepancies in both magnitude and energy dependence of the two-body current contributions. While part of the disagreement is likely to originate from using fully-relativistic and nonrelativistic MEC, as well as from the non relativistic nature of the GFMC calculations, the large interference contribution in the region of the quasi elastic peak observed in figure 3 may well arise from interference between amplitudes involving the one- and two-body currents and 1p1h final states.

Acknowledgements

The work of O.B. and N.R. was supported by INFN under grant MANYBODY. The work of AL, SCP, SG, JC and RS a was supported by the U.S. Department of Energy, Office of Science, Office of Nuclear Physics, under contracts DE-AC02- 06CH11357 (AL and SCP), DE-AC52-06NA25396 (SG. and JC), DE-AC05-06OR23177 (RS). The work of AL, SCP, SG and JC was also supported by the NUCLEI SciDAC program. Under an award of computer time provided by the INCITE program, this research used resources of the Argonne Leadership Computing Facility at Argonne National Laboratory, which is supported by the Office of Science of the U.S. Department of Energy under contract DE-AC02-06CH11357.

References

[1] LBNE Collaboration, C. Adams, D. Adams, T. Akiri, T. Alion, K. Anderson, C. Andreopoulos, M. Andrews, I. Anghel, J.C. Costa dos Anjos et al., ArXiv e-prints (2013), 1307.7335
[2] A. A. Aguilar-Areval et al. (MiniBooNE Collaboration), Phys. Rev. Lett. **100**, 032301 (2008)
[3] M. Martini, M. Ericson, G. Chanfray, J. Marteau, Phys. Rev. C **80**, 065501 (2009)
[4] J. Nieves, I. Ruiz Simo, M. Vicente Vacas, Phys. Lett. **B707**, 72 (2012), 1106.5374
[5] O. Benhar, D. Day, I. Sick, Rev. Mod. Phys. **80**, 189 (2008), nucl-ex/0603029
[6] M.H. Kalos, Phys. Rev. **128**, 1791 (1962)
[7] R. Grimm, R. Storer, Journal of Computational Physics **7**, 134 (1971)
[8] D.O. Riska, Phys. Rep. **181**, 207 (1989)
[9] L.E. Marcucci, R. Schiavilla, M. Viviani, A. Kievsky, S. Rosati, Phys. Rev. Lett. **84**, 5959 (2000)
[10] L.E. Marcucci, M. Viviani, R. Schiavilla, A. Kievsky, S. Rosati, Phys. Rev. C **72**, 014001 (2005)
[11] M. Pervin, S.C. Pieper, R.B. Wiringa, Phys. Rev. C **76**, 064319 (2007)
[12] L.E. Marcucci, M. Pervin, S.C. Pieper, R. Schiavilla, R.B. Wiringa, Phys. Rev. C **78**, 065501 (2008)
[13] R. Schiavilla, R.B. Wiringa, Phys. Rev. C **65**, 054302 (2002)
[14] L.E. Marcucci, M. Piarulli, M. Viviani, L. Girlanda, A. Kievsky, S. Rosati, R. Schiavilla, Phys. Rev. C **83**, 014002 (2011)
[15] R.B. Wiringa, R. Schiavilla, Phys. Rev. Lett. **81**, 4317 (1998)
[16] L.E. Marcucci, D.O. Riska, R. Schiavilla, Phys. Rev. C **58**, 3069 (1998)
[17] M. Viviani, R. Schiavilla, B. Kubis, R. Lewis, L. Girlanda, A. Kievsky, L.E. Marcucci, S. Rosati, Phys. Rev. Lett. **99**, 112002 (2007)
[18] A. Lovato, S. Gandolfi, J. Carlson, S.C. Pieper, R. Schiavilla, Phys. Rev. **C91**, 062501 (2015), 1501.01981
[19] J. Jourdan, Nuclear Physics A **603**, 117 (1996)
[20] J. Carlson, R. Schiavilla, Phys. Rev. Lett. **68**, 3682 (1992)
[21] R. Bryan, European Biophysics Journal **18**, 165 (1990)
[22] M. Jarrell, J. Gubernatis, Physics Reports **269**, 133 (1996)
[23] J. Carlson, J. Jourdan, R. Schiavilla, I. Sick, Phys. Rev. C **65**, 024002 (2002)
[24] O. Benhar, A. Fabrocini, S. Fantoni, Nuclear Physics A **505**, 267 (1989)
[25] O. Benhar, A. Fabrocini, S. Fantoni, I. Sick, Nuclear Physics A **579**, 493 (1994)
[26] O. Benhar, N. Rocco, Adv. High Energy Phys. **2013**, 912702 (2013)
[27] O. Benhar, A. Lovato, N. Rocco, Phys. Rev. **C92**, 024602 (2015), 1502.00887
[28] M.J. Dekker, P.J. Brussaard, J.A. Tjon, Phys. Rev. C **49**, 2650 (1994)

[29] A. De Pace, M. Nardi, W. Alberico, T. Donnelly, A. Molinari, Nuclear Physics A **726**, 303 (2003)

EPJ Web of Conferences **113**, 01011 (2016)
DOI: 10.1051/epjconf/201611301011

Short-range correlations and their implications for isospin-dependent modification of nuclear quark distributions

John Arrington[1,a]

[1] *Physics Division, Argonne National Laboratory, Argonne, IL 60439*

Abstract. The past decade has provided a much clearer picture of the structure of high-momentum components in nucleons, associated with hard, short-distance interactions between pairs of nucleons. Recent Jefferson Lab data on light nuclei suggest a connection between these so-called 'short-range correlations' and the modification of the quark structure of nucleons in the nuclear environment. In light of this discovery that the detailed nuclear structure is important in describing the nuclear quark distributions, we examine the potential impact of the isospin-dependent structure of nuclei to see at what level this might yield flavor-dependent effects in nuclear quark distributions.

1 Introduction

A small but important part of the Jefferson Lab (JLab) program involves studies of the high-momentum structure of nuclei, which requires high-energy interactions to probe cleanly [1]. After the initial observation of identical structure in the high-momentum components of nuclei at SLAC [2], measurements at JLab have mapped out the kinematic region where SRCs dominate [3, 4], mapped out the contribution of SRCs in various light and heavy nuclei relative to the deuteron [5, 6], and determined that the SRCs are dominated (at the ~90% level) by neutron-proton pairs [7–10].

Because these high-momentum components are associated with short-distance pairs, and thus high-density configurations in nuclei, it is natural to think that there could be a connection between SRCs and the nuclear dependence of nuclear parton distribution functions (pdfs). First it was noted [11] that the common scaling of SRCs and the EMC effect with density might suggest a connection between the two phenomena. Precise measurements of the EMC effect could be extended at JLab, taking advantage of the extended scaling in nuclei [12, 13]. Measurements of both the EMC effect [14] and SRCs [6] in light nuclei both showed that ^{9}Be deviated from the simple density dependence observed in heavier nuclei. While it seemed unlikely that the simple density-dependent or mass-dependent fits to data on heavy nuclei would apply to light nuclei, previous data on ^{4}He and ^{9}Be did not have sufficient precision to show deviations from these simple scaling models. For both the EMC effect and contribution of SRCs, ^{9}Be behaved like a rather dense nucleus, even though its average density is quite low [15, 16] due to its significant cluster structure. This led to more detailed examinations of the correlation, suggesting the possibility that the EMC effect may be related to either the large momenta or high density of SRCs in nuclei [17–19].

[a] e-mail: johna@anl.gov

Many calculations of the EMC effect include a binding contribution which explains a significant portion of the effect [20]. The average separation energy enters into these binding calculations, and a large part of this is associated with SRCs. It has also been argued that off-shell effects in these high-momentum nucleons could yield additional nucleon modification [21–24]. There have been direct measurements using electron scattering from a deuterium target, using a high-momentum spectator proton to tag scattering from high-momentum neutrons [25]. Existing data show effects consistent with such a modification of the neutron structure have been observed, but the interpretation is difficult due to potential final-state interactions. Future measurements are planned at Jefferson Lab [26]. More recently, such tagged measurements focusing on low-momentum nucleons have allowed for the extraction of the free neutron structure function [27, 28], and have been used to make the first direct extraction of the nuclear effects in the deuteron relative to the free proton and neutron [29]. At present, these measurements do not have sufficient precision or kinematic reach to differentiate between different models, but future measurements may provide further insight into the origin of the EMC effect [30]. One can also try to isolate SRCs by examining scattering at $x > 1$, while going to Q^2 values where one expects to be sensitive to the parton distributions in the region dominated by SRCs. Existing data suggest an onset of scaling at relatively low Q^2 for nuclei [12, 13, 31], and an experiment has been approved to extend such measurement to the maximum Q^2 available with the 12 GeV upgrade at Jefferson Lab [32].

Another channel for examining off-shell protons is the comparison of the proton form factors, extracted by recoil polarization, to the same polarization ratio for proton knockout from a nucleus [33, 34]. Again, questions about potential final-state interactions have been raised [35], but other polarization observables have some sensitivity to these effects [36] and future measurements will be able to map out the behavior in more detail [37]. For both the binding and off-shell contributions, the large momenta and energies, associated mainly with nucleons in short-range correlations, are the direct cause of the EMC effect, yielding the observed correlation between the EMC effect and presence of SRCs.

If the EMC effect is driven by the 'local density' observed by the struck nucleon, e.g. is due to direct quark exchange between nucleons with substantial overlap, then all short-distance configurations - nn, np, and pp pairs - will contribute. In this case, the presence of these short-distance configurations is the source of both the EMC effect (due to quark modification in the overlapping nucleons) and the high-momentum nucleons in SRCs (due to the strong interaction of the short-range components of the N-N potential. In this case, the EMC-SRC correlation occurs because both effects have a common origin in these small, dense, two-body configurations in nuclei. This implies a slightly different prediction for the detailed EMC-SRC correlation, as the EMC effect is sensitive to all short-distance NN pairs while SRCs are generated predominantly in np pairs. In this case, the connection between the EMC effect and SRCs in the nucleus will also depend on the number of potential nn, np, and pp pairs in the nucleus, implying an A- and Z-dependent correction to the simple linear correlation assumed in [11]. The detailed scaling of the EMC effect and SRCs were examined in Ref. [17], along with a comparison of the two pictures above, where the correlation is the result of either local-density or high-virtuality (HV) effects. Including a simple pair-counting correction to account for the difference in nn, np, and pp pairs, the correlation was found to be better (reduced χ^2 value and improved extrapolation to the deuteron value) under the local density hypothesis. However, the difference is not large and the χ^2 test is of somewhat limited use given the potential for correlated errors between the measurements on different nuclei, so it is far from a definitive test.

One new direction that has not yet been studied in detail is the flavor-dependence of the EMC effect. The connection between the EMC effect and presence of SRCs, combined with the isospin structure of SRCs suggests the potential for significant flavor-dependent effects in nuclear pdfs. This provides a new observable that can be used to examine the EMC effect and to test models of the nuclear modification. While this idea has been discussed [1, 38], exisint data on the EMC effect are been insufficient to claim an flavor-dependent effect.

2 Flavor dependence of the EMC effect

While the nature of the correlation between the EMC effect and the presence of SRCs is not understood, most explanations, conventional or exotic, imply that the isospin structure of the correlations would impact the flavor dependence of the EMC effect. In conventional binding calculations, the fact that the high-momenta nucleons are generated mainly in n-p pairs implies that a nucleus with significant neutron excess will have a larger *fraction* of it's protons at high momenta, yielding a larger separation energy and thus a larger binding effect for protons. The same is true if the correlation is due to off-shell effects in the high-momentum nucleons of the SRCs. If the EMC effect is driven by local-density effects, then a similar isospin dependence is natural in nuclei with a large neutron skin. Nuclei with significant neutron excess typically have a larger neutron radius, meaning that the lower density surface region will have more neutrons, again yielding a larger average EMC effect for protons. Thus, any of these explanations, as well as recent QCD-based models of nuclear parton distributions [39], all predict the same effect: an enhanced EMC effect for protons (and up-quarks) in neutron rich nuclei.

In the past, the EMC effect was generally assumed to scale with simple bulk properties of the nucleus, either mass or average nuclear density. The data on light nuclei [14] demonstrate that these models are not sufficient, and details of the nuclear structure must be accounted for. We examine here a variety of simple assumptions for the A dependence which account for the nuclear structure.

The observed EMC-SRC correlation has generally been interpreted based on the idea that the EMC effect is driven by either local density effects, where overlapping nucleon pairs can have direct quark exchange or contributions from more exotic states (hidden color or six-quark bags) [40–42], or by the presence of high-momentum nucleons, due to binding or off-shell effects. These yield slightly different predictions for the quantitative nature of the relation between the EMC effect and SRCs, as all NN pairs can contribute to high-density configurations, while the high-momentum components are dominated by np pairs [1], and will also yield different predictions for the relative effect on protons and neutrons in the nucleus. All of these yield an increased EMC effect for protons for nuclei with a neutron excess.

The same enhancement in the minority nucleons occurs in a calculation of nuclear pdfs through the quark-meson coupling arising in the NJL model [39, 43]. The model results are obtained for asymmetric nuclear matter, using a Z/N ratio that matches finite nuclei as a rough estimate for finite nuclei. The nuclear matter results are likely overestimates of the isospin dependence in finite nuclei, although these are mean-field calculations which do not include realistic nuclear structure for the finite nuclei. So for example, it does not include any effect connected to short-range correlations or the isospin-dependent density in a nucleus with a large neutron skin discussed above. Because all of the effects increase the EMC effect for up-quarks in neutron-rich nuclei, there could be an additional contribution if any of the nuclear structure effects are also present.

We examine a total of four updated scaling assumptions for the A dependence of the EMC effect:

- High-momentum fraction: We take the nucleon momentum distribution, $n(k)$, and assume that the EMC effect scales with the fraction of the distribution above 300 MeV/c.
- Average kinetic energy: We take the average kinetic energy of the nucleons, $\langle k^2 \rangle / 2M$, evaluated from $n(k)$.

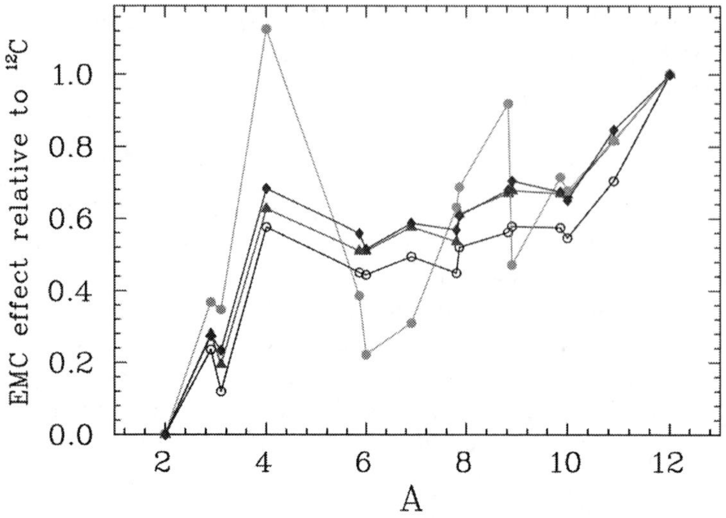

Figure 1. A dependence of the EMC effect for the four scaling assumptions: Fraction of momentum distribution above 300 MeV/c (black open circles), average kinetic energy (red triangles), average density (green circles), and probability to be within 1 fm of another nucleon (blue diamonds). The results are shown relative to Carbon, and a slight offset in A is added to differentiate nuclei with different N/Z ratios.

- Average density: We evaluate the average density of the nucleus as seen by an individual proton or neutron, using the one-body densities for the proton and neutrons
- Overlap probability: We determine the probability of a proton or neutron being within 1 fm of another nucleon, based on the two-body densities.

The first two quantities, high-momentum fraction and average kinetic energy are related to the idea that some combination of off-shell effects and the separation energy in conventional binding calculations are the source of the EMC effect through the presence of high-momentum nucleons. The average density is the traditional simple model used for scaling of the EMC effect in nuclei. The overlap probability is one method of examining the impact of local density, using the probability of two nucleons having significant overlap as an overall measure of the possible contribution to pdf modification due to the possible contribution from direct quark exchange between the nucleons or more exotic 6-quark or hidden color configurations. This set of quantities can be calculated from *ab initio* calculations of nuclear structure for light nuclei, providing predictions for the EMC effect in light nuclei, which will be measured in an approved experiment at Jefferson Lab [44]. We can also examine the difference between the EMC effect for protons and neutrons in asymmetric nuclei for each of these predictions, which can be examined in the comparison of ^{40}Ca and ^{48}Ca [44], as well as in parity-violating electron scattering measurements from asymmetric nuclei such as ^{9}Be and ^{48}Ca, where the parity-violating asymmetry, related to the ratio of weak to electromagnetic deep inelastic scattering, is insensitive to any flavor-independent rescaling of the nuclear pdfs.

2.1 Results

For each of the scaling models discussed in the previous section, we calculate the quantity of interest using the Quantum Monte Carlo calculation [45] for all particle-stable nuclei. The scaling parameter is determined independently for protons and neutrons, and for the conventional EMC effect, we combine these with a weight for the number of protons and neutrons and the relative e-p and e-n cross sections for deep-inelastic scattering at large x. The results are all plotted relative to ^{12}C, as only the A-dependence and not the overall scale is predicted by these simple scaling assumptions.

Figure 1 shows the prediction for the scaling of the EMC effect in light nuclei, taking the cross section and N/Z weighted average of the proton and neutron predictions. The assumption of a simple

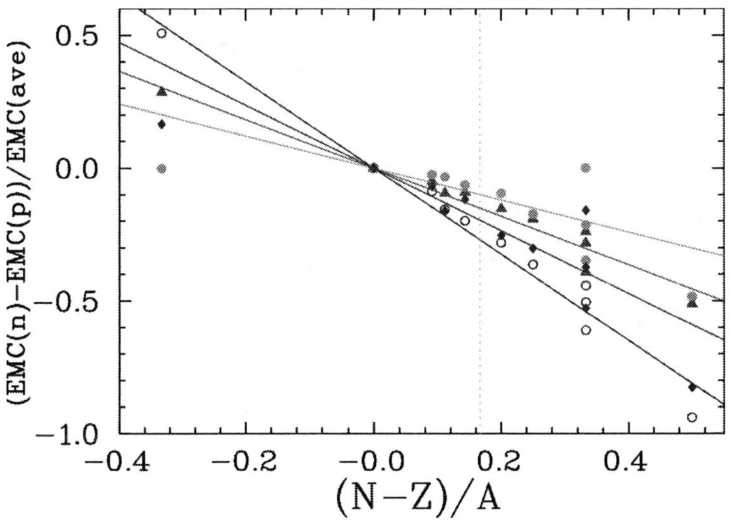

Figure 2. Isospin dependence of the EMC effect vs. fractional neutron excess of the nucleus for the four scaling models (same models as Fig. 1). The lines are simple unweighted linear fits. The short-dashed line shows the N/Z value corresponding to ^{48}Ca.

density-dependent scaling is significantly different from the others, and is ruled out by the existing data on ^{3}He, ^{4}He, ^{9}Be, and ^{12}C [14]. This is not surprising, as the simple density dependence picture was already excluded based on these data, but this demonstrates that accounting for the difference between proton and neutron distributions does not yield a significant improvement in the prediction. The other models all show a very similar A dependence for the EMC effect, and are qualitatively consistent with the observation that the modification is large and similar for ^{4}He, ^{9}Be, and ^{12}C. There is a small difference between the predictions, in particular for the first scaling model (fraction of high-momentum nucleons), but they are in generally good qualitative agreement.

Figure 2 shows the fractional difference between the proton and neutron EMC effect versus the fractional neutron excess of the nuclei. The lines are simple unweighted linear fits which provide a rough guide as to the overall size of the effects, but note that the behavior is not linear with fractional neutron excess, in particular for ^{3}H and ^{3}He. The predicted isospin dependence shows a significantly larger spread in the predictions for different nuclei, with very large differences between the EMC effect in protons and neutrons. The largest neutron excess shown in this figure are not easy to access experimentally while ^{3}H and ^{3}He, which have $|(N-Z)/A| = 1/3$, could be measured, but there are technical difficulties using a ^{3}H target and the sensitivity will be limited because the EMC effect in A=3 nuclei is small, making it difficult to extract even a relatively large fractional effect. Measurements on nuclei such as ^{9}Be and ^{48}Ca are straightforward, and these nuclei have a significant EMC effect as well as the potential to have significant sensitivity to different models via their isospin-dependence.

An approved experiment at Jefferson Lab [44] will measure the EMC effect for light nuclei (^{3,4}He, ^{6,7}Li, ^{9}Be, ^{10,11}B, and ^{12}C) to examine the sensitivity of the EMC effect to the detailed nuclear structure in these well-understood nuclei. In addition, it will have some sensitivity to the isospin dependence through the comparison of light non-isoscalar nuclei and the comparison of ^{40}Ca and ^{48}Ca. However, the EMC effect has limited to the isospin dependence, even where it is large. This is in because there are two effect which partially cancel in neutron rich nuclei. Increasing the number of neutrons increases the EMC effect for the protons (up quarks) which dominate the cross section for isoscalar nuclei, but it also increases the fraction of neutrons (down quarks) which have a decreased EMC effect. Thus, the sensitivity to a change in N/Z is enhanced when going to proton-rich nuclei and suppressed for neutron-rich nuclei, as seen in the calculations of Ref. [43].

A much more sensitive test of the isospin dependence of the EMC effect is party-violating deep-inelastic electron scattering from non-isoscalar nuclei [39]. The parity-violating asymmetry is sensitive to the interference between Z- and photon-exchange amplitudes relative to the photon-exchange amplitude squares [46, 47], making it sensitive to the ratio of the weak charge to electromagnetic charge of the quarks. From Ref. [39], the a_2 parity-violating asymmetry has the following simplified form when expanding around the $u(x) = d(x)$ limit:

$$a_2(x) = \frac{9}{5} - 4\sin^2(\theta_W) - \frac{12}{25}\frac{u(x) - d(x)}{u(x) + d(x)}. \tag{1}$$

An isoscalar nucleus will have a constant a_2 asymmetry, while a non-isoscalar nucleus will have a small x dependence which is independent of any isospin-independent EMC effect. The asymmetry expected from any nucleus in the absence of a flavor-dependent EMC effect can therefore be calculated with minimal uncertainty, based on the small difference between the up- and down-quark distributions in the excess protons or neutrons. This is relatively well known at low-to-modest values of x [27, 48–51], with further measurements planned to improve our knowledge at larger x values. Any deviation of the measured asymmetry from the predicted value is sensitive to an isospin-dependent EMC effect that would modify the ratio of up and down quarks [39].

3 Conclusions

Given what we have learned about the isospin structure of short-range correlations, combined with the importance of detailed nuclear structure for both SRCs and the EMC effect, it is now difficult to believe that the EMC effect could be flavor independent, in particular for non-isoscalar nuclei. As such, a detailed understanding of the EMC effect requires an understanding of the effect in terms of both A-dependent and isospin-dependent effects.

Measurements of the EMC effect in light nuclei (including several non-isoscalar nuclei), along with comparisons of ^{3}H and ^{3}He, along with ^{40}Ca and ^{48}Ca will provide some sensitivity to flavor-dependent effects. The comparison of ^{3}H and ^{3}He depends on having reliable understanding of the d/u ratio of the nucleon pdf, and is limited in sensitivity by the small size of the nuclear effects for A=3. The comparison of the Calcium isotopes should provide more sensitivity to isospin-dependent effects, but relies on comparison to a theoretical expectation for the difference between the EMC effect for A=40 and A=48 in the absence of isospin-dependent effects. This was assumed to be well understood based on earlier assumptions that the EMC effect had a simple scaling with the nuclear mass or average density. Because the recent data on light nuclei [14] demonstrated the impact of more complicated aspects of nuclear structure, including both clustering and isospin-structure, the comparison may be more sensitive to the assumed scaling in heavy nuclei than previously assumed.

Pion-induced Drell-Yan scattering has also been proposed as a way to study the flavor dependence of the EMC effect [52]. Such measurements would be possible at COMPASS or J-PARC, but there are no formal plans to make these measurements at present. The cleanest and most precise way to isolate flavor-dependent effects on the nuclear parton distributions is through measurements of the parity-violating DIS scattering from nuclei [39]. The initial estimates for the isospin dependence of the EMC effect presented here, along with similar calculations for medium-heavy nuclei, will allow for an optimization of the program, providing a separation of the A-dependence and isospin-dependence of the EMC effect. This will also provide the best approach for discriminating between different assumptions about the physics behind the EMC effect, as illustrated in Fig. 2.

Acknowledgements

This work was supported by the U.S. Department of Energy, Office of Science, Office of Nuclear Physics, under contract DC-AC02-06CH11357. The author thanks R. Wiringa, I. Cloet, S. Riordan, P. Souder, and R. Beminiwattha for useful discussions.

References

[1] J. Arrington, D. Higinbotham, G. Rosner, M. Sargsian, Prog. Part. Nucl. Phys. **67**, 898 (2012)
[2] L.L. Frankfurt, M.I. Strikman, D.B. Day, M. Sargsian, Phys. Rev. C **48**, 2451 (1993)
[3] K.S. Egiyan et al. (CLAS), Phys. Rev. C **68**, 014313 (2003)
[4] K.S. Egiyan et al. (CLAS), Phys. Rev. Lett. **96**, 082501 (2006)
[5] J. Arrington et al., Phys. Rev. C **64**, 014602 (2001)
[6] N. Fomin et al., Phys. Rev. Lett. **108**, 092502 (2012)
[7] R. Shneor et al., Phys. Rev. Lett. **99**, 072501 (2007)
[8] R. Subedi et al., Science **320**, 1476 (2008)
[9] I. Korover et al., Phys.Rev.Lett. **113**, 022501 (2014)
[10] O. Hen et al., Science **346**, 614 (2014)
[11] L. Weinstein, E. Piasetzky, D. Higinbotham, J. Gomez, O. Hen, R. Shneor, Phys. Rev. Lett. **106**, 052301 (2011)
[12] J. Arrington et al., Phys.Rev.Lett. **82**, 2056 (1999)
[13] J. Arrington, R. Ent, C.E. Keppel, J. Mammei, I. Niculescu, Phys. Rev. C **73**, 035205 (2006)
[14] J. Seely et al., Phys. Rev. Lett. **103**, 202301 (2009)
[15] J. Carlson et al. (2014), 1412.3081
[16] Z.T. Lu et al., Rev.Mod.Phys. **85**, 1383 (2013)
[17] J. Arrington et al., Phys. Rev. C **86**, 065204 (2012)
[18] O. Hen, D. Higinbotham, G. Miller, E. Piasetzky, L. Weinstein, Int. J. Mod. Phys. E. 22, **1330017** (2013)
[19] S. Malace, D. Gaskell, D.W. Higinbotham, I. Cloet, Int.J.Mod.Phys. **E23**, 1430013 (2014)
[20] S.A. Kulagin, R. Petti, Nucl. Phys. **A765**, 126 (2006)
[21] L.L. Frankfurt, M.I. Strikman, Phys. Rept. **160**, 235 (1988)
[22] F. Gross, S. Liuti, Phys. Rev. **C45**, 1374 (1992)
[23] W. Melnitchouk, A.W. Schreiber, A.W. Thomas, Phys. Lett. **B335**, 11 (1994)
[24] G.V. Dunne, A.W. Thomas, Phys. Rev. **D33**, 2061 (1986)
[25] A.V. Klimenko et al. (CLAS), Phys. Rev. **C73**, 035212 (2006)
[26] S. Gilad, O. Hen, L. Weinstein, S. Wood, spokespersons, Jefferson lab experiment E12-11-107
[27] N. Baillie et al. (CLAS), Phys. Rev. Lett. **108**, 142001 (2012), [Erratum: Phys. Rev. Lett.108,199902(2012)]
[28] S. Tkachenko et al. (CLAS), Phys. Rev. **C89**, 045206 (2014), [Addendum: Phys. Rev.C90,059901(2014)]
[29] K.A. Griffioen et al., Phys. Rev. **C92**, 015211 (2015)
[30] S. Bueltmann, M.E. Christy, H. Fenker, K. Griffioen, C. Keppel, S. Kuhn, W. Melnitchouk, V. Tvaskis, spokespersons, Jefferson lab experiment E12-06-113
[31] N. Fomin et al., Phys. Rev. Lett. **105**, 212502 (2010)
[32] J. Arrington, D. Day, N. Fomin, P. Solvignon-Slifer, spokespersons, Jefferson lab experiment E12-06-105

[33] S. Strauch et al. (Jefferson Lab E93-049), Phys. Rev. Lett. **91**, 052301 (2003)
[34] M. Paolone et al., Phys. Rev. Lett. **105**, 072001 (2010)
[35] R. Schiavilla, O. Benhar, A. Kievsky, L.E. Marcucci, M. Viviani, Phys. Rev. Lett. **94**, 072303 (2005)
[36] S.P. Malace et al., Phys. Rev. Lett. **106**, 052501 (2011)
[37] E. Brash, G. Huber, R. Ransome, S. Strauch, spokespersons, Jefferson lab experiment E12-11-002
[38] M.M. Sargsian, AIP Conf. Proc. **1560**, 480 (2013), arXiv:1209.2477
[39] I. Cloet, W. Bentz, A. Thomas, Phys.Rev.Lett. **109**, 182301 (2012)
[40] M.M. Sargsian et al., J. Phys. **G29**, R1 (2003)
[41] C.E. Carlson, T.J. Havens, Phys. Rev. Lett. **51**, 261 (1983)
[42] G.A. Miller, Phys. Rev. **C89**, 045203 (2014)
[43] I.C. Cloet, W. Bentz, A.W. Thomas, Phys. Rev. Lett. **102**, 252301 (2009)
[44] J. Arrington, A. Daniel, D. Gaskell, spokespersons, Jefferson lab experiment E12-10-008
[45] R. Wiringa, R. Schiavilla, S.C. Pieper, J. Carlson, Phys.Rev. **C89**, 024305 (2014)
[46] D. Wang et al. (PVDIS), Nature **506**, 67 (2014)
[47] D. Wang et al., Phys. Rev. **C91**, 045506 (2015)
[48] J. Arrington, F. Coester, R. Holt, T.S. Lee, J. Phys. **G36**, 025005 (2009)
[49] A. Accardi et al., Phys. Rev. D **81**, 034016 (2010), 0911.2254
[50] A. Accardi et al., Phys. Rev. D **84**, 014008 (2011)
[51] J. Arrington, J. Rubin, W. Melnitchouk, Phys. Rev. Lett. **108**, 252001 (2012)
[52] D. Dutta, J.C. Peng, I.C. Cloet, D. Gaskell, Phys. Rev. C **83**, 042201 (2011)

EPJ Web of Conferences **113**, 01012 (2016)
DOI: 10.1051/epjconf/201611301012

Experimental progress and status on nucleon electromagnetic form factors

H. Gao[1,2,a], C. Peng[1,b], and Z.W. Zhao[1,c]

[1] *Duke University, Durham, NC27708, U.S.A.*
[2] *Duke Kunshan University, Kunshan, Jiangsu, China 215316*

Abstract. Nucleon electromagnetic form factors are the fundamental quantities that are related to a two-dimensional view of the charge and magnetization distribution in the nucleon. They are essential in understanding the nucleon electromagnetic structure. In this article, we briefly summarize the recent experimental progress on this topic, and introduce several experiments related to the nucleon electromagnetic form factors in the near future.

1 Introduction

Nucleons are known as the fundamental building blocks of the visible matter. In 1933, Otto Stern discovered that the magnetic momentum of proton disagreed with Dirac's prediction $\mu = \frac{e}{2M}\hbar$ for a structureless spin-$\frac{1}{2}$ particle. The discovery firstly indicated the internal structure of nucleons. In 1950s, the electron-proton elastic scattering experiments by Hofstadter and others confirmed the spatial charge distribution of the nucleon. A decade later, the experimental efforts on deep-inelastic scattering of electrons on protons pioneered by Friedman, Kendall and Taylor at Stanford Linear Accelerator Center (SLAC) revealed the point-like constituent particles inside the proton, and led to the establishment of the quark model.

Quantum Chromodynamics (QCD) is the fundamental theory that describes the strong interaction between quarks and gluons. Although it was well tested in the high energy region, where the quarks are asymptotically free and perturbative QCD is applicable, the predictions and tests regarding the confinement and hadron structure in the non-perturbative regime are extremely challenging. Thus, the experimental efforts in revealing the internal structure of nucleons are essential for understanding QCD and the fundamental interactions at the level of quarks and gluons.

Among fundamental quantities related to the internal structure of the nucleon, the electromagnetic form factors describe the spatial distribution of charge and magnetization within the nucleon. The recent measurements of nucleon form factors at high Q^2 inspired many topics regarding the nucleon structure: the possible relations between G_E^p/G_M^p and the quark orbital angular momentum [1], the behaviors of individual quark-flavor contributions to the nucleon form factors [2], and the role of form factors in the unified view of nucleon structure based on the Wigner distributions. Besides, the

[a] e-mail: gao@phy.duke.edu
[b] e-mail: cp121@phy.duke.edu
[c] e-mail: zwzhao@jlab.org

so-called "proton charge radius puzzle" [3] has raised many interests to re-evaluate the proton charge radius value extracted from the electric form factor of the proton at low Q^2 region and to carry out new experiments to access previously unexplored lower Q^2 region.

The rest of this paper will introduce the experimental techniques and the recent experimental progress related to the nucleon form factors, as well as provide a perspective on the future experiments regarding this topic. This proceedings paper focuses on nucleon form factors in the space-like region only.

2 Experimental techniques

2.1 Rosenbluth separation

In the one-photon exchange approximation, the differential cross section of unpolarized elastic electron-proton scattering can be written as the Rosenbluth formula [4]:

$$\frac{d\sigma}{d\Omega} = \frac{\sigma_M}{\epsilon}\frac{\tau}{1+\tau}\left(\frac{\epsilon}{\tau}G_E^{p\,2} + G_M^{p\,2}\right), \text{ and } \sigma_{\rm M} = \frac{\alpha {\rm E}'\cos^2\frac{\theta}{2}}{4{\rm E}^3\sin^4\frac{\theta}{2}}, \tag{1}$$

,where $\tau = Q^2/(4M_p^2)$, $\epsilon = \left(1 + 2(1+\tau)\tan^2\frac{\theta}{2}\right)^{-1}$ is the virtual photon longitudinal polarization, and σ_M is the Mott cross section describing the scattering from a point-like target.

In Rosenbluth method, the proton electric and magnetic form factor squared $G_E^{p\,2}$ and $G_M^{p\,2}$ are separated by measuring the cross-section with varied beam energies and electron scattering angles but fixed Q^2. The reduced cross section $\sigma_R = \frac{\epsilon}{\tau}G_E^{p\,2} + G_M^{p\,2}$ at fixed Q^2 only depends on ϵ (or the scattering angle θ). $G_E^{p\,2}$ and $G_M^{p\,2}$ hence can be extracted as the slope and the intercept of $\sigma_R(\epsilon)$, respectively. Since the contribution from G_M term dominates the cross-section in the high Q^2 region, while G_E term becomes dominant as the Q^2 decreases, it is difficult to extract the G_E at large Q^2 and the G_M at low Q^2 using Rosenbluth method.

2.2 Recoil proton polarization measurement

Since the elastic scattering cross-section is much less sensitive to the electric form factor at high Q^2, the relative uncertainties of the G_E^p data determined by the Rosenbluth method increases drastically at $Q^2 > 1$ GeV^2. Nevertheless, experimental techniques using the polarization degrees of freedom in electron scattering provide an access to the interference term of the form factor $G_E G_M$, which does not suffer from the Q^2 dependent sensitivity. As one of such methods, the experiment based on the elastic scattering of longitudinally polarized electrons from unpolarized protons can obtain the ratio of form factors G_E/G_M by simultaneously measuring the transverse (P_l) and longitudinal (P_t) polarizations of the recoil proton:

$$\frac{G_E^p}{G_M^p} = -\frac{P_t}{P_l}\frac{E+E'}{2M}\tan\frac{\theta}{2}. \tag{2}$$

The recoil proton polarizations are extracted from the azimuthal distribution after recoil protons scattering on certain analyzer material, which is typically carbon. The experimental observable from recoil proton polarization measurement directly determines the ratio of form factors. It does not require several measurements at the same Q^2, and thus significantly reduces the systematical errors regarding the change of the kinematic settings. This method cancels many factors that would introduce experimental uncertainties, such as the beam polarization and luminosity, the detector efficiency, and the analyzing power of proton polarimeter. Combining cross-section measurements and recoil proton polarization measurements, one can precisely determine both form factors regardless of their sensitivities to Q^2.

2.3 Asymmetry super ratio method

Another method to extract the ratio of proton form factors is to measure the asymmetry from the elastic scattering of longitudinally polarized electrons from polarized protons. The experimental asymmetry can be written as:

$$A_{\text{exp}} = P_b P_t \frac{-2\tau\nu_{T'}\cos\theta^* {G_M^p}^2 + 2\sqrt{2\tau(1+\tau)}\nu_{TL'}\sin\theta^*\cos\phi G_M^p G_E^p}{(1+\tau)\nu_L {G_E^p}^2 + 2\tau\nu_T {G_M^p}^2}, \tag{3}$$

where θ^* and ϕ^* are the polar and azimuthal angles of the proton polarization with respect to the three-momentum transfer of the virtual photon and the scattering plane, P_b and P_t are beam and target polarizations, respectively, $\nu_{T'}$, $\nu_{TL'}$, ν_T and ν_L are kinematic factors [5]. The ratio of proton form factors is thus determined by the "super ratio" when one uses a symmetric detector system [6]:

$$R_A = \frac{A_1}{A_2} = \frac{a_1 - b_1 G_E^p/G_M^p}{a_2 - b_2 G_E^p/G_M^p}. \tag{4}$$

A_1 and A_2 are the measured asymmetries in the two symmetric detectors with the proton polarization fixed in the laboratory frame. This technique also benefits from the cancellation of detector efficiency, beam and target polarizations, and luminosity. Its systematic uncertainties are different with those of the recoil polarization measurements.

2.4 Neutron form factor measurement

The measurement of neutron form factors is challenging due to the lack of free neutron targets. Therefore, unpolarized elastic or quasi-elastic electron-deuteron scattering, coupled with theoretical deuteron calculations, is employed to extract the neutron form factors. However, a strong model dependence is involved with this technique, and the precision is worse in comparison to the proton form factors measurements.

The direct measurement of neutron electric form factor G_E^n is even more difficult from the unpolarized cross-section measurements, due to the fact that the contribution from magnetic form factor G_M^n dominates the cross-sections. Alternatively, the ratio of neutron form factor is extracted from the polarization observables. The experimental technique can be either achieved by measuring the recoil neutron polarization from the process $d(\vec{e}, e'\vec{n})$, or by measuring the spin-dependent asymmetry from the process $\vec{d}(\vec{e}, e'n)$.

An alternative target for the spin-dependent quasi-elastic scattering measurements is polarized ^{3}He. As an effective neutron target, the ground state of ^{3}He is dominated by the S wave, in which the spins of two protons cancel out, and the unpaired neutron carries the ^{3}He nuclear spin. The polarized ^{3}He target is typically utilized in two types of experiments: inclusive $^3\vec{\text{He}}(\vec{e}, e')$ measurement in which only the scattered electron is detected, and coincidence $^3\vec{\text{He}}(\vec{e}, e'n)$ measurement in which both the scattered electron and knockout neutron are detected to minimize the asymmetry contribution from protons.

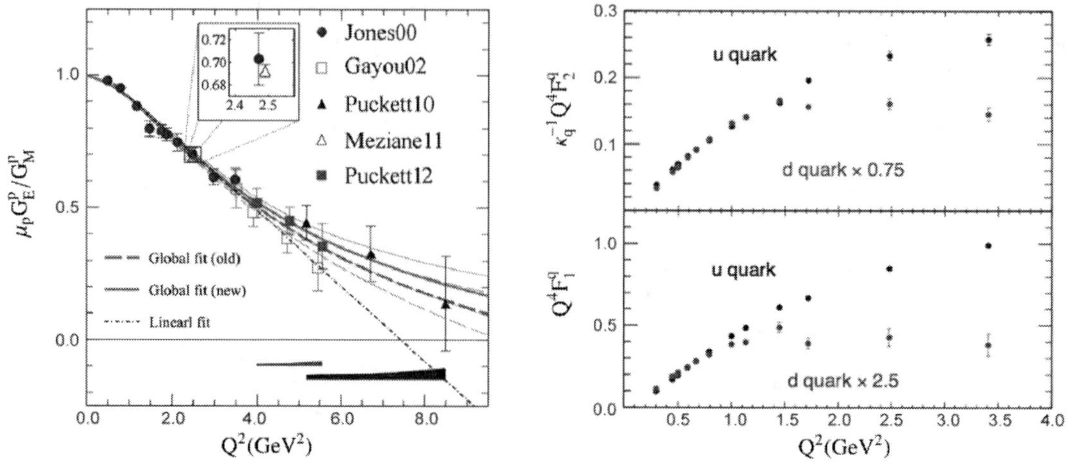

Figure 1. The left plot is the proton electromagnetic form factor ratio $\mu_p G_E^p/G_M^p$ from [9]. The right plot is the discomposed up and down quark contributions to the proton form factors from [2].

3 Progress on nucleon form factors

3.1 Relation to quark orbital angular momentum

The prediction of nucleon form factors at large Q^2 values was given by Brodsky and Farrar [7] based on the dimensional counting rules, implying the Dirac form factor $F_1^p \propto Q^{-4}$, and the Pauli form factor F_2^p has an additional factor of Q^{-2} due to helicity conservation. Thus, $Q^2 F_2^p/F_1^p$ (or G_E^p/G_M^p) should become constant at sufficiently large Q^2. This prediction was then confirmed with the perturbative QCD analysis given by Brodsky and Lepage [8]. However, as shown in Fig 1 (left), the current experimental data of the electromagnetic form factors ratio G_E^p/G_M^p decrease as Q^2 gets larger.

The scaling factor was then revisited by the pQCD analysis of nucleon's Pauli form factor [10] considering both leading- and subleading-twist light-cone wave functions, the latter describes quarks with one unit of orbital angular momentum. Its result predicts that $F_2^p/F_1^p \sim \ln^2(Q^2/\Lambda^2)/Q^2$, where Λ is a soft scale parameter related to the nucleon size. With a selection of $\Lambda = 0.2 - 0.3$ GeV^2, this prediction has a good agreement with the experimental data at $Q^2 > 1$ GeV^2.

Many relativistic constituent quark models (rCQM) [11–16] predict that G_E^p/G_M^p decreases with the increase of Q^2. This is related to the fact that these models, by imposing Poincaré invariance, lead to substantial violation of hadron helicity conservation, i.e., they include non-zero quark orbital angular momentum. A calculation based on Dyson Schwinger Equation (DSE) and Faddeev equation from Argonne [17] also predicts the decreasing behavior of G_E^p/G_M^p over Q^2. The calculation involves many interesting features, such as the dynamically generated quark mass, the contribution of quark orbital angular momentum, and the incorporation of diquark degrees of freedom.

3.2 Flavor decomposition of nucleon form factors

With experimental data on both G_E^p and G_E^n, the contributions from individual quark flavor can be extracted by assuming the charge asymmetry. Such a flavor decomposition was recently carried out for Q^2 values up to 3.4 GeV^2 [2]. The relations between nucleon form factors and the up or down quark contributions are:

$$F^u_{1(2)} = 2F^p_{1(2)} + F^n_{1(2)}, \text{ and } \mathrm{F}^d_{1(2)} = 2\mathrm{F}^n_{1(2)} + \mathrm{F}^p_{1(2)}, \tag{5}$$

where $F_{1(2)}$ denotes the Dirac (Pauli) form factor. As shown in Figure 1 (right), the d-quark contribution seems to scale as Q^{-4} at $Q^2 > 1$ GeV^2, while the u-quark contribution times Q^4 continually increases in the available data range. Several calculations describe the existing data reasonably well, including the DSE/Faddeev calculation [17] in which the diquark degrees of freedom accounts for the decreased ratio F_1^d/F_1^u and F_2^d/F_2^u at large Q^2, and a quark-diquark model with a pion cloud [18], as well as parameterizations of Generalized Parton Distributions (GPDs) [19, 20]. The future experiment [21] will allow the flavor decomposition at Q^2 up to 10 GeV^2, where the behaviors of the individual quark contributions as well as the calculations will be further tested.

3.3 Unified view of nucleon structure

Together with the Parton Distribution Functions (PDFs), the nucleon form factors have provided important information about the structure of nucleon. While a more comprehensive understanding of the nucleon structures, especially the orbital angular momentum of quarks and gluons, require information from multi-dimensions. An unified view of nucleon structure is possible by introducing the Wigner distributions $W_p^u(x, k_T, \vec{r}_T)$ [22], though its notion contradicts the uncertainty principle that forbids the simultaneous determination of momentum and position. However, by integrating the Wigner distributions over the parton transverse momentum k_T or position r_T, one can obtain the three-dimensional Generalized Parton Distributions (GPDs) or the Transverse Momentum Dependent Parton Distributions (TMDs), which provides key information on the spatial and/or momentum distribution of the partons inside a nucleon. In addition, the PDFs and form factors are also encompassed in the Wigner distributions. Fourier transformations of the nucleon electromagnetic form factors which describe the nucleon's transverse spatial distribution of charge and magnetization can be obtained by integrating the GPDs over Bjorken scaling variable x.

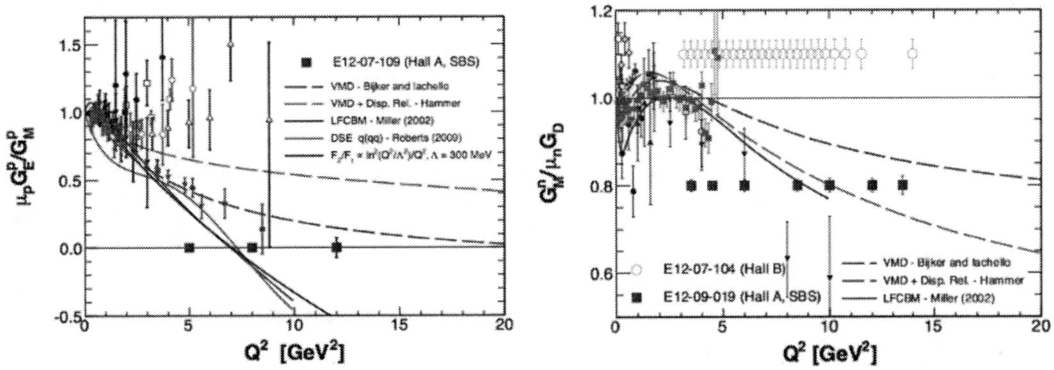

Figure 2. The projections of E12-07-109 [25] and E12-09-019 [26] experiments proposed in Hall A that will utilize the Super Bigbite.

3.4 Upcoming 12 GeV program at Jefferson Lab

The Continuous Electron Beam Accelerator Facility (CEBAF) at Jefferson Lab is under an energy upgrade to 12 GeV [23], which is expected to be completed in 2017. The upgraded instrumentation enables new experiments that explore the nucleon structures in an unprecedented kinematic coverage. Among those experiments, the Super Bigbite (SBS) projects proposed in Hall A [24] aim to precisely determine the nucleon form factor ratios G_E^p/G_M^p, G_E^n/G_M^n and G_M^n/G_M^p at Q^2 up to 10 – 14.5 GeV^2, which is far greater than the existing data. The Super Bigbite itself refers to a set of components which can be configured to fulfill the requirements of the specified measurement on the form factor G_E^n, G_M^n and G_E^p. By utilizing the Gas Electron Multiplier (GEM) technology, all the configurations of SBS have several similar advantages compared with the existing spectrometers: large geometry acceptance, capabilities to detect high rate events and to reach forward angles. Figure 2 shows the projections for two of the proposed SBS experiments.

4 Proton radius puzzle and PRad experiment

4.1 Proton radius puzzle

In the low Q^2 region, the "proton radius puzzle" recently triggered major efforts to reevaluate our understanding of the proton charge radius, which was thought as a well known fundamental quantity. Two measurements of the muonic hydrogen Lamb shift at PSI [27, 28] extracted the most precise value on the proton charge radius, but they were 7σ smaller than the value of CODATA 2012 compilation [29], which is a weighted average of world data from hydrogen spectroscopic measurements and elastic electron-proton scattering experiments. Figure 3 shows the discrepancy between PSI values and other experiment or analysis results.

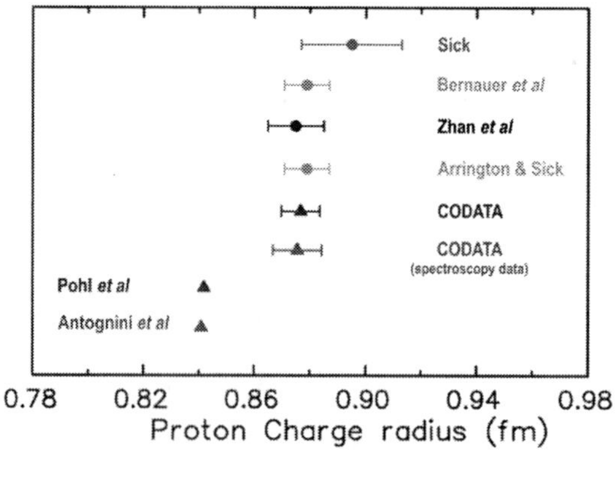

Figure 3. A comparison between the proton charge radius from PSI, CODATA, as well as the elastic *ep* scattering measurements [30–32] and analyses [33, 34].

This puzzle motivated intensive revisit of the QED calculations of the muonic hydrogen Lamb shift, including the evaluation of the proton polarizability contribution [35–37], yet no terms that can account for this discrepancy were found. The refined calculations were summarized in [38, 39]. New physics that may explain the puzzle was also explored, such as new particles [40, 41], new parity violating muonic force [42], and quantum gravity at the Fermi scale [43], just to name a few. In addition, some explanations in terms of the dispersion relations [44] or frame transformation [45, 46] were proposed to resolve the puzzle, but they remained controversial or unconfirmed. After all, the understanding of the "proton radius puzzle" awaits for input from new dedicated experiments.

4.2 Recent proton charge radius measurements

The absolute frequency of H energy levels has been measured with an accuracy of 1.4 part in 10^{14} via comparison with an atomic cesium fountain clock as a primary frequency standard. Such a precise measurement with the state-of-the-art QED calculations that include corrections for the finite size of the proton can indirectly determine the proton charge radius. The precision of this method is significantly enhanced by measuring the Lamb shift of muonic hydrogen. The contribution from the proton finite size term to the Lamb shift is much larger in muonic hydrogen than that in atomic hydrogen due to the fact that the muon mass is about 206 times of the electron mass. Such measurement was achieved at PSI in 2010 [27], and its result was reinforced by the same group in 2013 [28], which reported that $r_p = 0.84087 \pm 0.00039$. These measurements obtained the most precise value of the proton charge radius; its uncertainty was less than 0.05%. However, the PSI value is 4% less than the CODATA value, as well as the results from the recent *ep* scattering measurements.

Alternatively, the proton charge radius can be directly determined based on the elastic *ep* scattering experiments. Since the proton electric form factor G_E^p can be interpreted as the Fourier transform of the spatial charge distribution, the proton charge radius is extracted by fitting the slope of $G_E^p(Q^2)$ to $Q^2 \to 0$:

$$\langle r^2 \rangle = -6 \left. \frac{dG_E^p\left(Q^2\right)}{dQ^2} \right|_{Q^2=0} . \tag{6}$$

Thus, the G_E^p data at low Q^2 are essential for this method.

The recent experiment at Mainz [30, 31] measured the cross sections of elastic *ep* scattering within $Q^2 = 0.003 - 1.0$ GeV^2. Three spectrometers were utilized to cover a large amount of overlapping data sets. The statistical uncertainties of the measured cross-sections are lower than 0.2%. Both the electric and magnetic form factors of the proton were extracted from fitting the cross-section data. The recoil proton polarization experiment at Jefferson Lab [32] measured the elastic form factor ratio $\mu_p G_E/G_M$ of proton for $Q^2 = 0.3 - 0.7$ GeV^2. They extracted the proton charge radius by fitting the global data at $Q^2 < 0.5$ GeV^2. The two experimental results are consistent with each other. However, they are $3 - 5\sigma$ larger than the PSI value.

4.3 PRad experiment at Jefferson Lab

To investigate the proton charge radius puzzle, a precise experiment based on the *ep* elastic scattering with totally different systematic errors is necessary to check the result from previous *ep* scattering experiments. Thus, the proton charge radius (PRad) experiment is approved with A rating in Hall B, at Jefferson Lab.

The PRad experiment aims to precisely determine the proton charge radius based on the cross-sections measurement of elastic *ep* scattering with a calorimetry method. As shown in Figure 4, the experiment utilizes a high resolution, high efficiency calorimeter (HyCal) as the main detector. The calorimetric setup enables the experiment to reach the very forward angles and thus the very low Q^2 region $\left(\geq 2 \times 10^{-4} \text{ GeV}^2\right)$ that was never achieved before by magnetic spectrometer based measurements. As supplemental detectors, two GEM chambers at the upstream of HyCal can significantly increase the position resolution and reduce the systematic errors in the Q^2 determination. The hydrogen target used in PRad is a windowless gas-flow target. By removing the target windows, this experiment will not suffer from the typical background source of all previous *ep* experiments in order to reach the most forward scattering angles. The scattered electron from the elastic *ep* scattering and Møller scattering will be simultaneously detected. The cross-sections of elastic *ep* scattering hence will be normalized to those of the Møller process, which can be precisely calculated in QED.

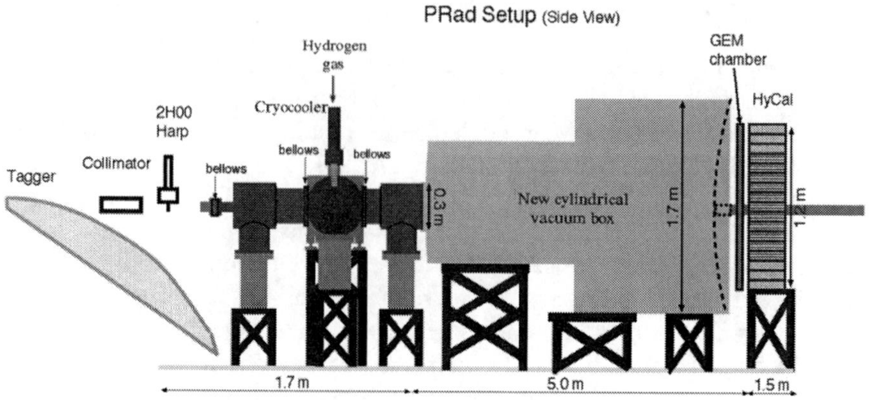

Figure 4. The layout of the PRad experiment, it is not to scale.

The extracted proton charge radius is expected to have a sub-percent precision. The systematic errors will be totally different compared with previous spectrometer based experiments. Its result will evaluate the discrepancy between the *ep* scattering experiments and the Lamb shift measurements of muonic hydrogen. Together with the μp elastic scattering experiment at PSI [47] and the Initial State Radiation (ISR) experiments at Mainz [48], these high precision experiments will have direct impacts on the "proton radius puzzle".

5 Summary

The nucleon electromagnetic form factors remain at the forefront of studies of the nucleon structure. On one hand, the 12-GeV CEBAF enables experiments that explore nucleon form factors at higher momentum transfers squared. On the other hand, the "proton radius puzzle", which is related to the proton electric form factor at low momentum transfers squared, remains unresolved, and awaits new experimental results.

Acknowledgements

This work is supported in part by U.S. Department of Energy under contract number DE-FG02-03ER41231 and NSF MRI award PHY-1229153. We thank J. Arrington, G. Cates, D. Dutta, R. Ent, A. Gasparian, R. Holt, M. Khandaker, M. Meziane, J.W. Qiu, and C. Roberts for useful discussions.

References

[1] C. F. Perdrisat, V. Punjabi, M. Vanderhaeghen, Prog. Part. Nucl. Phys. **59**, 694 (2007).
[2] G. D. Cates *et al.*, Phys. Rev. Lett. **106**, 252003 (2011).
[3] C. Carlson, Prog. Part. Nucl. Phys. **82**, 59 (2015).
[4] M. N. Rosenbluth, Phys. Rev. **79**, 615 (1950).
[5] T. W. Donnelly and A. S. Raskin, Ann. Phys. **169**, 247 (1986).
[6] C. Crawford, Phys. Rev. Lett. **98**, 052301 (2007).
[7] S. J. Brodsky and G. R. Farrar, Phys. Rev. D **11**, 1309 (1975).

[8] G. P. Lepage and S. J. Brodsky, Phys. Rev. D **22**, 2157 (1980).
[9] A. J. R. Puckett *et al.*, Phys. Rev. C **85**, 045203 (2012).
[10] A. V. Belitsky, X. Ji and F. Yuan, Phys. Rev. Lett. **91**, 092003 (2003).
[11] G. A. Miller and M. R. Frank, Phys. Rev. C **65**, 065205 (2002).
[12] S. Boffi *et al.*, Eur. Phys. J. A **14**, 17 (2002).
[13] F. Cardarelli *el al.*, Phys. Lett. B **357**, 267 (1995).
[14] F. Cardarelli and S. Simula, Phys. Rev. C **62**, 65201 (2000).
[15] P. L. Chung and F. Coester, Phys. Rev. D **44**, 229 (1991).
[16] F. Gross and P. Agbakpe, Phys. Rev. C **73**, 015203 (2006).
[17] I. C. Cloët *et al.*, Few Body Syst. **46**, 1 (2009).
[18] I. C. Cloët and G. A. Miller, Phys. Rev. C **86**, 015208 (2012).
[19] M. Guidal *et al.*, Phys. Rev. D **72**, 054013 (2005).
[20] M. Diehl and P. Kroll, Eur. Phys. J. C **73**, 2397 (2013).
[21] G. Cates, S. Riordan, and W. Bogdan (spokespersons), Experiment E12-09-016 at Jefferson Lab.
[22] X. Ji, Phys. Rev. Lett. **91**, 062001 (2003).
[23] Dudek *et al.*, arXiv:1208.1244 [hep-ex] (2012); https://www.jlab.org/12GeV.
[24] K. de Jager Int. J. Mod. Phys. E **19**, 844 (2010).
[25] L. P. Pentchev (contact person) *et al.*, Experiment E12-07-109 at Jefferson Lab.
[26] B. Wojtsekhowski (contact person) *et al.*, Experiment E12-09-019 at Jefferson Lab.
[27] R. Pohl *et al.*, Nature **466**, 213 (2010).
[28] A. Antognini *et al.*, Science **339**, 417 (2013).
[29] P.J. Mohr, B.N. Taylor, and D.B. Newell, Rev. Mod. Phys. **84**, 1527 (2012).
[30] J.C. Bernauer *et al.*, Phys. Rev. Lett. **105**, 242001 (2010).
[31] J. Bernauer *et al.*, Phys. Rev. C **90**, 015206 (2014).
[32] X. Zhan *et al.*, Phys. Rev. B **705**, 59 (2011).
[33] I. Sick, Phys. Lett. B **576**, 62 (2003).
[34] J. Arrington and I. Sick, J. Phys. Chem. Ref. Data **44**, 031204 (2015).
[35] M. C. Birse and J. A. McGovern *et al.*, Eur. Phys. J. A **48**, 120 (2012).
[36] G. A. Miller, Phys. Lett. B **718**, 1078 (2013).
[37] J. M. Alarcón, V. Lensky, and V. Pascalutsa, Eur. Phys. J. C **74**, 2852 (2014).
[38] U.D. Jentschura, Ann. Phys. **326**, 500 (2011).
[39] A. Antognini *et al.*, Ann. Phys. **331**, 127 (2013).
[40] V. Barger *et al.*, Phys. Rev. Lett. **106**, 153001 (2011).
[41] C. Carlson and B. Rislow, Phys. Rev. D **86**, 035013 (2012).
[42] B. Batell, D. Mckeen and M. Pospelov Phys. Rev. Lett. **107**, 011803 (2011).
[43] R. Onofrio, Europhys. Lett. **104**, 20002 (2013).
[44] I. T. Lorentz, H.-W. Hammer, Ulf-G. Meißner, Eur. Phys. J. A **48**, 151 (2012).
[45] D. Robson Int. J. Mod. Phys. E **23**, 14500906 (2014).
[46] T. W. Donnelly, D. K. Hasell, and R. G. Milner, arXiv:1505.04723 [nucl-ex] (2015).
[47] R. Gilman *et al.*, arXiv:1303:2160v1 [nucl-ex] (2013).
[48] M. Mihovilovic *et al.*, EPJ Web Conf. **72**, 00017 (2014).

EPJ Web of Conferences **113**, 01013 (2016)
DOI: 10.1051/epjconf/201611301013

Recent results on the nucleon resonance spectrum and structure from the CLAS detector

V. I. Mokeev[1,2,a], I. G. Aznauryan[3], V. D. Burkert[1], and R. W. Gothe[4] for the CLAS Collaboration

[1] *Thomas Jefferson National Accelerator Facility, Newport News, Virginia 23606, USA*
[2] *Skobeltsyn Nuclear Physics Institute and Physics Department at Moscow State University, Moscow 119899, Lenninskie gory, Russia*
[3] *Yerevan Physics Institute, 375036, Yerevan, Armenia*
[4] *University of South Carolina, Columbia, South Carolina 29208, USA*

Abstract. The CLAS detector at Jefferson Lab has provided the dominant part of all available worldwide data on exclusive meson electroproduction off protons in the resonance region. New results on the $\gamma_v pN^*$ transition amplitudes (electrocouplings) are available from analyses of the CLAS data and will be presented. Their impact on understanding of hadron structure will be discussed emphasizing the credible access to the dressed quark mass function that has been achieved for the first time by a combined analysis of the experimental results on the electromagnetic nucleon elastic and $N \to N^*$ transition form factors. We will also discuss further convincing evidences for a new baryon state $N'(1720)3/2^+$ found in a combined analysis of charged double pion photo- and electroproduction cross sections off the protons.

1 Introduction

Studies of the excited nucleon state (N^*) spectrum and structure from results on resonance electroexcitation amplitudes ($\gamma_v pN^*$ electrocouplings) at different photon virtualities (Q^2) represent an important direction in the broad efforts on the exploration of the non-perturbative strong interaction and its emergence from QCD [1, 2]. They are the only source of information on many facets of the non-perturbative strong interaction in the generation of N^*-states with different quantum numbers as bound systems of an infinite amount of the QCD current quarks and gauge gluons [2]. For the first time $\gamma_v pN^*$ electrocouplings of most well established nucleon resonances with masses below 1.8 GeV have been extracted from the data on exclusive meson electroproduction off protons measured with CLAS at Jefferson Lab at photon virtualities up to 5.0 GeV^2 [3, 5, 7]. In these proceedings we will present new results on $\gamma_v pN^*$ electrocouplings from CLAS and will discuss their impact on our understanding of the N^* structure, and in particular, the opportunity to map-out the dressed quark mass function which encodes the emergence of hadron masses and quark-gluon confinement from QCD. Furthermore, we will present new results from a combined analysis of the charged double pion photo- and electroproduction off protons that strongly support the existence of a new baryon state $N'(1720)3/2^+$.

[a] e-mail: mokeev@jlab.org

2 Evaluation of the resonance parameters from the CLAS data

Nucleon resonance electroexcitations are described by two transverse ($A_{1/2}(Q^2)$, $A_{3/2}(Q^2)$) and one longitudinal ($S_{1/2}(Q^2)$) electrocoupling amplitudes. These electrocouplings offer access to the resonance structure. The CLAS detector has contributed the lion's share of the world data on all essential exclusive meson electroproduction channels in the resonance excitation region including $N\pi$, ηp, KY, and $\pi^+\pi^-p$ electroproduction off protons with nearly complete coverage of the final hadron phase space [3]. The observables measured with the CLAS detector are stored in the CLAS Physics Data Base [8].The available observables in the resonance excitation region are listed in Table 1.

So far, most of the results on $\gamma_v pN^*$ electrocouplings have been extracted from independent analyses of π^+n, π^0p, and $\pi^+\pi^-p$ exclusive electroproduction data off the protons. A total of nearly 160,000 data points (d.p.) on unpolarized differential cross sections, longitudinally polarized beam asymmetries, and longitudinal target and beam-target asymmetries for $N\pi$ electroproduction off protons were obtained with the CLAS detector at $W < 2.0$ GeV and 0.2 GeV^2 $< Q^2 <$ 6.0 GeV^2. The data have been analyzed within the framework of two conceptually different approaches: a unitary isobar model (UIM) and dispersion relations (DR) [4, 5]. The UIM describes the $N\pi$ electroproduction amplitudes as a superposition of N^* electroexcitations in the s-channel, non-resonant Born terms and ρ- and ω- t-channel contributions. The latter are reggeized, which allows for a better description of the data in the second- and third-resonance regions. The final-state interactions are treated as πN rescattering in the K-matrix approximation [4]. In the DR approach, dispersion relations relate the real to the imaginary parts of the invariant amplitudes that describe the $N\pi$ electroproduction. Both approaches provide good and consistent description of the $N\pi$ data in the range of $W < 1.7$ GeV and $Q^2 < 5.0$ GeV^2, resulting in χ^2/d.p. < 2.9.

The $\pi^+\pi^-p$ electroproduction data from CLAS [9, 10] provide for the first time information on nine independent single-differential and fully-integrated cross sections binned in W and Q^2 in the mass range $W < 2.0$ GeV and at photon virtualities of 0.25 GeV^2 $< Q^2 <$ 1.5 GeV^2. The analysis of the data allowed us to develop the JM reaction model [6, 11] with the goal of extracting resonance electrocouplings as well as $\pi\Delta$ and ρp hadronic decay widths. This model incorporates all relevant reaction mechanisms in the $\pi^+\pi^-p$ final-state channel that contribute significantly to the measured electroproduction cross sections off protons in the resonance region, including the $\pi^-\Delta^{++}$, $\pi^+\Delta^0$, $\rho^0 p$, $\pi^+N(1520)\frac{3}{2}^-$, $\pi^+N(1685)\frac{5}{2}^+$, and $\pi^-\Delta(1620)\frac{3}{2}^+$ meson-baryon channels as well as the direct production of the $\pi^+\pi^-p$ final state without formation of intermediate unstable hadrons. The contributions from well established N* states in the mass range up to 2.0 GeV were included into the amplitudes of $\pi\Delta$ and ρp meson-baryon channels by employing a unitarized version of the Breit-Wigner ansatz [6]. The current analysis of the preliminary $\pi^+\pi^-p$ photoproduction data from CLAS at 1.6 GeV < W < 2.0 GeV [12] revealed the presence of a three-body final state interaction that is parameterized and fit to the nine single-differential cross sections. The JM model provides a good description of $\pi^+\pi^-p$ differential cross sections at $W < 1.8$ GeV and 0.2 GeV^2 $< Q^2 <$ 1.5 GeV^2 with χ^2/d.p. < 3.0, and the preliminary $\pi^+\pi^-p$ photoproduction data are also described successfully at W up to 2.0 GeV [12]. The achieved quality of the CLAS data description allows us to isolate the resonant contributions and to determine both resonance electrocouplings and πN, $\pi\Delta$, and ρN decay widths fitting them to the measured observables.

Resonance electrocouplings have been obtained from various CLAS data in the exclusive channels: π^+n and π^0p at $Q^2 <$ 5.0 GeV^2 in the mass range up to 1.7 GeV, ηp at $Q^2 <$ 4.0 GeV^2 in the mass range up to 1.6 GeV, and $\pi^+\pi^-p$ at $Q^2 <$ 1.5 GeV^2 in the mass range up to 1.8 GeV [3, 7]. For the first time photocouplings and $\pi\Delta$ and ρN hadronic decay widths of all well established resonances in the mass range from 1.6 GeV to 2.0 GeV that decay preferentially to the N$\pi\pi$ final states have become available in the analysis of the preliminary CLAS $\pi^+\pi^-p$ photoproduction data [12]. The resonance

Table 1. Observables for exclusive meson electroproduction off protons that have been measured with the CLAS detector in the resonance excitation region and stored in the CLAS Physics Data Base [8]: CM-angular distributions for the final mesons ($\frac{d\sigma}{d\Omega}$); beam, target, and beam-target asymmetries ($A_{LT'}$, A_t, A_{et}); and recoil hyperon polarizations (P', P^0).

Hadronic final state	W-range GeV	Q^2-range GeV^2	Measured observables
$\pi^+ n$	1.10-1.38	0.16-0.36	$\frac{d\sigma}{d\Omega}$
	1.10-1.55	0.30-0.60	$\frac{d\sigma}{d\Omega}$
	1.10-1.70	1.70-4.50	$\frac{d\sigma}{d\Omega}$, $A_{LT'}$
	1.60-2.00	1.80-4.50	$\frac{d\sigma}{d\Omega}$
$\pi^0 p$	1.10-1.38	0.16-0.36	$\frac{d\sigma}{d\Omega}$
	1.10-1.68	0.40-1.15	$\frac{d\sigma}{d\Omega}$, $A_{LT'}$, A_t, A_{et}
	1.10-1.39	3.00-6.00	$\frac{d\sigma}{d\Omega}$
ηp	1.50-2.30	0.20-3.10	$\frac{d\sigma}{d\Omega}$
$K^+\Lambda$	1.62-2.60	1.40-3.90	$\frac{d\sigma}{d\Omega}$
	1.62-2.60	0.70-5.40	P', P^0
$K^+\Sigma^0$	1.62-2.60	1.40-3.90	$\frac{d\sigma}{d\Omega}$
	1.62-2.60	0.70-5.40	P'
$\pi^+\pi^- p$	1.30-1.60	0.20-0.60	Nine single-differential
	1.40-2.10	0.50-1.50	cross sections

electrocouplings obtained from these $\pi^+\pi^- p$ photoproduction data are consistent with the published RPP results [13] from analyses of $N\pi$ photoproduction, confirming the reliability of the updated JM model in extracting resonance parameters.

3 Highlights of the recent results on the N^* spectrum and structure from the CLAS data

3.1 Impact of the new CLAS results on studies of the N^* structure

The studies of the $N(1440)1/2^+$ and $N(1520)3/2^-$ resonances with the CLAS detector [4, 6] have provided the dominant part of the worldwide available information on their electrocouplings in a wide range of photon virtualities 0.25 GeV^2 < Q^2 < 5.0 GeV^2. Currently the $N(1440)1/2^+$ and $N(1520)3/2^-$ states, together with the $\Delta(1232)3/2^+$ and $N(1535)1/2^-$ resonances [3], represent the most explored excited nucleon states. Furthermore, results on the $\gamma_v pN^*$ electrocouplings for the high-lying $N(1675)5/2^-$, $N(1680)5/2^+$, and $N(1710)1/2^+$ resonances have recently been determined for the first time from the CLAS $N\pi$ data at 1.5 GeV^2 < Q^2 < 4.5 GeV^2 [5].

Figure 1 shows the mentioned electrocoupling results for the $N(1440)1/2^+$, $N(1520)3/2^-$, and $N(1675)5/2^-$ resonances together with the preliminary results on the $N(1440)1/2^+$ and $N(1520)3/2^-$ electrocouplings from the CLAS $\pi^+\pi^- p$ electroproduction off protons at 0.5 GeV^2 < Q^2 < 1.5 GeV^2 [7]. The $N \rightarrow N(1440)1/2^+$ Dirac (F_1^*) and Pauli (F_2^*) electromagnetic transition form factors, as computed from these data, are shown in Fig. 2. Consistent results for the $\gamma_v pN^*$ electrocouplings of the $N(1440)1/2^+$ and $N(1520)3/2^-$ resonances, that have been determined in independent analyses of the dominant meson electroproduction channels, $N\pi$ and $\pi^+\pi^- p$, demonstrate that the extraction of these fundamental quantities is reliable, since good data description is achieved in the major electroproduction channels, that have quite different background contributions. This consistency also

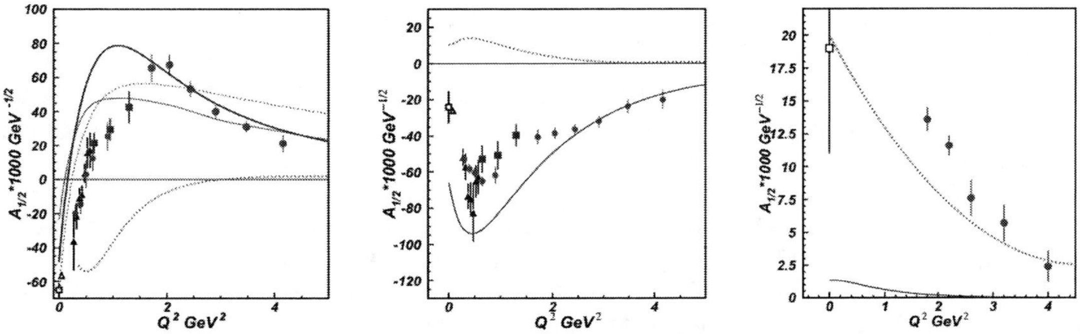

Figure 1. (Color Online) $A_{1/2}$ $\gamma_v pN^*$ electrocouplings of the $N(1440)1/2^+$ (left), $N(1520)3/2^-$ (center), and $N(1675)5/2^-$ (right) resonances from analyses of the CLAS electroproduction data off protons in the $N\pi$ - [4, 5] (red circles) and $\pi^+\pi^-p$ channels [6] (black triangles) with new preliminary results from $\pi^+\pi^-p$ channel [7] (blue squares). The electrocoupling results are shown in comparison with the DSEQCD - [16] (blue thick solid) and constituent quark model calculations [18] (thin red solid), [19] (thin red dashed), and [20] (thin black solid). The meson-baryon cloud contributions are presented by the magenta thick dashed lines. In case of the $N(1440)1/2^+$ resonance (left) they are based on the calculated DSEQCD results and the extracted electrocoupling data, whereas the absolute values at the resonance poles taken from Argonne-Osaka coupled channel analysis [22] are shown for $N(1520)3/2^-$ (center) and $N(1675)5/2^-$ (right). Photocouplings are taken from RPP [13] (black open squares) and the CLAS data analysis [14] of $N\pi$ photoproduction.

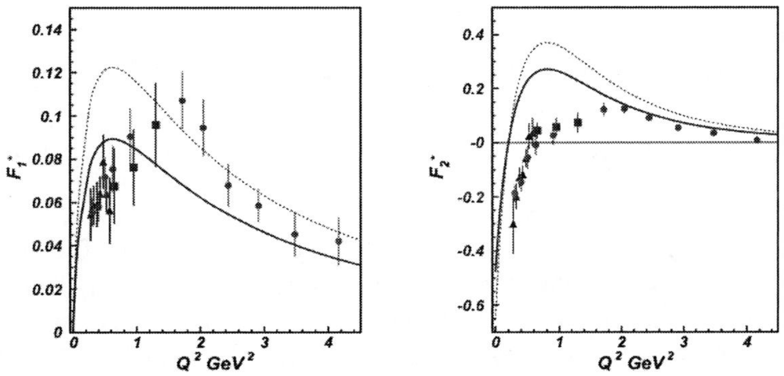

Figure 2. (Color Online) The F_1^* and F_2^* $p \to N(1440)1/2^+$ electromagnetic transition form factors. Experimental results from analyses of the CLAS electroproduction data off protons on $N\pi$ [4] (red circles) and on $\pi^+\pi^-p$ [6] (black triangles) and new preliminary results from $\pi^+\pi^-p$ channel (blue squares). These data are shown in comparison with DSEQCD evaluations [16] that start from the QCD Lagrangian and incorporate the contributions from dressed quarks only (black dashed line) and the dressed quark contributions after accounting for the admixture from meson-baryon cloud as described in Section 3 (blue solid line).

strongly suggests that the reaction models, described in Sec. 2, provide the basis for reliable evaluations of the $\gamma_v pN^*$ electrocouplings, and it is therefore possible to determine these electrocouplings for the majority of N^* resonances that decay preferentially to the $N\pi$ and/or $N\pi\pi$ final states.

Due to the rapid progress in the field of DSEQCD (Dyson-Schwinger Equations of QCD) studies of excited nucleon states [2, 15], the first evaluations of the electromagnetic transition $p \rightarrow N(1440)1/2^+$ Dirac F_1^* and Pauli F_2^* form factors starting from the QCD Lagrangian have recently become available [16]. They are shown in Fig. 2 by the black dashed lines and were computed accounting only for the contributions from three dressed quarks. However, we have to account for the fraction of meson-baryon degrees of freedom in the wave functions of the ground and excited nucleon states. We choose to estimate this by multiplying the $p \rightarrow N(1440)1/2^+$ electromagnetic transition form factors computed within the DSEQCD approach [16] by a common Q^2-independent factor that is fit to the data for $Q^2 > 3.0$ GeV^2, where the quark core is expected to be the biggest contributor. The $p \rightarrow N(1440)1/2^+$ electromagnetic transition form factors obtained in this manner are shown in Fig. 2 by the solid blue lines. They offer a good description of the experimental results for $Q^2 > 1.5$ GeV^2. Discrepancies at smaller Q^2 are related to meson-baryon-cloud contributions, which are beyond the scope of the current DSEQCD approach [15–17]. The dressed quark mass function used in the DSEQCD computations of the $p \rightarrow N(1440)1/2^+$ electromagnetic transition form factors [16] is *exactly the same* as the one employed in the previous evaluations of the electromagnetic nucleon form factors - and the $p \rightarrow \Delta$ transition form factors [15, 17]. This success strongly supports: a) the relevance of dynamical dressed quarks, with properties predicted by the DSEQCD approach [2], as constituents of the quark core in the structure of the ground and excited nucleon states; and b) the capability of the DSEQCD approach [15, 16] to map out the dressed quark mass function from the experimental results on the Q^2-evolution of the nucleon elastic - and $p \rightarrow N^*$ electromagnetic transition form factors or rather $\gamma_v pN^*$ electrocouplings. Studies of the dressed quark mass function will address the most challenging and still open problems of the Standard Model on the nature of the dominant part of the hadron mass, quark-gluon confinement, its emergence from QCD, and its relation to dynamical chiral symmetry breaking which is expected to be a source of more than 98% of the hadron mass in universe [2].

Recent advances in the development of constituent quark models make it possible to extend the Q^2 range in comparison with DSEQCD approaches where a reasonable description of the $\gamma_v pN^*$ electrocouplings is achieved by taking both the contributions, the quark core and meson-baryon cloud, into account. The two models [18, 19], that account for the meson-baryon cloud contributions, allowed us to considerably improve the description of $N(1440)1/2^+$ electrocouplings at $Q^2 < 2.0$ GeV^2, confirming the relevance of the meson-baryon degrees of freedom for the $N(1440)1/2^+$ structure at these distances, as it is shown in Fig. 1 (left). The credible DSEQCD evaluation of the quark core contributions to the electrocouplings of the $N(1440)1/2^+$ allows us to evaluate the meson-baryon cloud contributions as the difference between the fit of the experimental data at higher Q^2 and the quark core electroexcitation amplitudes from DSEQCD [16]. The meson-baryon cloud contributions extracted in this manner are shown in Fig. 1 (left). The CLAS results on the $\gamma_v pN^*$ electrocouplings of the $N(1520)3/2^-$ resonance are shown in Fig. 1 (center). The quark core contributions to these electrocoupling results have been explored within the framework of the two conceptually different approaches: a) the hypercentral constituent quark model [20], and b) the Bethe-Salpeter approach that employs structureless constituent quarks with momentum-independent quark mass and an instanton quark-quark interaction [21]. Both approaches provide a reasonable description of the experimental results at $Q^2 > 1.0$ GeV^2 as shown in Fig. 1 (center). At smaller photon virtualities they are unable to reproduce the CLAS experimental results. The absolute values of the meson-baryon cloud contributions from Argonne-Osaka analysis [22], also shown in Fig. 1 (center), are maximal at small photon virtualities, where the discrepancies between the quark model expectations and the experimental data are largest. The absolute value of the meson-baryon dressing amplitudes for the $A_{1/2}$ electrocouplings of the $N(1520)3/2^-$ are much smaller than those for the $S_{1/2}$ and $A_{3/2}$ electrocouplings as well as for the $A_{1/2}$ electrocoupling of the $N(1440)1/2^+$. This makes the $A_{1/2}$ electrocoupling of the $N(1520)3/2^-$ particularly attractive for the studies of quark degrees of freedom within the structure of the $N(1520)3/2^-$ resonance. On the other hand, the results on the electrocouplings of the $N(1675)5/2^-$ resonance are already reasonably described by only accounting for the meson-baryon cloud contributions from Argonne-Osaka approach [22], see Fig. 1 (right), since the quark core electroexcitation amplitudes vanish in the limit of the contribution from only the leading SU(6)-spin-flavor configuration and remain almost negligible accounting for the quark configuration mixing, see black solid line in Fig. 1 (right) [20]. These results offer for the first time an almost direct access to meson-baryon cloud contributions to the N^* structure [23] because of strong suppression for the quark core electroexcitations.

Analyses of the CLAS results strongly suggest that the structure of nucleon resonances for $Q^2 < 5.0$ GeV^2 is determined by a complex interplay between the inner core of three dressed quarks bound into a system with the quantum numbers of the nucleon resonance and the external meson-baryon cloud which also depends substantial on the quantum numbers of the excited nucleon state. The studies of resonance electrocouplings over the full spectrum of excited nucleon states of different quantum numbers are hence critical in order to explore different components in the N^* structure.

3.2 Further evidence for the existence of a new $N'(1720)3/2^+$ baryon state

The combined analysis of $\pi^+\pi^-p$ electroproduction and preliminary photoproduction data from CLAS provides further evidence for the existence of a new $N'(1720)3/2^+$ baryon state. For the first time signals from the $N'(1720)3/2^+$ candidate state have been observed in analyses of the CLAS $\gamma_v p \rightarrow \pi^+\pi^-p$ exclusive electroproduction data [9], which shows a pronounced structure in the W-dependence of fully integrated cross sections at $W \approx 1.7$ GeV and in all Q^2 bins covered by the measurements. Analyses of a very limited amount of the single-differential cross sections (three of nine available) carried out in [9] within the framework of the oversimplified initial version of the JM model [24, 25] suggested two possible ways to describe the structure observed at W $\approx$ 1.7 GeV: a) by a dominating decay of the $N(1720)3/2^+$ resonance to the $\pi\Delta$ final state accounting only for contributions from conventional resonances, or b) by implementing a new $N'(1720)3/2^+$ candidate state with parameters derived from the CLAS data fit, in which case the hadronic decay widths of all known resonances remain inside the ranges reported in the PDG03 [26].

Further studies of the CLAS $\gamma_v p \rightarrow \pi^+\pi^-p$ electroproduction cross sections [9] have been carried out within the framework of the current version of the JM model [6, 7, 11] outlined in Sec. 2. All nine single-differential cross sections were included in the data fit. In the fits we varied simultaneously electrocouplings, hadronic decay widths to the $\pi\Delta$ and ρN final states for all resonances that contribute to the region of the structure at $W \approx 1.7$ GeV. We also simultaneously varied the non-resonant parameters of the JM model. The hadronic decay width of each resonance remains the same in all Q^2-bins for the electroproduction data. The preliminary CLAS photoproduction data [12] and the CLAS electroproduction data [9] were fit independently, allowing us to examine the consistency of the resonance hadronic parameters extracted from the photo- and electroproduction data independently.

Two fits of the nine single-differential $\gamma_{r,v}p \rightarrow \pi^+\pi^-p$ cross sections have been carried out: a) assuming only the contribution from conventional resonances (fit A), or b) a $N'(1720)3/2^+$ candidate state was implemented in addition to the contributions from the conventional resonances (fit B). Both fit A and fit B provide a good description of the CLAS $\gamma_{r,v}p \rightarrow \pi^+\pi^-p$ photo- and electroproduction cross sections at W from 1.61 to 1.81 GeV with the $\chi^2/d.p.$ less than 3.03 and less than 2.80, respectively. A reliable description of the resonant content implies a valuable restriction on the resonance hadronic decay widths inferred at different values of photon virtuality Q^2. The hadronic decay widths of each resonance should be Q^2-independent. The hadronic decays widths of the conventional

Table 2. N(1720)3/2$^+$ hadronic decays determined from the independent fits to the data on charged double pion photo- [12] and electroproduction [9] off protons accounting only for contributions from conventional resonances.

Resonance states	N* total width MeV	Branching fraction for decays to $\pi\Delta$	Branching fraction for decays to ρN
N(1720)3/2$^+$			
electroproduction	126.0 ± 14.0	64% - 100%	< 5%
photoproduction	160.0 ± 65.0	14% - 60%	19% - 69%

Table 3. Hadronic decays into the $\pi\Delta$ and ρN final states of the resonances in the third resonance region with major decays to the N$\pi\pi$ final states determined from the fits to the data on charged double pion photo- [12] and electroproduction [9] implementing a new $N'(1720)3/2^+$ baryon state.

Resonance states	N* total width MeV	Branching fraction for decays to $\pi\Delta$	Branching fraction for decays to ρN
Δ(1700)3/2$^-$			
electroproduction	288.0 ± 14.0	77% - 95%	3% - 5%
photoproduction	298.0 ± 12.0	78% - 93%	3% - 6%
N(1720)3/2$^+$			
electroproduction	116.0 ± 7.0	39% - 55%	23% - 49%
photoproduction	112.0 ± 8.0	38% - 53%	31% - 46%
N$'$(1720)3/2$^+$			
electroproduction	119.0 ± 6.0	47% - 64%	3% - 10%
photoproduction	120.0 ± 6.0	46% - 62%	4% - 13%

$N(1720)3/2^+$ resonance obtained from fit A of the charged double pion photoproduction data [12] are listed in the Table 2 in comparison with those inferred in fit A of the electroproduction cross sections [9].

The branching fractions for the $N(1720)3/2^+$ decays into the ρN final state inferred by fit A of the data on charged double pion photo- and electroproduction off protons differ by more than a factor of four. Moreover, the branching fractions for the $N(1720)3/2^+$ decays into the $\pi\Delta$ final states obtained fitting the photo- and electroproduction data are also rather different. This makes it impossible to describe both the charged double pion photo- and electroproduction cross sections when only contributions from conventional resonances are taken into account. By implementing a new $N'(1720)3/2^+$ baryon state , a successful description of all nine single-differential $\gamma_{r,v}p \rightarrow \pi^+\pi^-p$ photo- and electroproduction cross sections has been achieved. Furthermore, the hadronic decay widths of all resonances in the third resonance region as inferred from the fits at different Q^2 remain Q^2-independent in the entire range of photon virtualities up to 1.5 GeV^2 that is covered by the CLAS measurements [9, 12] (Table 3.)

The implementation of a new $N'(1720)3/2^+$ baryon state represents the only way that allows us to achieve a good description of the CLAS data on charged double pion photo- and electroproduction in all four Q^2 bins covered by the CLAS measurements, centered at 0 GeV^2, 0.65 GeV^2, 0.95 GeV^2, and 1.30 GeV^2, with Q^2-independent hadronic decay widths of each relevant resonance. This success provides sound evidence for the first observation of the $N'(1720)3/2^+$ baryon state in the CLAS

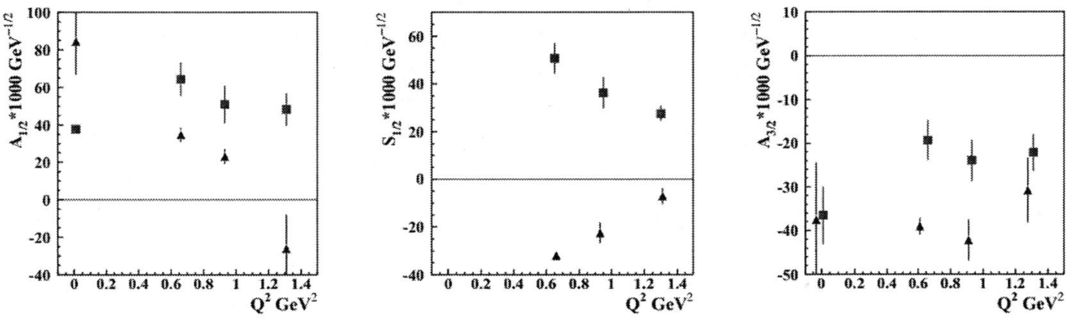

Figure 3. (Color online) Photo- and electrocouplings of the new $N'(1720)3/2^+$ (blue squares) and conventional $N(1720)3/2^+$ (black triangle) states as determined from the analysis of the CLAS charged double pion photo- and electroproduction data off protons [9, 12] within the framework of JM15 model [6, 7, 11] outlined in Sec. 2.

data [9, 12]. The electrocouplings of the two close-lying resonances, the new $N'(1720)3/2^+$ and the conventional $N(1720)3/2^+$, are superimposed in Fig. 3.

The analysis of the CLAS data within the framework of the recent JM model version [6, 11] revealed the contribution of a conventional $N(1720)3/2^+$ and a new $N'(1720)3/2^+$ baryon state to the $\pi^+\pi^-p$ photo- and electroproduction cross sections at $W \approx 1.7$ GeV. The two states have close masses (1.743 GeV to 1.753 GeV for $N(1720)3/2^+$ and 1.715 GeV to 1.735 GeV for $N'(1720)3/2^+$) and the same spin-parities $3/2^+$, but their hadronic decay widths of the $\pi\Delta$, ρN final states and Q^2-evolution of their electrocouplings (Fig. 3) are distinctively different. All this further strengthens the claim of a new $N'(1720)3/2^+$ baryon state found in the CLAS charged double pion photo- and electroproduction data.

Acknowledgements

We would like to acknowledge the outstanding efforts of the staff of the Accelerator and the Physics Divisions at Jefferson Lab. We are grateful to I.C. Clöet, M.M. Giannini, C.D. Roberts, E. Santopinto, J. Segovia for their theoretical support. This work was supported in part by the U.S. Department of Energy, the National Science Foundation, the Skobeltsyn Institute of Nuclear Physics, the Physics Department at Moscow State University, the University of South Carolina, and Yerevan Physics Institute. The Southeastern Universities Research Association (SURA) operates the Thomas Jefferson National Accelerator Facility for the United States Department of Energy under the contract DE-AC05-84ER40150.

References

[1] I. G. Aznauryan et al., Int. J. Mod. Phys. **E22**, 1330015 (2013).
[2] I. C. Clöet and C. D. Roberts, Prog. Part. Nucl. Phys. **77**, 1 (2014).
[3] I. G. Aznauryan and V. D. Burkert, Prog. Part. Nucl. Phys. **67**, 1 (2012).
[4] I. G. Aznauryan et al., CLAS Collaboration, Phys. Rev. C **80**, 055203 (2009).
[5] K. Park et al., CLAS Collaboration, Phys. Rev. C **91**, 045203 (2015).
[6] V. I. Mokeev et al., CLAS Collaboration, Phys. Rev. C **86**, 055203 (2012).

[7] V. I. Mokeev and I. G. Aznauryan, Int. J. Mod. Phys. Conf. Ser. **26**, 146080 (2014).
[8] CLAS Physics Data Base, http://clas.sinp.msu.ru/cgi-bin/jlab/db.cgi.
[9] M. Ripani et al., CLAS Collaboration, Phys. Rev. Lett. **91**, 022002 (2003).
[10] G. V. Fedotov et al., CLAS Collaboration, Phys. Rev. C **79**, 015204 (2009).
[11] V. I. Mokeev et al., Phys. Rev. C **80**, 045212 (2009).
[12] V.I.Mokeev, Invited talk at The International Workshop on Partial Wave Analysis for Hadron Spectroscopy, PWA 8 / ATHOS 3, http://physics.columbian.gwu.edu/program-and-contributions.
[13] J. Beringer et al., Review of Particle Physics, Phys. Rev. D **86**, 1 (2012).
[14] M. Dugger et al., CLAS Collaboration Phys. Rev. C **79**, 065206 (2009).
[15] J. Segovia *et al.*, Few Body Syst. **55**, 1185 (2014).
[16] J. Segovia *et al.*, arXiv:1504.04386 [nucl-th].
[17] I.C. Clöet, C.D. Roberts and A.W. Thomas, Phys. Rev. Lett. **111**, 101803 (2013).
[18] I.G. Aznauryan and V.D. Burkert, Phys. Rev. C **85**, 055202 (2012).
[19] I.T. Obukhovsky *et al.*, Phys. Rev. D **89**, 0142032 (2014).
[20] E. Santopinto and M.M. Giannini, Phys. Rev. C **86**, 065202 (2012).
[21] M. Ronninger and B. Ch. Metsch, Eur. Phys J. A **49**, 8 (2012).
[22] B. Julia-Diaz *et al.*, Phys. Rev. C **77**, 045205 (2008).
[23] I.G. Aznauryan and V.D Burkert, Phys. Rev. C **92**, 015203 (2015).
[24] M. Ripani et al., Nucl. Phys. A **672**, 220 (2000).
[25] V. I. Mokeev et al., Phys. Atom. Nucl. **64**, 6 (2001).
[26] D. E. Groom et al., Eur. Phys, J. C **15**, 1 (2000).

EPJ Web of Conferences **113**, 01014 (2016)
DOI: 10.1051/epjconf/201611301014

Impressions of the Meson Spectrum: Hybrids & Exotics, present and future

M.R. Pennington[1,a]

[1] *Theory Center, Thomas Jefferson National Accelerator Facility, 12000 Jefferson Avenue, Newport News, VA 23606, U.S.A.*

Abstract. It has long been expected that the spectrum of hadrons in QCD would be far richer and extensive than experiment has so far revealed. While there have been experimental hints of this richness for some time, it is really only in the last few years that dramatic progress has been seen in the exploration both experimentally and in calculations on the lattice. Precision studies enabled by new technology both with detectors and high performance computations are converging on an understanding of the spectrum in strong coupling QCD. These methodologies are laying the foundation for a decade of potential discovery that electro and photoproduction experiments at Jefferson Lab, which when combined with key results on B and charmonium decays from both e^+e^- and pp colliders, should turn mere impressions of the light meson spectrum into a high definition picture.

1 Introduction

The Chicago Art Institute is one of the great galleries of the world. It contains some of the best known Impressionist paintings of the 19th century: Monet's series of "waterlilies" and "haystacks". This talk is also an impression: an impression of the meson spectrum. The details may be fuzzy, but nevertheless there is a picture to be seen in a certain light. That light is the development of a detailed understanding of the workings of strong coupling QCD.

The spectrum of hadrons is, of course, a direct reflection of the degrees of freedom that underly hadronic states, and the forces between these. For the ground state baryon amd meson families, these appear to be just the 3 flavors of light quarks. For excited mesons, those with the maximum spin for a given mass, the $J^{PC} = 1^{--}, 2^{++}, 3^{--}, \cdots$ *i.e.* the ρ, ω, K^*, ϕ, $f_2(1270), a_2(1320), K_2^*(1430), f_2'(1525)$, $\cdots$ families, all appear strongly in hadro-production processes through their decays to $\pi\pi$, $K\pi$, 3π and $\overline{K}K$. All belong to nearly ideally mixed quark model multiplets. These states are said to lie along the leading Regge trajectories. They have set the paradigm for our idea of mesons. Their properties have defined the development of constituent quark models. Of course, QCD suggests that the hadronic states we observe are color singlets: the simplest being a quark and antiquark or three quarks. However QCD predicts there should be a far richer spectrum, with states made predominantly of glue, we call glueballs, tetraquark states made of two quarks and two antiquarks, some of which may like to rearrange themselves for good dynamical reasons as hadronic molecules, or states in

[a] e-mail: michaelp@jlab.org

which glue contributes to their spatial quantum numbers — states we call hybrids, with analogs to be found in the baryon sector too. For almost forty years we have been searching for these additional states. Indeed we may well have observed some of these, but there is little certainty of what has been found. They are in general more difficult to identify than the ideally mixed $\overline{q}q$ states. They have lower spins than the maximum at a given mass, and tend to be broad or only produced with smaller couplings, or both. Consequently one has to have data of high statistics with near perfect angular coverage to be able to perfom a robust Amplitude Analysis that can separate out the smaller partial waves with precision. It is only now with data on hadroproduction from COMPASS@CERN, from BaBar@SLAC, Belle@KEK, BESIII@IHEP and LHCb@CERN, and shortly with GlueX and CLAS12@JLab, that we have (and will continue to have) such data.

Historically the focus for a richer spectrum was on the pure glue sector [1]. It is there where lattice calculations can most easily be done in a world with no quarks. These [2–4] predicted the lightest glueball would be a scalar around 1.6-1.7 GeV, with a tensor higher up above 2 or 2.2 GeV. These were sought in what were thought to be glue-rich environments like J/ψ radiative decays, or $\overline{p}p$ annihilation. Of course, glueballs decay to flavorless combinations of pions and kaons. This means they inevitably mix with $\overline{q}q$ states of the same quantum numbers that decay to the very same channels. Though there are indeed more isoscalar scalars, below 1.8 GeV, than can fit into two nonets, which (if any) is the closest to a glueball is matter of debate and modeling [5–7]. Attention consequently turned to multi-quark states, some of which might indeed be the light scalar mesons. However, it is in the heavy flavor sectors of charmonium and bottomonium that mesons that are more than just $\overline{q}q$ have been most spectacularly seen.

2 The Richness of the Hadron Spectrum

While the very lightest mesons and baryons are stable as far as isospin conserving strong interactions are concerned, all excited states decay. Indeed, they decay very quickly. While they may be seeded by configurations with the minimum number of quarks, they almost immediately radiate gluons that create additional $\overline{q}q$ pairs. It is through these that the decay products emerge. Decays are an integral part of the lifecycle of all excited hadrons. The ρ^+ is not simply $u\overline{d}$, but composed of additional $u\overline{u}$

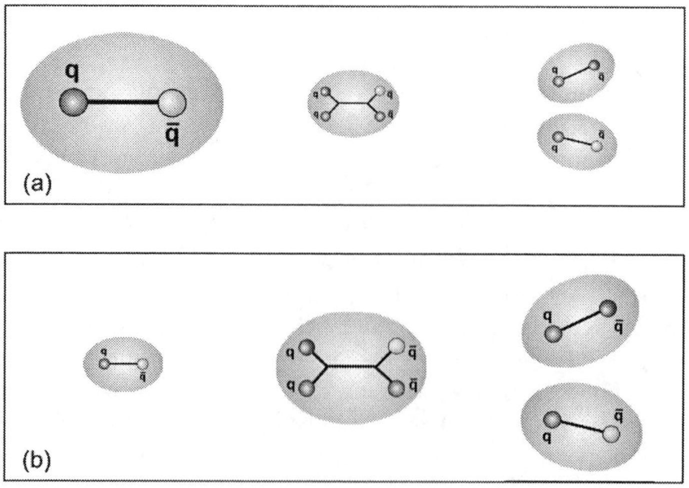

Figure 1. Cartoon of the Fock space of two types of mesons. In the upper graph (a) is shown a predominantly $\overline{q}q$ meson with smaller four quark and di-meson components, and in the lower (b) a possible orthogonal state, which though *seeded* by $\overline{q}q$ configurations, may be dominated by tetraquark or di-hadron components [8–10].

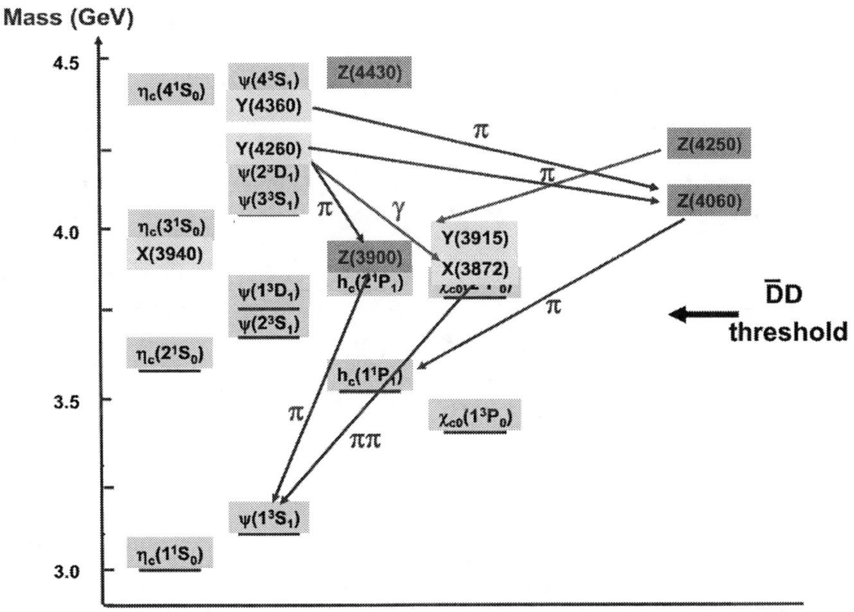

Figure 2. The standard charmonium spectrum with states (in light blue) labeled by their $n^{2S+1}L_J$ quantum numbers, overlaid with the states of the *new X, Y, Z* spectroscopy, showing some of their interconnections implied by their observed decay patterns [11].

and $d\bar{d}$ pairs. These rearrange themselves into a $\pi^+\pi^0$ final state. Thus as illustrated in Fig. 1a, the Fock space of the ρ^+ is not simply $u\bar{d}$. Nevertheless, this component dominates: the coupling to the $\pi\pi$ system is inhibited by being a P-wave interaction. Indeed all the states on the leading Regge trajectory, the easiest to be seen in experiment, are dominated by their $q\bar{q}$ configurations. However, lower lying states, like the chiral partner of the ρ, the $a_1(1260)$, spends more of its time in a multi-quark configuration, first $\rho\pi$ and then 3π. Ideal mixing is less apparent. As soon as one recognizes that the degrees of freedom are not just $\bar{q}q$, but include multiquark or multihadron configurations too, one understands that if the dynamics is favorable, there should also exist orthogonal states, in which the $\bar{q}q$ *seed* might be small and the dihadron components large, as illustrated in Fig. 1b. Just such a possibility appears to occur with the X, Y, Z states (or at least some of them). The opening of decay channels of the heavier charmonium states shifts them from where they would naively be in a potential model, and at the same time new states appear, Fig. 1. The longest known is the $X(3872)$ seen in B-decay to KX, where $X \to J/\psi\pi\pi$. While this state is 140 MeV above $\overline{D}D$ threshold, it lives 100 times longer than expected (with a width of less than 1.2 MeV). It sits within a "whisker" of $\overline{D^0}D^{*0}$ threshold and 8 MeV below the corresponding charged D threshold. This dynamics may make it favorable that one pion exchange binds the system [12]. Indeed, it may well be an example of a state with the Fock space shown in Fig. 1b — see, for instance [13].

Since the discovery of the $X(3872)$, a whole host of these unusual mesons have been found. Not every bump need be due to a state in the spectrum. Consequently the new "discoveries" have really to be shown to give rise to poles in the complex energy plane, which is the true confirmation of a state in the spectrum. Nevertheless, there are too many to be accidental, Fig. 2. All may well be

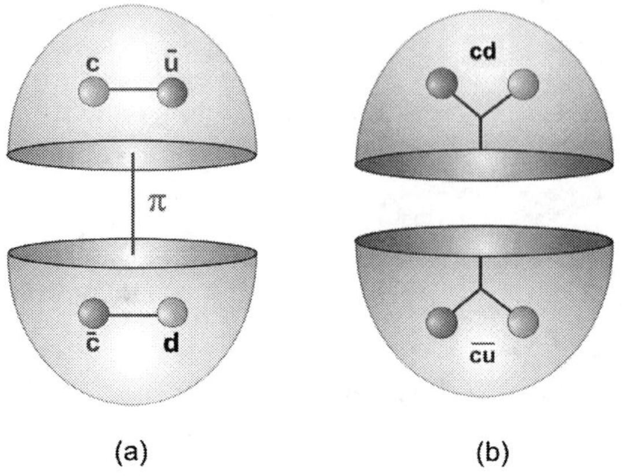

Figure 3. Representation of the $Z_C^-(4430)$ (a) as a molecule bound by interhadron forces, or (b) as a tetraquark state bound by interquark forces.

multi-quark states sitting close to a nearby open hadronic channel to which they couple in an S-wave. They may then be hadronic molecules or tetraquarks, Fig. 3. Indeed the appearance of the charged $Z_C(4430)$ points to just such configurations, definitely something more than $\bar{c}c$ as in Fig. 3. Two remarkable things hint to a relation between all these many states that Cheng-Ping Shen [14] and Eric Braaten [15] will discuss in much more detail. First they seem to be connected to each other (Fig. 2), one decays into another by emitting a photon or a pion, depending on quantum numbers. The second thing is that there are similarly narrow states in the bottomonium sector too. We don't yet have a working theoretical understanding of *all* of these states, though for some [16, 17], as Eric Braaten will discuss, this may be closer. Nevertheless, when we do understand why and how such systems bind, this will surely teach us about the workings of strong coupling QCD. There are even hints that similar states may occur in the hidden strange sector with a $\phi f_0(980)$ bound system [18]. Searches for $\overline{K}K^*$ systems, akin to $\overline{D}D^*$ of the $X(3872)$ will be a natural target for the future.

Indeed, it could well be that the nine lightest scalars, the $\sigma/f_0(500)$, $\kappa/K_0^*(800)$, $a_0(980)$ and $f_0(980)$, are of a similar type, tetraquark configurations, seeded by the heavier 0^{++} $\overline{q}q$ nonet underneath the corresponding 2^{++} states, but dominated in their Fock space by their di-meson components, as in Fig, 1b. In many respects [10], the $f_0(980)$ behaves as a $\overline{K}K$ molecule, probably the a_0 too (with the a_0^- being like Fig. 3b with s quarks replacing c's), while the σ and κ are loosely bound $\pi\pi$ and $K\pi$ systems, respectively. If so, any simple quark mixing scheme for them is quite misleading [6, 19, 20]. "Mixing" depends not just on the complex masses of the bound states, but on the masses of the pseudoscalar pairs to which they couple, and this has little (or nothing) to do with the intrinsic $\overline{q}q$, or even glue, they each likely have in their Fock space.

Results from lattice QCD are having a direct impact on experimental studies now underway. The Hadron Spectrum Collaboration [21] several years ago computed the spectrum of mesons built from quark-antiquark operators with both isovector and isoscalar quantum numbers at a pion mass of 396 MeV. As seen in Fig. 4, this accords well with the observed spectrum, just shifted by having a heavy pion [21, 22]. Including gluonic operators enlarges the spectrum to include $\overline{q}gq$ hybrids. As well as producing additional (but heavier) mesons with "conventional" quantum numbers on the left in Fig. 4, it also creates states (seen on the right) which are "exotic" in the quark model. The lightest of these hybrids, known as the π_1, has $J^{PC} = 1^{-+}$ in the mass range around 2 GeV, accessible to the 12 GeV campaign at Jefferson Lab [23]. Once such calculations include physical decay channels, with

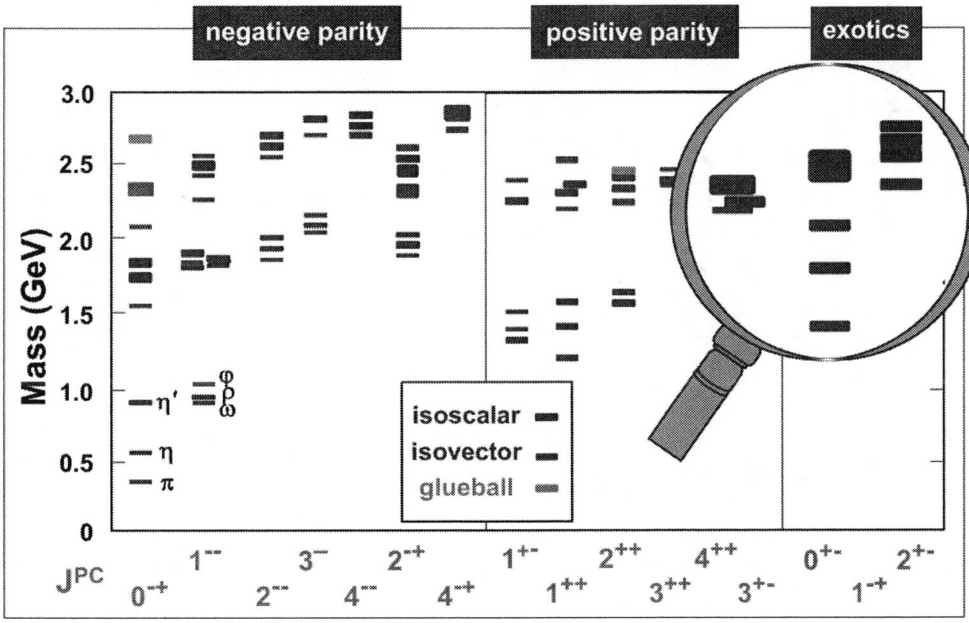

Figure 4. The masses of the negative and positive parity mesons generated in QCD by calculating quark-antiquark operators on the lattice at a pion mass of 396 MeV, in columns labeled by their J^{PC} values from the Hadron Spectrum Collaboration [21, 22]. The inclusion of gluonic operators adds hybrids with conventional quantum numbers on the left, roughly 1 GeV heavier than those with just quark operators, as well as generating states with "exotic" quantum numbers that are shown magnified on the right around 2 GeV in mass.

140 MeV pions, the masses of the mesons are expected to be shifted, generally lower. To be certain that any state discovered is really a gluonic hybrid, it is essential to identify several such hybrid states with different flavor and spatial quantum numbers — ideally whole multiplets. This is what the GlueX experiment at JLab will put under the magnifying glass, Fig. 4. To aid this further, the aim over the next few years is to predict using lattice QCD the decays and radiative transition amplitude of such hybrids. This follows the recent successful benchmarking of such quantities for well-known mesons like the ρ, calculating its decay to $\pi\pi$ and its radiative transition from a pion [24–27].

Previously experiments at CERN and Serpukhov with the GAMS detectors [28, 29], at Brookhaven with BNL-E852 [30, 31] and later at Protvino with VES [32, 33] were aimed at finding states with gluonic components. Indeed, they each found hints of mesons with 1^{-+} quantum numbers in $\pi\eta$ and $\pi\eta'$, but one could never be certain these were really resonances, rather than simple enhancements. It is in the 3π channel where larger statistics have been accumulated. So for instance in Fig. 5, we show the $\pi^+\pi^-\pi^-$ mass spectrum in $\pi^- p$ collisions at 18 GeV/c from BNL-E852 [31]. There the high lying states, the $a_2(1320)$ and $\pi_2(1670)$, are easily seen. However, the much smaller 1^{-+} $\rho\pi$ P-wave signal was found to be highly sensitive to leakage from the higher waves, and to treatment of the large $\pi\pi$ S-wave interaction, and so claims of a $\pi_1(1400)$ came and went to be replaced by a possible $\pi_1(1600)$ [30, 32, 33, 35, 36]. To prove the magnitude and relative phase variations arise from a pole in the complex energy plane requires more events with a hermetic detector and more detailed analysis.

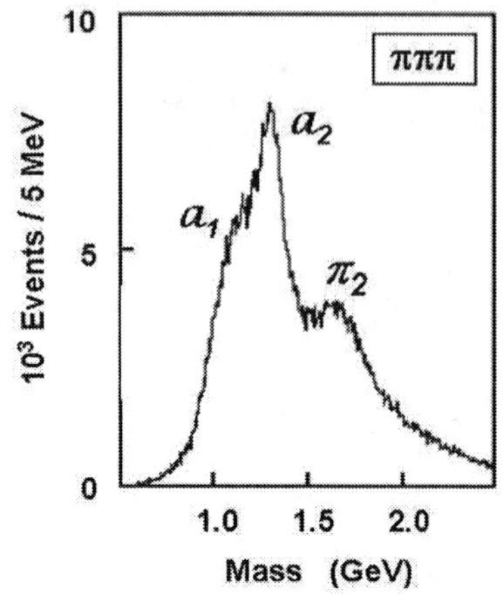

Figure 5. The $\pi^+\pi^-\pi^-$ mass spectrum found in π^-p collisions at 18 GeV/c by BNL-852 experiment [31]. The dominant resonances are indicated. The COMPASS experiment at 190 GeV/c has 400 times the statistics on the same channel with a very similar, but inevitably smoother, 3π spectrum [34].

3 Challenges for the near future

An attempt to use high statistics to completely clean up the light hadron spectrum is provided by combining results from several different production mechanisms to which new technology allows precision access. Hadroproduction with pion beams at COMPASS, heavy flavor decays from BaBar, Belle, BESIII and LHCb, and photoproduction (with real and virtual photons) at Jefferson Lab with GlueX and CLAS12. Here is where a whole suite of precision analysis tools are required to match the precision of the existing data, or that to come, if we are to truly extract definitive physics results from these data. COMPASS is the experiment at 190 GeV/c that provides the world's largest sample of three pion production with almost 100 Million events, four hundred times that of BNL-E852. The dominant partial waves (magnitudes and relative phases), which encode the $a_2(1320)$ and $\pi_2(1670)$, are robustly determined. Indeed, with such statistics, not only is the $a_1(1260)$ readily seen, as in Fig 4 in its $\rho\pi$ mode, but a new $a_1(1420)$ structure observed in the hundred times smaller $f_0(980)\pi$ channel [37] – see also [38]. It is at this level that waves with "exotic" quantum numbers appear: they are only a tiny fraction of the integrated cross-section.

To extract partial waves which contribute less than a few percent, one needs to be certain that one understands all of the other waves to even better accuracy. This is a major challenge. Small partial waves are most readily affected by artificially limiting the number of waves contributing. Something one had to do with earlier lower statistics data. COMPASS analyses [39] have dramatically illustrated the effect on the 1^{-+} $\rho\pi$ P-wave of different truncations by showing how fitting their data with restricted wavesets reproduces the results of analyses by BNL-E865 and by the VES group, matching the number of partial waves included by the BNL and VES experiments. The possible $\pi_1(1400)$ and $\pi_1(1600)$ broaden and change shape, as the number of contributing waves is altered [39]. To ensure this is not an issue, COMPASS includes a far bigger set of waves: 88 in total, a set that is deliberately "over-complete" [34].

In all these studies, the right theoretical foundations are essential. The Joint Physics Analysis Center (*JPAC*) project [40] at Jefferson Lab has taken on the challenge to coordinate these devlopments. The S-Matrix has been reloaded. Basic concepts of analyticity, crossing and unitarity well known for over 60 years provide the foundation for precision analysis of precison data. These are coupled with studies of the relevant reaction mechanisms, seen for instance in Fig. 6a, for hadro- and photo-production. These are key to avoiding the artificial truncation of partial waves. Cross-channel exchanges, seen in Fig. 6b, can provide a more efficient and effective way [41–43] of handling higher partial wave amplitudes. Similarly, we know precision data demand more than a crude isobar picture

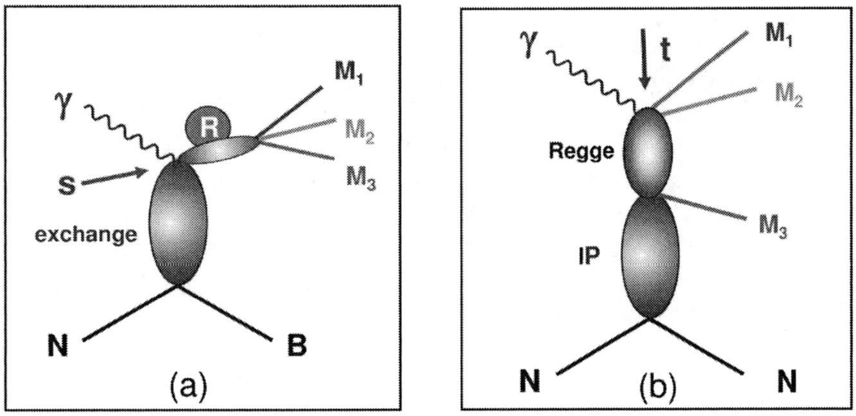

Figure 6. Examples of reaction mechanisms for multi-particle production. On the left (a) shows the peripheral mechanism for high energy photo-production of a state, with quantum numbers denoted by the circled R, which then decays to meson 1-3. This process is controlled by Pomeron and/or other Regge exchanges. The sum over channels of many quantum numbers R (circled) in γ-Reggeon scattering is expected to have a simpler Regge exchange description, illustrated on the right (b). Such a description provides a more economical representation of the higher waves than simply summing those on the left. Fig. (b) also includes the case when the upper Regge exchange and meson M_3 are pions. This is commonly called the *Deck mechanism* [41].

of pairwise interactions. Yes, the a_1 (and even a π_1) may decay to $\rho\pi$ and then the ρ decay to $\pi\pi$, but rescattering involving all three final state pions, Fig. 7, has to be taken into account to preserve unitarity. This is achieved employing the Khuri-Treiman equation [44]. *JPAC* calculations of the analogous graphs for the η, ω and ϕ [45–47] where high statistics data already exist, reveal rescattering and 3-body interactions to be small for these decays. This is not surprising given the limited phase space. Nevertheless this benchmarking is essential before tackling heavier states, like the a_1's and π_1, where such unitarity constraints become more complex as the number and energy range of waves in each interaction increases. Almost for the first time, data demand this level of detail.

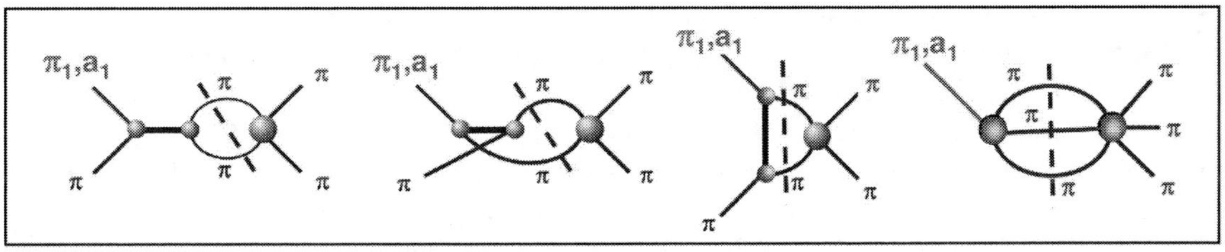

Figure 7. An example of the three pion decay of a parent hadron, like the a_1 or a putative π_1. This has contribution from the pairwise interaction of the pions, and their rescattering, as well as the contribution of direct three body decay. All are required to ensure probability is conserved. The dashed lines indicate the 2 and 3-body contributions to the unitarity relation. Recent calculations, using the Khuri-Treiman equation [44], of the analogous graphs for the η, ω and ϕ [45–47] reveal rescattering and 3-body interactions to be small, because of the limited phase space for interactions any more complex than low energy $\pi\pi$ in an S or P-wave. Calculations for the heavier states shown are underway.

The richest source of three body decays is from B and charm factories, from BaBar, Belle and BES, with rapidly appearing additions from LHCb. Again these are the channels with the largest statistics required to determine the closure of the weak unitarity triangle. The largest uncertainties in the CKM matrix parameters come from strong interaction effects from such multi-body decays. Thus the same technology of analysis required for understanding the hadron spectrum can inform and be informed by studying data from these machines. Cooperation and collaboration of theory and experiment, with data from different facilities is essential to progress. It is only by such planned coordination that we can turn impressions of the meson spectrum, of hybrids and exotics, into a definitive picture that translates the confinement dynamics of strong coupling QCD into the detailed pattern of hadrons we observe. That is the challenge for the next decade.

Acknowledgements

It is a pleasure to thank the conference organizers, particularly Charlotte Elster, Daniel Phillips and Craig Roberts, for inviting me to share my impressions of the meson spectrum. This material is based upon work supported by the U.S. Department of Energy, Office of Science, Office of Nuclear Physics under contract DE-AC05-06OR23177."

References

[1] E. Klempt and A. Zaitsev, Phys. Rept. **454**, 1 (2007) [arXiv:0708.4016 [hep-ph]].
[2] J. Sexton, A. Vaccarino and D. Weingarten, Phys. Rev. Lett. **75**, 4563 (1995) [hep-lat/9510022].
[3] C. J. Morningstar and M. J. Peardon, Phys. Rev. D **60**, 034509 (1999) [hep-lat/9901004].
[4] C. Michael and M. Teper, Nucl. Phys. B **314**, 347 (1989).
[5] F. E. Close and A. Kirk, Phys. Lett. B **483**, 345 (2000) [hep-ph/0004241].
[6] W. Ochs, J. Phys. G **40**, 043001 (2013) [arXiv:1301.5183 [hep-ph]].
[7] S. Janowski, F. Giacosa and D. H. Rischke, Phys. Rev. D **90**, no. 11, 114005 (2014) [arXiv:1408.4921 [hep-ph]].
[8] E. van Beveren, T. A. Rijken, K. Metzger, C. Dullemond, G. Rupp and J. E. Ribeiro, Z. Phys. C **30**, 615 (1986) [arXiv:0710.4067 [hep-ph]].
[9] M.F.M. Lutz and E.E. Kolomeitsev, *Nucl. Phys. A* **755**, 29 (2005).
[10] M. R. Pennington, Acta Phys. Polon. Supp. **8**, 9 (2015) [arXiv:1411.7902 [nucl-th]].
[11] R. E. Mitchell, PoS Hadron **2013**, 038 (2013).
[12] N. A. Tornqvist, Phys. Lett. B **590**, 209 (2004) [hep-ph/0402237].
[13] S. Coito, G. Rupp and E. van Beveren, Eur. Phys. J. C **73**, 2351 (2013) [arXiv:1212.0648 [hep-ph]].
[14] Cheng-Ping Shen, Proceedings of Few Body 21, Chicago, June 2015.
[15] E. Braaten, Proceedings of Few Body 21, Chicago, June 2015.
[16] E. Braaten, C. Langmack and D. H. Smith, Phys. Rev. D **90**, 014044 (2014) [arXiv:1402.0438 [hep-ph]].
[17] E. Braaten, Phys. Rev. Lett. **111**, 162003 (2013) [arXiv:1305.6905 [hep-ph]].
[18] M. Ablikim *et al.* [BES collaboration], Phys. Rev. Lett. **100**, 102003 (2008);
C.P. Shen *et al.* [Belle collaboration], Phys. Rev. **D80**, 031101 (2009).
[19] A. H. Fariborz, R. Jora and J. Schechter, Phys. Rev. D **76**, 014011 (2007) [hep-ph/0612200].
[20] D. Black, A. H. Fariborz and J. Schechter, AIP Conf. Proc. **541**, 100 (2000).

[21] J. J. Dudek, R. G. Edwards, M. J. Peardon, D. G. Richards and C. E. Thomas, Phys. Rev. Lett. **103**, 262001 (2009) [arXiv:0909.0200 [hep-ph]].

[22] J. J. Dudek, R. G. Edwards, B. Joo, M. J. Peardon, D. G. Richards and C. E. Thomas, Phys. Rev. D **83**, 111502 (2011) [arXiv:1102.4299 [hep-lat]].

[23] J. Dudek *et al.*, Eur. Phys. J. A **48**, 187 (2012) [arXiv:1208.1244 [hep-ex]].

[24] J. J. Dudek *et al.* [Hadron Spectrum Collaboration], Phys. Rev. Lett. **113**, 182001 (2014) [arXiv:1406.4158 [hep-ph]].

[25] D. J. Wilson, J. J. Dudek, R. G. Edwards and C. E. Thomas, Phys. Rev. D **91**, 054008 (2015) [arXiv:1411.2004 [hep-ph]].

[26] C. J. Shultz, J. J. Dudek and R. G. Edwards, Phys. Rev. D **91**, 114501 (2015) [arXiv:1501.07457 [hep-lat]].

[27] R. A. Briceno, J. J. Dudek, R. G. Edwards, C. J. Shultz, C. E. Thomas and D. J. Wilson, arXiv:1507.06622 [hep-ph].

[28] D. Alde *et al.* [IHEP-Brussels-Los Alamos-Annecy(LAPP) Collaboration], Phys. Lett. B **205**, 397 (1988).

[29] D. Alde *et al.* [IFVE-Brussels-Annecy-Los-Alamos Collaboration], Sov. J. Nucl. Phys. **48**, 1035 (1988) [Yad. Fiz. **48**, 1727 (1988)].

[30] S-U. Chung *et al.* [BNL-E852], Phys. Rev. D**60**, 092001 (1999) [hep-ex/9902003].

[31] G. S. Adams *et al.*, Nucl. Phys. A **680**, 335 (2000).

[32] G. M. Beladidze *et al.* [VES Collaboration], Phys. Lett. B **313**, 276 (1993).

[33] A. Zaitsev [VES Collaboration], Nucl. Phys. A **675**, 155C (2000).

[34] F. Krinner [COMPASS Collaboration], EPJ Web Conf. **96**, 01021 (2015) [arXiv:1501.00133 [hep-ex]].

[35] A. R. Dzierba *et al.*, Phys. Rev. D **73**, 072001 (2006) [hep-ex/0510068].

[36] M. Alekseev *et al.* [COMPASS Collaboration], Phys. Rev. Lett. **104**, 241803 (2010) [arXiv:0910.5842 [hep-ex]].

[37] C. Adolph *et al.* [COMPASS Collaboration], arXiv:1501.05732 [hep-ex].

[38] J. L. Basdevant and E. L. Berger, Phys. Rev. Lett. **114**, 192001 (2015) [arXiv:1504.05955].

[39] Florian Haas, "Two dimensional partial wave analysis of exclusive 190 GeV $\pi^- p$ scattering into the $\pi^-\pi^-\pi^+$ final state at COMPASS", Ph.D. thesis submitted to Technische Universität München, CERN-Thesis-2013-277.

[40] See https://jpac.jlab.org

[41] R. T. Deck, Phys. Rev. Lett. **13**, 169 (1964).

[42] M. G. Bowler, M. A. V. Game, I. J. R. Aitchison and J. B. Dainton, Nucl. Phys. B **97**, 227 (1975).

[43] J. L. Basdevant and E. L. Berger, Phys. Rev. D **16**, 657 (1977).

[44] N. N. Khuri and S. B. Treiman, Phys. Rev. **119**, 1115 (1960).

[45] I. V. Danilkin, C. Fernández-Ramírez, P. Guo, V. Mathieu, D. Schott, M. Shi and A. P. Szczepaniak, Phys. Rev. D **91**, 094029 (2015) [arXiv:1409.7708 [hep-ph]].

[46] P. Guo, I. V. Danilkin and A. P. Szczepaniak, arXiv:1409.8652 [hep-ph].

[47] P. Guo, I. V. Danilkin, D. Schott, C. Fernández-Ramírez, V. Mathieu and A. P. Szczepaniak, arXiv:1505.01715 [hep-ph].

EPJ Web of Conferences **113**, 01015 (2016)
DOI: 10.1051/epjconf/201611301015

Towards a Theoretical Understanding of the XYZ Mesons from QCD

Eric Braaten[1,a]

[1] *Department of Physics, Ohio State University, Columbus OH 43210 USA*

Abstract. The XYZ mesons are mesons that contain a heavy quark and antiquark but have properties that seem to require additional constituents. Some of them are electrically charged, so they must be tetraquark mesons whose additional constituents are a light quark and antiquark. The list of XYZ mesons has grown to about two dozen over the last decade. A promising approach to understanding these mesons within QCD is the Born-Oppenheimer approximation, which reduces the problem to the solution of the Schrödinger equation in potentials that can be calculated using lattice QCD. The Born-Oppenheimer approximation has not yet revealed the pattern of the XYZ mesons, but it provides a compelling framework for understanding them from the fundamental theory.

1 Introduction

The XYZ mesons are more than two dozen new $c\bar{c}$ and $b\bar{b}$ mesons that have been discovered since 2003 [1]. They contain a heavy quark and antiquark, but they have properties that seem to require additional constituents. Some of the more remarkable XYZ mesons are

- $X(3872)$, discovered by the Belle Collaboration in 2003 [2]. It has comparable branching fractions into $J/\psi\,\rho$ and $J/\psi\,\omega$, implying a severe violation of isospin symmetry.
- $Y(4260)$, discovered by the BaBar Collaboration in 2005 [3]. It has J^{PC} quantum numbers 1^{--}, but it is produced very weakly in e^+e^- annihilation.
- $Z^+(4430)$, discovered by the Belle Collaboration in 2007 [4]. It decays into $\psi(2S)\,\pi^+$, which implies that it must be a tetraquark meson with constituents $c\bar{c}u\bar{d}$.
- $Y(4140)$, discovered by the CDF Collaboration in 2009 [5]. It decays into $J/\psi\,\phi$, which suggests that it could be a tetraquark meson with constituents $c\bar{c}s\bar{s}$.
- $Z_b^+(10610)$ and $Z_b^+(10650)$, discovered by the Belle Collaboration in 2011 [6]. They both decay into $\Upsilon\,\pi^+$, which implies that they must be tetraquark mesons with constituents $b\bar{b}u\bar{d}$.
- $Z_c^+(3900)$, discovered by the BESIII Collaboration in 2013 [7]. It decays into $J/\psi\,\pi^+$, which implies that it must be a tetraquark meson with constituents $c\bar{c}u\bar{d}$.

Many of the XYZ mesons are surprisingly narrow. They present a major challenge to our understanding of the QCD spectrum!

[a] e-mail: braaten@mps.ohio-state.edu

2 Models for XYZ Mesons

Most of the theoretical work on the XYZ mesons has been carried out using constituent models. These models fall into three basic categories: (1) *conventional quarkonium*, (2) *quarkonium hybrids*, (3) *quarkonium tetraquarks*.

In a conventional quarkonium, the only constituents are a heavy quark Q and antiquark $\bar{Q}$. There is a well-developed phenomenology for conventional quarkonium based on quark potential models. These models are accurate below the open-heavy-flavor threshold, and they give well-defined predictions above the threshold. They imply that the heavy quarkonium states have energy levels labeled by radial and orbital-angular-momentum quantum numbers nL. Each energy level consists of a heavy-quark-spin multiplet that includes a spin-singlet state and one or three spin-triplet states with specific J^{PC} quantum numbers. For example, an S-wave multiplet consists of the two states $\{0^{-+}, 1^{--}\}$, a P-wave multiplet consists of the four states $\{1^{+-}, (0,1,2)^{++}\}$, and a D-wave multiplet consists of the four states $\{2^{-+}, (1,2,3)^{--}\}$. There are some exotic quantum numbers that are not allowed for quarkonium. If the only constituents are a quark and an antiquark, the forbidden quantum numbers are 0^{--}, 0^{+-}, $1^{-+}, 2^{+-}, 3^{-+}, \ldots$.

In a quarkonium hybrid, the Q and $\bar{Q}$ are accompanied by a gluonic excitation. There are various models for the gluonic excitation. In a *constituent gluon model*, it is a massive spin-1 particle. In a *flux tube model*, it is a vibrational state of a color flux tube extending between the heavy quark and antiquark. In a *gluelump model*, it is gluon fields bound to a compact $Q\bar{Q}$ pair in a color-octet state.

In a quarkonium tetraquark, the Q and $\bar{Q}$ are accompanied by a light quark q and a light antiquark $\bar{q}$. There are many ways the four colored constituents can be clustered in the meson. In a *compact tetraquark*, they are all in overlapping orbitals. In a *meson molecule*, there are two color-singlet clusters $Q\bar{q}$ and $\bar{Q}q$. In a *diquarkonium*, the constituents are clustered into color-triplet diquarks Qq and $\bar{Q}\bar{q}$. In a *hadro-quarkonium*, the $Q\bar{Q}$ pair forms a compact color-singlet core to which the q and $\bar{q}$ are bound. In a *quarkonium adjoint meson*, the $Q\bar{Q}$ pair forms a compact color-octet core to which the q and $\bar{q}$ are bound.

The various constituent models for XYZ mesons make little connection with the fundamental theory QCD. The constituents are plausible degrees of freedom from QCD, but the interactions between them are purely phenomenological. They have had some success in describing individual XYZ mesons, but they have failed to reveal the pattern of XYZ mesons. These models can appropriately be compared to blind-folded physicists studying an elephant. Depending on what aspect of the elephant they focus on, whether it be an ear or a tusk or the trunk or the tail, they arrive at very different conclusions about the nature of an elephant. All of these conclusion have some element of truth, but none of them comes close to capturing the true nature of an elephant.

In order to understand the true nature of the elephant, it is essential to approach the problem from within QCD. The starting point should be the fundamental fields, which are quarks and gluons. It should be expressed in terms of the parameters of QCD, which are the coupling constant α_s and the quark masses. Three approaches that fulfill these requirements are *QCD sum rules*, *lattice QCD*, and the *Born-Oppenheimer approximation*. I will not discuss QCD sum rules, because they seem to be too blunt a method to be very useful for this problem. I will summarize what is known from lattice QCD. I will then turn to the Born-Oppenheimer approximation, which I believe is the key to understanding the XYZ mesons.

3 Lattice QCD

The most extensive calculations of the $c\bar{c}$ meson spectrum using lattice gauge theory were pioneered by Dudek, Edwards, Mathur, and Richards [8] and extended by the Hadron Spectrum Collaboration

[9]. They used lattice QCD on an anisotropic lattice with $24^3 \times 128$ sites, but with light quarks that were too heavy. Their pion mass was 400 MeV, which is about 3 times its physical value. The most important caveats on their results are that they made no extrapolation to the physical up and down quark masses and they made no extrapolation to zero lattice spacing. Their results were impressive nonetheless. They determined the masses of charmonium and charmonium hybrids from the cross-correlators of many operators. They identified 46 statistically significant states with various J^{PC} quantum numbers, with spin J as high as 4 and masses as high as 4.6 GeV. They were also able to discriminate between charmonium and charmonium hybrids based on how strongly the states coupled to operators that had a factor of the gluon field strength. They identified 14 charmonium hybrid candidates. The 4 lowest candidates fill out a complete heavy-quark-spin multiplet: $\{1^{--}, (0,1,2)^{-+}\}$. One of these quantum numbers is exotic, namely 1^{-+}. Their other charmonium hybrid candidates were 3 spin singlets, $(0,1,2)^{++}$, and 7 spin triplets, $(0,1,1,1,2,2,3)^{+-}$. Two of these quantum numbers are exotic, namely 0^{+-} and 2^{+-}.

Lattice gauge theory has also been applied to the $b\bar{b}$ meson spectrum. Since the b quark is three times as heavy as the c quark, it is more practical to use a nonrelativstic effective field theory called NRQCD. The most extensive calculations of the $b\bar{b}$ meson spectrum using lattice NRQCD are still the pioneering calculations by Juge, Kuti, and Morningstar back in 1999 [14]. They used quenched lattice NRQCD on an anisotropic lattice with $15^3 \times 45$ sites. The adjective "quenched" implies that there were no light quarks at all, which necessarily introduces large systematic errors. They identified 4 bottomonium hybrid states, all spin singlets with quantum numbers 1^{--}, 1^{++}, 0^{++}, and a second 1^{--}. The quantum numbers 1^{--} of the lowest state match those of the spin-singlet member of the lowest multiplet of charmonium hybrid candidates in the lattice QCD calculations of the Hadron Spectrum Collaboration. The quantum numbers 1^{++} and 0^{++} of the next two states match those of two of the other three spin-singlet charmonium hybrid candidates.

There have been some applications of lattice gauge theory to charmonium tetraquark mesons, which have the flavor of a light quark and a light antiquark. Prelovsek, Leskovec, and collaborators used lattice QCD on an anisotropic lattice with $16^3 \times 32$ sites, but with two light quarks that were too heavy [10–12]. Their pion mass was 270 MeV, which is about twice its physical value. They studied only a few specific J^{PC} channels, and used a rather small set of interpolating operators. In the $J^{PC} = 1^{++}$ channel, they found a candidate for the $X(3872)$, but it is probably just the $2P$ charmonium state $\chi_{c1}(2P)$. In the $J^{PC} = 1^{+-}$ channel, they found no signal for a Z_c^+ tetraquark.

There have not yet been any applications of lattice gauge theory to bottomonium tetraquark mesons.

4 Born-Oppenheimer Approximation for Quarkonium Hybrids

The Born-Oppenheimer approximation in atomic physics provides a framework for understanding molecules and the low-energy scattering of atoms [13]. The Born-Oppenheimer approximation for QCD was pioneered by Juge, Kuti, and Morningstar, who applied it to quarkonium and quarkonium hybrids [14]. It is based on the fact that the heavy-quark mass m_Q is much larger than the energy scale for gluons. Because they are heavy, the Q and $\bar{Q}$ in a quarkonium ,or in a quarkonium hybrid, move nonrelativistically. The gluons can respond almost instantaneously to the motion of the Q and $\bar{Q}$. Given the positions of the Q and $\bar{Q}$, the gluon fields will be in a stationary state in the presence of static Q and $\bar{Q}$ sources at those positions. As the positions of the Q and $\bar{Q}$ change, the gluon fields will usually remain adiabatically in that stationary state. The energy of a stationary state of gluon fields in the presence of static Q and $\bar{Q}$ sources separated by the distance R defines a Born-Oppenheimer potential $V(R)$. In the simplest Born-Oppenheimer approximation, the motion of the $Q\bar{Q}$ pair is described by the Schrödinger equation in the potential $V(R)$.

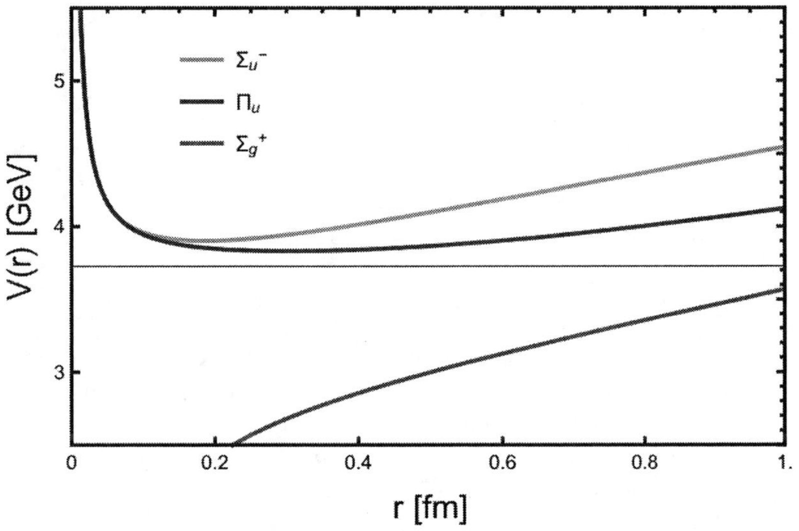

Figure 1. The deepest flavor-singlet Born-Oppenheimer potentials: the quarkonium potential Σ_g^+ and the quarkonium hybrid potentials Π_u and Σ_u^-. An additive constant in the potentials has been chosen so that the energy levels are those of $c\bar{c}$ mesons. The thin horizontal line is the charm-meson-pair threshold. For $b\bar{b}$ mesons, the potentials are the same except for an additional energy offset of about 6.8 GeV.

The stationary states for gluon fields in the presence of static Q and $\bar{Q}$ sources can be labeled by the quantum numbers that are conserved in the presence of the sources. A convenient set of quantum numbers are

- the absolute value $|\hat{\boldsymbol{R}} \cdot \boldsymbol{J}_{\rm light}|$ of the component of the angular momentum of the light fields along the axis defined by the sources. It has integer values $0, 1, 2, \ldots$, but they are traditionally labeled by upper case Greek letters $\Sigma, \Pi, \Delta, \ldots$,
- the product $(CP)_{\rm light}$ of the charge conjugation and parity of the light fields. Its values are $+1$ or -1, but they are traditionally labeled by a subscript g or u,
- in the case of Σ, an additional quantum number for a reflection through a plane containing the sources. Its values are $+1$ or -1, but they are traditionally labeled by a superscript $+$ or $-$ on the Σ.

Thus the Born-Oppenheimer potentials are labeled Λ_η, where $\Lambda = \Sigma^+, \Sigma^-, \Pi, \Delta, \ldots$ and $\eta = g, u$.

Many of the Born-Oppenheimer potentials were calculated by Juge, Kuti, and Morningstar in 1999 using quenched lattice QCD on an anisotropic lattice with $10^3 \times 30$ sites [14]. The deepest Born-Oppenheimer potentials are Σ_g^+, then Π_u, and then Σ_u^-. These potentials are illustrated in Figure 1. Juge, Kuti, and Morningstar also solved the Schrödinger equation in the Born-Oppenheimer potentials. The energy levels are labeled by nL, where $n = 1, 2, 3, \ldots$ is a radial quantum number and $L = 0, 1, 2, \ldots$ (or $S, P, D, \ldots$) is an orbital-angular-momentum quantum number. The energy levels in the Σ_g^+ potential are quarkonium states. The energy levels in the excited potentials, including Π_u and Σ_u^-, are quarkonium hybrids. Juge, Kuti, and Morningstar found that the predictions of the Born-Oppenheimer approximation for the energy splittings between the bottomonium hybrid energy levels were in semiquantitative agreement with their direct calculations using quenched lattice NRQCD [14]. This convinced me back in 1999 that quarkonium hybrids are real states in the QCD spectrum.

The lowest charmonium hybrid energy levels predicted by the Born-Oppenheimer approximation are

- the $\Pi_u(1P)$ energy level, which consists of the two heavy-quark-spin multiplets $\{1^{--}, (0,1,2)^{-+}\}$ and $\{1^{++}, (0,1,2)^{+-}\}$,
- the $\Pi_u(1D)$ energy level, which consists of the two multiplets $\{2^{--}, (1,2,3)^{-+}\}$ and $\{2^{++}, (1,2,3)^{+-}\}$,
- the $\Sigma_u^-(1S)$ energy level, which consists of the multiplet $\{0^{++}, 1^{+-}\}$.

The lowest multiplet $\{1^{--}, (0,1,2)^{-+}\}$ matches the lowest multiplet of charmonium hybrid candidates in the lattice QCD calculations of the Hadron Spectrum Collaboration [9]. Their other 10 charmonium hybrid candidates have quantum numbers that match all the states in three of the next 4 highest Born-Oppenheimer energy levels. Thus lattice QCD seems to reproduce the pattern of states predicted by the Born-Oppenheimer approximation, even when the light quarks are too heavy.

We proceed to discuss the Born-Oppenheimer potentials at small R. The lowest hybrid potentials are Π_u and then Σ_u^-. At small R, these potentials increase roughly like $1/R$. The Q and $\bar{Q}$ sources behave like a local color-octet source and therefore like a gluino, which is a hypothetical heavy particle predicted by supersymmetry that has the same color-octet charge as the gluon. The stationary state of gluon fields bound to a color-octet source (or a gluino) is called a *gluelump*. The potential at small R is the repulsive Coulomb potential between the Q and $\bar{Q}$ plus the energy of the gluelump. The gluelump spectrum has been calculated using lattice QCD. The most recent calculations were by Marsh and Lewis using an anisotropic lattice with $28^3 \times 56$ sites, but with light quarks that were too heavy [15]. Their pion mass is $m_\pi = 480$ MeV, which is about 3 times its physical value. The lowest-energy gluelump has quantum numbers 1^{+-}. The next gluelump has quantum numbers 1^{--} and energy larger by about 300 MeV. The energy of the 1^{+-} gluelump determines the common limit as $R \to 0$ of the Π_u and Σ_u^- potentials. Thus, just by knowing the quantum numbers 1^{+-} of the lowest-energy gluelump, one can infer correctly that the deepest hybrid Born-Oppenheimer potentials are probably Π_u and Σ_u^-.

We now discuss the Born-Oppenheimer potentials at large R. If there are no light quarks, all the Born-Oppenheimer potentials increase linearly with R at large R. The corresponding stationary state is a color flux tube extending between the Q and $\bar{Q}$ sources. The asymptotic behavior of the potential at large R is $V(R) \to \sigma R$, where σ is the energy per length of the color flux tube. If there are light quarks, the lowest-energy stationary state at large R actually consists of 2 static mesons, one consisting of a light antiquark bound to the Q source and the other consisting of a light quark bound to the $\bar{Q}$ source. In this case, the potential approaches a constant at large R that is equal to twice the energy of a static meson. One might expect that the linearly-rising quarkonium and hybrid potentials cease to exist above the energy of the static-meson-pair threshold. However stationary states with linearly-rising potentials do exist at higher energies, but they have avoided crossings with the threshold. This has been shown convincingly for the quarkonium Σ_g^+ potential by the SESAM Collaboration using lattice QCD [16]. They used an anisotropic lattice with $24^3 \times 40$ sites, but with light quarks that are too heavy. Their pion mass was 540 MeV, which is about 4 times its physical value. This work demonstrated that avoided crossings between Born-Oppenheimer potentials can be calculated using lattice QCD.

5 Born-Oppenheimer Approximation for Quarkonium Tetraquarks

Those XYZ mesons that have an electric charge are definitely quarkonium tetraquarks whose constituents include a $Q\bar{Q}$ pair and a light quark and antiquark. Some of the neutral XYZ mesons may also be quarkonium tetraquarks. Quarkonium tetraquarks can be treated using the Born-Oppenheimer approximation just like quarkonium hybrids [17]. This is because light quarks will respond almost instantaneously to the motion of the heavy quarks, just like gluon fields. The difference is that the stationary state of gluons and light quarks in a quarkonium tetraquark not only has Born-Oppenheimer quantum numbers, but also light-quark+antiquark flavors. Unfortunately there are as yet no lattice QCD calculations of tetraquark Born-Oppenheimer potentials.

There is one hint from lattice QCD about the tetraquark Born-Oppenheimer potentials, and that comes from calculations of the *adjoint meson* spectrum. An adjoint meson consists of light-quark and gluon fields with light-quark+antiquark flavor that are bound to a color-octet source (or a gluino). The adjoint meson spectrum was calculated by Foster and Michael using quenched lattice QCD on an anisotropic lattice with $24^3 \times 48$ sites [18]. They found that the quantum numbers of the lowest-energy adjoint meson was 1^{--} or 0^{-+}. The energy of the 1^{--} adjoint meson determines the common limit as $R \to 0$ of the Π_g and Σ_g^+ potentials. The energy of the 0^{-+} adjoint meson determines the limit as $R \to 0$ of the Σ_u^- potential. Thus, if the quenched lattice QCD calculations have correctly identified the quantum numbers of the lowest-energy adjoint mesons, one can infer that the deepest tetraquark Born-Oppenheimer potentials are probably Π_g and Σ_g^+ or Σ_u^-. The Born-Oppenheimer potentials may be different for each of three light-quark+antiquark flavors: isospin 1, isospin 0 and $s\bar{s}$.

The energy levels in the tetraquark Born-Oppenheimer potentials are labeled by quantum numbers nL. If the quenched lattice QCD calculations have correctly identified the the quantum numbers of the lowest-energy adjoint mesons, some of the lowest energy levels are likely to be

- the $\Pi_u(1P)$ energy level, which consists of the heavy-quark-spin multiplets $\{1^{-+}, (0,1,2)^{--}\}$ and $\{1^{+-}, (0,1,2)^{++}\}$,
- the $\Sigma_g^+(1S)$ energy level, which consists of the multiplet $\{0^{-+}, 1^{--}\}$,
- the $\Sigma_u^-(1S)$ energy level, which consists of the multiplet $\{0^{++}, 1^{+-}\}$.

There will be Born-Oppenheimer energy levels such as these for each of three light-quark+antiquark flavors: isospin 1, isospin 0 and $s\bar{s}$. Thus there are many possibilities for the J^{PC} and flavor quantum numbers of the quarkonium tetraquarks.

6 XYZ Mesons from the Born-Oppenheimer Perspective

The various constituent models for the XYZ mesons are based on different assumptions about their nature. What is their nature from the Born-Oppenheimer perspective? Are they compact tetraquarks? Are they diquarkonium? Are they adjoint mesons? Are they meson molecules? The answer is all of the above! Each of these possibilities describes some region of the Born-Oppenheimer wavefunction. In the small-R region where the hybrid potentials are repulsive Coulomb potentials, the $Q\bar{Q}$ pair form a compact color-octet source, so the configuration is like a gluelump or an adjoint meson. In the large-R region where the potential increases linearly, there is a color flux tube connecting a diquark consisting of the light quark bound to the Q and another diquark consisting of the light antiquark bound to the $\bar{Q}$, so the configuration is like a diquarkonium. In the intermediate region, so the configuration is like a compact tetraquark. In the region of R beyond an avoided crossing, the configuration is two separated mesons, so it is like a meson molecule.

The Born-Oppenheimer approach implies that quarkonium hybrids and quarkonium tetraquarks definitely exist as states in the QCD spectrum. Whether they can be observed in experiments depends on how narrow they are, how easily they can be produced, and whether they have favorable decay modes. The Born-Oppenheimer approximation has not yet revealed a compelling pattern for the XYZ mesons There are too many unknown Born-Oppenheimer potentials and too few XYZ mesons with known J^{PC} quantum numbers. Selection rules for hadronic transitions between Born-Oppenheimer configurations have been derived [19, 20]. They provide useful constraints, but these constraints have proven to be insufficient to reveal the pattern of the XYZ mesons.

7 Conclusions

The discoveries of the XYZ mesons have revealed a serious gap in our understanding of the QCD spectrum. Constituent models for the XYZ mesons have not presented a compelling pattern, and they make little contact with QCD. The Born-Oppenheimer approximation has not yet provided a compelling pattern for the XYZ mesons, but it is based firmly on QCD.

To fully develop the Born-Oppenheimer approximation would require a lot of information from lattice QCD. One simple problem is the calculation of the spectrum of gluelumps and adjoint mesons, which determines the small-R behavior of the Born-Oppenheimer potentials. It would be useful to have calculations of the $b\bar{b}$ hybrid meson spectrum with quality comparable to that of the $c\bar{c}$ hybrid meson spectrum calculated by the Hadron Spectrum Collaboration. Some more challenging problems are calculations of the avoided crossings between the Born-Oppenheimer potentials and meson-pair thresholds and calculations of the tetraquark Born-Oppenheimer potentials.

There are also many things that are needed from non-lattice theory. In the absence of lattice QCD calculations of the Born-Oppenheimer potentials, it might be possible to infer them from data on XYZ mesons. After all, the quarkonium potential was inferred quite accurately from very limited data on the charmonium spectrum way back in 1975 [21]. It would be useful to develop quantitative phenomenological models for hadronic transitions between quarkonium, quarkonium hybrids, and quarkonium tetraquarks, because most of the XYZ mesons have been observed through hadronic transitions. Finally, it would be desirable to develop an effective field theory in which the Born-Oppenheimer approximation arises as a first approximation, so that corrections can be calculated systematically.

Theory has so far failed to reveal the pattern of the XYZ mesons. Fortunately, future experimental progress on the XYZ mesons is guaranteed. There are ongoing e^+e^- experiments at the charm factory in China, and upcoming e^+e^- experiments at the bottom factory Super KEK-B in Japan. In the ongoing pp collision experiments at the LHC, the ATLAS and CMS detectors can contribute to the study of XYZ mesons and the LHCb detector is especially well-suited for this purpose. Finally, on the horizon, there is the PANDA experiment, which will use $p\bar{p}$ annihilation at resonance to produce charmonium, charmonium hybrids, and charmonium tetraquarks. What is needed from experiment is more J^{PC} quantum numbers, more transitions (hadronic and radiative), and more XYZ mesons. With enough new information and with the Born-Oppenheimer approximation as a guiding principle, the solution to the puzzle of the XYZ mesons is inevitable.

Acknowledgements

This work was supported in part by the Department of Energy, by the National Science Foundation, and by the Simons Foundation.

References

[1] G.T. Bodwin, E. Braaten, E. Eichten, S.L. Olsen, T.K. Pedlar and J. Russ, arXiv:1307.7425.

[2] S.K. Choi *et al.* [Belle Collaboration], Phys. Rev. Lett. **91**, 262001 (2003) [hep-ex/0309032].

[3] B. Aubert *et al.* [BaBar Collaboration], Phys. Rev. Lett. **95**, 142001 (2005) [hep-ex/0506081].

[4] S.K. Choi *et al.* [Belle Collaboration], Phys. Rev. Lett. **100**, 142001 (2008) [arXiv:0708.1790].

[5] T. Aaltonen *et al.* [CDF Collaboration], Phys. Rev. Lett. **102**, 242002 (2009) [arXiv:0903.2229].

[6] A. Bondar *et al.* [Belle Collaboration], Phys. Rev. Lett. **108**, 122001 (2012) [arXiv:1110.2251].

[7] M. Ablikim *et al.* [BESIII Collaboration], Phys. Rev. Lett. **110**, 252001 (2013) [arXiv:1303.5949].

[8] J.J. Dudek, R.G. Edwards, N. Mathur and D.G. Richards, Phys. Rev. D **77**, 034501 (2008) [arXiv:0707.4162].
[9] L. Liu *et al.* [Hadron Spectrum Collaboration], JHEP **1207**, 126 (2012) [arXiv:1204.5425].
[10] S. Prelovsek and L. Leskovec, Phys. Rev. Lett. **111**, 192001 (2013) [arXiv:1307.5172].
[11] S. Prelovsek and L. Leskovec, Phys. Lett. B **727**, 172 (2013) [arXiv:1308.2097].
[12] S. Prelovsek, C.B. Lang, L. Leskovec and D. Mohler, Phys. Rev. D **91**, 014504 (2015) [arXiv:1405.7623].
[13] M. Born and J.R. Oppenheimer, Annalen der Physik **389**, 457 (1927).
[14] K.J. Juge, J. Kuti and C.J. Morningstar, Phys. Rev. Lett. **82**, 4400 (1999) [hep-ph/9902336].
[15] K. Marsh and R. Lewis, Phys. Rev. D **89**, 014502 (2014) [arXiv:1309.1627].
[16] G.S. Bali *et al.* [SESAM Collaboration], Phys. Rev. D **71**, 114513 (2005) [hep-lat/0505012].
[17] E. Braaten, Phys. Rev. Lett. **111**, 162003 (2013) [arXiv:1305.6905].
[18] M. Foster *et al.* [UKQCD Collaboration], Phys. Rev. D **59**, 094509 (1999) [hep-lat/9811010].
[19] E. Braaten, C. Langmack and D.H. Smith, Phys. Rev. Lett. **112**, 222001 (2014) [arXiv:1401.7351].
[20] E. Braaten, C. Langmack and D.H. Smith, Phys. Rev. D **90**, 014044 (2014) [arXiv:1402.0438].
[21] E. Eichten, K. Gottfried, T. Kinoshita, J.B. Kogut, K.D. Lane and T.M. Yan, Phys. Rev. Lett. **34**, 369 (1975).

EPJ Web of Conferences **113**, 020 01 (2016)
DOI: 10.1051/epjconf/201611302001

Studies of HeH: Dissociative Excitation

E. Ertan[1,a], Å. Larson[1,b], and A. E. Orel[2,c]

[1] *Department of Physics, Stockholm University, SE-106 91 Stockholm, Sweden*
[2] *Department of Chemical Engineering and Materials Science, University of California, Davis, USA*

Abstract. We have used structure and scattering calculations to determine the potential energy curves, non-adiabatic couplings and autoionization widths for the HeH system. These will be used to study a variety of processes ranging from dissociative recombination to mutual neutralization. As an example, we present our results on the direct dissociative excitation of HeH^+ by electron impact via excitation to the two lowest excited states of the ion. The results are found to be in good agreement with experiment.

1 Introduction

The collision of an electron with a molecular ion such as HeH^+ results in a number of processes. These range from elastic scattering to inelastic processes, such as vibrational and rotational excitation, and at higher energies, electron-impact excitation to a dissociative electronic state can result in direct dissociative excitation. In addition, there are a number of resonant processes. Every state of the ion can serve as a parent ion for a series of neutral states. Below the ground state, these are a Rydberg series of neutral states converging to the ion. However, if the parent state is an excited state of the ion, these states are doubly excited (Feshbach) resonances that lie below the direct dissociation threshold, and are formed when an incoming electron excites the target ion and attaches to a Rydberg orbital. Capture into these states can initiate a number of resonant processes. As the neutral molecule evolves in time, the system can autoionize, meaning the electron can be re-emitted, returning the molecule to its original electronic state. If the ion does not have enough energy to dissociate to products, the target is left in some state of vibrational excitation. However, if the ion has enough energy to dissociate, the process is resonant dissociative excitation and it provides an efficient path to dissociation at energies below the direct excitation level. If, while evolving in time, the neutral fragments reach an internuclear separation beyond which autoionization is no longer possible, the states are considered electronically bound and the result is dissociative recombination. Depending on the electron affinity and ionization potential of the final products, it may be energetically possible to have ion-pair formation.

The study of these processes requires both the accurate treatment of the electron scattering processes, but must also include an accurate representation of the potential energy curves, both for electronically bound states and the resonant state. In addition, the couplings between these states, both the coupling between the resonant states and the scattering continuum (the autoionization width) and the

[a] e-mail: emelie.ertan@fysik.su.se
[b] e-mail: aasal@fysik.su.se
[c] e-mail: aeorel@ucdavis.edu

non-adiabatic coupling between all states are needed to complete describe the cross section including the branching ratios into final states. These same curves and couplings mediate another series of collision processes such as Penning ionization, associative ionization or mutual neutralization.

We have focused on the determination of accurate potential energy curves for the ion, Rydberg and resonant states for the HeH system, as well as autoionization widths and non-adiabatic coupling elements between all neutral states including the autoionizing states. These calculations included structure calculations as well as electron scattering calculations, using the complex Kohn variational method [1, 2] to obtain the autoionization widths and T-matrix elements (used for the direct excitation cross section calculations). Details on these structure and scattering calculations can be found in [3]. As an example, we will present the results of our calculations on direct dissociative excitation.

One of the earliest published experimental studies on dissociative excitation (DE) of HeH^+ is the work of F. B. Yousif and J. B. A. Mitchell from 1989 [4]. In this work the dissociative recombination (DR) and DE processes of HeH^+ were studied using a merged beam method. The cross sections for DE were reported in the $0-40$ eV energy range. The results showed an excitation energy threshold at about 20 eV for the low extraction conditions, where the ions are believed to be mainly in the ground electronic state. Series of sharp and very narrow peaks in the cross section were detected in the $20-26$ eV energy region. The narrowness of the peaks was suggested to originate from a process where the electron is trapped instantaneously into doubly excited neutral resonant states.

The findings of Yousif and Mitchell prompted a theoretical study [5]. In this work the DE of HeH^+ was studied in the $20-26$ eV energy region using the complex Kohn variational method. Excitation cross sections for the $X^1\Sigma \rightarrow a^3\Sigma^+$ transition were computed in overall $^2\Sigma^+$ and $^2\Pi$ symmetries as well as the total cross section at the equilibrium separation ($R_0 = 0.77$ Å). The calculation of the fixed-nuclei cross section resulted in a series of sharp peaks on a quite flat background. Closer inspection showed that most of the peaks were Feshbach resonances associated with energetically closed Rydberg states in this energy region. One of the peaks, situated at 24 eV, did not belong to the above mentioned category but proved to be a core-excited shape resonance. Further, it was shown that an autoionization process from an doubly excited state, as suggested by Yousif and Mitchell, was not a viable explanation of the narrowness of the peaks observed in the experiment. The computations in $^2\Sigma^+$ symmetry were also performed at $R = R_0 \pm 0.05$ Å in order to investigate how the cross section responds to changes in the internuclear distance. The results from these calculations showed that the widths of the resonance peaks and the value of the background cross section remained almost unchanged. The positions of the peaks were shifted with the excitation energy of the $X^1\Sigma - A^1\Sigma$ transition. The direct DE cross section was computed by integrating the fixed nuclei inelastic cross section over the square of the vibrational wave function of the target ion. The sharp peaks observed in the fixed-nuclei cross section were then smoothed out.

A second experimental study of the DE of HeH^+ was performed by C. Strömholm *et al.* [6]. In this work the DR and DE processes for HeH were studied and the absolute cross sections were determined for energies below 40 eV. The experiments were performed using CRYRING ion storage ring at the Manne Siegbahn Laboratory at Stockholm University. Contrary to the results of the cross section obtained by Yousif and Mitchell, it was found here that the absolute cross section for the direct DE process was basically constant in the $21-37$ eV energy region. Furthermore, it was found that there was an alternate DE pathway with an energy threshold already at 10 eV. In the reaction the electron is captured into a neutral doubly excited state which auto-ionizes into $He + H^+$. This reaction is resonant dissociative excitation which competes with the DR process.

The results of the direct DE cross section for the HeH^+ of the above mentioned studies are displayed in Fig. 1.

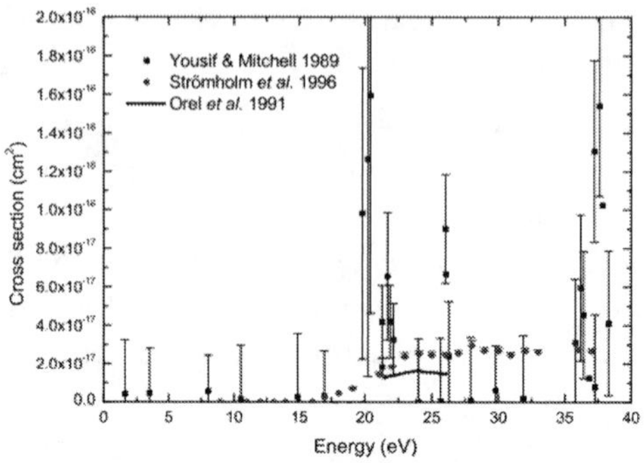

Figure 1. Previous results for the cross section of the direct DE of HeH^+ obtained from the experimental works of Yousif and Michell [4], Strömholm *et al.* [6] and a theoretical study of Orel *et al.* [5]

2 Theoretical formulation

In the work Orel *et al.* [5] a time-independent expression for a total cross section was derived by means of a delta-function approximation and using the fixed-nuclei excitation cross sections. We will now give a more detailed description of this expression.

The fixed-nuclei excitation cross sections are given by [5]

$$\tilde{\sigma}^{\Lambda}_{nn'}(E,R) = \frac{2\pi}{E} \sum_{\ell_0 \ell m_0 m} |T^{\Lambda,nn'}_{\ell_0\ell m_0 m}(E,R)|^2. \tag{1}$$

Here $T^{\Lambda,nn'}_{\ell_0\ell m_0 m}(E,R)$ is the fixed nuclei T-matrix element on the energy shell.

We will assume that a time-independent expression for the excitation cross section of DE can be formed by applying the adiabatic-nuclei approximation [7, 8],

$$\sigma^{\Lambda}_{nn'}(E) \approx \frac{2\pi}{E} \sum_{\ell_0 \ell m_0 m} \int_{E_0}^{E} |\langle \psi_{E'}(R)| T^{\Lambda,nn'}_{\ell_0\ell m_0 m}(E,R)|\chi_{v_0}(R)\rangle|^2 dE'. \tag{2}$$

Here $\psi_{E'}$ is an energy-normalized continuum function. Λ refers the overall symmetry of the scattering and $n, n' = 0, 1, 2$ are the electronic states of the target. In this work $0 \rightarrow 1$ and $0 \rightarrow 2$ scattering is studied. E refers to the scattering energy and E' is the energy of the dissociative nuclear state. Hence, the energy of the ejected electron is given by $E - E'$. E_0 is the asymptotic energy of the repulsive potential energy curve. Further, we also approximate the energy-normalized continuum wave function with the Dirac-delta function (details of this approximation can be found in [9] and references therein),

$$\psi_{E'}(R) \approx \left(\sqrt{\frac{dU}{dR}\Big|_{R_{E'}}} \right)^{-1} \delta(R - R_{E'}), \tag{3}$$

where $U(R)$ is the potential energy curve excited ionic state and $R_{E'}$ is the classical turning point at energy E'. Inserting Eq. (3) into Eq. (2) yields

$$\sigma^{\Lambda}_{nn'}(E) \approx \frac{2\pi}{E} \int_{E_0}^{E} \left(\frac{dU}{dR}\Big|_{R_{E'}} \right)^{-1} \sum_{\ell_0 \ell m_0 m} |T^{\Lambda,nn'}_{\ell_0\ell m_0 m}(E, R_{E'})|^2 [\chi_{v_0}(R_{E'})]^2 dE'. \tag{4}$$

Hence, we can write Eq. (4) as

$$\sigma_{nn'}^{\Lambda}(E) = \int_{E_0}^{E} \left(\frac{dU}{dR}\Big|_{R_{E'}} \right)^{-1} \tilde{\sigma}_{nn'}^{\Lambda}(E, R_{E'})[\chi_{v_0}(R_{E'})]^2 dE'. \tag{5}$$

Making the following change of variables,

$$\begin{aligned} U(R_{E'}) &= E' \\ dE' &= \frac{dU}{dR_{E'}} dR_{E'}, \end{aligned} \tag{6}$$

yields the following expression

$$\sigma_{nn'}^{\Lambda}(E) = \int_{R_E}^{\infty} \tilde{\sigma}_{nn'}^{\Lambda}(E, R)[\chi_{v_0}(R)]^2 dR. \tag{7}$$

From Eq. (7) we see that the total fixed-nuclei excitation cross section is multiplied by the square of the vibrational wave function of the initial state of the target and integrated over the internuclear distance, R. This is the formula used by Orel *et al.* [5].

3 Results

In order to calculate the total cross section using Eq. (7), the fixed-nuclei inelastic excitation cross section for a certain internuclear distance, $\tilde{\sigma}_{nn'}^{\Lambda}(E, R)$ is needed. These calculations were carried out using the complex Kohn variational method [1, 2]. The calculations used the aug-cc-pVQZ [10] basis set for He and the aug-cc-pVTZ [11] basis set for H. One extra diffuse d-functions was also added on He, resulting in a total of 106 functions. Using these basis sets a SCF calculation on the ionic ground state was performed. Then a full CI calculation was preformed on the three lowest excited states of the ion. From the full CI, natural orbitals are computed. All the possible excitations of the three electrons within the ten lowest natural orbitals form the reference configurations for the MRCI calculation. Additional single external excitations are also included. From each calculation the total fixed-nuclei elastic and inelastic scattering cross sections, T-matrix elements etc. are obtained. In the present study scattering with partial wave with $\ell \leq 6$, $|m| \leq 4$ are included. The calculations were done on a grid of internuclear distances R and $\tilde{\sigma}_{nn'}^{\Lambda}(E, R)$ is computed for the complete E-grid for each R. The fixed nuclei inelastic cross sections at 1.45 a_0 (near equilibrium distance) for A_1 symmetry is displayed in Fig. 2 as the solid lines.

As discussed above, along with the direct DE, which dissociates into He^++H there are simultaneous resonant processes in which the excited electron is temporarily captured into a Rydberg state from which it eventually autoionizes to He+H^+. The output of the scattering calculations includes both the direct and the resonant process. Thus, we would have to include both direct DE and resonant processes, as well as any interactions of the Rydberg states with the continuum, to exactly describe the full reaction mechanism. However, since the experiment will only measure the direct DE cross section, the contribution from the resonant states can be removed. There are more and less sophisticated methods to deal with the resonances. Here, a somewhat "brute force" method, consisting of removing any data points in the scattering output where there are resonance behavior, is employed. The data set without the resonances are then splined onto the same R-grid as the original data. These results at 1.45 a_0 (near equilibrium distance) for A_1 symmetry are displayed in Fig. 2 as the dashed lines. Using the fixed-nuclei excitation cross sections, $\tilde{\sigma}_{nn'}^{\Lambda}(E, R)$, calculated with the full T-matrix obtained from the complex Kohn variational method, we compute an averaged fixed-nuclei cross section using Eq. (7)

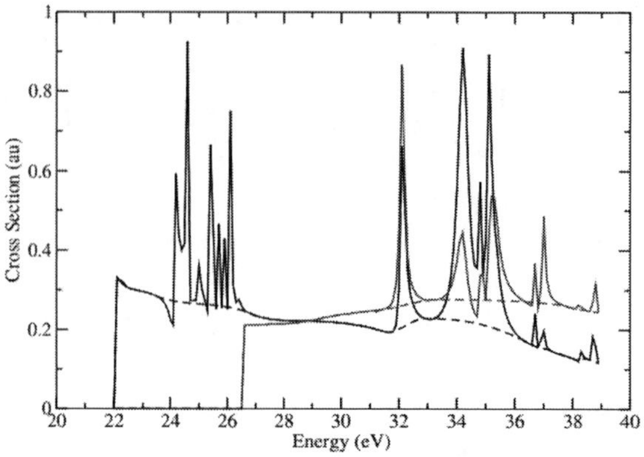

Figure 2. The fixed-nuclei inelastic cross sections at internuclear distance 1.45 a_0. Excitation to the $^3\Sigma$ and the $^1\Sigma$ states are shown with the black and red lines, respectively. The dashed lines show the cross sections when contributions from resonances are removed.

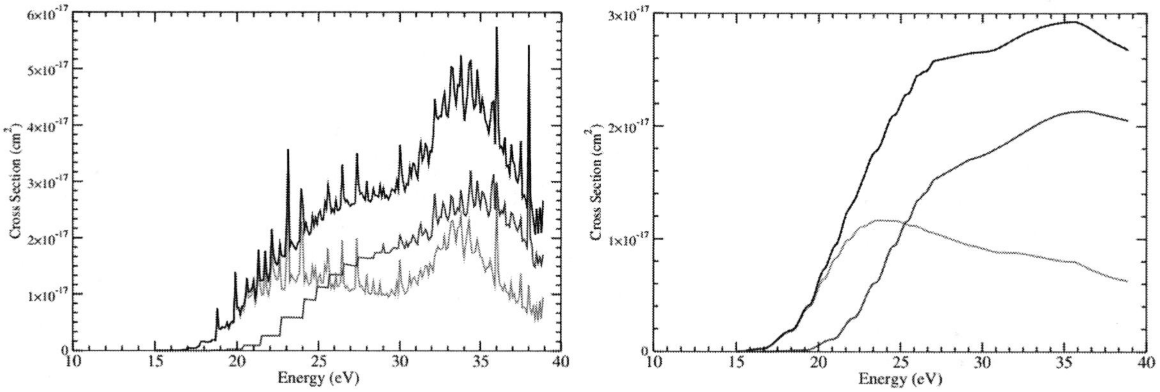

Figure 3. The direct DE cross section obtained using Eq. (7) for excitation to the $^3\Sigma$ state, green line; $^1\Sigma$, red line; and the total, black line. Left: with resonances; Right: With resonances removed.

both with and without removing the resonances. These results are shown in Fig. 3. There are sharp structures from the resonant states over the entire energy interval. The removal of the resonances as described above produces a smooth total cross section.

The total cross section obtained in the above described manner is compared with the experimental result of Strömholm *et al.* [6] in Fig 4. We can see that our theoretical result shows quite good agreement with the experimental result.

4 Conclusions

Our results on direct dissociative excitation of HeH^+ by electron impact via excitation to the two lowest excited states of the ion show good agreement with experiment. Future calculations to explore other processes in the system such as resonant dissociative excitation, dissociative recombination and mutual neutralization are planned.

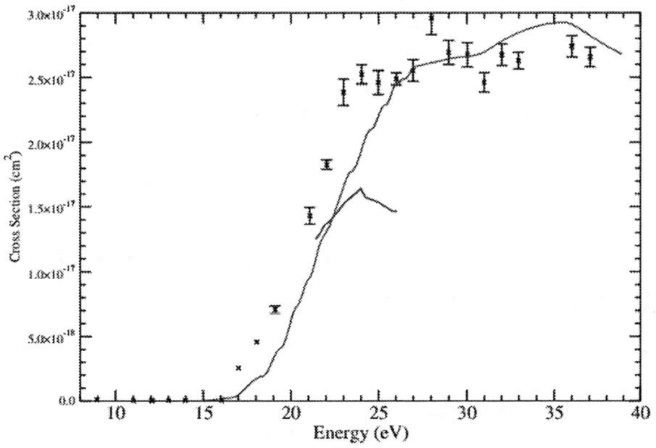

Figure 4. The total direct DE cross section (red line) is compared with the experimental result of Strömholm *et al.* [6] (black symbols) and the theoretical result of Orel *et al.* [5] (blue line).

Acknowledgements

This work is supported by the National Science Foundation, Grant No PHY-11-60611. In addition some of this material is based on work done while A.E.O. was serving at the NSF. Å.L. acknowledges support from the Swedish Research council, Grant No. 2014-4164, and the Carl Trygger Foundation.

References

[1] T. N. Rescigno, C. W. McCurdy, A. E. Orel and B. H. Lengsfield III, *Computational Methods for Electron-Molecule Collisions*, ed. by W. M. Huo and F. A. Gianturco (Plenum Press, New York 1995).
[2] B. I. Schneider and T. N. Rescigno, Phys. Rev. A **37**, 3749 (1988).
[3] Å. Larson, S. Nkambule, E. Ertan, J. Söder and A. E. Orel, EPJ Conf. **84**, 03001 (2015).
[4] F. B. Yousif and J. B. A. Mitchell, Phys. Rev. A **40**, 4318 (1989).
[5] A. E. Orel, T. N. Rescigno and B. H. Lengsfield III, Phys. Rev. A **44**, 4328 (1991).
[6] C. Strömholm, J. Semaniak, S. Rosén, H. Danared, S. Datz, W. van der Zande and M. Larsson, Phys. Rev. A **54**, 3086 (1996).
[7] D. M. Chase, Phys. Rev. **104**, 838 (1956).
[8] M. Shugard and A. U. Hazi, Phys. Rev. A **12**, 1895 (1975).
[9] T. F. O'Malley, Phys. Rev. **150**, 14 (1966).
[10] D. E Woon and T. H. Dunning, Jr., J. Chem. Phys. **100**, 2975 (1994).
[11] T. H. Dunning, Jr., J. Chem. Phys. **90**, 1007 (1989).

EPJ Web of Conferences **113**, 02002 (2016)
DOI: 10.1051/epjconf/201611302002

HD + *ortho-/para*-H_2 rotational energy transfer non-reactive scattering at low astrophysical temperatures: 2 K $\lesssim$ T $\lesssim$ 300 K

Renat A. Sultanov[1,a] and Dennis Guster[1,b]

[1] *Department of Information Systems, BCRL & Integrated Science and Engineering Laboratory Facility (ISELF) at St. Cloud State University, St. Cloud, MN, USA*

Abstract. State-resolved integral cross sections and corresponding thermal rate coefficients are computed for the HD + *ortho-/para*-H_2 rotational energy transfer collision at low temperatures of astrophysical interest: 2 K $\lesssim$ T $\lesssim$ 300 K. A recent original full-dimensional H_2-H_2 potential energy surface (PES) was adopted and appropriately modified for the current non-symmetrical four-atomic system.

1 Introduction

One of the most important few-body systems in the field of atomic and molecular physics is, probably, the non-reactive scattering problem between two hydrogen molecules, i.e. H_2+H_2. The four-atomic symmetrical system, H_2+H_2, represents the two simplest and lightest identical diatomic molecules. In astrophysical applications H_2 can play a significant role, since it is the most abundant chemical species in the universe's molecular clouds. Consequently, the H_2-H_2 interaction and collisions have always attracted significant theoretical and experimental attention. The system has been considered in many previous works, see for example [1–5]. After H_2, the second most abundant primordial molecule is HD - deuterated hydrogen, where D is the hydrogen isotope ^{2}H - deuterium. Just like the H_2 + H_2 collision, the HD + H_2 energy transfer collision is also of significant importance in the astrophysics of the early Universe and in the modeling of pre-galactic clouds and planetary atmospheres. The HD molecule plays an important role in the chemistry of the interstellar medium [6] as well as the rotational and possibly vibrational energy transfer collisions between H_2 molecules. Further, collisions between H_2 and HD are important for the cooling of primordial gas and in the formation of first stars [7, 8]. In order to compute elastic/in-elastic scattering outputs, one needs to construct a PES between colliding molecules first. The HD-H_2 potential can be obtained from the initial H_2-H_2 potential by shifting the original H_2 molecule's center of mass (c.m.) to the coordinate of the c.m. of the HD molecule. Once the exchange symmetry is broken in H_2+H_2 by replacing the H by the D atom in one H_2 one has the new HD-H_2 PES. Thereby, one can derive the HD-H_2 PES from the original H_2-H_2 PES employing all parts of the full HD-H_2 interaction including the HD's dipole moment. Using the HD-H_2 PES derived from the H_2-H_2 PES of Hinde [4] we carry out full quantum calculations for collisions of rotationally excited HD and H_2 molecules, i.e. the process:

$$\mathrm{HD}(j_1) + \mathrm{H}_2(j_2) \rightarrow \mathrm{HD}(j_1') + \mathrm{H}_2(j_2'). \tag{1}$$

[a] e-mail: rasultanov@stcloudstate.edu
[b] e-mail: dcguster@stcloudstate.edu

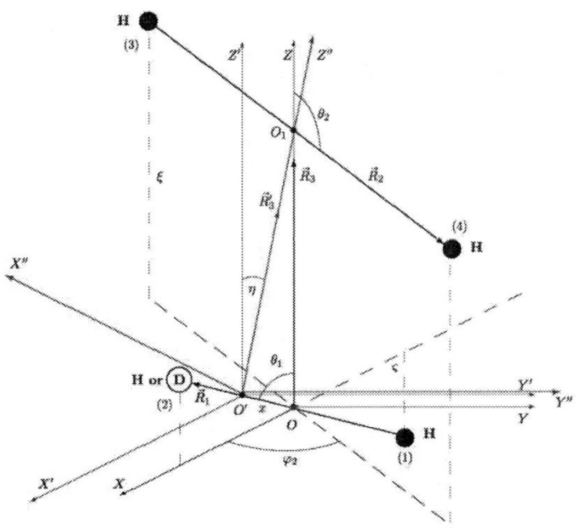

Figure 1. (Color online) Four-atomic system (1, 2, 3, 4) or H-D-H-H is shown together with its *few-body Jacobi coordinates* $\{\vec{R}_1, \vec{R}_2, \vec{R}_3\}$. The original Cartesian coordinate system is $OXYZ$. The center of mass of the original H_2 molecule lies at point O. $O'X'Y'Z'$ is a system which was shifted in parallel from the original system, O' lies in the center of mass of the actual HD molecule. The close-coupling equations are solved using the space-fixed coordinate system $O'X''Y''Z''$, which is obtained by rotation over the axis OY'. The vector $\vec{R}'_3$ connects the center of masses of the HD and H_2 molecules. HD is the first molecule with a rotational constant $B_e(1)$=44.7 cm^{-1} and quantum angular momentum of j_1, H_2 is the second molecule with rotational constant of $B_e(2)$=60.8 cm^{-1} and uses momentum j_2 in the system; $R_1 = 0.7631$ Å and $R_2 = 0.7668$ Å are fixed interatomic distances in each hydrogen molecule HD and H_2 respectively.

The scattering cross sections and their corresponding thermal rate coefficients are computed using a non-reactive quantum-mechanical close-coupling approach.

2 Quantum-mechanical method and results

1. Dynamical equations. The details of the quantum-mechanical approach used in the current work has been described in few previous papers. See, for example, preprint [9] and references therein. Briefly, a four-atomic Schrödinger equation is numerically solved in the framework of the quantum-mechanical close-coupling approach [2]. The HD and H_2 molecules are treated as linear rigid rotors. In our calculations, with the use of two PESs from works [3, 4] the bond length was fixed at 1.449 a.u. or 0.7668 Å for the H_2 molecule and 1.442 a.u. for HD which is 0.7631 Å. The Schrödinger equation for the (12) + (34) collision in the c.m. frame, where (12) and (34) are diatomic molecules is:

$$\left[\frac{\hat{P}^2_{\vec{R}_3}}{2\mathcal{M}_{12}} + \frac{\hat{L}^2_{\hat{R}_1}}{2\mu_1 R_1^2} + \frac{\hat{L}^2_{\hat{R}_2}}{2\mu_2 R_2^2} + V(R_1, \hat{R}_1, R_2, \hat{R}_2, \vec{R}_3) - E\right]\Psi(\hat{R}_1, \hat{R}_2, \vec{R}_3) = 0. \tag{2}$$

Here $\hat{P}_{\vec{R}_3}$ is the momentum operator of the kinetic energy of the collision, $\vec{R}_3$ is the collision coordinate, whereas $\vec{R}_1$ and $\vec{R}_2$ are relative vectors between atoms in the two diatomic molecules, and $\hat{L}_{\hat{R}_{1(2)}}$ are the quantum-mechanical rotation operators of the rigid rotors, μ_1 and μ_2 are the reduced masses of the HD and H_2 molecules and $\mathcal{M}_{12}$ is the reduced mass of the two molecules. Also, $V(R_1, \hat{R}_1, R_2, \hat{R}_2, \vec{R}_3)$ is the PES of the four-atomic system (1234), E is the total energy in the c.m. system, and $\hat{R}_{1(2)}$ are the angles of orientation for rotors (12) and (34), respectively, it is shown in Fig. 1. The use and modification of the original H_2-H_2 PESs [3, 4], $V(\vec{R}_1, \vec{R}_2, \vec{R}_3)$, is discussed below and in greater detail in [9, 10]. The cross sections for excitation from HD(j_1,m_1)+H_2(j_2, m_2) to HD($j'_1m'_1$)+H_2($j'_2m'_2$) are summed over the final angular momentum projections ($m'_1m'_2$) and averaged over the initial projections (m_1m_2) of the HD and H_2 molecules of angular momenta j_1 and j_2: $\sigma(j'_1, j'_2; j_1 j_2, \varepsilon) = \pi/[(2j_1+1)(2j_2+1)k_{\alpha\alpha'}]\sum_{Jj_{12}j'_{12}LL'}(2J+1)|\delta_{\alpha\alpha'} - S^J_{\alpha\alpha'}(E)|^2$. The kinetic energy is $\varepsilon = E - B_1 j_1(j_1+1) - B_2 j_2(j_2+1)$, where $B_1 = 44.7$ cm^{-1} and $B_2 = 60.8$ cm^{-1} are the rotational

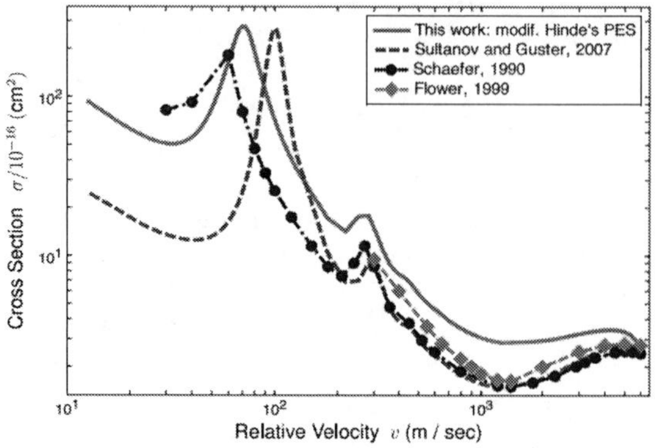

Figure 2. (Color online) The total cross section of the $HD(1) + H_2(0) \rightarrow HD(0) + H_2(0)$ inelastic rotational energy transfer collision. The numbers in parentheses are the rotational quantum numbers of the two-atomic molecules.

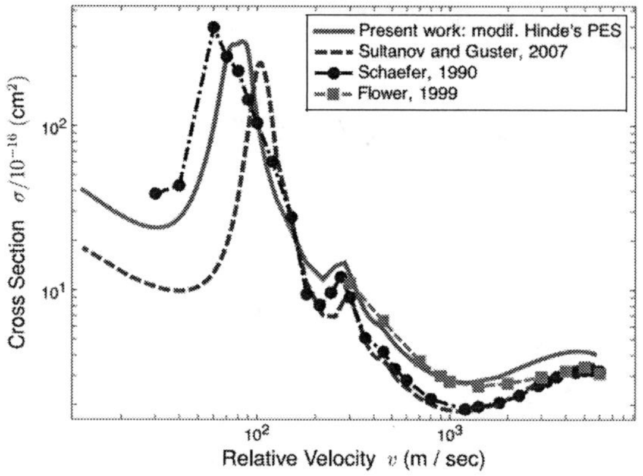

Figure 3. (Color online) The total cross section of the $HD(2) + H_2(0) \rightarrow HD(1) + H_2(0)$ inelastic rotational energy transfer collision. The numbers in parentheses are the rotational quantum numbers of the two-atomic molecules.

constants of rigid rotors (12) and (34) respectively. Further, J is total angular momenta of the four-atomic system, $\alpha \equiv (j_1 j_2 j_{12} L)$, where $j_1 + j_2 = j_{12}$ and $j_{12} + L = J$, $k_{\alpha\alpha'} = \sqrt{2M_{12}(E + E_\alpha - E_{\alpha'})}$ is the channel wavenumber and $E_{\alpha(\alpha')}$ are rotational channel energies.

2. Interaction potentials. Based on the Born-Oppenheimer idea [11] any quantum-mechanical, semi-classical or, say, quasi-classical trajectory treatment of the $HD+H_2$, D_2+HD or, say, D_2+D_2 four-atomic systems should begin with consideration of the original H_2-H_2 PES. The full six-dimensional function comprises a symmetrical surface over the OZ coordinate axis. In spherical coordinates the surface can be described by six variables: $R_1, \theta_1, R_2, \theta_2, R_3$, and φ_2 [4]. The variables are shown in Fig. 1. In work [3] an H_2-H_2 potential was formulated in a Cartesian coordinate system. In both works the H_2-H_2 potential was derived in the framework of the Born-Oppenheimer model [11] and can basically be considered as a symmetrical interaction field. When considering non-reactive scattering problems with participation of hydrogen molecules, one needs to solve the Schrödinger Eq. (2) with the H_2-H_2 potential $V(\vec{R}_1, \vec{R}_2, \vec{R}_3)$. The solution/propagation runs over the $\vec{R}_3$ Jacobi vector, which can be directed over the OZ axis, this is shown in Fig. 1. Therefore, in the case of the symmetrical H_2+H_2 and D_2+D_2 collisions, one can use the original H_2-H_2 PES as it is, i.e. without transformations. However, in the case of the non-symmetrical (symmetry-broken) HD+H2/D2 or HD+HD scattering systems, one should also apply the original H_2-H_2 interaction field or PES, but in this case the propagation (solution) of the Schrödinger equation should run over the corrected Jacobi vector $\vec{R}'_3$ which is directed over the new $O'Z''$ axis. This is shown in Fig. 1. In the current work, and in the preprint [9], we used two modified potentials from the papers [3, 4].

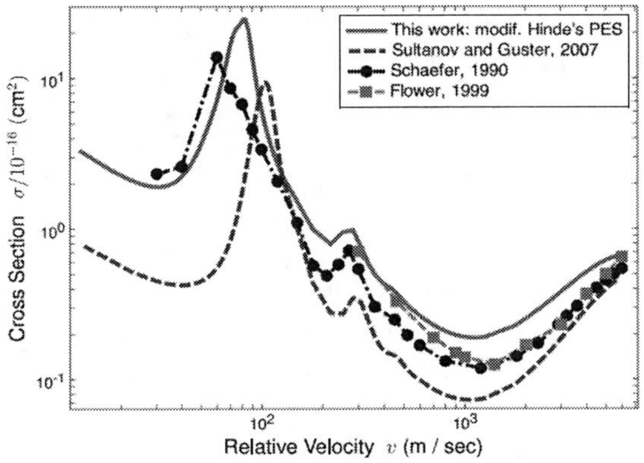

Figure 4. (Color online) The total cross section of the $HD(2) + H_2(0) \to HD(0) + H_2(0)$ inelastic rotational energy transfer collision. The numbers in parentheses are the rotational quantum numbers of the two-atomic molecules.

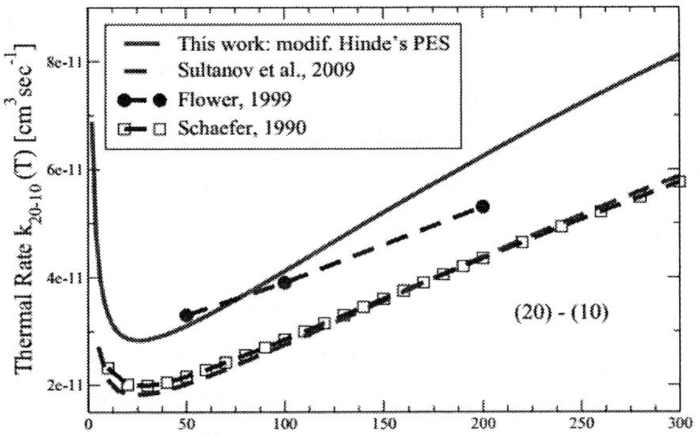

Figure 5. (Color online) The rotational de-excitation thermal rate coefficient $k_{ij \to i'j'}$ (T) for the following collision: $HD(2) + H_2(0) \to HD(1) + H_2(0)$.

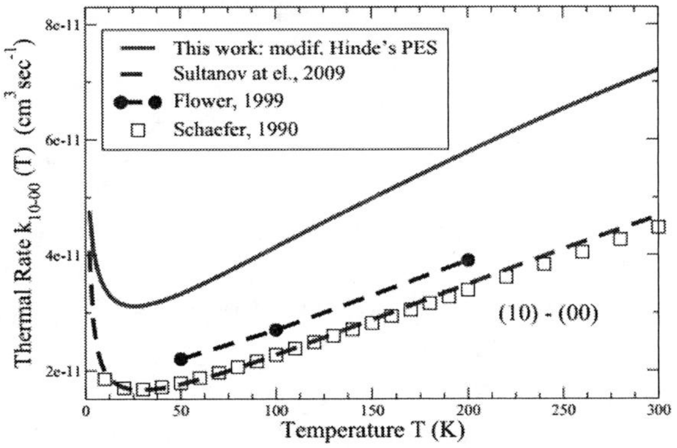

Figure 6. (Color online) The rotational de-excitation thermal rate coefficient $k_{ij \to i'j'}$ (T) for the following collision: $HD(1) + H_2(0) \to HD(0) + H_2(0)$.

3. Results. In order to solve the Schrödinger Eq. (2) the MOLSCAT program [12] has been used in our calculations. Because the $HD+H_2$ total rotational energy transfer cross sections have shape resonances at the low energy sector, a large number of energy points were needed. We used up to 250 energy points in each computation for each specific rotational transition in the HD and H_2 molecules considered. Convergence was obtained for the integral cross sections, $\sigma(j_1', j_2'; j_1 j_2, \varepsilon)$, with respect to the variation of the variables utilized in all considered collisions at different energies. For the intermolecular distance R_3, we used from $R_{3min} = 3.0$ a.u. to $R_{3max} = 30.0$ a.u. Our new rotational energy transfer integral cross sections are shown only for three different quantum-state transitions in the HD and H_2 molecules. Specifically, $HD(1) + H_2(0) \to HD(0) + H_2(0)$, $HD(2) + H_2(0) \to HD(1) + H_2(0)$ and $HD(2) + H_2(0) \to HD(0) + H_2(0)$ are presented in figures 2, 3 and 4. However, in the recent preprint [9] more results for different rotational transitions in the HD and H_2 molecules are available together with the important thermal rate coefficients at different temperatures: $k_{j_1 j_2 \to j'_1 j'_2}(T)$ [9]. The shape resonances in the low velocity region are reproduced, which are important in the cooling of the astrophysical media [6, 7]. Two thermal rate coefficients $k_{j_1 j_2 \to j'_1 j'_2}(T)$ [9] are presented in figures 5 and 6 as examples. Again, one can find more results in [9]. The new results are compared with the older quantum-dynamical calculations of Schaefer [7] and Flower [8] . The computational results from the older work [10] are also shown. These data have been obtained by using a PES obtained from a modification of the BMKP H_2-H_2 PES [3]. One can see that the new cross sections obtained with the Hinde's PES have the same structure and shape, but also have substantially larger ($\sim$ 60%) values at medium energies ($\upsilon \gtrsim 200$ m/sec) when compared with the results obtained from the modified BMKP PES [3] and the older result from Schaefer's calculations [7]. The cross sections from Ref. [8] are available only for higher collision velocities, i.e. $\upsilon \geq 300$ m/s. In summary, state-resolved close-coupling quantum-mechanical calculations for rotational de-excitation cross-sections and corresponding thermal rate coefficients of the HD+o-/p-H_2 collisions are computed using a linear rigid rotor model for the HD and H_2 molecules. The symmetrical H_2-H_2 PES of Ref. [4] has been appropriately adopted for the current non-symmetrical HD+H_2 system by appropriate translation and rotation. These geometrical operations lead to a new set of angle variables θ'_1, θ'_2 and φ'_2 for the Jacobi few-body coordinates, a new length of the intermolecular distance $\vec{R}'_3$ and, as a result, to a new HD-H_2 PES [9]. The new results can be used for subsequent consideration (at different astrophysical conditions) the astrophysical HD-cooling function, $\Lambda_{HD}(T)$, which is important in the astrophysics of the early universe [6]. Also, the substantially larger cross sections and rates in HD+o-/p-H_2 found in the current work and in [9] with the use of the modified Hinde PES are consistent with previous results for H_2+H_2 [13]. It has also been shown [13, 14] that full-dimensional dynamics which employ the Hinde PES [4] generally provide fairly good agreement with experiments for the low-lying vibrational states of H_2. Therefore, it can be recommended that the new results presented in [13, 14] for H_2+H_2, and in [9] for HD+H_2, which employ the Hinde PES, can be adopted in the astrophysical models.

References

[1] K. Takayanagi, Adv. At. Mol. Phys. 1, 149 (1965); G. Zarur and H. Rabitz, J. Chem. Phys. **60**, 2057 (1974); P. Diep and J.K. Johnson, J. Chem. Phys. **112**, 4465 (2000).

[2] S. Green, J. Chem. Phys. **62**, 2271 (1975).

[3] A.I. Boothroyd, P.G. Martin, W.J. Keogh, and M.J. Peterson, J. Chem. Phys. **116**, 666 (2002).

[4] R.J. Hinde, J. Chem. Phys. **128**, 154308 (2008).

[5] S. Montero and J. Perez Rios, J. Chem. Phys. **141**, 114301 (2014).

[6] A. Dalgarno and R. McCray, Ann. Rev. Astron. Astrophys. **10**, 375 (1972).

[7] J. Schaefer, Astron. Astrophys. Suppl. Ser. **85**, 1101 (1990).

[8] D.R. Flower, J. Phys. B **32**, 1755 (1999).

[9] R.A. Sultanov, D. Guster, and S.K. Adhikari, arXiv:1506.02161.

[10] R.A. Sultanov and D. Guster, Chem. Phys. Lett. **436**, 19 (2007).

[11] M. Born and R. Oppenheimer, Ann. Phys. **84**, 457 (1927).

[12] J. M. Hutson and S. Green, *MOLSCAT Computer Code version 14*, (Distributed by Collaborative Comp. Proj. 6 of the Engineering and Physical Sciences Research Council, Daresbury Lab., 1994).

[13] N. Balakrishnan, G. Quèmèner, R. C. Forrey, R. J. Hinde and P. C. Stancil, J. Chem. Phys. 134, 014301 (2011).

[14] S. Fonseca dos Santos, N. Balakrishnan, S. Lepp, G. Quèmèner, R. C. Forrey, R. J. Hinde and P. C. Stancil, J. Chem. Phys. 134, 214303 (2011).

EPJ Web of Conferences **113**, 02003 (2016)
DOI: 10.1051/epjconf/201611302003

On the Appearance of Families of Efimov States in the Spinor Three-Body Problem

V. E. Colussi[1,a], Chris H. Greene[2], and J. P. D'Incao[1,3]

[1] *Department of Physics, University of Colorado, Boulder, Colorado 80309-0440, USA*
[2] *Department of Physics, Purdue University, West Lafayette, Indiana 47907-2036, USA*
[3] *JILA, University of Colorado and NIST, Boulder, Colorado 80309-0440, USA*

Abstract. Few-body systems with access to multiple internal levels exhibit richness beyond that typically found in their single-level counterparts. One example is that of Efimov states in strongly-correlated spinor three-body systems. In [V. E. Colussi, C. H. Greene, and J. P. D'Incao, Phys. Rev. Lett. **113**, 045302 (2014)] this problem was analyzed for spinor condensates finding a complex level structure as in an early work [Bulgac and Efimov, Sov. J. Nucl. Phys. 22, 153 (1976)] in nuclear physics, and the impact of Efimov physics on the general form of the scattering observables was worked out. In this paper we discuss the appearance of novel families of Efimov states in the spinor three-body problem.

1 Introduction

The spinor three-body problem has received renewed interest recently due to the flourish of activity in the cold-atom community focused on the creation and manipulation of Bose-condensed gases of alkali atoms with access to the full multiplet of hyperfine ground states (see Ref. [1] and citations within.) These spinor condensates exhibit the rich interplay between superfluidity and magnetism, displaying interesting static and nonequilibrium many-body phenomenon from spin textures and spin domains to complex spin mixing dynamics [1]. This scenario also displays richness on the few-body level, where recent work has focused on the strongly-correlated regime [2, 3]. In spinor condensate experiments with alkali atoms to date, the interactions have typically been weak (with the exception of ^{85}Rb [3, 4] and possibly also ^{133}Cs and ^{7}Li), and so the strongly-correlated regime remains largely unexplored. In the strongly correlated regime, the s-wave scattering lengths associated with each two-body spin state exceeds the typical range of interatomic interactions, and Efimov physics becomes important [5–7]. In the single-level case, this leads to the formation of Efimov states whose level spectrum has an exponential structure and whose presence strongly affects the scattering observables, characterized by the geometric scaling factor $e^{\pi/|s_0|}$, where s_0 ($\approx 1.00624i$ for identical bosons) is a universal constant that relates to the strength of the three-body interaction. One of the major differences of the spinor three-body problem is the presence of multiple length scales in the problem associated with the s-wave scattering lengths for each of the two-body scattering channels. This not only modifies the energy spectrum but also increases the number of pathways three-body collisional processes like three-body

[a]e-mail: colussi@colorado.edu

recombination, atom-dimer relaxation and three-body spin-exchange, can occur [2]. In addition to that, the spin physics leads to the appearance of different values for the Efimov s_0 parameter, leading to the formation of novel families of Efimov states [2, 3].

In this paper we present a simple toy model which illuminates how these novel Efimov states arise in spinor systems. We consider a model where each boson has access to two degenerate internal states. By considering the full parameter space of scattering lengths and couplings between the two-body scattering channels, the mechanism by which the novel Efimov states arise is characterized. Our results echo the findings of Ref. [2, 3, 8], although the toy model presented here is the first to probe the parameter space of scattering lengths and couplings in the case of degenerate internal levels.

2 Background

On the one-body level the two internal states are labeled $|1\rangle$ and $|2\rangle$. Pairs of atoms interact via a multichannel generalization of the Fermi pseudopotential [9–11], which (in a.u.) has the form:

$$\hat{v}(r) = \frac{4\pi\hat{A}}{m}\delta^3(\vec{r})\frac{\partial}{\partial r}r = \frac{4\pi}{m}\left(\sum_{\sigma\sigma'}|\sigma\rangle A_{\sigma\sigma'}\langle\sigma'|\right)\delta^3(\vec{r})\frac{\partial}{\partial r}r, \tag{1}$$

where $\delta^3(\vec{r})$ is the three-dimensional Dirac delta-function, m is the atomic mass, and $\hat{A}$ is the scattering length matrix. The matrix $\hat{A}$ is a multichannel generalization of the scattering length where the diagonal elements represent the background scattering lengths in each channel and the off-diagonal elements are the couplings. The two-body basis is constructed by requiring them to be eigenkets of the symmetrization operator $\{|\sigma\rangle\} = \{|11\rangle, |12\rangle_S, |22\rangle\}$ [12].

We solve the three-body problem in the adiabatic hyperspherical representation, using the Green's function method developed in Refs. [9, 10]. The three-body system is characterized by the hyperradius R which sets the overall size of the system and a set of five hyperangles collectively denoted Ω which describe the internal motion. Treating R as an adiabatic parameter, the three-body wave-function can be written as

$$\Psi(R,\Omega) = \sum_{\nu} F_\nu(R) \sum_{\Sigma} \Phi_\nu^\Sigma(R;\Omega)|\Sigma\rangle, \tag{2}$$

where $F(R)$ are the hyperradial wave functions, $\Phi(R;\Omega)$ the channel functions, and $\{|\Sigma\rangle\} = \{|111\rangle, |112\rangle_S, |211\rangle, |122\rangle, |221\rangle_S, |222\rangle\}$ the basis set labeling the internal levels for each three-body configuration [12]. The task of solving the full three-body Schrödinger equation is reduced to solving the fixed-R hyperangular equation

$$\sum_{\Sigma'}\left[\frac{\hat{\Lambda}^2(\Omega)+15/4}{2\mu R^2}\delta_{\Sigma\Sigma'} + \langle\Sigma'|\hat{V}(R,\Omega)|\Sigma\rangle\right]\Phi_\nu^\Sigma(R;\Omega) = U_\nu(R)\Phi_\nu^\Sigma(R;\Omega), \tag{3}$$

where μ is the three-body reduced mass, $\hat{\Lambda}$ is the grand angular momentum operator [13], and $\hat{V}$ is the sum of pairwise interactions. The three-body potential $U_\nu(R)$ is obtained by solving Eq. 3 for fixed R values, where the problem reduces to solving a transcendental equation whose roots $s_\nu(R)$ enter as

$$U_\nu(R) = \frac{s_\nu(R)^2 - 1/4}{2\mu R^2}. \tag{4}$$

In the single-level case for three identical bosons in the limit $R/a \to 0$, solving Eq. 3 yields a single imaginary root $s_0 \approx 1.00624i$, which when inserted in Eq. 4 gives an attractive $1/R^2$ potential which supports an infinite amount of bound trimers (Efimov states) characteristic of the Efimov effect. For two identical and one dissimilar bosons $s_0 \approx 0.41370i$. In Ref. [2, 3], the first few $s_\nu(R)$ for spin-1, 2, and 3 are tabulated over all relevant length scales. The results demonstrate novel imaginary roots that lie in between the results from the analysis of single-level three-body systems as described in Ref. [8]. In the next section we illustrate with a simple toy model how to understand the appearance of these novel roots.

3 Toy Model

We will discuss first a toy model which explores the parameter space of the matrix $\hat{A}$. The toy model will then be generalized and a connection with the spinor three-body problem and the appearance of novel Efimov roots made. In our two-level toy model $\hat{A}$ is a real-valued symmetric 3×3 matrix with eigenvalues a_α, a_β, and a_γ. The full parameter space of $\hat{A}$ is large, but by considering a simple case the structure becomes more apparent. Consider the scenario where $A_{1,3} = A_{2,3} = 0$ with eigenvalues

$$a_{\alpha/\beta} = \left(\frac{A_{1,1} + A_{2,2}}{2}\right) \pm \left[A_{1,2}^2 + \left(\frac{A_{1,1} - A_{2,2}}{2}\right)^2\right]^{1/2} \quad \text{and} \quad a_\gamma = A_{3,3} \tag{5}$$

and eigenvectors

$$|\sigma_\alpha\rangle = \cos\theta|11\rangle + \sin\theta|12\rangle_S, \quad |\sigma_\beta\rangle = \sin\theta|11\rangle - \cos\theta|12\rangle_S, \quad \text{and} \quad |\sigma_\gamma\rangle = |22\rangle \tag{6}$$

parameterized by the lone parameter

$$\tan\theta = \frac{2A_{1,2}}{\left[(A_{1,1} - A_{2,2}) + \left[4A_{1,2}^2 + (A_{1,1} - A_{2,2})^2\right]^{1/2}\right]}. \tag{7}$$

By construction, variation of the angle $\theta \in [0, \pi/2]$ ensures that the scattering lengths a_α, a_β, and a_γ remain constant while the eigenvectors $|\sigma_\alpha\rangle$ and $|\sigma_\beta\rangle$ describe admixtures of the basis states $|11\rangle$ and $|12\rangle_S$.

In Figure 1, the results from solving Eq. 3 for the first few roots $s_\nu(R)$ in the hyperradial region $a_\alpha \ll R \ll a_\beta$ are shown. When $\theta = 0$, there are two imaginary roots $s_0 = s_1 \approx 0.41370i$, associated with the three-body configurations $|221\rangle$, $|122\rangle$, and all other configurations obtained by label permutation. When $\theta = \pi/2$, there is only one imaginary root $s_0 \approx 1.00624i$ associated with the three-body configuration $|111\rangle$. For intermediate values of θ the values s_ν are not assigned to any of the single-level results. In fact, the s_0 and s_1 are associated with a three-body configuration which is a linear combination of the configurations of the extreme $\theta = 0$ and $\pi/2$ cases and producing novel families of Efimov states.

The toy model is a result of a unitary transformation, $\hat{A} \to \hat{U}^\dagger \hat{A} \hat{U}$. Equivalently, it is a result of a transformation (rotation) of the basis $\{|11\rangle, |12\rangle_S, |22\rangle\} \to \{|\sigma_\alpha\rangle, |\sigma_\beta\rangle, |22\rangle\}$. What is then the class of unitary transformations which produce novel families of Efimov states? The eigenvector of $\hat{A}$ attached to the resonant channel must mix two-body product states under the transformation, which cannot arise from a unitary transformation on the one-body level.

There is a simple connection between these results and those from Ref. [2, 3] for spin-1, 2, and 3. When the internal levels are composed of the hyperfine spin multiplet there is the additional requirement that the two-body eigenstates be simultaneous eigenstates of both the symmetrization operator and spin operator $\hat{F}^2$. The unitary matrix that diagonalizes the matrix $\hat{F}^2$ carries the product basis $|m_{f_1}\rangle \otimes |m_{f_2}\rangle$ into the $|F_{2\text{B}}, m_{F2\text{b}}\rangle$ basis, and the elements of this matrix are Clebsch-Gordan coefficients, which is also equivalent to a unitary transformation of the matrix $\hat{A}$. In the toy model, the

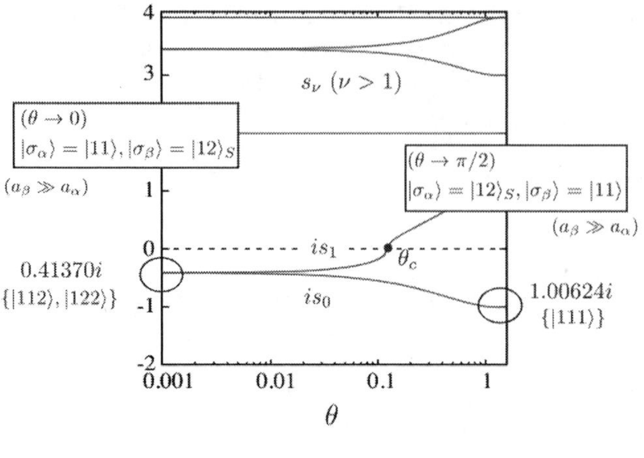

Figure 1. The first few roots s_ν obtained from solving Eq. 3 for the toy model in the hyperradial region $a_\alpha \ll R \ll a_\beta$. The roots are plotted versus the angle θ which controls the amount of admixture in the system.

admixture was set by the angle θ and treated as tunable for exploring the full parameter space. In the spinor case, this admixture is set by the values of the Clebsch-Gordan coefficients. As in the toy model, this admixture produces roots s_ν which are associated with three-body spin configurations that can be composed of a linear combination of product state spin functions and are capable of producing novel families of Efimov states.

Acknowledgements

This work was supported by the U.S. National Science Foundation, grant numbers PHY-1125844, PHY-1307380 and PHY-1306905.

References

[1] D. M. Stamper-Kurn and M. Ueda, Rev. Mod. Phys. **85**, 1191 (2013).

[2] V. E. Colussi, C. H. Greene, and J. P. D'Incao, Phys. Rev. Lett. **113**, 045302 (2014).

[3] V. E. Colussi, C. H. Greene, and J. P. D'Incao, arXiv:1507.08954.

[4] N. N. Klausen, J. L. Bohn, and C. H. Greene, Phys. Rev. A **64**, 053602 (2001).

[5] V. Efimov, Sov. J. Nucl. Phys. **10**, 62 (1970); **29**, 546 (1979).

[6] E. Braaten and H.-W. Hammer, Phys. Rep. **428**, 259 (2006).

[7] Y. Wang, J. P. D'Incao, and B. D. Esry, Advances in Atomic, Molecular, and Optical Physics **62**, 1 (2013).

[8] A. Bulgac and V. Efimov, Sov. J. Nucl. Phys. **22**, 153 (1976)

[9] N. P. Mehta, S. T. Rittenhouse, J. P. D'Incao, and C. H. Greene, Phys. Rev. A **78**, 020701 (2008).

[10] S. T. Rittenhouse, N. P. Mehta, and C. H. Greene, Phys. Rev. A **82**, 022706 (2010).

[11] O. I. Kartavtsev and J. H. Macek, Few-Body Syst. **31**, 249 (2002); J. H. Macek, ibid.. **31**, 241 (2002).

[12] The ket $|11\rangle$ is shorthand for the tensor product $|1\rangle \otimes |1\rangle$. The '$S$' subscript in $|12\rangle_S$ is shorthand for the symmetrization $(|12\rangle + |21\rangle)/\sqrt{2}$. Similarly, for the three-body ket $|112\rangle_S = (|112\rangle + |121\rangle)/\sqrt{2}$.

[13] J. Avery, *Hyperspherical Harmonics: Applications in Quantum Theory* (Kluwer Academic, Boston, 1989).

EPJ Web of Conferences **113**, 02004 (2016)
DOI: 10.1051/epjconf/201611302004

Reactivity in ion-neutral high density media

J. Pérez-Ríos[1,a] and C. H. Greene[1,b]

[1] *Department of Physics and Astronomy, Purdue University, West Lafayette, 47907 IN*

Abstract. The recent development of hybrid traps technology has revolutionized the field of cold chemistry. Recently, new trap configurations have been built in order to explore more involve chemical processes, among them is the three-body recombination involving ions and neutrals. Different ion-neutral-neutral reactions at low collision energies: 0.1 mK - 10 mK, are studied by means of a very recent method for classical trajectory calculations in hyperspherical coordinates for treating three-body collisions [J. Pérez-Ríos, S. Ragole, J. Wang and C. H. Greene, J. Chem. Phys. **140**, 044307 (2014)]. On the other hand a classically derived threshold law is obtained and numerically confirmed. The derived threshold law predicts molecular ions should dominate over molecular neutrals as the most products formed.

1 Introduction

The development of hybrid traps technology has brought the opportunity to study different kinds of chemical reactions involving neutrals and charged particles at cold temperatures ~ 1 mK. Hybrid trap technology relies on the combination of neutral atom traps and ion traps, in order to make a dual trap region where ions and neutrals overlap in space. On the other hand, by switching among different kinds of neutral traps or ion traps configurations, a broad spectra of ion-neutral chemical reactions can be explored. In particular, one interesting hybrid trap configuration is done by an optical dipole trap for neutrals, whereas a four-pole Paul trap is employed for trapping ions [1]. The scheme of this trap is shown in Fig.1, where also it is shown the specific Paul trap configuration employed by Härter *et al* [1], as well as the typical potential that an ion feels in the Paul trap. This hybrid trap configuration allows to study the same ion-neutral chemical reactions in a quite broad range of densities, and hence exploring the reactivity of the reactants as a function of the density.

In cold ion-neutral chemistry, most of the theoretical and experimental efforts have been devoted to the study of charge transfer reactions between ions and atoms, and more recently in molecular ion-atom systems [2]. Here, however we will focus on the study of three-body recombination (TBR) processes. TBR is a non-radiative few-body process in which three atoms collide, and eventually leading to the formation of a molecule as a product state, *i.e.*, $A + A + A \rightarrow A_2 + A$. Nevertheless, in ion-neutral-neutral TBR the reaction has two different channels: first, $B^+ + A + A \rightarrow A_2 + B^+$, and second, $B^+ + A + A \rightarrow AB^+ + A$. These two different reaction channels are studied for $^{87}Rb^+$ - ^{87}Rb - ^{87}Rb and $^{138}Ba^+$ - ^{87}Rb - ^{87}Rb, by means of a recent method based on classical trajectory calculations

[a] e-mail: jperezri@purdue.edu
[b] e-mail: chgreene@purdue.edu

(CTC) in hyperspherical coordinates [3, 4]. On the other hand, it is reported a classical threshold law for ion-neutral-neutral TBR, which is numerically tested, as well as it has been experimentally corroborated [5].

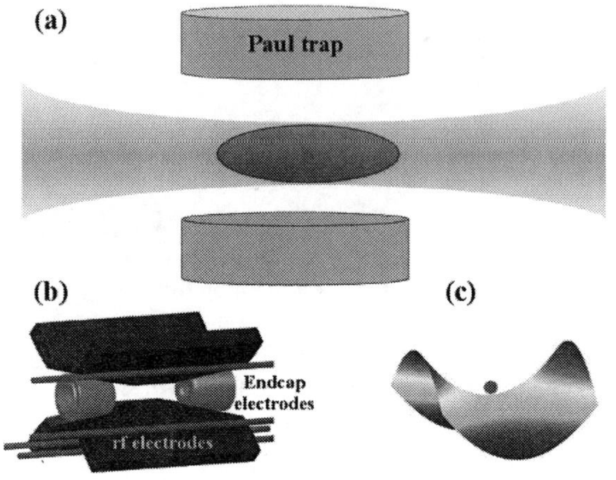

Figure 1. Hybrid trap. Panel (a); schematic representation of a hybrid trap involving a dipole trap for neutrals, represented by the orange beam and a Paul trap for ions, which is pictorial represented by the grey disks. The ultracold cloud of neutrals is represented by the blue ellipse, whereas the ion in this scheme is represented by the red dot. Panel (b): representation of the Paul trap configuration employed in the experimental results discussed here. The blue electrodes are the radio frequency electrodes, the endcap electrodes are DC electrodes to guarantee axial confinement of the lion along the z axis. The green conductors provide extra DC fields to compensate the micromotion of the ion in the Paul trap, which leads to a very high control on the kinetic energy of the ion. Panel (c): typical potential that an ion sees in a Paul trap at a given time.

2 Theoretical Approach

A recently developed method for the calculation of TBR cross sections based on classical trajectories [3] has been adapted for the study of ion-atom-atom recombination. The method relies on mapping the three-body problem into a 6-dimensional space (after separating out the trivial center of mass motion), where the cross section emerges as a generalization of the well-known two-body cross section, since the impact parameter can be defined as a 5-dimensional vector imbedded in the 6-dimensional space [3]. This approach is totally general and in principle it is suitable for solving the N-body problem. The position and momentum vectors in the 6-dimensional space is represented in hyperspherical coordinates [6, 7], which leads to a very efficient sampling of the phase-space.

The validity of a classical approach for ion-neutral-neutral collisions at cold temperatures may be questioned. However, the long-range ion-neutral interaction is dominated by the charge induced dipole moment interaction $-C_4/r^4$, where $C_4 = \alpha/2$, and α is the neutral atom polarizability. This long-range interaction is more attractive than its analog for neutral-neutral interactions, *i.e.*, the van der Waals interaction. Thus, the typical length scale of ion-neutral systems will be larger than its equivalent for neutral-neutral interactions. As a consequence, the energy in which the dynamics will be totally quantal is lower in ion-neutral than in neutral-neutral interactions. A more detailed analysis of this point can be found in [4]. In order to have a more quantitative argument, the angular momentum distributions of the TBR process for $^{138}Ba^+$ - ^{87}Rb - ^{87}Rb have been calculated at different collision energies, and the results are shown in figure 2. In this figure, it is noticed that even at collision energies $\sim 100\ \mu$ K, still many partial waves will contribute to the collision processes. Therefore, classical trajectory calculations might be an adequate approach for describing ion-neutral-neutral three-body recombination.

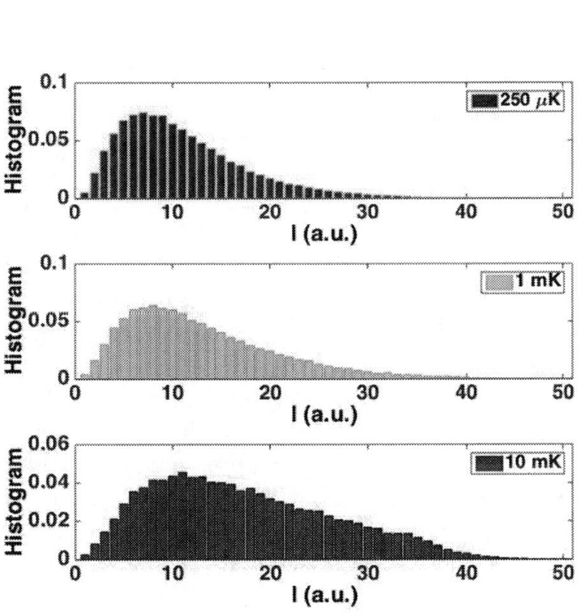

Figure 2. Total angular momentum distribution associated with the TBR for $^{138}Ba^+$ - ^{87}Rb - ^{87}Rb at different collision energies. The total angular momentum l is given in a.u., whereas the collision energy is given in K.

3 Results

The TBR cross section for $^{87}Rb^+$ -^{87}Rb -^{87}Rb and $^{138}Ba^+$ -^{87}Rb -^{87}Rb has been computed by running 10^5 trajectories per collision energy. Details about the numerical method employed to solve Hamilton's equations of motion, in conjunction with the sampling of the initial conditions, can be found elsewhere [3]. All of the calculations have assumed that neutral-neutral interactions, as well as ion-neutral interactions occur along one single potential energy curve. Concretely, Rb - Rb collisions are assumed to occur through the $^3\Sigma$ potential, *i.e.*, the spin flip transitions have been neglected [4]. On the other hand, the ion-neutral potential is described by the model potential $-C_4(1-(r_m/r)^4/2)/r^4$, where C_4 is taken as C_4=160 a.u., and r_m represents the position of the minimum of the potential [4].

Table 1. Classical threshold law for the TBR cross section. A power law dependence of the TBR cross section as a function of the collision energy is assumed and used as a fitting function for the CTC numerical results. The error on the fitting parameters are associated with a confidence interval of 95 %.

System	β (dimensionless)
$^{87}Rb^+$ - ^{87}Rb - ^{87}Rb	-1.27 ± 0.13
$^{138}Ba^+$ - ^{87}Rb - ^{87}Rb	-1.18 ± 0.07
Classical threshold law	-1.25

From a long-range perspective, the ion-atom interaction is of course more attractive than the atom-atom interaction. For any orientation, energy and impact parameter, the collision begins when the trajectories of the atoms start to deviate from uniform motion or free motion. This happens when the interaction potential where the moving atoms are currently located is comparable to the collision energy, *i.e*, $E_k \approx C_4/r^4$, where E_k represents the collision energy. Let us assume that this value of the radius defines the maximum impact parameter associated to a TBR event, *i.e.* $b_{max}(E_k) = (C_4/E_k)^{1/4}$. In our formalism the TBR cross section is defined as [3]

$$\sigma(E_k) \propto \int_0^{b_{max}(E_k)} b^4 db, \tag{1}$$

where a unit opacity function has been assumed. Finally, substituting the value of $b_{max}(E_k)$ in Eq. 1, $\sigma(E_k) \propto E_k^{-5/4}$ it is found. This prediction is tested by fitting the CTC numerical results. In particular, the fitting function employed has the form $\sigma(E_k) = \gamma E_k^{\beta}$, and the results are shown in table 1. In table 1, it is observed that energy scaling law for the TBR cross section numerically obtained for both systems, ^{87}Rb$^+$ - ^{87}Rb - ^{87}Rb and ^{138}Ba$^+$ - ^{87}Rb - ^{87}Rb, are in a good agreement with the predicted energy scaling law associated with the derived threshold law. This suggests that the derived threshold law is satisfied in different systems under different dynamical conditions. On the other hand, the presented numerical results for ^{138}Ba$^+$ - ^{87}Rb - ^{87}Rb have been corroborated in hybrid trap experiments [5]. The classically derived and numerically confirm threshold law has important implications on the final product state distributions. In particular, it predicts that for ion-neutral-neutral recombination the dominant product state will be the channel associated to molecular ions, *i.e.*, $B^+ + A + A \rightarrow AB^+ + A$.

4 Conclusions

The study of TBR for ^{87}Rb$^+$ -^{87}Rb -^{87}Rb and ^{138}Ba$^+$ -^{87}Rb -^{87}Rb has been performed by using a very recent classical trajectory method in hyperspherical coordinates. On the other hand, a classically derived threshold law for ion-neutral-neutral recombination has been obtained and corroborated numerically. Finally, this threshold is consistent with experimental results of the TBR rate for ^{138}Ba$^+$ -^{87}Rb -^{87}Rb [5].

Chemistry is maybe one of the most important playgrounds for few-body physics, where many important reactions involve the collision of very few molecules or atoms. However, chemistry has been receiving a very little of consideration by the few-body physics community. The present work is an example of how few-body physics can elucidate some of the intriguing chemical reactions . Therefore, we hope that our work serves as a motivation to go beyond the *apparent* limits of few-body physics.

Acknowledgements

This work was supported by the Department of Energy, Office of Science, under Award Number DE-SC0010545. The authors acknowledge Francis Robicheaux, Johannes Hecker Denschlag and Artjom Krükow for many fruitful discussions.

References

[1] A. Härter, A. Krükow, A. Brunner, W. Schnitzer, S. Schmid and J. Hecker Denschlag, Phys. Rev. Lett. **109**, 123201-4 (2012)

[2] F. H. J. Hall and S. Willistch, Phys. Rev. Lett. **109**, 233202-4 (2012)

[3] J. Pérez-Ríos, S. Ragole, J. Wang and C. H. Greene, J. Chem. Phys. **140**, 044307-12 (2014)

[4] J. Pérez-Ríos and C. H. Greene, J. Chem. Phys. **143**, 0441105-3 (2015)

[5] A. Krükow, A. Härter, M. Mohammadi, J. Hecker Denschlag, J. Pérez-Ríos and C. H. Greene, arXiv:1510.04938 (2015)

[6] J. Avery, *Hyperspherical Harmonics: Applications in Quantum Theory* (Kluwer,1986)

[7] J. M. Launay and M. Le Dourneuf J. Chem. Phys. Lett. **163**, 178-188 (1989)

EPJ Web of Conferences **113**, 02005 (2016)
DOI: 10.1051/epjconf/201611302005

Induced two-body scattering resonances from a square-well potential with oscillating depth

D. Hudson Smith[1,a]

[1] *Department of Physics, The Ohio State University, Columbus, OH 43210, USA*

Abstract. In systems of ultracold atoms, pairwise interactions can be resonantly enhanced by a new mechanism which does not rely upon a magnetic Feshbach resonance. In this mechanism, interactions are controlled by tuning the frequency of an oscillating parallel component of the magnetic field close to the Bohr frequency for the transition to a two-atom bound state. The real part of the s-wave scattering length a has a resonance as a function of the oscillation frequency near the Bohr frequency. The resonance parameters can be controlled by varying the amplitude of the oscillating field. The amplitude also controls the imaginary part of a which arises predominantly because the oscillating field converts atom pairs into molecules. For the case of a shallow bound state in the scattering channel, the dimensionless resonance parameters are universal functions of the dimensionless oscillation amplitude.

1 Introduction

A unique feature of ultracold atomic physics is the ability to precisely control the interactions among particles all the way from zero interactions to infinitely attractive or repulsive interactions. This tunability has led to many breakthroughs in few- and many-body physics. In most current experiments, interactions are controlled by exploiting a *magnetic Feshbach resonance* (MFR), where an external magnetic field is tuned near the value B_0 where a pair of unbound atoms becomes degenerate with a two-atom bound state [1]. For ultracold atoms, the strength of interactions is determined by the s-wave scattering length a. Near B_0, a is a simple function of the magnetic field B:

$$\frac{1}{a(B)} = \frac{1}{a_{\rm bg}} \frac{B - B_0}{B - B_0 - \Delta} + i\gamma, \tag{1}$$

where $a_{\rm bg}$ is the background scattering length, Δ is the width of the resonance, and γ is non-zero only if the colliding atoms have a spin-relaxation scattering channel.

In this letter, we examine a new mechanism, *modulated-magnetic Feshbach resonance* (MMFR), for resonantly enhancing the scattering length in ultracold gases. This mechanism is related to *modulated-magnetic spectroscopy* or *wiggle spectroscopy* which has been used to measure molecular binding energies and other properties for several alkali-metal atoms [2–6]. A MMFR is induced

[a]e-mail: smith.7991@osu.edu

by an oscillating magnetic field, $B(t) = \bar{B} + \widetilde{B}\cos(\omega t)$, that is parallel to the spin-quantization axis of the atoms and near the Bohr frequency for the transition between an atom pair and a bound state. Near a MFR, the Bohr frequency can be decreased by adjusting $\bar{B}$ to bring the resonance molecule closer to the threshold. Previous theoretical treatments involving oscillating magnetic fields have focused on molecule formation [8–10]. In constrast, the present work focuses on the effect of the oscillating magnetic field on the elastic scattering properties of the atoms.

For ω near the Bohr frequency ω_B, the scattering length is a simple function of ω:

$$\frac{1}{a(\omega)} = \frac{1}{\bar{a}} \frac{\omega - \omega_0}{\omega - \omega_0 - \delta} + i\gamma, \tag{2}$$

where $\bar{a}$ is the scattering length in the absence of the modulated field, ω_0 is close to ω_B, and δ is the width of the resonance. The inverse scattering length has been given a frequency-independent positive imaginary part γ that is only important very close to the resonance. The parametrization in Eq. (2) ensures that $\mathrm{Im}\,(a) \leq 0$ for all ω, as required by unitarity. The maximum value of $|\mathrm{Re}\,(a)|$ is $1/2\gamma$. The imaginary part of a arises from collisions in which a pair of low-energy atoms emits or absorbs one or more quanta of frequency ω and forms a molecule or an excited scattering state. Near resonance, the imaginary part is dominated by molecule formation.

We calculate $a(\omega)$ for a short-range potential with an oscillating depth. We show the existence of a resonance near $\omega = \omega_B$, and we extract the dependencies of the resonance parameters δ, $\Delta\omega_0 = \omega_0 - \omega_B$, and γ on the amplitude of the oscillating part of the potential. For small amplitudes, the resonance parameters each scale as the square of the amplitude. We examine the special case where the resonance molecule is a shallow dimer in the scattering channel. In this case, the dimensionless resonance parameters δ/ω_B, $\Delta\omega_0/\omega_B$, and $\gamma\bar{a}$ are universal functions of a dimensionless oscillatory magnetic field variable that is proportional to $\widetilde{B}$.

2 Single-channel scattering model

We consider a single-channel model for the effects of the oscillating magnetic field. For atomic separations r greater than r_0, the atoms are non-interacting. For $r < r_0$, the atoms interact through a square-well potential with an oscillating depth. The s-wave scattering properties for atoms with mass m are extracted from the solution to the time-dependent Schrödinger equation for $u(r,t) = rR(r,t)$, where $R(r,t)$ is the radial wave-function for the relative position of the atom pair with an interaction potential $(\bar{V} + \widetilde{V}\cos(\omega t))\theta(r_0 - r)$. The time-averaged depth of the potential, $\bar{V}$, and the range, r_0, are chosen such that the time-averaged potential supports a bound state with Bohr frequency ω_B. The term proportional to $\widetilde{V}$ represents the time-dependent shift in the binding energy of the resonance molecule relative to the scattering threshold. For a shallow bound state in the scattering channel, this energy shift is the change in the binding energy of the molecule when the magnetic field is shifted by $\widetilde{B}$. For a bound state in another hyperfine channel, the energy shift is $-\delta\mu\,\widetilde{B}\cos(\omega t)$, where $\delta\mu$ is the difference between the magnetic moment of the bound state and the magnetic moment of two atoms.

The Schrödinger equation can be solved by using Floquet's theorem as in Ref. [7]. This theorem asserts that for time-periodic potentials the wavefunction can be rewritten $u(r,t) = e^{-iE_F t}\phi(r,t)$, where the Floquet eigenvalue E_F is determined by the boundary conditions, and $\phi(r,t)$ has the same periodicity as the time-dependent part of the potential. The wavefunction for $r > r_0$ contains an incoming mode with energy k^2/m, where k is the small relative momentum of the atom pair. The requirement that $u(r,t)$ be continuous at $r = r_0$ implies that $E_F = k^2/m + j\pi\omega$ where j is any integer (taken to be

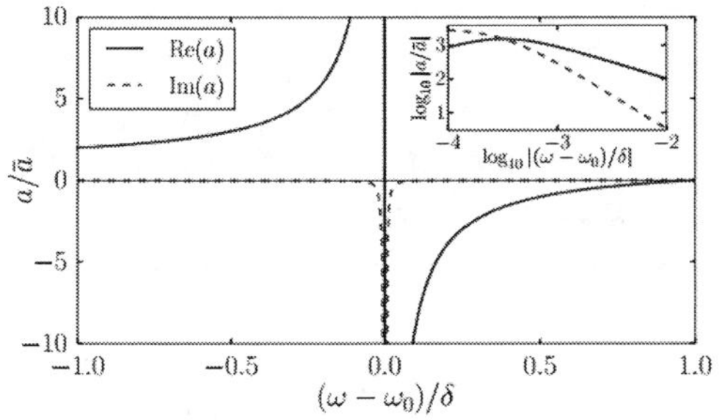

Figure 1. The real and imaginary parts of a as functions of ω for a MMFR where the resonance molecule is a shallow dimer in the scattering channel. The value of $\tilde{b}$ is 0.05. The inset shows the absolute values of the same quantities on a logarithmic scale.

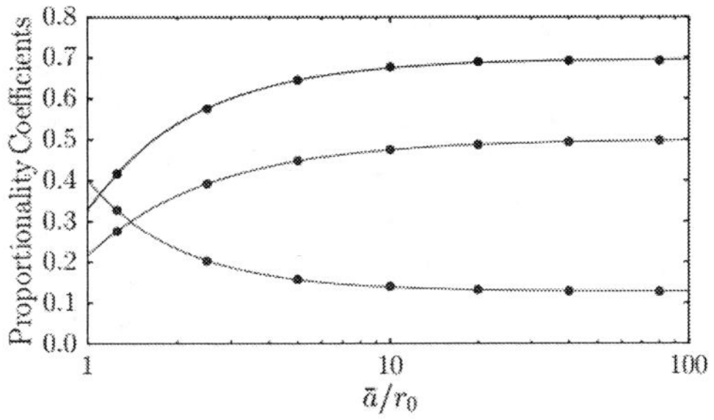

Figure 2. Numerical results for the coefficients of $\tilde{b}^2$ for the dimensionless resonance parameters $\Delta\omega_0/\omega_\mathrm{B}$, δ/ω_B, and $\gamma\bar{a}$ (top to bottom) as functions of $\bar{a}$ along with power-law fits to guide the eye. In the large $\bar{a}$ limit, the coefficients approach the universal numbers given in Eqs. (4).

zero below for simplicity). The Schrödinger equation can be solved analytically giving

$$u(r,t) = \sum_{n=-\infty}^{\infty} \begin{cases} 2ia_n \sin(q_n r) \exp\left[-i(k_n^2/m)t + i\widetilde{V}\sin(\omega t)/\omega\right] & r < r_0, \\ (A_n^{\mathrm{out}} e^{ik_n r} + A_n^{\mathrm{in}} e^{-ik_n r}) \exp\left[-i(k_n^2/m)t\right] & r \geq r_0, \end{cases} \tag{3}$$

where $k_n = [k^2 + mn\omega]^{1/2}$ and $q_n = [k^2 + m(\bar{V} + n\omega)]^{1/2}$. The solution for $r > r_0$ is a superposition of freely propagating modes. The coefficients A_n^{out} and A_n^{in} represent the amplitudes of outgoing and incoming modes, respectively. The $n = 0$ mode corresponds to the low-energy scattering state with energy k^2/m. The additional so-called *Floquet modes* with energies that differ by integer multiples of ω are necessary to satisfy the boundary conditions at $r = r_0$. Physically speaking, incident particles scattering with energy k^2/m can absorb (emit) quanta from (to) the oscillating field. They will, therefore, emerge as a superposition of modes with energies that differ from k^2/m by integer multiples of ω. The negative energy modes are exponentially damped for $r > r_0$, and they do not propagate. Both $u(r,t)$ and $\partial u(r,t)/\partial r$ must be continuous at $r = r_0$. These boundary conditions can be used to eliminate the a_n coefficients and find the S-matrix which relates the amplitudes of the incoming and outgoing modes. The scattering length is then extracted from the S-matrix element corresponding to low-energy elastic scattering (For a more detailed derivation, see Ref. [11])

Figure 1 plots the real and imaginary parts of a as functions of frequency, demonstrating the familiar Feshbach resonance shape. Molecule formation limits the maximum value of the real part of a. For small $\widetilde{V}$ the resonance parameters δ, $\Delta\omega_0$, and γ are proportional to $\widetilde{V}^2$ with coefficients that depend upon the parameters of the potential.

3 Shallow dimer in the scattering channel

The single-channel model is well-suited for describing an MMFR for which the resonance molecule is a shallow dimer in the scattering channel. The dimer's binding energy is $\omega_\mathrm{B} = 1/m\bar{a}^2$ where $\bar{a} = a(\bar{B})$ is given by Eq. (1) with $\gamma = 0$. We choose $\bar{V}$ such that the binding energy of the shallowest molecule $E(\bar{V})$ is ω_B. We choose $\widetilde{V}$ such that the change in $E(\bar{V})$ from a small modification $\widetilde{V}$ to the depth of the potential equals the change in ω_B from a small modification $\widetilde{B}$ to the amplitude of the magnetic field. Equating these energy shifts, $E'(\bar{V})\widetilde{V} = -2\omega_\mathrm{B}\tilde{b}$, where we have introduced the dimensionless oscillatory magnetic field variable $\tilde{b} = [a'(\bar{B})/\bar{a}]\widetilde{B}$. Inserting the value of $\widetilde{V}$ determined by equating the energy shifts into Eq. (3), we find the change in the scattering length resulting from a small oscillation of the magnetic field.

We now extract the dependencies of the resonance parameters on $\tilde{b}$ by fitting the numerical results for a extracted from Eq. (3) with the analytic parametrization in Eq. (2) for different values of $\tilde{b}$. The resonance parameter $\bar{a}$ is set equal to $a(\bar{B})$. The parameters δ and ω_0 are determined by fitting the real part of the inverse scattering length to the real part of Eq. (2). The parameter γ is then determined by calculating a at $\omega = \omega_0$. For small $\tilde{b}$, the dimensionless resonance parameters $\delta/\omega_{\rm B}$, $\Delta\omega_0/\omega_{\rm B}$, and $\gamma\bar{a}$ are proportional to $\tilde{b}^2$. In the limit $\bar{a} \gg r_0$ the proportionality coefficients are universal numbers that are independent of the short-range potential:

$$\begin{aligned} \delta/\omega_{\rm B} &= 0.50\,\tilde{b}^2, \\ \Delta\omega_0/\omega_{\rm B} &= 0.69\,\tilde{b}^2, \\ \gamma\bar{a} &= 0.13\,\tilde{b}^2. \end{aligned} \tag{4}$$

Figure 2 plots the proportionality coefficients as functions of $\bar{a}$. From Im(a), we can predict the on-resonance dimer formation rate. This rate agrees with the result from Fermi's Golden Rule [9].

4 Conclusion

We have examined a new mechanism for resonantly enhancing the s-wave scattering length a by tuning the frequency of an oscillating magnetic field that is parallel to the spin-quantization axis. Along with enhancing the real part of a, the oscillating field also generates an imaginary part associated with molecule formation. The real and imaginary parts of a can be controlled by the frequency and amplitude of the oscillating magnetic field. A simple model was used to determine how the resonance parameters scale with the amplitude of the oscillating field. For the special case where the resonance molecule is a shallow dimer in the scattering channel, the dimensionless resonance parameters are universal functions of $\tilde{b}$.

Acknowledgements

This research was supported by the National Science Foundation under grant PHY-1310862. Many thanks are due to Dr. Cheng Chin whose insightful question sparked this research topic and to Logan Clark for helpful discussions. I also thank Dr. Eric Braaten for invaluable feedback during the writing of this paper.

References

[1] C. Chin, R. Grimm, P. Julienne, and E. Tiesinga, Rev. Mod. Phys. **82**, 1225 (2010)
[2] S. T. Thompson, E. Hodby, and C. E. Wieman, Phys. Rev. Lett. **95**, 190404 (2005)
[3] C. Weber, G. Barontini, J. Catani, G. Thalhammer, M. Inguscio, and F. Minardi, Phys. Rev. A **78**, 061601(R) (2008)
[4] A. D. Lange, K. Pilch, A. Prantner, F. Ferlaino, B. Engeser, H.-C. Nägerl, R. Grimm, and C. Chin, Phys. Rev. A **79**, 013622 (2009)
[5] O. Machtey, Z. Shotan, N. Gross, and L. Khaykovich Phys. Rev. Lett. **108**, 210406 (2012)
[6] P. Dyke, S.E. Pollack, and R.G. Hulet, Phys. Rev. A **88**, 023625 (2013)
[7] W. Li and L. E. Reichl, Phys. Rev. **60**, 15732 (1999).
[8] T.M. Hanna, T. Koehler, and K. Burnett, Phys. Rev. A **75**, 013606 (2007)
[9] C. Langmack, D. H. Smith, and E. Braaten, Phys. Rev. Lett. **114**, 103002 (2015)
[10] A. Mohapatra and E. Braaten, Phys. Rev. A **92**, 013425 (2015)
[11] D. H. Smith, arXiv:1503.02688 (2015)

EPJ Web of Conferences **113**, 02006 (2016)
DOI: 10.1051/epjconf/201611302006

Few-body, hyperspherical treatment of the quantum Hall effect

R. E. Wooten[1,a], K. M. Daily[1,2], and Chris H. Greene[1,b]

[1] *Purdue University Department of Physics and Astronomy 525 Northwestern Avenue, West Lafayette, Indiana 47907-2036*
[2] *Wolfram Research, Inc, 100 Trade Center Dr., Champaign, IL 61820*

Abstract. The quantum Hall effect arises from the quantum behavior of two-dimensional, strongly-interacting electrons exposed to a strong, perpendicular magnetic field [1, 2]. Conventionally treated from a many-body perspective, we instead treat the system from the few-body perspective using collective coordinates and the hyperspherical adiabatic technique developed originally for atomic systems [3]. The grand angular momentum K from K-harmonic few-body theory, is shown to be an approximate good collective quantum number in this system, and is shown to correlate with known fractional quantum Hall (FQH) states at experimentally observed filling factors.

1 Introduction

When electrons are confined to a finite, two-dimensional plane in the presence of a strong, perpendicular magnetic field, the measurements of the Hall resistance R_H are quantized to values of $h/\nu e^2$ when the filling factor ν takes on integer or certain rational fraction values. The filling factor ν is the number of occupied Landau levels in the non-interacting ground state of a finite-area system. The integer values of ν are well understood to arise from Landau level quantization, but the rational fraction values of ν arise entirely from electron correlations and remain a rich area of study.

In conventional treatments, the N-particle Hamiltonian is diagonalized in a single-particle defined Slater determinant basis, but in our treatment, we address the problem by transforming first to a set of collective coordinates and then transforming to a hyperspherical representation [4]. While its use in condensed matter physics has been limited, the adiabatic hyperspherical representation has been used successfully in a wide range of problems in few-body physics, including nuclear structure and reactivity, few-electron atoms, positron-electron systems, and Bose condensates. We assert that the technique can also effectively be applied to the many body quantum Hall system, and that it introduces an interesting and distinct framework for interpreting the system.

We begin by noting that the center of mass is separable in this system. The non-interacting relative coordinate Hamiltonian in units of the magnetic energy, $\hbar\omega_c = \hbar eB/m_e$, is

$$H_{\text{rel}} = -\frac{1}{2\mu}\sum_{j=1}^{N_{\text{rel}}} \nabla_j^2 + \frac{\mu}{8}\sum_{j=1}^{N_{\text{rel}}} \rho_j^2 + \frac{1}{2\hbar}\sum_{j=1}^{N_{\text{rel}}} L_{z_j}^{\text{rel}}, \tag{1}$$

[a] e-mail: wootenr@purdue.edu
[b] e-mail: chgreene@purdue.edu

where m_e is the effective electron mass in the material (typically low temperature gallium arsenide), B is the magnetic field, $\mu = (1/N)^{(1/N_{rel})}$ is a dimensionless mass scaling factor, N_{rel} is the number of relative coordinates, and the ρ_j are the relative coordinate lengths, which have been scaled by the magnetic length, $\lambda_0 = \sqrt{\hbar/m_e\omega_c}$.

2 Hyperspherical treatment

The relative coordinate Hamiltonian can be recast in terms of the hyperradius, R and a set of internal coordinates called the hyperangles (collectively referred to as $\mathbf{\Omega}$) that can be defined a number of arbitrary ways (e.g. see [5] or Appendix 4 of [6]) . The hyperradius defines a single length scale for the relative coordinates $R^2 = \sum_{j=1}^{N_{\rm rel}} \rho_j^2$, where the ρ_j are the Jacobi coordinates. In hyperspherical coordinates, the Hamiltonian is

$$H_{\rm rel} = -\frac{1}{2\mu}\nabla^2_{R,\mathbf{\Omega}} + \frac{\mu}{8}R^2 + \frac{1}{2\hbar}L_z^{\rm rel,tot} + \kappa\frac{C(\mathbf{\Omega})}{R}, \tag{2}$$

where $\kappa = e^2/(4\pi\epsilon\lambda_0\hbar\omega_c)$ is the ratio of the magnetic to the Coulomb energy, ϵ is the dielectric constant of the material, and $C(\mathbf{\Omega})$ is the hyperangular form of the Coulomb interaction. $\nabla^2_{R,\mathbf{\Omega}}$ is the Laplacian in hyperspherical coordinates,

$$\nabla^2_{R,\mathbf{\Omega}} = \frac{1}{R^{2N_{\rm rel}-1}}\partial_R R^{2N_{\rm rel}-1}\partial_R - \frac{\hat{\mathbf{K}}^2}{R^2}, \tag{3}$$

where $\hat{\mathbf{K}}$ is the grand angular momentum operator. The eigenstates of $\hat{\mathbf{K}}^2$ are the hyperspherical harmonics $\Phi_{Ki}(\mathbf{\Omega})$, where

$$\hat{\mathbf{K}}^2\Phi^{(M)}_{Ki}(\mathbf{\Omega}) = K(K + 2N_{\rm rel} - 2)\Phi^{(M)}_{Ki}(\mathbf{\Omega}). \tag{4}$$

The i subscript here is a label indicating distinct, orthogonal functions that share K and M..

In the absence of the Coulomb interaction, each eigenfunction of Eq. 2 is separable into a product of a hyperradial function $F^{(M)}_{n_R K}(R)$ times an antisymmetrized hyperspherical harmonic, $\Phi^{(M)}_{Ka}(\mathbf{\Omega})$. Both the relative azimuthal angular momentum M and the grand angular momentum K are good quantum numbers in the non-interacting system, n_R is a hyperradial quantum number, and the index a distinguishes antisymmetric linear combinations of hyperspherical harmonic basis functions within the same K, M manifold. The non-interacting Schödinger equation simplifies to a one-dimensional, uncoupled ordinary differential equation,

$$\left\{-\frac{1}{2\mu}\frac{d^2}{dR^2} + U^{(M)}_K(R) - E\right\} F^{(M)}_{n_R K}(R) = 0, \tag{5}$$

where the noninteracting potentials $U^{(M)}_K(R)$ are given by

$$U^{(M)}_K(R) = \frac{(K + N_{\rm rel} - 1/2)(K + N_{\rm rel} - 3/2)}{2\mu R^2} + \frac{\mu}{8}R^2 + \frac{1}{2}M. \tag{6}$$

The non-interacting potential curves for the four particle system are shown in Fig. 1, where the lowest-lying curves each have $K = |M|$ and are the potentials for the lowest Landau level. These non-interacting solutions can be directly connected to the central parameter of the quantum Hall problem, the filling factor ν. In a typical 3-dimensional system, the hyperradius defines a characteristic "size" of a particle distribution, but in the case of particles on a plane, it defines an approximate area for a few

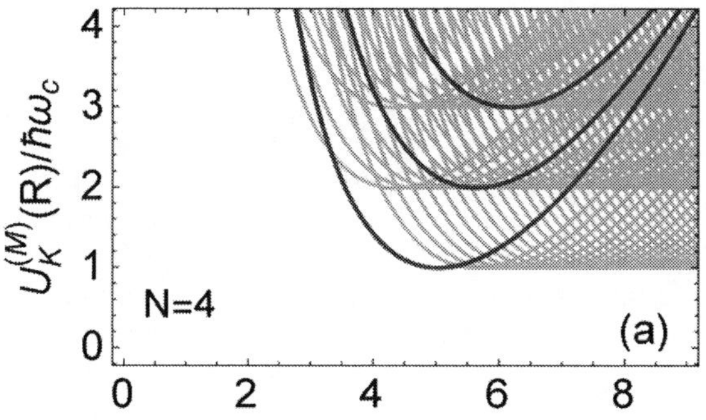

Figure 1. Light solid lines show the adiabatic potential curves of the noninteracting four-body system whose minima lie below $4\hbar\omega_c$. The lowest group ($K = |M|$) represents the lowest Landau level. The next and next-next higher groups are for $K = |M| + 2$ and $K = |M| + 4$, respectively. The dark solid lines are for $M = -6$, the lowest of which supports the integer quantum Hall state.

particle distribution. The hyperradial maxima of the ground states of the non-interacting hyperradial solutions, $F^{(M)}_{n_R K}(R)$ with $n_R = 0$, give characteristic hyperradii, and thus characteristic areas of the ground state solutions as a function of K. The characteristic areas leads to an expression for the hyperspherical filling factor in terms of only the particle number and the grand angular momentum,

$$\nu = \frac{N(N-1)}{2K}, \tag{7}$$

which connects the hyperspherical picture to the observations in the quantum Hall system.

With the introduction of the Coulomb interaction, the solutions become only approximately separable, and although M remains a good quantum number, K is only an approximate quantum number. We assume the solutions to the Schödinger equation that depend parametrically on the hyperradius, R,

$$\Psi(R, \mathbf{\Omega}) = R^{-N_{\rm rel}+1/2} \sum_{\chi} F^{(M)}_{E\chi}(R) \Phi^{(M)}_{\chi}(R; \mathbf{\Omega}), \tag{8}$$

where the subscript χ labels each channel, and the channel functions $\Phi^{(M)}_{\chi}(R; \mathbf{\Omega})$ are orthonormal for a fixed hyperradius. The potentials, $U^{(M)}_{\chi}(R)$, are found by diagonalizing the adiabatic Hamiltonian,

$$H_{ad} \Phi^{(M)}_{\chi}(R; \mathbf{\Omega}) = U^{(M)}_{\chi}(R) \Phi^{(M)}_{\chi}(R; \mathbf{\Omega}), \tag{9}$$

while treating the hyperradius as an adiabatic parameter and expanding the hyperangular channel functions at each hyperradius in terms of the antisymmetrized hyperspherical harmonics,

$$\Phi^{(M)}_{\chi}(R; \mathbf{\Omega}) = \sum_{Ka} c_{Ka}(R) \Phi^{(M)}_{Ka}(\mathbf{\Omega}). \tag{10}$$

As an example, the adiabatic potential curves for four particles with $M = -18$ and $\kappa = 1$ are shown in Fig. 2.

While an exact solution would include all adiabatic coupling between differing channel functions, in this system, coupling between different manifolds is weak, as shown for four particles in Fig.3 by the scaled adiabatic potentials $g^{(M)}_{\chi}(R)$,

$$g^{(M)}_{\chi}(R) = R^2 \left(U^{(M)}_{\chi}(R) - \left[\frac{\mu}{8} R^2 - \frac{M}{2} \right] \right). \tag{11}$$

With the scaling shown in Fig. 3, the avoided crossings indicating inter-channel coupling can be seen, but most appear diabatic in nature. In addition, the avoided crossings appear only at much higher values of the hyperradius ($\kappa R \gtrsim 30\lambda_0$), whereas the noninteracting hyperradial wave function

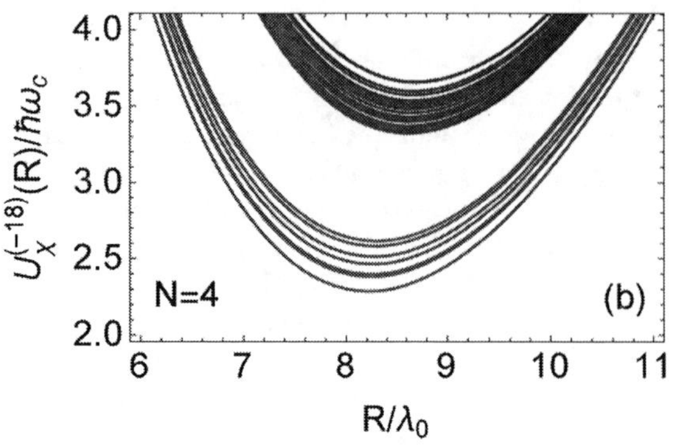

Figure 2. Adiabatic potentials $U_X^{(-18)}(R)$ for $N = 4$ particles, $M = -18$ and $\kappa = 1$. The larger gaps (on the order of $\hbar\omega_c$) indicate magnetic excitations, while the smaller splittings are due to Coulomb interactions. The separate clusters of curves indicate different K manifolds, with $K = |M|$ for the lowest grouping of curves and K increasing from bottom to top.

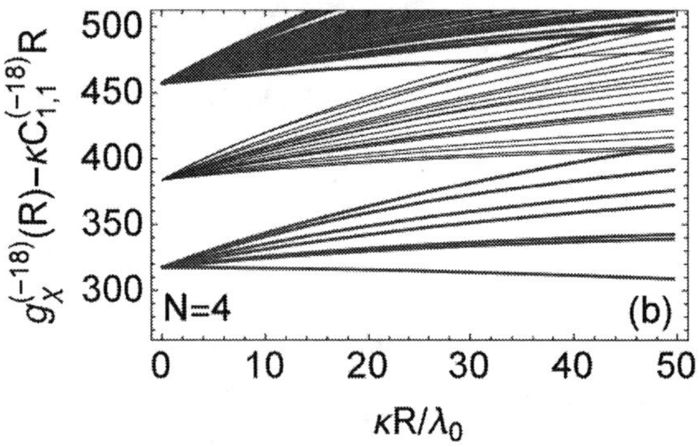

Figure 3. Bottom: $N = 4$ particles scaled adiabatic potentials $g_X(R)$ shifted by the smallest eigenvalue $C_{1,1}^{18}$ from diagonalizing the Coulomb interaction in the $K = |M| = 18$ manifold. Each bundle of curves originating from a higher point on the $\kappa R/\lambda_0$ axis corresponds to a higher K manifold. Avoided crossings indicating inter-channel coupling can be seen, but most are diabatic in nature and appear at hyperradii much larger than the typical scale of the system. That the majority of curves lack apparent adiabatic coupling indicates that the adiabatic approximation is particularly good in this system.

is contained within $R < 10.25\lambda_0$. Because the non-adiabatic coupling between manifolds is weak for the predominant hyperradial region, neglecting coupling between different K, M manifolds is a good approximation in this system.

Another indicator that the adiabatic approximation is particularly good is the fact that the upper and lower bounds in the hyperspherical representation differ by less than 10^{-4} in tested cases. By comparison, the ground state energies of the hydrogen negative ion differed by 1.7% in the upper and lower bounds in this approximation [7]. The impressive precision in the quantum Hall problem arises because the charged particles do not experience any attractive potentials, in contrast to the H^- system, where the attractive potentials cause potential valleys with much stronger deviations from adiabaticity.

In the absence of inter-K, M manifold coupling, the ground state energies can be found in a two-step process by first diagonalizing the Coulomb interaction within a hyperangular K, M basis then solving the hyperradial Schrödinger's equation numerically for varying magnetic fields. The ground state energies for the lowest Landau level, $K = |M|$ manifolds in a 6-particle system are shown in Fig. 4. The left-most curve corresponds to the 6-particle integer quantum Hall state, and the filling factor decreases towards the right as the magnetic field increases until slightly beyond the $\nu = 1/3$ Laughlin filling. The ground states at $\nu = 1$ and $\nu = 1/3$ filling are shown in red while ground states with $\nu = 2/3$ and $2/5$ are highlighted in blue. Other K, M manifold ground states that are also a ground state of the complete system at some values of the magnetic field are shown in green, and the remaining ground states are shown in black. The low energy states in this system other than the integer, Laughin, and Jain states are not identified, and their nature and properties are not predicted here.

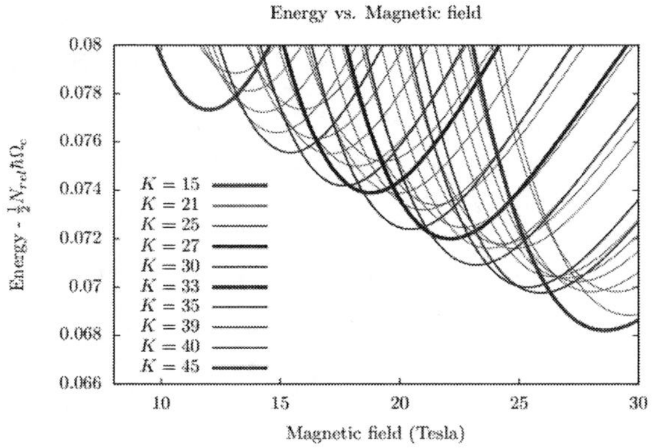

Figure 4. The ground state energies of K, M manifolds in the lowest Landau level (where $K = |M|$) as a function of the magnetic field, using experimental parameters matching a cold gallium arsenide system. These energies are found by first diagonalizing the Coulomb interaction in the hyperangular $K = |M| =$ bases, then solving the hyperradial Schrödinger's equation numerically with a restricted maximum hyperradius. The zero-point energies have been subtracted for scale. The curves for $K = 15, 27, 33, 45$ give the ground state energies of the $\nu = 1, 2/3, 2/5, 1/3$ quantum Hall states, respectively. Other states that are ground states of the system at some magnetic fields, but are not identified FQH states, are shown in purple, and the remaining higher energy states are shown in grey.

3 Exceptional degeneracy

One benefit of describing the system in hyperspherical coordinates is that the set of antisymmetrized hyperspherical harmonics in any K, M manifold forms a complete basis in the absence of interactions, and to a good approximation, is uncoupled to other K, M manifolds even in the presence of moderately strong Coulomb interactions. As such, it is interesting to consider the consequences of degeneracy on the individual manifolds without regard to the specific form of the interparticle interactions. As is well known from perturbation theory, introducing interactions into a degenerate manifold lowers the energy of the ground state relative to higher energy states, and the presence of additional degeneracy in the system strengthens this effect. As a result, it should be expected in this system that K, M manifolds with exceptionally high degeneracy relative to neighboring manifolds should also exhibit an enhanced energy gap between the ground state and excited states when interactions are introduced. As a dramatic energy gap is one of the primary requirements for a quantum Hall state, we predict that K, M manifolds with exceptional degeneracy are also likely to support quantum Hall states.

In the lowest Landau level ($K = |M|$), the degeneracy of each manifold can be determined analytically using group theoretical considerations. The degeneracies of the manifolds generally increase with $|M|$, but the relative change in degeneracy can be measured by enclosing the degeneracy function with an upper and lower envelope. Measuring the absolute degeneracy in comparison to the local height between the envelope functions gives a value for the relative degeneracy. In Fig. 5 we show a plot of the relative degeneracies of high degeneracy manifolds for a 6 particle system. The plot labels manifolds that support Laughlin and Jain states with their filling factors. All known Laughlin and Jain states in the 6-body system within the displayed $|M|$ range exist in manifolds with exceptionally high degeneracy, are labeled on this plot. The Laughlin and Jain fillings also display prominent degeneracy at significantly values of M, as well, and this trend has held true for systems with fewer particles, as well.

As can be seen in Fig. 5, while the integer, Laughlin, and Jain states display high relative degeneracy, this is not necessarily the case with other low energy states, and as such, the correlation between exceptional degeneracy and low energy ground states, which though suggestive, is not strict.

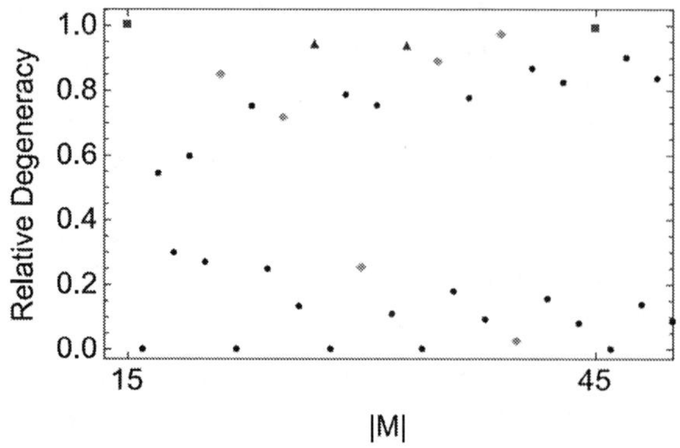

Figure 5. Relative degeneracies for the six-body system are shown as a function of $|M|$. Red squares show the integer quantum Hall effect $\nu = 1$ and the Laughlin $\nu = 1/3$ supporting manifolds; blue triangles show the Jain manifolds ($\nu = 2/3, 2/5$) that can exist in a 6 particle system; the green diamonds highlight manifolds that are ground states in Fig. 4, and the remaining black circles show the remaining K, M manifolds' relative degeneracies.

4 Conclusions

In this early study, we demonstrate that the adiabatic hyperspherical approximation from few-body theory is an appropriate framework from which to examine the complex many-body behavior of the quantum Hall system. In addition, the adiabatic hyperspherical approximation provides a new and different interpretation of the problem. In addition to the description of exceptional degeneracy, we expect that this approach will lead to other insights into the system's behavior that have been overlooked in the conventional many-body approach.

Acknowledgements

This work has been supported in part by a Purdue University Research Incentive Grant from the Office of the Vice President for Research. Some calculations were performed under resources from an NSF XSEDE allocation.

References

[1] D. C. Tsui, H. L. Stormer, A. C. Gossard, Phys. Rev. Lett. **48**, 1559 (1982).

[2] H. L. Stormer, D. C. Tsui, and A. C. Gossard, Rev. Mod. Phys. **71**, S298 (1999).

[3] J. Macek, J. Phys. B **1**, 831 (1968).

[4] K. M. Daily, R. E. Wooten, and Chris H. Greene, Phys. Rev. B **92**, 125427 (2015)

[5] Y. F. Smirnov and K. V. Shitikova, Sov. J. Part. Nucl. **8**, 44 (1977)

[6] S. T. Rittenhouse, J. von Stecher, J. P. D'Incao, N. P. Mehta, and C. H. Greene, Journal of Physics B: Atomic Molecular and Optical Physics **44**, 172001 (2011).

[7] H. Klar and M. Klar, Phys. Rev. A **17**, 1007 (1978).

EPJ Web of Conferences **113**, 02007 (2016)
DOI: 10.1051/epjconf/201611302007

Five is More: Comments on Symmetry, Integrability, and Solvability for a Few Particles in a One-Dimensional Trap

N.L. Harshman[1,a]

[1] *Department of Physics, American University, 4400 Massachusetts Ave. NW, Washington, DC, USA 20016-8058*

Abstract. This contributed conference proceeding reviews some results about a system of a few identical particles with spin trapped in one-dimensional potentials and experiencing two-body interactions. The focus is on how symmetry, integrability, and solvability depend on the trap shape, two-body interaction, the number of particles, and the number of spin components. A series of comments are presented that characterize the minimal symmetries possible for a composite system constructed from interacting single particles, with special focus on the contact interaction. For five and more particles with internal components like spin, a kind of universality called algebraically solvability is lost.

1 Introduction

This brief conference proceedings considers the symmetries of a model of N particles with J spin components in one-dimensional trap. This model has attracted much interest recently because many-body [1] and few-body [2] experiments with ultracold atoms in highly-elongated traps and interacting via tunable Feshbach resonances are well-modeled as particles in one-dimension with contact interactions [3]. This motivates examining the symmetries and solvability of the following N-body Hamiltonian:

$$\hat{H} = \hat{H}_0 + \sum_{\langle i,j \rangle} \hat{V}_{ij} \tag{1}$$

where the non-interacting part is the sum of one-particle Hamiltonians

$$\hat{H}_0 = \sum_{i=1}^{N} \hat{H}_i^1 = \sum_{i=1}^{N} \left(\frac{1}{2m} \hat{P}_i^2 + V^1(\hat{X}_i) \right) \tag{2}$$

and the two-body interactions are spin-independent functions of interparticle separation $\hat{V}_{ij} = V^2(|\hat{X}_i - \hat{X}_j|)$.

This model Hamiltonian also has a long history as a touchstone of theoretical and mathematical physics applied to few-body and many-body systems [4]. Depending on the shape of the trap V^1, the interaction potential V^2, the number of particles N, and number of internal particle components J, the system may be integrable, analytically solvable, or even algebraically solvable [5]. These properties have intrinsic interest to mathematical physicists, and they also have ramifications for physical

[a] e-mail: harshman@american.edu

properties like thermalization [6], universality [7], and entanglement [8], and practical matters like precision calculation and experimental control (c.f. recent reviews [9]).

The following comments are intended to provide a cross-section of results about the symmetries of the Hamiltonian. Some of these results are old and familiar, but some are from the recent wave of analysis inspired by current experiments. Others are new results presented at the conference for the first time and culled from my recent two-part article "One-Dimensional Traps, Two-Body Interactions, Few-Body Symmetries" [10, 11]. For a detailed set of references to classic and recent work, and for a systematic development of the results summarized in these comments, see those two articles.

2 Eight Comments

1. One-particle systems in one dimension are Liouville integrable.

   For any trap potential V^1, the one-particle Hamiltonian H^1 is in involution with itself, and so there are as many invariants as degrees of freedom, namely, one! Classically this is enough to solve for the equation of motion; quantum mechanically it means that a single observable H^1 and its spectrum is enough to completely separate the spatial Hilbert space $\mathcal{K}$ into subspaces characterized by energy. In addition to being integrable, some potentials are solvable, and some are even algebraically solvable. For truly trapped systems, the spectrum σ_1 is discrete, countably-infinite, and non-degenerate, i.e. $\sigma_1 = \{\epsilon_0, \epsilon_1, \epsilon_2, \cdots\}$. The associated kinematic symmetry K_1 is time translation T_t represented on the spatial Hilbert space by the exponentiation of the Hamiltonian. If the trap is parity symmetric, then the kinematic symmetry is isomorphic to $\mathrm{K}_1 \sim \mathrm{O}(1) \times \mathrm{T}_t$ and there is another integral of the motion, the parity operator $\hat{\Pi}$. For harmonic traps, the one-particle kinematic symmetry is $\mathrm{K}_1 \sim \mathrm{U}(1) \times \mathrm{T}_t$ because of the rotational invariance of phase space for a harmonic oscillator.

2. The Hamiltonian $\hat{H}^N$ has at least the minimal kinematic symmetry $\mathrm{K}_N \supseteq \mathrm{P}_N \times \mathrm{K}_1$.

   Particle permutation symmetry $\mathrm{P}_N \sim \mathrm{S}_N$ is clear from the Hamiltonian and the factor of K_1 is a consequence of the Galilean-invariance of the two-body interactions. This minimal symmetry is not enough to integrate the problem for general traps and interactions, but it does imply that (unless there are emergent symmetries or accidental degeneracies) every energy level should have a degeneracy corresponding to the dimension of an irreducible representation of S_N. In the case of a harmonic traps, both total parity and relative parity are invariants, and so there is an extra factor of O(1) in $\mathrm{K}_N \sim \mathrm{S}_N \times \mathrm{O}(1) \times \mathrm{U}(1) \times \mathrm{T}_t$. This is enough extra symmetry for the case with $N = 2$ to be integrable for any $\hat{V}_{12}$, and solvable for contact interactions [12].

3. The non-interacting Hamiltonian $\hat{H}_0^N$ has at least the minimal kinematic symmetry $\mathrm{K}_N^0 \supseteq \mathrm{P}_N \ltimes \mathrm{K}_1^{\times N}$.

   For the non-interacting system, all of the single-particle symmetries and invariants still commute with the total Hamiltonian, including single-particle time translations. That is what the N-fold direct product of one-particle symmetries $\mathrm{K}_1^{\times N}$ encapsulates. The semidirect product $\ltimes$ captures that fact that particle permutations act as homomorphisms on the normal abelian subgroup $\mathrm{K}_1^{\times N}$, e.g. for the unitary representation $\hat{U}(ij)$ of the transposition $(ij) \in \mathrm{P}_N$, one finds that $\hat{U}(ij)\hat{H}_i = \hat{H}_j\hat{U}(ij)$. This symmetry is enough for every system to be integrable for any N and J. A complete set of commuting observables for the spatial Hilbert space is $\{\hat{H}_1^1, \hat{H}_2^1, \ldots\}$ and this is equivalent to diagonalizing the $\mathrm{K}_1^{\times N}$ subgroup. For symmetrizing identical particles and perturbation theory, diagonalizing the P_N subgroup is more useful. For $N = 2$ and any trap, the system is superintegrable in a sense because $\hat{H}_0^2$ and $\hat{U}(12)$ are also sufficient to separate all of $\mathcal{K}$. For harmonic traps, $\hat{H}_0^N$ is maximally superintegrable because $\mathrm{K}_N^0 \sim \mathrm{U}(N)$ [13].

4. Symmetry breaking from K_N^0 to K_N for weak interactions is not algebraically universal for $N > 4$.

   Symmetry breaking is calculated using first-order perturbation theory aided by symmetrization of the subgroup P_N subgroup common to both K_N^0 and K_N. For $N = 2$, the magnitude of the first-order splitting energies depends on the specific properties of V^1 and V^2, but the form is algebraically universal, i.e. the same algebraic equation for every trap and every two-body interaction, and not, for example, a set of transcendental equations. Further, the expressions for the first-order eigenstates does not depend on V^1 or V^2. For $N = 3$ and $N = 4$, there are still algebraically-universal expressions for the first-order energy shifts and corresponding eigenstates. However, some universality is lost because now both expressions depend on the trap and interaction for mixed-symmetry states. These states that are neither totally symmetric or totally antisymmetric are relevant when $J > 1$. For $N > 4$, there are no algebraically universal expressions for first-order perturbation theory because solutions in mixed symmetry subspaces require solving polynomial equations of order quintic or higher. There are of course specific solutions for specific potentials and traps, e.g. harmonic interaction in harmonic traps [14].

5. Two-body matrix elements of the contact interaction potential have state permutation symmetry.

   Consider the two-body matrix elements $\langle\alpha\beta|\hat{V}_{12}|\gamma\zeta\rangle$ of the two body interaction, where $|\gamma\zeta\rangle = |\gamma\rangle \otimes |\zeta\rangle$ is the tensor product of the two one-particle states satisfying $\hat{H}^1|\gamma\rangle = \epsilon_\gamma|\gamma\rangle$. From Galilean invariance and hermiticity (and choosing phases so that the one-dimensional trap eigenstates are all real) the two-body matrix elements of the the contact interaction $V^2(|x_i - x_j|) = g\delta(x_i - x_j)$ have the property that all 24 permutations of the four states in $\langle\alpha\beta|\hat{V}_{12}|\gamma\zeta\rangle$ will have the same two-body matrix element. Since totally-antisymmetric spatial states are odd under permutations of state or particles, this symmetry explains why identical one-component fermions do not feel the contact interaction. For a flat potential with hard boundaries, i.e. the infinite square well, the two-body matrix elements are not only state permutation invariant, but they also do not depend any state property of $|\alpha\rangle$, $|\beta\rangle$, etc. except the for the degeneracy pattern of the quantum numbers in the the composition of the four states. The infinite square well with contact interactions is also integrable and solvable (but not algebraically solvable) for any value of g, number of particles N, and number of spin components J using the Bethe ansatz [15].

6. In the unitary limit of contact interactions $g \to \infty$, ordering permutation symmetry emerges as a symmetry of the Hamiltonian $\hat{H}_\infty^N$.

   The coincidence manifold $\mathcal{V}^N$ is the union of all the $(N-1)$-dimensional hyperplanes defined by $x_i = x_j$. The configuration space $\mathcal{X}^N$ is sectioned into $N!$ sectors $\mathcal{X}_s$ by the manifold $\mathcal{V}^N$, one for every permutation $s = \{s_1 s_2 s_3 \cdots s_N\} \in S_N$. Each sector $\mathcal{X}_s$ is the set of configurations with $x_{s_1} < x_{s_2} < \cdots < x_{s_N}$. In the unitary limit of the contact interaction, the manifold $\mathcal{V}^N$ is a nodal surface for all finite-energy states and there is no tunneling between sectors [16]. Particle permutations $p \in P_N$ act on $\mathcal{X}^N$ as orthogonal transformations and map sectors onto sectors like $p\mathcal{X}_s = \mathcal{X}_{sp^{-1}}$. Ordering permutations $\mathfrak{o} \in \mathfrak{O}_N$ rearrange the order of the positions, without regard to which particle is in which position. They are not linear transformations on $\mathcal{X}^N$ but they do map sectors onto sectors naturally: $\mathfrak{o}\mathcal{X}_s = \mathcal{X}_{\mathfrak{o}s}$. The set of ordering permutations $\mathfrak{O}_N$ is therefore also isomorphic to S_N, but distinct from and commuting with $P_N \sim S_N$.

7. The unitary-limit Hamiltonian $\hat{H}_\infty^N$ always has at least the minimal kinematic symmetry $K_N^\infty \supseteq P_N \times \mathfrak{O}_N \times K_1$ when restricted to the subspace of finite energy states.

   The finite-energy spectrum of $\hat{H}_\infty^N$ is the same as the spectrum of $\hat{H}_\infty^0$ restricted to totally antisymmetric spatial states. Therefore, for every distinct set of N one-particle states, there is an energy level of $\hat{H}_\infty^N$ that is $N!$-fold degenerate before symmetrization. A complete set of commuting operators for N distinguishable particles with spin s ($J = 2s + 1$) is provided by each particle's spin operator, e.g. $\hat{S}_i^z$, the total Hamiltonian $\hat{H}_\infty^N$, and a complete set of conjugacy class operators for canonical subgroup chains $\mathrm{P}_N \supset \mathrm{P}_{N_1} \supset \cdots \supset \mathrm{P}_2$ and $\mathfrak{O}_N \supset \mathfrak{O}_{N_1} \supset \cdots \supset \mathfrak{O}_2$. For indistinguishable fermions or bosons, these subgroup chains can be used (or in some cases modified to account for parity or spin composition) to select the state with the correct symmetrization properties [17]. The degeneracy of allowed states can be calculated algebraically using combinatorics.

8. Symmetry breaking from K_N^∞ to K_N for near-unitary interactions is not algebraically universal for $N > 4$.

   In the near-unitary limit, $\mathfrak{O}_N$ symmetry is broken by tunneling between adjacent sectors but $\mathrm{P}_N \times \mathrm{K}_1$ is preserved. The level-splitting can again be calculated by degenerate perturbation theory [18]. This time, the irreducible representations of $\mathfrak{O}_N \sim \mathrm{S}_N$ must be diagonalized. For $N = 2$ and $N = 3$ in a symmetric trap, the specific splitting is proportional to $1/g$, and the form of the splitting is the same for every trap shape. For three particles in an asymmetric trap, four particles, and five particles in a symmetric trap, the specific splitting depends two or three different tunneling parameters, all proportional to $1/g$, but whose relative ratios depend on the trap shape. Therefore, there is less universality in this case, but the equations are still algebraically solvable and take the same form in terms of the tunneling parameters for every case. For five particles in an asymmetric trap or six or more particles, there cannot be algebraic universality for the same reason at is was lost for weak interactions: quintic equations (or worse) arise. Note that for all cases, the totally antisymmetric state is universal, and the splitting of the totally symmetric state is universal. The failure of algebraic universality is only relevant for mixed symmetry states.

3 Concluding Remarks

One direction these comments on symmetry and integrability point is towards computational algorithms with increased efficiency. Another direction is implementation of quantum control protocol that exploits symmetry-protected subspaces of the Hamiltonian, perhaps for information processing. For example, if experimental control is sufficient, the symmetries of trapped ultracold gases in one-dimensional can be mapped on the symmetries of other quantum systems, enabling more robust simulation. And to conclude, if more is different, then five is more.

Acknowledgements

Thanks to the organizers and participants of FB21 for a stimulating conference, especially colleagues in the Friday session on Low-Dimensional Systems for their stimulating presentations.

References

[1] T. Kinoshita, T. Wenger, and D.S. Weiss, Nature **440**, 900-903 (2006); B. Paredes, et al. Nature **429**, 277 (2004); E. Haller, et al., Science **325**, 1224 (2009).

[2] F. Serwane, et al., Science **332**, 3362011 (2011); G. Zürn, et al., Phys. Rev. Lett. **108**, 075303 (2012); G. Zürn, et al., Phys. Rev. Lett. **111**, 175302 (2013); S. Murmann, et al., arXiv:1507.01117, (2015).

[3] M. Olshanii, Phys. Rev. Lett. **81**, 938 (1998); M.D. Girardeau and M. Olshanii, Phys. Rev. A **70**, 023608 (2004); Z. Idziaszek and T. Calarco, Phys. Rev. A **71**, 050701(R) (2005).

[4] M. Girardeau, J. Math. Phys. **1**, 516-523 (1960); E.H. Lieb and W. Liniger, Phys. Rev. **130**, 1605 (1963).

[5] B. Sutherland B., *Beautiful Models: 70 Years of Exactly Solved Quantum Many-Body Problems*, (World Scientific, Singapore, 2004); S. Albeverio, et al., *Solvable Models in Quantum Mechanics*, 2nd Ed., (AMS, Providence, 2004).

[6] M. Rigol, Phys. Rev. Lett. **103**, 100403 (2009); T. Langen, et al., Science **348**, 207-211 (2015).

[7] D.M. Gangardt, J. Phys. A **37**, 9335-9356 (2004); A. Farrell A. and B.P. van Zyl, J. Phys. A **43**, 015302 (2010).

[8] N.L. Harshman, AIP Conference Proceedings **1508**, 386-390 (2012).

[9] M.A. Cazalilla, Rev. Mod. Phys. **83**, 1405-1466 (2011); D. Blume, Rep. Prog. Phys. **75**, 046401 (2012); X.-W. Guan, M.T. Batchelor, and C. Lee, Rev. Mod. Phys. **85**, 1633-1691 (2013).

[10] N.L. Harshman, Few-Body Systems, First online (08 Sep 2015).

[11] N.L. Harshman, Few-Body Systems, First online (09 Sep 2015).

[12] M.P. Avakian, et al., Phys. Lett. A **124**, 233-236 (1987); T. Busch et al., Found. Physics **28**, 549-559 (1998); S. Jonsell, Few-Body Sys. **31**, 255-260 (2002).

[13] G.A. Baker, Phys. Rev. **103**, 1119 (1956); J.D. Louck, J. Math. Phys. **6**, 1786 (1965).

[14] J.R. Armstrong et al., Physica Scripta **T151**, 014061 (2012).

[15] N. Oelkers et al, J. Phys. A **39**, 1073-1098 (2006); Y. Hao, et al., Phys. Rev. A **79**, 033607 (2009).

[16] C.G. Bao, Few-Body Syst. **54**, 1957-1971 (2013).

[17] M.D. Girardeau and A. Minguzzi, Phys. Rev. Lett. **99**, 230402 (2007); F. Deuretzbacher et al., Phys. Rev. Lett. **100**, 160408 (2008); C.N. Yang et al., Chin. Phys. Lett. **26**, 120504 (2009); L. Guan et al., Phys. Rev. Lett. **102**, 160402 (2009); Z.-Q. Ma, et al., J. Phys. A **42**, 385210 (2009); M.D. Girardeau, Phys. Rev. A **82**, 011607(R) (2010); M.D. Girardeau and G.E. Astrakharchik, Phys. Rev. A **81**, 061601(R) (2010); M.D. Girardeau, Phys. Rev. A **83**, 011601(R) (2011); Fang et al., Phys. Rev. A **84**, 023626 (2011).

[18] A.G. Volosniev et al., Nat. Comm. **5**, 5300 (2014); F. Deuretzbacher et al., Phys. Rev. A **90**, 013611 (2014); J. Levinsen, et al., Science Advances **1**, e1500197 (2015); S.E. Gharashi, X.Y. Yin, and D. Blume, Phys. Rev. A **91**, 013620 (2015).

EPJ Web of Conferences **113**, 03001 (2016)
DOI: 10.1051/epjconf/201611303001

Weakly bound states with spin-isospin symmetry

A. Kievsky[1,a] and M. Gattobigio[2,b]

[1] *Istituto Nazionale di Fisica Nucleare, Largo Pontecorvo 3, 56127 Pisa, Italy*
[2] *Université de Nice-Sophia Antipolis, Institut Non-Linéaire de Nice, CNRS, 1361 route des Lucioles, 06560 Valbonne, France*

Abstract. We discuss weakly bound states of a few-fermion system having spin-isospin symmetry. This corresponds to the nuclear physics case in which the singlet, a_0, and triplet, a_1, $n-p$ scattering lengths are large with respect to the range of the nuclear interaction. The ratio of the two is about $a_0/a_1 \approx -4.31$. This value defines a plane in which a_0 and a_1 can be varied up to the unitary limit, $1/a_0 = 0$ and $1/a_1 = 0$, maintaining its ratio fixed. Using a spin dependant potential model we estimate the three-nucleon binding energy along that plane. This analysis can be considered an extension of the Efimov plot for three bosons to the case of three 1/2-spin-isospin fermions.

1 Introduction

In a series of papers [1, 2] V. Efimov has shown that three identical bosons interacting through a two-body short-range potential have a resonant spectrum at the unitary limit, $1/a = 0$, with a the two-body scattering length. In this limit a geometrical series of bound states appears whose energies accumulate to zero. The ratio between the energies of two consecutive states is constant and, remarkably it results independent of the particular form of the interaction. This behavior has been denoted Efimov effect and its characteristic of being universal has given to this effect a particular relevance. In fact in the last two decades an enormous amount of work has been dedicated to study this effect in different fields as molecular, atomic, nuclear and particle physics.

The spectrum of the three-boson system close to the unitary limit is described by the Efimov equations (or Efimov radial law) that can be put in the following form

$$\begin{aligned} E_3^n/(\hbar^2/ma^2) &= \tan^2\xi \\ \kappa_* a &= \mathrm{e}^{(n-n^*)\pi/s_0}\frac{\mathrm{e}^{-\Delta(\xi)/2s_0}}{\cos\xi}\,, \end{aligned} \tag{1}$$

where $\Delta(\xi)$ is an universal function. A parametrization of this function in the range $-\pi < \xi < -\pi/4$ can be found in Ref. [3]. The quantity $s_0 \approx 1.00624$ is a universal number and κ_* defines the energy $\hbar^2\kappa_*^2/m$ of the level $n = n^*$ at the unitary limit. Furthermore, knowing the value of κ_* the complete spectrum is determined as a function of a. It should be noticed however that the above equations have been derived in the zero-range limit. In this limit the two-body energy results to be $E_2 = \hbar^2/ma^2$ and,

[a] e-mail: kievsky@pi.infn.it
[b] e-mail: mario.gattobigio@inln.cnrs.fr

in addition, all the two-body low energy scattering observables can be written in terms of a, making this quantity a control parameter (see for instance Ref. [3] and references therein). When a two-body potential with finite range is considered, the Efimov radial law can be modified as it was discussed by the authors in Refs. [4, 5]. In the following we extend the discussion to the case in which the particles are 1/2-spin-isospin fermions [6].

2 Efimov plot with 1/2-spin-isospin fermions

The zero-range energy spectrum given by Eq.(1) is unbounded from below as has been shown first by Thomas [7]. Finite-range potentials cure this pathology modifying the energy spectrum. For values of a much bigger than the range of the interaction Eq.(1) is still valid for the highest levels, however the ground and first excited levels show finite-range effects. In order to take into account these corrections the authors have proposed a modification to the Efimov radial low [4, 5],

$$
\begin{aligned}
E_3^n/E_2 &= \tan^2\xi \\
\kappa_n^3 a_B + \Gamma_n^3 &= \frac{e^{-\Delta(\xi)/2s_0}}{\cos\xi} .
\end{aligned}
\tag{2}
$$

Here there are two corrections, one coming from the two-body system and it can be taken into account by substituting a with a_B, defined by $E_2 = \hbar^2/ma_B^2$, with E_2 the two-body binding energy if $a > 0$, or the two-body virtual-state energy in the opposite case, $a < 0$. The second correction is the introduction of a finite-range parameter, Γ_n^3, depending on the energy level, which produces a shift in the product $\kappa_* a_B$. Recently it has been shown that the value of the shift is almost the same for very different potentials in range and scale as the LM2M2 interaction of Aziz [8], a gaussian potential reproducing the values of E_2 and a given by that interaction and a combination of yukawians, the MT-III interaction [9], describing the s-wave triplet $n-p$ state. Interestingly the value of the shift for the ground state in the three cases results to be almost the same, $\Gamma_0^3 \approx 0.8$ [10].

In the following we would like to discuss the case of fermions having 1/2 isospin symmetry in which the interaction is different in the singlet and triple state as for the $n-p$ system. In this case the experimental data are $a_0 = -23.740 \pm 0.020$ fm and $a_1 = 5.424 \pm 0.003$ fm. To describe these data we use a gaussian potential

$$
V(r) = V_S e^{-r^2/r_S^2} ,
\tag{3}
$$

with different strength in the singlet ($S = 0$) and triplet ($S = 1$) and having the same range (in the following we consider $r_0 = r_1 = 1.65$ fm). The values $V_0 = -37.90$ MeV and $V_1 = -60.575$ MeV approximate the experimental data and, for $S = 1$, the deuteron binding energy is approximated too. Varying the strengths V_S it is possible to explore the three-nucleon binding energy as the values of a_S move toward the unitary limit. Among different possibilities we fix the ratio $a_0/a_1 = -4.31$ to its experimental value and calculate the energy of the ground and first excited states. In Fig.1 we show the different planes (the three-body energy defines the third axis not shown in the figure) determined by the values of a_0 and a_1. When the two scattering lengths are equal, $a_0/a_1 = 1$, the corresponding plane is the boson plane for three equal particles very well studied (see for instance Ref. [4]). In the figure this is given by the (red) solid line. The (blue) solid line corresponds to what we call the nuclear plane since it contains the physical nuclear point, the continuation to the fourth quadrant given by the (blue) dashed line does not give new information since the exchange of the singlet and triplet potentials gives the same binding energy. The two axes correspond to planes in which one of the scattering lengths is on the unitary limit. The energy values of E_3^n are different in the different planes however they collapse to the boson spectrum when the two scattering lengths are at the unitary limit. Using the gaussian potential given above the ground state energy in this limit results $E_3^0 \approx 3.6$ MeV.

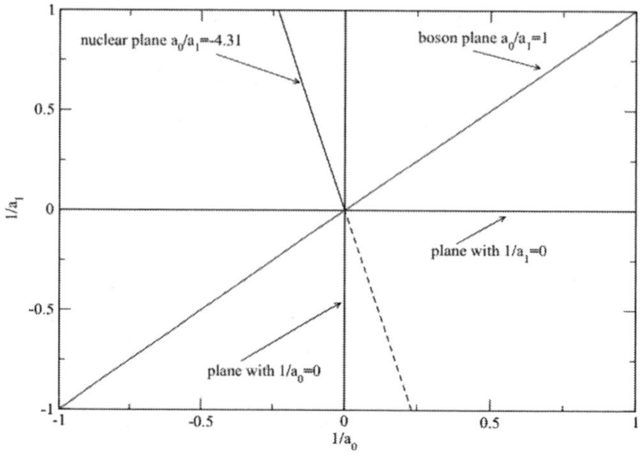

Figure 1. Different planes as function of the inverse of a_0 and a_1

The main results of the present study is given in Fig.2 where the binding momentum, defined as $(\hbar^2/m)K_3^2 = E_3$, for the ground and excited states are shown as a function of a_1 (in units of r_*, the value of the effective range at the unitary limit) maintaining the ratio a_0/a_1 fixed around -4.31. The three-fermion binding momenta are given by the (blue) solid line for the ground state and by the (blue) dot-dashed line for the first excited state. For comparison the binding momentum values in the case of three bosons are given too by the (black) solid and dashed lines. The (red) solid line is the two-body biding momentum. K_* is the momentum at the unitary limit and both, the fermion and boson values, are the same. The excited state in this limit has a value of $K_*/22.9$ showing a slight finite-range correction (the zero-range theory predicts the value $K_*/22.7$). The (negative) values of the triplet scattering lengths, a_-^0 and a_-^1, at which the three-boson ground and excited states disappear into the three-nucleon continuum are shown in the figure. Their ratio of about 17.6 is well below the prediction of 22.7 given by the zero-range theory. More interesting is the fact that the excited state of the three-fermion system disappears into the $n-d$ continuum before matching the nuclear physical point (at $a_*^1 \approx 20$ fm). This is not the case for the three-boson system, the excited state remains bound and always below E_2. This fact has been studied before in the case of a system of three helium atoms. The present analysis shows that in the case of three nucleons the difference in the single and triple potential strengths is such that the excited state disappear into the $n-d$ continuum before reaching the physical point. Evidence of the presence of this state embedded in the $n-d$ continuum has been given in the literature [11, 12] and the curvature of the $n-d$ effective range function close to the threshold is one of this [13, 14]. Finally, at the physical value, the two- and three-body energies are explicitly displayed. As it is well known, in order to reproduce the triton binding energy a three-body force has to be included.

The spectrum of the three-fermion system can be still described by Eq.(2) with a value of the shift depending on the ratio a_0/a_1. In the present analysis we found for the ground state level $\Gamma_3^0 \approx -0.2$, a negative value. If we compare this value to the value of about 0.8 obtained in the case of three bosons we can conclude that there is a particular value of the ratio at which the shift $\Gamma_3^0 \approx 0$. In this case Eq.(2) will be equivalent to the zero-range equation. However this would valid for the ground state level, the shift depends on the level and, as we have shown, the ratio with the excited state at the unitary limit is 22.9 (not 22.7) for a gaussian potential independently of the shift value.

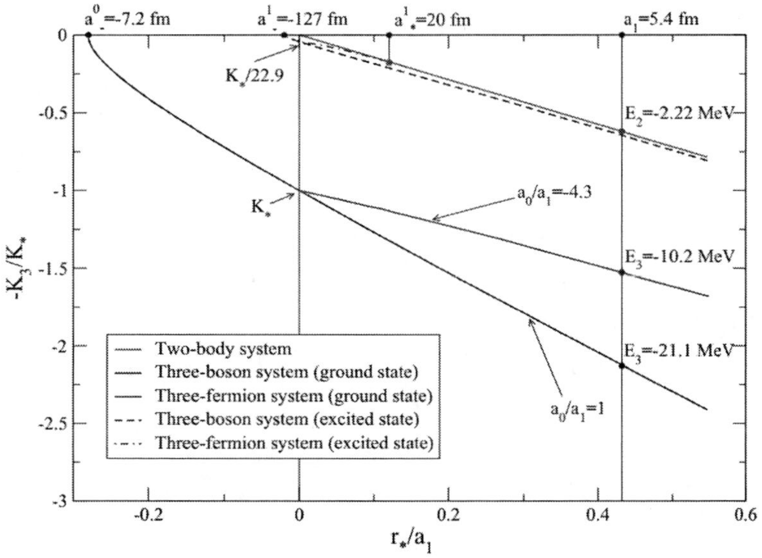

Figure 2. The binding momentum as a function of the inverse of a_1

3 Conclusions and Outlook

The present analysis focussed on the particularities of the Efimov plot in the case of $1/2$-spin-isospin fermions. We have selected a particular way of constructing the Efimov plot in which the ratio of the singlet and triplet scattering lengths has been kept fixed. In this way the binding momentum and a_1 define a plane similar to the boson case, in particular the values at the unitary limit coincide. The present work has to be consider a first step in the study of the Efimov plot for three nucleons. At present the role of the three-body force and the predictions on systems with $A > 3$ are underway.

References

[1] V. Efimov, Phys. Lett. B **33** (1970), 563
[2] V. Efimov, Sov. J. Nucl. Phys. **12** (1971), 589 [Yad. Fiz. 12, 10801090 (1970)]
[3] E. Braaten and H.-W. Hammer, Phys. Rep. **428** (2006), 259
[4] M. Gattobigio and A. Kievsky, Phys. Rev. **A90**, 012502 (2014)
[5] A. Kievsky, N.K. Timofeyuk and M. Gattobigio, Phys. Rev. **A90**, 032504 (2014)
[6] V. Efimov, Nuc. Phys. A **362** (1981), 45
[7] L.H. Thomas, Phys. Rev. **47**, 903 (1935)
[8] R.A. Aziz and M.J. Slaman, J. Chem. Phys. **94**, 8047 (1991)
[9] R.A. Malfliet and J.A. Tjon, Nuc. Phys. **A 127**, 161 (1969)
[10] A. Kievsky and M. Gattobigio, (2015) arXiv:1507.03402
[11] S.K. Adhikari, A.C. Fonseca and L. Tomio, Phys. Rev. **C 26**, 77 (1982)
[12] T. Frederico, I.D. Goldman and A. Delfino, Phys. Rev. **C 37**, 497 (1988)
[13] C.R. Chen, G.L. Payne, J.L. Friar and B.F. Gibson, Phys. Rev. C **39**, 1261 (1989)
[14] A. Kievsky and M. Gattobiggio, Phys. Rev. **A87**, 052719 (2014)

EPJ Web of Conferences **113**, 03002 (2016)
DOI: 10.1051/epjconf/201611303002

Resonance model for the three-body states of the A=6 reactions

Mark W. Paris[1,a] and Gerald M. Hale[1,b]

[1] *Theoretical Division (T-2), Los Alamos National Laboratory, Los Alamos NM 87545, USA*

Abstract. We present an R-matrix-based model for three-body final states that has aspects of the Faddeev approach to three-particle scattering. The model is applied to describing the nucleon spectra for breakup reactions in the $A = 6$ systems. Calculations using a charge-symmetric parametrization agree fairly well with the experimental data, although they indicate larger contributions from the ^{5}He or ^{5}Li ground state are necessary.

1 Introduction

There has been much interest and experimental activity lately at ICF facilities, such as OMEGA at the U. of Rochester, and the NIF at Livermore, to measure the particle spectra from three-body breakup reactions initiated by the laser implosion of capsules containing tritium ($t = {}^3$H), ^{3}He, or a mixture of the two. Here we describe an R-matrix-based model for fitting these spectra that maintains some aspects of the Faddeev method for three-particle scattering. The next section gives a brief introduction to the concepts of R-matrix theory and summarizes the resonance model, in which the T matrix is expressed as the sum of three Faddeev-like components. When specialized to the case of single-level terms, these expressions form the basis of our present model, which is similar to that of Brune *et al.* [1], but differs from it in detail.

The application of these formulas to calculate the spectra of nucleons from the 3-body breakup of the reactions induced by $t + t$, ^{3}He+^{3}He, and t+^{3}He is described in Section 3. These calculations use charge-symmetric (CS) parameters for the quantities dominated by nuclear forces, but take into account the Coulomb differences in the resonant sub-systems. This CS approach clearly has more justification for the first two (pure $T = 1$) mirror reactions than it does for the third, which is an isospin mixture of $T = 1$ and $T = 0$. The conclusions from this study thus far and the outlook for future work are given in Section 4.

2 R-matrix based resonance model

R-matrix theory asserts that, due to the short range of nuclear forces, there exists a finite region of coordinate space beyond which the scattering wave function for an A-nucleon system assumes its known asymptotic form in terms of T-matrix amplitudes. Through the device of the Bloch operator,

[a] e-mail: mparis@lanl.gov
[b] e-mail: ghale@lanl.gov

$\mathcal{L}_B$, acting at the surface enclosing that region, a complete set of eigenfunctions of the operator $H+\mathcal{L}_B$ can be used to expand the scattering wave function inside it as

$$\phi_{\mathbf{q}}^{+}(\mathbf{x}) = \sum_{\lambda} c_{\lambda}(\mathbf{q})\phi_{\lambda}(\mathbf{x}), \tag{1}$$

$$c_{\lambda}(\mathbf{q}) = \frac{1}{\sqrt{\pi}} \sum_{\lambda'} A_{\lambda\lambda'}\gamma_{\lambda'} O^{-1}(q)\mathcal{Y}_{sl}^{j\pi}(\hat{\mathbf{q}}). \tag{2}$$

In Eq. (2), $A_{\lambda\lambda'}$ is the level matrix, $\gamma_{\lambda'}$ is the reduced-width amplitude, O is the outgoing-wave solution, and $\mathcal{Y}_{sl}^{j\pi}$ is the spin-angle eigenfunction of total angular momentum and parity.

In the resonance model, a two-body initial state reacts to give a three-body final state. The assumed Jacobi coordinates and potentials in the final state are shown in Fig. 1. The pair-wise interactions in the final state are treated as in the Faddeev approach to three-particle scattering, but in these applications to the $A = 6$ systems, the initial state cannot be formed from a bound state of any pair of particles in the final state, as is required by the Faddeev theory. Hence, we call it "Faddeev-like."

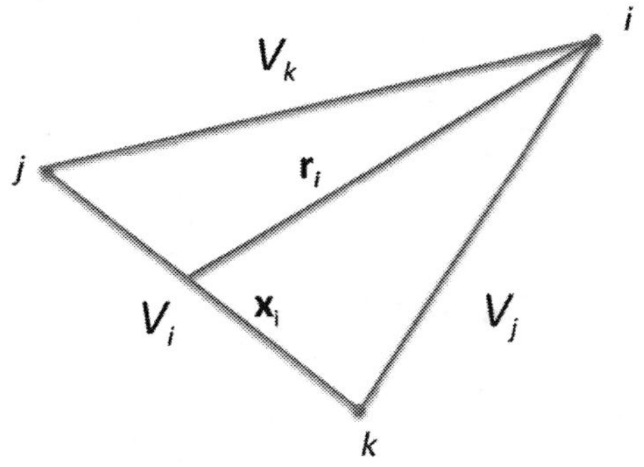

Figure 1: Jacobi coordinates and pair-wise potentials for describing a three-body final state.

The three-body breakup transition amplitude can be written as a sum of Faddeev-like components

$$T^{(3)} = \sum_{i=1}^{3} T^{(3)}_{\mathbf{q}_i\mathbf{k}_i\mathbf{k}_0}, \tag{3}$$

$$T^{(3)}_{\mathbf{q}_i\mathbf{k}_i\mathbf{k}_0} = -\pi\langle\chi_{\mathbf{q}_i}\chi_{\mathbf{k}_i}|V_i|\psi^{+}_{\mathbf{k}_0}\rangle, \tag{4}$$

where $\psi^{+}_{\mathbf{k}_0}$ is the scattering solution in the initial two-body state, and $\chi_{\mathbf{q}_i}$ and $\chi_{\mathbf{k}_i}$ are the plane-wave solutions for the momenta $\mathbf{q}_i$ and $\mathbf{k}_i$ conjugate to $\mathbf{x}_i$ and $\mathbf{r}_i$, respectively. Relating the plane-wave solution $\chi_{\mathbf{q}_i}$ on the left side of the matrix element to the two-body scattering solution $\phi^{-}_{\mathbf{q}_i}$ through the Lippmann-Schwinger equation gives Eq. (4) as the distorted-wave T-matrix element

$$T^{(3)}_{\mathbf{q}_i\mathbf{k}_i\mathbf{k}_0} = -\pi\langle\phi^{-}_{\mathbf{q}_i}\chi_{\mathbf{k}_i}|V_i + V_iG^{+}_{0i}V_i|\psi^{+}_{\mathbf{k}_0}\rangle. \tag{5}$$

The R-matrix phenomenology is introduced by making in each of the Jacobi arrangements the expansion of Eqs. (1) and (2) for $\phi^{-}_{\mathbf{q}_i}$, giving for the 3-body breakup T matrix,

$$T^{(3)} = \sum_{i=1}^{3}\sum_{\lambda_i} c^{(i)}_{\lambda_i}(\mathbf{q}_i)\tilde{T}^{(2)}_{\lambda_i\mathbf{k}_i\mathbf{k}_0}, \tag{6}$$

with

$$\begin{aligned}\tilde{T}^{(2)}_{\lambda_i \mathbf{k}_i \mathbf{k}_0} &= -\pi\langle \phi^{(i)}_{\lambda_i}\chi_{\mathbf{k}_i}|V_i + V_i G^{+}_{0i} V_i|\psi^{+}_{\mathbf{k}_0}\rangle \\ &= \sum_{Js'l'sl} \mathcal{Y}_{\lambda J s' l'}(\hat{\mathbf{k}}) O^{-1}_{l'}(k) R^{\bar{L}}_{\lambda s' l' sl}(\varepsilon) O^{-1}_{l}(k_0) \mathcal{Y}^{*}_{Jsl}(\hat{\mathbf{k}}_0), \end{aligned} \tag{7}$$

a pseudo two-body off-energy-shell T-matrix element expressed in terms of the outgoing-wave R-matrix amplitudes $R^{\bar{L}}_{\lambda s' l' sl}(\varepsilon)$. In the applications described below, all the expansions are single-level ($\lambda = \lambda' = 1$), the resonance parameters in the c_λ are determined by outside data, and only the magnitudes of the R-matrix amplitudes are adjusted to fit the experimental spectra.

3 Applications to A = 6 spectra

The single-level resonance parameters for the ground and first-excited states of ^{5}He and ^{5}Li were taken from analyses of $n + \alpha$ and $p + \alpha$ scattering. Similarly, those for the singlet $n - n$ and $p - p$ resonances were taken from a charge-independent analysis of $N + N$ scattering. The R-matrix amplitudes were first adjusted by hand to fit the $t + t$ spectra of Wong *et al.* [2], as shown in Fig. 2. Then those same amplitudes were used to predict the proton spectrum for the ^{3}He+^{3}He reaction shown in Fig. 3, and for the t+^{3}He reaction shown in Fig. 4.

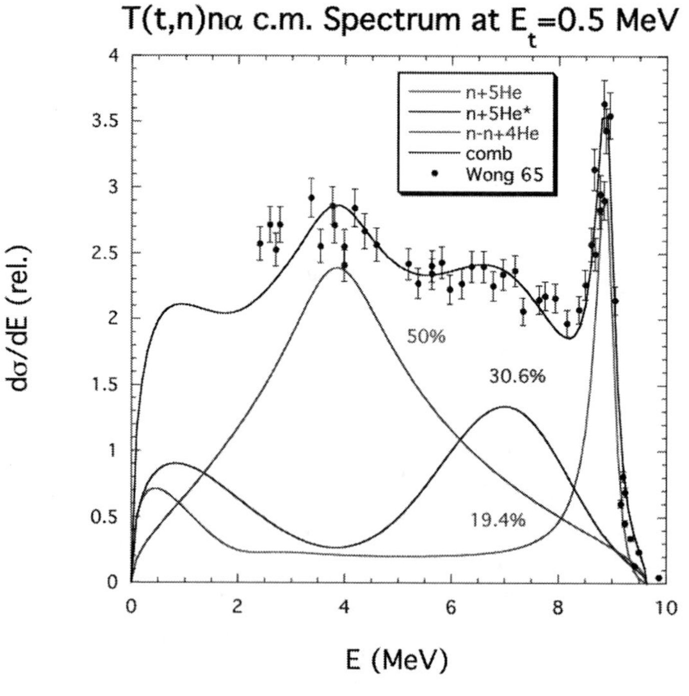

Figure 2: Neutron spectrum from the $t + t$ reaction at $E_t = 0.5$ MeV. The data are from Ref. [2], and the colored lines give the 3 Jacobi components of the spectrum with their branching ratios.

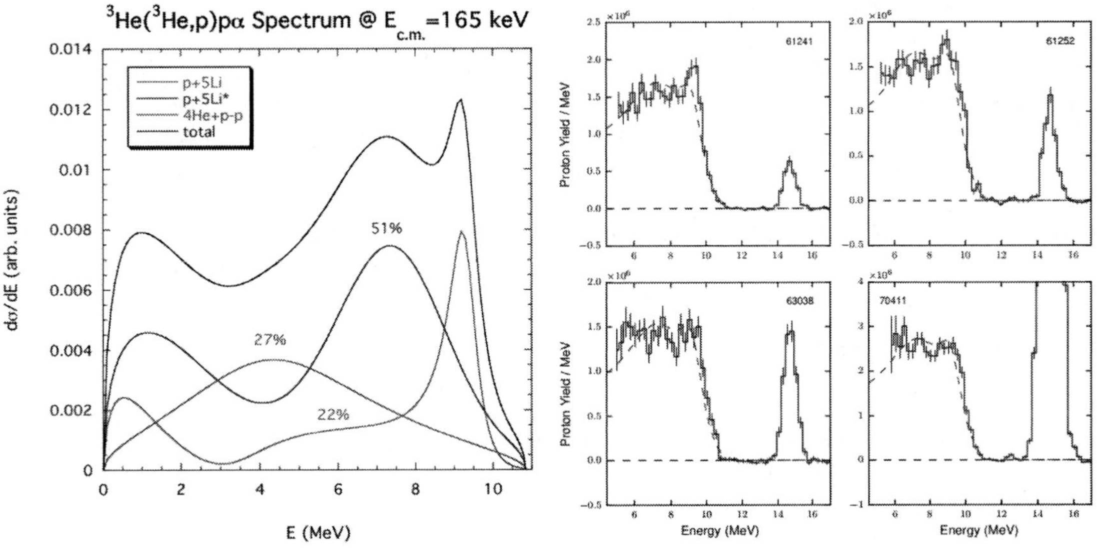

Figure 3: Proton spectra from the ^{3}He+^{3}He reaction at $E_{c.m.}$ = 165 keV. The left side of the figure shows the unbroadened spectrum with its 3 components, and the right side shows the instrumentally broadened spectrum compared with data from four different OMEGA shots [3, 4].

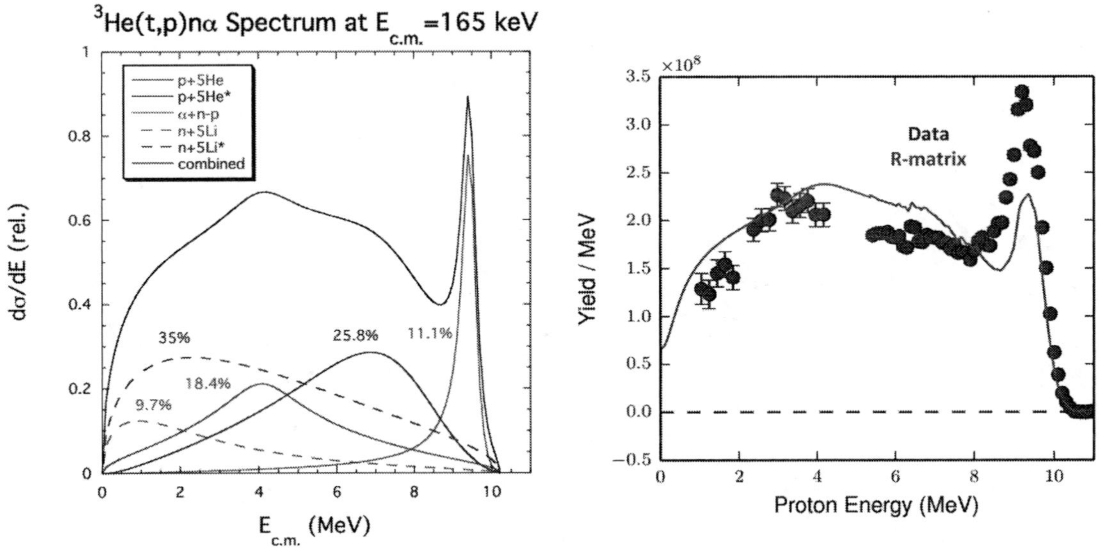

Figure 4: Proton spectra from the t+^{3}He reaction at $E_{c.m.}$ = 165 keV. The left side of the figure shows the unbroadened spectrum with its 5 components, and the right side shows the instrumentally broadened spectrum compared with OMEGA data [3, 4].

4 Conclusions and outlook

The agreement with experimental $A = 6$ spectra is reasonable. Coulomb differences in the resonances are important, and lead to different branching ratios for the components of the spectra even with CS amplitudes. However, the peaks at the ends of the spectra coming from the ground states of ^{5}He and ^{5}Li are under-predicted, and the details of the interference terms could be improved. Future plans are to make the fitting automated, include the angular distribution and interference effects with fewer approximations, make the kinematics relativistic, and to determine the energy dependence of the R^L amplitudes by linking them to R-matrix calculations of the integrated break-up cross sections.

Acknowledgements

We are indebted to our colleagues Andrew Bacher, Carl Brune, Johan Frenje, and Alex Zylstra for valuable discussions about theory and ICF experiments, and for folding our calculations with instrumental response functions in order to compare with their experimental data. This work was supported under the auspices of the US Department of Energy.

References

[1] C.R. Brune *et al.*, Phys. Rev. C **92** 014003 (2015)
[2] C. Wong, J.D. Anderson and J.W. McClure, Nucl. Phys. **71**, 106-112 (1965)
[3] A. B. Zylstra *et al.*, to be submitted to Phys. Rev. Lett. (2015)
[4] J.A. Frenje *et al.*, plenary talk, this conference (2015)

EPJ Web of Conferences **113**, 03003 (2016)
DOI: 10.1051/epjconf/201611303003

$T(T,2n)^4He$ and $^3He(^3He,2p)^4He$: The Reaction Mechanism from Solar Energies to 10 MeV*

A.D. Bacher[1], C.R. Brune[2], D.B. Sayre[3], G.M. Hale[4], J.A. Frenje[5], and M. Gatu Johnson[5]

[1] *Department of Physics, Indiana University, Bloomington, IN 47405, USA*
[2] *Ohio University, Athens OH 45701 USA*
[3] *Lawrence Livermore National Laboratory, Livermore CA 94550 USA*
[4] *Los Alamos National Laboratory, Los Alamos, NM 87545 USA*
[5] *Massachusetts Institute of Technology, Cambridge, MA 02139 USA*

Abstract. We have studied the energy dependence of the reaction mechanism of the $T(t,2n)^4He$ reaction at stellar energies and of its charge symmetric analog reaction $^3He(^3He,2p)^4He$ at energies up 10 MeV. We find that the reaction mechanism changes dramatically over this energy range in part due to the interference of the two identical fermions in the three-body final state.

This contribution describes a study of the Charge Symmetric reactions T+T and $^3He+^3He$ carried out at the laser facilities OMEGA and the National Ignition Facility (NIF) and at the tandem accelerator facility at Cal Tech. The $^3He(^3He,2p)^4He$ reaction reaction is of scientific interest because it completes over 98% of the p-p reaction chain in the Sun. Describing and measuring these reactions is complicated by the three body final states $^4He+2n$ and $^4He+2p$ which contain identical fermions. At low energies the reaction mechanism is dominated by the $L = 0$ (s-wave) partial wave. The shapes of the energy spectra of the neutrons and protons are determined by the stronger nucleon-^{4}He interaction and the weaker nucleon-nucleon interaction.

At OMEGA and the NIF the neutron spectra were determined by very fast Neutron Time-of-Flight detectors (TOF) and by a Medium Resolution Spectrometer (MRS) with a CD_2 foil to convert the neutrons into deuterons which could be momentum-analyzed in the magnetic field. Track detectors positioned along the MRS focal could be analyzed after the run. The properties of these detectors have been determined by previous measurements of n+D and D+T reactions. At Cal Tech the proton energy spectra were determined by a pair of solid-state detectors that used standard particle identification techniques to separate the charge-one protons from the charge-two alpha particles. While the measurements at the laser facilities measured the total cross section, the measurements at the accelerator laboratory could be measured at different angles with respect to the beam allowing an angular distribution of the protons and α particles to be determined at each beam energy. The targets were also very different. At OMEGA and the NIF the high pressure gases (at about 8 atm) were contained in small (3-mm diameter) capsules. At the NIF these were positioned inside a holram to enhance the reaction yield. The targets were positioned at the center of a large reaction chamber and lasers impinged on the targets creating fusion as the capsules collapsed and heated rapidly. At Cal Tech the ^{3}He beam entered the gas-filled scattering chamber through a thin (2500 Angstrom) foil. It exited through

*This contribution is dedicated to the memory of Tom Tombrello, my Ph.D. advisor at Cal Tech, who died in 2014.

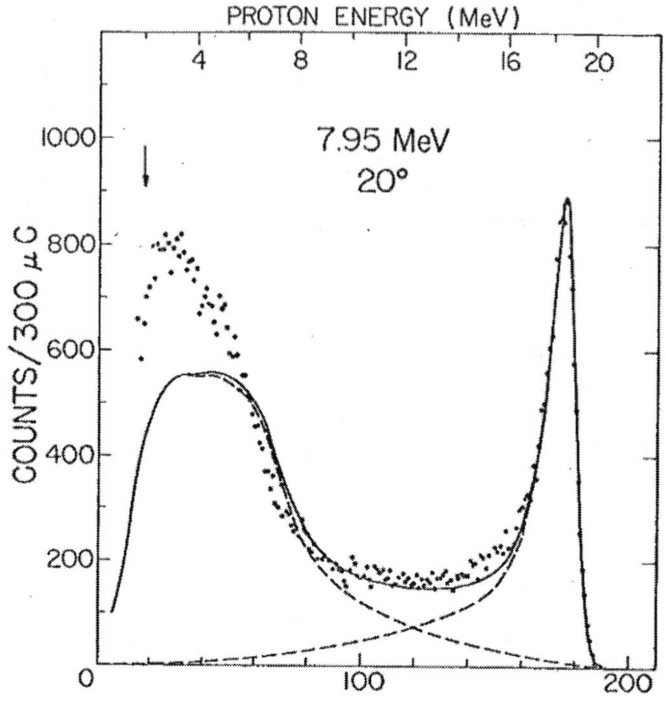

Figure 1. Proton energy spectrum measured at 20° and for a ^{3}He energy of 7.95 MeV [3].

a thicker foil so that the total charged could be integrated with a calibrated Faraday Cup, thus allowing the cross section for $^{3}He+^{3}He$ to be determined at each beam energy. The counter telescope could be positioned at angles with respect to the beam between 15° and 150°. State-of-the art detectors (in 1965) were used to stop the high-energy protons.

In addition to the s-wave initial state at the laser facilities, the energy spectra of the final state neutrons and protons are dominated by the strong nucleon-^{4}He final state interaction. This interaction is p-wave ($L = 1$) and has two components. The $3/2^{-}$ component corresponds to the ^{5}He ground state and the $1/2^{-}$ component corresponds to the ^{5}He excited state. Of less importance is the weaker n-n or p-p final state interaction. A new reaction model has been developed by Carl Brune at Ohio University, Dan Sayre at Lawrence Livermore National Laboratory, and by Gerry Hale at Los Alamos National Laboratory [1]. It is a based on an extended R-matrix which can handle both the three body final states and the presence of identical fermions.

Studies of the T+T reaction at solar energies provide an excellent account of the new measurements which were published recently in Physical Review Letters [2]. At solar energies the $1/2^{-}$ state is more strongly populated than the $3/2^{-}$ state by a factor of about 2:1. Fitting the low energy portion of the neutron spectrum also shows the importance of the oriented decay of the $3/2^{-}$ state. Because of the large energy available in the three-body final state, it is also possible that p-waves ($L = 1$) may play a role.

Gerry Hale has developed new reaction models to extend this work to higher energies. For spectra and calculations above 2 MeV we find that the $3/2^{-}$ state is more strongly populated than the $1/2^{-}$

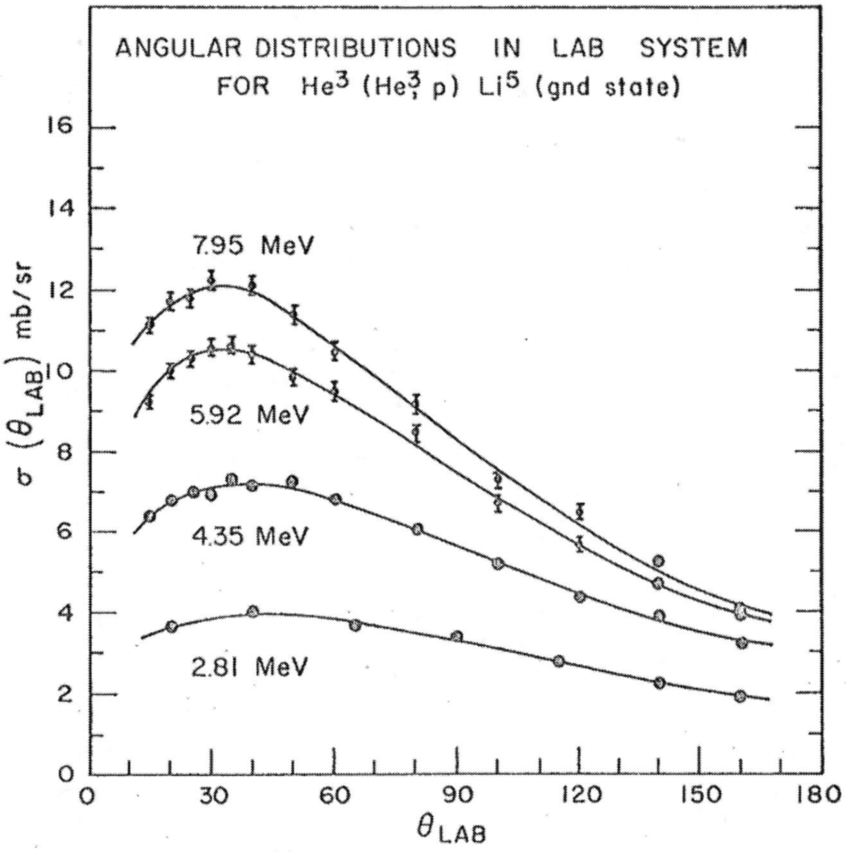

Figure 2. The evolution of the ^{3}He(^{3}He, p)^{5}Li(g.s) angular distribution with the incident ^{3}He energy [3].

state by a factor of about 2:1. This important difference from the factor at Solar energies is the result of including the interference terms for the identical fermions in the three-body final state. First we show several results from my Ph.D. thesis at Cal Tech [3]. In Figure 1, the proton spectrum at 7.95 MeV and 20° is fitted with a sequential decay model including the oriented decay of the recoiling ^{5}He nucleus which affects the shape of the proton spectrum below 6 MeV. The fit to the high energy portion of the spectrum is excellent. Next we show in Figure 2 angular distributions of this sequential decay model from 2.81 MeV up to 7.95 MeV. It is clear that the p-wave contribution is increasing with increasing energy. Finally, in Figure 3, we show a two dimensional plot of proton energy spectra at 20° from between 4 MeV and 18 MeV. Note how the overall yield increases dramatically above 12 MeV. There are some deuterons above 12 MeV due to the d+d final state.

Studies of the Charge Symmetric Reactions T+T and ^{3}He+^{3}He have shown how recent improvements in both experiment and theory have led to a new understanding of these reactions at Solar energies. For the ^{3}He+^{3}He reaction above 2 MeV the significant change in the reaction mechanism is

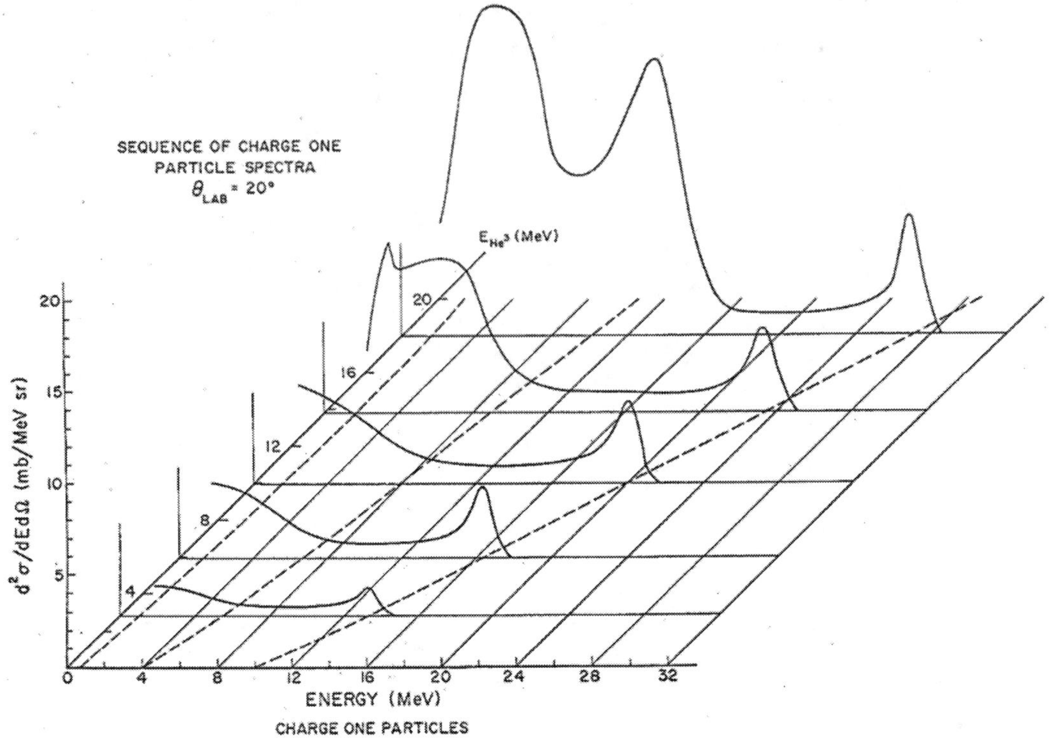

Figure 3. The evolution of the ^{3}He + ^{3}He proton spectrum at $\theta_{lab} = 20°$ with the incident ^{3}He energy [3].

understood to result from the interference between the two identical fermions in the final state. Less well understood is the increasing importance of p-wave in the spectra above 2 MeV. In the future we hope to make observations of the ^{3}He+^{3}He reaction at Solar energies with approved time at the NIF facility.

References

[1] C.R. Brune, J.A. Caggiano, D.B. Sayre, A.D. Bacher, G.M. Hale, M.W. Paris, Phys. Rev. C **92**, 014003 (2015)

[2] D.B. Sayre, C.R. Brune, J.A. Caggiano, V.Y. Glebov, R. Hatarik, A.D. Bacher, D.L. Bleuel, D.T. Casey, C.J. Cerjan, M.J. Eckart et al., Phys. Rev. Lett. **111**, 052501 (2013)

[3] A.D. Bacher, Ph.D. thesis, California Institute of Technology (1967), `http://resolver.caltech.edu/CaltechTHESIS:10262015-155317223`

EPJ Web of Conferences **113**, 03004 (2016)
DOI: 10.1051/epjconf/201611303004

Advances in the *ab initio* description of nuclear three-cluster systems

Carolina Romero-Redondo[1,a], Sofia Quaglioni[1], Petr Navrátil[2], and Guillaume Hupin[3]

[1] *Lawrence Livermore National Laboratory, P.O. Box 808, L-414, Livermore, California 94551, USA*
[2] *TRIUMF, 4004 Wesbrook Mall, Vancouver, British Columbia, V6T 2A3, Canada*
[3] *Institut de Physique Nucléaire, Université Paris-Sud, IN2P3/CNRS, F-91406 Orsay Cedex, France*

Abstract. We introduce the extension of the *ab initio* no-core shell model with continuum to describe three-body cluster systems. We present results for the ground state of ^{6}He and show improvements with respect to the description obtained within the no-core shell model and the no-core shell model/resonating group methods.

1 Introduction

The *ab initio* no-core shell model/resonating group method (NCSM/RGM) was presented in [1, 2] as a technique that is able to describe both structure and reactions in light nuclear systems. Within this approach, the wave function is expanded in a continuous cluster basis using the resonating group method with realistic interactions and a consistent *ab initio* description of the nucleon clusters.

The method was first introduced in detail for two-body cluster bases and has been shown to work efficiently in different systems [1–4]. Later, the expansion of the method for three-cluster systems was introduced in [5, 6]. The capability of *ab initio* methods to properly describe three-body cluster states is essential for the study of nuclear systems that present such configuration. This type of systems appear, e.g., in structure problems of two-nucleon halo nuclei such as ^{6}He and ^{11}Li, resonant systems such as ^{5}H, and reactions with three fragments in their final state such as ^{3}H(^{3}H,2n)^{4}He or ^{3}He(^{3}He,2p)^{4}He.

Despite the success of the NCSM/RGM in describing the asymptotic behavior of the wave functions, it has been shown that it has limitations when it comes to accurately describe systems at short to medium ranges (up to about 5 fm for the ^{6}He case). This is due to the fact that, in order to account for all many-body correlations, several excited states of the nuclear clusters must be included in the basis, resulting in an increase of the problem size that goes beyond current computational capabilities. This limitation has been overcome by introducing the *ab initio* no-core shell model with continuum (NCSMC). With this method, the wave function is written as a superposition of both continuous NCSM/RGM cluster states and discrete eigenstates of the compound system obtained with the no-core shell model (NCSM). The latter eigenstates compensate for missing cluster excitations, improving the description of the wave function at short to medium range.

The NCSMC was first introduced in [7, 8] for binary systems. Its expansion to three-cluster systems was recently achieved and we show here the first results for the ^{6}He ground state (g.s).

[a] e-mail: romeroredond1@llnl.gov

Table 1. Energy (in MeV) for the NCSM ^{4}He g.s. and the ^{6}He g.s. using the NCSM/RGM, NCSM and NCSMC approaches in terms of the absolute HO model space size $N_{\rm tot} = N_0 + N_{\rm max}$, where N_0 is the number of oscillator quanta shared by the nucleons in their lowest configuration. For the NCSM, we also show the extrapolated value to $N_{max} \to \infty$ (the extrapolation was performed with an exponential fit).

	^{4}He		^{6}He	
$N_{\rm tot}$	NCSM	NCSM/RGM	NCSM	NCSMC
8	−28.17	−28.62	−28.95	−29.69
10	−28.22	−28.72	−29.45	−29.86
12	−28.22	−28.70	−29.66	−29.86
Extrapolation	−28.230(5)	—	−29.84(4)	—

2 Formalism

In the NCSMC, the ansatz for the three-cluster many-body wave function is given by

$$|\Psi^{J^\pi T}\rangle = \sum_\lambda c_\lambda |A\lambda J^\pi T\rangle + \sum_\nu \iint dx\, dy\, x^2 y^2\, G_\nu^{J^\pi T}(x,y)\, \hat{\mathcal{A}}_\nu\, |\Phi^{J^\pi T}_{\nu xy}\rangle\,, \tag{1}$$

where c_λ and $G_\nu^{J^\pi T}(x,y)$ are, respectively, discrete and continuous variational amplitudes, $|A\lambda J^\pi T\rangle$ are NCSM eigenstates of the compound nucleus labeled by the set of quantum number λ, $\hat{\mathcal{A}}_\nu$ is an appropriate intercluster antisymmetrizer introduced to exactly preserve the Pauli exclusion principle, and

$$\begin{aligned}|\Phi^{J^\pi T}_{\nu xy}\rangle =& \Big[\Big(|A-a_{23}\ \alpha_1 I_1^{\pi_1} T_1\rangle \left(|a_2\, \alpha_2 I_2^{\pi_2} T_2\rangle |a_3\, \alpha_3 I_3^{\pi_3} T_3\rangle\right)^{(s_{23}T_{23})}\Big)^{(ST)} \left(Y_{\ell_x}(\hat\eta_{23}) Y_{\ell_y}(\hat\eta_{1,23})\right)^{(L)}\Big]^{(J^\pi T)} \\ &\times \frac{\delta(x-\eta_{23})}{x\eta_{23}} \frac{\delta(y-\eta_{1,23})}{y\eta_{1,23}}\,,\end{aligned} \tag{2}$$

are three-body cluster channels of total angular momentum J, parity π and isospin T, where ν represents a set of quantum numbers that describes the channel within the cluster basis. Here, $|A-a_{23}\ \alpha_1 I_1^{\pi_1} T_1\rangle$, $|a_2\, \alpha_2 I_2^{\pi_2} T_2\rangle$ and $|a_3\, \alpha_3 I_3^{\pi_3} T_3\rangle$ denote the microscopic (antisymmetric) wave functions of the three nuclear fragments calculated within the NCSM. The Jacobi coordinates describing the relative positions of the clusters are denoted by $\vec\eta_{23} = \eta_{23}\hat\eta_{23}$ and $\vec\eta_{1,23} = \eta_{1,23}\hat\eta_{1,23}$.

We calculate the unknowns of the NCSMC wave function [c_λ and $G_\nu^{J^\pi T}(x,y)$] by solving the orthogonalized coupled equations obtained by projecting the Schrödinger equation on the model space spanned by NCSM eigenstates and the NCSM/RGM basis $|\Phi^{J^\pi T}_{\nu xy}\rangle$. Those equations are solved by means of the microscopic R-matrix method on a Lagrange mesh [9]. Details on the procedure will be available in [10].

3 Application to ^{6}He

The lightest Borromean nucleus is ^{6}He [11, 12], formed by an ^{4}He core and two halo neutrons. It is, therefore, an ideal first candidate to be studied within a three-body formalism. Hence, it was used as a test case when the NCSM/RGM formalism for three-cluster dynamics was introduced in [5, 6] and here is studied again in order to perform a benchmark with such results. We describe the ^{4}He core by its g.s. wave function and couple the three-cluster basis with the ^{6}He g.s. eigenstate obtained through the NCSM.

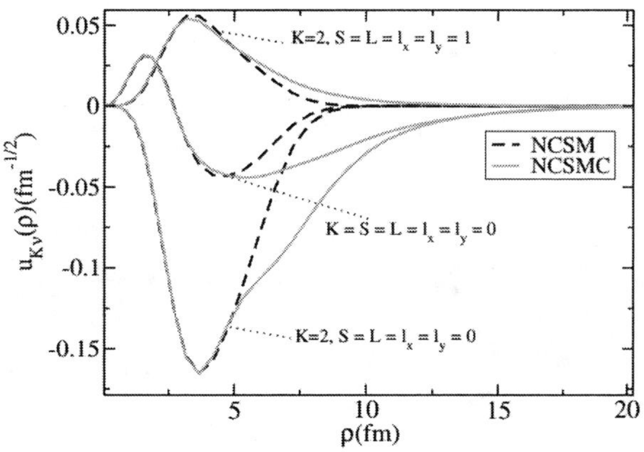

Figure 1. Most relevant hyperradial contributions to the ^{6}He g.s. wave function. Both the contribution from the NCSM wave function and the total NCSMC wave function are shown for a $N_{max} = 6$ model space. The figure shows how the addition of the three-cluster basis within the NCSMC compensates for the limitations of the NCSM to obtain an extended wave function characteristic of two-neutron halo nuclei. The hyperradial wave functions $u_{K\nu}(\rho)$ are the coefficients of the wave function when expanded in the hyperspherical basis, where K represents the hypermomentum.

We use the same potential as in [5, 6], i.e., the similarity-renormalization-group (SRG) [13, 14] evolved potential obtained from the chiral N^{3}LO NN interaction [15] with $\Lambda_{\rm SRG} = 1.5$ fm^{-1}. With this potential, the variational minimum for the NCSM ^{6}He g.s. is found at an HO frequency of around $\hbar\Omega = 14$ MeV, which is used in this work for all calculations. With this soft potential the binding energy can be accurately computed by extrapolating (through an exponential fit) the NCSM results to $N_{\rm max} \to \infty$, hence providing a good benchmark for the newly implemented NCSMC.

From Table 1, we can see that the NCSMC ^{6}He g.s. energy quickly converges to the NCSM extrapolated value, unlike in the NCSM/RGM, i.e., using only the basis (2) in the expansion of the wave function. This is due to the fact that the ^{6}He NCSM eigenstate takes into account the six-body correlations and ^{4}He core polarization that are missing when considering the cluster basis alone. It is also important to note that, in contrast to the behavior offered by the NCSM, the NCSMC presents the correct extended asymptotic behavior of the wave function. In Fig 1 such comparison is shown in a preliminary calculation at an $N_{max} = 6$ model space.

Finally, we can also compare the probability densities from the ^{6}He g.s. obtained with the NCSM/RGM and the NCSMC. In Fig. 2, such comparison is shown and it is interesting to find that while the two main configurations (di-neutron and cigar) appear to have the same probability within the NCSM/RGM, the di-neutron probability is enhanced when using the NCSMC. This asymmetry in the strength of the probability peaks is known to be a characteristic of ^{6}He and these results show that it is a consequence of the aditional six-body correlations.

4 Conclusions

The NCSMC uses an ansatz wave function that includes both an expansion in a continuous three-cluster basis and in a discrete basis of NCSM eigenstates. This provides a foundation that is capable of describing both short and long range characteristics of three-cluster systems. In the case of the ^{6}He g.s., we could see that this approach provides both the correct binding energy and extended asymptotic behavior unlike the NCSM, that does provide the correct binding energy, but not the correct

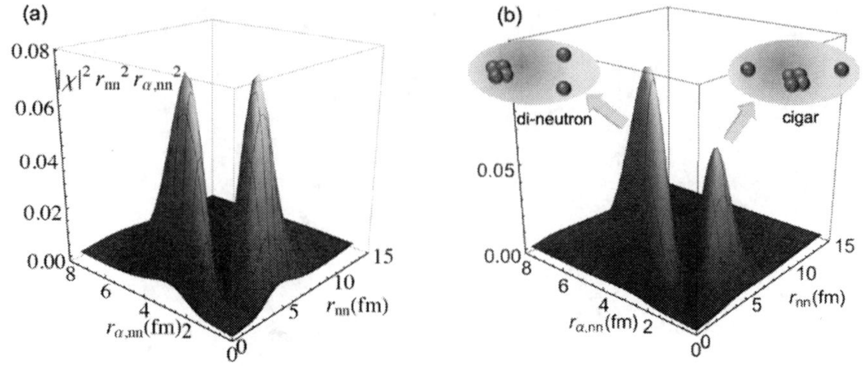

Figure 2. Probability distribution of the ^{6}He g.s. wave function in terms of the relative distance between the neutrons (r_{nn}) and the distance between the center of mass of the neutrons and the ^{4}He ($r_{\alpha,nn}$). The di-neutron and cigar configurations appear to have the same probability within the NCSM/RGM (a), while the di-neutron probability is enhanced when using the NCSMC (b).

asymptotics, or the NCSM/RGM that does the opposite. Calculations in larger model spaces for both g.s. and continuum states of ^{6}He are underway.

Acknowledgements

Prepared in part by LLNL under Contract DE-AC52-07NA27344. This material is based upon work supported by the U.S. Department of Energy, Office of Science, Office of Nuclear Physics, under Work Proposal Number SCW1158. Support from the NSERC Grant No. 401945-2011 is acknowledged. TRIUMF receives funding via a contribution through the Canadian National Research Council.

References

[1] S. Quaglioni, P. Navrátil, Physical Review Letters **101**, 092501 (2008)
[2] S. Quaglioni, P. Navrátil, Physical Review C **79**, 044606 (2009)
[3] P. Navrátil, S. Quaglioni, Physical Review C **83**, 044609 (2011)
[4] P. Navrátil, S. Quaglioni, Physical Review Letters **108**, 042503 (2012)
[5] S. Quaglioni, C. Romero-Redondo, P. Navrátil, Phys.Rev. **C88**, 034320 (2013)
[6] C. Romero-Redondo, S. Quaglioni, P. Navrátil, G. Hupin, Phys. Rev. Lett. **113**, 032503 (2014)
[7] S. Baroni, P. Navrátil, S. Quaglioni, Phys. Rev. Lett. **110**, 022505 (2013)
[8] S. Baroni, P. Navrátil, S. Quaglioni, Phys. Rev. C **87**, 034326 (2013)
[9] M. Hesse, J.M. Sparenberg, F.V. Raemdonck, D. Baye, Nucl. Phys. **A640**, 37 (1998)
[10] C. Romero-Redondo, S. Quaglioni, P. Navrátil, G. Hupin, In preparation (2015)
[11] I. Tanihata, J. Phys. **G22**, 157 (1996)
[12] I. Tanihata, H. Hamagaki, O. Hashimoto, Y. Shida, N. Yoshikawa, K. Sugimoto, O. Yamakawa, T. Kobayashi, N. Takahashi, Phys. Rev. Lett. **55**, 2676 (1985)
[13] S.K. Bogner, R.J. Furnstahl, R.J. Perry, Phys. Rev. C **75**, 061001 (2007)
[14] R. Roth, S. Reinhardt, H. Hergert, Phys. Rev. C **77**, 064003 (2008)
[15] D.R. Entem, R. Machleidt, Phys. Rev. C **68**, 041001 (2003)

EPJ Web of Conferences **113**, 03005 (2016)
DOI: 10.1051/epjconf/201611303005

The ^{3}H(d,γ)^{5}He Reaction for $E_{c.m.} \leq 300$ keV

C.E. Parker[1,a], C.R. Brune[1], T.N. Massey[1], J.E. O'Donnell[1,b], A.L. Richard[1], and D.B. Sayre[2]

[1]*Ohio University, Athens, OH 45701, USA*
[2]*Lawrence Livermore National Laboratory, Livermore, CA 94550, USA*

Abstract. The ^{3}H$(d,\gamma)^5$He reaction has been measured using a 500-keV pulsed deuteron beam incident on a stopping titanium tritide target at Ohio University's Edwards Accelerator Laboratory. The time-of-flight (TOF) technique has been used to distinguish the γ-rays from neutrons detected in the bismuth germinate (BGO) γ-ray detector. A stilbene scintillator and an NE-213 scintillator have been used to detect the neutrons from the ^{3}H$(d,n)^4$He reaction using both the pulse-shape discrimination and TOF techniques. A newly-designed target holder with a silicon surface barrier detector to simultaneously measure α-particles to normalize the neutron count was incorporated for subsequent measurements. The γ-rays have been measured at laboratory angles of 0°, 45°, 90°, and 135°. Information about the γ-ray energy distribution for the unbound ground state and first excited state of ^{5}He can be obtained experimentally by comparing the BGO data to Monte Carlo simulations. The ^{3}H(d,γ)/^{3}H(d,n) branching ratio has also been determined.

1 Introduction

The ^{3}H$(d,n)^4$He reaction has a very large cross section and provides the easiest approach to achieving fusion in the laboratory. However, about 1 in 10^4 times in a deuterium-tritium environment, the ^{3}H$(d,\gamma)^5$He reaction can occur. This reaction has gained importance in the scientific community because of inertial confinement fusion (ICF) studies, particularly at the National Ignition Facility. One way to investigate the fusion burn in the deuterium-tritium fuel is by detecting γ-rays above about 10 MeV. The γ-rays are useful for diagnostic purposes because they come out of the implosion nearly unscattered, as opposed to the neutrons. The γ-rays also travel faster, and therefore arrive sooner, than the neutrons, allowing them to be detected without interference from the neutrons.

In addition to the ICF application, studying the reaction can lead to a better understanding of the properties of the unbound ^{5}He nucleus. Figure 1 shows the two low-lying ^{5}He states, which are unbound with respect to ^{4}He + n, leading to both states having significant width when compared to the $\frac{3}{2}^+$ resonance state. There are γ-ray transitions from the resonance state to the $\frac{3}{2}^-$ ground state and the $\frac{1}{2}^-$ first excited state, also known as γ_0 and γ_1, respectively. A consequence of the non-negligible widths of the final states, the γ-rays from these transitions are not monoenergetic, but rather have an energy distribution which lies primarily in the 10- to 17-MeV range.

---

[a]e-mail: cp260509@ohio.edu
[b]*deceased*

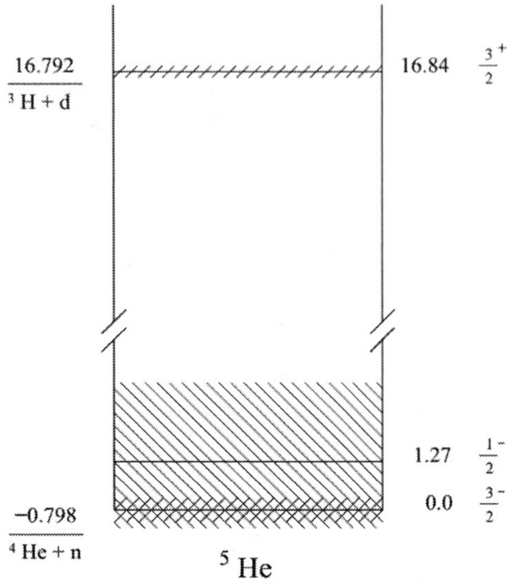

Figure 1. Selected levels in ^{5}He. In addition to the significant width in the lower-lying levels, they overlap, making it more difficult to experimentally determine their contributions to the overall lineshape.

Due to its importance, the ^{3}H(d,γ)/^{3}H(d,n) branching ratio has also been previously measured, however there is a discrepancy regarding the order of magnitude of the ratio. This discrepancy between experimental results for similar deuteron energies are likely due to undetected systematic errors, such as the manner in which the neutron background was addressed. Some reported data contains only measurements from the γ_0 branch, while others contain measurements from both the γ_0 and γ_1 branches. A summary of the branching ratio results from the last 50 years can be seen in Figure 2.

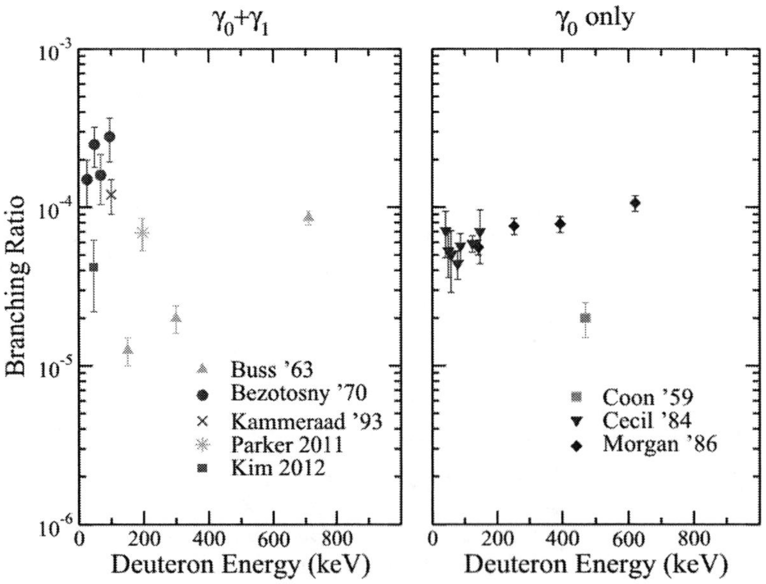

Figure 2. Summary of the branching ratios of ^{3}H(d,γ)/^{3}H(d,n). See references [1] and [2] for references to the other previous measurements.

2 Experimental Technique

The experiments were performed using the 4.5-MV Tandem Pelletron Accelerator at the John E. Edwards Accelerator Laboratory on the Ohio University campus. The beam swinger and concrete-shielded TOF tunnel [3] were crucial for the experimental set-up. Another important aspect of the experiments was the pulsed and bunched deuteron beam tuned to deliver beam packets with 200 ns between pulses to utilize the TOF technique. Sending the beam to the target in packets provides a means for a time correlation between when the packet hits the target and when the γ-rays or neutrons reach the detector. As γ-rays can also be produced from natural radioactivity or from the capture of thermal neutrons, these events would be seen as "out of time" on the TOF spectrum, indicating that they are not from the desired reaction. The neutrons will appear in the spectrum at different times, based on their energy. As higher-energy neutrons will reach the detector sooner than lower-energy neutrons, due to the difference in their velocities, the different events can be distinguished.

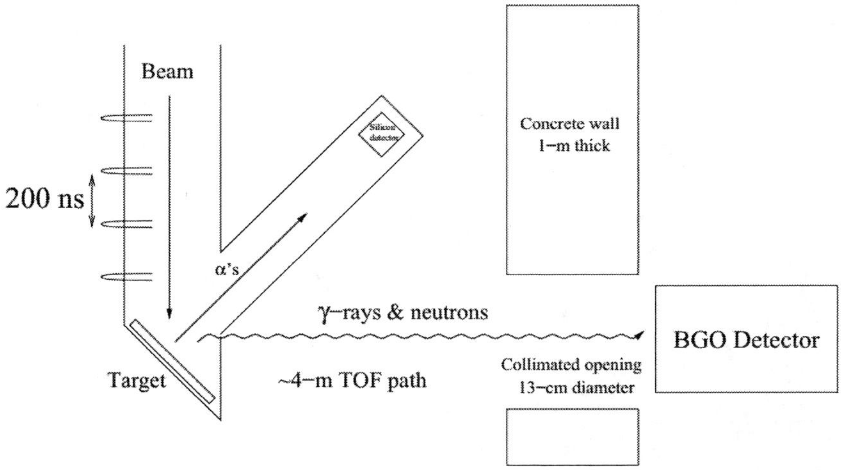

Figure 3. On-axis γ-rays and neutrons will go along the path toward the collimated tunnel entrance, and ultimately to the BGO γ-ray detector. Figure not drawn to scale.

The shielding and collimation provided by the tunnel, as seen in Figure 3, are key in the set-up. The collimated opening is useful because it provides shielding against nearly all of the γ-rays produced away from the target and scattered neutrons from being seen by the BGO. The distance of approximately 4 m is very beneficial to the TOF aspect of the experiment. By placing the BGO at that distance, the arrival time of the 14-MeV neutrons is about 75 ns after the γ-rays, giving a very good peak separation in the spectrum.

The main detector for the experiments was a 10.16-cm-diameter, 10.16-cm-thick BGO for the purpose of detecting the γ-rays. A BGO was chosen for the experiments because of the increased efficiency for γ-rays over other detector materials such as NaI. The high density and large atomic number of the bismuth helps the BGO have a large probability per unit volume for pair production, which is highly desirable for γ-rays with energies above 10 MeV.

In order to investigate the neutrons produced in the reaction, multiple detectors were implemented. A stilbene detector, which has excellent pulse shape discrimination, was utilized to look at the main flux of 14-MeV neutrons in the swinger area. An NE-213, located inside the tunnel area, was included to confirm the information in the stilbene spectrum and to provide an absolute neutron measurement, which can be done with a separate calibration reaction. As α-particles will be produced in a 1:1 ratio with the 14-MeV neutrons, a method to detect them was included in later measurements by placing a silicon surface-barrier detector inside the stainless steel target holder, vacuum sealed to the tritium disk holder. Beyond the simple cross-check of detecting α-particles in addition to neutrons, the method provides an absolute normalization of the neutron detector efficiency.

The tritium target for the measurements was of the solid metal tritide type, with a tritium to titanium ratio guaranteed to be $\geq$ 1.5, while the maximum achieved is known to be about 2.0. The solid target was chosen over the gas cell due to the increased safety precautions and recycling needs the gas cell would call for. There is also additional energy loss that takes place in the foil that seals a gas target that is not present with the solid one. Additional targets for detector calibration purposes were also used. A boron carbide target was utilized for the $^{11}B(d, n\gamma_{15.1})^{12}C$ reaction to provide a monoenergetic γ-ray to give an energy calibration for the BGO above standard offline sources, as well as provide input for the response function. To characterize and calibrate the silicon detector, protons were scattered off of a polished copper target at multiple energies.

3 Analysis and Outlook

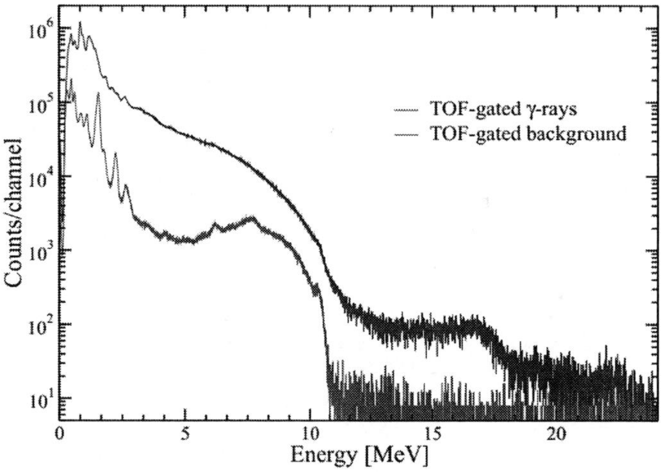

Figure 4. BGO spectrum for approximately 48 hours of beamtime at a laboratory angle of 90°. The γ-rays are gated by the TOF spectrum with an equal number of channels per gate for the same data set.

Analysis is currently underway for data at each laboratory angle. A sample of the γ-ray spectrum from the BGO detector can be seen in Figure 4. The γ-rays of interest for the $^3H(d, \gamma)^5He$ reaction are in the broad peak around 17 MeV; the higher-energy peak is presumably from fast neutron capture on the stainless steel target holder. GEANT4 will be used to determine the efficiency of the BGO for the target-to-detector distance of the set-up for the γ-ray energy range of interest. The relative γ_0 and γ_1 branches will be determined by adjusting the input parameters of the R-matrix calculation-based γ-ray distribution. The 14-MeV neutron yield will be determined using the α-particle yield in the silicon detector for the cross-normalization.

Acknowledgements

This work is supported in part by Lawrence Livermore National Laboratory, and the U.S. Department of Energy (National Nuclear Security Administration) through grant No. DE-NA0001837. The authors would also like to thank the staff of Edwards Accelerator Laboratory and the Ohio University Physics Machine Shop for their invaluable assistance in the completion of these experiments.

References

[1] C. E. Parker, Master's thesis, Ohio University, Athens, Ohio, Nov. 2011.
[2] Y. Kim *et. al.*, Phys. Rev. C, **85**, 061601 (2012).
[3] R. W. Finlay *et. al.*, Nucl. Instr. and Meth., **198**, 197 (1982).

EPJ Web of Conferences **113**, 03006 (2016)
DOI: 10.1051/epjconf/201611303006

Understanding the proton radius puzzle: Nuclear structure effects in light muonic atoms

Chen Ji[1,a], Oscar Javier Hernandez[1,2,b], Nir Nevo Dinur[3,c], Sonia Bacca[1,2,d], and Nir Barnea[3,e]

[1] *TRIUMF, 4004 Wesbrook Mall, Vancouver, BC V6T 2A3, Canada*
[2] *Department of Physics and Astronomy, University of Manitoba, Winnipeg, MB R3T 2N2, Canada*
[3] *Racah Institute of Physics, The Hebrew University of Jerusalem, Jerusalem 91904, Israel*

Abstract. We present calculations of nuclear structure effects to the Lamb shift in light muonic atoms. We adopt a modern *ab-initio* approach by combining state-of-the-art nuclear potentials with the hyperspherical harmonics method. Our calculations are instrumental to the determination of nuclear charge radii in the Lamb shift measurements, which will shed light on the proton radius puzzle.

1 Introduction

The proton radius extracted from recent high-precision measurements of the Lamb shift (2S-2P atomic transitions) in muonic hydrogen atom μH [1, 2] displays 7 standard deviations from results determined from *e*H spectroscopy or *e*-*p* elastic scattering. This discrepancy raises intriguing questions about lepton universality and challenges theories of standard model physics. To resolve the proton radius puzzle, the CREMA collaboration at Paul Scherrer Institute have performed new measurements of the Lamb shifts in other light muonic atoms (μD, μ^3He$^+$ and μ^4He$^+$) [3], with results yet to be published. These measurements aim to extract the nuclear charge radii with extremely high accuracy, which will be benchmarked with current and future spectroscopy measurements in electronic atoms [4–6] and electron-nucleus scattering data [7, 8].

The accuracy of the determined nuclear charge radius depends not only on the experimental precision of the Lamb shift, but also on the theoretical calculations. The theories relate the Lamb shift $\delta_{LS} \equiv E(2S) - E(2P)$ in a muonic atom/ion $\mu X^{(Z-1)+}$ to the charge radius $\langle r_{ch}^2 \rangle$ of a nucleus $^Z_A X$ by

$$\delta_{LS} = \delta_{\rm QED} + \mathcal{A}_{\rm OPE} \langle r_{ch}^2 \rangle + \delta_{\rm TPE}, \tag{1}$$

where $\delta_{\rm QED}$ is the contribution from the quantum-electrodynamics (QED) including the photon vacuum polarization, muon self energy, and relativistic recoil corrections. The other two terms in Eq. (1) are corrections due to nuclear structure. The term linear to $\langle r_{ch}^2 \rangle$ is dominated by one-photon exchange

[a] e-mail: jichen@triumf.ca
[b] e-mail: javierh@triumf.ca
[c] e-mail: nir.nevo@mail.huji.ac.il
[d] e-mail: bacca@triumf.ca
[e] e-mail: nir@phys.huji.ac.il

between the muon and the nucleus (Fig. 1), where $\mathcal{A}_{\rm OPE} \approx m_r^3(Z\alpha)^4/12$ with $m_r = m_\mu M_X/(m_\mu + M_X)$ denoting the reduced mass in the muon (m_μ) and nucleus (M_X) system. $\delta_{\rm TPE}$ represents the two photon-exchange (TPE) contribution (Fig. 1), which is a combination of elastic ($\delta_{\rm Zem}$) and inelastic ($\delta_{\rm pol}$) parts. $\delta_{\rm Zem}$ is proportional to the third electric Zemach moment $\langle r_{ch}^3\rangle_{(2)}$ by [9]

$$\delta_{\rm Zem} = -m_r^4(Z\alpha)^5\langle r_{ch}^3\rangle_{(2)}/24\,. \qquad (2)$$

The inelastic part δ_{pol} is the nuclear polarizability effect, where the nucleus in the muonic atom is virtually excited in the intermediate states by exchanging two photons with the muon.

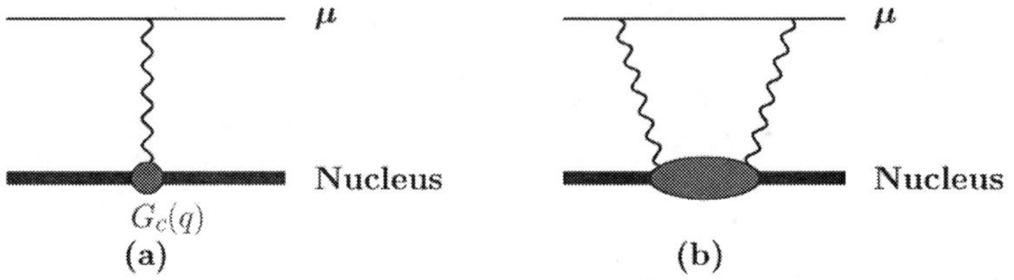

Figure 1. The muon-nucleus one-photon (a) and two-photon (b) exchange processes: The bulb in (a) indicates the insertion of nuclear electric form factor. The bulb in (b) denotes the excitation of nucleus in the intermediate states between two photons.

The bottleneck in accurately extracting the nuclear charge radius in Lamb shift measurements is the theoretical uncertainty in determining $\delta_{\rm TPE}$, especially $\delta_{\rm pol}$. To determine $\langle r_{ch}^2\rangle$ of ^{3}He and ^{4}He from measurements of their muonic ions to a 3×10^{-4} accuracy, $\delta_{\rm pol}$, which strongly relies on our knowledge of nuclear structures and dynamics, has to be known within a 5% accuracy [3].

$\delta_{\rm pol}$ can be represented as a combination of energy-dependent sum rules of a series of nuclear response functions. One way to calculate $\delta_{\rm pol}$ is to use inputs from measurements of photo-absorption cross sections and quasielastic scattering data. However, the applications of this approach in μD [10] and $\mu\,^4$He$^+$ [11–13] do not provide the accuracy desired by the Lamb shift measurements. The high-accuracy requirement can be achieved alternatively through calculations using state-of-the-art nuclear potentials. For example, the studies of nuclear structure effects in μD using Argonne v_{18} (AV18) potentials and pionless effective field theory improved the accuracy to a $\sim$ 1% level [14–16], which is one order of magnitude more accurate than previous methods.

We calculated the nuclear structure effects in μD using chiral effective field theory (χEFT) potentials [17]. Combining the modern inter-nucleon potentials with few-body methods, *i.e.*, effective interaction hyperspherical harmonics expansion (EIHH) [18] and Lanczos sum rules (LSR) [19], we also calculated $\delta_{\rm TPE}$ in μ^4He$^+$ [20, 21], μ^3He$^+$ and μ^3H [22]. By comparing the results using different potential models, we show that the uncertainty in $\delta_{\rm TPE}$ is dominated by nuclear physics. Our results are key to the determination of $\langle r_{ch}^2\rangle$ from future muonic atom measurements.

2 Nuclear structure effects in μD

$\delta_{\rm TPE}$ in μD was calculated by Pachucki [14] with the implementation of AV18 nuclear potentials [23]. The dominant contribution is the sum rule of the electric-dipole response function, with corrections from the higher-multipole response function sum rules, relativistic and Coulomb distortion effects, *etc.* This calculation achieved a $\sim$ 1% accuracy by systematically analyzing neglected higher-order effects in atomic physics. Our recent work [17] provided an independent derivation of the formalism for δ_{TPE}, which is consistent with Ref. [14].[1] The later work by Pachucki *et al.* [16] carefully analyzed the relativistic-recoil effects in δ_{TPE}, which indicate, overall, a few parts per mil correction.

To fully analyze the uncertainty in nuclear physics, we used in Ref. [17] not only the AV18 potential but also χEFT with a series of parameterizations [24, 25]. We studied the order-by-order chiral convergence and the regulation-cutoff dependence. Our study shows a 0.6% uncertainty from potential variations, which must be included in the total uncertainty of δ_{TPE} in μD. With a further investigation, we include an additional correction $\delta^{hadr}_{\text{Zem}}$ to our results in Ref. [17] to be consistent with Refs. [15, 16]. This term is proportional to the proton Zemach moment, its contribution to δ_{TPE} in $\mu\text{X}^{(Z-1)+}$, is related to the elastic TPE effects in muonic hydrogen $\delta_{\text{Zem}}(\mu\text{H})$ by

$$\delta^{hadr}_{\text{Zem}}(\mu\text{X}) = -\left[Zm_r(\mu\text{X})\right]^4 \alpha^5 \langle r_p^3\rangle_{(2)}/24 = \left[Zm_r(\mu\text{X})/m_r(\mu H)\right]^4 \delta_{\text{Zem}}(\mu\text{H}) \tag{3}$$

The hadronic polarizability in μX can be approximated from its effects in μH by

$$\delta^{hadr}_{\text{pol}}(\mu\text{X}) \approx Z^3(Z+N)\left[m_r(\mu\text{X})/m_r(\mu H)\right]^3 \delta^{hadr}_{\text{pol}}(\mu\text{H})\,. \tag{4}$$

However, this estimate is based on the hadronic polarizability effect involving a single nucleon, but does not include the interference with nucleon-nucleon interactions. A more accurate calculation of $\delta^{hadr}_{\text{pol}}$ in μD is done by Carlson *et al.* by using high-Q^2 part of the deuteron scattering data [10]. The combination of $\delta^{hadr}_{\text{Zem}}(\mu\text{X})$ and $\delta^{hadr}_{\text{pol}}(\mu\text{X})$ yields a total hadronic TPE contributions in μD to be $\delta^{hadr}_{\text{TPE}} = -47(10)\ \mu$eV. Through collaborative communications among independent groups, δ_{TPE} in μD is agreed to be 1.709 ± 0.020 meV, which is adopted in a recent review by the CREMA collaboration [26] and will be used for extracting the deuteron charge radius in their Lamb shift measurement.

3 Nuclear structure effects in $\mu^4\text{He}^+$, $\mu^3\text{He}^+$ and $\mu^3\text{H}$

To extend the calculations to other light muonic atoms, we applied the EIHH method to expand the nuclear Hamiltonians and wave functions in the hyperspherical harmonic basis. This method is suitable for Hamiltonians with local (AV18) or non-local (χEFT) potentials. The energy-dependent sum rules of the response functions are obtained by implementing the LSR method [19].

We calculated the nuclear structure effects to the Lamb shift in $\mu^4\text{He}^+$, including the inelastic part of δ_{TPE} due to nuclear excitations ($\delta^{nucl}_{\text{pol}}$) [20] and the elastic part due to nuclear charge distributions (δ_{Zem}) [21]. We have recently improved the treatment of nucleon size effects and we show in Table 1 updated results of $\delta^{nucl}_{\text{pol}}$ for μD and $\mu^4\text{He}^+$.[2] For the hadronic TPE effects, rigorous calculations using scattering data are not yet available. Therefore, we estimate such effects in $\mu^4\text{He}^+$ using Eqs. (3) and (4). We find $\delta^{hadr}_{\text{pol}}(\mu^4\text{He}^+) \approx -0.34$ meV and $\delta^{hadr}_{\text{Zem}}(\mu^4\text{He}^+) \approx -0.54(12)$ meV with the error dominated by the uncertainty in neutron polarizability. δ_{TPE} in μD and $\mu^4\text{He}^+$ are summarized in Table 1.

Table 1. Two-photon exchange contributions to the Lamb shift in light muonic atoms in units of meV.

	δ_{Zem}	$\delta^{nucl}_{\text{pol}}$	atomic error	$\delta^{hadr}_{\text{TPE}}$	δ_{TPE}
μD	-0.424(3)	-1.245(5)	±0.018	-0.047(10)	-1.716(20)
$\mu^4\text{He}^+$	-6.29(28)	-2.36(10)	±0.10	-0.89(12)	-9.54(34)

[1] Small discrepancies between Refs. [14, 17] appear in the reduced-mass–dependent coefficients of higher-multipole contributions. These are resolved later by Pachucki *et al.* in Ref. [16], which agrees with Ref. [17].

[2] The *np* correlated nucleon-size corrections to TPE were calculated in [20] with an inaccurate assumption that proton and neutron behave the same except for Coulomb. With a more rigorous calculation, we provide updated results.

We have recently extended our studies of TPE effects to μ^3He^+ and μ^3H. These muonic atoms of mirror isotopes, which contain unequal neutrons and protons, may provide additional information for the puzzle by probing the differences between muon-proton and muon-neutron interactions. The calculations of δ_{TPE} in μ^3He^+ and μ^3H will be detailed in an upcoming paper [22].

4 Conclusion

We illustrate *ab-initio* calculations of TPE contributions to the Lamb shift in light muonic atoms originated from nuclear structure. We provide the results for different muonic hydrogen and helium isotopes, which are important for determining the nuclear charge radii in Lamb shift measurements.

Acknowledgements

We thank Judith McGovern, Michael Birse, Aldo Antognini, Misha Gorchtein, and Krzysztof Pachucki for useful discussions. This work was supported in parts by the Natural Sciences and Engineering Research Council (NSERC), the National Research Council of Canada, the Israel Science Foundation (Grant number 954/09), and the Pazy foundation.

References

[1] R. Pohl *et al.*, Nature **466**, 213 (2010).
[2] A. Antognini *et al.*, Science 339, 417 (2013).
[3] A. Antognini *et al.*, Can. J. Phys. **89**, 47 (2011).
[4] C. G. Parthey *et al.*, Phys. Rev. Lett. **104**, 233001 (2010).
[5] R. van Rooij *et al.*, Science **333**, 196 (2011).
[6] P. Cancio Pastor *et al.*, Phys. Rev. Lett. **108**, 143001 (2012).
[7] P. Achenbach *et al.* (A1 Collaboration), MAMI Proposal A1/01-12 (2012).
[8] I. Sick, Phys. Rev. C **90**, 064002 (2014).
[9] J. Friar, Annals of Physics **122**, 151 (1979).
[10] C. E. Carlson, M. Gorchtein, and M. Vanderhaeghen, Phys. Rev. A **89**, 022504 (2014).
[11] J. Bernabeu and C. Jarlskog, Nucl. Phys. B **75**, 59 (1974).
[12] G. A. Rinker, Phys. Rev. A **14**, 18 (1976).
[13] J. L. Friar, Phys. Rev. C **16**, 1540 (1977).
[14] K. Pachucki, Phys. Rev. Lett. **106**, 193007 (2011).
[15] J. L. Friar Phys. Rev. C **88**, 034003 (2013).
[16] K. Pachucki and A. Wienczek, Phys. Rev. A **91**, no. 4, 040503 (2015).
[17] O. J. Hernandez, C. Ji, S. Bacca, N. N. Dinur and N. Barnea, Phys. Lett. B **736**, 344 (2014).
[18] N. Barnea, W. Leidemann, G. Orlandini, Phys. Rev. C **61**, 054001 (2000).
[19] N. Nevo Dinur, C. Ji, S. Bacca, N. Barnea, Phys. Rev. C **89**, 064317 (2014).
[20] C. Ji, N. Nevo Dinur, S. Bacca, N. Barnea, Phys. Rev. Lett. **111**, 143402 (2013).
[21] C. Ji, N. Nevo Dinur, S. Bacca, N. Barnea, Few Body Syst. **55**, 917 (2014).
[22] N. Nevo Dinur, C. Ji, S. Bacca, and N. Barnea, *in preparation*.
[23] R. B. Wiringa, V. G. J. Stoks, and R. Schiavilla, Phys. Rev. C **51**, 38 (1995).
[24] E. Epelbaum, W. Glöckle, and U.-G. Meißner, Nucl. Phys. A **747**, 362 (2005).
[25] D. R. Entem and R. Machleidt, Phys. Rev. C **68**, 041001 (2003).
[26] J. J. Krauth *et al.*, arXiv:1506.01298 [physics.atom-ph].

EPJ Web of Conferences **113**, 03007 (2016)
DOI: 10.1051/epjconf/201611303007

Proton Charge Radius (PRad) Experiment at Jefferson Lab

C. Peng[1,a] and H. Gao[1,2,b]

[1] *Duke University, Durham, NC27708, U.S.A.*
[2] *Duke Kunshan University, Kunshan, Jiangsu, China 215316*

Abstract. The puzzle of proton charge radius was recently raised by the measurement of muonic hydrogen Lamb shift at Paul Scherrer Institute (PSI), whose results were seven standard deviations smaller than the CODATA recommended value. To investigate this discrepancy, the PRad experiment was proposed and approved at Thomas Jefferson National Accelerator Facility (JLab). The experiment will extract the proton charge radius with a sub-percent accuracy by measuring the cross-sections of unpolarized electron-proton elastic scattering in an unprecedented low Q^2 region $\left(2 \times 10^{-4}\ \mathrm{GeV}^2/\mathrm{c}^2\right)$.

1 Introduction

The proton charge radius is an important quantity to understand the structure of proton. It is not only required for high precision tests of quantum electrodynamics (QED) calculations based on the hydrogen Lamb shift, but also crucial for understanding how quantum chromodynamics (QCD) works in the non-perturbative region. Recently, measurements of muonic hydrogen Lamb shift at PSI reported the most precise values of proton charge radius in 2010 [1] and 2013 [2], which were:

$$r_p = 0.84184 \pm 0.00067 \text{ fm [1]};$$

$$r_p = 0.84087 \pm 0.00039 \text{ fm [2]}.$$

However, the PSI results were 7σ away from the CODATA 2012 value [3], which is a compilation of the results from hydrogen Lamb shift measurements and elastic electron-proton (*ep*) scattering experiments. As shown in Figure 1, a large discrepancy exists between the PSI values and the results from the recent *ep* scattering experiments or data reanalyses, including the analysis of world *ep* elastic scattering data by Sick [4], the result extracted from the elastic *ep* scattering measurements at Mainz Microtron by Bernauer *et al.* [5], the updated global fit of the form factors data by Zhan *et al.* [6], and the world data average suggested by Arrington [7].

This discrepancy is still not understood and is known as the "proton radius puzzle". It is of great interest to atomic, particle, and nuclear physicists. Many efforts have been made to understand the discrepancy: calculations of muonic hydrogen energy levels were refined, and were summarized in [8, 9]; a muon-proton scattering measurement was approved at PSI [10] to search the possible difference between the muon and electron measurements. A high precision elastic *ep* scattering experiment

[a] e-mail: cp121@phy.duke.edu
[b] e-mail: gao@phy.duke.edu

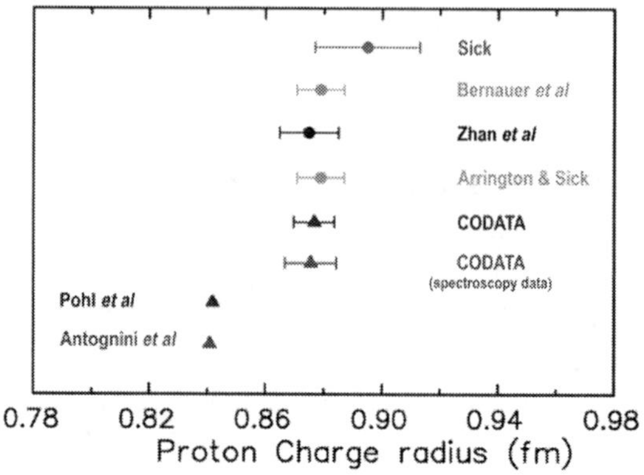

Figure 1. A compilation of values of the proton charge radius extracted from recent measurements and analysis.

with different systematical uncertainties is also necessary to evaluate the discrepancy, hence we have proposed the "high precision measurement of the proton charge radius" (PRad experiment) [11] in Hall B at Thomas Jefferson National Accelerator Facility (JLab). This report will introduce the experimental setup and its current status.

2 Experimental Setup

In 2012, the PRad experiment was approved at JLab with A rating and 15 days of 1.1 and 2.2 GeV beam. As shown in Figure 2, this experiment utilizes a non-magnetic calorimetric setup. It will measure the cross-section of unpolarized elastic *ep* scattering at very low four momentum transfer square (Q^2), and extract the electric form factors from it.

The PRad target is a windowless hydrogen gas-flow target, which removes the typical background source of all previous *ep* experiments – the elastic or quasi-elastic scattering off the nuclei in the target windows. In addition, a five meters long, two-stages vacuum box connects the target chamber to the detectors, hence minimizes any possible background sources between these two subsystems.

The detector package includes a Hybrid Calorimeter (HyCal) and two Gas Electron Multiplier (GEM) chambers. The HyCal consists of 1152 $PbWO_4$ crystal modules as the central part and 576 lead glass modules as the outer part. The energy and position resolutions of the $PbWO_4$ modules are $2.6\%/\sqrt{E(\text{GeV})}$ and $2.5\ \text{mm}/\sqrt{E(\text{GeV})}$, respectively, while the resolutions of the Pb-glass modules are approximately doubled. Two GEM chambers will be combined and mounted on the HyCal box to cover the entire geometrical acceptance of HyCal. The GEM detector improves the position resolution to better than 0.1 mm, and significantly reduces the systematical uncertainty in Q^2 determination. The detection angle range is 0.8° to 8°, corresponding to a kinematic coverage of $Q^2 = 2\times10^{-4} - 0.09\ \text{GeV}^2/c^2$. The lowest Q^2 region was never achieved before by magnetic spectrometer based measurements.

Scattered electrons from both Møller and *ep* elastic scattering will be simultaneously measured during the experiment. These two types of events within the kinematic coverage of this experiment can be easily separated by their energies. The cross sections of *ep* elastic scattering thus can be nor-

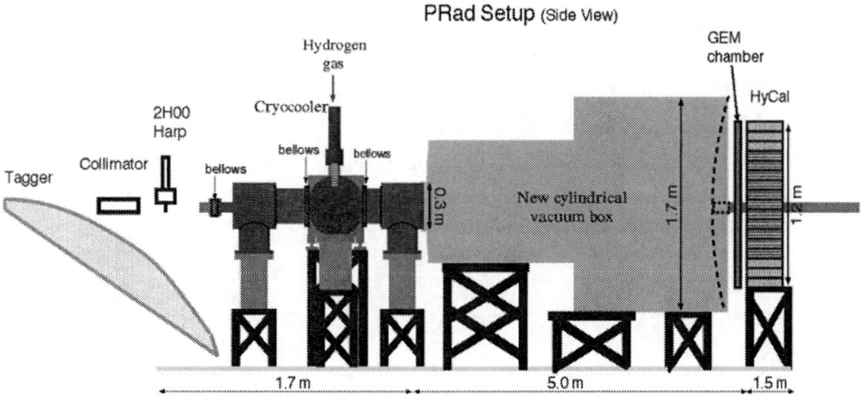

Figure 2. The proposed experimental layout (Not in scale). [11]

malized to those of the well known Møller process. This method significantly reduces the systematical uncertainty on determining the luminosity.

The experiment will eventually extract the electric form factor (G_E) from the measured cross-section data. According to the Rosenbluth formula,

$$\sigma_R = \left(\frac{d\sigma}{d\Omega}\right)_{exp} / \left(\frac{\sigma_{Mott}}{\epsilon}\frac{\tau}{1+\tau}\right) = \frac{\epsilon}{\tau} G_E^{p\,2} + G_M^{p\,2} . \tag{1}$$

, where $\tau = Q^2/(4M_p^2)$, and $\epsilon = \left(1 + 2(1+\tau)\tan^2\frac{\theta}{2}\right)^{-1}$, the contribution from magnetic form factor (G_M) strongly depends on Q^2. Within the low Q^2 region that is only achievable by 1.1 GeV beam, the G_M contribution is only 0.015% to 0.06%, and hence negligible. At higher Q^2, as the contribution of G_M becomes noticeable, we can either separate the two form factors using Rosenbluth method, or correct the G_M contribution by the existing world data.

3 Simulation Results

A full simulation program has been developed for the PRad experiment based on GEANT4 toolkit [12, 13]. The primary background source was found to be the electrons in the beam halo scattered off the nuclei in the target cell structures. To suppress the background, we enlarged the cell diameter, and planned to perform several empty target runs for background subtraction. The simulation including the empty target runs shows that the statistical uncertainties on the cross-sections will be less than 0.5% within the approved beam time, and the systematical uncertainty due to the subtraction can be less than 0.4%.

In the simulation program, two independent event generators with radiative effects were used. One is based on the complete set of calculations in [14], which goes beyond the ultra-relativistic approximations (URA). The other one is ESEPP [15], whose first order bremsstrahlung is calculated without soft-photon approximation or URA. The study with these two event generators shows the systematical uncertainty due to the radiative corrections is less than 0.3%.

The detector acceptance and Q^2 determination were also studied in the simulation. The implementation of GEM significantly improves the position resolution and hence the determination of Q^2, which helps control the systematical uncertainties of the acceptance. Reconstruction of tracking between the HyCal and GEM can further reduce the background from the target cell structures. Based on this simulation and the previous behavior of HyCal in the PrimEx experiment [16], a total uncertainty of 0.4% regarding the detector acceptance was assumed.

In addition, we have studied the rejection of neutral particles and the contamination of pions. Both uncertainties are less than 0.1%. A full simulation including all the above factors was performed. The proton charge radius extracted by a dipole fitting from simulation data indicates a total uncertainty of less than 0.7%.

4 Summary

The large experimental discrepancy on one of the fundamental quantities, the proton charge radius, has puzzled the communities since 2010. The PRad experiment aims to investigate this puzzle by extracting the proton charge radius with a sub-percent precision from the elastic electron-proton scattering measurement. Its magnetic-free calorimetric setup will reach the unprecedented low Q^2 region, and also significantly improve the systematical uncertainties that were typical in previous magnet spectrometer experiments. With these advantages, results from PRad experiment will directly impact on the "proton radius puzzle".

Acknowledgements

We thank members of the entire PRad collaboration. This project is supported in part by U.S. Department of Energy under contract number DE-FG02-03ER41231 and NSF MRI award PHY-1229153.

References

[1] R. Pohl *et al.*, Nature **466**, 213 (2010).
[2] A. Antognini *et al.*, Science **339**, 417 (2013).
[3] P.J. Mohr, B.N. Taylor, and D.B. Newell, Rev. Mod. Phys. **84**, 1527 (2012).
[4] I. Sick, Phys. Lett. B **576**, 62 (2003).
[5] J.C. Bernauer *et al.*, Phys. Rev. Lett. **105**, 242001 (2010).
[6] X. Zhan *et al.*, Phys. Lett. B **705**, 59 (2011).
[7] J. Arrington and I. Sick, arXiv:1505.02680v1 [nucl-ex] (2015).
[8] U.D. Jentschura, Ann. Phys. (N.Y.) **326**, 500 (2011).
[9] A. Antognini *et al.*, arXiv:1208:2637v2 [atom-ph] (2012).
[10] R. Gilman *et al.*, arXiv:1303:2160v1 [nucl-ex] (2013).
[11] A. Gasparian *et al.*, JLab experiment proposal E12-11-106 (2011). Spokespersons: D. Dutta, H. Gao, A. Gasparian (contact), M. Khandaker; A. Gasparian *et al.*, update to proposal C12-11-106 (2012).
[12] S. Agostinelli *et al.*, NIM A **506**, 3 (2003).
[13] J. Allison *et al.*, IEEE Trans. Nucl. Sci **53**, 1 (2006).
[14] I. Akushevich, H. Gao, A. Ilyichev, and M. Meziane Eur. Phys. J. A **51**, 1 (2015).
[15] A. V. Gramolin *et al.*, J. Phys. G. **41**, 115001 (2014).
[16] M. Kubantsev *et al.*, AIP Conf. Proc. **867**, 51 (2006).

EPJ Web of Conferences **113**, 03008 (2016)
DOI: 10.1051/epjconf/201611303008

Electron Shake-up and Shake-off Following ^{6}He Beta Decay

Eva Schulhoff[1,a] G.W.F. Drake[1,b]

[1]*Department of physics, University of Windsor, Windsor, Ontario, Canada N9E 3M7*

Abstract. Probabilities for electron shake-up and shake-off are calculated as relaxation processes following the beta decay of ^{6}He to form ^{6}Li, including corrections due to nuclear recoil. Within the sudden approximation, it is found that the correction due to nuclear recoil is nearly an order of magnitude less than that measured by Carlson *et al.* Phys. Rev. **129**, 2220 (1963).

1 Introduction

There are currently several experiments in progress to search for new physics beyond the Standard Model by high precision studies of angular correlations in the $E_{\rm max} = 3.508$ MeV β decay of the helium isotope $^6{\rm He} \to {}^6{\rm Li} + e^- + \bar{\nu}_e$ [1, 2]. An essential part of the analysis is to understand the energy distribution and spectra of the recoil ions. After the β decay event, the atomic electrons suddenly find themselves in a $^6{\rm Li}^+$ environment with nuclear charge $Z = 3$. The electrons redistribute themselves over all possible states of the $^6{\rm Li}^+$ ion, including the continuum leading to $^6{\rm Li}^{++}$ and $^6{\rm Li}^{3+}$. Evidence for new physics beyond the Standard Model would reveal itself by an additional tensor coupling contribution to the weak interaction, in addition to the simple Gamow-Teller axial-vector mechanism [1–3]. We present here a brief summary of a more extensive discussion to be published elsewhere [4].

2 Theory and Calculations

The maximum β-particle momentum is $P_{\rm max} = 1070$ a.u., which is relativistic. However, because of its larger mass, the maximum recoil velocity of the $^6{\rm Li}^+$ nucleus is only $v_{\rm rec} = 0.09749$ a.u., which is nonrelativistic. If the initial state of the atom is taken to be at rest in the laboratory frame, then the atomic electrons with position vectors $\mathbf{r}_i$ relative to the nucleus have a maximum recoil momentum $K \equiv m_e v_{\rm rec} = 0.09749$ a.u. after β-decay. Since the operator $\exp(i\mathbf{K}\cdot\mathbf{r})$ generates the transformation to a moving frame of reference, Born's rule for the transition probability from an initial atomic state i of ^{6}He to a final atomic state f of $^6{\rm Li}^+$ with recoil momentum $\mathbf{K}$ is

$$P_{i\to f}(\mathbf{K}) = \left|\left\langle \Psi_i(^6{\rm He}) \left| e^{i\mathbf{K}\cdot\mathbf{r}} \right| \Psi_f(^6{\rm Li}) \right\rangle\right|^2 \qquad (1)$$

[a]e-mail: schul1@uwindsor.ca
[b]e-mail: gdrake@uwindsor.ca

where $\mathbf{r} = \mathbf{r}_1 + \mathbf{r}_2$. Then by closure,

$$\sum_f P_{i\to f}(\mathbf{K}) = 1 \tag{2}$$

independent of $\mathbf{K}$, where the sum over f includes all bound and continuum states of Li^{++} and Li^{3+}, as well as Li^+. This sum rule will provide an important check on the final results.

Since $K = |\mathbf{K}|$ is small, it is convenient to use the multipole expansion (with z-axis in $\hat{\mathbf{K}}$ direction)

$$\begin{aligned} e^{i\mathbf{K}\cdot\mathbf{r}} &= e^{iKz} \\ &= 1 - \frac{(Kr)^2}{6} + iKz - \frac{(Kr)^2}{3}P_2(\cos\theta) + \cdots \end{aligned} \tag{3}$$

where $P_2(\cos\theta)$ is a Legendre polynomial inducing quadrupole transitions, and by definition $r^2 = |\mathbf{r}_1 + \mathbf{r}_2|^2 = r_1^2 + r_2^2 + 2\mathbf{r}_1\cdot\mathbf{r}_2$. The quadrupole term can be neglected because it is of order K^4 after squaring, but both the monopole and dipole terms contribute recoil corrections of order K^2 to the transition probabilities. The dominant contributions up to order K^2 are thus

$$\begin{aligned} P_{1S-nS} &= \left|\langle {}^6\mathrm{He}(1s^2\,{}^1S)\,|1 - (Kr)^2/6|\,{}^6\mathrm{Li}^+(n\,{}^1S)\rangle\right|^2 \\ &= \left|\langle {}^6\mathrm{He}(1s^2\,{}^1S)\,|\,{}^6\mathrm{Li}^+(n\,{}^1S)\rangle\right|^2 \\ &\quad - \frac{K^2}{3}\langle {}^6\mathrm{He}(1s^2\,{}^1S)\,|\,{}^6\mathrm{Li}^+(n\,{}^1S)\rangle\langle {}^6\mathrm{Li}^+(n\,{}^1S)\,|r^2|\,{}^6\mathrm{He}(1s^2\,{}^1S)\rangle \end{aligned} \tag{4}$$

$$P_{1S-nP} = K^2\left|\langle {}^6\mathrm{He}(1s^2\,{}^1S)|z|{}^6\mathrm{Li}^+(n\,{}^1P)\rangle\right|^2 \tag{5}$$

such that

$$\sum_n (P_{1S-nS} + P_{1S-nP}) = 1 + O(K^4) \tag{6}$$

For brevity, we will write $P_{S-S} = A_{S-S} + K^2 B_{S-S}$ and $P_{S-P} = K^2 B_{S-P}$. The recoil correction is then $K^2(B_{S-S} + B_{S-P})$. As will be seen, the two terms are of opposite sign, and so there is extensive cancellation. After summing over all states and ionization stages, the unitarity condition (6) requires $\sum_n (B_{1S-nS} + B_{1S-nP}) = 0$ [5]. Also, one can see by closure that $\sum_n B_{1S-nP} = \langle (z_1 + z_2)^2\rangle = \frac{1}{3}\langle 2r_1^2 + 2r_2^2 - r_{12}^2\rangle$ for S-states. For the ground state of He, the tabulated matrix elements [6] yield $\frac{1}{3}(4 \times 1.193\,482\,995 - 2.516\,439\,313) = 0.752\,497\,66$, in agreement with the tables to follow.

We employ Stieltjes imaging techniques [7, 8] in Hylleraas coordinates to represent the infinity of bound states for shake-up processes, and the continuum of scatttering states, including autoionizing ststes, for shake-off processes. In this approach, each pseudostate represents a block of physical scattering states over a narrow range of energies, and especially the additional recoil accompanying the emission of the shake-off electrons.

3 Results

The results of the calculations are presented in the tables. In Table 1, the initial state is assumed to be the ground $1s^2\ {}^1S$ state of helium, and the table shows the transition probabilities into energy bins corresponding to Li^+, Li^{++}, and Li^{3+} respectively. The results are expressed in the form

$$P_{i\to f} = A_{i\to f} + K^2 B_{i\to f} \tag{7}$$

Table 1. Singlet transition probability coefficients A and B from ^{6}He($1s^2\ ^1S$), summed over pseudostate energy bins for $^6Li^+$, $^6Li^{++}$, and $^6Li^{3+}$, where $P = A + K^2B$.

Ion	A_{S-S}	A_{S-P}	$A_{\rm Total}$	$A_{\rm Exp't.}$ [5]
Li^+	0.8903(3)	0.0	0.8903(2)	0.899(2)
Li^{++}	0.097(1)	0.0	0.097(1)	0.101(2)
Li^{3+}	0.012(1)	0.0	0.012(1)	0.00018(15)
Total	0.999(1)	0.0	0.999(1)	1.000(2)
Ion	B_{S-S}	B_{S-P}	$B_{\rm Total}$	$B_{\rm Exp't.}$ [5]
Li^+	−0.6650(2)	0.5730(2)	−0.0920(3)	−0.67(11)
Li^{++}	−0.078(2)	0.165(1)	0.087(2)	0.63(11)
Li^{3+}	−0.009(1)	0.014(1)	0.005(2)	0.049(20)
Total	−0.752(2)[1]	0.752(2)[1]	0.000(2)	-0.01(11)

[1] $\mp 0.752\,497\,66$ by closure. See text following Eq. (6).

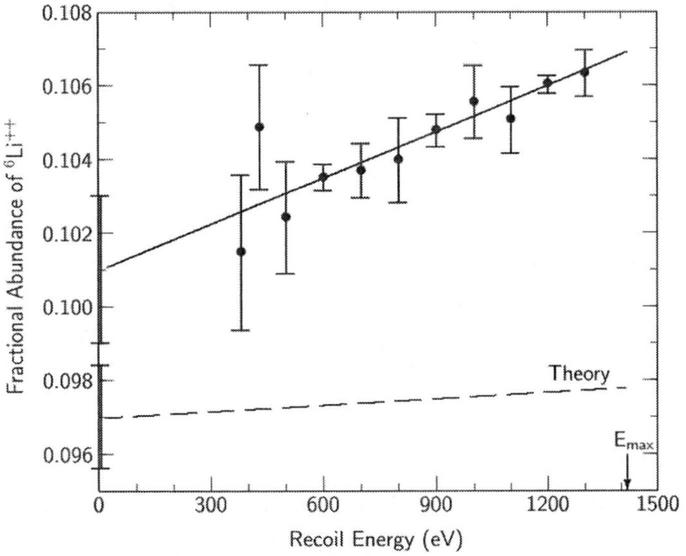

Figure 1. Comparison of theory and experiment for the fractional abundance of $^6Li^{++}$ as a function of recoil energy. The plotted points are from Carlson *et al.* [5], and the solid line is the least-squares fit. The dashed line is the theoretical probability $P(^6Li^{++}) = A_{S-S} + (B_{S-S} + B_{S-P})K^2$ with $K^2 = E/(149.2 \times 10^3\,\text{eV})$. The maximum recoil energy is $E_{\max} = 1418$ eV. The error bars on the vertical axis compare theory and experiment for A_{S-S}.

where $A_{i\to f}$ is the nonrecoil contribution, and $B_{i\to f}$ is the recoil contribution such that $149.2K^2 = E_{\rm rec}$ is the ion recoil energy in keV. The upper part of the table shows that the transition probabilities sum to unity when summed over all ionization states, and the lower part shows that the recoil coefficients $B_{i\to f}$ sum to zero, as expected. It is clear that the $S - P$ recoil term is canceled by the $S - S$ recoil contribution. We present similar results for the case of a He($1s2s\ ^3S$) initial state in Table 2.

Table 2. Triplet transition probability coefficients A and B from ^{6}He($1s2s\ ^3S$), summed over pseudostate energy bins for $^6Li^+$, $^6Li^{++}$, and $^6Li^{3+}$, where $P = A + K^2B$.

Ion	A_{S-S}	A_{S-P}	$A_{\rm Total}$
Li^+	0.88711(3)	0.0	0.88711(3)
Li^{++}	0.0942(7)	0.0	0.0942(7)
Li^{3+}	0.0186(7)	0.0	0.0186(7)
Total	0.9999(7)	0.0	0.9999(7)
Ion	B_{S-S}	B_{S-P}	$B_{\rm Total}$
Li^+	−6.718(0)	6.553(0)	−0.165(0)
Li^{++}	−0.741(7)	0.896(1)	0.155(1)
Li^{3+}	−0.143(7)	0.154(1)	0.011(1)
Total	−7.602(10)	7.603(2)	0.001(2)

In summary, this work provides the first calculations of the recoil term for the ^{6}Li ions formed by the β decay of neutral ^{6}He atoms, and the validity of a sum rule is directly demonstrated. The total result $0.005(1)K^2$ for Li^{3+} is in reasonable accord with the value $0.0000412E_{\rm rec} = 0.00615K^2$ obtained by Couratin *et al.* [9] for the corresponding one-electron $^6He^+$ case. However, the comparison with experiment in Table 1 shows that the calculated recoil contributions are nearly an order of magnitude smaller than the measured values from the experiment of Carlson *et al.* [5]. The disagreement between theory and experiment corresponds to the markedly different slopes in Fig. 2. Since the theoretical results are tightly constrained by the sum rules, it is difficult to see what change could be made to bring theory and experiment into agreement. Renewed experimental work would therefore be of great interest to resolve the disagreement between theory and experiment.

Research support by the Natural Sciences and Engineering Research Council of Canada, and by SHARCNET, are gratefully acknowledged.

References

[1] O. Naviliat-Cuncic, Rev. Mod. Phys. **78**, 991 (2006); Phys. Scr. T150, 014027 (2012).

[2] A. Knecht, Z. T. Alexander, Y. Bagdasarova, T. M. Cope, B . G. Delbridge, X. Fléchard, A. García, R. Hong, E. Liénard P. Mueller, O. Naviliat-Cuncic, A. S. C. Palmer, R. G. H. Robertson, D. W. Storm, H. E. Swanson, S. Utsuno, F. Wauters, W. Williams, C. Wrede and D. W. Zumwalt, ArXiv:1208.6433v2 [nucl-ex] 19 Oct (2013).

[3] C. H. Johnson, F. Pleasonton, and T. A. Carlson, Phys. Rev. **132**, 1149 (1963).

[4] E. E. Schulhoff and G. W. F. Drake, Phys. Rev. A, to be published.

[5] T. A. Carlson, F. Pleasonton, and C. H. Johnson, Phys. Rev. **129**, 2220 (1963).

[6] G. W. F. Drake in *Handbook for Atomic, Molecular, and Optical Physics*, (Springer, New York, 2006), p. 208.

[7] C. Winstead and P. W. Langhoff, J. Chem. Phys. **95**, 3107 (1991), and many earlier references therein.

[8] S. P. Goldman and G. W. F. Drake, Can. J. Phys. **61**, 198 (1983).

[9] C. Couratin, Ph. Velten, X. Fléchard,, E. Liénard, G. Ban, A. Cassimi, P. Delahaye, D. Durand, D. Hennecart, F. Mauger, A. Méry, O. Naviliat-Cuncic, Z. Patyk, D. Rodríguez, K. Siegień-Iwaniuk, and J-C. Thomas, Phys. Rev. Lett. **108**, 243201 (2012).

EPJ Web of Conferences **113**, 03009 (2016)
DOI: 10.1051/epjconf/201611303009

Strong interaction studies with kaonic atoms

J. Marton[1,a], M. Bazzi[2], G. Beer[3], C. Berucci[1,2], D. Bosnar[11], A.M. Bragadireanu[1,5], M. Cargnelli[1], A. Clozza[2], C. Curceanu[2], A. d'Uffizi[2], C. Fiorini[4], F. Ghio[6], C. Guaraldo[2], R. Hayano[7], M. Iliescu[2], T. Ishiwatari[1], M. Iwasaki[8], P. Levi Sandri[2], S. Okada[8], D. Pietreanu[2,5], K. Piscicchia[2], T. Ponta[5], R. Quaglia[4], A. Romero Vidal[10], E. Sbardella[2], A. Scordo[2], H. Shi[2], D.L. Sirghi[2,5], F. Sirghi[2,5], H. Tatsuno[12], O. Vazquez Doce[9], E. Widmann[1], and J. Zmeskal[1]

[1]*Stefan-Meyer-Institut für subatomare Physik, Boltzmanngasse 3, 1090 Wien, Austria*
[2]*INFN, Laboratori Nazionali di Frascati, C.P. 13, Via E. Fermi 40, I-00044 Frascati (Roma), Italy*
[3]*Dep. of Physics and Astronomy, University of Victoria, P.O.Box 3055, Victoria B.C. Canada V8W3P6*
[4]*Politecnico di Milano, Dip. di Elettronica e Informazione, Piazza L. da Vinci 32, I-20133 Milano, Italy*
[5]*Institutul National pentru Fizica si Inginerie Nucleara Horia Hulubei, Reactorului 30, Magurele, Romania*
[6]*INFN Sez. di Roma I and Instituto Superiore di Sanita I-00161, Roma, Italy*
[7]*University of Tokyo,7-3-1, Hongo, Bunkyo-ku,Tokyo, Japan*
[8]*RIKEN, Institute of Physical and Chemical Research, 2-1 Hirosawa, Wako, Saitama 351-0198 Japan*
[9]*Excellence Cluster Universe, Tech. Univ. München, Boltzmannstraße 2, D-85748 Garching, Germany*
[10]*Universidade de Santiago de Compostela, Casas Reais 8, 15782 Santiago de Compostela, Spain*
[11]*Department of Physics, Faculty of Science, University of Zagreb, Croatia*
[12]*High Energy Accelerator Research Organization (KEK), Tsukuba, 305-0801, Japan*

Abstract. The strong interaction of antikaons (K^-) with nucleons and nuclei in the low-energy regime represents an active research field connected intrinsically with few-body physics. There are important open questions like the question of antikaon nuclear bound states - the prototype system being K^-pp. A unique and rather direct experimental access to the antikaon-nucleon scattering lengths is provided by precision X-ray spectroscopy of transitions in low-lying states of light kaonic atoms like kaonic hydrogen isotopes. In the SIDDHARTA experiment at the electron-positron collider DAΦNE of LNF-INFN we measured the most precise values of the strong interaction observables, i.e. the strong interaction on the 1s ground state of the electromagnetically bound K^-p atom leading to a hadronic shift ϵ_{1s} and a hadronic broadening Γ_{1s} of the 1s state. The SIDDHARTA result triggered new theoretical work which achieved major progress in the understanding of the low-energy strong interaction with strangeness. Antikaon-nucleon scattering lengths have been calculated constrained by the SIDDHARTA data on kaonic hydrogen. For the extraction of the isospin-dependent scattering lengths a measurement of the hadronic shift and width of kaonic deuterium is necessary. Therefore, new X-ray studies with the focus on kaonic deuterium are in preparation (SIDDHARTA2). Many improvements in the experimental setup will allow to measure kaonic deuterium which is challenging due to the anticipated low X-ray yield. Especially important are the data on the X-ray yields of kaonic deuterium extracted from a exploratory experiment within SIDDHARTA.

[a]e-mail: johann.marton@oeaw.ac.at

1 Introduction

Hadronic atoms like pionic and kaonic atoms are extremely valuable systems for the investigation of the strong interaction in the low-energy domain.

Especially interesting is the strong interaction involving the strange quark which plays a peculiar role. The strange quark belongs to the light quarks but it is with a mass of about 100 MeV/c^2 much heavier than the up and down quarks which have masses in the order of few MeV/c^2. There are data on kaonic atoms available from past experiments (for a review see [1]), however the most simple kaonic atoms like kaonic hydrogen and deuterium are challenging and precision data were obtained in more recent experiments [2–4] using X-ray spectroscopy to unveil the low-energy strong interaction with high precision in these simple systems.

The most basic exotic atom with strangeness represents kaonic hydrogen (K^-p) which is an electromagnetically bound exotic atom consisting of a proton and an antikaon (K^-). In this system one can study the explicit and spontaneous chiral symmetry breaking in a fairly direct way by spectroscopy of x-rays emitted in the transitions to the 1s ground state, in this way the threshold data can be deduced. The K-p interaction at threshold is strongly influenced by the sub-threshold resonance $\Lambda(1405)$ which is an interesting hadronic object with a non-trivial nature [5]. In order to deduce the experimental values of isospin-separated antikaon-nucleon scattering lengths one needs the hadronic shift ϵ_{1s} and width Γ_{1s} in kaonic hydrogen and kaonic deuterium which can be extracted from the X-ray transitions to the 1s ground state in both kaonic systems.

2 Kaonic hydrogen

The SIDDHARTA experiment at the DAΦNE electron-positron collider succeeded to measure the X-ray spectrum of kaonic hydrogen by using an array of silicon drift detectors. From the K transitions the experimental values of ϵ_{1s} and width Γ_{1s} were determined. The energy shift ϵ_{1s} is given by the deviation of the measured K (np $\rightarrow$ 1s) transition energy $E^{meas.}_{np\rightarrow 1s}$ from the calculated electromagnetic value $E^{em}_{np\rightarrow 1s}$.

$$\epsilon_{1s} = E^{meas.}_{np\rightarrow 1s} - E^{em}_{np\rightarrow 1s} \tag{1}$$

With a modified Deser formula Equ.2 [6] the antikaon-nucleon scattering length a_p can be calculated which is the averaged sum of the a_0 (isospin I=0) and a_1 (isospin I=1) scattering lengths ($a_p = \frac{1}{2}(a_0+a_1)$).

$$\epsilon_{1s} + \frac{i}{2}\Gamma_{1s} = 2\alpha^3\mu_c^2 a_p(1 - 2\alpha\mu_c(ln\alpha - 1)a_p), \tag{2}$$

It is clear that one has to study the antikaon-neutron interaction by using kaonic deuterium to determine the a_0 and a_1.

3 Kaonic deuterium

The case of kaonic deuterium is more challenging than kaonic hydrogen mainly due to the larger widths of the K lines and the lower X-ray yield expected. In Fig.1 kaonic deuterium values of ϵ_{1s} and Γ_{1s} calculated in different theoretical approaches are displayed.

Experimentally the case of kaonic deuterium is still open. SIDDHARTA measured the X-ray spectrum with a pure deuterium filling but due to the limited statistics and the background condition

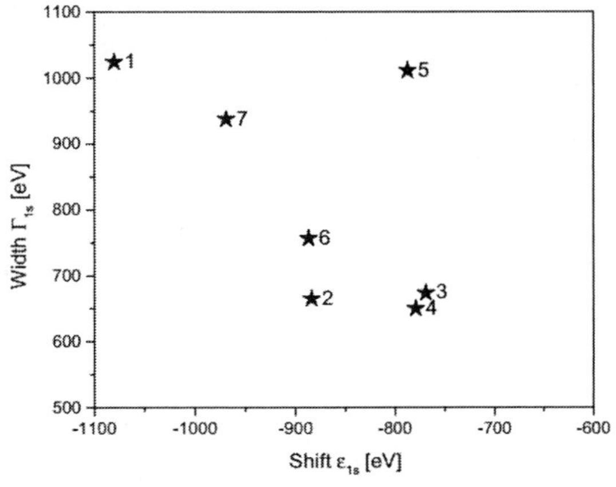

Figure 1. Predicted values of ϵ_{1s} and Γ_{1s} obtained in theoretical studies of kaonic deuterium indicated by the numbers 1-7: 1 ref.[7], 2 ref.[6], 3 ref.[8], 4 ref.[9], 5 ref.[10], 6 ref.[11], 7 ref.[12].

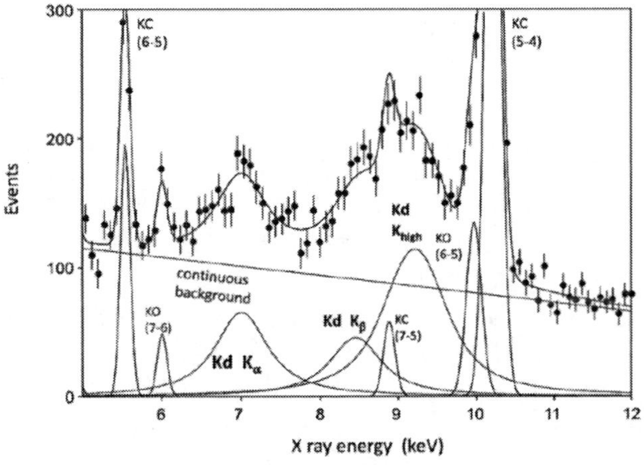

Figure 2. Monte-Carlo calculated X-ray spectrum of kaonic deuterium assuming ϵ_{1s} = -805 eV and Γ_{1s} = 750 eV [14]. With this values one gets an estimated precision of 70 eV in the shift and 150 eV in the width.

the determination of ϵ_{1s} and Γ_{1s} was impossible. An upper limit for the X-ray yield of the K lines could be extracted from the data: total yield <0.0143, Kα yield < 0.0039 [13].

A new experiment SIDDHARTA2 is planned which is based on a strongly improved apparatus. The improvements include an optimized geometry, deuterium gas density, discrimination of K^+, active shielding and better SDD timing performance by cooling. According to Monte Carlo studies one expects an X-ray energy spectrum shown in Fig.2.

4 Summary and Outlook

The SIDDHARTA experiment provided solid data for the antikaon-nucleon interaction at threshold and thus important constraints for the theory of strong interaction with strangeness at low energies. Now a clear description based on an effective field theory with coupled channels is available which is consistent with the information on the antikaon-nucleon interaction [15]. After SIDDHARTA the next important step is the study of the kaonic deuterium X-ray spectrum in order to deduce the isospin separated antikaon-nucleon scattering lengths. This measurement and related studies [16] are crucial for the understanding of strangeness and will provide stringent tests of the theoretical description.

Acknowledgements

We thank C. Capoccia, G. Corradi, B. Dulach, and D. Tagnani from LNF-INFN; and H. Schneider, L. Stohwasser, and D. Stückler from Stefan-Meyer-Institut, for their fundamental contribution in designing and building the SIDDHARTA setup. We thank as well the DAΦNE staff for the excellent working conditions and permanent support. Part of this work was supported by the European Community- Research Infrastructure Integrating Activity "Study of Strongly Interacting Matter" (HadronPhysics2, Grant Agreement No. 227431, and HadronPhysics3 (HP3) Contract No. 283286) under the Seventh Framework Programme of EU; HadronPhysics I3 FP6 European Community program, Contract No. RII3-CT-2004- 506078; Austrian Science Fund (FWF) (P24756-N20); Austrian Federal Ministry of Science and Research BMBWK 650962/0001 VI/2/2009; Croatian Science Foundation under Project No. 1680; Romanian National Authority for Scientific Research, Contract No. 2-CeX 06-11-11/2006; and the Grant-in-Aid for Specially Promoted Research (20002003), MEXT, Japan.

References

[1] E. Friedman, A. Gal, Phys. Rep. **452**, 89 (2007), C.J. Batty, E. Friedman, A. Gal, Phys. Rep. **287**, 385 (1997).

[2] M. Iwasaki et al., Phys. Rev. Lett. **78**, 3097(1997), T. M. Ito, et al., Phys. Rev. C **58**, 2366 (1998).

[3] G. Beer et al., Phys. Rev. Lett. **94**, 212302 (2005).

[4] M. Bazzi et al., Phys. Lett. B **704**, 113 (2011), Bazzi, M., Beer, G., Bombelli, L., Bragadireanu, A. M., Cargnelli, M. et al. (SIDDHARTA collaboration), Nucl. Phys. A **881**, 88 (2012).

[5] T. Hyodo and D. Jido, Prog.Part.Nucl.Phys. **67** (2012) 55-98 arXiv:1104.4474 [nucl-th].

[6] U.-G. Meißner, U. Raha, A. Rusetsky, Eur. Phys. J. C **47**, 473 (2006).

[7] S.S. Kamalov, E. Oset, A. Ramos, Nucl. Phys. A **690** 494 (2001).

[8] A. Gal, Int. J. Mod. Phys. A **22**, 226 (2007).

[9] M. Döring, U.-G. Meißner, Phys. Lett. B **704**, 663 (2011).

[10] N.V. Shevchenko, Nucl. Phys. A **890–891**, 50 (2012).

[11] T. Mizutani, C. Fayard, B. Saghai, K. Tsushima, Phys. Rev. C **87**, 035201 (2013).

[12] W. Weise, Hyp. Int. **233**, 131 (2015).

[13] M. Bazzi [SIDDHARTA], Nucl. Phys. A **907**, 69 (2013).

[14] M. Cargnelli, private communication.

[15] Y. Ikeda, T. Hyodo and W. Weise, Phys. Lett. B **706**, 63 (2011).

[16] C. Curceanu et al., Nucl. Phys. A **914**, 251 (2013).

EPJ Web of Conferences **113**, 03010 (2016)
DOI: 10.1051/epjconf/201611303010

Relativistic few-body methods

W. N. Polyzou[1,a]

[1] *Department of Physics and Astronomy, the University of Iowa*

Abstract. I discuss the role of relativistic quantum mechanics in few-body physics, various formulations of relativistic few-body quantum mechanics and how they are related.

1 Introduction

The observable degrees of freedom in any experiment are particles. Particles appear as initial and final states in scattering experiments. They are also the elementary building blocks of complex systems from nuclei to stars. While interactions do not conserve particle number, if the S-matrix is unitary then all of the reaction products remain particles. This observation implies that it must be possible to reformulate mathematical models of any physical system directly in terms of particle degrees of freedom. Since relativistic invariance is a fundamental symmetry of physical systems, it follows that it must be possible to represent mathematical models of physical systems as relativistically invariant quantum theories of particles.

If the underlying theory is a quantum field theory that describes interacting particles, it must have a representation as a relativistically invariant quantum theory of particles. The most interesting case is QCD, where the field theory involves non-observable quark and gluon degrees of freedom, but the theory ultimately describes reactions where the observable initial and final states are baryons and mesons. When expressed in terms of mesons and baryon degrees of freedom, QCD has the structure of a relativistic quantum theory of interacting particles. In the particle representation the degrees of freedom are experimentally observable, but the interactions necessarily become complicated. One important challenge is to construct the interactions between particles directly from QCD. If these interactions could be determined, it is likely that the particle representation of QCD would be a more efficient representation for calculating scattering observables, since the particle degrees of freedom are directly related to experiment. For this reason it also makes sense model these interactions by combining fundamental principles with experimental constraints.

The general principles of relativistic quantum theory provide representation independent constraints on these interactions. These constraints are non-trivial and may not be compatible with some truncations of the field theory. In addition, these constraints provide guidance in constructing realistic models of few-hadron systems that remain valid at relativistic energies.

I discuss various representations of relativistic quantum theories of particles and their relation to each other and to different representations of quantum field theory. One of the challenges of this subject is that there are may different representations. Some are more suited to computation while

[a] e-mail: polyzou@uiowa.edu

others are more directly related to an underlying quantum field theory. The equivalence between different representations is an important tool for understanding the structure of realistic models.

2 Relativistic invariance - general considerations

Relativistic invariance is a symmetry of a quantum theory which requires that quantum probabilities for equivalent experiments performed in different inertial frames are identical. Mathematically this requires that there is a unitary representation of the Poincaré group [1] $U(\Lambda, a)$

$$U(\Lambda_2, a_2)U(\Lambda_1, a_1) = U(\Lambda_2\Lambda_1, \Lambda_2 a_1 + a_2) \tag{1}$$

(the semidirect product of the Lorentz group with the spacetime translation group) on the Hilbert space of the quantum theory. Here Λ is a Lorentz transformation connected to the identity and a is a space-time translation parameter.

In addition, the ability to test special relativity by performing measurements on isolated subsystems requires that the unitary representation of the Poincaré group asymptotically approaches a tensor product of subsystem representations when the subsystems are asymptotically separated:

$$\lim_{|r_i - r_j| \to \infty} U(\Lambda, a) = U_i(\Lambda, a) \otimes U_j(\Lambda, a). \tag{2}$$

The smallest of these isolated subsystems are single-particle subsystems, which are the building blocks of the particle representation. The constraints (2) are fundamental constraints on acceptable models or truncations of field theories to a computable number of degrees of freedom.

3 Hilbert space representations

While there are many ways to construct unitary representations of the Poincaré group, the common feature of all of these representations is that they can always be decomposed into direct integrals of irreducible representations. This decomposition is the relativistic analog of diagonalizing the Hamiltonian in non-relativistic quantum mechanics. The diagonalization is more complicated in the relativistic case because it is necessary to simultaneously diagonalize the mass and spin Casimir operators. Both of these operators involve interactions in relativistic theories that satisfy (2).

The structure of the physical Hilbert space has a significant effect on the structure of the dynamical unitary representation of the Poincaré group. Covariant representations, which are typical of field theories, have Hilbert space inner products with non-trivial kernels, where the dynamics is encoded in the kernel. In explicit particle representations the dynamics appears in the unitary representation of the Poincaré group. The three most common representation are discussed below.

1. Local field theory. In this representation vacuum expectation values of products of fields appear in the kernel the Hilbert space scalar product:

$$W_n(x) = W_n(x_1, \cdots, x_n) := \langle 0|\phi(x_1)\cdots\phi(x_n)|0\rangle \tag{3}$$

$$\langle f|g\rangle = \sum_{mn} \int f_m^*(x) W_{m+n}(x, y) g_n(y) d^{4m}x d^{4n}y. \tag{4}$$

2. Euclidean field theory. In this representation the Euclidean Green functions,

$$S_n(x) = S_n(x_1 \cdots x_n) = \int D[\phi] d^{-A[\phi]} \phi(x_1)\cdots\phi(x_n), \tag{5}$$

which are formally moments of a Euclidean path integral, appear in the kernel of the physical Hilbert space inner product [2]

$$\langle f|g\rangle = \sum_{mn} \int f_m^*(Rx) S_{m+n}(x;y) g_n(y) d^{4m}x d^{4n}y \tag{6}$$

where R is the Euclidean time reversal operator and the test functions f have support for positive relative Euclidean times. The collection of Euclidean Green functions $\{S_n\}$ are called reflection positive if $\|f\|^2 = \langle f|f\rangle$ above is non-negative.

3. Direct interaction representations. The Hilbert space is the direct sum of tensor products of mass m spin j irreducible representation spaces $\mathcal{H}_{mj}$:

$$\mathcal{H} = \oplus(\otimes\mathcal{H}_{mj}). \tag{7}$$

This is the Hilbert space for a collection of free relativistic particles. Explicit dynamical representations of the Poincaré group or Lie algebra are constructed on this space.

4. Schwinger-Dyson - quasipotential representations. In this representation the dynamical quantities are time-ordered Green functions

$$G_n(x_1, \cdots, x_n) = \langle 0|T(\phi(x_1)\cdots\phi(x_n))|0\rangle \tag{8}$$

where T is the time ordering operator.

While these functions do not have a direct relation to a Hilbert space inner product, they can be used to calculate matrix elements of any operator between physical particle states [3]. In principle both the time-ordered Green functions, G_n, and the Wightman functions, W_n, are related to the Euclidean Green functions, S_n, by different analytic continuations. It follows that both have the same dynamical content.

Quasipotential methods replace the equations for the time-ordered Green functions by equivalent equations for quasipotentials and constrained time ordered Green functions. The simplest example is the Schwinger-Dyson equation for the four-point function, G_4, which is the Bethe-Salpeter equation:

$$G_4 = G_0 + G_0KG_4 \qquad G_0 = (G_2G_2)_{symmetrized} \tag{9}$$

Here K is the Bethe-Salpeter kernel which is formally defined by (9). The four-point function can be expressed in terms of the Bethe-Salpeter T by

$$G_4 = G_0 + G_0TG_0 \qquad \text{where} \qquad T = K + KG_0T. \tag{10}$$

In the quasi-potential formulation G_0 is replaced by a simpler g_0 with the same singularities. The equation for T is replaced by the pair of equations:

$$T = U + Ug_0T \qquad U = K + K\Delta gU \qquad \text{with} \qquad G_0 = g_0 + \Delta g \tag{11}$$

Quasipotential methods provide an alternative set of equations to calculate the time-ordered Green functions.

4 Unitary representations of the Poincaré group

In the case of quantum field theory, the covariance condition

$$W_n(x_1, \cdots, x_n) = W_n(\Lambda x_1 + a, \cdots, \Lambda x_n + a) \tag{12}$$

implies that the mapping

$$f_k(x_1, \cdots, x_k) = f_k(\Lambda x_1 + a, \cdots, \Lambda x_k + a) \tag{13}$$

preserves the inner product (4) and is thus defines a unitary representation of the Poincaré group.

In the Euclidean case the Euclidean time reversal operator R in (4) converts a unitary representation of the Euclidean group to a representation of a subgroup of the complex Poincaré group.

The Hamiltonian and rotationless boost generators in this representation are

$$Hg_k(x) = \sum_j \frac{\partial g_k}{\partial \tau_j}(x) \qquad K^i g_k(x) = \sum_j (\tau_j \frac{\partial g_k}{\partial x_{ij}}(x) - x_{ij}\frac{\partial g}{\partial \tau_j}(x)). \tag{14}$$

These are self-adjoint with respect to the inner product (6) and along with the generators of translations and rotations satisfy the commutation relations of the Poincaré Lie algebra.

For particle representations the problem is to construct a set of Poincaré generators on the particle Hilbert space. Once the mass and spin operators are simultaneously diagonailzed the structure of the irreducible representations is fixed by group representation theory. On each irreducible subspace, with vectors labeled by components of the three momentum and the projection of canonical spin on an axis, the transformation has the form

$$U(\Lambda, 0)|(m, j)p, \sigma\rangle_c = \sum |(m, j)\Lambda p, \sigma'\rangle_c \sqrt{\frac{\omega_m(\mathbf{\Lambda p})}{\omega_m(\mathbf{p})}} D^j_{\sigma'\sigma}[R_w(\Lambda, p)] \tag{15}$$

where $R_w(\Lambda, p)$ is a Wigner rotation and $\omega_m(\mathbf{p})$ is the energy of a particle of mass m and momentum $\mathbf{p}$. Dirac suggested a number of simplifying representations [4], called the instant, point and front-form of dynamics. They are distinguished by the choice of subgroup of the Poincaré group that is free of interactions. The different choices, called Dirac's forms of dynamics are all related by S-matrix preserving unitary transformations [5].

The connection between all three representations is easily seen for a single spinless free particle of mass m. In this case the two-point functions $W_2(x - y)$ and $S_2(x - y)$ have well known forms. The wave function in the Minkowski and Euclidean covariant representations are related to the wave function in the particle representation by

$$f_p(\mathbf{p}) = \frac{1}{(2\pi)^{3/2}} \int \frac{e^{-i\mathbf{p}\cdot\mathbf{x}+i\omega_\mu(\mathbf{p})t}}{\sqrt{2\omega_\mu(\mathbf{p})}} f_m(t, \mathbf{x}) d^4x = \frac{1}{(2\pi)^{3/2}} \int \frac{e^{-i\mathbf{p}\cdot\mathbf{x}-\omega_\mu(\mathbf{p})\tau}}{\sqrt{2\omega_\mu(\mathbf{p})}} f_e(\tau, \mathbf{x}) d^4x. \tag{16}$$

The simplest way to understand the origin of these relations is to observe that the kernel of the inner product (2) is a positive Hermetian quadratic form. If it was a matrix it could be written as the square of the positive square root of the matrix. These square roots could then be absorbed into the wave functions, thus removing the kernel of the quadratic form. What is actually done is to insert a complete set of intermediate particle states between the fields in the initial and final parts of the scalar product. This leads to an expression like

$$\langle f|g\rangle = \sum\!\!\!\!\!\!\int f^*(x_1, \cdots, x_n)\langle \phi(x_1) \cdots \phi(x_n)|(m, l, d)\mathbf{p}, \mu_l\rangle d\mathbf{p} \times$$

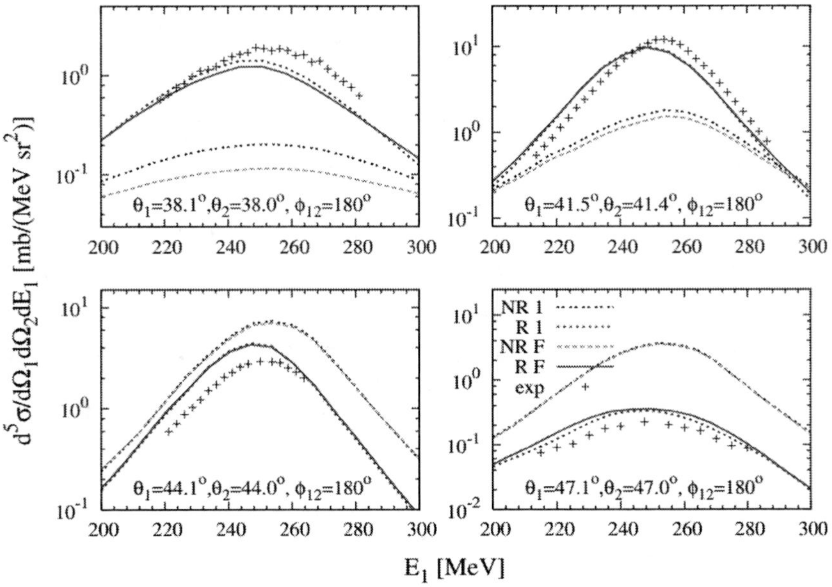

Figure 1. breakup - symmetric outgoing protons: solid curves - full relativistic calculation, long dashes - full non-relativistic calculation, short dashes - first order relativsitic, dot-dot-space - first order non-relativistic.

$$\langle (m,l,d)\mathbf{p},\mu_l|\phi(y_m)\cdots\phi(y_1)|0\rangle g(y_1,\cdots,y_m)d^{4m}yd^{4n}x \tag{17}$$

where the d represents degeneracy labels. The quantities

$$f((m,l)d,\mathbf{p},\mu_l) := \int \langle (m,l,d)\mathbf{p},\mu_l|\phi(y_m)\cdots\phi(y_1)|0\rangle f^*(y_1,\cdots,y_m)d^{4m}y \tag{18}$$

are wave functions of the particle representation. The amplitudes

$$\langle \phi(x_1)\cdots\phi(x_n)|(m,l,d)\mathbf{p},\mu_l\rangle$$

intertwine finite dimensional representation of the the Lorentz group with unitary irreducible representations of the Poincaré group:

$$\langle \phi(\Lambda x_1)\cdots\phi(\Lambda x_n)|(m,l,d)\mathbf{p},\mu_l\rangle = \langle \phi(x_1)\cdots\phi(x_n)|(m,l,d)\mathbf{\Lambda}p,\nu_l\rangle D^l_{\nu_l\mu_l}[R_w(\Lambda,p)] \tag{19}$$

where $R_w(\Lambda,p)$ is a Wigner rotation. This is a generalization of the standard intertwining properties of the $u(p)$ and $v(p)$ Dirac spinors.

5 Applications

Relativistic quantum mechanics is a powerful framework for investigation systems of light quarks. Relativistic quantum mechanical models have been used to look at spectral properties, lifetimes, and electroweak observables. They have also been used in few-hadron systems to study bound state, scattering and electromagnetic observables.

One illustrative example is $p-d$ scattering at the few GeV scale. At this scale one expects that the reaction might be sensitive to sub-nucleon degrees of freedom that cannot be easily modeled in a meson-exchange picture. In examining data it is important to be able to separate relativistic effects from potential sub-nucleon effects. Figure one shows the result of a relativistic direct interaction p-d calculation [6] scattering calculations with a Malfliet Tjon interaction at 500 MeV. The plots show

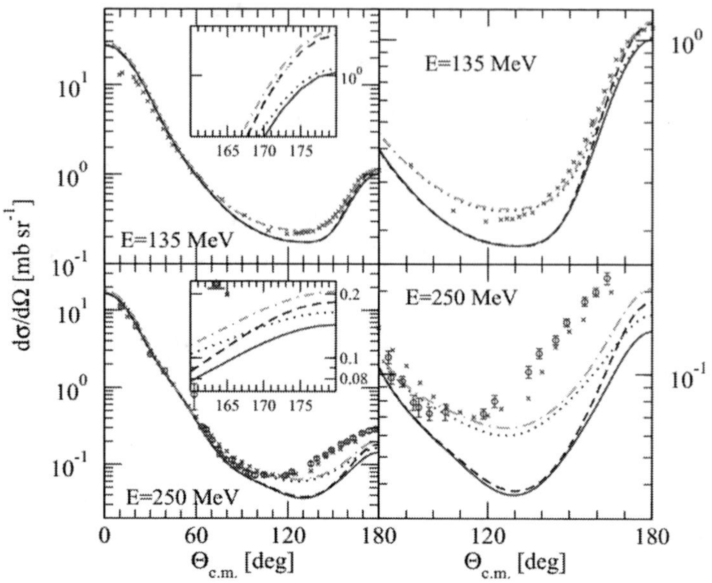

Figure 2. elastic pd scattering: solid - non-relativisitic no 3BF, long dash -relativistic no 3BF, dots - non-relativistic with 3BF, dot-dash - relativistic with 3BF.

the breakup cross section as a function of the energy of one of the protons when the final two protons make equal angles with the beam line. The solid curves are for the relativistic calculation, the long dashes are for the non-relativistic calculation. The relativistic and non-relativistic calculations are constrained to give the same two-body CM cross sections. The data is from [7]. This observable illustrates strong relativistic effects, as the non-relativistic and relativistic curves change places as the angle increases.

Figure 2 shows the elastic p-d scattering cross section at 135 and 250 MeV[8] [9](erratum). The dash-dot curve is a full relativistic calculation with realistic two (CD Bonn) and three-nucleon (TM99) forces, the long dashed curve is a relativistic calculation with only two-nucleon forces, the dotted curve is a non-relativistic calculation with two and three-nucleon forces, and the solid curve is a non-relativistic calculation with two nucleon forces. The 135 MeV calculations show no significant relativistic effects. Agreement with data [10] for scattering angles larger than 60 degrees is explained by the inclusion a three-nucleon force. At 250 MeV there are enhancements at back angles due to both relativistic effects and three-nucleon force effect, but the combined effect of both of these are almost an order of magnitude too small to explain the discrepancy with the data [11][12]. Back angles correspond to harder scattering and it is where one might expect to begin to see departures from meson-exchange models. It is interesting that the three-body force effects give the largest enhancement to the back-angle cross section, where one might expect hard scattering off of single nucleons to dominate. Since the energy is still below the pion production threshold, this is a clear indication of missing physics in standard realistic meson-exchange nuclear forces.

The application of relativistic quantum methods to few-nucleon problems at the few-GeV scale provides powerful tools to study nuclear reactions in the energy region where one expects to see a transition from the dominance by hadronic degrees of freedom to one dominated by QCD degrees of freedom.

I would like to acknowledge the may scientists with whom I have had the pleasure of collaborating with on various problems related to relativistic quantum mechanics. These include W. Glöckle (Bochum), T. Allen, P. L. Chung, F. Coester, H. C. Jean, W. Klink, P. Kopp, M. Herrmann, S. Kuthini Kunhammed, G. L. Payne, Y. Huang (Iowa), J. Golak, R. Skibínski, K. Topolnicki, H. Witała (Jagiellonian U) H. Kamada (Kyushu), B. Keister (NSF), Ch. Elster, T. Lin, M. Hadizadeh (Ohio). This research was supported by U.S. DOE contract No. DE-FG02-86ER40286 and NSF contract No. NSF-PHY-1005501 .

References

[1] E.P. Wigner, Annals Math. **40**, 149 (1939)
[2] K. Osterwalder, R. Schrader, Commun. Math. Phys. **31**, 83 (1973)
[3] K. Huang, H.A. Weldon, Phys. Rev. **D11**, 257 (1975)
[4] P.A.M. Dirac, Rev. Mod. Phys. **21**, 392 (1949)
[5] S.N. Sokolov, A.N. Shatnyi, Theor. Math. Phys. **37**, 1029 (1979)
[6] T. Lin, C. Elster, W.N. Polyzou, W. Glockle, Phys. Lett. **B660**, 345 (2008)
[7] V. Punjabi et al., Phys. Rev. **C38**, 2728 (1988)
[8] H. Witała, J. Golak, R. Skibiński, W. Glöckle, H. Kamada, W.N. Polyzou, Phys. Rev. C **83**, 044001 (2011)
[9] H. Witała, J. Golak, R. Skibiński, W. Glöckle, H. Kamada, W.N. Polyzou, Phys. Rev. C **88**, 069904 (2013)
[10] K. Sekiguchi et al., Phys. Rev. **C65**, 034003 (2002)
[11] K. Hatanaka et al., Phys. Rev. C **66**, 044002 (2002)
[12] Y. Maeda et al., Phys. Rev. C **76**, 014004 (2007)

EPJ Web of Conferences 113, 03011 (2016)
DOI: 10.1051/epjconf/201611303011

The Relativistic Three-Body Bound State in Three-Dimensions

M. R. Hadizadeh[1,a], Ch. Elster[1], and W. N. Polyzou[2]

[1] Institute of Nuclear and Particle Physics and Department of Physics and Astronomy, Ohio University, Athens, OH 45701, USA
[2] Department of Physics and Astronomy, The University of Iowa, Iowa City, IA 52242, USA

Abstract. Studying of the relativistic three-body bound state in a three-dimensional (3D) approach is a necessary first step in a process to eventually perform scattering calculations at GeV energies, where partial-wave expansions are not useful. To this aim we recently studied relativistic effects in the binding energy and for the first time, obtained the relativistic 3B wave function [1]. The relativistic Faddeev integral equations for the bound state are formulated in terms of momentum vectors, and relativistic invariance is incorporated within the framework of Poincaré invariant quantum mechanics.

1 Introduction and Formalism

While most three-nucleon calculations utilize partial-wave (PW) representations, at laboratory energies above about $\simeq 300$ MeV those become inefficient. For applications in the GeV regime one needs to consider vector formulations to implement the dynamics [2–4]. Convergence in the GeV regime has been demonstrated [2] in a relativistic one channel model using this approach. In this work we review the formulation of the 3B bound state problem in vector variables, which demonstrates that the relativistic problem can be treated using a straightforward generalization of methods that have been successfully used in the non-relativistic three-body problem.

In a Faddeev formulation, the relativistic bound state of three particles with mass m interacting with pairwise forces, is described by

$$|\psi\rangle = (M_t - M_0)^{-1}\, T(M_t)\, P\, |\psi\rangle, \tag{1}$$

where $M_t = E_t + 3m$ is the $3B$ mass eigenvalue, M_0 is the non-interacting $3B$ mass operator, $T(z) := V + V(z - M)^{-1} V$ is the boosted two-body (2B) t−matrix embedded in the 3B Hilbert space, and $P = P_{ij}P + P_{ik}P$ is the permutation operator for three identical particles. Projecting Eq. (1) on relativistic momentum states $|\mathbf{p}, \mathbf{k}\rangle$ leads to the integral equation,

$$\langle \mathbf{p}, \mathbf{k}|\psi\rangle = \int d\mathbf{p}'\, d\mathbf{k}'\, d\mathbf{p}''\, d\mathbf{k}'' \frac{\delta(\mathbf{k}-\mathbf{k}')}{M_t - M_0(p,k)}\, T\Big(\mathbf{p}, \mathbf{p}'; M_t - \omega_m(k)\Big) \langle \mathbf{p}', \mathbf{k}'|P|\mathbf{p}'', \mathbf{k}''\rangle \langle \mathbf{p}'', \mathbf{k}''|\psi\rangle. \tag{2}$$

The relativistic basis states contain two Jacobi momentum vectors **p** and **k**. The Jacobi momentum **k** is the relative momentum between the spectator and the pair and **p** is the relative momentum between the

[a] e-mail: hadizadm@ohio.edu

two particles in the pair in the two-body c.m. frame. For explicit construction of the Jacobi momentum vectors **p** and **k** in the three-body rest frame see Ref. [1]. In the kernel of relativistic Faddeev integral equation one needs the fully off-shell boosted 2B t−matrix $T(z)$ which can be calculated from the non-relativistic t−matrix t_{nr} by a two-step process, which guarantees that the nonrelativistic and relativistic 2B observables and CM wave functions remain unchanged. Those steps are

1. The relativistic right-half-shell 2B t−matrix is related to the non-relativistic right-half-shell by a multiplicative function F from [5],

$$T\Big(\mathbf{p},\mathbf{p}';\sqrt{m_0^2(\mathbf{p}')+k^2}+i0^+\Big) = F(p,p',k)\,t_{nr}\Big(\mathbf{p},\mathbf{p}';\frac{p'^2}{m}+i0^+\Big). \tag{3}$$

2. The relativistic fully off-shell t-matrix is obtained from the relativistic right-half-shell t−matrix by solving a first resolvent type equation,

$$T(z_j) = T(z) + T(z_j)\Big(g_0(z_j) - g_0(z)\Big)T(z). \tag{4}$$

   Here $g_0(z)$ is the relativistic two-body propagator, with z being the right-half-shell energy and z_j the fully-off-shell energy. This procedure was proposed by Keister and Polyzou [6] and implemented for the first time by Lin et al. in relativistic Faddeev equations for 3B scattering [2].

By following this strategy one can directly obtain the fully off-shell boosted t−matrix from nonrelativistic one.
The permutation operator P can be evaluated in relativistic basis states as

$$\begin{aligned}\langle\mathbf{p}',\mathbf{k}'|P|\mathbf{p}'',\mathbf{k}''\rangle &= {}_1\langle\mathbf{p}',\mathbf{k}'|\mathbf{p}'',\mathbf{k}''\rangle_2 + {}_1\langle\mathbf{p}',\mathbf{k}'|\mathbf{p}'',\mathbf{k}''\rangle_3\\ &= N(\mathbf{k}',\mathbf{k}'')\Big\{\delta^3\Big(\mathbf{p}'-\mathbf{k}''-\frac{1}{2}\mathbf{k}'C(\mathbf{k}',\mathbf{k}'')\Big)\delta^3\Big(\mathbf{p}''+\mathbf{k}'+\frac{1}{2}\mathbf{k}''C(\mathbf{k}'',\mathbf{k}')\Big)\\ &+ \delta^3\Big(\mathbf{p}'+\mathbf{k}''+\frac{1}{2}\mathbf{k}'C(\mathbf{k}',\mathbf{k}'')\Big)\delta^3\Big(\mathbf{p}''-\mathbf{k}'-\frac{1}{2}\mathbf{k}''C(\mathbf{k}'',\mathbf{k}')\Big)\Big\},\end{aligned} \tag{5}$$

where

$$N(\mathbf{k}',\mathbf{k}'') = \mathcal{N}^{-1}(-\mathbf{k}'-\mathbf{k}'',\mathbf{k}'')\,\mathcal{N}^{-1}(-\mathbf{k}'-\mathbf{k}'',\mathbf{k}'). \tag{6}$$

Here $\mathcal{N}$ is the square root of the Jacobian of the basis change,

$$\mathcal{N}(\mathbf{k},\mathbf{k}') = \left|\frac{\partial(\mathbf{k},\mathbf{k}')}{\partial(\mathbf{p},\mathbf{k}+\mathbf{k}')}\right|^{1/2} = \left[\frac{\omega_m(p)+\omega_m(p)}{\omega_m(k)+\omega_m(k')}\frac{\omega_m(k)\omega_m(k')}{\omega_m(p)\omega_m(p)}\right]^{1/2} \neq 1. \tag{7}$$

The function C is given by

$$C(\mathbf{k},\mathbf{k}') = 1 + \frac{\omega_m(k') - \omega_m(|\mathbf{k}+\mathbf{k}'|)}{\omega_m(k')+\omega_m(|\mathbf{k}+\mathbf{k}'|)+\sqrt{\Big(\omega_m(k')+\omega_m(|\mathbf{k}+\mathbf{k}'|)\Big)^2 - k^2}}, \tag{8}$$

where $\omega_m(k) = \sqrt{m^2+k^2}$. In the large mass limit, i.e. $m \gg |\mathbf{p}|, |\mathbf{k}|$, the permutation coefficient C and the Jacobian function N approach the nonrelativistic limit and are equal to one.

After evaluating the permutation operator and the matrix elements of the fully off-shell two-body t−matrix, the relativistic Faddeev integral equation reads,

$$\psi(\mathbf{p}, \mathbf{k}) = \frac{1}{M_t - M_0(p,k)} \int d\mathbf{k}'\, N(\mathbf{k},\mathbf{k}')\, T^{sym}\Big(\mathbf{p}, \tilde{\pi}; M_t - \omega_m(k)\Big)\, \psi(\pi, \mathbf{k}'), \tag{9}$$

where $T^{sym}(\mathbf{p},\mathbf{p}';\epsilon) = T(\mathbf{p},\mathbf{p}';\epsilon) + T(-\mathbf{p},\mathbf{p}';\epsilon)$ is symmetrized boosted two-body t-matrix and the shifted momentum arguments are given by

$$\tilde{\pi} = \mathbf{k}' + \frac{1}{2}C(\mathbf{k},\mathbf{k}')\,\mathbf{k}, \quad \pi = \mathbf{k} + \frac{1}{2}C(\mathbf{k}',\mathbf{k})\,\mathbf{k}'. \tag{10}$$

The relativistic and non-relativistic interaction models are defined so that they have the same two-body center-of-momentum scattering cross sections and wave functions. In the three-body space they are fixed by S-matrix cluster properties and Poincaré or Galilean invariance, respectively. Differences between the relativistic and nonrelativistic Faddeev equations arise from:

1. the Jacobian function N, representing from the change of the 3B basis states,
2. the coefficient C, which appears in the shifted arguments of the momenta,
3. the relation between the relativistic and non-relativistic t-matrices defined by function F,
4. the relations between the relativistic and non-relativistic free Green's functions.

2 Results and Discussion

For our numerical analysis we used two models of a spin-independent Malfliet-Tjon type potential, MT-V [3] and MT-Vc [4]. The model MT-Vc contains an additional cutoff function of dipole type. The relativistic Faddeev integral equation is solved using a Lanczos technique [7]. The calculated relativistic and nonrelativistic 3B binding energies for MT-V potential are $E_r = -7.4825$ (MeV) and $E_{nr} = -7.7382$ (MeV), which indicates a reduction of about $(E_r - E_{nr})/E_{nr} = 3.3\,\%$ in the 3B binding energy. This reduction is consistent with the partial wave calculation of Ref. [8], truncated to s-wave, with about 2.70 % reduction in the 3B binding energy. The similar behavior can also be seen for MT-Vc potential, where the relativistic effects leads to about 3.2 % reduction in 3B binding energy. The obtained nonrelativistic and relativistic 3B binding energies for MT-Vc potential are $E_{nr} = -7.7177$ (MeV) and $E_r = -7.4732$ (MeV), respectively.

While our calculations indicate that the overall relativistic effect is small, this results from large cancellations in the four relativistic corrections mentioned above. In Fig. 1, we show the contribution of each correction to the 3B relativistic binding energy of MT-V potential, where E_{appr} is the 3B binding energy when one of the ingredients is replaced by the corresponding non-relativistic quantity. By setting the Jacobian function N to one in our relativistic calculations, the 3B binding energy has a small increase of about 0.8 %, while setting the permutation coefficient C to one leads to a small reduction of about 0.6 %. Replacing the relativistic right-half-shell t−matrix with the non-relativistic one, i.e. setting $F = 1$ in Eq. (3), leads to about 2.4 % increase in the 3B binding energy. Finally, replacing the relativistic free propagator with the nonrelativistic one, the 3B binding energy is reduced by about 1.84 %. By setting both Jacobian functions, N and C, to one in our relativistic calculations, the net contribution of the dynamic ingredients to the 3B binding energy is a small increase of about 0.15 %. Similarly, the net contribution of the kinematic ingredients obtained by setting $F = 1$ and replacing the relativistic free propagator with the nonrelativistic one, leads to an increase in the 3B binding energy of about 0.22 %.

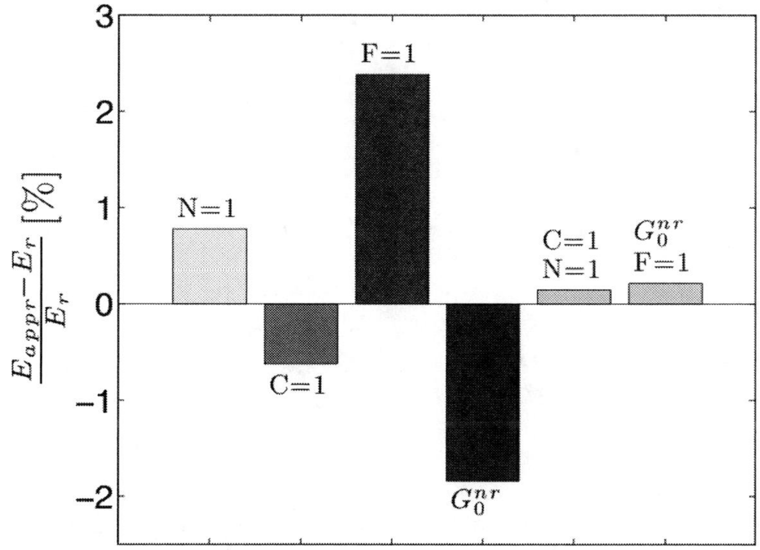

Figure 1. Contributions of different relativistic corrections to the 3B binding energy of MT-V potential. (1) $N = 1$: the Jacobian of the permutation operator (2) $C = 1$ relativistic Jacobi momenta → NR Jacobi momenta, (3) $F = 1$ the relativistic half-shell t-matrix → NR half-shell t-matrix (4) G_0: free resolvent $G_0 \to G_0^{NR}$.

The extension of the formulation of the relativistic Faddeev integral equations based on realistic nucleon-nucleon (NN) interactions is in progress. The input here is an operator representation of spin-dependent NN interaction in a helicity representation, which depends on the total spin and the relative momentum in the NN subsystems. Since the 3D approach automatically considers all partial waves the number of equations is fixed. The successful implementation in $3N$ bound state will pave the way to $3N$ scattering in the few-GeV energy range.

Acknowledgements

This work was performed under the auspices of the National Science Foundation under Contract No. NSF-PHY-1005587 with Ohio University and Contract No. NSF-PHY-1005501 with the University of Iowa. Partial support was also provided by the U. S. Department of Energy, Office of Science, Office of Nuclear Physics, under Contract No. DE-FG02-93ER40756 with Ohio University, and Contract No. DE-FG02-86ER40286 with the University of Iowa. We thank the Ohio Supercomputer Center (OSC) for the use of their facilities under Grant No. PHS0206.

References

[1] M. R. Hadizadeh, Ch. Elster, W. N. Polyzou, Phys. Rev. C **90**, 054002 (2014).
[2] T. Lin, C. Elster, W. Polyzou, and W. Glöckle, Phys. Rev. C **76**, 014010 (2007).
[3] Ch. Elster, W. Schadow, A. Nogga, W. Glöckle, Few Body Syst. **27**, 83 (1999).
[4] H. Liu, Ch. Elster, W. Glöckle, Few Body Syst. **33**, 241 (2003).
[5] F. Coester, Steven C. Pieper, F. J. D. Serduke, Phys. Rev. C **11**, 1 (1975).
[6] B. D. Keister and W. N. Polyzou, Phys. Rev. C **73**, 014005 (2006).
[7] A. Stadler, W. Glöckle, and P. Sauer, Phys. Rev. C **44**, 2319 (1991).
[8] W. Glöckle, T. S. H. Lee, and F. Coester, Phys. Rev. C **33**, 709 (1986).

EPJ Web of Conferences **113**, 03012 (2016)
DOI: 10.1051/epjconf/201611303012

Direct Bethe-Salpeter solutions in Minkowski space

J. Carbonell[1,a] and V.A. Karmanov[2,b]

[1]*Institut de Physique Nucléaire, Université Paris-Sud, IN2P3-CNRS, 91406 Orsay Cedex, France*
[2]*Lebedev Physical Institute, Leninsky Prospekt 53, 119991 Moscow, Russia*

Abstract. We review a method to directly solve the Bethe-Salpeter equation in Minkowski space, both for bound and scattering states. It is based on a proper treatment of the many singularities which appear in the kernel and propagators. The off-mass shell scattering amplitude for spinless particles interacting by a one boson exchange was computed for the first time. Using our Minkowski space solutions for the initial (bound) and final (scattering) states, we calculate elastic and transition (bound $\to$ scattering state) electromagnetic form factors. The conservation of the transition electromagnetic current $J \cdot q = 0$, verified numerically, confirms the validity of our solutions.

1 Introduction

Bethe-Salpeter (BS) approach, formulated originally in the Minkowski space, provides a self-consistent description of relativistic systems based on the field theory. The main reason for being interested in the Minkowski solutions is the fact that the Euclidean ones are not able to provide some fundamental observables. While the purely Euclidean solutions can be used to obtain the binding energy, the on-shell scattering observables (phase shifts) require already coupled equations with some particular Minkowski amplitudes[1, 2], As we have shown in [3, 4], Wick rotation is not directly applicable for computing electromagnetic (e.m.) form factors when the BS amplitudes are known numerically. Finally, the computation of the off-shell BS scattering amplitude – mandatory for finding e.g. the transition e.m. form factor – is possible only using a full Minkowski solution.

For bound states, the Minkowski solutions were found in [5], by representing the BS amplitude via Nakanishi integral [6]. A similar approach combined with the light-front projection was proposed in [7]. A modified formalism to the one developed in [7] aimed to compute the scattering states was developed in [8]. It has been tested for bound states [9] and, more recently, for computing the – zero-energy – scattering lengths [10].

We present here the off- and on-shell solutions the BS equation in Minkowski space found by a direct method [2] based on a proper treatment of the many singularities which appear in the kernel and propagators. It is applicable to bound and scattering states, and valid in any energy domain.

[a]e-mail: carbonell@ipno.in2p3.fr
[b]e-mail: karmanov@sci.lebedev.ru

2 BS amplitude and phase shifts

We have first computed the bound state solutions. The binding energies thus obtained coincide, within four-digit accuracy, with the ones calculated in our previous work [7].

For the scattering states, in c.m.-frame $\vec{p} = 0$, the BS off-shell partial wave amplitude F^{off}, for a given incident momentum p_s (corresponding to given invariant mass of the scattering state $M' = 2\sqrt{m^2 + p_s^2}$) depends on two variables k_0, k: $F^{off} = F(k_0, k; p_s)$. The BS amplitude F is factorized so that the strongest singularities are included in a known analytical prefactor, and the remaining regular part f is expanded in a spline basis. In this way, the BS equation is transformed in a numerically solvable form and the off-mass shell scattering amplitude for spinless particles interacting by a one boson exchange (with the mass μ) was computed for the first time. The detail of this transformation are given in [2].

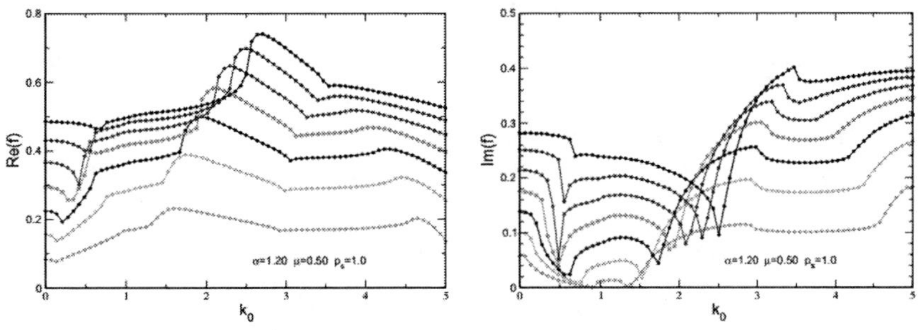

Figure 1. Real (left panel) and imaginary (right panel) parts of the regular part $f_0(k, k_0)$ of the off-shell scattering amplitude vs. k_0 for a few fixed values of k and $\mu = 0.5$ (in the units of mass m).

As an example, we display in Fig. 1 the regular part $f_0(k, k_0)$ of the S-wave off-shell scattering amplitude as a function k_0 for different values of k, for the coupling constant $\alpha = g^2/(16\pi m^2) = 1.2$. As one can see, this function is no longer singular, although it has highly non-trivial behavior with several sharp structures and cusps both in its real and imaginary parts. When the incident momentum p_s exceeds the threshold of the boson creation, the amplitude has a cusp and the phase shift automatically obtains an imaginary part. Some of these sharp behaviors can be understood analytically. In these – rare – cases, the numerical results are in close agreement with the analytical predictions giving thus a strong confidence on the validity of our approach (see for instance Fig. 12 in Ref. [2]).

The on-mass shell condition $k_1^2 = k_2^2 = m^2$, in the relative variables k_0, k, obtains the form $k_0 = 0, k = p_s$. The corresponding on-shell amplitude $F^{on} = F(k_0 = 0, k = p_s; p_s) = F(p_s)$ is the physical scattering amplitude related to the phase shift δ as:

$$F^{on} = \frac{M'}{4ip_s}\left[\exp(2i\delta) - 1\right]$$

Figure 2 (left panel) shows the real part of the phase shifts calculated via the BS equation (solid line) as a function of the scattering momentum p_s. They are compared to the non-relativistic (NR) values (dashed lines) provided by the Schrödinger equation with the Yukawa potential. For the value of $\alpha = 1.2$ there exists a bound state and, according to the Levinson theorem, the phase shift starts at 180°. One can see that the difference between relativistic and non-relativistic results is considerable even for relatively small incident momentum.

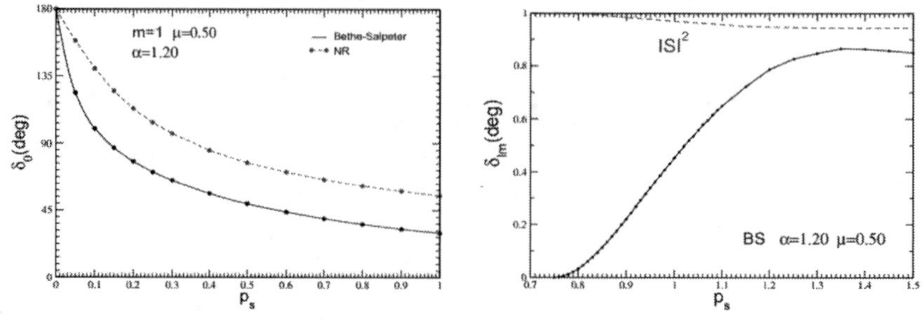

Figure 2. Left panel: real part of the phase shift (degrees) for $\alpha = 1.2$ and $\mu = 0.50$ calculated via BS equation (solid) compared to the non-relativistic results (dashed). Right panel: imaginary part of the phase shift (degrees) calculated via BS equation (solid) and the inelasticity parameter $\eta = |S|^2 = \exp[-2\mathrm{Im}(\delta)]$ (dashed).

The right panel shows the imaginary part of the phase shift. It appears starting from the first inelastic meson-production threshold $p_s^{(1)} = 0.75$ and in vicinity of $p_s = p_s^{(1)}$ displays the expected quadratic behavior $\mathrm{Im}[\delta_0] \sim (p_s - p_s^{(1)})^2$. Simultaneously the modulus squared of the S-matrix (dashed line) starts differ from unity. We reproduce the phase shifts found in [1]. The low energy parameters

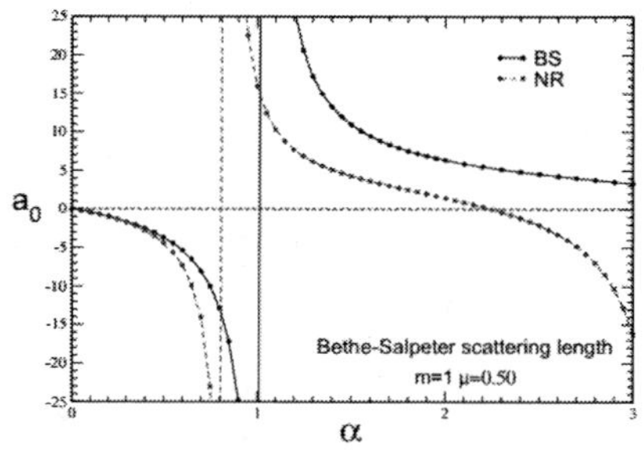

Figure 3. BS scattering length a_0 versus α (solid), compared to the non-relativistic results (dashed) for $\mu = 0.5$.

were computed directly at $p_s = 0$ and found to be consistent with a quadratic fit to the effective range function $k \cot \delta_0(k) = -\frac{1}{a_0} + \frac{1}{2} r_0 k^2$. The BS scattering length a_0 as a function of the coupling constant α is given in Fig. 3 for $\mu = 0.50$. It is compared to the non-relativistic (NR) values. The singularities correspond to the appearance of the first bound state at $\alpha_0 = 1.02$ for BS and $\alpha_0 = 0.840$ for NR. One can see that the differences between relativistic and non-relativistic treatments is substantially large even in processes involving zero energy, especially in presence of a bound state.

Our results for the scattering length coincide with good precision with ones found in [10] by a different method. Some discrepancy for small μ, indicated in [10], disappeared after taking the limit $p_s \to 0$ in our equations analytically.

The method can be easily generalized to fermions, since boson and fermion propagators have same singularities. This opens the way for studying the NN system in the BS framework.

3 E.M. form factors

Using our Minkowski space solutions for the initial (bound) and final (scattering) states, we calculate the elastic form factor [3] and the transition one for electrodisintegration of a two-body system [11].

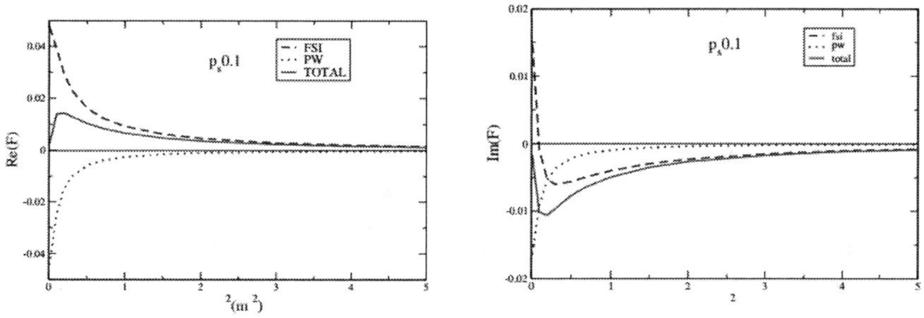

Figure 4. Transition EM form factor $F(Q^2)$ as a function of Q^2. Initial (bound) state corresponds to the binding energy $B = 0.01\,m$; final (scattering) state corresponds to a relative momentum $p_s = 0.1\,m$ (final state mass $M' = 2.00998\,m$). FSI contribution is shown by the dashed curve and the PW one by a dotted curve. Full form factor is shown by the solid curve. Left panel is the real part of form factor and right panel is the imaginary part.

We show in Fig. 4 the transition form factor for the initial (bound) state binding energy $B = 0.01\,m$ (the initial state mass $M = 1.99\,m$) and for the final (scattering) state relative momentum $p_s = 0.1\,m$ with corresponding final state mass values $M' = 2\sqrt{m^2 + p_s^2} \approx 2.00998\,m$. It is the sum of two contributions: final state interaction (FSI) and plane wave (PW). In contrast to the elastic scattering, the inelastic transition form factor is complex. Its real and imaginary parts as a function of Q^2 for $p_s = 0.1\,m$ are shown in Fig. 4, at left and right panels correspondingly. One can see that at relatively small momentum transfer $Q^2 < 1$ both contributions – FSI and PW – are equally important and they considerably cancel each other.

The gauge invariance, i.e. the conservation of the transition e.m. current $J \cdot q = 0$, is verified in [11] numerically. It results from full cancellation between the PW and FSI contributions to $J \cdot q$ which takes place only if the bound and scattering state amplitudes, and the e.m. current are strictly consistent with each other. This cancellation takes place numerically with rather high precision, that confirms the validity of our calculations.

The results shown above – the off-shell amplitude, the phase shifts, the scattering length and e.m. form factors – represent the exhaustive set of calculations for a two-body system. They put on the agenda the similar treatment, in the BS framework, of the realistic systems (nucleons and hadrons).

References

[1] M.J. Levine, J.A. Tjon, J. Wright, Phys. Rev. Lett. **16**, 962 (1966);
M.J. Levine, J. Wright, J.A. Tjon, Phys. Rev. **154**, 1433 (1967)
[2] J. Carbonell, V.A. Karmanov, Phys. Rev. D **90**, 056002 (2014)
[3] J. Carbonell, V.A. Karmanov, Few-Body Syst. **49**, 205 (2011)
[4] J. Carbonell, V.A. Karmanov, Few-Body Syst. **49**, 205 (2011)
[5] K. Kusaka, A. G. Williams, Phys. Rev. D **51**, 7026 (1995);
K. Kusaka, K. Simpson, A.G. Williams, Phys. Rev. D **56**, 5071 (1997)
[6] N. Nakanishi, Phys. Rev. **130**, 1230 (1963)
[7] V.A. Karmanov, J. Carbonell, Eur. Phys. J. A **27**, 1 (2006);
J. Carbonell, V.A. Karmanov, Eur. Phys. J. A **27**, 11 (2006)
[8] T. Frederico, G. Salmè, M. Viviani, Phys. Rev. D **85**, 036009 (2012)
[9] T. Frederico, G. Salmè, M. Viviani, Phys. Rev. D **89**, 016010 (2014)
[10] T. Frederico, G. Salmè, M. Viviani, Eur. Phys. J. C **75**, 398 (2015); arXiv:1504.01624
[11] J. Carbonell, V.A. Karmanov, Phys. Rev. D **91**, 076010 (2015)

EPJ Web of Conferences **113**, 03013 (2016)
DOI: 10.1051/epjconf/201611303013

Quarkonia and heavy-light mesons in a covariant quark model

Sofia Leitão[1,a], Alfred Stadler[2,1,b], M. T. Peña[1,3,c], and Elmar P. Biernat[1,d]

[1] *Centro de Física Teórica de Partículas, Instituto Superior Técnico (IST), Universidade de Lisboa, 1049-001 Lisboa, Portugal*
[2] *Departamento de Física, Universidade de Évora, 7000-671 Évora, Portugal*
[3] *Departamento de Física, Instituto Superior Técnico (IST), Universidade de Lisboa, 1049-001 Lisboa, Portugal*

Abstract. Preliminary calculations using the Covariant Spectator Theory (CST) employed a scalar linear confining interaction and an additional constant vector potential to compute the mesonic mass spectra. In this work we generalize the confining interaction to include more general structures, in particular a vector and also a pseudoscalar part, as suggested by a recent study [1]. A one-gluon-exchange kernel is also implemented to describe the short-range part of the interaction. We solve the simplest CST approximation to the complete Bethe-Salpeter equation, the one-channel spectator equation, using a numerical technique that eliminates all singularities from the kernel. The parameters of the model are determined through a fit to the experimental pseudoscalar meson spectra, with a good agreement for both quarkonia and heavy-light states.

1 Introduction

When applied to the description of hadrons, QCD is a highly complex nonperturbative theory that cannot be solved directly. Numerical simulations of a space-time discretized version of QCD, lattice QCD [2], are promising, but still have limitations. Continuum methods using the Bethe-Salpeter/Dyson-Schwinger equations [3–5] leave out some possibly important features of QCD, and so far do not describe all decay properties of the meson spectrum and in particular of the heavy-light sector. Moreover, both lattice QCD and Dyson-Schwinger calculations are performed in Euclidean and not in the physical Minkowski space. This simplification allows the calculation of mass spectra, but it is not clear how the obtained amplitudes can be reliably related to the physical ones.

This work is inserted in a larger project which proposes to implement the following: 1) a calculation of quark-antiquark bound states in Minkowski space; 2) dynamical quark mass generation, due to the self-interaction of quarks calculated consistently from the quark-quark interaction kernel and satisfying chiral symmetry constraints; 3) a confining term in the interaction kernel that in the nonrelativistic limit reduces to a linear potential. The parameters of the interaction kernel will be determined by fits of the meson spectra, and are constrained by the quark mass functions obtained in

[a] e-mail: sofia.leitao@tecnico.ulisboa.pt
[b] e-mail: stadler@uevora.pt
[c] e-mail: teresa.pena@tecnico.ulisboa.pt
[d] e-mail: elmar.biernat@tecnico.ulisboa.pt

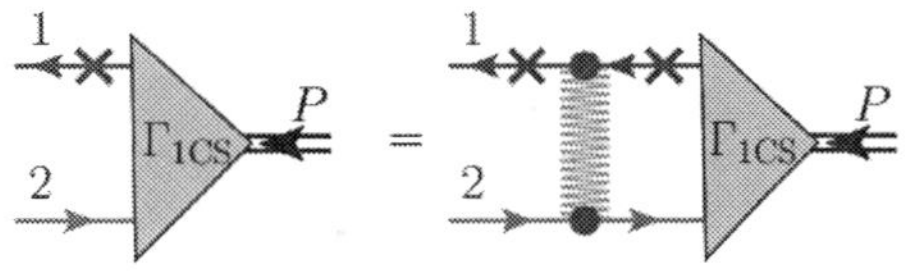

Figure 1. Diagrammatic representation of the 1CSE. The cross indicates that particle 1 is on its mass-shell. P is the total momentum and Γ_{1CS} the vertex function of the bound-state.

lattice QCD. This will also provide information on the Lorentz structure of the confining interaction, which is not well known at present. The dressed electromagnetic quark current will be calculated consistently, using the same kernel.

Once the quark current is known, many more applications become possible, such as hadron form factors and decay properties. We single out the calculation of hadronic contributions to light-by-light scattering, which is the most uncertain part of the theoretical prediction of the muon's anomalous magnetic moment. This uncertainty needs to be reduced in order to find out if the observed 3σ discrepancy between theory and measurement is a sign of new physics beyond the Standard Model [6].

2 Pseudoscalar meson spectra with the one-channel CST equation

We use the Covariant Spectator Theory (CST) [7], which can be viewed as a reorganization of the complete Bethe-Salpeter equation (BSE), to develop a dynamical quark model in Minkowski space that can describe the structure and the mass spectrum of both heavy and light quark systems.

The simplest version of the CST equation, called the one-channel CST equation (1CSE) and shown diagrammatically in Fig. 1, is obtained when only the positive-energy pole contribution from the heavier quark propagator in the energy loop integration is kept.

An important property of the 1CSE is that, unlike the BSE in ladder approximation, it has a smooth nonrelativistic limit, and therefore in calculations of heavy quarkonia its results can be compared directly to the ones obtained with the Schrödinger equation (SE).

To find an appropriate way to treat a linear confining interaction in momentum space, and to test the required numerical methods, in [8] we used the SE to calculate the two-body bound states of a linear potential. Since in S waves these solutions are known exactly, this is an ideal test case. It turned out that the singularities arising in the kernel can be treated efficiently, yielding accurate and stable numerical solutions. In this work, the same methods were applied for the first time to solve the relativistic 1CSE.

The kernel employed in our calculations with the 1CSE consists of a covariant generalization of the linear (L) confining potential used in [8], a one-gluon exchange (OGE), and a constant (C) interaction,

$$\mathcal{V}(p,k) = \mathcal{V}_{\rm L}(p,k) + \mathcal{V}_{\rm OGE}(p,k) + \mathcal{V}_{\rm C}(p,k)\,, \tag{1}$$

where

$$\mathcal{V}_{\rm L}(p,k) = -8\sigma\pi\left[\frac{1}{q^4} - \frac{E_{p_1}}{m_1}(2\pi)^3\delta^3(\mathbf{q})\int\frac{d^3k'}{(2\pi)^3}\frac{m_1}{E_{k'_1}}\frac{1}{q'^4}\right]\left[(1-y)(\mathbf{1}\otimes\mathbf{1}+\gamma^5\otimes\gamma^5) - y(\gamma^\mu\otimes\gamma_\mu)\right],$$

$$\mathcal{V}_{\rm OGE}(p,k) = -\frac{4\pi\alpha}{q^2}(\gamma^\mu\otimes\gamma_\mu)\,, \qquad \mathcal{V}_{\rm C}(p,k) = (2\pi)^3\frac{E_{k_1}}{m_1}C\delta^3(\mathbf{q})(\gamma^\mu\otimes\gamma_\mu)\,, \tag{2}$$

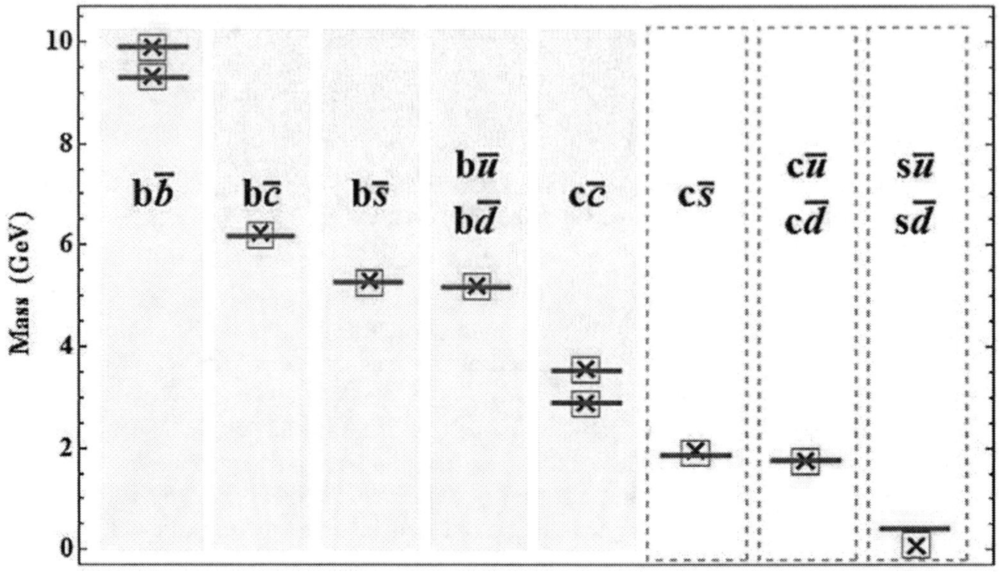

Figure 2. Mass spectra computed with the 1CSE for the $b\bar{b}$, $c\bar{c}$ quarkonia, and for the heavy-light $b\bar{s}$, $b\bar{u}/b\bar{d}$, $b\bar{c}$, $c\bar{u}/c\bar{d}$, $s\bar{u}/s\bar{d}$ pseudoscalar states. The horizontal bars represent the experimental $J^{PC} = 0^{--}$ states, the centers of the crosses and open squares indicate the calculated states for models 1 and 2, respectively. The states fitted to the data are those inside the shaded regions, whereas the remaining states inside the dashed rectangles are predictions.

Table 1. Fitted parameters for kernel models 1 and 2.

	σ [GeV^2]	α	C [GeV]	y
Model 1	0.22	0.38	0.337	8.60×10^{-7}
Model 2	0.21	0.37	0.311	0.38×10^{-7}

$q = p - k$ is the transferred four-momentum, and $E_{p_1} = (m_1^2 + \vec{p}^2)^{1/2}$ is the energy of the heavier quark with mass m_1. The mixing parameter y allows to dial continuously between a scalar-plus-pseudoscalar structure, suggested recently in [1] due to chiral-symmetry constraints, and a vector structure, while preserving the same nonrelativistic limit. The precise Lorentz structure of the confining interaction is not known, and by fitting the y parameter from the mesonic spectra some further information can be gained. The three coupling strengths σ, α and C, were treated as free model parameters.

For this preliminary work, we used constant masses for the constituent quarks, with the values

$$m_b = 4.793 \text{ GeV}, \quad m_c = 1.530 \text{ GeV}, \quad m_s = 0.400 \text{ GeV}, \quad \text{and} \quad m_u = m_d = 0.258 \text{ GeV}. \tag{3}$$

The 1CSE—with retardation included—requires regularization. We used a Pauli-Villars regularization with a cut-off parameter proportional to the average mass of the two constituent quarks, $\bar{m}$: $\Lambda = 1.7\bar{m}$.

The pseudoscalar meson spectra were fitted with two distinct models with different Lorentz structure of the confining interaction: whereas model 2 has exactly the equally weighted scalar-pseudoscalar component, mixed with a vector structure, as given in Eq. (2), model 1 omits the pseudoscalar part. The obtained results are given in Figure 2 and Table 1.

We find that good fits of the pseudoscalar meson spectra can be obtained with either model, and that the predictions for the mesons containing lighter quarks are also very reasonable. The observed slightly larger difference between the model predictions and the data for the lightest systems is in part due to a pole in the propagator of the lighter quark whose importance increases with decreasing bound-state mass, but which is omitted in the 1CSE. This pole will be included in future work.

It turns out that both model fits favor essentially pure Lorentz scalar structures of the confining interaction. It will be interesting to see if this feature persists once the fits are extended to include also vector mesons.

The inclusion of a pseudoscalar part in the confining interaction leads to a marginally better fit. However, because the heavier mesons are almost nonrelativistic systems for which the pseudoscalar interaction is strongly suppressed, the difference between models 1 and 2 becomes noticable only for the lighter states. In these cases, on the other hand, it may become necessary to replace the 1CSE by a more complete version of the CST bound-state equations, namely the two-channel (2CSE) or—in the case of the pion—the complete four-channel spectator equation (4CSE) [9].

3 Conclusions and Outlook

The numerical techniques developed so far allowed us to solve the 1CSE and perform the first fit with a realistic linear-plus-Coulomb-plus-constant kernel to a wider range of quarkonia and heavy-light pseudoscalar states.

As an immediate continuation of this work, we plan to extend the fit to other mesons (most importantly vector mesons), and to investigate the effect of replacing the constant constituent quark masses by quark mass functions that are calculated in a fully self-consistent fashion from the quark-antiquark kernel. The next goal will be to solve the 2CSE and 4CSE to obtain reliable solutions also for the light sector.

Meanwhile, based on the results determined so far, we can already state that the calculational methods we developed to treat a linear confining interaction in momentum space and tested in the nonrelativistic SE, also work in the relativistic case of the 1CSE in Minkowski space. These techniques will be useful in future studies of the mesonic bound-state problem using the CST.

Acknowledgements

This work received financial support from Fundação para a Ciência e Tecnologia (FCT) under Grant No. CFTP-FCT (PEst-OE/FIS/U/0777/2013).

References

[1] E. P. Biernat, M. T. Peña, J. E. Ribeiro, A. Stadler, and F. Gross, Phys. Rev. D **90**, 096008 (2014)
[2] C. Gattringer, C. B. Lang, *Quantum Chromodynamics on the Lattice* (Springer, 2010)
[3] I. C. Cloët, C. D. Roberts, Nucl. Phys. **77**, 69 (2014)
[4] R. Alkofer, L. von Smekal, Phys. Rept. **353**, 281 (2001)
[5] C. S. Fischer., J. Phys. **G32**, R253 (2006)
[6] F. Jegerlehner, A. Nyffeler., Phys. Rep. **477**, 1 (2009)
[7] F. Gross, *Relativistic quantum mechanics and field theory*, revised ed. (New York, USA: Wiley-VCH, 1999).
[8] S. Leitão, A. Stadler, M. T. Peña, E. P. Biernat, Phys. Rev. D **90**, 096003 (2014).
[9] E. P. Biernat, F. Gross, M. T. Peña, A. Stadler, Phys. Rev. D **89**, 016005 (2014).

EPJ Web of Conferences **113**, 03014 (2016)
DOI: 10.1051/epjconf/201611303014

Clothed Particles in Quantum Electrodynamics and Quantum Chromodynamics

Alexander Shebeko[a]

[1] *National Science Center "Kharkov Institute of Physics & Technology", Ukraine*

Abstract. The notion of clothing in quantum field theory (QFT), put forward by Greenberg and Schweber and developed by M. Shirokov, is applied in quantum electrodynamics (QED) and quantum chromodynamics (QCD). Along the guideline we have derived a novel analytic expression for the QED Hamiltonian in the clothed particle representation (CPR). In addition, we are trying to realize this notion in QCD (to be definite for the gauge group $SU(3)$) when drawing parallels between QCD and QED.

1 Introductory Remarks

As before in [1] - [4], by using the instant form of relativistic dynamics we start from a total Hamiltonian $H = H(\overset{\circ}{\alpha}) = H_0(\overset{\circ}{\alpha}) + V_0(\overset{\circ}{\alpha})$, where unperturbed Hamiltonian $H_0(\overset{\circ}{\alpha})$ and interaction term $V_0(\overset{\circ}{\alpha})$ depend on the creation and destruction operators of "bare" particles (e.g., bosons and fermions) with trial masses and coupling constants. Here $\overset{\circ}{\alpha}$ denotes the set of all such operators. Following [2] we introduce specific auxiliary transformations, reminiscent of the canonical transformations (in particular, the Bogoliubov ones) in the theory of superfluidity and superconductivity, that convert the primary bare bosons and fermions into some intermediate-level particles in a new α - representation with observable (in general, *a priori* given [1]) masses (cf. the terminology adopted in our survey [1]). As a result, the original Hamiltonian gets the form $H \equiv H(\alpha) = H_F(\alpha) + M_{ren}(\alpha) + V(\alpha) \equiv H_F(\alpha) + H_I(\alpha)$, where we meet the free Hamiltonian H_F expressed in terms of operators α and the operator M_{ren} which is contained the corresponding mass counterterms. Note that transition $\overset{\circ}{\alpha} \rightarrow \alpha$ remains the primary interaction intact and exemplifies one of nonequivalent representations of canonical commutation relations in QFT.

Just this form is subject to the similarity transformation $\alpha = W(\alpha_c)\alpha_c W^\dagger(\alpha_c)$ with the operator $W(\alpha_c) = W(\alpha) = \exp R$, $R = -R^\dagger$ that connects the bare particles representation (BPR) α and the clothed particles representation α_c. In its turn, transition $\alpha \rightarrow \alpha_c$ is a key point of the method of unitary clothing transformations (shortly, UCT method) for different approximations in our calculations with the Hamiltonian $H = W(\alpha_c)H(\alpha_c)W^\dagger(\alpha_c)$, the Lorentz boosts, the Nöther currents, etc. Physical constraints imposed upon W and mathematical aspects of the UCT method have been exposed in [1] and [4].

[a]e-mail: shebeko@kipt.kharkov.ua

[1]Some extension of our approach for field models such as QCD

2 Clothed Particles in QED

In the Coulomb gauge (CG) the interaction Hamiltonian of the spinor QED is given by

$$V_{qed} = \int d\vec{x} V_{qed}(\vec{x}) = -\int d\vec{x} j^k(\vec{x}) a_k(\vec{x}) + V_{Coul}, \tag{1}$$

with the electron-positron current $j^\mu(\vec{x}) = e\bar{\psi}(\vec{x})\gamma_\mu\psi(\vec{x})$ and the Coulomb part,

$$V_{Coul} = \frac{1}{2}\int d\vec{x}\int d\vec{y} \frac{j^0(\vec{x}) j^0(\vec{y})}{4\pi|\vec{x}-\vec{y}|}. \tag{2}$$

Evidently, the corresponding interaction density $V_{qed}(x)$ in the Dirac(D) picture is not the Lorentz scalar and we cannot use the so-called Belinfante ansatz to construct the boost generator $\vec{N}$, i.e., put $\mathbf{N}_{qed} = -\int \mathbf{x} V_{qed}(\mathbf{x}) d\mathbf{x}$. Therefore one has to seek other ways to provide the relativistic invariance (RI) in Dirac sense (see, e.g., [4]).
Our departure point is the particle number representation with the Fourier expansions

$$a^\rho(\boldsymbol{x}) = \int \frac{d\boldsymbol{k}}{\sqrt{2(2\pi)^3|\boldsymbol{k}|}} \sum_\sigma [e^\rho(\boldsymbol{k},\sigma) c(\boldsymbol{k},\sigma) + e^{\rho *}(\boldsymbol{k},\sigma) c^\dagger(\boldsymbol{k},\sigma)] exp(i\boldsymbol{k}\boldsymbol{x}), \tag{3}$$

$$\psi(\boldsymbol{x}) = \int d\boldsymbol{p}\sqrt{\frac{m}{(2\pi)^3 E_{\boldsymbol{p}}}} \sum_\mu [\bar{u}(\boldsymbol{p},\mu) b(\boldsymbol{p},\mu) + v(-\boldsymbol{p},\mu) d^\dagger(-\boldsymbol{p},\mu)] exp(i\boldsymbol{k}\boldsymbol{x}), \tag{4}$$

where $E_{\boldsymbol{p}} = \sqrt{\boldsymbol{p}^2 + m^2}$ fermion energy on its mass shell and the creation(destruction) operators meet commutation relations $\{b(\boldsymbol{p},\mu), b^\dagger(\boldsymbol{p}',\mu')\} = \{d(\boldsymbol{p},\mu), d^\dagger(\boldsymbol{p}',\mu')\} = \delta_{\mu\mu'}\delta(\boldsymbol{p}-\boldsymbol{p}')$, $[c(\boldsymbol{k},\sigma), c^\dagger(\boldsymbol{k}',\sigma')] = \delta_{\sigma\sigma'}\delta(\boldsymbol{k}-\boldsymbol{k}')$.
After the first clothing transformation $W^{(1)} = \exp[R^{(1)}]$ ($R^{(1)\dagger} = -R^{(1)}$), which eliminates from V_{qed}, the primary interaction $V^{(1)}$, with its generator $R^{(1)}$ being determined by the equation

$$V^{(1)} + [R^{(1)}, H_F] = 0 \tag{5}$$

we find following [1] the clothed-particle interaction operators responsible for all relevant 2→2 processes (stemmed from $[R^{(1)}, V^{(1)}]$ commutator), the operators responsible for all 2→3 processes (from $[[R^{(1)}, [R^{(1)}, V^{(1)}]]$ commutator) and so on. I would like to show one of them.

2.1 $e^-e^- \to e^-e^-$ process

It is the case where we need to evaluate the $[R^{(1)}, V^{(1)}]$ commutator. The corresponding interaction operator $V(ee \to ee)$ for the two clothed electrons is

$$V(ee \to ee) = \frac{1}{2}[R^{(1)}, V^{(1)}](e^-e^- \to e^-e^-) + V_{Coul}(e^-e^- \to e^-e^-), \tag{6}$$

where the first term is the relevant part of $[R^{(1)}, V^{(1)}]$, and after a simple algebra one can get

$$V(ee \to ee) = \int \sum_\mu V(1',2';1,2) b^\dagger(1') b^\dagger(2') b(1) b(2) d\boldsymbol{p}'_1 d\boldsymbol{p}'_2 d\boldsymbol{p}_1 d\boldsymbol{p}_2, \tag{7}$$

$$V(1',2';1,2) = \frac{e^2}{(2\pi)^3} \frac{{m_e}^2}{\sqrt{E_{\vec{p}'_1} E_{\vec{p}'_2} E_{\vec{p}_1} E_{\vec{p}_2}}} v(1',2';1,2)\delta(\vec{P}' - \vec{P}), \tag{8}$$

$$v(1',2';1,2) = \frac{1}{2}\frac{\bar{u}(1')\gamma^{\mu}u(1)\bar{u}(2')\gamma_{\mu}u(2)}{(p_1'-p_1)^2} + \frac{1}{2}\frac{\bar{u}(1')(\not{P}'-\not{P})u(1)\bar{u}(2')(\not{P}'-\not{P})u(2)}{k^2}\frac{1}{(p_1'-p_1)^2} + \\ + \frac{1}{2}\frac{\bar{u}(1')\gamma^0 u(1)\bar{u}(2')\gamma^0 u(2)}{k^2}\frac{(p_1'-p_1)^2-(p_2'-p_2)^2}{(p_1'-p_1)^2}, \quad (9)$$

where $P' = p_1' + p_2'$, $P = p_1 + p_2$ and $\boldsymbol{k} = \vec{p}_1' - \vec{p}_1 = \vec{p}_2 - \vec{p}_2'$. In eq. (7) the symbol $\sum_{\mu}$ denotes the summation over fermion polarization projections so $1 = \{\vec{p}_1, \mu_1\}$, etc. For brevity, we are omitting the lower index c at operators in the r.h.s. of the expression (7). One should note that the contribution from the noncovariant Coulomb interaction to the matrix elements by eqs. (8) - (9) on the energy shell $P_0' = P_0$ is canceled with the respective counterpart in the r.h.s. of eq. (6). An important point is that in the CPR such a cancellation (cf. our results [3] in mesodynamics) takes place directly in the Hamiltonian .

3 QCD Hamiltonian in CG. Some Parallels.

At the beginning , trying to extend our approach for describing different processes in quark-gluon systems we are following a remarkable survey [5], where a close similarity of QCD to QED when introducing the Yukawa-type vertex couplings between the interacting fields becomes popular even for pedestrians including myself. As a result, we find there the QCD Hamiltonian

$$H_{qcd} \equiv P^0_{qcd} = \int d\vec{x}(F_a^{0\lambda}F^a_{\lambda 0} + \frac{1}{4}F_a^{\mu\nu}F^a_{\mu\nu} + \frac{1}{2}[i\bar{q}\gamma_0\tilde{D}^0 q + h.c.]) \quad (10)$$

with the tensor $F_a^{\mu\nu} = \partial^{\mu}A_a^{\nu} - \partial^{\nu}A_a^{\mu} + igf^{ars}A_r^{\mu}A_s^{\nu} \equiv \overset{\circ}{F}{}_a^{\mu\nu} + g(A^{\mu}\times A^{\nu})_a$ of color-vector (gluon) potentials A_a^{μ} (gluon indices $a(rs)$ change from 1 to $n_c^2 - 1$) and the covariant color-derivative $n_c \otimes n_c$ matrices $\tilde{D}^0_{cc'} = \delta_{cc'}\partial^0 - ig\tilde{A}^0_{cc'}$ (color indices c and c' run from 1 to n_c) versus the QED Hamiltonian

$$H_{qed} \equiv P^0_{qed} = \int d\vec{x}(F^{0\lambda}F_{\lambda 0} + \frac{1}{4}F^{\mu\nu}F_{\mu\nu} + \frac{1}{2}[i\bar{\psi}\gamma_0 D^0\psi + h.c.]) \quad (11)$$

with the colorless counterparts. Recall also that in case of the $SU(3)$ model with the conserved color currents

$$J_a^{\mu} = j_{\mu}^a + gf^{ars}F_r^{\mu\lambda}A_{\lambda}^s, \quad (12)$$

where the quark current $j_{\mu}^a = g\bar{q}\gamma_{\mu}T_a q$, the $3\otimes 3$ matrices T_a are related to the Gell-Mann matrices $T_a = \frac{1}{2}\lambda_a$.

The operators H_{qcd} and H_{qed} are gauge independent, i.e., remain intact under local gauge transformations by formula (2.22) in [5], that has much in common with the Fock-Weyl criterion. From practical viewpoint it is convenient to employ the CG in which $\partial^k A_k^a = 0$ so Gauss' law takes place for each of the gluon field components. Then Hamiltonian H_{qcd} can be expressed through the longitudinal and transverse color electric fields to get the division $H_{qcd} = H_0(\alpha) + V(\alpha) +$ mass and vertex counterterms, $V(\alpha) = \int d\vec{x}V_{qcd}(\vec{x})$ with the interaction density

$$V_{qcd}(\vec{x}) = -j_a^k(\vec{x})A_k^a(\vec{x}) - \frac{1}{2}F_{k0}^{a(l)}F_a^{k0(l)} + \frac{g}{2}\overset{\circ}{F}{}^a_{kn}(A^k\times A^n)^a + \frac{g^2}{4}(A_k\times A_n)^a(A^k\times A^n)^a, \quad (13)$$

and the constraint equation

$$\partial^k F_{k0}^{a(l)} + g(A^m\times F_{m0}^{(l)})^a = -j_0^a + g(E_m\times A^m)^a. \quad (14)$$

At this point, we prefer to proceed with the set α composed of the creation and destruction operators (cf., [5]) in the expansions

$$A_a^\mu(\vec{x}) = \int \frac{d\vec{k}}{\sqrt{2(2\pi)^3|\vec{k}|}} \sum_\sigma \left[e^\mu(\vec{k},\sigma) c_a(\vec{k},\sigma) + e^{\mu^*}(-\vec{k},\sigma) c_a^\dagger(-\vec{k},\sigma) \right] \exp(i\vec{k}\vec{x}), \tag{15}$$

$$q_{fc}(\vec{x}) = \int d\vec{p} \sqrt{\frac{m}{(2\pi)^3 E_{\vec{p}}}} \sum_\mu \left[\bar{u}(\vec{p}\mu) b_{fc}(\vec{p}\mu) + v(-\vec{p}\mu) d_{fc}^\dagger(-\vec{p}\mu) \right] \exp(i\vec{p}\vec{x}), \tag{16}$$

with the flavor-color label fc, if we want, and canonical commutations

$$[c_a(\vec{k},\sigma), c_{a'}^\dagger(\vec{k}',\sigma')] = \delta(\vec{k}-\vec{k}')\delta_{\sigma\sigma'}\delta_{aa'},$$

$$\{b_{fc}(\vec{p}\mu), b_{f'c'}^\dagger(\vec{p}'\mu')\} = \{d_{fc}(\vec{p}\mu), d_{f'c'}^\dagger(\vec{p}'\mu')\} = \delta(\vec{p}-\vec{p}')\delta_{\mu\mu'}\delta_{ff'}\delta_{cc'}.$$

these preliminaries are aimed at reformulating the field model in terms of the "clothed " quarks whose masses, being some adjustable parameters, can be adopted, for instance, from the constituent quark models.

For the first time our current explorations are focused on studying the properties of positronium (quarkonium) states with instantaneous interactions between the clothed electrons and positrons (quarks and antiquarks). The corresponding 2→2 operators have the structure (in particular, for the annihilation processes $e^-e^+ \to \gamma\gamma$ and $q\bar{q} \to$ two gluons) similar to that by Eq. (7) with the c - number coefficients $V(1',2';1,2)$ that determine properly symmetrized kernels of the relevant sets of coupled equations for the states mentioned above (cf., Subsec 4,4 in [1]). By developing a recursive algebraic technique proposed in [2] we continue to calculate higher-order corrections to these coefficients that stem from the multiple commutations of UCT generators with the so-called bad terms in a primary Hamiltonian H that prevent the vacuum and clothed one-particle states to be the H eigenstates. As before (see. e.g., [4]), we work with nonlocal extensions of the available field interactions by introducing some covariant cutoff factors into their vertices to handle finite intermediate quantities for a permanent cancellation of charge and mass renormalization terms with their counterparts from more and more complex commutators in question. In this context, a separate challenge for us is to compare our results in mesodynamics, QED and QCD with another dressing procedure based on some renormalization group ideas (see surveys [6] and [7] and refs. therein) and do it in the instant! form of relativistic dynamics. I am very grateful to referee for this possibility to become familiar (unfortunately, for a very short time) with explorations in this direction.

Acknowledgements

I thank my collaborator, Adam Arslanaliev, for his assistance in preparation of this talk.

References

[1] Phys. Part. Nucl., A. V. Shebeko and M. I. Shirokov, **32** (2001) 31.
[2] Ann. Phys., V. Korda, L. Canton and, A. Shebeko, **322** (2007) 736.
[3] Few-Body Syst., I. Dubovyk and A. Shebeko, **48** (2010) 109.
[4] Few Body Syst., A. V. Shebeko and P. A. Frolov, **52** (2012) 125.
[5] Phys. Rep., S. J. Brodsky, H.-Ch. Pauli and S. S. Pinsky, **301** (1998) 299.
[6] Phys. Rep., F. J. Wegner, **348** (2001) 77.
[7] S. D. Glazek , arXiv:hep-th/9712188.

EPJ Web of Conferences **113**, 03015 (2016)
DOI: 10.1051/epjconf/201611303015

Quantum Scattering Theory in a Discrete Representation

O.A. Rubtsova[1,a], V.I. Kukulin[1,b], and V.N. Pomerantsev[1,c]

[1] *Skobeltsyn Institute of Nuclear Physics, Moscow State University, 119991, GSP-1, Leninskie gory 1(2), Moscow, Russia*

Abstract. The approach to solving few-body scattering problems in a discrete representation of the stationary wave packets is described briefly. By projecting into the wave-packet basis, all the operators and wave functions are represented with finite matrices and vectors, so that the integral equations of scattering theory are reduced to their matrix analogs. In such a discrete representation, it is easy to construct the matrix analogs for any complicated operators such as total resolvent and also effective interactions between composite particles. Using a special spectral shift function formalism, multichannel scattering problem can be solved in a discrete representation without any scattering equations at all. The approach is illustrated by examples of multichannel and three-body scattering.

1 Introduction

The correct solution for few-body scattering problems has been done, as is well known, many years ago by Faddeev and Yakubovsky [1] which gave rise to an extensive few-body activity worldwide both in theory and experiment. However, when treating realistic interactions, there are some additional problems related to account for internal degrees of freedom of the colliding particles, complex optical interactions of composite particles, correct account of a long-range Coulomb interaction, scattering of particles in resonance states etc., so that a few-body scattering problem with fully realistic interparticle interactions still remains to be a hard numerical task in spite of a great progress in computational facilities. There are few approaches employing the sets of L_2 functions which are adopted for treating the realistic interactions in few-nucleon systems (see the recent review [2]).

Nearly a decade ago, our group in Moscow State University have developed an original approach based on the complete few-body continuum discretization and formulation of the problem in a finite L_2 basis which corresponds to the treatment of a scattering problem in a discrete representation [4–6]. As a result of such a projection one arrives at a fully algebraic scheme for solving the initial problem in which all the operators and wave functions are represented by matrices and vectors, while the scattering equations are reduced to matrix ones. As an appropriate basis set we have taken the stationary wave packets, which have been introduced by Herman Weyl under a title of eigendifferentials [3]. The detailed description of the whole wave-packet approach and the main results attained can be found in our very recent review paper [4]. Here we briefly outline the basic results related to a construction

[a] e-mail: rubtsova-olga@yandex.ru
[b] e-mail: kukulin@nucl-th.sinp.msu.ru
[c] e-mail: pomeran@nucl-th.sinp.msu.ru

of a stationary wave-packet basis in few-body case, solving the realistic $3N$ scattering problem on ordinary PCs using the Graphics Processing Unit (GPU) technique. We present also a short description for the Discrete Spectral Shift formalism which allows to find the multichannel S-matrix without solving scattering equations at all [6]. The latter technique is realized in a discrete representation only and has close relations to the Lüscher finite volume approach, which is well known in the lattice QCD applications [7, 8].

2 Stationary wave-packet basis for solving scattering problems in a discrete representation

Let us consider some two-body Hamiltonian $h = h_0 + v$ where h_0 is a free Hamiltonian (kinetic energy operator) and v is an interaction, and divide the continuous spectrum of h into set of non-overlapping intervals $\{\mathfrak{D}_i \equiv [\mathcal{E}_{i-1}, \mathcal{E}_i]\}_{i=1}^{N}$. The stationary wave packets (WPs) are constructed as integrals of exact scattering wave functions $|\psi_p\rangle$ over such intervals:

$$|z_k\rangle = \frac{1}{\sqrt{B_k}} \int_{\mathfrak{D}_k} f(p)|\psi_p\rangle dp, \quad B_k \equiv \int_{\mathfrak{D}_k} |f(p)|^2 dp, \tag{1}$$

where $p = \sqrt{2mE}$ are relative momenta, m is the reduced mass of the system, B_k and $f(p)$ are normalization factors and weight functions respectively which are interrelated to each other.

The states (1) are well known as the Weyl's eigendifferentials. The integration over energy (or momentum) intervals is just enough to make normalized wave-functions for the continuum. Then the complete system of eigenfunctions for the Hamiltonian h is constructed from its bound states $|\psi_n\rangle$ and eigendifferentials (see the details in ref. [4]). The Hamiltonian matrix h and also the matrix of the resolvent $g(E) = [E + \mathrm{i}0 - h]^{-1}$ for the Hamiltonian h have explicit diagonal forms in a WP representation.

These properties are valid not only for short-range potentials. In fact, one can build the similar wave packets for a Hamiltonian which includes the long-range Coulomb interaction, for example, and then derive an analytical finite-dimensional representation for the Coulomb resolvent [4] as well.

Useful special examples of stationary wave packets are free WP states $|x_k\rangle$, which are defined just for the free Hamiltonian h_0. In momentum representation, the free WP states take the form of step-like functions:

$$\langle p|x_k\rangle = \frac{f(p)\theta(p \in \mathfrak{D}_k)}{\sqrt{B_i}}, \quad \theta(p \in \mathfrak{D}_k) = \begin{cases} 1, & p \in \mathfrak{D}_k, \\ 0, & p \notin \mathfrak{D}_k, \end{cases} \tag{2}$$

where θ is the Heavyside-type theta-function. In few-body and multidimensional cases the free WP bases are constructed as direct products of two-body ones, so that the few-body model space can be treated as a multidimensional lattice.

In the discrete representation, the wave packets corresponding to different Hamiltonians (including the Coulomb one) can be approximated very conveniently as superpositions of free WPs by using simple rotation matrices. In contrast, in order to find the scattering wave functions in the initial continuous momentum representation, one needs to solve an integral equation of the Lippmann–Schwinger type[1].

If the few-body Hamiltonian can be written in the form of a direct sum of two-body ones:

$$H_M = h_1 \oplus h_1 \oplus \ldots \oplus h_M, \tag{3}$$

[1]Also the discrete pseudostates of the Hamiltonian h obtained in some finite L_2 basis can be treated as approximations just for the WP states. In our practical calculations, we have employed the harmonic oscillator basis and also the basis of Gaussians and complex Gaussians as well (see [4] and the refs. therein).

the WP basis states for H_M can be constructed straightforwardly as direct products of the two-body ones:

$$|Z_{i_1 i_2 \ldots i_M}\rangle = |z_{i_1}\rangle \otimes |\bar{z}_{i_2}\rangle \otimes \ldots \otimes |\bar{\bar{z}}_{i_M}\rangle, \quad (4)$$

where we use bar marks up the z-function to distinguish states corresponding to different subsystems. In the basis (4), the matrix of the Hamiltonian H_M is diagonal and the matrix of the resolvent $G_M(E) = [E + i0 - H_M]^{-1}$ also has an explicit diagonal form.

When studying scattering in a system of three particles, 1,2 and 3, the useful examples for Hamiltonians of a type (3) are the free Hamiltonian $H_0 = h_{0p} \oplus h_{0q}$ and also the channel Hamiltonian $H_1 = (h_{0p} + v_1) \oplus h_{0q}$ defined for a given Jacobi partition (e.g. $\{23\}1$) with momenta (p, q), where h_{0p} and h_{0q} are kinetic energy operators and v_1 is the interaction between particles 2 and 3. Then one can introduce free WP states $|X_{ij}\rangle$ and channel WP states $|Z_{kj}\rangle$ from two-body ones (with account of necessary spin-angular couplings) and also connect the latter with the former by a simple rotation [4] (all the necessary quantum numbers have to be taken into account)[2].

3 GPU-optimization for a practical solution of 3N scattering problem

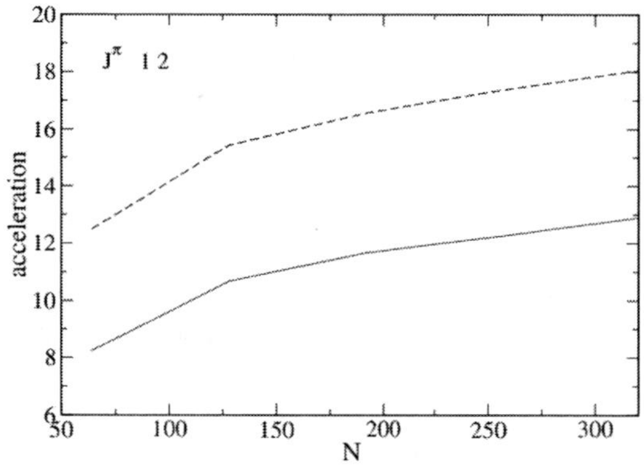

Figure 1. The dependence of the GPU acceleration for the whole solution (solid curve) and the permutation matrix calculation (dashed curve) on the dimension of the basis N (total dimension is N^2) when evaluating the elastic *nd* scattering amplitude for $J = \frac{1}{2}^+$.

One of the main advantages of the discrete WP approach is that the scattering problem can be solved directly in the basis space corresponding to *the channel Hamiltonian* instead of the three-body free Hamiltonian space. So that, an elastic *nd* scattering amplitude and a breakup amplitude can be found from the matrix analog of the Faddeev-type equation (in the AGS form) for the transition matrix $\mathbb{U}$ defined in the channel WP basis [4, 5]. The main computational effort in the numerical scheme for solving the *nd* scattering problem is spent on the calculation of the matrix elements of the particle permutation $\mathbb{P}$ in a channel WP basis. Because all these elements are calculated with the same code and fully independently from each other, the algorithm is very suitable for parallelization and implementation on multiprocessor systems, in particular on a Graphics Prosessing Unit (GPU).

In the Fig. 1, the rate of acceleration η (the ratio of ordinary CPU runtime to the GPU-optimized one) for different steps of the whole numerical scheme when calculating $\frac{1}{2}^+$ partial elastic *nd* amplitude for the Nijmegen I *NN* potential (with account of 18 spin-angular channels) is given.

It is evident from the Figure, that GPU optimization allows to accelerate the numerical solution nearly 10 times. The details of the procedure can be found in Ref. [5].

The obvious advantage of the above numerical scheme is that the permutation matrix does not depend on the incident energy. So that, when one needs scattering observables at many energies

[2]Also this scheme for the basis construction can be generalized to the charged particle case

(see e.g. the Fig. 2), the entire computation time will not noticeably increase because the most time-consuming part of the code is carried out only once for many energy points. The results of calculation

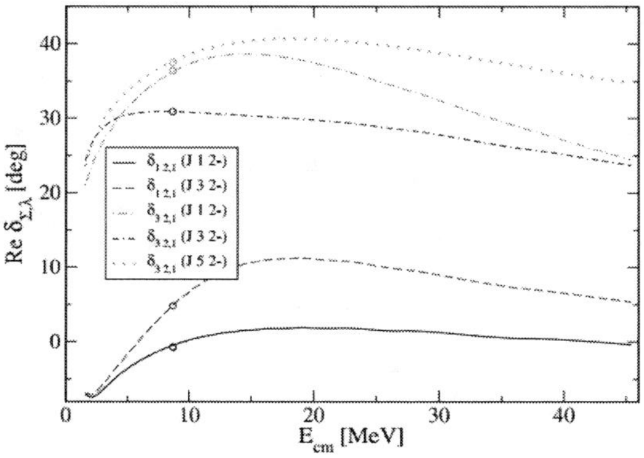

Figure 2. The P-wave partial phase shifts for the elastic *nd* scattering as functions of the neutron incident energy obtained within the wave-packet approach (solid curves) and within the standard Faddeev calculations (circles) [9]. Here Σ, J, π and λ are the total spin, total angular momentum, parity and the neutron orbital momentum respectively.

of elastic *nd* differential cross sections at different neutron incident energies can be found in our papers [4, 5].

4 Discrete spectral shift function formalism

The discrete representation for scattering theory objects opens new possibilities in practical solving scattering problems. Below we briefly report the method based on the spectral shift function formalism, which allows to find multichannel scattering matrix using spectral properties of a free and total Hamiltonian only.

The spectral shift function (SSF) $\xi(E) \equiv \xi(E; h, h_0)$ is an important object in the general spectral theory of perturbations which defines a spectral difference for two Hermitian operators h_0 and $h = h_0 + v$ (e.g., free and total Hamiltonians) both, in discrete and continuous parts of the total spectrum [10, 11]. There are no restrictions that the perturbation v has to be small. It is only assumed that the operator v has a finite trace. The most essential result of the SSF formalism for physical applications is the famous Birman–Krein formula [10], which relates the SSF with the determinant of the scattering operator S: $\det S(E) = \exp(-2\pi i \xi(E))$. In single-channel scattering (e.g. at a fixed angular momentum), this formula implies that the SSF, within a factor $(-\pi)$, is equal directly to a partial phase shift $\delta(E) = -\pi\xi(E)$.

At negative energies, the SSF is a counting function which changes by one unit when crossing each bound state energy [6]. Thus, in conventional scattering theory, the SSF is considered just as some generalized phase shift function. However, the discrete representation for the SSF opens new possibilities in a solution of scattering problems.

To define the SSF in a discretized representation, one has to use a concept of quasi-continuous spectrum introduced by I.M. Lifshits [12] (see also details in ref. [6]). He considers a family of Hermitian operators $\{h_0^{(\alpha)}\}$, where each operator depends on a small parameter α and has a purely discrete spectrum of eigenvalues (EVs) $\{E_j^0(\alpha)\}$ which can be approximated by a single continuous monotonous function $g(u)$:

$$E_j^0(\alpha) = g(j\alpha) + O(\alpha), \quad \text{and} \quad D_j^{(\alpha)} = E_{j+1}^0(\alpha) - E_j^0(\alpha) = \alpha \left[\left. \frac{\mathrm{d}g}{\mathrm{d}u} \right|_{u=j\alpha} + O(\alpha) \right]. \tag{5}$$

When α gets smaller, all the differences $D_j^{(\alpha)}$ get smaller as well and the quasi-continuous spectrum becomes more and more dense. Thus, in the limit $\alpha \to 0$ one has the limiting operator h_0 with a continuous spectrum. By adding the perturbation v to $h_0^{(\alpha)}$ operators, one gets a family of total Hamiltonians $h^{(\alpha)}$ with shifted EVs $\{E_j(\alpha)\}$. One or several EVs of the perturbed spectrum may occur to be below the threshold and thus correspond to the bound states of h, while the rest belongs to the quasi-continuous spectrum of this operator. The following relation between perturbed and unperturbed EVs in quasi-continuous spectrum takes place [6, 12]:

$$E_j(\alpha) = E_j^0(\alpha) + D_j^{(\alpha)}\xi_j + o(\alpha), \tag{6}$$

where $D_j^{(\alpha)}$ is defined in Eq. (5) and $\xi_j = \xi(E_j(\alpha))$ is the spectral shift function defined at discrete energy values. The formula (6) is the basic for the Discrete Spectral Shift (DSS) method. It results in a very simple approximate expression for the partial phase shift:

$$\delta(E_j(\alpha)) = -\pi\xi(E_j(\alpha)) \approx -\pi\frac{E_j(\alpha) - E_j^0(\alpha)}{D_j^{(\alpha)}}. \tag{7}$$

This method may be applied for any continuum discretization procedure. For example, when one considers a particle scattering in a box and the box size R increases to infinity ($\alpha \sim \frac{1}{R}$) [6]. Another useful case is the solution of the scattering problem in some finite L_2 basis when the parameter α is decreasing with increasing basis dimension N [6].

In a multichannel case one can evaluate *eigen phase shifts* from the differences of free and total Hamiltonian eigenvalues [6]. Here the discretized spectrum of the multichannel free Hamiltonian has to be degenerated with the multiplicity corresponding to the number of open channels at the current energy. In that case, the spectrum of the total multichannel Hamitonian **h** will consist of series of split levels [6]. Then one can define the different spectral shift functions in the eigenchannel representation (each for a separate eigen channel) and relate them with eigen phases of the multichannel problem. Finally, these eigen phases are defined again from the discrete spectral shifts:

$$\delta^{(\varkappa)}(E_j) \approx -\pi\frac{E_j^{(\varkappa)} - E_j^0}{D_j}, \quad \varkappa = 1,\ldots,d, \tag{8}$$

where $\varkappa$ is the eigenchannel index, E_j^0 is the eigenvalue of the free Hamiltonian $\mathbf{h}_0$ with the multiplicity d and $E_j^{(\varkappa)}$ is the eigenvalue of the total Hamiltonian **h**.

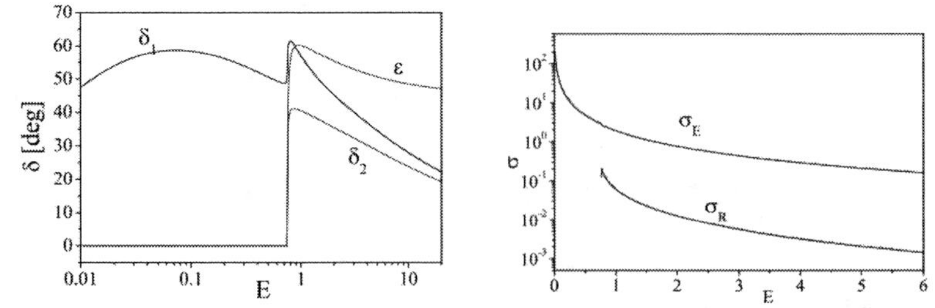

Figure 3. The eigen phases $\delta_{1,2}$, the mixing angle ϵ (left), and the elastic and reaction cross sections (right) found via the DSS method for the two-channel $e - H$ scattering (solid curves). The dash-dotted curves on the right panel reflect results of Ref. [13].

In Fig. 3, our results for the model two-channel $e + H$ scattering problem[3] found in the DSS approach in comparison with the results of Ref. [13] are given[4].

[3]Here we used a coupled-channel potential including $1S - 2S$ excitation of the hydrogen atom. The model parameters of interaction has been taken from Ref. [13].

[4]It is clear that the DSS technique represents the reaction cross section very well in a wide energy region except only one or two energy points just near the threshold of the second channel.

The detailed description of the DSS approach and numerical examples for multichannel scattering can be found in Refs. [6].

5 Summary

We have briefly shown that the discrete WP representation is a very useful tool for solving realistic few-body scattering problems. All details of the approach and the results can be found in the review paper [4].

It should be emphasized here that the discrete spectral shift approach is rather similar to the finite volume approaches and particulary to the Lüscher approach [7, 8]. However, in the DSS, there are no restrictions according to a small values of a perturbation or to low momentum. Also we do not need a special form of a finite volume boundary. Moreover, the discrete SSF approach is based on some fundamental spectral properties of operators and does not depend on the concrete representation. So that, the finite volume relations for partial phase shifts can be considered as some partial realization of the DSS approach. This fact might be important for the multichannel case, because finite volume approaches may use the DSS formalism for this case. The detailed analysis of the above aspects will be published elsewhere.

Acknowledgements

This work has been supported partially by the Russian Foundation for Basic Research, grant No. 13-02-00399.

References

[1] L.D. Faddeev, Sov. Phys.- JETP **12**, 1014 (1961); O.A. Yakubovsky, Sov. J. Nuclear. Phys. **5**, 937 (1967).

[2] J. Carbonell, A. Deltuva, A.C. Fonseca, R. Lazauskas, Progr. Part. Nucl. Phys. **74**, 55 (2014).

[3] W. Greiner, *Quantum Mecanics: An introduction* (Fourth Edition, Springer, 2001).

[4] O.A. Rubtsova, V.I. Kukulin, V.N. Pomerantsev, Ann. Phys. **360**, 613 (2015).

[5] V.N. Pomerantsev, V.I. Kukulin, O.A. Rubtsova, Phys. Rev. C **89**, 064008 (2014).

[6] O.A. Rubtsova et al., Phys. Rev. C **81**, 064003 (2010); O.A. Rubtsova et al., Phys. At. Nucl. **77**, 486 (2014).

[7] M. Lüscher, Nucl. Phys. **B354**, 531 (1991).

[8] Peng Guo et al., Phys. Rev. D **88**, 014501 (2013).

[9] W. Glöckle, H. Witała, D.Hüber, H. Kamada, J. Golack, Phys. Rep. **274**, 107 (1996).

[10] D.R. Yafaev, *Mathematical scattering theory: General theory*,(Amer. Math. Soc., Providence, RI, 1992).

[11] M.S.Birman, A.B. Pushnitsky, Integr. Equ. Oper. Theory **30**, 191 (1998).

[12] I.M. Lifshits, Sov. Phys.-JETP **17**, 1076 (1947) (in Russian).

[13] B.H. Brandsen, A.T. Stelbovics, J. Phys. B **17**, 1874 (1984).

EPJ Web of Conferences **113**, 03016 (2016)
DOI: 10.1051/epjconf/201611303016

Towards a Faddeev-AGS description of (d, p) reactions with heavy nuclei: Regularizing integrals with Coulomb functions.

V. Eremenko[1,5,a], L. Hlophe[1], Ch. Elster[1,b], F. M. Nunes[2], I. J. Thompson[3], G. Arbanas[4], and J. E. Escher[3]

TORUS Collaboration[†] (http://reactiontheory.org)

[1] *INPP and Department of Physics and Astronomy, Ohio University, Athens, Ohio 45701, USA*
[2] *NSCL and Dept. of Physics and Astronomy, Michigan State University, East Lansing, Michigan 48824, USA*
[3] *Lawrence Livermore National Laboratory, L-414, Livermore, California 94551, USA*
[4] *Reactor and Nuclear Systems Division, Oak Ridge National Laboratory, Oak Ridge, TN 37831, USA*
[5] *SINP M.V. Lomonosov Moscow State University, Moscow, 119991, Russia*

Abstract. The repulsive Coulomb force poses severe challenges when describing (d, p) reactions for highly charged nuclei as a three-body problem. Casting Faddeev-AGS equations in a Coulomb basis avoids introducing screening of the Coulomb force. However, momentum space partial-wave t-matrix elements need to be evaluated in this basis. When those t-matrices are separable, the evaluation requires the folding of a form factor, depending on one momentum variable, with a momentum space partial-wave Coulomb function, which has a singular behavior at the external momentum q. We developed an improved regularization scheme to calculate Coulomb distorted form factors as the integral over the Coulomb function and complex nuclear form factors.

1 Introduction

Direct reactions involving rare isotopes offer an important tool for understanding the structure of such nuclei. Deuteron induced reactions are attractive from the experimental perspective, since deuterated targets are readily available. Theoretically they are attractive, since the scattering problem can be reduced to an effective three-body problem, which can be solved exactly using Faddeev techniques. The momentum space Faddeev equations (here in the Alt-Grassberger-Sandhas form) have successfully been solved for (d, p) reactions involving light nuclei [1]. However, the screening technique employed to handle the Coulomb force encounters technical difficulties, when applied to (d, p) reactions with heavier nuclei [2]. An alternative to the screening procedure is a solution of the Faddeev-AGS equations in the Coulomb basis. This was suggested and carried out in Ref. [3] using real two-body transition operators in separable form, neglecting spin degrees of freedom.

Casting Faddeev-AGS equations in the Coulomb basis requires the evaluation of momentum-space partial-wave matrix elements of two-body transition operators,

$$t_l^C(q', q, E) = \int dp'\, dp\, \psi^C_{l,q',\eta'}(p')^\dagger\, t_l(p', p, E)\, \psi^C_{l,q,\eta}(p). \tag{1}$$

[a] e-mail: to@vsl.name
[b] e-mail: elster@ohio.edu

If $t_l(E)$ is separable, one needs to consider

$$t_l^C(q,q',E) = \sum_{zy} u_{l,z}^C(q)\,\lambda_{l,zy}(E)\,u_{l,y}^C(q')^\dagger. \tag{2}$$

In this case only the folding of a form factor $u_l(p)$, depending on one momentum variable, with partial-wave Coulomb function $\psi_{l,q}^C(p)$, for which we developed the numerical procedures in Ref. [6], is required:

$$u_{l,z}^C(q) = \int \frac{dp\,p^2}{2\pi^2} u_{l,z}(p)\psi_{l,q,\eta}^C(p)^*, \qquad \text{and} \qquad u_{l,y}^C(q')^\dagger = \int \frac{dp\,p^2}{2\pi^2} u_{l,y}(p)\psi_{l,q',\eta}^C(p). \tag{3}$$

Both integrals have an oscillatory singularity in the point $p = q$, where q is the external momentum. The partial-wave Coulomb functions are given by

$$\psi_{l,q,\eta}^C(p) \overset{p\to q}{=} \psi_{l,q,\eta}^C(\chi) \overset{\chi\to 0}{=} \mathcal{A}(q,l,\eta)\left[\frac{\mathcal{B}(\chi,q,l,\eta)}{(\chi+i0)^{1+i\eta}} - \frac{\mathcal{B}(\chi,q,l,\eta)^*}{(\chi-i0)^{1-i\eta}}\right], \qquad \chi \equiv p-q. \tag{4}$$

To evaluate $u_{l,z}^C(q)$, one needs to regularize the integral in Eq. (3). In Ref. [3] this regularization is performed by using a Gel'fand-Shilov (see Ref. [4]) technique for real form factors. When working with form factors describing complex nucleon-nucleus potentials, the regularization scheme must be constructed for complex form factors.

2 Regularization Procedure

Here we present a regularization scheme for calculating Coulomb distorted form factors as integrals over a Coulomb function and a complex form factor, which differs from the scheme presented in Ref. [5]. The key point of the work of Ref. [5] is that the integral is regularized only in a tiny neighborhood ($\Delta \sim 10^{-6}$ fm^{-1}) around q. That allows that the regularized singular part of the folding integral can be reduced to an analytic expression. However, the non-regularized parts of the integral must be calculated quite close to the value of q, which requires a lot of integration mesh points independent of the numerical integration quadrature used. For example, we used thousands of Gauss-Legendre quadrature points to obtain results converged to 3 significant figures.

To avoid using an excessive number of integration points, we developed here the new version of the regularization scheme, where we apply the regularization in a small, but finite neighborhood ($\Delta \sim 10^{-3}$ fm^{-1}) around q. Now the non-regularized integrals are converging much faster to an even better accuracy (see Fig. 1), while the regularized part is still mostly dominated by the analytical terms. Despite the fact that we need to calculate the regularized integral, we do not need to calculate it as accurately, as we had to calculate the non-regularized parts in Ref. [5]. Thus, the required amount of CPU time is greatly reduced.

Similar to Ref. [5], we start by splitting the folding integral into regularized and non-regularized parts,

$$u_{l,y}^C(q')^\dagger = I_0^{q'-\Delta} + I_{q'-\Delta}^{q'+\Delta} + I_{q'+\Delta}^{\infty}, \qquad I_a^b = \int_a^b \frac{dp\,p^2}{2\pi^2} u_{l,y}(p)\psi_{l,q',\eta}^C(p). \tag{5}$$

We first focus on the regularized part $I_{q-\Delta}^{q+\Delta}$, which we rewrite by substituting Eq. (4) into Eq. (5) as

$$I_{q'-\Delta}^{q'+\Delta} \equiv \mathcal{I}_{-\Delta}^{\Delta} = \frac{\mathcal{A}(q',l,\eta)}{2\pi^2}\left(\mathcal{I}_+ - \mathcal{I}_-\right), \qquad \mathcal{I}_\pm \equiv \int_{-\Delta}^{\Delta} d\chi \frac{\phi_\pm(\chi,q',l,\eta)}{(\chi \pm i0)^{1\pm i\eta}}, \tag{6}$$

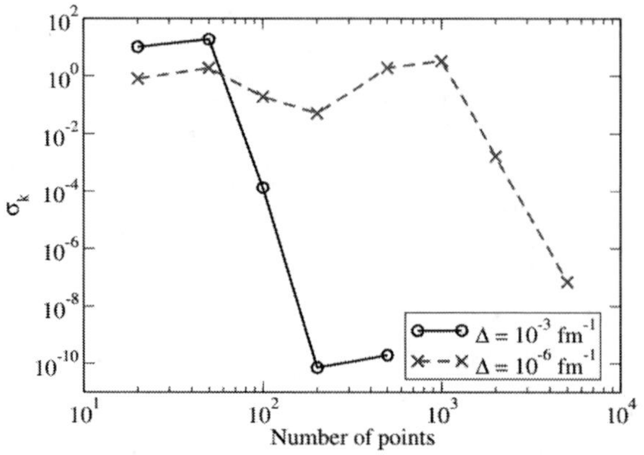

Figure 1. The error plot of the non-regularized integrals I_a^b with the same lower limit $a = 0.4$ fm^{-1}, but different upper limits: [solid black] $b = 0.499$ fm^{-1} (i.e. $\Delta = 1 \cdot 10^{-3}$ fm^{-1}), [dashed red] $b = 0.499999$ fm^{-1} ($\Delta = 1 \cdot 10^{-6}$ fm^{-1}). The parameters are: $q = 0.5$ fm^{-1}, $l = 0$, $\eta = 5.3$. These values of q and η correspond to the $p + {}^{208}$Pb system with $E_{c.m.} = 6$ MeV. See Sects. 2 and 3 for discussion and definition of the relative error σ_k.

where

$$\phi_+(\chi, q', l, \eta) \equiv u_{l,y}(q' + \chi)(q' + \chi)^2 \mathcal{B}(\chi, q', l, \eta), \tag{7}$$

$$\phi_-(\chi, q', l, \eta) \equiv u_{l,y}(q' + \chi)(q' + \chi)^2 \mathcal{B}(\chi, q', l, \eta)^*. \tag{8}$$

To apply the regularization scheme, we split the integrals $\mathcal{I}_\pm$ and take the limit $\pm i0$,

$$\mathcal{I}_\pm = \int_0^\Delta d\chi \frac{\phi_\pm(\chi, q', l, \eta)}{\chi^{1\pm i\eta}} + e^{\pi\eta} \int_0^\Delta d\chi \frac{\phi_\pm(-\chi, q', l, \eta)}{\chi^{1\pm i\eta}}. \tag{9}$$

While taking the limit, we used the following property,

$$(\chi \pm i0)^{-1\mp i\eta} = |\chi|^{-1\mp i\eta}(e^{\pm i\pi})^{-1\mp i\eta} = -e^{\pi\eta}|\chi|^{-1\mp i\eta}. \tag{10}$$

By applying the Gel'fand-Shilov regularization technique from Ref. [4], we obtain

$$\mathcal{I}_\pm = \int_0^\Delta d\chi \frac{\widetilde{\phi}_\pm(\chi, q', l, \eta)}{\chi^{1\pm i\eta}} + e^{\pi\eta} \int_0^\Delta d\chi \frac{\widetilde{\phi}_\pm(-\chi, q', l, \eta)}{\chi^{1\pm i\eta}} \mp \frac{\phi_\pm(0, q', l, \eta)}{i\eta}\left(1 - e^{\pi\eta}\right)\Delta^{\mp i\eta} + \frac{\phi'_\pm(0, q', l, \eta)}{1 \mp i\eta}\left(1 + e^{\pi\eta}\right)\Delta^{1\mp i\eta}, \tag{11}$$

where

$$\widetilde{\phi}_\pm(\chi, q', l, \eta) \equiv \phi_\pm(\chi, q', l, \eta) - \phi_\pm(0, q', l, \eta) - \phi'_\pm(0, q', l, \eta)\chi, \tag{12}$$

$$\phi'_\pm(0, q', l, \eta) = \left.\frac{d\phi_\pm(\chi, q', l, \eta)}{d\chi}\right|_{\chi=0}. \tag{13}$$

To calculate $\phi'_\pm(0, q', l, \eta)$, we also need $du_{l,y}(p)/dp$. Taking the limit $\Delta \to +0$, one arrives at the expressions from Ref. [5].

By using the expressions from Eq. (11) and Eq. (6), we can compute the regularized part $I_{q-\Delta}^{q+\Delta}$ for any given finite Δ. Then, using Eq. (5) we evaluate $u_{l,y}^C(q')^\dagger$, while $u_{l,z}^C(q)$ is given by

$$u_{l,z}^C(q) = \left[\int \frac{dp\, p^2}{2\pi^2} u_{l,z}(p)^* \psi_{l,q,\eta}^C(p)\right]^*. \tag{14}$$

To calculate $\phi'_{\pm}(0, q', l, \eta)$, the expression for $d\mathcal{B}/d\chi$ is required. By using Eq. (B9) from Ref. [5], since $p = q' + \chi$, Eq. (15.5.1) from Ref. [7], and by utilizing the property of the hypergeometric function ${}_2F_1(\ldots;0) = 1$, we obtain

$$\left.\frac{d\mathcal{B}(\chi, q, l, \eta)}{d\chi}\right|_{\chi=0} = e^{-i\sigma_l}\Gamma(1+i\eta)(-3+i\eta)2^{2l-2+i\eta}q^{l-3+i\eta}. \tag{15}$$

Here we substituted q' by q to simplify the expression.

3 Results, Discussions and Outlook

To prevent the unreasonably large computational times for calculating Coulomb distorted form factors, we improved the regularization scheme of Ref. [5]. This new scheme converges much faster, as the number of integration points increases. To demonstrate it, we present the error plot on Fig. 1. The integral I_a^b (see caption for a and b) was calculated by using Gauss-Legendre quadrature with 10, 20, 50, 100, 200, 500, 1 000, 2 000, and 5 000 points. Starting from the second step ($k = 2$, i.e. 20 points), the relative discrepancy was calculated,

$$\sigma_k \equiv \left(\frac{|\operatorname{Re}(I_a^b)_k - \operatorname{Re}(I_a^b)_{k-1}|}{|\operatorname{Re}(I_a^b)_k|} + \frac{|\operatorname{Im}(I_a^b)_k - \operatorname{Im}(I_a^b)_{k-1}|}{|\operatorname{Im}(I_a^b)_k|}\right)\Big/ 2, \tag{16}$$

where $(I_a^b)_k$ is the value of I_a^b, obtained on the k-th step. For the sake of illustration, here we were using simple Yamaguchi-style real form factor with constant 2.7 fm^{-1}. It should be noted, that for this large value of $\eta = 5.3$ and small value of $q = 0.5$ fm^{-1} (the corresponding system is $p + {}^{208}$Pb at $E_{c.m.} = 6$ MeV) the final result is being calculated by subtraction of two large numbers, and the integral I_a^b must be computed with $\sigma_k \lesssim 10^{-4}$ to get even the first digit of the final result. This new regularization scheme will be essential for successful progress in the numerical implementation of the Faddeev-AGS equations in Coulomb basis including spin degrees of freedom.

Acknowledgements

This work was performed in part under the auspices of the US Department of Energy, Office of Science of Nuclear Physics, under the topical collaborations in nuclear theory program No. DE-SC0004087 (TORUS Collaboration), and under Contract DE-FG02-93ER40756 with Ohio University.

References

[1] A. Deltuva, Phys. Rev. C **79**, 054603 (2009).
[2] N. J. Upadhyay *et al.* Phys. Rev. C **85**, 054621 (2012).
[3] A. M. Mukhamedzhanov *et al.* Phys. Rev. C **86**, 034001 (2012).
[4] I. Gel'fand and G. Shilov, *Generalized Functions, Volume 1: Properties and Operations* (Academic, New York, 1964) Chapter 1, Sections 1.7, 2.2 (Example 3), 3.2, 3.8.
[5] N. J. Upadhyay *et al.* (the TORUS Collaboration), Phys. Rev. C **90**, 014615 (2014).
[6] V. Eremenko *et al.* (the TORUS Collaboration), Comp. Phys. Comm. **187**, 195 (2015).
[7] *NIST Digital Library of Mathematical Functions* (`http://dlmf.nist.gov/`, Release 1.0.10 of 2015-08-07, online companion to [8]).
[8] F. W. J. Olver, D. W. Lozier, R. F. Boisvert, and C. W. Clark, editors, *NIST Handbook of Mathematical Functions* (Cambridge University Press, New York, NY, 2010, Print companion to [7]).

EPJ Web of Conferences **113**, 03017 (2016)
DOI: 10.1051/epjconf/201611303017

The general configuration-space Faddeev formalism for studying pd scattering

Vladimir M Suslov[1,a], Mikhail A Braun[2], Igor Filikhin[1], Ivo Slaus[3], and Branislav Vlahovic[1]

[1] *North Carolina Central University, 1801 Fayetteville St., Durham, NC, USA*
[2] *Dep. of High Energy physics, University of St. Petersburg, St. Petersburg, 199034 Russia*
[3] *Rudjer Boskovich Institute, Zagreb, 10000 Croatia*

Abstract. The configuration-space Faddeev equations are derived for p-d scattering taking into account the difference in interaction between the participant particles. Appropriate modifications have been made in the well-known configuration-space equations for n-d scattering. To show the effect of these modifications, the s-wave calculations are performed for bound state and scattering problems. We model the charge symmetry breaking effect for ^{3}H and ^{3}He with a modified Malfliet-Tjon MT I-III potential. Results obtained for elastic n-d and p-d scattering at E_{lab}=14.1 MeV are compared with our prediction (Ref. [1]) and those of the Los-Alamos/Iowa group (Ref. [2]) .

1 Introduction

The isotopic formalism was developed for the study of neutron-deuteron scattering in the framework of the configuration space Faddeev equations (Ref. [3]). Charge-independence breaking in the three-nucleon system was investigated in elastic neutron-deuteron and breakup prosesses (Ref. [4]). Here we study proton-deuteron scattering. Presence of the electromagnetic interaction requires one to consider the neutron and proton to be different particles and precludes literal use of the isotopic formalim of (Ref. [3]). So the FNNM equations used in (Ref. [5]) have to be changed.

Taking the neutron as particle 1 and protons as particles 2 and 3 we have the requirement $\Psi(1,2,3) = -\Psi(1,3,2)$. To satisfy this condition, we present Ψ in terms of Faddeev components as

$$\Psi(1,2,3) = \Phi_1(1,2,3) + \Phi_2(2,3,1) - \Phi_2(3,2,1), \tag{1}$$

where it is understood that in $\Phi(i,k,l)$ particles are grouped as $i + (kl)$ and Φ_1 is antisymmetric in the last pair of arguments: $\Phi_1(1,2,3) = -\Phi_1(1,3,2)$. As function $\Phi_2(2,3,1)$ has no definite properties under interchange $3 \leftrightarrow 1$ we encounter permutations which are not cyclic $P_{12}(231) = (321)$, $P_{13}(123) = (321)$. In terms of these operators and operators $P^{\pm}$ we obtain for the independent components Φ_1 and Φ_2 a system

$$\begin{aligned}
\big(E - \Delta - v_1(2,3)\big)\Phi_1(1,2,3) &= v_1(2,3)\big(P^{-}\Phi_2(1,2,3) - P_{13}\Phi_2(1,2,3)\big),\\
\big(E - \Delta - v_2(3,1)\big)\Phi_2(2,3,1) &= v_2(3,1)\big(P^{+}\Phi_1(2,3,1) - P_{12}\Phi_2(2,3,1)\big).
\end{aligned} \tag{2}$$

[a] e-mail: vsuslov@nccu.edu

Here $v_1(2,3)$ is a sum of the Coulomb and nuclear potentials and $v_2(3,1)$ is the pure nuclear potential:

$$v_1(2,3) = v_c(2,3) + v_{pp}(2,3), \quad v_2(3,1) = v_{pn}(3,1). \tag{3}$$

The final pair of equations is essentially the same as for the *nd* case with changes described above.

$$\Big[E + \frac{\hbar^2}{m}(\partial^2_{x_1} + \partial^2_{y_1}) - v^{\lambda l}_{\alpha}(x_1,y_1)\Big]\Phi^{\lambda_0,s_0,M_0}_{1,\alpha}(x_1,y_1) = \sum_\beta \Big[v_c(x_1) + v_{pp}(x_1)\Big]_{\alpha\beta}\Big[\Phi^{\lambda_0 s_0,M_0}_{1\beta}(x_1,y_1)+$$

$$\frac{1}{2}\int_{-1}^{1} du \sum_\gamma \Big(g^{(-)}_{\beta\gamma}(y_1/x_1,u)\Phi^{\lambda_0,s_0,M_0}_{2,\gamma}(x_2,y_2) - g^{(13)}_{\beta\gamma}(y_1/x_1,u)\Phi^{\lambda_0,s_0,M_0}_{3,\gamma}(x_3,y_3)\Big), \tag{4}$$

where $(x_2,y_2) = P^-(x_1,y_1)$ and $(x_3,y_3) = P_{13}(x_1,y_1)$.

$$\Big[E + \frac{\hbar^2}{m}(\partial^2_{x_2} + \partial^2_{y_2}) - v^{\lambda l}_{\alpha}(x_2,y_2)\Big]\Phi^{\lambda_0,s_0,M_0}_{2,\alpha}(x_2,y_2) = \sum_\beta \Big[v_{pn}(x_2)\Big]_{\alpha\beta}\Big[\Phi^{\lambda_0 s_0,M_0}_{2\beta}(x_2,y_2)+$$

$$\frac{1}{2}\int_{1}^{1} du \sum_\gamma \Big(g^{(+)}_{\beta\gamma}(y_2/x_2,u)\Phi^{\lambda_0,s_0,M_0}_{1,\gamma}(x_1,y_1) - g^{(12)}_{\beta\gamma}(y_2/x_2,u)\Phi^{\lambda_0,s_0,M_0}_{3,\gamma}(x_3 y_3)\Big), \tag{5}$$

where $(x_1,y_1) = P^+(x_2,y_2)$ and $(x_3,y_3) = P_{12}(x_2,y_2)$. In these formulas the multi index $\alpha = \{l,\sigma,j,s,\lambda\}$, $g^{(\pm)}_{\alpha\alpha'}$ and $g^{(ik)}_{\alpha\alpha'}$ are representatives of the operators $2P^\pm$ and $2P_{ik}$ in the MGL basis (Ref. [6]).

2 s-wave approach. Elastic scattering

In the *s*-wave approach there exists a single equation in the spin-quartet case for quantum state $\alpha = \{0,1,1,3/2,0\}$ and our new results for n-d and p-d elastic amplitudes at E_{lab}=14.1 MeV calculated with the Malfliet-Tjon MT-I-III potential do not practically differ from our predictions (Ref. [1]) and those of the Los-Alamos/Iowa group (Ref. [2]) and we do not present them here. However in the spin-doublet case there exist three equations for quantum states $\alpha_1 = \{0,0,0,1/2,0\}$ and $\alpha_2 = \{0,1,1,1/2,0\}$, one for Φ_{1,α_1} and two for Φ_{2,α_i}, (i=1,2). For the ppn system, a set of equations is writen as:

$$\Big[E + \frac{\hbar^2}{2m}(\partial^2_{x_1} + \partial^2_{y_1}) - v^{00}_q\Big]\Phi_{1,\alpha_1}(x_1,y_1) = [v^c(x_1) + v_{pp}(x_1)]_{\alpha_1\alpha_1}\Big[\Phi_{1,\alpha_1}(x_1,y_1) + \frac{1}{2}\int_{-1}^{1} du h^0_{0000}(y_1/x_1,u)$$

$$\times\Big[\Big(-\frac{1}{2}\Phi_{2,\alpha_1}(x_2,y_2) - \frac{1}{2}\Phi_{2,\alpha_1}(x_3,y_3)\Big) + \Big(\frac{\sqrt{3}}{2}\Phi_{2,\alpha_2}(x_2,y_2) + \frac{\sqrt{3}}{2}\Phi_{2,\alpha_2}(x_3,y_3)\Big)\Big]\Big] \tag{6}$$

$$\Big[E + \frac{\hbar^2}{2m}(\partial^2_{x_2} + \partial^2_{y_2}) - v^{00}_q\Big]\Phi_{2,\alpha_1}(x_2,y_2) = [v_{pn}(x_2)]_{\alpha_1\alpha_1}\Big[\Phi_{2\alpha_1}(x_2,y_2) + \frac{1}{2}\int_{-1}^{1} du h^0_{0000}(y_2/x_2,u)$$

$$\Big(-\frac{1}{2}\Phi_{1,\alpha_1}(x_1,y_1) - \frac{1}{2}\Phi_{2,\alpha_1}(x_3,y_3) - \frac{\sqrt{3}}{2}\Phi_{2,\alpha_2}(x_3,y_3)\Big)\Big] \tag{7}$$

$$\Big[E + \frac{\hbar^2}{2m}(\partial^2_{x_2} + \partial^2_{y_2}) - v^{00}_q\Big]\Phi_{2,\alpha_2}(x_2,y_2) = [v_{pn}(x_2)]_{\alpha_2\alpha_2}\Big[\Phi_{2\alpha_2}(x_2,y_2)$$

$$+\frac{1}{2}\int_{-1}^{1} du h^0_{0000}(y_2/x_2,u)\Big(\frac{\sqrt{3}}{2}\Phi_{1,\alpha_1}(x_1,y_1) - \frac{\sqrt{3}}{2}\Phi_{2,\alpha_1}(x_3,y_3) + \frac{1}{2}\Phi_{2,\alpha_2}(x_3,y_3)\Big)\Big]. \tag{8}$$

In the s-wave approach the functions $g^{(\pm)}_{\alpha\alpha'}$ and $g^{(ik)}_{\alpha\alpha'}$ are reduced to functions $h^0_{0000}(y_i/x_i, u) = x_i y_i/(x_k y_k)$. Here (x_k, y_k) are the coordinates of the integrand components $\Phi_{1,2}$ and $(k \neq i)$.

In the deuteron domain (x_2 finite, $y_2 \to \infty$) the asymptotic condition for the component Φ_2 corresponding to elastic channel:

$$\Phi^{0,1/2,1/2}_{2,\alpha_2}(x_2, y_2) \sim \left\{\delta_{\sigma 1}\delta_{J1} e^{i\Delta^c_0} F^c_0(qy_2) + e^{-i\Delta^c_0}\left(G^c_0(qy_2) + iF^c_0(qy_2)\right) a^{1/2}_{01/2,01/2}\right\}\psi_l(x_2). \tag{9}$$

In formula (9) $F_0(qy_2)$ and $G_0(qy_2)$ are the regular and irreqular Coulomb functions and $\psi_l(x_2)$ is the s - wave component of the deuteron wave function ($l = 0$). Amplitudes Φ_{1,α_1} and Φ_{2,α_1} have zero initial conditions and zero elastic asymptotics.

In the breakup domain we have for Φ_{1,α_1} and $\Phi_{2,\alpha_{1,2}}$ the asymptotics:

$$\Phi^{0\frac{1}{2}\frac{1}{2}}_{1,\alpha_1} \sim e^{-W_1(\theta_1)} A^{0\frac{1}{2}\frac{1}{2}}_{1,\alpha_1}(\theta_1),\ \Phi^{0\frac{1}{2}\frac{1}{2}}_{2,\alpha_{1,2}} \sim \mathcal{E}_0 A^{0\frac{1}{2}\frac{1}{2}}_{2,\alpha_{1,2}}(\theta_2),\ \mathcal{E}_0 = 2\pi \int_{-1}^{1} du e^{iW_2(\hat{\mathbf{x}}_2,\hat{\mathbf{y}}_2)},\ \ u = cos(\hat{\mathbf{x}}_2\hat{\mathbf{y}}_2). \tag{10}$$

In these formulas the Coulomb distorted phases W_1 and W_2 are as following

$$W_1(\theta) = -\frac{1}{2\sqrt{E}}\frac{me^2}{\hbar^2}\frac{\ln(2\sqrt{E}X)}{\cos\theta},\quad \cos\theta = \frac{x_1}{X}, \tag{11}$$

and

$$W_2(\hat{\mathbf{x}}_2, \hat{\mathbf{y}}_2) = -\frac{1}{2\sqrt{E}}\frac{me^2}{\hbar^2}\frac{X}{|\mathbf{x}_2/2 + \sqrt{3}\mathbf{y}_2/2|}\ln(2\sqrt{E}X),\quad X = \sqrt{x_i^2 + y_i^2}. \tag{12}$$

The two doublet amplitudes are given as follows

$$\mathcal{A}_{\alpha_1}(\theta_2) = A_{2,\alpha_1}(\theta_2) + \frac{1}{2}\int_{-1}^{1} du h^0_{0000}(y_2/x_2, u)$$
$$\left(g_{11}A_{1,\alpha_1}(\theta_1) - g^{(12)}_{11}A_{2,\alpha_1}(\theta_3) - g^{(12)}_{12}A_{2,\alpha_2}(\theta_3)\right) \tag{13}$$

$$\mathcal{A}_{\alpha_2}(\theta_2) = A_{2,\alpha_2}(\theta_2) + \frac{1}{2}\int_{-1}^{1} du h^0_{0000}(y_2/x_2, u)$$
$$\left(g_{21}A_{1,\alpha_1}(\theta_1) - g^{(12)}_{21}A_{2,\alpha_1}(\theta_3) - g^{(12)}_{22}A_{2,\alpha_2}(\theta_3)\right). \tag{14}$$

Equations (6 - 8) violate the isospin symmetry because of distinctions between nn, pp and np forces. The charge asymmetry is obtained by allowing the strengths of the central *nn* and *pp* forces to be different from *np* one. We modify MT-I-III potentials to reproduce singlet n-n and p-p scattering data by scaling the *pn* potential by factors 0.982 and 0.9745 for *nn* and *pp*, respectively. While MT-I-III uses for a_{np}, a_{nn} and a_{pp} the value of scattering length is -23.5 fm, we modify the potential to produce their scattering lenghts to agreement with experimental data (Ref. [10]).

The accuracy of this adjustment procedure is checked by calculating binding energies for ^{3}H and ^{3}He using MT-I-III potential and its modifications. Our new results and previous ones from Ref. [7] and Ref. [8] obtained in isospin formalism are given in table 1. Our new results calculated applying a new set of three Faddeev equations with modified MT-I-III NN potentials are in a good agreement with those from Ref. [7] and Ref. [8]. Results for our Coulomb energy ΔB_c and CSB energy ΔB(CBS) are given in table 1.

In the *s*-wave approach the value of ΔB_c is 661 keV slightly different from the result of 693 keV (Ref. [9]). Our result for the charge-symmetry breaking energy ΔB(CSB) is 61 keV close to 71 keV evaluated by Miller et al. [10]. Our new results for phase shifts and elasticities for n-d

Table 1. $B(^3H)$ and $B(^3He)$ binding energies (in MeV), the Coulomb energy ΔB_c (in keV), CSB effect for energy ΔB(CSB) (in keV). The results of Ref. [7] (Ref. [8]) are given in brackets (square brackets). m_n (m_p) is the mass of neutron (proton).

		m_n	m_p	$B(^3H)$	$B(^3He)$	ΔB_c	ΔB(CSB)
MT-I-III		939.0	939.0	8.545			
		–	–	(8.535)	(7.868)		
		–	–	[8.54]	[7.88]		
		939.565	938.272	8.548	7.882		
Modified MT-I-III	nn	–	–	8.396	7.735	661	
	pp	–	–		7.674	–	61

Table 2. Spin-doublet case. n-d and p-d elastic shifts and inelasticities at E_{lab}=14.1 MeV.

	n - d			p - d		
	MT-I-III	[1]	modified MT-I-III	MT-I-III	[1]	modified MT-I-III
$\delta(deg)$	106.16	105.47	105.56	111.05	108.06	110.76
η	0.4653	0.4649	0.4744	0.533	0.4929	0.536

and p-d breakup scattering at E_{lab}=14.1 MeV calculated assuming the neutrons and protons to be distinguishable particles are presented in table 2. One concludes that CSB is visible in phase shifts and inelasticities for neutron-deuteron and proton-deuteron scattering at E_{lab}=14.1 MeV. However these results have been obtained applying partly artificial procedure for constructing s-wave singlet nn and pp components of MT-I-III potential. Therefore we are currently extending our studies using the charge dependent AV18 NN potential.

Acknowledgements

This work is supported by the NSF (HRD-1345219).

References

[1] V.M. Suslov and B. Vlahovic, Phys. Rev. C **69** (2004) 044003.

[2] J.L. Friar, G.L. Payne, W. Glöckle, D. Hüber, and H. Witala, Phys. Rev. C **51** (1995) 2356.

[3] A.A. Kvitsinsky, Yu.A. Kuperin, S.P. Merkuriev, A.K. Motovilov, S.L. Yakovlev, EChaYa, 17 (1986) 267-317.

[4] H. Witala, W. Glöckle, and H. Kamada, Phys. Rev. C **43** (1991) 1619.

[5] V.M. Suslov, M.A. Braun, I. Filikhin, B. Vlahovic, I. Slaus, Few-Body Syst. (2011) 50; 267.

[6] S.P. Merkuriev, C. Gignoux, A. Laverne, Annals of Phisics **99**, 30-71 (1976).

[7] J.L. Friar, B.F. Gibson, G.L. Payne, Phys. Rev. C **24** (1981) 2279.

[8] J. Bernabeu, V.M. Suslov, T.A. Strizh, S.I. Vinitsky, Hyperfine interaction 110/102 (1996) 391-399

[9] S.A. Coon and R.C. Barrett, Phys. Rev. C **36** (1987) 2189.

[10] G.A. Miller and W.T.H. van Oers, arXiv:nucl-th/9409013

EPJ Web of Conferences **113**, 03018 (2016)
DOI: 10.1051/epjconf/201611303018

Effective Screened Coulomb Potential from e^-- e^- to ^{208}Pb -^{208}Pb Systems

Takashi Watanabe[1,a], Yasuhisa Hiratsuka[1,b], and Shinsho Oryu[1,c]

[1] *Department of Physics, Faculty of Science and Technology, Tokyo University of Science, Yamazaki, Noda City, Chiba 278-8510 Japan*

Abstract. We confirmed that the Coulomb phase shifts can not be given with a screened Coulomb potential by a monotonic increase of the range R. However, several screening "discrete range-bands" reproduce the Coulomb phase shift in a wide energy region. If the range-band of a proton-proton system for example is obtained, then the range-bands are generalized by three universal arguments: the universal range $\mathcal{R}(k)(\equiv kR)$, the Sommerfeld parameter $\eta(k)(\equiv Z_1Z_2e^2\nu/k)$, and the universal asymptotic phase $\mathcal{F}(k)(\equiv 2kr_c)$. Finally, the phase shifts of any other systems from e^-–e^- to heavy ions such as ^{208}Pb–^{208}Pb can be reproduced automatically by an individual range $R = \mathcal{R}(k)/k$ on the universal range-bands $\mathcal{R}$(k).

1 Introduction

The screened Coulomb potential can not reproduce the Coulomb *phase shift* by the monotonic increase-range method, although the screened Coulomb potential itself reaches the Coulomb potential in the infinite range [1][2]. However, the method is still adopted as the second best way in a wide field from the electron-electron system to the heavy ion-systems. In this paper, we would like to present a method to avoid the Coulomb singularity by using a universal "range-bands".

2 A Phase Function in the Asymptotic Region

The Coulomb phase shift is analytically obtained in configuration space and expressed by

$$\sigma_L(k) = \arg\Gamma(L + 1 + i\eta(k)), \tag{1}$$

where $\eta(k) = Z_1Z_2e^2\nu/k$ is the Sommerfeld parameter, L the two-body angular momentum, and $\nu = m_1m_2/(m_1 + m_2)$ reduced mass, respectively.
The phase shift appears in the asymptotic Coulomb wave function,

$$w_L^{(\pm)}(kr) = \exp\Big[\pm i\{kr - \pi L/2 - \eta(k)\ln 2kr + \sigma_L(k)\}\Big], \tag{2}$$

[a] e-mail: watanabe@ph.noda.tus.ac.jp
[b] e-mail: yasuhisa1059@gmail.com
[c] e-mail: oryu@rs.noda.tus.ac.jp

where the r-dependent phase: $\eta(k)\ln 2kr$ differs from the one obtained from short-ranged potentials,

$$h_L^{(\pm)}(kr) = \exp\Big[\pm i\{kr - \pi L/2 + \delta_L^R(k)\}\Big], \tag{3}$$

where $\delta_L^R(k)$ is the phase shift with short range potential.

Let us define a phase function by the ratio between both wave functions in the region $r \geq r_c$, where r_c is a starting point of the asymptotic region,

$$\mathcal{Y}_L^{\pm}(kr) \equiv \ln y_L^{\pm}(kr) \equiv \ln\Big[\frac{h_L^{(\pm)}(kr)}{w_L^{(\pm)}(kr)}\Big] = \pm i\big(\eta(k)\ln 2kr + \delta_L^R(k) - \sigma_L(k)\big). \tag{4}$$

The phase function satisfies a differential equation,

$$2i\Big(k - \frac{\eta(k)}{r}\Big)\frac{d\mathcal{Y}_L(kr)}{dr} = V^R(r) - V^C(r), \tag{5}$$

where $V^C(r)$ and $V^R(r)$ are the Coulomb and the screened Coulomb potentials, respectively.
The solution of the equation is obtained by the integral from r_c to ∞ with a Yukawa-type screened Coulomb potential

$$\exp\Big[-\eta(k)/\mathcal{R}(k)\Big]Ei(-t_c) - \ln\Big|\frac{\mathcal{F}(k) - 2\eta(k)}{\mathcal{F}(k)}\exp\Big[\pm 2m\pi/\eta(k)\Big]\Big| = 0, \quad (m = 0, 1, 2, \dots) \tag{6}$$

with the integral exponential function, and the universal asymptotic phase $\mathcal{F}(k) \equiv 2kr_c$,

$$Ei(-t_c) \equiv -\int_{t_c}^{\infty}\exp(-t)\frac{dt}{t}\,; \qquad t_c = \frac{\mathcal{F}(k) - 2\eta(k)}{2\mathcal{R}(k)}, \tag{7}$$

where $\mathcal{F}(k)$ is a variational parameter to minimize $\mathcal{R}(k)$ at a fixed $\eta(k)$ in Eq.(6). Therefore, a unique screened Coulomb potential is given by the permitted range $\mathcal{R}(k) = \mathcal{R}_n(k)$: $(n = 0,\ 1,\ 2, \dots)$ where n indicates phase shift regions: $n = 0$ for $\sigma_0 < 0$, and $(n-1)\pi \leq \sigma_0 < n\pi$ for $1 \leq n$.
However, the minimization procedure is very sensitive in the numerical calculation, therefore we will refer the analytic phase shift of Eq.(1) and obtain the $\mathcal{R}(k) - \eta(k)$ relation with the universal variables, instead of $R - k$ relation, which is shown for the $L = 0$ case in Figure 1.
Fortunately, the σ_L cases for $1 \leq L$ are automatically obtained by the $\mathcal{R}(k) - \eta(k)$ relation.

3 Numerical Results

We found five discrete $\mathcal{R}(k) - \eta(k)$-bands which reproduce the Coulomb phase shift for $\sigma_0 \leq 4\pi$.
1) The lowest $\mathcal{R}(k) - \eta(k)$-band (black circle in Figure 1) represents $\sigma_0(k) < 0$ in the higher energies.
2) The second band (black nabla sign) recreates $0 \leq \sigma_0(k) \leq \pi$.
3) The third band (black square sign) gives $\pi \leq \sigma_0(k) \leq 2\pi$.
4) The fourth band (white circle sign) offers $2\pi \leq \sigma_0(k) \leq 3\pi$.
5) The fifth band (black delta sign) displays $3\pi \leq \sigma_0(k) \leq 4\pi$.

In order to represent $\sigma_0(k) = n\pi$ $(n = 0, 1, 2, \cdots)$, the zero range potential $V^{R=0}(r)$ is required, however, it seems to be unusual. Therefore, we should jump to another nearby band before arriving at $\sigma_0(k) = n\pi$ at which the nearby band has a finite range (see Figure 1).

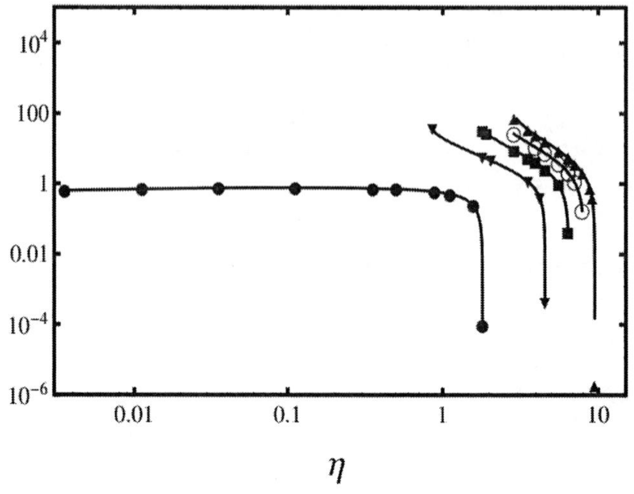

Figure 1. Arrowed universal ranges $\mathcal{R}(k)$ are denoted by five different signs: black circle, black nabla, black square, white circle, and black delta. The universal ranges $\mathcal{R} = \mathcal{R}(k) = \mathcal{R}_n(k)$ $(n = 0, 1, 2, 3, 4)$ are fitted as the functions of $\eta = \eta(k)$.
1) $\mathcal{R}_0 = 0$ at $\sigma_0(k) = 0$, $\mathcal{R}_0 = (37.283 + 5708.9\eta - 3166.9\eta^2)/(64.616 + 7062.0\eta - 2564.0\eta^2)$,
2) $\mathcal{R}_1 = 0$ at $\sigma_1(k) = \pi$,
$\mathcal{R}_1 = (-259.88 + 447.042\eta - 142.14\eta^2 + 12.414\eta^3)/(10.301 - 38.589\eta + 36.964\eta^2 - 5.8202\eta^3)$,
3) $\mathcal{R}_2 = 0$ at $\sigma_2(k) = 2\pi$,
$\mathcal{R}_2 = (42.289 - 42.199\eta + 26.9977\eta^2 - 3.3228\eta^3)/(0.050814 + 0.32437\eta - 0.91936\eta^2 + 0.57813\eta^3)$,
4) $\mathcal{R}_3 = 0$ at $\sigma_3(k) = 3\pi$,
$\mathcal{R}_3 = (516.55 - 217.77\eta + 40.497\eta^2 - 2.6540\eta^3)/(-22.633 + 19.254\eta - 4.6868\eta^2 + 0.50235\eta^3)$,
5) $\mathcal{R}_4 = 0$ at $\sigma_0(k) = 4\pi$,
$\mathcal{R}_4 = (-520.31 + 230.76\eta - 18.004\eta^2 - 0.056522\eta^3)/(15.524 - 11.918\eta + 2.1584\eta^2 + 0.034126\eta^3)$.
All bands become zero at $\sigma_0(k) = n\pi$ $(n = 0,\ 1,\ 2,\ \cdots)$. Such regions should be replaced by the finite range upper nearby-band. Each band can derive the Coulomb phase shifts very well (see Figure 2).

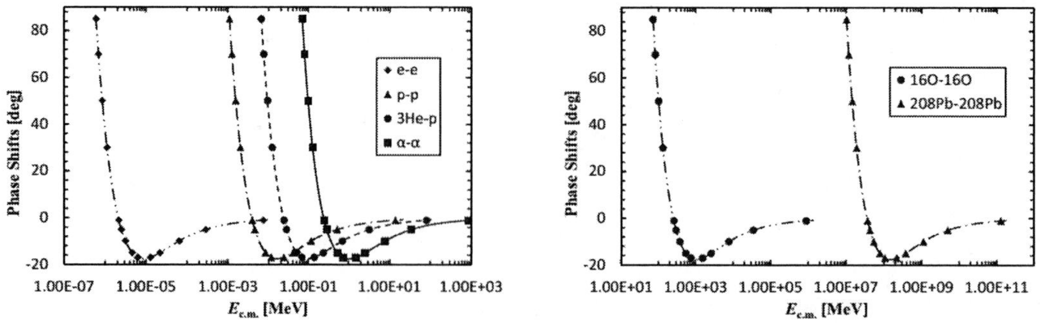

Figure 2. The Coulomb phase shifts in the S-wave are illustrated for several systems from e^--e^- to ^{208}Pb-^{208}Pb. These are calculated by the universal ranges which are given in Figure 1. Present screening range is only available for the Yukawa-type screened Coulomb potential which can reproduce the Coulomb phase shift not only for the wide energy range but also for any other system.

Table 1. $\sigma_0(k) - \eta(k) - R_n(k)$ relation which is seen in Figure 1. The Coulomb phase shift, $\sigma_0(k)$[deg], for the proton-proton system, with respect to the screening range bands $R_n(k)$[fm] ($n = 0, 1, 2, \cdots$), and the universal range bands $\mathcal{R}_n(k)$ (no dimension). The Sommerfeld parameter $\eta(k)$ as a function of the momentum, $k = \sqrt{2\nu E}$. E_{pp}[MeV] is the proton-proton c.m. energy. The energy for $\eta(k) = 1.81$ leads to $\sigma_0(k) = 0$[deg], the energy for $\eta(k) = 4.56$ shows almost $\sigma_0(k) = \pi$, $\eta(k) = 6.41$ corresponds nearly to $\sigma_0(k) = 2\pi$, $\eta(k) = 7.90$ means $\sigma_0 = 3\pi$, and $\eta(k) = 9.457$ gives practically $\sigma_0 = 4\pi$, respectively.

σ_0[deg]	−0.370	−3.66	−17.6	0.00	179.5
$\eta(k)$	0.0112	0.112	0.884	1.81	4.56
E_{pp}[MeV]	1.00×10^{2}	1.00×10^{0}	1.60×10^{-2}	3.81×10^{-3}	6.00×10^{-4}
R_0[fm]	4.60×10^{-1}	5.00×10^{-1}	3.10×10^{1}	1.04×10^{-4}	
$\mathcal{R}_0$	7.14×10^{-1}	7.76×10^{-1}	6.09×10^{-1}	9.61×10^{-6}	
R_1[fm]			1.84×10^{3}	6.08×10^{2}	1.10×10^{-1}
$\mathcal{R}_1$			3.72×10^{1}	5.84×10^{0}	4.18×10^{-4}
R_2[fm]			6.75×10^{4}	3.45×10^{3}	6.75×10^{2}
$\mathcal{R}_2$			1.37×10^{3}	3.32×10^{1}	2.57×10^{0}

σ_0[deg]	−1.01	167.6	0.00
$\eta(k)$	6.41	7.90	9.457
E_{pp}[MeV]	3.03×10^{-4}	2.00×10^{-4}	1.40×10^{-4}
R_2[fm]	1.60×10^{0}		
$\mathcal{R}_2$	4.33×10^{-2}		
R_3[fm]	7.76×10^{2}	8.00×10^{1}	
$\mathcal{R}_3$	9.16×10^{-5}	1.76×10^{-1}	
R_4[fm]	2.05×10^{3}	9.30×10^{2}	1.00×10^{-3}
$\mathcal{R}_4$	5.54×10^{0}	2.04×10^{0}	1.83×10^{-6}

4 Summary

One of the aims of this paper is to show that the usual screened Coulomb method is not correct, but the proper screening range-band exists. Second, if we adopt the universal range, the Sommerfeld parameter, and the universal asymptotic phase, then the Coulomb phase shifts of any two-charged particle systems can be obtained by the same universal range, and phase shifts of the higher partial wave are automatically given. An extension of this method to define the off-shell Coulomb amplitude will be done by using a so-called *Lemma* (see ref. [2]), where the universal range should be re-defined to satisfy an auxiliary phase shift, $\phi(k) \equiv \sigma(k) - \delta^R(k) = 0$, by solving the Schrödinger equation with an auxiliary potential $V^\phi(r) \equiv V^C(r) - V^R(t)$.

References

[1] S. Oryu, Y. Hiratsuka, S. Nishinohara, S. Chiba, J. Phys. G: Nucl. Part. Phys. **39** 045101 (2012), *ibid*, Phys. Rev. **C75**, 021001 (2007).

[2] S. Oryu, Phys. Rev. **C73**, 054001 (2006), *ibid*, **C76**, 069901 (2007).

EPJ Web of Conferences **113**, 03019 (2016)
DOI: 10.1051/epjconf/201611303019

Universal noise and Efimov physics

Amy N. Nicholson[1,a]

[1] *Department of Physics, University of California, Berkeley, CA, USA*

Abstract. Probability distributions for correlation functions of particles interacting via random-valued fields are discussed as a novel tool for determining the spectrum of a theory. In particular, this method is used to determine the energies of universal N-body clusters tied to Efimov trimers, for even N, by investigating the distribution of a correlation function of two particles at unitarity. Using numerical evidence that this distribution is log-normal, an analytical prediction for the N-dependence of the N-body binding energies is made.

Probability distributions are often studied in the context of Monte Carlo calculations, where a finite set of field configurations is used to estimate the value of an observable and its associated statistical uncertainty. Previously, it has been argued that the mean and variance of the distribution can often be estimated based on physical arguments, in efforts to better understand and control the signal-to-noise ratios in statistically noisy calculations [1, 2]. Therefore, it seems plausible that this argument may be turned on its head, such that knowledge of the distribution can be used to make predictions about the theory. As an example, probability distributions of correlation functions in certain cases may prove to be useful tools for calculations of the spectrum of the theory.

The concept of probability distributions arises naturally in lattice calculations. As an example, I will consider a Euclidean theory of non-relativistic particles interacting via a two-particle interaction, which will be mediated on the lattice by a Hubbard-Stratonovich field, ϕ. The discretized action of the theory may be written $S = \psi^\dagger K \psi$, where,

$$
\begin{aligned}
K_{\mathbf{p},\mathbf{p}'}(\tau,\tau') &= \delta_{\tau,\tau'} D_{\mathbf{p},\mathbf{p}'} + \delta_{\tau,\tau'-1} X_{\mathbf{p},\mathbf{p}'}(\tau)\,, \\
D_{\mathbf{p},\mathbf{p}'} &= \delta_{\mathbf{p},\mathbf{p}'} e^{\mathbf{p}^2/2M}\,, \qquad X_{\mathbf{p},\mathbf{p}'}(\tau) = \delta_{\mathbf{p},\mathbf{p}'} + C^{1/2}(\mathbf{p}-\mathbf{p}')\tilde{\phi}_{\mathbf{p}-\mathbf{p}'}(\tau)\,,
\end{aligned} \tag{1}
$$

[3, 4]. Here τ denotes Euclidean time and $\tilde{\phi}_{\mathbf{p}}$ represents the Fourier transform of the Z_2-valued auxiliary field ϕ. The limit of unitarity may be seen as a non-trivial UV fixed point in the β-function for the two-particle coupling, corresponding to a fine-tuning of the couplings in the lattice theory. In particular, one may define a tower of couplings which depend on powers of the momentum transfer, $C(\mathbf{p}) = \frac{4\pi}{M}\sum_{n=0}^{N_O-1} C_{2n}O_{2n}(\mathbf{p})$, up to some cutoff N_O, and tune these to the unitary point, effectively removing the first N_O orders of the effective range expansion for the scattering phase shift [3].

In this formulation, the auxiliary fields are chosen to live along the temporal links of the lattice, so that, along with the choice of open temporal boundary conditions (valid only for zero temperature calculations), the matrix K takes on an upper tridiagonal form. One important consequence of these choices is that the determinant, $\det K$ is simply the determinant of a product of free kinetic operators, D, which is independent of ϕ, and therefore makes no contribution to the probability measure.

[a] e-mail: anicholson@berkeley.edu

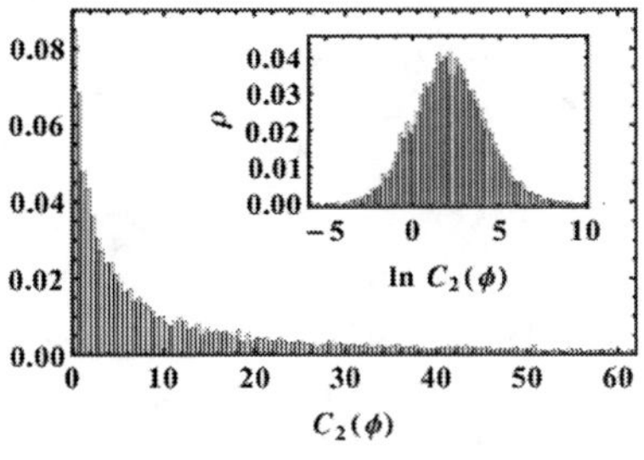

Figure 1. Histogram of the two-body correlator, C_2, at large Euclidean time, τ. Inset: Histogram of the logarithm of the two-body correlator at the same τ.

Furthermore, propagators may be written as iterative operations in time,

$$K^{-1}(\tau) = D^{-1}X(\tau-1)K^{-1}(\tau-1) \tag{2}$$

with $K^{-1}(0) = D^{-1}$, which, for large Euclidean time, act as filters for the ground state,

$$\sum_{\phi}\langle\Psi|K^{-1}[\phi,\tau]|\Psi\rangle = \sum_{n}\langle\Psi|n\rangle e^{-E_n\tau}\langle n|\Psi\rangle \underset{\tau\to\infty}{\longrightarrow} \mathcal{Z}e^{-E_0\tau}\,, \tag{3}$$

where $\sum_\phi$ denotes a sum over all field configurations ϕ, and E_n are single particle energies corresponding to the eigenstates $|n\rangle$ that have non-zero overlap onto the wavefunction Ψ (the overlap factor onto the ground state is denoted by $\mathcal{Z}$).

Given a finite set of field configurations, the probability distribution for an observable may be sampled by calculating the observable on each configuration (as is done in Eq. 2 prior to performing the sum). The values of the observable on each configuration may be plotted as a histogram, with the shape of the histogram representing the distribution. For the two-particle correlator tuned to unitarity and calculated at large Euclidean time using the lattice theory described above, one finds the histogram shown in Fig. 1. The inset shows the histogram obtained by calculating the logarithm of the observable, rather than the observable itself, on each sample. This distribution looks remarkably Gaussian, characteristic of an observable which obeys the log-normal distribution.

While the lattice provides a natural tool for sampling probability distributions, the distributions themselves are not simply an artifact of lattice calculations, but may be defined field-theoretically in the continuum. Assuming we have a theory of particles interacting through a field, ϕ, with probability measure $e^{-S[\phi]}$, then the probability distribution, P, and characteristic function, Φ, for a functional $C_N[\phi]$, may be defined as

$$P(x) \propto \int[d\phi]e^{-S[\phi]}\delta(C_N[\phi]-x)\,, \quad \Phi(s) \propto \int[d\phi]e^{-S[\phi]+isC_N[\phi]}\,, \tag{4}$$

The characteristic function is the Fourier transform of the probability distribution, and is itself a partition function for a theory that includes the observable in the Lagrangian, and may be amenable to perturbative treatments [5, 6]. In particular, $W(s) = -\ln\Phi(s)$ is the generating functional for the cumulants, κ, of the distribution,

$$W(s) = -\sum_{n=1}^{\infty}\frac{(is)^n}{n!}\kappa_n\,. \tag{5}$$

Thus, if $C_N = \langle C_N[\phi]\rangle$ is an N-body correlation function (angle brackets denote an average over the field ϕ according to the appropriate probability measure for ϕ) containing N external propagators, then κ_n may be identified as a sum of connected diagrams containing $n\times N$ external legs.

In this field theoretic language, the central limit theorem may be described in terms of renormalization group (RG) flow, the idea being that by successively block averaging random numbers drawn from a given distribution characterized by a set of cumulants, κ_n, one finds that the nth cumulant becomes rescaled as $\kappa_n \longrightarrow N_b^{1-n/2}\kappa_n$ where N_b is the number of block averaging operations (see, e.g., [7]). Thus, cumulants with $n > 2$ act as irrelevant operators, vanishing in the limit $N_b \longrightarrow \infty$.

The log-normal distribution follows the same RG arguments, however it applies to products of random numbers rather than sums. Thus, the log-normal distribution may be described as an RG fixed-point of products of random numbers. As described previously, propagators are given by τ products of $D^{-1}X[\phi]$, where the fields ϕ on each time slice are independent, random fields. Since we are interested in the limit $\tau \longrightarrow \infty$, some RG argument along these lines may be applicable to this case. However, recall that $D^{-1}X[\phi]$ is a matrix, and little is known about products of random matrices above two-dimensions [8], so at this time a rigorous mathematical treatment along these lines is not possible.

One may understand the distribution, however, on physical grounds by examining the long-time behavior of the moments of the N-body correlation function. In the path integral language, the nth moment of the distribution, like the nth cumulant, is given by diagrams containing $n \times N$ external legs. The moments, however, include both connected and disconnected diagrams, and may be shown to correspond to $n \times N$-body correlation functions. As an example, if we begin with a two-body correlation function composed, for simplicity of this argument, of single particle propagators $\mathcal{S}[\phi,\tau]$,

$$C_2(\tau) = \langle (\mathcal{S}[\phi,\tau])^2 \rangle \,, \tag{6}$$

then the variance of the probability distribution of this correlation function is given by the product of four propagators, averaged over a set of background fields,

$$\sigma^2_{C_2}(\tau) = \langle (\mathcal{S}[\phi,\tau])^4 \rangle \,, \tag{7}$$

This quantity is, in turn, a four-body correlation function. Following this argument, one may easily verify that the Nth moment corresponds to a $2N$-body correlation function.

At this point it is important to note that in order for all moments to correspond to physical correlation functions, the formulation must be such that the particle content specified by the probability measure is not at odds with the particle content in the operator. For example, many theories include a fermion (or boson) determinant as part of the probability measure, with the power of the determinant determining the number of degenerate "flavors" of particle in the theory. As noted previously, determinants in this formulation do not contribute to the probability measure, and the particle content of the theory is fully determined by the operator (the set of propagators) for a given moment. In a non-relativistic theory this is in a sense natural because at zero temperature and zero chemical potential there are no particle/antiparticle pairs arising from the vacuum, so the only information about particle content must be encoded in the operators measured.

Finally, because the long-time behavior of correlation functions is dominated by the ground-state energy of the system, the long-time behavior of the Nth moment of the two-particle correlator, $\mathcal{M}_N(\tau)$. is then

$$\mathcal{M}_N(\tau) \underset{\tau\to\infty}{\longrightarrow} \mathcal{Z}_{2N} e^{-E_0^{(2N)}\tau} \,, \tag{8}$$

where $E_0^{(2N)}$ and $\mathcal{Z}_{2N}$ are the ground-state energy and wavefunction overlap of the $2N$-body system, respectively. It should be noted that the set of propagators in all cases are inherently symmetric under particle interchange, and will have overlap with both identical bosons and distinguishable particles.

At unitarity, the energies of N-body bosonic systems have been shown to be tied to the spectrum of Efimov trimers [9, 10], whose binding energies are known to obey a discrete scaling symmetry, for at least $N \lesssim 16$ [11–15]. One method for showing this is to study the ratio of some N-body

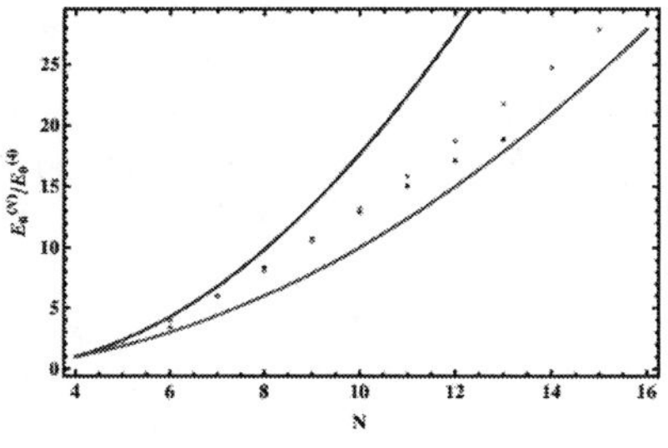

Figure 2. Ratio of energies $E_0^{(N)}/E_0^{(4)}$ predicted by the log-normal distribution (green), compared to direct numerical calculations of the energies, (blue) [12], (red) [18], and (gray) [15], as well an ansatz (purple band) for the N dependence of the energies based on direct numerical calculations [13, 14]. The red point shows the only calculation involving an excited state, where non-universal effects are expected to be smaller than for ground-state calculations.

binding energy to that of a trimer, while varying the short-distance behavior of the theory. If the ratio remains constant, this proves that there is no relevant n-body scale (for $3 < n \leq N$), meaning the three-body scale is the only scale determining the physics. This implies that for each trimer there must be a corresponding N-body state with the measured binding energy ratio, due to the discrete scale invariance. It also implies that this ratio is universal, in the sense that it is insensitive to any UV physics. These ratios have been calculated directly using numerical methods (see data in Fig. 2), however, at this time there is still notable disagreement among the various calculations.

It is now clear from the discussion that the energies of N-body bound states at unitarity should have the form $E_0^{(N)} = -a_N \Lambda_*$, where Λ_* is a three-body scale that may be set using the energy of any of the Efimov trimers, and a_N is a universal N-body energy ratio. Thus, the moments of the two-body correlation function may be shown to have the following form at large Euclidean time,

$$\mathcal{M}_N \underset{\tau \to \infty}{\longrightarrow} \mathcal{Z}_{2N} e^{a_{2N} \Lambda_* \tau} \,. \tag{9}$$

The moments of the log-normal distribution are known to be,

$$\mathcal{M}_N^{(\mathrm{LN})} = e^{N\mu + \frac{1}{2}N^2\sigma^2} \longrightarrow \mathcal{M}_N^{(\mathrm{LN})} = e^{\frac{1}{2}N(N-1)\sigma^2} \,, \tag{10}$$

where μ and σ are the mean and standard deviation of the distribution of the logarithm of the random variable, and on the right I have used the constraint that the first moment is a constant, corresponding to a correlation function with zero ground state energy. This form implies that the log-normal distribution corresponds to correlation functions whose moments are controlled by a single scale, since correlation functions decay as the exponential of an energy scale times Euclidean time. This is precisely what is expected from the form of the moments of the two-body correlation function at unitarity, Eq. 9. Furthermore, by comparing the log-normal moments with those of the two-body correlator at unitarity, one may predict the N-dependence of the universal constants, a_{2N}, giving the following energy ratios:

$$E_0^{(2N)}/E_0^{(4)} = \frac{1}{2}N(N-1) \,. \tag{11}$$

One should note that the expression above is not expected to hold for arbitrarily large N, because as the states become increasingly bound, eventually they will become affected by non-universal cutoff effects. For the (zero effective range) lattice theory which is used to sample the distribution, the energy cutoff imposed by the lattice spacing is found to be a factor of ~ 335 times larger than the energy of the lowest Efimov trimer, allowing a large separation of scales in which to find universal behavior.

The prediction based on the log-normal distribution is plotted in Fig. 2 (lower solid line), along with a comparison to other works. The upper solid line is an analytic prediction from [13, 14], using data produced from a finite range potential, coupled with a procedure for removing the finite effective range effects. This conjecture confirmed the leading N^2 dependence predicted by this work, but with a significantly different prefactor. The points represent direct numerical calculations of the N-body binding energies using potential models with varying dependences on non-universal effects such

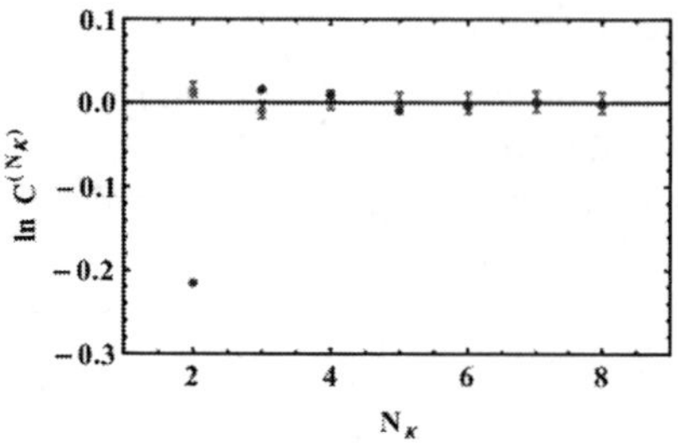

Figure 3. Cumulant expansion (Eq. 12) for the two-body correlator at unitarity cut off at N_κ. The green points represent the lattice data on a $16^3 \times 1000$ lattice, while the blue points are from a mock distribution created by expanding around unitarity in such a way as to reproduce moments corresponding to the energies found in [12].

as non-zero effective range and three- and higher-body cutoffs. Note that all of these calculations (including this work) are based on ground-state calculations, except for the single red point. For this point the binding energy of a 6-body state tied to the first excited Efimov trimer was calculated, and is therefore expected to be the least affected by non-universal UV effects.

Finally, note that for odd numbers of particles the correlator is not positive so an exactly log-normal distribution would be somewhat unexpected. However, in practice it is found that the distributions for odd-numbered correlation functions are also approximately log-normal with only a tiny negative contribution. Because all direct numerical calculations of the N-body binding energies in other works show smooth behavior with N, I have interpolated the prediction for even N in this plot.

To determine quantitatively how close the probability distribution for the two-body correlation function at unitarity is to a log-normal distribution, we may turn to the cumulant expansion [16], which relates the logarithm of the (field averaged) correlation function to the sum of the cumulants of the distribution of $\ln C$,

$$\ln\langle C\rangle = \sum_{n=1}^{\infty} \frac{\kappa_n(\ln C)}{n!} . \tag{12}$$

Recall that the cumulants of a distribution act like connected n-point operators with increasing dimension. Thus the cumulant expansion may be regarded in an effective field theory context: as a distribution is driven toward log-normal, cumulants $\kappa_n(\ln C)$ with $n > 2$ become increasingly irrelevant [17]. For the log-normal distribution, cumulants for $n > 2$ are exactly zero; any deviation from log-normal may be quantified by the relative size of the third and higher cumulants.

In Fig. 3 the cumulant expansion for the logarithm of the two-body correlation function at large time is shown versus the number of cumulants included in the sum, N_κ. Comparing the value of the cumulant expansion cut off at $N_\kappa = 2$ to the asymptotic value, we see that the distribution deviates by approximately 2% from the log-normal distribution. This small discrepancy is likely due to sensitivity to lattice artifacts for large moments of C_2. In the same figure an attempt has been made to quantitatively determine how much variations in the energies affect the cumulants of the distribution. An example of a mock distribution generated by expanding around log-normal and fitting the coefficients to produce the energies calculated in [12] is shown by the blue points. Using different parameterizations of the expansion around log-normal and fitting different combinations of energies, the discrepancy from log-normal (given by the value of the cumulant expansion at third order) for the energies from [12] varies between $\sim 17\% - 30\%$, which is approximately the same as the largest discrepancy between the energies from [12] and those predicted by an exactly log-normal distribution. Using similar expansions around log-normal and allowing for the $\sim 2\%$ deviation seen in the lattice data results in energies that differ from the log-normal predictions by no more than a few percent.

I would like to emphasize that the log-normal distribution itself implies that these $2N$-body energies are universal, because all moments depend on a single energy scale, and are therefore not particularly sensitive to any non-universal effects. Thus, the lattice data, which shows a log-normal distribution, gives further confidence to the claim that there are universal N-body states tied to Efimov trimers. The log-normal distribution implies other universal properties as well. For example, if the two-body correlation function is log-normal, then by extension the correlation functions for all $2N$-body Efimov states must also be log-normal, but with rescaled parameters μ and σ^2. The logarithm of the nth moment of a $2N$-body correlator corresponds to $E_0^{(2Nn)}T = \frac{1}{2}Nn(Nn-1)E_0^{(4)}T$. Comparing this to the moments of the log-normal distribution (Eq. 10), gives $\mu = \frac{1}{2}NE_0^{(4)}T$, $\sigma^2 = -N^2E_0^{(4)}T$ for the distribution of the $2N$-body correlation function. Thus, the distribution itself may be seen as universal for all correlation functions of states tied to Efimov trimers.

It is compelling that as the beta function for the two-particle coupling approaches an RG fixed point, the distribution for the two-particle correlator appears to as well. It may be interesting to explore an effective field theory type (or possibly some perturbative) expansion around the log-normal distribution [17], much in the way that the KSW expansion [19, 20] for nuclear effective field theory expands around the unitary fixed point, either analytically using Eq. 4 or numerically using lattice data. Such a program, in addition to giving further insight into the connection between the log-normal distribution and unitarity, would allow calculations of non-universal effects at large N, as well as any possible deviations from log-normal within the universal regime.

Acknowledgements

The author would like to thank M. Endres, D. B. Kaplan, J.-W. Lee, J. E. Drut, P. Bedaque, T. Cohen, E. Berkowitz, K. McElvain, and T. Kurth for helpful discussions on this work. This work was supported by the U.S. Department of Energy under grants DE-FG02-93ER-40762, DE-SC00046548.

References

[1] G. Lepage, *The Analysis of Algorithms for Lattice Field Theory* (1989)
[2] P.E. Gibbs, *Understanding Finite Baryonic Density Simulations in Lattice QCD* (1986)
[3] M.G. Endres, D.B. Kaplan, J.W. Lee, A.N. Nicholson, Phys.Rev. **A84**, 043644 (2011)
[4] M.G. Endres, D.B. Kaplan, J.W. Lee, A.N. Nicholson, Phys. Rev. **A87**, 023615 (2013)
[5] D. Grabowska, D.B. Kaplan, A.N. Nicholson, Phys. Rev. **D87**, 014504 (2013)
[6] A.N. Nicholson, D. Grabowska, D.B. Kaplan (2012), [J. Phys. Conf. Ser.432,012032(2013)]
[7] D. Sornette, *Critical phenomena in natural sciences: chaos, fractals, selforganization, and disorder: concepts and tools* (Berlin:Springer, 2004)
[8] A.D. Jackson, B.E. Lautrup, P. Johansen, M. Nielsen, Phys. Rev. **E66**, 066124 (2002)
[9] V.N. Efimov, Physics Letters B **33**, 563 (1970)
[10] V.N. Efimov, Sov. J. Nucl. Phys. **12**, 589 (1971)
[11] D. Blume, K. Daily, Comptes Rendus Physique **12**, 86 (2011)
[12] J. von Stecher, J.Phys.B.At.Mol.Opt.Phys. **43**, 101002 (2010)
[13] M. Gattobigio, A. Kievsky, Phys. Rev. **A90**, 012502 (2014)
[14] A. Kievsky, N.K. Timofeyuk, M. Gattobigio, Phys. Rev. **A90**, 032504 (2014)
[15] Y. Yan, D. Blume, ArXiv e-prints (2015), **1508.00081**
[16] M.G. Endres, D.B. Kaplan, J.W. Lee, A.N. Nicholson, Phys. Rev. Lett. **107**, 201601 (2011)
[17] M.G. Endres, D.B. Kaplan, J.W. Lee, A.N. Nicholson (2011)
[18] J. von Stecher, Phys. Rev. Lett. **107**, 200402 (2011)
[19] D.B. Kaplan, M.J. Savage, M.B. Wise, Phys. Lett. **B424**, 390 (1998)
[20] D.B. Kaplan, M.J. Savage, M.B. Wise, Nucl. Phys. **B478**, 629 (1996)

EPJ Web of Conferences **113**, 04001 (2016)
DOI: 10.1051/epjconf/201611304001

Recent results on deuteron–proton scattering from RIKEN

Kimiko Sekiguchi[1,a]

[1] *Department of Physics, Tohoku University, Sendai, 980-8578, JAPAN*

Abstract. Nucleon-deuteron (*Nd*) scattering, for which a rigorous formulation in terms of Faddeev equations exists and exact solutions of these equations for any dynamical input can be obtained, offers a good opportunity to study the dynamical aspects of 3NFs, such as momentum and spin dependences. Since the first indication of 3NF effects in *Nd* elastic scattering around 100 MeV/nucleon, precise measurements of proton–deuteron / neutron–deuteron scattering have been extensively performed at 60–300 MeV/nucleon. Direct comparison between the data and the Faddeev calculations based on realistic nucleon–nucleon forces plus 2π–exchange three nucleon forces draws the following conclusions. (1) The 3NF is definitely needed in *Nd* elastic scattering. (2) The spin dependent parts of the 3NFs may be deficient. (3) The shorter-range components of the 3NFs are probably required for the cross section as well as the spin observables at backward angles with increasing an incident energy.

1 Introduction

One of the main interests in nuclear physics is understanding the forces acting between nuclear constituents. A hot topic in the study of nuclear forces is to clarify the roles of three nucleon forces (3NFs) in nuclei, and to describe various phenomena of nuclei by explicitly taking into account nucleon–nucleon (NN) interactions combined with 3NFs. The 3NFs arise naturally in the standard meson exchange picture [1] as well as in the framework of chiral effective field theory (χEFT) which has a link to QCD [2, 3].

The first evidence for a 3NF was found in the three-nucleon bound states, ^{3}H and ^{3}He [4, 5]. The binding energies of these nuclei are not reproduced by exact solutions of three-nucleon Faddeev equations employing modern NN forces only, *i.e.* AV18 [6], CD Bonn [7], Nijmegen I, II [8]. The underbinding of ^{3}H and ^{3}He is explained by adding a 3NF, mostly based on 2π-exchange, acting between three nucleons [4, 5, 9]. The importance of 3NFs has been further supported by the binding energies of light mass nuclei, and by the empirical saturation point of symmetric nuclear matter. *Ab initio* microscopic calculations of light mass nuclei, such as Green's Function Monte Carlo [10] and no-core shell model calculations [11], highlight the necessity of including 3NFs to explain the binding energies and low-lying levels of these nuclei. As for the density of symmetric nuclear matter, it has been reported that all NN potentials provide saturation at too high density, and a short-range repulsive 3NF is one possibility to shift the theoretical results to the empirical point [12]. In the past decade, low energy scattering, binding energies of light [13] and medium mass nuclei [14, 15] and nuclear

[a]e-mail: kimiko@lambda.phys.tohoku.ac.jp

matter [16] have been extensively studied also in the framework of chiral effective field theory (χEFT). In all these investigations, it became evident that 3NFs are taken as one key element to understand various nuclear phenomena. Therefore, they should be investigated in a wide momentum region to understand their properties in detail.

In order to study the dynamical aspects of 3NFs, such as momentum, spin, and iso-spin dependences, the three-nucleon scattering system is an attractive probe because various kinematical conditions are allowed to measure not only differential cross sections but also a rich set of polarization observables. The importance of 3NFs in three-nucleon scattering was shown in the *Nd* elastic scattering for the first time in Ref. [17]. Clear signals from 3NFs were found around the cross section minimum occurring at c.m. angle $\theta_{c.m.} \approx 120°$ for incident energies above 60 MeV/nucleon. Since then the *pd*/*nd* scattering at 60–200 MeV/nucleon have been performed at the facilities, e.g. RIKEN, RCNP, KVI, and IUCF, providing precise data of the cross sections as well as various types of the spin observables [19]. At RIKEN we performed the measurements of the cross sections and the spin observables with the polarized deuteron beams at the incident energies up to 135 MeV/nucleon [20]. Recently we have extended the measurements at the RIKEN RI Beam Factory (RIBF) with the polarized deuteron beams at 250, and 300 MeV/nucleon which are slightly above the pion emission threshold energy 210 MeV [21, 22].

In the following sections the recent achievements on study of 3NFs via measurements for the *dp* scattering at the RIKEN RIBF are discussed.

2 Experiment at RIKEN

The schematic view of the experimental setup is shown in Fig. 1.

At the RIBF the vector and tensor polarized deuteron beam was provided by the polarized ion source [31] and was accelerated by the AVF, RRC, and SRC up to 250 (294) MeV/nucleon through the Intermediate-stage Ring Cyclotron (IRC) bypass transport beam line [23]. The measurement for elastic *dp* scattering was performed with the detector system BigDpol which was installed at the extraction beam line of the SRC. A polyethylene (CH_2) target with a thickness of 330 mg/cm^2 was used as a hydrogen target. In the BigDpol four pair of plastic scintillators coupled with photomultiplier tubes were placed symmetrically in the directions of azimuthal angles to left, right, up and down. Scattered deuterons and recoil protons were detected in a kinematical coincidence condition by each pair of the detectors. The measured angles in the center of mass system are $\theta_{\rm c.m.} = 40°$–$160°$. In the experiment the deuteron beams were stopped in the Faraday cup which was installed at the focal plane F0 of the BigRIPS spectrometer.

The beam polarizations were monitored continuously with a beam line polarimeter Dpol prior to acceleration by the SRC using the reaction of elastic *dp* scattering at 90 (100) MeV/nucleon. The analyzing powers for this reaction have been calibrated in the previous measurement by using the ^{12}C$(d,\alpha)^{10}$B* [2^+] reaction, the $A_{zz}(0°)$ of which is exactly -1 because of parity conservation [24]. At the RIKEN RIBF the single-turn extractions were available for all the cyclotrons used for the experiments. Therefore depolarizations were expected to be small during beam acceleration. In the measurement typical values of the beam polarizations were 80% of the theoretical maximum values.

3 Results and Discussions

In Fig. 2 some representative experimental results for *dp* and *nd* elastic scattering are compared with the Faddeev calculations with and w/o 3NFs. The red (blue) bands are the calculations with (without) Tucson-Melbourne99 (TM99) 3NF [25], which is a version of the Tucson-Melbourne 3NF [26] more

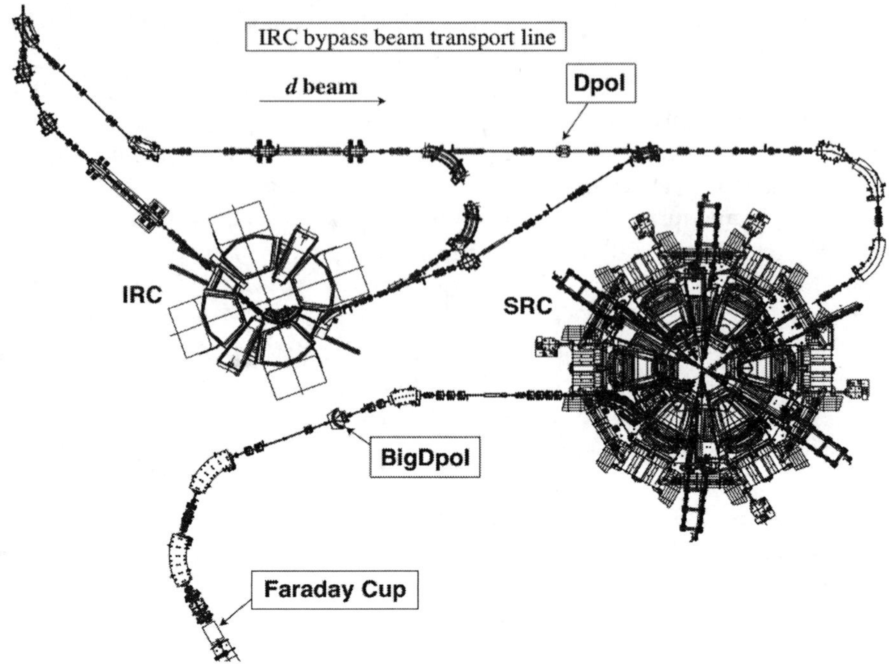

Fig. 1. Schematic view of the experimental setup for *dp* elastic scattering with polarized deuteron beams at RIKEN RI Beam Factory.

consistent with chiral symmetry [27, 28], based on the modern NN potentials, *i.e.* CD Bonn, AV18, Nijmegen I and II. The solid lines are the calculations based on the AV18 potential with including the Urbana IX 3NF [29].

For the cross section specific features are seen depending on the scattering angles in the center of mass system $\theta_{c.m.}$. At the forward angles $\theta_{c.m.} \lesssim 80^\circ$, the theoretical calculations based on the various NN potentials are well converged and the predicted 3NF effects are very small. The experimental data are well described by the calculations except for the very forward angles. This discrepancy comes from that fact that the calculations shown in the figure do not take into account the Coulomb interactions between protons [30]. At the angles $\theta_{c.m.} \gtrsim 80^\circ$ the clear discrepancies between the data and the calculations based on the NN potentials are found. They become larger as an incident energy increases. At the angles around $\theta_{c.m.} = 80^\circ$–120° the discrepancies are explained by taking into account the 2π exchange type 3NF models (TM99, and Urbana IX). However, at the backward angles $\theta_{c.m.} \gtrsim 120^\circ$ the differences appear between the experimental data and the calculations even including the 3NF potentials with increasing an incident energy, which lie to the very backward angle $\theta_{c.m.} \sim 180^\circ$ at a higher energy 250 MeV/nucleon. Since these features are clearly seen as going to higher energies [31, 32], the relativistic effects have been estimated by using the Lorentz boosted NN potentials with the TM99 [33]. However the relativistic effects have turned out to be small and only slightly alter the cross sections (see Fig. 3).

As for the polarization observables the energy dependence of the predicted 3NF effects and the difference between the theory and the data is not always similar to that of the cross section. The deuteron vector analyzing power iT_{11} has features similar to those of the cross section. Meanwhile the tensor analyzing power T_{22} reveals different energy dependence from that of iT_{11}. Starting from $\sim$ 100

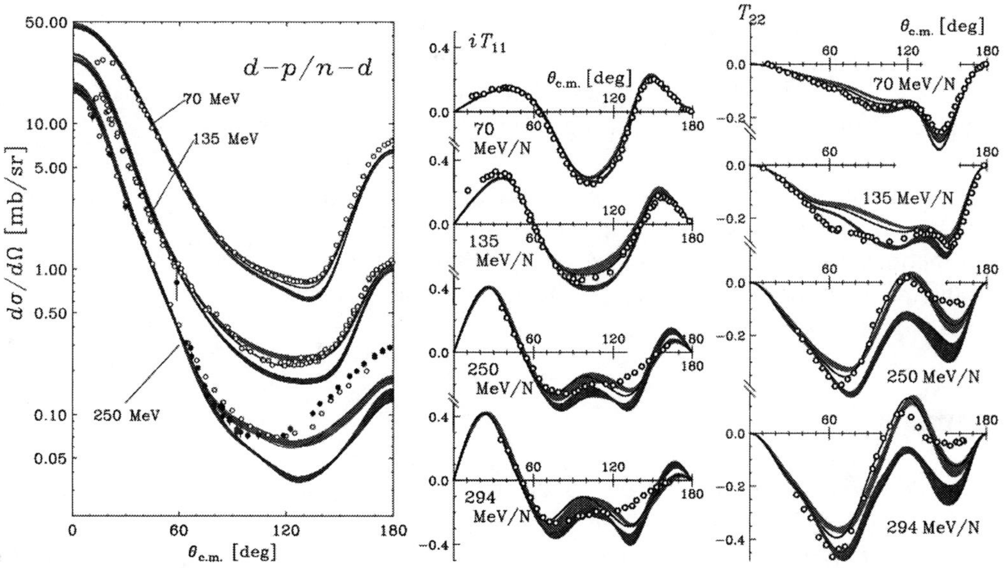

FIG. 2. Differential cross sections and deuteron analyzing powers iT_{11}, T_{22} for elastic *Nd* scattering at 70–294 MeV/nucleon (MeV/N). The red (blue) bands are the calculations with (w/o) TM99 3NF based on the modern NN potentials, namely CD Bonn, AV18, Nijmegen I and II. The solid lines are the calculations with including Urbana IX 3NF based on AV18 potential. For the cross sections the open circles at 70 and 135 MeV/nucleon are the *dp* data in Refs. [20]. The open and solid circles at 250 MeV/nucleon are the *pd* and *nd* data, respectively in Refs. [31]. For the deuteron analyzing powers the data at 70 and 135 MeV/nucleon are from Refs. [20]. The data at 250 and 294 MeV/nucleon are taken at the RIBF [21, 22].

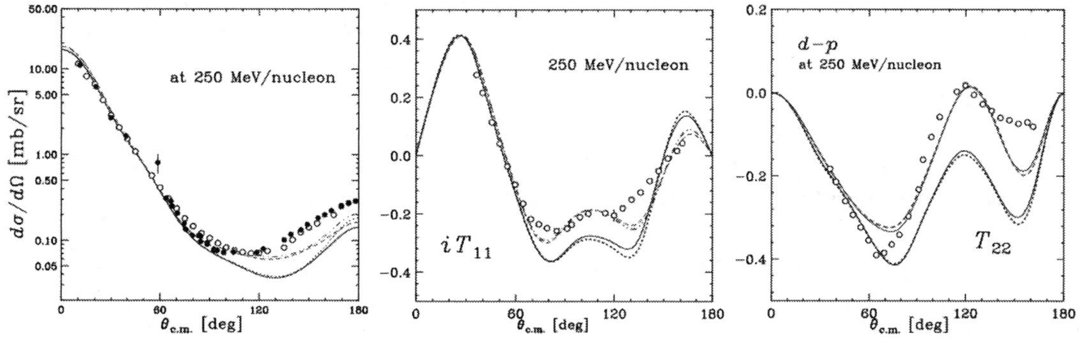

FIG. 3. Differential cross sections and deuteron analyzing powers iT_{11}, T_{20} for *dp* elastic scattering at 250 MeV/nucleon. The solid (blue) and dashed (red) curves show the results of nonrelativistic Faddeev calculations with the CD Bonn potential alone and combined with the TM99 3NF, respectively. The relativistic calculations based on the CD Bonn potential without Wigner spin rotations are shown with blue (dotted) curves. The red (double-dotted-dashed) curves show the relativistic calculations with the TM99 3NF included.

MeV/nucleon large 3NF effects are predicted. At 135 MeV/nucleon and below adding 3NFs worsens the description of data in a large angular region. It is contrary to what happens at the highest energies above 250 MeV/nucleon, where large 3NF effects are supported by the T_{22} data. The relativistic effects are estimated to be small also for these polarization observables for *Nd* elastic scattering (see Fig. 3) [22].

The results obtained for *Nd* elastic scattering draw the following conclusions; (1) the 3NF is definitely needed in *Nd* elastic scattering, (2) the spin dependent parts of the 3NF may be deficient, (3) the compornents other than the 2π-exchange 3NFs, i.e. the shorter-range components of the 3NFs are probably required for the cross section as well as the spin observables at backward angles as going to higher energies.

4 Summary

3NFs are now accepted as key elements in understanding various nuclear phenomena, e.g. properties of light mass nuclei and equation of state for nuclear matter. The *Nd* scattering data provide rich sources to explore the properties of 3NFs such as momentum and spin dependence. In this talk the experiments performed with polarized deuteron beams at RIKEN are presented and the recent achievements of study of 3NFs via *dp* scattering at 100–300 MeV/nucleon are discussed. The energy and angular dependent results of the cross sections as well as the polarization observables show that (1) clear signatures of the 3NF effects are found in the cross section, (2) the spin dependent parts of the 3NFs may be deficient, and (3) the shorter-range components of the 3NFs are probably required for the cross sections as well as the spin observables at backward angles with increasing an incident energies.

As the next step of 3NF study in the few nucleon scattering it would be interesting to see how theoretical approaches, *e.g.* inclusion of 3NFs other than 2π–exchange types. Recently the calculations based on the χ EFT potentials are becoming available for the *Nd* scattering up to 200 MeV/nucleon in which the NN forces up to the next-to-next-to-next-to-next leading order are taken into account [34]. Calculated results show possible signatures of 3NF effects at backward angles. Theoretical analysis with 3NFs which include not only the 2π–exchange types but other various diagrams of the 3NFs are now in progress. Together with this, it should also be mentioned that careful treatments for the effects of *pi*-emission might be necessary for *Nd* scattering around and above the π threshold energy. So far we expect that cross sections for π emission are quite small in analogy of total cross section of *pp* scattering [35].

Experimentally, it is interesting to measure spin correlation coefficients as well as polarization transfer coefficients for elastic *dp* scattering at higher energies 200–400 MeV/nucleon. Various kinematic configurations of the exclusive *pd* breakup reactions should also be measured in order to study the properties of 3NFs as well as relativistic effects. As a first step from few to many body systems it is interesting to extend the measurements to 4N scattering systems, e.g. $p+{}^3$He scattering, which would provide a valuable source of information on 3NFs including their iso-spin dependences.

Acknowledgments

The author would like to thank the collaborators for the experiments performed with the polarized deuteron beams at RIKEN RI Beam Factory. She is also grateful to the strong supports from the theorists, H. Witała, W. Glöckle, H. Kamada, J. Golak, A. Nogga, R. Skibiński, P. U. Sauer, A. Deltuva, and A. C. Fonseca.

References

[1] J. Fujita, and H. Miyazawa, Prog. Theor. Phys. **17**,360 (1957).
[2] U. van Kolck, Phys. Rev. C **49**, 2932 (1994).
[3] E. Epelbaum, H.-W. Hammer, U.-G. Meißner, Rev. Mod. Phys. **81**, 1773 (2009).
[4] C. R. Chen *et al.*, Phys. Rev. C **33**, 1740 (1986).
[5] T. Sasakawa and S. Ishikawa, Few-Body Syst. **1**, 3 (1986).
[6] R. B. Wiringa *et al.*, Phys. Rev. C **51**, 38 (1995).
[7] R. Machleidt, Phys. Rev. C **63**, 024001 (2001).
[8] V. G. J. Stoks *et al.*, Phys. Rev. C **49**, 2950 (1994).
[9] A. Nogga *et al.*, Phys. Rev. C **65**, 054003 (2002).
[10] S. C. Pieper *et al.*, Phys. Rev. C. **64**, 014001 (2001).
[11] P. Navrátil and W. E. Ormand, Phys. Rev. C **68**, 034305 (2003).
[12] see for example, A. Akmal *et al,*, Phys. Rev. C **58**, 1804 (1998).
[13] E. Epelbaum *et al.*, Phys. Rev. C **66**, 064001 (2002).
[14] S. Binder *et al.*, Phys. Rev. C **87**, 021303 (2013).
[15] T. A. Lähde *et al.* arXv.1311.0477 [nucl-th].
[16] K. Hebeler *et al.*, Phys. Rev. C **83**, 031301 (2011).
[17] H. Witała *et al.*, Phys. Rev. Lett. **81**, 1183 (1998).
[18] S. Nemoto *et al.*, Phys. Rev. C **58**, 2599 (1998).
[19] See references in N. Kalantar-Nayestanaki *et al.*, Rep. Prog. Phys. **75**, 016301 (2012).
[20] N. Sakamoto *et al.*, Phys. Lett. B **367**, 60 (1996); H. Sakai *et al.*, Phys. Rev. Lett. **84**, 5288 (2000); K. Sekiguchi *et al.*, Phys. Rev. C **65**, 034003 (2002); K. Sekiguchi *et al.*, Phys. Rev. C **70**, 014001 (2004); K. Sekiguchi *et al.*, Phys. Rev. Lett **95**, 162301 (2005).
[21] K. Sekiguchi *et al.*, Phys. Rev. C **83**, 061001 (2011).
[22] K. Sekiguchi *et al.*, Phys. Rev. C **89**, 064007 (2014).
[23] Y. Watanabe *et al.*, RIKEN Accel. Prog. Rep. **42**, 111 (2009).
[24] K. Suda *et al.*, Nucl. Instrum. Methods Phys. Res. A. **572**, 745 (2007).
[25] S. A. Coon and H. K. Han, Few Body Syst., **30**, 131 (2001).
[26] S. A. Coon and W. Glöckle, Phys. Rev. C**23**, 1790 (1981).
[27] J. L. Friar *et al.*, Phys. Rev. C **59**, 53 (1999).
[28] D. Hüber *et al.*, Few-Body Syst. **30**, 95 (2001).
[29] B. S. Pudliner *et al.*, Phys. Rev. C**56**, 1720 (1997).
[30] A. Deltuva *et al.*, Phys. Rev. C **71**, 054005 (2005).
[31] K. Hatanaka *et al.*, Phys. Rev. C **66**, 044002 (2002); Y. Maeda *et al.*, Phys. Rev. C **76**, 014004 (2007).
[32] A. Ramazani-Moghaddam-Arani *et al.*, Phys. Rev. C **78**, 014006 (2008).
[33] H. Witała *et al.*, Phys. Rev. C **83**, 044001 (2011); H. Witała *et al.*, Phys. Rev. C **88**, 069904(E) (2013).
[34] E. Epelbaum, proceedings in this conference; S. Binder *et al.*, arXiv:1505.07218 (2015).
[35] See for example, K. A. Olive *et al.* (Particle Data Group), Chin. Phys. C, **38**, 090001 (2014).

EPJ Web of Conferences **113**, 04002 (2016)
DOI: 10.1051/epjconf/201611304002

Studies of three-nucleon systems with improved chiral forces

R.Skibiński[1,a], J.Golak[1], K.Topolnicki[1], H.Witała[1], and E.Epelbaum[2]

[1] *M.Smoluchowski Institute of Physics, Jagiellonian University, PL-30348 Kraków, Poland*
[2] *Institut für Theoretische Physik II, Ruhr-Universität Bochum, D-44780 Bochum, Germany*

Abstract. Recently developed chiral two-nucleon (2N) potentials are applied to various processes in three-nucleon (3N) systems within the approach based on Faddeev equations. The predictions for the cross sections in elastic nucleon-deuteron (Nd) scattering and for radiative proton-deuteron (pd) capture are similar to results based on the AV18 potential. The very good convergence with respect to the chiral expansion and a weak dependence on a regularization parameter are observed. This gives hope that the Hamiltonian combining these 2N potentials with consistent 3N forces will become an important tool to study nuclear structure and processes.

1 Introduction

Recently, the improved 2N forces up to next-to-next-to-next-to-next-to leading order (N^4LO) were derived by the Bochum-Bonn group [1, 2]. Among important new features of these potentials, one should mention especially a new way of fixing values of some low energy constants (LECs) and a new regularization method. Namely, the values of LECs which occur in the long-range part of the interactions are taken directly from the pion-nucleon data without any fine tuning as it was done in older versions of the chiral NN potential. The second important change is the improved regularization scheme for the long-range part of the potential. The non-local regulator in momentum space, which was used in the previous versions, is now replaced by a local regulator in coordinate space. This allows one for a better control of regularization artifacts. In particular, such a procedure makes the long-rang part of the interaction less affected by the regularization procedure and removes unwanted short-range parts of the two-pion exchange potential. As a consequence, no spectral function regularization is needed.

The 2N interactions of Ref. [1, 2] have been recently applied not only to 2N scattering but also to the Nd scattering reactions [3]. This very first application proves usefulness of the chiral forces [1, 2] in describing properties and processes in the 3N systems. These results will be briefly reported in Section 2. In Section 3 we apply the same forces to radiative proton-deuteron capture. Also for this process new results are very promising. Especially, the weak dependence on the value of the regularization parameter increases the predictive power of the new forces compared to the older ones [4].

[a] e-mail: roman.skibinski@uj.edu.pl

2 Elastic neutron-deuteron scattering

Elastic Nd scattering is the simplest process in 3N systems in which the nuclear forces can be tested. Already for the cross section, strong 3N force effects have been found at intermediate energies. However, a careful study of 3NF effects requires detailed knowledge of the 2N interaction and a very weak cut-off dependence of theoretical predictions. To test the quality of the new chiral potentials, we calculate, neglecting the 3N forces, the differential cross section for elastic Nd scattering. We solve in momentum space the Faddeev equation [5] for the state $T|\phi\rangle$,

$$T|\phi\rangle \;=\; tP|\phi\rangle \;+ tPG_0T|\phi\rangle \tag{1}$$

with the off-the-energy shell 2N t-matrix t, the sum P of a cyclic and anti-cyclic permutation of three nucleons, the free 3N propagator G_0, and the initial state $|\phi\rangle$, composed of the deuteron and a momentum eigenstate of the projectile nucleon. The 2N interaction enters Eq.1 via the t-matrix. The transition amplitude for elastic scattering with a nucleon-deuteron final state $|\phi'\rangle$ is then given by [5]

$$\langle\phi'|U|\phi\rangle = \langle\phi'|PG_0^{-1}|\phi\rangle + \langle\phi'|PT|\phi\rangle. \tag{2}$$

The first test of usefulness of the new model of the chiral force is the convergence of results with respect to the order of chiral expansion. This is shown in the left panel of Fig. 1 for the differential cross section at the relatively high nucleon laboratory energy of E= 200 MeV and for the regularization parameter R = 0.9 fm. We clearly see, that going from next-to-leading order (NLO) to next-to-next-to-leading order (N^2LO) and subsequently to N^3LO causes a shift of predictions. The predictions at N^4LO overlap with these at N^3LO for most of the scattering angles. Only below $\theta \approx 20°$, a tiny difference between N^3LO and N^4LO predictions is seen. Finally, we observe a nice agreement with predictions based on the AV18 2N force [6].

Next we check the cut-off dependence shown in Fig. 1 (the right panel) for the same energy and at N^4LO. That dependence of the cross section is very weak at this energy, however, it is stronger than at, not shown here, lower energies. The prediction obtained for R = 1.2 fm lies further away from the other four curves what confirms the indication from 2N scattering that such a value of the R parameter generates a large amount of cut-off artifacts. While it is instructive to study cut-off dependence of the predictions, we want to stress that not all values of the cut-off parameter used here are equally preferred. In particular, the values of R = 0.9 fm and R = 1.0 fm lead to the χ^2/data ratio close to 1.0 for high energy 2N scattering. It remains to check if the same values of the cut-off parameter will be preferred in 3N reactions or nuclear structure calculations. We also want to emphasize that the cut-off dependence of predictions does not provide a proper way to estimate theoretical uncertainties in calculations of this kind. In Refs. [1, 3] we formulated an approach to estimate truncation errors, that is the theoretical uncertainty due to neglected contributions from higher orders of the chiral expansion. This novel approach was justified for the case of R = 0.9 fm in the Bayesian analysis of Ref. [7].

3 The radiative proton-deuteron capture reaction

The radiative pd capture reaction is another process to test the new chiral 2N forces. Compared to elastic scattering, a new dynamical ingredient, namely the electromagnetic current operator j_μ, enters the calculation. The transition amplitude for radiative capture can be obtained via time reversal symmetry from the transition amplitude N_μ^{Nd} for two-body photodisintegration of the corresponding 3N bound state, $\gamma +^3$ He $\to p + d$ in our case [8, 9].

The corresponding Faddeev equation, if one neglects the 3N interaction, is given as

$$|U\rangle \;=\; tG_0(1+P)j_\mu|\Phi^{3N}\rangle \;+ tG_0P|U\rangle \tag{3}$$

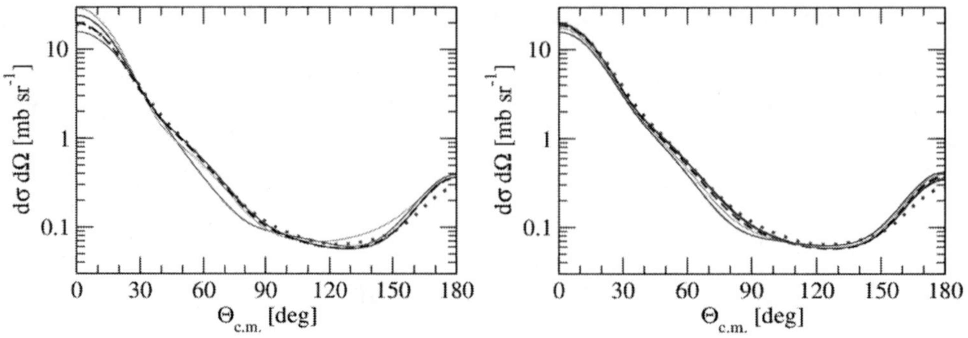

Figure 1. The neutron-deuteron elastic scattering differential cross section $d\sigma/d\Omega$ as a function of the center-of-mass angle $\theta_{c.m.}$ at the neutron energy E_{lab}=200 MeV. On the left, the convergence of the cross section is shown for $R = 0.9$ fm. The red, green, blue and dashed black curves correspond to the predictions at NLO, N^2LO, N^3LO and N^4LO, respectively. On the right, the cut-off dependence at N^4LO is shown. The solid black, red, dashed black, green and blue curves correspond to the predictions with $R = 0.8, 0.9, 1.0, 1.1$ and 1.2 fm, respectively. In both panels the dotted violet line shows the AV18 predictions.

and the transition amplitude N_μ^{Nd} can be calculated by

$$N_\mu^{Nd} = \langle\phi'|(1+P)j_\mu|\Phi^{3N}\rangle + \langle\phi'|P|U\rangle, \tag{4}$$

where the state $|\Phi^{3N}\rangle$ is the ^{3}He wave function. In the following, we use the Siegert theorem to include implicitly many-body electromagnetic current operators (see [8, 9] for more details).

In Fig. 2 we show the differential cross section for proton-deuteron radiative capture at the incoming deuteron laboratory energy of E= 95 MeV. The left panel again shows the convergence with respect to the chiral order. While leading order predictions are far from others at forward and backward photon emission angles, all other curves are overlapping in this area. At the maximum of the cross section, the predictions at NLO and N^2LO come close together, however there is a shift in predictions when moving to N^3LO. The predictions at N^4LO overlap with the N^3LO ones. At this energy, the cut-off dependence is seen only in the region of the maximum of the cross section, see the right panel of Fig. 2). The cross section increases smoothly with the increasing value of the regularization parameter, however, the differences between the biggest and the lowest values of the cross section amount only to about 10%, which is factor of two smaller than the corresponding spread of results for the older version of the chiral 2N potential [10].

In summary, the improved chiral 2N interactions of Refs. [1, 2] have been tested in elastic Nd scattering and in radiative pd capture. For both processes, the new forces show very promising behavior (reasonable values of observables, good convergence with respect to the chiral order and weak cut-off dependence). These findings are very promising in view of the ongoing work towards the inclusion of consistent chiral 3N forces and electromagnetic currents.

Acknowledgements

This work was done as a part the LENPIC project and was supported from the resources of the National Science Center (Poland) under grant DEC-2013/10/M/ST2/00420. It was also supported in part by DFG (SFB/TR 16, "Subnuclear Structure of Matter") and ERC project 259218 NUCLEAREFT. Numerical calculations have been performed on the supercomputer clusters of the JSC, Jülich, Germany.

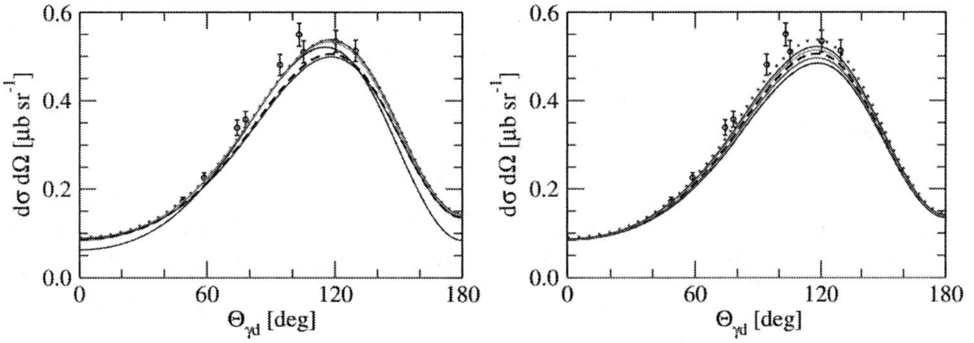

Figure 2. The differential cross section $d\sigma/d\Omega$ for proton-deuteron radiative capture as a function of the photon emission angle $\theta_{\gamma,d}$ at the deuteron laboratory energy E_{lab}=95 MeV. The panels and curves as in Fig.1, except for dotted curve which corresponds to predictions based on the AV18 and the UrbanaIX [11] interactions. Data are from [12].

References

[1] E. Epelbaum, H. Krebs, U.-G.Meißner, Eur. Phys. J. **A5**, 53 (2015).

[2] E. Epelbaum, H. Krebs, U.-G.Meißner, arXiv:1412.4623 [nucl-th] (2014).

[3] S.Binder et al., arXiv:1505.07218.

[4] H. Witała, J. Golak, R. Skibiński, K. Topolnicki, *J. Phys. G: Nucl. Part. Phys.* **41**, 094011 (2014).

[5] W. Glöckle, H. Witała, D. Hüber, H. Kamada and J. Golak, *Phys. Rep.* **274**, 107 (1996).

[6] R.B. Wiringa, V.G.J. Stoks, and R. Schiavilla, Phys. Rev. C **51**, 38 (1995).

[7] R.J.Furnstahl, N.Klco, D.R.Phillips, S.Wesolowski, arXiv.1506.01343.

[8] J. Golak et al., Phys. Rev. **C62**, 054005 (2000).

[9] J. Golak et al., Phys. Rept. **415**, 89 (2005)

[10] R.Skibiński et al. Acta Phys.Polon. **B37** 2905 (2006).

[11] B.S.Pudliner et al., Phys. Rev. **C56**, 1720 (1997).

[12] W.K. Pitts et al., Phys. Rev. **C37**, 1 (1988).

EPJ Web of Conferences **113**, 04003 (2016)
DOI: 10.1051/epjconf/201611304003

Experimental Studies of the Coulomb Force Effects in Deuteron-Proton Break-up Reaction at Medium Energy Regime

I. Ciepał[1,a], W. Parol[2], N. Kalantar-Nayestanaki[3], G. Khatri[2], St. Kistryn[2], B. Kłos[4], A. Kozela[1], P. Kulessa[1], J. Messchendorp[3], I. Skwira-Chalot[5], E. Stephan[4], and B. Włoch[6]

[1] *Institute of Nuclear Physics PAS, PL-31342, Kraków, Poland*
[2] *Institute of Physics, Jagiellonian University, PL-30348, Kraków, Poland*
[3] *KVI-CART, University of Groningen, NL-9747 AA Groningen, The Netherlands*
[4] *Institute of Physics, University of Silesia, PL-40007, Katowice, Poland*
[5] *Faculty of Physics, University of Warsaw, PL-02093, Warsaw, Poland*
[6] *AGH – University of Science and Technology, Faculty of Physics and Applied Computer Science, al. A. Mickiewicza 30, PL-30059 Kraków, Poland*

Abstract. A set of differential cross-section data of the ^{1}H*(d, pp)n* breakup reaction at 130 and 160 MeV deuteron beam energies has been measured in the forward polar angles domain. The data were collected with the use of the Germanium Wall (FZ Jülich) and BINA (KVI Groningen) detectors. This part of the phase-space is special with respect to the dominant Coulomb force influence on the system dynamics. The data are compared with the theoretical calculations based on the Argonne V18 potential supplemented with the long-range electromagnetic component. The predictions also include the Urbana IX three nucleon force model. The strongest Coulomb effects are found in regions where the relative energy of the two protons is the smallest.

1 Introduction

Deuteron break-up in collision with one proton, leading to a final state of three-nucleon (3N) continuum is one the simplest systems to study, providing a good testing ground for the nuclear Hamiltonian. Such a process offers a rich kinematics which makes it more selective regarding the interaction. One of the most extensively studied 3N observables is the differential cross section, very sensitive to different pieces of the system dynamics like three-nucleon force (3NF), Coulomb interaction or relativistic effects. One can investigate their influence on the observables at various parts of the phase space, simply by choosing different geometries of the outgoing nucleons.

In the very forward angular range, investigated with the Germanium Wall and BINA detectors (see Sec. 2), the dynamics is dominated with the Coulomb force influence. For the first time, such significant effects were observed in the *dp* breakup differential cross-section data measured at 130 MeV [1]. Then, the Coulomb force effects were studied in more details in a dedicated experiment at FZJ [2–4].

The first calculations which included the Coulomb interaction were done with CD Bonn and CD

[a] e-mail: izabela.ciepal@ifj.edu.pl

Bonn + Δ potentials [5, 6] for elastic proton-deuteron scattering. Afterwards, they were developed for the *pd* breakup process [7] and the existing experimental data [1] strongly supported the significant effects predicted by the calculations. The dedicated experiment provided the possibility to test them at the new level of accuracy [4]. Recently, a consistent theoretical treatment of a phenomenological 3NF (Urbana IX) and the Coulomb force has been achieved also for the Argonne V18 (AV18) potential [8] that allows to investigate the role of both effects to a high level of accuracy.
In the present work the extensive data sets obtained in a series of the experiments are used to investigate the Coulomb force influence on the differential cross section.

2 Experiments

The experiments at FZJ was conducted with the Germanium Wall (GeWall) setup [4] at the deuteron beam energy of 130 MeV. GeWall consisted of three high-purity germanium position sensitive detectors. Two different types of the detectors were used: a thin transmission detector *Quirl* with an excellent spatial resolution for determining the position and energy loss (ΔE detector) of the passing charged particles, and two thick detectors $E1$ and $E2$ for measurement of particle energies with an excellent resolution. The angular acceptance of the apparatus was $5° - 14°$ for the polar and 2π for the azimuthal angles.

With the BINA detector and the deuteron beam energy of 160 MeV the new rich contribution to the database of the differential cross section was obtained [9, 10]. The apparatus is composed of the two main parts called Wall and Ball. Wall, covering the angular range from 15° to 37°, is built of the MWPC for momenta reconstruction, ΔE (24 strips) and E (10 slabs) scintillator detectors. The ΔE and E create virtual matrix of 120 telescopes. The backward part is ball-shaped and consists of 149 phoswich detectors which cover polar angles between 40° and 160°. The Ball plays two roles: of the particle detector and scattering chamber.

3 Studies of Coulomb force effects

In the investigated part of the phase space, the electromagnetic interaction plays a crucial role. Therefore, calculations which take into account both dynamical components, i.e., the 3NF and the Coulomb force, are able to properly reproduce the data. Adding the electromagnetic force into the calculations does not essentially change the quality of the data description at large values of the invariant mass of the protons emerging from the breakup reaction $\sqrt{s_{pp}}$. However, at small $\sqrt{s_{pp}}$ values the observed discrepancies are almost totaly removed [1, 4]. The inclusion of 3NF and Coulomb effects into the calculations does not remove completely the disagreement between data and theoretical predictions. There are still some discrepancies at large $\sqrt{s_{pp}}$ [3] which can be interpreted as indication that parts of the dynamics are missing, either relativistic effects or unresolved problems in our understanding of the 3NF structure.

The experimental investigations at FZJ [4] were focusing on a very forward and narrow part of the phase space where the predicted Coulomb force effects are significant. Very recently also a new set of the breakup cross section at 160 MeV was obtained [9, 10], allowing one to search for the effects but at slightly higher energy. The new data were evaluated for angular configurations on a grid defined by the polar angles of the two protons θ_1, θ_2 given in a range of $17° - 38°$ with a step of 2°, and their relative azimuthal angle φ_{12} ($20° - 180°$), with a step of 20°. Sample preliminary distributions of the differential cross section are presented in Fig. 1. The importance of the Coulomb interaction for correct description of the data at low φ_{12} (left panel) is quite clear.

To check in a systematic way how the electromagnetic processes influence the data, the cross sec-

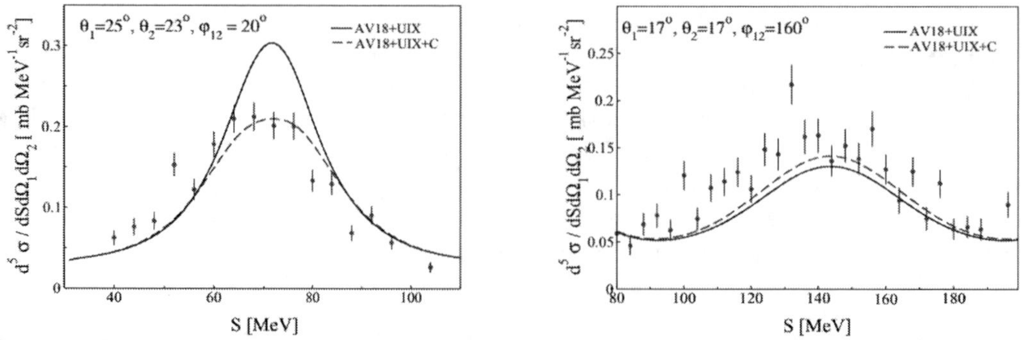

Figure 1. (Color online) Differential cross-section data for the dp breakup reaction at 160 MeV at two different kinematical configurations (specified in the panels). In the configuration presented on the left panel significant Coulomb force effects are observed. The data are compared to the AV18-based calculations and are described in the legend.

tions were integrated over S and studied as a function of the φ_{12} variable. The same integration, limited by the experimental energy threshold of 30 MeV, was done for the model predictions of AV18+UIX and AV18+UIX+C. The two closest in θ-distance geometries originating from the 130 and 160 MeV datasets were investigated, see Fig. 2.

The calculations which neglect the Coulomb force usually overestimate the data at the lowest relative azimuthal angle ($\varphi_{12} < 40°$) for both the studied datasets. Such a behavior is consistent with the intuitive understanding (neglecting any interference of the dynamical ingredients) - the Coulomb repulsion decreases the number of protons in configurations with small relative energies. This corresponds to a situation when the protons stay close together for a relatively long time and the repulsion can lead to a significant change of their initial trajectories. For higher φ_{12}, one observes opposite situation at 130 MeV, whereas at 160 MeV, only very small Coulomb influence is seen.

With growing φ_{12} the discrepancy changes sign and at a certain point the theoretical curves intersect. Such points are visible at $\varphi_{12} = 60°$ and $\varphi_{12} = 40°$ at 130 and 160 MeV, respectively. This refers

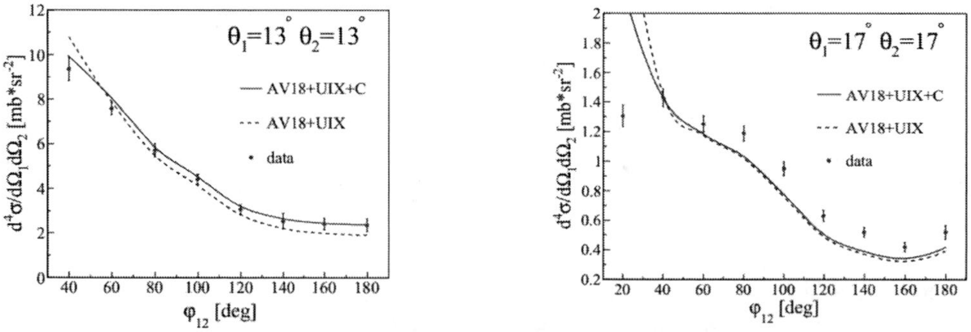

Figure 2. Differential, integrated over energy, cross-section values presented as a function of the relative azimuthal angle φ_{12}, for a given θ_1, θ_2 combination. The presented data were obtained at 130 MeV (left panel) and at 160 MeV (right panel) deuteron beam energies. The data are compared with the calculations indicated in the legend.

to a situation when the number of "incoming" and "outgoing" protons is balanced and the net effect of the Coulomb force is equal to zero. In a range of the higher relative angles the number of protons is increased. This, in turn, is caused by the protons originating from the region where the repulsion is strong.

The results obtained at 130 MeV revealed a quite strong influence of the Coulomb force effects on the breakup reaction. The quantitative comparisons with the available predictions for the 3N systems demonstrates a crucial role of the electromagnetic component in the data description [4]. The data evaluated at 160 MeV are still preliminary and one also needs to estimate the possible systematic effects to draw final conclusions on the quality of description provided by the calculations.

4 Summary and conclusions

Precise and systematic studies of the breakup reaction in a large part of the phase space are very important for understanding of the interaction between nucleons in few-nucleon systems. Currently available theoretical approaches which try to model the interaction need very precise and large experimental database to be verified and further developed. Within these predictions different pieces of the dynamics can be studied separately and also their mutual interplay can be investigated. Moreover, there is still strong need to have possibly complete theoretical treatments including all ingredients of the 3N system dynamics (3NF, Coulomb interaction, relativistic effects).

New experiments to study 3N system dynamics are planned, including also investigations of the three-body system in the four-body environment. The data obtained with the WASA detector at FZJ at higher energies are under analysis. They will allow one to unambiguously fix a relevance of the 3NF. A new scientific program concentrated on investigation of the few-nucleon systems dynamics was proposed to be carried out with the use of the BINA detector at the Cyclotron Center Bronowice (CCB) in Cracow. Moreover, an experimental apparatus which will be used in measurements with a polarized ^{3}He target is under construction. Such studies planned at CCB will allow one to fill a data-deficient sector of 4N systems.

Acknowledgements

This work was partially supported by the Polish National Science Center under Grants No. 2012/05/E/ST2/02313 (2013-2016) and No.DEC-2013/10/M/ST2/00420.

References

[1] St. Kistryn et al., Phys. Lett. B **641**, 23-27 (2006)
[2] I. Ciepał et al., EPJ Web of Conferences **37**, 09011 (2012)
[3] I. Ciepał et al., Few-Body Syst. **601**, 1-5 (2013)
[4] I. Ciepał et al., Few-Body Syst., http://dx.doi.org/10.1007/s00601-015-1014-8, 1-26 (2015) doi=10.1007/s00601-015-1014-8
[5] A. Deltuva, A.C. Fonseca, P.U. Sauer, Phys. Rev. C **71**, 054005 (2005)
[6] A. Deltuva, A.C. Fonseca, P.U. Sauer, Phys. Rev. C **72**, 054004 (2005)
[7] A. Deltuva, A.C. Fonseca, P.U. Sauer, Phys. Rev. C **73**, 057001 (2006)
[8] A. Deltuva, Phys. Rev. C **80**, 064002 (2009)
[9] W. Parol et al., Acta Physica Polonica B, Vol. **45**, No. 2, 527-532 (2013)
[10] W. Parol et al., EPJ Web of Conferences **81**, 06007 (2014)

EPJ Web of Conferences **113**, 04004 (2016)
DOI: 10.1051/epjconf/201611304004

Experimental study of Three-Nucleon Dynamics in the dp breakup reaction

B. Kłos[1,a], I. Ciepał[2] , G. Khatri[3], S. Kistryn[3], A. Kozela[2], A. Magiera[3], W. Parol[3], I. Skwira-Chalot[4], E. Stephan[1], K.Szpik[1], and the WASA-at-COSY Collaboration

[1] *University of Silesia, PL-40007 Katowice, Poland*
[2] *Institute of Nuclear Physics PAN, PL-31342 Kraków, Poland*
[3] *Jagiellonian University, PL-30059 Kraków, Poland*
[4] *Faculty of Physics, University of Warsaw, Warsaw, Poland*

Abstract.
An experiment to investigate the ^{1}H(*d,pp*)n breakup reaction using a deuteron beam of 340, 380 and 400 MeV and the WASA detector has been performed at the Cooler Synchrotron COSY-Jülich. The main goal was a detailed study of various aspects of few-nucleon dynamics like the three nucleon force (3NF), the long-range Coulomb interaction or relativistic effects in the medium energy region. The relativistic effects and their interplay with 3NF become more important with increasing available energy in the three nucleon system. The almost 4π geometry of the WASA detector provides an unique possibility to study various aspects of dynamics.
The studies of the 3N system dynamics in the breakup reaction with BINA detector are continued in the Cyclotron Center Bronowice.

1 Introduction

Few-nucleon system is ideal laboratory to study nuclear forces. The system composed of three nucleons (3N) is the simplest nontrivial environment, in which various models of the nucleon-nucleon (NN) interaction can be tested. Three-nucleon system dynamics can be investigated quantitatively by comparing observables calculated with the use of Faddeev equations with precise measurements. The breakup observables can be predicted using modern realistic pairwise nucleon-nucleon NN interactions, combined with model of 3N forces [1]. Moreover, the two- and three-nucleon interactions can be modeled within the coupled-channel (CC) framework by an explicit treatment of the Δ-isobar [2]. Alternatively, the dynamics is generated by the Chiral Perturbation Theory (ChPT), so far at the next-to-next-to-leading order with all relevant NN and 3N contributions taken into account [3]. The modern theoretical calculations include different pieces of nucleon-nucleon dynamics like the above mentioned three nucleon force but also the long-range Coulomb interaction or relativistic effects. Differential cross section in the region of medium energies is very sensitive to all these effects which reveal their influence in different regions of the phase space. In recent years the relativistic treatment of the breakup reaction in the 3N system has been developed using the NN potential [4]

[a] e-mail: barbara.klos@us.edu.pl

and this approach has also been extended for calculations including 3NF [5]. It was shown that in some particular regions of the breakup phase space, relativistic effects can increase or decrease the calculated breakup cross sections by up to 60%. At the same time the effects of 3NF may change certain observables by a similar factor. The relativistic effects and their interplay with 3NF become more important with increasing available energy in the three nucleon system. Therefore the energy range of interest is possibly to close pion production threshold (0.4 GeV for dp breakup reaction). We measured the differential cross section of the ^{1}H*(d,pp)n* breakup reaction at energies of 340, 380 and 400 MeV. The investigations will enable to study the evolution of the relativistic and 3NF effects in this energy range. This study will put strong constraints on the theoretical calculations.

2 Experiment

The experiment using the $^{1}\mathrm{H}(\vec{d}, pp)n$ breakup reaction at 340 MeV, 380 MeV and 400 MeV deuteron beam energy has been performed in January 2013 at the Cooler Synchrotron COSY-Jülich with the WASA-at-COSY detector [6], [7]. The WASA (Wide Angle Shower Apparatus) detector, covering almost full solid angle, consists of four main components: Central Detector (CD), Forward Detector (FD), Pellet Target Device and the Scattering Chamber (see Figure 1). Such a geometry of WASA gives an unique possibility to study various aspects of dynamics of the three nucleon system, which influence the differential cross section in different phase space regions.

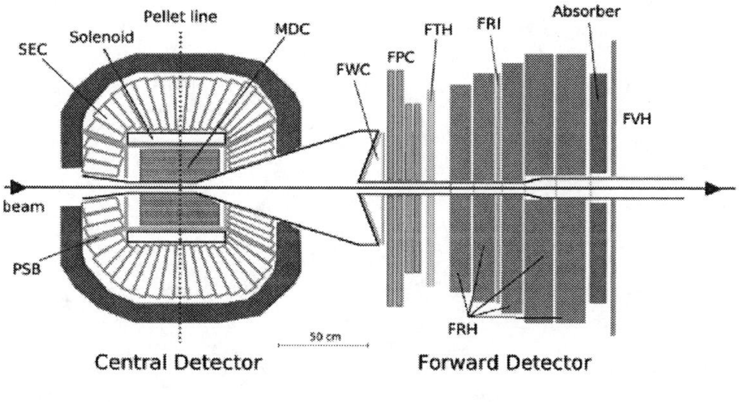

Figure 1. Schematic view of the detection system

The target components are situated on top of the platform of the CD. The target provides a narrow stream of frozen hydrogen or deuteron pellets with diameters of about 20μm. The CD surrounds the interaction region and covers the region of the polar angles from 20° to 169°. It consists of four elements: Mini Drift Chamber (MDC), Plastic Scintillator Barrel (PSB), Scintillator Electromagnetic Calorimeter (SEC) and Super Conduction Solenoid (SCS). The FD covers the region of the polar angles from deuterons of 3° to 18°. It consists of a set of plastic scintillators for the identification of charged hadrons and track reconstruction: Forward Window Counter (FWC), Forward Proportional Chamber (FPC), Forward Trigger Hodoscope (FTH), The Forward Range Hodoscope (FRH) and Forward Veto Hodoscope (FVH).

3 Data Analysis

The first step of data analysis is the identification of interesting events, *i.e.* two protons from the breakup process and deuteron-proton pairs from elastic scattering channel in the Forward Detector (range of the polar angles is from 3° to 18°). The particle identification is based on the ΔE-E technique.

After selection of the proton-proton coincidences and having performed the energy calibration, the differential cross section of the breakup reaction can be determined for a chosen kinematic configuration. The configuration is defined by the emission angles of the two outgoing protons: two polar angles θ_1 and θ_2 and the relative azimuthal angle ϕ_{12} (reconstructed from FPC in Forward Detector and MDC in Central Detector).

For a chosen configuration a kinematical spectrum (E_1 vs E_2) is built and the events are projected onto a kinematical curve in order to obtain the distribution as a function of the arc-length S measured along the kinematics. An example of the preliminary non-normalized experimental breakup rate (not corrected for efficiency and acceptance) obtained for the chosen kinematical configurations are presented in Fig. 2. In the right panel in this figure the theoretical calculations [1] [8] are also shown. In the studied angular region, close to FSI configuration of a proton pair (two protons flying close to each other), Coulomb force plays an important role, decreasing locally the cross section. Shape of the experimental distribution confirms that prediction. Although only 10% of collected data has been analysed, the statistical uncertainty per data point is already between 5%-10% in configuration registred in Forward Detector.

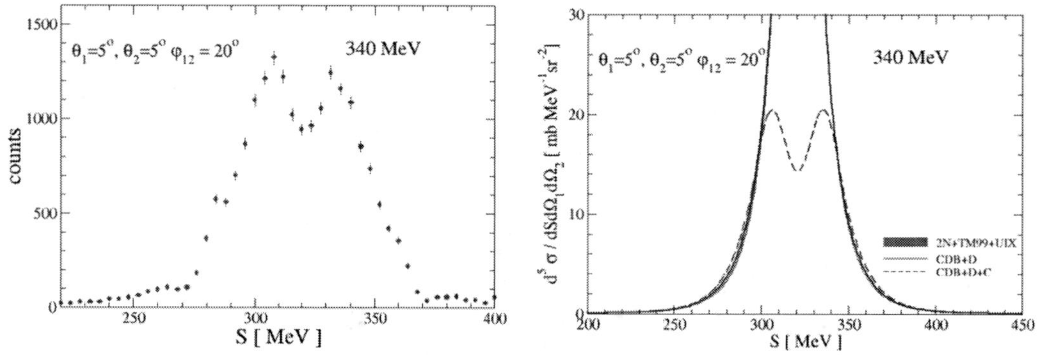

Figure 2. An example of the preliminary non-normalized experimental breakup event rate (not efficiency and acceptance corrected) obtained for chosen kinematical configurations as a function of the S value (left panel). Magenta bands represent calculations based on realistic potentials: 2N complemented with the TM99 3NF and the realistic AV18 potential combined with the Urbana IX. The calculations within the coupled-channel approach with the CD Bonn+ Δ potential without and with the Coulomb force included are represented by blue solid line and green dashed line, respectively (right panel).

4 Summary and outlook

The analysis is continued with the aim to determine the differential cross sections for the the deuteron breakup process in the d+p system at energies of 340, 380 and 400 MeV. The theoretical calculations using modern realistic nucleon-nucleon (NN) interactions, combined with a suitable model of 3N forces [1] and to the calculations within the coupled-channel approach with the CD Bonn+Δ

potential and with the Coulomb force included [8] exist and can be compared with experimental data after normalization. The calculations including relativistic effects are also been prepared [10].

Currently, the analysis is focused on obtaining normalized experimental results of the cross sections for the different combinations of kinematical configurations of the polar angles θ_1 and θ_2 and the relative azimuthal angle ϕ_{12}. The luminosity will be determined using measured angular distribution of the elastically-scattered deuterons and the known cross section [9]. Preliminary selection of registered deuterons has been done.

The studies of the 3N system dynamics in the breakup reaction are continued with BINA detector in Cyclotron Center of Bronowice (CCB). The Proteus C-235 cyclotron at CCB provides proton beams with energy from 70 MeV to 230 MeV with energy spread $\Delta E/E < 0.7\%$ and beam current in the range of 500 nA to 0.1 nA. BINA detector has been instalated in CCB (adapted to beam line and infrastructure) and commissioned. First test run with all detectors using ZnS, CD_2 and CH_2 targets has been done. We plan the measurement of the cross sections of *p+d* elastic scattering for 108, 135 and 160 MeV and, subsequently, the differentional cross sections of *p+d* breakup reaction at 108 MeV.

Acknowledgements

This work has been supported by FFE funds of the Forschunszentrum Jülich; by the European Commision under the 7th Framework Programme through the "Research Infrastructures" action of the Capacities Programme (Call: FP7-INFRASTRUCTURES-2008-1), the Polish 2013-2015 science funds as research Project No. 2012/05/E/ST2/02313 and by Polish National Science Center from grant DEC-2012/05/B/ST2/02556.

References

[1] W. Glöckle , H. Witała, D. Hüber, H. Kamada, J. Golak, Phys. Rep. **274**, 107 (1996)
[2] A. Deltuva, K. Chmielewski and P.U. Sauer, Phys. Rev. **C67**, 034001 (2003);
A. Deltuva, R. Machleidt and P.U. Sauer, Phys. Rev. **C68**, 024005 (2003)
[3] E. Epelbaum, Rep. Prog. Nucl. Phys. **57**, 654 (2006)
[4] H. Witała, R. Skibinski and J. Golak, The European Physical Journal A **30**, 369 (2006)
[5] H. Witała *et al.*, Phys. Rev. C **83** 044001 (2011)
[6] Ch. Bargholtz *et al.*, Nucl. Instrum. Methods in Physics Research A **594**, 339-350 (2008)
[7] H. H. Adam *et al.*, arXiv:nucl-ex/0411038 (2004)
[8] A. Deltuva *et al.*, Phys. Rev. C **72**, 054004 (2005)
[9] K. Ermisch *et al.*, Phys. Rev. C **68** 051001(R) (2003)
[10] H. Witała, privite communication.

EPJ Web of Conferences **113**, 04005 (2016)
DOI: 10.1051/epjconf/201611304005

Recent Deuteron Compton Scattering Results and Extracted Neutron Polarizabilities

L.S. Myers[1,2,a]

[1] *Thomas Jefferson National Accelerator Facility, Newport News, VA 23606, USA*
[2] *Presented on behalf of the COMPTON@MAX-lab collaboration*

Abstract. The COMPTON@MAX-lab collaboration has recently published a new measurement of elastic photon scattering from deuterium using tagged photons at the MAX IV Laboratory [1]. The experiment utilized the Tagged Photon Facility at MAX IV and three of the largest NaI(Tl) detectors in the world. Correction terms to the cross section were determined via Monte Carlo simulations [2, 3] and were confirmed by comparisons to the well-known $^{12}C(\gamma,\gamma)^{12}C$ reaction [4]. These results represent the most extensive data on deuteron Compton scattering ever measured and effectively double the world data set. In addition, the energy range overlaps previous experiments and extends nearly 20 MeV higher where the sensitivity to the polarizabilities is enhanced. As a result, we have obtained the neutron polarizabilities as α_n=[11.55 ± 1.25(stat) ± 0.2(BSR) ± 0.8(th)] × 10^{-4} fm^3 and β_n=[3.65 ∓ 1.25(stat) ± 0.2(BSR) ± 0.8(th)] × 10^{-4} fm^3, which represents a 30% reduction in the statistical uncertainty.

1 Introduction

The electric (α) and magnetic (β) polarizabilities of the proton have recently been extracted with a statistical uncertainty of 0.35×10^{-4} fm^3 [5]. This precision is a result of the large number of experimental measurements of $p(\gamma,\gamma)p$ conducted over the last twenty years (see [6] for a review of the experiments). By comparison, in 2012, the uncertainty of the neutron polarizabilities was 1.8×10^{-4} fm^3 [6] – more than five terms larger than for the proton. The primary reason for this discrepancy is the lack of data for Compton scattering from the deuteron which is used to extract α_n and β_n.

The previous data for $d(\gamma,\gamma)d$ come from three measurements: Lucas et al. at Illinois [7], Hornidge et al. at Saskatoon [8], and Lundin et al. at Lund [9]. The results of these experiments are summarized in table 1. The goals of this measurement were therefore

1. to produce more than twenty additional data points with
2. statistical (systematic) uncertainties comparable to those from Illinois (Saskatoon) while
3. the energy bin width $\Delta E_\gamma \approx 8$ MeV and
4. overlapping with the low energy data near 70 MeV and extending up to ~115 MeV.

Data that meets these goals would effectively double the world database for $d(\gamma,\gamma)d$ and significantly reduce the uncertainties of α_n and β_n.

[a] e-mail: lmyers@jlab.org

Table 1. Comparison of reported results from prior $d(\gamma,\gamma)d$ measurements.

Ref.	E_γ [MeV]	ΔE_γ [MeV]	Data Points	$\delta\frac{d\sigma}{d\Omega}$ Stat	Syst
[7]	49, 69	~7	6	4–13%	~4%
[8]	95	20	5	5–10%	~5%
[9]	55, 66	10	18	8–24%	7–14%

2 Experimental Setup

This measurement was performed at the Tagged Photon Facility at MAX IV (see figure 1). Over the course of two run periods, electron beams with energy of 144(165) MeV were used to produce and tag photons with energies of 65–97(81–115) MeV. The electrons were detected in a focal-plane (FP) hodoscope, which is used for both energy determination and cross-section normalization. The bremsstrahlung photons scattered off the target (either deuterium or carbon) and were detected by the large-volume NaI detectors. These detectors consist of NaI cylinders with an outside diameter of 50 cm and a length greater than 50 cm. The detectors also have a cosmic-ray veto shield and an energy resolution of ~2% at 100 MeV. Further details of the facility and detectors can be found in references [4, 10].

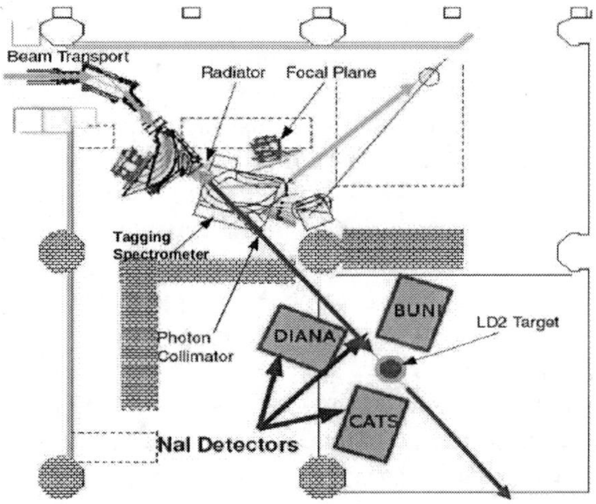

Figure 1. Schematic diagram of the experimental setup at the MAX IV facility. The electron beam impinges on an Al radiator producing bremsstrahlung photons. The recoil electrons are momentum-analyzed by the tagging spectrometer magnet and detected in the focal-plane hodoscope. The photons interact with the target and scatter into the NaI detectors.

2.1 Experimental Challenges

Compton scattering off the deuteron has been difficult to measure precisely. Unlike $p(\gamma,\gamma)p$, the potential for contamination of elastic deuteron data with events from $d(\gamma,\gamma')np$ requires (1) that detectors have sufficient energy resolution to separate the elastic photons or (2) that simulations be used to estimate the photon yield from $d(\gamma,\gamma')np$. Secondly, the $d(\gamma,\gamma)d$ cross section is ~15 nb/sr at these energies. Photon intensity at the target is typically on the order of 1–2 MHz/MeV which necessitates long run periods to collect sufficient statistics while also dealing with a large, untagged background over the same energy range. The approach to handling these issues are to have detectors which good energy resolution and to collect as much data as possible – both of which were accomplished in this experiment.

A third challenge arose during data collection and analysis that is unique to this experiment. As seen in figure 5 of reference [10], the beam suffered from a significant amount of structure. This structure, which was eventually connected to an incomplete filling of the electron ring and the frequency of the extraction shaker, led to large ($\geq$ 4 MHz) instantaneous electron rates in the FP hodoscope. As a result, the cross sections suffered from rate-dependent effects (see sect. 3). The collaboration developed and benchmarked a Monte Carlo simulation [2, 3] to calculate the correction factors and estimate the uncertainties.

3 Data Analysis

A detailed description of the data analysis and simulations can be found in [10] so only a very brief overview is given here. First, cosmic-ray events were filtered from the raw data. From the reduced data, a missing energy spectra (E_{NaI} - E_{tagged}) was filled for events within the prompt timing peak. A similar histogram was also filled for events in the accidental window. A GEANT4 simulation was employed to determine the detector response lineshape for the elastic photons. Finally, the prompt data are fit with the sum of the accidental spectrum (scaled) and the simulated elastic spectrum. From this fit it is possible to subtract the accidental contribution and extract the photon yield.

The elastic photons were integrated over a range from –2 to +2 MeV in missing energy. This region was deliberately chosen to minimize any contamination from $d(\gamma,\gamma')np$. The analysis determined that the contamination is small and statistically consistent with zero, in part because the elastic lineshape describes the data very well over the integrating region (χ^2/d.o.f. $\approx$ 1). Total yields were on the order of 10^3 events in each energy/angle bin, with statistical uncertainties of 6–16%.

Normalization factors for the cross section included the number of photons incident on the target, the thickness (areal density) of the target, and the solid angle subtended by each NaI detector. For the deuterium target, data was also collected using an empty cell in order to subtract any scattering contribution originating from the cell.

Monte Carlo simulations were used to determine correction factors arising from event filters and the extended geometry of the setup (GEANT4), the effective target thickness, and rate-dependent effects in the FP hodoscope. The systematic uncertainties on these terms was typically on the order of a few percent.

4 Results

The analysis steps from the previous section were first applied to data collected using a carbon target. As shown in [4], the cross sections are in excellent agreement with prior measurements. These results indicate that the corrections and their uncertainties are accurately described by the Monte Carlo simulations. As such, it is reasonable to apply the same method to the data collected with the deuterium target.

The final cross sections for $d(\gamma,\gamma')d$ are shown in figure 2. The error bars shown are statistical only. The systematic (normalization) uncertainties are ~5% for all the data points. There is an additional uncertainty of 3–4% correlated for all points at a given angle arising from differences in detection efficiency and solid angle for each NaI. The open symbols correspond to previous measurements at similar kinematics and indicate the consistency of the world data set.

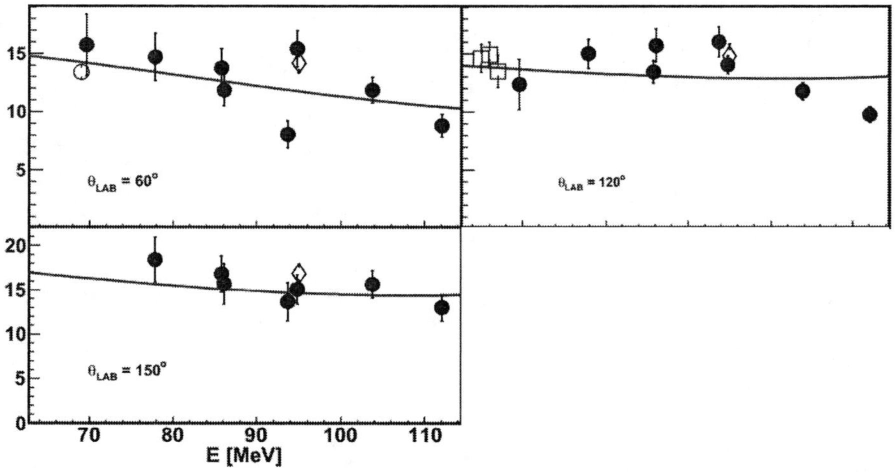

Figure 2. Cross sections from the experiment presented here (•). Open symbols are results from previous measurements, and the solid line represents the best fit from reference [1].

5 Conclusions

The results from the COMPTON@MAX-lab collaboration represent the addition of 23 new data points to the world data set with uncertainties comparable to the previous data. The energy range for the Compton scattering cross section has been extended by ~20 MeV. As a result of these data, a new extraction of the neutron polarizabilities [1] yields $\alpha_n = 11.55 \pm 1.25_{\text{stat}}$ and $\beta_n = 3.65 \mp 1.25_{\text{stat}}$, where the statistical uncertainties are reduced by ~30%. This is an important first step in developing a deuteron Compton scattering database comparable to the one for the proton. Future data from MAX IV, as well as upcoming measurements at HIγS and Mainz, should continue to add to the database.

Acknowledgements

The author acknowledge the support of the COMPTON@MAX-lab collaboration, the staff of the MAX IV Laboratory, and the Data Management and Software Centre, a Danish contribution to the European Spallation Source ESS AB, for generously providing access to their computations cluster. This work was sponsored in part by the US National Science Foundation under Award Number 0855569.

References

[1] L. S. Myers et al., Phys. Rev. Lett. **113**, 262506 (2014).
[2] L. S. Myers et al., Nucl. Inst. Meth. A **729**, 707 (2013).
[3] M. F. Preston et al., Nucl. Inst. Meth. A **744**, 17 (2014).
[4] L. S. Myers et al., Phys. Rev. C **89**, 035202 (2014).
[5] J. A. McGovern, D. R. Phillips, and H. W. Griesshammer, Eur. Phys. J. A **49**, 12 (2013).
[6] H. W. Griesshammer, J. A. McGovern, D. R. Phillips, and G. Feldman, Prog. Part. Nucl. Phys. **67**, 841 (2012).
[7] M. A. Lucas, Ph.D. thesis, University of Illinois at Urbana-Champaign, unpublished (1994).

[8] D. L. Hornidge et al., Phys. Rev. Lett. **84**, 2334 (2000).
[9] M. Lundin et al., Phys. Rev. Lett. **90**, 192501 (2003).
[10] L. S. Myers et al., Phys. Rev. C **92**, 025203 (2015).

EPJ Web of Conferences **113**, 04006 (2016)
DOI: 10.1051/epjconf/201611304006

News on Compton Scattering $\gamma X \to \gamma X$ in Chiral EFT

Harald W. Grießhammer[1,a], Judith A. McGovern[2], and Daniel R. Phillips[3]

[1] *Institute for Nuclear Studies, Dept. of Physics, George Washington University, Washington DC 20052, USA*
[2] *School of Physics and Astronomy, The University of Manchester, Manchester M13 9PL, UK*
[3] *Dept. of Physics and Astronomy, Inst. of Nucl. and Part. Physics, Ohio University, Athens OH 45701, USA*

Abstract. We review theoretical progress and prospects to understand the nucleon's static dipole polarisabilities from Compton scattering on few-nucleon targets, including new values; see Refs. [1–5] for details and a more thorough bibliography.

Why Compton Scattering? Since the electromagnetic field of a real photon induces radiation multipoles by displacing charged constituents and currents, the energy- and angle-dependence of the emitted radiation elucidates the distribution, symmetries and dynamics of the charges and currents which constitute the low-energy degrees of freedom inside the nucleon and nucleus; see e.g. a recent review [1]. *Energy-dependent polarisabilities* parametrise the stiffness of the nucleon N (spin $\frac{\vec{\sigma}}{2}$) against transitions $Xl \to Yl'$ of given photon multipolarity at fixed frequency ω ($l' = l \pm \{0; 1\}$; $X, Y = E, M$; $T_{ij} = \frac{1}{2}(\partial_i T_j + \partial_j T_i)$; $T = E, B$). Up to 500 MeV in photon energy, the relevant terms are:

$$\mathcal{L}_{\rm pol} = 2\pi\, N^\dagger\, [\alpha_{E1}(\omega)\, \vec{E}^2 \;+\; \beta_{M1}(\omega)\, \vec{B}^2 \;+\; \gamma_{E1E1}(\omega)\, \vec{\sigma}\cdot(\vec{E}\times\dot{\vec{E}}) \;+\; \gamma_{M1M1}(\omega)\, \vec{\sigma}\cdot(\vec{B}\times\dot{\vec{B}})$$
$$-\; 2\gamma_{M1E2}(\omega)\, \sigma^i\, B^j\, E_{ij} \;+\; 2\gamma_{E1M2}(\omega)\, \sigma^i\, E^j\, B_{ij} \;+\; \dots \text{(photon multipoles beyond dipole)}]\; N \quad (1)$$

Two scalar polarisabilities $\alpha_{E1}(\omega)/\beta_{M1}(\omega)$ encode electric/magnetic dipole transitions; 4 spin-polarisabilities $\gamma_{E1E1}(\omega)$, $\gamma_{M1M1}(\omega)$, $\gamma_{E1M2}(\omega)$ and $\gamma_{M1E2}(\omega)$ the response of the nucleon's spin-structure. Intuitively interpreted, the electromagnetic field associated with the spin degrees causes bi-refringence in the nucleon (Faraday-effect).

Polarisabilities test our understanding of the subtle interplay between electromagnetic and strong interactions and enter in theoretical determinations of the proton-neutron mass difference, and in the two-photon-exchange contribution to the Lamb shift in muonic hydrogen; see e.g. the introduction of [5]. Finally, nuclear targets provide an opportunity to study not only neutron polarisabilities, but indirectly also the nuclear force, since the photons couple to the charged pion-exchange currents in the nucleus. The values $\alpha_{E1}(\omega = 0)$ etc. are often called "the (static) polarisabilities"; they compress the richness of information extrapolated from data in a wide range of energies between about 70 MeV and the Δ resonance region into just a few numbers. In the canonical 10^{-4} fm^3, Chiral Effective Field Theory (χEFT, see below) gives (Fig. 1 and [2, 4]):

$$\alpha_{E1}^{(\rm p)} = 10.65 \pm 0.35_{\rm stat} \pm 0.2_{\rm Baldin} \pm 0.3_{\rm th}\;,\quad \beta_{M1}^{(\rm p)} = 3.15 \mp 0.35_{\rm stat} \pm 0.2_{\rm Baldin} \mp 0.3_{\rm th}$$
$$\alpha_{E1}^{(\rm n)} = 11.55 \pm 1.25_{\rm stat} \pm 0.2_{\rm Baldin} \pm 0.8_{\rm th}\;,\quad \beta_{M1}^{(\rm n)} = 3.65 \mp 1.25_{\rm stat} \pm 0.2_{\rm Baldin} \mp 0.8_{\rm th} \quad (2)$$

[a] e-mail: hgrie@gwu.edu; presenter.

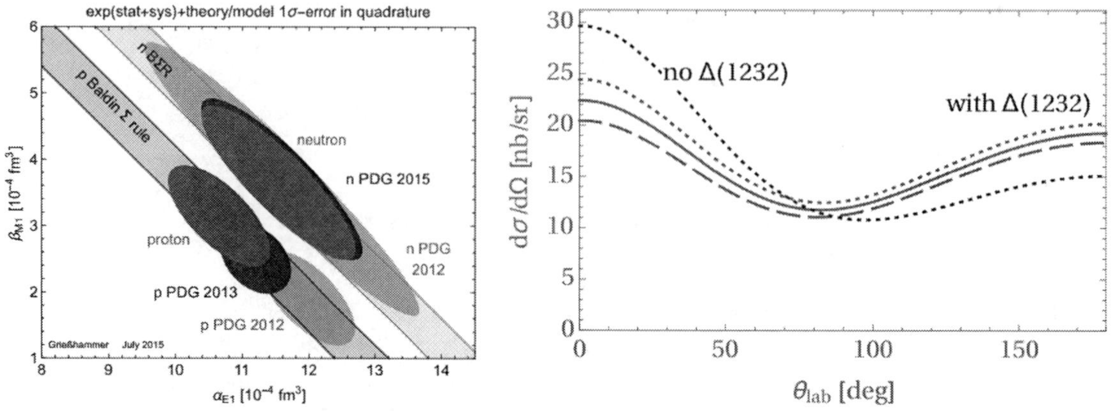

Figure 1. *Left:* Static scalar polarisabilities in our fits (red: proton; blue: neutron); old (green) and new (black) PDG values. 1σ errors, with statistic, systematic and theory error added in quadrature. *Right:* ^{3}He Compton scattering at 120 MeV with $\Delta(1232)$ (red solid), and without (black dotted); blue dashed/green dotted: $\alpha_{E1}^{(n)} \pm 2$.

We also predicted the spin values [2, 5], prior to the pioneering MAMI results [6] (in 10^{-4} fm^4):

	γ_{E1E1}	γ_{M1M1}	γ_{E1M2}	γ_{M1E2}	
χEFT neutron	$-4.0 \pm 1.9_{\text{th}}$	$1.3 \pm 0.5_{\text{stat}} \pm 0.5_{\text{th}}$	$-0.1 \pm 0.6_{\text{th}}$	$2.4 \pm 0.5_{\text{th}}$	(3)
χEFT proton	$-1.1 \pm 1.9_{\text{th}}$	$2.2 \pm 0.5_{\text{stat}} \pm 0.6_{\text{th}}$	$-0.4 \pm 0.6_{\text{th}}$	$1.9 \pm 0.5_{\text{th}}$	
proton MAMI 2014	-3.5 ± 1.2	3.2 ± 0.9	-0.7 ± 1.2	2.0 ± 0.3	

M. Ahmed and L. Myers reported on the concerted ongoing and planned efforts at HIγS [7], MAX-Lab [4, 8], and MAMI [9]. Interpreting such data requires of course commensurate theory support; cf. the open letter by theorists from a variety of backgrounds Ref. [10]. One must carefully evaluate data-consistency for hidden systematic errors; subtract binding effects in few-nucleon systems; extract the polarisabilities; identify their underlying mechanisms and relate them to QCD – and all that with reproducible theoretical uncertainties and minimal theoretical bias.

Indeed, χEFT, the low-energy theory of QCD and extension of Chiral Perturbation Theory to few-nucleon systems, has been quite successful in proton and few-nucleon Compton scattering. With $\delta \equiv \frac{\Delta_M}{\Lambda} \approx \left(\frac{m_\pi}{\Lambda}\right)^{1/2} < 1$ [11], we recently derived single-nucleon Compton amplitudes from zero energy to about 400 MeV. For $\omega \lesssim m_\pi$, they contain all contributions at $O(e^2\delta^4)$ (N^{4}LO, accuracy $\delta^5 \lesssim 2\%$), and for $\omega \sim \Delta_M$ all at $O(e^2\delta^0)$ (NLO, accuracy $\delta^2 \lesssim 20\%$) [1, 2]. A reproducible and systematically improvable *a priori* estimate of the theoretical accuracies of observables, like in eqs. (2) and (3) or Fig. 5, is essential to uniquely disentangle chiral dynamics from data; see Ref. [5] and Refs. therein.

Figure 2 lists examples of the 3 classes of contributions in few-nucleon systems. Charged exchange currents and rescattering often dominate over the nucleonic structure contributions. Compton photons therefore also provide direct, non-trivial benchmarks how accurately the chiral expansion accounts order-by-order for nuclear binding and its mesonic contributions. For the deuteron, our results are complete at $O(e^2\delta^3)$ from the Thomson limit up to about 120 MeV, including the $\Delta(1232)$ [1].

In the preceding talk, L. Myers eloquently described the difficulties and successes of adding 22 deuteron points at MAXlab [4, 8]. This first new data in over a decade effectively doubled the deuteron's world dataset. Our analysis shows that is well consistent with and within the world dataset ($\chi^2 = 45.2$ for 44 degrees of freedom), and with the Baldin sum rule. These data alone slashed the statistical error by 30% . Just to illustrate the data quality, Figure 3 shows that the χ^2 distribution of the new world dataset agrees quite well with the analytic expectation.

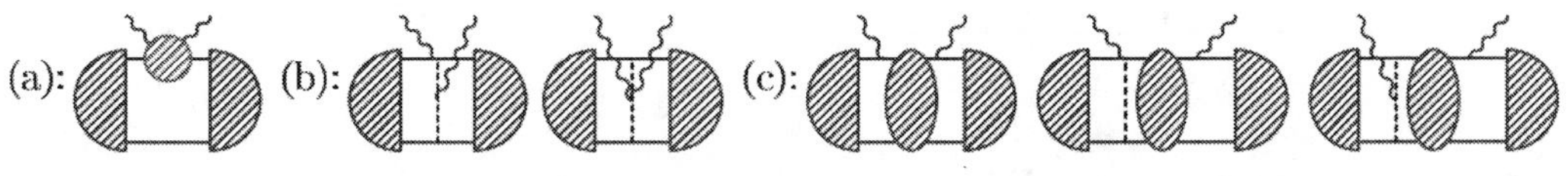

Figure 2. Contributions to deuteron Compton scattering. Ellipse: NN S-matrix:(a): single-nucleon; (b) photon coupling to charged exchange currents which bind the nucleus as dictated by chiral symmetry; (c) rescattering between emission and absorption restores the low-energy Thomson limit and guarantees current conservation.

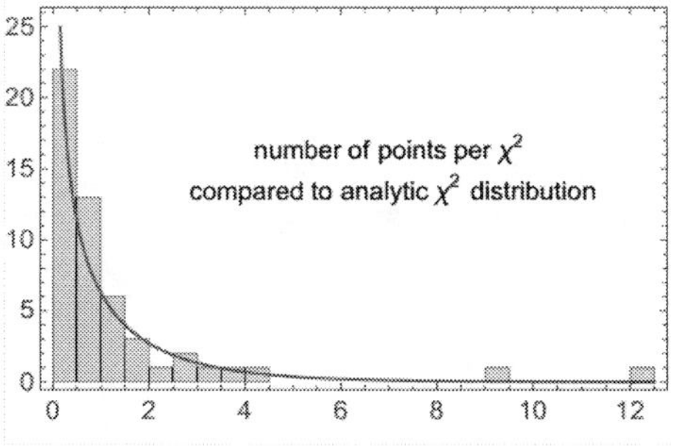

Figure 3. Histogram of the number of deuteron Compton data with a given χ^2, overlaid with the predictions of an ideal, statistically consistent set. The two data with $\chi^2 \geq 9$ are pruned by statistical-likeliness criteria; including them does not have a significant impact on the neutron values. We add point-to-point and angle-dependent systematic errors in quadrature to the statistical error, and subsume overall systematic errors into a floating normalisation. The norm of each dataset floats by $\leq 5\%$ and within the respective quoted normalisation errors of the data.

The future lies in un-, single- and double-polarised experiments of high accuracy and theoretical analyses with reproducible systematic uncertainties. To understand the subtle differences of the pion clouds around the proton and neutron induced by explicit chiral symmetry breaking in QCD, we need the neutron polarisabilities with uncertainties comparable to those of the proton – eq. (2) shows that this is mostly an issue of better data. Therefore, MAMI, MAXlab and HIγS aim for deuteron data with statistical and systematic uncertainties of better than 5%, and plan extensions to ^{3}He. In general, heavier nuclei are experimentally better to handle and provide count rates which scale at least linearly with the target charge when photons scatter incoherently from the protons, i.e. for $\omega \gtrsim 100$ MeV. But a theoretical description of their energy levels with adequate accuracy is involved. For the proton, amplitudes on the $\lesssim 2\%$-level are available; for deuteron and ^{3}He, we now extend descriptions with similar accuracies into the Delta resonance region. Around ^{3}He-^{4}He-^{6}Li may well be the "sweet-spot" between needs and wants of theorists and experimentalists. Figure 1 shows that not including the $\Delta(1232)$ can lead to false signals in high-accuracy extractions of neutron polarisabilities [12].

Since the 8 spin-polarisabilities are a top priority of experiment and theory alike, we performed a series of sensitivity studies; see Fig. 4 and summary in [1, Sec. 6.1]. Recently, the deuteron cross section and asymmetry with arbitrary photon and target polarisations has for example been parametrised via 18 independent observables [3]. Particularly attractive are some asymmetries which are sensitive to only 1 or 2 polarisabilities. For spin polarisabilities with an error of $\pm 2 \times 10^{-4}$ fm^4, asymmetries should be measured with an accuracy of $\gtrsim 10^{-2}$, with differential cross sections of a dozen nb/sr at 100 MeV or a few dozen nb/sr at 250 MeV. Relative to single-nucleon Compton scattering, interference with the deuteron's D wave and pion-exchange current increases the sensitivity to the "mixed" spin polarisabilities γ_{E1M2} and γ_{M1E2}. A *Mathematica* file for $\omega < 120$ MeV is available from `hgrie@gwu.edu` (see screen-shot in Fig. 4), and is finalised for the proton and ^{3}He.

Finally, χEFT connects data with emerging lattice-QCD simulations by reliable extrapolations from numerically less costly, heavier pion masses within the χEFT regime to the physical point and circumvents a direct lattice simulation of Compton scattering – that would be highly nontrivial. The lattice, in turn, tests to which extent χEFT adequately captures the m_π-dependence of the low-energy dynamics, and may predict short-distance (fit) parameters from QCD . W. Detmold's plenary showed intriguing lattice results for the proton's and neutron's β_{M1} at $m_\pi = 806$ MeV; see Fig. 5 and Ref. [5].

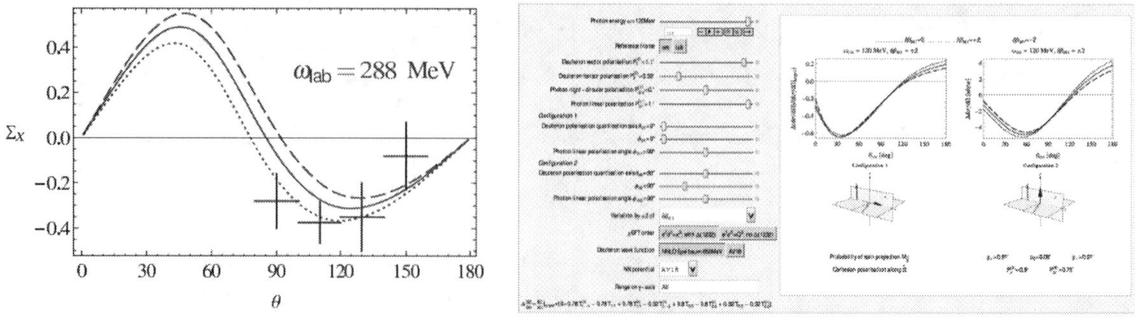

Figure 4. *Left:* χEFT prediction and MAMI data [6] for the double-polarisation observable Σ_{2x} on the proton. Solid: $\gamma_{E1E1} = -1.1$ (predicted); dashed/dotted: theory uncertainty from eq. (3) and Ref. [5]. *Right: Mathematica* notebook screenshot for deuteron Compton scattering with arbitrary polarisations [3].

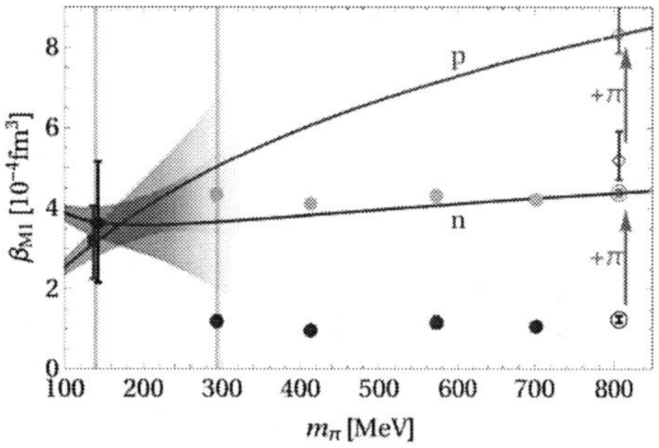

Figure 5. Comparison of our χEFT *predictions* to lattice simulations of β_{M1}: • (neutron) Hall et al.; ⋄ (proton) and ∘ (neutron) NPLQCD. Gray "ghost points" found by shifting all lattice results by $+\pi \times 10^{-4}$ fm^3 [*sic!*]. The *difference* $\beta_{M1}^{(p)} - \beta_{M1}^{(n)}$ is nearly identical to the chiral result even well beyond the range in which χEFT should be applicable. This suggests that the experimental finding $\beta_{M1}^{(p)} \approx \beta_{M1}^{(n)}$ is something of a coincidence. The agreement with simulations for α_{E1} is even better [5]. Corridors: theoretical uncertainties in the regime where χEFT can be expected to converge [5].

Acknowledgements

HWG cordially thanks the organisers for a stimulating atmosphere and incessant promotion of Chicago's highlights of popular comics culture. This work was supported in part by UK Science and Technology Facilities Council grants ST/J000159/1 and ST/L005794/1 (JMcG), by the US Department of Energy under contracts DE-FG02-93ER-40756 (DRP) and DE-FG02-95ER-40907 (HWG), and by the Dean's Research Chair programme of the Columbian College of Arts and Sciences of The George Washington University (HWG).

References

[1] H.W. Grießhammer, J.A. McGovern, D.R. Phillips, G. Feldman, Prog.Part.Nucl.Phys.**67**,841(2012).
[2] J. A. McGovern, D. R. Phillips and H. W. Grießhammer, Eur. Phys. J. A **49** 12 (2013).
[3] H. W. Grießhammer, Eur. Phys. J. A **49** 100 (2013).
[4] L. S. Myers *et al.* [COMPTON@MAX-lab Collaboration], Phys. Rev. Lett. **113**, 262506 (2014).
[5] H. W. Grießhammer, J. A. McGovern and D. R. Phillips, arXiv:1511.01952.
[6] P. P. Martel *et al.* [A2 Collaboration], Phys. Rev. Lett. **114**, 112501 (2015).
[7] H. R. Weller *et al.*, Prog. Part. Nucl. Phys. **62** 257 (2009).
[8] L. Myers *et al.*, Phys. Rev.**C92** 025203 (2015).
[9] G. M. Huber and C. Collicott, arXiv:1508.07919 [nucl-ex].
[10] H. W. Grießhammer, A. I. L'vov, J. A. McGovern, V. Pascalutsa, B. Pasquini and D. R. Phillips, arXiv:1409.1512 [nucl-th].
[11] V. Pascalutsa and D. R. Phillips, Phys. Rev. C **67**, 055202 (2003).
[12] H. W. Grießhammer, J. A. McGovern, D. R. Phillips, D. Shukla and B. Strandberg, forthcoming.

EPJ Web of Conferences **113**, 04007 (2016)
DOI: 10.1051/epjconf/201611304007

Electric dipole polarizability: from few- to many-body systems

Mirko Miorelli[1,2,a], Sonia Bacca[1,3,b], Nir Barnea[4], Gaute Hagen[5,6], Giuseppina Orlandini[7,8], and Thomas Papenbrock[5,6]

[1] *TRIUMF, 4004 Wesbrook Mall, Vancouver, British Columbia, Canada V6T 2A3*
[2] *Department of Physics and Astronomy, University of British Columbia, Vancouver, British Columbia, Canada V6T 1Z4*
[3] *Department of Physics and Astronomy, University of Manitoba, Winnipeg, Manitoba, Canada R3T 2N2*
[4] *Racah Institute of Physics, Hebrew University, 91904 Jerusalem, Israel*
[5] *Physics Division, Oak Ridge National Laboratory, Oak Ridge, Tennessee 37831, USA*
[6] *Department of Physics and Astronomy, University of Tennessee, Knoxville, Tennessee 37996, USA*
[7] *Dipartimento di Fisica, Universitá di Trento, Via Sommarive 14, I-38123 Trento, Italy*
[8] *Istituto Nazionale di Fisica Nucleare, TIFPA, Via Sommarive 14, I-38123 Trento, Italy*

Abstract. We review the Lorentz integral transform coupled-cluster method for the calculation of the electric dipole polarizability. We benchmark our results with exact hyperspherical harmonics calculations for ^{4}He and then we move to a heavier nucleus studying ^{16}O. We observe that the implemented chiral nucleon-nucleon interaction at next-to-next-to-next-to-leading order underestimates the electric dipole polarizability.

1 Introduction

The electric dipole polarizability is a fundamental quantity to understand nuclear dynamics and is related to the response of the nucleus to an external electric field. Typically, it is measured via photo-absorption reactions [1], Compton scattering [2] or hadronic processes, such as (p, p') reactions [3]. Ab-initio calculations of this observable were traditionally available only for very light nuclei. It is our goal to extend such studies to the medium mass regime using coupled-cluster theory.

2 Review of the method

The electric dipole polarizability can be calculated from the dipole response function $R(\omega)$ as an inverse energy weighted sum rule

$$\alpha_D = 2\alpha \int_{\omega_{th}}^{\infty} \frac{R(\omega)}{\omega} d\omega. \tag{1}$$

In the coupled-cluster formalism [4] the dipole response function is

$$R(\omega) = \sum_n \langle 0_L | e^{-\hat{T}} \Theta^\dagger e^{\hat{T}} | n_R \rangle \langle n_L | e^{-\hat{T}} \Theta e^{\hat{T}} | 0_R \rangle \delta(E_n - E_0 - \omega), \tag{2}$$

[a] e-mail: mmiorelli@triumf.ca
[b] e-mail: bacca@triumf.ca

where $\hat{\Theta}$ is the dipole excitation operator, $\hat{T}$ is the cluster operator defined in coupled-cluster theory [5], and $\langle 0_L|$, $|0_R\rangle$ ($\langle n_L|$, $|n_R\rangle$) are the left and right reference ground states (excited states), respectively. Using Eq. (1) we can then rewrite the electric dipole polarizability as

$$\alpha_D = 2\alpha\langle 0_L|\overline{\Theta}^\dagger|\tilde{\Psi}_R\rangle, \tag{3}$$

where $\overline{\Theta} = e^{-\hat{T}}\Theta e^{\hat{T}}$ is the similarity transformed operator and $\tilde{\Psi}_R$ is the solution of the Schrödinger-like equation

$$(\overline{H} - \Delta E_0)|\tilde{\Psi}_R\rangle = \overline{\Theta}|0_R\rangle, \tag{4}$$

where ΔE_0 is the correlation energy of the ground state. This equation is obtained using the Lorentz integral transform method [6] in the coupled-cluster formulation [4, 7]. Similarly to what was done in [8, 9], the polarizability can be expressed as a continued fraction of the Lanczos coefficients [10]

$$\alpha_D = 2\alpha\langle 0_L|\overline{\Theta}^\dagger\overline{\Theta}|0_R\rangle\left\{\frac{1}{a_0 - \frac{b_0^2}{a_1 - \frac{b_1^2}{a_2 - \cdots}}}\right\}. \tag{5}$$

If one is able to calculate the response function of Eq. (2), then the polarizability could be obtained using the definition in Eq. (1). However, using Eq. (5) allows us to calculate the polarizability directly from the Lanczos coefficients, thus avoiding further uncertainties coming from the inversion of the transform. Furthermore, due to the nature of the continued fraction, convergence is reached with less than 100 Lanczos steps. Eq. (5) is an exact result, free of any approximation, if the cluster expansion is performed up to *A-particle-A-hole* excitations, *A* being the total number of nucleons. However, in practical calculations the cluster expansion is truncated since a full expansion is not feasible due to the very demanding computational cost. In the results we present here we truncated the cluster operators at the singles-and-doubles excitations level and dub this approximation scheme LIT-CCSD.

3 Results and Conclusions

3.1 Benchmark in ^{4}He

While the LIT-CCSD method has been already benchmarked with the effective interaction hyperspherical harmonic (EIHH) method [11] for the dipole response function [4], here we show a benchmark for the electric dipole polarizability calculated using Eq. (5) with a chiral nucleon-nucleon force derived at next-to-next-to-next-to-leading order [12].

In Figure 1, we show the electric dipole polarizability of ^{4}He as a function of the model space size N_{max}. The green band represents the combined experimental values from [13, 14] which have been extracted by fitting to experimental photo-absorption data [15, 16]. The continuous line (brown) is the polarizability obtained with the EIHH method. The lines with scatter-symbols are the results from the LIT-CCSD method obtained for different values of the frequency $\hbar\Omega$ of the harmonic oscillator basis used to expand the potential. As expected, the polarizability is independent of the oscillator frequency as one increases N_{max}, and converges rapidly to a value $\alpha_D = 0.0815$ fm^3, which compares fairly well with the EIHH [9] result of $\alpha_D = 0.0831$ fm^3 and $\alpha_D = 0.0822(5)$ fm^3 obtained using the no-core-shell-model with the same interaction [17]. Moreover, the convergence with respect to the number of Lanczos coefficients used in Eq. (5) is much faster than the one observed for the response function in [4]. Already with 15 Lanczos coefficients the error with respect to the converged value is 0.3%, which shows the advantage of using directly Eq. (5). The small difference between the LIT-CCSD and EIHH results can thus be attributed to the truncation to singles-and-doubles excitations only.

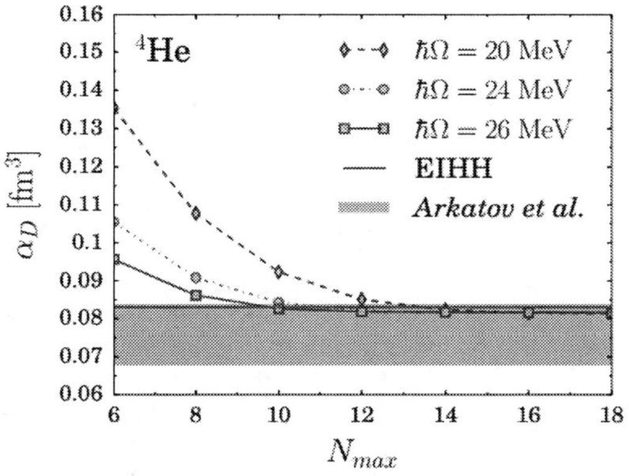

Figure 1. The electric dipole polarizability in ^{4}He as a function of the model space size N_{max} for different values of the harmonic oscillator frequency $\hbar\Omega$ (black, red, blue). The converged value at N_{max} = 18 is compared with the result obtained from the EIHH method (brown) and the experimental data from [13–16] (green band).

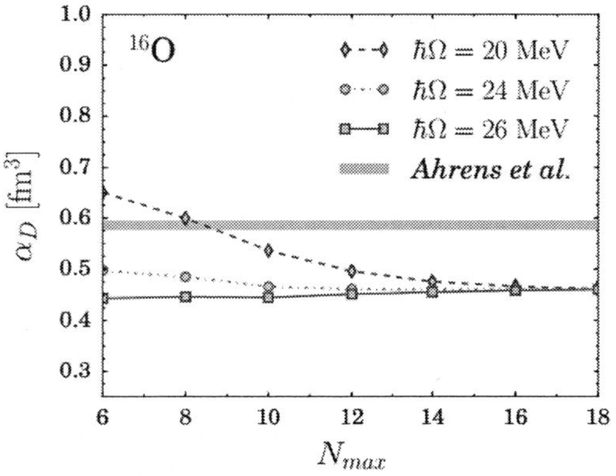

Figure 2. Comparison of the electric dipole polarizability in ^{16}O as a function of the model space size N_{max} for three different values of $\hbar\Omega$ = 20, 24 and 26 MeV (black, red, blue) with the experimental data from Ahrens *et al.* [1] (green).

3.2 Application to ^{16}O

The very nice agreement in the benchmark for the ^{4}He polarizability suggests that we can extend the calculation of α_D via this method to heavier nuclei. We then consider ^{16}O, whose response function has been already studied previously [4, 7]. In Figure 2, we study the convergence of the polarizability as a function of the model space size. Again, the convergence is very fast and the final value α_D = 0.461 fm^3 is independent of the underlying harmonic oscillator frequency. However, the converged value is rather low if compared to the experimental one of α_D^{exp} = 0.585(9) [1]. This trend is observed also in heavier nuclei such as ^{40}Ca and ^{48}Ca [4, 10], and the discrepancy is even larger when the mass number is increased. As shown in Figure 1, the effect of the cluster expansion truncation is tiny and, because of the size extensive nature of coupled-cluster theory, we expect the effect to be very small in heavier systems as well. Discrepancies between the calculated values and the experimental data are mainly attributed to deficiencies of the employed Hamiltonian.

3.3 Conclusions

We used the LIT-CCSD method to extend the calculation of the electric dipole polarizability from few- to many-body nuclei using nucleon-nucleon chiral interactions. We have shown the reliability of the method with benchmarks in ^{4}He where both EIHH and LIT-CCSD calculations and the experimental data are in agreement. Moving to heavier systems, *e.g.*, ^{16}O, we observe disagreement between calculated and measured values. This points to the inadequacy of chiral interactions at the two-body level in accounting for the dipole polarizability in heavier nuclei, which we also observed for the radii [4, 10]. The use of newly developed Hamiltonians with three-nucleon forces [18] calibrated on radii of finite nuclei can help solving this problem. Work in this direction is underway.

Acknowledgements

This work was supported in parts by the Natural Sciences and Engineering Research Council (NSERC), the National Research Council of Canada, the US-Israel Binational Science Foundation (Grant No. 2012212), the Pazy Foundation, the U.S. Department of Energy (Oak Ridge National Laboratory), under Grant Nos. DEFG02-96ER40963 (University of Tennessee), de-sc0008499 (NUCLEI Sci-DAC collaboration), and the Field Work Proposal ERKBP57.

References

[1] J. Ahrens, H. Borchert, K.H. Czock, H.B. Eppler, H. Gimm, H. Gundrum, M. Kröning, P. Riehn, G. Sita Ram, A. Zieger, and B. Ziegler, Nuclear Physics A **251** 479-492 (1975)

[2] H.W. Griesshammer, J.A. McGovern, D.R. Phillips, G. Feldman, Prog. Part. Nucl. Phys., **67**, 841-897 (2012)

[3] A. Tamii, P. von Neumann-Cosel, and I. Poltoratska, Eur. Phys. J. A **50** 28 (2014)

[4] S. Bacca, N. Barnea, G. Hagen, M. Miorelli, G. Orlandini and T. Papenbrock, Phys. Rev. C **90**, 064619 (2014)

[5] I. Shavitt, and R. J. Bartlett, *Many-body Methods in Chemistry and Physics* (Cambridge University Press, Cambridge UK, 2009) 255-256

[6] Efros, V.D., Leidemann, W., Orlandini, G.: Phys. Lett. B 338, 130 (1994)

[7] S. Bacca, N. Barnea, G. Hagen, G. Orlandini and T. Papenbrock, Phys. Rev. Lett. **111**, 122502 (2013)

[8] N. Nevo Dinur, J. Chen, S. Bacca and N. Barnea, Phys. Rev. C **89**, 064317 (2014)

[9] R. Goerke, S. Bacca, and N. Barnea, Phys. Rev. C **86** 064316 (2012)

[10] M. Miorelli, S. Bacca, N. Barnea, G. Hagen, G. Orlandini and T. Papenbrock, in preparation.

[11] N. Barnea, W. Leidemann, and G. Orlandini, Phys. Rev. C **61**, 054001 (2000)

[12] D. R. Entem, and R. Machleidt, Phys. Rev. C **68** 041001 (2003)

[13] J. L. Friar, Phys. Rev. C **16** 1540-1548 (1977)

[14] K. Pachucki, and A. M. Moro, Phys. Rev. A **75** 032521 (2007)

[15] Y. M. Arkatov, P. I. Vatset, V. I. Voloshuchuk, V. A. Zolenko, I. M. Prokhorets, and V. I. Chimil', Sov. J. Nucl. Phys. **19** 598 (1980)

[16] Y. M. Arkatov, P. I. Vatset, V. I. Voloshuchuk, V. A. Zolenko, and I. M. Prokhorets, Sov. J. Nucl. Phys. **31** 726 (1980)

[17] I. Stetcu, S. Quaglioni, J. L. Friar, A. C. Hayes, and P. Navrátil Phys. Rev. C **79** 064001 (2009)

[18] A. Ekström, G. R. Jansen, K. A. Wendt, G. Hagen, T. Papenbrock, B. D. Carlsson, C. Forssén, M. Hjorth-Jensen, P. Navrátil, and W. Nazarewic, Phys. Rev. C **91** 051301(R) (2015)

EPJ Web of Conferences **113**, 04008 (2016)
DOI: 10.1051/epjconf/201611304008

Few-Nucleon Research at TUNL: Probing Two- and Three-Nucleon Interactions with Neutrons

C.R. Howell[1,a], W. Tornow[1,b], and H. Witała[2,c]

[1] *Department of Physics, Duke University and Triangle Universities Nuclear Laboratory, Durham, NC, USA*
[2] *M. Smoluchowski Institute of Physics, Jagiellonian University, Cracow, Poland*

Abstract. The central goal of few-nucleon research at the Triangle Universities Nuclear Laboratory (TUNL) is to perform measurements that contribute to advancing ab-initio calculations of nuclear structure and reactions. The program aims include evaluating theoretical treatments of few-nucleon reaction dynamics through strategically comparing theory predictions to data, determining properties of the neutron-neutron interaction that are not accessible in two-nucleon reactions, and searching for evidence of long-range features of three-nucleon interactions, e.g., spin and isospin dependence. This paper will review studies of three- and four-nucleon systems at TUNL conducted using unpolarized and polarized neutron beams. Measurements of neutron-induced reactions performed by groups at TUNL over the last six years are described in comparison with theory predictions. The results are discussed in the context of the program goals stated above. Measurements of vector analyzing powers for elastic scattering in A=3 and A=4 systems, differential cross sections for neutron-deuteron elastic scattering and neutron-deuteron breakup in several final-state configurations are described. The findings from these studies and plans for the coming three years are presented in the context of world-wide activities in this front, in particular, research presented in this session.

1 Introduction

The theoretical treatment of few-nucleon systems and light nuclei using realistic models of interactions between nucleons provides a formalism bridge for connecting phenomena in nuclear structure and reactions to quantum chromodynamics (QCD). Studies of bound-state properties and reaction dynamics in these systems allow for assessments of two- and three-nucleon force models over a wide range of phase space. The theory tool kit includes rigorous few-nucleon calculations, ab-initio structure calculations of light nuclei, phenomenological nuclear potential models, effective field theory calculations and lattice QCD computations. Studies of three- and four-nucleon systems at TUNL are discussed with emphasis on measurements sensitive to long-range three-nucleon interactions and to features of the nucleon-nucleon (NN) interaction that are not accessible using two-nucleon systems.

The mechanism by which the residual strong nuclear force is produced by the color interactions in QCD is one of nature's great mysteries. Because hadrons are color neutral the static strong force

[a] e-mail: howell@tunl.duke.edu
[b] e-mail: tornow@tunl.duke.edu
[c] e-mail: ufwitala@cyf-kr.edu.pl

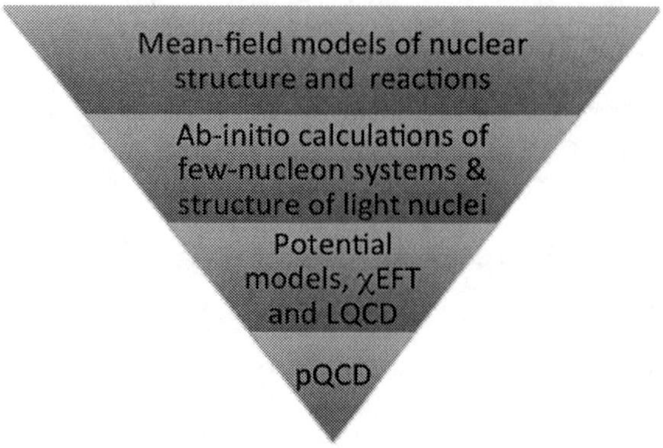

Figure 1. Schematic diagram for coherent theoretical treatment of nuclear systems starting from high energies where perturbative QCD can be applied going to low-energy nuclear phenomena where mean-field potential models are most efficient.

between nucleons is zero. Therefore, the strong nuclear force must be of a dynamical nature similar to Van der Waals interactions between charge-neutral atoms or molecules. For example, the spatial features of the NN potential are qualitatively similar to the London dispersive potential [1], which is due to the induced polarization of the atomic electron cloud relative to the nucleus. In this analogy the NN interaction is associated with the distortion of the pion cloud about the massive core. Qualitatively both potentials are attractive with effective ranges of distances several times the diameter of the interacting bodies and are repulsive at distances where the objects overlap. The dynamical nature of the nuclear force provides for rich phenomena and complexity that makes analysis of the systems interesting and challenging.

Over the last decade substantial progress has been made on developing formalism and computational methods that contribute to having theoretically coherent descriptions of nuclear phenomena with origins in QCD. An ultimate goal would be to describe nuclear matter over wide distance and energy scales with a QCD Lagrangian. Achieving this aim will likely require work done over generations of scientists, similar to other grand challenges in science and technology. A schematic diagram is shown in Fig. 1 of a plausible hierarchy of organizing the theoretical treatments of nuclear systems spanning a variety of phenomena. The scheme starts at ultra high energies, where the most fundamental degrees of freedom are quarks and gluons, and progressively evolves in complexity. This diagram is intended only to represent coarse features that should be included in a coherent picture of strongly interacting matter. In this framework the unifying concept is the residual strong force between nucleons. At the top of the diagram, mean-field potentials that describe nuclear structure properties, collective motion of nuclei and nuclear reactions should be derived from residual strong interactions between nucleons. Ab-initio calculations of the structure of light nuclei [2–5] and few-nucleon reaction dynamics [6] enable refinement of two-nucleon and multi-nucleon interactions using effective degrees of freedom. Current theory tools for describing the strong nuclear force (two- and three-nucleon interactions) include semi-empirical potential models, e.g., Refs. [7–11], effective field theory formulations of two-nucleon (2N) and three-nucleon (3N) interactions [12, 13], and Lattice QCD (LQCD) calculations of few-nucleon systems, e.g., see Refs. [14, 15]. Descriptions of the collective properties of nucleons in terms of effective field theories, e.g., Ref. [16] and LQCD, e.g., Ref. [17], are steps toward bridging gaps between QCD and theoretical treatments of few-nucleon systems.

In this paper we discuss the applications of few-nucleon systems to probe features of three-nucleon interactions (3NIs) and to evaulate current formulations of the neutron-neutron (*nn*) $^1S_0$ interaction. The general approach is to use ab-initio calculations of few-nucleon systems or light nuclei to search for observables that are sensitive to 3N force (3NF) effects or to the $^1S_0$ *nn* interaction. Such theory studies are useful for guiding experimental efforts. However, experiment planning should not be

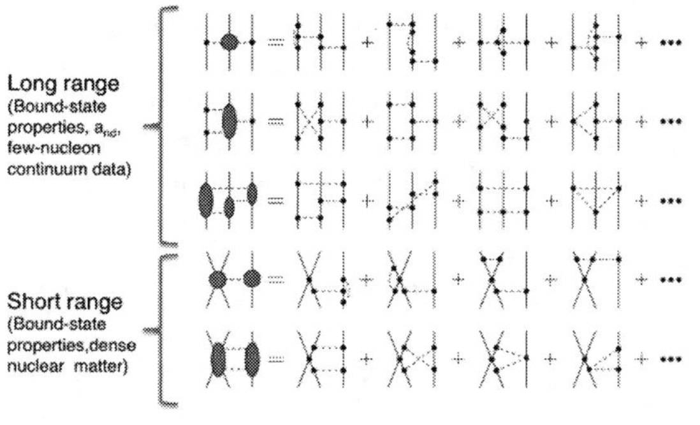

Figure 2. Diagrams of additional 3NI terms when increasing the χEFT expansion from N^2LO to N^3LO. This figure is from Ref. [13]. Expanding to N^3LO results in increasing the number of interaction terms. However, there are no additional LECs when adding these higher order terms to the 3NI expansion. The diagrams can be divided into five topologies with the upper three lines contributing to the long-range component of the force, and the lower two lines providing short-range interactions.

strictly limited by them; there should be a purely exploratory component to provide opportunities for discovering effects not predicted by current theory. The sections that follow provide an overview of few-nucleon research at TUNL with focus on 3NF effects and the *nn* interaction.

2 Three-nucleon Interactions

There is extensive evidence illustrating the importance of 3NIs in ab-initio theoretical treatment of nuclei. Examples include the triton-binding energy discrepancy [8, 10] and systematic studies of the ground and excited states of light nuclei using Monte-Carlo Green function simulations [2] and no-core shell model calculations [3–5]. Also, in the calculation of the nuclear equation of state, the energy per nucleon as a function of density, is highly sensitive to the inclusion of 3NFs, especially at high nuclear densities relevant to the interior of neutron stars [18]. The indications for 3NF effects in low-energy scattering data are less definitive.

There are two types of 3NIs used in modern ab-initio nuclear calculations, semi-empirical potential, e.g., Refs. [10, 11] and effective field theory formulations, e.g., Refs. [12, 19–21]. The lowest order in Chiral Effective Field Theory (χEFT) where 3NI occur is next-to-next-to-leading order (N^2LO). In this order there are two low-energy constants (LECs) associated with 3NI diagrams, often referred to as c_D and c_E [21]. These parameters can be determined by simultaneously fitting, e.g., the triton binding energy and $^2a_{nd}$. In this session, König [22] described calculations of the quartet proton-deuteron (*pd*) scattering length as potentially being another observable for constraining the values of LECs in EFT interactions.

The additional non-vanishing 3NI diagrams added when increasing the expansion to next-to-next-to-next-to-leading order (N^3LO) are shown in Fig. 2. The terms in N^3LO 3NIs can be grouped according to their effective range. The features of the high-momentum components of nuclear wavefunctions and high-density nuclear matter, e.g., the interior of neutron stars as mentioned above, are sensitive to the short-range part of the 3NI. The long-range component of the 3NI, which is the theme of this session, can be probed by studying low-momentum components of nuclear wavefunctions via low-energy few-nucleon reactions. Three presentations in this session are examples of such investigations. Sagara *et al.* [23] describe the energy dependence of the space-star (SST) in *pd* breakup at proton energies below 20 MeV. They found that the ratio of the cross-section data to calculations is nearly constant with the data being about 8% lower than theory. In contrast, as shown in Figs. 3 and 4, neutron-deuteron (*nd*) data for the SST are about 30% larger than theory. The difference between the data and theory decreases with increasing neutron energy. In Fig. 3 the cross section is shown as a function of the arc-length S of the kinematical curve at incident nucleon energy of 13.0 MeV. The cross-section curves are Faddeev calculations using the CD-Bonn [7] NN potential and the Tucson-Melbourne 3N potential model [10]. The *pd* data [26] plotted with the star symbols are about 10%

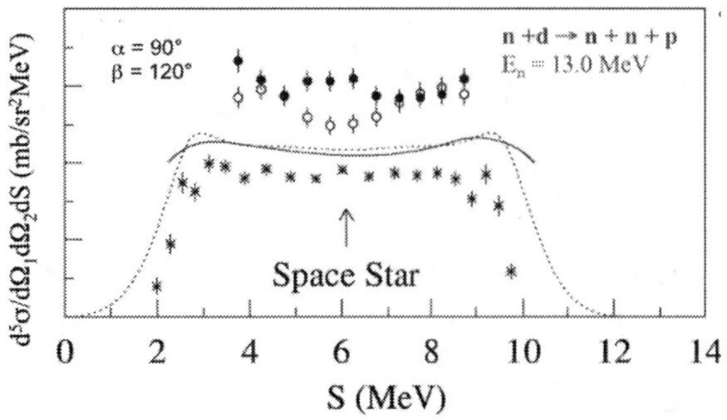

Figure 3. Cross sections for the space-star configuration in *nd* breakup at an incident neutron energy of 13.0 MeV. The average angle of the detected neutrons are $\theta_{n1} = \theta_{n2} = 50.3°$ and $\phi_{12} = 120°$. This figure is from Ref. [24]. The data are *nd* breakup from Ref. [24] (solid circles), nd breakup from Ref. [25] (open circles) and *pd* breakup from Ref. [26] (stars). The curves are predictions of ab initio *nd* breakup calculation with 3NI. The dotted curve is a point-geometry calculation, and the solid curve is a simulation for the TUNL experiment [24].

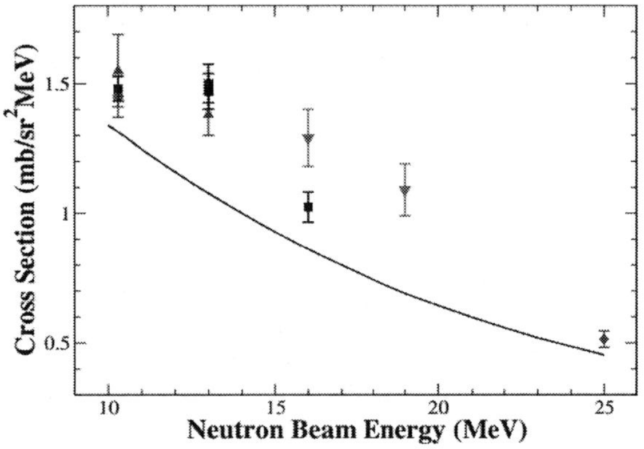

Figure 4. Cross sections for the space-star configuration in *nd* breakup as a function of an incident neutron energy. This figure is from Ref. [27]. The data are *nd* breakup from Refs. [28–30] (squares), [27] (inverted triangles), [25, 31] (triangles), [32] (circles), and [33] (diamond).

lower than the theoretical predictions; the findings reported by Sagara at this conference [23] are consistent with these results. Long-range 3NIs that are not included in current calculations is a possible cause of the discrepancy between data and theory for the SST in nucleon-deuteron breakup.

Also in this session, calculations presented by Girlanda [34] illustrate that a solution to the long-standing "A_y puzzle" in 3N and 4N systems might be provided through the fitting flexibility enabled by the additional LECs for 3NIs in N^4LO EFT interactions. Recent measurements made at TUNL of the vector analyzing power (A_y) for neutron-^{3}He elastic scattering [35] are shown in Fig. 5. As in the case of the 3N system, theory underpredicts the value of $A_y(\theta)$ at the maximum for n^{3}He elastic scattering. One approach to interpreting the A_y issue in the 4N system is through the lens of isospin. Data for both mixed isospin (T=0,1) and pure isospin (T=1) states exist for proton scattering from ^{3}H and ^{3}He, respectively. However, only data for the mixed isospin state (n^{3}He) is available for neutron scattering. Esterline *et al.* [35] found that the energy dependence of the fractional difference between data and theory was similar for the mixed isospin (T = 0, 1) scattering systems of p^{3}H and n^{3}He and has a slope in the energy range studied opposite to that for the pure T=1 p^{3}He scattering. Measurements of $A_y(\theta)$ for n^{3}H elastic scattering are planned using the pulsed polarized neutron beam at TUNL. The goal of this experiment is to provide data that will complement the p^{3}He (T=1) data.

3 Studies of the Neutron-neutron Interaction at TUNL

In the two-nucleon system the values for the $^1S_0$ NN scattering length a_{NN} and effective range parameter r_{NN} are tightly constrained by limits on charge dependence and charge-symmetry violation in the strong NN interaction. This feature along with the high sensitivity of these parameters to the strength of the S-wave NN interaction provide means for probing few-nucleon systems and light nuclei for

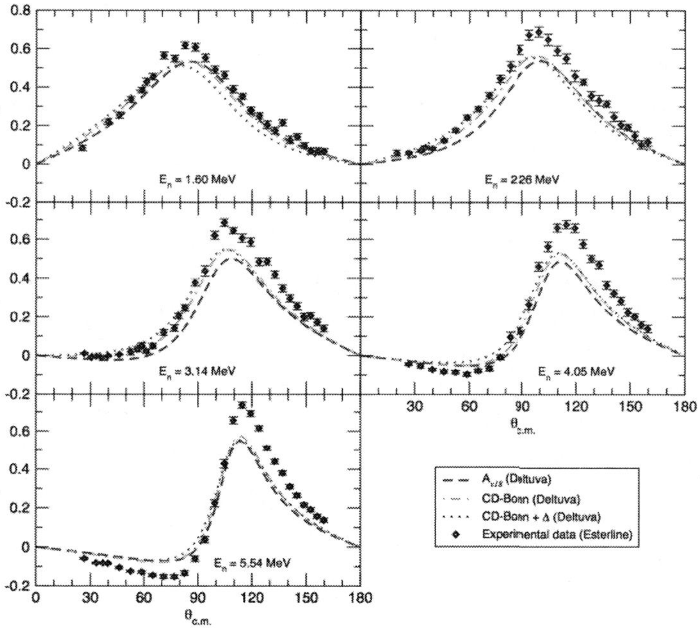

Figure 5. $A_y(\theta)$ for n-^{3}He elastic scattering at several energies. This figure is adapted from Ref. [35]. The data are measurements from TUNL, and the curves are ab-initio calculations with the attributes indicated in the legend.

3NF effects. For example, discrepancies in the values of a_{nn} from $\pi^- d$ capture data and those obtained from some *nd* breakup measurements could indicate 3NF effects that are not included in current *nd* breakup calculations [36] . A more recent observation of effects that could be due to 3NIs that are not included in current *nd* calculations was reported by Witała and Glöckle in their analysis of *nn* quasi-free scattering (QFS) cross-section data in *nd* breakup [37]. The resolution of the about 20% discrepancy between data and theory using current 3NIs required substantial charge-symmetry breaking in the NN interactions beyond what is generally accepted; the value of r_{nn} in their calculation was about 12% larger than allowed by charge symmetry.

There are two experiments underway at TUNL to further investigate these issues in the *nd* continuum. One experiment is to measure the cross section for *nn* QFS scattering in *nd* breakup. This experiment is being carried out at two beam energies using the pulsed neutron beam facility at TUNL. Details and preliminary results for measurements at 10 MeV are presented by Malone *et al.* [38] in this session. The other experiment is a measurement of the differential cross section for exclusive photodisintegration of ^{3}H in the kinematic region of the *nn* FSI. These data will be used to determine a value of a_{nn} to investigate 3NF effects in tritium as probed by the photodisintegration reaction at gamma-ray energies below 20 MeV. The conceptual design of the experiment is described in the poster presentations by Friesen *et al.* [39] and Han *et al.* [40].

Acknowledgements

This work is supported in part by the U.S. Department of Energy under grant No. DE-FG02-97ER41033 and by the Polish National Science Center under Grant No. DEC-2013/10/M/ST2/00420.

References

[1] F. London, Transactions of the Faraday Society **33**, 8–26 (1937); doi:10.1039/tf937330008b
[2] S.C. Pieper and V.R. Pandharipande, Annu. Rev. Nucl. Part. Sci. **51**, 53 (2001)

[3] P. Navratil *et al.*, Phys. Rev. Lett. **99**, 042501 (2007)
[4] P. Navratil *et al.*, Few-Body Syst. **43**, 129 (2008)
[5] A. Nogga, P. Navratil, B.R. Barrett and J.P. Vary, Phys. Rev. C **73**, 064002 (2006)
[6] W. Glöckle *et al.*, Phys. Rep. **274**, 107 (1996)
[7] R. Machleidt, Phys. Rev. C **63**, 024001 (2001)
[8] R.B. Wiringa, V.G.J. Stoks and R. Schiavilla, Phys. Rev. C **51**, 38 (1995)
[9] V.G.J. Stoks, R.A.M. Klomp, C.P.F. Terheggen and J.J. de Swart, Phys. Rev. C **49**, 2950 (1994)
[10] S.A. Coon, M.D. Scadron and B.R. Barrett, Nucl. Phys. A **242**, 467 (1975)
[11] S.C. Pieper, V.R. Pandharipande, R.B. Wiringa and J. Carlson, Phys. Rev. C **64**, 014001 (2001)
[12] R. Machleidt and D.R. Entem, Phys. Rep. **503**, 1 (2011)
[13] E. Epelbaum, H.-W. Hammer and Ulf-G. Meissner, Rev. Mod. Phys. **81**, 1773 (2009)
[14] E. Epelbaum, H Krebs, D. Lee and Ulf-G. Meissner, Phys. Rev. Lett. **106**, 192501 (2011)
[15] S.R. Beane *et al.*, Phys. Rev. Lett. **115**, 132001 (2015)
[16] H.W. Griesshammer, J.A. McGovern, D.R. Phillips and G. Feldman, Prog. Part. Nucl. Phys. **67**, 841 (2012)
[17] S. Beane *et al.*, Phys. Rev. Lett. **113**, 252001 (2014)
[18] P. Danielewicz, AIP Conf. Proc. **597**, 24 (2001)
[19] R. Higa and M.R. Robilotta, Phys. Rev. C **68**, 024004 (2003)
[20] R. Higa, M.R. Robilotta and C.A. da Rocha, Phys. Rev. C **69**, 034009 (2004)
[21] E. Epelbaum *et al.*, Phys. Rev. C **66**, 064001 (2002)
[22] S. König and H. Hammer, Proc. Int. Conf. on Few-Body Problems in Physics (2015)
[23] K. Sagara *et al.*, Proc. Int. Conf. on Few-Body Problems in Physics (2015)
[24] H.R. Setze *et al.*, Phys. Lett. B **388**, 229 (1996)
[25] J. Strate *et al.*, Nucl. Phys. A **501**, 51 (1989)
[26] G. Rauprich *et al.*, Nucl. Phys. A **535**, 313 (1991)
[27] A.H. Couture *et al.*, Phys. Rev. C **85**, 054004 (2012)
[28] H.R. Setze *et al.*, Phys. Rev. C **71**, 034006 (2005)
[29] A.S. Crowell, Ph.D. thesis, Duke University, 2001
[30] R.A. Macri, Ph.D. thesis, Duke University, 2004
[31] K. Gebhardt *et al.*, Nucl. Phys. A **561**, 232 (1993)
[32] M. Stephan *et al.*, Phys. Rev. C **39**, 2133 (1989)
[33] Z. Zhou *et al.*, Nucl. Phys. A **684**, 545c (2001)
[34] L. Girlanda *et al.*, Proc. Int. Conf. on Few-Body Problems in Physics (2015)
[35] J.H. Esterline *et al.*, Phys. Rev. Lett. **110**, 152503 (2013)
[36] C.R. Howell, Proc. X Intern. Seminal. on Interactions of Neutrons with Nuclei, Dubna, Russia, ISINN-10 (2003), p. 71; arXiv:0805.1177
[37] H. Witała and W. Glöckle, Phys. Rev. C **83**, 034004 (2011)
[38] R.C. Malone *et al.*, Proc. Int. Conf. on Few-Body Problems in Physics (2015)
[39] F.Q.L. Friesen *et al.*, Proc. Int. Conf. on Few-Body Problems in Physics (2015)
[40] Z. Han *et al.*, Proc. Int. Conf. on Few-Body Problems in Physics (2015)

EPJ Web of Conferences **113**, 04009 (2016)
DOI: 10.1051/epjconf/201611304009

Tuning the $3N$ force from $3N$ scattering data

Luca Girlanda[1,2,a], Alejandro Kievsky[3,b], Michele Viviani[3,c], and Laura Elisa Marcucci[3,4,d]

[1] *Department of Mathematics and Physics, University of Salento, I-73100 Lecce, Italy*
[2] *INFN Sez. di Lecce, I-73100 Lecce, Italy*
[3] *INFN Sez. di Pisa, I-56127 Pisa, Italy*
[4] *Department of Physics, University of Pisa, I-56127 Pisa, Italy*

Abstract. We report on our progress in the inclusion of the subleading $3N$ contact operators for the elastic $p-d$ scattering below the deuteron break-up. We find that, with a nuclear interaction consisting of the AV18 $2N$ potential, supplemented by the leading and subleading $3N$ contact operators, existing discrepancies between theory and experiment, like the well-known A_y puzzle, are substantially reduced.

1 Introduction

Despite long-lasting efforts in the determination of a realistic three-nucleon ($3N$) force, none of the presently available models leads to a satisfactory description of bound and scattering states of the $A = 3$ system [1]. It seems natural to ascribe the above situation to the fact that these models include a very small number of adjustable parameters, compared to the two-nucleon ($2N$) interaction case. For example, in the framework of the chiral expansion, only 2 low-energy constants (LECs) enter up to and including N3LO [2].

In Ref. [3] we have classified all subleading $3N$ contact operators, which involve two powers of nucleon momenta. These terms would show up at the N4LO of the low-energy expansion, but they represent the first correction to the $3N$ force in the pionless version of the effective theory. A matching procedure to the pionful theory at N3LO allows to express them in terms of pion-nucleon LECs [4]. The resulting three-nucleon potential, which depends on 10 LECs ($E_{1,...,10}$) whose values should be fitted to experimental data, can be cast in a local form in coordinate space,

$$
\begin{aligned}
V = \sum_{i\neq j\neq k} \quad & (E_1 + E_2\boldsymbol{\tau}_i\cdot\boldsymbol{\tau}_j + E_3\boldsymbol{\sigma}_i\cdot\boldsymbol{\sigma}_j + E_4\boldsymbol{\tau}_i\cdot\boldsymbol{\tau}_j\boldsymbol{\sigma}_i\cdot\boldsymbol{\sigma}_j)\left[Z_0''(r_{ij}) + 2\frac{Z_0'(r_{ij})}{r_{ij}}\right]Z_0(r_{ik}) \\
& +(E_5 + E_6\boldsymbol{\tau}_i\cdot\boldsymbol{\tau}_j)S_{ij}\left[Z_0''(r_{ij}) - \frac{Z_0'(r_{ij})}{r_{ij}}\right]Z_0(r_{ik}) \\
& +(E_7 + E_8\boldsymbol{\tau}_i\cdot\boldsymbol{\tau}_k)(\mathbf{L}\cdot\mathbf{S})_{ij}\frac{Z_0'(r_{ij})}{r_{ij}}Z_0(r_{ik}) \\
& +(E_9 + E_{10}\boldsymbol{\tau}_j\cdot\boldsymbol{\tau}_k)\boldsymbol{\sigma}_j\cdot\hat{\mathbf{r}}_{ij}\boldsymbol{\sigma}_k\cdot\hat{\mathbf{r}}_{ik}Z_0'(r_{ij})Z_0'(r_{ik})
\end{aligned}
\qquad (1)
$$

a. e-mail: girlanda@le.infn.it
b. e-mail: kievsky@pi.infn.it
c. e-mail: viviani@pi.infn.it
d. e-mail: marcucci@pi.infn.it

where S_{ij} and $(\mathbf{L}\cdot\mathbf{S})_{ij}$ are respectively the tensor and spin-orbit operators for particles i and j, and the function $Z_0(r)$, which also depends on the cutoff Λ, is the Fourier transform of the cutoff function,

$$Z_0(r) = \int \frac{d\mathbf{p}}{(2\pi)^3} e^{i\mathbf{p}\cdot\mathbf{r}} F(\mathbf{p}^2;\Lambda). \tag{2}$$

chosen of gaussian type, $F(\mathbf{p}^2;\Lambda) = \exp -(\mathbf{p}^2/\Lambda^2)$. The 10 additional LECs parametrize the short-range component of the $3N$ force, and are unconstrained by chiral symmetry. As such, they could provide the necessary flexibility to arrive at a truly realistic model for the $3N$ interaction.

2 Numerical procedure

In order to put such conjecture to quantitative scrutiny, we examine to which extent the AV18 $2N$ interaction [5], supplemented by the leading and subleading contact-range $3N$ force, provides a satisfactory fit to the high-precision low-energy $p-d$ scattering data of Ref. [7].

We performed the calculation at $E = 3$ MeV proton energy. The three-body problem was solved with the Hyperspherical Harmonics method, reviewed in Ref. [6]. The $p-d$ scattering wave function with relative orbital angular momentum L , total spin S and total angular momentum J, is written as the sum of an internal and an asymptotic part

$$\Psi_{LSJJ_z} = \Psi_{\mathrm{C}} + \Psi_{\mathrm{A}}, \tag{3}$$

where the internal part is expanded on the Hyperspherical Harmonics basis,

$$\Psi_C = \sum_{\mu} c_\mu \Phi_\mu \tag{4}$$

μ denoting a set of quantum numbers necessary to completely specify the basis element, while the asymptotic part describes the relative motion between the proton and the deuteron at large separation, which takes the form of a linear combination of the regular and irregular solutions of the Coulomb $p-d$ Schroedinger equation at relative momentum q, duly regulated at small distance, $\Omega^{\lambda}_{LSJJ_z}$ with $\lambda = R, I$ respectively,

$$\Psi_A^{LSJJ_z} = \Omega^R_{LSJJ_z} + \sum_{L'S'} \mathcal{R}^J_{LS,L'S'}(q)\Omega^I_{L'S'JJ_z}. \tag{5}$$

The weights $\mathcal{R}^J_{LS,L'S'}$ of the irregular solution relative to the regular one are related to the K matrix whence the scattering phaseshifts and mixing parameters can be determined. They can be found, together with the coefficient c_μ in Eq. (4), from the Kohn variational principle, which requires the functional

$$\left[\mathcal{R}^J_{LS,L'S'}(q)\right] = \mathcal{R}^J_{LS,L'S'}(q) - \langle\Psi_{L'S'JJ_z}|H-E|\Psi_{LSJJ_z}\rangle \tag{6}$$

to be stationary under changes of the variational parameters in Ψ_{LSJJ_z}, normalized such that

$$\langle\Omega^R_{LSJJ_z}|H-E|\Omega^I_{LSJJ_Z}\rangle - \langle\Omega^I_{LSJJ_z}|H-E|\Omega^R_{LSJJ_Z}\rangle = 1. \tag{7}$$

This leads to a linear system

$$\sum_{L''S''} \mathcal{R}^J_{LS,L''S''}(q) X_{L'S',L''S''} = Y_{LS,L'S'}, \tag{8}$$

with the matrices

$$X_{LS,L'S'} = \langle\Omega^I_{LS} + \Psi^I_C|H-E|\Omega^I_{L'S'}\rangle, \quad Y_{LS,L'S'} = \langle\Omega^R_{LS} + \Psi^R_C|H-E|\Omega^I_{L'S'}\rangle, \tag{9}$$

and the $\Psi_C^{R/I}$ solutions of

$$\sum_{\mu'} c_\mu \langle \Phi_\mu | H - E | \Phi_{\mu'} \rangle = -\langle \Phi_\mu | H - E | \Omega_{LS}^{R/I} \rangle. \tag{10}$$

The interest of such a procedure lies in that the matrices can be computed once and for all, for each one of the operators entering the contact potential, and kept on disk. Then, linear combinations depending on the actual values of the LECs E_i can be formed to obtain the corresponding phaseshifts and mixing angles. From the phaseshifts we compute the observables by adding all partial waves up to $J = 15/2$ and $L = 6$. We checked that such truncation does not lead to appreciable differences.

3 Results

We performed multiparameter fits of the $3N$ LECs to the ^{3}H binding energy, the doublet and quartet $N-d$ scattering lengths and to $p-d$ data on the total differential cross section and vector and tensor polarization observables reported in Ref. [7]. We used the POUNDerS algorithm [8] developed at Argonne. It should be noticed that, in the isospin limit, $N-d$ scattering is only sensitive to the isospin $T = 1/2$ component of the $3N$ force. There exists only one combination of the subleading $3N$ contact operators which projects on isospin $T = 3/2$ [1], therefore we can at most determine 9 out of the 10 subleading contact LECs. It could be, however, that $p-d$ scattering alone cannot determine all of them, and only entails correlations among them. This would be apparent if no sensible improvement in the χ^2 is obtained when a parameter is added. This starts to be the case from more than 6 parameters. In Fig. 1 we show the results of a 6-parameter fit for the cutoff $\Lambda = 300$ MeV, showing that the χ^2 can be lowered to 2.0 per degree of freedom.

The description of the polarization observables is much improved, in particular the well-known A_y and T_{11} discrepancies are much decreased. The LECs have been rescaled by inserting appropriate powers of F_π and Λ according to naive dimensional analysis, so that they become adimensional and should be of order 1, if natural. This turns out to be the case for most of them. Similar results are obtained for other values of the cutoff $\Lambda = 200$, 400 and 500 MeV. The exploration of the parameter space is not complete yet, and there is room for a further decrease of the reduced χ^2. Work is under way to check whether the improved description is mantained at different energies.

We also checked, along the lines of Ref. [9], the constraints from the large-N_c limit of QCD. Excluding one of the LECs (e.g. E_8) to account for the $T = 1/2$ projection, and absorbing its effect in the remaining LECs, these constraints take the form

$$E_2 = \frac{2}{3}(E_3 - E_5) + O(1/N_c), \quad E_9 = 3E_5 + O(1/N_c). \tag{11}$$

It is interesting to notice that these constraints are qualitatively satisfied by the fit outcome, while a strict application of them would conflict with the aimed accuracy.

References

[1] A. Kievsky, M. Viviani, L. Girlanda and L. E. Marcucci, Phys. Rev. C **81**, 044003 (2010).

[2] E. Epelbaum, A. Nogga, W. Gloeckle, H. Kamada, U. G. Meissner and H. Witala, Phys. Rev. C **66**, 064001 (2002).
V. Bernard, E. Epelbaum, H. Krebs and U. -G. Meissner, Phys. Rev. C **77** (2008) 064004, Phys. Rev. C **84** (2011) 054001.

1. Strictly speaking this is only true at the level of the momentum space operators, neglecting cutoff effects.

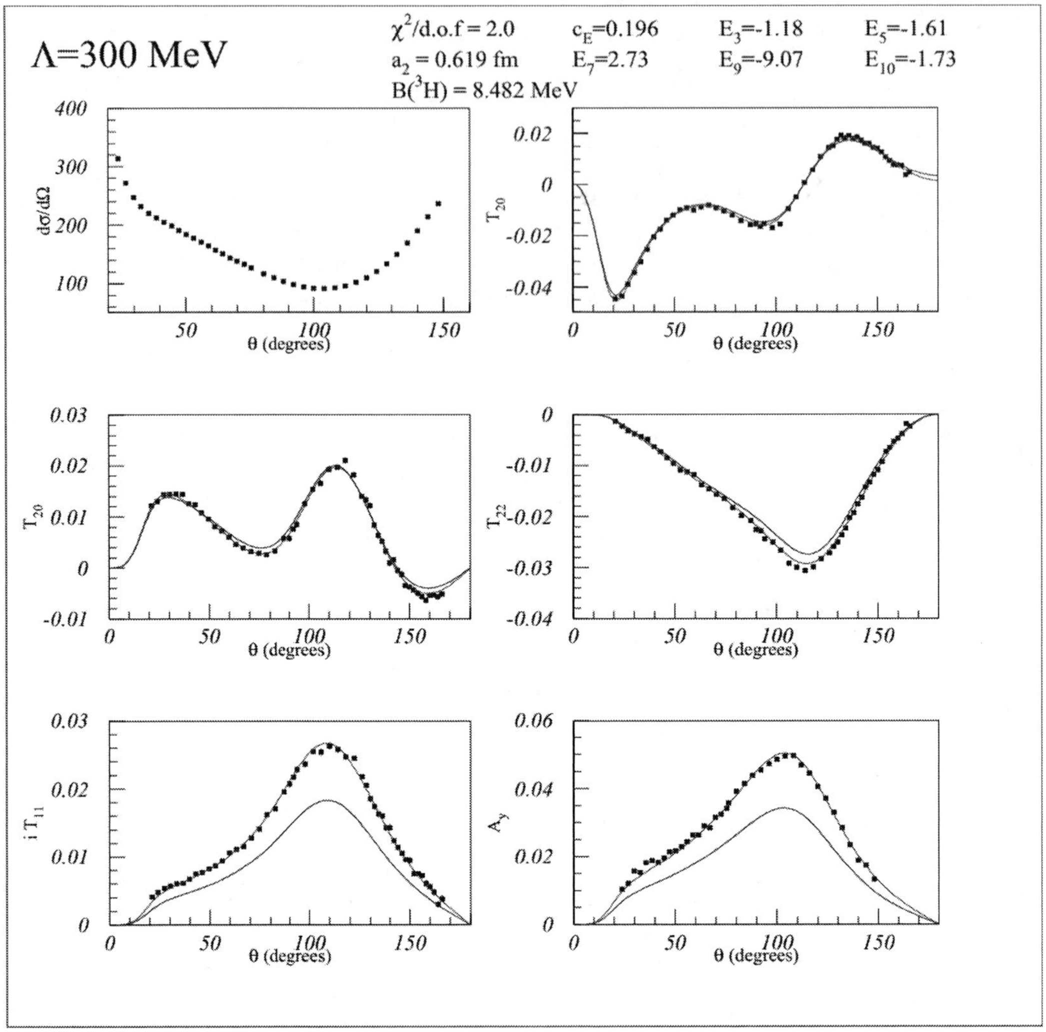

Figure 1. Results of a 6-parameter fit to $p - d$ scattering observables at 3 MeV laboratory energy. Blue lines denote the AV18 prediction, while the red lines include a subset of the contact-range $3N$ force, with the cutoff $\Lambda = 300$ MeV. Also fitted are the triton binding energy $B(^3\text{H})$ and the doublet scattering length a_2.

[3] L. Girlanda, A. Kievsky and M. Viviani, Phys. Rev. C **84** (2011) 014001.
[4] L. Girlanda, A. Kievsky and M. Viviani, PoS QNP **2012**, 142 (2012).
[5] R. B. Wiringa, V. G. J. Stoks and R. Schiavilla, Phys. Rev. C **51**, 38 (1995).
[6] A. Kievsky, S. Rosati, M. Viviani, L. E. Marcucci and L. Girlanda, J. Phys. G **35**, 063101 (2008).
[7] S. Shimizu *et al.*, Phys. Rev. C **52**, 1193 (1995).
[8] T. Munson, J. Sarich, S. Wild, S. Benson, L. McInnes, TAO 2.0 Users Manual, Technical Report ANL/MCS-TM-322, http ://www.mcs.anl.gov/tao
[9] D. R. Phillips and C. Schat, Phys. Rev. C **88**, no. 3, 034002 (2013).

EPJ Web of Conferences **113**, 04010 (2016)
DOI: 10.1051/epjconf/201611304010

Neutron-neutron quasifree scattering in *nd* breakup at 10 MeV

R.C. Malone[1,a], B. Crowe[2], A.S.Crowell[1], L.C. Cumberbatch[1], J.H. Esterline[1], B.A. Fallin[1], F.Q.L. Friesen[1], Z. Han[1], C.R. Howell[1], D. Markoff[2], D. Ticehurst[1], W. Tornow[1], and H. Witała[3]

[1] *Department of Physics, Duke University and Triangle Universities Nuclear Laboratory, Durham, NC, USA*
[2] *Department of Physics, North Carolina Central University and Triangle Universities Nuclear Laboratory, Durham, NC, USA*
[3] *M. Smoluchowski Institute of Physics, Jagiellonian University, Kraków, Poland*

Abstract.
The neutron-deuteron (*nd*) breakup reaction provides a rich environment for testing theoretical models of the neutron-neutron (*nn*) interaction. Current theoretical predictions based on rigorous ab-initio calculations agree well with most experimental data for this system, but there remain a few notable discrepancies. The cross section for *nn* quasifree (QFS) scattering is one such anomaly. Two recent experiments reported cross sections for this particular *nd* breakup configuration that exceed theoretical calculations by almost 20% at incident neutron energies of 26 and 25 MeV [1, 2]. The theoretical values can be brought into agreement with these results by increasing the strength of the $^1S_0$ *nn* potential matrix element by roughly 10%. However, this modification of the *nn* effective range parameter and/or the $^1S_0$ scattering length causes substantial charge-symmetry breaking in the nucleon-nucleon force and suggests the possibility of a weakly bound di-neutron state [3].
We are conducting new measurements of the cross section for *nn* QFS in *nd* breakup. The measurements are performed at incident neutron beam energies below 20 MeV. The neutron beam is produced via the $^2\mathrm{H}(d,n)^3\mathrm{He}$ reaction. The target is a deuterated plastic cylinder. Our measurements utilize time-of-flight techniques with a pulsed neutron beam and detection of the two emitted neutrons in coincidence. A description of our initial measurements at 10 MeV for a single scattering angle will be presented along with preliminary results. Also, plans for measurements at other energies with broad angular coverage will be discussed.

1 Introduction

Despite the success of rigorous ab-initio calculations in predicting most three-nucleon (3N) data, a few notable exceptions remain. One discrepancy between theory and data is the neutron-neutron quasifree scattering (*nn* QFS) cross section in neutron-deuteron (*nd*) breakup. In the quasifree configuration, the projectile transfers momentum to only one of the target nucleons. The other nucleon behaves like a spectator, i.e., its momentum is unchanged by the collision. Measurements of the neutron-proton QFS

[a] e-mail: rm216@phy.duke.edu

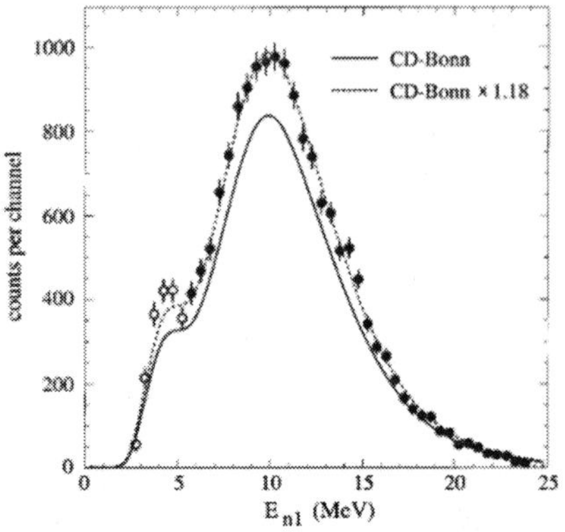

Figure 1. Experimental data (circles) for *nn* QFS scattering in *nd* breakup at 26 MeV. The solid curve is the theoretical prediction from a Monte-Carlo simulation using cross sections calculated with CD-Bonn nucleon-nucleon potential, and the dashed curve represents the simulation scaled by a factor of 1.18. The open circles represent data below the detector energy threshold [1].

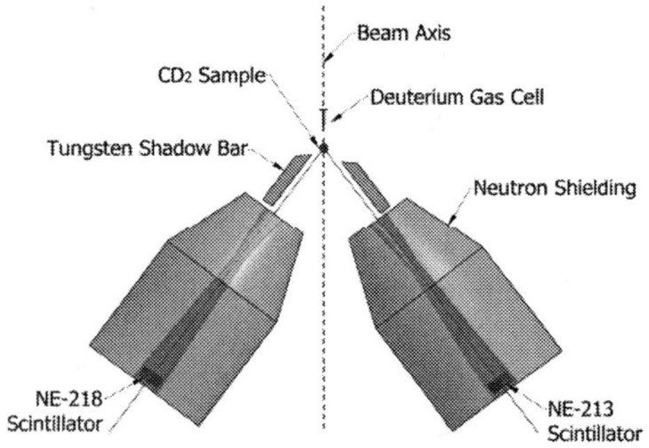

Figure 2. Schematic diagram of the experimental setup for the measurement of the *nn* QFS cross section in *nd* breakup at TUNL (not to scale). The two shielded neutron detectors are placed at equal angles of 36.7° on either side of the beam axis. The two detectors used to monitor the neutron flux are not shown.

cross section are in good agreement with theoretical calculations. However, recent measurements of the *nn* QFS cross section exceed theoretical predictions by almost 20% [1, 2] (see Fig. 1). The theory can be brought into agreement with the data by scaling the $^1S_0$ matrix element by a factor of 1.08. This scaling results in a sign change of the *nn* scattering length, suggesting the existence of a weakly bound di-neutron state. The discrepancy can also be resolved by decreasing the effective range parameter by approximately 10%, implying greater charge-symmetry- and charge-independence-breaking than is expected [3]. Because of the serious implications of this discrepancy, we have undertaken an independent measurement of this cross section.

2 Experimental Setup

Our measurement of the *nn* QFS cross section in *nd* breakup utilized neutron time-of-flight (TOF) techniques. A pulsed neutron beam with a period of 400 ns and pulse width of about 2 ns was used. An uncollimated beam of 10 MeV neutrons produced via the ^{2}H$(d,n)^3$He reaction bombarded a cylindrical deuterated polyethylene target (⌀ = 28.3 mm and h = 36.4 mm). The target was oriented such that its central axis was perpendicular to the incident neutron beam axis. Scattered neutrons were detected by two heavily-shielded liquid scintillators placed at 36.7° on opposite sides of the beam axis (see Fig. 2). Two additional neutron detectors were used to monitor the neutron beam flux.

A minimum detector bias was used to reduce backgrounds from low-energy neutrons and pulse shape discrimination techniques reduced background events from gamma rays. A coincidence between the two neutron detectors was required to identify *nd* breakup events. A two-dimensional coincidence TOF spectrum collected during our measurement is shown in Fig. 3. The raw coicidence

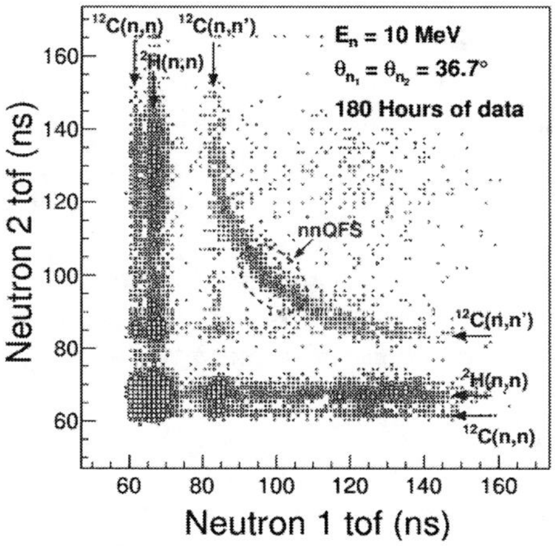

Figure 3. (Color online) Two-dimensional *nd* breakup coincidence spectrum from our measurement. The kinematic locus is clearly visible with the *nn* QFS region circled with a (red) dashed line. Regions with accidental coincidences due to elastic and inelastic scattering are indicated by the (blue) arrows.

TOF spectrum contains both true *nd* breakup events and events due to accidental coincidences. The kinematic locus from breakup events is clearly visible and the quasifree region is well separated from accidental-coincidence backgrounds.

Accidental coincidences are caused by neutrons from two independent events involving neutron scattering from deuterium and carbon. Accidental events were identified using the difference in TOF of the two detected neutrons. The accidental-coincidence backgrounds were measured by forming coincidences of neutrons from two consecutive beam pulses. The net coincidence TOF spectrum is obtained by subtracting the accidental-coincidence spectrum from the raw coincidence spectrum.

3 Preliminary Results

The breakup yields were measured by placing a two-dimensional gate around the experimental kinematic locus and projecting the counts inside this region onto the ideal kinematic locus. The cross section is given by equation 1:

$$\frac{d^5\sigma}{d\Omega_1 d\Omega_2 dS} = \frac{Y_{bu}}{N_n \cdot N_D \cdot \varepsilon(E_1) \cdot \varepsilon(E_2) \cdot \alpha(E_0) \cdot \alpha(E_1) \cdot \alpha(E_2) \cdot d\Omega_1 \cdot d\Omega_2 \cdot dS} \tag{1}$$

where Y_{bu} are the breakup yields, $N_n \cdot N_D$ is the integrated beam-target luminosity, $\varepsilon(E)$ is the neutron detection efficiency, $\alpha(E)$ is the neutron transmission through the sample, and $d\Omega$ is the detector solid angle (the subscript 0 refers to the incoming neutron; the subscripts 1 and 2 refer to the outgoing neutrons). The efficiency factor is for a pulse-height threshold setting of 238.5 keVee ($\frac{1}{2}\times$Cs edge). The luminosity was determined in-situ by normalizing to neutron elastic scattering off of deuterium. Equation 2 was used to calculate the luminosity using the elastic yields (Y_{el}) and the well-established *nd* elastic cross section $\frac{d\sigma}{d\Omega}$ [4]:

$$N_n \cdot N_D = \frac{Y_{el}}{\frac{d\sigma}{d\Omega} \cdot \varepsilon(E_{el}) \cdot \alpha(E_0) \cdot \alpha(E_{el}) \cdot d\Omega} \tag{2}$$

The luminosity measured by the two detectors is then averaged.

The preliminary results of our measurement are shown in Fig. 4, which compares about half of the accumulated experimental data to the theoretical cross section (calculated for a point geometry). There is a systematic uncertainty of ±7% (not shown in the plot) due to uncertainties in the absolute

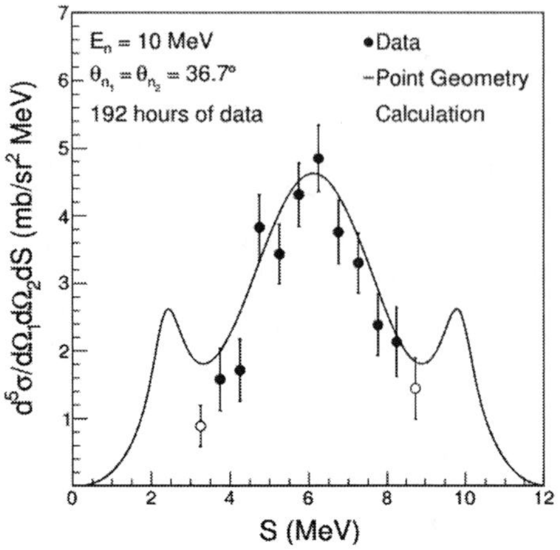

Figure 4. Preliminary results for the *nn* QFS cross section as a function of S (arclength along the kinematic locus). The open circles represent points near the detection energy threshold and have much larger uncertainties than represented on the plot. The theoretical point-geometry cross section was calculated using the CD-Bonn potential. The error bars represent statistical uncertainty only. Systematic uncertainties are discussed in the text.

detector efficiency (5%), the shape of the efficiency curve (3%), neutron attenuation in the sample (2%), and normalization to the *nd* elastic scattering cross section (4%).

A Monte-Carlo simulation is under development to average the theoretical cross section over the finite geometry of the experiment. We do not expect the theoretical prediction to change significantly after implementing the simulation; however, the Monte Carlo will help account for backgrounds associated with the *nd* elastic scattering yields and allow for more accurate determination of the luminosity, which is currently overestimated.

4 Future Work

Data collection at 10 MeV was finished in June. Data analysis is on schedule to be completed by October 2015. A Monte-Carlo simulation is being developed to aid in analysis of current data and to plan future measurements at higher energies. These measurements will be conducted at 13 and 16 MeV at multiple angles using an open detector geometry with a shielded neutron source. Brief tests using the shielded neutron source at TUNL indicate that the signal-to-noise ratio is acceptable to conduct these measurements using this setup. Further tests will be conducted in the fall of 2015.

Acknowledgements

This work is supported in part by the U.S. Department of Energy under grant Nos. DE-FG02-97ER41033 and DE-SC0005367 and by the Polish National Science Center under grant No. DEC-2013/10/M/ST2/00420.

References

[1] A. Siepe, J. Deng, V. Huhn, L. Wätzold, C. Weber, W. von Witsch, H. Witała, W. Glöckle, Phys. Rev. C **65**, 034010 (2002)

[2] X.C. Ruan, Z.Y. Zhou, X. Li, J. Jiang, H.X. Huang, Q.P. Zhong, H.Q. Tang, B.J. Qi, J. Bao, B. Xin et al., Phys. Rev. C **75**, 057001 (2007)

[3] H. Witała, W. Glöckle, Phys. Rev. C **83**, 034004 (2011)

[4] M. Chadwick, M. Herman, P. Obložinský et al., Nuclear Data Sheets **112**, 2887 (2011), special Issue on ENDF/B-VII.1 Library

EPJ Web of Conferences **113**, 04011 (2016)
DOI: 10.1051/epjconf/201611304011

The proton–deuteron scattering length in pionless EFT

Sebastian König[1] and Hans-Werner Hammer[2,3]

[1] *Department of Physics, The Ohio State University, Columbus, Ohio 43210, USA*
[2] *Institut für Kernphysik, Technische Universität Darmstadt, 64289 Darmstadt, Germany*
[3] *ExtreMe Matter Institute EMMI, GSI Helmholtzzentrum für Schwerionenforschung GmbH, 64291 Darmstadt, Germany*

Abstract. We present a fully perturbative calculation of the quartet-channel proton–deuteron scattering length ($^4a_{p\text{–}d}$) up to next-to-next-to-leading order (NNLO) in pionless effective field theory. In particular, we use a framework that consistently extracts the Coulomb-modified effective range function for a screened Coulomb potential in momentum space. We find a natural convergence pattern as we go to higher orders in the EFT expansion. Our NNLO result of (10.9 ± 0.4) fm agrees with older experimental determinations but deviates from more recent calculations, which find values around 14 fm. To resolve this discrepancy, we discuss the scheme dependence of Coulomb subtractions in a three-body system.

1 Introduction

The quartet-channel proton–deuteron scattering length $^4a_{p\text{–}d}$ is a fundamental observable in the nuclear three-body system. The most recent determination of this quantity, carried out by Black *et al.* [1], extracts a value of $^4a_{p\text{–}d} = (14.7 \pm 2.3)$ fm. While this falls in line with theoretical extractions of the quantity based on potential-model calculations [2–4] that find values close to about 13.8 fm, it deviates quite significantly from older experimental results for $^4a_{p\text{–}d}$ between 11 and 12 fm [5, 6]. In this work, we present a theoretical calculation of $^4a_{p\text{–}d}$ using pionless effective field theory (pionless EFT). This EFT is tailored specifically for the systematic description of few-nucleon systems at very low energies. It only includes short-range contact interactions between nucleons [7, 8] and is constructed to reproduce the effective range expansion [9] in the two-body system. Observable are determined as an expansion in the parameter $Q/\Lambda_{\not\pi}$, where $Q \sim \gamma_d \approx 45$ MeV is the typical momentum scale set by the deuteron binding momentum, and $\Lambda_{\not\pi} = O(m_\pi)$ is the natural breakdown scale set by the left-out pion physics; it can alternatively be written in terms of the $N\text{–}N$ scattering lengths and effective ranges.

We obtain here a fully perturbative result up to next-to-next-to-leading order (N^{2}LO). The key feature of our approach is a consistent numerical calculation of the Coulomb-modified effective range function that takes into account the screening of the Coulomb interaction by introducing a small photon mass in the momentum-space Skorniakov-Ter-Martirosian (STM) equation. This method can also be applied to other systems of charged particles, *e.g.*, to halo nuclei, using the EFT constructed to describe such systems [10, 11]. We argue that the Coulomb-modified scattering length in a three-body system like *p–d* depends on the convention used for the subtraction of Coulomb effects. If we use a field-theoretical Coulomb subtraction scheme based on diagrammatic methods, our result agrees quite well with the older experimental determinations. With a two-body subtraction scheme that mimics the approach taken by configuration-space potential model calculations, we can achieve good agreement with 13.8 fm. For a more detailed discussion of the results presented here we refer the reader to Ref. [12].

2 Formalism

For the purpose of this paper, the relevant part of the pionless EFT Lagrangian is

$$\mathcal{L} = N^\dagger\left(\mathrm{i}D_0 + \frac{\boldsymbol{D}^2}{2M_N}\right)N - d^{i\dagger}\left[\sigma_d + \left(\mathrm{i}D_0 + \frac{\boldsymbol{D}^2}{4M_N}\right)\right]d^i + y_d\left[d^{i\dagger}\left(N^T P^i_d N\right) + \text{h.c.}\right] + \mathcal{L}_{\text{photon}}\,, \tag{1}$$

including a nucleon field N (doublet in spin- and isospin space) and a single auxiliary dibaryon field d^i corresponding to the deuteron-channel ($^3S_1$). The electromagnetic sector is included via the covariant derivative $D_\mu = \partial_\mu + \mathrm{i}eA_\mu\hat{Q}$ (with the charge operator $\hat{Q}$ and the photon field A_μ) and a kinetic term for the photons included in $\mathcal{L}_{\text{photon}}$. It suffices here to keep only so-called *Coulomb photons*, which simply correspond to a static potential between charged particles. More details of the formalism can be found, for example, in Ref. [13] and earlier references therein.

The full leading-order dibaryon propagator $\Delta_d(p_0, \mathbf{p})$ is obtained by dressing the bare expression with nucleon bubbles to all orders and matching $-y_d^2\Delta_d(p_0 = k^2/M_N, \mathbf{p} = \mathbf{0})$ to the effective range expansion around the deuteron pole, $k\cot\delta_d = -\gamma_d + \frac{\rho_d}{2}(k^2 + \gamma_d^2) + \cdots$, for renormalization [7, 8]. For the parameters, we use $\gamma_d = \sqrt{M_N E_d} = 45.7022(1)$ MeV [14] and $\rho_d = 1.765(4)$ fm [15]. To perform a strictly perturbative calculation up to N^{2}LO as described in Ref. [16], it is convenient to define $D_d(E;q) \equiv (-\mathrm{i})\cdot\Delta_d\left(E - q^2/(2M_N), q\right)$ and expand this as

$$D_d(E;q) = D_d^{(0)}(E;q) + D_d^{(1)}(E;q) + D_d^{(2)}(E;q) + \cdots = -\frac{4\pi}{M_N y_d^2}\frac{1}{-\gamma_d + \sqrt{3q^2/4 - M_N E - \mathrm{i}\varepsilon}}$$
$$\times\left[1 + \frac{\rho_d}{2}\frac{\left(3q^2/4 - M_N E - \gamma_d^2\right)}{-\gamma_d + \sqrt{3q^2/4 - M_N E - \mathrm{i}\varepsilon}} + \left(\frac{\rho_d}{2}\frac{\left(3q^2/4 - M_N E - \gamma_d^2\right)}{-\gamma_d + \sqrt{3q^2/4 - M_N E - \mathrm{i}\varepsilon}}\right)^2 + \cdots\right]. \tag{2}$$

The deuteron wavefunction renormalization is defined as the residue of Δ_d at the bound-state pole, and its perturbative expansion can be read off from Eq. (2).

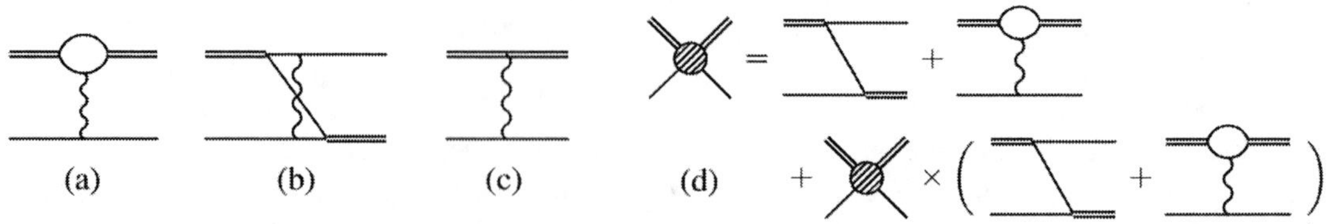

Figure 1. $\mathcal{O}(\alpha)$ diagrams involving Coulomb photons (a,b,c), and integral equation for the full (*i.e.* strong + Coulomb) scattering quartet-channel amplitude $\mathcal{T}_{\text{full}}$ (d). A single line represents a nucleon whereas the deuteron is expressed as a double line.

In Fig. 1(a-c) we show the $\mathcal{O}(\alpha)$ Feynman diagrams that are relevant for our p–d scattering calculation. With the operation $A \otimes B \equiv \frac{1}{2\pi^2}\int_0^\Lambda \mathrm{d}q\, q^2 A(\ldots, q)$, where Λ denotes a momentum cutoff used for the numerical solution, the integral equation represented by Fig. 1(d) can be written as

$$\mathcal{T}_{\text{full}} = -M_N y_d^2\left(K_s - \tfrac{1}{2}K_{\text{bub}}\right) + \mathcal{T}_{\text{full}} \otimes \left[M_N y_d^2 D_d\left(K_s - \tfrac{1}{2}K_{\text{bub}}\right)\right], \tag{3}$$

where $K_s(E;k,p)$ corresponds to the one-nucleon exchange diagram and $K_{\rm bub}(E;k,p)$ is given by Fig. 1(a). Explicit expressions for these kernel functions, as well as for $K_{\rm box}(E;k,p)$ and $K_{\rho_d}(E;k,p)$, represented by Figs. 1(b) and (c), respectively, are given in Ref. [12]. The latter two play a role as we go to higher orders in the perturbative expansion, as discussed below. The equation for the pure Coulomb amplitude $\mathcal{T}_{\rm c}$ is obtained by omitting K_s in Eq. (3). From the on-shell amplitudes $\mathcal{T}_{\rm full}(E_k;k,k)$ and $\mathcal{T}_{\rm c}(E_k;k,k)$ with $E_k = 3k^2/(4M_N) - \gamma_d^2/M_N$ we obtain the full and pure Coulomb phase shifts, respectively, and then calculate the Coulomb-subtracted $\delta_{\rm diff}(k) = \delta_{\rm full}(k) - \delta_{\rm c}(k)$. This is then used in the Coulomb-modified effective range expansion (ERE),

$$C^2_{\eta,\lambda}\,k\cot\delta_{\rm diff}(k) + \gamma_{p\text{-}d}\,h_\lambda(\eta) = -\frac{1}{{}^4a_{p\text{-}d}} + O(k^2)\ ,\ \eta = \frac{\gamma_{p\text{-}d}}{2k}\ ,\ \gamma_{p\text{-}d} = 4\alpha M_N/3\,, \tag{4}$$

to determine the scattering length ${}^4a_{p\text{-}d}$. The λ here denotes a small photon mass that is used to regulate the otherwise singular photon-exchange diagrams. It has to be taken to zero in order to obtain physical results. A key feature of our approach is that we keep a nonzero photon mass throughout the calculation and consistently calculate the screened Coulomb functions as

$$C^2_{\eta,\lambda}(k) = \left|1 + \frac{2M_N}{3\pi^2}\int_0^\Lambda \frac{\mathrm{d}p\,p^2\,Z_0\mathcal{T}_{\rm c}(E;p,k)}{p^2-k^2-\mathrm{i}\varepsilon}\right|^2\ ,\ h_\lambda(\eta) = \frac{3k^2}{2\alpha M_N}\frac{1}{\pi}\,\mathrm{P}\int_0^\Lambda \mathrm{d}q\,\frac{C^2_{\eta,\lambda}(q)}{(q+k)(q-k)}\,, \tag{5}$$

and then evaluate ${}^4a_{p\text{-}d}$ at $\lambda = 0$ by extrapolation. This approach is based on the general modified ERE derived by van Haeringen and Kok [17].

All quantities discussed here have perturbative expansions. Higher order corrections are generated primarily by the effective range, $\gamma_d\rho_d \sim Q/\Lambda_{\not\pi}$, as shown in Eq. (2). Similarly, Fig. 1(c) (K_{ρ_d}) is formally a range correction. The diagram shown in Fig. 1(b), $K_{\rm box}$, was counted as a higher-order correction by Rupak and Kong [18]. We present here results in this "RK" scheme as well as for an alternative "$O(\alpha)$ counting that includes this diagram already at LO, *i.e.*, iterating it together with the other diagrams shown in Fig. 1(d) [13].

3 Results and discussion

In Fig. 2 we show our results for ${}^4a_{p\text{-}d}$ as a function of the regulating photon mass λ. In the left panel, we use exactly the procedure described in the preceding section and find that the two counting schemes are practically indistinguishable at N^{2}LO. The very weak λ-dependence allows us to unambiguously extrapolate to $\lambda = 0$ and obtain, at N^{2}LO, ${}^4a_{p\text{-}d} = (10.9\pm0.4)$ fm. This agrees with older experimental determinations [5, 6] but deviates from the more recent determination of Black *et al.* [1] and potential-model calculations [2–4].

The conclusion, however, is not to dismiss the latter, but to carefully analyze where the discrepancy might come from. We note that in our EFT framework, it is natural to define the pure Coulomb sector by keeping only the diagrams without strong interaction between the proton and the deuteron, *i.e.*, Figs. 1(a) and (c). The leading diagram, Fig. 1(a), has some contributions from the short-range substructure of the deuteron because the photon only couples to a nucleon bubble. This is a three-body effect. Configuration-space potential-model calculations, on the other hand, subtract Coulomb effects purely at the two-body level [19]. We can mimic this scheme by defining the pure-Coulomb sector through a simple Yukawa potential, which keeps the photon-mass screening but contains no three-body effects. As shown in the right panel of Fig. 2, this procedure gives a result compatible with ${}^4a_{p\text{-}d} \approx 13.8$ fm, *i.e.*, good agreement with potential-model calculations, but it is somewhat unsatisfactory from the EFT perspective. We conclude that the Coulomb-modified scattering length should be called a subtraction-scheme-dependent quantity, at least for the scattering of composite particles.

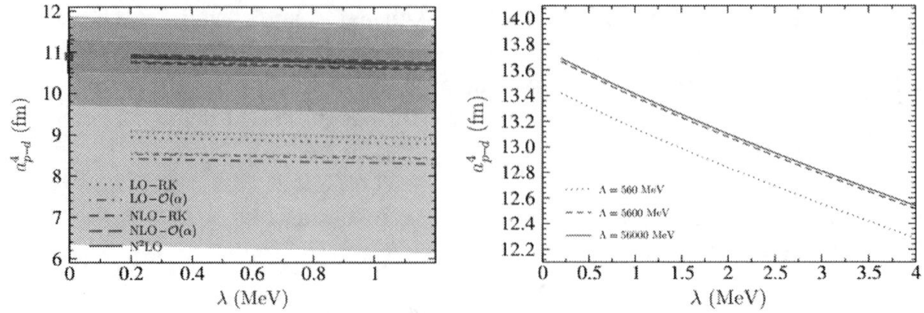

Figure 2. Photon-mass dependence and extrapolation of $^4a_{p\text{-}d}$. **Left panel:** Result with EFT subtraction scheme. Dotted lines: LO result, dashed lines: NLO result, solid lines: N^{2}LO result. Each calculation was performed at three different cutoffs, $\Lambda = 140$ MeV (thick lines), $\Lambda = 280$ MeV (medium lines), and $\Lambda = 560$ MeV (thin lines). "RK" and "$\mathcal{O}(\alpha)$" indicate the different Coulomb-counting schemes (see text). The bands (shown for the "RK" results only) reflect the expected EFT expansion uncertainty. **Right panel:** Result with a simple two-body Yukawa subtraction at N^{2}LO in the "$\mathcal{O}(\alpha)$" counting scheme.

Acknowledgements

We thank our collaborators for many useful discussions. This research was supported in part by the NSF under Grant No. PHY–1306250 and the NUCLEI SciDAC Collaboration under DOE Grant DE-SC0008533 (SK), as well as by the BMBF under contracts 05P12PDFTE and 05P15RDFN1 and by the Helmholtz Association under contract HA216/EMMI (HWH).

References

[1] T. C. Black *et al.*, Phys. Lett. B **471** (1999) 103.

[2] G. H. Berthold and H. Zankel, Phys. Rev. C **34** (1986) 1203.

[3] C. R. Chen, G. L. Payne, J. L. Friar and B. F. Gibson, Phys. Rev. C **44** (1991) 50.

[4] A. Kievsky *et al.*, Phys. Lett. B **406** (1997) 292.

[5] J. Arvieux, Nucl. Phys. A **221** (1973) 253.

[6] E. Huttel *et al.*, Nucl. Phys. A **406** (1983) 443.

[7] D. B. Kaplan, M. J. Savage and M. B. Wise, Phys. Lett. B **424** (1998) 390.

[8] U. van Kolck, Nucl. Phys. A **645**, (1999) 273.

[9] H. A. Bethe, Phys. Rev. **76** (1949) 38.

[10] C. A. Bertulani, H.-W. Hammer, and U. van Kolck, Nucl. Phys. A **712** (2002) 37.

[11] P. F. Bedaque, H.-W. Hammer, and U. van Kolck, Phys. Lett. B **569** (2003) 159.

[12] S. König and H.-W. Hammer, Phys. Rev. C **90** (2014) 3, 034005.

[13] S. König, H. W. Grießhammer and H.-W. Hammer, J. Phys. G **42** (2015) 045101.

[14] C. van der Leun and C. Anderliesten, Nucl. Phys. A **380** (1982) 261.

[15] J. J. de Swart, C. P. F. Terheggen and V. G. J. Stoks, arXiv:nucl-th/9509032.

[16] J. Vanasse, Phys. Rev. C **88** (2013) 044001.

[17] H. van Haeringen and L. P. Kok, Czech. J. Phys. B **32** (1982) 307.

[18] G. Rupak and X. Kong, Nucl. Phys. A **717** (2003) 73.

[19] C. R. Chen, G. L. Payne, J. L. Friar and B. F. Gibson, Phys. Rev. C **39** (1989) 1261.

EPJ Web of Conferences **113**, 04012 (2016)
DOI: 10.1051/epjconf/201611304012

Reactions in the four-nucleon system above breakup threshold

A. Deltuva[1,a] and A. C. Fonseca[2,b]

[1] *Institute of Theoretical Physics and Astronomy, Vilnius University, A. Goštauto 12, LT-01108 Vilnius, Lithuania*
[2] *Centro de Física Nuclear da Universidade de Lisboa, P-1649-003 Lisboa, Portugal*

Abstract. Microscopic calculations of four-body scattering become very challenging in the energy regime above the threshold for four free particles. We consider mixed-isospin four-nucleon reactions initiated by the proton-^{3}H, neutron-^{3}He, or deuteron-deuteron collisions. We solve the Alt, Grassberger, and Sandhas equations for the four-nucleon transition operators in the momentum-space framework. The complex-energy method with special integration weights is applied to deal with the complicated singularities in the kernel of AGS equations. Results for the differential cross section and spin observables in elastic, charge-exchange, transfer, and breakup reactions using realistic potentials are presented.

1 Introduction

The four-nucleon (4N) continuum is an important "theoretical laboratory" for a quantitative test of the two-nucleon (2N) and three-nucleon (3N) interaction models. However, its theoretical description is technically and computationally highly challenging. Exact four-particle scattering equations have been formulated 50 years ago by Faddeev and Yakubovsky (FY) [1] and by Alt, Grassberger, and Sandhas (AGS) [2, 3]. However, their numerical solution using realistic force models only became possible many years later. In the last decade accurate numerical calculations for low-energy nucleon-trinucleon elastic scattering have been performed using both coordinate-space and momentum-space rigorous approaches, namely, the hyperspherical harmonics (HH) expansion method [4, 5], the Faddeev-Yakubovsky (FY) equations [1] for the wave function components [6], and the Alt, Grassberger and Sandhas (AGS) equations [3] for transition operators [7, 8]. The reliability of all these methods was confirmed in a benchmark calculation [9] for neutron-^{3}H (n-^{3}H) and proton-^{3}He (p-^{3}He) elastic scattering observables.

Above the four-body breakup threshold the calculations become even more difficult owing to singularities in the momentum-space or complicated boundary conditions in the coordinate-space framework. In this regime rigorous and realistic results have been obtained so far by only two methodologies. These are the complex-energy method in the framework of momentum-space integral equations [10] and the complex scaling method in the framework of coordinate-space differential equations [11]. Only the former method has been applied to reactions involving the coupled proton-^{3}H (p-^{3}H),

[a] e-mail: arnoldas.deltuva@tfai.vu.lt
[b] e-mail: fonseca@cii.fc.ul.pt

neutron-^{3}He (n-^{3}He) and deuteron-deuteron (d-d) systems. The scattering process in these systems resembles a typical nuclear reaction where elastic, charge exchange, transfer, and breakup reactions take place simultaneously. The present contribution summarizes the achievements in the calculation of $4N$ reactions initiated by $p + {}^3$H, $n + {}^3$He, or $d + d$ collisions above the breakup threshold.

2 Theory

Four-nucleon scattering process is described exactly using the Alt, Grassberger, and Sandhas (AGS) equations [3]. The symmetrized version for the four-particle transition operators $\mathcal{U}_{\beta\alpha}$ is derived in ref. [7], where the nucleons are treated as identical particles in the isospin formalism, i.e.,

$$\mathcal{U}_{11} = -(G_0\, t\, G_0)^{-1} P_{34} - P_{34} U_1 G_0\, t\, G_0\, \mathcal{U}_{11} + U_2 G_0\, t\, G_0\, \mathcal{U}_{21}, \tag{1a}$$

$$\mathcal{U}_{21} = (G_0\, t\, G_0)^{-1}(1 - P_{34}) + (1 - P_{34}) U_1 G_0\, t\, G_0\, \mathcal{U}_{11}, \tag{1b}$$

$$\mathcal{U}_{12} = (G_0\, t\, G_0)^{-1} - P_{34} U_1 G_0\, t\, G_0\, \mathcal{U}_{12} + U_2 G_0\, t\, G_0\, \mathcal{U}_{22}, \tag{1c}$$

$$\mathcal{U}_{22} = (1 - P_{34}) U_1 G_0\, t\, G_0\, \mathcal{U}_{12}. \tag{1d}$$

Here, $\alpha = 1$ corresponds to the $3+1$ partition (12,3)4, whereas $\alpha = 2$ corresponds to the $2+2$ partition (12)(34). The free resolvent

$$G_0 = (Z - H_0)^{-1} \tag{2}$$

is taken with a complex energy parameter $Z = E + i\varepsilon$, where H_0 is the free Hamiltonian H_0; the operator

$$t = v + vG_0 t \tag{3}$$

is the pair (12) transition matrix resulting from the potential v, and

$$U_\alpha = P_\alpha G_0^{-1} + P_\alpha t\, G_0\, U_\alpha \tag{4}$$

are the 3+1 or 2+2 subsystem transition operators. For the four-nucleon system the basis states are antisymmetric under exchange of two particles in the subsystem (12) and, in the 2 + 2 partition, also in the subsystem (34). The full antisymmetry is ensured by the permutation operators P_{ab} of particles a and b with $P_1 = P_{12} P_{23} + P_{13} P_{23}$ and $P_2 = P_{13} P_{24}$.

The scattering amplitudes for two-cluster reactions at available energy $E = \epsilon_\alpha + p_\alpha^2/2\mu_\alpha = \epsilon_\beta + p_\beta^2/2\mu_\beta$ are obtained from the on-shell matrix elements $\langle \mathbf{p}_\beta | T_{\beta\alpha} | \mathbf{p}_\alpha \rangle = S_{\beta\alpha} \langle \phi_\beta | \mathcal{U}_{\beta\alpha} | \phi_\alpha \rangle$ in the limit $\varepsilon \to +0$. Here $|\phi_\alpha\rangle$ is the Faddeev component of the asymptotic two-cluster state in the channel α, characterized by the bound state energy $\epsilon_\alpha < 0$, the relative momentum $\mathbf{p}_\alpha$, and the reduced mass μ_α; $S_{\beta\alpha}$ are the symmetrization factors [7]. The AGS equations (1) are solved in the momentum-space partial-wave framework. Two different types of basis states $|k_x k_y k_z \nu\rangle_\alpha$ with $\alpha = 1$ and 2 are employed, where k_x, k_y, and k_z denote magnitudes of the Jacobi momenta, and ν abbreviates all discrete quantum numbers. The reduced masses associated with Jacobi momenta k_x and k_y in the partition α are denoted by $\mu_{\alpha x}$ and $\mu_{\alpha y}$, respectively.

In the solution process the integrals are discretized using Gaussian quadrature rules [12] turning Eqs. (1) into a system of linear equations as described in ref. [7]. However, in the limit $\varepsilon \to +0$ needed for the calculation of the observables the kernel of the AGS equations contains integrable singularities: At $E + i\varepsilon - \epsilon_\alpha - k_z^2/2\mu_\alpha \to 0$ the subsystem transition operator in the bound state channel has the pole

$$G_0 U_\alpha G_0 \to \frac{P_\alpha |\phi_\alpha\rangle S_{\alpha\alpha} \langle \phi_\alpha | P_\alpha}{E + i\varepsilon - \epsilon_\alpha - k_z^2/2\mu_\alpha}. \tag{5}$$

At $E + i\varepsilon - \epsilon_d - k_y^2/2\mu_{\alpha y} - k_z^2/2\mu_\alpha \to 0$ the two-nucleon transition matrix in the channel with the deuteron quantum numbers for the pair (12) has the pole

$$t \to \frac{v|\phi_d\rangle\langle\phi_d|v}{E + i\varepsilon - \epsilon_d - k_y^2/2\mu_{\alpha y} - k_z^2/2\mu_\alpha}, \tag{6}$$

with $|\phi_d\rangle$ being the pair (12) deuteron wave function and ϵ_d being internal deuteron energy. At $E+i\varepsilon - k_x^2/2\mu_{\alpha x} - k_y^2/2\mu_{\alpha y} - k_z^2/2\mu_\alpha \to 0$ the free resolvent (2) becomes singular as well. The interplay of all these singularities with permutation operators and basis transformations leads to a very complicated singularity structure of the AGS equations.

This difficulty can be formally avoided by performing calculations for a set of finite $\varepsilon > 0$ values where the kernel contains no singularities and then extrapolating the results to the $\varepsilon \to +0$ limit. This idea was proposed in Refs. [13–15] and the point method [13] was suggested for the extrapolation. However, this extrapolation method as well as alternative choices are only precise if ε is small enough. However, for in that case the kernel of the AGS equations, although formally being non-singular, exhibits a quasi-singular behavior and therefore requires dense grids for the numerical integration. In practical calculations with large number of partial waves necessary for the convergence it is highly desirable to keep the number of integration grid points as small as possible. Thus, a more sophisticated integration method is needed.

An important technical improvement when calculating $\mathcal{U}_{\beta\alpha}$ at finite ε was introduced in Ref. [10]. The method of special weights for numerical integration involving any of the above-mentioned quasi-singularities is used, i.e.,

$$\int_a^b \frac{f(x)}{x_0^n + iy_0 - x^n} dx \approx \sum_{j=1}^N f(x_j) w_j(n, x_0, y_0, a, b). \tag{7}$$

The quasi-singular factor $(x_0^n + iy_0 - x^n)^{-1}$ is separated and absorbed into the special integration weights $w_j(n, x_0, y_0, a, b)$. The set of N grid points $\{x_j\}$ where the remaining smooth function $f(x)$ has to be evaluated is chosen the same as for the standard Gaussian quadrature. However, while the standard weights are real [12], the special weights $w_j(n, x_0, y_0, a, b)$ are complex. They are chosen such that for a set of N spline functions $S_j(x)$ the result (7) is exact. The corresponding special weights are

$$w_j(n, x_0, y_0, a, b) = \int_a^b \frac{S_j(x)}{x_0^n + iy_0 - x^n} dx. \tag{8}$$

Further details as well as examples demonstrating the reliability of this method are given in Refs. [10, 16].

3 Results and summary

As an example of our work we consider $p + {}^3$H scattering and in Figs. 1 - 4 we show the differential cross section and selected spin observables for elastic, charge-exchange, and transfer reactions. Further examples can be found in Refs. [16, 26–28]. We use several realistic interaction models, i.e., the charge-dependent Bonn potential (CD Bonn) [29], its coupled-channel extension CD Bonn + Δ [30], and the inside-nonlocal outside-Yukawa (INOY04) potential by Doleschall [6, 31]. The latter yields nearly the experimental value for the ^{3}He and ^{3}H binding energy without an additional three-nucleon force. The overall description of the available experimental data is reasonably good.

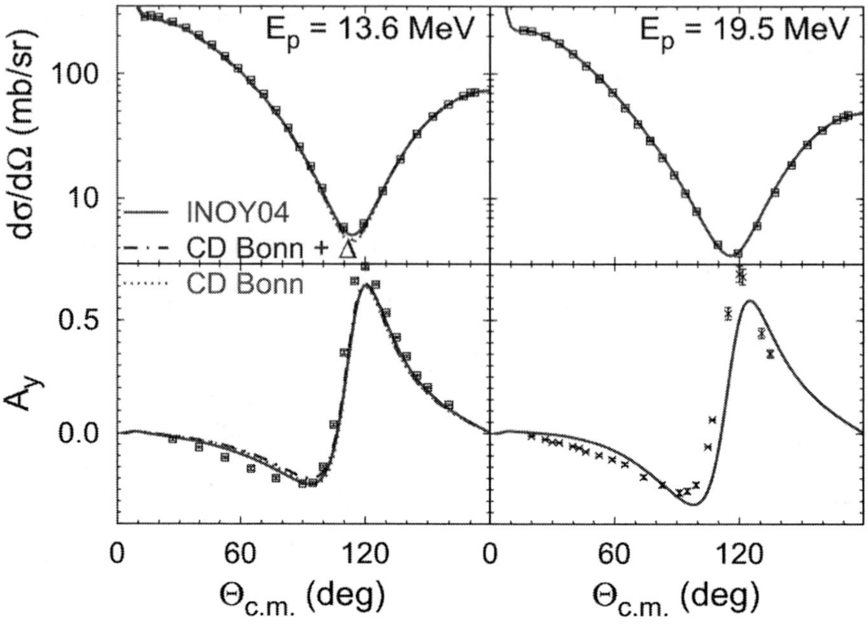

Figure 1. Differential cross section and proton analyzing power of elastic p-^{3}H scattering at 13.6 and 19.5 MeV proton energy. Results obtained with potentials INOY04 (solid curves), CD Bonn (dotted curves), and CD Bonn + Δ (dashed-dotted curves) are shown. The data are from Refs. [17, 18].

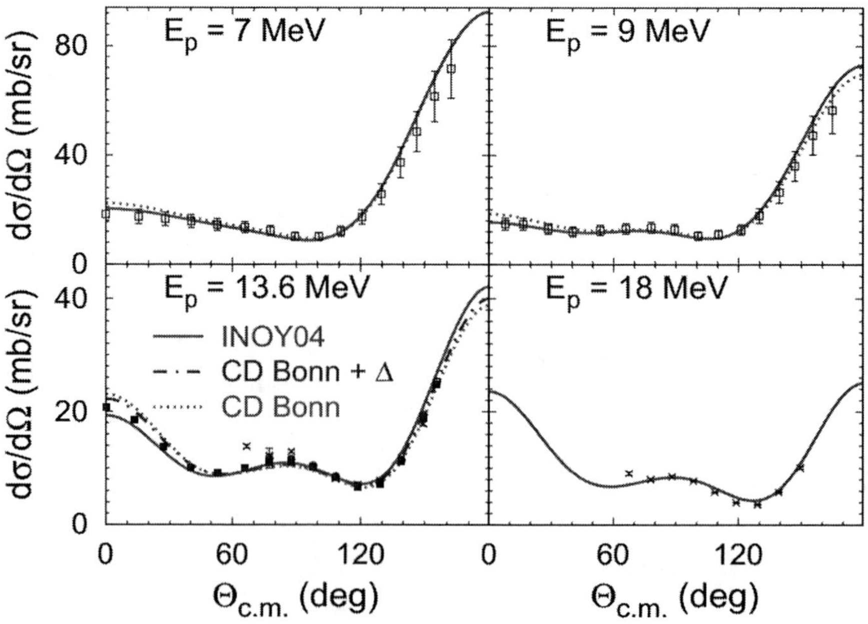

Figure 2. Differential cross section of ^{3}H$(p,n)^3$He reaction at 7, 9, 13.6, and 18 MeV proton energy. Curves as in Fig. 1. The data are from Refs. [19–22].

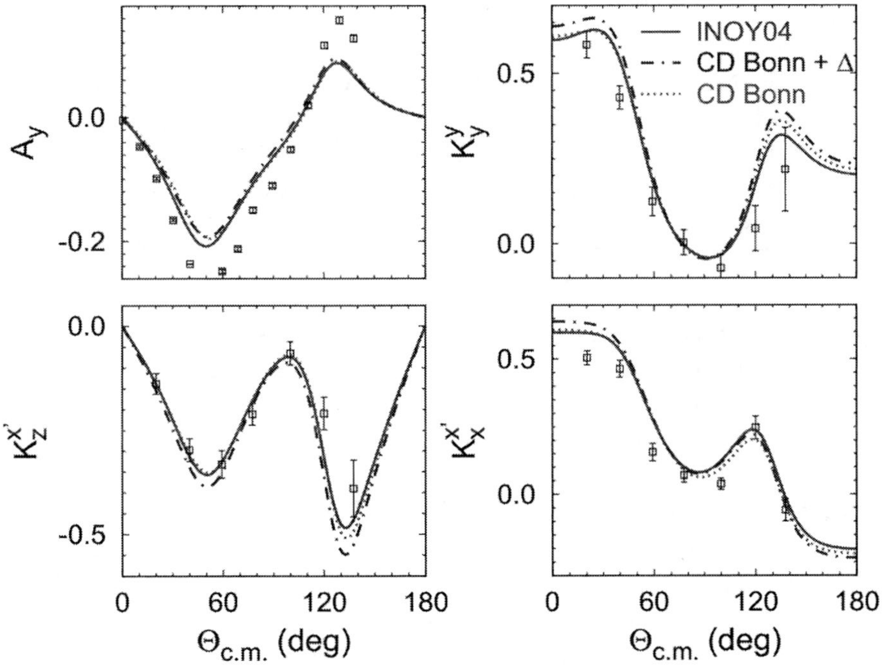

Figure 3. Proton analyzing power and proton to neutron polarization transfer coefficients of $^3H(p,n)^3He$ reaction at 13.6 MeV proton energy. Curves as in Fig. 1. The data are from Refs. [23, 24].

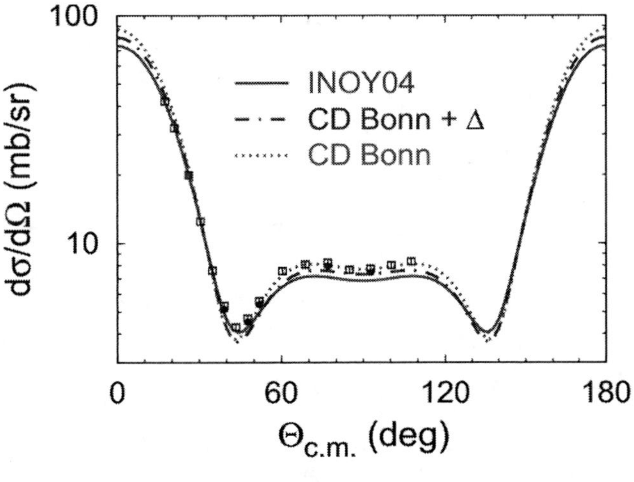

Figure 4. Differential cross section of $^3H(p,d)^2H$ reaction at 13.6 MeV proton lab energy. Curves as in Fig. 1. The data are from Refs. [17, 25].

References

[1] O.A. Yakubovsky, Yad. Fiz. **5**, 1312 (1967) [Sov. J. Nucl. Phys. **5**, 937 (1967)]

[2] E.O. Alt, P. Grassberger, W. Sandhas, JINR report **E4-6688**, 1 (1972)

[3] P. Grassberger, W. Sandhas, Nucl. Phys. **B2**, 181 (1967)

[4] M. Viviani, A. Kievsky, S. Rosati, E.A. George, L.D. Knutson, Phys. Rev. Lett. **86**, 3739 (2001)

[5] A. Kievsky, S. Rosati, M. Viviani, L.E. Marcucci, L. Girlanda, J. Phys. G **35**, 063101 (2008)

[6] R. Lazauskas, J. Carbonell, Phys. Rev. C **70**, 044002 (2004)

[7] A. Deltuva, A.C. Fonseca, Phys. Rev. C **75**, 014005 (2007)

[8] A. Deltuva, A.C. Fonseca, Phys. Rev. Lett. **98**, 162502 (2007)

[9] M. Viviani, A. Deltuva, R. Lazauskas, J. Carbonell, A.C. Fonseca, A. Kievsky, L.E. Marcucci, S. Rosati, Phys. Rev. C **84**, 054010 (2011)

[10] A. Deltuva, A.C. Fonseca, Phys. Rev. C **86**, 011001(R) (2012)

[11] R. Lazauskas, Phys. Rev. C **91**, 041001 (2015)

[12] W.H. Press, B.P. Flannery, S.A. Teukolsky, W.T. Vetterling, *Numerical Recipes* (Cambridge University Press, Cambridge, 1989)

[13] L. Schlessinger, Phys. Rev. **167**, 1411 (1968)

[14] F.A. McDonald, J. Nuttall, Phys. Rev. A **4**, 1821 (1971)

[15] H. Kamada, Y. Koike, W. Glöckle, Prog. Theor. Phys. **109**, 869L (2003)

[16] A. Deltuva, A.C. Fonseca, Phys. Rev. C **91**, 034001 (2015)

[17] J.L. Detch, R.L. Hutson, N. Jarmie, J.H. Jett, Phys. Rev. C **4**, 52 (1971)

[18] R. Darves-Blanc, N. Sen, J. Arvieux, A. Fiore, J. Gondrand, G. Perrin, Lett. Nuovo Cimento **4**, 16 (1972)

[19] W.E. Wilson, R.L. Walter, D.B. Fossan, Nucl. Phys. **27**, 421 (1961)

[20] M. Drosg, Nucl. Sci. Eng. **67**, 190 (1978)

[21] N. Jarmie, J.H. Jett, Phys. Rev. C **16**, 15 (1977)

[22] R.G. Allas, L.A. Beach, R.O. Bondelid, E.M. Diener, E.L. Petersen, J.M. Lambert, P.A. Treado, I. Slaus, Phys. Rev. C **9**, 787 (1974)

[23] J.J. Jarmer, R.C. Haight, J.E. Simmons, J.C. Martin, T.R. Donoghue, Phys. Rev. C **9**, 1292 (1974)

[24] R.C. Haight, J.E. Simmons, T.R. Donoghue, Phys. Rev. C **5**, 1826 (1972)

[25] N. Jarmie, J.H. Jett, Phys. Rev. C **10**, 54 (1974)

[26] A. Deltuva, A.C. Fonseca, Phys. Rev. Lett. **113**, 102502 (2014)

[27] A. Deltuva, A.C. Fonseca, Phys. Rev. C. **90**, 044002 (2014)

[28] A. Deltuva, A. Fonseca, Phys. Lett. B **742**, 285 (2015)

[29] R. Machleidt, Phys. Rev. C **63**, 024001 (2001)

[30] A. Deltuva, R. Machleidt, P.U. Sauer, Phys. Rev. C **68**, 024005 (2003)

[31] P. Doleschall, Phys. Rev. C **69**, 054001 (2004)

EPJ Web of Conferences **113**, 04013 (2016)
DOI: 10.1051/epjconf/201611304013

Description of the four-nucleon collisions by including breakup

Rimantas Lazauskas[1,a]

[1] *IPHC, IN2P3-CNRS/Université de Strasbourg BP 28, F-67037 Strasbourg Cedex 2, France*

Abstract. Four-nucleon reactions above the breakup threshold are described by solving Faddeev-Yakubovsky equations for the realistic nuclear Hamiltonians. Complex-scaling method is applied in order to simplify the boundary conditions.

1 Introduction

The accurate description of nuclear interaction remains one of the most important challenges of nuclear physics. The success of this enterprize strongly depends on the ability to provide accurate numerical tools, which could test modern nuclear interaction models in describing nuclear reactions. A proper formalism of three-particle reactions has been introduced by Faddeev in 1961 [1] and few years later generalized by Yakubovsky [2] to any number of particles. Regardless the fact that this formalism enables to handle binary collisions at relative ease, the description of the breakup process into three or more clusters constitutes an important technical and formal challenge. This complexity is mostly due to the complex behavior of the system's wave function in the far asymptotes.

It is quite obvious that a direct approach based on the explicit treatment of boundary conditions becomes overcomplex already for the systems containing more than three particles. Recently a formalism [3], which combines Faddeev-Yakubovsky (FY) equations [2] and complex scaling method [4], has been developed in order to solve three- and four-body scattering problems above the three-particle breakup threshold. This formalism relies on trivial boundary conditions, is easy to handle but still provides accurate results [3].

In this manuscript the latest developments in the four-nucleon sector are reviewed. This includes calculations for n-^{3}H and p-^{3}He scattering above the three- and four-nucleon breakup thresholds.

2 The formalism

A detailed overview of complex scaling method has been recently presented in [5]. This method, introduced by Nuttall and Cohen [4], might be employed to solve Scrödinger equation directly. Nevertheless, in this study configuration space FY equations [2] are used. FY equations are built to separate the asymptotes of the binary channels and therefore are advantageous in describing systems, where binary collisions are dominant. After separation of the incoming plane wave, FY equations formulated for a system of four identical fermions reads:

$$K(\vec{x},\vec{y},\vec{z}) = K^{out}(\vec{x},\vec{y},\vec{z}) + K^{in}(\vec{x},\vec{y},\vec{z}) \tag{1}$$

$$H(\vec{x},\vec{y},\vec{z}) = H^{out}(\vec{x},\vec{y},\vec{z}) + H^{in}(\vec{x},\vec{y},\vec{z}). \tag{2}$$

[a] e-mail: rimantas.lazauskas@iphc.cnrs.fr

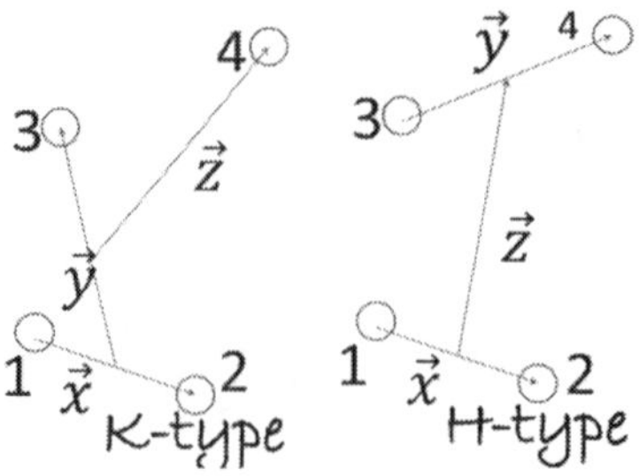

Figure 1. The FY components $K^{4}_{12,3}$ and H^{34}_{12} for a given particle ordering. As $z \to \infty$, the K components absorb asymptotes of 3+1 particle channels, whereas the H components absorb asymptotes of 2+2 channels.

Four-body FY equations involve components of two types, K and H. In their $z \to \infty$ asymptotes these components separate respectively outgoing (K^{out}, H^{out}) and incoming (K^{in}, H^{in}) waves of (3+1,2+2) particle channels, see fig. 1. These components are the solutions of the inhomogeneous FY equations:

$$(E - H_0 - V_{12})\, K^{out} \;-\; V_{12}(P^+ + P^-)\left[(1 + Q)K^{out} + H^{out}\right] = V_{12}(P^+ + P^-)\left[H^{in} + QK^{in}\right], \quad (3)$$

$$(E - H_0 - V_{12})\, H^{out} \;-\; V_{12}\tilde{P}\left[(1 + Q)K^{out} + H^{out}\right] = V_{12}\tilde{P}\left[(1 + Q)K^{in}\right], \quad (4)$$

here $P^- = P_{23}P_{12}$, $P^+ = P_{12}P_{23}$, $Q = P_{34}$ and $\tilde{P} = P_{13}P_{24}$ are particle permutation operators.

Then the complex scaling method is applied. I.e. one acts on the last set of equations with a complex-scaling operator:

$$\widehat{S} = e^{i\theta r \frac{\partial}{\partial r}} = e^{i\theta(x\frac{\partial}{\partial x} + y\frac{\partial}{\partial y} + z\frac{\partial}{\partial z})}, \quad (5)$$

where parameter θ is often referred as complex scaling angle. The complex scaled FY equations are solved to determine the transformed FY components $\tilde{K}^{out} = \widehat{S}K^{out}$ and $\tilde{H}^{out} = \widehat{S}H^{out}$. In particular, the complex scaling operator transforms outgoing waves into square integrable functions $\tilde{K}^{out}$ (or $\tilde{H}^{out}$) if a complex scaling angle is chosen in the interval $0 < \theta < \pi$. At the same time functions $\tilde{K}^{in}$ (or $\tilde{H}^{in}$) become exponentially divergent. Nevertheless these functions appear only in inhomogeneous term of the FY equations (3-4) premultiplied with potential energy operator, and thus if interactions are exponentially bound the kernel of the complex scaled equations becomes compact [1].

The transformed FY equations are solved using standard techniques, developed in our preceding studies, see [3] and references therein. Namely, the spin-isospin as well as angular dependence of the FY components is expanded using partial-waves. The radial dependence of the complex-scaled FY components $\tilde{K}^{out}$ and $\tilde{H}^{out}$ is expanded on the Lagrange-Laguerre basis and a system of integro-differential equations is transformed into linear algebra problem by using Lagrange-mesh method [6]. For neutron energies below 30 MeV numerical convergence is achieved by including partial-waves with angular momenta $max(l_x, l_y, l_z) \leq 4$, also 3D Lagrange-mesh of $\sim 30^3$ points is required to describe radial dependence of the FY components $(\tilde{K}^{out}, \tilde{H}^{out})$. This leads to linear systems of $\sim 10^8$ equations, which are solved using iterative linear algebra methods. Scattering amplitudes, including breakup ones, are calculated from transformed solutions $\tilde{K}^{out}$ (or $\tilde{H}^{out}$) by employing integral relations [3, 5].

[1] One should note that when considering scattering in $A > 2$ particle system additional condition is implied on complex scaling angle in order to guarantee compactness of the kernel [5].

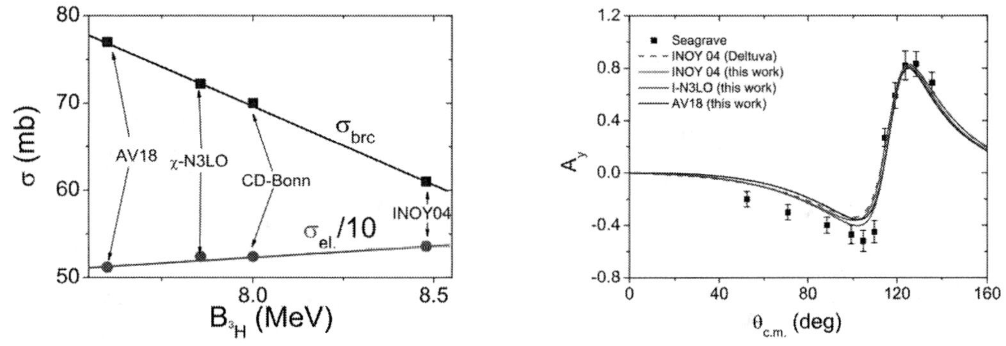

Figure 2. Correlation of n-^{3}H total elastic and breakup cross sections with the model predicted ^{3}H binding energy (left panel). Neutron-^{3}H analyzing powers A_y (right panel). Calculations have been performed for incident neutrons with laboratory energy of 22.1 MeV. Experimental results on analyzing powers are from J. Seagrave *et al.* [9].

3 Selected results

The complex scaling method enables to consider 4N system above the breakup threshold employing fully realistic NN interaction models. First this method was applied to explore neutron scattering on ^{3}H nucleus, a system which does not involve Coulomb interaction. In particular, the region of neutron laboratory energies from 14 to 22 MeV has been explored, using INOY04, χN3LO and AV18 NN interaction models [3, 5]. Few years earlier the same system had been examined by A. Deltuva *et al.* [7], within the momentum space formulation of the complex-energy method. An excellent agreement is found between two very different approaches, reaching three-digit accuracy for the calculated observables. By analyzing different model predictions, a correlation pattern has been established between both, the total elastic and the total break up cross section, and the predicted ^{3}H binding energy (see fig. 2). Moreover, once trinucleon binding energy is reproduced, theoretical prediction for the total cross section complies with the experimental data. Neutron analyzing powers also agree reasonably well with the experimental data, nevertheless the theoretical predictions systematically underestimate experimental data at the minima.

In the previous studies on 3-body system a possibility to implement the repulsive Coulomb interaction has been demonstrated, when describing nuclear reactions above the breakup threshold [3, 5]. The formalism essentially remains the same, only free in(out)going waves should be replaced with the Coulomb ones. As discussed in [5], small approximation is however involved in this case, which consist in neglecting long-range Coulomb polarization terms between the separating fragments.

The inclusion of the repulsive Coulomb interaction in the complex-scaling formalism allows to consider proton scattering on ^{3}He. This reaction has been explored for the protons with energy from 8.5 MeV and 19.4 MeV, using realistic Hamiltonian based on AV18 and INOY04 NN interaction models. The obtained results are compared with the calculations of A. Deltuva *et al.* [8] and the experimental data from ref. [10]. This time the agreement between the two calculations is slightly worse than for a Coulomb-free case, small discrepancy of order 5% between two theoretical predictions is observed at the backward angles. It is difficult to trace the origin of this discrepancy due to the fact that the final cross section is a result of the strong interference of separate partial waves. Nevertheless, a good overall agreement is observed with the existent experimental cross sections for the INOY04 Hamiltonian. On contrary the AV18 interaction model slightly underestimates the differential cross

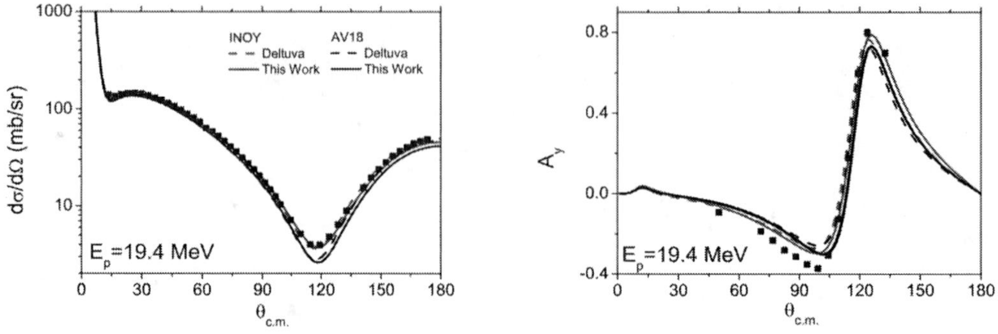

Figure 3. Calculated p-^{3}He elastic differential cross sections and analyzing powers for the protons of laboratory energy 19.4 MeV. Calculated values are compared with ones from ref. [8] and with the experimental data from ref. [10].

section at the minima, this is visibly related with ^{3}He nucleus being underbound by this model. The proton-^{3}He analyzing powers are similar to the neutron-^{3}H ones, in particular the theoretical predictions underestimate experimental data at the minima.

Acknowledgements

This work was granted access to the HPC resources of TGCC and IDRIS under the allocation 2015-x2015056006 made by GENCI. We thank the staff members of the TGCC and IDRIS for their constant help.

References

[1] L. D. Faddeev, Zh. Eksp. Teor. Fiz. **39**, 1459 (1960). [Sov. Phys. JETP **12**, 1014 (1961)].
[2] O. A. Yakubovsky, Sov. J. Nucl. Phys. **5**, 937 (1967).
[3] R. Lazauskas and J. Carbonell, Phys. Rev. C **84**, 034002 (2011); R. Lazauskas, Phys. Rev. C **91**, 041001(R) (2015).
[4] J. Nuttall and H. L. Cohen, Phys. Rev. **188**, 1542 (1969).
[5] A. Deltuva *et al.*, PPNP **74**, 55 (2014).
[6] D. Baye, Phys. Rep. **565** (2015) 1.
[7] A. Deltuva and A. C. Fonseca, Phys. Rev. C **86**, 011001 (2012).
[8] A. Deltuva and A. C. Fonseca, Phys. Rev. **C 87**, 054002 (2013).
[9] J. Seagrave *et al.*, Annals of Physics **74**, 250 (1972).
[10] S. D. Baker, et al., Nucl. Phys. **A160**, 428 (1971).

EPJ Web of Conferences **113**, 04014 (2016)
DOI: 10.1051/epjconf/201611304014

Resolution of resonances with the Lorentz integral transform

Winfried Leidemann[1,2,a]

[1] *Dipartimento di Fisica, Università di Trento, I-38123 Trento, Italy*
[2] *Istituto Nazionale di Fisica Nucleare, TIFPA, I-38123 Trento, Italy*

Abstract. The energy resolution of cross sections calculated with the Lorentz integral transform (LIT) method is investigated. It is shown that a crucial role is played by the density of LIT states obtained in the calculation. The situation is illustrated for the simple case of deuteron photodisintegration in unretarded dipole approximation. In a next step the ^{4}He isoscalar monopole resonance is considered. It is pointed out that in a specific energy interval the density of LIT states may depend on the A-body basis chosen for the solution of the LIT equation. Using a central NN potential and choosing a proper A-body basis set it is shown that the width of the ^{4}He isoscalar monopole resonance can be determined. A value of 180(70) keV is obtained, which is close to the experimental value of 270(50) keV.

1 Introduction

The LIT approach [1] is an ab initio technique that reduces an A-body continuum problem to an A-body bound-state like problem. The method is usually applied to perturbation induced inclusive breakup reactions of an A-body nucleus, where the corresponding cross section can be described in terms of response functions. The LIT represents an integral transform of such a response function and is just given by the norm of a localized function $\tilde{\Psi}$. The latter is determined by the solution of the LIT equation. Finally, the response function, and thus also the cross section, is obtained via the inversion of the transform. Since the LIT method is an approach with a controlled resolution, one knows in a given LIT calculation up to which limit one is able to resolve structures in the reaction cross section. In most of the past LIT applications the relevant structures, e.g. in electron scattering the quasi-elastic peak, were resolved without greater problems. An exception is the recent LIT-EIHH [2] application for the ^{4}He inelastic isoscalar monopole response, where it was realized that the width of the narrow ^{4}He isoscalar monopole resonance could not be resolved due to a too low density of LIT states in the resonance region. The aim of the present work is to describe how this problem can be overcome.

This work is organized as follows. In Sect. 2 the LIT approach is briefly summarized and the term "LIT state" is explained, in addition, the effect of the density of LIT states is illustrated for a simple example (deuteron photodisintegration in unretarded dipole approximation). In Sect. 3 it is discussed that the LIT state density can be increased by choosing a proper basis for the expansion of the localized function $\tilde{\Psi}$. Results with such a basis are presented, in fact they lead to a determination of the width of the ^{4}He inelastic isoscalar monopole resonance [3].

[a] e-mail: winfried.leidemann@unitn.it

2 LIT method

The inclusive response function $R(E_f)$ induced by an excitation operator θ is given by

$$R(E_f) = \int dE_f |\langle f|\theta|0\rangle|^2 \delta(E_f - E_0 - \omega) , \qquad (1)$$

where $|0\rangle$ and $|f\rangle$ are ground-state and final-state wave functions of an Hamiltonian H of the A-body system under consideration, E_0 and E_f are the corresponding eigenenergies, while $\omega = E_f - E_0$ is the energy transferred to the system. Instead of calculating the response function R via Eq. (1), where however a knowledge of $|f\rangle$ is necessary, one can determine R via the inversion of the LIT L, defined as

$$L(\sigma = \sigma_R + i\sigma_I) = \int dE_f \frac{R(E_f)}{(E_f - \sigma_R)^2 + \sigma_I^2} , \qquad (2)$$

where σ_R controls the position and σ_I the width of the Lorentzian kernel. Because of the adjustable width, the LIT is a transform with a controlled resolution.

The basic LIT equations for an inclusive reaction induced by an excitation operator θ read

$$(H - \sigma)\tilde{\Psi} = \theta\,|0\rangle , \qquad L(\sigma) = \langle\tilde{\Psi}|\tilde{\Psi}\rangle . \qquad (3)$$

The solution $\tilde{\Psi}$ is localized and can thus be determined via bound-state techniques. In fact, often an expansion of $\tilde{\Psi}$ is made on a complete localized A-body basis, where the number of basis states N is increased up to the point that convergence is obtained for the relevant quantities. For a given N one determines the spectrum H for the so defined basis thus finding N eigenstates ϕ_n with eigenenergies E_n. The energies E_n define the positions of the above mentioned LIT states. Furthermore, the LIT solution assigns to any E_n a Lorentzian with strength $S_n = |\langle\phi_n|\theta|0\rangle|^2$ and width σ_I. The LIT is then simply given by a sum of LIT states positioned at $\sigma_R = E_n$ with strength S_n and width $\Gamma = 2\sigma_I$:

$$L(\sigma) = \sum_{i=1}^{N} \frac{S_n}{(\sigma_R - E_n)^2 + \sigma_I^2} . \qquad (4)$$

The result should resemble a smooth function without that a single LIT state is sticking out. In other words to any value of σ_R one should have significant contributions from more than one of the N Lorentzians. In case of a rather large σ_I a few scattered LIT states will be sufficient. However with such a LIT one can only resolve rather broad structures. In order to disentangle finer structure one needs a smaller σ_I, comparable to the size of the structure to resolve, and thus a higher density of LIT states.

As example for the considerations above the case of deuteron photodisintegration in unretarded dipole approximation is discussed in the following. The total cross section is given by

$$\sigma_{\rm unret}(\omega) = 4\pi^2\,\alpha\,\omega\,R_{\rm unret}(E_f = E_0 + \omega) , \qquad (5)$$

where ω denotes the photon energy and α is the fine structure constant. The transition operator θ for the response function $R_{\rm unret}$ is the z-component of the dipole operator.

In Fig. 1 the LIT of $R_{\rm unret}$ is shown [3] (as final state only the $^3P_1$ NN partial wave is included). One sees for the σ_I value of 10 MeV that at lower σ_R a smooth LIT is obtained with only eleven basis functions. At higher σ_R the LIT result starts to oscillate showing the underlying structure, in fact one sees single Lorentzians sticking out. The oscillations vanish if the number of basis functions N, i.e. the LIT state density, is increased. The situation is similar for $\sigma_I = 2.5$ MeV, but oscillations already start at rather low σ_R and still remain even for $N = 51$. To obtain also in this case a smoother LIT one needs to increase the LIT state density further.

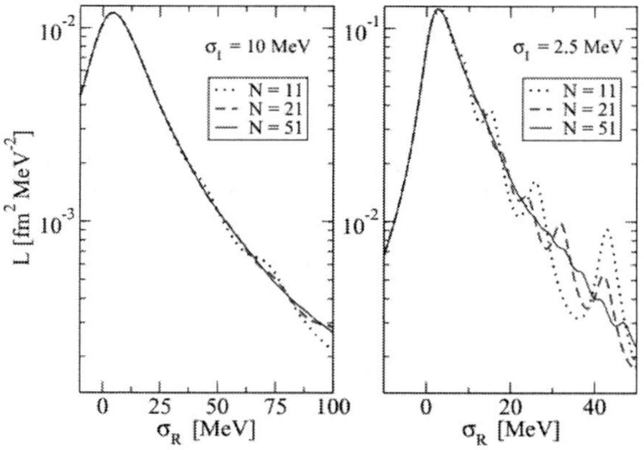

Figure 1. LIT $L(\sigma_R, \sigma_I)$ of the response function R_{unret} to a specific final state, namely the ${}^3P_1$ NN partial wave. The number of basis functions N used for the expansion of $\tilde{\Psi}$ to solve the LIT equation is as follows: $N = 11, 21, 51$ (calculational details are given in [3]). Left panel: σ_I=10 MeV. Right panel: σ_I=2.5 MeV.

3 LIT approach for narrow resonances

From the example of the deuteron case above it is evident that the possibility to determine the shape of a narrow cross section structure with the LIT technique is correlated to the density of LIT states. It is important to realize that, in particular at low energy, the density of LIT states depends on the A-body basis chosen for the expansion of $\tilde{\Psi}$. In [2] a hyperspherical harmonics (HH) was used and it was found that the LIT state density was not sufficiently high to determine the width of the ^{4}He isoscalar monopole resonance. In fact the resonance is located at a rather low energy, where only a two-body breakup of ^{4}He is possible (bound three-nucleon system plus free fourth nucleon). As a matter of fact an increase of the number of the HH basis states leads mainly to an increase of the LIT state density beyond the three-body breakup threshold, whereas the energy region below is much less affected. In [3] it was shown that it is much better to use a different A-body basis such that the dynamical variable of a two-body breakup explicitly appears. For the ^{4}He case considered here it corresponds to the vector pointing from the center-of-mass of the bound three-nucleon system to the position of the fourth nucleon (in Eq. (6) below this variable is called $\mathbf{r}'_4$).

It is worthwhile to mention that in principle the LIT result should not depend on the chosen A-body basis. However, this is only the case for a completely converged calculation. Such a complete convergence would imply that the LIT is converged for any value of σ_I even if the value is extremely small. On the other hand in the limit $\sigma_I \to 0$ one is in principle back to the continuum problem and it is evident that a complete convergence cannot be the aim of a LIT calculation. Nonetheless it is interesting to note that in such a hypothetical fully converged LIT calculation all the necessary information is also encoded in the LIT results with very large σ_I values. However, the information on a small cross section structure like the shape of narrow resonance would be hidden in tiny contributions. Using such a LIT with large σ_I would make the inversion unnecessarily difficult. Conversely this means that one should always work with the LIT with the smallest σ_I that is sufficiently converged. Then one also has a control about the size of structures that can be resolved. In conclusion, in principle in a complete calculation any A-body basis leads to the same LIT result, but in reality one has always to work with a limited number of basis states and then it is quite possible that in a given energy interval the density of LIT states depends strongly on the chosen basis.

For the four-body calculation presented here the basis consists of a three-body HH basis with additional two-body short-range correlations (CHH basis of [4]), whereas the basis for the fourth nucleon is given by

$$|\Phi(\mathbf{r}'_4)\rangle = \sum_{l_4 m_{s_4} m_{t_4}} \phi_{l_4}(r'_4)\, |l_4 m_{l_4}\rangle\, |s_4 m_{s_4}\rangle\, |t_4 m_{t_4}\rangle\,, \tag{6}$$

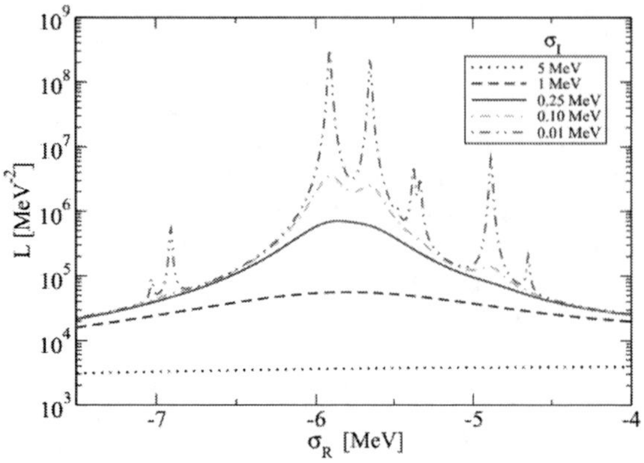

Figure 2. Low-energy part of the LIT $L(\sigma_R, \sigma_I)$ for the ^{4}He isoscalar monopole response function with various values of σ_I. The number of basis functions used for the expansion of $\phi_{l_4}(r'_4)$ is twenty (further details are given in [3]).

where l_4, $s_4 = 1/2$ and $t_4 = 1/2$ are the orbital angular momentum, spin, and isospin quantum numbers, respectively, and m_{l_4}, m_{s_4} and m_{t_4} denote the corresponding projections. The explicit form of the radial wave function $\phi_{l_4}(r'_4)$ can be found in [3]. The CHH part of the so defined four-body basis provides antisymmetric states for the first three nucleons, whereas the product of the CHH basis states and the single-particle states has to be antisymmetrized in order to have totally antisymmetric basis states for all four particles. The NN potential used in the present study is a central spin dependent NN potential (TN potential) that had been previously employed in the very first LIT applications (electromagnetic break-up of ^{4}He by electrons and real photons [5, 6]). As discussed in greater detail in [3] the TN potential model leads to a ^{4}He binding energy of 31.41(5) MeV.

Concerning the ^{4}He isoscalar monopole resonance one has to consider the isoscalar monopole response function $M(q,\omega)$, where the operator θ of Eq. (1) becomes q-dependent and takes the form

$$\theta(q) = \frac{G_E^s(q^2)}{2} \sum_{i=1}^{A} j_0(qr_i)\,, \tag{7}$$

where $G_E^s(q^2)$ is the nucleon isoscalar electric form factor, $\mathbf{r_i}$ is the position of nucleon i, and j_0 is the spherical Bessel function of 0th order.

In Fig. 2 the low-energy part of the LIT of $M(q,\omega)$ is shown, various σ_I values are taken. In the resonance region the LIT exhibits a peak which becomes more pronounced for decreasing σ_I. For $\sigma_I \geq 0.1$ MeV the LIT state density is high enough to lead to a sufficiently smooth LIT, whereas for the smallest σ_I of 0.01 MeV the single LIT states are clearly sticking out. Via LIT-inversions with $\sigma_I = 0.1$ and 0.25 MeV the width Γ of the resonance can be determined, a value of Γ=180(70) keV is obtained [3], which agrees quite well with the experimental result of 270(50) keV [7].

References

[1] V. D. Efros, W. Leidemann, and G. Orlandini, Phys. Lett. B **338**, 130 (1994);
V. D. Efros, W. Leidemann, G. Orlandini and N. Barnea, J. Phys. G **34**, R459 (2007)

[2] S. Bacca, N. Barnea, W. Leidemann, and G. Orlandini, Phys. Rev. Lett. **110**, 042503 (2013)

[3] W. Leidemann, Phys. Rev. C **91**, 054001 (2015)

[4] V. D. Efros, W. Leidemann, and G. Orlandini, Phys. Lett. B **408**, 1 (1997)

[5] V. D. Efros, W. Leidemann, and G. Orlandini, Phys. Rev. Lett. **78**, 432 (1997)

[6] V. D. Efros, W. Leidemann, and G. Orlandini, Phys. Rev. Lett. **78**, 4015 (1997)

[7] Th. Walcher, Phys. Lett. B, **31**, 442 (1970)

EPJ Web of Conferences **113**, 04015 (2016)
DOI: 10.1051/epjconf/201611304015

Properties of ^{4}He and ^{6}Li with improved chiral EFT interactions

P. Maris[1,a], S. Binder[2], A. Calci[3], E. Epelbaum[4], R.J. Furnstahl[5], J. Golak[6], K. Hebeler[7], H. Kamada[8], H. Krebs[4], J. Langhammer[7], S. Liebig[9], U.-G. Meißner[9,10,11], D. Minossi, A. Nogga[9], H. Potter[1], R. Roth[7], R. Skibiński[6], K. Topolnicki[6], J.P. Vary[1], and H. Witała[6]

[1]*Department of Physics and Astronomy, Iowa State University, Ames, IA 50011, USA*
[2]*Physics Division, Oak Ridge National Laboratory, Oak Ridge, TN 37831, USA*
[3]*TRIUMF, 4004 Wesbrook Mall, Vancouver, British Columbia, V6T 2A3, Canada*
[4]*Institut für Theoretische Physik II, Ruhr-Universität Bochum, D-44780 Bochum, Germany*
[5]*Department of Physics, The Ohio State University, Columbus, OH 43210, USA*
[6]*M. Smoluchowski Institute of Physics, Jagiellonian University, PL-30348 Kraków, Poland*
[7]*Institut für Kernphysik, Technische Universität Darmstadt, D-64289 Darmstadt, Germany*
[8]*Dept. of Physics, Faculty of Engineering, Kyushu Institute of Technology, Kitakyushu 804-8550, Japan*
[9]*Institut für Kernphysik, Institute for Advanced Simulation and Jülich Center for Hadron Physics, Forschungszentrum Jülich, D-52425 Jülich, Germany*
[10]*Helmholtz-Institut für Strahlen- und Kernphysik and Bethe Center for Theoretical Physics, Universität Bonn, D-53115 Bonn, Germany*
[11]*JARA - High Performance Computing, Forschungszentrum Jülich, D-52425 Jülich, Germany*

Abstract. We present recent results for ^{4}He and ^{6}Li obtained with improved NN interactions derived from chiral effective field theory up to N^{4}LO. The many-body calculations are performed order-by-order in the chiral expansion. At N^{3}LO and N^{4}LO additional renormalization using the Similarity Renormalization Group is adopted to improve numerical convergence of the many-body calculations. We discuss results for the ground state energies, as well as the magnetic moment and the low-lying spectrum of ^{6}Li.

1 Nuclear potential from chiral EFT

Chiral effective field theory (EFT) allows one to derive nuclear forces (and corresponding electoweak currents) in a systematical way [1]. In particular, the leading order (LO) and next-to-leading order (NLO) contributions to the interaction are given solely by nucleon-nucleon (NN) operators while three-nucleon forces (3NFs) appear first at next-to-next-to-leading order (N^{2}LO) [2] in the chiral expansion. Four-nucleon forces are even more suppressed and start contributing at N^{3}LO. The chiral power counting thus provides a natural explanation of the observed hierarchy of nuclear forces.

In Refs. [3, 4] a new generation of chiral EFT NN potentials up to N^{4}LO was developed, using local regulators for the long-range terms and with reduced finite-cutoff artefacts by employing a novel ultraviolet regularization scheme; in addition, a procedure to estimate the theoretical uncertainties due to the truncation of the chiral expansion was proposed (see also [5]). Here we present results for ^{4}He and ^{6}Li obtained with these improved chiral NN potentials using a regulator of $R = 1.0$ fm [6].

[a]e-mail: pmaris@iastate.edu

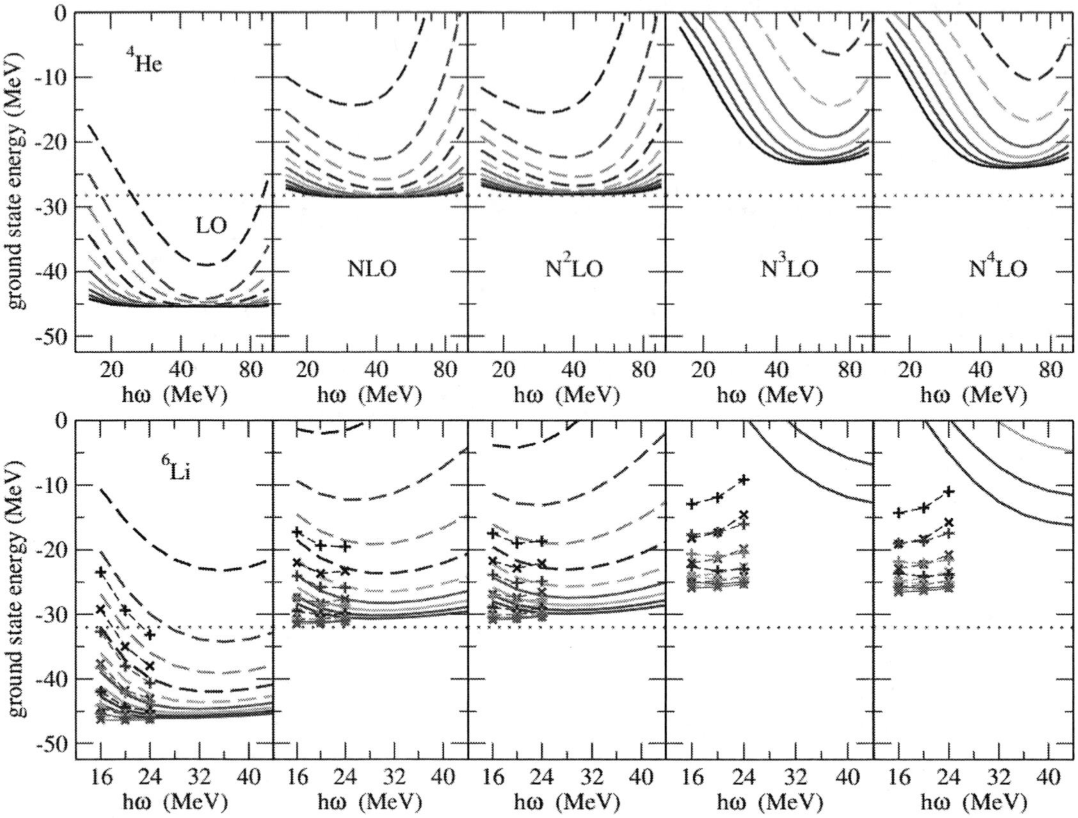

Figure 1. Ground state (gs) energy of ^{4}He (top) and ^{6}Li (bottom) as function of the HO parameter $\hbar\omega$ for $N_{\max} = 2$ (dashed black line) to $N_{\max} = 20$ (solid black line, ^{4}He) and to $N_{\max} = 18$ (solid red line, ^{6}Li) at LO, NLO, N^{2}LO, N^{3}LO, and N^{4}LO. For ^{6}Li we also show results with SRG renormalization (crosses and plusses) from $N_{\max} = 2$ (black) to $N_{\max} = 12$ (magenta). The horizontal dotted green line indicates the experimental values.

2 No-Core Configuration Interaction calculations

In No-Core Configuration Interaction (NCCI) calculations of nuclei [7] the wavefunction Ψ of a nucleus consisting of A nucleons is expanded in an A-body basis of Slater determinants Φ_k of single-particle wavefunctions $\phi_{nljm}(\vec{r})$. Conventionally, one uses a harmonic oscillator (HO) basis with energy parameter $\hbar\omega$ for the single-particle wavefunctions. A convenient and efficient truncation of the complete (infinite-dimensional) basis is a truncation on the total number of HO quanta: the basis is limited to many-body basis states with $\sum_A N_i \leq N_0 + N_{\max}$, with N_0 the minimal number of quanta for that nucleus and $N_{\max}$ the truncation parameter. Thus the many-body Schrödinger equation becomes a large sparse matrix eigenvalue problem with even (odd) values of $N_{\max}$ providing results for natural (unnatural) parity. Convergence toward the exact results is obtained with increasing $N_{\max}$, and is marked by approximate $N_{\max}$ and $\hbar\omega$ independence.

3 Results for ^{4}He and ^{6}Li

Figure 1 clearly illustrates convergence patterns of NCCI calculations: as $N_{\max}$ increases, we see a compression of successive $N_{\max}$ curves, and the dependence on $\hbar\omega$ near the variational minimum

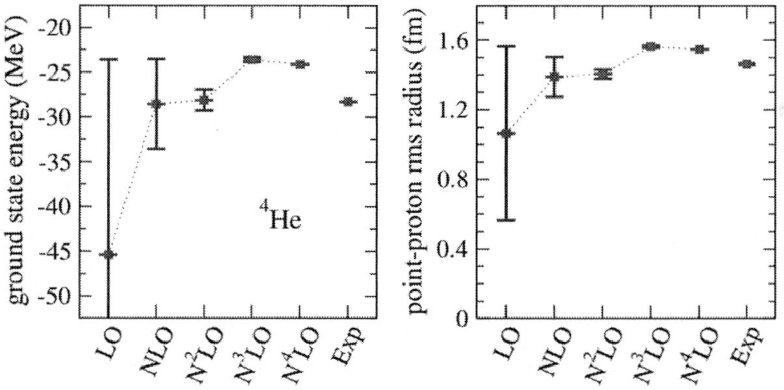

Figure 2. Results for ^{4}He: gs energy and point-proton rms radius (r_p) at different chiral order, with both theoretical (chiral) uncertainty estimates (blue) and many-body numerical uncertainties (red), with experimental values in green.

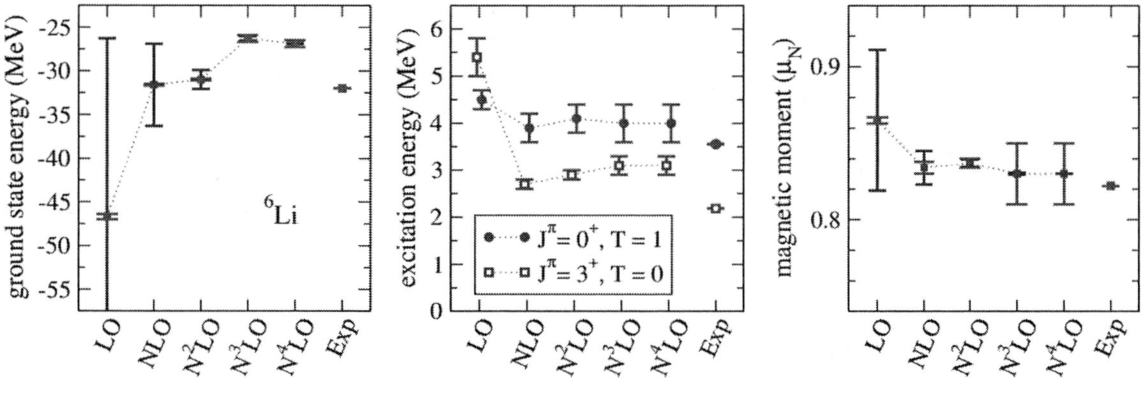

Figure 3. Results for ^{6}Li: gs energy, excitation energies of the two lowest excited states, and gs magnetic moment at different chiral order.

weakens. We also see a dramatic difference in convergence rate: the strongly bound nucleus ^{4}He converges more rapidly than the weakly bound nucleus ^{6}Li; furthermore, up to N^{2}LO convergence is rapid, but at N^{3}LO and N^{4}LO our results for ^{6}Li are far from being converged, even at $N_{\max} = 18$.

In order to improve the convergence of the many-body calculations we apply the Similarity Renormalization Group (SRG) at the three-body level to 'soften' the chiral NN interaction [8–11]. **Indeed, at SRG evolution values of $\alpha = 0.04\,\mathrm{fm}^4$ and $\alpha = 0.08\,\mathrm{fm}^4$ we do find rapid convergence of the many-body calculation, and, including induced 3N interactions, only very weak dependence on the SRG evolution. Up to N^{2}LO, the SRG evolution produces results for ^{6}Li to within a fraction of a percent of those without SRG; at N^{3}LO and N^{4}LO the results with the SRG evolution are significantly better converged than, and within the extrapolation uncertainties of, the unconverged results without SRG evolution.** Finally, as a cross-check we also confirm that, to within our estimated numerical accuracy, our results for ^{4}He agree with results obtained in the Faddeev–Yakubovsky framework [6].

In figures 2 and 3 we summarize our results at different orders in the chiral expansion. In addition to the estimated numerical uncertainties in the many-body calculation, we also display the estimated theoretical chiral uncertainties following [3–6]. The chiral uncertainties decrease with increasing chiral order (as they should). However, the many-body numerical uncertainty increases with increasing chiral order, and at N^{3}LO and N^{4}LO our results for ^{6}Li are dominated by the many-body uncertainties.

Starting at N^2LO there are 3NFs in the chiral expansion, which we have not taken into account in the present calculations – this is the main cause for the 'jumps' between N^2LO and N^3LO in figures 2 and 3. We are confident that incorporating consistent 3NFs, starting at N^2LO, will reduce (or even eliminate) these jumps.

The patterns for the energies are very similar to each other: we find large overbinding at LO, close to the experimental data at NLO and N^2LO, and at N^3LO and N^4LO a modest underbinding, indicative of the need for 3NFs. The radius r_p in ^{4}He is correlated to its binding: far too small at LO, and slightly too large at N^3LO and N^4LO. Similar patterns are also found for the gs energy of ^{3}H and the Nd total cross section at 10 MeV [6]. Up to N^2LO our results for ^{4}He are also consistent, to within the estimated chiral uncertainties, with similar results obtained using Quantum Monte Carlo methods in combination with local chiral EFT NN potentials [12, 13] – **any differences can be attributed to differences in the NN potentials (e.g. different choices for the regulators).**

The binding energies of the lowest two excited states in ^{6}Li follow the same pattern as the gs energy. Hence their excitation energies, depicted in figure 3, depend much less on the chiral order than one might naıvely expect based on the theoretical uncertainties in the binding energies. At present we are, however, unsure about the best way to estimate the chiral uncertainties in the excitation energies.

Finally, the gs magnetic moment of ^{6}Li, calculated using the canonical one-body M1 operator, is reasonable well converged at N^2LO. Calculations incorporating meson-exchange currents [14] suggest that higher order chiral contributions to the M1 operator could very well lead to a correction of a few percent. The derivation and implementation of electroweak current operators consistent with these improved chiral EFT interactions is currently in progress.

Acknowledgements

This work was performed by the LENPIC collaboration with support from: BMBF (contract 06DA7047I); DFG and NSFC (CRC 110); DFG (SFB 634, SFB/TR 16); ERC projects 259218 NUCLEAREFT and 307986 STRONGINT; EU (HadronPhysics3, Grant 283286); HIC for FAIR; Polish National Science Center DEC-2013/10/M/ST2/00420; US DOE DESC0008485, DE-FG-02-87ER40371, DE-SC0006758, DE-SC0008533; US NSF PHY-0904782, PHY-1306250. Supercomputer usage included: JSC-Jülich; PAS0680-Ohio SC; Edison-NERSC.

References

[1] S. Weinberg, Phys. Lett. B **251**, 288 (1990).
[2] E. Epelbaum, H. W. Hammer and U.-G. Meißner, Rev. Mod. Phys. **81**, 1773 (2009).
[3] E. Epelbaum, H. Krebs and U.-G. Meißner, Eur. Phys. J. A **51**, 53 (2015).
[4] E. Epelbaum, H. Krebs and U.-G. Meißner, Phys. Rev. Lett. **115**, 122301 (2015).
[5] E. Epelbaum, these proceedings.
[6] S. Binder *et al.*, arXiv:1505.07218 [nucl-th].
[7] B. R. Barrett, P. Navratil and J. P. Vary, Prog. Part. Nucl. Phys. **69**, 131 (2013).
[8] S. K. Bogner *et al.*, Nucl. Phys. A **801**, 21 (2008).
[9] S. K. Bogner, R. J. Furnstahl and A. Schwenk, Prog. Part. Nucl. Phys. **65**, 94 (2010).
[10] E. D. Jurgenson, P. Navratil and R. J. Furnstahl, Phys. Rev. C **83**, 034301 (2011).
[11] R. Roth, A. Calci, J. Langhammer and S. Binder, Phys. Rev. C **90**, 024325 (2014).
[12] J. E. Lynn *et al.*, Phys. Rev. Lett. **113**, 192501 (2014).
[13] J. E. Lynn, arXiv:1510.05450 [nucl-th] (these proceedings).
[14] S. Pastore, S. C. Pieper, R. Schiavilla and R. B. Wiringa, Phys. Rev. C **87**, 035503 (2013).

EPJ Web of Conferences **113**, 04016 (2016)
DOI: 10.1051/epjconf/201611304016

A theoretical approach to electromagnetic reactions in light nuclei

Tianrui Xu[1,2], Mirko Miorelli[1,2], Sonia Bacca[1,3,a], and Gaute Hagen[4,5]

[1] *TRIUMF, 4004 Wesbrook Mall, Vancouver, British Columbia, Canada V6T 2A3*
[2] *Department of Physics and Astronomy, University of British Columbia, Vancouver, British Columbia, Canada V6T 1Z4*
[3] *Department of Physics and Astronomy, University of Manitoba, Winnipeg, Manitoba, Canada R3T 2N2*
[4] *Physics Division, Oak Ridge National Laboratory, Oak Ridge, Tennessee 37831, USA*
[5] *Department of Physics and Astronomy, University of Tennessee, Knoxville, Tennessee 37996, USA*

Abstract. We briefly review the theory for electromagnetic reactions in light nuclei based on the coupled-cluster formulation of the Lorentz integral transform method. Results on photodisintegration reactions of ^{22}O and ^{40}Ca are reported and preliminary calculations on the Coulomb sum rule for ^{4}He are discussed.

1 Introduction

The investigation of nuclear reactions from first principles is fundamental in bridging nuclear physics with the underlying quantum chromo-dynamics regime. Nuclear reactions induced by electromagnetic probes are very useful as the electromagnetic current is well known and a clean comparison with experimental data can be performed [1]. Nowadays this valuable information is not only accessible for the lightest nuclei, but novel theoretical approaches are being developed to tackle nuclei with a larger number of nucleons. Below, we briefly review the Lorentz integral transform (LIT) approach in its coupled-cluster (CC) theory formulation [2]. Thereafter, we discuss some recent results regarding photoabsorption and electron-scattering reactions.

2 Theoretical Formulation

The key ingredient to study reactions induced by external electromagnetic probes, like photons or electrons, is the nuclear response function

$$R(\omega, q) = \sum_n |\langle n| O(q) |0\rangle|^2 \, \delta(E_n - E_0 - \omega)\,, \tag{1}$$

where $O(q)$ is the excitation operator, which will depend specifically on the external probe and on the momentum-transfer q. The nuclear response function is a dynamical observable which requires knowledge of the whole spectrum of the nucleus, being $|0\rangle$ and $|n\rangle$ the ground- and excited-state,

[a] e-mail: bacca@triumf.ca

respectively. The exact solutions of the many-body problem for the excited-states, typically in the continuum, are limited in mass number and in energy ω. In particular, it is difficult to calculate when many channels are open. However, one can use the LIT approach [3] to reduce the continuum problem to the solution of a bound-state equation. In particular, recently this method was reformulated using coupled-cluster theory [4], obtaining a new approach to extend *ab-initio* studies of electromagnetic observables to heavier and neutron-rich nuclei [2, 5].

In coupled-cluster theory one introduces the similarity transformed Hamiltonian

$$\overline{H} = \exp(-T)\hat{H}_N \exp(T), \tag{2}$$

where H_N is normal-ordered with respect to a chosen Slater determinant $|\Phi_0\rangle$. Correlations are introduced through the cluster operator T which is expanded into particle-hole (ph) excitation operators, *i.e.*, $T = T_1 + T_2 + \ldots$, with the $1p$-$1h$ excitation operator T_1, the $2p$-$2h$ excitation operator T_2, and so on. In the coupled-cluster formalism [2] the response function in Eq. (1) becomes

$$R(\omega, q) = \sum_n \langle 0_L|\bar{\Theta}^\dagger(q)|n_R\rangle\langle n_L|\bar{\Theta}(q)|0_R\rangle\delta(E_n - E_0 - \omega), \tag{3}$$

where $\bar{\Theta}^{(\dagger)}(q) = e^{-\hat{T}}O^{(\dagger)}(q)e^{\hat{T}}$ is the similarity transformed (adjoint) excitation operator and $\langle 0_L|$, $|0_R\rangle$ ($\langle n_L|$, $|n_R\rangle$) are the left and right reference ground-states (excited-states), respectively (see also Ref. [6]). The utilization of the LIT method requires to find the solution $\tilde{\Psi}_R$ of the following bound-state equation [2, 5]

$$(\overline{H} - \Delta E_0)|\tilde{\Psi}_R\rangle = \overline{\Theta}(q)|0_R\rangle, \tag{4}$$

where ΔE_0 is the ground-state correlation energy and $|0_R\rangle = |\Phi_0\rangle$. By assuming that $|\widetilde{\Psi}_R\rangle = \mathcal{R}|\Phi_0\rangle$ and expanding $\mathcal{R}$ in ph excitation operators, similarly to what done with T, one can see that Eq. (4) is equivalent to an equation-of-motion [4] with a source in the right-hand-side. In Ref. [2, 5] we have implemented this method with a truncation of both the operators T and $\mathcal{R}$ up to the $2p$-$2h$ excitation level, namely up to single and double (SD) excitations. Such a scheme is being named here the LIT-CCSD method.

By solving Eq. (4) one finds the LIT of the response function, integrated from threshold ω_{th} to infinity

$$L(\omega_0, \Gamma) = \frac{\Gamma}{\pi}\int_{\omega_{th}}^{\infty} d\omega \frac{R(\omega, q)}{(\omega - \omega_0)^2 + \Gamma^2} = \langle\tilde{\Psi}_L|\tilde{\Psi}_R\rangle, \tag{5}$$

where ω_0 and Γ are auxiliary parameters and $\langle\tilde{\Psi}_L|$ is the solution of the equivalent to Eq. (4) for the left-hand-side. To obtain the response function $R(\omega, q)$ we then invert Eq. (5) using the method outlined in Refs. [7, 8], which looks for the regularized solution of the integral transform equation.

Below we highlight some of our recent calculations obtained for photoabsorption and electron scattering reactions using as only ingredient an Hamiltonian H that contains, besides kinetic energy, a two-nucleon force derived in chiral effective field theory at next-to-next-to-next-to-next-to-leading (N^3LO) order [9].

3 Photodisintegration reactions

Photodisintegration reactions have been widely studied in the '70s with experiments on a variety of stable nuclei. The main observed feature of the measured cross sections was a very pronounced peak, referred to as the giant dipole resonance, located at excitation energies of about 10–30 MeV. First theoretical interpretations were given in terms of collective models [10, 11] and until very recently, most

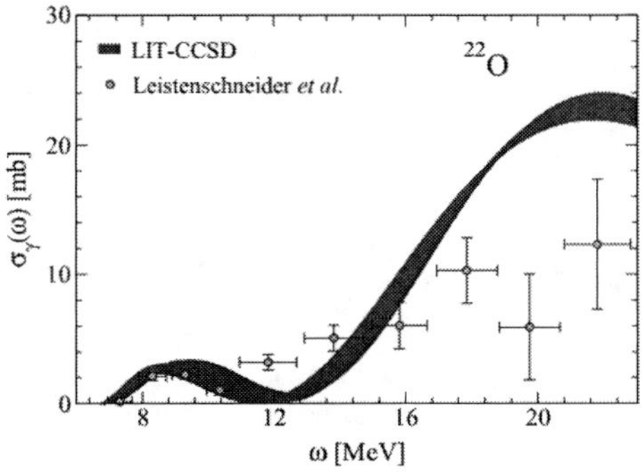

Figure 1. ^{22}O photodisintegration cross section compared with data from Ref. [13]. The curve is calculated with the N^{3}LO nucleon-nucleon force and is shifted to the experimental threshold.

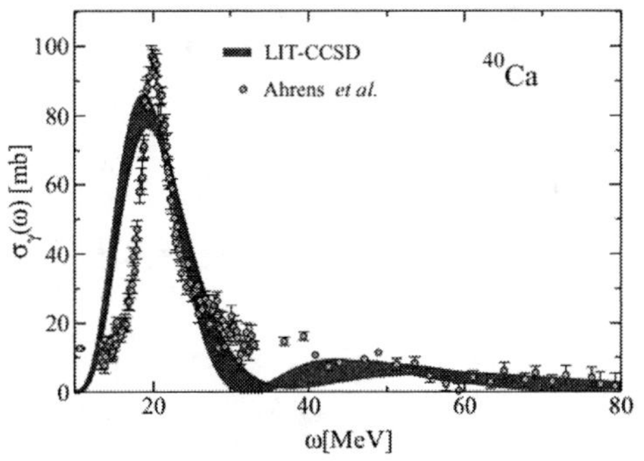

Figure 2. ^{40}Ca photodisintegration cross section compared to data from Ref. [14]. The curve is calculated with the N^{3}LO nucleon-nucleon force and is shifted to the experimental threshold.

microscopic calculations were obtained in terms of phenomenological approaches and interactions, see, *e.g.*, Ref. [12]. The LIT-CCSD approach introduced above, offers the opportunity to investigate photodisintegration reactions from an *ab-initio* point of view using nucleon-nucleon potentials, which reproduce two-nucleon scattering data [9]. Moreover, it allows one to study equivalent processes in neutron-rich nuclei, which have been more recently investigated with Coulomb-excitations experiments at rare isotope facilities, see *e.g.* Ref. [13].

In the long-wavelength limit, the photodisintegration cross section can be written as

$$\sigma_\gamma(\omega) = 4\pi^2\alpha\,\omega\,R^D(\omega)\,, \tag{6}$$

where α is the electromagnetic coupling constant and $R^D(\omega)$ is the dipole response function, namely Eq. (1) with the the dipole operator as excitation operator

$$D_z = \sum_k^A (z_k - Z_{cm})\left(\frac{1+\tau_k^3}{2}\right). \tag{7}$$

Here, z_k is the z-coordinate of the k-th nucleon in the lab-frame, while Z_{cm} denotes the z−component of the center of mass of the nucleus and τ_k^3 is the third component of the isospin of the k-th nucleon.

In Refs. [2, 5] we implemented the one-body operator of Eq. (7) in the solution of Eq. (4) and investigated the photodisintegration of a variety of light and medium-mass nuclei. After having benchmarked the LIT-CCSD method with exact hyperspherical harmonics [15] on ^{4}He, where we have observed that higher-order corrections to the T and $\mathcal{R}$ operator are very small, we have proceeded to study heavier nuclei. In particularly, here we review our results for the neutron-rich ^{22}O isotope in Fig. 1 and the stable ^{40}Ca nucleus in Fig. 2.

For ^{22}O one notices the appearance of a small peak at low energy, also named pigmy dipole resonance and experimentally observed in a variety of neutron-rich nuclei, which agrees very nicely with data from Leistenschneider *et al.* [13]. It is worth noticing that nothing in the interaction has been adjusted to ^{22}O and that no cluster structure has been imposed a priori in the calculation. This is not the first time that the LIT approach suggests the existence of a low-energy dipole mode, as in Ref. [16] a similar situation was found in ^{6}He.

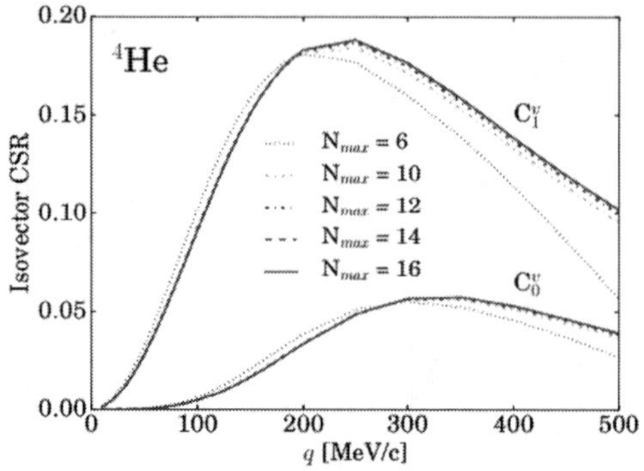

Figure 3. LIT-CCSD sum rules from C_0^v and C_1^v for ^{4}He for different model space sizes N_{max} with the N^{3}LO nucleon-nucleon chiral interaction.

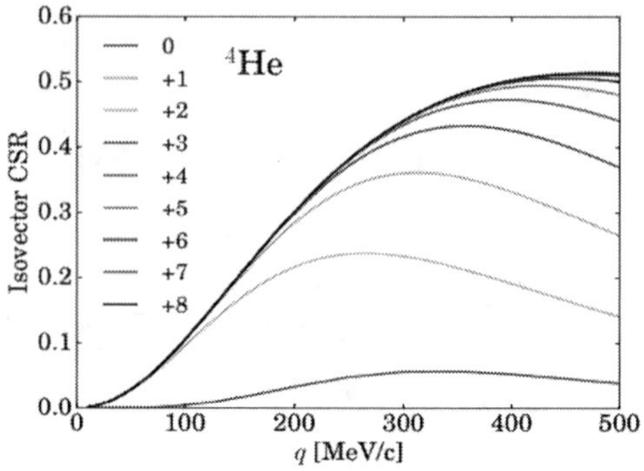

Figure 4. Recursive sum of isovector CSR for ^{4}He calculated using the chiral N^{3}LO nucleon-nucleon potential in the LIT-CCSD scheme.

Because the computational cost of the coupled-cluster method scales mildly with respect to the mass number, we can extend our studies to the photonuclear cross-section of ^{40}Ca. Fig. 2 shows a comparison of the LIT-CCSD calculations with data by Ahrens *et al.* [14]. We clearly see a giant resonance, even though the theoretical prediction is slightly broader and lower in strength. Both in Fig. 1 and Fig. 2 the band in the curve is obtained by inverting LIT with different Γ parameters in Eq. (5) and can be seen as a lower estimate of the theoretical error bar.

4 Electron-scattering reactions: Coulomb sum rule

Electron scattering reactions have been already investigated within the LIT method, see, *e.g.*, [17]. With the objective in mind to eventually tackle neutrino-nucleus interactions relevant for long-baseline neutrino experiments, we initiated a study of electron-scattering reactions within coupled-cluster theory. The latter has in fact the potential to be applied to ^{16}O and other medium-mass nuclei, which are used as detector materials in neutrino experiments. Following the work done with the Green's function Monte Carlo method in Ref. [18], we will first investigate the Coulomb sum rule.

The longitudinal Coulomb sum rule (CSR) is defined as

$$S_L(q) = \frac{1}{Z}\int_{\omega_{th}^+}^{\infty} d\omega \frac{R_L(\omega, q)}{G_E^{p\,2}(Q^2)}, \tag{8}$$

where $R_L(\omega, q)$ is the longitudinal response function, *i.e.*, Eq. (1) with $O(q) = \rho(q)$ and a recoil term $\frac{\mathbf{q}^2}{2M}$ in the delta function, while $G_E^p(Q^2)$ is the proton electric form factor as a function of the square of the four-momentum Q^2. The charge operator can be written as

$$\rho(q) = \sum_k^A \left(\frac{1+\tau_k^3}{2} G_E^p(Q^2) + \frac{1-\tau_k^3}{2} G_E^n(Q^2) \right) e^{iqz_k}, \tag{9}$$

with $G_E^n(Q^2)$ being the neutron electric form factor. Similarly to what done in Ref. [17] we perform a multipole expansion of Eq. (9) in terms of Coulomb multipoles, separating isoscalar (s) and isovector (v) contributions as

$$\rho(q) = \sum_J \left(C_J^s(q) + C_J^v(q) \right). \tag{10}$$

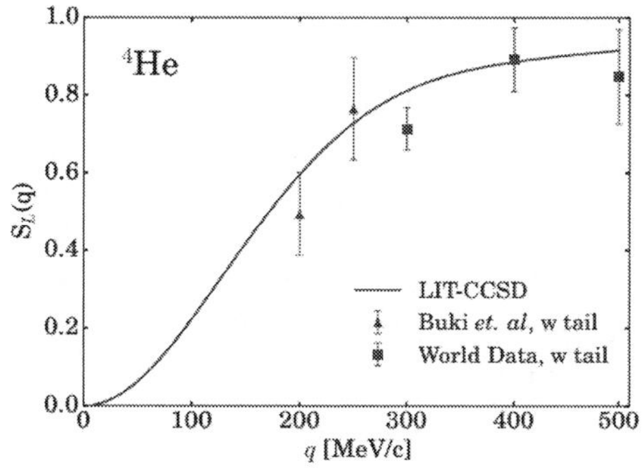

Figure 5. Preliminary results for the ^{4}He CSR from coupled-cluster theory compared with experimental CSR obtained from data from Buki *et. al.* (red triangle) [19] and from Ref. [20] (green square), as described in the text.

Each of the Coulomb multipole is calculated separately solving Eq. (4) for various fixed values of the momentum transfer q. Below, we show LIT-CCSD results for ^{4}He obtained using the N^{3}LO nucleon-nucleon chiral interaction.

In Fig. 3 we test the convergence of our calculations in terms of the model space size $N_{max} = 2n + l$, determined by the number of single-particle shells used in the ph expansion. For the isovector monopole C_0^v and isovector dipole C_1^v operator, one observes that very good convergence is reached already for $N_{max} = 14$. For higher order multipoles, slightly larger model spaces are needed. With respect to exact hyperspherical harmonics [15], C_1^v calculations from coupled-cluster theory agree very nicely up to about $q = 200$ MeV/c, but are 20% smaller at higher momenta. Note that at energies of about 300 MeV and higher, the chiral interaction of Ref. [9] is at the limit of its applicability. Work is in progress to understand the role of center of mass contamination in our calculation, given that the operator in Eq. (9) is not translational invariant.

In Fig. 4 we compare the recursive sum of the isovector multipoles that make up the CSR. One can readily see that, while at momentum transfer below $q = 100$ MeV/c two or three multipoles suffice, at the highest momentum transfer considered $q = 500$ MeV/c, nine multipoles need to be calculated before reaching convergence. A similar situation is found for the isoscalar multipoles. Finally, in Fig. 5 we compare the total CSR calculated from coupled-cluster theory with experimental data. Our calculation still misses the contribution of the isoscalar monopole term. Measurements of the longitudinal response functions at intermediate momentum transfer have been performed in the past and are collected in Ref. [20], while low momentum data at $q = 200$ and 250 MeV/c are taken from Buki *et. al.* [19]. Since finite maximal values of the energy transfer ω_{max} are measured in experimental data, the experimental CSR is obtained as

$$S_L(q) = \frac{1}{Z}\int_{\omega_{th}^+}^{\omega_{max}} d\omega \frac{R_L(\omega, q)}{G_E^{p2}(Q^2)} + S_{L,\,\mathrm{tail}} \tag{11}$$

where $S_{L,\,\mathrm{tail}}$ is taken from the theoretical calculations of ^{4}He response functions of Ref. [17].
As shown in Fig. 5, our calculation agrees well with the experimental data from low- to intermediate-momenta. It is known from Ref. [17] that three-body forces are not very pronounced on the CSR at the values of momentum transfer where data are available. Work is being presently directed to benchmark these studies with hypersherical harmonics, with the prospect of tackling ^{16}O.

5 Conclusions

After a short review of the coupled-cluster formulation of LIT method, we report on our recent calculations of the photodisintegration of the neutron-rich ^{22}O nucleus and the stable ^{40}Ca medium-mass nucleus. We then show our preliminary calculation of the Coulomb sum rule for ^{4}He, which describe rather well the available experimental data. A benchmark with data is essential for any theory that is going to be used to model the neutrino-nucleus interaction taking place within the detectors of neutrino long-baseline experiments. Using the LIT-CCSD theory, our long term goal is to study neutrino scattering off ^{16}O, which is relevant for the T2K experiment.

Acknowledgements

We thank N. Barnea and T. Papenbrock for useful discussions. This work was supported in parts by the Natural Sciences and Engineering Research Council (NSERC), the National Research Council of Canada, the Canadian Institute of Nuclear Physics (CINP), the U.S. Department of Energy (Oak Ridge National Laboratory), under Grant Nos. DEFG02-96ER40963 (University of Tennessee), de-sc0008499 (NUCLEI Sci-DAC collaboration), and the Field Work Proposal ERKBP57.

References

[1] S. Bacca and S. Pastore, J. Phys. G: Nucl. Part. Phys. **41** 123002 (2014)

[2] S. Bacca, N. Barnea, G. Hagen, G. Orlandini and T. Papenbrock, Phys. Rev. Lett. **111**, 122502 (2013)

[3] V. D. Efros, W. Leidemann, G. Orlandini, and N. Barnea, J. Phys. G: Nucl. Part. Phys. **34** R459 (2007)

[4] I. Shavitt, and R. J. Bartlett, *Many-body Methods in Chemistry and Physics* (Cambridge University Press, Cambridge UK, 2009) 255-256

[5] S. Bacca, N. Barnea, G. Hagen, M. Miorelli, G. Orlandini and T. Papenbrock, Phys. Rev. C **90**, 064619 (2014)

[6] M. Miorelli, S. Bacca, N. Barnea, G. Hagen, G. Orlandini, T. Papenbrock, FB21 proceedings.

[7] V.D. Efros, W. Leidemann, G. Orlandini, Few-Body Systems **26**, 251-269 (1999)

[8] D. Andreasi, W. Leidemann, C. Reiss, M. Schwamb, Eur. Phys. J. A **24**, 361-372 (2005).

[9] D. R. Entem, and R. Machleidt, Phys. Rev. C **68** 041001 (2003)

[10] M. Goldhaber, E. Teller, E., Phys. Rev. **74**, 1046–1049 (1948)

[11] H. Steinwedel, J.H.D Jensen, Z. Naturforsch. **5A**, 413 (1950)

[12] J. Erler, P. Kluepfel, P-G Reinhard, J. Phys. G: Nucl. Part. Phys. **38**, 033101 (2011)

[13] A. Leistenschneider *et al.*, Phys. Rev. Lett. **86**, 5442, (2001)

[14] J. Ahrens *et al.*, Nuclear Physics A **251** 479-492 (1975)

[15] N. Barnea, W. Leidemann, and G. Orlandini, Phys. Rev. C **61**, 054001 (2000)

[16] S. Bacca, M.A. Marchisio, N. Barnea, W. Leidemann, G. Orlandini, Phys. Rev. Lett., **89** 052502 (2002)

[17] S. Bacca, N. Barnea, W. Leidemann and G. Orlandini, Phys. Rev. C **80** 064001 (2009)

[18] A. Lovato *et al.* Phys. Rev. Lett., **111**, 092501 (2013)

[19] A. Yu. Buki, I. S. Timchenko, N. G. Shevchenko and I. A. Nenko, Phys. Lett. B **641**, 156 (2006)

[20] J. Carlson, J. Jourdan, R. Schiavilla, and I. Sick, Phys. Rev. C **65**, 024002 (2002)

EPJ Web of Conferences **113**, 04017 (2016)
DOI: 10.1051/epjconf/201611304017

Deuteron form factor measurements at low momentum transfers

B. S. Schlimme[1,a], P. Achenbach[1], J. Beričič[2], R. Böhm[1], D. Bosnar[3], L. Correa[4], M. O. Distler[1], A. Esser[1], H. Fonvieille[4], I. Friščić[3], K. A. Griffioen[5], Y. Huan[5], S. Kegel[1], Y. Kohl[1], H. Merkel[1], M. Mihovilovič[1], J. Müller[1], U. Müller[1], J. Pochodzalla[1], M. Schoth[1], F. Schulz[1], C. Sfienti[1], S. Širca[2,6], S. Štajner[2], M. Thiel[1], and A. Weber[1]

[1] *Institut für Kernphysik, Johannes Gutenberg-Universität Mainz, D-55099 Mainz, Germany*
[2] *Jožef Stefan Institute, SI-1000 Ljubljana, Slovenia*
[3] *Department of Physics, University of Zagreb, HR-10002 Zagreb, Croatia*
[4] *Clermont Université, Université Blaise Pascal, F-63000 Clermont-Ferrand, France*
[5] *College of William and Mary, Williamsburg, VA 23187, USA*
[6] *Department of Physics, University of Ljubljana, SI-1000 Ljubljana, Slovenia*

Abstract. A precise measurement of the elastic electron-deuteron scattering cross section at four-momentum transfers of $0.24\,\text{fm}^{-1} \leq Q \leq 2.7\,\text{fm}^{-1}$ has been performed at the Mainz Microtron. In this paper we describe the utilized experimental setup and the necessary analysis procedure to precisely determine the deuteron charge form factor from these data. Finally, the deuteron charge radius r_d can be extracted from an extrapolation of that form factor to $Q^2 = 0$.

1 Introduction

Electromagnetic form factors of light nuclei can be measured in elastic electron scattering. The results can be compared to predictions obtained by different theoretical approaches (for instance, see [1]) and therefore offer a unique opportunity to test our understanding of nuclear dynamics. In addition, charge radii can be determined from precise measurements of the charge form factors at low momentum transfers Q.

Assuming single photon exchange, the cross section for unpolarized electron-deuteron scattering can be written as

$$\frac{d\sigma}{d\Omega} = \sigma_{NS}\left[A(Q^2) + B(Q^2)\tan^2\frac{\theta_e}{2}\right], \tag{1}$$
$$A(Q^2) = G_C^2(Q^2) + \frac{8}{9}\eta^2 G_Q^2(Q^2) + \frac{2}{3}\eta G_M^2(Q^2),$$
$$B(Q^2) = \frac{4}{3}\eta(1+\eta)G_M^2(Q^2),$$

where σ_{NS} is the Mott differential cross section multiplied by the deuteron recoil factor, θ_e the electron scattering angle and $\eta = \frac{Q^2}{4M^2}$ with the deuteron mass M. G_C, G_Q and G_M are the the charge, quadrupole

[a]e-mail: schlimme@kph.uni-mainz.de

and magnetic form factors, respectively. In the kinematic regime investigated within the experiment, the cross section is dominated by the contribution of the deuteron charge form factor $G_C(Q^2)$. Taking into account corrections due to the small contributions of the quadrupole and the magnetic form factors, the cross section measurements allow for a determination of the charge form factor as a function of the momentum transfer. From the extrapolation of $G_C(Q^2)$ to $Q^2 = 0$, the deuteron radius can then be extracted from the slope $\mathrm{d}G_C/\mathrm{d}Q^2$.

The interest in the deuteron radius is renewed in the context of the proton radius puzzle, pointing out the inconsistency between proton radius values obtained using different experimental techniques: The recommended deuteron radius value from the CODATA group is dominated by the determined value for the proton radius combined with isotope shift data, relating the proton and deuteron radii. Including the result from muonic hydrogen measurements [2] results in a drastic change of r_d. An overview of deuteron charge radius results is illustrated in Fig. 1.

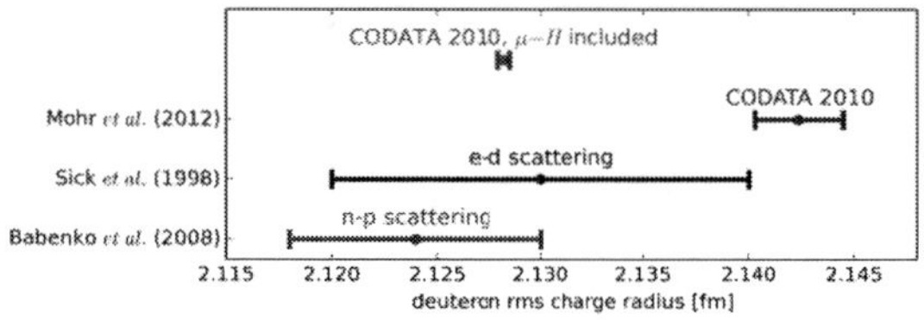

Figure 1. Various determinations of the deuteron charge radius r_d [3–5]. The uncertainty of the radius determination from electron-deuteron scattering is too large to reasonably distinguish between the two quoted CODATA values with and without inclusion of the muonic hydrogen result.

2 Experiment

Deuteron form factor measurements have been carried out by the A1 Collaboration at the spectrometer facility [6] at MAMI [7] to considerably improve the current form factor data base and to provide an extraction of the deuteron radius with a significant smaller uncertainty. Data were taken in 2014 during three weeks of beam time. Cross section measurements of the elastic $d(e,e')d$ reaction have been performed for three different beam energies (E_e = 180, 315, and 450 MeV). The scattered electrons were detected with three high resolution magnetic spectrometers simultaneously (spectrometers A, B, and C) to maximize the covered angular range and the internal redundancy in the data. The spectrometers have a relative momentum resolution of $\sim 10^{-4}$ and an angular resolution at the target of $\sim$ 3 mrad [6]. The angular acceptance of the spectrometers is well defined by collimators. The central electron scattering angles were varied in the range between 15° and 107° by changing the positions of the spectrometers. The chosen kinematic settings reach down in Q^2 as far as possible with the available setup (down to $Q = 0.24\,\mathrm{fm}^{-1}$, similar to lowest published data), and have high redundancy in the region $0.5\,\mathrm{fm}^{-1} \leq Q \leq 1.2\,\mathrm{fm}^{-1}$ where sensitivity to the rms-radius is high [1]. Altogether, almost 200 different kinematic settings have been investigated during the beam time.

The electron beam impinged on a liquid deuterium target with beam currents between $\sim$ 1 nA and 4 μA, which were measured with Foerster probes and a pA-meter. Vertical drift chambers (VDCs) were used for tracking, scintillation detectors for trigger and timing purposes, and a threshold-gas-Cherenkov detector for electron identification in the offline analysis. Due to the large cross sections, the low background, and the relatively high data taking rate of about 500 Hz per spectrometer, all kinematic settings could be measured with reasonable statistics during the available time. Hence, the statistical errors of the resulting cross section determinations are well below the 1% level for most of the settings. Usually only one spectrometer angle was varied at a time, allowing to monitor changes in the luminosity with the other two spectrometers. The beam delivery was generally not stopped during angle changes, feasible through the use of a remote spectrometer movement control, to minimize possible beam parameter changes.

3 Data analysis

For each recorded event the raw data of the individual detectors are considered to reconstruct kinematic quantities like the angle and the momentum of the detected particle and the reaction vertex. The VDC and scintillator data is used for track reconstruction of the detected particles in the focal plane of the spectrometers. The well known optics are used to reconstruct the particle tracks at the target position. Identification of elastic electron-deuteron scattering events is then accomplished via comparison of the detected energy E'_{exp} with the energy calculated from the detected scattering angle, $E'(\theta_{\mathrm{exp}})$, see Fig. 2. Comparison of the experimental spectra with the spectra from a simulation then yields the

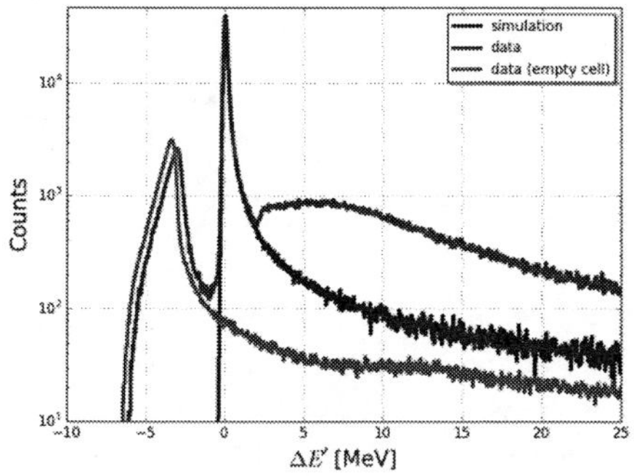

Figure 2. Identification of elastic electron-deuteron scattering events. Data (blue) and simulation (black) for spectrometer A at $\theta_e = 23.6°$ and $E_e = 315$ MeV. The additional large contribution for $\Delta E' = E'(\theta_{\mathrm{exp}}) - E'_{\mathrm{exp}} > 2.2$ MeV from the deuteron breakup reaction can be clearly separated from the elastic peak, which is located around zero. Furthermore, there is a contribution from electrons originating from the foils surrounding the deuterium target. Empty cell data, taken for the purpose of background subtraction, are shown in green.

experimental cross sections relative to the cross sections employed in the simulation. The simulation accounts for leading radiative processes beyond the one photon exchange approximation including the generation of the radiative tail, energy loss and straggling of the electrons in the target and window materials. Not all of the necessary corrections have been applied, yet. For instance, there is a contribution from background electrons originating from the target foils beneath the elastic deuteron peak. That contribution is in general larger for smaller angles. Since the deuteron charge radius will be determined from the slope of the extracted charge form factor with respect to the momentum transfer, a careful study of this kind of systematic effects is mandatory. Empty cell data have been taken for the purpose of background subtraction. Due to a slightly different energy loss situation when measuring on an empty cell without deuterium gas inside, there is a slight mismatch between the empty cell spectrum and the spectrum measured with the deuterium target which needs to be accounted for. To avoid an erroneous background subtraction, a simulation of the target wall contribution, including the contribution from excited states, is being developed to study that effect in detail. To illustrate the statistical quality of the data, a preliminary cross section result for one particular data set is shown in Fig. 3.

4 Outlook

The extraction of the individual cross section values is clearly the central part of the analysis. Several crucial corrections have not been applied yet, for instance Coulomb corrections or the mentioned background subtraction. Also, the modelling of the spectrometer resolutions, needed for a precise comparison between data and simulation, has to be finalized. Moreover, the determination of the luminosity is actually solely based on the measured target parameters and the beam current, and can be improved by the comparison of the count rates of the spectrometers, which were not moved between a setting change of the third spectrometer. Once all the necessary corrections and cross-checks have been performed, the obtained cross sections can be used to extract the charge form factor of the deuteron, $G_C^2(Q^2)$. An appropriate fitting procedure will be applied for this purpose. From

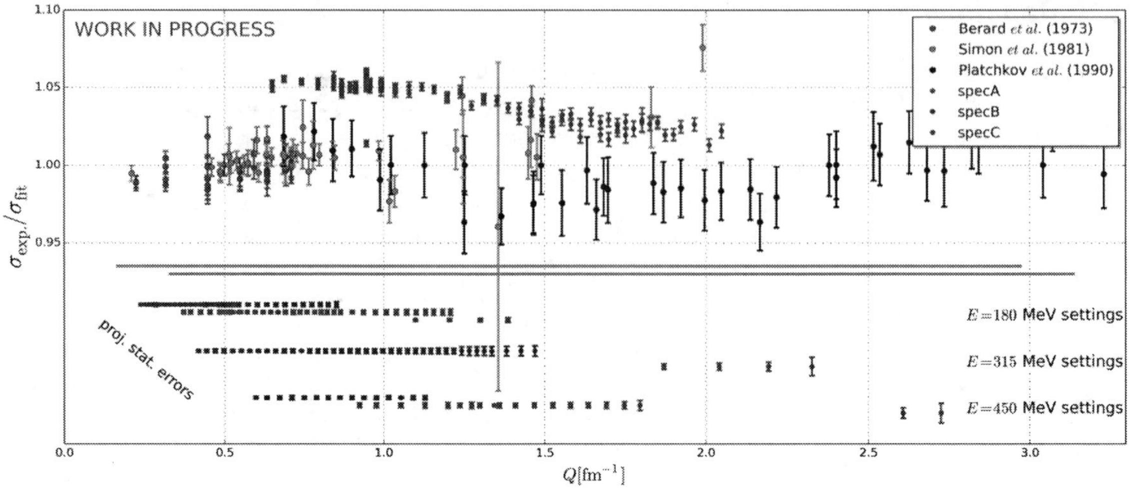

Figure 3. Top: Experimental cross sections divided by the cross sections calculated with the form factor parametrization of [8] (Sum-of-Gaussians) for one particular data set (red circles) at $E_e = 315$ MeV (statistical errors only). The reevaluated (see [4]) most precise existing low-Q data sets [9–11] are shown for comparison. The beam current, which enters the luminosity determination, was measured with a pA-meter, which has not been calibrated at this stage of the analysis. This can result in a systematic shift of our data compared to the previous data, as can be observed in the figure. Finally, the absolute normalization will be fixed by the known values of the form factors at $Q^2 = 0$. There is an additional rise of our data at smaller momentum transfers Q essentially caused by background produced at the target foils, which has not been subtracted, yet. Bottom: Projected statistical errors of the other data sets of this experiment.

the slope of $G_C^2(Q^2)$ at $Q^2 = 0$, the deuteron radius can then be determined, with the aim to provide additional insight to the proton radius puzzle.

Acknowledgements

This work was supported by the Deutsche Forschungsgemeinschaft with the Collaborative Research Center 1044: *The Low-Energy Frontier of the Standard Model. From Quarks and Gluons to Hadrons and Nuclei.*

References

[1] L.E. Marcucci *et al.*, arXiv:1504.05063 [nucl-th]
[2] R. Pohl *et al.*, Nature **466**, 213 (2010)
[3] P.J. Mohr *et al.*, Rev. Mod. Phys. **84**, 1527 (2012)
[4] I. Sick and D. Trautmann, Nucl. Phys. A **637**, 559 (1998)
[5] V.A. Babenko and N.M. Petrov, Phys. Atom. Nucl. **71**, 1730 (2008)
[6] K.I. Blomqvist *et al.*, Nucl. Instrum. Meth. A **403**, 263 (1998)
[7] H. Herminghaus *et al.*, Nucl. Instrum. Meth. **138**, 1 (1976)
[8] D. Abbott *et al.*, Eur. Phys. J. A **7** (2000) 421
[9] R.W. Berard *et al.*, Phys. Lett. B **47**, 355 (1973)
[10] G.G. Simon *et al.*, Nucl. Phys. A **364**, 285 (1981)
[11] S. Platchkov *et al.*, Nucl. Phys. A **510**, 740 (1990)

EPJ Web of Conferences **113**, 04018 (2016)
DOI: 10.1051/epjconf/201611304018

The contribution of the $^3\vec{He}(\vec{\gamma}, n)pp$ reaction to the GDH integrand below pion production threshold

G. Laskaris[1,2,a]

[1] *Triangle Universities Nuclear Laboratory, Durham, North Carolina 27708, USA*
[2] *Department of Physics, Duke University, Durham, North Carolina 27708, USA*

Abstract. The first measurements of the three-body phototdisintegration of ^{3}He polarized parallel and anti-parallel to a circularly polarized γ-ray beam were carried out at the High Intensity γ-ray Source (HIγS) facility located at Triangle Universities Nuclear Laboratory (TUNL). A high pressure ^{3}He target, polarized via spin-exchange optical pumping with alkali metals, was used in the experiments. The neutrons from the three-body photodisintegration were detected with sixteen 12.7 cm diameter liquid scintillator detectors. The spin-dependent cross sections and the contributions from the three-body photodisintegration to the ^{3}He Gerasimov-Drell-Hearn sum rule integrand were extracted and compared with state-of-the-art three-body calculations at the incident photon energies of 12.8, 14.7, and 16.5 MeV. The calculations, which include the Coulomb interaction are in good agreement with the results of the measurements at 12.8 and 14.7 MeV but deviate from the results at 16.5 MeV.

1 Introduction

Sum rules involving the spin structure of the nucleon and nuclei offer an important opportunity to study QCD. Among spin sum rules, the Gerasimov-Drell-Hearn (GDH) sum rule [1] is particularly interesting. It connects long distance behavior of the nucleon (nucleus), i.e. static property, to the entire excitation spectrum. The GDH sum rule relates the energy-weighted difference of the spin-dependent total photo-absorption cross sections σ^P (for target spin and beam helicity parallel) and σ^A (for target spin and beam helicity anti-parallel) to the anomalous magnetic moment of the target nucleus/nucleon as follows:

$$I^{GDH} = \int_{\nu_{thr}}^{\infty} (\sigma^P - \sigma^A)\frac{d\nu}{\nu} = \frac{4\pi^2\alpha}{M^2}\kappa^2 I, \qquad (1)$$

where ν is the photon energy, ν_{thr} is the pion production/photodisintegration threshold on the nucleon/nucleus, κ is the anomalous magnetic moment, M is the mass and I is the spin of the nucleon or the nucleus. This sum rule is based on fundamental principles such as Lorentz and Gauge invariance, crossing symmetry, causality and unitarity, and an assumption of unsubtracted dispersion relation. A detailed derivation of the GDH sum rule based on the aforementioned principles can be found in Ref. [2].

[a]Currently at Stanford University, Stanford, CA 94305, USA, e-mail: laskaris@stanford.edu

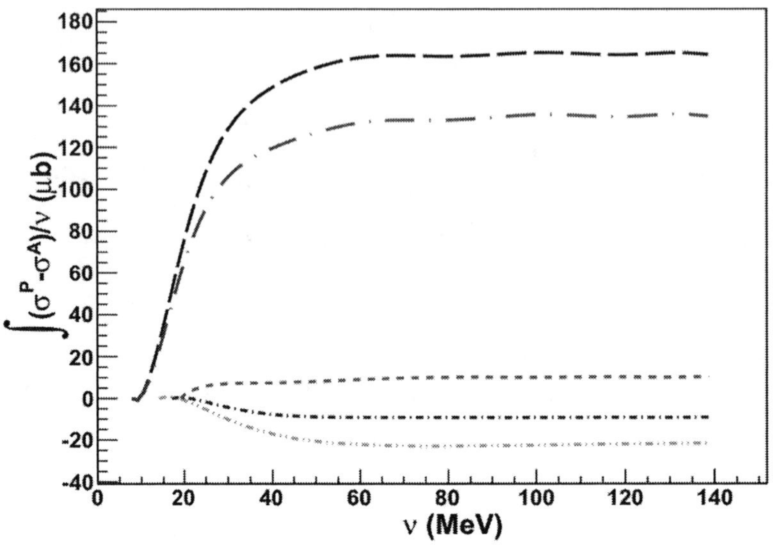

Figure 1. (Color online) Theoretical predictions of two- and three-body channel contributions to the ^{3}He GDH Integral below pion production threshold. The curves from top to bottom are: (i) long-dashed (black) curve: three-body channel contribution by Deltuva *et al.* (ii) long-dashed dotted (red) curve: three-body channel contribution by Skibiński *et al.* (iii) short-dashed (purple) curve: two-body channel contribution by Skibiński *et al.* (explicit MEC) (iv) short-dashed dotted (blue) curve: two-body channel contribution by Skibiński *et al.* (implicit MEC via Siegert theorem) (v) short-dashed double dotted (green) curve: two-body channel contribution by Deltuva *et al.*

The sum rule for ^{3}He can be written as the sum of two terms:

$$496\mu b = \int_{\nu_{thr}}^{\infty} GDH_{^3He} = \int_{\nu_{thr}}^{\nu_\pi} GDH_{^3He} + \int_{\nu_\pi}^{\infty} GDH_{^3He} \tag{2}$$

where the first term is from the two-body breakup threshold of ^{3}He ($\sim$ 5.5 MeV) to the pion production threshold ($\sim$ 140 MeV) and the second term is for the energy region from the pion production threshold to infinity. The first term can be measured experimentally by carrying out double-polarized two-body and three-body photodisintegration experiments in this energy region and it can also be estimated based on the state-of-the-art three-body calculations.

The state-of-the-art three–body calculations which can estimate the first part of the integral in Eq. (2) are performed mainly through the machinery of Faddeev [6] and Alt-Grassberger-Sandhas equations (AGS) [7], and have been carried out for both two-body and three-body photodisintegration of ^{3}He with double polarizations. These calculations [8, 9] use a variety of nucleon-nucleon (NN) potentials like Argonne V18 (AV18) [10] or charge dependent (CD) Bonn [11, 12] and three-nucleon forces (3NFs) like Urbana IX (UIX) [13] or CD Bonn + Δ [8], with the latter yielding an effective 3NF through the Δ-isobar excitation. The calculations by Deltuva *et al.* are based on AGS equations and employ the CD Bonn+Δ [8] with the corresponding single-baryon and meson-exchange electromagnetic currents plus relativistic single-nucleon charge corrections [8]. The proton-proton Coulomb force is included using the method of screening and renormalization [8]. Skibiński *et al.* solve the Faddeev equations by using the AV18 potential and the UIX 3NF accounting for single-nucleon currents and the two most important meson-exchange electromagnetic currents, the seagull and pion-in-flight terms [9].

Figure 1 shows the GDH integral for the two- and three-body photodisintegration of ^{3}He below pion production threshold. The short-dashed double dotted (green) and the long-dashed (black) curves are the estimations of the GDH integral for the two- and three-body photodisintegration from Deltuva *et al.* Their sum amounts to 142 μb. The short-dashed (purple) and short-dashed dotted (blue) curves are two different estimations by Skibiński *et al.* of the two-body channel contribution to the GDH integral taking into account explicitly or implicitly the MEC via the Siegert theorem. The long-dashed dotted curve is the estimation of the three-body contribution to the GDH integral coming from the same group. The maximum estimation of the GDH integral below pion threshold according to Skibiński *et al.* is 145 μb. Both theoretical calculations predict that the contribution to the GDH sum rule below pion production threshold is 140–150 μb which is an important part of the overall GDH integral (496 μb). This contribution lays mainly below the incident photon energy of 40 MeV. Therefore, a spin-dependent study of $^3\vec{He}(\vec{\gamma},n)pp$ below 40 MeV not only provides a stringent test of modern three-body calculations, but also serves as an important step towards an experimental test of the GDH sum rule on the ^{3}He nucleus.

2 The Experiment

The first experiments [14, 15] on the three–body photodisintegration of ^{3}He using a longitudinally polarized ^{3}He target and a circularly polarized γ-ray beam took place at the HIγS facility [17] of TUNL at the incident photon energies of 12.8, 14.7 and 16.5 MeV. A nearly mono-energetic, ~100% circularly-polarized pulsed γ-ray beam was used. The beam was collimated using a 12 mm diameter collimator resulting in on-target intensities of (0.7-2.0)$\times 10^8 \gamma/s$ and an energy spread of $\Delta\nu/\nu$ ~3.0-5.0%. The on-target intensity of the beam was determined using the well-known d(γ,n)p cross section [18] and two BC501A liquid scintillator neutron detectors mounted at a scattering angle of 90° degrees downstream of the ^{3}He target.

Upstream of the flux monitor, the polarized γ-beam was incident on a polarized ^{3}He cell. A N_2 reference cell was used for background subtraction. Details concerning their technical characteristics and the spin exchange optical pumping technique used to polarize the ^{3}He target can be found in Refs. [14, 15, 19–21]. The spin of the ^{3}He target was flipped every 15 mins in order to extract the spin-dependent cross sections and the GDH integrand, $(\sigma^P - \sigma^A)/\nu$. The polarization was measured using the nuclear magnetic resonance-adiabatic fast passage [22] technique calibrated by electron paramagnetic resonance [23]. The latter can measure the absolute ^{3}He target polarization which was found to be between 33% and 45%.

An array of sixteen liquid scintillator BC-501A counters was used to detect the neutrons from the $^3\vec{He}(\vec{\gamma},n)pp$ reaction. The detectors were placed at the horizontal plane every 15°, symmetrically on each side of the beam, at laboratory scattering angles from 30° to 165°. No detectors were placed at the laboratory angles 60° and 120° due to the proximity to a pair of Helmholtz coils which provided the holding field for the polarized ^{3}He target.

Three quantities were recorded for each event: the pulse height (PH), the time-of-flight (TOF) and the pulse shape discrimination (PSD) signals. Initially, a PH cut was applied at 0.162 MeVee to set the detector efficiency. The correlations between the PSD, PH and TOF were utilized to extract the neutron events and to remove the γ-ray events and two-dimensional cuts were applied on these histograms. The same cuts were used for the data taken with the N_2 reference cell to subtract the background contributions. The outgoing neutron energy was determined using the measured TOF of the neutrons assuming they were emitted from the center of the ^{3}He target cell. More details about this analysis can be found in Refs. [14, 15].

3 Results and Discussion

The spin-dependent double-, single-differential and total cross sections, and the contributions of the three–body photodisintegration of ^{3}He to the GDH integrand were obtained [14–16] and compared with the state-of the-art three-body calculations [8, 9] at all incident energies. Fig. 2 [16] shows the GDH integral results including the statistical and systematic uncertainties compared with the theoretical calculations. Although a very good agreement is observed between the measurements [14, 15] and the calculations based on Ref. [8] at 12.8 and 14.7 MeV, a difference between the measured GDH integrand and the calculations can be seen at the incident photon energy of 16.5 MeV. The measured GDH integrand at 16.5 MeV is found to be slightly more than one standard deviation larger than the maximum calculated value based on Ref. [8].

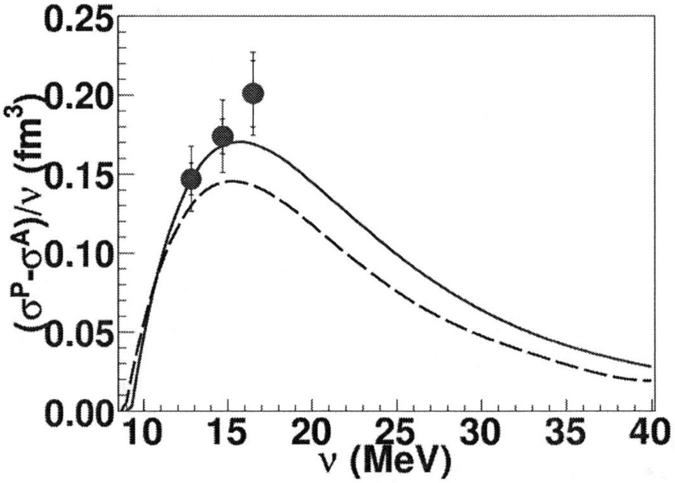

Figure 2. (Color online) The combined GDH integrand results as can be seen in Ref. [16] compared with the theoretical predictions of Ref. [8] (solid-blue curve) and Ref. [9] (dashed-black curve). The inner error bars of the data points represent the statistical uncertainties while the outer include both the statistical and systematic uncertainties added in quadrature.

To investigate whether the larger than expected GDH integrand value at 16.5 MeV is due to statistics and to further quantify the three-body contribution to the GDH integral, measurements above 16.5 MeV for the three-body photodisintegration channel are necessary. These measurements combined with the recently acquired data from the two-body photodisintegration channel [15] will constrain the contribution to the GDH integral for ^{3}He below the pion-threshold.

Acknowledgements

The author wish to thank the members of the Medium Energy Physics group and the Capture group of Duke University, A. Deltuva, A.C. Fonseca, J. Golak, P. U. Sauer, R. Skibiński and H. Witała for providing theoretical calculations, Michael Souza of Princeton University for making the target cells, Gordon Cates' group from

University of Virginia for providing the Rb/K ampules and for Sol-Gel coating of the target cells, T. Averett for filling the cell and the TUNL personnel for the technical support of this experiment. This work is supported by the U.S. Department of Energy under contract number DE-FG02-03ER41231 and Duke University.

References

[1] S.D. Drell and A.C. Hearn, Phys. Rev. Lett. **16**, 908 (1966); S.B. Gerasimov, Yad. Fiz. **2**, 598 (1965) (Sov. J. Nucl. Phys. **1**, 430 (1966)).
[2] K. Helbing, Progress in Particle and Nuclear Physics **57**, 405 (2006).
[3] J.L. Friar, B.F. Gibson, G.L. Payne, A.M. Bernstein, and T.E. Chupp, Phys. Rev. C **42**, 2310 (1990).
[4] P. Aguar Bartolomé *et al.*, Phys. Lett. B **723**, 71 (2013).
[5] S. Costanza, Eur. Phys. J. A **50**, 173 (2014).
[6] L.D. Faddeev, Zh. Eksp. Theor. Fiz. **39** 1459 (1960); Sov. Phys. JETP **12**, 1041 (1961).
[7] E.O. Alt, P. Grassberger, and W. Sandhas, Nucl. Phys. **B2**, 167 (1967).
[8] A. Deltuva, L.P. Yuan, J. Adam, A.C. Fonseca, P.U. Sauer, Phys. Rev. C **69**, 034004 (2004); A. Deltuva, A.C. Fonseca, and P.U. Sauer, Phys. Rev. C **72**, 054004 (2005); Ann. Rev. Nucl. Part. Sci. **58**, 27 (2008); A. Deltuva, A.C. Fonseca, and P.U. Sauer, Phys. Rev. C **80**, 064004 (2009).
[9] R. Skibiński, J. Golak, H. Witała, W. Glöckle, A. Nogga, and H. Kamada, Phys. Rev. C **72**, 044002 (2005); R. Skibiński, J. Golak, H. Witała, W. Glöckle, H. Kamada and A. Nogga, Phys. Rev. C **67**, 054002 (2003); R. Skibiński, J. Golak, H. Kamada, H. Witała, W. Glöckle and A. Nogga, Phys. Rev. C **67**, 054001 (2003).
[10] R. B Wiringa, V. G. J. Stoks, R. Schiavilla. Phys. Rev. C **51**, 38 (1995).
[11] R. Machleidt and K. Holinde and Ch. Elster, Phys. Rep. **149**, 1 (1987).
[12] R. Machleidt, Phys. Rev. C **63**, 024001 (2001).
[13] J. Carlson, V.R. Pandharipande, and R.B. Wiringa, Nucl. Phys. **A 401**, 59 (1983).
[14] G. Laskaris *et al.*, Phys. Rev. Lett. **110**, 202501 (2013); Phys. Rev. C **89**, 024002 (2014).
[15] G. Laskaris, Ph.D. thesis, Duke University (2015).
[16] G. Laskaris *et al.*, Phys. Lett. B **750C**, 547 (2015);
[17] H.R. Weller M.W. Ahmed, H. Gao, W. Tornow, Y.K. Wu, M. Gai, and R. Miskimen, Prog. Part. Nucl. Phys. **62**, 257 (2008).
[18] D.M. Skopik, Y.M. Shin, M.C. Phenneger, and J.J. Murphy II, Phys. Rev. C **9**, 531 (1974); Y. Birenbaum, S. Kahane, and R. Moreh, Phys. Rev. C **32**, 1825 (1985); R. Bernabei *et al.*, Phys. Rev. Lett. **57**, 1542 (1986); A. De Graeve *et al.*, Phys. Rev. C **45**, 860 (1992).
[19] W. Happer Rev. Mod. Phys. **42**, 169 (1972).
[20] K. Kramer, X. Zong, R. Lu, D. Dutta, H. Gao, X. Qian, Q. Ye, X. Zhu, T. Averett, and S. Fuchs, Nucl. Instrum. Methods Phys. Res., Sect. A **582**, 318 (2007).
[21] Q. Ye, G. Laskaris, W. Chen, H. Gao, W. Zheng, X. Zong, T. Averett, G.D. Cates, and W.A. Tobias, Eur. Phys. J. A **44**, 55 (2010).
[22] W. Lorenzon, T.R. Gentile, H. Gao, R.D. McKeown, Phys. Rev. A **47**, 468 (1993).
[23] M.V. Romalis and G.D. Cates, Phys. Rev. A **58**, 3004 (1998), and the references therein.

EPJ Web of Conferences **113**, 04019 (2016)
DOI: 10.1051/epjconf/201611304019

Results on Double-polarization Asymmetries in Quasielastic Scattering from Polarized ^{3}He

Vincent A. Sulkosky[1,a]

[1] *Physics Department, University of Virginia, 382 McCormick Rd, PO Box 400714, Charlottesville, VA 22904, USA.*

Abstract. The ^{3}He nucleus has become extremely important in the investigation of the neutron's spin structure. When polarized, ^{3}He acts as an effective polarized neutron target and hence facilitates our understanding of the neutron's internal structure. However, to be used in this manner, our understanding of the internal structure of ^{3}He is of extreme importance. As the precision of experiments has improved, the extraction of polarized neutron information from ^{3}He leads to an ever larger share of the systematic uncertainty for these experiments. In these proceedings, I present a precise measurement of beam-target asymmetries in the $^3\vec{\mathrm{He}}(\vec{e}, e'd)$ and $^3\vec{\mathrm{He}}(\vec{e}, e'p)$ reactions. The former process is a uniquely sensitive probe of hadron dynamics in ^{3}He and the structure of the underlying electromagnetic currents. The measurements have been performed around the quasi-elastic peak at $Q^2 = 0.25$ (GeV/c)^2 and 0.35 (GeV/c)^2 for recoil momenta up to 270 MeV/c. The experimental apparatus, analysis and results were presented together with a comparison to state-of-the art Faddeev calculations.

1 Introduction

Protons and neutrons, collectively referred to as nucleons, comprise the nuclei of atomic elements. While many of the characteristics and underlying dynamics of the proton have been reasonably measured, the neutron is not understood to a desirable accuracy in particular regarding its charge, magnetization and spin distributions. Direct measurements on the neutron are difficult due to its short half-life outside the nucleus of an atom. Physicists have had to rely on indirect measurements using the appropriate nuclear targets to unveil the internal structure of the neutron. The two most common nuclei used to investigate the neutron are the deuteron (proton + neutron) and ^{3}He (two protons + neutron). However, detailed knowledge on the structure of light nuclei is crucial for extracting precise information on neutron structure. In these proceedings, we will focus on the ^{3}He nucleus.

^{3}He nuclei are interesting because theoretical predictions of their nuclear structure can be compared with data to an increasingly accurate degree. Understanding the ground-state spin aids in experiments that attempt to extract neutron information from polarized ^{3}He targets such as the inelastic structure functions g_1^n and g_2^n. However, the statistical precision of the experimental measurements on ^{3}He is now similar to the systematic uncertainty, which is related to our imperfect knowledge of the polarization of the nucleons within the ^{3}He nucleus. Theoretical calculations have predicted that there

[a] On Behalf of the Jefferson Lab Hall A Collaboration. e-mail: vasulk@jlab.org

are three main components in the ^{3}He ground-state wave functions, and our knowledge of the small components aids our understanding of the "standard model" of the few-body system.

A complete understanding of the ground-state wave function, reaction-mechanism effects such as final-state interactions (FSI) and meson-exchange currents (MEC) along with the roles of two- and three-nucleon forces are necessary to obtain good agreement between theory and experimental data. The experimental data reported here are compared with two independent full Faddeev calculations from the Bochum-Krakow (BK) [1, 2] and Hannover-Lisbon (HL) [3–6] groups. Additionally, calculations from the Pisa [7] group, though not Faddeev calculations, are expected to account for all relevant reaction mechanisms. The calculations have shown that the $^3\vec{\mathrm{He}}\,(\vec{e}, e'd)$ scattering process is strongly sensitive to the subleading components of the ^{3}He ground-state wave function and that the components have different signatures as functions of the missing momentum, $p_m = |\mathbf{q} - \mathbf{p_d}|$, which is the difference between the transferred and the detected deuteron momentum.

2 Experimental Details

During the experiment, the ^{3}He nucleus was examined by electron-induced knock-out of protons and deuterons. In the reactions, the electrons with incident energy E transfer energy ω and three-momentum $\mathbf{q}$ to the nucleus. These processes are measured as a function of the magnitude of the particle's missing momentum, which corresponds to the momentum of the recoiled particle. The experiment [8] was conducted at Jefferson Lab in Hall A [10] using the polarized electron beam with an energy of 2.425 GeV and currents > 10 μA. The beam polarization was measured by a Møller polarimeter to be ~ 84%, and the helicity of the beam was flipped at 30 Hz using a pseudorandom sequence. The electron beam was scattered from a polarized ^{3}He target.

The polarized ^{3}He target consisted of a 40-cm long glass cell filled with ^{3}He gas to approximately 9 atm and was polarized using spin-exchange optical pumping of a Rb-K alkali admixture [9]. The orientation of the target polarization was maintained by two pairs of Helmholtz coils, and the target polarization was measured by nuclear magnetic resonance and electron paramagnetic resonance throughout the experiment.

The quasi-elastically scattered electrons and knocked-out deuterons (protons) were detected by a High Resolution Spectrometer (HRS) [10] and the BigBite Spectrometer (using a hadron detector package) [11], respectively. The HRS was located at an electron scattering angle of either 12.5° or 14.5°, which translates into an average four-momentum transfer squared $Q^2 = \mathbf{q}^2 - \omega^2$ of 0.25 (GeV/c)^2 or 0.35 (GeV/c)^2. Additional experimental details are available in Ref. [8].

The measured asymmetries are given by the expression:

$$A\,(\theta^*, \phi^*) = \frac{(d\sigma/d\Omega)_+ - (d\sigma/d\Omega)_-}{(d\sigma/d\Omega)_+ + (d\sigma/d\Omega)_-}, \tag{1}$$

where the subscript on the differential cross section $(d\sigma/d\Omega)$ indicates the helicity of the beam. These asymmetries were measured for both the $^3\vec{\mathrm{He}}\,(\vec{e}, e'd)$ and $^3\vec{\mathrm{He}}\,(\vec{e}, e'p)$ processes, though only preliminary results are available for the latter reaction. The orientation of the target polarization vector is defined by the the polarization angles θ^* and ϕ^*, where the z axis is along the $\mathbf{q}$-vector and the y axis is defined by the cross product of the incident electron and scattered electron vectors. During the experiment, the Helmholtz coils were used to orient the ^{3}He spin vector to measure the asymmetries: $A\,(160°, 0°)$ and $A\,(71°, 0°)$. The values of the spin angles are averages over the spectrometer acceptance. The measured asymmetries are corrected for helicity-dependent effects such as the beam charge asymmetry and dead time. The physics asymmetries are then determined by correcting for the beam and target polarizations. Radiative corrections were also applied.

3 Results and Comparisons to Theory

Figure 1 shows the asymmetries versus missing momentum (left panel) and ω (right panel). The top panels provide the results on $A(71°, 0°)$, whereas the bottom panels illustrate the results for $A(160°, 0°)$. The error bars on the data represent the statistical (inner) and total uncertainties (outer), where the total uncertainty is the statistical and systematic uncertainties added linearly. The total systematic uncertainty is 7% relative, in which the systematic uncertainties have been added in quadrature.

In Fig. 1, the results of the state-of-the-art calculations of the BK [1, 2], HL [3–6] and Pisa [7] groups are shown with the data. Due to the spread in θ^* and ϕ^* caused by the finite acceptances of the spectrometers, the theory calculations were acceptance averaged in the procedure detailed in Ref. [12]. The shaded (light blue in color) bands represent the uncertainties caused by the acceptance-averaging procedure. In the figure, they have been placed on the HL curves, since that calculation was averaged over the finest mesh. It is clear from the comparison that neither of the theoretical curves matches the data well, though improvement is seen when a cut is placed on $\omega < 140$ MeV to isolate the quasielastic region (thin solid curve). This improvement is expected, since all present calculation are known to preform better in the region of the quasielastic peak. This is caused by the production of pions and the influence of the resonances, which have not been included in the calculations. The high precision of these data reveal deficiencies in the theories, which may require further refinement in the treatment of their two-and/or three-body dynamics.

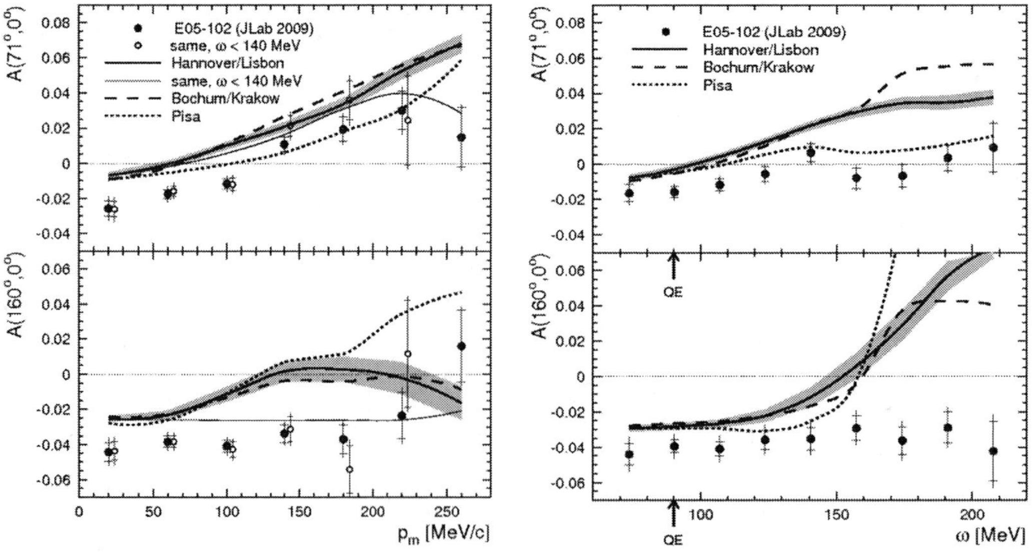

Figure 1. The asymmetries $A(71°, 0°)$ (top) and $A(160°, 0°)$ (bottom) in the quasielastic $^3\vec{He}(\vec{e}, e'd)$ process versus missing momentum (left panel) and energy transfer (right panel). The arrows indicate the approximate location in ω of the quasielastic peak. Reproduced from Ref. [8]. See text for additional details.

In the right-side panel of Fig. 1, the asymmetries versus the energy transfer are shown along with the theoretical curves. At large ω, the measured asymmetries deviate substantially from the calculations, which indicates that the input of the reaction mechanisms in the dip region is incomplete. At lower ω, an absolute systematic offset between 1% and 2% is seen, though the trend of the data is reasonably reproduced by the theories.

Preliminary results for the $^3\vec{\mathrm{He}}\,(\vec{e}, e'p)$ reaction for $Q^2 = 0.25\ (\mathrm{GeV}/c)^2$ and $0.35\ (\mathrm{GeV}/c)^2$ are shown in Fig. 2. Due to the resolution of the relative momentum of the BigBite spectrometer ($\approx 2\%$), the two and three body-break up regions in this reaction are difficult to separate experimentally, and hence, the preliminary asymmetries are shown by including both regions. Analysis is continuing to provide comparisons of these data to the theoretical predictions. The depiction in between the graphs shows the ^{3}He spin vector orientation with respect to the incident electron beam momentum and $\mathbf{q}$.

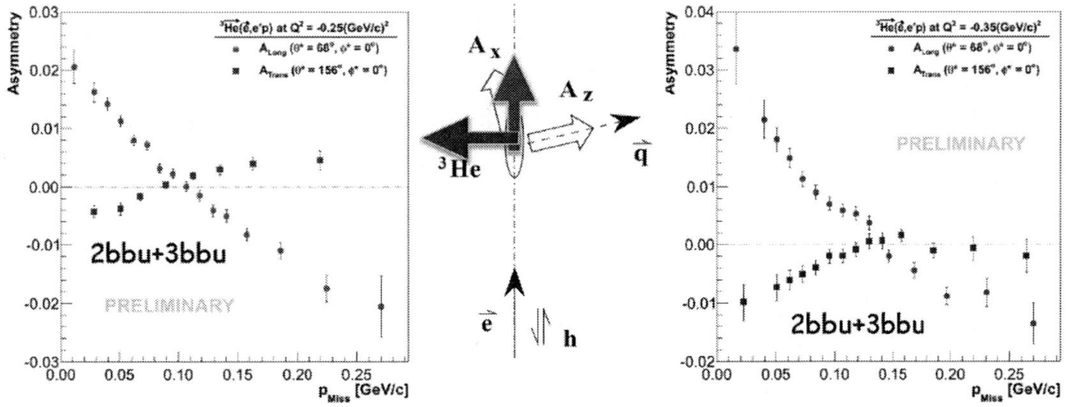

Figure 2. The asymmetries $A\,(68^\circ, 0^\circ)$ (circles) and $A\,(156^\circ, 0^\circ)$ (squares) in the quasielastic $^3\vec{\mathrm{He}}\,(\vec{e}, e'p)$ process versus missing momentum at $Q^2 = 0.25\ (\mathrm{GeV}/c)^2$ (left panel) and $0.35\ (\mathrm{GeV}/c)^2$ (right panel).

4 Conclusion

In these proceedings, the published results on the $^3\vec{\mathrm{He}}\,(\vec{e}, e'd)$ process were presented at $Q^2 = 0.25\ (\mathrm{GeV}/c)^2$ versus the missing momentum and energy transfer. This process is a sensitive probe of hadron dynamics in ^{3}He and the structure of the underlying electromagnetic currents. Asymmetries show a systematic offset compared to state-of-the-art theory predictions, but they are in fair agreement in terms of their functional dependencies with respect to p_m and ω. Data on the $^3\vec{\mathrm{He}}\,(\vec{e}, e'p)$ reaction were also presented, though the final comparison with the available theories is ongoing. Finally, data at $Q^2 = 0.35\ (\mathrm{GeV}/c)^2$ also exist for both reactions, though no calculations are currently available to compare against them. The new data along with the theoretical calculations will significantly advance our knowledge of the three-nucleon system.

Acknowledgements

This work was supported by the U.S. Department of Energy, Office of Science, Office of Nuclear Physics under Contract No. DE-AC05-06OR23177. This work was also supported in part by the Polish National Science Center under Grant No. DEC-2013/10/M/ST2/00420. The numerical calculations of the Bochum-Krakow group were partly performed on the supercomputer cluster of the JSC, Jülich, Germany.

References

[1] J. Golak *et al.*, Phys. Rev. C **65** 064004 (2002).

[2] J. Golak *et al.*, Phys. Rev. C **72** 054005 (2005).
[3] L. P. Yuan *et al.*, Phys. Rev. C **66** 054004 (2002).
[4] A. Deltuva *et al.*, Phys. Rev. C **69** 034004 (2004).
[5] A. Deltuva *et al.*, Phys. Rev. C **70** 034004 (2004).
[6] A. Deltuva *et al.*, Phys. Rev. C **71** 054005 (2005).
[7] L. E. Marcucci *et al.*, Phys. Rev. C **72** 014001 (2005).
[8] M. Mihovilovic *et al.*, Phys. Rev. Lett. **113**, 232505 (2014).
[9] E. Babcock *et al.*, Phys. Rev. Lett. **91** 123003 (2003).
[10] J. Alcorn *et al.*, Nucl. Instrum. Methods A **522** 294 (2004).
[11] M. Mihovilovic *et al.*, Nucl. Instrum. Methods Phys. Res., Sect. A **686**, 20 (2012).
[12] M. M. Dalton *et al.*, arXiv:1402.7028 25 (2014).

EPJ Web of Conferences **113**, 04020 (2016)
DOI: 10.1051/epjconf/201611304020

Chiral nuclear forces: Recent developments

Evgeny Epelbaum[1,a]

[1] *Institut für Theoretische Physik II, Ruhr-Universität Bochum, D-44780 Bochum, Germany*

Abstract. In this talk, I review recent progress towards developing accurate and precise nuclear forces in the framework of chiral effective field theory. The new generation of NN potentials up to fifth order in the chiral expansion is shown to provide excellent description of phase shifts and to yield a convergent pattern for low-energy NN observables. I also address the issue of uncertainty quantification and discuss some recent results in nucleon-deuteron scattering.

1 Introduction

Chiral effective field theory (EFT) provides a well established and nowadays most commonly used approach to low-energy nuclear reactions and light nuclei, see [1, 2] for recent review articles. It relies on the symmetries of QCD and allows one to derive nuclear forces and the corresponding current operators within a systematically improvable perturbative expansion, the so-called chiral expansion [3]. In the standard formulation based on pions and nucleons as the only explicit degrees of freedom, the expansion parameter Q is defined as $Q \in \{M_\pi/\Lambda_b, \ |\vec{p}\,|/\Lambda_b\}$ with M_π, $\vec{p}$ and Λ_b referring to the pion mass, typical three-momenta of the nucleons and the breakdown scale to be specified below, respectively.

In the past two and a half decades, this approach was employed by different groups to derive nuclear forces up to fourth order Q^4 in the chiral expansion (N^{3}LO) [4–8] and to perform *ab initio* calculations of few-nucleon reactions and light nuclei. For a related recent work on the electromagnetic currents see [9–12]. Given that the numerical implementation of the N^{3}LO three-nucleon force (3NF) contributions represents a non-trivial task, which could be accomplished only recently [13, 14], most of the available *ab initio* calculations utilized the nuclear Hamiltonian consisting of the N^{3}LO nucleon-nucleon (NN) potentials of [4, 5] accompanied by the leading, i.e. order-Q^3 (N^{2}LO), 3NF. These studies have revealed the important role of the chiral 3NF yielding many promising results, see [15] and references therein. For a progress report on the discretized formulation of nuclear chiral EFT see [16]. Still, the well-known discrepancies between state-of-the-art theoretical calculations and experimental data in the three-nucleon continuum at intermediate energies of $E_N \sim 50 \ldots 100$ MeV and higher [17] could so far not be resolved. Further developments are needed to tackle the long-standing 3NF problem in the framework of chiral EFT. First, given that the 3NF is itself a small correction to the dominant NN force, *accurate* and *precise* chiral NN potentials are required to pin down the 3NF effects. Secondly, it was argued that the chiral expansion of the 3NF is not yet converged at N^{3}LO [18] and should, therefore, be pushed to at least N^{4}LO (i.e. Q^5) in the standard approach [19–21] or

[a] e-mail: evgeny.epelbaum@rub.de

to N^{3}LO in the formulation with explicit $\Delta(1232)$ degrees of freedom. Last but not least, it is crucial to be able to reliably quantify the theoretical uncertainty due to truncation of the chiral expansion for all calculated observables, see [22] for a general discussion. In this talk I discuss recent developments along these lines focusing mainly on the first and the last topics.

2 A new generation of chiral NN forces up to N^{4}LO

Given that the chiral expansion of the 3NF is not yet converged at N^{3}LO, the expansion for the nuclear Hamiltonian needs to be pushed to N^{4}LO. For the NN force, the pertinent long-range contributions due to 2π- and 3π-exchange have been derived and applied to peripheral NN scattering in Ref. [23]. Notice that no new isospin-conserving NN contact interactions appear at this order in the chiral expansion.

In [24, 25], we have introduced a new generation of NN potentials up to N^{4}LO in the chiral expansion. In addition to pushing the chiral expansion to a higher order, these new potentials feature a number of important improvements as compared to the first-generation chiral N^{3}LO NN forces. First, we used a coordinate-space regularization procedure for the long-range components $V_{\rm long}$,

$$V_{\rm long}(\vec{r}) \to V_{\rm long}^{\rm Reg.}(\vec{r}) = V_{\rm long}(\vec{r})\, F\left(\frac{r}{R}\right) \quad \text{with} \quad F(x) = \left(1 - \exp(-x^2)\right)^6, \tag{1}$$

where the short-range cutoff R is chosen in the range of $R = 0.8\ldots1.2$ fm in agreement with the estimated breakdown distance of ~ 0.8 fm of the chiral expansion for the pion-exchange potential [26]. The new regulator maintains, per construction, the analytic structure of the amplitude at low momenta and allowed us to substantially reduce the amount of finite-cutoff artefacts. Moreover, since the new regulator cuts off precisely the undesired *short-range* components of the pion exchange contributions instead of their *large-momentum* parts as the old non-local regulator does, no additional spectral-function regularization (SFR) [27] used e.g. in [4] is required. This is a particularly welcome feature in view of the ongoing and upcoming 3NF studies, in which the implementation of the SFR would be highly non-trivial. Notice that while the authors of [5] have managed to avoid the introduction of the SFR at N^{3}LO, this was achieved at the cost of employing a nonlocal regulator, whose functional form was adjusted from partial wave to partial wave, see Table F.2 of [2].

Secondly, the new potentials of [24, 25] adopt the values of the πN low-energy constants (LECs) determined in πN scattering without any fine tuning. In particular, at N^{4}LO, we used the LECs taken from the order-Q^4 fit to the Karlsruhe-Helsinki partial wave analysis (PWA) of [19], see [28] for a recent extraction from the subthreshold parameters and [29] for a related work.

With the short-range contact interactions being determined from a fit to the Nijmegen PWA (NPWA) of neutron-proton (np) and proton-proton (pp) scattering [30], see [24, 25] for more details, our results for S-, P- and D-waves and the mixing angles $\epsilon_{1,2}$ at different chiral orders are visualized in Fig. 1 for $R = 0.9$ fm. One observes a very good convergence pattern of the chiral expansion in the whole energy range up to $E_{\rm lab} = 300$ MeV. To get more quantitative insights, we show in Table 1 the χ^2/datum for the description of the Nijmegen np and pp phase shifts. Here, we estimated the uncertainty of the NPWA for a given phase shift and mixing angle from the deviations to the results of the Nijmegen I, II and Reid93 potentials [31], which are regarded as alternative PWA. Notice, however, that such an estimation does *not* allow for a statistical interpretation of the χ^2 values shown in table 1. Still, the χ^2 defined as described above is a useful tool to quantify the accuracy of the fits.

As expected, one observes the improved description of the phase shifts with increasing orders of the chiral expansion. It is particularly encouraging to see a strong reduction in the χ^2 when going from NLO (Q^2) to N^{2}LO (Q^3) and from N^{3}LO (Q^4) to N^{4}LO (Q^5), which comes entirely from the

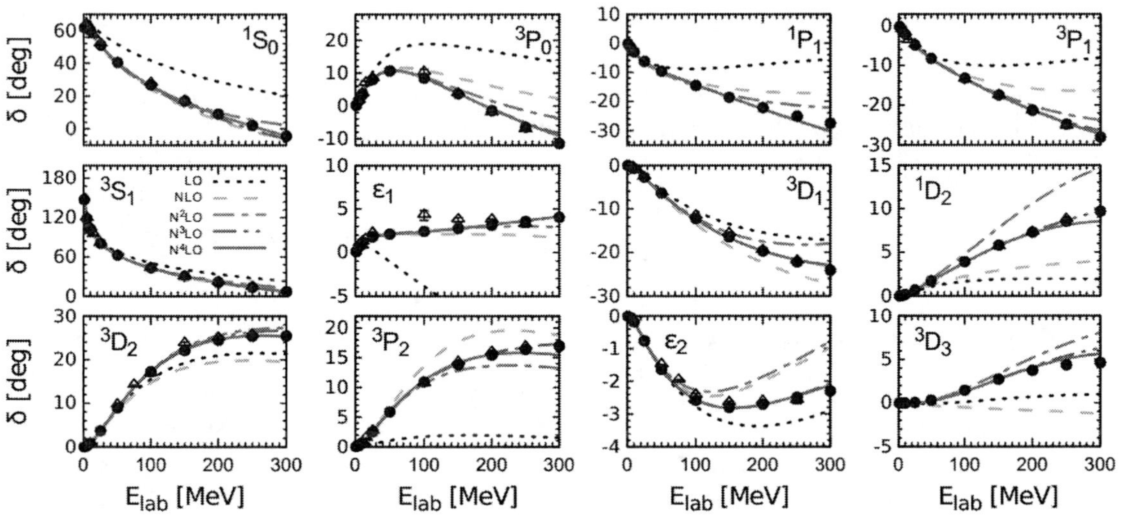

Figure 1. Chiral expansion of the NN phase shifts in comparison with the NPWA (solid dots) and the GWU single-energy np partial wave analysis (open triangles). Dotted (black), dashed (brown), dash-dotted (blue), dash-double-dotted (red) and solid (violet) lines show the results at LO, NLO, N^{2}LO, N^{3}LO and N^{4}LO, respectively, calculated using the cutoff $R = 0.9$ fm

Table 1. χ^2/datum for the description of the Nijmegen np and pp phase shifts [30] at different orders in the chiral expansion for the cutoff $R = 0.9$ fm. The numbers in the brackets give the number of adjustable isospin-invariant contact interactions at the corresponding order. Only those channels are included which have been used in the N^{3}LO/N^{4}LO fits, namely the S-, P- and D-waves and the mixing angles ϵ_1 and ϵ_2.

$E_{\rm lab}$ bin	LO (2)	NLO (9)	N^{2}LO (9)	N^{3}LO (24)	N^{4}LO (24)
neutron-proton phase shifts					
0–100	360	31	4.5	0.7	0.3
0–200	480	63	21	0.7	0.3
proton-proton phase shifts					
0–100	5750	102	15	0.8	0.3
0–200	9150	560	130	0.7	0.6

corresponding two-pion exchange components without invoking new parameters.[1] These results constitute an important consistency check of the theoretical approach and provide a nice illustration of the importance of the chiral 2π-exchange potential, which is determined entirely by the chiral symmetry of QCD and experimental information on πN scattering. Furthermore, the obtained results suggest – fully in line with the Weinberg power counting [3] – that the theoretical uncertainties at NLO and N^{3}LO are dominated by the neglected 2π-exchange contributions at orders N^{2}LO and N^{4}LO, respectively. Indeed, if certain order-Q^4 and order-Q^6 contact interactions would have to be promoted to lower orders in violation with naive dimensional analysis, see e.g. [32–34], the inclusion of the 2π-

[1] At N^{4}LO, there is one additional isospin-breaking contact interaction in the neutron-proton $^1S_0$ channel.

exchange contributions alone at N^{2}LO and N^{3}LO would not result in the improved accuracy of the fits.

We have also tested the convergence of the chiral expansion for various observables. For example, for the np total cross section at various energies, we obtain using $R = 0.9$ fm

$$\begin{aligned}
\sigma_{\rm tot}(50\,{\rm MeV}) &= 183.6_{\rm LO} - 17.1_{\rm NLO} + 0.5_{\rm N^2LO} - 0.2_{\rm N^3LO} + 0.7_{\rm N^4LO} = 167.5\,{\rm mb}\,,\\
\sigma_{\rm tot}(96\,{\rm MeV}) &= 84.8_{\rm LO} - 9.7_{\rm NLO} + 3.2_{\rm N^2LO} - 0.8_{\rm N^3LO} + 0.5_{\rm N^4LO} = 78.0\,{\rm mb}\,,\\
\sigma_{\rm tot}(143\,{\rm MeV}) &= 52.5_{\rm LO} - 3.4_{\rm NLO} + 5.1_{\rm N^2LO} - 0.5_{\rm N^3LO} + 0.2_{\rm N^4LO} = 53.9\,{\rm mb}\,,\\
\sigma_{\rm tot}(200\,{\rm MeV}) &= 34.9_{\rm LO} + 1.0_{\rm NLO} + 6.7_{\rm N^2LO} + 0.6_{\rm N^3LO} - 0.5_{\rm N^4LO} = 42.7\,{\rm mb}\,. \qquad (2)
\end{aligned}$$

Notice that the natural size of the N^{2}LO, N^{3}LO and N^{4}LO contributions at $E = 50$ MeV is estimated to be ~ 3 mb, ~ 0.8 mb and ~ 0.2 mb, respectively [24]. The somewhat large N^{4}LO contribution at this energy reflects, in fact, the adopted fitting protocol, see [25] for more details. Similarly, the NLO contributions at the two highest energies come our rather small compared with their expected size of ~ 10 mb. Generally, the convergence pattern of the chiral expansion for the np cross section is very satisfactory. We also found rapid convergence for the deuteron properties [24, 25].

3 Uncertainty quantification

Uncertainty quantification is key for performing a meaningful comparison between theory and data. In the past, theoretical uncertainty of nuclear chiral EFT due to truncation of the chiral expansion was, at best, estimated via a variation of the cutoff in a certain range. As pointed out already in [4], such an approach suffers from certain drawbacks. First, the residual cutoff dependence of a calculated observable measures the impact of neglected contact interactions which contribute only at *even* orders Q^{2n}, $n = 0, 1, 2, \ldots$. Secondly, the available cutoff range appears, in practice, to be rather limited. As a result, such "cutoff bands" can generally not be expected to provide a realistic estimation of the theoretical uncertainty, see [4, 24, 25] for more detail.

In [24], we proposed an alternative approach to quantify the uncertainty of a chiral EFT prediction by directly estimating the size of neglected higher-order contributions to a given observable. To this aim, we assume the expansion parameter Q to be $Q = \max\{M_\pi/\Lambda_b,\ |\vec{p}\,|/\Lambda_b\}$. The breakdown scale of the chiral expansion Λ_b was estimated in [24] to be $\Lambda_b \sim 600$ MeV for $R = 0.8 \ldots 1.0$ fm, $\Lambda_b \sim 500$ MeV for $R = 1.1$ fm and $\Lambda_b \sim 400$ MeV for $R = 1.2$ fm. The accuracy of a chiral EFT prediction for the observable of interest can then be estimated by multiplying the leading-order result with the appropriate power of Q. In order to have as reliable as possible estimation of the uncertainty, we proposed to also employ the information on the actual size of higher-order contributions (if available). The precise description of the suggested algorithm for error analysis can be found in [24, 25], see also [35] for a related discussion. Here we just mention that our new approach can be straightforwardly applied (i) to any observable of interest and (ii) for any specific choice of the regulator R.

As a representative example, consider the chiral EFT predictions for the np total cross section in Eq. (2). Performing the error analysis as described above, our final predictions for $R = 0.9$ fm at N^{4}LO take the form $\sigma_{\rm tot}(50\,{\rm MeV}) = 167.5 \pm 0.2$ mb, $\sigma_{\rm tot}(96\,{\rm MeV}) = 78.0 \pm 0.2$ mb, $\sigma_{\rm tot}(143\,{\rm MeV}) = 53.9 \pm 0.4$ mb and $\sigma_{\rm tot}(200\,{\rm MeV}) = 42.7 \pm 0.9$ mb. These predictions are in excellent agreement with the NPWA results of $\sigma_{\rm NPWA}(50\,{\rm MeV}) = 167.8 \pm 0.2$ mb, $\sigma_{\rm NPWA}(96\,{\rm MeV}) = 78.1 \pm 0.5$ mb, $\sigma_{\rm NPWA}(143\,{\rm MeV}) = 53.6 \pm 0.5$ mb and $\sigma_{\rm NPWA}(200\,{\rm MeV}) = 42.35 \pm 0.20$ mb. Last but not least, our quoted theoretical uncertainties were found in [36] to be consistent with the 68% degree-of-belief intervals for EFT predictions.

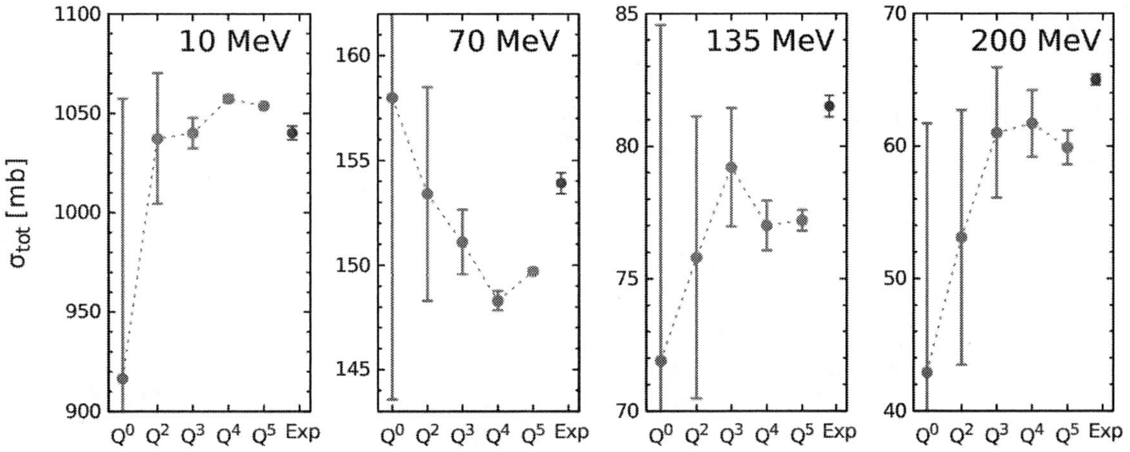

Figure 2. Predictions for Nd total cross section at laboratory energies of 10, 70, 135 and 200 MeV based on the NN potentials of [24, 25] for $R = 0.9$ fm without including the 3NF.

4 Beyond the NN system

With all the necessary tools being developed as described in the previous sections, one is now well equipped for testing chiral EFT and especially the novel chiral 3NFs in three- and more-nucleon systems. As a first step along this line, the new generation of the chiral NN potentials has been recently applied to nucleon-deuteron scattering and selected low-energy observables in ^{3}H, ^{4}He and ^{4}Li [35]. With a detailed account of this study being given in the talks by Maris [37] and Skibinski [38], I only briefly address here the results for nucleon-deuteron (Nd) total cross section visualized in Fig. 2. While these calculations can, starting from N^{2}LO, not be regarded as complete due to the missing 3NFs, they do reveal a number of interesting insights. First, one observes a convergent pattern at all considered energies in spite of the fact that the calculations are actually incomplete. The predicted results for the cross section at N^{4}LO are in a clear disagreement with experimental values with the discrepancies being much larger than the experimental and estimated theoretical uncertainties. This provides truly unambiguous evidence for missing 3NF within the employed theoretical framework. The estimated accuracy of the chiral EFT results at N^{4}LO suggests that Nd scattering at intermediate and even higher energies will be a very promising testing ground for the chiral 3NF. Furthermore, it is comforting to see that the deviations between our N^{3,4}LO predictions and experimental data are similar in size to the NLO (i.e. Q^2) error bars, which give the estimated size of N^{2}LO contributions. Indeed, this is precisely the chiral order at which the first nonvanishing 3NF appear according to the Weinberg power counting.

To summarize, the newly developed improved chiral NN potentials up to N^{4}LO, along with a new analysis of the theoretical uncertainties, open the way for precision tests of chiral 3NFs in few-nucleon systems. Work along these lines is in progress by the LENPIC collaboration [39].

Acknowledgements

I am grateful to my collaborators Hermann Krebs, Ulf-G. Meißner and to all members of the LENPIC collaboration for sharing their insights into the discussed topics. This work was supported in part by DFG (SFB/TR 16, "Subnuclear Structure of Matter") and ERC project 259218 NUCLEAREFT.

References

[1] E. Epelbaum, H. W. Hammer and U.-G. Meißner, Rev. Mod. Phys. **81**, 1773 (2009).
[2] R. Machleidt and D. R. Entem, Phys. Rept. **503**, 1 (2011).
[3] S. Weinberg, Phys. Lett. B **251**, 288 (1990).
[4] E. Epelbaum, W. Glockle and U.-G. Meißner, Nucl. Phys. A **747**, 362 (2005).
[5] D. R. Entem and R. Machleidt, Phys. Rev. C **68**, 041001 (2003).
[6] V. Bernard, E. Epelbaum, H. Krebs and U.-G. Meißner, Phys. Rev. C **77**, 064004 (2008).
[7] V. Bernard, E. Epelbaum, H. Krebs and U.-G. Meißner, Phys. Rev. C **84**, 054001 (2011).
[8] E. Epelbaum, Eur. Phys. J. A **34**, 197 (2007).
[9] S. Kölling, E. Epelbaum, H. Krebs and U.-G. Meißner, Phys. Rev. C **80**, 045502 (2009).
[10] S. Kölling, E. Epelbaum, H. Krebs and U.-G. Meißner, Phys. Rev. C **84**, 054008 (2011).
[11] S. Pastore, R. Schiavilla and J. L. Goity, Phys. Rev. C **78**, 064002 (2008).
[12] S. Pastore *et al.*, Phys. Rev. C **80**, 034004 (2009).
[13] J. Golak *et al.*, Eur. Phys. J. A **43**, 241 (2010).
[14] K. Hebeler *et al.*, Phys. Rev. C **91**, no. 4, 044001 (2015).
[15] H. W. Hammer, A. Nogga and A. Schwenk, Rev. Mod. Phys. **85**, 197 (2013).
[16] D. Lee, these proceedings.
[17] N. Kalantar-Nayestanaki *et al.*, Rept. Prog. Phys. **75**, 016301 (2012).
[18] R. Machleidt and D. R. Entem, J. Phys. G **37**, 064041 (2010).
[19] H. Krebs, A. Gasparyan and E. Epelbaum, Phys. Rev. C **85**, 054006 (2012).
[20] H. Krebs, A. Gasparyan and E. Epelbaum, Phys. Rev. C **87**, no. 5, 054007 (2013).
[21] E. Epelbaum, A. M. Gasparyan, H. Krebs and C. Schat, Eur. Phys. J. A **51**, no. 3, 26 (2015).
[22] R. J. Furnstahl, D. R. Phillips and S. Wesolowski, J. Phys. G **42**, no. 3, 034028 (2015).
[23] D. R. Entem, N. Kaiser, R. Machleidt and Y. Nosyk, Phys. Rev. C **91**, no. 1, 014002 (2015).
[24] E. Epelbaum, H. Krebs and U.-G. Meißner, Eur. Phys. J. A **51**, no. 5, 53 (2015).
[25] E. Epelbaum, H. Krebs and U.-G. Meißner, arXiv:1412.4623 [nucl-th].
[26] V. Baru *et al.*, Eur. Phys. J. A **48**, 69 (2012).
[27] E. Epelbaum, W. Glöckle and U.-G. Meißner, Eur. Phys. J. A **19**, 125 (2004).
[28] M. Hoferichter, J. R. de Elvira, B. Kubis and U.-G. Meißner, arXiv:1507.07552 [nucl-th].
[29] K. A. Wendt, B. D. Carlsson and A. Ekström, arXiv:1410.0646 [nucl-th].
[30] V. G. J. Stoks *et al.*, Phys. Rev. C **48**, 792 (1993).
[31] V. G. J. Stoks *et al.*, Phys. Rev. C **49**, 2950 (1994).
[32] A. Nogga, R. G. E. Timmermans and U. van Kolck, Phys. Rev. C **72**, 054006 (2005).
[33] M. C. Birse, PoS CD **09**, 078 (2009).
[34] B. Long and C. J. Yang, Phys. Rev. C **85**, 034002 (2012).
[35] S. Binder *et al.*, arXiv:1505.07218 [nucl-th].
[36] R. J. Furnstahl *et al.*, Phys. Rev. C **92**, no. 2, 024005 (2015).
[37] P. Maris, these proceedings.
[38] R. Skibinski, these proceedings.
[39] Low Energy Nuclear Physics International Collaboration (LENPIC), *www.lenpic.org*.

EPJ Web of Conferences **113**, 04021 (2016)
DOI: 10.1051/epjconf/201611304021

The Falsification of Nuclear Forces

R. Navarro Perez[1,a], J. E. Amaro[2,b], and E. Ruiz Arriola[2,c]

[1] *Nuclear and Chemical Science Division, Lawrence Livermore National Laboratory Livermore, California 94551, USA*
[2] *Departamento de Física Atómica, Molecular y Nuclear and Instituto Carlos I de Física Teórica y Computacional, Universidad de Granada E-18071 Granada, Spain*

Abstract. We review our work on the statistical uncertainty analysis of the NN force. This is based on the Granada-2013 database where a statistically meaningful partial wave analysis comprising a total of 6713 np and pp published scattering data from 1950 till 2013 below pion production threshold has been made. We stress the necessary conditions required for a correct and self-consistent statistical interpretation of the discrepancies between theory and experiment which enable a subsequent statistical error propagation and correlation analysis

1 Introduction

Error propagation and uncertainty quantification have recently become a central topic in nuclear physics [1–6]. In the particular field of phenomenological Nucleon-Nucleon (NN) interactions uncertainties can be classified as statistical or systematic[1]. Statistical uncertainties are the result of unavoidable random fluctuations during the experimental process. Most NN scattering measurements consist of counting events which corresponds to a Poisson distribution. If the number of events is large enough then the distribution can be safely approximated as a normal one. This allows to fix the parameters of a phenomenological potential via the usual chi square minimization process to reproduce the collection of NN scattering data. As a consequence the statistical uncertainty of the experimental data propagates into the fitting parameters in the form of a confidence region in which the parameters are allowed to vary and still give an accurate description of the data. Such confidence region can be easily determined by the parameters' covariance matrix if the assumption of normally distributed residuals is fulfilled. Systematic uncertainties are a consequence of our lack of knowledge of the actual form of the NN potential and the assumptions that have to be made in order to give a representation of the NN interaction. Some potentials are separable in momentum space while others are not, some are energy dependent and others range from fully local to different types non-localities in coordinate space. Even though most of the potentials are fitted to the same type of experimental NN scattering

[a] e-mail: navarroperez1@llnl.gov
[b] e-mail: amaro@ugr.es
[c] e-mail: earriola@ugr.es

[1] Numerical uncertainties are also present but can be made small enough to be dominated by the other two.

data, their predictions of unmeasured scattering observables or nuclear structure properties can sometimes be incompatible. This residual incompatibilities beyond statistical consistency and equivalence is what we refer as systematic uncertainties. In this contribution we review our determination of the NN interaction statistical uncertainties along with the necessary conditions for such type of analysis.

2 Coarse graining and the delta-shell potential

Coarse graining embodies the Wilsonian renormalization [7] concept and represents a very reliable tool to simplify the description of pp and np scattering data while still retaining all the relevant information of the interaction up to a certain energy range set by the de Broglie wavelength of the most energetic particle considered. The $V_{\mathrm{low}k}$ potentials in momentum space are a good example of an implementation of coarse graining by removing the high-momentum part of the interaction [8, 9]. In 1973 Aviles introduced the delta-shell (DS) potential in the context of NN interactions [10]. In [11] the DS potential was used to implement coarse graining in coordinate space via a local potential that samples the np interaction at certain concentration radii r_i by

$$V(r) = \sum_{i=1}^{n} \frac{\lambda_i}{2\mu} \delta(r - r_i), \tag{1}$$

where μ is the reduced mass and λ_i are strength coefficients. After fitting the λ_i parameters to np phase-shifts the properties and form factors of the deuteron were calculated. A variational method with harmonic oscillator wave functions was used to calculate upper bounds to the binding energy of the double-closed shell nuclei ^{4}He, ^{16}O and ^{40}Ca [11].

3 Description of NN scattering data

In order to quantify the statistical uncertainties of the NN interaction a fit to experimental data becomes necessary. The usual first step to fit a phenomenological potential to reproduce scattering phase-shifts is not sufficient to get an accurate description of the actual experimental data. As was recently shown in [6], the local chiral effective potential of [12] fitted to phase-shifts yields a significantly large χ^2/d.o.f. value when compared to experimental scattering data. However it is possible that small readjustments of the potential parameters have a significant impact in lowering the total χ^2. Given the wide applicability of this type of interactions [13, 14] a full fledged fit to NN experimental scattering data would be of great interest.

Historically, a successful description of the complete database with $\chi^2/\mathrm{d.o.f.} \lesssim 1$ has never been possible. The potentials and PWA were gradually improved over time by explicitly including different physical effects like OPE in the long range part, charge symmetry breaking in the central channel and electromagnetic interactions among others. The first PWA with $\chi^2/\mathrm{d.o.f.} \lesssim 1$ was obtained in 1993 when the Nijmegen group introduced the 3σ criterion to exclude over 1000 inconsistent data[15]. The 3σ criterion deals with possible over and underestimations of the statistical uncertainties by excluding data sets with improbably high or improbably low values of χ^2 (for a clear description of this process see [16]). However, this method identifies only inconsistencies between individual data sets and a model trying to describe the complete database. To improve on this method, so that inconsistencies between each data set and the rest of the database can be found, the 3σ criterion was applied iteratively to the *complete* database and the potential parameters were refitted to the *accepted* data sets until no more data are excluded or recovered [17]. The self-consistent data base obtained with this procedure contains 6713 experimental points and recovers 300 of initially discarded data with the usual 3σ

criterion [18]. Although the 300 extra data do not significantly change the potential parameters, their inclusion can only improve the estimate of statistical errors. A simultaneous fit to pp and np scattering data was made representing the short range part of the interaction with a DS potential and OPE for the long range part. The fit requires a total of 46 parameters and yields $\chi^2/\mathrm{d.o.f.} = 1.04$ to the self-consistent data base [17, 19].

4 Description of NN scattering errors

On a more fundamental level, any chi-squared distribution can be used to test goodness of fit, provided that the experimental data can be assumed to have a normal distribution [20, 21]. If the residuals defined as

$$R_i = \frac{O_i^{\mathrm{exp}} - O_i^{\mathrm{theor}}}{\Delta O_i^{\mathrm{exp}}} \tag{2}$$

a theoretical model correctly describes the data if they follow the standard normal distribution $N(0,1)$. [2] This self-consistency condition, which can only be checked *a posteriori* and entitles legitimate error propagation, has usually been overlooked in the NN literature. In [22] a few of these tests are reviewed along with a recently proposed Tail-Sensitive test [23] and it was found that the three potentials DS-OPE [17], DS-χTPE [24] and Gauss-OPE [22] have standard normal residuals. The three more recent potentials DS-Born, Gauss-χTPE and Gauss-Born also were found to have normally distributed residuals [25]. A simple and straightforward recipe to apply the Tail-Sensitive test to any set of empirical data with a sample size up to $N = 9000$ was developed in [26]. Thus, this six new potentials are the first to qualify for error estimation in nuclear physics. A direct application of normality is the re-sampling of experimental data via Monte Carlo techniques, as is noted in [27], for a robust analysis of possible asymmetries on the potential parameters distribution. Most recently a Monte Carlo method was used to calculate a realistic statistical uncertainty of the Triton binding energy stemming from NN scattering data [28].

5 Conclusions

In this contribution we outline the two main requirements for a correct quantification of the NN statistical uncertainties; a correct description of the experimental data and the reproduction of experimental errors. Although the first one may seem obvious, some widely used interactions in nuclear structure and nuclear reaction calculations are fitted to phase-shifts instead of experimental data and as shown in [6] those two descriptions are not entirely equivalent. The second requirement can be easily reformulated into positively testing for the normality of residuals. Normality of residuals provides a criterion of "falsibility" to distinguish those NN interactions that can come in conflict with observation and those that cannot. Interactions that fail the criterion [5, 29] come in conflict with observation, even if they apparently give a good description of the data, because the chi-squared distribution cannot be used to test goodness of fit, and require more complex statistical techniques to analyze the data. Even though the description here is for statistical uncertainties, a certain glance into the systematic uncertainty can already be perceived by looking into the differences in phase-shifts and scattering amplitude predictions given by the different realistic potentials with $\chi^2_{\mathrm{min}}/\mathrm{d.o.f.} \sim 1$. In particular the six interactions fitted to the Granada self-consistent database, give similar statistical uncertainties but present inconsistent phase-shifts at low angular momentum and high energy. The discrepancies between different potentials, accounting for the systematic uncertainty, are usually an order of magnitude larger than the statistical error bars.

[2] More generally, if the sufficient goodness of fit condition $\chi^2/\nu = 1 \pm \sqrt{2/\nu}$ is not fulfilled one can scale globally the errors $\Delta O_i^{\mathrm{exp}} \to \alpha \Delta O_i^{\mathrm{exp}}$ by a common Birge factor α *provided* the residuals follow a scaled normal distribution $N(0,\alpha)$.

Acknowledgements

This work is supported by Spanish DGI (grant FIS2014-59386-P) and Junta de Andalucía (grant FQM225). Work performed under the auspices of the U.S. Department of Energy by Lawrence Livermore National Laboratory under Contract No. DE-AC52-07NA27344. Funding was also provided by the U.S. Department of Energy, Office of Science, Office of Nuclear Physics under Award No. DE-SC0008511 (NUCLEI SciDAC Collaboration)

References

[1] D.G. Ireland, W. Nazarewicz, J. Phys. **G42**, 030301 (2015)
[2] J. Dobaczewski, W. Nazarewicz, P.G. Reinhard, J. Phys. **G41**, 074001 (2014), 1402.4657
[3] J.D. McDonnell, N. Schunck, D. Higdon, J. Sarich, S.M. Wild, W. Nazarewicz, Phys. Rev. Lett. **114**, 122501 (2015), 1501.03572
[4] R.J. Furnstahl, D.R. Phillips, S. Wesolowski, J. Phys. **G42**, 034028 (2015), 1407.0657
[5] B.D. Carlsson, A. Ekström, C. Forssén, D.F. Strömberg, O. Lilja, M. Lindby, B.A. Mattsson, K.A. Wendt (2015), 1506.02466
[6] R. Navarro Pérez, J.E. Amaro, E.R. Arriola, Phys. Rev. **C91**, 054002 (2015), 1411.1212
[7] K. Wilson, J.B. Kogut, Phys.Rept. **12**, 75 (1974)
[8] S. Bogner, T. Kuo, A. Schwenk, Phys.Rept. **386**, 1 (2003), nucl-th/0305035
[9] S. Bogner, R. Furnstahl, A. Schwenk, Prog.Part.Nucl.Phys. **65**, 94 (2010), 0912.3688
[10] J. Aviles, Phys.Rev. **C6**, 1467 (1972)
[11] R. Navarro Pérez, J.E. Amaro, E. Ruiz Arriola, Prog.Part.Nucl.Phys. **67**, 359 (2012), 1111.4328
[12] A. Gezerlis, I. Tews, E. Epelbaum, M. Freunek, S. Gandolfi et al. (2014), 1406.0454
[13] J.E. Lynn, J. Carlson, E. Epelbaum, S. Gandolfi, A. Gezerlis, A. Schwenk, Phys. Rev. Lett. **113**, 192501 (2014), 1406.2787
[14] S. Gandolfi, A. Lovato, J. Carlson, K.E. Schmidt, Phys. Rev. **C90**, 061306 (2014), 1406.3388
[15] V. Stoks, R. Kompl, M. Rentmeester, J. de Swart, Phys.Rev. **C48**, 792 (1993)
[16] F. Gross, A. Stadler, Phys.Rev. **C78**, 014005 (2008), 0802.1552
[17] R. Navarro Pérez, J.E. Amaro, E. Ruiz Arriola, Phys.Rev. **C88**, 064002 (2013), 1310.2536
[18] R. Navarro Perez, J. Amaro, E. Ruiz Arriola, *2013 Granada Database*, http://www.ugr.es/~amaro/nndatabase/ (2013), accessed: 2015-08-15
[19] R. Navarro Pérez, J.E. Amaro, E. Ruiz Arriola, Phys.Rev. **C88**, 024002 (2013), 1304.0895
[20] M. Evans, J. Rosenthal, *Probability and Statistics: The Science of Uncertainty* (W. H. Freeman, 2009), ISBN 9781429224628
[21] F. James, W. Eadie, *Statistical Methods in Experimental Physics* (World Scientific, 2006), ISBN 9789812567956
[22] R. Navarro Pérez, J.E. Amaro, E. Ruiz Arriola, Phys.Rev. **C89**, 064006 (2014), 1404.0314
[23] S. Aldor-Noiman, L.D. Brown, A. Buja, W. Rolke, R.A. Stine, Am. Statist. **67**, 249 (2013)
[24] R. Navarro Pérez, J.E. Amaro, E. Ruiz Arriola, Phys.Rev. **C89**, 024004 (2014), 1310.6972
[25] R. Navarro Pérez, J. Amaro, E.R. Arriola (2014), 1410.8097
[26] R. Navarro Pérez, J.E. Amaro, E. Ruiz Arriola, J. Phys. **G42**, 034013 (2015), 1406.0625
[27] R. Navarro Pérez, J.E. Amaro, E. Ruiz Arriola, Phys.Lett. **B738**, 155 (2014), 1407.3937
[28] R. Navarro Pérez, E. Garrido, J. Amaro, E. Ruiz Arriola, Phys.Rev. **C90**, 047001 (2014)
[29] A. Ekström, G. Baardsen, C. Forssén, G. Hagen, M. Hjorth-Jensen et al., Phys.Rev.Lett. **110**, 192502 (2013), 1303.4674

EPJ Web of Conferences **113**, 04022 (2016)
DOI: 10.1051/epjconf/201611304022

Minimally non-local nucleon-nucleon potentials with chiral two-pion exchange including Δ's

Maria Piarulli[1,a]

[1] *Department of Physics, Old Dominion University, Norfolk, VA 23529, USA*

Abstract. A coordinate-space nucleon-nucleon potential is constructed in chiral effective field theory (χEFT) retaining pions, nucleons and Δ-isobars as explicit degrees of freedom. The calculation of the potential is carried out by including one- and two-pion-exchange contributions up to next-to-next-to-leading order (N2LO) and contact interactions up to next-to-next-to-next-to-leading order (N3LO). The low-energy constants multiplying these contact interactions are fitted to the 2013 Granada database in the laboratory-energy range 0–300 MeV.

1 Introduction

The recent history of nuclear physics has witnessed the tremendous development of nuclear chiral effective field theory (χEFT), originally proposed by Weinberg in a series of papers in the early 1990's [1]. The (approximate) chiral symmetry exhibited by the underlying theory of QCD in the low-energy regime severely restricts the form of the strong interactions among nucleons, Δ-isobars, and pions, as well as the electroweak interactions of these hadrons with external (electroweak) fields. In the specific case of two nucleons, the requirements imposed by χEFT can be incorporated into a non-relativistic quantum mechanical potential, constructed by a perturbative matching, order by order in the chiral expansion, between the on-shell scattering amplitude and the solution of the Schrödinger equation (see, for example, the review paper by Machleidt and Entem [2]). By its own nature, χEFT needs to be organized within a given power counting scheme and the resulting chiral potentials are conveniently organized in powers of Q/Λ_χ, where $Q \ll \Lambda_\chi$ is a general low-momentum scale entering the theory and $\Lambda_\chi \sim 1$ GeV specifies the chiral-symmetry breaking scale.

The objective of this presentation is to construct a coordinate-space chiral nucleon-nucleon (NN) potential derived up to next-to-next-to-next-to leading order (N3LO or Q^4) in the chiral expansion, including pions, nucleons and Δ-isobars degrees of freedom. The calculation of the potential is carried out by including one- and two-pion-exchange (OPE and TPE) contributions up to next-to-next-to-leading order (N2LO or Q^3) and contact interactions up to N3LO. While the OPE and TPE potentials represent the long-range part of the NN interaction, the contact terms, instead, encode the short-range physics, and their strength are specified by unknown low-energy constants (LEC's), which are then determined by fits to experimental data. The inclusion of Δ-isobars in the TPE component of the NN interaction is dictated from phenomenological considerations which explain the important role

[a] e-mail: mpiar001@odu.edu

of Δ isobars in nuclear structure and reactions [3]. The necessity to derive a coordinate-space chiral potential, whose natural formulation is in momentum-space, is related to the fact that many computational techniques utilized to calculate properties of nuclei and nuclear matter such as Quantum Monte Carlo (QMC) methods [4] require a local coordinate-space representation of the nuclear interactions. However, available momentum-space chiral potentials have the feature of being strongly non-local meaning that, upon Fourier transformation, they lead to non-local interactions (or **p**-dependent interactions, where $\mathbf{p} \longrightarrow -i\boldsymbol{\nabla}$ is the relative momentum operator) in coordinate-space. The sources of non-localities in χEFT are mostly due to contact interactions that depend not only on the momentum transfer $\mathbf{k} = \mathbf{p}' - \mathbf{p}$ but also on $\mathbf{K} = (\mathbf{p}' + \mathbf{p})/2$ ($\mathbf{p}$ and $\mathbf{p}'$ are the initial and final relative momenta of the two nucleons), and also to specific choices of cutoff functions. It is for this reason that we construct a chiral potential as local as possible by minimizing the number of non-localities due to contact interactions and removing those due to the choice of regulator functions. In order to make the short-range part as local as possible, we use Fierz identities to remove terms which in coordinate-space would lead to powers higher than two in the relative momentum operator **p**. However, while this chiral potential is local at N2LO [5, 6], terms proportional to p^2 still persist at N3LO [6]. To avoid non-localities due to regulators, we choose cutoff functions that depend only on the relative distance between the two nucleons.

2 Potentials

The NN potential presented in this work includes a strong interaction component derived from χEFT up to N3LO and denoted as v_{12}, and an electromagnetic interaction component, including up to terms quadratic in the fine structure constant α, and denoted as $v_{12}^{\rm EM}$. The $v_{12}^{\rm EM}$ component is the same as that adopted in the Argonne v_{18} (AV18) potential [7]. The component induced by the strong interaction is conveniently separated into long- and short-range parts, labeled, respectively, $v_{12}^{\rm L}$ and $v_{12}^{\rm S}$. The $v_{12}^{\rm L}$ part includes the OPE and TPE contributions up to N2LO, illustrated in Fig. 1: panel (a) represents the static OPE contribution at leading order (LO or Q^0); panels (b)–(g) represent the TPE contributions at next-to leading (NLO or Q^2) without and with Δ-isobars in the intermediate states; lastly, panels (h)–(m) represent sub-leading TPE contributions at N2LO. The open circles denote πN and $\pi N\Delta$ couplings from the sub-leading chiral Lagrangians $\mathcal{L}_{\pi N}^{(2)}$ [8] and $\mathcal{L}_{\pi N\Delta}^{(2)}$ [9]. The NLO and N2LO loop corrections contain ultraviolet divergencies, which are isolated in dimensional regularization and then reabsorbed into contact interactions by renormalization of the associated LEC's [10, 11]. Note that one-loop as well as two-loop TPE and three-pion exchange contributions at N3LO have been neglected, since their effect on the peripheral phase shifts would have negligible influence on the fit to the NN data (see, for example, Ref. [2] for a discussion on this issue). Furthermore it is the LEC's at Q^4 that are critical for a good reproduction of phase shifts in lower partial waves, particularly D-waves, and a good fit to the NN database in the 0–300 MeV range of energies considered in the present study. The LO and NLO terms depend on the the pion decay amplitude $F_\pi = 184.80\,$MeV, and the nucleon and N-to-Δ axial coupling constants, respectively $g_A = 1.29$ and $h_A = 3\,g_A/\sqrt{2}$. The sub-leading N2LO terms also depend on the LEC's c_1, c_2, c_3, and c_4 and the combination of LEC's $(b_3 + b_8)$, respectively from the second order πN and $\pi N\Delta$ chiral Lagrangians $\mathcal{L}_{\pi N}^{(2)}$ [8] and $\mathcal{L}_{\pi N\Delta}^{(2)}$ [9]. The values of these LEC's are determined by fits to πN scattering data and are given in [9]. Note that we retained the LEC $(b_3 + b_8)$ in our fit, even though it is redundant at this order [12].

The coordinate-space expressions for the TPE terms are obtained by using the spectral function representation [13], however with no spectral cutoff [14]. Therefore, the potential $v_{12}^{\rm L}$ reads [6]

$$v_{12}^{\rm L} = \left[\sum_{l=1}^{6} v_{\rm L}^{l}(r)\, O_{12}^{l}\right] + v_{\rm L}^{\sigma T}(r)\, O_{12}^{\sigma T} + v_{\rm L}^{tT}(r)\, O_{12}^{tT}\,, \tag{1}$$

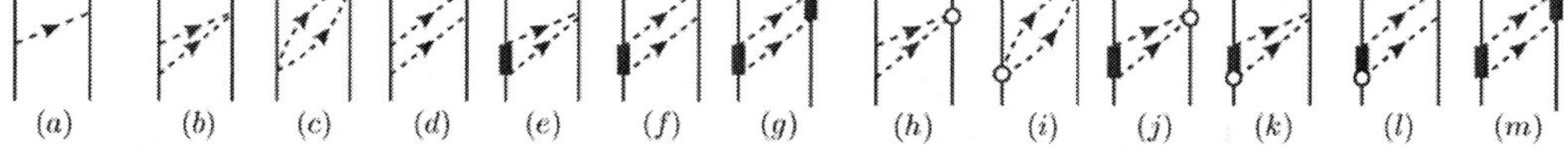

Figure 1. OPE and TPE contributions at LO [(a)], NLO [(b)–(g)], and N2LO [(h)–(m)]. Nucleons, Δ isobars, and pions are denoted, respectively, by the solid, thick-solid, and dashed lines; both direct and crossed box contributions are retained in diagrams (d), (f)–(g), (l)–(m). The open circles denote πN and $\pi N\Delta$ couplings from the sub-leading chiral Lagrangians $\mathcal{L}^{(2)}_{\pi N}$ [8] and $\mathcal{L}^{(2)}_{\pi N\Delta}$ [9].

with $O_{12}^{l=1,\ldots,6} = [\mathbf{1}\,,\,\boldsymbol{\sigma}_1\cdot\boldsymbol{\sigma}_2\,,\,S_{12}]\otimes[\mathbf{1}\,,\,\boldsymbol{\tau}_1\cdot\boldsymbol{\tau}_2]$ denoted as $c, \tau, \sigma, \sigma\tau, t, t\tau$, and $O_{12}^{\sigma T} = \boldsymbol{\sigma}_1\cdot\boldsymbol{\sigma}_2\, T_{12}$, and $O_{12}^{tT} = S_{12}\, T_{12}$, and $T_{12} = 3\,\tau_{1z}\tau_{2z} - \boldsymbol{\tau}_1\cdot\boldsymbol{\tau}_2$ is the isotensor operator. The terms proportional to T_{12} account for the charge-independence breaking induced by the difference between the neutral and charged pion masses in the only OPE. The radial functions $v^l_{\rm L}(r)$, $v^{\sigma T}_{\rm L}(r)$ and $v^{tT}_{\rm L}(r)$ are listed in [6]. They are singular at the origin (they behave as $1/r^n$ with n taking on values up to $n = 6$), and each is regularized by a cutoff of the form

$$C_{R_{\rm L}}(r) = 1 - \frac{1}{(r/R_{\rm L})^6\, e^{(r-R_{\rm L})/a_L} + 1}\,, \tag{2}$$

where three values for the radius R_L are considered $R_L = (0.8, 1.0, 1.2)$ fm with the diffuseness a_L fixed at $a_L = R_L/2$ in each case.

The potential $v^{\rm S}_{12}$ includes charge-independent (CI) contact interactions at LO, NLO and N3LO, and charge-dependent (CD) ones at LO and NLO, in momentum-space $v^{\rm S}_{12}(\mathbf{k},\mathbf{K}) = v^{\rm S,CI}_{12}(\mathbf{k},\mathbf{K}) + v^{\rm S,CD}_{12}(\mathbf{k},\mathbf{K})$, and their expressions are given in [6]. In the NLO and N3LO contact interactions terms proportional to K^2 and K^4, which would lead to p^2 and p^4 operators in coordinate space, have been removed by a Fierz rearrangement as discussed in Refs. [5, 6]. However, mixed terms of the type $k^2 K^2$ or $\mathbf{K}\times\mathbf{k}$ cannot be Fierz-transformed away. In the potential $v^{\rm S,CD}_{12}(\mathbf{k},\mathbf{K})$ only terms up to NLO, involving charge-independence breaking (proportional to T_{12}) and charge-symmetry breaking (proportional to $\tau_{1z} + \tau_{2z}$), are accounted for. However, the LEC's associated with $\tau_{1z} + \tau_{2z}$, while providing some additional flexibility in the data fitting (especially the LO term which is necessary for reproducing the singlet nn scattering length), are not well constrained. The potential $v^{\rm S}_{12}$ in coordinate-space reads (see Ref. [6])

$$v^{\rm S}_{12} \;=\; \left[\sum_{l=1}^{19} v^l_{\rm S}(r)\, O^l_{12}\right] + \{\, v^p_{\rm S}(r) + v^{p\sigma}_{\rm S}(r)\,\boldsymbol{\sigma}_1\cdot\boldsymbol{\sigma}_2 + v^{pt}_{\rm S}(r)\, S_{12} + v^{pt\tau}_{\rm S}(r)\, S_{12}\,\boldsymbol{\tau}_1\cdot\boldsymbol{\tau}_2\;,\; \mathbf{p}^2\,\}\,, \tag{3}$$

where $O_{12}^{l=1,\ldots,6}$ have been defined above, $O_{12}^{l=7,\ldots,11} = \mathbf{L}\cdot\mathbf{S}\,,\,\mathbf{L}\cdot\mathbf{S}\,\boldsymbol{\tau}_1\cdot\boldsymbol{\tau}_2\,,\,(\mathbf{L}\cdot\mathbf{S})^2\,,\,\mathbf{L}^2\,,\,\mathbf{L}^2\,\boldsymbol{\sigma}_1\cdot\boldsymbol{\sigma}_2$, referred to as b, $b\tau$, bb, q, $q\sigma$, and $O_{12}^{l=12,\ldots,19} = [\mathbf{1}\,,\,\boldsymbol{\sigma}_1\cdot\boldsymbol{\sigma}_2\,,\,S_{12}\,,\,\mathbf{L}\cdot\mathbf{S}]\otimes\left[T_{12}\,,\,\tau^z_1+\tau^z_2\right]$ referred to as T, τz, σT, $\sigma\tau z$, tT, $t\tau z$, bT, $b\tau z$. The four additional terms, denoted as p, $p\sigma$, pt, and $pt\tau$, in the anti-commutator of Eq. (3) are $\mathbf{p}^2$-dependent. The radial functions $v^l_{\rm S}(r)$, $v^p_{\rm S}(r)$, $v^{p\sigma}_{\rm S}(r)$, $v^{pt}_{\rm S}(r)$ and $v^{pt\tau}_{\rm S}(r)$ are listed in [6]. The potential $v^{\rm S}_{12}$ is regularized via a Gaussian cutoff,

$$C_{R_{\rm S}}(r) = \frac{1}{\pi^{3/2} R^3_{\rm S}}\, e^{-(r/R_{\rm S})^2}\,, \tag{4}$$

and we consider, in combination with $R_L = (0.8, 1.0, 1.2)$ fm, $R_s = (0.6, 0.7, 0.8)$ fm, corresponding to typical momentum-space cutoffs $\Lambda_{\rm S} = 2/R_{\rm S}$ from about 660 MeV down to 500 MeV.

3 Data analysis and results

The NN potential discussed in the previous section involves 34 unknown LEC's associated with the charge-independent contact interactions entering at LO, NLO and N3LO, and the charge-dependent contact interactions entering the LO and NLO. These contact parameters are determined by fitting the 2013 Granada database [15], consisting of 2309 pp and 2982 np data in the laboratory-energy range $E_{\text{lab}} = 0 - 300\,\text{MeV}$, as well as the deuteron binding energy. In the optimization procedure, as discussed in Ref. [6], we fit first phase shifts, then we refine the fit by minimizing the χ^2 obtained from a direct comparison with the database. In fact, sizable changes in the total χ^2 are found when passing from phase shifts to observables, so this refining is absolutely necessary to claim reasonable fits to data. This is a general feature which is often found, and reflects the different weights in the χ^2 contributions of the two different fitting schemes. Indeed, the initial guiding fit to phase shifts chooses a prescribed energy grid arbitrarily, which *does not* correspond directly to measured energies, nor necessarily samples faithfully the original information provided by the experimental data. Moreover, there are different PWA's which describe the same data but yield different phase shifts with significantly larger discrepancies than reflected by the inferred statistical uncertainties [15].

In Ref. [6] we report results for the potentials $v_{12} + v_{12}^{\text{EM}}$ corresponding to three different choices of cutoffs $(R_{\text{L}}, R_{\text{S}})$: model a with $(1.2, 0.8)$ fm, model b with $(1.0, 0.7)$ fm, and model c with $(0.8, 0.6)$ fm. The $\chi^2(pp)$/datum and $\chi^2(np)$/datum obtained by fitting the Granada database up to lab energies of 300 MeV are about 1.48, 1.48, 1.52 and 1.20, 1.19, 1.23 for models a, b, and c, respectively; the corresponding global $\chi^2(pp + np)$/datum are 1.33, 1.33, 1.37. Errors for pp data are significantly smaller than for np, thus explaining the consistently higher $\chi^2(pp)$/datum. The fitted values of the LEC's corresponding to models a, b, and c are listed in [6].

The S-wave, P-wave, and D-wave phase shits for np (in $T = 0$ and $T = 1$) and pp are reported in Ref. [6] where the calculated phases are compared to those obtained in partial-wave analyses (PWA's) by the Nijmegen [16], Granada [15], and Gross-Stadler [17] groups. There, we also provide tables of the pp, np and nn effective range parameters and of deuteron properties, including a figure of the deuteron S and D waves.

References

[1] S. Weinberg, Phys. Lett. **B251**, 288 (1990); Nucl. Phys. **B363**, 3 (1991); Phys. Lett. **B295**, 114 (1992).
[2] R. Machleidt and D.R. Entem, Phys. Rep. **503**, 1 (2011).
[3] R. Schiavilla *et al.*, Phys. Rev. C **45**, 2628 (1992).
[4] J. Carlson *et al.*, arXiv:1412.3081[nucl-th].
[5] A. Gezerlis *et al.*, Phys. Rev. Lett. **111**, 032501 (2013); Phys. Rev. C **90**, 054323 (2014).
[6] M. Piarulli *et al.*, Phys. Rev. C **91**, 024003 (2015).
[7] R. B. Wiringa, V. G. J. Stoks, and R. Schiavilla, Phys. Rev. C **51**, 38 (1995).
[8] N. Fettes *et al.*, Ann. Phys. **283**, 273 (2000).
[9] H. Krebs, E. Epelbaum, and Ulf.-G. Meißner, Eur. Phys. J. A **32**, 127 (2007).
[10] E. Epelbaum, W. Glöckle, and U.-G. Meissner, Nucl. Phys. **A637**, 107 (1998).
[11] S. Pastore *et al.*, Phys. Rev. C **80**, 034004 (2009).
[12] B. Long and V. Lensky, Phys. Rev. C **83**, 045206 (2011).
[13] E. Epelbaum, W. Glöckle, and U.-G. Meißner, Eur. Phys. J. A **19**, 125 (2004).
[14] N. Kaiser, R. Brockmann, and W. Weise, Nucl. Phys. A **625**, 758 (1997); Nucl. Phys. A **637**, 395 (1998).

[15] R. Navarro Pérez, J.E. Amaro, and E. Ruiz Arriola, Phys. Rev. C **88**, 064002 (2013); Phys. Rev. C **89**, 024004 (2014); Phys. Rev. C **89**, 064006 (2014).

[16] V.G.J. Stoks *et al.*, Phys. Rev. C **48**, 792 (1993); Phys. Rev. C **49**, 2950 (1994).

[17] F.L. Gross and A. Stadler, Phys. Rev. C **78**, 014005 (2008).

EPJ Web of Conferences **113**, 04023 (2016)
DOI: 10.1051/epjconf/201611304023

Nucleon-nucleon scattering at small angles, measured at ANKE-COSY

Z. Bagdasarian[1,2,3,a] for the ANKE collaboration

[1] *Forschungszentrum Jülich*
[2] *Ivane Javakhishvili Tbilisi State University*
[3] *University of Cologne*

Abstract. The most accepted approach to describe nucleon-nucleon (NN) interaction is the partial wave analysis (PWA), which translates various experimental observables to the common language of the partial waves. The reliable analysis relies not only on the quality experimental data, but also on the measurements of scattering observables over preferably the full angular range. Small angle scattering has been measured for six beam energies between 0.8 and 2.4 GeV using polarized proton beam incident on both proton and deuteron unpolarized targets at COSY-ANKE. This proceeding will report on the published and preliminary results for both *pp* and *pn* scattering from this and other recent experiments at ANKE. This study aims to provide the valuable observables to the SAID group in order to improve the phenomenological understanding of the nucleon-nucleon interaction.

1 Introduction

The interactions between nucleons and their constituent quarks and gluons, are mainly governed by the strong interaction. However, Quantum Chromodynamics, the theory of strong interaction, which perfectly describes the phenomena at high energies, cannot give us quantitative description of the strong interaction at lower energies. Hence the only feasible approach to describe nucleon interaction at intermediate energies is phenomenological one. That is why, the partial wave decomposition of the proton-proton and proton-neutron scattering data is a very important issue in nuclear physics.

The SAID data base and analysis (Scattering Analysis Interactive Dialin) [1] have proved to be a truly invaluable tool over many years for researchers working in this area. Such an analysis is based on the measurements of various NN scattering observables at different energies over the full angular range. Many accelerators around the world included the NN study into their research program, however even after many years of studies, there are still many gaps in our knowledge. As one can see on the Fig. 1, even in the data base of the most basic reaction of proton-proton elastic scattering, there has been a significant gap at the small angles ($\theta_{cm} < 30°$) above 1 GeV. The COSY-ANKE facility was designed for the investigation of the small angle region and is thus well suited to cover this significant gap in the database. The situation is far less advanced for the isoscalar channel where the much poorer neutron-proton data base only permits the I = 0 phase shifts to be evaluated up to at most 1.3 GeV but with significant ambiguity above about 800 MeV.

[a] e-mail: z.bagdasarian@fz-juelich.de

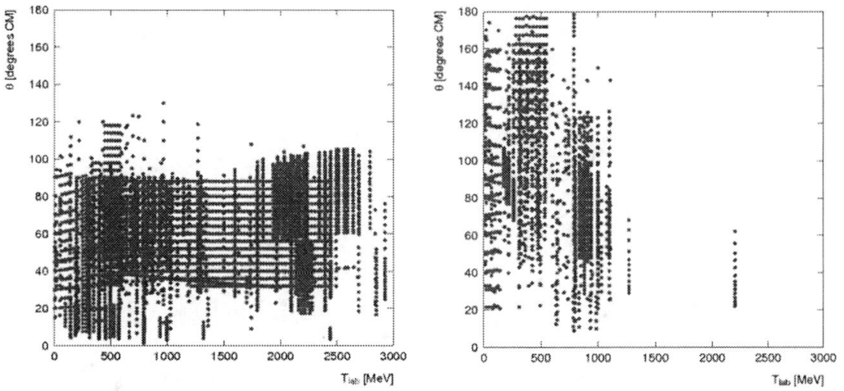

Figure 1. Abundance plot of c.m. scattering angle (θ_{cm}) versus beam energy (T_{lab}) for experiments on the analysing power for proton proton elastic scattering (left) and for proton neutron quasi-elastic scattering (right). *Source: nn-online.org*

2 Experiment

The experiment was carried out using the ANKE magnetic spectrometer positioned inside the storage ring of the COoler SYnchrotron (COSY) of the Forschungszentrum Jülich. The ANKE spectrometer can provide precise data on the differential cross section and spin observables in the energy range up to 2.8 GeV for center-of-mass angles $5° < \theta_{cm} < 30°$. The ANKE forward detector (FD) comprises a set of multiwire proportional and drift chambers (MWCs) and a two-plane scintillation hodoscope. The two silicon tracking telescopes (STT) were placed symmetrically inside the vacuum chamber, to the left and right of the beam near the unpolarised cluster-jet target. Each telescope consists of three double-sided silicon strip detectors of 70 μm, 300 μm, and 5 mm thickness.

The ANKE experiment used a vertically polarised beam incident on an unpolarised target so that the preparation and the measurement of the beam polarisation are critical. The H^- ions from the polarised ion source were accelerated to 45 MeV in the cyclotron JULIC before being stripped of their electrons and injected into COSY.

3 Results

3.1 Proton-proton scattering

The results of all the ANKE measurements of A_y for pp elastic scattering are shown for the six energies in Fig. 2 [3]. For the five of these energies the measurements have been performed for the first time in the angular range from 4° to 28°. The agreement between the STT and FD data, which involved completely independent measurements of the final state, is remarkably good. Although the analysing power results agree all with the many published data at 796 MeV, and also with the most recent partial wave solution (SP07) at this energy, the ANKE data at higher energies lie well above the predictions of this solution at small angles. An updated phase shift analysis that uses the ANKE results together with the World data leads to a much better description of these new measurements.

The experiment measuring the differential cross-sections $\frac{d\sigma}{d\Omega}$ has been performed using an unpolarized proton beam at eight different beam energies, T_p= 1.0, 1.6, 1.8, 2.0, 2.2, 2.4, 2.6, and 2.8 GeV, interacting with a hydrogen cluster-jet target. The preliminary results of this experiment are presented in Fig. 3.

3.2 Proton-neutron scattering

The detailed studies of $dp \to \{pp\}_s n$ were undertaken by the ANKE collaboration at energies per nucleon of $T_N = 0.6, 0.8, 0.9$, and 1.135 GeV [9][10]. At the three lower energies the predictions of

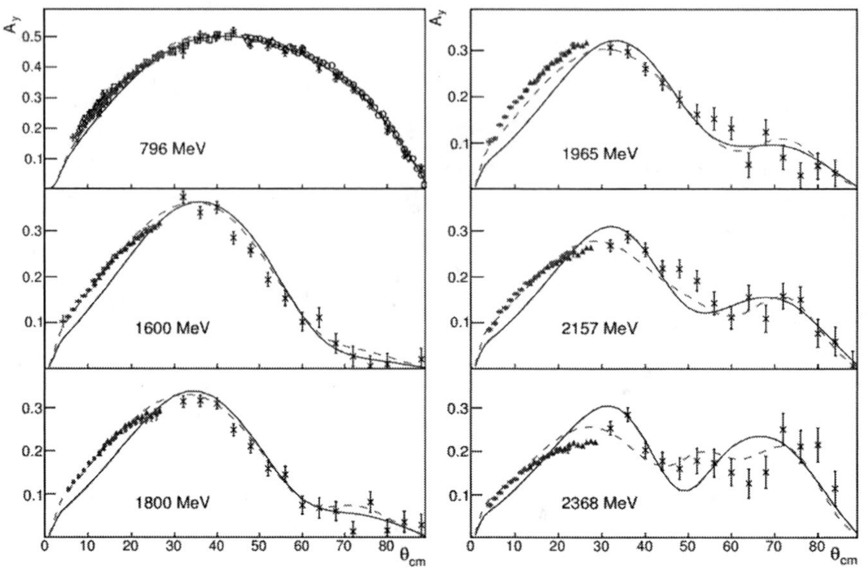

Figure 2. Comparison of the ANKE measurements of the proton analysing power in *pp* elastic scattering using the STT (red filled circles) and FD (blue filled triangles) systems with the curves corresponding to the SAID SP07 (solid black line) and the revised fit (dashed red) solutions [1]. Only statistical errors are shown so that the systematic uncertainties arising, for example, from the calibration of the EDDA polarimeter have not been included. Also shown are selected results from EDDA (black crosses) [2] at the energies different by no more than 7 MeV and, at 796 MeV, LAMPF [4–6], and SATURNE [7] (black open symbols).

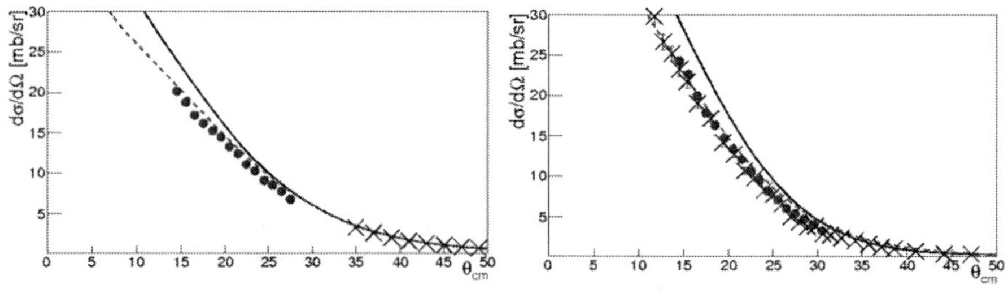

Figure 3. Comparison of the preliminary ANKE measurements of the differential cross section in *pp* elastic scattering (blue filled circles) systems with the curves corresponding to the SAID SP07 (solid black line) and the revised fit (dashed red) solutions [1] at 2.0 GeV (left) and 2.8 GeV (right).

the impulse approximation model describe the data very well on the basis of np input taken from the SAID SP07 partial wave solution [1]. Deviations were, however, noted in the 2.27 GeV data [10] that were ascribed to an overestimate of the strength of the *np* spin-longitudinal amplitude at 1135 MeV.

To continue the studies at COSY to higher energies, where there is great uncertainty in the neutron-proton amplitudes, the experiments had to be carried out in inverse kinematics, with a proton beam incident on a polarized deuterium target. The study of the charge exchange at low momentum transfers would then require the measurement of two low energy protons recoiling from the target. The results of the first measurement for the $pd \to n\{pp\}_s$ charge-exchange at 600 MeV that extends the earlier deuteron beam data out to larger values of the momentum transfer q, are reported in [11]. The ANKE

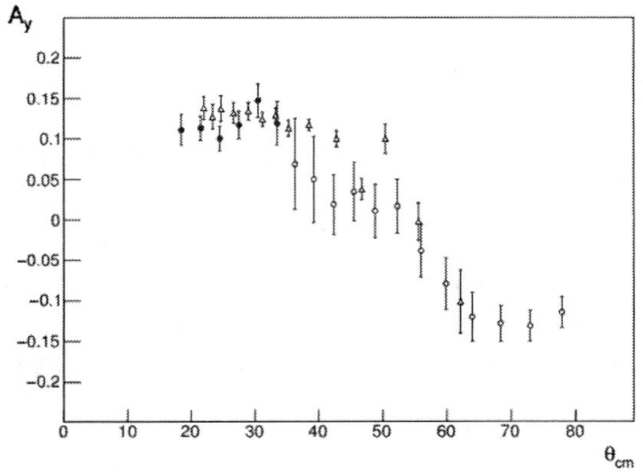

Figure 4. Comparison of the preliminary ANKE measurements of the analysing power in *pn* quasi-elastic scattering (blue filled circles) with the results from Argonne Narional Laboratory at 2.157 GeV [12] [13].

measurements of A_y for *pn* quasi-elastic scattering have been performed using polarised proton beam at the six energies and unpolarised deuterium target. The SAID SP07 solution [1] describes our results very well at the lowest energy of 796 MeV. The solution is valid, however, only up to 1.3 GeV kinetic energy The results at 2.157 GeV are shown in Fig. 4 in order to show how well our measurements agree with the existing data from Argonne National Laboratory. The measurements at other energies have been performed for the first time at small angles.

Acknowledgements

This work and the conference trip has been supported by the Shota Rustaveli Science Foundation Grant #DO/30/6-200/14.

References

[1] R.A. Arndt, W.J. Briscoe, I.I. Strakovsky, R.L. Workman, Phys. Rev. C 76 025209 (2007); `http://gwdac.phys.gwu.edu`.
[2] M. Altmeier et al., Phys. Rev. Lett. 85 1819 (2000); M. Altmeier et al., Eur. Phys. J. A 23 351(2005).
[3] Z. Bagdasarian et al., Phys. Lett. B 739 (2014) 152 .
[4] M.W. McNaughton et al., Phys. Rev. C 23 1128 (1981).
[5] F. Irom, G.J. Igo, J.B. McClelland, C.A. Whitten, Phys. Rev. C 25 373 (1982).
[6] P.R. Bevington et al., Phys. Rev. Lett. 41384 (1978).
[7] C.E. Allgower et al., Nucl. Phys. A 637 231 (1998).
[8] D.Bugg and C. Wilkin, Nucl. Phys. A 467 575 (1987).
[9] D. Chiladze et. al., Phys. Rev. St Accel. Beams 9 050101 (2006).
[10] D. Mchedlishvili et al., Eur. Phys. J. A 49 49 (2013).
[11] B. Gou al., Phys. Lett. B 741 305 (2015).
[12] Y. Makdisi et.al., Phys. Rev. Lett. 45 1529 (1980).
[13] R. Diebold et. al., Phys. Rev. Lett. 35 632 (1975).

EPJ Web of Conferences **113**, 04024 (2016)
DOI: 10.1051/epjconf/201611304024

Nucleon-nucleon scattering in the $^1S_0$ partial wave in the modified Weinberg approach

A. M. Gasparyan[1,2,a], E. Epelbaum[1], J. Gegelia[1,3], and H. Krebs[1]

[1] *Institut für Theoretische Physik II, Fakultät für Physik und Astronomie, Ruhr-Universität Bochum, 44780 Bochum, Germany*
[2] *SSC RF ITEP, Bolshaya Cheremushkinskaya 25, 117218 Moscow, Russia*
[3] *Tbilisi State University, 0186 Tbilisi, Georgia*

Abstract. Nucleon-nucleon scattering in the $^1S_0$ partial wave is considered in chiral effective field theory within the recently suggested renormalizable formulation based on the Kadyshevsky equation. Contact interactions are taken into account beyond the leading-order approximation. The subleading contact terms are included non-perturbatively by means of subtractive renormalization. The dependence of the phase shifts on the choice of the renormalization condition is discussed. Perturbative inclusion of the subleading contact interaction is found to be justified only very close to threshold. The low-energy theorems are reproduced significantly better compared with the leading order results.

1 Introduction

Recently a novel formulation of the NN scattering problem in EFT has been suggested in Refs. [1–4]. It is based on a modification of the original Weinberg's approach [5]. In the new scheme one solves the relativistic Kadyshevsky equation instead of the Lippmann-Schwinger equation. One of the advantages of such an approach is that ultraviolet divergences reproduce the ones that appear in covariant Feynman diagrams, i.e. there exist only logarithmic or even-power divergences. The odd-power divergences present in iterations of the Lippmann-Schwinger equation are artifacts of the non-relativistic approximation. In particular, the integral equation based on the leading-order potential, which consists of the momentum- and energy-independent contact interaction and the one-pion-exchange potential, possesses only logarithmic divergences and is therefore renormalizable (there are exceptions from such a naive picture, e.g. the case of $^3P_0$ partial wave, see Ref. [1]). Higher-order contributions are included perturbatively in this scheme without spoiling the renormalizability feature. The present approach is alternative to the standard non-relativistic EFT approach based on the Lippmann-Schwinger equation [6, 7] where one necessarily has to introduce a cut off regulator of the order of the hard scale [8, 9].

In the $^1S_0$ channel a large discrepancy between the leading order EFT results and the phase shift of the Nijmegen partial wave analysis [10] is observed, which starts already at rather low energies. This indicates that at least some parts of the higher-order contributions to the effective potential, in contrast with the original formulation, have to be included non-perturbatively. A similar observation

[a] e-mail: ashot.gasparyan@rub.de

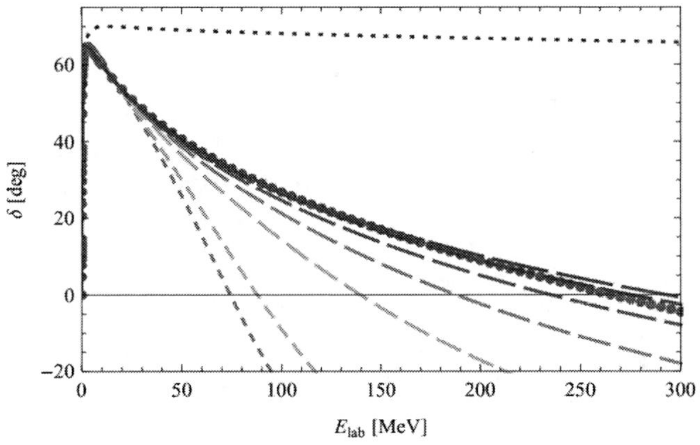

Figure 1. Neutron-proton ${}^1S_0$ phase shifts versus the energy in the laboratory frame. Circles (red) correspond to the Nijmegen PWA [10]. Dotted line represents the LO result. Curves with different dashing (colors) correspond to non-perturbative inclusion of the NLO contact interaction potential for $\mu = 50, 100, 300, 500, 700, 850$ and 900 MeV respectively.

was made in Ref. [11] in the scheme with an explicit dibaryon field. In this talk we report on the results for the ${}^1S_0$ phase shifts calculated both with perturbative and non-perturbative inclusion of the next-to-leading-order (NLO) contact interaction and make the comparison between the two options.

2 Formalism

We assume that only the short range part of the next-to-leading-order (NLO) potential (in standard Weinberg power counting [5]) needs to be treated non-perturbatively. It involves only the contact interaction terms quadratic in momenta and the pion mass. This makes it possible to perform the subtractive renormalization explicitly in non-perturbative expressions [12].

The NLO ${}^1S_0$ partial wave NN scattering amplitude can be obtained by extracting the ${}^1S_0$ component from the solution to the integral equation for the off-shell amplitude T

$$T(p_0, \vec{p}\,', \vec{p}) = V(\vec{p}\,', \vec{p}) + \frac{m^2}{2} \int \frac{d^3k}{(2\pi)^3} \frac{V\left(\vec{p}\,', \vec{k}\right) T\left(p_0, \vec{k}, \vec{p}\right)}{\left(\vec{k}^2 + m^2\right)\left(p_0 - \sqrt{\vec{k}^2 + m^2} + i\epsilon\right)}, \tag{1}$$

with the potential given by

$$V(\vec{p}\,', \vec{p}) = \left[C + C_2\left(\vec{p}\,'^2 + \vec{p}\,^2\right)\right] - \frac{g_A^2 M_\pi^2}{4F_\pi^2} \frac{1}{(\vec{p}\,' - \vec{p})^2 + M_\pi^2}. \tag{2}$$

After some algebraic transformations it is possible to rewrite Eq.(1) in terms of two finite renormalized low-energy constants $\tilde{C}^R(\mu)$ and $C_2^R(\mu)$, which depend on the the subtraction momentum μ [12].

3 Results

The renormalized integral equations for the scattering amplitude have been solved numerically and the values of the low-energy constants $\tilde{C}^R(\mu)$ and $C_2^R(\mu)$ were determined from a fit to the phase shifts of the Nijmegen partial-wave analysis [10] for different choices of the subtraction point μ. The resulting phase shifts for different choices of μ are plotted in Fig. 1. The description of the data is significantly improved as compared to the leading order result. On the other hand one can observe some residual μ-dependence of the predicted phase shifts which gets strongly enhanced if one chooses μ of the order or smaller than the pion mass. This supports the fact that the appropriate choice of the subtraction

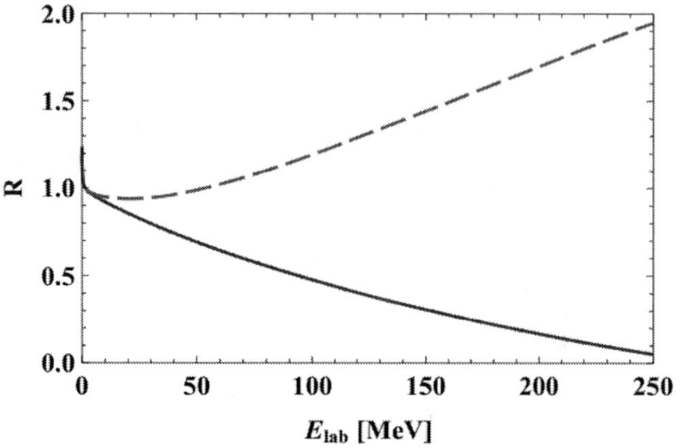

Figure 2. The ratio R of the absolute values of the T-matrices calculated in the leading and next-to-leading order approximation as described in text versus the energy in the laboratory frame. The solid (dashed) line shows the result based on T^{NLO} corresponding to the non-perturbative (perturbative) inclusion of the subleading contact interaction. In both cases, the subtraction scale is set to μ = 850 MeV.

scale for the non-perturbative treatment of C_2^R should be $\mu \sim \Lambda$ (see e.g. Refs. [8, 9]), where the hard scale Λ can be estimated to be of the order of $\Lambda \sim$ 600-700 MeV (masses of the sigma and rho mesons which phenomenologically are known to yield the most important short-range contribution to the nucleon-nucleon potential [13]).

We also address the question of perturbativeness of the subleading contact interaction within our scheme. The ratio $R(E_{\text{lab}}) = \frac{|T^{\text{NLO}}(E_{\text{lab}})|}{|T^{\text{LO}}(E_{\text{lab}})|}$ for the cases when the subleading contact term is included perturbatively and non-perturbatively is plotted in Fig. 2. Note that the ratio R is not equal to 1 at threshold, because in our subtraction scheme diagrams containing NLO contact interactions but no overall divergences are not subtracted. Looking at this figure or based on explicit calculations one can conclude that it is advantageous to include the subleading contact interaction non-perturbatively for energies of about $E_{\text{lab}} \sim$ 50 MeV and higher. This conclusion is similar to the one made in Ref. [11] but in a different regularization and renormalization scheme.

Finally, we found a clear improvement when going from leading order to next-to-leading order for the coefficients in the effective range expansion, i.e. for the low-energy theorems (LETs) (see Table 1). It is instructive to compare the convergence pattern for the LETs in our approach with the one in the scheme with perturbative pions suggested by Kaplan, Savage and Wise (KSW) [14], where convergence is, at best, very slow as follows from Table 1. On the other hand, the μ dependence of the low-energy coefficients appears to be rather weak because they are mostly determined by the energy region very close to threshold. By the same reason perturbative inclusion of the subleading contact terms leads to nearly the same values for the low-energy coefficients.

4 Summary

We extended the modified Weinberg scheme for the NN-interactions of Ref. [1] by including the subleading contact interaction both perturbatively and non-perturbatively. The perturbative treatment appears to be reasonable for energies below $E_{\text{lab}} \sim$ 50 MeV whereas the non-perturbative approach leads to a good description of the data up to $E_{\text{lab}} \sim$ 300 MeV for appropriate choices of the subtraction point. The results of our work allow one to proceed with inclusion of higher-order terms not taken into account in the present work—such as two-pion exchange loop contributions—perturbatively according to the original scheme.

Table 1. Predictions for the coefficients in the effective range expansion of the $^1S_0$ phase shifts (low-energy theorems) at LO and NLO in the modified Weinberg approach in comparison with the NLO and NNLO KSW predictions of Ref. [15] and empirical numbers extracted from the Nijmegen PWA [16, 17]. For the NLO Weinberg results, we show the predictions corresponding to the variation of the subtraction point in the range of $\mu = 500\ldots900$ MeV. The errors quoted for the LO predictions refer to the uncertainty in the numerical extraction of the coefficients [1].

	a [fm]	r [fm]	v_2 [fm^3]	v_3 [fm^5]	v_4 [fm^7]
LO, Ref. [1]	fit	1.50	-1.9	8.6(8)	$-37(10)$
NLO, nonperturb. C_2	fit	fit	$-0.55\ldots-0.61$	$5.1\ldots5.5$	$-29.6\ldots-30.8$
NLO, perturbative C_2	fit	fit	$-0.51\ldots-0.57$	$4.5\ldots4.7$	$-28.8\ldots-29.8$
NLO KSW, Ref. [15]	fit	fit	-3.3	19	-117
NNLO KSW, Ref. [15]	fit	fit	-3.3	2.9	-0.7
Nijmegen PWA	-23.7	2.67	-0.5	4.0	-20

Acknowledgements

This work is supported by the Deutsche Forschungsgemeinschaft (SFB/TR 16, "Subnuclear Structure of Matter"), by the European Community Research Infrastructure Integrating Activity "Study of Strongly Interacting Matter" (acronym HadronPhysics3, Grant Agreement n. 283286) under the 7th Framework Programme of the EU, by the European Research Council (acronym NuclearEFT, ERC-2010-StG 259218).

References

[1] E. Epelbaum, J. Gegelia, Phys.Lett. **B716**, 338 (2012), 1207.2420
[2] E. Epelbaum, J. Gegelia, Few Body Syst. **54**, 1473 (2013), 1210.3964
[3] E. Epelbaum, J. Gegelia, PoS **CD12**, 090 (2013), 1301.6134
[4] E. Epelbaum, A.M. Gasparyan, J. Gegelia, M.R. Schindler, Eur. Phys. J. **A50**, 51 (2014), 1311.7164
[5] S. Weinberg, Phys. Lett. **B251**, 288 (1990)
[6] R. Machleidt, D. Entem, Phys.Rept. **503**, 1 (2011), 1105.2919
[7] E. Epelbaum, U.G. Meissner, Ann. Rev. Nucl. Part. Sci. **62**, 159 (2012), 1201.2136
[8] J. Gegelia, Phys. Lett. **B429**, 227 (1998)
[9] E. Epelbaum, J. Gegelia, Eur. Phys. J. **A41**, 341 (2009), 0906.3822
[10] V. Stoks, R. Klomp, M. Rentmeester, J. de Swart, Phys.Rev. **C48**, 792 (1993)
[11] B. Long, Phys. Rev. **C88**, 014002 (2013), 1304.7382
[12] E. Epelbaum, A.M. Gasparyan, J. Gegelia, H. Krebs, Eur. Phys. J. **A51**, 71 (2015), 1501.01191
[13] E. Epelbaum, U.-G. Meißner, W. Glöckle, C. Elster, Phys. Rev. **C65**, 044001 (2002), nucl-th/0106007
[14] D.B. Kaplan, M.J. Savage, M.B. Wise, Phys. Lett. **B424**, 390 (1998), nucl-th/9801034
[15] S. Fleming, T. Mehen, I.W. Stewart, Nucl. Phys. **A677**, 313 (2000), nucl-th/9911001
[16] V.G.J. Stoks, R.A.M. Klomp, C.P.F. Terheggen, J.J. de Swart, Phys. Rev. **C49**, 2950 (1994), nucl-th/9406039
[17] M. Pavon Valderrama, E. Ruiz Arriola, Phys. Rev. **C72**, 044007 (2005)

EPJ Web of Conferences **113**, 04025 (2016)
DOI: 10.1051/epjconf/201611304025

Three-Body Nuclear Systems in Pionless Effective Field Theory

Jared Vanasse[1,a]

[1] *Department of Physics, Duke University, Durham, NC 27708, USA*

Abstract. New perturbative techniques for three-body systems with contact interactions are discussed. Their application to pionless effective field theory ($\mathrm{EFT}_{\not\pi}$) for *nd* scattering is shown, and their extension to bound states addressed. With the extension to bound states a leading-order $\mathrm{EFT}_{\not\pi}$ calculation of the triton charge radius and novel treatments of three-body forces are discussed.

1 Introduction

At low energies ($E < m_\pi^2/M_N$) nuclear physics can be described by a theory of short range interactions between nuclei known as pionless effective field theory ($\mathrm{EFT}_{\not\pi}$) (see e.g. [1]). This has been used to great success in the two-body sector, with precision calculations (<1%) being performed for $np \to d\gamma$ (see e.g. [2]). Progress has also been made in three-body sector with *nd* scattering but not at the level of precision in the two-body sector (see e.g. [3, 4]). Previously, calculations in the three-body sector of $\mathrm{EFT}_{\not\pi}$ have been performed using the partial resummation technique (see e.g. [5]). This technique calculates all diagrams to the order one is working but also includes an infinite subset of higher order diagrams and therefore is not strictly perturbative in the $\mathrm{EFT}_{\not\pi}$ power counting. A technique that is strictly perturbative and no more numerically expensive than the partial resummation technique is discussed below and further detail given in Ref. [4]. Its extension to bound state calculations, and the future prospects for precision three-body calculations are also discussed.

2 Scattering

The leading-order (LO) *nd* scattering amplitude in $\mathrm{EFT}_{\not\pi}$ is given by an infinite sum of diagrams, which can be represented by an integral equation that is solved numerically. The numerical solution of the half off-shell LO *nd* scattering amplitude is less numerically expensive than solving the full off-shell scattering amplitude. Higher order corrections to the *nd* scattering amplitude in principle require the full off-shell LO *nd* scattering amplitude. Calculations of higher order corrections using the full off-shell LO scattering amplitude have been performed for cold atoms systems but not for nuclear systems in $\mathrm{EFT}_{\not\pi}$ [6]. To circumvent the need for the full off-shell scattering amplitude the partial resummation technique was created [5]. This technique gives all contributions up to the order one is working but includes an infinite subset of higher order diagrams. Thus this method is not strictly

[a] e-mail: jjv9@phy.duke.edu

perturbative. Also in this method the imaginary part of the quartet (total spin $S = \frac{3}{2}$) S-wave phase shift has a negative sign. This is unphysical, and its origin is currently not understood.

A method to calculate higher order corrections strictly perturbatively, that is no more numerically expensive than solving the half off-shell LO scattering amplitude and gives the correct sign for the imaginary part of the quartet S-wave phase shift was developed in Ref. [4]. This method was slightly improved upon in Ref. [7], in which the numerical technique for corrections to dibaryon propagators is simplified. In this improved perturbative technique the general nth order correction to the scattering amplitude in the quartet channel is given by the integral equation represented in Fig. 1. The final diagram, in Fig. 1, is the kernel of the integral equation and is the same kernel for the LO nd scattering amplitude. All preceding diagrams are the inhomogeneous term of the integral equation and depend on lower-order scattering amplitudes. In this technique the kernel of the integral equation is exactly the same at each order, and this simplifies the numerical solution of higher order corrections. Also in this improved technique the diagrams with deuteron corrections become the deuteron wavefunction renormalization corrections in the on-shell limit. Thus in the on-shell limit this improved technique gives the full nth order correction including corrections due to the deuteron wavefunction renormalization. Higher order corrections in the doublet channel are given by a similar set of coupled integral equations involving a spin-singlet dibaryon propagator. Additional complications occur in the doublet S-wave channel at all orders due to the necessity of three-body forces.

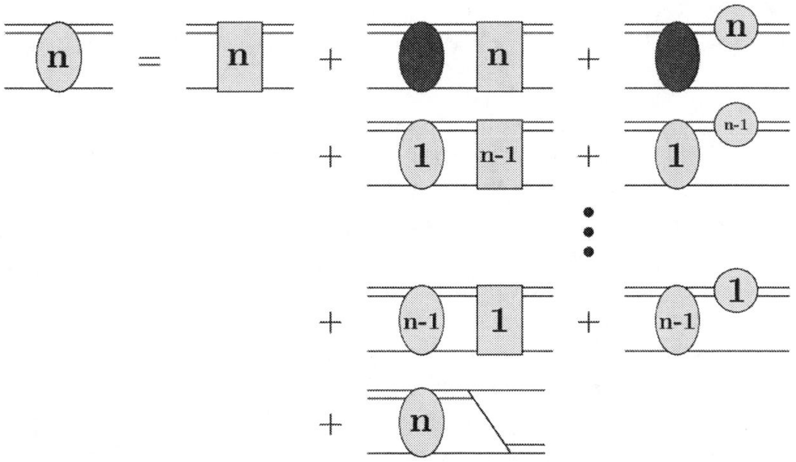

Figure 1. Integral equation for the general nth order correction to the quartet channel nd scattering amplitude. The single (double) lines are nucleon (deuteron) propagators, the solid red oval the LO nd scattering amplitude, and the n-oval is the nth order correction to the nd scattering amplitude. The n-circles attached to the deuteron propagators are nth order corrections to the deuteron propagator, and the n-rectangles are nth order "three-body" corrections between neutron and deuteron.

Using these strict perturbative techniques nd scattering has been calculated to next-to-next-to-leading-order (N^{2}LO) in EFT$_{\not\pi}$ [4]. At next-to-leading-order (NLO) and N^{2}LO there are effective range corrections to the dibaryon propagators, and at N^{2}LO there is a two-body SD-mixing term, which is important since it gives the first non-zero contribution to mixing angles and polarization observables. There is one three-body force at LO with no derivatives that receives corrections at each order and a new energy dependent three-body force at N^{2}LO. It was found that at N^{2}LO in comparison to potential model calculations that the eigen-phases and mixing angles are mostly reproduced well. However, significant differences were found for the ϵ_J^π mixing angles (mixes different total spin angular momentum but not orbital angular momentum) and the ^4P_J eigen-phase shifts.

3 Bound States

The new perturbative method in Ref. [7] can also be applied to three-body bound states, by introducing a triton auxiliary field, $\hat{\psi}$, that interacts with a spin-triplet (spin-singlet) dibaryon field $\hat{t}_i$ ($\hat{s}_a$) and nucleon field $\hat{N}$ via the Lagrangian

$$\mathcal{L}_3 = \hat{\psi}^\dagger\left(\Omega - h_2(\Lambda)\left(i\partial_0 + \frac{\vec{\nabla}^2}{2M_N} + \frac{\gamma_t^2}{M_N}\right)\right)\hat{\psi} + \sum_{n=0}^{2}\omega_0^{(n)}\hat{\psi}^\dagger\left(\sigma_i\hat{N}\hat{t}_i - \tau_a\hat{N}\hat{s}_a\right) + \text{H.c.} \tag{1}$$

Here, $\gamma_t = 45.7025$ MeV, is the deuteron binding momentum. From this formalism the LO triton vertex function [8] can be calculated via an integral equation entirely analogous to the integral equation for *nd* scattering in the doublet S-wave channel with no three-body force. The only difference between these integral equations is the inhomogeneous term. Higher order corrections to the triton vertex function are calculated by integral equations exactly the same as those for perturbative corrections to *nd* scattering in the doublet S-wave channel with no three-body forces, except the *n*th order *nd* scattering amplitude is replaced by the *n*th order triton vertex function. Using the LO triton vertex function the triton charge radius has been calculated using methods found in Refs. [7, 9] yielding a value of [1.41 ± 0.42] fm [7], which is in agreement with the experimental value of [1.755 ± 0.086] fm [10]. There is also a EFT$_{\not\pi}$ calculation using wavefunction methods that obtains a value of [2.1±0.6] fm [11].

This new formalism also offers a novel approach to the treatment of three-body forces. The LO on-shell *nd* scattering amplitude is given by

$$T_{\text{LO}}(k) = Z_{\text{LO}}\, t_{0,Nt\to Nt}^{\ell=0}(k,k) + H_{\text{LO}}\frac{1}{1 - H_{\text{LO}}\,\Sigma_0(E)}\pi Z_{\text{LO}}\left(G_{0,\psi\to Nt}(E,k)\right)^2, \tag{2}$$

where k is the center of mass (c.m.) momentum of the *nd* system. The function $t_{0,Nt\to Nt}^{\ell=0}(k,k)$ is the *nd* scattering amplitude in the doublet S-wave channel with no three-body force, and $G_{0,\psi\to Nt}(E,k)$ is the LO triton vertex function, with $E = (\frac{3}{4}k^2 - \gamma_t^2)/M_N$ being the total energy of the system. Z_{LO} is the LO deuteron wavefunction renormalization, H_{LO} the LO three-body force, and $\Sigma_0(E)$ a function calculated from integrating over the LO triton vertex function with a LO dibaryon propagator. The novel feature of this expression is that the three-body force is factored out of all numerically determined expressions. Therefore, the three-body force can be fit algebraically using numerically determined quantities yielding

$$H_{\text{LO}} = \frac{x}{1 + x\Sigma_0(-\gamma_t^2)}, \tag{3}$$

where

$$x = \frac{-\left(\frac{3\pi a_{nd}}{M_N} + Z_{\text{LO}}t_{0,Nt\to Nt}^{\ell=0}(0,0)\right)}{\pi Z_{\text{LO}}(G_{0,\psi\to Nt}(-\gamma_t^2,0))^2}. \tag{4}$$

Here the three-body force has been fit to the doublet S-wave scattering length, $a_{nd} = .65$ fm.

4 Conclusions

Using these new perturbative methods, *nd* scattering has been calculated to N^{2}LO in EFT$_{\not\pi}$ including the SD-mixing term. Comparing with potential model calculations good agreement was found with most eigen-phases and mixing angles, with the exception of the ϵ_J^π mixing angles and ${}^4P_J$ eigen-phases. As a result poor agreement was found with low-energy polarization observables in *nd* scattering. Thus a higher order calculation of *nd* scattering is necessary to properly address polarization observables. Such a calculation will include shape parameter corrections and two-body *P*-wave contact interactions. Investigating polarization observables using EFT$_{\not\pi}$ is important as there exists the longstanding A_y puzzle. This is a discrepancy between theoretical calculations and experimental data for the vector analyzing power, A_y, which is worse at low energies. However, EFT$_{\not\pi}$ is ideally suited for low energies and offers the prospect of understanding this longstanding puzzle.

Using a triton auxiliary field, a LO EFT$_{\not\pi}$ calculation of the triton charge radius, within the expected LO error, yields agreement with the experimental value. The extension of the perturbative techniques to bound states using the triton auxiliary field gives a straightforward method for calculating higher order corrections to the triton charge radius. This extension also offers a novel numerical treatment and understanding of three-body forces. It allows the LO three-body force to be determined algebraically from numerically determined quantities. Previous methods were more numerically expensive as they were essentially root finding algorithms and required multiple iterations.

Finally, these perturbative techniques work equally well for processes involving external currents and systems with charged particles such as *pd*. Therefore, calculations of *pd* scattering can be performed, for which there is much more experimental data for polarization observables than *nd*. Investigation of low energy Compton scattering off of the triton and Helium-3 is also possible. For astrophysical applications *pd* and *nd* capture at low energies can be considered. Beta decay of the triton can also be addressed and gives a way to fit the $L_{1,A}$ coupling that is an important unknown in *pp* fusion in EFT$_{\not\pi}$ [12]. In short there is a plethora of three-body nuclear physics at low energies that can benefit from these new perturbative techniques and make precision three-body physics in EFT$_{\not\pi}$ feasible.

Acknowledgements

This material is based upon work supported by the U.S. Department of Energy Office of Science, Office of Nuclear Physics, under Award Number DE-FG02-05ER41368. I would also like to thank Roxanne Springer and Thomas Mehen for useful conversations during the course of this work.

References

[1] S.R. Beane, P.F. Bedaque, W.C. Haxton, D.R. Phillips, M.J. Savage (2000), in *At the Frontier of Particle Physics*, edited by M. Shiftman, Vol. 1 (World Scientific, River Edge, NJ, 2001).
[2] G. Rupak, Nucl. Phys. A **678**, 405 (2000)
[3] H.W. Grießhammer, Nucl. Phys. A **744**, 192 (2004)
[4] J. Vanasse, Phys. Rev. C **88**, 044001 (2013)
[5] P.F. Bedaque, G. Rupak, H.W. Grießhammer, H.W. Hammer, Nucl. Phys. A **714**, 589 (2003)
[6] C. Ji, D.R. Phillips, Few-Body Syst. **54**, 2317 (2013)
[7] J. Vanasse, in preparation
[8] R.D. Amado, Phys. Rev. **141**, 902 (1966)
[9] P. Hagen, H.W. Hammer, L. Platter, Eur.Phys.J. **A49**, 118 (2013)
[10] A. Amroun et al., Nucl. Phys. **A579**, 596 (1994)
[11] L. Platter, H.W. Hammer, Nucl. Phys. **A766**, 132 (2006)
[12] M. Butler, J.W. Chen, Phys. Lett. **B520**, 87 (2001)

EPJ Web of Conferences **113**, 04026 (2016)
DOI: 10.1051/epjconf/201611304026

Hadronic parity violation in few-body systems

M. R. Schindler[1,a]

[1] *Department of Physics and Astronomy, University of South Carolina, Columbia, SC 29208, USA*

Abstract.
Efforts to describe hadronic parity violation in few-nucleon systems using effective field theories are described. Particular emphasis is given to observables in two- and three-nucleon systems and their relation to ongoing experimental efforts. The role of parity-violating three-nucleon interactions is also discussed.

1 Introduction

Parity-violating (PV) interactions of hadrons are the manifestation of an interplay of strong and weak interactions between quarks inside the hadrons (for recent reviews see, e.g., Refs. [1, 2]). For the case of nucleon-nucleon (NN) interactions, the PV component is expected to be suppressed by a factor of about 10^{-6} to 10^{-7} compared to the dominant parity-conserving (PC) part. These small effects can be isolated by considering observables that would vanish if parity was conserved. Examples of such observables include longitudinal and angular asymmetries in scattering with polarized beams or targets. Parity-violating effects can be enhanced by up to several orders of magnitude in more complex nuclei with closely-spaced energy levels. However, the theoretical description and interpretation in terms of two- and three-nucleon interactions becomes difficult and can introduce uncontrolled errors. The development of high-intensity sources and advances in the control of systematics have led recent efforts to be focussed on the detection of parity violation in few-nucleon systems. For these systems, the relation to the underlying nucleon-nucleon interactions is much easier to establish. Most of the corresponding experiments are related to processes at very low energies, e.g., the capture of cold neutrons. At these energies, effective field theories (EFT) provide a model-independent framework to analyze and interpret data, and to give theoretical uncertainty estimates.

Parity-violating interactions between nucleons originate in weak quark-quark interactions inside nucleons. While the weak interactions at the quark level are well-understood, their manifestation at the hadronic level is complicated by nonperturbative QCD confining the quarks into hadrons. Because the weak interactions between quarks are very short-ranged, the PV nucleon forces are sensitive to short-distance quark-quark correlations inside the nucleon. Hadronic parity violation can therefore be used to gain a better understanding of nonperturbative strong interactions. While the EFT analysis to be discussed does not directly relate hadronic parity violation to standard model degrees of freedom and their interactions, any quark-level calculation of hadronic parity violation must be consistent with the EFT results. One possible approach is given by lattice QCD [3].

[a] e-mail: mschindl@mailbox.sc.edu

2 Parity violation in pionless EFT

At energies well below the pion mass, it is possible to construct a theory in which the only dynamical degrees of freedom are nucleons, while pions and all heavier particles are "integrated out" and do not appear explicitly. In this so-called pionless EFT (EFT($\not\pi$)), the interactions between nucleons are described in terms of contact terms with an increasing number of derivatives and nucleon fields. Each contact term is accompanied by a so-called low-energy constant (LEC) that encodes the short-distance details that have been integrated out. Observables are calculated order-by-order in an expansion in a small parameter that is based on a ratio of scales. Pionless EFT has been very successfully applied in the PC sector, for reviews see, e.g., Refs. [4–6].

At leading order (LO) in EFT($\not\pi$), parity violation can be described in terms of five operators that parameterize S- to P-wave transitions [7–9]. The corresponding Lagrangian can be written as [10]

$$\begin{aligned}\mathcal{L}_{PV} = -\Big[& g^{({}^3S_1-{}^1P_1)} d_t^{i\dagger} \left(N^T \sigma_2 \tau_2 \, i\overleftrightarrow{D}_i N\right) + g^{({}^1S_0-{}^3P_0)}_{(\Delta I=0)} d_s^{A\dagger} \left(N^T \sigma_2 \, \vec{\sigma} \cdot \tau_2 \tau_A \, i\overleftrightarrow{D} N\right) \\ & + g^{({}^1S_0-{}^3P_0)}_{(\Delta I=1)} \, \epsilon^{3AB} \, d_s^{A\dagger} \left(N^T \sigma_2 \, \vec{\sigma} \cdot \tau_2 \tau^B \overleftrightarrow{D} N\right) + g^{({}^1S_0-{}^3P_0)}_{(\Delta I=2)} \, \mathcal{I}^{AB} \, d_s^{A\dagger} \left(N^T \sigma_2 \, \vec{\sigma} \cdot \tau_2 \tau^B \, i\overleftrightarrow{D} N\right) \\ & + g^{({}^3S_1-{}^3P_1)} \, \epsilon^{ijk} \, d_t^{i\dagger} \left(N^T \sigma_2 \sigma^k \tau_2 \tau_3 \overleftrightarrow{D}^j N\right)\Big] + \text{h.c.},\end{aligned} \tag{1}$$

where d_t and d_s represent auxiliary fields for two-nucleon states in the ${}^3S_1$ and ${}^1S_0$ channels, respectively [11], $a\mathcal{O}\overleftrightarrow{D}_i b = a\mathcal{O}D_i b - (D_i a)\mathcal{O}b$, with $\mathcal{O}$ some spin-isospin-operator, and $\mathcal{I} = \text{diag}(1, 1, -2)$. The five LECs $g^{(X-Y)}$ encode the short-distance details of the interactions and cannot be determined within the EFT framework. In principle, they can be calculated in terms of standard-model parameters. However, this requires a nonperturbative QCD calculation, which has not been feasible so far. As an alternative, the LECs can be determined from data once a sufficient number of experiments and corresponding calculations has been performed.

3 Two-nucleon systems

Two-nucleon systems provide the theoretically simplest cases to study PV nucleon-nucleon interactions. The PV longitudinal asymmetry A_L in scattering of a polarized beam on an unpolarized target is defined as

$$A_L = \frac{\sigma_+ - \sigma_-}{\sigma_+ + \sigma_-}\,, \tag{2}$$

where $\sigma_\pm$ denotes the total cross section for scattering with beam helicity ± 1. At LO in EFT($\not\pi$), the results for various beam and target nucleons, adjusted to the conventions of Eq. (1), are [9]

$$A_L^{nn} = -\sqrt{\frac{32M}{\pi}}\, p \left(g^{({}^1S_0-{}^3P_0)}_{(\Delta I=0)} - g^{({}^1S_0-{}^3P_0)}_{(\Delta I=1)} + g^{({}^1S_0-{}^3P_0)}_{(\Delta I=2)}\right), \tag{3}$$

$$A_L^{pp} = -\sqrt{\frac{32M}{\pi}}\, p \left(g^{({}^1S_0-{}^3P_0)}_{(\Delta I=0)} + g^{({}^1S_0-{}^3P_0)}_{(\Delta I=1)} + g^{({}^1S_0-{}^3P_0)}_{(\Delta I=2)}\right), \tag{4}$$

$$\begin{aligned}A_L^{np} = & -\sqrt{\frac{32M}{\pi}}\, p \frac{\frac{d\sigma^{{}^1S_0}}{d\Omega}}{\frac{d\sigma^{{}^1S_0}}{d\Omega} + 3\frac{d\sigma^{{}^3S_1}}{d\Omega}} \left(g^{({}^1S_0-{}^3P_0)}_{(\Delta I=0)} - 2g^{({}^1S_0-{}^3P_0)}_{(\Delta I=2)}\right) \\ & -\sqrt{\frac{32M}{\pi}}\, p \frac{\frac{d\sigma^{{}^3S_1}}{d\Omega}}{\frac{d\sigma^{{}^1S_0}}{d\Omega} + 3\frac{d\sigma^{{}^3S_1}}{d\Omega}} \left(g^{({}^3S_1-{}^1P_1)} + 2g^{({}^3S_1-{}^3P_1)}\right),\end{aligned} \tag{5}$$

where M denotes the nucleon mass and p is the nucleon momentum in the center-of-mass frame. The result of Eq. (4) neglects Coulomb effects. However, as shown in Ref. [9], they only amount to a correction of about 3% for the lowest energy that has been experimentally considered.

Parity-violating interactions induce a rotation of the nucleon spin in the transmission of a perpendicularly polarized beam through an unpolarized target. In the case of a neutron beam and a hydrogen target, the rotation angle per unit length at NLO is given by [12]

$$\frac{1}{\rho}\frac{d\phi^{np}_{\rm PV}}{dl} = 4\sqrt{2\pi M}\left(\frac{2g^{(^3S_1-^3P_1)} + g^{(^3S_1-^1P_1)}}{\gamma_t}\frac{Z_t+1}{2} + \frac{g^{(^1S_0-^3P_0)}_{(\Delta I=0)} - 2g^{(^1S_0-^3P_0)}_{(\Delta I=2)}}{\gamma_s}\frac{Z_s+1}{2}\right). \tag{6}$$

Here, ρ denotes the target density, $\gamma_{t/s}$ are the poles in the NN scattering amplitude in the S-wave spin-triplet/spin-singlet channel, and $Z = (1-\gamma_{t/s}r_{t/s})^{-1}$ with $r_{t/s}$ the effective range in the corresponding channel.

Additional observables are available when interactions with photons are considered. In the radiative capture of polarized neutrons on protons ($\vec{n}p \to d\gamma$), the direction of the outgoing photon can be used to define an angular asymmetry A_γ. Determination of A_γ has been the goal of the NPDGamma collaboration at the Spallation Neutron Source (SNS) at Oak Ridge National Laboratory [13]. With data taking being completed, the result of this measurement is eagerly being awaited. In EFT($\not{\pi}$) at LO, the result is given by [10, 14]

$$A_\gamma = \frac{4}{3}\sqrt{\frac{2}{\pi}}\frac{M^{\frac{3}{2}}}{\kappa_1(1-\gamma_t a_s)}g^{(^3S_1-^3P_1)}, \tag{7}$$

with a_s the scattering length in the $^1S_0$ channel and κ_1 the isovector nucleon anomalous magnetic moment.

The capture of unpolarized neutrons provides an independent and complementary observable in the induced circular polarization P_γ of the outgoing photons. For reversed kinematics, it is identical to the PV asymmetry A_L^γ in the deuteron breakup reaction with polarized photons ($\vec{\gamma}d \to np$),

$$A_L^\gamma = \frac{\sigma_+ - \sigma_-}{\sigma_+ + \sigma_-}, \tag{8}$$

where $\sigma_\pm$ stands for the total breakup cross section for photon helicities ± 1. A measurement of A_L^γ would require a high-intensity photon source with good control of the beam. The possibility of performing this experiment at an upgraded HIGS facility at the Triangle Universities Nuclear Laboratory has been under consideration. The PV asymmetry was calculated to NLO in EFT($\not{\pi}$) in Ref. [15]. Because the PV LECs have not been reliably determined, their values were related to different model estimates (see Ref. [15] for details). The energy dependence of A_L^γ using three different estimates for the LECs is shown in Fig. 1. The asymmetry is maximum at threshold, where the denominator in Eq. (8) vanishes. To determine at which energy to best perform the experiment, a very simplistic figure of merit of the form

$$f = \left(A_L^\gamma\right)^2 \times \sigma(\gamma d \to np) \tag{9}$$

was considered. As shown in Fig. 2, for the chosen values of the PV LECs, the figure of merit is maximized for photon energies in the range $2.259\,\text{MeV} < k < 2.264\,\text{MeV}$. It is important to note, though, that because of the uncertainty in the estimation of the LECs, the shown results are only representations of the possible values of A_L^γ and should only be considered as order-of-magnitude estimates.

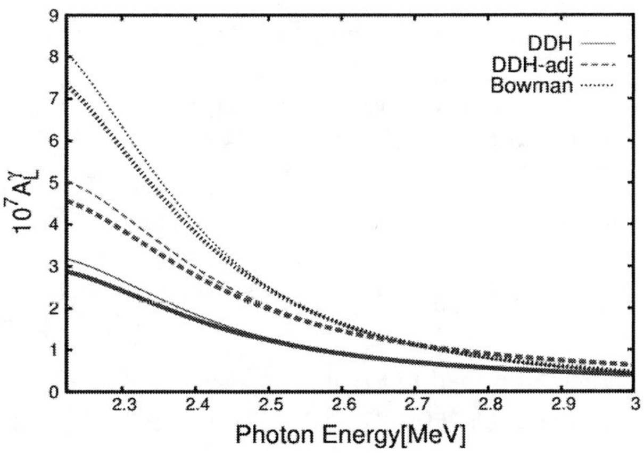

Figure 1. Energy dependence of the PV asymmetry A_L^γ for various LEC estimates from Ref. [15]. The thin and thick lines correspond to different EFT($\not\pi$) parametrizations and can be viewed as an indication of the theoretical uncertainty.

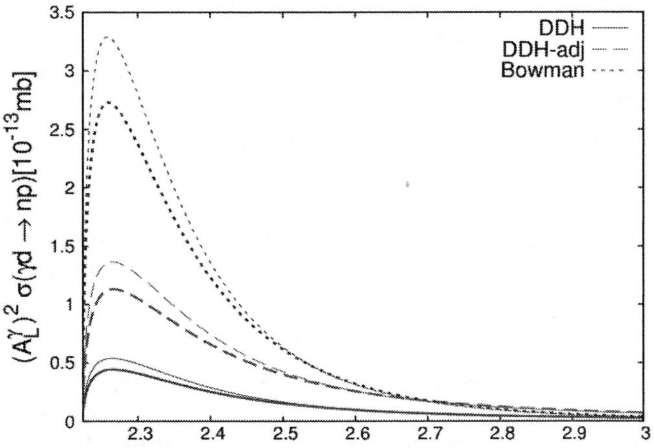

Figure 2. Figure of merit f for various LEC estimates from Ref. [15]. The thin and thick lines correspond to different EFT($\not\pi$) parametrizations and can be viewed as an indication of the theoretical uncertainty.

4 Three-nucleon systems

In the PC sector of pionless EFT, a direct application of the power-counting rules does not predict three-nucleon forces at LO. However, as shown in Ref. [16], an unphysical cutoff dependence occurs in the scattering amplitude for spin-doublet neutron-deuteron scattering. This cutoff dependence can be removed by introducing a three-nucleon contact term already at LO. The corresponding LEC can be determined from the experimental value of, e.g., the three-nucleon binding energy or the neutron-deuteron scattering length. A similar possible enhancement of a PV three-nucleon interaction to LO would require an additional experimental result to determine the corresponding LEC. Given the scarcity of data in the few-nucleon sector, this would complicate the analysis of hadronic parity violation considerably. However, the analysis of PV nucleon-deuteron scattering of Ref. [17] showed that such an enhancement of a PV three-nucleon interaction does not occur at LO or NLO. Therefore, up to an accuracy of roughly 10%, it is sufficient to analyze observables using only PV two-nucleon interactions.

The spin rotation angle for the transmission of a polarized neutron through a deuteron target was calculated in EFT($\not\pi$) to NLO in Ref. [12]. The result is

$$\frac{1}{\rho}\frac{d\phi^{nd}_{\mathrm{PV}}}{dl} = \Big([16 \pm 1.6] g^{(^3S_1-^1P_1)} + [34 \pm 3.4] g^{(^3S_1-^3P_1)} + [4.6 \pm 1.0](3g^{(^1S_0-^3P_0)}_{(\Delta I=0)} - 2g^{(^1S_0-^3P_0)}_{(\Delta I=1)})\Big)\frac{\mathrm{rad}}{\mathrm{MeV}^{\frac{1}{2}}}. \tag{10}$$

The theoretical uncertainties correspond to the most conservative of three error estimates based on power counting arguments, cutoff dependence, and different inputs used to fix the PC three-nucleon interaction.

5 Few-nucleon systems

A measurement of a PV asymmetry in the charge-exchange reaction $\vec{n}\,^3\mathrm{He} \to p\,^3\mathrm{H}$ is currently underway at the SNS. The corresponding calculation using an EFT with explicit pion degrees of freedom was performed in Ref. [18]. Experimental results also exist for the PV asymmetry in $\vec{p}\,^4\mathrm{He}$ scattering [19] and the neutron spin rotation angle in a $^4\mathrm{He}$ target [20]. No calculations in terms of two- and three-nucleon interactions have so far been performed, but will form an important component in improving our understanding of hadronic parity violation.

6 Conclusions and outlook

Because of their origin in an interplay between short-distance weak interactions and nonperturbative QCD, PV interactions between nucleons provide a unique probe of our understanding of the standard model. Considering PV observables in few-nucleon systems avoids possible complications from nuclear structure. At very low energies, EFT($\not\pi$) provides a model-independent framework for the analysis and interpretation of hadronic parity violation. At LO, the PV nucleon-nucleon interactions can be described by five independent operators. The corresponding LECs encode the short-distance details of the underlying interactions. Once the LECs have been determined from comparison with experimental results, their values provide a nontrivial target that any calculation in the nonperturbative regime has to reproduce. A preliminary calculation of a PV pion-nucleon coupling using lattice QCD was performed in Ref. [3], and further work in this direction will be of great importance.

Acknowledgements

This work was supported by the US Department of Energy under Grant No. DE-SC0010300. I am grateful to H. W. Grießhammer, D. R. Phillips, R. P. Springer, and J. Vanasse for their collaboration and many stimulating discussions.

References

[1] W.C. Haxton, B.R. Holstein, Prog.Part.Nucl.Phys. **71**, 185 (2013), 1303.4132
[2] M.R. Schindler, R.P. Springer, Prog.Part.Nucl.Phys. **72**, 1 (2013), 1305.4190
[3] J. Wasem, Phys.Rev. **C85**, 022501 (2012), 1108.1151

[4] S.R. Beane, P.F. Bedaque, W.C. Haxton, D.R. Phillips, M.J. Savage, In *Shifman, M. (ed.): At the frontier of particle physics, vol. 1* pp. 133–269. (2000), nucl-th/0008064
[5] P.F. Bedaque, U. van Kolck, Ann. Rev. Nucl. Part. Sci. **52**, 339 (2002), nucl-th/0203055
[6] L. Platter, Few Body Syst. **46**, 139 (2009), 0904.2227
[7] S.L. Zhu, C.M. Maekawa, B.R. Holstein, M.J. Ramsey-Musolf, U. van Kolck, Nucl.Phys. **A748**, 435 (2005), nucl-th/0407087
[8] L. Girlanda, Phys. Rev. **C77**, 067001 (2008), 0804.0772
[9] D.R. Phillips, M.R. Schindler, R.P. Springer, Nucl. Phys. **A822**, 1 (2009), 0812.2073
[10] M.R. Schindler, R.P. Springer, Nucl. Phys. **A846**, 51 (2010), 0907.5358
[11] D.B. Kaplan, Nucl. Phys. **B494**, 471 (1997), nucl-th/9610052
[12] H.W. Grießhammer, M.R. Schindler, R.P. Springer, Eur.Phys.J. **A48**, 7 (2012), 1109.5667
[13] M.T. Gericke, R. Alarcon, S. Balascuta, L. Barron-Palos, C. Blessinger et al., Phys.Rev. **C83**, 015505 (2011)
[14] M.J. Savage, Nucl. Phys. **A695**, 365 (2001), nucl-th/0012043
[15] J. Vanasse, M.R. Schindler, Phys. Rev. **C90**, 044001 (2014), 1404.0658
[16] P.F. Bedaque, H.W. Hammer, U. van Kolck, Nucl. Phys. **A676**, 357 (2000), nucl-th/9906032
[17] H.W. Grießhammer, M.R. Schindler, Eur. Phys. J. **A46**, 73 (2010), 1007.0734
[18] M. Viviani, A. Baroni, L. Girlanda, A. Kievsky, L. Marcucci et al. (2014), 1403.2267
[19] J. Lang, T. Maier, R. Muller, F. Nessi-Tedaldi, T. Roser et al., Phys.Rev.Lett. **54**, 170 (1985)
[20] W.M. Snow, C.D. Bass, T.D. Bass, B.E. Crawford, K. Gan et al., Phys.Rev. **C83**, 022501 (2011)

EPJ Web of Conferences **113**, 04027 (2016)
DOI: 10.1051/epjconf/201611304027

Proton-Deuteron Scattering and Test of Time-Reversal Invariance

Yu.N. Uzikov[1,2,a]

[1] *Laboratory of Nuclear Problems, JINR, Dubna, Moscow reg., 141980 Russia*
[2] *Department of Physics, Moscow State University, Moscow, 119991 Russia*

Abstract. The integrated proton-deuteron scattering cross section $\widetilde{\sigma}$ for transversely polarized protons (P_y^p) and tensor polarized deuterons (P_{xz}) constitutes a null test signal for time- reversal invariance violating but P-parity conserving effects. This cross section will be measured at COSY. Using the generalized optical theorem and Glauber theory we study the null-test observable $\widetilde{\sigma}$ for different types of T-odd P-even NN-interactions.The formalism includes full spin dependence of elementary pN-amplitudes and S- and D-components of the deuteron wave function.

1 Introduction

Time-invariance-violating (T-odd) P-parity conserving (P-even) (TVPC) interactions do not arise at the fundamental level within the standard model. This type of interaction can be generated by radiative corrections to the T-odd P-odd interaction discovered in the physics of kaons and B-mesons. However, in this case its intensity is too low to be observed in experiments at present [1]. Thus, observation of TVPC effects would be considered as an indication of physics beyond the standard model.

As was shown in Ref. [2], the total polarized cross section $\widetilde{\sigma}$ of the proton-deuteron scattering with vector polarization of the proton p_y^p and tensor polarization of the deuteron P_{xz} constitutes a null-test observable for TVPC effects. The dedicated experiment is planned at COSY [3] at proton beam energy 135 MeV. The first analysis of the TVPC null-test signal [4] was done within the nonmesonic deuteron breakup channel $pd \to ppn$ estimated in the single scattering approximation. Recently we used the spin-dependent formalism [5] of the Glauber theory to calculate the cross section $\widetilde{\sigma}$ [6] and "null-combinations" of some differential spin observables of the *pd* elastic scattering [7] which deviate from zero if the TVPC effects occur. The formalism includes full spin dependence of elementary pN-amplitudes and S- and D-components of the deuteron wave function. This formalism allows one to explain existing data on the non-polarized differential cross section and spin observables of the elastic *pd* scattering at 135 MeV [8]. Here we consider some qualitative arguments concerning the ρ-meson contribution to $\widetilde{\sigma}$ and briefly explain the role of the deuteron D-wave.

[a] e-mail: uzikov@jinr.ru

2 Elements of formalism

Time-reversal symmetry conserving and P-parity conserving (TCPC or T-even P-even) interactions lead to the following transition amplitude of the elastic pd scattering at 0 degree [9]

$$e'^*_\beta M(0)^{TCPC}_{\alpha\beta} e_\alpha = g_1[\mathbf{e}\,\mathbf{e}'^* - (\mathbf{me})(\mathbf{me}'^*)] + g_2(\mathbf{me})(\mathbf{me}'^*) +$$
$$ig_3\{\sigma[\mathbf{e}\times\mathbf{e}'^*] - (\sigma\mathbf{m})(\mathbf{m}\cdot[\mathbf{e}\times\mathbf{e}'^*])\} + ig_4(\sigma\mathbf{m})(\mathbf{m}\cdot[\mathbf{e}\times\mathbf{e}'^*]), \quad (1)$$

where $\mathbf{e}$ ($\mathbf{e}'$) is the polarization vector of the initial (final) deuteron, $\mathbf{m}$ is the unit vector along the beam momentum, σ is the Pauli matrix, g_i ($i = 1,\ldots,4$) are complex amplitudes. To the right-hand side of Eq.(1) one can add the TVPC (T-odd P-even) term in a very general form

$$e'^*_\beta M(0)^{TVPC}_{\alpha\beta} e_\alpha = \widetilde{g}\{(\sigma\cdot[\mathbf{m}\times\mathbf{e}])(\mathbf{m}\cdot\mathbf{e}'^*) + (\sigma\cdot[\mathbf{m}\times\mathbf{e}'^*])(\mathbf{m}\cdot\mathbf{e})\}, \quad (2)$$

where $\widetilde{g}$ is the TVPC transition amplitude. The matrix elements of the operators (1), (2) are

$$<\mu' = \frac{1}{2}, \lambda' = 1|M^{TCPC}|\mu = \frac{1}{2}, \lambda = 1 >= g_1 + g_4, \quad (3)$$

$$<\mu' = \frac{1}{2}, \lambda' = -1|M^{TCPC}|\mu = \frac{1}{2}, \lambda = -1 >= g_1 - g_4, \quad (4)$$

$$\langle\mu|' = \frac{1}{2}, \lambda' = 0|M^{TCPC}|\mu = \frac{1}{2}, \lambda = 0 >= g_2, \quad (5)$$

$$<\mu' = \frac{1}{2}, \lambda' = 0|M^{TCPC} + M^{TVPC}|\mu = -\frac{1}{2}, \lambda = 1 >= \sqrt{2}g_3 + i\sqrt{2}\widetilde{g}, \quad (6)$$

$$<\mu' = \frac{1}{2}, \lambda' = -1|M^{TCPC} + M^{TVPC}|\mu = -\frac{1}{2}, \lambda = 0 >= \sqrt{2}g_3 - i\sqrt{2}\widetilde{g}. \quad (7)$$

where μ (μ') and λ (λ') are spin projections of the initial (final) proton and deuteron on the beam direction, respectively. All diagonal matrix elements of the M^{TVPC} operator are zeros.

The total cross section of the pd scattering has the form [8]

$$\sigma_{tot} = \sigma_0 + \sigma_1\mathbf{p}^p\cdot\mathbf{p}^d + \sigma_2(\mathbf{p}^p\cdot\mathbf{m})(\mathbf{p}^d\cdot\mathbf{m}) + \sigma_3 P_{zz} + \widetilde{\sigma} p_y^p P_{xz}^d, \quad (8)$$

where $\mathbf{p}^p$ ($\mathbf{p}^d$) is the vector polarization of the initial proton (deuteron) and P_{zz} and P_{xz} are the tensor polarizations of the deuteron. The OZ axis is directed along the proton beam momentum $\mathbf{m}$, OY$\uparrow\uparrow$ $\mathbf{p}^p$, OX $\uparrow\uparrow$ $[\mathbf{p}^p\times\mathbf{m}]$. In Eq. (8) the terms σ_i with $i = 0, 1, 2, 3$ are non-zero only for T-even P-even interactions corresponding to Eq. (1) and the last term $\widetilde{\sigma}$ constitutes a null-test signal of T-invariance violation with P-parity conservation. Using the generalized optical theorem we find $\widetilde{\sigma} = -4\sqrt{\pi} Im \frac{2}{3}\widetilde{g}$.

Hadronic amplitudes of pN scattering are taken as [5]

$$M_N(\mathbf{p},\mathbf{q};\sigma,\sigma_N) = A_N + C_N\sigma\hat{\mathbf{n}} + C'_N\sigma_N\hat{\mathbf{n}} + \sum_{l=n,q,k} B^l_N(\mathbf{q})(\sigma\hat{\mathbf{l}})(\sigma_N\hat{\mathbf{l}}), \quad (9)$$

where $\hat{\mathbf{q}}$, $\hat{\mathbf{k}}$ and $\hat{\mathbf{n}}$ are defined as unit vectors along the vectors $\mathbf{q} = \mathbf{p} - \mathbf{p}'$, $\mathbf{k} = \mathbf{p} + \mathbf{p}'$ and $\mathbf{n} = [\mathbf{k}\times\mathbf{q}]$, respectively; $\mathbf{p}$ ($\mathbf{p}'$) is the initial (final) proton momentum; σ_N is the Pauli matrix acting on the spin state of the nucleon N. We consider the following terms of the TVPC NN interaction which were under discussion in Ref. [4]:

$$t_{pN} = h_N[(\sigma\cdot\mathbf{k})(\sigma_N\cdot\mathbf{q}) + (\sigma_N\cdot\mathbf{k})(\sigma\cdot\mathbf{q}) - \frac{2}{3}(\sigma_N\cdot\sigma)(\mathbf{k}\cdot\mathbf{q})]/m_p^2 + \quad (10)$$
$$+g_N[\sigma\times\sigma_N]\cdot[\mathbf{q}\times\mathbf{k}]/m_p^2 + g'_N(\sigma - \sigma_N)\cdot i\,[\mathbf{q}\times\mathbf{k}][\tau\times\tau_N]_z/m_p^2.$$

Here σ (σ_N) is the Pauli matrix acting on the spin state of the proton (nucleon $N = p, n$), τ (τ_N) is the corresponding matrix acting on the isospin state; m_p is the proton mass. In the framework of the phenomenological meson exchange interaction the term g' corresponds to ρ-meson exchange, and h-term provides the axial meson h_1 exchange.

2.1 g'-term

The g' term contributes only to the charge exchange transitions, because the non-zero matrix elements of the isospin-operator connected with the g' term in Eq. (10) are the following

$$< n,p|[\tau \times \tau_N]_z|p,n >= -i2, \quad < p,n|[\tau \times \tau_N]_z|n,p >= i2. \tag{11}$$

The isospin matrix element of the C-odd isospin operator $\mathcal{T}_z = [\tau \times \tau_N]_z$ in Eq. (11) changes the sign under replacement $p \leftrightarrow n$. In contrast, the similar matrix elements for T-even P-even (strong) NN-interaction are equal one to other. This difference is one cause for the vanishing of the amplitude $\widetilde{g}$ for the double scattering mechanism of the process $pd \to pd$. As was shown in Ref. [6], the g'-term gives zero contribution to $\widetilde{g}$ within the Glauber model. Below we discuss this observation briefly.

The TVPC charge-exchange Glauber operator of the double scattering has a form [6]

$$O^c_{TVPC} = -\frac{1}{2}[M_{np\to pn}(\mathbf{q}_2)t_{pn\to np}(\mathbf{q}_1) + t_{np\to pn}(\mathbf{q}_2)M_{pn\to np}(\mathbf{q}_1)], \tag{12}$$

where $\mathbf{q}_1 = \mathbf{q}/2+\mathbf{q}'$ is the transferred momentum in the first and $\mathbf{q}_2 = \mathbf{q}/2-\mathbf{q}'$ in the second collision and $\mathbf{q}$ is the total transferred momentum. For the next step one has to calculate the matrix element of the operator (12) over the deuteron states $\psi(\mathbf{r})$ with the factor $\exp(i\mathbf{q}'\mathbf{r})$ and integrate over $\mathbf{q}'$. Under the sign of this integral the operator (12) is not changed after the substitution $\mathbf{q}_1 \leftrightarrow \mathbf{q}_2$ [5, 6]. Therefore, one may add to the right side of Eq. (12) the term $O^c_{TVPC}(1 \leftrightarrow 2)$ and divide the obtained sum by a factor of 2. In collinear kinematics ($\mathbf{q} = 0$), this symmetry and linear dependence of g'-term on $[\mathbf{q}\times\mathbf{k}]$ lead to cancellation of the spin-independent term A_N in the transition operator (12). The same is true for the B_N terms in Eq. (9). Thus, only C_N and C'_N terms may contribute as [6]

$$O^c_{TVPC} = \frac{g'}{\Pi}\Big[C_n(\sigma\cdot\hat{\mathbf{n}}_1)(\sigma_n-\sigma_p)\cdot\mathbf{n}_1 - C'_n(\sigma_n\cdot\hat{\mathbf{n}}_1)(\sigma_p\cdot\mathbf{n}_1) + C'_n\mathbf{n}_1\hat{\mathbf{n}}_1 +$$
$$C_p(\sigma\cdot\hat{\mathbf{n}}_1)(\sigma_p-\sigma_n)\cdot\mathbf{n}_1 - C'_p(\sigma_p\cdot\hat{\mathbf{n}}_1)(\sigma_n\cdot\mathbf{n}_1) + C'_p\mathbf{n}_1\hat{\mathbf{n}}_1\Big], \tag{13}$$

where $\mathbf{n}_1 = [\mathbf{k}\times\mathbf{q}']$, $\hat{\mathbf{n}}_1 = \mathbf{n}_1/|\mathbf{n}_1|$ and $\mathbf{q}' = \mathbf{q}_2 = -\mathbf{q}_1$; Π is a constant. In Eq. (13) the C'_N-terms do not contain the proton beam spin σ. We can show that this is a consequence of Eqs. (11). According to Eqs. (6), (7), it means that the contribution of the $C'_N\times g'$ term to the amplitude $\widetilde{g}$ is zero. Furthermore, due to Eqs. (11) the remaining terms with C_N in Eq. (13) contain the difference $\sigma_n - \sigma_p$, but not the sum. These terms can be rewritten as $\mathbf{V}_p\sigma_p + \mathbf{V}_n\sigma_n = 2(\mathbf{V}_p + \mathbf{V}_n)(\sigma_p + \sigma_n) \equiv 0$ [6]. Thus, the contribution of the operator (13) to the amplitude $\widetilde{g}$ vanishes and this fact is directly connected with Eqs. (11).

Strong suppression of the contribution of the ρ-meson as compared to the axial h_1 meson was found numerically in the Faddeev calculations [10] of the null-test signal for the *nd* scattering at 100 keV, but no explanation of this result was offered. We suppose that the cause for this suppression is the same spin-isospin structure of the scattering amplitude which leads to the vanishing ρ-meson contribution within the Glauber approach.

2.2 h- and g-terms

For the single scattering mechanism the amplitude $\widetilde{g}$ vanishes within the Glauber theory. Using Eqs. (3)-(7) for the double scattering mechanism with pN-amplitudes (9) and (10) we find for the h- and g-terms in Eq.(10) that all T-even P-even amplitudes of the pd-scattering are zeros: $g_1 = g_2 = g_3 = g_4 = 0$. Furthermore, we find for the TVPC amplitude

$$\widetilde{g} = \frac{i}{4\pi m_p}\int_0^\infty dq q^2[S_0^{(0)}(q) - 2\sqrt{2}S_2^{(1)}(q)][C'_n(q)(g_p-h_p) + C'_p(q)(g_n-h_n)], \tag{14}$$

where $S_0^{(0)} = \int_0^\infty dru^2(r)j_0(qr)$ and $S_2^{(1)}(q) = 2\int_0^\infty dru(r)w(r)j_2(qr)$ are the elastic form factors of the deuteron, and $u(r)$ and $w(r)$ are the S-wave and D-wave $w(r)$ of the deuteron, respectively, [6]. We can show that the S-D wave interference, not considered in Ref.[6], considerably diminishes the null-test signal $\widetilde{\sigma}$ at the energies of the planned COSY experiment [3] ~100 MeV as compared to the pure S-wave contribution and provides an enhancement at 700-800 MeV.

3 Summary

In contrast to Ref. [4] we show, using the optical theorem, that within the single scattering approximation the null-test observable $\widetilde{\sigma}$ is zero. Our result obtained within the Glauber theory is formulated by Eq. (14). Only the amplitude C'_N appears in Eq.(14) whereas other T-even P-even pN amplitudes, which were found in Ref. [4] to contribute to the TVPC null-test signal, are absent in Eq. (14). Furthermore, we find the deuteron D-wave gives a valuable contribution to the null-test signal for the case of the h- and g-type of interaction. The g'-term caused by the $\rho-$ meson exchange in the TVPC NN-interaction makes a zero contribution to $\widetilde{g}$ and this result is true in the case when both the S- and D-components of the deuteron wave function are taken into account. We discuss some symmetry arguments to clarify a cause for the vanishing contribution of the g'-term. The g'-optical potential [11] and the corresponding coupling constant of the $\rho-$meson to the nucleon $\bar{g}_\rho$ is widely used as a measure of intensity of the TVPC effects [12, 13]. Since the g'-term gives zero contribution to $\widetilde{\sigma}$ within the Glauber theory, this parameter cannot be applied straightforwardly for the nucleon-deuteron scattering as a scale of the TVPC interactions at large enough energies. However, the g'-term can give contribution to the null-test signal $\widetilde{\sigma}$ if this interaction is included into the deuteron bound state [6]. One-pion exchange is excluded from the TVPC NN-interaction [12], however, two-pion exchange probably contributes similarly to the P-violating pp-interaction [14]. Finally, TVPC NN forces can contribute to the electromagnetic p-d interaction due to the toroidal quadrupole form factor of the deuteron [15].

References

[1] I. B. Khriplovich, Nucl. Phys. B **352**, 385 (1991).
[2] H. E. Conzett, Phys. Rev. C **48**, 423 (1993).
[3] COSY Proposal N 215,
`[http://apps.fz-juelich.de/pax/paxwiki/images/8/8c/215-TRI_Prop_sum.pdf]`
[4] M. Beyer, Nucl. Phys. A **560**, 895 (1993).
[5] M. N. Platonova, V. I. Kukulin, Phys. Rev. C **81**, 014004 (2010).
[6] Yu. N. Uzikov, A. A. Temerbayev, Phys.Rev. C **C 92**, 014002 (2015).
[7] A. A. Temerbayev, Yu. N. Uzikov, Izv. Ross. Akad. Nauk Ser. Fiz. **79**, N4, 535 (2015) [Bull. Russ. Acad. Sci. Phys. **79**,N4, 493 (2015)].
[8] A. A. Temerbayev, Yu. N. Uzikov, Yad. Fiz. **78**, 38 (2015).
[9] M. P. Rekalo N. M. Piskunov, I. M. Sitnik, Few-Body Syst. **23**, 187 (1998).
[10] Y. H. Song, R. Lazauskas and V. Gudkov, Phys. Rev. C **84**, 025501 (2011).
[11] J. Engel, C. R. Gould, V. Hnizdo, Phys. Rev. Lett. **73**, 3508 (1994).
[12] M. Simonius, Phys. Rev. Lett. **78**, 4161 (1997).
[13] P. R. Huffman *et al.*, Phys. Rev. C **55**, 2684 (1997).
[14] J.de Vries, Ulf-G. Meissner, E. Epelbaum, N. Keiser, Eur. Phys. J. A **49**, 149 (2013).
[15] E. Mereghetti, J. de Vries, R.G.E. Timmermans, U. van Kolck, Phys. Rev. C **88**, 034001 (2013).

EPJ Web of Conferences **113**, 04028 (2016)
DOI: 10.1051/epjconf/201611304028

Neutrino pion production off deuteron

F. Myhrer[1,a] and S. Pastore[1,2,b]

[1] *Dept. Physics and Astronomy, Univ. South Carolina, Columbia, SC 29208*
[2] *Theoretical Division, Los Alamos National Laboratory, Los Alamos, NM 87545, USA*

Abstract. Experimental investigations of neutrino properties, using neutrino beams generated at accelerators facilities, necessitate a detailed and precise knowledge of neutrino-nucleus reaction mechanisms. In the energy region of nuclear quasi-elastic scattering, pion-production reactions constitute an important background process. A theoretical understanding of these processes is then required in order to correctly determine the produced neutrino energy spectrum. In the first stage of our research project, we study neutrino induced pion-production off deuterons. The choice of the deuteron minimizes the complications of the nuclear dynamics associated with larger nuclear systems. We evaluate the pion-production reaction near threshold using heavy baryon chiral perturbation theory.

1 Introduction

Neutrino beams generated in accelerator-based neutrino experiments [1] are not mono-energetic. Their energy spectrum spans from a few hundred MeV to several GeV and the neutrino energy spectrum is reconstructed from final state observations of neutrino-nucleus reactions. The reconstruction procedure relies on an accurate theoretical understanding of the involved neutrino-nucleus reactions along with a quantitative evaluation of the associated cross-sections. The relevant reactions occurring in this energy regime are nuclear quasi-elastic scattering, pion-productions and deep inelastic scattering. A detailed and precise knowledge of the neutrino-nucleus reaction mechanisms is required in order to determine the neutrino energy spectrum with sufficient accuracy [2].

Theoretical studies of neutrino scattering off light nuclei (A $\leq$ 16) in the quasi-elastic regime [3–6], show that many-body terms in both the nuclear Hamiltonian and electro-weak currents provide sizable, non-negligible contributions to the evaluated cross sections. An important background process to the quasi-elastic reactions is generated by neutrino-nucleus pion-production reactions. In the following, we focus on the neutrino induced pion-production reactions off light nuclei near threshold, where an effective field theory approach is applicable. In order to minimize nuclear dynamics effects, we consider pion-production rections in neutrino-scattering off deuterons and focus on charge current reactions $\nu_l + d \to l + \pi + N + N$.

[a] e-mail: myhrer@physics.sc.edu
[b] e-mail: saori@lanl.gov

2 The chiral effective field theory approach

We use heavy baryon chiral perturbation theory (ChPT) to evaluate the transition amplitudes entering pion-production in neutrino scattering off deuteron. ChPT allows a systematic evaluation of high order corrections to the leading order (LO) amplitudes. In this theory, which is applicable for long wave length probes, the short distance hadronic structure and dynamics are parametrized by low-energy constants (LECs). ChPT provides effective interactions that preserve the symmetries and pattern of symmetry breaking exhibited by the underlying theory of QCD. We use an effective hadronic Lagrangian, $\mathcal{L}$, which is written in terms of pion and nucleon fields. It can be expressed as an expansion in $(Q/\Lambda_\chi)^n \ll 1$, where Q represents a typical four-momentum transfer in the process under study, or the pion mass, m_π, and $\Lambda_\chi \sim 1$ GeV is the chiral symmetry-breaking energy scale. The $\mathcal{L}^{(n)}$ terms of order $(Q/\Lambda_\chi)^n$ entering the effective Lagrangian can be arranged according to their relevance as:

$$\mathcal{L} = \mathcal{L}_{\pi\pi}^{(2)} + \mathcal{L}_{\pi N}^{(1)} + \mathcal{L}_{\pi N}^{(2)} + \mathcal{L}_{\pi N}^{(3)} + \cdots . \qquad (1)$$

This calculational scheme has been utilized to study electro-induced pion-production rections (see, *e.g.*, [7–9]). Here, we evaluate the neutrino-induced pion-production transition amplitude including up to next-to-next-to-leading order (NNLO) contributions in the chiral expansion. Leading order (LO) terms—or impulse approximation terms—-where pions are produced off single nucleons, contain the LO vertices constructed from the the $\mathcal{L}_{\pi\pi}^{(2)}$ and $\mathcal{L}_{\pi N}^{(1)}$ terms in the chiral Lagrangian, for which we take the expressions given in Ref. [10]. NLO and NNLO corrections to the LO one-body operator are evaluated using higher order terms in the chiral Lagrangian, namely $\mathcal{L}_{\pi N}^{(2)}$ and $\mathcal{L}_{\pi N}^{(3)}$, which are taken from Refs. [10, 11]. Physically, this corresponds to accounting for nucleon recoil corrections of order m_N^{-2}. At NNLO one-body currents include also loop corrections to the LO diagrams. The ultraviolet divergences of these loops are regulated by LECs contained in $\mathcal{L}_{\pi N}^{(3)}$. In addition to renormalizing the nucleon mass etc., these loop-diagrams naturally generate lowest order vertex form factors. For example, they generate the nucleon isovector charge form factor expanded to order q^2—here q is the momentum transferred to the nucleon—-where the finite part of the corresponding LEC is given by the isovector charge r.m.s. radius of the nucleon.

Two-nucleon currents enter at NNLO. Hadronic axial currents contributing to the pion-production reaction are evaluated from tree-level diagrams—some of which are illustrated in the panels of Fig. 1—while tree-level hadronic vector currents are derived from the diagrams illustrated in the panels of Fig. 2. It is to be noted that at this order, i.e., NNLO, all the LECs are determined and the theory has predictive power.

The next step in this project, is to evaluate the calculated transition operators sandwiched between initial deuteron wave functions and nucleon-nucleon final scattering states.

Acknowledgements

This work is supported in part by NSF Grant No. PHY-1068305.

References

[1] The LBNE Collaboration, arXiv:1307.7335 (2013).

[2] H. Gallagher, G. Garvey and C.P. Zeller, Annu. Rev. Nuc. Part. Sci., **61**, 355 (2011); G. Garvey, talk at Fermi National Laboratory (2011), http://minerva-docdb.fnal.gov/cgi-bin/ShowDocument?docid=5878 .

[3] A. Lovato, S. Gandolfi, J. Carlson, S. Pieper and R. Schiavilla, Phys. Rev. C **91**, 062501 (2015).

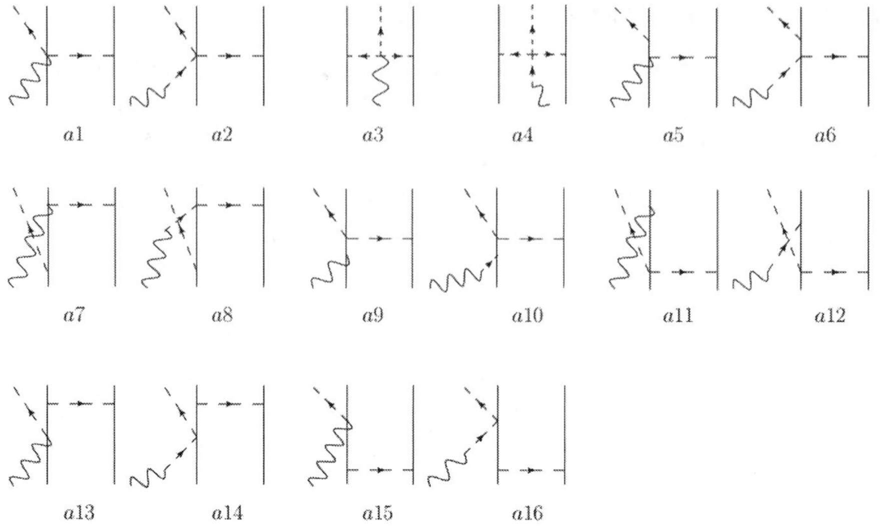

Figure 1. Sample of diagrams contributing to the two-nucleon axial current. The wiggly line represent incoming axial probes, while pions and nucleons are indicated by dashed and solid lines, respectively. The verticies entering the associated amplitudes are derived from $\mathcal{L}_{\pi\pi}^{(2)}$, $\mathcal{L}_{\pi N}^{(1)}$ and $\mathcal{L}_{\pi N}^{(2)}$ [10].

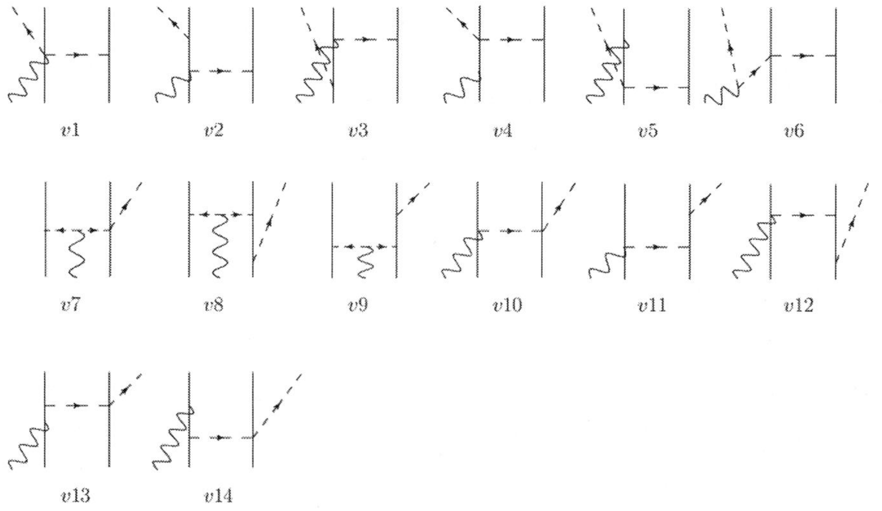

Figure 2. Diagrams contributing to the two-nucleon vector current. Notation is as in Fig. 1

[4] M. Martini, M. Ericson, G. Chanfray and J. Marteau, Phys. Rev. C **80**, 065501 (2009); M. Martini and M. Ericson, Phys. Rev. C **90**, 025501 (2014).
[5] G.D. Megias *et al.*, Phys. Rev. D **91**, 073004 (2015).
[6] R. Gran, J. Nieves, F. Sanchez, M.J. Vicente Vacas, Phys. Rev. D **88**, 113007 (2013)
[7] S.R. Beane, C.Y. Lee and U. van Kolck, Phys.Rev. C **52**, 2914 (1995).
[8] S.R. Beane, V. Bernard, T.-S.H. Lee, Ulf-G. Meißner and U. van Kolck, Nucl.Phys. A **618**, 381 (1997).
[9] H. Krebs, V. Bernard, and Ulf-G. Meißner, Nucl.Phys. A **713** 405 (2003).
[10] V. Bernard, N. Kaiser and U.-G. Meißner, Int. J. Mod. Phys. E **4**, 193 (1995).
[11] N. Fettes, U.-G. Meißner and S. Steininger, Nucl. Phys. A **640**,199 (1998).

EPJ Web of Conferences **113**, 04029 (2016)
DOI: 10.1051/epjconf/201611304029

Break-up channels in muon capture on ^{3}He

J. Golak[1,a], R. Skibiński[1,b], H. Witała[1,c], K. Topolnicki[1,d], A. E. Elmeshneb[1,e], H. Kamada[2,f], A. Nogga[3,g], and L. E. Marcucci[4,h]

[1] *M. Smoluchowski Institute of Physics, Jagiellonian University, PL-30059 Kraków, Poland*
[2] *Department of Physics, Faculty of Engineering, Kyushu Institute of Technology, Kitakyushu 804-8550, Japan*
[3] *IKP-3, IAS-4, JCHP and JARA-HPC, Forschungszentrum Jülich, D-52425 Jülich, Germany*
[4] *Department of Physics, University of Pisa, IT-56127 Pisa, Italy and INFN-Pisa, IT-56127 Pisa, Italy*

Abstract. The $\mu^- +^3 \mathrm{He} \rightarrow \nu_\mu + n + d$ and $\mu^- +^3 \mathrm{He} \rightarrow \nu_\mu + n + n + p$ capture reactions are studied under full inclusion of final state interactions with the AV18 nucleon-nucleon potential, augmented by the Urbana IX three-nucleon force, and employing the single nucleon weak current operator. We give first realistic estimates of the total capture rates: 544 s^{-1} and 154 s^{-1} for the $n + d$ and $n + n + p$ channels, respectively. Our results are compared with the most recent experimental data, finding a rough agreement for the total capture rates, but failing to reproduce the differential capture rates.

1 Introduction

Muon capture reactions on light nuclei have been studied intensively both experimentally and theoretically for many years and information on earlier achievements can be found in Refs. [1–4]. Recent theoretical efforts, presented for example in Refs. [5–8], focused on the $\mu^- +^2 \mathrm{H} \rightarrow \nu_\mu + n + n$ and $\mu^- +^3 \mathrm{He} \rightarrow \nu_\mu +^3 \mathrm{H}$ reactions. These calculations were based on various dynamical inputs, representing the so-called phenomenological approach, the "hybrid" chiral effective field theory (χEFT) approach and also the "non-hybrid" χEFT approach. A good agreement between the results obtained within different approaches was found, as well as between theoretical predictions and available experimental data.

In Ref. [9] we joined our expertise: in momentum space treatment of electromagnetic processes [10, 11] and in the potential model approach developed in Ref. [5] to perform a systematic study of all the $A = 2$ and $A = 3$ muon capture reactions. We compared results of new calculations carried out in the momentum space for the $\mu^- +^2 \mathrm{H} \rightarrow \nu_\mu + n + n$ and $\mu^- +^3 \mathrm{He} \rightarrow \nu_\mu +^3 \mathrm{H}$ reactions with

[a] e-mail: jacek.golak@uj.edu.pl
[b] e-mail: roman.skibinski@uj.edu.pl
[c] e-mail: henryk.witala@uj.edu.pl
[d] e-mail: kacper.topolnicki@uj.edu.pl
[e] e-mail: alaa.elmeshneb@uj.edu.pl
[f] e-mail: kamada@mns.kyutech.ac.jp
[g] e-mail: a.nogga@fz-juelich.de
[h] e-mail: marcucci@df.unipi.it

those of Ref. [5], finding a very good agreement not only for the predictions obtained with the single nucleon current operator but also for the case, when the meson-exchange currents from Ref. [12] (Eqs. (4.16)–(4.39), without Δ-isobar contributions) were included. These calculations employed the AV18 nucleon-nucleon (NN) potential [13] supplemented with the Urbana IX three-nucleon (3N) force [14]. Results for the break-up channels in muon capture on ^{3}He obtained with the same forces will be demonstrated in the next section.

Recently, improved chiral NN forces were published in Refs. [15, 16] and thus we would like to present here *preliminary* results for the $\mu^- +^2 \mathrm{H} \rightarrow \nu_\mu + n + n$ and $\mu^- +^3 \mathrm{He} \rightarrow \nu_\mu +^3 \mathrm{H}$ reactions. The calculations used the single nucleon current operator [9], since no consistent current operators are available for these new potentials at the moment. Even if these reactions are not in the focus of the present paper and the calculations are not yet consistent, we believe they might be interesting for the reader. The spread of the results due to the different regulator parameters gets very narrow for the N4LO predictions and we obtain $\Gamma_d^{1/2} \in [384.6, 386.3]\,\mathrm{s}^{-1}$ for the total doublet capture rate in the case of the $\mu^- +^2 \mathrm{H} \rightarrow \nu_\mu + n + n$ reaction and $\Gamma \in [1285, 1308]\,\mathrm{s}^{-1}$ for the total capture rate in the case of the $\mu^- +^3 \mathrm{He} \rightarrow \nu_\mu +^3 \mathrm{H}$ process. Future calculations will combine the presently included dynamical ingredients with consistent two-body currents and three-nucleon forces, to produce complete chiral EFT calculations at high orders. The convergence pattern of such calculations can then be used to assess the theoretical uncertainty in predictions for the muon capture rates.

2 Results for the $\mu^- +^3 \mathrm{He} \rightarrow \nu_\mu + n + d$ and $\mu^- +^3 \mathrm{He} \rightarrow \nu_\mu + n + n + p$ reactions

In Ref. [9] we provided, for the first time, predictions for the total and differential capture rates of the reactions $\mu^- +^3 \mathrm{He} \rightarrow \nu_\mu + n + d$ and $\mu^- +^3 \mathrm{He} \rightarrow \nu_\mu + n + n + p$. To this aim we first considered kinematically allowed regions for these two reactions (see Fig. 1) and made sure that the non-relativistic kinematics could be used consistently with the corresponding non-relativistic dynamics. Our essential results for the differential capture rates are shown in Fig. 2. We see that inclusion of the 3N force reduces the peak heights by about 20 %. It is also clear that the main contributions to the total decay rates come from the regions with highest neutrino energies.

The corresponding total decay rates for the two break-up channels are displayed in Table 1. The 3N force effects for the two reactions are comparable and amount to about 10 %. In the same table we collect also the experimental results from Refs. [17–20] and find a fair agreement with most of them.

In Ref. [9] we analyzed also most recent data on more differential capture rates published in Ref. [20]. The data in Ref. [20] were evaluated by means of two methods (denoted by "md I" and "md II" in Table 1). In Fig. 3 we show the second set of data points for the averaged capture rates and our theoretical results obtained without and with inclusion of 3N force effects. For the $\mu^- +^3 \mathrm{He} \rightarrow \nu_\mu + n + d$ reaction our predictions in the whole range of the deuteron energies clearly underestimate the data by nearly a factor of 2. In the case of the $\mu^- +^3 \mathrm{He} \rightarrow \nu_\mu + n + n + p$ process we underestimate the data for smaller proton energies and overshoot them for the higher proton energies.

It is very important to realize that *only these data points* were used to obtain the total capture rates by means of simple extrapolations. If we combine information from Figs. 1 and 2, we see that the interval $E_d \in [13, 30]$ MeV corresponds to the interval $E_\nu \in [10, 75]$ MeV. Thus one misses the bulk of the contribution to the total capture rate. The situation for the $\mu^- +^3 \mathrm{He} \rightarrow \nu_\mu + n + n + p$ process is very similar. Our results raise doubts if the extrapolations performed in [20] lead to correct results for the total capture rates.

Undeniably, further theoretical work and new precision measurements in the whole kinematical region are needed to improve our knowledge about break-up channels in muon capture on ^{3}He.

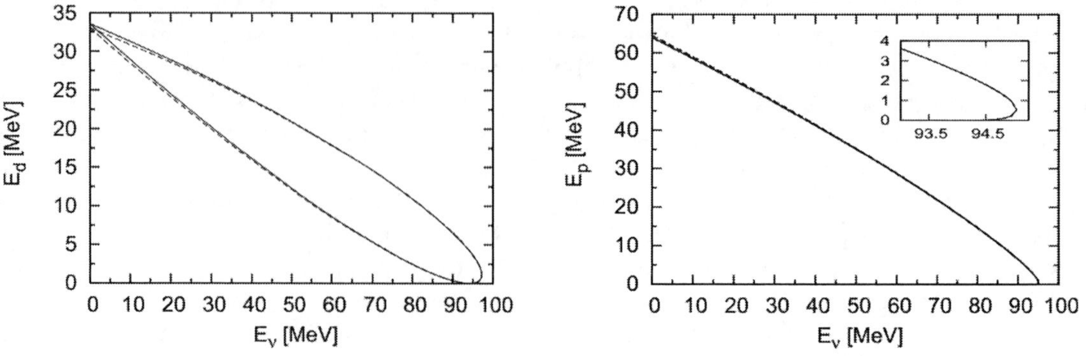

Figure 1. The kinematically allowed regions in the $E_\nu - E_d$ plane for the $\mu^- +^3$ He $\longrightarrow \nu_\mu + n + d$ process (left panel) and in the $E_\nu - E_p$ plane for the $\mu^- +^3$ He $\longrightarrow \nu_\mu + n + n + p$ process (right panel) calculated relativistically (solid curve) and nonrelativistically (dashed curve).

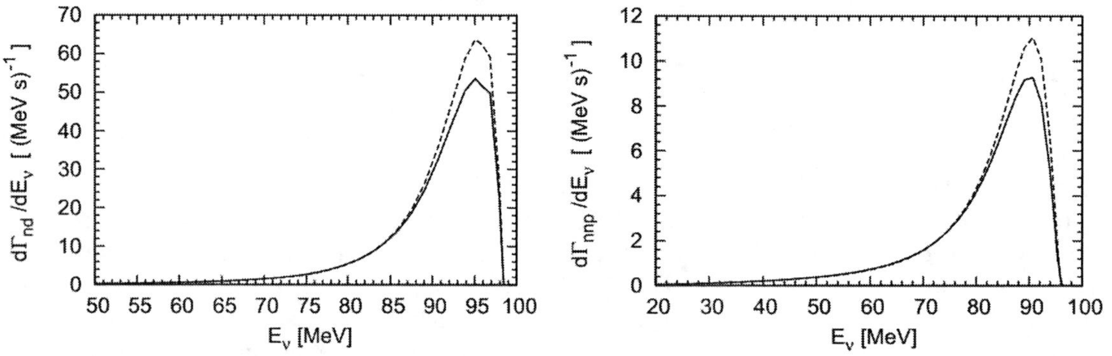

Figure 2. The differential capture rates $d\Gamma_{nd}/dE_\nu$ for the $\mu^- +^3$ He $\longrightarrow \nu_\mu + n + d$ process (left panel) and the differential capture rates $d\Gamma_{nnp}/dE_\nu$ for the $\mu^- +^3$ He $\longrightarrow \nu_\mu + n + n + p$ reaction (right panel) calculated with the single nucleon current operator without (dashed curve) and with the 3N force (solid curve). The calculations are based on the AV18 nucleon-nucleon potential [13] and the Urbana IX 3N force [14].

Table 1. Capture rates in s^{-1} for the $\mu^- +^3$ He $\longrightarrow \nu_\mu + n + d$ (Γ_{nd}) and $\mu^- +^3$ He $\longrightarrow \nu_\mu + n + n + p$ (Γ_{nnp}) processes and their sum $(\Gamma_{br} \equiv \Gamma_{nd} + \Gamma_{nnp})$ calculated with the AV18 [13] nucleon-nucleon potential and the Urbana IX [14] 3N force, using the single nucleon current and including final state interaction effects. Experimental data are from Refs. [17–20].

	AV18	AV18+Urb. IX	[17]	[18]	[19]	[20]-md I	[20]-md II
Γ_{nd}	604	544				491 ± 125	497 ± 57
Γ_{nnp}	169	154				187 ± 11	190 ± 7
Γ_{br}	773	698	660 ± 160	$665\ ^{+170}_{-430}$	720 ± 70	678 ± 126	687 ± 60

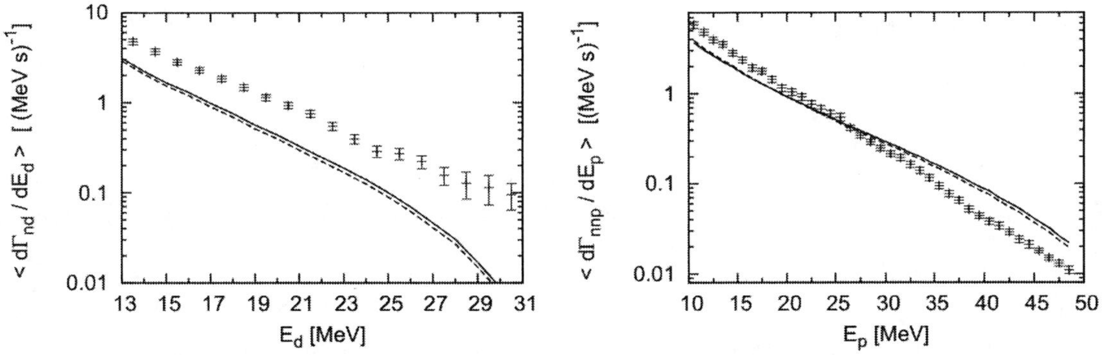

Figure 3. The capture rates $\langle d\Gamma_{nd}/dE_d \rangle$ for the $\mu^- + ^3\text{He} \rightarrow \nu_\mu + n + d$ process (left panel) and the capture rates $\langle d\Gamma_{nnp}/dE_p \rangle$ for the $\mu^- + ^3\text{He} \rightarrow \nu_\mu + n + n + p$ process (right panel) averaged over 1 MeV deuteron or proton energy bins compared with the experimental data given in Tables VI and V of Ref. [20] evaluated by means of method II. Curves are the same as in Fig. 2.

Acknowledgements

This work was supported from the resources of the National Science Center (Poland) under grant DEC-2013/10/M/ST2/00420. Numerical calculations have been performed on the supercomputer clusters of the JSC, Jülich, Germany.

References

[1] R. Skibiński, J. Golak, H.Witała, and W. Glöckle, Phys. Rev. C **59**, 2384 (1999)
[2] D.F. Measday, Phys. Rep. **354**, 243 (2001)
[3] T. Gorringe and H.W. Fearing, Rev. Mod. Phys. **76**, 31 (2004)
[4] P. Kammel and K. Kubodera, Annu. Rev. Nucl. Part. Sci. **60**, 327 (2010)
[5] L.E. Marcucci *et al.*, Phys. Rev. C **83**, 014002 (2011)
[6] L.E. Marcucci, Int. J. Mod. Phys. A **27** 1230006 (2012)
[7] L.E. Marcucci *et al.*, Phys. Rev. Lett. **108**, 052502 (2012)
[8] J. Adam, Jr. *et al.*, Phys. Lett. B **709**, 93 (2012)
[9] J. Golak *et al.*, Phys. Rev. C **90**, 024001 (2014)
[10] J. Golak *et al.*, Phys. Rept. **415**, 89 (2005)
[11] R. Skibiński, J. Golak, H. Witała, W. Glöckle, and A. Nogga, Eur. Phys. J. A **2**4, 11 (2005)
[12] L.E. Marcucci *et al.*, Phys. Rev. C **63**, 015801 (2000)
[13] R.B. Wiringa, V.G.J. Stoks, and R. Schiavilla, Phys. Rev. C **51**, 38 (1995)
[14] B.S. Pudliner *et al.*, Phys. Rev. C **56**, 1720 (1997)
[15] E. Epelbaum, H. Krebs, U.-G.Meißner, Eur. Phys. J. A **5**, 53 (2015)
[16] E. Epelbaum, H. Krebs, U.-G.Meißner, arXiv:1412.4623 [nucl-th] (2014)
[17] O.A. Zaĭmidoroga *et al.*, Phys. Lett. **6**, 100 (1963)
[18] L.B. Auerbach *et al.*, Phys. Rev. **138**, B127 (1965)
[19] E.M. Maev *et al.*, Hyp. Interact. **101/102**, 423 (1996)
[20] V.M. Bystritsky *et al.*, Phys. Rev. A **69**, 012712 (2004)

EPJ Web of Conferences **113**, 05001 (2016)
DOI: 10.1051/epjconf/201611305001

A perspective on Dyson-Schwinger equation: toy model of Pion

Lei Chang[1,a]

[1] *CSSM, School of Chemistry and Physics University of Adelaide, Adelaide SA 5005, Australia*

Abstract. We suggest a model representation of pion leading Bethe-Salpeter amplitude which involves a particular momentum dependence. Combining the constituent-quark propagator approximation we discuss the twist-2 parton distribution amplitude, parton distribution function, pion elastic form factor and $\pi - \gamma$ transition form factor.

1 Introduction

Given its twofold roles as a conventional bound-state in quantum field theory and as the Goldstone mode associated with dynamical chiral symmetry breaking, the pion has been proven critical to explaining phenomena as diverse as the long-range nucleon-nucleon interactions. The study of the pion parton structure function and electromagnetic properties is of great interest as a fundamental test of our understanding of nonperturbative QCD.

We describe pion as bound state using the Bethe-Salpeter equation. This takes the abbreviated form:

$$\Gamma_\pi(k;P) = \int_{dq}^{\Lambda} K(q,k;P)\chi_\pi(q;P) \tag{1}$$

where $\int_{dq}^{\Lambda}$ is a Poincaré-invariant regularization of the four-dimensional integral, with Λ the ultraviolet regularization mass-scale; q and k are the relative momenta between the quark-antiquark pair, P is the pion's four momentum and

$$\chi_\pi(q;P) = S(q_+)\Gamma_\pi(q;P)S(q_-) \tag{2}$$

is the pion's Poincaré-covariant Bethe-Salpeter wave-function, with Γ_π the Bethe-Salpeter amplitude. Using isospin symmetry we label the dressed quark propagators $S(q_\pm)$, where $q_\pm = q \pm \frac{P}{2}$, without loss of generality. Explicitly, these take the form:

$$S^{-1}(k) = i\gamma \cdot kA(k^2) + B(k^2) \tag{3}$$

where the scalar functions A, B depend on both momentum and the choice of renormalization point. With some form of quark-gluon vertex the quark-antiquark scattering kernel K could be derived consistently within a symmetry preserving framework and then we can solve the quark gap equation and Bethe-Salpeter equation tegother. Some model-independent statements may also be made about the pion's Bethe-Salpeter amplitude and its relationship to the dressed quark propagator.

[a] e-mail: l.chang@adelaide.edu.au

In this contribution we do not try to summarize the development of Dyson-Schwinger equation apporach [1] on describing the pion properties but just rely on a simple model to cover some story of pion. We consider an algebraic model [2] where the quark propagator and the meson amplitude take of constituent mass form

$$S^{-1}(k) = i\gamma \cdot k + M \tag{4}$$

and

$$\Gamma_\pi(k;P) = i\gamma_5 \frac{12}{5}\frac{M}{f_\pi}\int_{-1}^{1} dz\rho(z)\frac{M^2}{k^2 + zk\cdot P + \Lambda_\pi^2}, \tag{5}$$

where M is a dressed-quark mass, f_π is the pion decay constant and we focus on the case of a massless pion, the scale in the amplitude can be set $\Lambda_\pi = M$ for simplify. The factor $12/5$ is the normalization constant needed to ensure the charge conservation. k is the relative momentum of the quarks in the pion and we choose the amplitude to behave like $1/k^2$ asymptotically, as this is the leading order result if one takes a one-gluon exchange interaction between the quark and antiquark. We introduce a specific Nakanishi-like representation [3] with weight function $\rho(z)$, which takes a form as

$$\rho(z) = \frac{1}{\pi}\frac{1}{\sqrt{1-z^2}} \tag{6}$$

In this special case we have the exact relation between M and f_π as $f_\pi = \frac{3M}{\sqrt{10}\pi}$ and M can be defined to produce the empirical f_π. In some more realistic cookings the mass scale in amplitude Λ_π could be different to the constituent quark mass.

2 Pion parton structure

The leading twist-2 projection of the pion's Bethe-Salpeter wave function on the light-front [4] can be written as

$$f_\pi\, \varphi_\pi(x) = \mathrm{tr}_{\mathrm{CD}} Z_2 \int_{dq}^{\Lambda} \delta(n\cdot q_+ - x\, n\cdot P)\,\gamma_5\gamma\cdot n\,\chi_\pi(q;P)\,, \tag{7}$$

where: the trace is over colour and spinor indices; $Z_2(\zeta,\Lambda)$ is the quark wave-function renormalisation constant, with ζ the renormalisation scale; n is a light-like four-vector, $n^2 = 0$.

With the algebraic model of quark propagator and Bethe-Salpeter amplitude we are able to compute the moments $\langle x^m\rangle = \int_0^1 dx x^m \varphi(x)$ of the parton distribution amplitude(PDA) which yields in the form

$$\langle x^m\rangle = \frac{1}{f_\pi (n\cdot P)^{m+1}} \mathrm{tr}_{\mathrm{CD}} Z_2 \int_{dq}^{\Lambda} (n\cdot k)^m \gamma_5\gamma\cdot n\,\chi_\pi(q;P) = \frac{4\Gamma[\frac{3}{2}+m]}{\sqrt{\pi}\Gamma[3+m]}, \tag{8}$$

with the Euler gamma function $\Gamma[z]$. The related PDA can be read from the moments immediately as

$$\varphi(x) = \frac{8}{\pi}\sqrt{x(1-x)} \tag{9}$$

This model for the PDA follows from the precise mapping of string amplitudes in Anti-deSitter space to the light-front wavefunction of a hadron in physical space-time using holographic methods [5]. And such PDA is practically indistinguishable from the contemporary DSE prediction [4]. Whereas the scale that should be associated with the model analyses is poorly known which would be issued when it is concerned.

In Ref. [6] a corrected, leading-order expression was given for the pion's valence-quark parton distribution function(PDF). This expression produces the model-independent result that quarks dressed via the RL truncation carry all of the pion's light-front momentum at a characteristic hadronic scale, even if the meson amplitude is momentum dependent. We quote the form of the quark distribution function in the RL truncation here:

$$q(x) = \mathrm{tr}_{\mathrm{CD}} \int_{dk}^{\Lambda} \delta(n \cdot k - xn \cdot P)\, \partial_k \left[\Gamma(k - \frac{P}{2}; -P) S(k)\right] \Gamma(k - \frac{P}{2}; P) S(k - P), \tag{10}$$

The present algebraic model produce a simple form of distribution function as in the form

$$q(x) = 6x(1 - x) \propto \varphi^2(x) \tag{11}$$

It should be emphasised that the related PDF has the power-law behavior, $1 - x$, near $x = 1$. That is in contrast to the QCD parton prediction.

Ezawa [7] predicted that the pion PDF would behave as $(1 - x)^{2\alpha}$ if the pion amplitude behaved as $\frac{1}{(k_1^2)^\alpha}$, where k_1 is the struck quark momentum. Let us take a look at asymptotic form of this model of amplitude. Introduce quarks momentum as $k_{1,2}$ with fixed pion mass $P^2 = (k_1 - k_2)^2$ and integrate weight function argument z we can get $\Gamma_\pi(k_1, k_2) \propto \frac{4}{\sqrt{s^2 - 4(k_1^2 - k_2^2)^2}}$ with $s = 2(k_1^2 + k_2^2 + 2M^2) - P^2$. The leading oder of k_1 would be $1/k_1$ if one fix k_2, $1/k_2^2$ if one fix the ratio k_2/k_1. Based on Ezawa's work one readily finds a PDF which behaves as $(1 - x)^{2(1-0.5)}$, which is consistent with the present model calculation.

To cover QCD parton prediction one should build model that amplitude behavior like $1/k_2^2$ asymptotically with the fixed k_1. A possible way to guarantee the asymptotic behavior $1/k_2^2$ with fixed k_2 but taking the same value of the second central moment of PDA as the present model caluclation is to soften the weight function at boundaries $z = \pm$. A promised candidate is the two-δ function like structure of weight function. The related PDA/PDF would take the power form $1 - x, (1 - x)^2$ respectively near $x = 1$ and behavior much flat in the central region of x. We will make a research on this possiblility throughly in future.

3 Pion electromagnetic properties

In this part we will take a look at the electromagnetic properties of pion in chiral limit whith the input the above model of amplitude. At leading order in the symmetry-preserving truncation scheme the pion form factor is given by

$$P_\mu F(Q^2) = -i\mathrm{tr}_{\mathrm{CD}} \int_{dk}^{\Lambda} \Gamma_\pi(k - \frac{P}{2} + \frac{Q}{4}; -P - \frac{Q}{2}) S(k + \frac{Q}{2}) \gamma_\mu S(k - \frac{Q}{2}) \Gamma_\pi(k - \frac{P}{2} - \frac{Q}{4}; P - \frac{Q}{2}) S(k - P) \tag{12}$$

where: we have repalced the dressed quark-photon vertex by its bared form due to the consistituent quark propagator approximation; Q denotes the incoming photon momentum and massless pion mass shell is defined by $(P - \frac{Q}{2})^2 = 0 = (P + \frac{Q}{2})^2$. Near $Q^2 = 0$ we can expand the form factor by Taylor series to order Q^2 as

$$F(Q^2) = 1 - \frac{189}{1000} \frac{1}{f_\pi^2 \pi^2} Q^2 \tag{13}$$

The pion charge radius is about $0.52\mathrm{fm}^2$ compared to the empirical value $0.44\mathrm{fm}^2$. When we take Q^2 to infinity the form factor behavior like

$$Q^2 F(Q^2) \rightarrow \frac{1}{\sqrt{Q^2}} \tag{14}$$

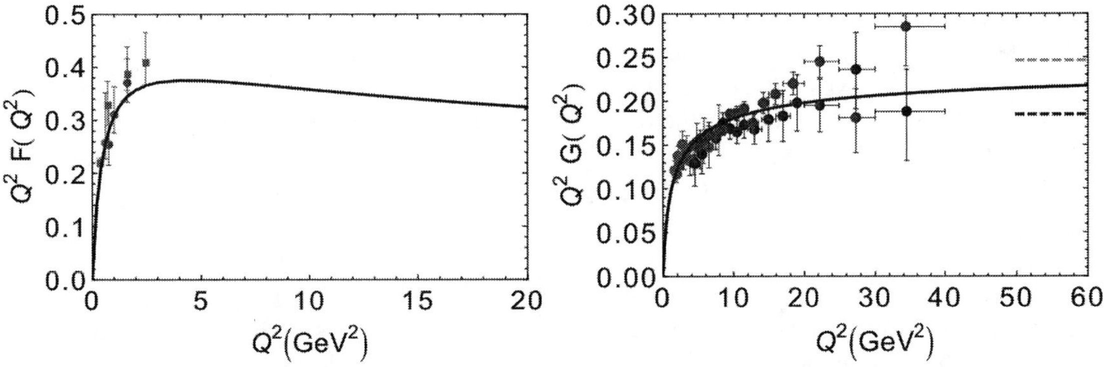

Figure 1. left panel-the momentum dependence of elastic form factor. solid line: the present model calculation and the filled circles are the data described in Ref [9]; right panel-the momentum dependence of transition form factor(divided by $4f_\pi\pi^2$). solid line: the model calculation and the filled circles are the data described in Ref [10](Red for CLEO; Blue for BaBar and Black for Belle). The black dashed line denote the limit with asymptotic PDA input and green dashed line is related to the limit with the present model PDA input.

which is in contrast to the QCD prediction for pion charge form factor at suitably large Q^2. It should be noted that if one take the weight function softening at boundaries the power behavior would be $1/Q^2$. It has been wildly known that large Q^2 QCD prediction would be rescured by including the pseudovector part of Bethe-Salpeter amplitude that is lacked in our model building. The model calculation of momentum dependence of $Q^2F(Q^2)$ is displayed in Fig. 1. Form factor within this simple model takes roughly same charactor as the complicate RL DS calculation [8] that there existed a maximum around 5GeV^2. The results match the data well in the region $Q^2 < 2\mathrm{GeV}^2$ and underestimate a little bit above. The numerical calculation could be improved by changing the value of the mass scale M.

For the $\pi - \gamma$ transition form factor we also suppose the bared vertex ansatz for both virtual and real photon coupling. At the leading order the triangular diagram for the transition form factor can be expressed as

$$\frac{1}{4\pi^2 f_\pi}\epsilon_{\mu\nu\rho\sigma}k_1^\rho k_2^\sigma G(k_1^2) = -2\mathrm{tr_D}\int_{dk}^{\Lambda}\gamma_\mu S(k-k_1+k_2)\gamma_\nu\chi_\pi(k-\frac{k_1}{2}+\frac{k_2}{2};-k_1-k_2) \tag{15}$$

where we set $k_1^2 = Q^2$ the virtual photon momentum and $k_2^2 = 0$ the real photon momentum. Near $Q^2 = 0$ we have

$$G(Q^2) = \frac{4}{5} - \frac{27}{250}\frac{1}{f_\pi^2\pi^2}Q^2 \tag{16}$$

It is not surprised that the present model and truncation can not produce chiral anormaly exactly because the relation between quark mass function and pion amplitude has not been preserved. A possible improvement in the model calculation can be done if we introduce the quark mass function as $M^3/(k^2+M^2)$ or rely on the exact consistent RL DSE calculation. We leave this topic for future.

In the large Q^2 limit we can get the analytical expression within this model as

$$\frac{1}{4\pi^2 f_\pi}Q^2G(Q^2\to\infty)\to 8f_\pi\int_{-1}^{1}dz\frac{\log(\frac{1+z}{2})}{3(z-1)}\rho(z) = 2f_\pi\int_0^1 dx\frac{\varphi(x)}{3(1-x)} \tag{17}$$

which is consistent to the QCD predication with the input of PDA related to $\rho(z)$. The model calculation of G is displayed in Fig. 1. With the increasing of momentum the form factor increases and overtakes the perturbative prediction limit $2f_\pi$. At very large momentum form factor goes to the limit $\frac{4}{3}2f_\pi$ from below. At $Q^2 = 60\text{GeV}^2$ the dressed triangular diagram suppress form factor about 12% compared to its large Q^2 limit.

4 Summary

We suggest and explain a model representation of pion amplitude which is simple but provides a powerful tool to analysis the parton structure and electromagnetic properties. The corresponding relation between weight function and parton distribution amplitude is straightforward in this approach that builds a bridge to study the effect of different parton distribution amplitude on the form factors. We also like to point that he improvements, such as preserving chiral Ward-Takahashi identity, including one gluon exchange in the form factor calculation and evoluting Bethe-Salpeter amplitude with running scale, would be necessary.

Acknowledgements

I am grateful for the financial support by Prof. Anthony W. Thomas.

References

[1] I. C. Cloet,C. D. Roberts Prog. Part. Nucl. Phys. 77 (2014) 1-69.
[2] L. Chang,A. W. Thomas arXiv: 1410.8250.
[3] N. Nakanishi, Phys. Rev. 130 (1963) 1230.
[4] L. Chang, *et al.*, Phys. Rev. Lett. 110 (2013) 132001.
[5] S. J. Brodsky, and G. F. de Teramond, Phys. Rev. Lett. 96 (2006) 201601. S. J. Brodsky, and G. F. de Teramond, Phys. Rev. D 77 (2008) 056007.
[6] L. Chang, *et al.*, Phys. Lett. B 737 (2014) 23.
[7] Z. F. Ezawa, IL NUOVO CIMENTO 23 (1974) 271.
[8] L. Chang, *et al.*, Phys. Rev. Lett. 111 (2013) 141802.
[9] G. Huber, *et al.*, Phys. Rev. C. 78 (2008) 045203.
[10] J. Gronberg, *et al.*, Phys. Rev. D 57 (1998) 33; B. Auberts, *et al.*, Phys. Rev. D 80 (2009) 052002; S. Uehara, *et al.*, Phys. Rev. D 86 (2012) 092007;

EPJ Web of Conferences **113**, 05002 (2016)
DOI: 10.1051/epjconf/201611305002

From continuum QCD to hadron observables

Daniele Binosi[1,a]

[1] *European Centre for Theoretical Studies in Nuclear Physics and Related Areas (ECT*) and Fondazione Bruno Kessler, Villa Tambosi, Strada delle Tabarelle 286, I-38123 Villazzano (TN) Italy*

Abstract. We show that the form of the renormalization group invariant quark-gluon interaction predicted by a refined nonperturbative analysis of the QCD gauge sector is in quantitative agreement with the one required for describing a wide range of hadron observables using sophisticated truncation schemes of the Schwinger-Dyson equations relevant in the matter sector.

1 Introduction

The quark gap equation is arguably one of the most important equations appearing in QCD. Solutions to this equation must describe the evolution of a (chiral) current quark of perturbative QCD into a constituent quark with a mass around 350 MeV. This nonperturbative effect thusly generates mass from nothing: being responsible for approximately 98% of the proton's mass, it represents the most important mass generating mechanism for visible matter in the Universe (the Higgs is almost irrelevant for light quarks) [1].

Writing for the quark propagator $S(p) = 1/[iA(p^2)\gamma\cdot p + B(p^2)]$, the gap equation reads

$$S^{-1}(p) = Z_2(i\gamma\cdot p + m_0) + \frac{4}{3}Z_1\int_q g^2\Delta_{\mu\nu}(k)\gamma^\mu S(q)\Gamma^\nu(q,p), \tag{1}$$

where $k = p - q$, m_0 is the quark bare mass, $Z_{1,2}$ the vertex and wave-function renormalization constants, $\Delta_{\mu\nu} = (g_{\mu\nu} - k_\mu k_\nu/k^2)\Delta(k^2)$ the (Landau gauge) gluon propagator and Γ^ν the quark gluon vertex. Owing to asymptotic freedom, the model input for realistic studies of this equation can be reduced to providing a statement on the nature of the gap equation kernel in the momentum region $k^2 \lesssim 2\ \mathrm{GeV}^2$, as above this region perturbation theory takes over. In practice one writes

$$Z_1 g^2\Delta_{\mu\nu}(k)\Gamma^\nu(q,p) = \mathcal{I}(k^2)\Delta^{\rm free}_{\mu\nu}(k)Z_2\Gamma^\nu_{\rm A}(q,p). \tag{2}$$

where $\mathcal{I}$ represents the so-called quark-gluon interaction strength, $\Delta^{\rm free}_{\mu\nu} = (g_{\mu\nu} - k_\mu k_\nu/k^2)/k^2$ is the tree-level gluon propagator and, finally, $\Gamma^\nu_{\rm A}$ represents a suitable vertex Ansatz reducing to $Z_2\gamma^\nu$ in the perturbative region $k^2 > 2\ \mathrm{GeV}^2$.

There are now two possible approaches that can be employed for determining the interaction strength $\mathcal{I}$. In a *bottom-up* approach, this quantity is obtained by fitting bound state properties data within a well-defined truncation scheme of the relevant equations (see, e.g., [2]). In a *top-down*

[a] e-mail: binosi@ectstar.eu

approach, one attempts instead to directly compute $\mathcal{I}$ by means of *ab-initio* studies of the QCD gauge and ghost sectors.

The widespread phenomenological success of the former approach, together with the significant progress made over the last decade in our understanding of the QCD infrared physics from first principles– in particular the consensus that the gauge sector is characterized by the dynamical generation of a gluon mass-scale [3, 4]–, raises an important question [5]: Is the interaction strength evaluated within these two approaches in qualitative and (possibly) quantitative agreement?

2 Bottom-up approach

As reviewed elsewhere [6–8] successful explanations and predictions of numerous hadrons observables can be obtained choosing

$$\mathcal{I}(k^2) = k^2\mathcal{G}(k^2); \quad \mathcal{G}(k^2) = \frac{8\pi^2}{\omega^4}De^{-k^2/\omega^2} + \frac{8\pi^2\gamma_m(1 - e^{-k^2/4m_t^2})}{k^2\ln[\tau + (1 + k^2/\Lambda^2_{\rm QCD})^2]}, \tag{3}$$

where $\gamma_m = 12/(33 - 2N_f)$ [typically, $N_f = 4$], $\Lambda_{\rm QCD} = 0.57$ GeV (in the momentum subtraction scheme); $\tau = e^2 - 1$, $m_t = 0.5$ GeV.

At a first glance the bottom-up interaction strength (3) depends on two parameters: D and ω. However, explicit computations have shown that as long as the two parameters are related through $D\omega = (\varsigma_{\rm G})^3 =$ const and $\omega \in [0.4, 0.6]$ GeV, one can reproduce a large body of observable properties of ground-state vector- and isospin-nonzero pesudoscalar mesons as well as numerous properties of the nucleon and Δ resonance [9]. The value of $\varsigma_{\rm G}$ is then fixed by the requirement of reproducing the correct value of the pion decay constant f_π, with the resulting value depending on the form of the vertex Ansatz employed.

In the case of the rainbow-ladder truncation, corresponding to $\Gamma^\nu_{\rm A} \sim \gamma_\nu$, one obtains $\varsigma_{\rm RL} = 0.87$ GeV, whereas the improved truncation scheme of [10] (which incorporates dynamical chiral symmetry breaking nonperturbatively in the bound-states integral equation through an accurate representation of the dressed quark-gluon vertex) gives $\varsigma_{\rm DB} = 0.55$ GeV.

Thus, the question we originally posed can be rephrased as follows: does a top-down approach produce a shape for the interaction strength qualitatively consistent with the one found in a bottom up approach? and if so, is there a value of $\varsigma_{\rm G}$ turning the agreement quantitative?

3 Top-down approach

The top-down interaction strength can be obtained within the so-called pinch-technique background field method framework [11–14] and is found to be [5]

$$\mathcal{I}(k^2) = k^2\widehat{d}(k^2); \qquad \widehat{d}(k^2) = \frac{\alpha_s(\mu^2)\Delta(k^2;\mu^2)}{[1 + G(k^2;\mu^2)]^2}, \tag{4}$$

with α_s the strong coupling evaluated at the renormalization point μ^2, and G is one of two possible form factors (the other one being L) of a certain function $\Lambda_{\mu\nu} = g_{\mu\nu}G + (k_\mu k_\nu/k^2)L$ describing the ghost-gluon dynamics and typically appearing in the PT-BFM framework. BRST symmetry implies then that, in the Landau gauge, these two form factors are related to the ghost dressing function F (defined as k^2 times the ghost propagator) through $F^{-1}(k^2) = 1 + G(k^2) + L(k^2)$ [15]. It is worth noticing that $\widehat{d}$ is a renormalization group invariant quantity [16], and does not depend on the valence-quark content of the corresponding Bethe-Salpeter equation; in addition Eq. (4) is parameter free.

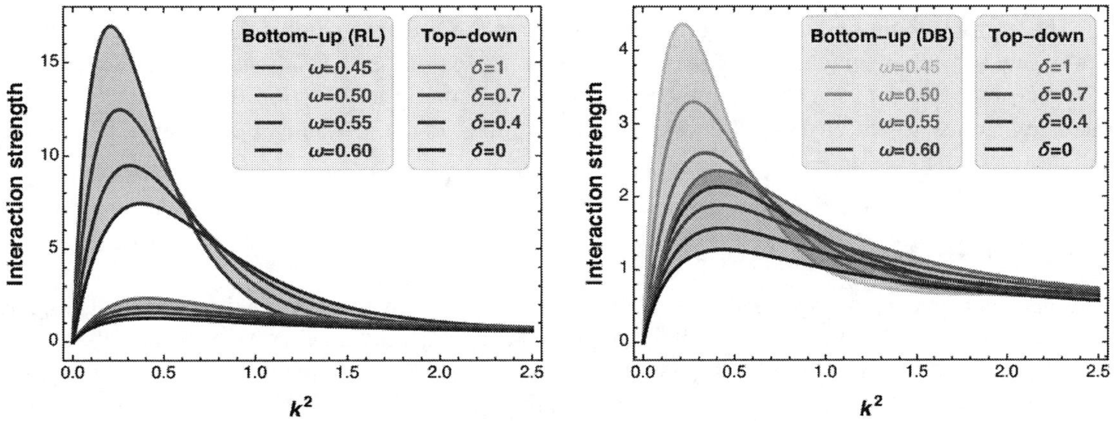

Figure 1. Comparison between top-down results for the gauge-sector interaction [Eq. (4), with those obtained using the bottom-up approach Eq. (3). *Left panel*: RL truncation, ($\varsigma_{\rm RL} = 0.87$ GeV) *Right panel*: DB truncation, ($\varsigma_{\rm DB} = 0.55$). The bands denote either the domain of constant ground-state physics $\omega \in [0.45, 0.6]$ GeV (bottom-up approach), or $\delta \in [0, 1]$ (top-down approach). Notice in particular that one can see in this latter approach that the often employed approximation $L(k^2) \approx 0$ is inaccurate, as L shows a sizeable maximum in the intermediate momentum region, resulting in an enhancement of the interaction strength and an overall better agreement with the bottom-up interaction strength.

The procedure of constructing the top-down interaction involves the following main steps [5, 16]. One begins with the lattice QCD results for the gluon propagator renormalized at a suitable value μ^2, and use them as input in the gap equation satisfied by the ghost dressing function F. The latter equation is then solved and the strong coupling α_s determined so that the resulting solution best matches the available lattice data. From the obtained F and α_s, one can next determine the function G (or, equivalently, L), and therefore construct the final quantity $\widehat{d}$. The renormalization group invariance can be checked by changing the initial renormalization point μ^2 of the gluon propagator, and observing that (*i*) the resulting curves for $\widehat{d}$ precisely overlap, and (*ii*) that this happens for the correct values of the strong coupling $\alpha_s(\mu^2)$ [5].

The major source of uncertainty in this procedure is related to the determination of G (or L) from the ghost dressing function F, as this entails a modelling of the ghost-gluon vertex. A (gross) estimate of this uncertainty can be obtained by first evaluating the form factor L from its Schwinger-Dyson equation, and then use the identity between F, G and L with the replacement $L \to \delta L$ and $\delta \in [0, 1]$. $\delta = 1$ corresponds to the $1 + G$ obtained through the solution of the corresponding Schwinger-Dyson equation, while $\delta = 0$ corresponds to the often employed approximation $1 + G \approx 1/F$ (L is known to be a subleading term with $L(0) = 0$).

4 Comparison

It is now possible to fully address the questions asked at the end of Sect. 2. Specifically, on the left (right) panel of Fig. 1 we compare the top-down interaction strength $\mathcal{I}$ with the one obtained from a bottom-up approach using the RL (DB) truncation.

It is immediately evident that while the RL interaction strength has the correct shape, it is way too large in the infrared. This is clearly an artefact of the truncation, as the bare vertex is too simple a description for capturing the complexity of hadronic phenomena: therefore, their satisfactory description can only be achieved in this case by artificially increasing the interaction strength.

The bottom-up interaction strength obtained within the DB truncation scheme is instead in both qualitative as well as quantitative agreement with the top-down one. This represents quite a remarkable fact, meaning that the two approaches capture the same representation of the gauge and ghost sector dynamics, much beyond a generic acknowledgment of the dynamical emergence of non-zero and finite gluon mass-scale.

5 Conclusions

Summarizing, our results shows that the interaction predicted by modern analyses of QCD's gauge sector is in near precise agreement with that required for a veracious description of measurable hadron properties using the most sophisticated matter-sector gap and Bethe-Salpeter kernels available today. This bridges a gap that had lain between nonperturbative continuum QCD studies and the *ab-initio* prediction of bound-state properties.

Acknowledgements

I thank the organizers of the "21st International Conference on Few-Body Problems in Physics" for their kind invitation and hospitality, and all participants for providing a most stimulating and pleasant atmosphere.

References

[1] C. D. Roberts, J. Phys. Conf. Ser. **630**, no. 1, 012051 (2015), arXiv:1501.06581 [nucl-th]
[2] G. Eichmann, I. C. Cloet, R. Alkofer, A. Krassnigg and C. D. Roberts, Phys. Rev. C **79**, 012202 (2009), arXiv:0810.1222 [nucl-th]
[3] J. M. Cornwall, Phys. Rev. D **26**, 1453 (1982),
[4] A. C. Aguilar, D. Binosi and J. Papavassiliou, Phys. Rev. D **78**, 025010 (2008), arXiv:0802.1870 [hep-ph]
[5] D. Binosi, L. Chang, J. Papavassiliou, C.D. Roberts, Phys.Lett. **B742**, 183 (2015), arXiv:1412.4782 [nucl-th]
[6] L. Chang, C.D. Roberts, P.C. Tandy, Chin.J.Phys. **49**, 955 (2011), arXiv:1107.4003 [nucl-th]
[7] A. Bashir, L. Chang, I.C. Cloet, B. El-Bennich, Y.X. Liu et al., Commun.Theor.Phys. **58**, 79 (2012), arXiv:1201.3366 [nucl-th]
[8] I.C. Cloet, C.D. Roberts, Prog. Part. Nucl. Phys. **77**, 1 (2014), arXiv:1310.2651 [nucl-th]
[9] S.x. Qin, L. Chang, Y.x. Liu, C.D. Roberts, D.J. Wilson, Phys. Rev. **C84**, 042202 (2011), arXiv:1108.0603 [nucl-th]
[10] L. Chang, C.D. Roberts, Phys. Rev. Lett. **103**, 081601 (2009), arXiv:0903.5461 [nucl-th]
[11] A.C. Aguilar, J. Papavassiliou, JHEP **12**, 012 (2006), hep-ph/0610040
[12] D. Binosi, J. Papavassiliou, Phys.Rev. **D77**, 061702 (2008), arXiv:0712.2707 [hep-ph]
[13] D. Binosi, J. Papavassiliou, JHEP **0811**, 063 (2008), arXiv:0805.3994 [hep-ph]
[14] D. Binosi, J. Papavassiliou, Phys. Rept. **479**, 1 (2009), arXiv:0909.2536 [hep-ph]
[15] D. Binosi, A. Quadri, Phys.Rev. **D88**, 085036 (2013), arXiv:1309.1021 [hep-th]
[16] A.C. Aguilar, D. Binosi, J. Papavassiliou, J. Rodriguez-Quintero, Phys. Rev. **D80**, 085018 (2009), arXiv:0906.2633 [hep-ph]

EPJ Web of Conferences **113**, 05003 (2016)
DOI: 10.1051/epjconf/201611305003

Contemporary continuum QCD approaches to excited hadrons

Bruno El-Bennich[1,2,a] and Eduardo Rojas[3,b]

[1] *Laboratório de Física Teórica e Computacional, Universidade Cruzeiro do Sul, Rua Galvão Bueno 868, 01506-000 São Paulo, SP, Brazil*
[2] *Instituto de Física Teórica, Universidade Estadual Paulista, Rua Bento Teobaldo Ferraz 271, 01140-070 São Paulo, SP, Brazil*
[3] *Instituto de Física, Universidad de Antioquia, Calle 70, no. 52-21, Medellín, Colombia*

Abstract. Amongst the bound states produced by the strong interaction, radially excited meson and nucleon states offer an important phenomenological window into the long-range behavior of the coupling constant in Quantum Chromodynamics. We here report on some technical details related to the computation of the bound state's eigenvalue spectrum in the framework of Bethe-Salpeter and Faddeev equations.

1 Introduction: excited states as an eigenvalue problem

A great deal of research activity in hadron physics is concerned with hadron structure and revolves around two fundamental questions: what constituents are the hadrons made of and how does Quantum Chromodynamics (QCD), the strong interaction component of the Standard Model, produce them? These are simple questions which, however, may not entail simple answers. To understand the measurable content of QCD, spectroscopy is a valuable and time-honored tool — suffice it to mention the inestimable progress made in the computation of atomic or molecular spectra and subsequent comparison with experiments that lead to a deeper understanding of Quantum Electrodynamics.

The same is true for QCD: if confinement is related to the analytic properties of QCD's Schwinger functions, then light-quark confinement ought to be understood by mapping out the infrared behavior of the theory's universal β function. Obviously, this cannot be possibly achieved in perturbation theory. A nonperturbative continuum approach to QCD is provided by Dyson-Schwinger equations (DSE) which relate the theory's β function to experimental observables [1]. Therefore, comparison between DSE predictions embedded in bound-state calculations and observations of the hadron mass spectrum as well as of elastic and transition form factors can be used to study the long-range behavior of QCD's interaction. As the properties of excited hadron states are considerably more sensitive to the long-range behavior of the strong interaction than those of ground states [2–4], excited mesons and nucleons [5] are an important source of information and complement our understanding of the strong interaction in light mesons [6–9] and in heavy-light mesons with disparate energy scales [10–13].

Of course, the properties of these hadrons can only be understood, and the functional behavior of the β function be inferred therefrom, by studying the quark's DSE in conjunction with quark-antiquark

[a] e-mail: bruno.bennich@cruzeirodosul.edu.br
[b] e-mail: eduardo.rojas@cruzeirodosul.edu.br

or three-quark bound-state equations, the Bethe-Salpeter and Faddeev equations, respectively. Both are treated as eigenvalue problems which we here exemplify with the relativistic bound-state equation of pseudoscalar $J^P = 0^-$ mesons. The homogeneous Bethe-Salpeter equation for this $q\bar{q}$ bound state with relative momentum p and total momentum P can be generally written as,

$$\Gamma_{0^-}(p,P) = \int_k^\Lambda \mathcal{K}(p,k,P)\,[S(k+\eta_+P)\,\Gamma_{0^-}(k,P)\,S(k-\eta_-P)] \,, \qquad (1)$$

where $S(k \pm \eta_\pm P)$ are dressed quark propagators with $\eta_+ + \eta_- = 1$ (NB: in a Poincaré invariant calculation numerical results are independent of the momentum partition parameter $\eta_\pm$). For the sake of simplicity, we omit flavor and Dirac indices, as the following discussion is independent of them. In the same spirit, we restrict ourselves to the rainbow-ladder truncation of the interaction kernel,

$$\mathcal{K}(p,k,P) = -\frac{Z_2^2\,\mathcal{G}(q^2)}{q^2}\,\frac{\lambda^a}{2}\gamma_\mu T_{\mu\nu}(q)\,\frac{\lambda^a}{2}\gamma_\nu \,, \qquad (2)$$

with the transverse projection operator $T_{\mu\nu}(q) := g_{\mu\nu} - q_\mu q_\nu/q^2$, $q = p - k$, Z_2 is the wave-function renormalization constant and λ^a are the SU(3) color matrices in the fundamental representation. Various model ansätze [14–17] have been proposed for the effective interaction, $\mathcal{G}(q^2)$, which emulates the combined effect of the gluon and quark-gluon vertex dressing functions and most recent efforts extend these models to include important transverse components of this vertex beyond the leading truncation [18]. Most importantly, Eq. (2) satisfies the axialvector Ward-Takahashi identity [19] and therefore ensures a massless pion in the chiral limit. As we shall see below, Eqs. (1) and (2) define an eigenvalue problem with physical solutions at the mass-shell points, $P^2 = -m^2$, where m is the bound-state mass.

The Bethe-Salpeter equation's Poincaré-invariant solutions can be cast in the form,

$$\Gamma_{0^-}(p,P) = \gamma_5\left[\, i\,\mathbb{I}_D E_{0^-}(p,P) + \not{P} F_{0^-}(p,P) + \not{p}(p\cdot P)\,G_{0^-}(p,P) + \sigma_{\mu\nu}\,p_\mu P_\nu\,H_{0^-}(p,P)\right] \,, \qquad (3)$$

Note that this Euclidean-metric basis, $\mathcal{A}^\alpha(p,P) = \gamma_5\{i\,\mathbb{I}_D,\ \not{P},\ \not{p}(p\cdot P),\ \sigma_{\mu\nu}p_\mu P_\nu\}$, is nonorthogonal with respect to the Dirac trace. The functions $\mathcal{F}^\alpha_{0^-}(p,P) = \{E_{0^-}(p,P), F_{0^-}(p,P), G_{0^-}(p,P), H_{0^-}(p,P)\}$ are Lorentz-invariant scalar amplitudes and are extracted from the Bethe-Salpeter amplitude (3) with appropriate projectors, $P^\alpha(p,P)$,[1]

$$\tfrac{1}{4}\,P^{\alpha\beta}(p,P)\,\mathrm{Tr}_\mathrm{D}\left[\mathcal{A}^\beta(p,P)\mathcal{A}^\gamma(p,P)\right] = \delta_{\alpha\gamma}\,; \qquad P^\alpha(p,P) = P^{\alpha\beta}(p,P)\,\mathcal{A}^\beta(p,P)\,, \qquad (4)$$

where $\alpha,\beta = 1,\ldots,4$ and the projection is given by:

$$\mathcal{F}^\alpha_{0^-}(p,P) = \tfrac{1}{4}\,\mathrm{Tr}_\mathrm{D}\left[P^\alpha(p,P)\,\Gamma_{0^-}(p,P)\right]. \qquad (5)$$

Using the Bethe-Salpeter equation, this leads to the eigenvalue problem,

$$\lambda(P^2)\,\mathcal{F}^\alpha_{0^-}(p,P) = \int_k^\Lambda \mathcal{K}^{\alpha\beta}(p,k,P)\,\mathcal{F}^\beta_{0^-}(k,P)\,, \qquad (6)$$

where $\mathcal{K}^{\alpha\beta}(p,k,P)$ stems from the projection of Eq. (1) using Eqs. (2) and (12):

$$\mathcal{K}^{\alpha\beta}(p,k,P) = -\frac{Z_2^2}{4}\frac{\mathcal{G}(q^2)}{q^2}\,T_{\mu\nu}(q)\,\mathrm{tr}_\mathrm{CD}\left[P^\alpha(p,P)\,\gamma_\mu\,\lambda^a S(k_+)\,\mathcal{A}^\beta(k,P)\,S(k_-)\,\gamma_\nu\,\lambda^a\right]. \qquad (7)$$

[1] see, e.g., Ref. [20] for the derivation of the change of basis coefficients $P^{\alpha\beta}(p,P)$.

In Eq. (6), $\lambda(P^2)$ is a scalar function and the eigenvalue equation has a solution for every value of P^2. Typically, iterative eigenvalue algorithms are employed as only under simplifying assumptions an inversion of the Bethe-Salpeter problem is possible [21–23]. To elucidate the iterative procedure, we simplify Eq. (6) (see also Ref. [24]), so that:

$$\lambda(P^2)\,|\Phi\rangle = \mathcal{K}(P^2)\,|\Phi\rangle\,. \tag{8}$$

The kernel, $\mathcal{K}(P^2)$, has a complete set of real eigenvectors ϕ_i with eigenvalues $\lambda_i(P^2)$ which are ordered as $\lambda_0(P^2) > \lambda_1(P^2) > \lambda_2(P^2) > \ldots. > \lambda_i(P^2)$. Thus, any solution can be written as a linear superposition,

$$|\Phi\rangle = \sum_{i=1}^{\infty} a_i\,|\,\phi_i\rangle\,, \tag{9}$$

where a_i are real constants and the vector must be normalized. To begin the iterative process, one may "guess" as solution, such as in Eq. (9), for a given value of P^2 and n successive actions of the kernel lead to,

$$|\,\phi_n\rangle := \mathcal{K}^n(P^2)\,|\Phi\rangle = \sum_{i=1}^{\infty} \lambda_i^n a_i\,|\,\phi_i\rangle = \lambda_0^n\left[a_0\,|\,\phi_0\rangle + \sum_{i=1}^{\infty}\left(\frac{\lambda_i}{\lambda_0}\right)^n a_i\,|\,\phi_i\rangle\right]\,. \tag{10}$$

Since $\lambda_0 > \lambda_i$, the coefficients of $|\,\phi_i\rangle$ converge to zero for sufficiently large values of n and therefore the amplitude $|\,\phi_n\rangle$ converges to the ground state amplitude $|\,\phi_0\rangle$:

$$|\,\phi_n\rangle \overset{n\to\infty}{=} \lambda_0^n\,a_0\,|\,\phi_0\rangle \simeq \lambda_0\,\mathcal{K}^{n-1}(P^2)\,|\Phi\rangle\,. \tag{11}$$

This is the most basic method of computing the largest eigenvalue $\lambda_0(P^2)$ and its associated eigenvector to any required accuracy and is referred to as power or *von Mises* iteration. The trajectory of this eigenvalue function for a range of P^2 values yields the ground-state meson mass, that is one finds $\lambda_0(P^2) = 1$ when $P^2 = -m_0^2$.

2 Gram-Schmidt orthogonalization

A common procedure to extract the wave function of excited states from the spectrum of the interaction kernel is based on the Gram-Schmidt orthogonalization process. As mentioned in Section 1, the homogeneous Bethe-Salpeter equation is an eigenvalue problem, where the largest eigenvalue the interaction kernel produces corresponds to the ground state. However, the eigenvalue spectrum is not limited to the ground state and excited states with smaller eigenvalues can be determined with the same iterative methods discussed above. As sketched, e.g. in Ref. [24], one chooses a mass that is larger than the ground state mass, $P^2 = -m^2 < -m_0^2$, and applies the iterative procedure outlined in Section 1:

- find the largest eigenvalue, λ_0, and the associated eigenvector for $m^2 > m_0^2$ and $\lambda_0(m^2) > \lambda_0(m_0^2)$. This is the unphysical "ground state" at the mass scale $P^2 = -m^2$.
- make again a guess for the Bethe-Salpeter amplitude *now projecting out* the eigenvector that pertains to the largest eigenvalue $\lambda_0(m^2)$ in a first iteration.
- use the eigenvector obtained in the previous step as input for the Bethe-Salpeter amplitude and project out *again* the eigenvector associated with $\lambda_0(m^2)$ in a second iteration; the resulting eigenvector must be projected as before in a third iteration and so on.
- the iteration converges after the nth projection which yields an eigenvalue, $\lambda_1(m^2)$, with an eigenvector orthogonal to the one associated with the "ground state" eigenvalue $\lambda_0(m^2) > \lambda_1(m^2)$.

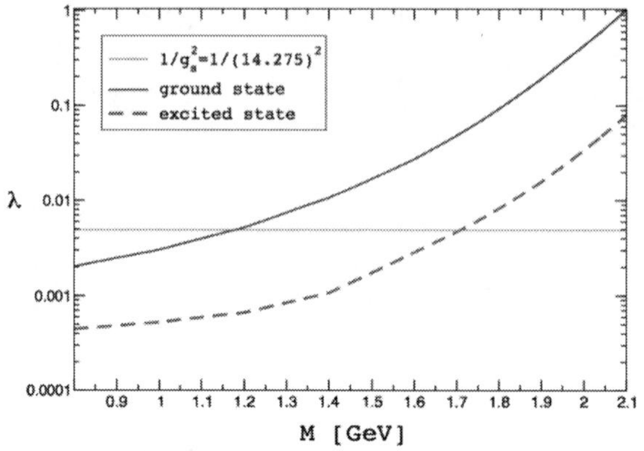

Figure 1. First two eigenvalue trajectories of the quark-diquark kernel used in Refs. [25, 26] as a function of $\sqrt{-P^2}$. Note that the eigenvector solutions are here differently normalized so that the eigenvalues are $\lambda_i = 1/g_s^2$ instead of $\lambda_i = 1$ for $P^2 = -m_i^2$; see Ref. [25] for details. The intersections of the solid horizontal line with the trajectories locates the mass-pole position of the nucleon and its first excited state identified with the *Roper*.

- by varying P^2 one obtains the mass evolution of the second largest eigenvalue $\lambda_1(P^2)$, which is exemplified in Figure 1. The solution for $\lambda_1(P^2) = 1$ yields the mass of the first excited state, $P^2 = -m_1^2$ where $m_1 > m_0$, at which the eigenvector is the state's Bethe-Salpeter amplitude.

The normalized projection within this Gram-Schmidt procedure is effected by,

$$|\tilde{\Phi}\rangle = |\Phi\rangle - \frac{\langle\phi_0\,|\,\Phi\rangle}{\langle\phi_0\,|\,\phi_0\rangle}\,|\phi_0\rangle\,, \tag{12}$$

where $|\Phi\rangle$ is the initial guess for the Bethe-Salpeter amplitude in Eq. (3) and $|\,\phi_0\rangle$ is the vector of the ground-state Bethe-Salpeter amplitude. In projecting out this ground-state, one must define a norm in Euclidean space via an inner product, as is evident from Eq. (12). Within Quantum Field Theory in rainbow-ladder approximation, this product is defined as:

$$\langle\Psi\,|\,\Phi\rangle := \mathrm{tr}_{\mathrm{CD}}\int_k^{\Lambda} \overline{\Psi}(k,-P)\,S(k+\eta_+P)\,\Phi(k,p)\,S(k-\eta_-P)\,. \tag{13}$$

Orthogonality is thus defined by the condition, $\langle\Psi|\Phi\rangle = 0$, which expresses the vanishing of the overlap amplitude at P^2. Note that Eq. (13) is not valid for computations beyond the leading approximation, i.e. the rainbow-ladder truncation.

3 Krylov subspace and Arnoldi iteration

After momentum discretization and Chebyshev expansion of Eq. (6), the numerical kernel of the Bethe-Salpeter equation (or the Faddeev equation for the three-body problem) is a non-symmetric matrix of large dimensions and with some of its eigenvalues close to zero. In general, non-symmetric matrices have eigenvalues and eigenvectors which are very sensitive to small changes in the matrix elements due to the lack of symmetries on which traditional methods rely to ensure numerical stability [27]. This is particularly true for radial excitations, where angular dependence encoded in higher Chebyshev moments contributes in a nontrivial manner. Therefore, we make use of the numerical ARPACK library [28] which is designed to compute a few eigenvalues and corresponding eigenvectors of a general $n \times n$ matrix by means of the implicitly restarted Arnoldi method (IRAM). The strength of the Arnoldi method [29] lies in the application of the stabilized orthogonalization algorithm in the Krylov space of a given matrix K to find its eigenvectors, as will be explained shortly. The most important feature of this algorithm is its ability to decouple the calculation of the eigenvectors corresponding to the eigenvalues with the largest absolute value from the eigenvectors whose eigenvalues have an absolute value close to zero. This is highly desirable as the calculation of the latter are plagued by numerical instabilities.

In algebra, the order-r Krylov subspace is generated by an operator whose representation is given by an $n \times n$ matrix $K = K_{ij}$, and a vector Φ of dimension n. It corresponds to the linear subspace spanned by actions of K on Φ:

$$\mathcal{S}_r := \left\{\Phi, K\Phi, K^2\Phi, K^3\Phi, ..., K^{r-1}\Phi\right\} . \tag{14}$$

The power iteration of K yields the sequence in Eq. (14) and converges to the eigenvector associated with the largest eigenvalue λ_0. However, hereby one only makes very limited use of the stored information, as only the final vector, $K^{r-1}\Phi$, is kept. On the other hand, the basis for the Krylov subspace is derived from the Cayley-Hamilton theorem which implies that the inverse of a matrix can be expressed as a linear combination of its powers. Thus, Krylov subspaces play an important role in contemporary iterative methods to obtain one or few eigenvalues of large sparse matrices or to solve large systems of linear equations. Instead of heavy matrix operations, these methods rely on successive multiplications of vectors by the matrix, thus forming a Krylov subspace, and then employ the resulting vectors.

The vectors of the Krylov space are initially not ortogonal and usually become almost linearly dependent due to the properties of the matrix power iteration. Nonetheless, an orthogonal matrix can be constructed from the basis vectors by means of the Gram-Schmidt process described in Section 2. This method proves to be unstable but its shortcoming can be overcome with the Arnoldi iteration [29] which uses the *stabilized* Gram-Schmidt process and can be applied to general, possibly non-Hermitian matrices. The Arnoldi method generalizes the Gram-Schmidt process by computing the eigenvalues of the orthogonal projection of K onto the Krylov subspace, where the projection is represented by the upper Hessenberg matrix H_r [28]. For Hermitian (symmetric) matrices, the Arnoldi iteration is analogous to the Lanczos iteration.

4 A note on orthogonality

In the context of Bethe-Salpeter and Faddeev equations, orthogonality is defined by Eq. (13). In vectorial form, using the shorthand $x = k^2$ and $y = p^2$, Eq. (6) can be written as,

$$\lambda(y)\, \boldsymbol{F}_R(y) = \int dx\, \mathcal{K}(x, y) \cdot \boldsymbol{F}_R(x) , \tag{15}$$

which in a numerical treatment is evaluated as the sum,

$$\lambda(x_j)\, \boldsymbol{F}_R(x_j) = \sum_i w_i\, \mathcal{K}(x_j, x_i) \cdot \boldsymbol{F}_R(x_i) , \tag{16}$$

and similarly,

$$\lambda(x_j)\, \boldsymbol{F}_L(x_j) = \sum_i w_i\, \boldsymbol{F}_L(x_i) \cdot \mathcal{K}(x_j, x_i) , \tag{17}$$

where x_i are the nodes of a given quadrature with weights w_i. Since the integral in Eq. (13) is over a charge-conjugate Bethe-Salpeter amplitude, $\bar{\Psi}(k, -P) := C\, \Psi^T(-k, -P) C^T$, we introduce in Eqs. (16)

and (17) "left" and "right" eigenvectors, $\boldsymbol{F}_L$ and $\boldsymbol{F}_R$, respectively. Multiplying Eq. (16) from the left by $\sum_j w_j \boldsymbol{F}'_L(x_j)$ we obtain,

$$\lambda \sum_j w_j \boldsymbol{F}'_L(x_j) \cdot \boldsymbol{F}_R(x_j) = \sum_{i,j} w_i w_j \boldsymbol{F}'_L(x_j) \cdot \mathcal{K}(x_j, x_i) \cdot \boldsymbol{F}_R(x_i) = \lambda' \sum_i w_i \boldsymbol{F}'_L(x_i) \cdot \boldsymbol{F}_R(x_i) , \quad (18)$$

where $\boldsymbol{F}'_L(x_j)$ is the eigenvector associated with the eigenvalue λ'. For $\lambda \neq \lambda'$ this relation implies,

$$\int dx\, \boldsymbol{F}'_L(x) \cdot \boldsymbol{F}_R(x) = 0 , \quad (19)$$

The left eigenvector is given by the Bethe-Salpeter wave function, $S(k - \eta_- P)\, \overline{\Phi}_{\lambda'}(k, -P) S(k + \eta_+ P)$, and the right eigenvector is the Bethe-Salpeter amplitude [30]. We thus deduce from Eq. (19),

$$\mathrm{tr_{CD}} \int_k S(k - \eta_- P)\, \overline{\Phi}_{\lambda'}(k, -P)\, S(k + \eta_+ P)\, \Phi_\lambda(k, P) = 0, \quad \text{for } \lambda' \neq \lambda , \quad (20)$$

and because the trace is cyclic this orthogonality condition is equivalent to that described in the paragraph below Eq. (13). Moreover, we verify that the Bethe-Salpeter equation spectrum of Eq. (16) is equal to that in Eq. (17). We stress that Eq. (19) is generally valid and no assumption was made about the kernel structure or about the eigenvectors.

5 Examples in hadron physics

The meson and nucleon resonance structure has been the object of a long history of studies and we here limit ourselves to the approaches based on the combined Dyson-Schwinger and Bethe-Salpeter (or Faddeev) equations, in particular their application to excited mesons and nucleons to compute their masses, weak decay constants and/or electromagnetic form factors making use of the techniques described in Sections 1–4.

Seminal studies on the ground state spectrum of light pseudoscalar mesons established that the $\pi(1300)$ can be described as the first radially excited state of the Goldstone boson and furthermore that the decay constant of excited states, P_n, vanishes identically in the chiral limit [31],

$$f_{P_n}^{\hat{m}=0}(\mu) \equiv 0 , \; n \geq 1 , \quad (21)$$

where $\hat{m}$ is the renormalization-group invariant current-quark mass. Electromagnetic properties of ground and excited state pseudoscalar mesons, making use of the the Gram-Schmidt process illustrated in Section 2, were studied in Ref. [32]. A first approach to computing the eigenvalues of unflavored light and heavy mesons and their corresponding Bethe-Salpeter amplitudes with the implicitly restarted Arnoldi factorization, as implemented in the ARPACK [28] library, is expounded in Refs. [30, 33] and has successively been applied to the light [35] and heavy [34, 35] quarkonia spectrum for pseudoscalar, vector and tensor states. A recent detailed analysis of the Maris-Tandy interaction [16] parameter space for ground and radially excited states is presented in Ref. [35], where the authors conclude that the preferred inverse effective range of the Maris-Tandy interaction in rainbow-ladder truncation, ω, is shorter for heavy quarkonia than for the light quarkonium spectrum: $\omega_{Q\bar{Q}} = 0.7$ GeV vs. $\omega_{q\bar{q}} = 0.5$ GeV.

The same library has then been employed to study radial excitations of flavor-singlet and flavored pseudoscalar mesons within the framework of the rainbow-ladder truncation [4, 36, 37]. A summary of the theoretical values for the pseudoscalar's ground- and excited-state masses and weak decay constants for flavor-singlet and nonsinglet $J^P = 0^-$ mesons is reproduced in Table 1 for two parameter

Table 1. Masses and decay constants for flavor singlet and nonsinglet $J^P = 0^-$ mesons; see Section 5.

	Model 1 [GeV]	Model 2 [GeV]	Reference
m_π	0.138	0.153	0.139 [38]
f_π	0.139	0.189	0.1304 [38]
$m_{\pi(1300)}$	0.990	1.414	1.30 ± 0.10 [38]
$f_{\pi(1300)}$	-1.1×10^{-3}	-8.3×10^{-4}	
m_K	0.493	0.541	0.493 [38]
f_K	0.164	0.214	0.156 [38]
$m_{K(1460)}$	1.158	1.580	1.460 [38]
$f_{K(1460)}$	−0.018	−0.017	
$m_{\bar{s}s}$	1.287	1.702	
$f_{\bar{s}s}$	−0.0214	−0.0216	
$m_{\eta_c(1S)}$	3.065	3.210	2.984 [38]
$f_{\eta_c(1S)}$	0.389	0.464	0.395 [39]
$m_{\eta_c(2S)}$	3.402	3.784	3.639 [38]
$f_{\eta_c(2S)}$	0.089	0.105	

sets of the interaction introduced by Qin et al. [17]. Namely, Model 1 and Model 2 correspond to the interaction parameters, $\omega = 0.4$ GeV, $\omega D = (0.8\text{ GeV})^3$ and $\omega = 0.6$ GeV, $\omega D = (1.1\text{ GeV})^3$; see Ref. [4] for details and Ref. [37] for a discussion of the effective interaction. Note that we explored a range of parameter combinations for $\omega D \neq$ const., yet could not find a unique set which produces theoretical observables that compare well with all data in Table 1. We therefore observe that we are unable to consistently and simultaneously describe both ground and excited state observables with the given interaction [2] in the rainbow-ladder truncation — whereas one parameter set successfully reproduces experimental ground state masses and decay constants but not the experimental data on excited states, the second parameter set only provides a reasonable description of the excited pseudoscalar mass spectrum.

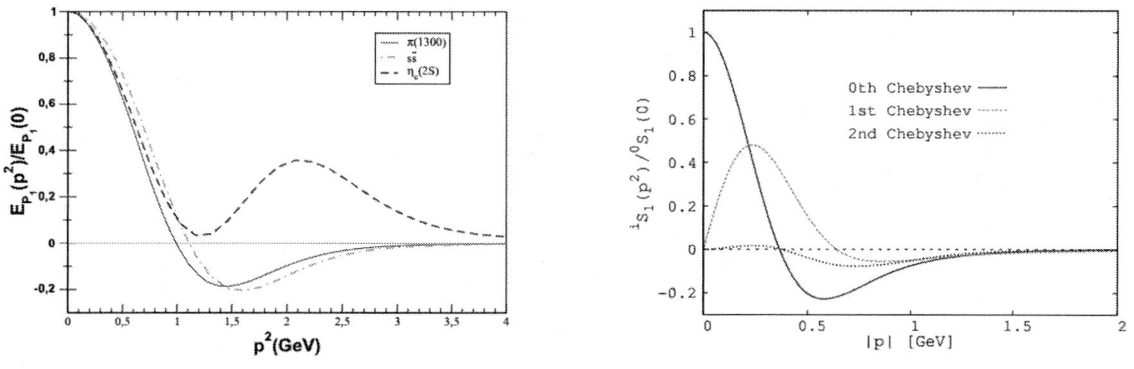

Figure 2. Left panel: Lowest Chebyshev moment, ${}^0E_{P_1}(p^2)$, associated with the leading Dirac structure $E_{P_1}(p^2)$ of the pseudoscalar's Bethe-Salpeter amplitude for the first radial excitations $P_1 = \pi(1300)$, $(\bar{s}s)_{n=1}$ and $\eta_c(2S)$. Right panel: First three normalized Chebyshev moments, ${}^iS_1(p^2)$, $i = 0, 1, 2$, of the leading S-wave component in the nucleon's first excited-state Faddeev amplitude.

As it becomes clear from the left graph of Fig. 2, for $p^2 \gtrsim 1$ GeV^2 the amplitude's lowest Chebyshev projection, ${}^0E_{P_1}(p^2)$ associated with the leading covariant $i\gamma_5$ of the pseudoscalar's Bethe-Salpeter amplitude, becomes negative definite in case of the first radial excitations, $\pi(1300)$ and $(\bar{s}s)_{n=1}$, whereas it remains positive for the ground states. It is remarkable that the behavior of the Bethe-Salpeter amplitudes parallels the pattern of wave functions in quantum mechanics, namely, the number of zeros can be associated with a principal quantum number n. As just mentioned, the ground state amplitude has no zeros and can thus be associated with $n = 0$. The amplitude of the next highest mass mesons possesses one zero and one assigns the quantum number $n = 1$ and so on. An analogous behavior is found for the nucleon's first excited state Faddeev amplitude for which we plot the first three Chebyshev moments of its leading S-wave component. Indeed, very recently a range of properties of the proton's radial excitation was predicted which strongly suggests that the nucleon's first radial excitation is the Roper resonance [26]. In particular, in Fig. 3 of Ref. [26] one can appreciate that the overlap amplitude between the nucleon and its first excited state described by the nucleon-Roper transition form factor $F_1^*(Q^2)$ vanishes at $Q^2 = 0$ and therefore satisfies orthogonality.

Acknowledgements

We thank the organizers of the *21 International Conference on Few-Body Problems in Physics* in Chicago for the opportunity to present our work at this successful event. The work of B. E. is supported in parts by a State of São Paulo Research Foundation (FAPESP) grant, an MCTI/CNPq/Universal grant and by the Brazilian agency CNPq with a visiting fellowship at the Institute for Theoretical Physics, State University of São Paulo (IFT-Unesp). The work of E. R. is supported by "*Patrimonio Autónomo Fondo Nacional de Financiamiento para la Ciencia, la Tecnología y la Innovación, Francisco José de Caldas*" and by "*Sostenibilidad-UDEA 2014-2015*".

References

[1] A. Bashir, L. Chang, I. C. Cloët, B. El-Bennich, Y. X. Liu, C. D. Roberts and P. C. Tandy, Commun. Theor. Phys. **58**, 79 (2012).
[2] S. X. Qin, L. Chang, Y. X. Liu, C. D. Roberts and D. J. Wilson, Phys. Rev. C **85**, 035202 (2012).
[3] E. Rojas, J. P. B. C. de Melo, B. El-Bennich, O. Oliveira and T. Frederico, JHEP **1310**, 193 (2013).
[4] E. Rojas, B. El-Bennich and J. P. B. C. de Melo, Phys. Rev. D **90**, 074025 (2014).
[5] I. G. Aznauryan *et al.*, Int. J. Mod. Phys. E **22**, 1330015 (2013).
[6] B. El-Bennich, J. P. B. C. de Melo, B. Loiseau, J.-P. Dedonder and T. Frederico, Braz. J. Phys. **38**, 465 (2008).
[7] L. Chang, I. C. Cloët, B. El-Bennich, T. Klähn and C. D. Roberts, Chin. Phys. C **33**, 1189 (2009).
[8] E. O. da Silva, J. P. B. C. de Melo, B. El-Bennich and V. S. Filho, Phys. Rev. C **86**, 038202 (2012).
[9] B. El-Bennich, J. P. B. C. de Melo and T. Frederico, Few Body Syst. **54**, 1851 (2013).
[10] B. El-Bennich, M. A. Ivanov and C. D. Roberts, Nucl. Phys. Proc. Suppl. **199**, 184 (2010).
[11] B. El-Bennich, M. A. Ivanov and C. D. Roberts, Phys. Rev. C **83**, 025205 (2011).
[12] B. El-Bennich, G. Krein, L. Chang, C. D. Roberts and D. J. Wilson, Phys. Rev. D **85**, 031502 (2012).
[13] B. El-Bennich, C. D. Roberts and M. A. Ivanov, POS(QCD-TNT-II) 018, arXiv:1202.0454 [nucl-th] (2012).
[14] P. Maris, C. D. Roberts and P. C. Tandy, Phys. Lett. B **420**, 267 (1998).
[15] P. Maris and C. D. Roberts, Phys. Rev. C **56**, 3369 (1997).
[16] P. Maris and P. C. Tandy, Phys. Rev. C **60**, 055214 (1999).

[17] S. x. Qin, L. Chang, Y. x. Liu, C. D. Roberts and D. J. Wilson, Phys. Rev. C **84**, 042202 (2011).

[18] D. Binosi, L. Chang, J. Papavassiliou and C. D. Roberts, Phys. Lett. B **742**, 183 (2015).

[19] A. Bender, C. D. Roberts and L. Von Smekal, Phys. Lett. B **380**, 7 (1996).

[20] J. J. Cobos-Martínez, *Static and dynamic properties of the pion from continuum model ling of strong QCD*, PhD Thesis, Durham University (2010).

[21] W. Lucha and F. F. Schöberl, Phys. Rev. D **87**, no. 1, 016009 (2013).

[22] T. Frederico, G. Salmè and M. Viviani, Phys. Rev. D **89**, 016010 (2014).

[23] M. S. Bhagwat, A. Höll, A. Krassnigg, C. D. Roberts and S. V. Wright, Few Body Syst. **40**, 209 (2007).

[24] A. Krassnigg and C. D. Roberts, Fizika B **13**, 143 (2004).

[25] I. C. Cloët, G. Eichmann, B. El-Bennich, T. Klähn and C. D. Roberts, Few Body Syst. **46**, 1 (2009).

[26] J. Segovia, B. El-Bennich, E. Rojas, I. C. Cloët, C. D. Roberts, S. S. Xu and H. S. Zong, Phys. Rev. Lett. **115**, no. 17, 171801 (2015).

[27] W. H. Press, S. A. Teukolsky, W. T. Vetterling and B. P. Flannery, ISBN-9780521430647.

[28] R. B. Lehoucq, D. C. Sorensen, C. Yang, *ARPACK Users' Guide: Solution of Large-Scale Eigenvalue Problems with Implicitly Restarted Arnoldi Methods*, Siam (1998).

[29] W. E. Arnoldi, Quarterly of Applied Mathematics, vol. 9, pp. 17-29 (1951).

[30] M. Blank, arXiv:1106.4843 [hep-ph] (2011).

[31] A. Höll, A. Krassnigg and C. D. Roberts, Phys. Rev. C **70**, 042203 (2004).

[32] A. Höll, A. Krassnigg, P. Maris, C. D. Roberts and S. V. Wright, Phys. Rev. C **71**, 065204 (2005).

[33] M. Blank and A. Krassnigg, Comput. Phys. Commun. **182**, 1391 (2011).

[34] T. Hilger, C. Popovici, M. Gomez-Rocha and A. Krassnigg, Phys. Rev. D **91**, no. 3, 034013 (2015).

[35] T. Hilger, M. Gomez-Rocha and A. Krassnigg, arXiv:1508.07183 [hep-ph].

[36] B. El-Bennich, E. Rojas, M. A. Paracha and J. P. B. C. de Melo, AIP Conf. Proc. **1625**, 80 (2014).

[37] E. Rojas, B. El-Bennich, J. P. B. C. De Melo and M. A. Paracha, Few Body Syst. **56**, no. 6-9, 639 (2015).

[38] J. Beringer *et al.* [Particle Data Group Collaboration], Phys. Rev. D **86**, 010001 (2012).

[39] C. T. H. Davies, C. McNeile, E. Follana, G. P. Lepage, H. Na and J. Shigemitsu, Phys. Rev. D **82**, 114504 (2010).

EPJ Web of Conferences **113**, 05004 (2016)
DOI: 10.1051/epjconf/201611305004

Quark model with chiral-symmetry breaking and confinement in the Covariant Spectator Theory

Elmar P. Biernat[1,a], M. T. Peña[1,2], J. E. Ribeiro[3], A. Stadler[4,1], and F. Gross[5]

[1] *CFTP, Instituto Superior Técnico, Universidade de Lisboa, 1049-001 Lisboa, Portugal*
[2] *Departamento de Física, Instituto Superior Técnico, Universidade de Lisboa, 1049-001 Lisboa, Portugal*
[3] *CeFEMA, Instituto Superior Técnico, Universidade de Lisboa, 1049-001 Lisboa, Portugal*
[4] *Departamento de Física, Universidade de Évora, 7000-671 Évora, Portugal*
[5] *Thomas Jefferson National Accelerator Facility (JLab), Newport News, Virginia 23606, USA*

Abstract. We propose a model for the quark-antiquark interaction in Minkowski space using the Covariant Spectator Theory. We show that with an equal-weighted scalar-pseudoscalar structure for the confining part of our interaction kernel the axial-vector Ward-Takahashi identity is preserved and our model complies with the Adler-zero constraint for π-π-scattering imposed by chiral symmetry.

1 Introduction

As the lightest quark-antiquark bound state the pion is of particular importance for our understanding of confinement and spontaneous chiral-symmetry breaking (SχSB). It emerges non-perturbatively from the strong interaction and it is at the same time identified with the Goldstone boson associated with SχSB. Various modern approaches have addressed the non-perturbative dynamics underlying such hadronic systems. For instance, lattice-QCD simulations [1, 2], light-front quantum field theory [3, 4], as well as models based on the Dyson-Schwinger–Bethe-Salpeter (DSBS) approach and the mass gap equation [5–10] have significantly contributed to an understanding of a wide range of hadronic phenomena. The framework we use is the Covariant Spectator Theory (CST) [11–16] — another modern field-theoretic approach that implements SχSB through the famous Nambu–Jona-Lasinio mechanism, similarly to DSBS. Whereas the latter is usually treated in a Euclidean formulation, CST is established in Minkowski space, an advantage for instance when computing form factors in the timelike region. Another distinct feature of CST is its capability of accommodating a Lorentz-scalar confining interaction kernel without destroying chiral symmetry, which is of particular importance in view of the approaches [17–19] suggesting the existence of a scalar component for the quark-antiquark interaction. In the present work we study to what extent such confining forces can be made consistent with SχSB. Our strategy is to start from the most general Lorentz structure for the interaction kernel and then determine the constraints imposed by chiral symmetry, similarly to what has been done in Ref. [20] for a different formalism. It turns out that a CST model with scalar confinement, together with an equal-weighted pseudoscalar counterpart satisfies the SχSB condition of the Adler consistency zero [21] in π-π scattering in the chiral limit.

[a] e-mail: elmar.biernat@tecnico.ulisboa.pt

2 Axial-vector Ward-Takahashi identity

Explicit and spontaneous chiral-symmetry breaking is expressed in quantum field theory through the axial-vector Ward-Takahashi identity (AV-WTI) involving the dressed quark propagators and the dressed axial-vector and pseudoscalar vertices. When dealing with strong quark form factors according to Gross and Riska [22–24] the AV-WTI reads

$$P_\mu \Gamma_R^{5\mu}(p',p) + 2m_0 \Gamma_R^5(p',p) = \tilde{S}^{-1}(p')\gamma^5 + \gamma^5 \tilde{S}^{-1}(p)\,, \tag{1}$$

where $\Gamma_R^{5\mu}(p',p)$ and $\Gamma_R^5(p',p)$ are the dressed axial-vector and pseudoscalar vertices, respectively, m_0 is the bare quark mass, p and p' are the incoming and outgoing quark momenta, respectively, $P = p' - p$ is the momentum flowing into the vertex, and $\tilde{S}(p)$ is the dressed quark propagator as introduced in Ref. [25]. The combination on the LHS of Eq. (1) is sometimes called the dressed axial vertex $\Gamma_R^A(p',p)$, which is the solution of an inhomogeneous CST Bethe-Salpeter equation (CST-BSE),

$$\Gamma_R^A(p',p) = \gamma_R^A(p',p) + \mathrm{i}\int_{k0} \mathcal{V}_R(p-k)\tilde{S}(k')\Gamma_R^A(k',k)\tilde{S}(k)\,, \tag{2}$$

where $\gamma_R^A(p',p)$ is the bare axial vertex, $\mathcal{V}_R(p-k)$ is the covariant $q\bar{q}$ interaction kernel depending only on the four-momentum transfer $p - k = p' - k'$, and "$k0$" indicates the charge-conjugation invariant CST prescription for performing the k_0 contour integration [15]. The most general structure of the CST generalization of the linear-confining potential, together with a vector–axial-vector remainder, is given by

$$\begin{aligned}\mathcal{V}_R(p-k) \;=\; & V_{LR}(p-k)\Big[\lambda_S(\mathbf{1}\otimes\mathbf{1}) + \lambda_P(\gamma^5\otimes\gamma^5) + \lambda_V(\gamma^\mu\otimes\gamma_\mu) + \lambda_A(\gamma^5\gamma^\mu\otimes\gamma^5\gamma_\mu) \\ & + \frac{\lambda_T}{2}(\sigma^{\mu\nu}\otimes\sigma_{\mu\nu})\Big] + V_{CR}(p-k)\Big[\kappa_V(\gamma^\mu\otimes\gamma_\mu) + \kappa_A(\gamma^5\gamma^\mu\otimes\gamma^5\gamma_\mu)\Big]\,,\end{aligned} \tag{3}$$

where V_{LR} and V_{CR} are the Lorentz-invariant momentum-dependent parts of the linear-confining and remaining kernels, respectively. The corresponding weight parameters λ_i and κ_i [with $i = S$ (scalar), P (pseudoscalar), V (vector), A (axial-vector), and T (tensor)] are arbitrary constants. Further, V_{LR} satisfies the CST generalization of the non-relativistic condition $V_L(r = 0) = 0$, given by

$$\int \frac{\mathrm{d}^3k}{E_k} V_{LR}(p \pm \hat{k}) = 0\,, \tag{4}$$

where $E_k = \sqrt{m^2 + \vec{k}^2}$, $\hat{k} = (E_k, \vec{k})$, and m is the dressed quark mass. For this kernel with $\lambda_S = \lambda_P$ it has been shown [25] that the AV-WTI (1) together with the CST-BSE (2) implies that $\tilde{S}(p)$ is the solution of the CST-Dyson Equation (CST-DE),

$$\tilde{S}^{-1}(p) = \tilde{S}_0^{-1}(p) - \mathrm{i}\int_{k0} \mathcal{V}_R(p-k)\tilde{S}(k)\,, \tag{5}$$

where $\tilde{S}_0$ is the bare quark propagator, which obeys an AV-WTI involving $\gamma_R^A(p',p)$. It turns out that $\gamma_R^A(p',p)$ vanishes in the chiral limit of vanishing bare quark mass, $m_0 \to 0$, and vanishing vertex momentum, $P \to 0$. In this limit, the CST-BSE (2) becomes homogeneous and identical to the zero-mass pion CST equation for the pion vertex function in the chiral limit, $\Gamma_{R\chi}^\pi$, which implies the relation

$$\Gamma_{R\chi}^A(p,p) \propto \Gamma_{R\chi}^\pi(p,p)\,. \tag{6}$$

Because of the condition (4), and imposing equal weights to the scalar and pseudoscalar terms in the interaction kernel only $\mathcal{V}_{CR}$ contributes to the chiral-limit pion equation and to the scalar part of the CST-DE (5). This corresponds to dynamical quark mass generation. Therefore, the linear-confinement part $\mathcal{V}_{LR}$ that includes the scalar, pseudoscalar and tensor structures in our model, entirely decouples from these equations [13].

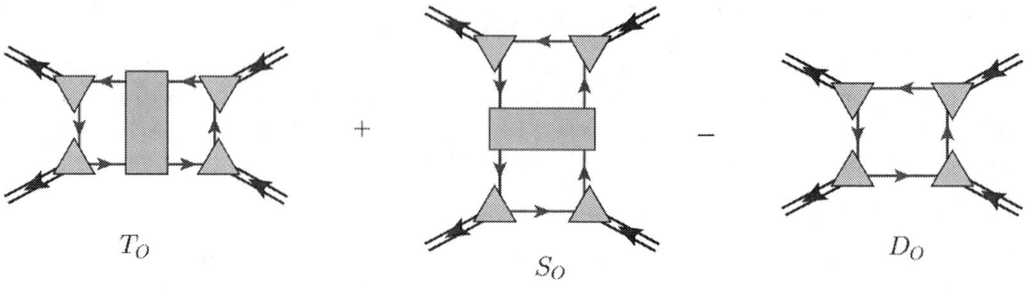

Figure 1. The O contributions to π-π scattering. The orange boxes denote the unamputated quark-quark scattering amplitudes.

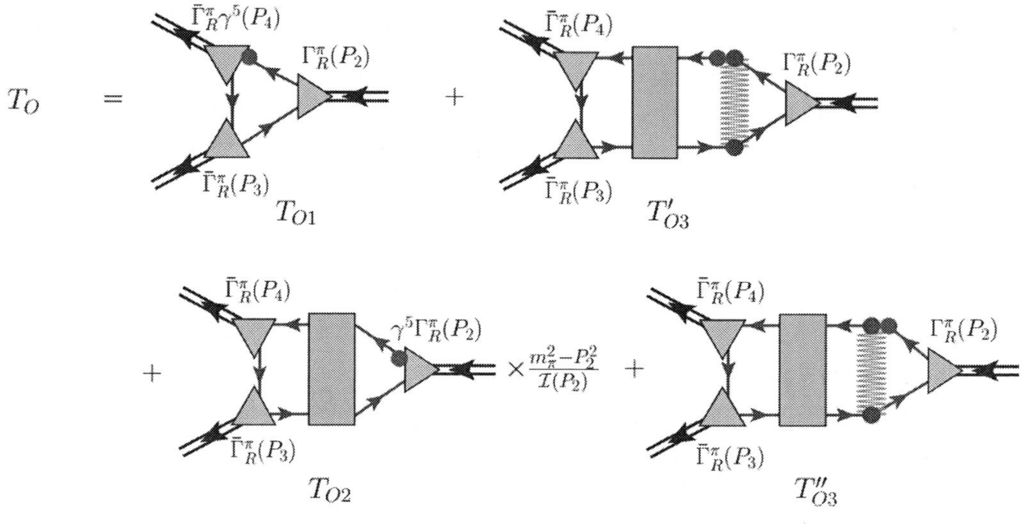

Figure 2. The T_O term. A purple blob denotes a γ^5 and $\mathcal{I}(P)$ is a non-vanishing normalization integral.

3 π-π scattering and Adler zero

Consistency with chiral symmetry implies the vanishing of the π-π scattering amplitude in the chiral limit [21]. This property, known as the Adler zero, is due to remarkable cancellations, which occur between different scattering diagrams that go beyond the (lowest-order) impulse approximation. In these diagrams intermediate-state interactions to all orders are included through the complete quark-quark ladder sum [26, 27], which involves nine different types of contributions. The proof of the cancellations between these contributions is rather lengthy. To illustrate the technical procedure we discuss here only the diagrams shown in Fig. 1. The same treatment applies to the other diagrams that have to be taken into account. The full proof is given in detail in Ref. [25]. In order to show that the sum of these diagrams vanishes in the chiral limit, one inserts an additional ladder sum at one pion vertex by using the spectral decomposition of the ladder sum. Then, by Eq. (6), $\Gamma^{\pi}_{R\chi}$ is replaced by $\Gamma^{A}_{R\chi}$, allowing the application of the AV-WTI (1) between two ladder sums. For T_O this results in 4 terms as depicted in Fig. 2. Because the sum $T'_{O3} + T''_{O3}$ is proportional to the anticommutator $\{\mathcal{V}_R, \gamma^5\}$, all vector and axial-vector contributions cancel. The scalar, pseudoscalar, and tensor structures from the confining potential integrate to zero in the chiral limit because of the decoupling property discussed in the previous section. The T_{O2} term vanishes as consequence of the fact that the pion does not couple to the scalar channel. Finally, the T_{O1} term, together with S_O cancel exactly the D_O term by applying again the AV-WTI.

Acknowledgements

This work received financial support from Fundação para a Ciência e a Tecnologia (FCT) under Grants No. PTDC/FIS/113940/2009 and No. CFTP-FCT (PEst-OE/FIS/U/0777/2013). The research leading to these results has received funding from the European Community's Seventh Framework Programme FP7/2007-2013 under Grant Agreement No. 283286. This work was also partially supported by Jefferson Science Associates, LLC, under U.S. DOE Contract No. DE-AC05-06OR23177.

References

[1] R.G. Edwards, N. Mathur, D.G. Richards, S.J. Wallace, Phys. Rev. D **87**, 054506 (2013)
[2] P. Guo, J.J. Dudek, R.G. Edwards, A.P. Szczepaniak, Phys. Rev. D **88**, 014501 (2013)
[3] S.J. Brodsky, H.C. Pauli, S.S. Pinsky, Phys. Rept. **301**, 299 (1998)
[4] J. Carbonell, B. Desplanques, V. Karmanov, J. Mathiot, Phys. Rept. **300**, 215 (1998)
[5] P.J. de A. Bicudo, J.E.F.T. Ribeiro, Phys. Rev. D **42**, 1635 (1990)
[6] A.V. Nefediev, J.E.F.T. Ribeiro, Phys. Rev. D **70**, 094020 (2004)
[7] R. Alkofer, L. von Smekal, Phys. Rept. **353**, 281 (2001)
[8] P. Maris, C.D. Roberts, Int. J. Mod. Phys. E **12**, 297 (2003)
[9] C.S. Fischer, J. Phys. G **32**, R253 (2006)
[10] E. Rojas, J. de Melo, B. El-Bennich, O. Oliveira, T. Frederico, J. High Energy Phys. **1310**, 193 (2013)
[11] F. Gross, Phys. Rev. **186**, 1448 (1969)
[12] F. Gross, J. Milana, Phys. Rev. D **43**, 2401 (1991)
[13] F. Gross, J. Milana, Phys. Rev. D **45**, 969 (1992)
[14] F. Gross, J. Milana, Phys. Rev. D **50**, 3332 (1994)
[15] C. Savkli, F. Gross, Phys. Rev. C **63**, 035208 (2001)
[16] E.P. Biernat, F. Gross, M.T. Peña, A. Stadler, Phys. Rev. D **89**, 016005 (2014)
[17] C. Michael, Phys. Rev. Lett. **56**, 1219 (1986)
[18] T.J. Allen, M.G. Olsson, S. Veseli, Phys. Rev. D **62**, 094021 (2000)
[19] G.S. Bali, K. Schilling, A. Wachter, Phys. Rev. D **56**, 2566 (1997)
[20] J.E. Villate, D.S. Liu, J.E. Ribeiro, P.J. de A. Bicudo, Phys. Rev. D **47**, 1145 (1993)
[21] S.L. Adler, Phys. Rev. **137**, B1022 (1965)
[22] F. Gross, D.O. Riska, Phys. Rev. C **36**, 1928 (1987)
[23] F. Gross, Y. Surya, Phys. Rev. C **47**, 703 (1993)
[24] Y. Surya, F. Gross, Phys. Rev. C **53**, 2422 (1996)
[25] E.P. Biernat, M.T. Peña, J.E. Ribeiro, A. Stadler, F. Gross, Phys. Rev. D **90**, 096008 (2014)
[26] P. Bicudo, S. Cotanch, F. Llanes-Estrada, P. Maris, E. Ribeiro, A. Szczepaniak, Phys. Rev. D **65**, 076008 (2002)
[27] P. Bicudo, Phys. Rev. C **67**, 035201 (2003)

EPJ Web of Conferences **113**, 05005 (2016)
DOI: 10.1051/epjconf/201611305005

Antihyperon-Hyperon production in antiproton-proton annihilations with PANDA at FAIR

Michael Papenbrock[1,a]
on behalf of the PANDA collaboration

[1] *Dept. of Physics and Astronomy, Uppsala University, Box 516, S-751 20 Uppsala, Sweden*

Abstract. Hyperon production is an excellent probe of QCD in the confinement domain, and spin observables are a powerful tool in understanding the underlying physics. For the Ω hyperon, seven polarisation parameters can be extracted from the angular distributions of its decay products with the future PANDA experiment at FAIR. Simulation studies reveal great prospects for strange and single charmed hyperon channels with PANDA. Software tools supporting these investigations are currently under development.

Hyperon production in $\overline{p}p \to \overline{Y}Y$ reactions gives important insights into strangeness and charm production. In particular, the production of single-, multi-strange and single-charmed hyperons from the light quarks in $\overline{p}p$ implies processes where they are replaced by heavier quarks. For any given process the relevant degrees of freedom are given by its energy scale, which for strangeness production is determined by the mass of the strange quark, $m_s \approx 100\,\mathrm{MeV}$. This is close to the QCD cut-off, $\Lambda_{QCD} \approx 200\,\mathrm{MeV}$ [1], where the coupling constant α_s grows too large for perturbative QCD to be applicable. Therefore, the production of strange hyperons presents an excellent tool to study QCD in the region below the perturbative regime, *i.e.* the confinement domain, about which very little is known so far. On the other hand, the scale of charm production is governed by the mass of the charm quark, $m_c \approx 1300\,\mathrm{MeV}$, which is more than ten times larger than m_s. In this region, the strong coupling constant becomes $\alpha \approx 0.3$, just barely enough for a perturbative treatment to be valid. Comparing the production of strange and charmed hyperons could give important insights into the differences in the production mechanism at these two energy scales. The production of hyperons in $\overline{p}p \to \overline{Y}Y$ is often described by theoretical models based on a constituent quark-gluon picture [2]. In case of strange hyperons, models based on the exchange of one or more t-channel kaons exist as well [3]. There have also been efforts to combine both approaches [4]. In order to discriminate between different models, spin observables are very powerful. For hyperons these can be related to the spin of the individual quarks. The Λ hyperon is often modelled as consisting of a spin 0 ud di-quark and an s-quark carrying the hyperon spin. Since the Λ_c^+ hyperon can be described in a similar way, comparing the spin observables of Λ and Λ_c^+ could give deeper insight into the role of spin degrees of freedom in the production of s- and c-quarks.

The density matrix of a particle of arbitrary spin j is given by

$$\rho = \frac{1}{2j+1} I + \sum_{L=1}^{2j} \frac{2j}{2j+1} \sum_{M=-L}^{L} Q_M^L r_M^L, \tag{1}$$

[a] e-mail: michael.papenbrock@physics.uu.se

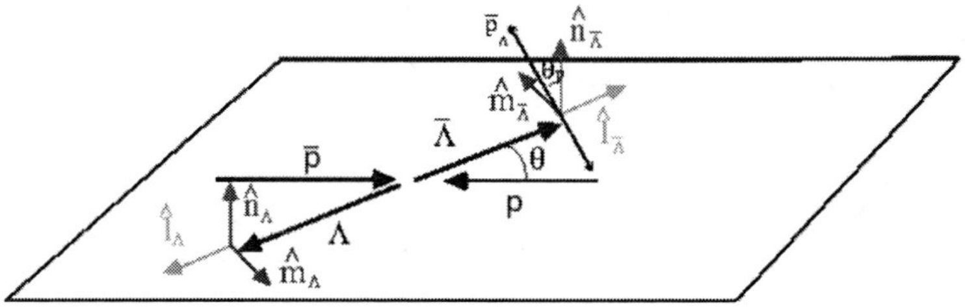

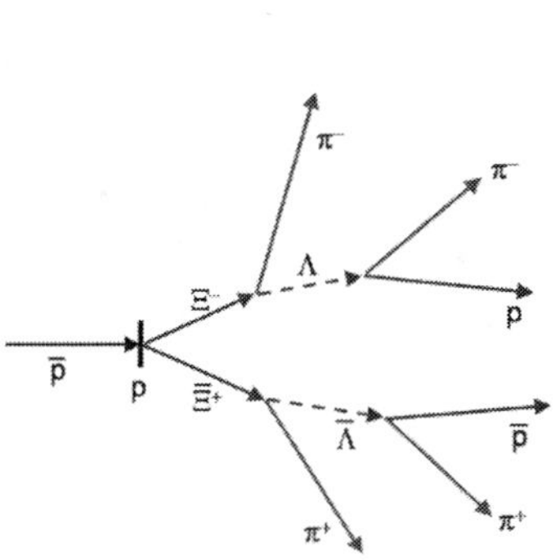

Figure 1. **(a)** The reference system of $\overline{p}p \rightarrow \overline{Y}Y$. **(b)** The $\overline{p}p \rightarrow \overline{\Xi}\Xi, \Xi \rightarrow \Lambda\pi, \Lambda \rightarrow p\pi$ decay.

in terms of the hermitian matrices Q^L_M and the polarisation parameters r^L_M [5]. Here, the first term represents the unpolarised differential cross section I and the second the polarised part with the r^L_M parameters, L being the angular momentum and M its third component. For spin $\frac{1}{2}$ particles the Q^L_M are the Pauli matrices and the polarisation parameters correspond to the vector polarisations P_l, P_m and P_n. The indices are defined in the coordinate system shown in the left panel of figure 1. In parity conserving processes, *e.g.* $\overline{p}p \rightarrow \overline{\Lambda}\Lambda$ with unpolarised beam and target, P_l and P_m equal zero due to symmetries of the spin density matrix. The only non-zero polarisation component is the one perpendicular to the production plane, namely P_n.

Since the hyperons have a weak, parity violating decay, the decay products are preferentially emitted along the direction of the spin of the parent hadron. For $\Lambda \rightarrow p\pi^-$, the polarisation is accessible by measuring the angular distribution of the proton, both of which are related by $I(\cos\theta_p) = \frac{1}{4\pi}(1 + \alpha_\Lambda P_n \cos\theta_p)$, with $\alpha_\Lambda = 0.64$ [6] being the asymmetry parameter. Some hyperons decay into other weakly decaying hyperons, *e.g.* $\Xi^- \rightarrow \Lambda\pi^-$ (see right panel of figure 1), in which case additional asymmetry parameters become accessible through the angular distribution of the protons [7].

The spin structure of spin $\frac{3}{2}$ hyperons is more complicated. Fifteen Q^L_M matrices with corresponding r^L_M parameters become relevant. Eight polarisation parameters were found to be zero using symmetries imposed by the strong interaction [8]. In the $\Omega^- \rightarrow \Lambda K^-$ decay the parameters r^2_0, r^2_1, and r^2_2 can be obtained from the angular distribution of the Λ [8], assuming that $\alpha_\Omega = 0$ according to previous measurements [6]. The combined angular distribution $I(\theta_\Lambda, \phi_\Lambda, \theta_p, \phi_p)$ of the Λ hyperons from Ω^- decay and the protons from the subsequent Λ decay grants access to the other four non-zero polarisation parameters r^1_{-1}, r^3_{-1}, r^3_{-2} and r^3_{-3}.

The major part of the existing data on $\overline{p}p \rightarrow \overline{Y}Y$ are collected with beam momenta below 2 GeV/c. In this region, the PS185 experiment provided a high-quality data set for Λ and Σ hyperons [9]. The spin structure of the $\overline{p}p \rightarrow \overline{\Lambda}\Lambda$ reaction was explored in detail, but neither quark-gluon models nor hadron models were in agreement with the data on the spin observables [10]. For multi-strange and charmed hyperons produced in $\overline{p}p$ annihilation the existing data are much more limited, with only a few bubble chamber events on $\overline{p}p \rightarrow \overline{\Xi}^+\Xi^-$ and no data on $\overline{p}p \rightarrow \overline{\Omega}^+\Omega^-$ nor $\overline{\Lambda_c}\Lambda_c^+$. New experimental data will serve as a guideline for the theoretical investigation which in turn could widen our understanding of the underling processes in this energy regime.

The foreseen PANDA experiment at FAIR provides unique possibilities for the investigation of hyperon physics. The HESR storage ring can be operated with antiprotons in a momentum range

Table 1. Estimated production rates of various hyperon channels [8, 13]. The cross sections marked with a * have a very large uncertainty. Details are given in the text.

Beam mom. (GeV/c)	Reaction	σ (μb)	Eff (%)	Decay	Rate at $1\cdot10^{31}\mathrm{cm}^{-2}\mathrm{s}^{-1}$
1.64	$\overline{p}p \to \overline{\Lambda}\Lambda$	64	11	$\Lambda \to p\pi^-$	29 s^{-1}
4	$\overline{p}p \to \overline{\Lambda}\Sigma^0$	$\approx$40	$\approx$30	$\Sigma^0 \to \Lambda\gamma$	30 s^{-1}
4	$\overline{p}p \to \overline{\Xi}^+\Xi^-$	$\approx$2	$\approx$20	$\Xi^- \to \Lambda\pi^-$	1.5 s^{-1}
12	$\overline{p}p \to \overline{\Omega}^+\Omega^-$	$\approx$0.002*	$\approx$30	$\Omega \to \Lambda K^-$	$\approx$4 h^{-1}
12	$\overline{p}p \to \overline{\Lambda}_c\Lambda_c^+$	$\approx$0.1*	$\approx$35	$\Lambda_c \to \Lambda\pi^+$	$\approx$2 d^{-1}

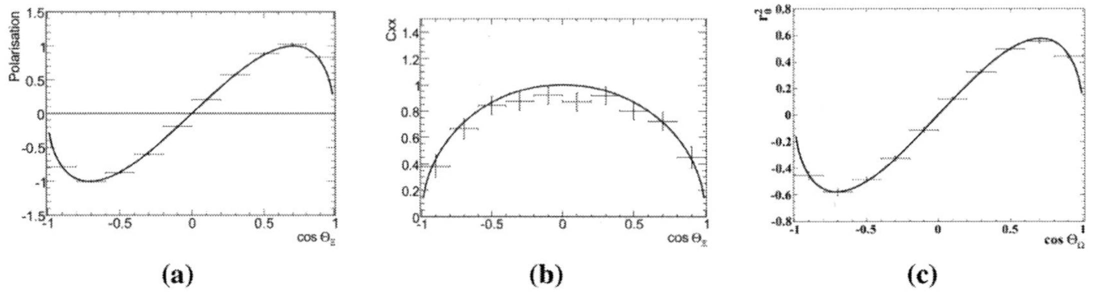

Figure 2. **(a)** The polarisation of the Ξ^- as a function of $\cos\theta_\Xi$. **(b)** The spin correlation C_{xx} in the x direction ($\hat{m}$ in Fig. 1) of the $\overline{\Xi}^+$ and the Ξ^-. **(c)** The polarisation parameter r_0^2 of the Ω as a function of $\cos\theta_\Omega$. The lines represent the input sine functions (see text) and the points and bars the reconstructed values for 10000 simulated events.

from 1.5 to 15 GeV/c, interacting with an internal hydrogen target. The PANDA experiment will have nearly full angular coverage with precise tracking and vertex reconstruction, particle identification, calorimetry, and will benefit from the high luminosity of HESR [11]. Simulation studies with a simplified Monte Carlo framework [12] show excellent prospects with high signal and low background rates and a good detection efficiency across the full phase space for all strange and single-charmed hyperon channels [8, 13]. The estimated signal rates in Table 1 are calculated using the day-1 luminosity mode of HESR of $1\cdot10^{31}\mathrm{cm}^{-2}\mathrm{s}^{-1}$. The only well known cross section is the one for the $\overline{\Lambda}\Lambda$ channel at 1.64 GeV/c, whereas the cross sections for the other strange channels are extrapolated from data at other energies. For the heavier hyperons, *i.e.* Ω and Λ_c, where no data exist from $\overline{p}p$ annihilation, the values are taken from theoretical predictions which can differ by an order of magnitude [14, 15].

Simulation studies of spin observables for Λ, Σ, Ξ, Ω, and Λ_c hyperons have been performed as well. Since there exist no data on multi-strange or charmed hyperon channels, a parametrisation based on sine functions was introduced at the event generation stage. These fulfil the requirement of the definition of the production plane (see figure 1) of being equal to zero at $\cos\theta_Y = \pm1$. Figure 2 shows some examples of spin observables of Ξ and Ω, parametrised by $P_{y,\Xi} = \sin 2\theta_\Xi$, $C_{xx,\Xi} = \sin\theta_\Xi$ and $r^2_{0,\Omega} = \sin 2\theta_\Omega/\sqrt{3}$. It is clear that these can be accurately reconstructed with PANDA.

The weak decay of the hyperons goes along with comparatively long lifetimes. Their decay vertices can easily be displaced by several tens of centimetres from the production point, setting high demands on precise track reconstruction. The PANDA experiment will include a variety of subdetectors designed for charged track reconstruction, including a Micro Vertex Detector (MVD), a Straw Tube Tracker (STT), and a Gas Electron Multiplier (GEM). The reconstruction can also be supported by a Scintillator Tile Hodoscope (SciTil). Multiple algorithms for pattern recognition and tracking are currently under development with different algorithms for different detectors or detector groups. A special focus is given here on the STT, which is a large barrel of 4636 straw tubes. Since its main purpose is to gather transverse information from the tracks, most of these are aligned parallel to the beam axis, while a few layers of straws are skewed by $\pm 2.9°$, allowing to gain longitudinal information as well. Simulation studies show that tracks may completely miss one or more of the aforementioned subdetectors, creating only hits within the STT. In these cases, the skewed straws of the STT are the only viable source for a longitudinal momentum information of the track. A tool to support a standalone track finder for the STT is under development which matches the isochrones from the skewed straws with axial information gained from the parallel straws to gather p_z information [16].

Using the high luminosity mode of HESR, the mean time between events can be as small as 100 ns while the drift times in the STT can be up to 250 ns, usually causing hits from multiple events to overlap. Events are reconstructed on a software level, using time gaps between hits in one or more of the faster subdetectors, *e.g.* the MVD. This procedure also yields a first approximation of the start time of each event, T_0, which in turn is required to determine the isochrones from the straw tube signals. With this first approximation, preliminary tracks will be propagated to detectors with excellent time resolution, *e.g.* the SciTil, in order to acquire a precise T_0.

References

[1] Ellis R K, Stirling W J and Webber B R, *QCD and Collider Physics* (Cambridge Monographs on particle physics, Nuclear Physics and Cosmology, Cambridge, 1996)

[2] Kohno M and Weise W, Phys. Lett. B **179**, 15 (1986); Rubinstein H R and Snellman H, Phys. Lett. B **165**, 187 (1985); Furui S and Faessler A, Nucl. Phys. A **486**, 669 (1987); Burkardt M and Dillig M, Phys. Rev. C **37**, 1362 (1988); Alberg M A et al., Z. Phys. A **331**, 207 (1988)

[3] Tabakin F and Eisenstein R A, Phys. Rev. C **31**, 1857 (1985); Kohno M and Weise W, Phys. Lett. B **179**, 15 (1986); La France P et al., Phys. Lett. B **214**, 317 (1988); Timmermanns R G E et al., Phys. Rev. D **45**, 2288 (1992); Haidenbauer J et al., Phys. Rev. C **46**, 2516 (1992)

[4] Ortega P G et al., Phys. Lett. B **696**, 352 (2011)

[5] Doncel M G et al., Nucl. Phys B **38**, 477 (1972); Doncel M G et al., Phys. Rev. D **7**, 815 (1973)

[6] Yao W M et al., J. Phys. G **33**, 1 (2006)

[7] Koch W ed. Nikolic M, *Analysis of scattering and decay* (Gordon and Breach, New York-London-Paris, 1968)

[8] Thomé E, *Multi-Strange and Charmed Antihyperon-Hyperon Physics for PANDA* (Ph. D. Thesis, Uppsala Univeryity, 2012)

[9] Johansson T, AIP Conf. Proc. 8th Int. Conf. on Low Energy Antiproton Physics, 95 (2003)

[10] Klempt E et al., Phys. Rep. **368**, 119 (2002)

[11] The PANDA collaboration, Technical Performance Report, (2005) https://panda.gsi.de/oldwww/archive/public/panda_tpr.pdf

[12] The PANDA collaboration, Physics Performance Report, 43 (2009) https://panda.gsi.de/oldwww/archive/public/panda_pb.pdf

[13] Grape S, *Studies on PWO Crystals and Simulations of the* $\bar{p}p \rightarrow \bar{\Lambda}\Lambda, \bar{\Lambda}\Sigma^0$ *Reactions for the PANDA experiment* (Ph. D. Thesis, Uppsala University, 2009)

[14] Kaidalov A B and Volkovitsky P E, Z. Phys. C **63**, 51 (1994)

[15] Titov A I and Kämpfer B 2008, Phys. Rev. C **78**, 025201 (2008); Gornitschnig A T et al., Eur. Phys. J A **42**, 43 (2009); He J et al., Phys. Rev. D **84**, 114010 (2011); Khodjamirian A et al., Eur. Phys. J A **48**, 31 (2012)

[16] Ikegami Andersson W K, *private communication*

EPJ Web of Conferences **113**, 05006 (2016)
DOI: 10.1051/epjconf/201611305006

The Pion-Cloud Contribution to the Electromagnetic Nucleon Form Factors

Daniel Kupelwieser[1,a] and Wolfgang Schweiger[1,b]

[1] *Institute of Physics, University of Graz, A-8010 Graz, Austria*

Abstract. We study the electromagnetic structure of the nucleon within a hybrid constituent-quark model that comprises, in addition to the $3q$ valence component, also a $3q$+π non-valence component. To this aim we employ a Poincaré-invariant multichannel formulation based on the point-form of relativistic quantum mechanics. With a simple 3-quark wave function for the bare nucleon, i.e. the $3q$-component, we obtain reasonable results for the nucleon form factors and predict the meson-cloud contribution to be significant only below $Q^2 \lesssim 0.5$ GeV^2 amounting to about 10% for $Q^2 \to 0$, in accordance with the findings of other authors.

1 Introduction

Pions, undoubtedly, play an important role in explaining the nucleon structure [1]. In the present contribution we try to substantiate the physical picture of a nucleon as consisting of a quark core that is surrounded by a pion cloud by means of a hybrid constituent-quark model in which a physical nucleon is a superposition of a $3q$ and a $3q$+π component. Thereby both, the constituent quarks and and the pion, are treated as dynamical degrees of freedom. Within our model pions couple directly to the constituent quarks which are, in addition, subject to an instantaneous confining force. Such kind of models have also been considered by other authors [2, 3]. These authors use the light-front formalism to account for relativity in a proper way, we rather adopt the point-form approach which has already turned out to be very convenient for form-factor calculations of mesonic systems [4–7].

2 Formalism

Following the same strategy as in Refs. [4–7] we calculate the invariant 1γ-exchange electron-nucleon scattering amplitude, extract the electromagnetic nucleon current, analyze the covariant structure of the current and identify the electromagnetic nucleon form factors. In order to derive the 1γ-exchange amplitude we employ a Poincaré invariant coupled-channel formulation which relies on the point form of relativistic quantum mechanics [5]. Typical for the point-form, all four components of the momentum operator are interaction dependent, whereas the generators of Lorentz transformations stay free of interactions. As a welcome consequence one has simple rotation and boost properties and

[a]e-mail: con-fuser43@web.de
[b]e-mail: wolfgang.schweiger@uni-graz.at

angular-momentum addition works like in non-relativistic quantum mechanics. Poincaré invariance is achieved by employing the Bakamjian-Thomas construction which allows to separate the overall motion of the system from the internal motion in a neat way:

$$\hat{P}^\mu = \hat{\mathcal{M}}\,\hat{V}^\mu_{\text{free}} = \left(\hat{\mathcal{M}}_{\text{free}} + \hat{\mathcal{M}}_{\text{int}}\right)\hat{V}^\mu_{\text{free}}\,, \tag{1}$$

i.e. the 4-momentum operator factorizes into an interaction-dependent mass operator $\hat{\mathcal{M}}$ and a free 4-velocity operator $\hat{V}^\mu_{\text{free}}$. Bakamjian-Thomas-type mass operators are most conveniently represented in a velocity-state basis. Velocity states $|V;\mathbf{k}_1,\mu_1;\mathbf{k}_2,\mu_2;\ldots;\mathbf{k}_n,\mu_n\rangle$ are characterized by the overall velocity V ($V_\mu V^\mu = 1$) of the system, the CM momenta $\mathbf{k}_i$ of the individual particles and their (canonical) spin projections μ_i [5].

We now want to calculate the 1γ-exchange amplitude for elastic electron scattering off a nucleon that consists of a $3q$ and a $3q+\pi$ component. Thereby not only the dynamics of electron and quarks, but also the dynamics of the photon and the pion should be fully taken into account. This is accomplished by means of a multichannel formulation that comprises all states which can occur during the scattering process (i.e. $|3q,e\rangle$, $|3q,\pi,e\rangle$, $|3q,e,\gamma\rangle$, $|3q,\pi,e,\gamma\rangle$). What one then needs, in principle, are scattering solutions of

$$\begin{pmatrix} \hat{M}^{\text{conf}}_{3qe} & \hat{K}_\pi & \hat{K}_\gamma & \hat{K}_{\pi\gamma} \\ \hat{K}^\dagger_\pi & \hat{M}^{\text{conf}}_{3q\pi e} & \hat{\hat{K}}_{\pi\gamma} & \hat{K}_\gamma \\ \hat{K}^\dagger_\gamma & \hat{\hat{K}}^\dagger_{\pi\gamma} & \hat{M}^{\text{conf}}_{3qe\gamma} & \hat{K}_\pi \\ \hat{K}^\dagger_{\pi\gamma} & \hat{K}^\dagger_\gamma & \hat{K}^\dagger_\pi & \hat{M}^{\text{conf}}_{3q\pi e\gamma} \end{pmatrix} \begin{pmatrix} |\psi_{3qe}\rangle \\ |\psi_{3q\pi e}\rangle \\ |\psi_{3qe\gamma}\rangle \\ |\psi_{3q\pi e\gamma}\rangle \end{pmatrix} = \sqrt{s} \begin{pmatrix} |\psi_{3qe}\rangle \\ |\psi_{3q\pi e}\rangle \\ |\psi_{3qe\gamma}\rangle \\ |\psi_{3q\pi e\gamma}\rangle \end{pmatrix} \tag{2}$$

which evolve from an asymptotic electron-nucleon in-state $|eN\rangle$ with invariant mass $\sqrt{s}$. The diagonal entries of this matrix mass operator contain, in addition to the relativistic kinetic energies of the particles in the particular channel, an instantaneous confinement potential between the quarks. The off-diagonal entries are vertex operators which describe the transition between the channels. In the velocity-state representation these vertex operators are directly related to usual quantum-field theoretical interaction-Lagrangean densities [5]. The 4-vertices $\hat{K}^{(\dagger)}_{\pi\gamma}$ and $\hat{\hat{K}}^{(\dagger)}_{\pi\gamma}$ show up only for pseudovector pion-quark coupling and vanish for pseudoscalar pion-quark coupling which we will concentrate on in the following.

At this point it is convenient to reduce Eq. (2) to an eigenvalue problem for $|\psi_{3qe}\rangle$ by means of a Feshbach reduction:

$$\left[\hat{M}^{\text{conf}}_{3qe} + \hat{K}_\pi(\sqrt{s} - \hat{M}^{\text{conf}}_{3q\pi e})^{-1}\hat{K}^\dagger_\pi + \hat{V}^{\text{opt}}_{1\gamma}(\sqrt{s})\right]|\psi_{3qe}\rangle = \sqrt{s}\,|\psi_{3q\pi e}\rangle\,, \tag{3}$$

where $\hat{V}^{\text{opt}}_{1\gamma}(\sqrt{s})$ is the 1γ-exchange optical potential. The invariant 1γ-exchange electron-nucleon scattering amplitude is now obtained by sandwiching $\hat{V}^{\text{opt}}_{1\gamma}(\sqrt{s})$ between (the valence component of) physical electron-nucleon states $|eN\rangle$, i.e. eigenstates of $[\hat{M}^{\text{conf}}_{3qe} + \hat{K}(\sqrt{s} - \hat{M}^{\text{conf}}_{3q\pi e})^{-1}\hat{K}^\dagger]$. The crucial point is now to observe that, due to instantaneous confinement, propagating intermediate states do not contain free quarks, they rather contain either physical nucleons N or bare baryons $\tilde{B}$, the latter being eigenstates of the pure confinement problem. This allows us to rewrite the scattering amplitude in terms of pure hadronic degrees of freedom with the quark substructure being hidden in vertex form factors. This is graphically represented in Fig. 1.

In a first step only nucleons are taken into account in the pion loop. To calculate the diagrams shown in Fig. 1 we thus need the electromagnetic and strong form factors of the bare nucleon and the the electromagnetic form factor of the pion. The microscopic expression (in terms of quark-degrees

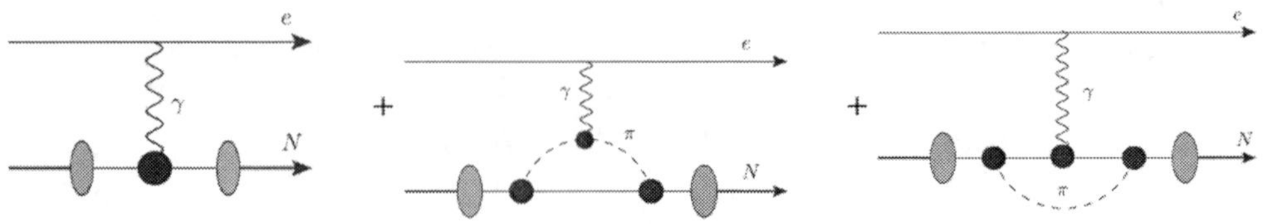

Figure 1. Diagrams representing the 1γ-exchange amplitude for electron scattering off a "physical" nucleon N, i.e. a bare nucleon dressed by a pion cloud. The time orderings of the γ-exchange are subsumed under a covariant photon propagator. Black blobs represent vertex form factors for the coupling of a photon or pion to the bare nucleon $\tilde{N}$. A vertex form factor is also assumed at the photon-pion vertex. The ovals represent the wave function (i.e. essentially the probability $P_{\tilde{N}/N}$) for finding the bare nucleon in the physical nucleon.

of freedom) for the first diagram in Fig. 1 gives the electromagnetic form factors of the bare nucleon.[1] What enters into the analytical expressions for these form factors (and also for the $\pi\tilde{N}\tilde{N}$ form factor) is the 3-q bound-state wave function of the bare nucleon. Instead of solving the bound-state problem for a particular confinement potential, we adopt a model for the 3-quark wave function which has already been used in corresponding front-form calculations [2, 3]:

$$\Phi_{\tilde{N}}\left(\tilde{\mathbf{k}}_i\right) = \frac{\mathcal{N}}{((\sum \tilde{\omega}_i)^2 + \beta^2)^\gamma}\,. \tag{4}$$

$\tilde{\mathbf{k}}_i$ and $\tilde{\omega}_i$ denote the quark momenta and energies in the rest frame of the nucleon and the parameters β, γ are taken from Ref. [3]. The normalization $\mathcal{N}$ has to be fixed such that the whole nucleon wave function, including the $3q$+π component, is normalized to one. The electromagnetic pion form factor is taken from Ref. [4], where it has been calculated within the same approach as here using a harmonic-oscillator model for the $u\bar{d}$ bound-state wave function of the π^+. An analogous calculation gives also the strong coupling and the strong form factor for the $\pi\tilde{N}\tilde{N}$ vertex [8]. Here we want to emphasize that, unlike the authors of Refs. [2, 3], who took a phenomenological $\pi\tilde{N}\tilde{N}$ form factor, we have calculated both, the electromagnetic form factors of the bare nucleon as well as the strong $\pi\tilde{N}\tilde{N}$ vertex form factor with the same microscopic input, namely the 3-quark bound-state wave function $\Phi_{\tilde{N}}$ given in Eq. (4).

3 Results and Outlook

With the model sketched above we achieve good agreement with the experimental data for proton electric and magnetic form factors. The corresponding results for the neutron electric and magnetic form factors are shown in Fig. 2. Whereas our neutron magnetic form factor is also in reasonable agreement with a parameterization of the corresponding experimental data, the reproduction of the neutron electric form factor seems to be less satisfactory. But here one has to notice that it is a rather small quantity and the error bars on the experimental data points are, in general, much larger than the indicated uncertainty of the parameterization. The size of the pionic contribution to all the nucleon form factors is comparable with the one found in Refs. [2, 3]. A significant effect of the $3q$+π component on the form factors is only observed for momentum transfers $Q^2 \lesssim 0.5$ GeV^2, where it leads to a welcome modification of the Q^2-dependence.

To improve the model one may first think of a more sophisticated 3-q wave function of the (bare) nucleon, containing, e.g., a mixed $SU(6)$ spin-flavor-symmetry component like in Ref. [3]. Further

[1] For details of their extraction and a discussion of the problems connected with wrong cluster properties associated with the Bakamjian-Thomas construction we refer to Refs. [4, 5, 7].

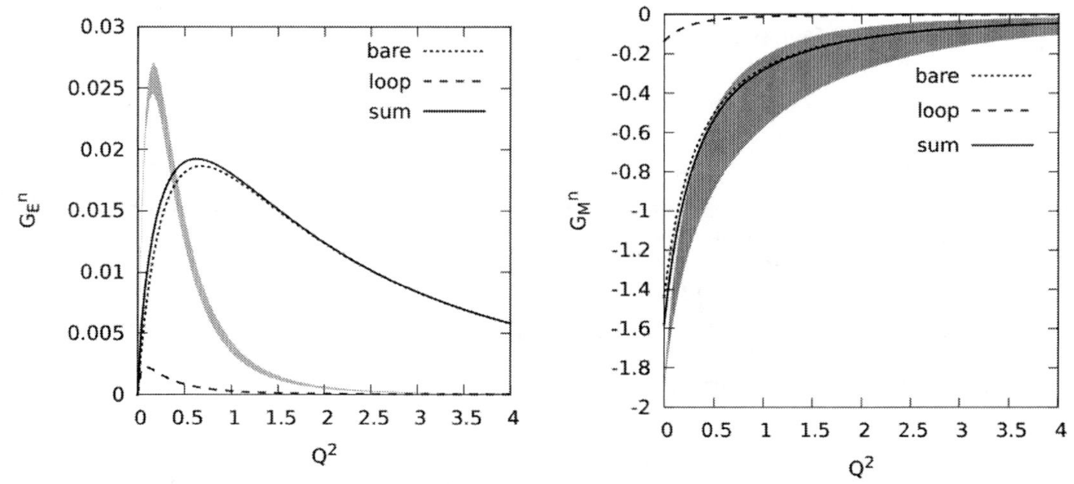

Figure 2. The neutron electric (left) and magnetic (right) form factors as predicted by our model (solid line). The $3q$ valence contribution is indicated by the dotted line, the contribution due to the $3q + \pi$ non-valence component by the dashed line. The shaded area is a parameterization of the experimental data (including uncertainties) [9].

improvements concern the replacement of the pseudoscalar by the pseudovector $\pi\tilde{N}\tilde{N}$ coupling, which guarantees correct properties in the chiral limit. This would entail additional pion-loop graphs with a $\gamma\pi\tilde{N}\tilde{N}$ vertex. A complete treatment of the pion-loop contribution will also require to take into account other baryons, different from the nucleon, within the loop, the lightest and most important of them being the Δ. We do not expect that this kind of refinements, which are the subject of ongoing work, will drastically change the size of the pion-loop contribution, but they hopefully will help to achieve an even better reproduction of the electromagnetic nucleon form factors.

Acknowledgements

D. Kupelwieser acknowledges the support of the "Fonds zur Förderung der wissenschaftlichen Forschung in Österreich" (FWF DK W1203-N16).

References

[1] A. W. Thomas and W. Weise, *The Structure of the Nucleon* (Wiley-VCH, Berlin, 2001)
[2] G. A. Miller, Phys. Rev. C **66**, 032201 (2002)
[3] B. Pasquini and S. Boffi, Phys. Rev. D **76**, 074011 (2007)
[4] E. P. Biernat, W. Schweiger, K. Fuchsberger and W. H. Klink, Phys. Rev. C **79**, 055203 (2009)
[5] E. P. Biernat, W. H. Klink and W. Schweiger, Few Body Syst. **49**, 149 (2011)
[6] M. Gomez-Rocha and W. Schweiger, Phys. Rev. D **86**, 053010 (2012)
[7] E. P. Biernat and W. Schweiger, Phys. Rev. C **89**, 055205 (2014)
[8] D. Kupelwieser and W. Schweiger, Few Body Syst. **55**, 881 (2014)
[9] J. J. Kelly, Phys. Rev. C **70**, 068202 (2004)

EPJ Web of Conferences **113**, 05007 (2016)
DOI: 10.1051/epjconf/201611305007

Exclusive π^- Electroproduction off the Neutron in Deuterium in the Resonance Region

Ye Tian[1,a] and Ralf W. Gothe[1,b] for the CLAS Collaboration

[1] *Department of Physics and Astronomy, University of South Carolina, Columbia, SC, USA*

Abstract. The goal of our research is to provide the exclusive $\gamma^*(n) \rightarrow p^+\pi^-$ reaction cross section from deuterium data using the correction factor that account for the final state re-scattering that can be determined from the data set itself. The "e1e" Jefferson Lab CLAS data set that we analyze includes both a hydrogen and deuterium target run period, which allows a combined analysis of pion electroproduction off the free proton, the bound proton, and the bound neutron under the same experimental conditions. Hence it will provide the experimentally best possible information about the off-shell and final state interaction effects in deuterium, which must be considered in order to extract the neutron information. This data set will provide results with a kinematic coverage for the hadronic invariant mass W up to 1.7 GeV and in the momentum transfer Q^2 range of $0.4 - 1.0\ GeV/c^2$. The cross section analysis of this data set is currently underway, which will considerably improve our knowledge of the Q^2 evolution of $\pi^- p$ electroproduction cross sections off bound neutron needed for the extraction of excited neutron state electrocouplings for the first time.

1 Introduction

The nucleon resonance (N^*) studies are crucial to our understanding of the structure and interaction of hadrons, and are poised to push the development of quark models and QCD-based calculations forward. QCD is very successful at large Q^2 within the perturbative regime with current quarks and gauge gluons as the fundamental degree of freedom, however when Q^2 drops down to the non-perturbative regime, there is a transition to completely different degrees of freedom, the dressed quarks and gluons as well as the mesons and nucleons, which requires the development of the based on QCD non-perturbitive approaches [1] for description of the phenomena corresponding to this scale such as the nucleon structure and its excitations. Therefore we need to accumulate sufficient and precise data on meson electroproduction reactions to pin down the distance dependent baryon structure and eventually to shed light on dynamics of non-perturbitive strong interaction from these experimental data. On the experimental side, although the low-lying excited states of the proton have been studied in greater detail, there is still very few data available on neutron excitations [2]. Because of the inherent difficulty in obtaining a free neutron target, a deuterium target is the best alternative.

[a] e-mail: tianye8001@gmail.com
[b] e-mail: gothe@sc.edu

2 Preliminary results

2.1 Accounting for the re-scattering in the final state

In order to get to the exclusive $\gamma^* n \rightarrow p\pi^-$ reaction cross section from the deuterium, the final-state-interaction (FSI) correction and the off-shell effect need to be studied well. The correction factor (R) related to the final state re-scattering can be extracted directly from data by using

$$R_{FSI} = \frac{(\frac{d\sigma^{quasi-free}}{d\Omega^*_{\pi^-}})}{(\frac{d\sigma^{full}}{d\Omega^*_{\pi^-}})}, \tag{1}$$

where the exclusive quasi-free process can be isolated by applying cuts on missing mass square of the spectator proton (m_s^2) and its missing momentum ($|\vec{P}_s|$), and the exclusive full process was obtained from cutting on the missing mass square of the spectator proton only. Figure 1 shows the spectator momentum distribution (black line) from data with the detector-smeared generated proton momentum distribution from the Bonn potential (blue line). For $|\vec{P}_s| < 200\,MeV$, the comparison between the two curves reveals that the quasi-free process is absolutely dominant in this region. When $|\vec{P}_s| > 200\,MeV$, the final state interaction becomes the dominant process. This shows that we can successfully separate the quasi-free process by cutting on $|\vec{P}_s|$ at $200\,MeV$. Preliminary results of R_{FSI} as a function of the polar angle of pion in the Center of Mass (CM) frame (θ^*_π) are shown in Fig. 2.

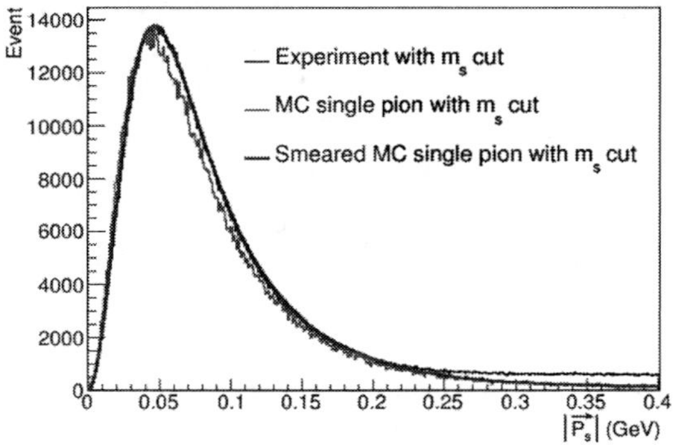

Figure 1. (Color online) The black line represents the missing momentum distribution of the unmeasured proton from data. Based on Bonn potential, the Monte Carlo simulated proton momentum distribution leads to the red line and the detector-smeared simulated distribution to the blue line.

2.2 Cross section

After particles identification, fiducial cuts, and the exclusive quasi-free event selection, the cross section for the quasi-free exclusive $\gamma^*(n) \rightarrow p^+\pi^-$ reaction with unpolarized electron beam and unpolarized deuteron target is given by

$$\frac{d^4\sigma}{dWdQ^2d\Omega_{\pi^-_*}} = \Gamma_v \frac{d\sigma}{d\Omega^*_{\pi^-}}, \tag{2}$$

where Γ_v is the virtual photon flux defined by Eq. (3), in which ϵ is the transverse polarization of the virtual photon, $\nu = E_{beam} - E_{scattered\ electron}$, and θ_e is the scattering angle of electron in the lab frame.

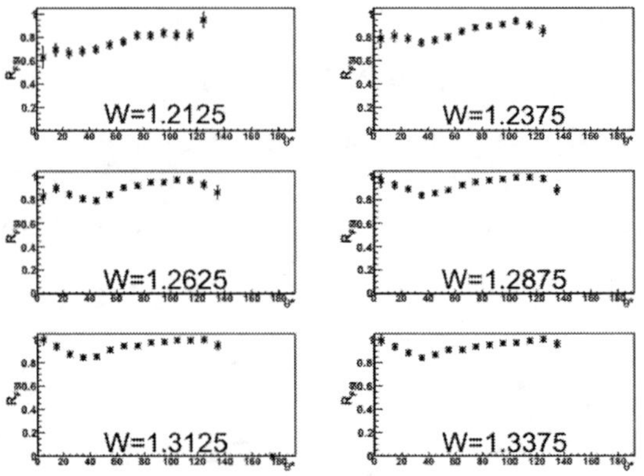

Figure 2. The FSI correction factor R_{FSI} as a function of (θ^*_π) is represented by the stars for different W bins at $Q^2 = 0.7\,GeV$.

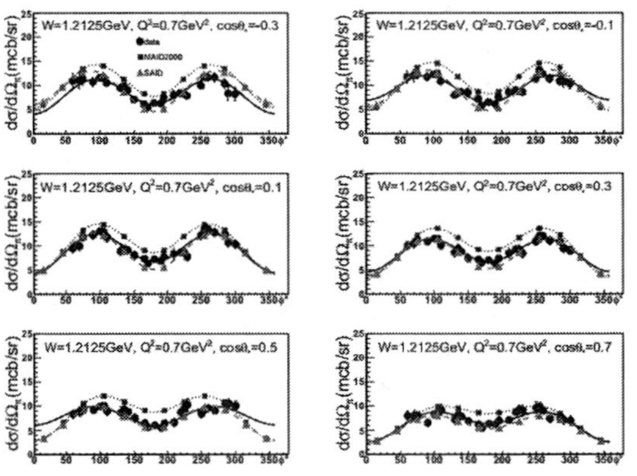

Figure 3. (Color online) A typical example of the CM frame $\phi^*_{\pi^-}$ dependent cross section for different $cos\theta^*_\pi$ values at $Q^2 = 0.7\,GeV$ and $W = 1.212\,GeV$ in comparison to MAID2000 (blue dotted line) and SAID (purple dashed line) predictions. Black solid line is the fit to extract structure functions.

$$\Gamma_\upsilon = \frac{\alpha}{4\pi}\frac{1}{E^2_{beam}M^2_n}\frac{W(W^2 - M^2_n)}{(1-\epsilon)Q^2}, \quad \epsilon = (1 + 2(1 + \frac{\nu^2}{Q^2})tan^2\frac{\theta_e}{2})^{-1} \tag{3}$$

Finally, quasi-free differential pion electroproduction cross section, after the final state re-scattering was taken off, is given by

$$\frac{d\sigma}{d\Omega^*_{\pi^-}} = \frac{1}{\Gamma_\upsilon}\frac{1}{R_{FSI}}\frac{d^4\sigma}{dWdQ^2d\Omega_{\pi^*_-}}. \tag{4}$$

Figure 3 shows the CM frame $\phi^*_{\pi^-}$ dependent differential cross section in the Δ resonance region as a typical subset of the available data. We also compare the cross section with two physics models SAID [4] and MAID2000 [3].

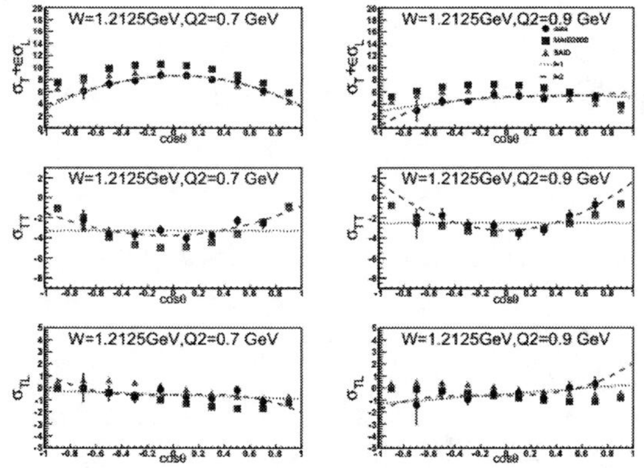

Figure 4. (Color online) $\sigma_T + \epsilon\sigma_L$, σ_{TT}, and σ_{TL} in the Δ resonance region for different Q^2 bins corresponding to top, middle, and bottom row (black points), which are compared to MAID2000 (blue squares) and SAID (purple triangles) predictions. The blue dotted and red dashed lines are Legendre polynomial fits to the black points for $l = 1$ and $l = 2$, respectively.

2.3 Structure function

The hadronic cross section $\frac{d\sigma}{d\Omega^*_{\pi^-}}$ is fit in terms of $cos\phi^*$ (Eq. (5)) to extract the structure function. The fitting function has three fit parameters a,b, and c, which corresponds to structure functions $\sigma_T + \epsilon\sigma_L$, σ_{TT}, and σ_{TL}, respectively.

$$\frac{d\sigma}{d\Omega^*_{\pi^-}} = a + b\cos 2\phi^* + c\cos\phi^*,\ a = \sigma_T + \epsilon\sigma_L,\ b = \epsilon\sigma_{TT},\ c = \sqrt{2\epsilon(1+\epsilon)}\sigma_{TL} \qquad (5)$$

An example of typical structure functions as a function of $cos\theta^*_\pi$ for $Q^2 = 0.7\,GeV$ and $0.9\,GeV$ at $W = 1.212GeV$ are shown in Fig. 4 in comparison to SAID [3] and MAID2000 [2] models . In order to gain some insight on the dominant partial wave contribution in this particular resonance region, we also perform a Legendre polynomial expansion of the structure functions up to $l = 2$, see Fig. 4.

Acknowledgements

The first author is very grateful to the second author for his kind support and encouragement. She would also like to thank Kijun Park, Gleb Fedotov, Evan Phelps, and Arjun Trivedi for many inspiring discussions on this work, Annalisa D'Angelo of the CSC committee for reviewing a preliminary version of this proceeding and offering helpful comments and suggestions, as well as the JSA Graduate Fellowship program committee and the NSF (under Grant PHY-1205782) for supporting this work.

References

[1] I.C. Clöet and C.D. Roberts, Prog. Part. Nucl. Phys. **77**, 1 (2014).
[2] J.V.Morris et al, Physics Letter B **73**, 495 (1978)
[3] D.Drechsel et al, Nuclear Physics A **645**, 145 (1999), and the full database can be accessed via website http://portal.kph.uni-mainz.de/MAID/.
[4] R.A.Arndt et al, Chinese Physics C **33**, 1063 (2009), and the full database can be accessed via website http://gwdac.phys.gwu.edu.

EPJ Web of Conferences **113**, 05008 (2016)
DOI: 10.1051/epjconf/201611305008

Pion's valence-quark GPD and its extension beyond DGLAP region

Cédric Mezrag[1,a], Hervé Moutarde[1,b], and José Rodríguez-Quintero[2,c]

[1] *Centre de Saclay, IRFU/Service de Physique Nucléaire, F-91191 Gif-sur-Yvette, France*
[2] *Departamento de Física Aplicada, Facultad de CCEE, Universidad de Huelva, Huelva E-21071, Spain*

Abstract. We briefly report on a recent computation, with the help of a fruitful algebraic model, sketching the pion valence dressed-quark generalized parton distribution and, very preliminary, discuss on a possible avenue to get reliable results in both Dokshitzer-Gribov-Lipatov-Altarelli-Parisi (DGLAP) and Efremov-Radyushkin-Brodsky-Lepage (ERBL) kinematial regions.

1 Introduction

Generalized Parton Distributions (GPDs) were introduced independently by Müller *et al.* [1], Ji [2] and Radyushkin [3]. They are related to hadron form factors by sum rules, and contain the usual Parton Distribution Functions (PDFs) as a limiting case. But they not only generalize the classical objects describing the static or dynamical content of hadrons; they also provide unique information about the structure of hadrons, including 3D imaging of its partonic components and access to the quark orbital angular momentum. GPDs have been the object of an intense theoretical and experimental activity ever since (see *e.g.* [4–9] or the reviews [10–15] and references therein).

Most of the constraints that apply to GPDs are fulfilled when the GPD is written as a double distribution (DD) [1, 16, 17], which is equivalent to expressing the GPD as a Radon transform [18]. In order to obtain insights into the nature of hadron GPDs, it has been common to model the Radon amplitudes, F, G, following Refs. [19]. This approach has achieved some phenomenological success (see *e.g.* [15, 20]); but more flexible parametrisations enable a better fit to data [21]. Such fits play a valuable role in establishing the GPD framework; However there is no known parameterization of GPDs relying on first principles only. Computing GPDs in a symmetry-preserving framework is a key ingredient for the *a priori* fulfillment of all GPD theoretical constraints. This observation is highlighted by experience drawn from the simpler case of the pion's valence-quark PDF [22]. In [23, 24], steps are made by following a different approach for the computation of hadron GPDs based on the example provided by the pion's valence-quark PDF. As sketched in [24], such a procedure only leaves with a reliable result near the so-called forward limit ($\xi = 0$), within the DGLAP region [25–28]. We will now preliminary discuss how, mainly by invoking the overlap representation and Radon-transform technology, the ERBL region [29, 30] can be reached.

[a] e-mail: cedric.mezrag@cea.fr
[b] e-mail: herve.moutarde@cea.fr
[c] e-mail: jose.rodriguez@dfaie.uhu.es

2 Computing the pion valence-quark GPD

A veracious description of the pion is only possible within a framework that faithfully expresses symmetries and their breaking patterns [31]. The Dyson-Schwinger equations (DSEs) fulfill this requirement [32, 33] and hence we employ that framework to compute pion properties in [24]. Despite its complexity, the pion bound-state is still a $J = 0$ system and hence there is only one GPD associated with a quark q in the pion ($\pi^{\pm}, \pi^0$), which is defined by the matrix element

$$H_\pi^q(x,\xi,t) = \int \frac{d^4z}{4\pi} e^{ixP\cdot z}\, \delta(n\cdot z)\, \delta^2(z_\perp)\, \langle \pi(P_+)|\bar{q}(-z/2)\, n\cdot\gamma\, \mathcal{W}(-z/2,z/2) q(z/2)\, |\pi(P_-)\rangle, \quad (1)$$

where k, n are light-like four-vectors, satisfying $k^2 = 0 = n^2$, $k\cdot n = 1$; $z_\perp$ represents that two-component part of z annihilated by both k, n; and $P_\pm = P \pm \Delta/2$ and $\mathcal{W}(-z/2,z/2)$ represents a Wilson line laid along a light-like path that joins the two listed vectors. In Eq. (1), $\xi = -n\cdot\Delta/[2n\cdot P]$ is the "skewness", $t = -\Delta^2$ is the momentum transfer, and $P^2 = t/4 - m_\pi^2$, $P\cdot\Delta = 0$. The GPD also depends on the resolving scale, ζ. Within the domain upon which perturbation theory is valid, evolution to another scale ζ' is described by the ERBL equations [29, 30] for $|x| < \xi$ and the DGLAP equations [25–28] for $|x| > \xi$, where $\xi \geq 0$.

As discussed at length in [24], the valence-quark piece of the GPD expressed by (1) can be first approximated by

$$H_\pi^{\rm v}(x,\xi,t) = \frac{1}{2} N_c {\rm tr}\int_{d\ell} \delta_n^{xP}(\ell)\, i\Gamma_\pi(\ell_+^{\rm R}; -P_+)\, S(\ell_+)\, in\cdot\Gamma(\ell_+,\ell_-)\, S(\ell_-) i\Gamma_\pi(\ell_-^{\rm R}; P_-)\,, \quad (2)$$

derived from the incomplete impulse-approximation for the so-called *handbag* diagram [22], and can be next corrected by the additional contribution

$$\begin{aligned} H_\pi^{\rm C}(x,0,-\Delta_\perp^2) &= \frac{1}{2} N_c {\rm tr}\int_{d\ell} \delta_n^{xP}(\ell) \Big[n\cdot\partial_{\ell_+^{\rm R}}\Gamma_\pi(\ell_+^{\rm R}; -P_+) S(\ell_P)\Gamma_\pi(\ell_-^{\rm R}; P_-)\, S(\ell_-) \\ &\quad + \Gamma_\pi(\ell_+^{\rm R}; -P_+) S(\ell_P) n\cdot\partial_{\ell_-^{\rm R}}\Gamma_\pi(\ell_-^{\rm R}; P_-) S(\ell_-)\Big]\,, \end{aligned} \quad (3)$$

within the non-skewed kinematical region, for the transverse momentum $\Delta_\perp^2$. In Eqs. (2,3), $\int_{d\ell} := \int \frac{d^4\ell}{(2\pi)^4}$ is a translationally invariant regularisation of the integral; $\delta_n^{xP}(\ell) := \delta(n\cdot\ell - xn\cdot P)$; the trace is over spinor indices; $\eta \in [0,1]$, $\bar{\eta} = 1-\eta$; $\ell_+^{\rm R} = \bar{\eta}\ell_+ + \eta\ell_P$, $\ell_-^{\rm R} = \eta\ell_- + \bar{\eta}\ell_P$, $\ell_\pm = \ell \pm \Delta/2$, $\ell_P = \ell - P$ (N.B. Owing to Poincaré covariance, no observable can legitimately depend on η; i.e., the definition of the relative momentum). In order to gain novel insights into pion structure, in ref. [24], we used the algebraic model of [34],

$$S(\ell) = [-i\gamma\cdot\ell + M]\Delta_M(\ell^2) \quad (4a)$$

$$\rho_\nu(z) = \frac{1}{\sqrt{\pi}} \frac{\Gamma(\nu+3/2)}{\Gamma(\nu+1)} (1-z^2)^\nu \quad (4b)$$

$$n_\pi \Gamma_\pi(\ell_\pm^{\rm R}; \pm P) = i\gamma_5 \int_{-1}^{1} dz\, \rho_\nu(z)\, \hat{\Delta}_M^\nu(\ell_{z\pm}^2) \quad (4c)$$

for the dressed-quark and pion elements in Eqs. (2,3); where $\Delta_M(\ell^2) = 1/(\ell^2 + M^2)$, M is a dressed-quark mass-scale; $\hat{\Delta}_M(\ell^2) = M^2\Delta_M(\ell^2)$; $\ell_{z\pm} = \ell_\pm^{\rm R} + (z \pm 1)P/2$ and we work in the chiral limit ($P^2 = 0 = \hat{m}$, where $\hat{m}$ is the current-quark mass); and n_π is the Bethe-Salpeter amplitude's canonical normalisation constant. We then applied the algebra and approximations described in [23, 24] to be finally left with the results displayed in Fig. 1. Notably, the so-computed non-skewed GPD shows

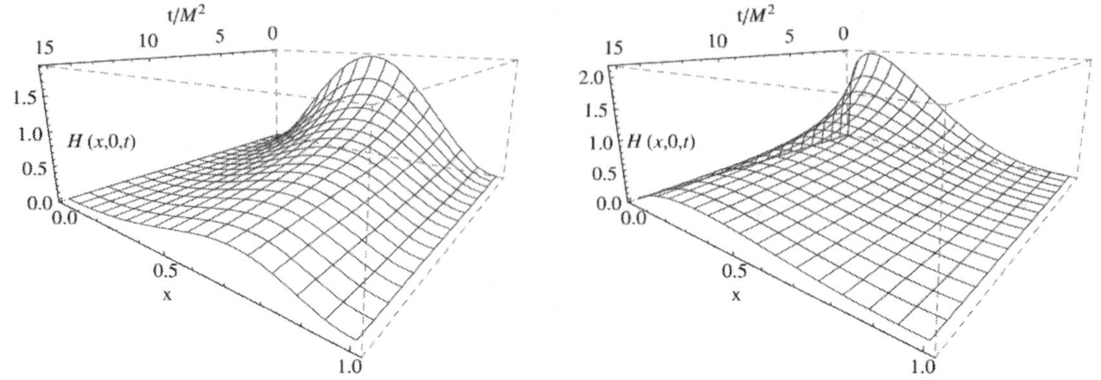

Figure 1. Pion valence dressed-quark GPD, $H^v_\pi(x,0,-\Delta^2_\perp)$, defined by the addition of Eqs. (2,3), obtained as explained in [24] and plotted as a function of $t/M^2 = \Delta^2_\perp/M^2$, where M is a mass scale for the dressed-quark in the algebraic model of [34]. *Left panel*.- Result obtained at the model scale, $\zeta_H = 0.51$ GeV. *Lower panel* – GPD evolved to $\zeta_2 = 2$ GeV using leading-order DGLAP equations.

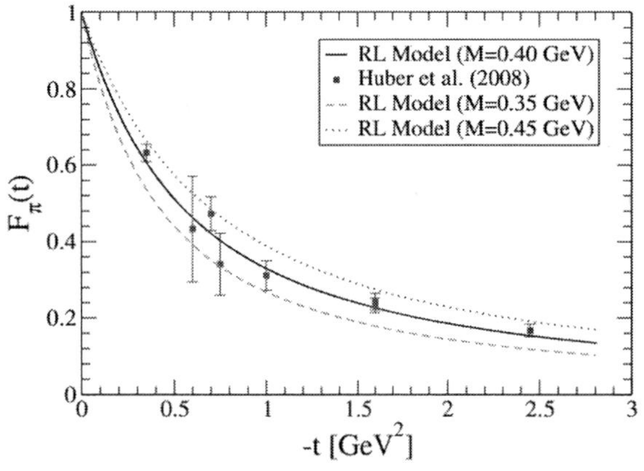

Figure 2. Pion electromagnetic form factor obtained from the integtation over x (sum rule) of $H^v_\pi(x,0,-\Delta^2_\perp)$, resulting from Eqs. (2), (4) and associated definitions, in Ref. [24]. The data are described in Ref. [38]. The most favourable comparison is obtained with $M = 0.40$ GeV in Eqs. (4) and the band shows results with $M = 0.40 \pm 0.05$ GeV.

all the properties expected on the basis of the GPD overlap representation [12, 35–37], discussed in [24], at both the model scale, $\zeta_H = 0.51$ GeV and, after being evolved using leading-order DGLAP equations, up to $\zeta_2 = 2$ GeV. In its forward limit, additionally, it naturally reproduces the pion valence dressed-quark distribution function found in [22]; and, integrated over x, it produces an estimate for the pion electromagnetic form factor which compares very well with experimental data (see Fig. 2).

3 Overlap representation, Radon transform and skewed GPD

The correction (3) appears to be well defined for a non-skewed GPD. In the aim of obtaining results beyond $\xi = 0$, both in DGLAP and ERBL kinematical regions, a sensible extension of (3) correcting (2) is far from being obvious. A different approach based on the representation of the pion GPD as overlap of light-cone wave functions (LCWF),

$$H^q_\pi(x,\xi,t)_{\xi\leq x\leq 1} = C^q \int d^2\mathbf{k}^2_\perp \Psi^*\left(\frac{x-\xi}{1-\xi}, \mathbf{k}_\perp + \frac{1-x}{1-\xi}\frac{\Delta_\perp}{2}; P_-\right)\Psi\left(\frac{x+\xi}{1+\xi}, \mathbf{k}_\perp - \frac{1-x}{1+\xi}\frac{\Delta_\perp}{2}; P_+\right), \quad (5)$$

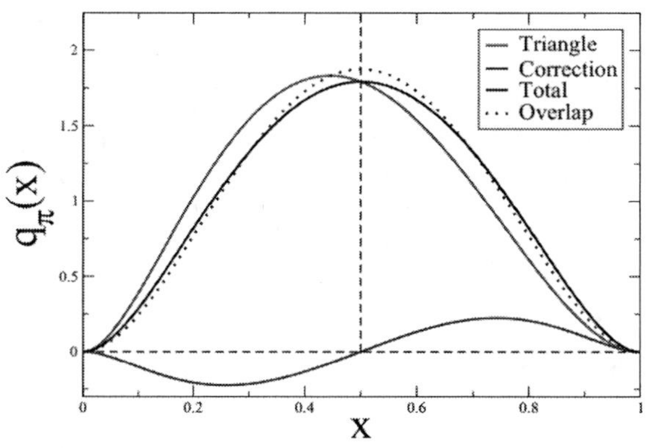

Figure 3. Pion dressed-quark distribution function (PDF) obtained from the GPD in the forward limit ($t = 0$, $\xi = 0$); solid red and blue lines stand for the results from (2) and (3), respectively, and black solid one is for their addition; the dotted line corresponds to the overlap result given by (7), obtained from Eqs. (5,6), for $\nu = 1$.

can be followed, where C^q is a normalization constant and the LCWF can be computed by integrating over k^- the pion Bethe-Salpeter wave function $\chi_\pi(k, P)$ projected onto $\gamma^+\gamma_5$

$$\Psi(k^+, \mathbf{k}_\perp; P) = -\frac{1}{2\sqrt{3}} \int \frac{dk^-}{2\pi} \mathrm{Tr}\left[\gamma^+\gamma_5 \chi_\pi(k, P)\right] . \tag{6}$$

If we obtain the Bethe-Salpeter wave function from (4), apply it to derive the LCWF with (6), and use then Eq. (5) to compute the GPD in the forward limit ($t = 0$, $\xi = 0$), we get

$$q_\pi(x) = H_\pi^q(x, 0, 0) = k_\nu x^{2\nu}(1 - x)^{2\nu} . \tag{7}$$

In the case $\nu = 1$, the normalization factor is $k_1 = 30$, and the result for the pion dressed-quark distribution function (PDF) has been proved to be numerically consistent with that obtained from Eqs. (2,3), also in the forward limit. This can be seen in Fig. 3, where both results appear depicted and, to the eye, can be barely distinguishable from each other. In ref. [22], the same result of (7) with $\nu = 1$ was introduced as an excellent and efficacious approximation to the pion's valence dressed-quark PDF, while also resulted, in [24], from a heuristic LCWF implemented in (5). Here, it appears as the natural result from the algebraic model of Eqs. (4) [34] and the overlap representation of pion's GPD.

Therefore, Eq. (5) provides with a natural extension for the GPD beyond the non-skewed limit, although only valid within the DGLAP region. Moreover, one can take also advantage of that, within the one-component DD (1CDD) scheme (see discussion in [20] and references therein), the GPD can be expressed as the Radon transform, $\mathcal{R}f$, of the distribution $f(\beta, \alpha, t)$,

$$\frac{\sqrt{1-\xi^2}}{x} H(x, \xi, t) = \mathcal{R}f(s, \varphi, t) = \int_{|\alpha|+|\beta|\leq 1} d\beta\, d\alpha\, \delta(s - \beta\cos\varphi - \alpha\sin\varphi)\, f(\beta, \alpha, t) , \tag{8}$$

where $s = x\cos\varphi$ and $\xi = \tan\varphi$. Then, the singular value decomposition of the Radon transform defined in (8), investigated within the context of the computerized tomography [39], allows for the GPD to be recast as

$$\mathcal{R}f(s, \varphi, t) = \sum_{m=0}^{\infty} \sum_{l=0}^{m} g_{ml}(t)\, e^{i\,(-m+2l)\,\varphi}\, C_m^\alpha(s) , \tag{9}$$

where $g_{ml}(t)$ are the coefficients for the expansion in terms of Geigenbauer polynomials, $C_m^\alpha(s)$. Therefore, identifying properly the coefficients $g_{ml}(t)$, one is also left with the natural extension to the ERBL kinematical region for the GPD, *via* (9). This might pave the way to build a model implementing, *a priori*, positivity and polynomiality properties.

4 Conclusions

We have briefly reported on the first steps recently made towards the computation of the pion's valence dressed-quark GPD, within the symmetry-preserving framework provided by DSEs. In addition, and very preliminary, we also discussed the open avenue, on the basis of the GPD representation as overlap of light-cone wave functions, and that can be exploited to extend previous results; in particular, from those for the GPD in DGLAP kinematical region and near the forward limit to non-zero skewness and to the ERBL domain.

Acknowledgements

This work has been partially supported by Spanish ministry projects FPA-2011-23781 and FPA-2014-53631-C2-2-P and French GDR 3034 PH-QCD "Cromodynamique Quantique et Physique des Hadrons" and ANR-12-MONU-0008-01 "PARTONS". We thank for their very valuable remarks to L. Chang, C.D. Roberts, F. Sabatié and S.M. Schmidt. H. M. and J. R-Q. are grateful for the chance to participate in the "21st Conference on Few-Body Problems in Physics", held in Chicago, and specially to C.D. Roberts for his warm hospitality at the Argonne National Laboratory.

References

[1] D. Mueller, D. Robaschik, B. Geyer, F. Dittes, J. Hořejši, Fortsch.Phys. **42**, 101 (1994), hep-ph/9812448
[2] X.D. Ji, Phys.Rev. **D55**, 7114 (1997), hep-ph/9609381
[3] A. Radyushkin, Phys.Rev. **D56**, 5524 (1997), hep-ph/9704207
[4] B. Tiburzi, G. Miller, Phys.Rev. **D67**, 113004 (2003), hep-ph/0212238
[5] L. Theussl, S. Noguera, V. Vento, Eur.Phys.J. **A20**, 483 (2004), nucl-th/0211036
[6] W. Broniowski, E. Ruiz Arriola, Phys.Lett. **B574**, 57 (2003), hep-ph/0307198
[7] C.R. Ji, Y. Mishchenko, A. Radyushkin, Phys.Rev. **D73**, 114013 (2006), hep-ph/0603198
[8] W. Broniowski, E. Ruiz Arriola, K. Golec-Biernat, Phys.Rev. **D77**, 034023 (2008), 0712.1012
[9] T. Frederico, E. Pace, B. Pasquini, G. Salme, Phys.Rev. **D80**, 054021 (2009), 0907.5566
[10] X.D. Ji, J.Phys. **G24**, 1181 (1998), hep-ph/9807358
[11] K. Goeke, M.V. Polyakov, M. Vanderhaeghen, Prog.Part.Nucl.Phys. **47**, 401 (2001), hep-ph/0106012
[12] M. Diehl, Phys.Rept. **388**, 41 (2003), hep-ph/0307382
[13] A. Belitsky, A. Radyushkin, Phys.Rept. **418**, 1 (2005), hep-ph/0504030
[14] S. Boffi, B. Pasquini, Riv.Nuovo Cim. **30**, 387 (2007), 0711.2625
[15] M. Guidal, H. Moutarde, M. Vanderhaeghen, Rept.Prog.Phys. **76**, 066202 (2013), 1303.6600
[16] A. Radyushkin, Phys.Rev. **D59**, 014030 (1999), hep-ph/9805342
[17] A. Radyushkin, Phys.Lett. **B449**, 81 (1999), hep-ph/9810466
[18] O. Teryaev, Phys.Lett. **B510**, 125 (2001), hep-ph/0102303
[19] I. Musatov, A. Radyushkin, Phys.Rev. **D61**, 074027 (2000), hep-ph/9905376
[20] C. Mezrag, H. Moutarde, F. Sabatié, Phys.Rev. **D88**, 014001 (2013), 1304.7645
[21] K. Kumericki, D. Mueller, K. Passek-Kumericki, Eur.Phys.J. **C58**, 193 (2008), 0805.0152
[22] L. Chang, C. Mezrag, H. Moutarde, C.D. Roberts, J. Rodríguez-Quintero, P.C. Tandy, Phys. Lett. **B737**, 23 (2014), 1406.5450
[23] C. Mezrag, H. Moutarde, J. Rodríguez-Quintero, F. Sabatié (2014), 1406.7425

[24] C. Mezrag, L. Chang, H. Moutarde, C.D. Roberts, J. Rodríguez-Quintero, F. Sabatié, S.M. Schmidt, Phys. Lett. **B741**, 190 (2014), 1411.6634
[25] Y.L. Dokshitzer, Sov. Phys. JETP **46**, 641 (1977), [Zh. Eksp. Teor. Fiz.73,1216(1977)]
[26] V.N. Gribov, L.N. Lipatov, Sov. J. Nucl. Phys. **15**, 438 (1972), [Yad. Fiz.15,781(1972)]
[27] L.N. Lipatov, Sov. J. Nucl. Phys. **20**, 94 (1975), [Yad. Fiz.20,181(1974)]
[28] G. Altarelli, G. Parisi, Nucl. Phys. **B126**, 298 (1977)
[29] A.V. Efremov, A.V. Radyushkin, Phys. Lett. **B94**, 245 (1980)
[30] G.P. Lepage, S.J. Brodsky, Phys. Rev. **D22**, 2157 (1980)
[31] S.X. Qin, C.D. Roberts, S.M. Schmidt, Phys. Lett. **B733**, 202 (2014), 1402.1176
[32] L. Chang, C.D. Roberts, P.C. Tandy, Chin. J. Phys. **49**, 955 (2011), 1107.4003
[33] A. Bashir, L. Chang, I.C. Cloet, B. El-Bennich, Y.X. Liu et al., Commun.Theor.Phys. **58**, 79 (2012), 1201.3366
[34] L. Chang, I. Cloet, J. Cobos-Martinez, C. Roberts, S. Schmidt et al., Phys.Rev.Lett. **110**, 132001 (2013), 1301.0324
[35] M. Burkardt, Phys.Rev. **D62**, 071503 (2000), hep-ph/0005108
[36] M. Diehl, T. Feldmann, R. Jakob, P. Kroll, Nucl. Phys. **B596**, 33 (2001), [Erratum: Nucl. Phys.B605,647(2001)], hep-ph/0009255
[37] M. Burkardt, Int. J. Mod. Phys. **A18**, 173 (2003), hep-ph/0207047
[38] G. Huber et al. (Jefferson Lab), Phys.Rev. **C78**, 045203 (2008), 0809.3052
[39] F. Natterer, *The Mathematics of Computerized Tomography* (Society for Industrial and Applied Mathematics, 2001), ISBN 9780898714937, https://books.google.fr/books?id=pZMdcI0WZxcC

EPJ Web of Conferences **113**, 05009 (2016)
DOI: 10.1051/epjconf/201611305009

Light-Front Quark Model Phenomenology Consistent with Chiral Symmetry

Ho-Meoyng Choi[1,a] and Chueng-Ryong Ji[2,b]

[1] *Department of Physics, Teachers College, Kyungpook National University, Daegu 702-701, Korea*
[2] *Department of Physics, North Carolina State University, Raleigh, NC 27695-8202, USA*

Abstract. We discuss the link between the chiral symmetry of QCD and the numerical results of the light-front quark model, analyzing both the twist-2 and twist-3 distribution amplitudes of a pion. In particular, we exhibit that the pion twist-3 distribution amplitude can flip the sign of curvature as the cutoff scale for the transverse momentum gets lower than a typical scale of around 1 GeV.

1 Introduction

Hadronic distribution amplitudes (DAs) provide essential information on the QCD interaction of quarks, antiquarks and gluons inside the hadrons and play an essential role in applying QCD to hard exclusive processes [1]. It has motivated many studies using various nonperturbative models and led to develop distinct phenomenological models over the past two decades. Among them, the light-front quark model (LFQM) appears to be one of the most efficient and effective tools in hadron physics as it takes advantage of the distinguished features of the LFD. The rational energy-momentum dispersion relation of LFD yields the sign correlation between the LF energy $k^-(= k^0 - k^3)$ and the LF longitudinal momentum $k^+(= k^0 + k^3)$ and leads to the suppression of vacuum fluctuations. This simplification is a remarkable advantage in LFD and facilitates the partonic interpretation of the amplitudes. Based on the advantages of the LFD, the LFQM has been developed and subsequently applied for various meson phenomenologies such as the meson mass spectra, the decay constants, DAs, form factors and GPDs [2, 3]. Despite these successes in reproducing the general features of the data, however, it has proved very difficult to obtain direct connection between the LFQM and QCD.

Through the recent analysis of the twist-2 and twist-3 DAs of pseudoscalar and vector mesons [4, 5], we discussed the link between the chiral symmetry of QCD and the numerical results of the LFQM. In particular, through the analysis of twist-3 DAs of π and ρ mesons, we observe that the LFQM with effective degrees of freedom represented by the constituent quark and antiquark may provide the view of effective zero-mode cloud around the quark and antiquark inside the meson. Consequently, the constituents dressed by the zero-mode cloud may be expected to satisfy the chiral symmetry of QCD. Our numerical results appear consistent with this expectation and effectively indicate that the constituent quark and antiquark in the LFQM may be considered as the dressed constituents including the zero-mode quantum fluctuations from the vacuum.

[a]e-mail: homyoung@knu.ac.kr
[b]e-mail: crji@ncsu.edu

2 Model Description

The twist-2(3) DAs $\phi^{\mathcal{A}(\mathcal{P})}_{2(3);\pi}$ for a pion are defined in the light-cone gauge as follows [6]:

$$\begin{aligned}\langle 0|\bar{q}(z)\gamma^\mu\gamma_5 q(-z)|\pi(P)\rangle &= if_\pi P^\mu \int_0^1 dx e^{i(2x-1)P\cdot z}\phi^{\mathcal{A}}_{2;\pi}(x),\\ \langle 0|\bar{q}(z)i\gamma_5 q(-z)|\pi(P)\rangle &= f_\pi\mu_\pi \int_0^1 dx e^{i(2x-1)P\cdot z}\phi^{\mathcal{P}}_{3;\pi}(x),\end{aligned} \tag{1}$$

where the normalization parameter μ_π is given by $\mu_\pi = -2\langle\bar{q}q\rangle/f_\pi^2$ from the Gell-Mann-Oakes-Renner relation. Defining the local matrix elements $\mathcal{M}_\alpha \equiv \langle 0|\bar{q}\Gamma_\alpha q|\pi(P)\rangle$ ($\alpha = \mathcal{A},\mathcal{P}$) for axial-vector ($\Gamma_A = \gamma^\mu\gamma_5$) and pseudoscalar ($\Gamma_P = i\gamma_5$) channels, we start from the exactly solvable manifestly covariant Bethe-Salpeter (BS) model:

$$\mathcal{M}_\alpha = N_c \int \frac{d^4k}{(2\pi)^4}\frac{H_0}{(p^2-m_q^2+i\varepsilon)(k^2-m_{\bar{q}}^2+i\varepsilon)}\mathrm{Tr}\left[\Gamma_\alpha\left(\not{p}+m_q\right)\gamma_5\left(-\not{k}+m_{\bar{q}}\right)\right], \tag{2}$$

where N_c denotes the number of colors, $m_{q(\bar{q})}$ is the constituent quark (antiquark) mass, and $p = P-k$. In order to regularize the covariant loop, we use the usual multipole ansatz $H_0(p^2,k^2) = g/(p^2 - \Lambda^2 + i\varepsilon)^2$ for the $q\bar{q}$ bound-state vertex function, where g and Λ are constant parameters. Since the vertex function H_0 may not be the realistic approximation of a $q\bar{q}$ bound state, we utilize it only as a tool to analyze the zero-mode complication in the exactly solvable BS model. Comparing the two results of Eq. (2) obtained from the manifestly covariant calculation and the LF one, we found that only the LF result for $\mathcal{M}_{\mathcal{A}}$ receives the zero modes. What is remarkable in our finding [4, 5] is that the nonvanishing zero-mode contributions as well as the instantaneous ones to $\mathcal{M}_{\mathcal{A}}$ that appeared in the covariant BS model now vanish explicitly when the phenomenological wave function such as the Gaussian wave function in the LFQM is used. In other words, the decay constants and the quark DAs of pseudoscalar mesons can be obtained only from the on-mass-shell valence contribution within the framework of the standard LFQM using the Gaussian radial wave function and they still satisfy the chiral symmetry consistent with the QCD. This observation is also applicable to the calculation of the pion elastic form factor [5]. The self-consistent correspondence relation between the covariant BS model and the LFQM can be found in [4, 5].

The explicit forms of the twist-2 and twist-3 DAs of the pion in the LFQM are given by [5]

$$\begin{aligned}\phi^{\mathcal{A}}_{2;\pi}(x) &= \frac{\sqrt{2N_c}}{f_\pi 8\pi^3}\int^{|\mathbf{k}_\perp|<\mu} d^2\mathbf{k}_\perp \frac{\phi_R(x,\mathbf{k}_\perp)}{\sqrt{\mathbf{k}_\perp^2+m_q^2}}m_q \equiv \int^{|\mathbf{k}_\perp|<\mu} d^2\mathbf{k}_\perp \psi^{\mathcal{A}}_{2;\pi}(x,\mathbf{k}_\perp),\\ \phi^{\mathcal{P}}_{3;\pi}(x) &= -\frac{\sqrt{2N_c}}{f_\pi\mu_\pi\cdot 16\pi^3}\int^{|\mathbf{k}_\perp|<\mu} d^2\mathbf{k}_\perp \frac{\phi_R(x,\mathbf{k}_\perp)}{\sqrt{\mathbf{k}_\perp^2+m_q^2}}\frac{\mathbf{k}_\perp^2+m_q^2}{x(1-x)} \equiv \int^{|\mathbf{k}_\perp|<\mu} d^2\mathbf{k}_\perp \psi^{\mathcal{P}}_{3;\pi}(x,\mathbf{k}_\perp),\end{aligned} \tag{3}$$

respectively. From the point of view of QCD, one should note that the quark DAs of a hadron depend on the scale μ that may separate nonperturbative and perturbative regimes. In our LFQM, we can associate μ with the transverse integration cutoff via $|\mathbf{k}_\perp| \leq \mu$. The dependence on the scale μ is then consistently given by the QCD evolution equation [1]. In our previous analysis [5] for the radial wave function $\phi_R(x,\mathbf{k}_\perp)$, we used the Gaussian or Harmonic Oscillator (HO) wave function, $\phi_R^{\rm HO} = N_{\rm HO}\exp(-\vec{k}^2/2\beta^2)$ where $\vec{k}^2 = \mathbf{k}_\perp^2 + k_z^2 = \mathbf{k}_\perp^2 + (x-1/2)^2(\mathbf{k}_\perp^2+m_q^2)/x(1-x)$ for the pion case. But, in this work, we also use the power-law (PL) wave function, $\phi_R^{\rm PL} = N_{\rm PL}(1+\vec{k}^2/\beta^2)^{-2}$ and check the sensitivity of the scale μ to $\phi^{\mathcal{A}}_{2;\pi}(x)$ and $\phi^{\mathcal{P}}_{3;\pi}(x)$.

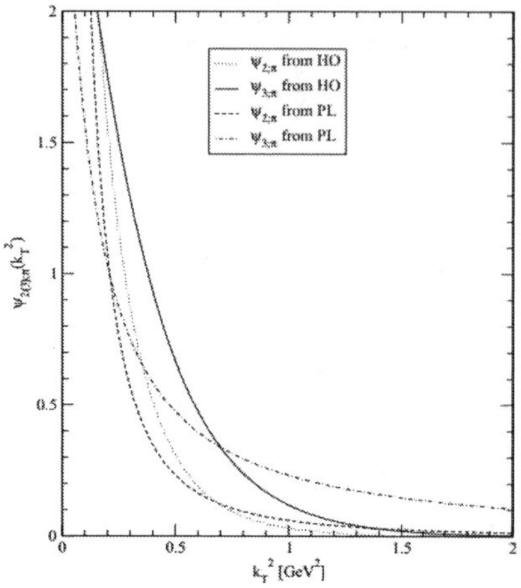

Figure 1. The $\mathbf{k}_\perp^2$ distributions $\psi^{\mathcal{A}(\mathcal{P})}_{2(3);\pi}(\mathbf{k}_\perp^2)$ obtained from HO and PL wave functions. The dotted, solid, dashed, dot-dashed lines represent $[\psi^{\mathcal{A}}_{2;\pi}]_{\rm HO}$, $[\psi^{\mathcal{P}}_{3;\pi}]_{\rm HO}$, $[\psi^{\mathcal{A}}_{2;\pi}]_{\rm PL}$, and $[\psi^{\mathcal{P}}_{3;\pi}]_{\rm PL}$, respectively.

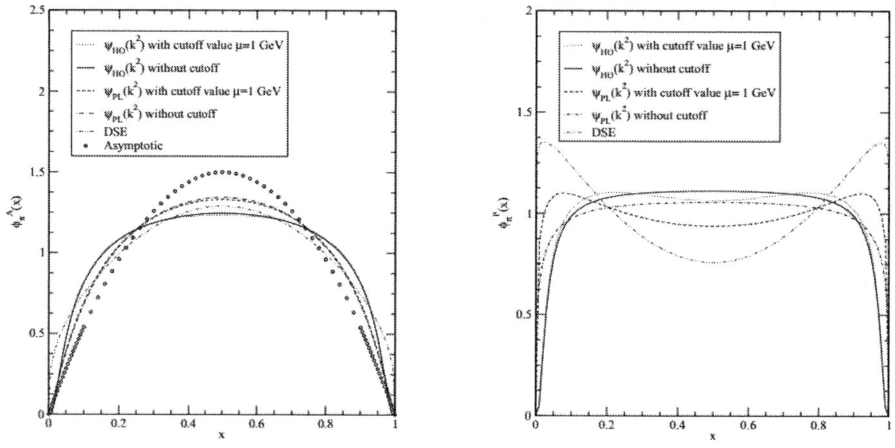

Figure 2. Twist-2 and-3 DAs from Gaussian (left) vs. Power-law (right) wave function.

3 Numerical Results

In our numerical calculations, we use the model parameters $m_{u(d)} = 0.22$ GeV and $\beta = 0.3659$ GeV for both $\phi_R^{\rm HO}$ and $\phi_R^{\rm PL}$. Both $\phi_R^{\rm HO}$ and $\phi_R^{\rm PL}$ are normalized without the momentum cutoff (i.e. $|\mathbf{k}_\perp| \longrightarrow \infty$) and the corresponding pion decay constants give the same value $f_\pi = 130$ MeV, which is in good agreement with the experimental data , $f_\pi^{\rm Exp.} = (130.41 \pm 0.03 \pm 0.20)$ MeV.

Defining the $\mathbf{k}_\perp^2$ distribution $\psi^{\mathcal{A}(\mathcal{P})}_{2(3);\pi}(\mathbf{k}_\perp^2)$ after integrating over x for the full wave function $\psi^{\mathcal{A}(\mathcal{P})}_{2(3);\pi}(x, \mathbf{k}_\perp)$ in Eq. (3), we show in Fig. 1 $\psi^{\mathcal{A}(\mathcal{P})}_{2(3);\pi}(\mathbf{k}_\perp^2)$ obtained from both HO and PL wave functions. As one can see from Fig. 1, $[\psi^{\mathcal{A}(\mathcal{P})}_{2(3);\pi}]_{\rm PL}$ have more high momentum tails than the corresponding $[\psi^{\mathcal{A}(\mathcal{P})}_{2(3);\pi}]_{\rm HO}$ for the cutoff values $\mu \geq 1$ GeV. Figure 2 shows $\phi^{\mathcal{A}}_{2;\pi}(x)$ (left panel) and $\phi^{\mathcal{P}}_{3;\pi}(x)$ (right panel) of a pion. We plotted for the two cases, i.e. without the cutoff (i.e. $\mu \longrightarrow \infty$) and with the cutoff value $\mu \longrightarrow 1$ GeV for both $\phi_R^{\rm HO}$ and $\phi_R^{\rm PL}$. As one see from Fig. 2, $\psi^{\mathcal{P}}_{3;\pi}(x)$ for both $\phi_R^{\rm HO}$ and $\phi_R^{\rm PL}$ are more sensitive to the cutoff scale than $\psi^{\mathcal{A}}_{2;\pi}(x)$. While $\psi^{\mathcal{P}}_{3;\pi}$ for both HO and PL wave functions show slight convex shapes when $\mu \longrightarrow \infty$, they flip the convex shapes to concave ones when the typical momentum cutoff scale $\mu = 1$ GeV or less is taken. The concave-shape $\psi^{\mathcal{P}}_{3;\pi}(x)$ was obtained in other models such as the QCD sum rules [6] and Dyson-Schwinger approach [7] (double-dot dashed lines) and our LFQM calculation can generate the concave-shape with a finite transverse momentum cutoff. This concave behaviors of $\psi^{\mathcal{P}}_{3;\pi}(x)$ are more pronounced for the smaller cutoff scale as one may expect from Fig. 1. This rebuts the remark made in Ref.[8] that our LFQM has curvature of the opposite sign on almost the entire domain of support in conflict with a model-independent prediction of QCD. We have shown in our recent works[4, 5] that our LFQM is indeed consistent with the nature of chiral symmetry in QCD. Without the momentum cutoff ($\mu \to \infty$), our results for both $\phi_R^{\rm HO}$ and $\phi_R^{\rm PL}$ reproduce the asymptotic result [9], $[\phi^{\mathcal{P}}_{3;\pi}]_{\rm as}(x) \to 1$, in the chiral symmetry ($m_q \to 0$) limit in our LFQM. Our LFQM result in the chiral symmetry limit is consistent with the conclusion drawn from our previous analysis [4] of the twist-2 ($\phi^{\parallel}_{2;\rho}(x)$) and twist-3 ($\phi^{\perp}_{3;\rho}(x)$) ρ meson DAs, where both DAs also remarkably reproduce the exact asymptotic DAs in the chiral symmetry limit. However, with the momentum cutoff, the asymptotic behavior of $\phi^{\mathcal{P}}_{3;\pi}(x)$ becomes concave shape following the results of the QCD sum rules [6] and Dyson-Schwinger approach [7].

4 Summary

We have discussed the link between the chiral symmetry of QCD and the numerical results of the LFQM, analyzing both the twist-2 and twist-3 DAs of a pion using two different model wave functions (HO vs. PL). While the twist-2 DA is rather insensitive to the transverse momentum cutoff $\mu \geq 1$ GeV, the twist-3 DA is sensitive to the cutoff value and thus the sign of curvature should be discussed discretely according to the scale taken consistently with the nature of chiral symmetry in QCD. The idea of our LFQM is to provide the nonperturbative wave functions at the momentum scale consistent with the use of constituent quark mass. The DAs determined from this nonperturbative wave functions can be fed into the QCD evolution equation to provide the shorter distance information of the corresponding hadrons. The DAs obtained without the cutoff should not be regarded as the fully evolved DAs but still be nonperturbative as they just mean that the cutoff dependence becomes marginal beyond a certain nonperturbative cutoff scale.

Acknowledgements

This work is supported by the Korean Research Foundation (No. NRF-2014R1A1A2057457). CRJ acknowledges partial support from the US Department of Energy (No. DE-FG02-03ER41260).

References

[1] G. P. Lepage and S. J. Brodsky, Phys. Rev. D **22**, 2157 (1980).

[2] H.-M. Choi and C.-R. Ji, Phys. Rev. D **59**, 074015 (1999); Phys. Lett. B **460**, 461 (1999); Phys. Rev. D **75**, 034019 (2007); Phys. Rev. D **75**, 073016 (2007); Phys. Rev. D **80**, 054016 (2009).

[3] J.P.B.C. de Melo and T. Frederico, Phys. Lett. B **708**, 87 (2012).

[4] H.-M. Choi and C.-R. Ji, Phys. Rev. D **89**, 033011 (2014).

[5] H.-M. Choi and C.-R. Ji, Phys. Rev. D **91**, 014018 (2015).

[6] P. Ball, V. M. Braun, and A. Lenz, J. High Energy Phys. 05, 004 (2006).

[7] L. Chang, C. D. Roberts, and S. M. Schmit, Phys. Lett. **B** 727, 255 (2013); L. Chang, I. C. Cloët, J. J. Cobos-Martinez, C. D. Roberts, S. M. Schmit, and P. C. Tandy, Phys. Rev. Lett. **110**, 132001 (2003).

[8] C. Shi et.al., Phys. Rev. D **92**, 014035 (2015).

[9] V. M. Braun and I. Filyanov, Z. Phys. C **48**, 239 (1990).

EPJ Web of Conferences **113**, 05010 (2016)
DOI: 10.1051/epjconf/201611305010

The ^{3}He spectral function in light-front dynamics

Matteo Rinaldi[1,a], Alessio Del Dotto[2], Leonid Kaptari[3], Emanuele Pace[4], Giovanni Salmè[5], and Sergio Scopetta[1]

[1] *Dipartimento di Fisica e Geologia, Università di Perugia and INFN, Sezione di Perugia, Italy*
[2] *Dipartimento di Fisica, Università di Roma Tre and INFN, Sezione Roma 3, Italy*
[3] *Bogoliubov Lab. Theor. Phys., 141980, JINR, Dubna, Russia*
[4] *Dipartimento di Fisica, Università di Roma "Tor Vergata" and INFN, Sezione Roma 2, Italy*
[5] *INFN Sezione di Roma*

Abstract. A distorted spin-dependent spectral function for ^{3}He is considered for the extraction of the transverse-momentum dependent parton distributions in the neutron from semi-inclusive deep inelastic electron scattering off polarized ^{3}He at finite momentum transfers, where final state interactions are taken into account. The generalization of the analysis to a Poincaré covariant framework within the light-front dynamics is outlined.

In the next few years, several experiments involving ^{3}He nuclear targets will be performed at JLab, with the aim of extracting information on the parton structure of the neutron. In particular, the three-dimensional neutron structure, described in terms of quark transverse momentum dependent parton distributions (TMDs) [1], will be probed through spin-dependent semi-inclusive deep inelastic scattering (SIDIS) processes off ^{3}He, where a high-energy pion is detected in coincidence with the scattered electron [2]. Hence, to clarify the flavour dependence of the TMDs, high precision experiments, involving both protons and neutrons, are needed. To obtain a reliable information from ^{3}He SIDIS one has to take into account: i) the nuclear structure of ^{3}He, ii) the interaction in the final state (FSI) between the observed pion and the remnant debris, and iii) the relativistic effects. Our efforts about the above issues are summarized here.

An initial study of this kind was reported in Ref. [3], where the plane wave impulse approximation (PWIA) was adopted to describe the reaction mechanism, treating the ^{3}He structure non-relativistically, using the AV18 interaction [4]. In facts, a polarized ^{3}He is an ideal target to study the neutron, since at a 90% level a polarized ^{3}He is equivalent to a free polarized neutron [5]. In Ref. [3] it was found, using a realistic ^{3}He spin-dependent spectral function, $P^{\tau}_{\sigma,\sigma'}(\vec{p}, E)$, with $\vec{p}$ the nucleon momentum in the laboratory frame and E the removal energy [6], that the formula

$$A_n \simeq \left(A_3^{exp} - 2p_p f_p\, A_p^{exp}\right)/(p_n f_n) \qquad (1)$$

can be safely adopted to extract the neutron Sivers and Collins asymmetries [7, 8] from ^{3}He and proton data. This formula has been already used by experimental collaborations [9, 10]. Nuclear effects are hidden in the proton and neutron "effective polarizations" (EPs), $p_{p(n)}$, and in the dilution factors, $f_{p(n)}$, properly defined [3]. With the AV18 interaction, one obtains $p_p = -0.023$, $p_n = 0.878$ [3].

[a] e-mail: matteo.rinaldi@pg.infn.it

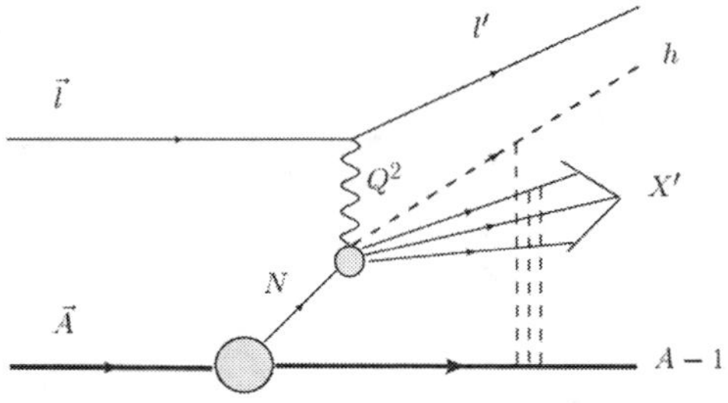

Figure 1. Interaction between the $(A-1)$ spectator system and the debris produced by the absorption of a virtual photon by a nucleon in the nucleus.

We are now taking into account FSI between the observed pion and the remnants. In a recent paper [11], the distorted spin dependent spectral function of ^{3}He has been introduced, and evaluated when the final $A-1$ system is a deuteron state. The evaluation considering the most general final state is in progress [12]. A few preliminary results are reported here. In SIDIS off ^{3}He, the relative energy between the spectator $(A-1)$ system and the system composed by the detected pion and the remnant debris (see Fig. 1) is a few GeV. FSI can be therefore treated through a generalized eikonal approximation (GEA) [11]. The GEA was already successfully applied to unpolarized SIDIS in Ref. [13]. In GEA, FSI effects are due to the propagation of the debris, formed after the γ^* absorption by a target quark, and the subsequent hadronization, both of them influenced by the presence of a fully interacting $(A-1)$ spectator system (see Fig. 1). The key quantity to describe FSI effects is the *distorted* spin-dependent spectral function, whose relevant part in the evaluation of single spin asymmetries (SSAs) is: $P^{PWIA(FSI)} = O^{IA(FSI)}_{\frac{1}{2}\frac{1}{2}} - O^{IA(FSI)}_{-\frac{1}{2}-\frac{1}{2}}$ with:

$$O^{IA}_{\lambda\lambda'}(\vec{p},E) = \sum\!\!\!\!\!\!\int d\epsilon^*_{A-1}\, \rho(\epsilon^*_{A-1})\, \langle S_A,\mathbf{P_A}|\Phi_{\epsilon^*_{A-1}},\lambda',\vec{p}\rangle\langle\Phi_{\epsilon^*_{A-1}},\lambda,\vec{p}|S_A,\mathbf{P_A}\rangle\, \delta(E-B_A-\epsilon^*_{A-1}) \quad (2)$$

and

$$O^{FSI}_{\lambda\lambda'}(\vec{p},E) = \sum\!\!\!\!\!\!\int d\epsilon^*_{A-1}\, \rho(\epsilon^*_{A-1})\, \langle S_A,\mathbf{P_A}|(\hat{S}_{Gl})\{\Phi_{\epsilon^*_{A-1}},\lambda',\vec{p}\}\rangle\langle(\hat{S}_{Gl})\{\Phi_{\epsilon^*_{A-1}},\lambda,\vec{p}\}|S_A,\mathbf{P_A}\rangle \times \delta(E-B_A-\epsilon^*_{A-1})\,, \quad (3)$$

where $\rho\left(\epsilon^*_{A-1}\right)$ is the density of the $(A-1)$-system states with intrinsic energy ϵ^*_{A-1}, and $|S_A,\mathbf{P_A}\rangle$ is the ground state of the A-nucleons nucleus with polarization S_A. $\hat{S}_{Gl}$ is the Glauber operator:

$$\hat{S}_{Gl}(\mathbf{r}_1,\mathbf{r}_2,\mathbf{r}_3) = \prod_{i=2,3}[1-\theta(z_i-z_1)\Gamma(\mathbf{b}_1-\mathbf{b}_i,z_1-z_i)] \quad (4)$$

and $\Gamma(\mathbf{b}_{1i},z_{1i})$ the generalized profile function:

$$\Gamma(\mathbf{b}_{1i},z_{1i}) = \frac{(1-i\,\alpha)\,\sigma_{eff}(z_{1i})}{4\,\pi\,b_0^2}\exp\left[-\frac{\mathbf{b}^2_{1i}}{2\,b_0^2}\right], \quad (5)$$

where $\mathbf{r}_{1i} = \{\mathbf{b}_{1i},\mathbf{z}_{1i}\}$ with $\mathbf{z}_{1i} = \mathbf{z}_1 - \mathbf{z}_i$ and $\mathbf{b}_{1i} = \mathbf{b}_1 - \mathbf{b}_i$. The models for the profile function, $\Gamma(\mathbf{b}_{1i},z_{1i})$, and for the effective cross section, $\sigma_{eff}(z_{1i})$, as well as the values of the parameters α and b_0, are the ones used in Ref. [13] to successfully describe deuteron JLab data [14]. $\sigma_{eff}(z_{1i})$ is the cross section for the interaction of the formed hadrons, α is the ratio of the real to the imaginary part of the forward scattering amplitude, b_0 is the slope parameter (cf Ref. [15] for details). It occurs that P^{PWIA} and P^{FSI} can be very different, but the relevant observables for the SSAs involve integrals, dominated by the low momentum region, where the FSI effects are minimized and the spectral function is large [13]. As a consequence, the EPs change from the *PWIA* values, i.e. $p_p = -0.023$ and $p_n = 0.878$, to $p_p^{FSI} = -0.026$, $p_n^{FSI} = 0.76$, where

$$p^{FSI}_{p(n)} = \int d\epsilon_S \int d\vec{p}\; Tr[\mathbf{S}_{He} * \sigma\, P^{p(n)}_{FSI}(\vec{p},\epsilon_S,S_{He})]\,. \quad (6)$$

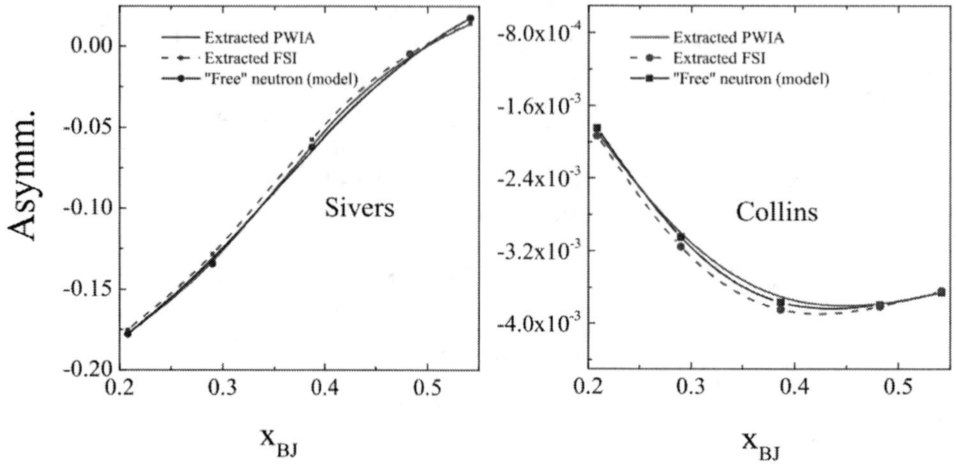

Figure 2. Effectiveness of the extraction procedure, Eq. (7), for the Sivers (left panel) and Collins (right panel) asymmetries, with and without FSI, in the actual kinematics of JLab [2]. Preliminary results to appear in [12].

Then $p_{p(n)}$ with and without the FSI differ by 10-15% . Actually, one has also to consider the effect of the FSI on dilution factors. We have found, in a wide range of kinematics typical for the experiments at JLAB [2], that the product of polarizations and dilution factors changes very little [16]. Indeed the effects of FSI on the dilution factors and on the EPs are found to compensate each other to a large extent and the usual extraction seems to be safe, viz. :

$$A_n \simeq \left(A_3^{exp} - 2p_p^{FSI} f_p^{FSI} A_p^{exp}\right)/(p_n^{FSI} f_n^{FSI}) \simeq \left(A_3^{exp} - 2p_p f_p\, A_p^{exp}\right)/(p_n f_n)\,, \tag{7}$$

as one can clearly see in Fig. 2.

We now address the relativistic description of the parton structure of ^{3}He, within a novel approach which, due to the high energies involved in the forthcoming experiments [2], is very timely [17, 18]. To study relativistic effects, we adopt the Light-Front (LF) form of the Relativistic Hamiltonian Dynamics (RHD) for an interacting system with a fixed number of on-mass-shell constituents. The Bakamijan-Thomas construction of the Poincaré generators leads to a fully Poincaré covariant RHD [17, 18]. The Light-Front form of RHD has a subgroup structure of the LF boosts, which allows a separation of the intrinsic motion from the global one, very important for the description of DIS, SIDIS and deeply virtual Compton scattering processes. We have not taken into account a possible violation of cluster separability in our approach [19]. The key quantity to consider in SIDIS processes is the LF relativistic spectral function, $\mathcal{P}^{\tau}_{\sigma'\sigma}(\tilde{\kappa}, \epsilon_S, S_{He})$, with $\tilde{\kappa}$ an intrinsic nucleon momentum and ϵ_S the energy of the two-nucleon spectator system. The LF spectral function is useful also for other studies (e.g., for nuclear generalized parton distributions, where final states have to be properly boosted). With respect to previous attempts to describe DIS processes off ^{3}He in a LF framework (see, e.g., the one in Ref. [20]), here a special care is devoted to the definition of the intrinsic light-cone variables as well as to the spin degrees of freedom in $\mathcal{P}^{\tau}_{\sigma'\sigma}(\tilde{\kappa}, \epsilon_S, S_{He})$. This PWIA LF spectral function is defined in terms of LF overlaps [18] between the ground state of a polarized ^{3}He and the cartesian product of an interacting state of two nucleons with energy ϵ_S and a plane wave for the third nucleon. Within a reliable approximation [18], it can be given in terms of the Melosh Rotations, $D^{\frac{1}{2}}[\mathcal{R}_M(\tilde{\kappa})]$, and the usual instant-form spectral function $P^{\tau}_{\sigma'_1\sigma_1}$:

Table 1. Proton and neutron EPs in PWIA, within the non relativistic appproach (NR) and within the LF RHD. First line: longitudinal EP; second line: transverse EP. Preliminary results.

	proton NR	*proton LF*	*neutron NR*	*neutron LF*
$\int d\epsilon_S d\tilde{\kappa}\ Tr(\mathcal{P}\sigma_z)_{\vec{S}_A=\widehat{z}}$	-0.02263	-0.02231	0.87805	0.87248
$\int d\epsilon_S d\tilde{\kappa}\ Tr(\mathcal{P}\sigma_y)_{\vec{S}_A=\widehat{y}}$	-0.02263	-0.02268	0.87805	0.87494

$$\mathcal{P}^{\tau}_{\sigma'\sigma}(\tilde{\kappa}, \epsilon_S, S_{He}) \propto \sum_{\sigma_1\sigma'_1} D^{\frac{1}{2}}[\mathcal{R}^{\dagger}_M(\tilde{\kappa})]_{\sigma'\sigma'_1}\ P^{\tau}_{\sigma'_1\sigma_1}(\mathbf{p}, \epsilon_S, S_{He})\ D^{\frac{1}{2}}[\mathcal{R}_M(\tilde{\kappa})]_{\sigma_1\sigma} \tag{8}$$

Preliminary results are quite encouraging, since, as shown in Tab. 1, LF longitudinal and transverse EPs change little from the ones obtained within the NR spectral function and weakly differ from each other. Then the usual extraction procedure should work well also within the LF approach. Concerning FSI, we plan to include in our LF description the FSI between the jet produced from the hadronizing quark and the two-nucleon spectator system through the GEA introduced in Ref. [13].

References

[1] V. Barone, A. Drago, P. G. Ratcliffe, Phys. Rept. **359**, 1 (2002)

[2] G. Cates et al., E12-09-018, JLAB approved experiment, *hallaweb.jlab.org/collab/PAC/PAC38//E12-09-018-SIDIS.pdf*

[3] S. Scopetta, Phys. Rev. D **75**, 054005 (2007)

[4] R. B. Wiringa, V. G. J. Stoks and R. Schiavilla, Phys. Rev. C **51**, 38 (1995)

[5] C. Ciofi degli Atti, S. Scopetta, E. Pace and G. Salmè, Phys. Rev. C **48**, 968 (1993)

[6] C. Ciofi degli Atti et al., Phys. Rev. C **46**, 1591 (1992)
A. Kievsky et al., Phys. Rev. C **56**, 64 (1997)

[7] D.W. Sivers, Phy. Rev. D **41**, 83 (1990)
D.W. Sivers, Phys. Rev. D **43**, 261 (1991)

[8] J.C. Collins, Nucl. Phys. B **396**, 161 (1993)

[9] X. Qian, Phys. Rev. Lett. **107**, 072003 (2011)

[10] K. Allada et al., Jefferson Lab Hall A Collaboration, Phys. Rev. C **89**, 042201 (2014)

[11] L.P. Kaptari, A. Del Dotto, E. Pace, G. Salmè, and S. Scopetta, Phys. Rev. C **89**, 035206 (2014)

[12] L.P. Kaptari, A. Del Dotto, E. Pace, G. Salmè, and S. Scopetta, in preparation.

[13] C. Ciofi degli Atti, L.P. Kaptari, Phys. Rev. C **83**, 044602 (2011)

[14] S. Kuhn, C. Keppel, W. Melnitchouk (spokespersons), et al., "*The structure of the free neutron via spectator tagging*", www.jlab.org/exp_prog/proposals/03prop.html (PR03-012.pdf); www.jlab.org/exp_prog/proposals/06/PR12-06-113.pdf

[15] C. Ciofi degli Atti, L. P. Kaptari and B. Z. Kopeliovich, Eur. Phys. J. A **19**, 145 (2004).

[16] A. Del Dotto, L. Kaptari, E. Pace, G. Salmè, S. Scopetta and E. Cisbani, EPJ Web Conf. **73**, 02019 (2014)

[17] E. Pace, G. Salmè, S. Scopetta, A. Del Dotto and M. Rinaldi, Few Body Syst. **54**, 1079 (2013)

[18] A. Del Dotto, E. Pace, G. Salmè, S. Scopetta, to be published

[19] B. D. Keister and W. N. Polyzou, Adv. Nucl. Phys. **20**, 225 (1991)

[20] U. Oelfke, P. U. Sauer and F. Coester, Nucl. Phys. A **518**, 593 (1990)

EPJ Web of Conferences **113**, 05011 (2016)
DOI: 10.1051/epjconf/201611305011

Quarks in Few Body Nuclei

Roy J. Holt[1,a]

[1] *Physics Division, Argonne National Laboratory, Argonne, Illinois 60439*

Abstract. Electron scattering at very high Bjorken x from hadrons provides an excellent test of models, has an important role in high energy physics, and from nuclei, provides a window into short range correlations. Light nuclei have a key role because of the relatively well-known nuclear structure. The development of a novel tritium target for Jefferson Lab has led to renewed interest in the mass three system. For example, deep inelastic scattering experiments in the light nuclei provide a powerful means to determine the neutron structure function. The isospin dependence of electron scattering from mass-3 nuclei provide information on short range correlations in nuclei. The program using the new tritium target will be presented along with a summary of other experiments aimed at revealing the large-x structure of the nucleon.

1 Introduction

During the past decades there has been an enormous effort to determine the parton distribution functions (PDFs) of the most stable hadrons: neutron, proton and pion [1]. The behavior of the PDFs on the far valence region (Bjorken-$x > 0.5$) is of particular interest because this domain defines a hadron. Recognizing the importance of the far valence region, a new generation [2] of experiments, focused on $x \gtrsim 0.5$, is planned at the Thomas Jefferson National Accelerator Facility (JLab), and under examination in connection with Drell-Yan studies at the Fermi National Accelerator Facility (FNAL) [3] and a possible Electron Ion Collider (EIC) [4]. One of the new experiments required the development of a tritium target for JLab, which led to new planned experiments. Theoretical calculations [5–12] of the pion structure function have demonstrated the success of the nonperturbative approach in QCD. Finally, the high-x region could impact high energy physics since low momentum transfer and high x evolves to high momentum transfer and low x. This talk will focus on three main areas of research at JLab: first, the planned measurements of the F_2^n/F_2^p and d/u ratios as well as present status; secondly, measurements of the longitudinal spin asymmetries for the proton and neutron, and third, the planned measurements that will make use of the new low-activity tritium target developed for JLab.

2 Status of high-x experiments

Two types of experiments will be extremely useful for understanding models of the nucleon. First, measurements of the F_2^n/F_2^p ratio at very high x are essential to constrain the models. Secondly, accurate, high-x, polarized, longitudinal asymmetries for both the proton and neutron are extremely important to understand the nucleon structure.

[a] e-mail: holt@anl.gov

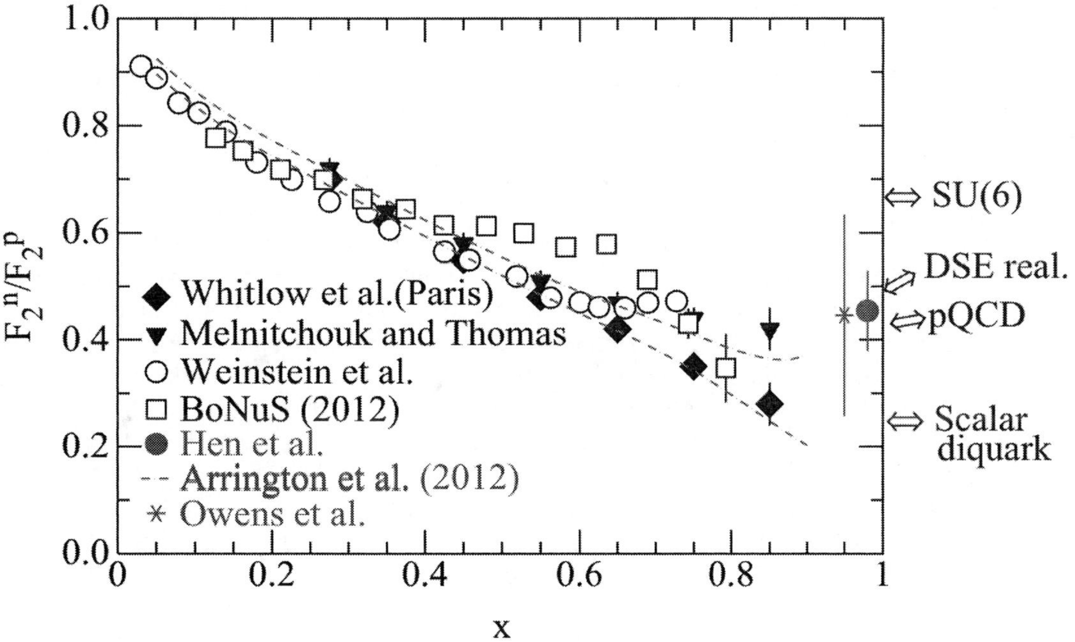

Figure 1. F_2^n/F_2^p as a function of x. Results from various extraction methods are shown [14–21] along with selected predictions. Adapted from ref. [5].

2.1 F_2^n/F_2^p ratio at high x

There is considerable theoretical interest in the ratio of F_2^n/F_2^p and d/u at very high x. The ratio as $x \to 1$ is sensitive to the model of the nucleon. One can see from the right side of Fig. 1 that the F_2^n/F_2^p or d/u ratio varies substantially among the models. Within the parton model at very high x:

$$\frac{F_2^n}{F_2^p} \overset{x\simeq 1}{=} \frac{1 + 4(d_v/u_v)}{4 + (d_v/u_v)}. \tag{1}$$

Thus a measurement of the neutron and proton structure functions at large-x provides a determination of the d_v/u_v ratio. However, while proton and deuteron DIS data are well measured at reasonably high x, the extraction of the neutron structure function at very high x from DIS data on the deuteron is problematic. The central difficulty is that the extraction of F_2^n/F_2^p at high x is sensitive to the poorly known high-momentum components of the deuteron wave function [13].

Many extractions of the neutron-proton structure function ratios have been performed [14–22]. They are summarised in Fig. 1, with recent inferences indicated by the points with error bars near $x = 1$: there is a large uncertainty in the ratio for $x \gtrsim 0.6$. (See also Fig. 25 in Ref. [1].) A very recent work [23] has found that the F_2^n/F_2^p ratio rises rapidly at large x, and consequently, has opened up the uncertainty range once again. New experimental methods are necessary in order to place tighter constraints on the far valence domain. A primary goal is to empirically eliminate or constrain the quite different theoretical predictions.

Two new experiments [24, 25] should provide data up to $x \approx 0.85$ with the upgraded beam energy at JLab. Since much of the uncertainty can be traced to the poorly known short-range part of the deuteron wave function, the JLab BoNuS Collaboration has performed [21] and will perform [24] an experiment where a very low energy spectator proton from the deuteron can be detected in coincidence with a DIS event from the neutron in the deuteron. In this way, one can restrict the data to a region where the well-known long-range part of the deuteron wave function dominates the process. The central difficulty with this experiment is detecting the very low energy proton of about 150 MeV/c or less. This requires extremely thin target and detector components. An interesting variant of this approach is to use an EIC with, e.g., an 8 GeV electron beam impinging on a deuteron beam of 30 GeV/nucleon. The forward going energetic proton would be detected at very small angles in coincidence with a DIS event from the neutron. Simulations suggest that this should be feasible [26].

Another method is to perform deep inelastic scattering from the mirror nuclei ^{3}He and ^{3}H over a broad range in x [25, 27–30]. Theoretical calculations indicate that nuclear effects cancel to a high degree in extracting the F_2^n/F_2^p ratio from these two nuclei. This experiment would also be useful in determining the EMC effect in the mass-three system [31]. Although providing a tritium target for JLab is straightforward [32], it is not trivial. The necessary tritium target has been developed for JLab and will be installed in 2016.

Finally, parity violating DIS can avoid the problem encountered with neutrons bound in nuclei. Parity-violating DIS from the proton is sensitive to the d/u ratio [33]. The proposed SoLID experiment plans to measure the d/u ratio up to an x of 0.7.

2.2 Longitudinally polarized deep inelastic scattering at high x

It is evident from the right side of Fig. 2 that measurements of the longitudinal asymmetries in DIS provide an important constraint on models of nucleon structure. Numerous experiments and extractions aimed at determining nucleon longitudinal spin structure functions have been performed [34].

Existing measurements of A_1^p are summarised in Fig. 2. Unfortunately, these data are not of sufficient accuracy and high enough x to discriminate among the models. As indicated in Fig. 2, however, a new JLab experiment [35] will extend the results up to $x \approx 0.8$ with a projected error that promises a significant constraint on the models.

The status of existing data for A_1^n is shown for polarized ^{3}He targets in Fig. 3. The data extend only to $x \approx 0.6$ and place little constraint on descriptions of the nucleon. New experiments proposed at JLab [36, 37] are expected to provide results up to $x \approx 0.75$, as indicated in Fig. 3. These new results should permit discrimination between the pQCD model and other predictions.

3 New studies in the mass-3 system

Three other experiments that make use of the new tritium target at JLab are: (i) $x > 1$ measurements [38], (ii) nucleon momentum distributions in three body nuclei [39] and (iii) the ^{3}H -^{3}He charge radius difference measurement [40]. It is well established [41] that $x > 1$ experiments are sensitive to short-range nucleon-nucleon correlations (SRC) in nuclei. An issue is the isospin dependence of the SRC. The ^{3}He - ^{3}H system represents the best opportunity to observe isospin differences since ^{3}He has the largest Z/N ratio available and ^{3}H has a 2:1 N/Z ratio. In fact, a simple estimate gives the ratio of ^{3}He to ^{3}H $x > 1$ cross section to be a factor of 1.4; whereas, for complete n-p dominance of the SRC, the ratio is expected to be unity. This experiment would be expected to extract the ratio of the $T = 1$ to $T = 0$ component to better than 4%.

An especially clever way to measure the ratio of the neutron to proton momentum distributions in nuclei is to use the mirror nuclei, ^{3}H and ^{3}He, and measure only (e,e'p) reactions for both nuclei [39]. This avoids the technically difficult problem of using only ^{3}He, say, and detecting both (e,e'n)

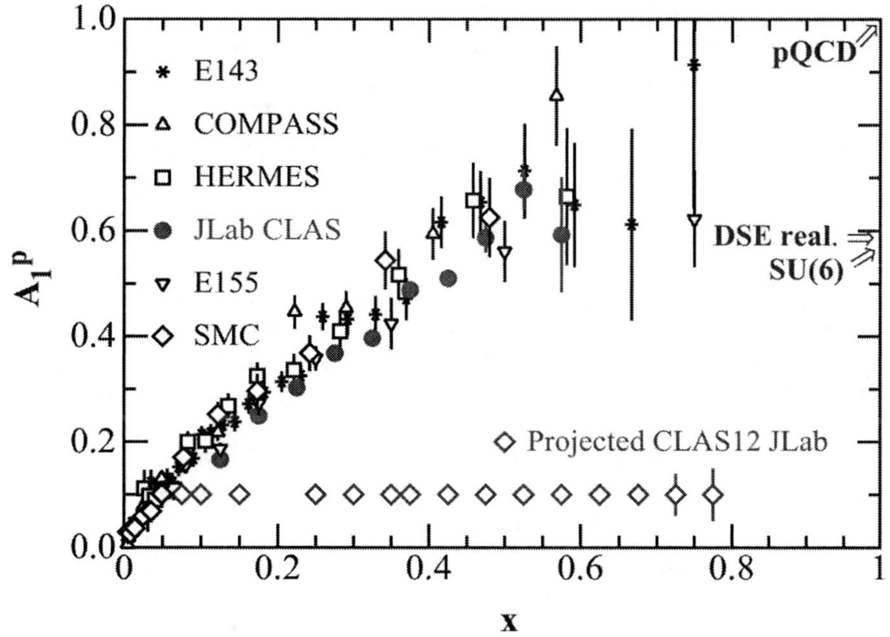

Figure 2. Existing and projected measurements of the proton's longitudinal spin asymmetry as a function of x (statistical errors only), along with selected predictions. Adapted from ref. [5].

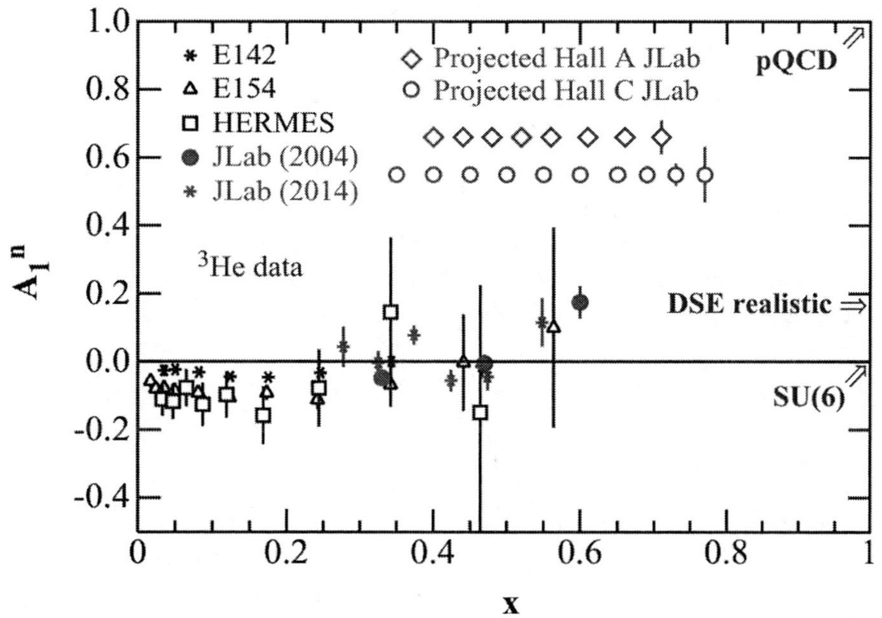

Figure 3. Existing and projected measurements of the neutron's longitudinal spin asymmetry as a function of x (statistical errors only), along with selected predictions. Here, only A_1^n data obtained from polarised ^{3}He targets are displayed. Adapted from ref. [5].

and (e,e'p) reactions. This experiment is especially sensitive to the N-N interaction in the nuclear medium. For example, under total n-p dominance of the N-N interaction, the ratio should be unity. This experiment is tentatively scheduled to run in conjunction with the other tritium experiments at JLab.

The third tritium experiment [40]is aimed at measuring the electric form factor ratio for 3He and $3H$ at low momentum transfer. The goal is to provide an accurate difference between the charge radii for these mirror nuclei.

4 Summary

Two and three-body systems have a key role in determining the quark structure of the nucleon. JLab at 12 GeV is an ideal facility for beginning the study of the far valence region of the nucleon, and the EIC should provide a large dynamic range in Q^2 and W for these studies. The advent of a tritium target at JLab presents new physics opportunities. These new experiments should provide benchmark data for the studies of the nucleon structure, the short range N-N interaction and the structure of 3-body nuclei. Possible experiments that would make use of a polarized tritium target at JLab should be evaluated.

Acknowledgements

I am grateful to C. D. Roberts and J. Arrington for fruitful discussions. This work was funded by the Department of Energy, Office of Science, Office of Nuclear Physics, contract no. DE-AC02-06CH11357.

References

[1] R.J. Holt, C.D. Roberts, Rev. Mod. Phys. **82**, 2991 (2010)
[2] R.J. Holt (2013), arXiv:1311.1527
[3] P. Reimer, D. Geesaman et al. (Fermilab E906/Drell-Yan), E906: Fermilab Proposal 906, 2006
[4] A. Accardi, J. Albacete, M. Anselmino, N. Armesto, E. Aschenauer et al. (2012), 1212.1701
[5] C.D. Roberts, R.J. Holt, S.M. Schmidt, Phys.Lett. **B727**, 249 (2013), 1308.1236
[6] L. Chang et al., Phys. Rev. Lett. **110**, 132001 (2013)
[7] I. Cloët, L. Chang, C. Roberts, S. Schmidt, P. Tandy, Phys.Rev.Lett. **111**, 092001 (2013), 1306.2645
[8] L. Chang, I. Cloët, C. Roberts, S. Schmidt, P. Tandy, Phys.Rev.Lett. **111**, 141802 (2013), 1307.0026
[9] M.B. Hecht, C.D. Roberts, S.M. Schmidt, Phys. Rev. C **63**, 025213 (2001)
[10] K. Wijesooriya, P.E. Reimer, R.J. Holt, Phys. Rev. C **72**, 065203 (2005)
[11] M. Aicher, A. Schafer, W. Vogelsang, Phys.Rev.Lett. **105**, 252003 (2010), 1009.2481
[12] T. Nguyen, A. Bashir, C.D. Roberts, P.C. Tandy, Phys. Rev. C **83**, 062201(R) (2011)
[13] J. Arrington, D.W. Higinbotham, G. Rosner, M. Sargsian, Prog. Part. Nucl. Phys. **67**, 898 (2012)
[14] L. Whitlow, E. Riordan, S. Dasu, S. Rock, A. Bodek, Phys.Lett. **B282**, 475 (1992)
[15] W. Melnitchouk, A.W. Thomas, Phys.Lett. **B377**, 11 (1996), nucl-th/9602038
[16] J. Arrington, F. Coester, R.J. Holt, T.S.H. Lee, J.Phys. **G36**, 025005 (2009), 0805.3116
[17] L. Weinstein, E. Piasetzky, D. Higinbotham, J. Gomez, O. Hen et al., Phys.Rev.Lett. **106**, 052301 (2011), 1009.5666

[18] O. Hen, A. Accardi, W. Melnitchouk, E. Piasetzky, Phys. Rev. D **84**, 117501 (2011), 1110.2419

[19] J. Arrington, J. Rubin, W. Melnitchouk, Phys. Rev. Lett. **108**, 252001 (2012), 1110.3362

[20] J. Owens, A. Accardi, W. Melnitchouk, Phys. Rev. D **87**, 094012 (2013)

[21] N. Baillie et al. (CLAS Collaboration), Phys. Rev. Lett. **108**, 199902 (2012), 1110.2770

[22] S. Tkachenko et al. (CLAS), Phys. Rev. **C89**, 045206 (2014), [Addendum: Phys. Rev.C90,059901(2014)], 1402.2477

[23] W. Cosyn, M.M. Sargsian (2015), 1506.01067

[24] S. Bueltmann et al. (BoNuS Collaboration), *The Structure of the Free Neutron at Large x-Bjorken*, JLab Experiment E12-06-113

[25] G. Petratos et al. (MARATHON Collaboration), *Measurement of the* F_2^n/F_2^p, d/u *Ratios and the* $A = 3$ *EMC Effect in Deep Inelastic Scattering off the Tritium and Helium Mirror Nuclei*, JLab Experiment E12-06-118

[26] A. Accardi, C. Keppel, R. Ent (2010), Presentation at the "Workshop on Nuclear Chromodynamics with an EIC", Argonne National Laboratory

[27] I.R. Afnan et al., Phys. Lett. B **493**, 36 (2000)

[28] F.R.P. Bissey, A.W. Thomas, I.R. Afnan, Phys. Rev. C **64**, 024004 (2001)

[29] I.R. Afnan et al., Phys. Rev. C **68**, 035201 (2003)

[30] R.J. Holt, G.G. Petratos, AIP Conf.Proc. **1369**, 106 (2011)

[31] O. Hen, D. Higinbotham, G. Miller, E. Piasetzky, L. Weinstein, Int. J. Mod. Phys. E. **22**, 1330017 (2013)

[32] B. Brajuskovic, T. O'Connor, R. Holt, J. Reneker, D. Meekin et al., Nucl.Instrum.Meth. **A729**, 469 (2013), 1306.6000

[33] P. Souder et al. (SoLID Collaboration), *Precision Measurement of Parity Violation in Deep Inelastic Scattering over a Broad Kinematic Range*, JLab Experiment E10-007

[34] C.A. Aidala, S.D. Bass, D. Hasch, G.K. Mallot, Rev.Mod.Phys. **85**, 655 (2013), 1209.2803

[35] S. Khun et al. (CLAS Collaboration), *The Longitudinal Spin Structure of the Nucleon*, JLab Experiment E12-06-109

[36] X. Zheng et al. (E12-06-110 Collaboration), *Measurement of Neutron Spin Asymmetry* A_1^n *in the Valence Quark Region using an 11 GeV Beam and a Polarized* ^{3}*He Target in Hall C*, JLab Experiment E12-06-110

[37] N. Liyanage et al. (E12-06-122 Collaboration), *Measurement of Neutron Spin Asymmetry* A_1^n *in the Valence Quark Region using 8.8 and 6.6 GeV Beam Energies and Bigbite Spectrometer in Hall A*, JLab Experiment E12-06-122

[38] P. Solvignon et al. (JLab E12-11-112 Collaboration), Jefferson Lab Proposal 12-11-112

[39] O. Hen, L. Weinstein, S. Gilad, W. Boeglin (2014), arXiv:1410.4451

[40] L.S. Myers et al. (JLab E12-14-009 Collaboration), Jefferson Lab Proposal 12-14-009

[41] J. Arrington, A. Daniel, D. Day, N. Fomin, D. Gaskell et al., Phys.Rev. **C86**, 065204 (2012), 1206.6343

EPJ Web of Conferences **113**, 05012 (2016)
DOI: 10.1051/epjconf/201611305012

Sea Quarks in the Proton

Paul E Reimer[1,a] and the Fermilab SeaQuest Collaboration

[1] *Physics Division, Argonne National Laboratory, Argonne, IL 60439 USA*

Abstract. The proton is a composite particle in which the binding force is responsible for the majority of its mass. To understand this structure, the distributions and origins of the quark-antiquark pairs produced by the strong force must be measured. The SeaQuest collaboration is using the Drell-Yan process to elucidate antiquark distributions in the proton and to study their modification when the proton is held within a nucleus.

Over the last half century, an understanding of the proton has developed in which it contains three valence quarks and an effervescing sea of antiquarks, quarks and gluons. The flavors of the three valence quarks determine many of the characteristics, but the strong force and the sea of antiquark-quark pairs are primarily responsible for its mass. Experimentally, this picture was established based on a variety of probes, one of the most important being deep inelastic scattering (DIS). DIS, however, lacks the ability to distinguish between antiquarks and quarks.

Measurements of the sea quarks are important, however, as they are manifestations of the effects of the strong force. The Drell-Yan reaction necessarily involves antiquarks. To leading order in the strong coupling constant, α_s, the Drell-Yan process is the annihilation of a quark from one hadron with an antiquark from another hadron into a massive virtual photon that decays into a detectable lepton-antilepton pair, a process first observed by J.H. Christenson *et al.* [1]. The features of the cross section were explained by S.D. Drell and T.-M. Yan [2] in terms of the parton model as a hard scattering of point-like particles multiplied by a convolution of the parton distributions of the interacting hadrons:

$$\frac{d^2\sigma}{dx_1 dx_2} = \frac{4\pi\alpha_e^2}{9sx_1x_2} \sum_{q\in\{u,d,s,\ldots\}} e_q^2 \left[\bar{q}_1(x_1)q_2(x_2) + q_1(x_1)\bar{q}_2(x_2)\right], \tag{1}$$

where $x_{1(2)}$ represent Bjorken-x, x_{Bj}; $q_i(x_i)$ is the parton distribution of quark of flavor q; e_q is the charge of quark flavor q; $\sqrt{s}$ is the center-of-mass energy; $\alpha_e \approx 1/137$ is the fine structure constant; and the sum is over all quark flavors. The e_q^2 weighting of the parton distributions implies that the cross section is primarily sensitive to the u- and $\bar{u}$-quark distributions. This expression is only leading order, and next-to-leading order (NLO) terms in α_s contribute up to half of the cross section. In a typical fixed-target experiment, the acceptance is biased toward large Feynman-x, $x_{\mathrm{F}} \approx x_1 - x_2$, and thus the beam parton is generally a large x_{Bj} valence parton. For a proton beam this implies that the interaction is between a valence beam quark and a lower-x_{Bj} target *anti*-quark.

The SeaQuest experiment was proposed to measure the distributions of sea quarks in the nucleon and the modifications of these distributions in nuclei. The experiment is designed to measure the $\mu^+\mu^-$ (dimuon) decay of the Drell-Yan virtual photon, using a 120 GeV proton beam extracted in a

[a] e-mail: reimer@anl.gov

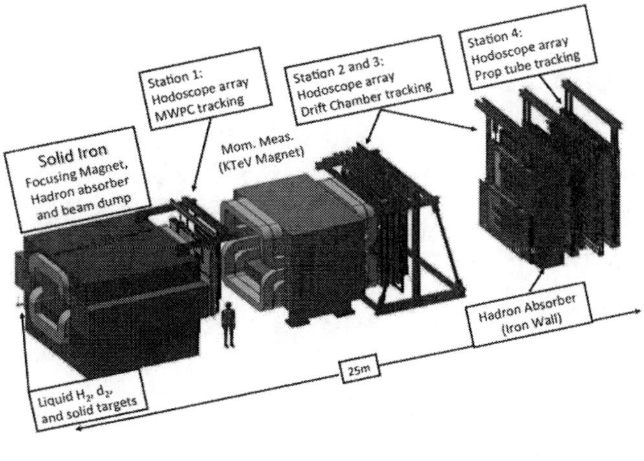

Figure 1. This figure schematically shows the SeaQuest spectrometer. The 120 GeV proton beam enters from the left, and the solid iron magnet also serves as an absorber for the beam that did not interact in a target.

4 s long, slow-spill from the Fermilab Main Injector. The spectrometer's basic concept is based on previous Fermilab Drell-Yan spectrometers that were used with 800 GeV extracted beams [3–5].To account for the difference in boost, the spectrometer was shortened significantly. Key features of the spectrometer include two large dipole magnets and four independent tracking/triggering stations. A schematic drawing of the spectrometer is shown in Fig. 1. More details about the spectrometer may be found in Ref. [6]. The experiment uses liquid hydrogen, liquid deuterium, carbon, iron and tungsten targets as well as an empty liquid target flask and a "no target" position for background subtraction. These targets are mounted on a horizontally movable table, so that only one target intercepts the beam during any given slow-extraction spill. The first dipole is a closed-aperture, solid iron magnet. The beam which does not interact in the targets is absorbed in the iron of the first magnet, allowing only (to a large extent) muons to traverse the remaining spectrometer. The downstream magnet is a large open-aperture magnet that was previously used in the Fermilab KTeV experiment. Each of the tracking/triggering stations consisted of a set of scintillator hodoscopes that provided fast information for an FPGA-based road look up and a drift chamber for fine-grained tracking.

Initially, it was believed that, at some very low energy scale, the proton could be described by empirically derived distributions of three valence quarks [7]; however, this approach disagreed with existing data [8]. Gluons spontaneously splitting into antiquark-quark pairs and the spontaneous recombination of antiquark-quark pairs into gluons was believed to be the process primarily responsible for the generation of the quark sea. Studying the flavor dependence of the antiquark sea can determine the relevant importance of this process. Nevertheless, the assumption that the sea was flavor symmetric was maintained until measurements by NMC [9] of the Gottfried sum rule [10] showed otherwise. Because of its sensitivity to the anti-quarks, Martin suggested that a comparison of proton-proton to proton-deuterium Drell-Yan would be an excellent method to study this difference [11]. Using the approximations that the Drell-Yan cross section is dominated by u_{beam} and $\bar{u}_{\mathrm{target}}$ annihilations and that the proton and neutron are charge symmetric, the approximate sensitivity is:

$$\frac{\sigma^{\mathrm{pd}}}{2\sigma^{\mathrm{pp}}} \approx \frac{1}{2}\left(1 + \frac{\bar{d}}{\bar{u}}\right), \tag{2}$$

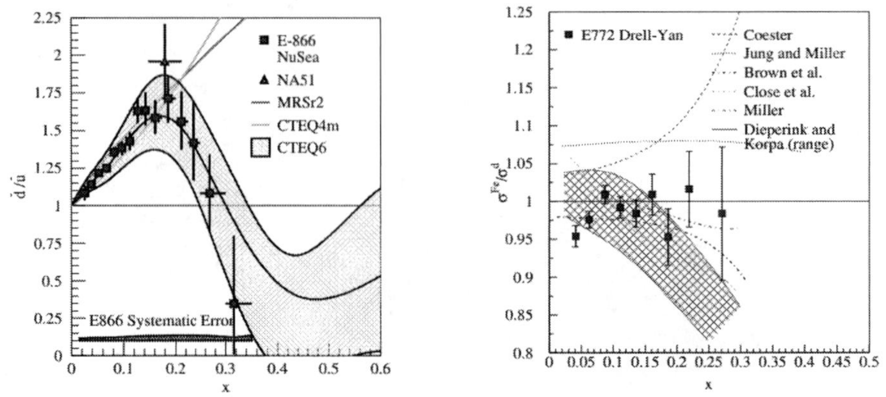

Figure 2. On the left, this figure shows the CERN NA51 [12] and Fermilab E866/NuSea [5] measurements of the $\bar{d}/\bar{u}$ ratio using Drell-Yan. The yellow band indicates a global fit by CTEQ including these data [13]. The right figure shows the lack of any nuclear dependence in the sea quarks as measured by Fermilab E-772, along with a model prediction [16] and several postdictions [17].

where $\sigma^{pp(d)}$ is the proton (deuterium) cross section and $\bar{d}$ ($\bar{u}$) is the antidown (antiup) quark distribution. This idea was seized upon by the CERN NA51 [12] and the Fermilab E-866/NuSea [5] experiments, whose results are shown in Fig. 2. The Fermilab NuSea data revealed a surprising trend in the flavor asymmetry–returning to unity as x_{Bj} increases–but with decreasing statistical precision at larger x_{Bj}. The Fermilab E-906/SeaQuest experiment will measure this ratio up to $x_{\mathrm{target}} \approx 0.45$. After engineering runs for both the SeaQuest spectrometer and the high-duty factor, slow-spill extraction of protons from the Main Injector, SeaQuest began collecting production data in the spring of 2014. SeaQuest anticipates to continue recording data until the summer of 2016.

The residual strong force is responsible for holding the positively-charged nucleus of an atom together. Evidence from DIS [14] has shown that the distributions of quarks and antiquarks within the individual nucleons are modified, presumably because of this. The modification of the sea quark distributions may be completely different, but DIS is much less sensitive to this. In Fermilab E-772 using the Drell-Yan reaction, no modifications to sea quark distributions were observed [4]. (See Fig. 2.) Models of nuclear binding that explain the EMC effect rely on the exchange of virtual mesons [15]. Based on these models, significant enhancements of the antiquark distributions in nuclei were expected [16]. The lack of sea-quark nuclear effects prompted a number of new models [17]. For $x > 0.2$, the E-772 statistical uncertainties allow significant freedom for these models.

The SeaQuest spectrometer is a general purpose muon spectrometer. Several future measurements using the SeaQuest spectrometer have been proposed to and approved by Fermilab. These include plans for a polarized hydrogen target to explore the Sivers asymmetry, $f_1^{\perp}(x, k_T)$, in the sea quarks and its relation to orbital angular momentum. The polarized target is being provided by Los Alamos collaborators and work on the beam line for better collimation is underway at Fermilab. The target will be installed immediately after the completion of the unpolarized experiment, starting in summer, 2016 [18]. An interesting property of the Sivers asymmetry is that its sign is probe dependent. That is, $f_1^{\perp}(x, k_T)\big|_{\mathrm{DIS}} = -f_1^{\perp}(x, k_T)\big|_{\mathrm{DY}}$. This property has yet to be tested, primarily due to a lack of polarized Drell-Yan data in a the same kinematic region as the existing DIS data. This can be remedied by using a polarized proton beam on an unpolarized target in SeaQuest, as proposed in Fermilab E-1027 [19].

A recent and very exciting addition to the SeaQuest physics program is a search for dark photons and dark Z's. If long-lived dark photons are produced by the proton-nucleus collisions in the iron of the first magnet, they may be identified by selecting dimuons with a vertex downstream of the first magnet and searching for a peak in the reconstructed dimuon mass spectrum [20].

The SeaQuest experiment was initially designed to measure unpolarized sea quark distributions to elucidate the origins of the sea quarks and the modifications of their distributions in a nuclear environment. The physics program has vastly expanded since then to include searches for dark sector particles and measurements of the Sivers asymmetry with polarized targets and beams. The experiment has been commissioned and is collecting data. The unpolarized phase is expected to be complete in the summer of 2016, after which the installation of a polarized target will begin.

Acknowledgements

This work is supported in part by the U.S. Department of Energy, Office of Science, Office of Nuclear Physics, under Contract No. DE-AC02-06CH11357.

References

[1] J.H. Christenson, et al. Phys. Rev. Lett. **25**, 1523 (1970); J.H. Christenson, et al. Phys. Rev. D **8**, 2016 (1973)

[2] S.D. Drell, T.M. Yan, Phys. Rev. Lett. **25**, 316 (1970); S.D. Drell, T.M. Yan, Phys. Rev. Lett. **25**, 902 (1970)

[3] G. Moreno, et al. Phys. Rev. D **43**, 2815 (1991)

[4] D.M. Alde, et al., Phys. Rev. Lett. **64**, 2479 (1990)

[5] E.A. Hawker et al. (Fermilab E866/NuSea), Phys. Rev. Lett. **80**, 3715 (1998); R.S. Towell et al. (Fermilab E866/NuSea), Phys. Rev. **D64**, 052002 (2001)

[6] *The SeaQuest spectrometer at Fermilab* (2016), in preparation

[7] M. Glück, R. Godbole, E. Reya, Z. Phys. C **41**, 667 (1989)

[8] M. Glück, E. Reya, A. Vogt, Z. Phys. C **53**, 127 (1992)

[9] P. Amaudruz, et al. Phys. Rev. Lett. **66**, 2712 (1991)

[10] K. Gottfried, Phys. Rev. Lett. **18**, 1174 (1967)

[11] A. Martin, W. Stirling, R. Roberts, Physics Letters B **308**, 377 (1993)

[12] A. Baldit et al. (NA51), Phys. Lett. **B332**, 244 (1994)

[13] J. Pumplin, et al., Journal of High Energy Physics **2002**, 012 (2002)

[14] J.J. Aubert et al. (European Muon), Phys. Lett. **B123**, 275 (1983)

[15] J. Carlson, R. Schiavilla, Rev. Mod. Phys. **70**, 743 (1998)

[16] F. Coester (2001), private communication

[17] H. Jung, G.A. Miller, Phys. Rev. **C41**, 659 (1990); G.E. Brown, M. Rho, Nucl. Phys. **A590**, 527c (1995); J.R. Smith, G.A. Miller, Phys. Rev. Lett. **91**, 212301 (2003) A.E.L. Dieperink, C.L. Korpa, Phys. Rev. **C55**, 2665 (1997)

[18] A. Klein, X. Jiang et al. (Fermilab E1039/SeaQuest) (2013), Fermilab Proposal 1039 http://www.fnal.gov/directorate/program_planning/June2013PACPublic/P-1039_LOI_polarized_DY.pdf

[19] W. Lorenzon, P.E. Reimer et al. (Fermilab E1027/SeaQuest) (2012), Fermilab Proposal 1027 https://seaquest-docdb.fnal.gov:cgi-bin/ShowDocument?docid=535

[20] S. Gardner, R.J. Holt, A.S. Tadepalli, Phys. Rev. D (2015), in preparation

EPJ Web of Conferences **113**, 05013 (2016)
DOI: 10.1051/epjconf/201611305013

Pseudoscalar mesons parton distribution amplitudes

Chen Chen[1,2,a] and Chao Shi[3,4,b]

[1] *Hefei National Laboratory for Physical Sciences at the Microscale, University of Science and Technology of China, Hefei, Anhui 230026, People's Republic of China*
[2] *Institute for Theoretical Physics and Department of Modern Physics, University of Science and Technology of China, Hefei, Anhui 230026, People's Republic of China*
[3] *Key Laboratory of Modern Acoustics, MOE, Institute of Acoustics, Nanjing University, Nanjing 210093, China*
[4] *Department of Physics, Nanjing University, Nanjing 210093, China*

Abstract. We compute all kaon and pion parton distribution amplitudes (PDAs) to twist-three and find that all of the pion's distributions are symmetric and all of the kaon's distributions are skewed in favour of the heavier s-quark, which are clear signals of dynamical chiral symmetry breaking (DCSB). And we can also notice that only the pseudotensor PDA can reasonably be approximated by its conformal limit. At any realistic energy scale, the pseudovector and pseudoscalar PDAs differ markedly from the corresponding functions in QCD's conformal limit. We can use these PDAs as the inputs to study hard exclusive processes and get trustable results.

1 Introduction

The PDAs of pion and kaon are very crucial to study hard exclusive processes[1, 2]. In the previous studies, people can only use the expressions of PDAs in the QCD's conformal limit[3, 4]. Now people can recalculate mesons' more realistic PDAs by using a new method. With this new method[5, 6], we calculate six PDAs of pion and kaon to twist-three.

2 Distribution amplitudes and Bethe-Salpeter wave functions

A pseudoscalar meson, $P_{\bar{g}f}(q)$, has three two-particle distribution amplitudes on the light front, they can be expressed as[7]

$$f_P\varphi_P(u) = N_c \mathrm{tr} Z_2 \int_{dk}^{\Lambda} \delta(n\cdot k_\eta - u\, n\cdot q)\gamma_5\gamma\cdot n\chi_P^q(k_\eta, k_{\bar{\eta}}) \quad (1a)$$

$$i\rho_P^\zeta\omega_P(u) = N_c \mathrm{tr} Z_4 \int_{dk}^{\Lambda} \delta(n\cdot k_\eta - u\, n\cdot q)\gamma_5\chi_P^q(k_\eta, k_{\bar{\eta}}) \quad (1b)$$

$$\frac{1}{4}\rho_P^\zeta v'(u) = N_c \mathrm{tr} Z_4 \int_{dk}^{\Lambda} \delta(n\cdot k_\eta - u\, n\cdot q)\gamma_5\sigma_{\mu\nu}q_\mu n_\nu\chi_P^q(k_\eta, k_{\bar{\eta}}) \quad (1c)$$

a. e-mail: chenchen.anl@gmail.com
b. e-mail: shichao0820@gmail.com

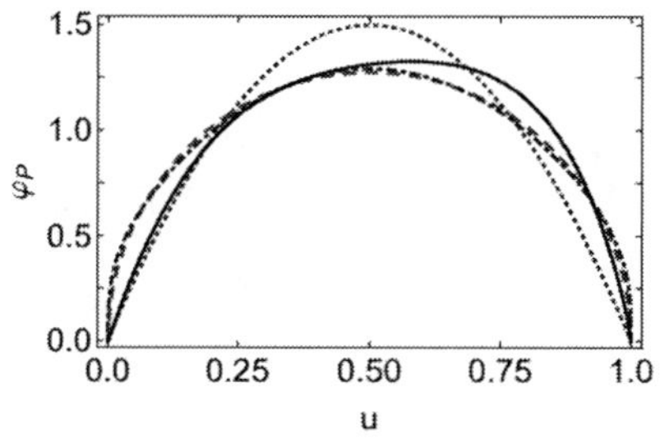

Figure 1. Two-particle twist-two PDAs, computed at ζ_2: dot-dashed curve (dark blue) – pion, $\varphi_\pi^{\rm DB}(u;\zeta_2)$; solid curve (black) – kaon, $\varphi_K^{\rm DB}(u;\zeta_2)$; dashed curve (dark green) – $\varphi_\pi^{\rm model}(u)$, Eq. (5); and dotted curve (indigo) – asymptotic distribution, Eq. (4).

where $N_c = 3$; the trace is over spinor indices; $\int_{dk}^{\Lambda}$ is a Poincaré-invariant regularisation of the four-dimensional integral, with Λ the ultraviolet regularisation mass-scale; $Z_{2,4}(\zeta,\Lambda)$ are renormalisation constants; ζ is the renormalisation scale; $\chi_P^q(k_\eta,k_{\bar\eta})$ is the Bethe-Salpeter wave function; and f_P, ρ_P^ζ are, respectively, the pseudovector and pseudoscalar projections of the meson's Bethe-Salpeter wave function onto the origin in configuration space. Every PDA should be normalized, *viz.*

$$\int_0^1 du\,\{\varphi_P(u;\zeta)\,,\,\omega_P(u;\zeta)\,,\,\upsilon_P(u;\zeta)\} = 1\,. \tag{2}$$

We can numerically solve the coupled integral equations[8–11], namely, QCD's gap and Bethe-Salpeter equations, to get the expressions of quark propagators and the mesons' Bethe-Salpeter amplitudes. Here we solve these equations in two symmetry-preserving truncations[12]: the rainbow-ladder (RL) truncation, which is the most widely used in the calculation; and the modern DCSB-improved (DB) kernels, which can give the most realistic predictions by now.

Then we can employ the procedure in previous works to calculate PDAs[5, 6, 13]. First we need to calculate PDAs' moments

$$\langle u_\Delta^m\rangle_\phi = \int_0^1 du\,(2u-1)^m \phi_P(u), \tag{3}$$

where $\phi = \varphi_P,\ \omega_P,\ \upsilon'_P$, with these moments, we can reconstruct the PDAs by using Gegenbauer polynomials of order α, $\{C_n^\alpha(2u-1)\,|\,n=0,\ldots,\infty\}$. Gegenbauer polynomials form a complete orthonormal set on $u\in[0,1]$ with respect to the measure $[u(1-u)]^{\alpha_-}$, $\alpha_- = \alpha - 1/2$. They therefore enable reconstruction of any function defined on $u\in[0,1]$.

3 Numerical results

3.1 Pseudovector

We depict our pseudovector results in Fig. 1, and compare them with the asymptotic two-particle distribution

$$\varphi^{\rm cl}(u) = 6u(1-u) \tag{4}$$

which is the asymptotic distribution in QCD's conformal limit, and with

$$\varphi_\pi^{\rm model}(u) = (8/\pi)\sqrt{u(1-u)}\,, \tag{5}$$

which is a model result, it is practically indistinguishable from the DSE prediction.

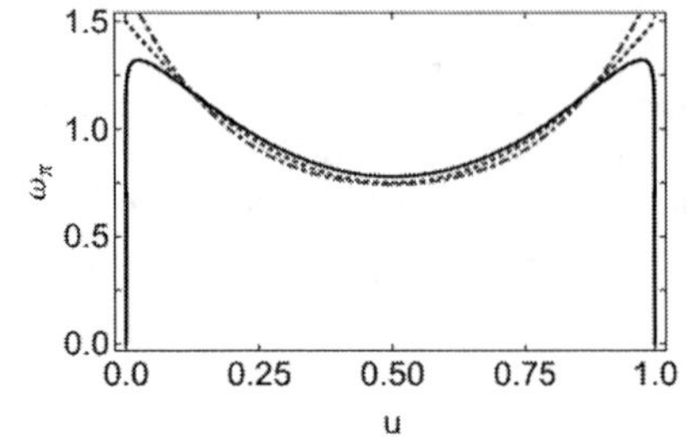

Figure 2. Pion two-particle, twist-three PDAs, computed at ζ_2. Solid curve (black), $\omega_\pi(u;\zeta_2)$; dot-dashed curve (dark green), QCD sum rules estimate [14]; and dotted curve (indigo), $\omega_P^{\rm asy}$ in Eq. (6).

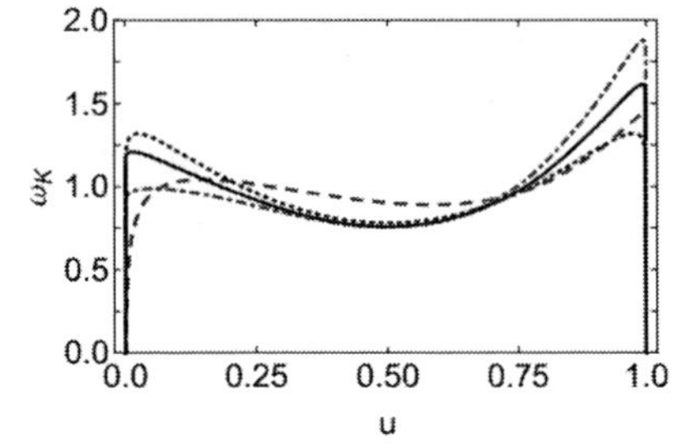

Figure 3. Kaon two-particle, twist-three PDAs, computed at ζ_2. Solid curve (black), $\omega_K^{\rm DB}(u;\zeta_2)$; dot-dashed curve (green), $\omega_K^{\rm RL}(u;\zeta_2)$; dashed curve (red), QCD sum rules estimate [14]; and dotted curve (blue), $\omega_\pi(u;\zeta_2)$.

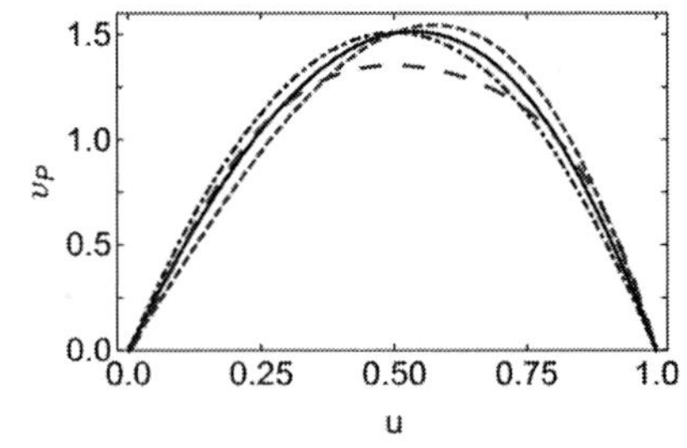

Figure 4. Pseudotensor two-particle, twist-three PDAs, computed at ζ_2: dot-dashed curve (dark blue) – pion, $\upsilon_\pi(u)$; solid curve (black) – kaon in DB truncation, $\upsilon_K^{\rm DB}(u)$; dashed curve (dark green) – kaon in RL truncation $\upsilon_K^{\rm RL}(u)$; and long-dashed curve (red) – QCD sum rules result for the kaon from Ref. [14]. We do not plot the asymptotic form because it is effectively indistinguishable from our prediction for the pion.

3.2 Pseudoscalar

We depict our pseudoscalar results in Fig. 2 and 3, and compare them with the asymptotic distribution

$$\omega_P^{\rm asy}(u) = 1 + \frac{1}{2}C_2^{(1/2)}(2u-1)\,. \tag{6}$$

3.3 Pseudotensor

We depict our pseudotensor results in Fig. 4, we do not plot the asymptotic form because it is effectively indistinguishable from our prediction for the pion..

4 Discussion and conclusion

We can get many important information from our results. First, we can see that all of the pion's distributions are symmetric and all of the kaon's distributions are skewed in favour of the heavier s-quark, which are clear signals of DCSB. In kaon's pseudoscalar case, the RL PDA is more skewed than the DB result, *viz.* the RL truncation allocates more fraction of the in-kaon condensate to its valence s-quark. The RL-kernels ignore DCSB in the quark-gluon vertex, meanwhile the DB-kernels put DCSB into the quark-gluon vertex, so the the DB-PDA is more realistic. In kaon's pseudovector and pseudotensor case, we can get similar observations. It can be shown that in all PDAs, the $SU(3)$ flavour-symmetry breaking is a 13% effect[5, 15]. And we can also notice that only the pseudotensor PDA can reasonably be approximated by its conformal limit. At any realistic energy scale, the pseudovector and pseudoscalar PDAs differ markedly from the corresponding functions in QCD's conformal limit.

Now we have calculated pion's and kaon's all the PDAs to twist-three, which are remarkably different from their expressions in the conformal limit except for the pseudotensor case. We can use these PDAs as the inputs to study hard exclusive processes and get trustable results.

Acknowledgements

The material described in this contribution is the result of an international collaborative effort that has involved, amongst others, L. Chang, G. Krein, S.-X. Qin, C. D. Roberts, J. Segovia, P.C. Tandy, S.-L. Wan and H.-S. Zong. This work was supported by the Fundamental Research Funds for the Central Universities Programme of China (Grant No. WK2030040050); the National Natural Science Foundation of China (Grants No. 11275097 and No. 11475085); and the National Basic Research Programme of China (Grant No. 2012CB921504).

References

[1] G. P. Lepage and S. J. Brodsky, Phys. Lett. B **87**, 359 (1979).
[2] A. V. Efremov and A. V. Radyushkin, Phys. Lett. B **94**, 245 (1980).
[3] S. J. Brodsky, Y. Frishman, G. P. Lepage and C. T. Sachrajda, Phys. Lett. B **91**, 239 (1980).
[4] V. Braun, G. Korchemsky and D. Mueller, Prog. Part. Nucl. Phys. **51**, 311 (2003).
[5] L. Chang *et al.*, Phys. Rev. Lett. **110**, 132001 (2013).
[6] F. Gao, L. Chang, Y.-X. Liu, C. D. Roberts and S. M. Schmidt, Phys. Rev. D **90**, 014011 (2014).
[7] V. M. Braun and I. Filyanov, Z. Phys. C **48**, 239 (1990).
[8] P. Maris and C. D. Roberts, Int. J. Mod. Phys. E **12**, 297 (2003).
[9] L. Chang, C. D. Roberts and P. C. Tandy, Chin. J. Phys. **49**, 955 (2011).
[10] A. Bashir *et al.*, Commun. Theor. Phys. **58**, 79 (2012).
[11] I. C. Cloët and C. D. Roberts, Prog. Part. Nucl. Phys. **77**, 1 (2014).
[12] L. Chang, C. D. Roberts and S. M. Schmidt, Phys. Rev. C **87**, 015203 (2013).
[13] C. Shi *et al.*, Phys. Lett. B **738**, 512 (2014).
[14] P. Ball, V. M. Braun and A. Lenz, JHEP **0605**, 004 (2006).
[15] C. Shi, C. Chen, L. Chang, C. D. Roberts, S. M. Schmidt and H.-S. Zong Phys. Rev. D **92**, 014035 (2015).

EPJ Web of Conferences **113**, 05014 (2016)
DOI: 10.1051/epjconf/201611305014

Leading Twist Parton Distribution Amplitudes in Heavy Vector Mesons

Fei Gao[1,2, a], Minghui Ding[1,2, b], Lei Chang[3, c], Yu-Xin Liu[1,2,4, d], and Craig D. Roberts[5, e]

[1] *Department of Physics and State Key Laboratory of Nuclear Physics and Technology, Peking University, Beijing 100871, China*
[2] *Collaborative Innovation Center of Quantum Matter, Beijing 100871, China*
[3] *CSSM, School of Chemistry and Physics, University of Adelaide, Adelaide SA 5005, Australia*
[4] *Center for High Energy Physics, Peking University, Beijing 100871, China*
[5] *Division of Physics, Argonne National Laboratory, Argonne, IL 60439, USA*

Abstract. We employed QCD's Dyson-Schwinger equations (DSEs) for heavy quarks and obtained the leading twist parton distribution amplitudes (PDAs) in heavy vector mesons J/Ψ and Υ. We found that all of the amplitudes are narrower than the asymptotic form, while they deviate from δ function. This indicates that the interaction between the two continient quarks are still important in the mesons consisted of charm and bottom quarks.

1 Introduction

The leading twist parton distribution amplitudes (PDAs) of mesons are defined by integrating out the transeverse momentum $k_\perp$ from the light front wave function [1]. They play an essential role in the hard exclusive processes [2–5], and moreover, the vector meson PDAs can be inferred [6, 7] from diffractive vector-meson production experiments at the Hadron Electron Ring Accelerator (HERA) [8–10]. Herein we calculate the leading twist PDAs in heavy vector mesons within DSEs approach.

The DSEs approach has been employed to calculate the light front information including the light flavour parton distribution amplitudes [11–13], parton distribution functions [14] and pion form factor [15] through projecting the Bethe-Salpeter wave function onto the light front. It is then natural to generalize the calculation to the heavy quarks case to obtain the PDAs for vector mesons J/Ψ and Υ. Our results show that for the mesons consisted of the charm and bottom quarks, their PDAs still deviate from δ function. Such heavy meson's PDAs will bring in the non-factorizable corrections in the hard exclusive processes [4].

Our presentation is organized as follows. In Sec. 2 we introduce some relevant definitions to calculate mesons' PDAs. In Sec. 3 we give our results of PDAs and some discussions.

a. e-mail: hiei@pku.edu.cn,speaker
b. e-mail: mhding@pku.edu.cn
c. e-mail: lei.chiong@gmail.com
d. e-mail: yxliu@pku.edu.cn
e. e-mail: c.d.roberts@anl.gov

2 PDAs and Bethe-Salpeter wave function

The leading twist PDAs could be defined as the following projection of Bethe-Salpeter wave function onto the light front[11, 12], at leading twist there are two independent PDAs for vector meson. They could be defined as following.

$$f_V^{\perp}\phi_{V\perp}(x,\zeta)m_V^2 = n\cdot P tr_{CD} Z_T(\zeta,\Lambda)\int_{dq}^{\Lambda}\delta(n\cdot q_+ - xn\cdot P)\sigma_{\mu\lambda}P_{\mu}\chi_{\lambda}(q;P)\,, \tag{1a}$$

$$f_V n\cdot P\phi_{V\parallel}(x,\zeta) = m_V tr_{CD} Z_2(\zeta,\Lambda)\int_{dq}^{\Lambda}\delta(n\cdot q_+ - xn\cdot P)n\cdot\gamma n_{\lambda}\chi_{\lambda}(q;P)\,, \tag{1b}$$

where the trace is over the color and spinor indices ; V is for vector meson ; $\perp$ is for the transversal PDA of vector meson and $\parallel$ is for the longitudinal part ; $Z_{2,T}(\zeta,\Lambda)$ is the quark wave function renormalization constant, with ζ the renormalization scale ; n is a lightlike four-vector ; P is the meson's four-momentum, with $P^2 = -m_V^2$ and $n\cdot P = -m_V$, with m_V being the meson's mass ; $f_V^{\perp}$ and f_V are decay constants which would normalize the distributions as $\int_0^1 dx\phi_{V\perp,V\parallel}(x) = 1$. The meson's Bethe-Salpeter wave function is defined as $\chi_{\lambda}(q;P) = S(q_+)\Gamma_{\lambda}(q,P)S(q_-)$ with Γ_{λ} the Bethe-Salpeter amplitude, S the dressed quark propagator and $q_+ = q+\eta P$, $q_- = q-(1-\eta)P$, $\eta\in[0,1]$. Using Eq. (1a-1b) and the numerical results of meson's Bethe-Salpeter amplitudes, one may get the moments of distribution $< x^m >= \int_0^1 dx x^m\phi(x)$, and then would fit the PDAs.

The PDAs in heavy flavour mesons are quite different from the light quark case. The PDA would tend to be δ function as the quark mass goes to infinity, and hence intuitively, it would be appropriate to fit the PDAs in heavy flavour mesons through an exponential distribution instead of the Gegenbauer polynomials used in the light quark case [1, 11, 12, 16]. Here we use the following form :

$$\phi(x) = 6ax(1-x)e^{-c(\frac{3}{2}-6x(1-x))}. \tag{2}$$

3 Results and Discussions

3.1 Meson properties in rainbow-ladder truncation

In our calculation, we solve the quark gap equation and Bethe-Salpeter equation obtained with the rainbow-ladder truncation of QCD's DSEs [17–19] and the interaction introduced in Ref. [20]. We employ two choices for the parameters in this interaction, $D\omega = (0.87\text{GeV})^3$ and $D\omega = (0.55\text{GeV})^3$. The current charm and bottom quark masses are chosen to be $m_c(\zeta = m_c) = 1.29$ GeV and $m_b(\zeta = m_b) = 4.16$ GeV which are renormalized at the respective current mass scale. After then we get the meson properties including the masses and decay constants exhibited in Table. 1.

Tableau 1. Mesons mass and decay constants (GeV)

	$m_{J/\Psi}$	m_{Υ}	$f_{J/\Psi}$	$f_{J/\Psi}^{\perp}$	f_{Υ}	$f_{\Upsilon}^{\perp}$
$D\omega = (0.87\text{GeV})^3$	3.17	9.46	0.483	0.386	0.602	0.506
$D\omega = (0.55\text{GeV})^3$	3.17	9.63	0.288	0.226	0.496	0.440

The results does not change a lot when we vary the parameters from $D\omega = (0.87\text{GeV})^3$ to $D\omega = (0.55\text{GeV})^3$. As the interaction strength becomes smaller, the decay constants decrease. Besides, the gap between $f^{\perp}$ and f becomes smaller, and these two decay constants will gradually become identical as in leading order approximation [21]. Besides, if we employ a naive quark model, the two decay constants for vector meson satisfy the relation [22] : $2m_q(\zeta)f_V = m_V f_V^{\perp}$, and thus the deviation for this relation indicates the dynamical chiral symmetry breaking. For light flavour mesons, this relation is sabotaged severely owing to the disparity between the large dynamical quark mass and the current quark mass, while for heavy flavour vector mesons, this relation is almost satisfied. This reveals that the effect of dynamical chiral symmetry breaking (DCSB) becomes less superior since the current quark mass is quite large compared to the dynamical mass from DCSB for heavy vector mesons and hence will lead to quite different behaviour compared to the light mesons.

3.2 Parton Distribution Amplitudes

In light quark case, we cannot handle the calculation for the moments in Euclidean space since the $(\frac{n\cdot q_+}{n\cdot P})$ is highly oscillated especially on the domain close to the ultraviolet cutoff, and thus we need to fit our numerical data into some analytical forms and then calculate in Minkowski space. However, the heavy quark mass offers the numerical stability and plausibility which make people to be able to directly calculate the moments from the numerical data without fitting procedure. The results show stability against the choices of the cut-off. We then obtain the moments shown in Table. 2.

Tableau 2. Mesons' moments with $D\omega = (0.87\text{GeV})^3$ *(left panel)* and $D\omega = (0.55\text{GeV})^3$ *(right panel)*

$<(2x-1)^m>$		m=2	4	$<(2x-1)^m>$		m=2	4
J/Ψ	$\perp$	0.0260	0.0020	J/Ψ	$\perp$	0.0276	0.0023
	$\parallel$	0.0049	7.23×10^{-5}		$\parallel$	0.0223	0.0015
Υ	$\perp$	0.0192	0.0011	Υ	$\perp$	0.0222	0.0015
	$\parallel$	0.0028	2.37×10^{-5}		$\parallel$	0.0165	8.17×10^{-4}

Tableau 3. Fitting parameters for PDAs with $D\omega = (0.87\text{GeV})^3$ *(left panel)* and $D\omega = (0.55\text{GeV})^3$ *(right panel)*

		a	c			a	c
J/Ψ	$\perp$	3.30	12.13	J/Ψ	$\perp$	3.20	11.34
	$\parallel$	7.59	67.23		$\parallel$	3.56	14.24
Υ	$\perp$	3.84	16.65	Υ	$\perp$	3.57	14.28
	$\parallel$	10.04	118.0		$\parallel$	4.14	19.48

We then fit the PDAs with Eq. (2), and show the obtained parameters in Table. 3. As the parameter c tends to be zero, the PDA becomes the asymptotic form, on the contrary, if c becomes infinite the PDA becomes δ function. All of the heavy flavour mesons' PDAs are narrower than the asymptotic form, while in light quark case, the PDAs are always broader than the asymptotic form. As the quark mass increases, the shape of the PDAs tends to be narrower, and it can be expected to be a δ function as the quark mass goes to infinity. For the same quark, the longitudinal PDA in vector mesons is narrower than the transversal PDA. However, the PDAs for J/Ψ and Υ meson still deviate from δ function. We also compare the results of different coupling strengths. The transversal PDA for the vector meson barely changes as the coupling strength changes, however, the longitudinal PDA gets closer to the transversal part as the interaction strength decreases.

In summary, we compute the heavy vector mesons' PDAs. The PDAs are all narrower than the asymptotic form and as the quark mass increases the form would be closer to δ function. However, the PDAs still deviate from δ function. We also compare the effects of different coupling constants. We find that as the coupling constant decreases, the longitudinal and transversal PDAs of vector meson get closer to each other.

Acknowledgments

The work was supported by the National Natural Science Foundation of China under Contracts No. 11435001 and No. 11175004 ; the National Key Basic Research Program of China under Contract No. G2013CB834400 ; the U.S. Department of Energy, Office of Nuclear Physics, Contract No. DE-AC02-06CH11357.

References

[1] P. Ball, V.M. Braun, Phys. Rev. **D54**, 2182 (1996), hep-ph/9602323

[2] G.P. Lepage, S.J. Brodsky, Phys. Lett. **B87**, 359 (1979)

[3] G.R. Farrar, D.R. Jackson, Phys.Rev.Lett. **43**, 246 (1979)

[4] M. Beneke, G. Buchalla, M. Neubert, C.T. Sachrajda, Phys. Rev. Lett. **83**, 1914 (1999), hep-ph/9905312

[5] M. Beneke, G. Buchalla, M. Neubert, C.T. Sachrajda, Nucl. Phys. **B591**, 313 (2000), hep-ph/0006124

[6] J. Forshaw, R. Sandapen, JHEP **1011**, 037 (2010), 1007.1990

[7] B. Cox, J. Forshaw, R. Sandapen, JHEP **0906**, 034 (2009), 0905.0102

[8] S. Chekanov, M. Derrick, S. Magill, B. Musgrave, D. Nicholass, J. Repond, R. Yoshida, M. Mattingly, M. Jechow, N. Pavel et al., PMC Phys. A **356**, 601 (2007)

[9] F. Aaron, M.A. Martin, C. Alexa, V. Andreev, B. Antunovic, A. Asmone, S. Backovic, A. Baghdasaryan, E. Barrelet, W. Bartel et al., JHEP **2010**, 1 (2010)

[10] C. Adloff, V. Andreev, B. Andrieu, V. Arkadov, A. Astvatsatourov, I. Ayyaz, A. Babaev, J. Bähr, P. Baranov, E. Barrelet et al., Phys. Lett. B **483**, 23 (2000)

[11] L. Chang, I. Cloet, J. Cobos-Martinez, C. Roberts, S. Schmidt et al., Phys. Rev. Lett. **110**, 132001 (2013), 1301.0324

[12] F. Gao, L. Chang, Y.X. Liu, C.D. Roberts, S.M. Schmidt, Phys. Rev. **D90**, 014011 (2014), 1405.0289

[13] I. Cloët, L. Chang, C. Roberts, S. Schmidt, P. Tandy, Phys. Rev. Lett. **111**, 092001 (2013), 1306.2645

[14] L. Chang, C. Mezrag, H. Moutarde, C.D. Roberts, J. Rodríguez-Quintero et al., Phys. Lett. **B737**, 23 (2014), 1406.5450

[15] L. Chang, I. Cloët, C. Roberts, S. Schmidt, P. Tandy, Phys. rev. lett. **111**, 141802 (2013)

[16] G.P. Lepage, S.J. Brodsky, Phys. Rev. **D22**, 2157 (1980)

[17] I.C. Cloet, C.D. Roberts, Prog. Part. Nucl. Phys. **77**, 1 (2014), 1310.2651

[18] P. Maris, C.D. Roberts, Int. J. Mod. Phys. **E12**, 297 (2003), nucl-th/0301049

[19] A. Bashir, L. Chang, I.C. Cloet, B. El-Bennich, Y.X. Liu et al., Commun. Theor. Phys. **58**, 79 (2012), 1201.3366

[20] S.x. Qin, L. Chang, Y.x. Liu, C.D. Roberts, D.J. Wilson, Phys. Rev. **C84**, 042202 (2011), 1108.0603

[21] H.Y. Cheng, K.C. Yang, Phys. Rev. D **63**, 074011 (2001)

[22] J. Chay, C. Kim, arXiv preprint hep-ph/0009244 (2000)

EPJ Web of Conferences **113**, 05015 (2016)
DOI: 10.1051/epjconf/201611305015

Light-quark mass behaviour of the X(3872) as a molecular state

V. Baru[1,2,a], E. Epelbaum[1,b], A.A. Filin[1,c], J. Gegelia[3,4,d], and A.V. Nefediev[2,5,6,e]

[1] *Institut für Theoretische Physik II, Ruhr-Universität Bochum, D-44780 Bochum, Germany*
[2] *Institute for Theoretical and Experimental Physics, B. Cheremushkinskaya 25, 117218 Moscow, Russia*
[3] *Forschungszentrum Jülich, Institute for Advanced Simulation, Institut für Kernphysik and Jülich Center for Hadron Physics, D-52425 Jülich, Germany*
[4] *Tbilisi State University, 0186 Tbilisi, Georgia*
[5] *National Research Nuclear University MEPhI, 115409, Moscow, Russia*
[6] *Moscow Institute of Physics and Technology, 141700, Dolgoprudny, Moscow Region, Russia*

Abstract.
Chiral extrapolations of the binding energy of the $X(3872)$ molecular state are investigated using an explicitly renormalizable framework free of finite cut-off artefacts. Insights into the binding mechanisms are discussed: if the X is less bound with the growing pion mass, its binding energy is governed by the explicit pion mass dependence from one-pion exchange; an opposite behaviour would indicate the importance of the pion-mass dependent short-range interactions, in addition to pionic effects. The important role of the three-body $D\bar{D}\pi$ dynamics is emphasised.

1 Introduction

The light-quark mass dependence of hadronic observables is of particular importance for understanding various phenomena in modern hadron and nuclear physics. For example, the knowledge of the light-quark or equivalently the pion mass dependence of the nucleon-nucleon (NN) scattering lengths is crucial for understanding certain fine-tunings in nuclear physics such as the formation of the life-important elements in connection to the anthropic principle, see a review [1] for a detailed discussion. In hadron physics the quark-mass dependence of the resonance poles could be useful to investigate the interplay of the long- and short-range forces in the formation of hadronic molecules. Furthermore, this information should be of practical value because it provides the connection between the lattice simulations which are still being carried out at relatively large unphysical pion masses and the physical point. One of the most interesting candidates is an enigmatic meson 1^{++} X(3872) discovered by the Belle collaboration in 2003, see Ref. [2] for a review. The existence of the X was predicted a long time ago [3, 4] based on the similarities between the $\bar{D}D^*$ and NN forces. In particular, it is assumed

[a] e-mail: vadim.baru@tp2.rub.de
[b] e-mail: evgeny.epelbaum@ruhr-uni-bochum.de
[c] e-mail: arseniy.filin@tp2.rub.de
[d] e-mail: jgegelia@hotmail.com
[e] e-mail: nefediev@itep.ru

[5–7] that the X is a bound (or virtual) state of $\bar{D}D^*$ because of its close proximity to the $\bar{D}D^*$ threshold and because of its appropriate quantum numbers. In analogy to the deuteron, the X is amenable to the effective field theory treatment, see [8–10] for various studies within pionless framework and [11–17] for pionful approaches.

2 Description of the approach

In ref. [15] the pion mass dependence of the X binding energy was investigated within the EFT framework based on the non-perturbative solution of the non-relativistic Lippmann-Schwinger-type integral equations for the $\bar{D}D^*$ problem with the leading-order (LO) potential consisting of the one-pion-exchange (OPE) and the pion-mass dependent contact interaction. Special care was taken to include non-perturbatively all scales from three-body $D\bar{D}\pi$ intermediate states which play an important role because of a close proximity of the $D^{\pm}D^{\mp}\pi^0$, $D^{\pm}\bar{D}^0\pi^{\mp}$, and $D^0\bar{D}^0\pi^0$ thresholds to the $X(3872)$ pole. Notice that the use of non-perturbative OPE for chiral extrapolations allows one to extend the domain of applicability of the approach to larger pion masses which is important for analysing the results of lattice QCD calculations.

In this contribution we briefly discuss the extension of the work of Ref. [15] to address another important issue which is related to nonperturbative renormalisation of the integral equations for the $\bar{D}D^*$ problem, see a detailed discussion in the full-length publication [18]. The standard nonrelativistic approach to heavy mesons leads to coupled-channel integral equations for the scattering amplitudes which, at leading order in the EFT expansion, are linearly divergent. As a consequence, iterations of the OPE potential within the integral equations generate an infinite series of ultraviolet (UV) divergent higher-order contributions to the amplitude which cannot be absorbed into a finite number of counter terms (contact interactions) included in the truncated potential. To suppress the unwanted higher-order contributions, one usually employs a finite UV cut-off of the order of a natural hard scale in the problem, as advocated in Ref. [19]. This strategy was followed, in particular, in Refs. [9, 10, 14, 15]. Recently, another solution of this problem was proposed within a novel renormalizable approach to NN scattering [20] where exactly the same problem occurs. The core of the problem lies in the non-commutativity of the non-relativistic expansion and the renormalisation procedure [21], although the effect of relativistic corrections at low energies is of course small after renormalisation. Therefore, we derive the system of 3-dimensional dynamical equations for the $\bar{D}D^*$ problem which satisfy the relativistic unitarity, as naturally comes out if one uses time-ordered perturbation theory. This approach is renormalizable at leading order since all iterations of the potential generate only logarithmically divergent contributions which can be absorbed in the redefinition of the contact terms, see Refs. [18, 20]. Then the higher-order contributions to the potential are included perturbatively.

While both approaches described above should agree in general, the explicitly renormalizable approach has an advantage that finite cut-off artefacts are removed which is an important pre-requisite for understanding the binding mechanism. For example, unlike the finite cut-off results of Ref. [15], the pion mass dependence of the binding energy in the explicitly renormalizable framework can be predicted at LO and is entirely governed by the pion mass dependence from OPE. The novel approach should therefore be very useful for carrying out chiral extrapolations and studying correlations in the scattering amplitude induced by the analytic structure of the long-range forces Refs. [22].

3 Discussion of the results

We start from the LO. As was explained above, the relevant potential at this order reads

$$V_{\rm LO} = V_{\rm OPE}(\boldsymbol{p}, \boldsymbol{p}', \xi) + C_0, \tag{1}$$

where the first term on the r.h.s. denotes the OPE potential and the contact term C_0 is m_π-independent at this order. Once C_0 is adjusted to reproduce the binding energy at the physical pion mass $m_\pi^{\rm ph}$ (for definiteness we set $E_B(m_\pi^{\rm ph}) = 0.5$ MeV), the scattering amplitude can be calculated for unphysical pion masses without loss of renormalizability of the LO equations. Therefore, at the LO of our EFT, the dependence $E_B(m_\pi)$ can be predicted in a parameter free way—see Fig. 1. At this order, the pion mass dependence of E_B originates only from the pionic effects in the OPE potential and in the renormalised selfenergy loops. The binding energy at the LO demonstrates a clear tendency to disappear with the growing m_π. Furthermore, the slope of the binding energy in m_π at the physical point, $(\partial E_B/\partial m_\pi)\big|_{m_\pi=m_\pi^{\rm ph}}$, exhibits a strong sensitivity to the 3-body $D\bar{D}\pi$ effects. In particular, neglecting the 3-body dynamics (the so-called static OPE) results in a much steeper fall of the binding energy—compare the dotted (blue) line versus the solid (black) in the left panel of Fig. 1.

Since no real experiment is possible for unphysical pion masses, the only source of information on the m_π-dependence of the X pole is provided by lattice simulations. In analogy to the deuteron (see e.g. Ref. [22] for the discussion), the first lattice calculations for the $X(3872)$ evidence a stronger bound X for $m_\pi > m_\pi^{\rm ph}$ [23–25], although these results still suffer from potentially large finite-range corrections, as pointed out in Ref. [13]. To understand this type of behaviour we proceed to the NLO potential which takes into account the short-range corrections quadratic in m_π

$$V_{\rm NLO} = V_{\rm LO} + D\,(\xi^2 - 1), \quad \xi = m_\pi/m_\pi^{\rm ph}. \tag{2}$$

The unknown coefficient D can be fixed to the slope of the binding energy at the physical pion mass, which is therefore considered as an additional input quantity. For example, in Fig. 1 we illustrate the behaviour of the binding energy for the slope of natural size $(\partial E_B/\partial m_\pi^2)\big|_{m_\pi=m_\pi^{\rm ph}} \approx E_B^{\rm ph}/m_\pi^{\rm ph\,2}$—see the dashed curve in the right panel. Interestingly, in a theory with the same polynomial behaviour of the contact operator but without pions one would observe a much flatter behaviour $E_B(m_\pi)$ for the same slope, as shown by the dashed-dotted curve. The difference between the two curves demonstrates the role of dynamical pions as an explicit long-range degree of freedom. As seen from Fig. 1, the contact interaction provides a smooth background for a rapidly varying pion mass dependence stemming from OPE. Therefore integrating out pions and the corresponding 3-body soft scales while still trying to at least partially compensate for neglecting these long-range effects, one would arrive at unnaturally large m_π-dependent coefficients accompanying short-range operators.

In summary, the explicitly renormalizable approach for $\bar{D}D^*$ scattering is developed and applied to explore the light-quark mass dependence of the $X(3872)$ pole. The latter appears to be strongly affected by the 3-body effects. Our approach should be of practical use to provide extrapolation of the lattice data to the physical limit.

Acknowledgements

This work is supported by the project HadronPhysics3 (Grant Agreement n. 283286), the ERC project 259218 NUCLEAREFT, the DFG and the NSFC through funds provided to the Sino-German CRC 110 "Symmetries and the Emergence of Structure in QCD", the Russian Science Foundation (Grant No. 15-12-30014) and by the Georgian Shota Rustaveli National Science Foundation (grant FR/417/6-100/14).

References

[1] U.-G. Meißner, Sci. Bull. (2015) **60**(1):43-54 [arXiv:1409.2959 [hep-th]].

[2] N. Brambilla *et al.*, Eur. Phys. J. C **71**, 1534 (2011).

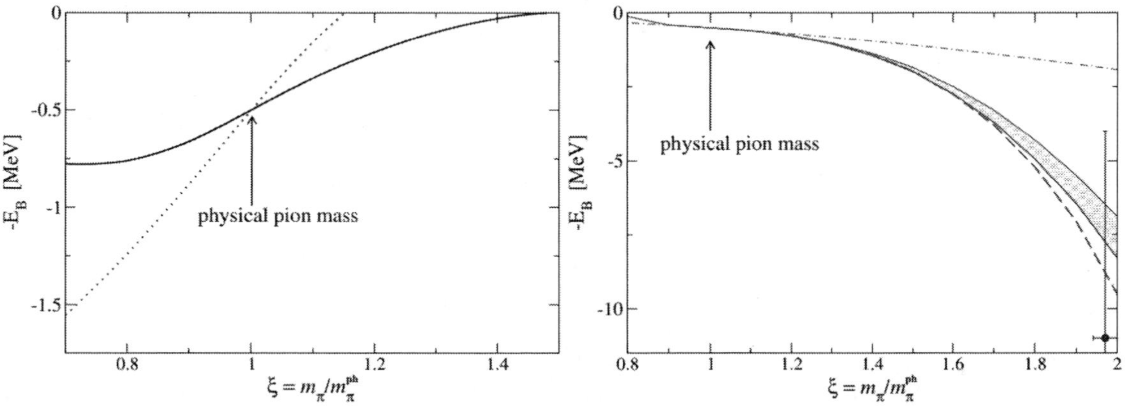

Figure 1. Pion mass dependence of the $X(3872)$ binding energy. Left panel: LO calculation; the results of the full dynamical theory with 3-body effects (black solid curve) vs. static OPE (blue dotted curve). Right panel: NLO calculation with an opposite slope used as input; perturbative inclusion of the contact operator D (red dashed curve) is in agreement with nonperturbative treatment of the same operator employing resonance saturation (red band). The dashed-dotted line corresponds to the calculation without pions, the blue data point is from Ref. [25].

[3] M. B. Voloshin, L. B. Okun, Pisma Zh. Eksp. Teor. Fiz. **23**, 369 (1976).
[4] N. A. Tornqvist, Phys. Rev. Lett. **67**, 556 (1991).
[5] M. B. Voloshin, Phys. Lett. B **579**, 316 (2004).
[6] E. Braaten and M. Lu, Phys. Rev. D **76**, 094028 (2007).
[7] Yu. S. Kalashnikova and A. V. Nefediev, Phys. Rev. D **80**, 074004 (2009).
[8] M. T. AlFiky, F. Gabbiani, and A. A. Petrov, Phys. Lett. B **640**, 238 (2006).
[9] F.-K. Guo, C. Hidalgo-Duque, J. Nieves, and M. P. Valderrama, Phys. Rev. D **88**, 054007 (2013).
[10] J. Nieves and M. P. Valderrama, Phys. Rev. D **86**, 056004 (2012).
[11] S. Fleming, M. Kusunoki, T. Mehen, and U. van Kolck, Phys. Rev. D **76**, 034006 (2007).
[12] M. Jansen, H.-W. Hammer, and Y. Jia, Phys. Rev. D **89**, 014033 (2014).
[13] M. Jansen, H.-W. Hammer, and Y. Jia, arXiv:1505.04099 [hep-ph].
[14] V. Baru et *al.*, Phys. Rev. D **84**, 074029 (2011).
[15] V. Baru, E. Epelbaum, A. A. Filin, C. Hanhart, U.-G. Meißner, and A. V. Nefediev, Phys. Lett. B **726**, 537 (2013).
[16] P. Wang and X. G. Wang, Phys. Rev. Lett. **111**, 042002 (2013).
[17] V. Baru, E. Epelbaum, A. A. Filin, F.-K. Guo, H.-W. Hammer, C. Hanhart, U.-G. Meißner and A. V. Nefediev, Phys. Rev. D **91**, no. 3, 034002 (2015).
[18] V. Baru, E. Epelbaum, A. A. Filin, G. Gegelia, and A. V. Nefediev, in preparation
[19] G. P. Lepage, nucl-th/9706029.
[20] E. Epelbaum and J. Gegelia, Phys. Lett. B **716**, 338 (2012).
[21] E. Epelbaum and J. Gegelia, Few Body Syst. **54**, 1473 (2013).
[22] V. Baru, E. Epelbaum, A. A. Filin, and J. Gegelia, Phys. Rev. C **92**, 014001 (2015).
[23] S. Prelovsek and L. Leskovec, Phys. Rev. Lett. **111**, 192001 (2013).
[24] S. H. Lee *et al.* [Fermilab Lattice and MILC Collaborations], arXiv:1411.1389 [hep-lat].
[25] M. Padmanath, C. B. Lang, and S. Prelovsek, Phys. Rev. D **92**, 034501 (2015)

EPJ Web of Conferences **113**, 05016 (2016)
DOI: 10.1051/epjconf/201611305016

Mesonic Decay of Charm Hypernuclei Λ_c^+

Sabyasachi Ghosh[1,a], Carlos E. Fontoura[1,b], Gastão Krein[1,c],

[1] *Instituto de Física Teórica, Universidade Estadual Paulista*
Rua Dr. Bento Teobaldo Ferraz, 271 - Bloco II, 01140-070 São Paulo, SP, Brazil

Abstract. Λ_c^+ hypernuclei are expected to have binding energies and other properties similar to those of strange hypernuclei in view of the similarity between the quark structures of the strange and charmed hyperons, namely $\Lambda(uds)$ and $\Lambda_c^+(udc)$. One striking difference however occurs in their mesonic decays, as there is almost no Pauli blocking in the nucleonic decay of a charm hypernucleus because the final-state nucleons leave the nucleus at high energies. The nuclear medium nevertheless affects the mesonic decays of charm hypernucleus because the nuclear mean fields modify the masses of the charm hyperon. In the present communication we present results of a first investigation of the effects of finite baryon density on different weak mesonic decay channels of the Λ_c^+ baryon. We found a non-negligible reduction of the decay widths as compared to their vacuum values.

1 Introduction

The possible existence of charm hypernuclei has been suggested [1] right after the discovery of the charm quark and a first explicit calculation [2] of their binding energies was performed using a meson exchange model with coupling constants determined by SU(4) flavor symmetry. More recently, calculations were performed [3] using the quark meson coupling model at the Hartree level — the predicted binding energies per baryon in Λ_c^+, Σ_c, Ξ_c hypernuclei ranging from O to Pb are quite substantial, on average of the order of 7.5 MeV. Such binding energies are of similar magnitude to those of J/Ψ binding in nuclear matter [4] and finite nuclei [5]. Vacuum effects [6] and Fock terms [7], although crucial for capturing pionic effects in matter [8], are not expected to have a great impact on the binding energies. Up to now there is no experimental evidence of the existence of a charm hypernucleus, apart from an emulsion experiment at Fermilab with 250 GeV protons that indicated that Λ_c^+ hypernuclei have been produced in the collisions [9]. New experimental possibilities were explored [10, 11] at the beginnings of the 1990's, but concrete hopes for the production of such hypernuclei comes with the completion of the FAIR facility (where $\Lambda_c^\pm$ can be produced in $\bar{p}p$ annihilation processes [12]) and the installation of a 50 GeV high-intensity proton beam at the J-PARC complex.

The study of charm hypernuclei is interesting for several reasons; among the most interesting in our view are: (1) they bring the opportunity to learn about the poorly known interactions of charmed baryons with nucleons [13, 14]; (2) they offer also the opportunity to learn about medium effects on

[a] e-mail: sabyaphy@ift.unesp.br
[b] e-mail: eduardo@ift.unesp.br
[c] e-mail: gkrein@ift.unesp.br

diquark correlations, as the masses of the light u and d quarks are affected by the medium while the mass of the heavy c quark is not affected; and (3) few-nucleon (typically one or two) charmed states have the potential of having a rich spectroscopy similar to the X,Y,Z mesons.

One particular observable that can be affected by the medium is the weak-decay lifetime of the charmed baryon because of the change in its mass. There are no Pauli blocking effects on the decaying baryon – for the case of Λ_c^+, for instance, the decays $\Lambda\pi^+$ and $\Sigma^+\pi^0$ are not blocked, and in the decay $p\bar{K}^0$ the proton leaves the nucleus with a large momentum. In the present communication we present the results of a first investigation [15] of the effects of finite baryon density on different weak mesonic decay channels of the Λ_c^+ baryon.

2 In-medium mesonic decays of Λ_c^+

We evaluate the decay rate in infinite nuclear matter via the imaginary part of the self-energy diagram depicted in Fig. 1. To evaluate the diagram, we employ the effective weak-decay Lagrangian density

$$\mathcal{L}_{\Lambda_c BM} = i g_{\Lambda_c BM} \bar{\psi}_{\Lambda_c} \left(A_{BM} + B_{BM}\gamma^5\right) \phi_M \psi_B + \text{h.c.}, \tag{1}$$

where $g_{\Lambda_c BM} = 10^{-2} G_F V_{ud} V_{cs}$, with $G_F = 1.16 \times 10^{-5}$ GeV^{-2} being the Fermi constant and $V_{ud} = 0.974$, $V_{cs} = 0.973$ relevant CKM matrix elements. For the density dependence of Λ_c^+ we use the results from the QMC model [3]. Finite nuclei can be treated in the local density approximation.

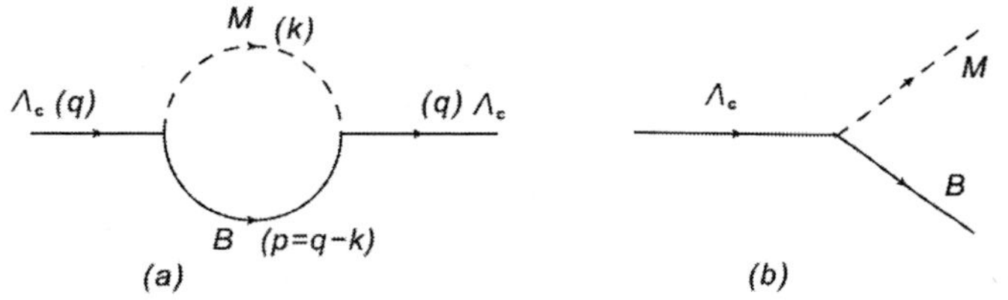

Figure 1. (a) One loop baryon-meson BM self-energy of Λ_c^+: $B = \Lambda, \Sigma^+, p$; $M = \pi^+, \pi^0, \bar{K}^0$. (b) decay diagram, the imaginary part of (a).

We start examining the role of the Pauli principle in the $\Lambda_c^+ \to p\bar{K}^0$ decay, neglecting all other possible medium effects. The in-medium decay width $\Gamma_T(q)$ for a generic baryon-meson (BM) decay $\Lambda_c^+ \to BM$, where $q = |\mathbf{q}|$ with $\mathbf{q}$ the momentum of the Λ_c^+, can be written as the difference $\Gamma_T(q) = \Gamma_V(q) - \Gamma_D(q)$, where $\Gamma_V(q)$ is vacuum decay width and $\Gamma_D(q)$ is a density dependent width due to Pauli-blocking:

$$\begin{pmatrix} \Gamma_V(q) \\ \Gamma_D(q) \end{pmatrix} = \frac{g^2_{\Lambda_c BM}}{16\pi q m_{\Lambda_c}} \left[(m_{\Lambda_c} + m_B)^2 - m_M^2\right] \left[|A_{BM}|^2 + \left(\frac{(m_{\Lambda_c} - m_B)^2 - m_M^2}{(m_{\Lambda_c} + m_B)^2 - m_M^2}\right) |B_{BM}|^2\right] \times \int_{\omega^-(q)}^{\omega^+(q)} \begin{pmatrix} 1 \\ \theta(\omega - \omega^{\text{th}}(q)) \end{pmatrix} d\omega, \tag{2}$$

where m_{Λ_c}, m_B and m_M are the masses of Λ_c^+, baryon B and meson M, respectively, and θ is the step function with

$$\omega^\pm(q) = \frac{R^2}{2m^2_{\Lambda_c}} \left[(q^2 + m^2_{\Lambda_c})^{1/2} \pm qW\right], \qquad \omega^{\text{th}}(q) = \left(q^2 + m^2_{\Lambda_c}\right)^{1/2} - \left(k_F^2 + m_B^2\right)^{1/2}, \tag{3}$$

where $R^2 = m_{\Lambda_c}^2 + m_M^2 - m_B^2$, $W = (1 - 4m_M^2 m_{\Lambda_c}^2/R^4)^{1/2}$, and k_F is the nuclear matter Fermi momentum given in terms of the baryon density ρ as $k_F = (3\pi^2/2\rho)^{1/3}$.

There is no Pauli blocking in the $\Lambda_c^+ \to \Lambda\pi^+$ and $\Lambda_c^+ \to \Sigma^+\pi^0$ decays and $\Gamma_D(q)$ is nonzero only in the nucleonic decay $\Lambda_c^+ \to p\bar{K}^0$. Notice that $\Gamma_D(q)$ vanishes for $\rho_B = 0$ as $\omega^{\rm th}(q) > \omega^+(q)$ in this case and the step function in Eq. (2) gives zero for the integral. The dot-dashed line in Fig. 2 presents the momentum dependence of the ratio $\Gamma_T/\Gamma_V = 1 - \Gamma_D/\Gamma_V$ for the $\Lambda_c^+ \to p\bar{K}^0$ decay channel in nuclear matter ($\rho = \rho_0 = 0.16$ fm^{-3}). Clearly, the effect of the Pauli principle is negligible. Physically, this is due to the fact that the momentum of the outgoing proton is of the order of, or larger than the Fermi momentum — this is transparent in the plot of $\omega^{\rm th}(q)$ that shows that it barely takes values within the region delimited by the curves of $\omega^+(q)$ and $\omega^-(q)$ as demanded by the theta function in Eq. (2) for a $\Gamma_D(q) \neq 0$.

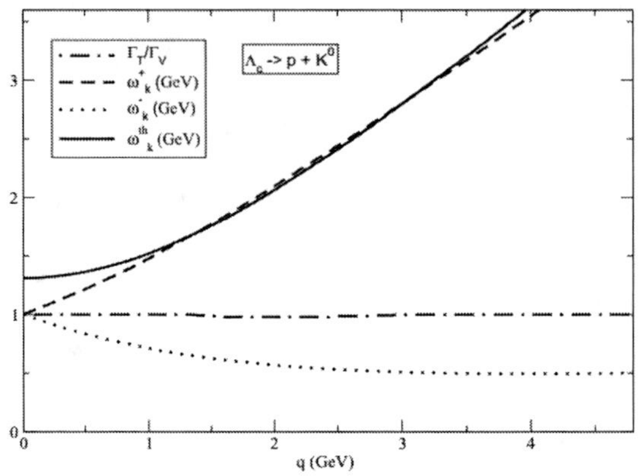

Figure 2. Momentum dependence of the ratio of the total to vacuum decay widths (dash-dotted line) for the $\Lambda_c \to p\bar{K}^0$ decay channel in nuclear matter at the saturation density ($\rho = \rho_0 = 0.16$ fm^{-3}). Also shown are the quantities $\omega^+(q)$ (dashed line), $\omega^-(q)$ (dotted line), and $\omega^{\rm th}(q)$ (solid line).

Next, we consider the effect of the in-medium mass shift of Λ_c^+ on the decay widths of the different channels — as the proton leaves the nucleus with high momentum, its mass shift is negligible. The density dependence on the mass is taken from the QMC model [3], which can be parametrized as

$$m_{\Lambda_c}^*(\rho)/m_{\Lambda_c} = a_0\, e^{-a_1\rho/\rho_0} + a_2, \tag{4}$$

with $a_0 = 0.121$, $a_1 = 0.565$ and $a_2 = 0.878$. The in-medium decay width is denoted by Γ_ρ and is given by the same expression as Γ_V in Eq. (2) but with m_{Λ_c} replaced by $m_{\Lambda_c}^*$. In Fig. 3 we present the ratio Γ_ρ/Γ_V as function of the baryon density, where Γ_ρ is evaluated at $q = m_{\Lambda_c}^*$ and Γ_V is evaluated at $q = m_{\Lambda_c}$. At the nuclear matter saturation density, $m_{\Lambda_c}^* = 2.146$ GeV. The in-medium reduction of the decay widths is not small. Specifically, at the normal nuclear matter density, the reductions are 14.5%, 12% and 6.5% for $\Lambda_c \to \Lambda\pi^+, \Sigma^+\pi^0$, and $p\bar{K}^0$ decay channels, respectively. A local density approximation leads to similar reductions for medium to heavy nuclei [15].

Several issues need careful examination before definite conclusions can be drawn. Specifically, SU(4) flavor symmetry breaking in meson-baryon couplings have been shown to be significant [16] and its impact on the results needs to be assessed — the subject is also of crucial importance for the DN interaction [14] in connection to D-mesic nuclei. The consequences of medium modifications on the mass of Λ_c^+ on non-mesonic weak decays of charmed hypernuclei is under investigation [17].

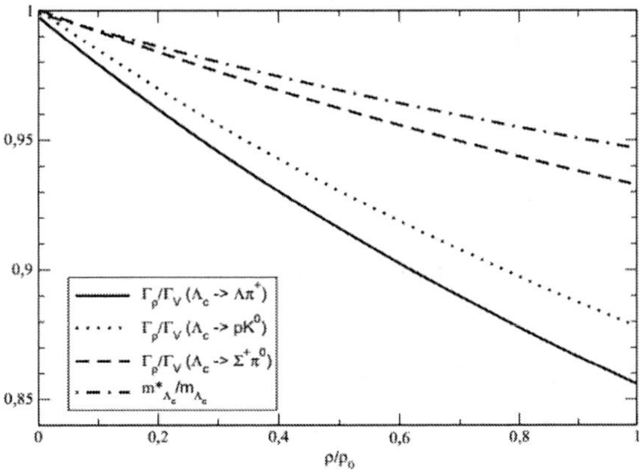

Figure 3. Γ_ρ/Γ_V for $\Lambda_c \to \Lambda\pi^+$ (solid line), $\Lambda_c \to \Sigma^+\pi^0$ (dashed line), $\Lambda_c \to p\bar{K}^0$ (dotted line) decay channels with ρ/ρ_0. Also shown is $m^*_{\Lambda_c}/m_{\Lambda_c}$ (dot-dashed line).

Acknowledgements

Work partially by Fundação de Amparo à Pesquisa do Estado de São Paulo-FAPESP, Grants 2012/16766-0 (S.G.) and No. 2013/01907-0 (G.K.), and Conselho Nacional de Desenvolvimento Científico e Tecnológico - CNPq, Grant No.305894/2009-9 (G.K.). The work of C.E.F. was supported by a scholarship from Universidade Estadual Paulista.

References

[1] A.A. Tyapkin, Sov. J. Nucl. Phys. **22**, 181 (1975).
[2] C.B. Dover and S.H. Kahana, Phys. Rev. Lett. **39**, 1506 (1977).
[3] K. Tsushima and F.C. Khanna, Phys. Rev. C **67**, 015211 (2003); *ibid.* J. Phys. G **30**, 1765 (2004).
[4] G. Krein, A. W. Thomas and K. Tsushima, Phys. Lett. B **697**, 136 (2011).
[5] K. Tsushima, D. H. Lu, G. Krein and A. W. Thomas, Phys. Rev. C **83**, 065208 (2011).
[6] G. Krein, M. Nielsen, R. D. Puff and L. Wilets, Phys. Rev. C **47**, 2485 (1993).
[7] G. Krein, A. W. Thomas and K. Tsushima, Nucl. Phys. A **650**, 313 (1999).
[8] M. E. Bracco, G. Krein and M. Nielsen, Phys. Lett. B **432**, 258 (1998).
[9] Yu. A. Batusov et al. JETP Lett. **33**, 56 (1981),
[10] T. Bressani and F. Iazzi, Nuovo Cimento A **102**, 597 (1989).
[11] S.A. Buyatov et al. Nuovo Cimento **104**, 1361 (1981).
[12] J. Haidenbauer and G. Krein, Phys. Lett. B **687**, 314 (2010).
[13] J. Haidenbauer, G. Krein, U. G. Meissner and A. Sibirtsev, Eur. Phys. J. A **33**, 107 (2007).
[14] C. E. Fontoura, G. Krein and V. E. Vizcarra, Phys. Rev. C **87**, 025206 (2013).
[15] S. Ghosh, C.E. Fontoura and G. Krein, in preparation.
[16] B. El-Bennich, G. Krein, L. Chang, C. D. Roberts and D. J. Wilson, Phys. Rev. D **85**, 031502 (2012).
[17] C. E. Fontoura, A. P. Galeão, G. Krein and F. Krmpotic, in preparation.

EPJ Web of Conferences **113**, 05017 (2016)
DOI: 10.1051/epjconf/201611305017

Recoil corrections in antikaon-deuteron scattering

Maxim Mai[1,a], Vadim Baru[2,3,b], Evgeny Epelbaum[2,c], and Akaki Rusetsky[1,d]

[1] *Helmholtz–Institut für Strahlen- und Kernphysik (Theorie) and Bethe Center for Theoretical Physics, Universität Bonn, D-53115 Bonn, Germany*
[2] *Institut für Theoretische Physik II, Fakultät für Physik und Astronomie, Ruhr-Universität Bochum, 44780 Bochum, Germany*
[3] *Institute for Theoretical and Experimental Physics, 117218, B. Cheremushkinskaya 25, Moscow, Russia*

Abstract. Using the non-relativistic effective field theory approach for K^-d scattering, it is demonstrated that a systematic *perturbative* expansion of the recoil corrections in the parameter $\xi = M_K/m_N$ is possible in spite of the fact that K^-d scattering at low energies is inherently non-perturbative due to the large values of the $\bar{K}N$ scattering lengths. The first order correction to the K^-d scattering length due to single insertion of the retardation term in the multiple-scattering series is calculated. The recoil effect turns out to be reasonably small even at the physical value of $M_K/m_N \simeq 0.5$.

1 Introduction

Antikaon-nucleon scattering is an excellent testing ground for understanding of the $SU(3)$ QCD dynamics at low energies in the one-baryon sector. Starting from the seminal paper [1], it is often described within the so-called unitarized Chiral Perturbation Theory (ChPT), which uses the chiral potential calculated at a certain order. The common feature of such approaches is a relatively large number of free parameters, which are fixed from the fit to the experimental data, see for recent developments, e.g., Refs. [2–5]. An essential part of the input is coming from the S-wave $\bar{K}N$ scattering lengths, which "nail down" the amplitudes at the $\bar{K}N$ threshold and thus impose stringent constraints both on the scattering in the $\bar{K}N$ channel as well as the sub-threshold behavior of the amplitudes.

The experiments with kaonic atoms have been carried out in order to extract the precise values of the S-wave $\bar{K}N$ scattering lengths. Recently, the energy shift and width of kaonic hydrogen were measured very accurately in the SIDDHARTA experiment at DAΦNE [6]. These two quantities can be related to the K^-p scattering lengths via the so-called modified Deser-type formula, see Refs. [7, 8]. The same experimental collaboration has made an attempt to measure the energy and the width of the ground state of the kaonic deuterium as well. What makes the experiments with kaonic deuterium extremely important is the fact that the S-wave $\bar{K}N$ scattering lengths are complex-valued. Therefore, extracting two complex scattering lengths a_0, a_1 corresponding to the total isospin $I = 0, 1$ in the $\bar{K}N$ system implies the determination of four real quantities and thus requires measurements of four

[a] e-mail: mai@hiskp.uni-bonn.de
[b] e-mail: vadimb@tp2.rub.de
[c] e-mail: evgeny.epelbaum@ruhr-uni-bochum.de
[d] e-mail: rusetsky@hiskp.uni-bonn.de

independent observables. Two observables are provided by the kaonic hydrogen, and the remaining two can come from, e.g., the kaonic deuterium. Therefore, it remains to derive an explicit relation between the $\bar{K}d$ and $\bar{K}N$ scattering lengths. This can be done assuming that the nucleons are infinitely heavy, leading to the type of Brueckner formula [9, 10]. However, there exists no *a priori* reason to believe that this is a good approximation in view of the fact that the mass ratio $M_K/m_N \simeq 0.5$ is not small. The goal of the present work is to continue the work of Ref. [11] with the main goal to formulate a procedure for including the recoil corrections *perturbatively* into the multiple-scattering series, in which the static interactions are summed up *to all orders*.

2 Framework

Our approach relies on the existence of two distinct momentum scales. The nucleon-nucleon and three-particle interactions are characterized by a low scale (of the order of the pion mass) and are described by non-local, energy-independent potentials $V_{NN}(\mathbf{p}, \mathbf{q})$ and $V_3(\mathbf{p}_1, \mathbf{p}_2, \mathbf{p}_3; \mathbf{q}_1, \mathbf{q}_2, \mathbf{q}_3)$, respectively. On the contrary, $\bar{K}N$ interactions are characterized by a heavier scale (numerically of the order of the mass of the $\rho, \omega, \cdots$ resonances). We shall describe these interactions by a tower of local terms in the Lagrangian with zero, two,... space derivatives. The couplings emerging in these terms are expressed through the $\bar{K}N$ scattering lengths, effective radii and so on in a standard manner. For this reason, a *perturbative* expansion in such an effective theory automatically yields the *multiple-scattering series*, known from the potential scattering framework.

A generic term in the multiple-scattering expansion contains the diagrams in which the kaons are exchanged between two nucleons as well as kaons hopping on the same nucleon (the self-energy-type diagrams). Since NN interactions are non-perturbative, they have to be included to all orders. This is normally done by solving the Lippmann-Schwinger type equations which yield the NN amplitudes at low-energies. The NN amplitude in its turn is to be included in each intermediate state of the $\bar{K}NN - \bar{K}NN$ Feynman diagrams. In the vicinity of the static limit ($\xi \to 0$), only diagrams where the kaon is exchanged between two nucleons survive. In this limit the series in the diagrams can be summed up to all orders. Taking the Fourier transform and folding the result with the deuteron wave function in the isospin limit, we finally arrive at

$$\mathcal{A}_{\text{st}} = \frac{1}{1+\xi/2} \int d^3\mathbf{r}\, |\Psi(\mathbf{r})|^2 \, \frac{3\tilde{a}_1 r^2 + r\tilde{a}_0(4\tilde{a}_1 + r)}{\tilde{a}_0(r - 2\tilde{a}_1) + r(2r - \tilde{a}_1)}\,, \tag{1}$$

where $r := |\mathbf{r}|$ and $\Psi(\mathbf{r})$ stands for the wave function of the deuteron, normalized to unity. The $\bar{K}N$ scattering lengths of isospin i enter the above equation as $\tilde{a}_i = (1+\xi)\, a_i$. Away from the static limit, the $\bar{K}NN$ Green functions g in the exchange diagram can be decomposed to a static part g_{st} and the recoil correction Δg as $g := g_{\text{st}} + \Delta g$. Let $\tilde{a}$ denote a generic $\bar{K}N$ scattering length which is related to the non-derivative coupling in the effective Lagrangian. The multiple-scattering series can be therefore written as

$$\tilde{a} + \tilde{a}g\tilde{a} + \tilde{a}g\tilde{a}g\tilde{a} + \cdots = \left\{\tilde{a} + \tilde{a}^2 g_{\text{st}} + \cdots\right\} + \left\{\tilde{a} + \tilde{a}^2 g_{\text{st}} + \cdots\right\} (\Delta g) \left\{\tilde{a} + \tilde{a}^2 g_{\text{st}} + \cdots\right\} + \cdots . \tag{2}$$

It is seen that the whole multiple-scattering series can be rearranged so that the static contributions are re-summed to all orders (the expression in the first curly bracket in the r.h.s of Eq. (2)), whereas the recoil corrections enter perturbatively, in the form of one, two, ... "recoil insertions", see the terms $\sim (\Delta g)^n$, with $n = 1, \cdots$. The contributions from the self-energy-type diagrams and those including NN interactions are parts of the recoil insertions, because they are absent in the static case. Moreover, in the original publication [12] we show that each insertion counts as $O(\xi^{1/2})$ (or, in some cases, as $O(\xi)$) and, consequently, to carry out calculations at a given order in the expansion parameter ξ, it suffices to consider a finite number of insertions.

3 Single recoil insertion

The antikaon-deuteron scattering length can be written as $\mathcal{A} = \mathcal{A}_{\rm st}+\mathcal{A}^{(1)}+\mathcal{A}^{(2)}+\ldots$. In this section, we present compact explicit expression for the one recoil insertion correction $\mathcal{A}^{(1)}$ as derived in Ref. [12]. At this order, one has two diagrams, where the kaon scatters either on the same nucleon or on different nucleons, as well as one diagram which takes into account the intermediate NN interactions to all orders. Following [12], it is advantageous to combine all three contributing diagrams to two blocks of a definite isospin of the intermediate NN pair, $\mathcal{A}^{(1)} = (\mathcal{A}_1 + \Delta\mathcal{A}_{\rm st,1}) + (\mathcal{A}_0 + \mathcal{A}^{(c)} + \Delta\mathcal{A}_{\rm st,0})$. The single terms read ($\mathbf{p}_{1/2} := \mathbf{p} \pm \mathbf{l}/2$)

$$
\begin{aligned}
\mathcal{A}_0 &= \frac{1}{2}\frac{1}{1+\xi/2}\int\frac{d^3\mathbf{p}d^3\mathbf{l}}{(2\pi)^6}\,G_r(\mathbf{p},\mathbf{l})\Big(\Phi^+(\mathbf{p}_1)\big(\Phi^+(\mathbf{p}_1)+\Phi^+(\mathbf{p}_2)\big)\Big),\\
\mathcal{A}_1 &= \frac{1}{2}\frac{1}{1+\xi/2}\int\frac{d^3\mathbf{p}d^3\mathbf{l}}{(2\pi)^6}\,G_r(\mathbf{p},\mathbf{l})\Big(\Phi^-(\mathbf{p}_1)\big(\Phi^-(\mathbf{p}_1)-\Phi^-(\mathbf{p}_2)\big)+2\Phi_x(\mathbf{p}_1)\big(\Phi_x(\mathbf{p}_1)-\Phi_x(\mathbf{p}_2)\big)\Big),\\
\mathcal{A}^{(c)} &= \frac{\xi}{8\pi m_N}\frac{1}{1+\xi/2}\int\frac{d^3\mathbf{p}d^3\mathbf{l}d^3\mathbf{q}}{(2\pi)^9}\,G(\mathbf{p},\mathbf{l})\,M_{NN}(\mathbf{p},\mathbf{q},\mathbf{l})\,G(\mathbf{q},\mathbf{l})\Big(\Phi^+(\mathbf{p}_1)\Phi^+(\mathbf{q}_1)\Big), \qquad (3)\\
\Delta\mathcal{A}_{\rm st,0} &= -\frac{\xi}{2(1+\xi)(1+\xi/2)}\int d^3\mathbf{r}\frac{\Psi^2(\mathbf{r})}{r}A^+(r)^2,\\
\Delta\mathcal{A}_{\rm st,1} &= \frac{\xi}{2(1+\xi)(1+\xi/2)}\int d^3\mathbf{r}\frac{\Psi^2(\mathbf{r})}{r}\left(A^-(r)^2+2A_x(r)^2\right).
\end{aligned}
$$

Here, the Fourier transform of $\Psi(\mathbf{r})A^\pm_{(x)}(r)$ is denoted by $\Phi^\pm_{(x)}$, where the rational functions $A^\pm(r) := A_p(r) \pm A_n(r)$ and $A_x(r)$ account for static contributions, resummed to all orders. Explicitly, they read

$$
A_p(r) = \frac{r^2a_1 + ra_0(2a_1+r)}{a_0(r-2a_1)+r(2r-a_1)},\quad A_n(r) = \frac{2ra_1(a_0+r)}{a_0(r-2a_1)+(2r-a_1)r},\quad A_x(r) = \frac{r^2(a_0-a_1)}{a_0(2a_1-r)+(a_1-2r)r}.
$$

The nucleon-nucleon amplitude M_{NN} is determined from a solution of the Lippmann-Schwinger equation for a given two-nucleon potential V_{NN}. For the explicit representation of this as well as Green functions $G_{(r)}(p,l)$ we refer the reader to Ref [12].

The above expression of the single recoil insertion depend on the variable ξ in a non-trivial way. In order to establish systematic power-counting rules, it is therefore necessary to perform an expansion of this expression in ξ. Further, considering such an expansion helps one to reveal the pattern of cancellations of the leading terms, which has been discussed already in Ref. [11]. The uniform expansion method, see Refs. [13, 14], allows one to perform such an expansion by identifying three different momentum regions according to the scales appearing in the problem. The integrand is expanded in each region before performing the integration, and the results are summed up. Utilizing this method and introducing a modified expansion parameter $\tilde{\xi} = \xi/(1+\xi/2)$, we find that the expansion of each contribution $\mathcal{A}_1$, $\mathcal{A}_0$ and $\mathcal{A}^{(c)}$ in powers of $\sqrt{\tilde{\xi}}$ indeed converges quite rapidly. For more details on the expansion, explicit results and discussion of cancellation patterns, the reader is referred to the original publication [12]. Before coming to the final *numerical* results, we also wish to note that these do not rely on the latter expansion of individual contributions.

Since no experimental results on kaon-deuteron scattering length are available yet, in our exploratory study we estimate the size of the recoil correction for a given set of the KN scattering lengths used as input in Ref. [15], i.e. $a_0 = -1.62 + i0.78$ fm, $a_1 = 0.18 + i0.68$ fm. Moreover, at this stage we use the phenomenological nucleon-nucleon Hulthén and PEST separable potentials instead of a more complicated treatment based completely on chiral EFT. The results of the calculations shown in the following table.

Hulthén	$\mathcal{A}_{\rm st}$		$-1.49 + i\,1.19$
	$\mathcal{A}^{(1)}$	$\mathcal{A}_1$	$-0.13 + i\,0.16$
		$\Delta\mathcal{A}_{\rm st,1}$	$+0.12 - i\,0.20$
		$\mathcal{A}_0$	$-0.27 + i\,0.87$
		$\Delta\mathcal{A}_{\rm st,0}$	$-0.12 + i\,0.33$
		$\mathcal{A}^{(c)}$	$+0.35 - i\,1.06$
	$\mathcal{A}_{\rm st} + \mathcal{A}^{(1)}$		$-1.52 + i\,1.27$

PEST	$\mathcal{A}_{\rm st}$		$-1.55 + i\,1.25$
	$\mathcal{A}^{(1)}$	$\mathcal{A}_1$	$-0.13 + i\,0.18$
		$\Delta\mathcal{A}_{\rm st,1}$	$+0.13 - i\,0.22$
		$\mathcal{A}_0$	$-0.29 + i\,0.97$
		$\Delta\mathcal{A}_{\rm st,0}$	$-0.11 + i\,0.34$
		$\mathcal{A}^{(c)}$	$+0.36 - i\,1.19$
	$\mathcal{A}_{\rm st} + \mathcal{A}^{(1)}$		$-1.59 + i\,1.32$

We conclude that in the isoscalar channel, the individual contributions, which still contain the dominant $O(\tilde{\xi}^{1/2})$ term, are very large, especially the imaginary parts thereof. However, they undergo significant cancellations, yielding only about a 10% net correction to the imaginary part of the static term. The resulting isovector recoil correction appears to be even smaller providing only about a 3% correction to the static term. Its smallness can be understood from the exact cancellation of the $O(\tilde{\xi}^{1/2})$ term along with some additional cancellations among higher-order terms. Furthermore, the net correction $\mathcal{A}^{(1)}$ which stems from one recoil insertion in the multiple scattering diagrams appears to be quite small, of order of $\simeq 6-8\%$ of $\mathcal{A}_{\rm st}$, despite the large value of ξ. An additional suppression is partly accounted for by cancellations of the isoscalar and isovector recoil corrections.

Finally, it is instructive to compare this results to the result of numerical estimations. In Ref. [15] the value $\mathcal{A} \simeq (-1.47 + i1.11)$ fm was obtained from the solution of the Faddeev equations with the one-channel energy-independent optical potential, which was adjusted to reproduce the same $\bar{K}N$ scattering lengths as used in our calculation. Despite the differences in two calculations, such as different NN models, or the use of the off-shell form factors, needed to regularize the Faddeev calculations, the differences between the static result and the solution of the Faddeev equation are, generally, of the same order as the full recoil correction $\mathcal{A}^{(1)}$.

References

[1] N. Kaiser, P. B. Siegel and W. Weise, Nucl. Phys. A **594** (1995) 325 [arXiv:nucl-th/9505043].

[2] Y. Ikeda, T. Hyodo and W. Weise, Nucl. Phys. A **881** (2012) 98 [arXiv:1201.6549 [nucl-th]].

[3] M. Mai and U.-G. Meißner, Nucl. Phys. A **900** (2013) 51 [arXiv:1202.2030 [nucl-th]].

[4] M. Mai and U.-G. Meißner, Eur. Phys. J. A **51** (2015) 3, 30 [arXiv:1411.7884 [hep-ph]].

[5] Z. H. Guo and J. A. Oller, Phys. Rev. C **87** (2013) 3, 035202 [arXiv:1210.3485 [hep-ph]].

[6] M. Bazzi, G. Beer, L. Bombelli, A. M. Bragadireanu, M. Cargnelli, G. Corradi, C. Curceanu (Petrascu) and A. d'Uffizi *et al.*, Phys. Lett. B **704** (2011) 113 [arXiv:1105.3090 [nucl-ex]].

[7] S. Deser, M. L. Goldberger, K. Baumann and W. E. Thirring, Phys. Rev. **96** (1954) 774.

[8] U.-G. Meißner, U. Raha and A. Rusetsky, Eur. Phys. J. C **35** (2004) 349 [hep-ph/0402261].

[9] R. Chand and R. H. Dalitz, Annals Phys. **20** (1962) 1.

[10] S. S. Kamalov, E. Oset and A. Ramos, Nucl. Phys. A **690** (2001) 494 [nucl-th/0010054].

[11] V. Baru, E. Epelbaum and A. Rusetsky, Eur. Phys. J. A **42** (2009) 111 [arXiv:0905.4249 [nucl-th]].

[12] M. Mai, V. Baru, E. Epelbaum and A. Rusetsky, Phys. Rev. D **91** (2015) 5, 054016 [arXiv:1411.4881 [nucl-th]].

[13] M. Beneke and V. A. Smirnov, Nucl. Phys. B **522** (1998) 321 [hep-ph/9711391].

[14] R. F. Mohr, R. J. Furnstahl, R. J. Perry, K. G. Wilson and H. -W. Hammer, Annals Phys. **321** (2006) 225 [nucl-th/0509076].

[15] N. V. Shevchenko, Phys. Rev. C **85** (2012) 034001 [arXiv:1103.4974 [nucl-th]].

EPJ Web of Conferences **113**, 05018 (2016)
DOI: 10.1051/epjconf/201611305018

Higher order effective interactions and effective bosonized model for 2-N particle states

Fabio L. Braghin[1,a]

[1] *Instituto de Física, Univ. Fed. de Goiás, P.B.131, Campus II, 74.001-970, Goiânia, GO, Brazil*

Abstract. In this work, an effective fermion model with particular higher order interactions given by: $I_{II} = \sum_n^N g_{2^n} (\bar{\psi}_a \psi_a)^{2^n}$, for finite N, is investigated by means of the auxiliary field method by taking into account $2n$-particle states as proposed in Ref. [1]. The bosonized model was found to exhibit an approximate symmetry when expanded for weak field fluctuations around the mean field solutions. In the present work, the role of the mean field solutions for the corresponding auxiliary fields is investigated. With the integration of fermions, the resulting determinant is expanded in a polynomial boson model in the weak field approximation and the approximate symmetry found for this series does not appear if the boson mean fields are considered to be zero.

1 Introduction

The auxiliary field method [2–5] is a powerful non perturbative method which reduces the non linearities of a theory at the expenses of introducing interacting auxiliary fields (a.f.). It has been successefully applied in different systems from quantum field, strong interacting and nuclear systems models to many body atomic and electronic systems [5–9]. N-body effective interactions can contribute in few and many body systems and it is one extremely difficult problem to take them into account. Lately it has been shown that these higher order interactions constitute important part of the Effective Field Theory approach (EFT) [10, 11]. Some new few body states and interactions have been observed in several systems in the last years such as in meson spectroscopy with the appearance of tetraquark states and in cold atoms. A particularly interesting observed feature is an apparent approximate degeneracy of 2-particle and 4-particle states in light scalar mesons (masses below/around 1 GeV) [12] and also in 2 and 3 or 4 particle states in cold atoms by means of their binding energies [13]. Even though this last systems are non relativistic one might search for general feature(s) of few body states built with few body interactions. In the present work a complementary calculation to Ref. [1] is performed based in a dynamical fermion effective model. In the present work, the particular series addressed in [1] given by $I_{II} = \sum_n^N g_{2^n} (\bar{\psi}_a \psi_a)^{2^n}$, for finite N, is analysed within the auxiliary field method if there are no mean field solutions. The logics of the work is the following. A dynamical model with higher order interactions is considered and it is bosonized by introducing a set of N auxiliary fields. Although from EFT power counting or renormalization group arguments the more relevant terms of a series of interactions are not those present in the series I_{II}, the idea of the present development is the following: to investigate the role of separate terms or classes of higher order effective interactions by switching

[a] e-mail: braghin@ufg.br

on and off the others to understand their roles in the bosonized resulting effective model, which describes sets of $2n-$ particles composite states. The minimum number of a.f. is introduced such as to cancel out all the interacting terms by means of minimum number of shifts, which are considered to be the simplest ones. Although this approach is strictly valid for weak fluctuations of the a.f. with respect to a coupling constant scale, it was shown in Ref. [1] that it yields the correct expressions if this weak field approximation is lifted. By integrating out the original fermion fields a bosonized model for a collection of N-particle states is obtained which can be expanded in polynomial boson effective interactions. The saddle point equations of the bosonized model are considered to have only trivial solutions without the formation of mean fields differently from Ref. [1]. The resulting polynomial effective model is found and compared to the case in which there is non trivial a.f. mean field (condensates). It will be shown that the new approximate symmetry between N-fermion states appear only if the condensates (mean fields) are formed, i.e. if mean fields are taken into account.

2 Auxiliary fields and bosonization

Consider the following generating funcional with a polynomial effective model:

$$Z = \int \mathcal{D}[\bar{\psi}_a, \psi_a] e^{\,i\int_x \mathcal{L}_{II}[\psi_a, \bar{\psi}_a]}, \tag{1}$$

where the following series of bilinears will be considered:

$$\mathcal{L}_{II} = \bar{\psi}_a(x)\left(i\partial\!\!\!/ - m_a\right)\psi_a(x) + \sum_{n=1}^{N} g_{2^n}\left(\bar{\psi}_a \Gamma \psi_a\right)^{2^n}, \tag{2}$$

where N even and $\Gamma = I$ will be addressed; where g_{2^n} are the effective coupling constants with dimension: $[g^{2^n}] = M^{d-(d-1)2^n}$, where M has dimension of mass, m_a are the masses for each of the fermion species and the index $a = 1...N_r$ stands for the fermion components. The auxiliary fields, with dimension $[M^2]$, are introduce by means of $N/2$ unity integrals that are given by:

$$N' \int \mathcal{D}[\varphi_n] e^{-i\int_x \frac{1}{2}\sum_n^N \frac{1}{d_n}\varphi_n^2(x)} = 1, \tag{3}$$

The simplest necessary shifts of the auxiliary fields that cancel out the interactions can be written as:

$$\varphi_n^2 \to (\varphi_n - \beta_n (\bar{\psi}_a \psi_a)^{2^{n-1}})^2, \tag{4}$$

where β_n are dimensionful parameters that are determined by imposing the corresponding cancelations of all polynomial interactions. There are other possible shifts in the auxiliary fields, however these are the simplest ones with the minimal number of auxiliary fields that cancel out all the polynomial interactions without introducing further previously non existing interations. This reduces eventual ambiguities in the procedure.

For an arbitrary n, these conditions can be written in the following form:

$$\begin{aligned} g_{2^n} &= -\frac{\beta_{n+1}}{d_{n+1}}\varphi_{n+1} + \frac{\beta_n^2}{2d_n}, \qquad \text{for n} < \text{N}, \\ g_{2^n} &= \frac{\beta_n^2}{2d_n}, \qquad\qquad\qquad\qquad \text{for n} = \text{N}. \end{aligned} \tag{5}$$

From these expressions the parameters β_n might become functions of the fields φ_{n+1} without changing the unity Jacobian. This minimal procedure is valid when all the fermion coupling constants, except

g_{2^N}, are quite strong and the subset of φ_{N-1} fields only assume positive values and/or these auxiliary fields are weak with respect to the mean field which are weaker than the (normalized) fermion coupling constants. This means that higher order auxiliary fields, which are introduced to cancel out progressively more irrelevant fermion interactions, are progressively weaker, i.e. $|\varphi_m \beta_m| < g_{2^{m-1}}$ for positive coupling constants.

The resulting effective action for fermions interacting with the auxiliary fields is given by:

$$S_{eff} = \int d^4x \left[\bar{\psi}_a \left(i\partial\!\!\!/ - m_a + \frac{\beta_1}{d_1}\varphi_1 \right) \psi_a - \sum_{n=1}^{N} \frac{1}{2d_n} \varphi_n^2 \right]. \tag{6}$$

The saddle point equations for these auxiliary fields provide relations between the ground state average of the auxiliary fields the progressively large power of bilinears, i.e. $\varphi_n \sim (\bar{\psi}_a \psi_a)^n$, with an implicit summation on the index a. Therefore the auxiliary fields represent corresponding scalar composite fermion states, basically as a sum of states. By integrating out fermions and by using that $\det A = \exp(Tr \ln A)$ it yields:

$$S_{eff} = -iTr \log\left(i\partial\!\!\!/ - m_a + \frac{\beta_1}{d_1}\varphi_1 \right) - \sum_{n=1}^{N} \int_x \frac{1}{2d_n} \varphi_n^2, \tag{7}$$

where Tr is the traces taken over all the internal indices and integration over space-time. According to expression (5), there is a recursive dependence on all the a.f. in the determinant because: $\beta_1[\varphi_n]$. Only the trivial zero solutions for the gap equation will be considered.

3 Expanded effective boson model

The first terms expansion corresponds to:

$$S_{eff} \simeq S_{eff,(0)}[\varphi_i^{(0)}] + \sum_i^N \sum_j^N \frac{1}{n_i! n_j!} \int_{x_1,x_2} \left. \frac{\delta^2 S_{eff}}{\delta\varphi_i(x_1)\delta\varphi_j(x_2)} \right|_{\varphi_i = \varphi_i^{(0)}} \varphi_i(x_1)\varphi_j(x_2) + h.o. \tag{8}$$

where $\int_{x_1,x_2} = \int dx_1 \int dx_2$, $h.o.$ stands for (even) higher order derivatives, n_i, n_j in the second derivative term and n_i, n_j, n_k in the third order are such that $n_i + n_j = 2$ and $n_i + n_j + n_k = 3$ respectively. The first derivative term is set to zero due to the stability condition.

By assembling the interaction terms, it yields for the first three auxiliary fields:

$$\mathcal{V}_I^{eff} = \sum_{j=2} \frac{c_j}{2} \varphi_1^j + \sum_{j,k=1} t_{j,k} \varphi_1^j \varphi_2^k + \sum_{j,k,i=1} t_{j,k,i} \varphi_1^j \varphi_2^k \varphi_3^i \tag{9}$$

where c_j are the resulting self interaction coupling constants for φ_1, $t_{j,k}$ for $j + k \geq 2$ and $t_{j,k,i}$ for $j + k + i \geq 3$. It is seen that only the first a.f. develops a mass term, and higher order fields only couple to successively lower order a.f., i.e. φ_n only develops couplings to all the a.f. of the type $(\prod_m^{n-1} \varphi_m \varphi_n)$. This is a result of the zero condensates case. For the sake of comparison the following expression obtained in Ref. [1] is reproduced here for the case of non zero condensates in the case the fields are suitably redefined to incorporate:

$$\begin{aligned} \mathcal{V}_I^{eff} = & \frac{1}{2}\left[c_2 X_1^2 + (c_2 + c_{2,2}) X_2^2 + (c_2 + c_{2,2} + c_{2,3}) X_3^2 + (c_2 + c_{2,2} + c_{2,3} + c_{2,4}) X_4^2 + ... \right] \\ & + \sum_{n \geq 3} \left[c_n X_1^n + (c_n + c_{n,2}) X_2^n + (c_n + c_{n,2} + c_{n,3}) X_3^n + ... \right] + \sum_{i,j,k} t_{i,j,k} X_1^i X_2^j X_3^k, \\ & t_{i,j,k} \quad \text{for } i + j + k \geq 2, \text{ at least two indices non zero,} \end{aligned} \tag{10}$$

where c_n and $c_{n,m}$ are the resulting self interaction coupling constants and contributions for masses, and also the couplings $t_{i,j,k}$ are those couplings between at least two different components, being that at least two indices are non zero, i.e. $i, j \neq 0$ or $i, k \neq 0$ and so on. From this expression the limit of progressively relatively large condensates yielded an effective potential with an approximate symmetry. By comparing expressions (9) and (10) it is seen there is a complete absence of the explicit effective and approximate symmetry that was found in the case of non zero (relatively large) higher order condensates [1].

4 Conclusions

In this work an effective bosonized model for n-fermion scalar states was built from a dynamical effective fermion model with very particular higher order interactions by means of the auxiliary field method in the case of zero mean fields. By integrating out fermions the resulting determinant was expanded in polynomial boson effective interactions. The approximate symmetry found in Ref. [1] in the case of relatively large higher order condensates (mean fields) was found to be completely absent in the case of zero scalar fields condensates. The resulting effective potential (9) exhibits a preferential role for the lowest order scalar a.f. (φ_1) which appears in all interaction terms.

Acknowledgements

The author thanks short discussion with P. Bedaque, and financial support from FAPEG-Goias and CAPES-MCT and CNPq, Brazil.

References

[1] F.L. Braghin, submitted to publication, arXiv:1505.01209 [hep-th].
[2] T. Kashiwa, Phys. Rev. **D 59**, 085002 (1999).
[3] D.J. Gross, A. Neveu, Phys. Rev. **D 10**, 3235 (1974).
[4] H. Kleinert, Electronic Journal of Theoretical Physics **8**, 57 (2011).
[5] J.W. Negele, H. Orland, Quantum Many-Particle Systems, Addison-Wesley, 1988.
[6] S.P. Klevansky, Rev. of Mod. Phys. **64**, 649 (1992).
[7] E. Baum, Nucl. Phys. **B 266**, 547 (1986).
[8] H. Kleinert, Lectures presented at the Erice Summer Institute 1976, in Understanding the Fundamental Constituents of Matter, Plenum Press, New York, ed. by A. Zichichi, 289 (1978).
[9] F.L. Braghin, F.S. Navarra, Phys. Rev. **D 91**, 074008 (2015).
[10] P.F. Bedaque, U. van Kolck, Annu. Rev. Nucl. Part. Sci. **52**, 339 (2002).
[11] E. Braaten, H.-W. Hammer, Phys. Rept. **428**, 259 (2006).
[12] For example in: K.A. Olive et al. (Particle Data Group), Chin. Phys. **C 38**, 090001 (2014). G. 't Hooft, G. Isidori, L. Maiani, A.D. Polosa, V. Riquer, Phys.Lett.B **662**, 424 (2008).
[13] For example in: J. H. Gurian, P. Cheinet, P. Huillery, A. Fioretti, J. Zhao, P. L. Gould, D. Comparat, and P. Pillet Phys. Rev. Lett. **108**, 023005 (2012). F. Ferlaino, S Knoop, M. Berninger, W. Harm, J. P. D'Incao, H.-C. Nagerl, R. Grimm Phys. Rev. Lett. **102**, 140401 (2009).

EPJ Web of Conferences **113**, 05019 (2016)
DOI: 10.1051/epjconf/201611305019

Towards the next QCD Frontier with the Electron Ion Collider

Abhay Deshpande[1,a], Zein-Eddine Meziani[2,b], and Jian-Wei Qiu[3,4,c]

[1] *Department of Physics and Astronomy, Stony Brook University, Stony Brook, NY 11794, USA*
[2] *Department of Physics, Temple University, Philadelphia, PA 19122, USA*
[3] *Physics Department, Brookhaven National Laboratory, Upton, NY 11973, USA*
[4] *C.N.Yang Institute for Theoretical Physics and Department of Physics and Astronomy, Stony Brook University, Stony Brook, NY 11794, USA*

Abstract. In this talk, we argue that the proposed Electron-Ion Collider (EIC) with its unique capability to collide polarized electrons with polarized protons and light ions at unprecedented luminosity, and with heavy nuclei at high energy, will be the most powerful tomographic scanner able to precisely image gluons and quarks inside the proton and nuclei. This precision microscope will allow us to "see" and explore the dynamics binding gluons and quarks together to form hadrons. The EIC will address the most compelling unanswered questions in QCD and hadron physics.

1 Introduction

Understanding the evolution of our universe from its origin (the Big Bang) to its current state, and the fundamental structure of all matters within it, is one of the most central goals of human scientific endeavor. This evolution is intimately tied to the properties of fundamental particles that existed in each of its phases as well the interactions that govern their dynamics. The discovery of the atomic structure and the nucleus by the Rutherford experiment over 100 years ago led to the discovery of quantum mechanics and the quantum world, which forever changed our view of our universe. It was the modern "Rutherford" experiment, namely the scattering between electrons and protons, performed at the Stanford Linear Accelerator Center (SLAC) in the sixties, that revealed *quarks*, the fundamental constituents of the proton, and led to the discovery of Quantum Chromo-Dynamics (QCD) – the theory of strong interaction between color charges that is responsible for confining colored quarks into color neutral hadrons through the exchange of *gluons*. In contrast to quantum electromagnetism where the force carriers (photons) are electrically neutral and non self-interacting, the force carriers (gluons) in QCD themselves carry a color charge causing them to interact among themselves. It is the non-linear gluon self-interaction that is responsible for both the *confinement* of color and the *asymptotic freedom* of the color force, two remarkable yet seemingly contradicting properties of QCD which make the strong interaction and hadron physics very rich, with plenty of dynamics mysteries.

[a] e-mail: abhay.deshpande@stonybrook.edu
[b] e-mail: meziani@temple.edu
[c] e-mail: jqiu@bnl.gov, Speaker

Despite the great experimental challenge of probing the internal structure of a hadron without being able to observe the quarks and gluons in isolation, QCD has been extremely successful in interpreting data from all high energy e^+e^-, lepton-hadron and hadron-hadron collisions with momentum transfer(s) $\gg \Lambda_{\rm QCD} \sim 1/{\rm fm}$. This was only possible due to the great theoretical development of QCD factorization formalisms which allows to connect the observed hadrons to the active quarks and gluons in high energy scattering. At the same time, the *ab initio* lattice QCD calculations have made tremendous progress and provided an impressive description of the mass spectrum of all known hadrons [1]. Today no one questions the validity of QCD as the fundamental theory of strong interactions in the Standard Model, and yet, our understanding of QCD remains incomplete. We know very little about the hadrons' quark-gluon structure and their emergence from quarks and gluons, a result closely connected to the defining property of QCD namely the *color confinement*.

The proposed Electron-Ion Collider (EIC) [2] will have highly polarized electrons, light ions and unpolarized nuclei (all the way to Uranium) colliding in the CoM energy range of 40-140 GeV, with a luminosity (e-nucleon equivalent) of $L_{e-N} = 10^{33-34}$ cm^{-2}sec^{-1}. The EIC with such parameters will be able to address many overarching questions about QCD and hadron physics, such as:

- How are the sea quarks and gluons distributed in space and momentum inside the nucleon, and how do they build up nucleon's intrinsic properties, such as the spin?
- Does the gluon density saturate to produce the matter of universal properties in all hadrons, and is there a simple boundary that separates the saturated from more diluted quark-gluon matter?
- How does the nuclear environment affect the distribution of quarks and gluons and their interactions in nuclei, and how do hadrons emerge from quarks and gluons?

2 The hadron structure

The relation between the observable physical states, i.e. hadrons and the basic degrees of freedom, quarks and gluons (collectively called partons), is most intriguing in QCD. Proton structure is dynamic, not static, rich in features beyond our current knowledge and intuition. Quarks and gluons interact strongly when they are far apart – *the confinement of color* – but weakly when they are closer – *the asymptotic freedom*; they appear and disappear almost continuously; and they are confined to form the proton. Complete understanding of the proton structure and its properties (such as its mass, spin, size, etc...), the emergent phenomena of QCD dynamics, is still beyond the capability of today's best minds.

2.1 Nucleon spin

Several decades of experiments of deep inelastic scattering (DIS) of electron or muon beams off nucleons have taught us how partons share the momentum of a fast-moving nucleon. They have not, however, completely resolved the question of how partons build up the nucleon's spin? Following an intensive and worldwide experimental program over the past two decades, including recent measurements at RHIC, along with state-of-the-art QCD global analyses, we learned that the spin of quarks and antiquarks is only responsible for $\sim$ 30% of the proton spin, while the contribution from the gluons' spin is about 20-30% from the x-range explored so far. EIC will greatly increase the kinematic coverage in x and Q^2, as shown in Fig. 1(Left), and hence, reduce the uncertainty of the gluonic contribution to the proton's spin in a very dramatic way, as shown by the red and yellow bands in Fig. 1(Right). The reduced uncertainties would definitively resolve the question of whether parton spin preferences alone can account for the overall proton spin, or whether an additional contribution from the total orbital angular momentum of partons in the nucleon is needed. Clearly, the EIC will make a profound impact on our knowledge of these quantities, unmatched by any other existing or anticipated facility.

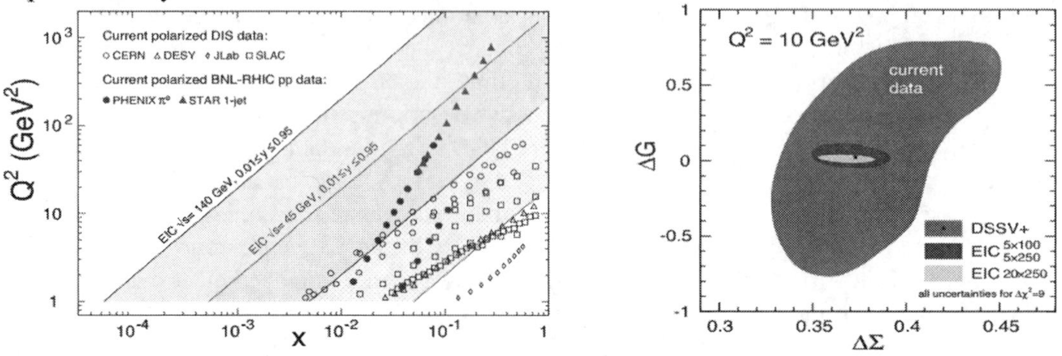

Figure 1. **Left:** The increase coverage in the proton momentum fraction x vs. the square of the momentum transferred by the electron to the proton Q^2 in $e + p$ collisions at the EIC. **Right:** The projected reduction in the uncertainties of gluon and quark helicity contributions to the spin of proton: ΔG and $\Delta\Sigma$.

2.2 Confined motion of quarks and gluons inside the nucleon

Semi-inclusive DIS (SIDIS) measurements have two independent momentum scales: the large momentum transfer from the electron beam that ensures the spatial resolution of the probe, and the momentum of the produced hadrons perpendicular to the direction of the momentum transfer that prefers a small value sensitive to the motion of confined partons. Remarkable theoretical advances over the past decade have led to a rigorous framework where information on the confined motion of the partons inside a fast-moving nucleon is matched to transverse-momentum dependent parton distributions (TMDs). TMDs allow us to investigate the full three-dimensional quark-gluon dynamics inside the proton, about which very little is known to date. In Fig. 2(Left), the probability distribution (the color code) to find the up quarks inside a proton moving in the z-direction (out of the page) with its spin polarized in the y-direction is plotted as a function of quark transverse-momentum. The anisotropy in transverse momentum is induced by the correlation between the proton's spin direction and the motion of its quarks and gluons. While the figure is based on a preliminary extraction of this distribution from current experimental data, nothing is known about the spin and momentum correlations of the gluons and sea quarks. The EIC will be crucial to initiate and realize such a program, and it will dramatically advance our knowledge of the motion of confined gluons and sea quarks in ways not achievable at any existing or proposed facility.

2.3 Spatial Imaging of quarks and gluons inside the nucleon

By choosing particular final states in electron+proton scattering without breaking the proton, the EIC with its unprecedented luminosity and detector coverage will be able to create detailed spatial images of quarks and gluons in the colliding proton. These tomographic images obtained from cross-sections and polarization asymmetries in exclusive processes are encoded in generalized parton distributions (GPDs), and are complementary to those obtained from the TMDs of quarks and gluons, revealing aspects of proton structure that are intimately connected with the dynamics of QCD at large distances. In Fig. 2(Right), the spatial distribution of gluons, as measured in the exclusive process: electron + proton $\rightarrow$ electron + proton + J/ψ, is plotted as a function of b_T, the transverse position of the gluons. No existing or planned facility will be able to perform such measurements.

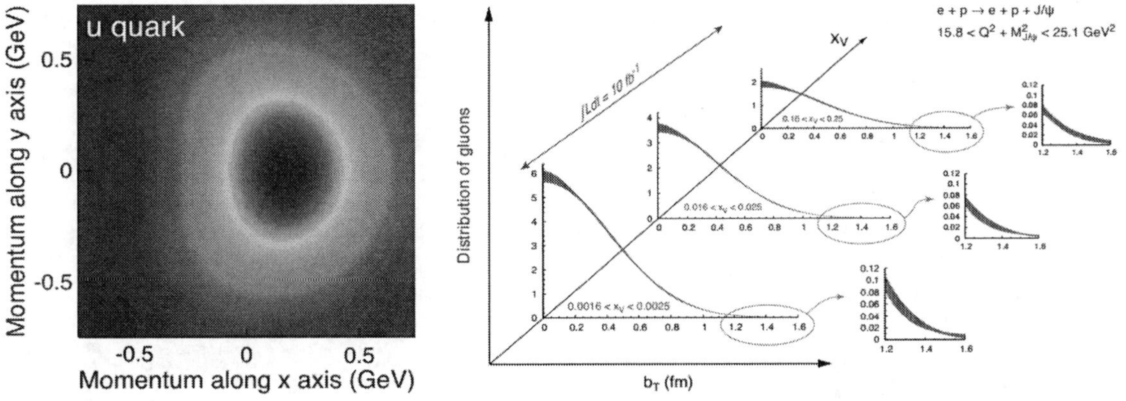

Figure 2. **Left:** The transverse-momentum distribution of an up quark with longitudinal momentum fraction $x = 0.1$ in a transversely polarized proton moving in the z-direction, while polarized in the y-direction. The color code indicates the probability of finding the up quarks. **Right:** The projected precision of the transverse spatial distribution of gluons as obtained from the cross-section of exclusive J/ψ production at the EIC.

3 The nucleus: A femtometer laboratory for QCD

The emergence of hadrons from almost massless quarks and gluons, or heavy quarks, is a necessary and critical process in the formation of our visible universe shortly after its birth. Color neutralization, is key to this emergent phenomenon of QCD and is likely to take place at the femtometer size, the color confining scale, has yet to be understood within QCD.

3.1 Quark and gluon landscape in a nucleus

The EMC experiment at CERN and experiments in the following two decades clearly revealed that the distribution of quarks in a fast-moving nucleus is not a simple superposition of their distributions within nucleons, as shown in Fig. 3. With its much wider kinematic reach in both x and Q^2, the EIC would allow measurements of the suppression of the structure functions to a much lower value of x, approaching the gluon dominated region, where the universal saturation phenomenon related to the gluons in nuclei, is predicted to exist. The EIC could for the first time reliably quantify the probability to find a soft gluon in this interesting regime.

If the nuclear effect in the DIS cross section, as shown in Fig. 3, is mainly due to the abundance of nucleons at the same impact parameter of the nucleus (proportional to $A^{1/3}$), while the elementary scattering is still relatively weak, one would expect the ratio of nuclear over nucleon structure functions to saturate when x goes below 0.01, as shown for example, by the upper line of the blue area extrapolated from the current data. On the other hand, if the soft gluons are a property of the whole nucleus and the coherence is strong, one would expect the ratio of the nuclear to nucleon structure function to fall continuously as x decreases, as sketched by the lower line of the blue band, and eventually, reach a constant when both nuclear and nucleon structure functions are in the saturation region. From the size of the purple error band in Fig. 3, which is the expected systematic uncertainty at the EIC, it is self-evident that (statistical uncertainty is expected to be much smaller) the EIC will easily distinguish between these two extreme scenarios and light on the nature of soft gluons and sea quarks in a nuclear environment.

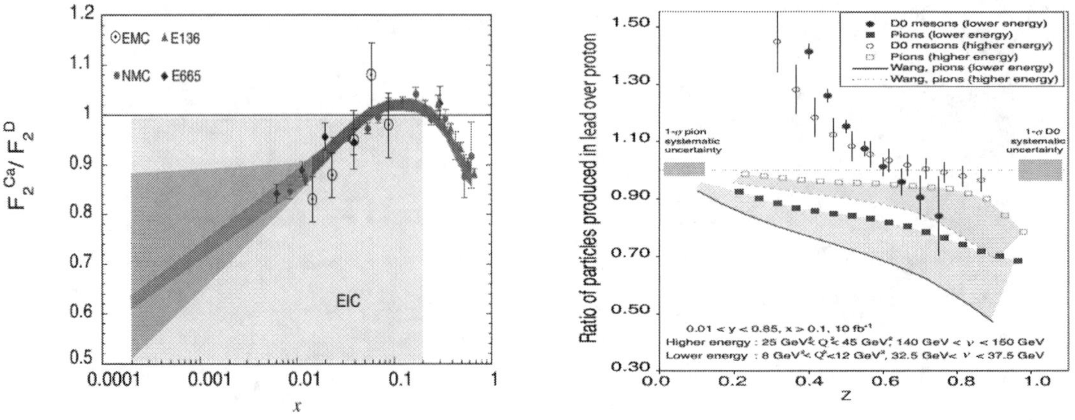

Figure 3. **Left:** The ratio of nuclear over nucleon F_2 structure function as a function of Bjorken x, with data from existing fixed target DIS experiments at $Q^2 > 1$ GeV^2, along with the QCD global fit from EPS09. Also shown is the expected kinematic coverage of the inclusive measurements at the EIC. The purple error band is the expected systematic uncertainty at the EIC assuming a 2% (a total of 4%) systematic error, while the statistical uncertainty is expected to be much smaller. **Right:** The ratio of the semi-inclusive cross-section for producing a pion (red) composed of light quarks, and a D^0 meson (blue) composed of heavy quarks in e+lead collisions to e+deuteron collisions, plotted as a function of z, the ratio of the energy carried by the produced hadron to that of the exchange virtual photon in the rest frame of the colliding proton or ion.

3.2 The color propagation and neutralization

The exact mechanism of how colored partons pass through colored media, both cold nuclear matter and the hot quark-gluon plasma (QGP), and their emergence as colorless hadrons, is not understood. At the EIC, a nucleus will provide a femtometer size filter to help determine the correct mechanism by which colored partons interact and hadronize in nuclear matter. By measuring pion and D^0 meson production in both $e+p$ and $e+A$ collisions, the EIC will provide the first measurement of the quark mass dependence of the response of nuclear matter to a fast moving quark. The dramatic difference between them, shown in Fig. 3 (Right), would be easily discernable at the EIC. The color bands reflect the limitation of our knowledge on hadronization – the emergence of a pion from a colored light quark. Enabling all such studies in one place, the EIC will be a true "femtometer size QCD Laboratory", unique of its kind in the world.

3.3 QCD matter at an extreme gluon density

When fast moving hadrons are probed by a high resolution local probe, the low-momentum gluons contained in their wave functions become experimentally accessible. The large soft-gluon density at small-x, especially in a large nucleus, enables the gluon-gluon recombination to limit the density growth. Such a self-regulation mechanism necessarily generates a dynamic scale from the interaction of high density massless gluons, known as the saturation scale, Q_s, at which gluon splitting and recombination reach a balance. At this scale, the density of gluons is expected to saturate, producing new hadronic matter of universal properties. The saturation scale Q_s separates the condensed and saturated soft gluonic matter from the dilute, but confined, quarks and gluons in a hadron, as shown in Fig. 4(Left). Such universal cold gluon matter is an emergent phenomenon of QCD dynamics and is of high scientific interest. With its wide kinematic reach, the EIC will be the first experimental facility capable of exploring the internal three-dimensional sea quark and gluon structure of a fast-moving nucleus, and the transition from a dilute to a dense gluon density to provide access to a thus far unconfirmed regime of matter where abundant gluons dominate its behavior.

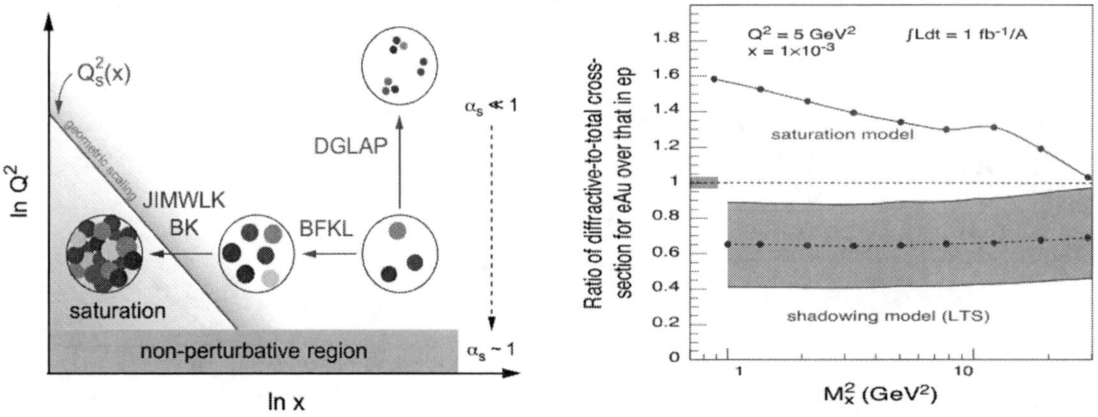

Figure 4. Left: The schematic probe resolution vs. energy landscape, including the regions of low and high saturated parton density, and the transition region between them. **Right:** The ratio of diffractive over total cross-section for DIS on gold normalized to DIS on proton for different values of mass squared of hadrons produced in the collisions with and without saturation.

The existence of such saturated soft gluon matter is a direct consequence of gluon self-interactions in QCD. With many more soft gluons at the same impact parameters in a large nucleus, enhanced by $A^{1/3}$ with the atomic weight A than that in a proton, the gluon saturation is much easier to be reached in $e + A$ than $e + p$ collisions. By measuring the ratio of diffractive cross-sections over the total DIS cross-sections in both $e + A$ and $e + p$ collisions, as shown in Fig. 4(Right), the EIC would provide the first unambiguous evidence for the novel QCD matter of saturated gluons. Furthermore, knowing the properties of this matter and its underlying QCD dynamics is critically important for understanding the dynamical origin of QGP formation from colliding two relativistic heavy ions.

4 Summary and outlook

The EIC is a ultimate machine to explore the three-dimensional quark-gluon structure of nucleons and nuclei, to search for hints and clues of the color confinement, to study the color neutralization, to probe the existence of the saturated gluonic matter and to explore it in detail. At extreme energy and luminosities, it may also discover the unexpected phenomena within and beyond the Standard Model. The EIC promises to unite and extend the scientific programs at CEBAF and RHIC in dramatic and fundamentally important ways, and to propel them to the next QCD frontier. Its realization will ensure US leadership in nuclear science research.

Acknowledgements: We thank all members of the EIC White Paper Writing Committee [2] for helpful discussions and the collaboration.

References

[1] N. Brambilla, S. Eidelman, P. Foka, S. Gardner, A. Kronfeld et al., Eur.Phys.J. **C74**, 2981 (2014), 1404.3723

[2] A. Accardi, J. Albacete, M. Anselmino, N. Armesto, E. Aschenauer et al. (2012), 1212.1701

EPJ Web of Conferences **113**, 05020 (2016)
DOI: 10.1051/epjconf/201611305020

Leading Twist Parton Distribution Amplitudes in Heavy Flavor Pseudoscalar Mesons From Dyson-Schwinger Equations of QCD

Minghui Ding[1,a], Fei Gao[1,b], Lei Chang[2,c], Yu-Xin Liu[1,d], and Craig D. Roberts[3,e]

[1] *Department of Physics and State Key Laboratory of Nuclear Physics and Technology, Peking University, Beijing 100871, China*
[2] *CSSM, School of Chemistry and Physics, University of Adelaide, Adelaide SA 5005, Australia*
[3] *Theory Group, Physics Division, Argonne National Laboratory, 9700 S. Cass Ave., Lemont, IL 60439, USA*

Abstract. We compute the valence-quark leading twist parton distribution amplitudes (PDAs) of heavy pseudoscalar mesons η_c and η_b and find that they are both broader than the δ-like function while narrower than the asymptotic one. The evolution of distribution amplitude with momentum scale is then considered and PDAs will turn to a asymptotic form when the momentum goes to infinity.

1 Introduction

The B_C meson had been studied under various approaches since its observation by the CDF and D0 group in different channels [1–3]. The theoretical researches include the non-relativistic QCD approach [4], QCD sum rules [5], full lattice QCD [6] approaches and so on. Among them QCD light-cone sum rules method for example, its valence-quark parton distribution amplitudes (PDAs) of c-meson or b-meson were introduced through an integration over light-cone wave functions, which were ansatz presented by Brodsky-Huang-Lepage [7]. Once the distribution amplitude was obtained, it will be widely applicable for QCD hard exclusive process. Numbers of the measurable quantities such as transition form factors and the branching ratio for particular decay channel could be computed from theory [8].

Recently pion's distribution amplitude has been obtained directly from a projection of the pion's Bethe-Salpeter wave function onto the light-cone [9], and the asymptotic form was also obtained naturally, $\phi_\pi^{asy}(x) = 6x(1-x)$. The distribution amplitude of pion was broader than the asymptotic one, which explained well the emergent phenomenon of dynamic chiral symmetry breaking. In order to get the Bethe-Salpeter wave function, the quark propagator was introduced after solving QCD's Dyson-Schwinger equations (DSEs), and the bound state appeared with Bethe-Salpeter equations (BSEs) [10].

a. e-mail: mhding@pku.edu.cn,speaker
b. e-mail: hiei@pku.edu.cn
c. e-mail: lei.chiong@gmail.com
d. e-mail: yxliu@pku.edu.cn
e. e-mail: c.d.roberts@anl.gov

Heavy flavor pseudoscalar meson leading twist PDAs are then planned to be studied after a clear image of pion. In the heavy flavour sector pseudoscalar mesons η_c and η_b are well established with the meson picture $\bar{c}c$ and $\bar{b}b$. PDAs of these two mesons are calculated in this document under the framework of DSEs following this new method that successfully applied to pion.

2 Distribution amplitude and Bethe-Salpeter wave function

The projection of Bethe-Salpeter wave function onto the light-cone can be written as follows

$$f_{\eta_c(\eta_b)}\phi(x) = tr_{CD} Z_2 \int_{dk}^{\Lambda} \delta(n\cdot k_+ - xn\cdot P)\gamma_5\gamma\cdot n\chi(k;P)\,, \tag{1}$$

where the trace is over colour and spinor indices ; $\int_{dk}^{\Lambda}$ is a four-dimensional integral over momentum, with Λ the ultraviolet regularization mass-scale ; $Z_2(\zeta,\Lambda)$ is the quark wave function renormalisation constant, with ζ the renormalisation scale ; $f_{\eta_c(\eta_b)}$ is the leptonic decay constant ; n is light-like four-vector, $n^2 = 0$; P is the four-momentum of the meson, $P^2 = -m^2_{\eta_c(\eta_b)}$ and $n\cdot P = -m_{\eta_c(\eta_b)}$, with m being the meson mass ; and the Bethe-Salpeter wave function

$$\chi(k;P) = S(k_+)\Gamma(k;P)S(k_-)\,, \tag{2}$$

with Γ the Bethe-Salpeter amplitude, S the dressed light quark propagator, and $k_+ = k + \eta P$, $k_- = k - (1-\eta)P$, $\eta \in [0,1]$. In detail, the dressed propagator satisfies the Gap equation

$$S(p)^{-1} = Z_2(i\gamma\cdot p + m^{bm}) + Z_1 \int_k^{\Lambda} g^2 D_{\mu\nu}(p-k)\frac{\lambda^a}{2}\gamma_\mu S(k)\frac{\lambda^a}{2}\Gamma_\nu(k;p), \tag{3}$$

where $D_{\mu\nu}$ is the gluon propagator ; Γ_ν, the quark-gluon vertex ; m^{bm}, the current-quark bare mass ; and Z_1 the quark vertex renormalisation constant.

Similarly the Bethe-Salpeter amplitude can be obtained from a homogeneous Bethe-Salpeter equation $[\Gamma(k;P)]_{tu} = \int_q^{\Lambda}[S(q+)\Gamma(q;P)S(q-)]_{sr}K_{tu}^{sr}(q,k;P)$, in which K is the full-amplitude quark-antiquark scattering kernel [11]. The axial-vector Ward-Takahashi identity enables the relation existing between the kernels in the gap equation and that in the BSE.

3 Nakanishi-like representation

We follow the approach introduced in Ref. [9] and use a Nakanishi-like representation of meson Bethe-Salpeter wave functions that are obtained via numerical solution of Bethe-Salpeter equation. Computation of the distribution amplitude is obtained by interpolating quark propagator and the Bethe-Salpeter amplitude separately. The $S(p)$ and $\Gamma(k;P)$ can be expressed in the following way :

$$S(p) = [-i\gamma\cdot p + M]\Delta_M(p^2)\,, \qquad \Gamma(k;P) = i\gamma_5\frac{M^3}{f_{\eta_c(\eta_b)}}\int_{-1}^{1} dz\rho_\nu(z)\Delta_M^\nu(k_{+z}^2)\,, \tag{4}$$

where $\Delta_M(s) = 1/[s+M^2]$, $\rho_\nu(z) = \frac{1}{\sqrt{\pi}}\frac{\Gamma(\nu+3/2)}{\Gamma(\nu+1)}(1-z^2)^\nu$, $k_{\pm z} = k - (1\mp z)P/2$. Inserting Eqs. (4) into Eq. (1), using a Feynman parametrisation to combine denominators, shifting the integration variable, one can get the moments of the parton distribution amplitude $\phi(x)$. Algebraical analysis shows that different spectral function $\rho_\nu(z)$ represents different meson structure in every limit case. When $\nu = 0$, it means a point-like particle, and if $\nu = 1$, this is the so-called QCD's asymptotic limit.

We present the quark propagator as functions with no poles on the real p^2-axis, which is consistent with the feature of confinement [12]. The real-p^2-axis mass pole splits into pairs of complex conjugate poles. Each scalar function in the Bethe-Salpeter amplitude is described in Nakanishi-like representation following the method in Ref. [9].

4 Calculated distribution amplitudes

We use a common kernel of Dyson-Schwinger equations and Bethe-Salpeter equations, rainbow-ladder truncation (RL) , which is widely used in DSE framework. This approximation gives most of the exact results : static properties of π- and ρ- mesons and those of the neutron and proton. It also provides predictions for experiment [13]. For the gluon sector, we use the interaction described in Ref. [14], and the infrared composition of this interaction is consistent with that determined in modern studies of QCD's gauge sector [15]. The quark mass function, $M(p^2) = B(p^2, \zeta^2)/A(p^2, \zeta^2)$, is independent of the renormalisation point, ζ. Basic meson properties we obtain is listed in the left panel of Table 1.

Furthermore, one can use the Gegenbauer polynomials to give a complete expression to the PDA

$$\phi(x;\tau) = \phi^{asy}(x)[1 + \sum_{j=2,4,...}^{\infty} a_j^{3/2}(\tau)C_j^{3/2}(2x-1)], \tag{5}$$

with $\{C_j^{3/2}(2x-1)|j = 0, ..., \infty\}$, which is a complete orthonormal set on $x \in [0, 1]$. Coefficients of the fitting that $j \geq 10$ is not required because $a_j^{3/2}|j \geq 10$ decreases rapidly.

The leading twist distribution amplitudes of η_c and η_b together with the asymptotic case are presented in the left panel of Fig. 1. They are symmetric and much narrower than the asymptotic one but broader than δ-like function, which is consistent with the trend that distribution amplitude of meson varies from light to heavy, and turn to a δ-like function in the heavy quark limit.

Once the distribution amplitude is obtained, the moments of it could be calculated, where $\langle\xi^n\rangle = \int_0^1 dx(2x-1)^n\phi(x)$ is the n_{th}-order moment of $\phi(x)$. The 0_{th}-order moment $\langle\xi^0\rangle = \int_0^1 dx\phi(x) = 1$ normalises $\phi(x)$. Comparison is made in moments of the distribution amplitude with a computation using QCD sum rules [16]. Results in the right panel of Table 1 show that our $\langle\xi^2\rangle$ is a little larger than that of Ref. [16], and this could be explained by that our distribution is broader so that the variance of it is bigger.

Then the evolution of leading twist PDA, that is the dependence of distribution amplitude on the momentum scale ζ or, equivalently, the length-scale $\tau = 1/\zeta$, is considered. The expansion coefficient $\{a_j^{3/2}, j = 1..., \infty\}$ evolve logarithmically with τ, and vanish as $\tau \to 0$ [7, 17]. The evolution of PDA for η_c in Eq. (5) from $\tau_{1.275} \to \tau_{100}$ is displayed in the right panel of Fig. 1. It is evident that PDA becomes broader with the increase of ζ, and goes to the asymptotic form when ζ is infinity finally .

Tableau 1. *Left panel* : Meson basic properties (Units in GeV). *Right panel* : Moments of PDAs $\langle\xi^n\rangle$.

	$(Dw)^3$	m	ζ	M	f
η_c	0.545	1.275	1.275	3.0	0.386
η_b	0.545	4.28	4.28	9.0	0.629

	η_c	η_c [16]	η_b	η_b [16]
$\langle\xi^2\rangle$	0.095	0.073	0.072	0.067
$\langle\xi^4\rangle$	0.026	0.014	0.017	0.011

In summary, we use a Rainbow-Ladder truncation of Dyson-Schwinger equations and Bethe-Salpeter equations in order to get the valence-quark parton distribution amplitude of heavy pseudoscalar mesons η_c and η_b. Bethe-Salpeter wave function is acquired by interpolating quark propagator and the Bethe-Salpeter amplitude separately. After projecting the Bethe-Salpeter wave function on the light-front, we obtain the moments of distribution amplitude numerically. Then the distribution amplitude is reconstructed through a best fitting of moments of them. The computed $\eta_{c(b)}$'s distribution amplitude shows to be much wider than a δ-like function, which is the heavy quark limit, while narrower than the asymptotic case, $\phi^{asy}(x) = 6x(1-x)$. Evolution of distribution amplitude of η_c with the momentum scale ζ is also investigated . It will become broader with the increase of ζ, and

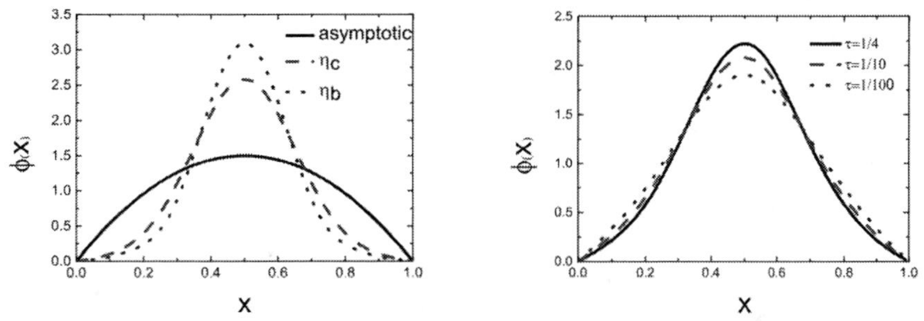

FIGURE 1. Left panel : Computed distribution amplitude. Curves : red dashed, PDA for η_c ; blue dotted, PDA for η_b ; black solid, asymptotic distribution. Right panel : Distribution amplitude evolution of η_c.

finally goes to asymptotic form when ζ is infinity. Study of distribution amplitude on other heavy pseudoscalar meson such as B_c may be good verification for the variation trend we get from light meson to heavy ones, and we are in the process of it.

Acknowledgement

The work was supported by the National Natural Science Foundation of China under Contracts No. 11435001 and No. 11175004 ; the National Key Basic Research Program of China under Contract No. G2013CB834400 ; the U.S. Department of Energy, Office of Nuclear Physics, Contract No. DE-AC02-06CH11357.

References

[1] F. Abe et al. (CDF Collaboration), Phys.Rev. **D58**, 092002 (1998)
[2] A. Abulencia et al. (CDF Collaboration), Phys.Rev.Lett. **96**, 082002 (2006)
[3] M. Corcoran (CDF Collaboration, D0 Collaboration), pp. 29–32 (2005), `hep-ex/0506061`
[4] G.T. Bodwin, E. Braaten, G.P. Lepage, Phys.Rev. **D51**, 1125 (1995)
[5] V. Kiselev, A. Kovalsky, A. Likhoded, Nucl.Phys. **B585**, 353 (2000)
[6] E.B. Gregory, C.T. Davies, I.D. Kendall, J. Koponen et al., Phys.Rev. **D83**, 014506 (2011)
[7] G.P. Lepage, S.J. Brodsky, Phys.Rev. **D22**, 2157 (1980)
[8] M. Beneke, G. Buchalla, M. Neubert, C.T. Sachrajda, Nucl. Phys. **B591**, 313 (2000)
[9] L. Chang, I. Cloet, J. Cobos-Martinez, C. Roberts et al., Phys.Rev.Lett. **110**, 132001 (2013)
[10] C.D. Roberts, A.G. Williams, Prog.Part.Nucl.Phys. **33**, 477 (1994)
[11] P. Maris, C.D. Roberts, P.C. Tandy, Phys.Lett. **B420**, 267 (1998)
[12] C.D. Roberts (2012), `1203.5341`
[13] P. Maris, C.D. Roberts, Int.J.Mod.Phys. **E12**, 297 (2003)
[14] S.x. Qin, L. Chang, Y.x. Liu, C.D. Roberts, D.J. Wilson, Phys.Rev. **C84**, 042202 (2011)
[15] P.O. Bowman, U.M. Heller, D.B. Leinweber, M.B. Parappilly, A.G. Williams, Phys. Rev. **D70**, 034509 (2004)
[16] T. Zhong, X.G. Wu, T. Huang (2014), `1408.2297`
[17] A. Efremov, A. Radyushkin, Phys.Lett. **B94**, 245 (1980)

EPJ Web of Conferences **113**, 05021 (2016)
DOI: 10.1051/epjconf/201611305021

The Proton Radius Puzzle

E. J. Downie[1,a]

[1] *Department of Physics, George Washington University, 725 21st Street NW, Washington DC 20052, USA.*

Abstract. The proton radius puzzle is the difference between the proton radius as measured with electron scattering and in the excitation spectrum of atomic hydrogen, and that measured with muonic hydrogen spectroscopy. Since the inception of the proton radius puzzle in 2010 by the measurement of Pohl et al.[1], many possible resolutions to the puzzle have been postulated, but, to date, none has been generally accepted. New data are therefore necessary to resolve the issue. We briefly review the puzzle, the proposed solutions, and the new electron scattering and spectroscopy experiments planned and underway. We then introduce the MUSE experiment, which seeks to resolve the puzzle by simultaneously measuring elastic electron and muon scattering on the proton, in both charge states, thereby providing new information to the puzzle. MUSE addresses issues of two-photon effects, lepton universality and, possibly, new physics, while providing simultaneous form factor, and therefore radius, measurements with both muons and electrons.

1 Introduction

The proton radius puzzle (PRP) concerns the difference between the radius of the proton as measured with electron scattering and atomic hydrogen spectroscopy, and that measured in muonic hydrogen. In 2010, the CREMA Collaboration published their measurement of the proton radius $R_p = 0.8409(4)$ fm, which was made by studying the Lamb shift in muonic hydrogen [1]. This result, although ten times more precise than, is completely incompatible with the CODATA value of $R_p = 0.8775(51)$ fm [2]. Until that point, the CODATA value was in good agreement with both scattering and spectroscopy results. Since 2010, various attempts have been made to resolve the PRP by looking closely at the experimental data and its interpretation: looking for errors in the theory of the radius extraction from the spectroscopy data, looking for errors in the data itself, and re-analyzing and refitting the electron scattering data, but all to no avail.

There were initially many suggestions as to what could theoretically cause this discrepancy in radius measurements, but, to date, none of the postulated solutions have been generally accepted as resolving the puzzle. Many have been ruled out. Remaining explanations include possible physics beyond the Standard Model (BSM), whereby the electron is measuring an electromagnetic radius, while the muon is measuring an electromagnetic radius modified by BSM effects. There are also suggestions that there might be a currently unconstrained correction in the proton polarizability. Neither is universally accepted but neither has been conclusively ruled out.

[a] On behalf of the MUSE Collaboration. e-mail: edownie@gwu.edu

The puzzle, if anything, has become more puzzling since its inception in 2010. Now the main focus for resolution of the PRP is on new measurements: atomic hydrogen spectroscopy measurements are being repeated; the CREMA collaboration is extending the range of muonic measurements to include D, ^{3}He and ^{4}He; new, lower Q^2, electron scattering measurements are planned and underway JLab and MAMI; and a simultaneous measurement of muon and electron scattering, of both charge states, will be performed at PSI by the MUSE collaboration. By measuring elastic scattering cross sections down to low Q^2, it is possible to test for the new particles postulated as part of the BSM explanation. By measuring in both charge states, MUSE can access two-photon exchange effects, postulated to be responsible for the breakdown of the Rosenbluth separation and which would be enhanced by the enhanced proton polarizability proposition.

2 Radius Measurements

2.1 Electron Scattering Measurements

The "radius" of the proton is defined as the slope of the electric form factor at $Q^2 = 0$. To date, the most comprehensive elastic electron scattering cross section measurement was performed at MAMI by Bernauer *et al.* [3]. They fit the data with a variety of functional forms and determined the proton electric radius to be $< r_E^2 >^{1/2} = 0.879(5)_{stat}(4)_{syst}(2)_{model}(4)_{group}$ fm. The data agree with the recent JLab polarized measurement [4] and the CODATA value. Other fits including the world data set, but deliberately excluding the Mainz data, [5], agree with the CODATA value, strengthening the puzzle. For a recent review, see Pohl *et al.* [6].

2.2 The Muonic Spectroscopy Measurements

The muonic hydrogen measurements [1, 7] were made as the knowledge of the proton radius limits precision in many electromagnetic physics measurements. Due to the finite size of the proton, the electron in atomic hydrogen actually has a finite probability to be found inside the proton, when occupying S-states. While inside the proton, some of its attractive potential is screened, thereby reducing the average attractive potential seen by the electron and perturbing the energy levels of the S-states. This perturbs the S-P transitions, so by measuring them very precisely, and applying some theoretical calculations, one can extract the radius of the proton.

The probability that the electron is found inside the proton is simply proportional to the volume of the proton relative to the volume of the atom. By replacing the electron with a muon, which is ~205 times as massive, one shrinks the size of the atom and thereby increases the strength of the perturbation, vastly improving the sensitivity of the measurement to the proton radius, and enabling a much more precise radius extraction.

The measurement by Pohl *et al.* of $r_p = 0.84184 \pm 0.00067$ fm [1] was highly unexpected. The result was confirmed in 2013 by the analysis of Antognini *et al.*, who measured the proton electric radius to be $r_p = 0.84087 \pm 0.00039$ fm. Interestingly, Antognini *et al.* produced a magnetic radius which agrees with that obtained from electron scattering, while still supporting the discrepant electric radius of Pohl *et al.*.

The CREMA collaboration has extended their measurements beyond hydrogen to deuterium, ^{3}He and ^{4}He. Preliminary results from muonic deuterium are consistent with the CODATA, electron-scattering and spectroscopy measured values for the deuterium radius. The ^{3}He measurements are also underway, and the ^{4}He data is under analysis.

3 New Measurements

It is commonly agreed that the only way to solve the proton radius puzzle is by new measurements. Some of these plan to confirm the old results with new, more precise measurements, some intend to extend their kinematic range, and some to measure completely new things, such as precise muon scattering or spectroscopy of new muonic ions.

3.1 Spectroscopy and Electron Scattering Measurements

There are current efforts underway to remeasure the proton radius via atomic hydrogen spectroscopy. In addition to this, as described above, the CREMA collaboration are extending their measurement series to a number of light nuclei. There are also new elastic electron scattering experiments, planned and underway, to extend the measurements at JLab and Mainz down to lower Q^2. To date, there have been experiments in spectroscopy involving electrons and muons and elastic scattering experiments involving electrons, but no elastic muon scattering experiment at a level of precision able to answer the proton radius puzzle. The MUSE experiment will change that.

3.2 The MUon proton Scattering Experiment (MUSE)

MUSE entails a simultaneous measurement of elastic scattering of muons and electrons on the proton, investigating both positive and negative charge states. The experiment will reach as low a Q^2 as the existing electron scattering measurements, with sub-percent precision. The experiment will use beam line Cherenkov, scintillating fiber and GEM detectors to determine the trajectory and species of each incoming beam particle. The scattered particles will be detected in Straw Tube Trackers and a double scintillator wall. This will allow a direct comparison of cross sections and form factors, and, ultimately, the proton radius, as extracted from muon and electron elastic scattering. The simultaneous measurement of electrons and muons allows very good systematic control and measurement of both charge states also allows a test of two-photon-exchange effects.

The MUSE experiment (see Fig. 1) will take place at the πM1 beam line of the Paul Scherrer Institute (PSI). The πM1 beam line is a secondary beam line and the beam consists of a mixture of pions, electrons and muons with a 50 MHz time structure. We will measure at three different beam momenta: $p_{in} \simeq 115$, 153 and 210 MeV/c^2. These momenta are chosen to allow good separation of the different particle types in time and give overlapping kinematic points for good systematic control while providing data in the range of kinematic interest for measurements sensitive to the proton radius.

Due to the large dispersion and the mixed particle nature of the beam, it is necessary to monitor beam timing and angle accurately. Precise timing will be achieved by tagging beam particles as they enter the experimental area at the two higher momenta using a Cherenkov detector, following the idea prototyped by Albrow *et al.* [8]. This will provide accurate timing of beam particles relative to the beam RF signals and time of flight. The Cherenkov detector will be the first detector in the beamline in the target area. At the lowest beam momentum (115 MeV), timing constraints are not as strict and multiple scattering is a much larger issue, so the Cherenkov will be replaced by a thin plastic scintillator.

The position and timing of beam particles for trigger purposes will then be measured using a scintillating fiber (SciFi) detector in the target area. This consists of four layers of forty fibres each, with double-ended maPMT readout. The SciFi detector is followed by a stack of three Gas Electron Multiplier (GEM) chambers, repurposed from OLYMPUS at DESY [9]. These provide accurate tracking for offline scattering angle reconstruction, but the faster SciFi signal is necessary for trigger-purposes. While all the other equipment will be custom-made for the experiment, the GEM chambers already

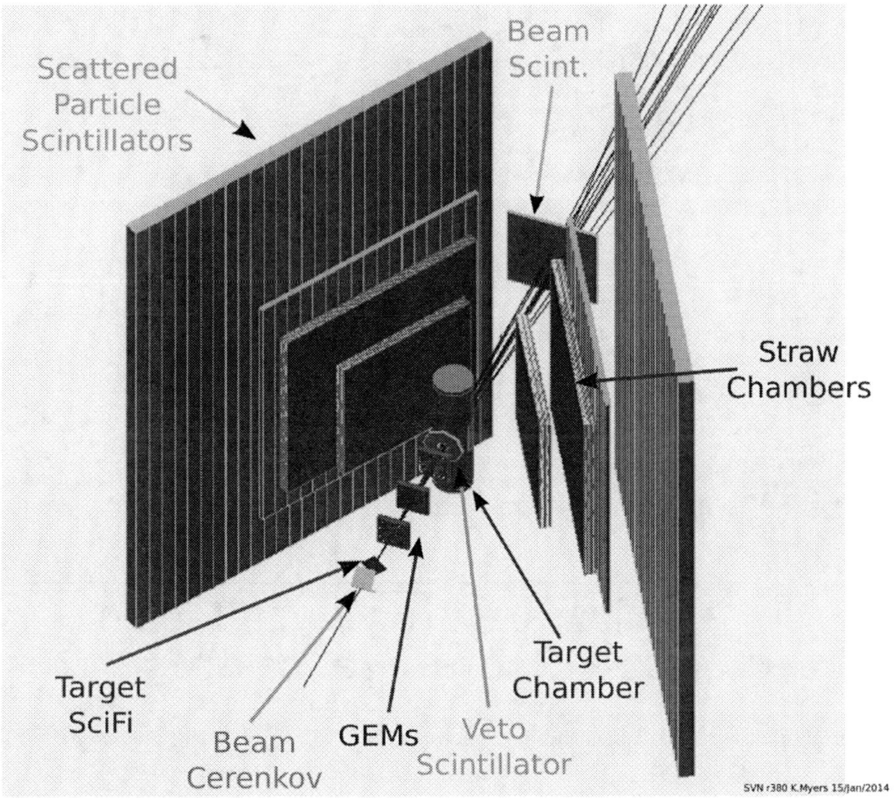

Figure 1. The MUSE setup: showing Cherenkov detector, SciFi, and GEM beamline detectors, and the scattered particle detector arms, with STTs and triggering scintillator walls.

exist, are already present in PSI and have been used for test beam times. The GEM chambers are, however, being reconfigured to shorten their stack length in order to minimize their impact on multiple scattering in the experiment. The SciFi and GEM chambers will be used in the offline analysis to determine the exact angle of the particles entering the target to enable accurate scattering angle definition.

In order to compensate for the low beam flux of the πM1 beam line, the scattered particle detector system employed by the MUSE experiment will take the form of a non-magnetic, large-acceptance tracking detector. There will be two identical sectors, one on each side of the beamline. Each side will consist of two ten-layer straw tube tracking chambers (STTs) with 1425 straws per side followed by two layers of fast scintillators (92 bars in total) for triggering and timing.

The STTs will be based on the PANDA design and will be read out for timing only using PADIWA discriminators and TRB3 boards which provide timing, scalars and trigger logic [10]. The scintillation counter signals will be sent into PADIWA discriminators and TRB3 boards for timing etc., but also into QDCs in order to allow for offline pulse-height timing corrections.

The experimental trigger will be based on the requirement of a particle firing the Cherenkov / beam line scintillation detector, the SciFi detector and a scattered particle with appropriate timing and position reaching the scattered particle scintillators. The trigger algorithm will be processed in the specially programmed FPGAs (Field-Programmable Gate Arrays) of the TRB3 boards used to

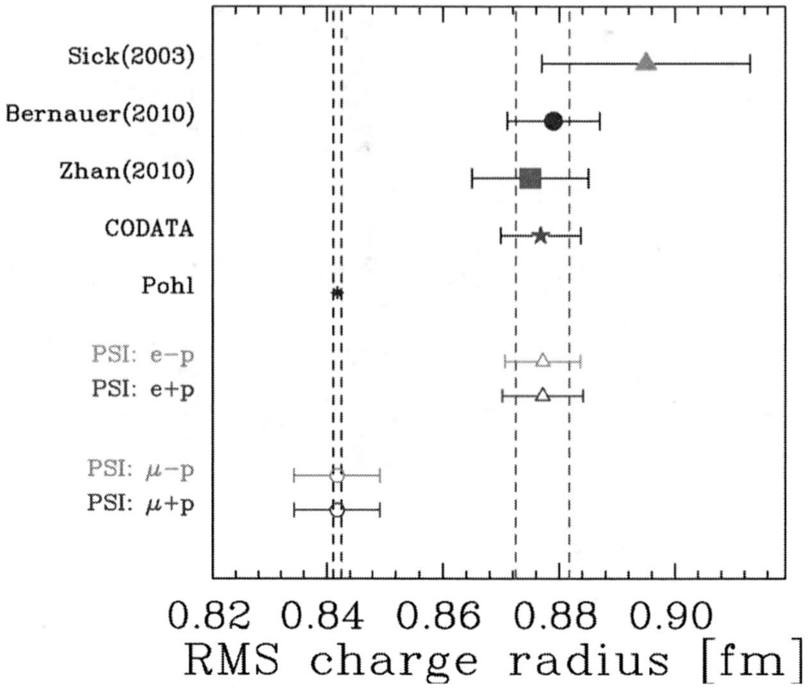

Figure 2. Current proton radius values and the anticipated MUSE measurement precision.

provide timing and scalar information for each chanel, with one TRB3 trigger master which will make the final trigger decision and distribute the trigger and trigger number to the TRB3s and non-TRB3 parts of the detector system.

In order to suppress backgrounds caused by muon decays in-flight, and scattering in the beamline, there will be an annular veto detector, comprising of eight scintillators, surrounding the entrance to the liquid hydrogen target scattering chamber. The signals from these detectors will be incorporated in the trigger system as a veto. These backgrounds will be further reduced in the offline analysis by studying the time of flight from the initial target Cherenkov or, at low energy, plastic scintillation detector to the scattered particle scintillator wall. In order to monitor the beam and ensure stability throughout the experiment, there will be twelve beam-monitor scintillators which will be used to monitor beam flux and position.

It is expected that MUSE will produce cross section measurements for the elastic scattering of $\mu^{+/-}$ and $e^{+/-}$ with sub-1% relative precision over a Q^2 range from 0.002 to 0.07 GeV^2. Using these data, form factors will be extracted and the resulting radii from muon and electron scattering, and from both charge states, will be compared. See Fig. 2 for the anticipated precision.

4 Conclusion

After five years of effort, the proton radius puzzle remains puzzling. There are multiple experimental efforts underway to find a solution to this very interesting, and challenging, conundrum. These include renewed hydrogen spectroscopy measurements, an extended study of muonic ^{2}H, ^{3}He and ^{4}He, and new, lower Q^2 elastic electron scattering measurements at JLab and MAMI. Complementary to these efforts, covering the one clear gap in the proton radius puzzle experimental information: MUSE will measure elastic muon scattering on the proton, simultaneously with elastic electron scattering in order to have a very firm grasp on possible systematic uncertainties. By measuring in both charge states, MUSE will also be able to access two-photon effects. In so doing, MUSE can access the signatures of the main postulated explanations for the proton radius puzzle. New particles would modify the scattering cross sections at low Q^2, and a correction to the polarizability would be clearly visible in enhanced two-photon-effects. Thus, with the new data from scattering and spectroscopy experiments, we hope to resolve the proton radius puzzle and identify any possible signatures of new physics.

Acknowledgements

This material is based upon work supported by the National Science Foundation under Grant No. PHY-1309130 and the U.S. Department of Energy, Office of Science, Office of Nuclear Physics program under contract number DE-SC0012485.

References

[1] R. Pohl, A. Antognini, F. Nez, F.D. Amaro, F. Biraben et al., Nature **466**, 213 (2010)
[2] P.J. Mohr, B.N. Taylor, D.B. Newell, Rev.Mod.Phys. **80**, 633 (2008), 0801.0028
[3] J. Bernauer et al. (A1 Collaboration), Phys. Rev. Lett. **105**, 242001 (2010), arXiv: [nucl-ex]1007.5076
[4] X. Zhan et al., Phys. Lett. **B705**, 59 (2011), arXiv: [nucl-ex]1102.0318
[5] I. Sick, Phys.Lett. **B576**, 62 (2003), nucl-ex/0310008
[6] R. Pohl, R. Gilman, G.A. Miller, K. Pachucki, Ann.Rev.Nucl.Part.Sci. **63**, 175 (2013), arXiv: [atom-ph]1301.0905
[7] A. Antognini et al., Science **339**, 417 (2013)
[8] M. Albrow, H. Kim, S. Los, E. Ramberg, A. Ronzhin et al., JINST **7**, P10027 (2012), arXiv: [ins-det]1207.7248
[9] R. Milner et al., Nuclear Inst. and Methods in Physics Research A **741**, 1 (2014), arXiv: [ins-det]1312.1730
[10] TRB homepage containing full specification and documentation: trb.gsi.de

EPJ Web of Conferences **113**, 05022 (2016)
DOI: 10.1051/epjconf/201611305022

Studies on η meson production in dp collisions at the ANKE spectrometer

Christopher Fritzsch[1,a], Daniel Guderian[1], Alfons Khoukaz[1], Malte Mielke[1], Michael Papenbrock[1], and Daniel Schröer[1] for the ANKE-Collaboration

[1] *Institut für Kernphysik, Westfälische Wilhelms-Universität Münster, Wilhelm-Klemm-Straße 9, 48149 Münster, Germany*

Abstract. Current investigations at the internal fixed target experiment ANKE at the storage ring COSY, located at the Forschungszentrum Jülich in Germany, focus on the behaviour of the η meson production process as well as on the study of angular distributions in the reactions $d + p \to {}^3\text{He} + \eta$ and $p + d \to d + \eta + p_{\text{sp}}$. The recent analysis status for both reactions will be presented and discussed.

1 Introduction

Investigations on the total cross sections of the reaction $d + p \to {}^3\text{He} + \eta$ are of special interest since they differ strongly from a pure phase space behaviour near threshold (cf. Figure 1 (left)) [1–5].

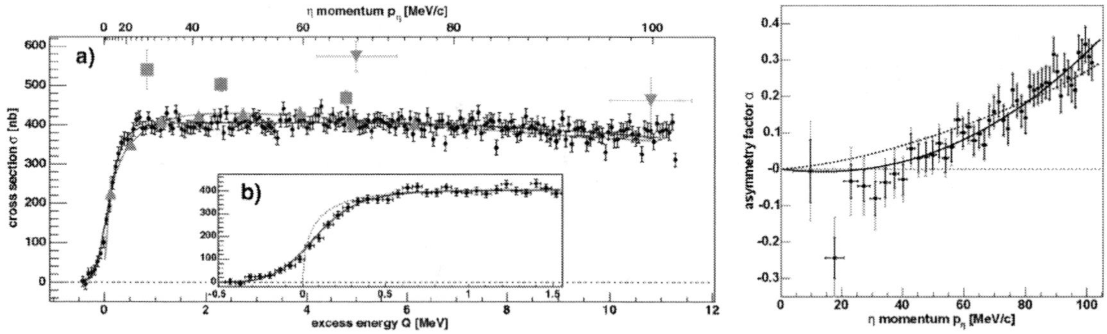

Figure 1. Left: a) Overview of the obtained total cross section of the reaction $d+p \to {}^3\text{He}+\eta$. The red solid line shows a fit to the ANKE data set from threshold up to an excess energy of $Q = 11$ MeV, which is represented by the black filled circles [1]. **b)** Enlargement of the near threshold region. The red dotted line shows the FSI ansatz corrected for the beam momentum spread. **Right:** Variation of the asymmetry factor α with the η momentum p_η. The solid line is a fit to the ANKE data [1] with regard to s- and p-wave interference and the dashed line without s- and p-wave interference.

[a] e-mail: c.fritzsch@uni-muenster.de

Furthermore, analyses of the asymmetry factor α [6] of the differential cross sections (cf. Figure 1 (right)) show a distinct effect of s- and p-wave interference with the η momentum p_η, which can be explained by a rapid variation of the phase. These effects are an indication for an unexpected strong final state interaction (FSI) between η mesons and ^{3}He nuclei which could lead to a quasi bound state of the η^3He-system [6]. New high precision data from the ANKE spectrometer enable the extraction of additional total and differential cross sections for the η production up to an excess energy of $Q = 15\,\text{MeV}$ with improved accuracy, which will clarify the behaviour of the asymmetry factor α with less uncertainties.
Assuming the FSI ansatz to be appropriate, the non pure phase space behaviour of the total cross section near threshold should be seen also in other production channels. This has been tested by different facilities [7–9]. Nevertheless, none of these measurements could clearly prove whether the η-nucleus system is in a bound or virtual state. In order to investigate this in more detail, studies on the properties of η-mesic nuclei with the reaction $p + d \to d + \eta + p_{\text{spec}}$ are in progress. The behaviour of the total cross sections near threshold will allow the determination of the scattering length $a_{d\eta}$ to gain new insights on the possible formation of η-mesic nuclei.

2 The ANKE spectrometer

The magnetic spectrometer **ANKE** ("**A**pparatus for Studies of **N**ucleon and **K**aon **E**jectiles") is an internal fixed target experiment at the accelerator and storage ring **COSY** ("**CO**oler **SY**nchrotron"), located at the Forschungszentrum Jülich in Germany, to study the properties and interaction of hadrons in a medium energy regime. For this purpose, COSY provides (un)polarized deuteron and proton beams with momenta between 0.3 GeV/c and 3.7 GeV/c. With its polarized storage-cell target and unpolarized cluster-jet target, the ANKE experiment focusses on the investigation of pp-, pd- and dd-collisions. The magnet spectrometer ANKE (cf. Figure 2) consists of three dipole magnets (D1 - D3), an internal target, and three detection systems (Nd-, Pd-, and Fd-system). Negatively (positively) charged ejectiles will be deflected by the second dipole magnet D2 into the Nd(Pd)-system. In addition to the Pd-system, the Fd-system was designed for heavy (^{3}He) or fast (p or d) positively charged ejectiles, which will be deflected by D2 under small laboratory scattering angles due to the large Lorentz boost in beam direction close to the beam pipe. The Fd-system is composed of one multiwire drift chamber and two multiwire proportional chambers used for track reconstruction and two layers of scintillator hodoscopes for energy loss and time-of-flight measurements. A special feature of the ANKE detector is its movable D2 magnet to optimize the detector acceptance for each reaction which will be investigated. The design of the ANKE spectrometer setup with its three dipole magnets is very

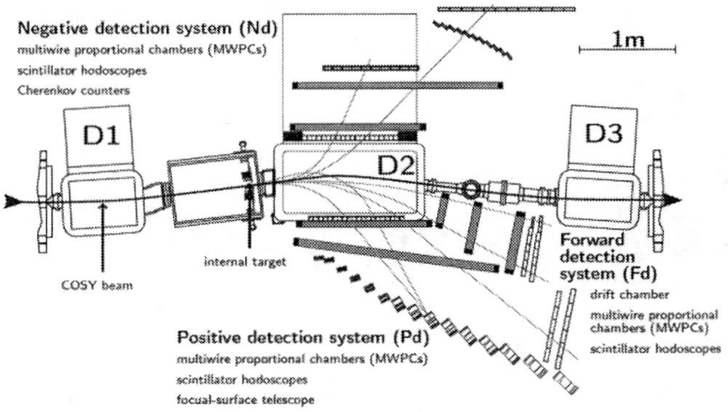

Figure 2. Schematic representation of the magnatic spectrometer ANKE consisting of three dipole magnets (D1 - D3), the internal target, and three detection systems (Nd-, Pd-, and Fd-system). The blue (red) lines illustrate negatively (positively) charged particles, which will be deflect by the second dipole magnet D2 into the Nd(Pd or Fd)-system.

well suited for the detection and reconstruction of charged particles generated even at small laboratory scattering angles in reactions, e.g. $d + p \to {}^3\text{He} + \eta$, close to threshold.

3 The η production channel $d + p \to {}^3\text{He} + \eta$

New data from the ANKE spectrometer allowed to determine the η meson mass with high precision using the $d{+}p \to {}^3\text{He}{+}\eta$ reaction [10] and to study the two pion production using $d{+}p \to {}^3\text{He}{+}\pi^+{+}\pi^-$ [11]. In order to determine the η mass, data has been analysed at 18 deuteron beam momenta in a range between $3120.17\ \text{MeV}/c \leq p_d \leq 3204.16\ \text{MeV}/c$ which could be extracted very accurately via the resonant depolarisation technique with a precision of $\Delta p_d/p_d < 6 \times 10^{-5}$ [10, 12]. Moreover, due to the high statistics of $> 10^5$ ${}^3\text{He}\eta$ events per energy in combination with full angle coverage these high precision ANKE data allow to investigate the total and differential cross sections of the reaction $d + p \to {}^3\text{He} + \eta$. Such data can be used to investigate the behaviour of the asymmetry factor α especially in the near threshold region (cf. Figure 1 (right)). To extract total and differential cross section values a careful luminosity determination was performed for each of the 18 beam momenta of the beam time via dp-elastic scattering with high precision, i.e. $\Delta L_{\text{stat}} = 1\%$ and $\Delta L_{\text{sys}} = 6\%$, which corresponds to an improvement by at least a factor of two compared to previous measurements. The identification of the reaction $d + p \to {}^3\text{He} + \eta$ is ensured by detecting the ${}^3\text{He}$ nucleus in the Fd-system. As shown in Figure 3 (left), it is straightforward to separate the reaction of interest from other reactions, e.g. pion production channels. Due to the fully reconstructed four momentum of the ${}^3\text{He}$ nuclei the usage of the missing mass method (cf. Figure 3 (right)) is feasible. A smooth model independent background description was realized by data taken below the η production threshold ($d + p \to {}^3\text{He} + \eta$) at a beam momentum of $p_d = 3120\ \text{MeV}/c$ which, after boosting them to the energy above threshold, perfectly fits the data. By this, a clean η signal remains, so that first results on total and differential cross sections will be available soon.

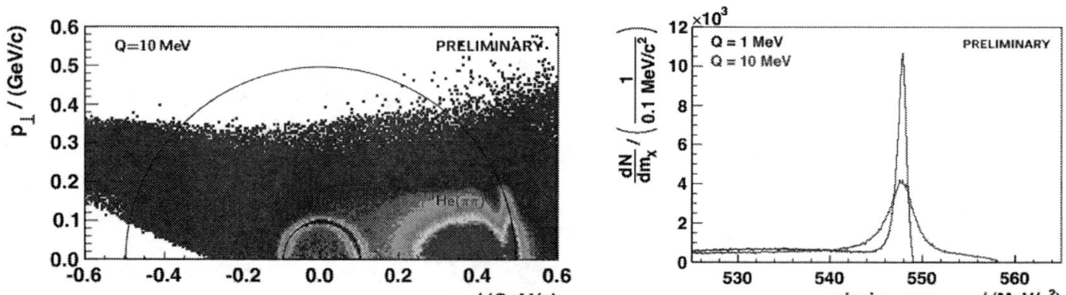

Figure 3. Left: The magnitude of the transverse momentum plotted against the longitudinal momentum in the center of mass system for $Q = 10\,\text{MeV}$. The solid black lines represent the calculated kinematical loci of the reactions $d + p \to {}^3\text{He} + \eta$ and $d + p \to {}^3\text{He} + \pi^0$ **Right:** Missing mass distributions for two different excess energies ($Q = 1\,\text{MeV}$ and $Q = 10\,\text{MeV}$).

4 The η production channel $p + d \to d + \eta + p_{\text{sp}}$

An additional beam time was performed at ANKE to investigate the properties of η-mesic nuclei with the reaction $p + d \to d + \eta + p_{\text{spec}}$, where the initial deuteron serves as an effective neutron target. The combination of two beam momenta $p_1 = 2.09\ \text{GeV}/c$ and $p_2 = 2.25\ \text{GeV}/c$ and the Fermi motion inside the deuteron allows to extract differential and total cross sections in a wide excess energy range between 0 – 100 MeV. Moreover, it enables the determination of the scattering length $\alpha_{d\eta}$ to validate the range of the s-wave FSI ansatz and the production of higher partial waves. For this purpose, the precise measurement and clear identification of both the spectator protons and the deuterons is necessary. The identification of the spectator protons is ensured via two Silicon Tracking Telescopes (acronym: "STT"), each consisting of three layers of segmented semiconductor for track reconstruction and energy loss measurements (cf. Figure 4 (left)). The reconstruction of the deuteron in the Fd-system is more challenging because of a similar energy loss of the huge proton background and the deuterons in this momentum range. Therefore, using a cut on the time-of-flight difference between pions in the Pd-system and particles in the Fd-system, energy loss cut conditions can be determined, so that the proton background can be suppressed and a clear deuteron band becomes visible, as shown in Figure 4 (right). With both charged ejectiles, i.e. spectator proton and deuteron, and an accurate calibration of the detection systems, the reaction can be identified via the missing mass technique. First results on the cross section will be available soon.

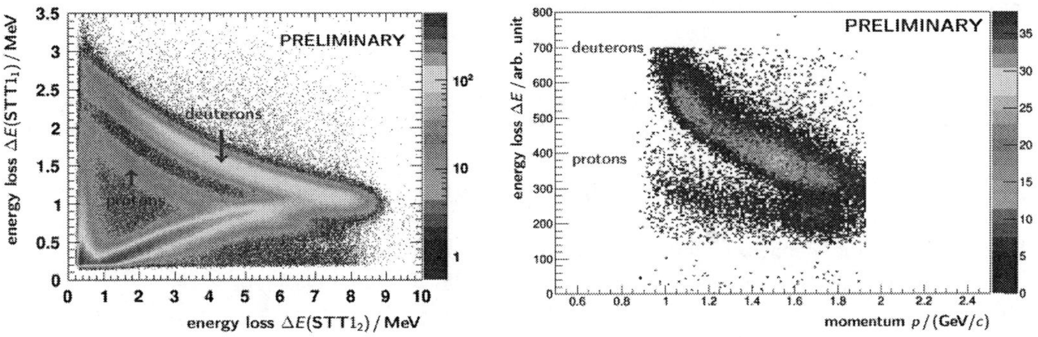

Figure 4. Left: Deposited energy of protons and deuterons in the first layer of one of the Silicon Tracking Telescopes plotted against the energy loss information of the second semiconductor layer of the same STT. **Right:** Energy loss measurements of particles detected in the Fd-system. By applying cuts on the time-of-flight difference between particles in the Fd-system and positively charged pions in the Pd-system, a clear deuteron band becomes visible.

Acknowledgements

This work has been supported by the COSY-FFE program of the Forschungszentrum Jülich.

References

[1] T. Mersmann *et al.*, Phys. Rev. Lett. **98**, 242301 (2007).
[2] J. Smyrski *et al.*, Phys. Lett. B **649**, 258 (2007).
[3] J. Berger *et al.*, Phys. Rev. Lett. **61**, 919 (1988).
[4] B. Mayer *et al.*, Phys. Rev. C **53**, 2068 (1996).
[5] H.-H. Adam *et al.*, Phys. Rev. C **75**, 014004 (2007).
[6] C. Wilkin *et al.*, Phys. Lett. B **654**, 92 (2007).
[7] F. Pheron *et al.*, Phys. Lett. B **709**, 21 (2012).
[8] A. Budzanowski *et al.*, Nucl. Phys. A **821**, 193 (2009).
[9] A. Budzanowski *et al.*, Phys. Rev. C **82**, 041001(R) (2010).
[10] P. Goslawski *et al.*, Phys. Rev. D **85**, 112011 (2012).
[11] M. Mielke *et al.*, Eur. Phys. J. A **50**, 102 (2014).
[12] P. Goslawski *et al.*, Phys. Rev. ST-AB **13**, 022803 (2010).

EPJ Web of Conferences **113**, 05023 (2016)
DOI: 10.1051/epjconf/201611305023

Studies on η meson production and the C-parity violating decay $\eta \to \pi^0 e^+ e^-$ with WASA-at-COSY

Nils Hüsken[1,a], Florian Bergmann[1], Kay Demmich[1], Alfons Khoukaz[1], and Karsten Sitterberg[1] on behalf of the WASA-at-COSY Collaboration

[1] *Institut für Kernphysik, Westfälische Wilhelms-Universität Münster, Wilhelm-Klemm-Straße 9, 48149 Münster, Germany*

Abstract. The WASA-at-COSY experiment is a near 4π detector, located at the Forschungszentrum Jülich (Germany). It is highly suited to study both the production of light mesons in hadron collisions and their subsequent decays. Current investigations include a study on the energy dependence of the production cross section and angular distributions of the η meson in the proton-deuteron fusion reaction $pd \to {}^3\text{He}\,\eta$, as well as a search for beyond standard model physics in the C-parity violating decay $\eta \to \pi^0 e^+ e^-$. The status of both these analyses and preliminary results are presented.

1 Introduction

In the very near threshold region, the $pd \to {}^3\text{He}\,\eta$ reaction has been studied in great detail [1–5], whereas at higher excess energies Q the amount of available data is limited [6–9]. Moreover, while the data from ANKE [6] and WASA/PROMICE [7, 8] suggest a plateau in the total cross section for higher Q-values, recent results from the WASA-at-COSY experiment [10] suggest an unexpected narrow variation of the total cross section at $Q = 48.8\,\text{MeV}$ as can be seen in Fig.1. Furthermore, theoretical model calculations available in the literature [11–13] based on either one- or two-step models fail to describe the strongly forward peaked angular distributions and total cross section in parallel. While new calculations based on a boson exchange model look promising [13], more high quality data at intermediate excess energies is needed to support and test new calculations in order to deepen our understanding of the underlying production mechanism.

Due to the rather large production cross section of η mesons in both pp- and pd-collisions, data taken with the WASA-at-COSY experiment also present a nice opportunity to investigate rare η meson decays. A first dataset consisting of approximately 3×10^7 $pd \to {}^3\text{He}\,\eta$ events is being used for the search of the C-parity violating decay $\eta \to \pi^0 e^+ e^-$ among others. The current experimental upper limit for this decay is $\text{BR}\left(\eta \to \pi^0 e^+ e^-\right) < 4 \times 10^{-5}$ (90% CL), whereas the standard model expectation for the C-parity conserving fourth order process $\eta \to \pi^0 \gamma^* \gamma^* \to \pi^0 e^+ e^-$ is of the order of 10^{-9} [14]. An observation of a significantly higher branching ratio would therefore indicate beyond standard model physics being involved in the decay chain, like for example a Z-like dark boson.

[a] e-mail: n_hues02@uni-muenster.de

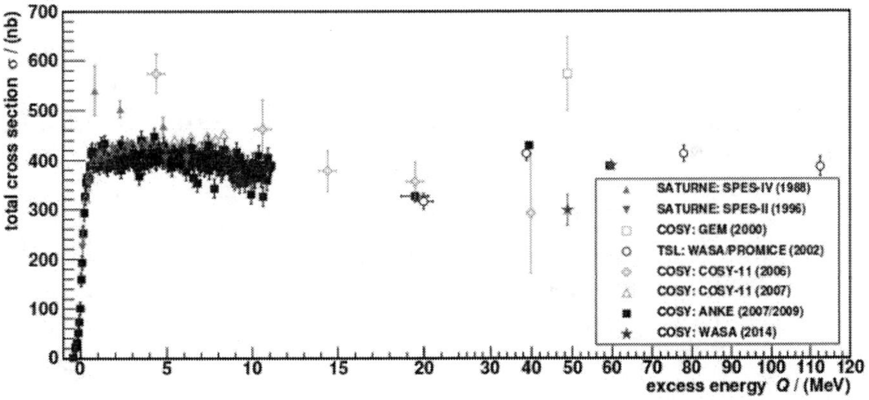

Figure 1. Total cross section data for the reaction $pd \rightarrow {}^3\mathrm{He}\,\eta$ obtained by Ref. [1–9] as a function of the excess energy Q. The uncertainties shown do not include systematic uncertainties resulting from an absolute normalization. The data from WASA- at-COSY (red stars) are scaled to the ANKE data point at $Q = 59.4\,\mathrm{MeV}$. Figure taken from Ref. [10].

2 The WASA-at-COSY experiment

The WASA-at-COSY experiment is a fixed target experiment located at the COoler SYnchroton (COSY). With COSY providing either polarized or unpolarized protons or deuterons with beam momenta of up to $p = 3.7\,\mathrm{GeV/c}$ and a frozen hydrogen or deuterium pellet-target, it is possible to investigate pp-, pd- and dd-collisions with WASA-at-COSY. The Central Detector system, consisting of a solenoid magnet, a drift chamber, a plastic scintillator barrel and an electromagnetic calorimeter, is used for the identification and reconstruction of light mesons and their decay particles, like pions, electrons and photons. Due to the fixed target geometry, all ejectiles experience a large boost in forward direction, so that heavier particles like protons, deuterons or helium nuclei are scattered under small polar angles and therefore measured in the Forward Detector system. The FD consists of various layers of thin scintillators, a proportional chamber and five layers of a thick scintillator hodoscope. This experimental setup offers the possibility to precisely reconstruct both the azimuthal and the polar scattering angles of forward going particles, as well as their energy. With its near 4π coverage and excellent detector systems, the WASA-at-COSY experiment allows for both inclusive and exclusive measurements.

3 η meson production in pd fusion to ${}^3\mathrm{He}\,\eta$

In May 2014, a beam time was conducted with the COSY storage ring providing unpolarized protons at 15 different beam momenta between $p_p = 1.60\,\mathrm{GeV/c}$ and $p_p = 1.74\,\mathrm{GeV/c}$, that were brought to collision with the deuterium pellet-target. With this new dataset, covering an excess energy range relative to the ${}^3\mathrm{He}\,\eta$ threshold from $Q \approx 13.6\,\mathrm{MeV}$ to $Q \approx 80.9\,\mathrm{MeV}$ with a step size of $\Delta Q \approx 4.8\,\mathrm{MeV}$, it is possible to extract precise total and differential cross sections of the $pd \rightarrow {}^3\mathrm{He}\,\eta$ reaction.

The ^{3}He-nuclei can be identified in the forward detector due to their energy loss ΔE by use of the ΔE-E technique. Combining the energy loss information with a measurement of both the azimuthal and the polar scattering angles φ and ϑ, the full four-momenta of the ^{3}He-nuclei can be reconstructed, so that the missing mass technique can be applied. Such a missing mass distribution is shown in

Fig.2. Performing a simultaneous fit of a polynomial for the background and a gaussian for the signal, the background can be subtracted to get a first estimate of more than 1.2×10^5 η mesons per excess energy. Applying the same background subtraction technique to missing mass spectra binned in $\cos \vartheta_\eta^{cm}$, first raw angular distributions can be extracted (see Fig.2). These distributions have yet to be corrected for the detector acceptance and cut efficiencies. Due to a full angular coverage and good statistics, the angular distributions can be examined in great detail in the future. The reaction $pd \to {}^3\mathrm{He}\,\pi^0$ will be used for normalization purposes, so that ultimately total and differential cross sections can be determined with an estimated point-to-point uncertainty of $\sim 8\,\%$, mainly resulting from the normalization procedure, over the whole excess energy range from $Q \approx 13.6\,\mathrm{MeV}$ to $Q \approx 80.9\,\mathrm{MeV}$. This will present a considerable extension of the current database, allowing a detailed investigation of theoretical models describing the underlying production mechanism of η mesons in pd fusion to ${}^3\mathrm{He}\,\eta$.

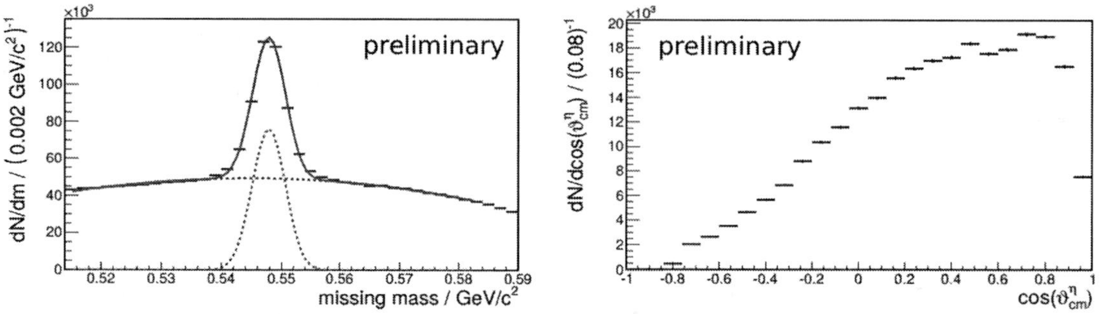

Figure 2. Left: Missing mass of $pd \to {}^3\mathrm{He}\,X$ for ^{3}He-nuclei stopped in the first layer of the Forward Range Hodoscope. The solid red line is a combined fit of a signal gaussian (dotted red line) and a background polynomial (dotted blue line). **Right:** Raw angular distribution of ${}^3\mathrm{He}\,\eta$ events. No geometrical acceptance and efficiency correction have been applied so far. Both spectra show data for $p_p = 1.70\,\mathrm{GeV/c}$ ($Q \approx 61.7\,\mathrm{MeV}$).

4 The C-parity violating decay $\eta \to \pi^0\, e^+\, e^-$

A huge dataset of approximately 3×10^7 $pd \to {}^3\mathrm{He}\,\eta$ events allows the search for (very) rare and forbidden η meson decays with the WASA-at-COSY experiment. The decay $\eta \to \pi^0\, e^+\, e^-$ is of particular interest, as the decay via a virtual photon would violate C-parity and is therefore forbidden. In the standard model, a fourth order process involving two virtual photons exists, that does not violate C-parity. However, the estimated branching ratio of this process is in the order of 10^{-9} [14]. With the current experimental upper limit at $\mathrm{BR}\left(\eta \to \pi^0\, e^+\, e^-\right) < 4 \times 10^{-5}$ (90% CL), there remain four orders of magnitude that can be probed for beyond standard model physics. To significantly reduce the existing upper limit, it is highly important to gain a very good understanding of the contributing background processes. Knowing these, cuts can be optimized in a way that the signal-to-background ratio is maximized. This procedure is applied to cuts on particle identification, the invariant mass of two neutral particles $\left(\pi^0 \to \gamma\gamma\right)$, the invariant mass of two neutral and two oppositely charged particles $\left(\eta \to \pi^0\, e^+\, e^-\right)$, a kinematic fit probability for the hypothesis of a ${}^3\mathrm{He}\,\gamma\,\gamma\,e^+\,e^-$ final state and finally a cut on the missing mass of the ^{3}He-nucleus in a blind analysis. An eventual signal of BSM physics could appear in the invariant mass distribution of the lepton pair, for example, in the case that the decay is mediated by a Z-like dark photon. Preliminary results of this analysis of roughly one third of the whole dataset can be seen in Fig.3. The observed distributions of both missing mass of ^{3}He and invariant mass of the lepton pair are consistent with the expectation from Monte-Carlo simulations for direct $^3\mathrm{He}\,\pi^0\pi^0$ production with one of the two pions performing a Dalitz decay and one photon escaping the detector.

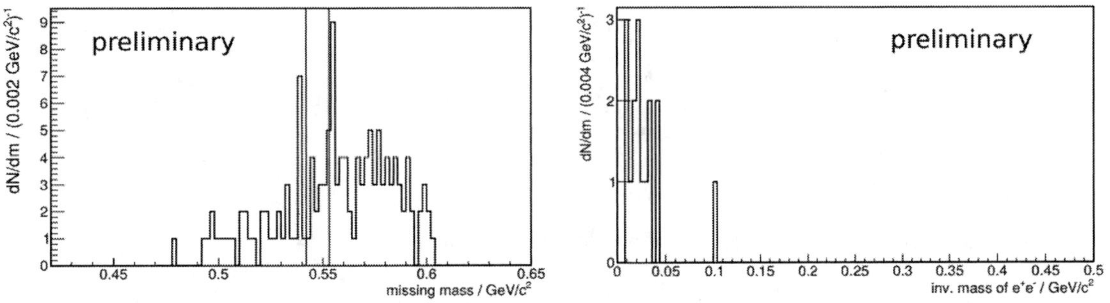

Figure 3. Left: Missing mass of ^{3}He after all cuts were applied. The blue lines represent the cut conditions on the missing mass of ^{3}He used for the invariant mass spectrum of the lepton pair. **Right:** Invariant mass of the lepton pair after all cuts were applied. Both distributions are consistent with direct $\pi^0\pi^0$ production with one pion performing a Dalitz decay. A sample of the whole dataset (from [16]) is displayed here.

The ongoing analysis of the full dataset will significantly reduce the existing upper limit for the C-parity violating η meson decay $\eta \to \pi^0 e^+ e^-$ and will therefore reduce the available parameter space for BSM physics in the near future.

Acknowledgements

Supported by COSY-FFE and the European Union Seventh Framework Programme (FP7/2007-2013) under grant agreement n 283286.

References

[1] T. Mersmann *et al.*, Phys. Rev. Lett. **98**, 242301 (2007).
[2] J. Smyrski *et al.*, Phys. Lett. B **649**, 258 (2007).
[3] J. Berger *et al.*, Phys. Rev. Lett. **61**, 919 (1988).
[4] B. Mayer *et al.*, Phys. Rev. C **53**, 2068 (1996).
[5] H.-H. Adam *et al.*, Phys. Rev. C **75**, 014004 (2007).
[6] T. Rausmann *et al.*, Phys. Rev. C **80**, 017001 (2009).
[7] R. Bilger *et al.*, Phys. Rev. C **65**, 044608 (2002).
[8] R. Bilger *et al.*, Phys. Rev. C **69**, 014003 (2004).
[9] M. Betigeri *et al.*, Phys. Lett. B **472**, 267 (2000).
[10] P. Adlarson *et al.*, Eur. Phys. J. A **50**, 100 (2014).
[11] K.P. Khemchandani *et al.*, Phys. Rev. C **76**, 069801 (2007).
[12] K.P. Khemchandani *et al.*, Phys. Rev. C **68**, 064610 (2003).
[13] N.G. Kelkar *et al.*, Rep. Prog. Phys. **76**, 066301 (2013).
[14] J.N. Ng *et al.*, Phys. Rev. D **47**, 4939 (1993).
[15] H.-H. Adam *et al.*, WASA-at-COSY Collaboration, arXiv:nucl-ex/0411038.
[16] F. Bergmann, PhD Thesis in Preparation, Westfälische Wilhelms-Universität Münster, Germany, (2015).

EPJ Web of Conferences **113**, 05024 (2016)
DOI: 10.1051/epjconf/201611305024

A systematic approach to sketch Bethe-Salpeter equation

Si-xue Qin[1,a]

[1] *Physics Division, Argonne National Laboratory, Argonne, Illinois 60439, USA*

Abstract. To study meson properties, one needs to solve the gap equation for the quark propagator and the Bethe-Salpeter (BS) equation for the meson wavefunction, self-consistently. The gluon propagator, the quark-gluon vertex, and the quark–anti-quark scattering kernel are key pieces to solve those equations. Predicted by lattice-QCD and Dyson-Schwinger analyses of QCD's gauge sector, gluons are non-perturbatively massive. In the matter sector, the modeled gluon propagator which can produce a veracious description of meson properties needs to possess a mass scale, accordingly. Solving the well-known longitudinal Ward-Green-Takahashi identities (WGTIs) and the less-known transverse counterparts together, one obtains a nontrivial solution which can shed light on the structure of the quark-gluon vertex. It is highlighted that the phenomenologically proposed anomalous chromomagnetic moment (ACM) vertex originates from the QCD Lagrangian symmetries and its strength is proportional to the magnitude of dynamical chiral symmetry breaking (DCSB). The color-singlet vector and axial-vector WGTIs can relate the BS kernel and the dressed quark-gluon vertex to each other. Using the relation, one can truncate the gap equation and the BS equation, systematically, without violating crucial symmetries, e.g., gauge symmetry and chiral symmetry.

1 Introduction

The visible mass of the universe is mainly contributed by hadrons – bound states of fundamental blocks, i.e., quarks and gluons. Their dynamics is described by quantum chromodynamics (QCD) – the strong interaction sector of the Standard Model. QCD has two fascinating features: dynamical chiral symmetry breaking (DCSB) and confinement, which are not apparent in its Lagrangian. It is believed that DCSB is responsible for the generation of mass from nothing (the mass of constituent light quarks is several hundreds MeV), and thus the Higgs mechanism has almost nothing to do with the origin of the mass of the visible matter (the mass of current light quarks is only several MeV). On the other hand, the fundamental degree of freedoms, i.e., quarks and gluons, are confined and cannot directly be detected. To understand confinement is one of the greatest challenges in modern physics. Neither DCSB nor confinement is understood perturbatively. The key to the whole story is to solve QCD non-perturbatively.

In order to solve QCD non-perturbatively, we try to study its equations of motion, i.e., Dyson-Schwinger equations (DSEs) [1]. Specifically, we need to deal with three categories of equations, i.e., the one-body gap equation for quarks, the two-body Bethe-Salpeter (BS) equation for mesons, and the

[a]e-mail: sqin@anl.gov

three-body Faddeev equation for baryons. In the rest of this paper, we will focus on the gap equation and the BS equation, i.e.,

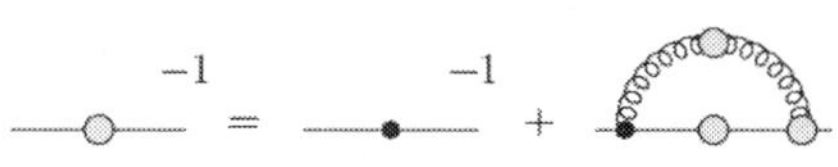

and

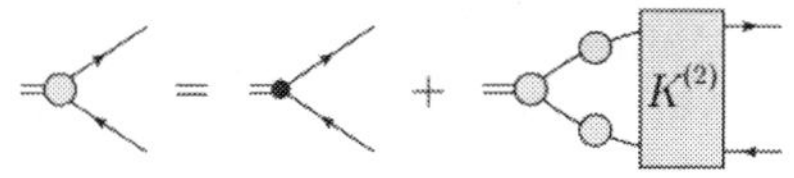

, (1)

where gray circular blobs denote dressed propagators and vertices, $K^{(2)}$ denotes the full quark–anti-quark scattering kernel, and black dots denote bare propagators or vertices. Once the gluon propagator, the quark-gluon vertex, and the scattering kernel are specified, the two equations can be solved, and thus meson properties can be studied accordingly.

2 Gluon propagator

In pure SU(3) Yang-Mills theory, the Lagrangian is gauge invariant and thus does not include a mass scale. Certainly, for the standard perturbative solution of QCD, the gluon is massless. However, as discussed in Ref. [2], the gluon can be non-perturbatively massive.

In Euclidean space and Landau-gauge, the gluon propagator is decomposed as

$$\mathcal{D}_{\mu\nu}(k) = \left(\delta_{\mu\nu} - \frac{k_\mu k_\nu}{k^2}\right) D(k^2), \quad D(k^2) = \frac{Z(k^2)}{k^2 + m_g^2(k^2)}, \tag{2}$$

where $D(k^2)$ is the dressing function of gluon and has a massive-type form according to lattice QCD [3] or DSE [4] calculations; $Z(k^2)$ is the wavefunction renormalization function, and $m_g(k^2)$ is the running gluon mass. Inspired by the result in one-loop perturbative QCD, one can parameterize $Z(k^2)$ and $m_g(k^2)$ as

$$Z(k^2) = z_0 \left(\ln \frac{k^2 + rM_g^2}{\Lambda^2}\right)^{-\gamma}, \quad m_g^2(k^2) = \frac{M_g^4}{k^2 + M_g^2}, \tag{3}$$

where γ is the gluon anomalous dimension; z_0 , r, Λ, and M_g are the fitting parameters. Using such a form, lattice data can be well fitted with a typical gluon mass scale $M_g \sim 700$ MeV [3].

Putting the non-perturbatively massive gluon into consideration, we proposed a realistic interaction model [5] as

$$g^2 \mathcal{D}_{\mu\nu}(k) = k^2 \mathcal{G}(k^2) \mathcal{D}^{\rm free}_{\mu\nu}(k) = [k^2 \mathcal{G}_{\rm IR}(k^2) + \tilde{\alpha}_{\rm pQCD}(k^2)] \mathcal{D}^{\rm free}_{\mu\nu}(k), \tag{4}$$

wherein $\mathcal{D}^{\rm free}_{\mu\nu}(k)$ is the free-gauge-boson propagator, $\mathcal{G}_{\rm IR}(k^2)$ is an ansatz which dominates the infrared interaction, and $\tilde{\alpha}_{\rm pQCD}(k^2)$ is a regular continuation of the perturbative-QCD running coupling constant. Explicitly,

$$\mathcal{G}(s) = \frac{8\pi^2}{\omega^4} D e^{-s/\omega^2} + \frac{8\pi^2 \gamma_m \mathcal{F}(s)}{\ln\left[\tau + (1 + s/\Lambda_{\rm QCD}^2)^2\right]} \approx \frac{4\pi\alpha(s)}{s + m_g^2(s)}, \tag{5}$$

where $\gamma_m = 12/25$, $\Lambda_{\rm QCD} = 0.234$ GeV; $\tau = e^2 - 1$; $\mathcal{F}(s) = [1 - \exp(-s)]/s$; and $\alpha(s)$ is the effective running coupling constant. In such a form, the one-loop renormalization-group behavior of QCD is preserved. The two parameters, i.e., the interaction strength D and the interaction width ω, shape the infrared interaction and may implicitly represent higher-order ultraviolet contributions. With the product $D\omega$ fixed, one can obtain a uniformly good description of pseudoscalar and vector mesons [5]. In the favorable parameter space, the values of M_g are typical. The infrared value of the coupling is, typically, $\alpha_{\rm RL}(0)/\pi \sim 10$ and $\alpha_{\rm DB}(0)/\pi \sim 2$, where RL stands for the simplest rainbow-ladder approximation and DB for the DCSB-improved approximation.

3 Quark-gluon vertex

In perturbation theory, the quark-gluon vertex can be calculated order by order in loop expansion. However, since QCD is non-perturbative, the dynamics dressing effect fundamentally alters the appearance of the vertex. In order to expose the structure of the quark-gluon vertex (or the fermion–gauge-boson vertex in general), we solve the familiar longitudinal WGTI and its less well known transverse counterparts together, where we work in the Abelian approximation to avoid dealing with ghost fields.

As a consequence of gauge invariance, the divergence of the dressed-fermion–gauge-boson vertex can be related to the dressed-fermion propagators ($q = k - p$ and $S(p) = 1/[i\gamma \cdot pA(p^2) + B(p^2)]$)

$$iq_\mu \Gamma_\mu(k, p) = S^{-1}(k) - S^{-1}(p) . \tag{6}$$

Combining the Lorentz transformation with the gauge and chiral transformations, one can derive the vector and axial-vector transverse WGTIs ($t = k + p$), respectively:

$$q_\mu \Gamma_\nu(k, p) - q_\nu \Gamma_\mu(k, p) = S^{-1}(p)\sigma_{\mu\nu} + \sigma_{\mu\nu} S^{-1}(k) + 2im\Gamma_{\mu\nu}(k, p) + t_\lambda \varepsilon_{\lambda\mu\nu\rho} \Gamma^A_\rho(k, p) + A^V_{\mu\nu}(k, p) , \tag{7}$$

$$q_\mu \Gamma^A_\nu(k, p) - q_\nu \Gamma^A_\mu(k, p) = S^{-1}(p)\sigma^5_{\mu\nu} - \sigma^5_{\mu\nu} S^{-1}(k) + t_\lambda \varepsilon_{\lambda\mu\nu\rho} \Gamma_\rho(k, p) + V^A_{\mu\nu}(k, p) , \tag{8}$$

where m is the fermion mass; $\mathrm{tr}[\gamma_5\gamma_\mu\gamma_\nu\gamma_\rho\gamma_\sigma] = -4\varepsilon_{\mu\nu\rho\sigma}$; $\sigma_{\mu\nu} = (i/2)[\gamma_\mu, \gamma_\nu]$, $\sigma^5_{\mu\nu} = \gamma_5\sigma_{\mu\nu}$; $\Gamma_{\mu\nu}(k, p)$ is an inhomogeneous tensor vertex; and $A^V_{\mu\nu}(k, p)$ and $V^A_{\mu\nu}(k, p)$ stands for contributions from high-order Green functions. The transverse WGTIs express curls of the vertices.

A first observation for those transverse WGTIs is that the identities for different vertices are coupled together. However, with the well-defined projection tensors [6]

$$T^1_{\mu\nu} = \frac{1}{2}\varepsilon_{\alpha\mu\nu\beta} t_\alpha q_\beta \mathbf{I}_\mathrm{D} , \quad T^2_{\mu\nu} = \frac{1}{2}\varepsilon_{\alpha\mu\nu\beta}\gamma_\alpha q_\beta , \tag{9}$$

one can decouple the identities and isolate the transverse WGTIs for the vector vertex as

$$q \cdot tt \cdot \Gamma(k, p) \quad = \quad T^1_{\mu\nu}\left[S^{-1}(p)\sigma^5_{\mu\nu} - \sigma^5_{\mu\nu} S^{-1}(k)\right] + t^2 q \cdot \Gamma(k, p) + T^1_{\mu\nu} V^A_{\mu\nu}(k, p) , \tag{10}$$

$$q \cdot t\gamma \cdot \Gamma(k, p) \quad = \quad T^2_{\mu\nu}\left[S^{-1}(p)\sigma^5_{\mu\nu} - \sigma^5_{\mu\nu} S^{-1}(k)\right] + \gamma \cdot tq \cdot \Gamma(k, p) + T^2_{\mu\nu} V^A_{\mu\nu}(k, p) . \tag{11}$$

Choosing a proper basis for the matrix-valued equations, one can obtain a unique solution for the vector vertex. The solution confirms the longitudinal Ball-Chiu vertex [7] as a precise piece of the full fermion–gauge-boson vertex and exposes the transverse ACM terms, e.g.,

$$\Gamma^{\mathrm{ACM5}}_\mu(k, p) = -\Delta_B(k^2, p^2)\sigma_{\mu\nu} q_\nu , \quad \Delta_B(k^2, p^2) = [B(k^2) - B(p^2)]/[k^2 - p^2] , \tag{12}$$

which is comparable with that in Ref. [8]. Apparently, the ACM strength is proportional to the magnitude of DCSB. Thus, it is expected that the ACM terms have significant consequences in observables.

4 Scattering kernel

It has been shown that, in the bare vertex approximation, the one-gluon exchange form for the BS scattering kernel fails to produce meson spectrum above 1 GeV, e.g., a_1, b_1 meson, radial excitation states of pion and rho meson [9, 10] . An exact kernel must be valid when the quark-gluon vertex is fully dressed.

The solutions of the inhomogeneous BS equations must satisfy the color-singlet vector and axial-vector WGTIs ($k_\pm = k \pm P$), respectively,

$$iP_\mu \Gamma_\mu(k, P) \quad = \quad S^{-1}(k_+) - S^{-1}(k_-) , \tag{13}$$

$$P_\mu \Gamma_{5\mu}(k, P) + 2im\Gamma_5(k, P) \quad = \quad S^{-1}(k_+)i\gamma_5 + i\gamma_5 S^{-1}(k_-) . \tag{14}$$

The vector WGTI guarantees current conservation, e.g., the pion form factor $F_\pi(Q^2 = 0) = 1$; and the axial-vector counterpart guarantees pion's twofold role: pseudo-scalar bound-state and Goldstone boson, i.e., pion must be massless in the chiral limit [11, 12]. A scattering kernel which is useful for studying meson properties must preserve such identities.

Inserting the inhomogeneous BS equations and the quark gap equation into the WGTIs, one can relate the scattering kernel to the quark-gluon vertex as

$$\int_q K^{(2)}{}_{\alpha\alpha',\beta'\beta}[S(q_-) - S(q_+)]_{\alpha'\beta'} = \int_q D_{\mu\nu}(k-q)\gamma_\mu[S(q_+)\Gamma_\nu(q_+,k_+) - S(q_-)\Gamma_\nu(q_-,k_-)]\,, \quad (15)$$

$$\int_q K^{(2)}{}_{\alpha\alpha',\beta'\beta}[S(q_+)\gamma_5 + \gamma_5 S(q_-)]_{\alpha'\beta'} = \int_q D_{\mu\nu}(k-q)\gamma_\mu[S(q_+)\Gamma_\nu(q_+,k_+)\gamma_5 - \gamma_5 S(q_-)\Gamma_\nu(q_-,k_-)]\,. \quad (16)$$

Using the relation and a proper ansatz, one can construct the scattering kernel for a specified vertex, and vice versa.

5 Epilogue

This paper explained a systematic approach to sketch BS equation: modeling the gluon propagator, investigating the structure of the quark-gluon vertex, and constructing the quark–anti-quark scattering kernel. Using this approach, the spectrum of light-flavor mesons including ground and radial excitation states can be well described. Moreover, the approach can potentially be extended to more applications, e.g., meson form factors and baryon properties in the diquark picture.

Acknowledgements

The author would like to thank C. D. Roberts for helpful discussions. The work was supported by the Office of the Director at Argonne National Laboratory through the Named Postdoctoral Fellowship Program – Maria Goeppert Mayer Fellowship.

References

[1] C.D. Roberts, A.G. Williams, Prog.Part.Nucl.Phys. **33**, 477 (1994)
[2] J.M. Cornwall, Phys. Rev. **D26**, 1453 (1982)
[3] O. Oliveira, P. Bicudo, J.Phys. **G38**, 045003 (2011)
[4] A.C. Aguilar, A.A. Natale, P.S. Rodrigues da Silva, Phys. Rev. Lett. **90**, 152001 (2003)
[5] S.x. Qin, L. Chang, Y.x. Liu, C.D. Roberts, D.J. Wilson, Phys.Rev. **C84**, 042202 (2011)
[6] S.x. Qin, L. Chang, Y.x. Liu, C.D. Roberts, S.M. Schmidt, Phys.Lett. **B722**, 384 (2013)
[7] J.S. Ball and T.-w. Chiu, Phys.Rev. **D22**, 2542 (1980)
[8] L. Chang, Y.x. Liu, C.D. Roberts, Phys.Rev.Lett. **106**, 072001 (2011)
[9] S.x. Qin, L. Chang, Y.x. Liu, C.D. Roberts, D.J. Wilson, Phys.Rev. **C85**, 035202 (2012)
[10] L. Chang, C.D. Roberts, Phys.Rev. **C85**, 052201 (2012)
[11] L. Chang, C.D. Roberts, Phys.Rev.Lett. **103**, 081601 (2009)
[12] S.x. Qin, C.D. Roberts, S.M. Schmidt, Phys. Lett. **B733**, 202 (2014)

EPJ Web of Conferences **113**, 05025 (2016)
DOI: 10.1051/epjconf/201611305025

Strong diquark correlations inside the proton

Jorge Segovia[1,a]

[1] *Instituto Universitario de Física Fundamental y Matemáticas (IUFFyM), Universidad de Salamanca, E-37008 Salamanca, Spain*

Abstract. Quantum Chromodynamics is thought to be the relativistic quantum field theory that describes the strong interaction of the Standard Model. This interaction produces mesons but it is also able to generate quark-quark (diquark) correlations inside baryons. In this work, we employ a continuum approach to QCD based on Dyson-Schwinger equations to calculate the electromagnetic form factors of the proton and analyze in a deeper way the consequences of having strong diquark correlations. Comparison with the experimental data reveals that the presence of strong diquark correlations within the proton is sufficient to understand empirical extractions of the flavour-separated form factors. The explained reduction of the ratios F_1^d/F_1^u and F_2^d/F_2^u at high Q^2 in the quark-diquark picture are responsible of the precocious scaling of the F_2^p/F_1^p observed experimentally.

1 Origin of diquarks

Quantum Chromodynamics (QCD) is the strong-interaction part of the Standard Model of Particle Physics and solving QCD presents a fundamental problem that is unique in the history of science. Never before we have been confronted by a theory whose elementary excitations are not those degrees of freedom readily accessible via experiment, that is, whose elementary excitations are confined. Moreover, QCD generates forces which are so strong that they produce about 98% of the mass of the proton and, consequently, most of the mass in the visible Universe. The underlying phenomenon which generates this mass – from nothing – is dynamical chiral symmetry breaking (DCSB). Neither confinement nor DCSB are apparent in QCD's Lagrangian and yet they play the dominant role in determining the observable characteristics of real-world QCD, *e.g.*, hadron masses and coupling.

An important consequence of DCSB is that any interaction capable of creating bound-states of a light dressed-quark and -antiquark will necessarily also generate strong colour-antitriplet correlations between any two dressed quarks contained within a nucleon. Although a rigorous proof within QCD cannot be claimed, this assertion is based upon an accumulated body of evidence, gathered in two decades of studying two- and three-body bound-state problems in hadron physics (see, for instance, Refs. [1–5]). No realistic counter examples are known; and the existence of such diquark correlations is also supported by simulations of lattice QCD [6, 7].

The properties of diquark correlations have been charted. Diquarks are confined. Additionally, owing to properties of charge-conjugation, a diquark with spin-parity J^P may be viewed as a partner to the analogous J^{-P} meson [1]. It follows that scalar, isospin-zero and pseudovector, isospin-one diquark correlations are the strongest; and whilst no pole-mass exists, the following mass-scales,

---

[a] e-mail: segonza@usal.es

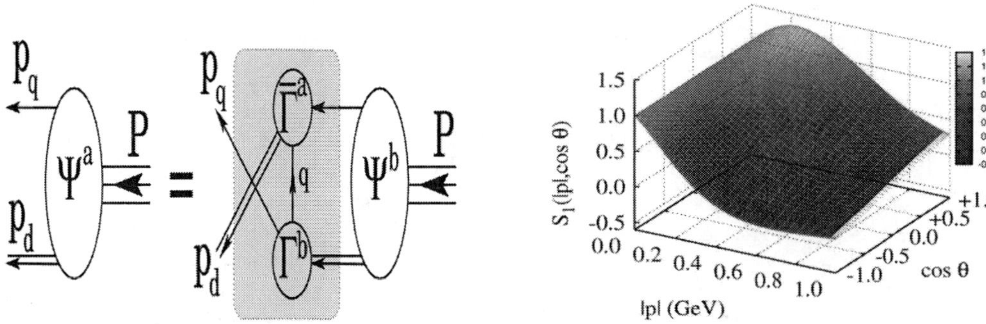

Figure 1. Left panel: Poincaré covariant Faddeev equation. Ψ is the Faddeev amplitude for a baryon of total momentum $P = p_q + p_d$, where $p_{q,d}$ are, respectively, the momenta of the quark and diquark within the bound-state. The shaded area demarcates the Faddeev equation kernel: *single line*, dressed-quark propagator; Γ, diquark correlation amplitude; and *double line*, diquark propagator. **Right panel:** Dominant piece in the nucleon's eight-component Poincaré-covariant Faddeev amplitude: $S_1(|p|, \cos\theta)$. In the nucleon rest frame, this term describes that piece of the quark-diquark relative momentum correlation which possesses zero *intrinsic* quark-diquark orbital angular momentum, *i.e.* $L = 0$ before the propagator lines are reattached to form the Faddeev wave function. Referring to Fig. 1, $p = P/3 - p_q$ and $\cos\theta = p \cdot P/\sqrt{p^2 P^2}$. The amplitude is normalised such that its U_0 Chebyshev moment is unity at $|p| = 0$.

which express the strength and range of the correlation and are each bounded below by the partnered meson's mass, may be associated with these diquarks [1, 6, 7]: $m_{[ud]_{0^+}} \approx 0.7 - 0.8\,\text{GeV}$, $m_{\{uu\}_{1^+}} \approx 0.9 - 1.1\,\text{GeV}$, with $m_{\{dd\}_{1^+}} = m_{\{ud\}_{1^+}} = m_{\{uu\}_{1^+}}$ in the isospin symmetric limit. Realistic diquark correlations are also soft. Finally, they possess an electromagnetic size that is bounded below by that of the analogous mesonic system, *viz.* [8, 9]: $r_{[ud]_{0^+}} \gtrsim r_\pi$, $r_{\{uu\}_{1^+}} \gtrsim r_\rho$, with $r_{\{uu\}_{1^+}} > r_{[ud]_{0^+}}$.

2 Diquarks inside the nucleon

The existence of tight diquark correlations considerably simplifies analyses of the three valence-quark scattering problem and hence baryon bound states. The nucleon is a compound system whose properties and interactions are primarily determined by the quark+diquark structure evident in the Poincaré covariant Faddeev equation depicted in the left panel of Fig. 1. While an explicit three-body term might affect fine details of baryon structure, the dominant effect of non-Abelian multi-gluon vertices is expressed in the formation of diquark correlations.

A nucleon (and kindred baryons) described by the left panel of Fig. 1 can be view as a Borromean bound-state, the binding within which has two contributions. One part is expressed in the formation of tight diquark correlations. That is augmented, however, by attraction generated by the quark exchange depicted in the shaded area of the left panel of Fig. 1. This exchange ensures that diquark correlations within the nucleon are fully dynamical: no quark holds a special place because each one participates in all diquarks to the fullest extent allowed by its quantum numbers. The continual rearrangement of the quarks guarantees, *inter alia*, that the nucleon's dressed-quark wave function complies with Pauli statistics.

It is important to highlight that both scalar-isoscalar and pseudovector-isotriplet diquark correlations feature within a nucleon. The relative probability of scalar versus pseudovector diquarks in a nucleon is a dynamical statement. Realistic computations predict a scalar diquark strength of approximately 60% [3–5]. As will become clear, this prediction can be tested by contemporary experiments.

The quark+diquark structure of the nucleon is elucidated in the right panel of Fig. 1, which depicts the leading component of its Faddeev amplitude: with the notation of Ref. [4], $S_1(|p|, \cos\theta)$, computed

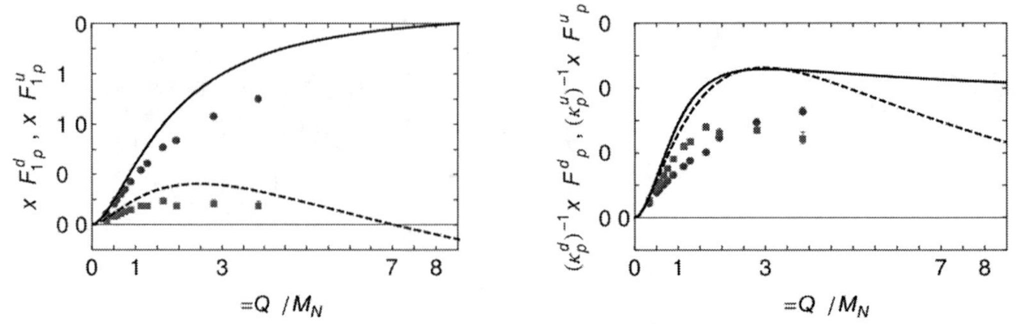

Figure 2. Left panel: Flavour separation of the proton's Dirac form factor as a function of $x = Q^2/M_N^2$. The solid-curve is the u-quark contribution, and the dashed-curve is the d-quark contribution. Experimental data taken from Ref. [10] and references therein: circles – u-quark; and squares – d-quark. **Right panel:** Same for Pauli form factor.

using the Faddeev kernel described therein. This function describes a piece of the quark+scalar-diquark relative momentum correlation. Notably, in this solution of a realistic Faddeev equation there is strong variation with respect to both arguments. Support is concentrated in the forward direction, $\cos\theta > 0$, so that alignment of p and P is favoured; and the amplitude peaks at ($|p| \simeq M_N/6, \cos\theta = 1$), whereat $p_q \approx P/2 \approx p_d$ and hence the *natural* relative momentum is zero. In the antiparallel direction, $\cos\theta < 0$, support is concentrated at $|p| = 0$, *i.e.* $p_q \approx P/3$, $p_d \approx 2P/3$.

3 Predictions of diquark clustering

A nucleon described by the Faddeev equation in Fig. 1 is constituted from dressed-quarks, any two of which are always correlated as either a scalar or pseudovector diquark. If this is a veracious description of Nature, then the presence of these correlations must be evident in many physical observables. We focus our attention on the flavour separated versions of the Dirac a Pauli form factors of the nucleon. The evaluation within our framework of the nucleon electromagnetic current is detailed in Ref. [4] and the results we describe herein are derived from that analysis.

Figure 2 displays the proton's flavour separated Dirac and Pauli form factors. The salient features of the data are: the d-quark contribution to F_1^p is far smaller than the u-quark contribution; $F_2^d/\kappa_d > F_2^u/\kappa_u$ on $x < 2$ but this ordering is reversed on $x > 2$; and in both cases the d-quark contribution falls dramatically on $x > 3$ whereas the u-quark contribution remains roughly constant. Our calculations are in semi-quantitative agreement with the empirical data.

It is natural to seek an explanation for the pattern of behaviour in Fig. 2. We have emphasised that the proton contains scalar and pseudovector diquark correlations. The dominant piece of its Faddeev wave function is $u[ud]$; namely, a u-quark in tandem with a $[ud]$ scalar correlation, which produces 62% of the proton's normalisation. If this were the sole component, then photon–d-quark interactions within the proton would receive a $1/x$ suppression on $x > 1$, because the d-quark is sequestered in a soft correlation, whereas a spectator u-quark is always available to participate in a hard interaction. At large $x = Q^2/M_N^2$, therefore, scalar diquark dominance leads one to expect $F^d \sim F^u/x$. Available data are consistent with this prediction but measurements at $x > 4$ are necessary for confirmation.

It is natural now to consider the proton ratio: $R_{21}(x) = xF_2(x)/F_1(x)$, $x = Q^2/M_N^2$, drawn in Fig. 3. The momentum dependence of $R_{21}(x)$ is principally determined by the scalar diquark component of the proton. Moreover, the rest-frame $L = 1$ terms are seen to be critical in explaining the data: the behaviour of the dashed (green) curve highlights the impact of omitting these components.

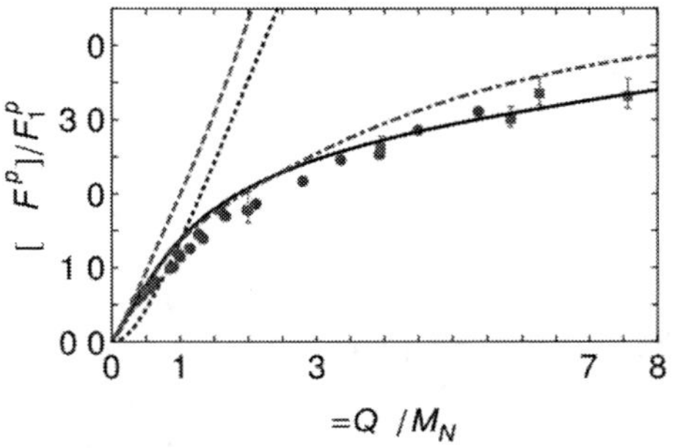

Figure 3. Proton ratio $R_{21}(x) = xF_2(x)/F_1(x)$, $x = Q^2/M_N^2$. Curves: solid (black) – full result, determined from the complete proton Faddeev wave function and current; dot-dashed (red) – momentum-dependence of the scalar-diquark contribution; dashed (green) – momentum-dependence of that component of the scalar diquark contribution to the proton's Faddeev wave function which is purely S-wave in the rest-frame; dotted (blue) – momentum-dependence of the pseudovector diquark contribution. The data are drawn from Ref. [10] and references therein.

4 Summary

We have explained that the same interaction which produces mesons is able to generate tight, dynamical colour-antitriplet quark-quark correlations inside baryons. The quark-diquark picture of a baryon should have numerous empirical consequences. Focusing on the flavour-separated proton's Dirac and Pauli form factors, we have seen that dominant scalar diquark correlations inside the proton is enough to observe that, at large Q^2, the d-quark contributions to both Dirac and Pauli proton form factors are reduced relative to the u-quark contributions. Moreover, the explained reduction of the ratios F_1^d/F_1^u and F_2^d/F_2^u at high Q^2 in the quark-diquark picture are responsible of the precocious scaling of the F_2^p/F_1^p observed experimentally.

Acknowledgements

A significant body of this work is drawn from Ref. [11]. I am grateful for insightful comments from V. Mokeev, I. C. Cloët, R. Gothe, and T.-S. H. Lee. I acknowledges financial support from a postdoctoral IUFFyM contract at *Universidad de Salamanca*, Spain.

References

[1] R.T. Cahill, C.D. Roberts, J. Praschifka, Phys. Rev. **D36**, 2804 (1987)
[2] A. Bender, C.D. Roberts, L. Von Smekal, Phys. Lett. **B380**, 7 (1996)
[3] I.C. Cloët, G. Eichmann, B. El-Bennich, T. Klahn, C.D. Roberts, Few Body Syst. **46**, 1 (2009)
[4] J. Segovia, I.C. Cloët, C.D. Roberts, S.M. Schmidt, Few Body Syst. **55**, 1185 (2014)
[5] J. Segovia, B. El-Bennich, E. Rojas, I.C. Cloët, C.D. Roberts, S.S. Xu, H.S. Zong, Phys. Rev. Lett. **115**, 171801 (2015)
[6] C. Alexandrou, P. de Forcrand, B. Lucini, Phys. Rev. Lett. **97**, 222002 (2006)
[7] R. Babich et al., Phys. Rev. **D76**, 074021 (2007)
[8] P. Maris, Few Body Syst. **35**, 117 (2004)
[9] H.L.L. Roberts et al., Phys. Rev. **C83**, 065206 (2011)
[10] G.D. Cates, C.W. de Jager, S. Riordan, B. Wojtsekhowski, Phys. Rev. Lett. **106**, 252003 (2011)
[11] J. Segovia, C.D. Roberts, S.M. Schmidt, Phys. Lett. **B750**, 100 (2015)

EPJ Web of Conferences **113**, 06001 (2016)
DOI: 10.1051/epjconf/201611306001

How well do we understand $^{7}\mathrm{Be} + p \to {}^{8}\mathrm{B} + \gamma$? An Effective Field Theory perspective

Xilin Zhang[1,2,a], Kenneth M. Nollett[3,4,2,b], and D. R. Phillips[2,c]

[1] *Physics Department, University of Washington, Seattle, WA 98195, USA*
[2] *Institute of Nuclear and Particle Physics and Department of Physics and Astronomy, Ohio University, Athens, OH 45701, USA*
[3] *Department of Physics, San Diego State University, 5500 Campanile Drive, San Diego, California 92182-1233, USA*
[4] *Department of Physics and Astronomy, University of South Carolina, 712 Main Street, Columbia, South Carolina 29208, USA*

Abstract. We have studied the $^{7}\mathrm{Be}(p,\gamma)^{8}\mathrm{B}$ reaction in the Halo effective field theory (EFT) framework. The leading order (LO) results were published in Ref. [1] after the isospin mirror process, $^{7}\mathrm{Li}(n,\gamma)^{8}\mathrm{Li}$, was addressed in Ref. [2]. In both calculations, one key step was using the final shallow bound state asymptotic normalization coefficients (ANCs) computed by *ab initio* methods to fix the EFT couplings. Recently we have developed the next-to-LO (NLO) formalism [3], which could reproduce other model results by no worse than 1% when the ^{7}Be-p energy was between 0 and 0.5 MeV. In our recent report [4], a different approach from that in Ref. [1] was used. We applied Bayesian analysis to constrain all the NLO-EFT parameters based on measured S-factors, and found tight constraints on the S-factor at solar energies. Our $S\,(E = 0\,\mathrm{MeV}) = 21.3 \pm 0.7$ eV b. The uncertainty is half of that previously recommended. In this proceeding, we provide extra details of the Bayesian analysis, including the computed EFT parameters' probability distribution functions (PDFs) and how the choice of input data impacts final results.

1 Introduction

The $^{7}\mathrm{Be}(p,\gamma)^{8}\mathrm{B}$ reaction cross section around zero energy directly affects the solar neutrino flux [5]. (An expanded discussion on the reaction's importance for astrophysics can be found in Ref [4].) The previously recommended $S(0) = 20.8 \pm 0.7 \pm 1.4$ eV b [6]. The first error comes from the capture experiment uncertainties. Due to the Coulomb repulsion barrier, the cross section is exponentially suppressed, so that the lowest energy of available data is above 0.1 MeV. Extrapolating the data down to zero energy based on various models gives rise to the second quoted $S(0)$ error. Therefore eliminating the model dependence is critical for improvement.

Fortunately different scales in this reaction are well separated, and thus the problem is "simple" enough to be treated in Halo EFT. The ^{7}Be and proton are considered as two fundamental particles

[a] e-mail: xilinz@uw.edu
[b] e-mail: nollett@mailbox.sc.edu
[c] e-mail: phillid1@ohio.edu

and ^8B as a shallow bound state of the two; the high momentum modes resolving the short distance (< 3 fm) physics are integrated out, and their effects are subsumed to contact terms in the lagrangian. Although the idea is close to the previous potential models', the EFT puts the scale separation on rigorous grounds and should increase in accuracy as terms of higher order in momentum are retained. The number of parameters increases while going to higher order, but it is kept minimal, and the error of truncation can be estimated. Moreover the important effect of the Coulomb interaction between ^7Be and proton can be treated exactly. The reaction is entirely an electric-dipole ($E1$) induced transition from the s- and d-wave initial states to p-wave configurations in the final ^8B. The resulting S-factor [3] is proportional to

$$S(E) \propto \sum_s C_s^2 \left[\left| \mathcal{S}_{\rm EC}(E; a_s, r_s) + \overline{L}_s \mathcal{S}_{\rm SD}(E; a_s, r_s) + \epsilon_s \mathcal{S}_{\rm CX}(E; a_s, r_s) \right|^2 + |\mathcal{D}_{\rm EC}(E)|^2 \right].$$

The proportionality factor is in Ref. [1]. The total spin s of ^{7}Be-p can be 1 or 2, defining two different reaction channels. The LO term $\mathcal{S}_{\rm EC}$ is the amplitude of the external direct capture to the ^{7}Be-p component in ^8B; the same capture but to the lowest excited core $^7Be^*$ plus proton component is a NLO contribution denoted as $\epsilon_s \mathcal{S}_{\rm CX}$; thus the dimensionless factor $|\epsilon_s|$ is small. $\overline{L}_s \mathcal{S}_{\rm SD}$ is another NLO term coming from the capture in short distance (< 3 fm); the coefficient $\overline{L}_s \sim 3$ fm. The three amplitudes in each channel depend on the s-wave scattering length a_s and effective range r_s due to the initial state interaction effect. $\mathcal{D}_{\rm EC}$ is the $E1$ transition from the d-wave initial state to the final bound state, which is not affected by the initial state interaction at LO. The overall factor in each channel, C_s, is the ANC of the two-particle wave function (with total spin s) in ^8B.

There are nine EFT parameters, $C_{1,2}^2$, $a_{1,2}$, $r_{1,2}$, $\overline{L}_{1,2}$, and ϵ_1 ($\epsilon_2 = 0$ based on angular momentum conservation). Their so-called posterior PDF can be extracted from direct capture data by using Bayesian analysis [7]. In our analysis, another five parameters $\xi_{1,2,3,4,5}$ were introduced to float the normalizations of the five capture data sets used. (In total 42 data points with energies up to 0.5 MeV were used; Fig. 2 provides some information about them.) A Markov chain Monte Carlo (MCMC) algorithm with Metropolis sampling [7] was used to compute the 14-dim posterior PDF [4]. Note prior knowledge about the parameters was incorporated by assigning a prior PDF to each of them. For $a_{1,2}$, it was a Gaussian distribution centered at the measured value [8] with the width equal to the experimental uncertainty; the other EFT parameters were given flat prior PDFs over finite value windows [4]. The ξ_i priors were also Gaussian distributions but centered at 0 (the data set's normalization factor is $1 + \xi_i$); the widths were the corresponding experimental common mode errors (CMEs), 2.7%, 2.3%, 11.25%, 5%, and 2.2% in the order of being mentioned in Fig. 2. (Ref. [9] contains mostly relative cross sections, so its CME is handled differently.)

2 EFT parameters

Fig. 1 shows a few interesting 1-dim parameter posterior PDFs. (Each PDF is the 14-dim PDF with the other 13 parameters marginalized, and was computed via drawing its histogram based on the MCMC samples [4].) We see in the first panel that the data prefer a large C_2^2 and a small C_1^2, and the sum is tightly constrained ($C_1^2 + C_2^2 = 0.564(23)\,{\rm fm}^{-1}$ [4]). The second panel shows the ^{7}Be-p s-wave effective ranges. Although we assigned positive flat priors for them, we also required that no s-wave resonance exists below 0.6 MeV [4]. The computed PDF of r_1 disfavors large r_1 value, while that of r_2's is almost flat. (The scattering lengths' PDFs are close to their priors.) In the third panel, each $\overline{L}_s$ distribution is broad but does have a significant peak around a few fm, overlapping with the $\overline{L}_{1,2} \sim 3$ fm region suggested by the EFT power counting. Thus we believe that we have detected the direct capture at short distance from the data, although the large distance capture dominates the

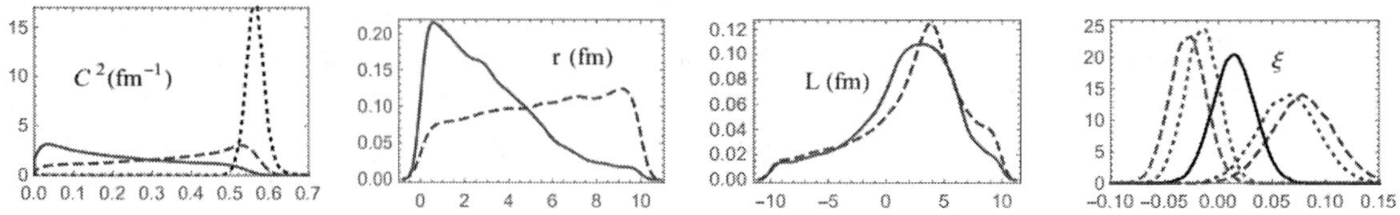

Figure 1. In the first panel, the red solid, blue dashed, and black dashed lines are the posterior PDFs of C_1^2, C_2^2, and $C_1^2 + C_2^2$; in the second (third) panel, the red solid and blue dashed lines are the PDFs of r_s ($\overline{L}_s$) in $s = 1$ and 2 channels, respectively; in the last panel, the red dashed (for the Junghans BE1 data set), red dotted (the Junghans BE3), blue dashed (Filippone), blue dotted (Hammache 1998), and black solid (Baby) lines are the PDFs of the corresponding data normalization variable ξ_i. The data sets are shown in Fig. 2.

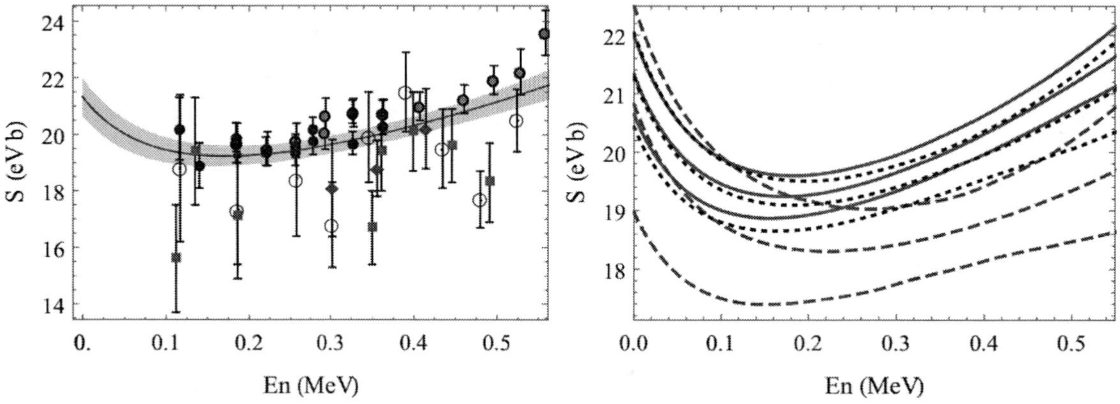

Figure 2. The left panel shows the S-factor median values (solid blue curve) and the 1-σ intervals (shaded in green) at different energies. The data used in our analysis and a few others above 0.5 MeV are shown, including Junghans *et.al.*, BE1 and BE3 [10] (filled black circle and filled grey circle), Filippone *et.al.*, [11] (open circle), Baby *et.al.*, [12, 13] (filled purple diamond), and Hammache *et.al.*, [9, 14] (filled red box). In the right panel, we show three different sets of the $S(E)$ median values and 1-σ bounds, as computed by using different combinations of the data sets. The blue dashed lines use the combination of Filippone, Hammache and Baby sets, the black dashed lines add the Junghans BE3 set, and the red solid lines use all the sets.

reaction. The last panel shows the PDFs of ξ_i. The corresponding data sets are plotted in Fig. 2, which shows minor tension between the recent Junghans data sets and the rest. This is also reflected in the ξ_i PDFs in the last panel: reducing the Junghans BE1 and BE3 normalizations is favored, the Filippone and Hammache normalizations need to be increased a bit, and Baby's is not modified significantly. However all floating norms (i.e. $1 + \xi_i$), while $\neq 1$, are entirely consistent with the quoted CMEs. It is worth mentioning that we also did χ^2 minimization based on the combined data sets, and found $\chi^2_{\rm min}/\text{dof} = 1.08$ (the corresponding $S(E)$ curve is in the 1-σ interval shown in Fig. 2). However, there are several local minima having similar values.

3 S factor

The left panel in Fig. 2 shows the computed 1-σ lower and upper bounds and the median value for the S factor. The three quantities were inferred from the S-factor distributions at individual energies, which were computed via making the histograms of S based on the MCMC samples. The data are plotted for comparison. At the shown energies, the S factor uncertainties $\approx$ 3.5% or less, and are correlated. A numerical table can be found in Ref. [4]. It is interesting to ask how the choice of data sets impacts our results; this is illustrated in the right panel. Without having the BE1 and BE3 sets in the analysis, the S factor uncertainty $\approx$ 9% at zero energy; adding either BE1 or BE3 narrows the error bands significantly within the bounds provided by other data. Note that all the three 1-σ intervals overlap with each other. We also see that combining BE1 with the other four sets does not change the constraints below 0.1 MeV much; at higher energy the BE1's impact shows up by constraining $S(E)$ further and systematically favoring larger $S(E)$.

We also repeated our analysis using different priors for the scattering lengths and introducing a N^2LO term into the capture amplitude, and we found negligible differences. We also found that the current 0.7% uncertainty of ^{8}B's proton separation energy could change $S(0)$ by about 0.75%.

4 Summary

In summary, the $^7\mathrm{Be}(p,\gamma)^8\mathrm{B}$ reaction has been studied in Halo EFT up to NLO. We applied Bayesian analysis to constrain the EFT parameters. The analysis gave a stringent constraint on the total ANC squared, and showed that positive short-distance-contribution parameters $\overline{L}_{1,2}$ were preferred. Minor tensions among the modern capture data sets were seen. We found that $S(E)$ is constrained to $\approx$ 3.5% between 0 and 0.5 MeV. The $S(0)$ results were robust against exclusion of either Junghans data set and against addition of an N^2LO term to the reaction amplitude.

Acknowledgements

X.Z. and D.R.P. acknowledge support from the US Department of Energy under grant DE-FG02-93ER-40756. X.Z. also acknowledges support from the US Department of Energy under grant DE-FG02-97ER-41014. K.M.N. acknowledges support from the Institute of Nuclear and Particle Physics at Ohio University, and from U.S. Department of Energy Awards No. DE-SC 0010 300 and No. DE-FG02-09ER41621 at the University of South Carolina.

References

[1] X. Zhang, K.M. Nollett, D. Phillips, Phys. Rev. C **89**, 051602 (2014), 1401.4482
[2] X. Zhang, K.M. Nollett, D. Phillips, Phys. Rev. C **89**, 024613 (2014), 1311.6822
[3] X. Zhang, K.M. Nollett, D. Phillips (2016), 16xx.xxxxx
[4] X. Zhang, K.M. Nollett, D.R. Phillips, Phys. Lett. **B751**, 535 (2015), 1507.07239
[5] W.C. Haxton et al., Ann. Rev. Astron. Astrophys. **51**, 21 (2013)
[6] E. Adelberger et al., Rev.Mod.Phys. **83**, 195 (2011), 1004.2318
[7] D. Sivia, *Data Analysis: A Bayesian Tutorial* (Oxford University Press, New York, 1996)
[8] C. Angulo, M. Azzouz, P. Descouvemont, G. Tabacaru et al., Nucl. Phys. **A 716**, 211 (2003)
[9] F. Hammache et al., Phys.Rev.Lett. **86**, 3985 (2001), nucl-ex/0101014
[10] A. Junghans et al., Phys.Rev. **C81**, 012801 (2010)
[11] B. Filippone, A. Elwyn, C. Davids, D. Koetke, Phys.Rev. **C28**, 2222 (1983)
[12] L. Baby et al. (ISOLDE), Phys. Rev. Lett. **90**, 022501 (2003), nucl-ex/0208005
[13] L. Baby et al. (ISOLDE), Phys.Rev. **C67**, 065805 (2003), nucl-ex/0212011
[14] F. Hammache et al., Phys.Rev.Lett. **80**, 928 (1998), nucl-ex/9712003

EPJ Web of Conferences **113**, 06002 (2016)
DOI: 10.1051/epjconf/201611306002

Towards an ab initio description of light-nuclei radiative captures

Jérémy Dohet-Eraly[1,a], Petr Navrátil[1], Sofia Quaglioni[2], Wataru Horiuchi[3], and Guillaume Hupin[2]

[1] *TRIUMF, 4004 Wesbrook Mall, Vancouver BC V6T 2A3, Canada*
[2] *Lawrence Livermore National Laboratory, P.O. Box 808, L-414, Livermore, California 94551, USA*
[3] *Department of Physics, Hokkaido University, Sapporo 060-0810, Japan*

Abstract. The ^{3}He$(\alpha,\gamma)^7$Be and ^{3}H$(\alpha,\gamma)^7$Li astrophysical S factors are evaluated at low collision energies (less than 2.5 MeV in the centre-of-mass frame) within the no-core shell model with continuum using a renormalized chiral nucleon-nucleon interaction.

1 Introduction

Radiative-capture processes play an important role in stellar nucleosynthesis. Of particular interest is the ^{3}He$(\alpha,\gamma)^7$Be cross section or astrophysical S factor at collision energies about 20 keV in the centre-of-mass (c.m.) frame, which is essential to evaluate the fraction of pp-chain terminations resulting in ^{7}Be neutrinos [1, 2]. The study of the ^{3}He$(\alpha,\gamma)^7$Be reaction at low energies, complemented by the study of its mirror reaction, the ^{3}H$(\alpha,\gamma)^7$Li radiative capture, is also needed to evaluate the primordial ^{7}Li abundance in the universe [3]. In spite of the interest in these capture reactions, measurements at astrophysical energies have not been possible because at such low energies, the capture cross sections are very small due to the Coulomb repulsion between the colliding nuclei. Theoretical models (or extrapolations) are thus needed to provide the ^{3}He$(\alpha,\gamma)^7$Be and ^{3}H$(\alpha,\gamma)^7$Li capture cross sections at solar energies.

The ^{3}He$(\alpha,\gamma)^7$Be astrophysical S factor has been measured for collision energies between ~0.1 and 3 MeV in the last decade [4–10] with significantly increased accuracy with respect to the previous experiments [1]. These recent experimental data provide a useful ground test to check the accuracy of theoretical models, which are then used to predict the astrophysical S factor at lower energies. In contrast, the ^{3}H$(\alpha,\gamma)^7$Li astrophysical S factor is known less accurately and has been only measured between ~50 keV and 1.2 MeV [11, 12].

Many theoretical models have been developed to describe these radiative capture processes: from simple external-capture models [13] to much more complicated microscopic approaches [3, 14–17]. However, no *ab initio* approach, i.e. a microscopic approach based on a realistic inter-nucleon interaction able to describe accurately bound states and continuum states in a consistent framework, has been developed, yet. In this proceeding, we present an application of the no-core shell model with continuum (NCSMC) approach [18], using a renormalized chiral nucleon-nucleon (NN) interaction, to the description of the ^{3}He$(\alpha,\gamma)^7$Be and ^{3}H$(\alpha,\gamma)^7$Li radiative-capture processes. This work is a first step towards a complete *ab initio* description of these radiative-capture reactions.

[a] e-mail: jdoheter@triumf.ca

2 No-core shell model with continuum approach

The ^{3}He$(\alpha,\gamma)^7$Be and ^{3}H$(\alpha,\gamma)^7$Li astrophysical S factors are calculated from the matrix elements of the electromagnetic transition multipole operators between initial scattering states and final bound states [19]. The initial and final states are obtained with the NCSMC approach [18], as solutions of a microscopic Schrödinger equation based on a realistic nucleon-nucleon interaction. In this approach, each partial wave function $|\psi^{J^\pi T}\rangle$ of total angular momentum J, parity π, and isospin T is described by a combination of $\alpha+{}^3$He or $\alpha+{}^3$H cluster states and eigenstates of the compound ^{7}Be or ^{7}Li systems obtained with the no-core shell model (NCSM) [20]. The latter eigenstates, denoted by $|^7\mathrm{Be}\lambda J^\pi T\rangle$ or $|^7\mathrm{Li}\lambda J^\pi T\rangle$ with λ the energy label, are fully antisymmetric and translational-invariant. They are built from linear combinations of products of harmonic oscillator (HO) functions with frequency Ω and up to $N_{\max}$ HO quanta above the lowest configuration. The cluster states, denoted by $|\Phi_{\nu r}^{J^\pi T}\rangle$, are also translational-invariant. They are built from the NCSM states of ^{4}He and ^{3}He or ^{3}H by means of the resonating group method (RGM) [21]. For the ^{3}He$(\alpha,\gamma)^7$Be study, they are given by

$$|\Phi_{\nu r}^{J^\pi T}\rangle = \left[\left[|^4\mathrm{He}\lambda_\alpha J_\alpha^{\pi_\alpha}T_\alpha\rangle|^3\mathrm{He}\lambda_\mathrm{h} J_\mathrm{h}^{\pi_\mathrm{h}}T_\mathrm{h}\rangle\right]^{(sT)} Y_\ell(\hat{r}_{\alpha\mathrm{h}})\right]^{(J^\pi T)} \frac{\delta(r-r_{\alpha\mathrm{h}})}{r r_{\alpha\mathrm{h}}}, \tag{1}$$

where $|^4\mathrm{He}\lambda_\alpha J_\alpha^{\pi_\alpha}T_\alpha\rangle$ is a NCSM state of ^{4}He with $\lambda_\alpha J_\alpha \pi_\alpha T_\alpha$ standing for the energy label, total angular momentum, parity, and isospin associated with ^{4}He, $|^3\mathrm{He}\lambda_\mathrm{h} J_\mathrm{h}^{\pi_\mathrm{h}}T_\mathrm{h}\rangle$ is a NCSM state of ^{3}He with similar definitions of the quantum numbers, s is the channel spin, $\boldsymbol{r}_{\alpha\mathrm{h}}$ is the relative coordinate between the centres of mass of ^{4}He and ^{3}He, ℓ is the relative orbital angular momentum between the clusters, and ν is a collective index for all relevant quantum numbers. An operator $\hat{\mathcal{A}}_{\alpha\mathrm{h}}$ is applied to enforce the antisymmetrization between the nucleons belonging to ^{4}He and those belonging to ^{3}He. As an example, for the ^{3}He$(\alpha,\gamma)^7$Be study the NCSMC partial wave can be written as

$$|\psi^{J^\pi T}\rangle = \sum_\lambda c_\lambda^{J^\pi T}|^7\mathrm{Be}\lambda J^\pi T\rangle + \sum_\nu \int dr r^2 \frac{\gamma_\nu^{J^\pi T}(r)}{r} \hat{\mathcal{A}}_{\alpha\mathrm{h}}|\Phi_{\nu r}^{J^\pi T}\rangle. \tag{2}$$

The $c_\lambda^{J^\pi T}$ coefficients and the $\gamma_\nu^{J^\pi T}$ functions are determined by inserting the ansatz (2) in a variational form of the seven-nucleon Schrödinger equation. The proper asymptotic behaviour of a bound state or a scattering state is imposed by using the microscopic R-matrix method (MRM) [22, 23]. From the MRM, the asymptotic normalization coefficients for bound states and the collision matrix for scattering states (or, equivalently, the phase shift for one-open-channel studies) are determined.

3 ^{3}He$(\alpha,\gamma)^7$Be and ^{3}H$(\alpha,\gamma)^7$Li astrophysical S factors

A chiral N^{3}LO NN interaction [24] softened via the similarity-renormalization-group (SRG) method [25–27], which reduces the influence of momenta higher than $\hbar\Lambda$, is considered. The SRG transformation induces three-body forces even if the initial interaction is limited to two-body terms. For computational reasons, in this calculation, we restrict ourselves to the two-body part of the renormalized interaction. We adjust the SRG parameter ($\Lambda = 2.15$ fm^{-1} is used) to approximately reproduce the separation energies of the ^{7}Be and ^{7}Li bound states for the considered model space, which is essential to get the correct tail of the wave function and hence the astrophysical S factors. This adjustement is required because three-nucleon forces are neglected and because the largest accessible model space is still too small to get converged results. Partial waves with $J^\pi = \{1/2^+, 3/2^+, 5/2^+, 1/2^-, 3/2^-\}$ and isospin $T = 1/2$ are considered. The HO frequency is set to $\Omega = 20$ MeV$/\hbar$ for both NCSM and NCSM/RGM states. Nine ^{7}Be or ^{7}Li NCSM states are included in expansion (2): one in partial wave

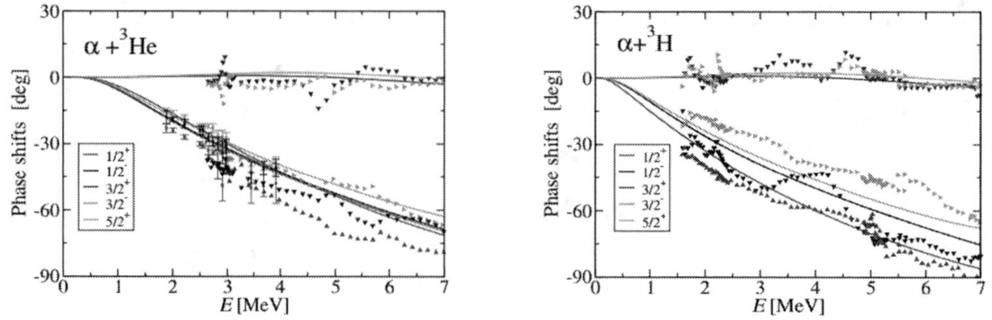

Figure 1. The $\alpha + {}^3$He and $\alpha + {}^3$H elastic phase shifts, obtained from the NCSMC and from experiments [28, 29]. See the text for model details.

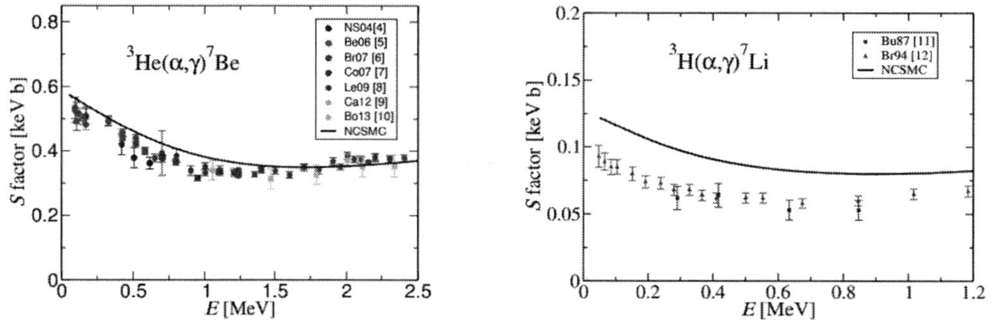

Figure 2. The ^{3}He$(\alpha,\gamma)^7$Be and ^{3}H$(\alpha,\gamma)^7$Li astrophysical S factors obtained from the NCSMC and from experiments [4–12]. See the text for model details.

$J^\pi = 5/2^+$ and two in the other partial waves. Positive-parity states are obtained with $N_{\rm max} = 11$ and negative-parity states with $N_{\rm max} = 10$. The NCSM/RGM states are obtained by coupling the NCSM ground state of ^{4}He, characterized by $(J^\pi T) = (0^+0)$, with the $(1/2^+1/2)$ NCSM ground state of ^{3}He or ^{3}H. Both are obtained with $N_{\rm max} = 12$.

The computed $\alpha + {}^3$He and $\alpha + {}^3$H elastic phase shifts for relative collision energies (E) up to 7 MeV are compared with experiment [28, 29] in Fig. 1. The agreement is not perfect, in particular for the $1/2^+$ wave. However, the accuracy of the experimental data is quite poor, making a quantitative comparison difficult. The ^{3}He$(\alpha,\gamma)^7$Be astrophysical S factors for collision energies up to 2.5 MeV and the ^{3}H$(\alpha,\gamma)^7$Li astrophysical S factors for collision energies up to 1.2 MeV are compared, in Fig. 2, with the most recent and accurate experimental data [4–12]. For these ranges of energies, the $E1$ transitions are dominant and thus, we neglect the other transitions. Qualitatively, the experimental ^{3}He$(\alpha,\gamma)^7$Be astrophysical S factor is rather well reproduced by the NCSMC approach but the ^{3}H$(\alpha,\gamma)^7$Li astrophysical S factor is overestimated. A similar mismatch was obtained by Neff with a microscopic appoach based on a NN realistic interaction [17]. A quantitative comparison requires the inclusion of three-body forces. By including them, we expect a reproduction of the experimental ^{7}Be and ^{7}Li separation energies without fine-tuning the SRG parameter Λ, involved in the renormalization of the interaction, and a smaller Λ-dependence of the results. Then, a smaller Λ value could be considered, which would lead to converged results with a smaller model space. Overall, the predictive power of the approach would be improved.

Acknowledgements

TRIUMF receives funding via a contribution through the National Research Council Canada. This work was supported in part by NSERC under Grant No. 401945-2011, by LLNL under Contract DE-AC52-07NA27344, by the U.S. Department of Energy, Office of Science, Office of Nuclear Physics, under Work Proposal Number SCW1158, and by JSPS KAKENHI Grant Numbers 25800121 and 15K05072. W.H. acknowledges Excellent Young Researcher Overseas Visit Program of JSPS that allowed him to visit LLNL (2009-2010). Computing support came from the LLNL institutional Computing Grand Challenge Program and from an INCITE Award on the Titan supercomputer of the Oak Ridge Leadership Computing Facility (OLCF) at ORNL.

References

[1] E.G. Adelberger et al., Rev. Mod. Phys. **70**, 1265 (1998)
[2] E.G. Adelberger et al., Rev. Mod. Phys. **83**, 195 (2011)
[3] K.M. Nollett, Phys. Rev. C **63**, 054002 (2001)
[4] B.S. Nara Singh, M. Hass, Y. Nir-El, G. Haquin, Phys. Rev. Lett. **93**, 262503 (2004)
[5] D. Bemmerer et al. (LUNA Collaboration), Phys. Rev. Lett. **97**, 122502 (2006)
[6] T.A.D. Brown et al., Phys. Rev. C **76**, 055801 (2007)
[7] F. Confortola et al., Phys. Rev. C **75**, 065803 (2007)
[8] A. Di Leva et al., Phys. Rev. Lett. **102**, 232502 (2009)
[9] M. Carmona-Gallardo et al., Phys. Rev. C **86**, 032801 (2012)
[10] C. Bordeanu et al., Nucl. Phys. A **908**, 1 (2013)
[11] S. Burzyński, K. Czerski, A. Marcinkowski, P. Zupranski, Nucl. Phys. A **473**, 179 (1987)
[12] C.R. Brune, R.W. Kavanagh, C. Rolfs, Phys. Rev. C **50**, 2205 (1994)
[13] T.A. Tombrello, P.D. Parker, Phys. Rev. **131**, 2582 (1963)
[14] T. Kajino, Nucl. Phys. A **460**, 559 (1986)
[15] T. Mertelmeier, H.M. Hofmann, Nucl. Phys. A **459**, 387 (1986)
[16] A. Csótó, K. Langanke, Few-Body Syst. **29**, 121 (2000)
[17] T. Neff, Phys. Rev. Lett. **106**, 042502 (2011)
[18] S. Baroni, P. Navrátil, S. Quaglioni, Phys. Rev. Lett. **110**, 022505 (2013); Phys. Rev. C **87**, 034326 (2013)
[19] D. Baye, P. Descouvemont, Nucl. Phys. A **407**, 77 (1983)
[20] P. Navrátil, J.P. Vary, B.R. Barrett, Phys. Rev. Lett. **84**, 5728 (2000); Phys. Rev. C **62**, 054311 (2000)
[21] S. Quaglioni, P. Navrátil, Phys. Rev. Lett. **101**, 092501 (2008); Phys. Rev. C **79**, 044606 (2009)
[22] D. Baye, P.-H. Heenen, M. Libert-Heinemann, Nucl. Phys. A **291**, 230 (1977)
[23] P. Descouvemont, D. Baye, Rep. Prog. Phys. **73**, 036301 (2010)
[24] D.R. Entem, R. Machleidt, Phys. Rev. C **68**, 041001 (2003)
[25] F. Wegner, Annalen der Physik **506**, 77 (1994)
[26] S.K. Bogner, R.J. Furnstahl, R.J. Perry, Phys. Rev. C **75**, 061001 (2007)
[27] E.D. Jurgenson, P. Navrátil, R.J. Furnstahl, Phys. Rev. Lett. **103** 082501 (2009); Phys. Rev. C **83**, 034301 (2011)
[28] R.J. Spiger and T.A. Tombrello. *Phys. Rev.* **163** (1967) 964.
[29] W.R. Boykin, S.D. Baker, and D.M. Hardy. *Nucl. Phys. A* **195** (1972) 241.

EPJ Web of Conferences **113**, 06003 (2016)
DOI: 10.1051/epjconf/201611306003

Deuteron-like correlation of valence nucleons for the $T = 0$ channel in ^{18}F

Hiroshi Masui[1,a] and Masaaki Kimura[2,b]

[1] *Information Processing Center, Kitami Institute of Technology, 165 Koen-Cho, Kitami, Japan*
[2] *Department of Physics, Faculty of Science, Hokkaido University, N10-W8, Kita-Ku, Sapporo, Japan*

Abstract. We study the deuteron-like correlation of the valence proton and neutron with the $T = 0$ channel in ^{18}F using an efficient calculational formalism; the cluster-orbital shell model approach. In this study, we show a particular configuration $[j_1 \otimes j_2]^J_T = [d_{5/2} \otimes d_{3/2}]^1_0$ is the key ingredient for the $T = 0$ pair formation. Also, the continuum contributions plays an important role of the spatial correlation of the proton and neutron.

1 Introduction

Deuteron is the lightest bound system. In the case for the proton and neutron in a nucleus, the spin-orbit interaction from the core nucleus separates the single-particle orbits of $j_\gtrless$, and the coupling scheme is changed from the free deuteron. It has been pointed out that the $[\,j_1 \otimes j_2]^J_T = [\,j_> \otimes j_<]^1_0$ configuration in the same major shell plays an important role for the $T = 0$ channel correlation [1–3]. In our study, sharing the basic aim with the previous works, we investigate how the coupling of the spin-orbit partners, $[\,j_> \otimes j_<]^1_0$ affects to the $T = 0$ channel correlation in ^{18}F. We also examine how the valence proton and neutron in ^{18}F are spatially localized with the presence of the $[\,j_> \otimes j_<]^1_0$ channel coupling and continuum states with higher partial-wave.

For the purpose of precise investigation to the valence nucleons including unbound states, an appropriate theoretical approach, which is capable to treat the unbound states within the same footing as the bound states, is needed. Because, for the nucleons above the ^{16}O core, only two orbits, $0d_{5/2}$ and $1s_{1/2}$ are bound, and all other states are unbound states. Therefore, we employ the cluster-orbital shell model (COSM) approach [4]. By using the ^{16}O+N potential which reproduces the bound and unbound states of ^{17}O and ^{17}F [5], we show how the $[j_> \otimes j_<]^1_0$ configuration and the continuum states enhance the correlation energy of the $T = 0$ channel and also discuss the deuteron-like spatial localization arisen due to the continuum contribution.

[a] e-mail: hgmasui@mail.kitami-it.ac.jp
[b] e-mail: masaaki@nucl.sci.hokudai.ac.jp

2 Model and formalism

We employ the cluster-orbital shell model (COSM) approach [4] to study ^{18}F in an ^{16}O+p+n three-body model. The Hamiltonian of COSM is formulated as follows:

$$\hat{H} = \sum_{i=1}^{2} \left(\hat{t}_i + \hat{v}_i^{\rm CN}\right) + \left(\hat{T}_{12} + \hat{V}_{12}\right), \tag{1}$$

where $\hat{T}_{12} = \boldsymbol{p}_1 \cdot \boldsymbol{p}_2 / M_C$ is the recoil term coming from the subtraction of the center of mass motion due to the finite mass M_C of the core nucleus. The basis set of COSM is constructed in the jj-coupling scheme using basis functions ϕ_α defined in the coordinates of the core+N subsystems as

$$\Phi_m \equiv \mathcal{A}\left\{[\phi_{\alpha_1} \otimes \phi_{\alpha_2}]^{JM}_{TM_T}\right\}_m = \mathcal{A}\left\{u_1(r_1)\,u_2(r_2) \cdot [j_1 \otimes j_2]_{JM} \cdot [\chi_1^\tau \otimes \chi_2^\tau]_{TM_T}\right\}_m, \tag{2}$$

where α_i denotes a set of the angular momentum and the isospin of the ith particle. The eigenstates of the Hamiltonian, $\hat{H}\Psi_k = E_k\Psi_k$ are described by the expansion using Φ_m as, $\Psi_k = \sum_m C_m^{(k)}\Phi_m$. For the radial part of the basis function, we apply the Gaussian expansion method [6],

$$\phi_{\alpha_i}(\boldsymbol{r}_i) = N_i r_i^{\ell_i} \exp(-\frac{1}{2}a_i r_i^2)\,|j_i\, m_i\rangle|\chi_i^\tau\rangle, \tag{3}$$

where a_i is determined in a variational procedure, and N_i is the normalization.

For the ^{16}O+N potential, we use the same interaction applied in Refs. [5, 7]. This potential reproduces the energy of lowest three states; $5/2^+$, $1/2^+$ and $3/2^+$ of ^{17}O and ^{17}F. For the two-body interaction part $\hat{V}_{12}$, we use an effective nucleon-nucleon interaction, *i.e.* the Minnesota potential [8]. We introduce an isospin dependence to $\hat{V}_{12}$ by multiplying a factor as $0.85\,(T = 0)$ and $1.50\,(T = 1)$ in order to adjust to the experimental binding and excited energies of ^{18}F.

3 Results and discussions

3.1 Physical quantities of ^{18}F

One of the way to examine the property of the ground (1^+) and first excited (0^+) states is to calculate the $M1$ transition strength $B(M1; 0^+ \to 1^+)$. Our calculation gives the large $B(M1)$-value as $18.40\,\mu_N^2$, which almost corresponds to the experimental one, $19.71\,\mu_N^2$ [9]. Other theoretical approaches give 15.18 [10] and $18.15\,(\mu_N^2)$ [3]. Here, the essential difference between Ref. [10] and [3] is the inclusion of the $[j_> \otimes j_<]$ component in the model space, and the former one does not include such the component. We consider this component is the key ingredient for the $T = 0$ channel deuteron-like correlation as discussed later and refer to the component as "cross-term" contribution. The magnetic moment of the ground state is not measured in experiments. Our calculation shows the large magnetic moment as $\mu = 0.810\,\mu_N$, and the result is consistent with other theoretical calculations, 0.834 [3] and $0.82\,(\mu_N)$ [11]. Since the magnetic moment is sensitive to the spin-coupling scheme of the wave function, we decompose to the orbital angular momentum and spin parts; μ^L and μ^S as follows:

$$\mu = \langle \sum_i g_l^{(i)} \boldsymbol{l}_i \rangle + \langle \sum_i g_s^{(i)} \boldsymbol{s}_i \rangle \equiv \mu^L + \mu^S\,. \tag{4}$$

If the ^{18}F ground state has a $S = 1$ channel dominance, μ^S becomes large. Our calculation follows such the situation, *i.e.* $\mu^L = 0.091$ and $\mu^S = 0.719\;(\mu_N)$. Hence, in the 1^+ state of ^{18}F, the orbital angular momentum part is small, and the $S = 1$ channel is dominant.

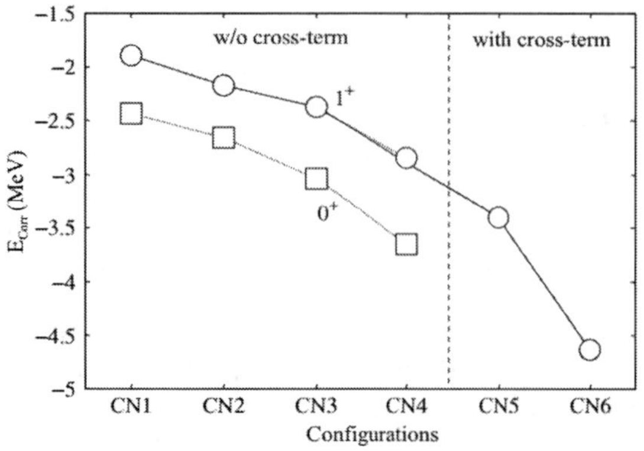

Figure 1. Configuration dependence of the correlation energies for the 1^+ $(T = 0)$ and $0^+(T = 1)$ state of ^{18}F. Open circles and squares correspond to the calculations for the 1^+ and 0^+ states, respectively. For the definition of the numbers of the configurations "CN#", please see text. The lines are to guide the eye.

3.2 Configuration dependence and deuteron-like correlation

In the previous section, we showed that the valence proton and neutron are correlated with the $S = 1$ channel dominance in the ground state of ^{18}F $(T = 0)$. Next, we discuss the importance of two ingredients for the deuteron-like correlation; the cross-term and the continuum contributions. To this end, we classify the set of basis functions into six types of configurations labeled as "CN#" and see the configuration dependence. For the series with the $(\ell_j)^2$-type coupling of the angular momentum part, we use CN1, CN2, CN3 and CN4, which are set of basis functions with $(d_{5/2})^2$, $(s_{1/2})^2$, $(d_{3/2})^2$ and $(pf)^2$. Here, for example, CN2 includes $(s_{1/2})^2$ in addition to CN1; $(d_{5/2})^2$, and other configurations, CN3 and CN4 are defined as the same manner. CN5 includes the "cross-term" $(d_{5/2})(d_{3/2})$ and $(s_{1/2})(d_{3/2})$ in addition to CN3, and CN6 is the full configuration up to pf-waves including the cross-term.

The configuration dependence of the correlation energies and magnetic moments are shown in Figs. 1 and 2. The correlation energy is defined as $E^{n\ell_j}_{\mathrm{Corr}}(J^\pi) \equiv E(J^\pi) - (\epsilon^\pi_{0d_{5/2}} + \epsilon^\nu_{0d_{5/2}})$, where $E(J^\pi)$ is the three-body energy. First, we focus on the results with the series of the $(\ell_j)^2$-type configurations (CN1, CN2, CN3 and CN4). The absolute values of the correlation energies for both 1^+ and 0^+ almost monotonically become large with increasing the number of the configurations, and the results of the 1^+ and 0^+ are similar each other. Moreover, for the $(\ell_j)^2$-type configurations, the change of the magnetic moment of the 1^+ state is small, and the orbital part μ^L is systematically larger than the spin part μ^S.

Next, we discuss the importance of the "cross-term" contribution in the 1^+ state (CN5) to gain the correlation energy and to change the coupling scheme of the valence nucleons. With the CN5 configuration, the correlation energy of the 1^+ state drastically increases as $E_{\mathrm{Corr}} = -3.41$ MeV and is even larger than that of CN4 (-2.85 MeV), which includes the $(pf)^2$ configurations to the model space. From CN3 to CN5, the magnetic moment increases form 0.613 to 0.779 (μ_N). The role of the angular momentum part μ^L and spin part μ^S are interchanged drastically. μ^S increases from 0.262 to 0.646 (μ_N), and μ^L decreases from 0.351 to 0.133 (μ_N). The large contribution of the spin-part μ^S at CN5 is an evidence of the importance of the cross-term to change the coupling scheme of the wave function from the jj-coupling to LS-coupling with $S = 1$.

The other important ingredient is the continuum contribution. Since all of the pf-waves with this potential model are obtained as the unbound states, the calculations with CN4 and CN6 correspond to the inclusion of the continuum contributions. With the CN6 configuration, the correlation energy is obtained as -4.64 MeV. The calculated value shows the continuum contribution is equally important to the cross-term one. However, the effect of the continuum contribution depends on the presence of the cross-term. This is because that without the cross-term (CN4), the correlation energy becomes less than half of the CN6 case.

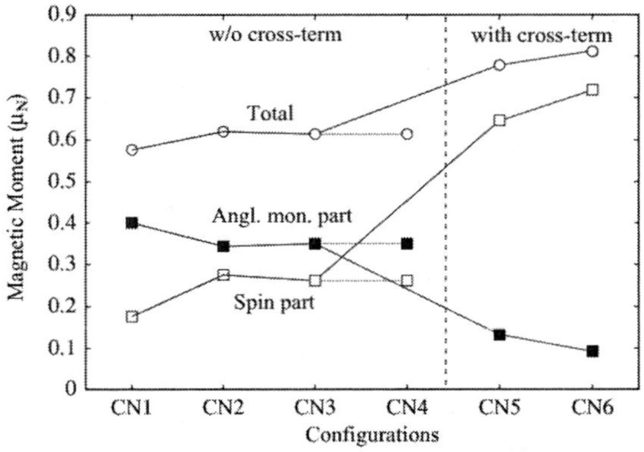

Figure 2. Configuration dependence of magnetic moment for ^{18}F(1^+). For the definition of the numbers of the configurations "CN#", please see text.

4 Summary and discussion

We study the deuteron-like correlation of the valence neutron and proton in ^{18}F using the COSM approach. From the analysis, we showed the essential ingredients of the deuteron-like correlation are the cross-term, which is the coupling of the spin-orbit partners $j_{\gtrless}$, and the continuum contributions. In the model space of the $(\ell_j)^2$-type configurations, the property of the correlation energies of the $T = 1$ and $T = 0$ channels becomes similar with the change of the number of basis functions. The cross-term component, $[d_{5/2} \otimes d_{3/2}]$ in ^{18}F changes the correlation energy drastically. The magnetic moment also increases, and its spin part becomes dominant by including the cross-term. The continuum contribution is the other important ingredient for the deuteron-like spatial correlation. The calculated opening angle of valence nucleons show the spatially localized deuteron-like correlation of the proton and neutron in ^{18}F.

Acknowledgements

The authors would like to thank the members of the nuclear theory group in Hokkaido University for their kind interest and discussions.

References

[1] A. Poves, G. Martines-Pinedo Phys. Lett. B **430**, 203 (1998).
[2] H. Sagawa, Y. Tanimura and K. Hagino, Phys. Rev. C **87**, 034310 (2013).
[3] Y. Tanimura, H. Sagawa and K. Hagino, Prog. Theor. Exp. Phys. **2014**, 053D02 (2014).
[4] Y. Suzuki and K. Ikeda, Phys. Rev. C **38**, 410 (1988).
[5] H. Masui, K. Katō and K. Ikeda, Phys. Rev. **C73**, 034318 (2006).
[6] E. Hiyama, Y. Kino and M. Kamimura, Prog. Part. Nucl. Phys. **51**, 223 (2003).
[7] H. Masui, K. Katō and K. Ikeda, Nucl. Phys. **A895**, 1 (2012).
[8] D. R. Thompson, M. LeMere, and Y. C. Tang, Nucl. Phys. **A286**, 53 (1977).
[9] National Nuclear Data Center, Chart of Nuclides (National Nuclear Data Center, New York, n.d.). (Available at: http://www.nndc.bnl.gov/chart, date last accessed April 11, 2014).
[10] A. F. Lisetskiy, R. V. Jolos, N. Pietralla, and P. von Bretano, Phys. Rev. C **60**, 064310 (1999).
[11] Y. Kanada-En'yo and F. Kobayashi, Phys. Rev. C **90**, 054332 (2014).

EPJ Web of Conferences **113**, 06004 (2016)
DOI: 10.1051/epjconf/201611306004

Low-lying ^{12}C continuum states in three-α model

Souichi Ishikawa[1,a]

[1] *Hosei University, 2-17-1 Fujimi, Chiyoda, Tokyo 102-8160, Japan*

Abstract. The electromagnetic monopole ($E0$) transition of the ^{12}C nucleus, the 0^+ ground state to three α-particles (3α) continuum states, is calculated by applying the Faddeev three-body formalism in coordinate space accommodating the long range Coulomb potential between α particles. Results show that there are two 0^+ states, ^{12}C(0_3^+) and ^{12}C(0_4^+) states at excitation energies around 10 MeV, which is consistent with results of recent semi-microscopic calculations and analysis of α-^{12}C inelastic scattering.

1 Introduction

Low-lying states in ^{12}C nucleus are interesting subjects to study as a 3α system. For example, it is widely known that the first 0^+ resonant state ^{12}C(0_2^+) at the excitation energy $E_x = 7.65$ MeV (so called the Hoyle state) plays an essential role in the synthesis of ^{12}C from a 3α continuum state (the triple-α process). However, there still have been uncertainties in the ^{12}C energy level structure at these energies ($E_x \approx 10$ MeV): While a 0^+ state was tentatively listed at $E_x = 10.3$ MeV in the compilation of experimental data [1], there is an experimental report that there exist two 0^+ states (0_3^+ and 0_4^+) [2, 3] near $E_x = 10$ MeV. Theoretically, the existing of two 0^+ states in this energy region was predicted by a semi-microscopic model, the orthogonal-condition model (OCM), combined with the complex scaling method [4].

In this paper, I will consider a transition of the ^{12}C ground state ^{12}C(0_1^+) to 3α 0^+ continuum states by the $E0$ operator and calculate the transition strength as a function of 3α energy E at the center of mass (c.m.) system above the 3α threshold. This is an extension of recent works [5, 6], in which 3α continuum states up to $E = 0.6$ MeV were studied to calculate the reaction rate of the triple-α process at stellar temperature $T \sim 10^9$ K, and clarify the decay mode of the Hoyle state.

In Sec. 2, I will give a short description of the formalism to calculate the strength function. In such a treatment, every complicacies arising from nucleon structure of the α-particle are assumed to be incorporated in interaction potentials among the α-particles, which usually consist of two-α potential (2αP) as well as three-α potential (3αP). Such models used in this work and results of the strength function will be presented in Sec. 3. Summary will be given in Sec. 4.

[a] e-mail: ishikawa@hosei.ac.jp

2 Formalism

We will consider a transition process of the $^{12}C(0_1^+)$ state $|\Psi_b\rangle$ to 3α continuum states induced by the $E0$ operator (see, *e.g.* [7]),

$$\hat{O}(E0) = \boldsymbol{x}^2 + \frac{4}{3}\boldsymbol{y}^2, \tag{1}$$

where $\boldsymbol{x}$ is the relative coordinate of α-α pair and $\boldsymbol{y}$ is the relative coordinate of the spectator α-particle with respect to the c.m. of the pair.

With the transition amplitude of the reaction, $\langle\Psi_{qp}^{(-)}|\hat{O}(E0)|\Psi_b\rangle$, where $|\Psi_{qp}^{(-)}\rangle$ is an eigen-state of 3α Hamiltonian H, and $\boldsymbol{q}$ and $\boldsymbol{p}$ are Jacobi momenta conjugate to the coordinates $\boldsymbol{x}$ and $\boldsymbol{y}$, respectively, the $E0$ transition strength function is defined by

$$\frac{dB(E0;E)}{dE} = \int d\boldsymbol{q}d\boldsymbol{p}\left|\langle\Psi_{qp}^{(-)}|\hat{O}(E0)|\Psi_b\rangle\right|^2 \delta\left(E-(E_q+E_p)\right), \tag{2}$$

where $E_q = \frac{\hbar^2}{m_\alpha}\boldsymbol{q}^2$ and $E_p = \frac{3\hbar^2}{4m_\alpha}\boldsymbol{p}^2$ with m_α being the α particle mass.

Let us define a wave function for the transition process by

$$\Psi(\boldsymbol{x},\boldsymbol{y}) = \left\langle \boldsymbol{x},\boldsymbol{y}\left|\frac{1}{E+\imath\epsilon-H}\hat{O}(E0)\right|\Psi_b\right\rangle, \tag{3}$$

which has the asymptotic form [5],

$$\Psi(\boldsymbol{x},\boldsymbol{y}) \underset{\substack{x\to\infty \\ y/x \text{ fixed}}}{\longrightarrow} \frac{e^{\imath(K+O(1/R))R}}{R^{5/2}}F^{(B)}(\hat{\boldsymbol{q}},\hat{\boldsymbol{p}},E_q), \tag{4}$$

where $K = \sqrt{\frac{m_\alpha}{\hbar^2}E}$, $R = \sqrt{x^2+\frac{4}{3}y^2}$, $O(1/R)$ expresses the effects from the long-range Coulomb interaction, and $F^{(B)}(\hat{\boldsymbol{q}},\hat{\boldsymbol{p}},E_q)$ is the breakup amplitude,

$$F^{(B)}(\hat{\boldsymbol{q}},\hat{\boldsymbol{p}},E_q) = e^{\frac{\pi}{4}\imath}\sqrt{\frac{\pi}{2}\frac{m_\alpha}{\hbar^2}}\left(\frac{4K}{3}\right)^{3/2}\langle\Psi_{qp}^{(-)}|\hat{O}|\Psi_b\rangle. \tag{5}$$

The wave function Ψ will be obtained by applying the Faddeev three-body formalism to solve integral equations in coordinate space with accommodating the long range Coulomb force effects. (See Refs. [5, 6, 8] for the details of the calculations.) This procedure provides the amplitude $F^{(B)}(\hat{\boldsymbol{q}},\hat{\boldsymbol{p}},E_q)$, and then the $E0$ strength function is calculated as

$$\frac{dB(E0;E)}{dE} = \frac{1}{2\pi K^3}\left(\frac{3}{4}\right)^2\int d\hat{\boldsymbol{q}}d\hat{\boldsymbol{p}}pqdE_q\left|F^{(B)}(\hat{\boldsymbol{q}},\hat{\boldsymbol{p}},E_q)\right|^2. \tag{6}$$

3 Results

In the present work, I used the model D of the Ali-Bodmer potential AB(D) [9] with the (point) Coulomb potential for the α-α interaction. In addition, 3αP of the following form is introduced:

$$\Delta(a) = W_\Delta \exp\left(-\frac{r_{12}^2+r_{23}^2+r_{31}^2}{a^2}\right) + [\text{c.p.}], \tag{7}$$

Table 1. The range and strength parameters of the 3αP models, Eq. (7), used in this work, and calculated energy of the ^{12}C ground state with respect to the 3α threshold, $E[^{12}\mathrm{C}(0_1^+)]$.

Model	a (fm)	W_Δ (MeV)	$E[^{12}\mathrm{C}(0_1^+)]$ (MeV)
Δ(2.4)	2.43	-143.7	-7.546
Δ(2.6)	2.61	-101.22	-7.584
Δ(3.0)	3.00	-52.3	-7.759
Δ(3.4)	3.39	-30.95	-7.789
Δ(4.0)	4.00	-16.18	-7.281
Exp.			-7.275

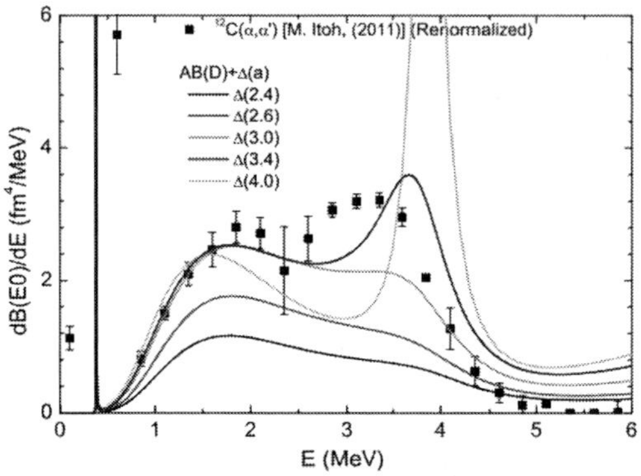

Figure 1. The $E0$ strength function for a transition from ^{12}C ground state to 3α 0^+ continuum states as a function of the final 3α energy in the c.m. system E. The black, red, green, blue, and aqua lines are calculations by Δ(2.4), Δ(2.6), Δ(3.0), Δ(3.4), and Δ(4.0), respectively. Experimental data are taken from Ref. [2, 11] renormalized according to Ref. [10] (see the text).

where [c.p.] denotes the cyclic permutations. Taking into account of uncertainty in the knowledge of 3αP, I examined some different values for the range parameter a from 2.4 fm to 4.0 fm in Eq. (7).

The strength parameter W_Δ for each of $\Delta(a)$ is determined to reproduce the energy of the Hoyle state, and results are shown in Table 1 along with calculated energy of $^{12}\mathrm{C}(0_1^+)$. In the table, calculated energies of $^{12}\mathrm{C}(0_1^+)$ differ from the empirical value by (at most) 10 %, which leaves the construction of a 3αP that reproduces both of bound and continuum states for a future work.

Fig. 1 shows calculated $E0$ transition strength functions for five different models of 3αP denoted as $\Delta(a)$ together with the AB(D) 2αP. In the figure, the experimentally extracted $E0$ strength from $^{12}\mathrm{C}(\alpha,\alpha')$ at $E_\alpha = 386$ MeV [2, 11] is also plotted, but it is renormalized, for a comparison, to reproduce the strength $B(E0, 0_3^+ \to 0_1^+) = |M(E0, 0_3^+ \to 0_1^+)|^2 = (2.9 \pm 0.3 \text{ fm})^2$, which was evaluated by a folding-mode analysis of the inelastic α + ^{12}C scattering data at medium energies [10].

Besides the sharp peak of the Hoyle state $^{12}\mathrm{C}(0_2^+)$, the calculated strength functions reveal two peaks at $E \approx 1.8$ MeV and at $E \approx 3.5$ MeV with rather large widths. These peaks may correspond to $^{12}\mathrm{C}(0_3^+)$ and $^{12}\mathrm{C}(0_4^+)$ resonant states, which were predicted in Ref. [4] and found in Ref. [2]. For the calculations with smaller value of a such as Δ(2.4), two-peak structure of the strength function is not clear, especially for the higher $^{12}\mathrm{C}(0_4^+)$ peak. As a increases, two-peak structure develops gradually, and the peak height of $^{12}\mathrm{C}(0_4^+)$ increases much faster than that of $^{12}\mathrm{C}(0_3^+)$. Comparing the heights of two peaks, calculations with Δ(3.4) and Δ(3.0) look favorable with the experimental data. It is remarkable that $a = 3.4$ fm corresponds to a configuration that two α particles are almost in touch.

Three particles in the final state spread by sharing momenta and energies in variety of manners as far as conservation laws are satisfied in principle. In Ref. [6], it was shown that the decay of the Hoyle state is dominated by a sequential decay (SD) process, in which the decay occurs in two steps: first 3 α-particles separate to 2-α resonant state, $^8\mathrm{Be}(0_1^+)$ $[E_{2\alpha,r} = 92 \text{ keV}, \Gamma_{2\alpha} = 6.8 \text{ eV}]$ and the rest

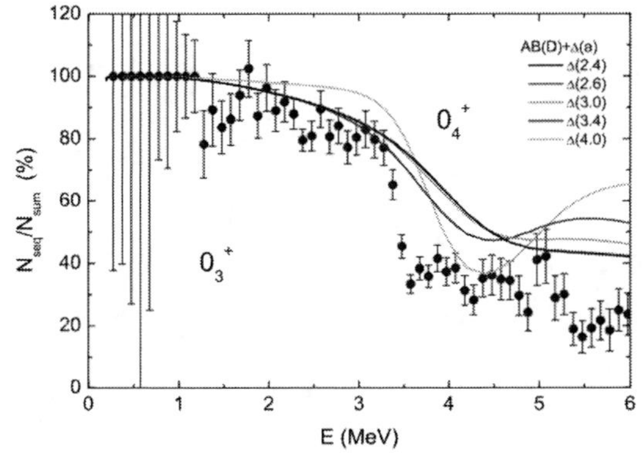

Figure 2. Ratio of the sequential decay component to the total flux for $E0$-disintegration of ^{12}C ground state as a function of the c.m. energy E. The meaning of the lines is the same as in Fig. 1. Experimental data are evaluated from the figure 2 of Ref. [3].

α-particle, and then ^{8}Be(0_1^+) decays to 2-α particles. In Ref. [3], from an analysis of semi-exclusive ^{12}C($\alpha, \alpha'\alpha''$) reaction, contributions of the SD mode were extracted. In the present work, the SD component is evaluated by limiting the E_q-integration in Eq. (6) as $E_{2\alpha,r} - \Delta E \leq E_q \leq E_{2\alpha,r} + \Delta E$, where ΔE is set to be a few times of the width $\Gamma_{2\alpha}$.

In Fig. 2, calculated fraction of the SD component to the total is plotted and compared to the data. While the 3α decay mode for the ^{12}C(0_3^+) state is dominated by the SD process via the ^{8}Be(0_1^+) state, that for the ^{12}C(0_4^+) state is not. The details of the 3α decay mode depend on the choice of the 3αP.

4 Summary

The 3α-model calculations in this work suggest that there are two peaks in the $E0$ strength function from ^{12}C ground state, which may be assigned as ^{12}C(0_3^+) and ^{12}C(0_4^+) states.

While the sequential decay mode is dominant one for ^{12}C(0_3^+) state, it is not for ^{12}C(0_4^+) one. Further and more precision studies of the $E0$ strength function as a function of E as well as those of the decay mode of ^{12}C(0_4^+) state will provide valuable information on the interaction model of α-particles, especially on 3α potential.

Acknowledgements

The author is grateful to Dr. Masatoshi Itoh for valuable discussions and providing his data.

References

[1] F. Ajzenberg-Selove, Nucl. Phys. A **506**, 1 (1990)
[2] M. Itoh *et al.*, Phys. Rev. C **84**, 054308 (2011)
[3] M. Itoh *et al.*, J. Phys.: Conf. Ser. **436**, 012006 (2013)
[4] C. Kurokawa and K. Kato, Phys. Rev. C **71**, 021301 (2005)
[5] S. Ishikawa, Phys. Rev. C **87**, 055804 (2013)
[6] S. Ishikawa, Phys. Rev. C **90**, 061604(R) (2014)
[7] T. Yamada *et al.*, Prog. Theor. Phys. **120**, 1139 (2008)
[8] S. Ishikawa, Phys. Rev. C **80**, 054002 (2009)
[9] S. Ali and A. R. Bodmer, Nucl. Phys. **80**, 99 (1966)
[10] D. C. Cuong *et al.*, Phys. Rev. C **88**, 064317 (2013)
[11] M. Itoh, *private communication*

EPJ Web of Conferences **113**, 06005 (2016)
DOI: 10.1051/epjconf/201611306005

Rotational character of the ^{12}C spectrum investigated through inelastic cross sections via photon emission

E. Garrido[1,a], A.S. Jensen[2], and D.V. Fedorov[2]

[1] *Instituto de Estructura de la Materia, IEM-CSIC, Serrano 123, E-28006 Madrid, Spain*
[2] *Department of Physics and Astronomy, Aarhus University, DK-8000 Aarhus C, Denmark*

Abstract. In this work the bremsstrahlung and photon dissociation cross sections for transitions between 0^+, 2^+, and 4^+ states in ^{12}C are computed. The nucleus is described within the three-alpha model, and the wave functions are computed by means of the hyperspherical adiabatic expansion method. The continuum states are discretized by imposing a box boundary condition. The transition strengths are obtained from the cross sections, and compared to schematic rotational model predictions. The computed results strongly suggest that the two lowest bands are made, respectively, by the states with angular momentum and parity $\{0_1^+, 2_1^+, 4_2^+\}$ and $\{0_2^+, 2_2^+, 4_1^+\}$. The transitions between the states in the first band are consistent with the rotational pattern corresponding to three alphas in an equilateral triangular structure. For the second band, the transitions are also consistent with a rotational pattern, but with the three alphas in an aligned distribution.

1 Introduction

The properties of the spectrum of ^{12}C have attracted a lot of attention along the years. In particular, the presence in the energy spectrum of at least two sequences of states with angular momentum and parity $\{0^+, 2^+, 4^+\}$ immediately suggests that these states could correspond to states in a rotational band. At least two rotational bands could exist in ^{12}C, one of them sitting on the ground state, and another one sitting on the Hoyle state.

The same rotational sequence has also been observed in ^{8}Be, where the ground state and the two first excited states follow as well the angular momentum and parity sequence 0^+, 2^+, and 4^+ (together with very broad 6^+ and 8^+ states, which are found numerically as poles of the S-matrix). All the states in the ^{8}Be spectrum are unbound (resonances), with the ground state only about 0.1 MeV above the threshold for emission of two alpha particles. This spectrum was recently investigated in [1–3] by computation of the electric quadrupole cross sections for bremsstrahlung emission after transitions between the ^{8}Be-states, which were described as two-body systems made of two alpha particles. These calculations required a careful treatment of the continuum wave functions and clarification of the definition of the cross section for transitions between states with a non well-defined energy.

In this work we extend the investigations to the spectrum of ^{12}C, which is described as a three-alpha system. In particular, we shall investigate whether the behavior of the $\{0_1^+, 2_1^+, 4_1^+\}$ and $\{0_2^+, 2_2^+, 4_2^+\}$ sequences of states agree with the predictions of the rotational model. This will be done

[a] e-mail: e.garrido@csic.es

by means of the the electric quadrupole γ-emission bremsstrahlung cross sections for the different transitions between the states. These cross sections contain the information about the corresponding transition strengths, which in the case of transitions between states within a rotational band must be given by very simple expressions. In particular, these expressions contain the intrinsic quadrupole moment, which should be the same for all the states in the band. The consistency of the intrinsic quadrupole moment with specific (linear or triangular) structures of the three-alpha system are also investigated.

2 Method

Let us consider the three-body system made of particles a, b, and c, and the continuum to continuum γ-emission process $A^* \to A + \gamma$, where A^* and A represent three-body continuum states with energies E and E' ($E > E'$), respectively. The cross section for this reaction is given by

$$\frac{d\sigma^{(\lambda)}}{dE'}(E) = \nu! \frac{2(2J'+1)}{(2J_a+1)(2J_b+1)(2J_c+1)} \frac{32\pi}{\kappa^5} \left(\frac{E_\gamma}{\hbar c}\right)^2 \frac{d\sigma_\gamma^{(\lambda)}}{dE'}(E_\gamma), \tag{1}$$

where $E_\gamma = E - E'$ is the photon energy, J and J' are the initial and final angular momenta, ν is the number of identical particles in the three-body system, J_a, J_b, and J_c are the total angular momenta of particles a, b, and c, respectively, and κ is the three-body momentum, which is defined as $\kappa = \sqrt{2mE/\hbar^2}$. The mass m is the normalization mass used to define the Jacobi coordinates. Finally, the differential photo-absorption cross section is given by:

$$\frac{d\sigma_\gamma^{(\lambda)}}{dE'}(E) = \frac{(2\pi)^3(\lambda+1)}{\lambda[(2\lambda+1)!!]^2} \left(\frac{E_\gamma}{\hbar c}\right)^{2\lambda-1} \frac{d\mathcal{B}^{(\lambda)}}{dEdE'}(J' \to J). \tag{2}$$

In this work the continuum spectrum is discretized by imposing a box boundary condition. When done, the differential transition strength contained in the equation above can be written as:

$$\frac{d\mathcal{B}^{(\lambda)}}{dEdE'}(J' \to J) = \frac{1}{2J'+1} \sum_{i,j} \left|\langle \Phi_J^{(i)} \| \hat{O}_\lambda \| \Phi_{J'}^{(j)} \rangle\right|^2 \delta(E - E_i)\delta(E' - E'_j), \tag{3}$$

where the indeces i and j run over the initial and final discrete continuum states with energies E_i and E'_j, respectively. The corresponding continuum wave functions are denoted by $\Phi_J^{(i)}$ and $\Phi_{J'}^{(j)}$. The electromagnetic operator with multipolarity λ is denoted by $\hat{O}_\lambda$. It is important to note that the summation over i and j in the Eq.(3) is not unrestricted, but limited to the initial and final energy ranges of experimental interest [4].

2.1 Transition strength

As shown in Ref.[3], in the case of continuum to continuum transitions, a direct calculation of the transition strength from Eq.(3) is not possible. It is in fact necessary to extract it directly from the cross section, which, as evident from Eq.(2), contains all the information about the transition strength. In particular, in this work two different methods will be used.

In the first method the strength is obtained from the total (integrated) cross section. More precisely, for a transition between some initial and final energy ranges, typically around resonances in the initial and final states, integration of Eq.(1) over E and E' within those two windows will provide the total cross section for the transition. If the photon energy, E_γ, were constant, one could immediately get the

Table 1. Experimental and computed energies E_R (in MeV) and widths Γ_R (in keV) of the ^{12}C resonances for different J^π and the Buck α-α potential. For each of the calculations, the strength S (in MeV) used in the gaussian effective three-body potential is also given (the range is taken to be $b = 6.0$ fm). The energies are referred to the three-alpha threshold.

J^π	$E_{R,\text{exp}}$ (MeV)	$\Gamma_{R,\text{exp}}$ (keV)	$E_{R,\text{Buck}}$ (MeV)	$\Gamma_{R,\text{Buck}}$ (keV)	S
0_1^+	$-7.275^{(a)}$	–	-7.27	–	-22.0
0_2^+	$0.380^{(a)}$	$0.009 \pm 0.001^{(a)}$	0.38	$\lesssim 0.05$	-18.0
0_3^+	$4.20 \pm 0.14^{(b)}$	$3440 \pm 220^{(b)}$	4.12	700	-18.0
2_1^+	$-2.8356 \pm 0.0003^{(a)}$	0.0	-2.85	–	-12.1
2_2^+	$2.76 \pm 0.11^{(c)}$	$800 \pm 130^{(c)}$	1.72/2.35	161/920	$-12.1/-3.0$
	$2.86 \pm 0.05^{(d)}$	$2080 \pm 300^{(d)}$			
2_3^+	$3.88 \pm 0.05^{(a)}$	$430 \pm 80^{(a)}$	5.04/5.48	570/3600	$-12.1/-3.0$
4_1^+	$6.0 \pm 0.2^{(e)}$	$1700 \pm 200^{(e)}$	5.48	3800	13.4
4_2^+	$6.81 \pm 0.02^{(a)}$	$258 \pm 15^{(a)}$	6.81	800	13.4
4_3^+			13.1	2060	13.4

(*a*) Ref.[5]; (*b*) Ref.[6]; (*c*) Ref.[7]; (*d*) Ref.[8], (*e*) Ref.[9]

total transition strength simply after division of the total cross section by the constants that multiply the transition strength. However, since this is not correct, we must use an average value for $E_\gamma^{2\lambda+1}$. In particular, the photon energy will be taken as $E_\gamma = E_c - E_c'$ where E_c is the energy of the cross section peak corresponding to the resonance in the initial state and E_c' is the center of the final energy window (usually the resonance energy in the final state). Of course, for this procedure to be applied, some initial information about the resonance positions is needed. Furthermore, this procedure can be sensitive to rather small variations around a chosen E_γ due to the $(2\lambda + 1)$ exponent for the $\mathcal{B}(E\lambda)$ transitions, see Ref.[1] for details.

The second method exploits the fact that in the vicinity of a resonance the cross section of the photo-dissociation reaction $A + \gamma \rightarrow a + b + c$ takes the form

$$\sigma_\gamma(E_\gamma) = \frac{2J+1}{2(2J_A+1)} \frac{\pi\hbar^2c^2}{E_\gamma^2} \frac{\Gamma_R\Gamma_\gamma}{(E-E_R)^2+\Gamma^2/4}, \tag{4}$$

where after the collision the particles a, b, and c are assumed to populate a resonance with angular momentum J, energy E_R, and width for decay into three particles Γ_R. J_A is the angular momentum of A and $\Gamma = \Gamma_R + \Gamma_\gamma$, where Γ_γ is the γ-decay width of the three-body resonance. In this method the value of Γ_γ in Eq.(4) is used as a parameter in order to fit the peak of the computed cross section corresponding to the three-body resonance at $E = E_R$. The Γ_γ value obtained in this way is then used to extract the transition strength by means of the well-known expression:

$$\Gamma_\gamma = \frac{8\pi(\lambda+1)}{\lambda[(2\lambda+1)!!]^2} \left(\frac{E_\gamma}{\hbar c}\right)^{2\lambda+1} \mathcal{B}^{(E\lambda)}(J \rightarrow J_A). \tag{5}$$

In the following, the transition strengths computing using these two methods will be denoted by $\mathcal{B}_\sigma$ and $\mathcal{B}_\gamma$, respectively.

3 The three-alpha system

The resonances of the three-alpha system are obtained by use of the hyperspherical adiabatic expansion combined with the the complex scaling method. The three-body resonances appear then formally

Table 2. $\mathcal{B}(E2)$ transition strengths in $e^2\mathrm{fm}^4$ for different transitions between ^{12}C states. For transitions from continuum states, $\mathcal{B}_\gamma$ and $\mathcal{B}_\sigma$ denote the two different methods used (see text). The second, third, and fourth columns give the known experimental values, and the results obtained with the microscopic cluster model calculation described in Ref.[10] and with the Antisymmetrized Molecular Dynamics (AMD) method.

Transition	Exp.	α-cluster	AMD	$\mathcal{B}(E2)$ ($e^2\mathrm{fm}^4$)	
$2_1^+ \to 0_1^+$	$7.6 \pm 0.4^{(a)}$	$9.16^{(a)}$	$8.4^{(d)}$	9.9	
				$\mathcal{B}_\gamma$	$\mathcal{B}_\sigma$
$4_2^+ \to 2_1^+$			$15.8^{(d)}$	12.5	9.9
				$\mathcal{B}_\gamma$	$\mathcal{B}_\sigma$
$2_2^+ \to 0_2^+$			$102^{(d)}$	158	136
$4_1^+ \to 2_2^+$			$595^{(d)}$	220	145
$2_1^+ \to 0_2^+$	$2.6 \pm 0.4^{(a)}$	$0.84^{(a)}$	$5.1^{(d)}$	0.96	
				$\mathcal{B}_\gamma$	$\mathcal{B}_\sigma$
$4_1^+ \to 2_1^+$				1.2	1.0
$2_2^+ \to 0_1^+$	$0.73 \pm 0.13^{(b)}$ $1.57 \pm 0.13^{(c)}$	$1.99^{(a)}$	$0.4^{(d)}$	2.3	2.2
$2_2^+ \to 2_1^+$				5.5	5.4
$4_2^+ \to 2_2^+$			$7.5^{(d)}$	7.5	4.7

(a) Ref.[10]; (b) Ref.[7]; (c) Ref.[8]; (d) Ref.[11]

as ordinary bound states with complex energy, whose real and imaginary parts give the resonance energy and half the width of the resonance, respectively. In order to reproduce the known experimental energies in the ^{12}C-spectrum a fine tuning with a short-range three-body force is required. This is done by an effective three-body potential. The results obtained with the α-α Buck potential are shown in Table 1.

The use of the complex-rotated three-body wave functions permits to compute very easily the complex expectation value $\rho_{rms} = \langle \rho^2 \rangle^{1/2}$, whose real part gives a feeling of the spatial distribution of the system (the imaginary part is related to the uncertainty). States arising as the result of a rotation of a reasonable rigid structure should show similar values of ρ_{rms}. It is remarkable that, on the one side, the 0_1^+ and 2_1^+ states have similar sizes, both of them in the vicinity of $\rho_{rms} \sim 6.8$ fm and $r_{rms} \sim 2.5$ fm, and, on the other side, the same happens with the 0_2^+ and 2_2^+ resonances, with a value of $\rho_{rms} \sim 10$ fm and $r_{rms} \sim 3.4$ fm. This may suggest that these two sets of 0^+ and 2^+ states could correspond to two different rotational bands, each of them with a reasonably well-"frozen" structure. It is interesting to note that for the 4_1^+ and 4_2^+ states the computed values of ρ_{rms} are precisely about 10 fm and 6 fm, respectively, which might indicate a crossing of the first and second 4^+ states, in such a way that the 4_1^+ state should belong to the second band and the 4_2^+ state to the first one. This is consistent with the fact that the 4_2^+ state becomes actually the first 4^+ and the 4_1^+ state becomes the second 4^+ when the repulsion in the three-body force is diminished [4].

In Table 2 we give the computed electric quadrupole transition strengths for the transitions between the 0^+, 2^+, and 4^+ states shown in Table 1. In those cases where the initial state corresponds to a continuum state we give in the table the two values, $\mathcal{B}_\sigma$ and $\mathcal{B}_\gamma$, corresponding to the two different computation procedures described in the previous section.

For transitions between states within a schematic rotational band, and assuming axial symmetry for the system, the quadrupole transition strength is given by

$$\mathcal{B}^{(E2)}(J \to J') = \frac{5}{16\pi} Q_0^2 \langle J0; 20|J'0\rangle^2, \tag{6}$$

Table 3. Absolute value of the intrinsic transition quadrupole moments $|Q_0|$ (in $e$fm^2) obtained from the $\mathcal{B}(E2)$ transition strengths given in Table 2 and Eq.(6). $|Q_0|_\sigma$ and $|Q_0|_\gamma$ denote the intrinsic quadrupole moments obtained from the transition strengths $\mathcal{B}_\sigma$ and $\mathcal{B}_\gamma$, respectively.

Transition	Exp.	α-cluster	AMD	$\|Q_0\|$ ($e$fm^2)	
$2_1^+ \to 0_1^+$	19.5 ± 0.5	21.46	20.5	22.3	
				$\|Q_0\|_\gamma$	$\|Q_0\|_\sigma$
$4_2^+ \to 2_1^+$			23.6	21.0	18.7
				$\|Q_0\|_\gamma$	$\|Q_0\|_\sigma$
$2_2^+ \to 0_2^+$			72	89	83
$4_1^+ \to 2_2^+$			145	88	71
$2_1^+ \to 0_2^+$	11.4 ± 0.9	6.50	16.0	6.9	
				$\|Q_0\|_\gamma$	$\|Q_0\|_\sigma$
$4_1^+ \to 2_1^+$				6.5	5.9
$2_2^+ \to 0_1^+$	6.03 ± 0.54	10.0	4.5	10.8	10.5
	8.88 ± 0.37				
$2_2^+ \to 2_1^+$				13.9	13.8
$4_2^+ \to 2_2^+$			16.2	16.2	12.9

where the projection, K, of the angular momentum on the intrinsic symmetry axis has been assumed to be zero. The intrinsic quadrupole moment Q_0 is given by:

$$Q_0 = \langle \sum_i q_i (2z_i^2 - x_i^2 - y_i^2) \rangle \, , \tag{7}$$

where i runs over all the charged particles with charge q_i and whose center of mass coordinates are denoted by (x_i, y_i, z_i), where the z-axis is chosen along the intrinsic symmetry axis. The expectation value is taken in the intrinsic body-fixed coordinate system. The quadrupole moment is a measure of the deformation and it has ideally one characteristic value for a sequence of states belonging to a given rotational band. From each of the calculated transition strengths given in Table 2, it is then easy, to obtain from Eq.(6) the absolute value of the intrinsic quadrupole moments. The $|Q_0|$-values obtained in this way are given in Table 3.

4 Discussion

The transition strengths obtained from the cross sections are in general consistent with each other. The results are similar no matter what method is used to extract the transition strengths ($\mathcal{B}_\sigma$ or $\mathcal{B}_\gamma$), especially when taking into account the uncertainties inherent to each of the methods. Although not shown, it has also been observed that the results are, to a large extent, independent of the potential used [4].

An important result is that the 4_1^+ and 4_2^+ states have actually crossed, in such a way that the 4_1^+ state belongs to the second band and the 4_2^+ to the first one. This is first suggested by the fact that the computed values of ρ_{rms}, which should be similar for systems having the same 'frozen' spatial structure, provides values for the 4_1^+ state similar to the ones of the 0_2^+ and 2_2^+ states, and values for the 4_2^+ state similar to the ones of the 0_1^+ and 2_1^+ states. Furthermore, investigating how the 4_1^+ and 4_2^+ resonances move in the complex energy plane when making the effective three-body force more and more attractive, we have seen that the first 4^+ state becoming bound would actually be the 4_2^+ state.

The assignment of the 4_1^+ state to the second band and the 4_2^+ to the first is also confirmed by the analysis of the transition strengths and the intrinsic quadrupole moments. The values of Q_0 are obtained assuming a rotational character for the two bands in ^{12}C under investigation, in such a way that the transition strength is basically the square of the intrinsic quadrupole moment multiplied by some geometrical factor depending on the initial and final angular momenta of the transition. Doing like this we have seen that for the $4_2^+ \to 2_1^+$ reaction the computed transition strength is clearly more consistent with the previous AMD calculation than the one corresponding to the $4_1^+ \to 2_1^+$. The large discrepancy still remaining between our result and the AMD calculation can be understood from the fact that, as shown in Ref.[4], the interference between the 4_1^+ and 4_2^+ resonances and the removal of the so-called soft-photon contribution introduce in the calculation a series of uncertainties that lead us to consider our result in Table 3 as a lower limit for the transition strength (and the intrinsic quadrupole moment). Furthermore, the intrinsic quadrupole moments derived from the transition strengths for the $2_1^+ \to 0_1^+$ and $4_2^+ \to 2_1^+$ reactions are rather similar and consistent with previous calculations and the experimental value. This consistency would disappear if the reaction $4_1^+ \to 2_1^+$ were the one taken into account. The same happens for the reactions in the second band. Although the computed transition strength for the $4_1^+ \to 2_2^+$ is even a factor of 3 smaller than the previous AMD calculation, this discrepancy is certainly less important than when considering the $4_2^+ \to 2_2^+$ reaction, and it can actually be understood from the uncertainties associated to the method used to extract the strength. Also, the intrinsic quadrupole moments for the $2_2^+ \to 0_2^+$ and $4_1^+ \to 2_2^+$ are reasonably stable, stability that would be clearly broken if assuming the $4_2^+ \to 2_2^+$ transition as belonging to the second band. Therefore, from the stability of the intrinsic quadrupole moments for each of the $\{0_1^+, 2_1^+, 4_2^+\}$ and $\{0_2^+, 2_2^+, 4_1^+\}$ bands we can conclude that each of the bands correspond to states having a rather well preserved rigid structure.

The rotational character of two bands is confirmed when comparing with the prediction obtained from an axially symmetric rotating structure made by three point-like alpha particles. Assuming values for ρ_{3b} similar to the ones computed numerically for the states in the first and second bands ($\rho_{rms} \sim 6.7$ fm and ~ 10 fm, respectively) we have seen that the Q_0 values previously computed for the states in the first band are consistent with a equilateral triangular structure rotating around an axis perpendicular to the plane holding the three particles. For the states in the second band, the computed Q_0 values are mostly consistent with a linear distribution of the particles. The same results are obtained from the analyses of the moments of inertia.

References

[1] E. Garrido, A.S. Jensen, D.V. Fedorov, Phys. Rev. C **86**, 064608 (2012)

[2] E. Garrido, A.S. Jensen, D.V. Fedorov, Phys. Rev. C **88**, 024001 (2013)

[3] E. Garrido, A.S. Jensen, D.V. Fedorov, Few-body Syst. **55**, 101 (2014)

[4] E. Garrido, A.S. Jensen, D.V. Fedorov, Phys. Rev. C **91**, 054003 (2015)

[5] F. Ajzenberg-Selove, Nucl. Phys. A **506** (1990) 1

[6] C. Aa. Diget, *Beta delayed particle emission: Probing the triple alpha continuum*, Ph.D. Thesis, Natural Science Faculty, University of Aarhus 2006, http://phys.au.dk/fileadmin/site_files/publikationer/phd/Christian_Aa_Diget.pdf.

[7] W.R. Zimmerman *et al.*, Phys. Rev. Lett **110**, 152502 (2013)

[8] W.R. Zimmerman, *Direct Observation of the Second 2$^+$ State in ^{12}C*, Ph.D. Thesis, University of Connecticut Graduate School 2013, http://digitalcommons.uconn.edu/cgi/viewcontent.cgi?article=6439&context=dissertations

[9] M. Freer *et al.*, Phys. Rev. C **83**, 034314 (2011)

[10] M. Chernykh *et al.*, Phys. Rev. Lett. **98**, 032501 (2007)

[11] D.C. Cuong, D.T. Khoa, Y. Kanada-En'yo, Phys. Rev. C **88**, 064317 (2013)

EPJ Web of Conferences **113**, 06006 (2016)
DOI: 10.1051/epjconf/201611306006

Search for $4n$ contributions in the reaction ^{14}Be(CH$_2$,X)^{10}He

M. D. Jones[1,2, a] and Z. Kohley,[1,3] T. Baumann,[1] G. Christian,[1,2, b] P. A. DeYoung,[4] J. E. Finck,[5] N. Frank,[6] R. A. Haring-Kaye,[7] A. N. Kuchera,[1] B. Luther,[8] S. Mosby,[1,2, c] J. K Smith,[1,2,b] J. Snyder,[1,2] A. Spyrou,[1,2] S. L. Stephenson,[9] M. Thoennessen[1,2]

[1] *National Superconducting Cyclotron Laboratory, Michigan State University, East Lansing, MI 48824, USA*
[2] *Department of Physics and Astronomy, Michigan State University, East Lansing, MI 48824, USA*
[3] *Department of Chemistry, Michigan State University, East Lansing, MI 48824, USA*
[4] *Department of Physics, Hope College, Holland, MI 49422-9000, USA*
[5] *Department of Physics, Central Michigan University, Mount Pleasant, MI 48859, USA*
[6] *Department of Physics and Astronomy, Augustana College, Rock Island, IL 61201, USA*
[7] *Department of Physics and Astronomy, Ohio Wesleyan University, Delaware, OH 43015, USA*
[8] *Department of Physics, Concordia College, Moorhead, MN 56562, USA*
[9] *Department of Physics, Gettysburg College, Gettysburg, PA 17325, USA*

Abstract. A previously published measurement of the ground state resonance of ^{10}He, populated by a reaction of a 59 MeV/u ^{14}Be beam on a deuterated polyethylene target, was further analyzed to search for $4n$ emission resulting from $2p$ removal. No evidence for $4n$ events was found. A lower limit of about 1 MeV was determined for a possible resonance in ^{12}He.

1 Introduction

There is some controversy over the location of the ground state of ^{10}He. Recently, a missing-mass measurement performed at JINR in Dubna populated ^{10}He via a (t,p) reaction on ^{8}He and found the ground state to be at 2.1(2) MeV [1] in contrast to another measurement done at GSI using proton knockout from ^{11}Li, which found the ground state to be lower at 1.54(11) MeV [2]. It has been suggested by Grigorenko and Zhukov that the observed peak in the 3-body spectrum of the GSI measurement could be shifted by the halo nature of the ^{11}Li beam [3], apparently reconciling the discrepancy between the Dubna and GSI results.

In addition, a recent measurement at the NSCL populated ^{10}He through 2p2n removal from ^{14}Be and found the 3-body resonance to be at 1.6(25) MeV [4]. This reaction was considered to be more dissipative and so the invariant mass spectrum should not be influenced by the proposed initial state effects – yet a lower energy was observed. In this experiment, the 2p2n removal could proceed multiple ways. Here we consider α-knockout and 2p removal to produce ^{10}He, the latter process populating ^{12}He. In this work we investigate possible evidence for direct cluster removal and search for a potential resonance in ^{12}He.

a. e-mail: jonesm@nscl.msu.edu
b. Present address :TRIUMF, 4004, Wesbrook Mall, Vancouver, British Columbia, V6T 2A3 Canada
c. Present address : LANL, Los Alamos, New Mexico 87545. USA

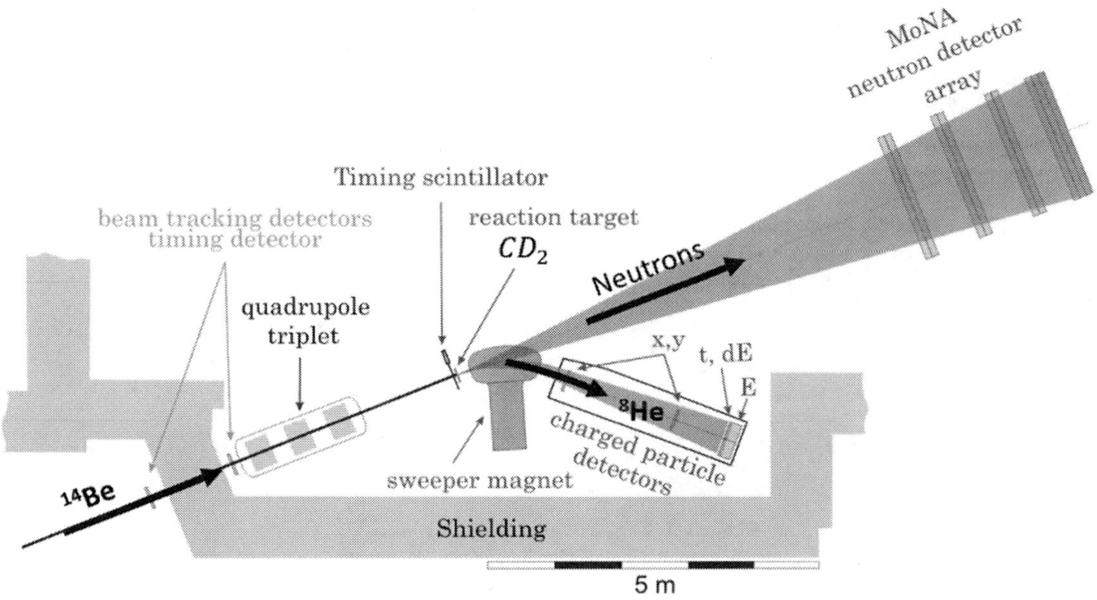

FIGURE 1: Schematic drawing of the experimental setup at the NSCL.

2 Experimental Method

The experiment was performed at the National Superconducting Cyclotron Laboratory where a 3196 mg/cm^2 ^{9}Be target was bombarded with an ^{18}O beam at 120 MeV/u. The A1900 fragment separator allowed for selection of the ^{14}Be secondary beam which continued on to impinge a 435 mg/cm^2 deuterated polyethylene target in the experimental area. The resulting charged fragments were bent 43.3° by a 4-Tm superconducting sweeper magnet [5] into a collection of charged-particle detectors, which allowed for element and isotope identification. The neutrons emitted in-flight traveled undisturbed by the magnet towards MoNA (Modular Neutron Array), which provided time-of-flight and position information. Together, MoNA and the Sweeper provide a full kinematic measurement of the neutrons and the charged fragments from which the N-body decay energies of ^{8}He + Xn can be determined. Additional experimental details can be found in Ref. [4]. Figure 1 shows a schematic of the experimental setup.

2.1 Analysis

Depending on the reaction path ($\alpha/2p2n$ or $2p$ removal), a different number of neutrons is emitted. Thus the analysis was extended from 2- to 5-body spectra corresponding to the decays of $^{9-12}$He. The N-body decay energy is defined as :

$$E_{\text{decay}} = M_{\text{Nbody}} - M_{^8\text{He}} - \sum_{i=1}^{i=N-1} m_{\text{n}},$$

where M_{Nbody} is the invariant mass of the N-body system, $M_{^8\text{He}}$ the mass of ^{8}He, and m_{n} the mass of a neutron. The invariant mass of an N-body system was calculated from the experimentally measured four-momenta of ^{8}He and the first $N-1$ time-ordered interactions in MoNA.

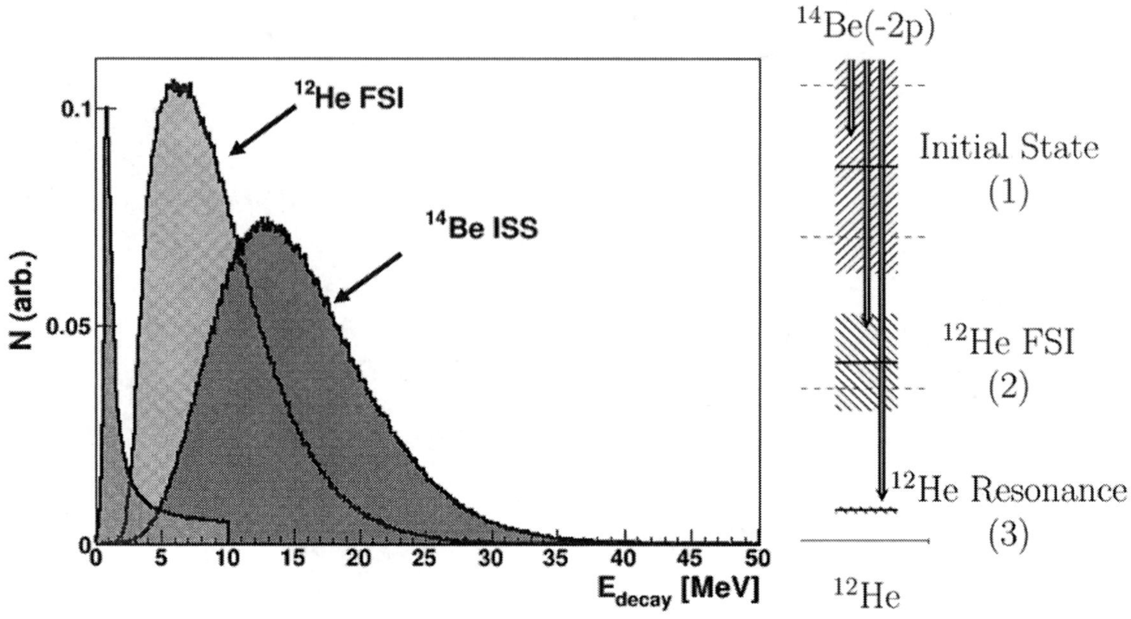

FIGURE 2: Input 5-body distributions for the ^{8}He + 4n system. (1) Shown in blue is a distribution influenced by the halo structure of ^{14}Be [10]. (2) In grey is a resonant final-state interaction [10]. (3) In red is a low-lying 1 MeV Breit-Wigner lineshape.

In order to extract information from the invariant mass spectra, detailed Monte-Carlo simulations were performed. These simulations include the beam characteristics, the reaction mechanism, and decay mechanism. Using GEANT4 [6] and MENATE_R [7], the resolution, efficiency and acceptances of MoNA and the charged particle detectors were fully incorporated into the simulations making them directly comparable to data.

Multi-neutron decay reconstruction is very sensitive to detector response effects, and multiple scattering within the array must be properly taken into account. In the case of 2n events (^{10}He), this is accomplished by applying causality cuts on relative distance and velocity as described in Ref. [4]. One extension to 4n events could be to apply causality cuts on each pair of interactions, of which there are six. However, a typical causality cut has an efficiency on the order of 30%, and applying this cut six-fold is not feasible. Instead, a 4n-signal can be indirectly searched for by a simultaneous χ^2 minimization on several decay energy spectra of varying multiplicities described in Ref. [8]. A similar method was used to search for 3n events coming from ^{15}Be in another study [9]. Since a 4n event will induce more scattering than a 2n event, a simultaneous fit across several multiplicities is sensitive to the number of neutrons emitted. Using this method, it was found in this analysis that there is no need to invoke a 4n model to describe the data [8], as the data can be completely described by 2n events coming from either α-knockout or 2p2n removal.

However, contributions from the 2p removal reaction could be present in the data up to some statistical limit. For this reason we consider several possibilities for the ^{12}He 5-body system illustrated in Figure 2. Three different cases for the population of ^{12}He were considered, of varying energy : (1) a distribution influenced by the initial halo structure of ^{14}Be (Initial State Structure [10]), (2) a resonant final-state interaction (^{12}He FSI) [10] peaking at roughly 6.5 MeV, and (3) a low-lying Breit-Wigner lineshape centered at 1 MeV. The 5-body spectrum for ^{8}He is most sensitive to the presence of a $4n$ component and so it is useful to track the log-likelihood ratio, $Ln[\lambda]$ of this spectrum. Since the data can be described entirely with $2n$ events, the best-fit parameters from α/2p2n-knockout are used as a baseline. The quality of the fit and the parameters used for ^{9}He and ^{10}He can be found in Ref. [8]. The $4n$ component from a hypothesis is then added to the spectrum and its amplitude is incremented. Plotting the log-likelihood as a function of the ratio of $4n$ to $2n$ events gives the statistical limits for each model. The result of this procedure is shown in Figure 3.

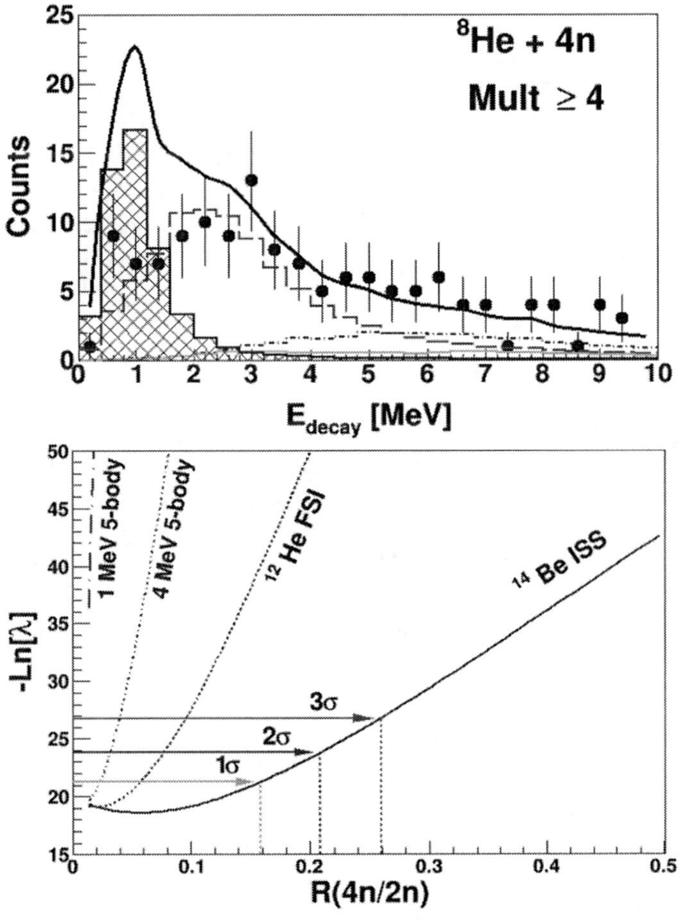

FIGURE 3: Figures from Ref. [8]. (Top) Five-body decay energy spectrum for all multiplicities ≥ 4. The hatched blue histogram is the contribution from a five-body breakup of ^{12}He at E = 1 MeV with $R(4n/2n)$ = 1.5%. The dash, dot-dashed, and solid lines are contributions from direct population of ^{10}He. (Bottom) Maximum likelihood for the five-body decay spectra (and multiplicity) as a function of the ratio of 2p to α or $2p2n$ removal for several possibilities in ^{12}He. For more details refer to Ref. [8]

Figure 3 (top panel) shows the result of including a low-lying resonance in ^{12}He with a strength of only 1.5% that of the net α or $2p2n$ components. A large excess at low E_{decay} is evident in the 5-body spectrum and the prediction is at odds with what is observed. Since one would expect a distinct resonance in ^{12}He to be populated from $2p$ removal, the data do not show evidence for a low-lying state below 1 MeV. The other scenarios, which cover a broad range of energies, do not minimize in favor of $2p$ removal. The ratio $R(4n/2n)$ does not exceed 30% for any case considered here while remaining within 3σ confidence – implying that the data are dominated by α/2p2n-knockout. The amount of contribution from $2p$ removal increases with the 5-body energy primarily due to a drop-off in efficiency for higher decay energies. It should be noted that for this experiment the sweeper magnet was set for lower rigidities and only the low-energy tail of the overall momentum distribution for ^{8}He was recorded. These events probably originate from more dissipative reactions which could bias the data towards α/2p2n removal relative to 2p removal.

3 Conclusions

A complete analysis of multi-neutron decay energy spectra is a tool to explore neutron unbound systems which decay by emission of three or four neutrons. Even if the statistics are not sufficient to extract spectra with clean identification of each neutron, the presence of additional neutrons can be inferred through the multiplicity and the relative intensities of the decay spectra. In low-energy cases where the experimental sensitivity is high a limit can be established. In the present case, the data are dominated by $\alpha/2p2n$-knockout and there is no evidence for the existence of a low-lying ($\leq$ 1 MeV) resonance in ^{12}He.

Acknowledgments

We would like to thank L.V. Grigorenko for thoughtful discussion, and for providing theoretical estimates of the 5-body decay energies of the ^{12}He system. This work was supported by the National Science Foundation under Grants No. PHY09-22335, PHY09-69058, PHY09-69173, PHY12-05357, PHY12-05537, and PHY11-02511. The material is also based upon work supported by the Department of Energy National Nuclear Security Administration under Award No. DE-NA0000979.

References

[1] S. Sidorchuk et al., Phys. Rev. Lett. **108**, 202502 (2012)
[2] H. Johansson et al., Nucl. Phys. A **842**, 15 (2010)
[3] L.V. Grigorenko, M.V. Zhukov, Phys. Rev. C **77**, 034611 (2008)
[4] Z. Kohley et al., Phys. Rev. Lett. **109**, 232501 (2012)
[5] M. Bird et al., IEEE Trans. Appl. Supercond. **15**, 1252 (2005)
[6] S. Agostinelli et al., Nucl. Instrum. Mehtods A **506**, 240 (2003)
[7] B. Roeder, EURISOL Design Study, Report No. 10-25-2008-006-In-beamvalidations.pdf pp. 31–44 (2008)
[8] M.D. Jones et al., Phys. Rev. C **91**, 044312 (2015)
[9] A.N. Kuchera et al., Phys. Rev. C **91**, 017304 (2015)
[10] L.V. Grigorenko, private communication (2014)

EPJ Web of Conferences **113**, 06007 (2016)
DOI: 10.1051/epjconf/201611306007

Four- and three-body channel coupling effects on ^{6}Li elastic scattering with CDCC

Shin Watanabe[1,a], Takuma Matsumoto[1], Kosho Minomo[2], Kazuyuki Ogata[2], and Masanobu Yahiro[1]

[1] *Department of Physics, Kyushu University, Fukuoka 812-8581, Japan*
[2] *Research Center for Nuclear Physics (RCNP), Osaka University, Ibaraki 567-0047, Japan*

Abstract. We investigate breakup dynamics in ^{6}Li elastic scattering on heavy targets (T = ^{209}Bi or ^{208}Pb) near the Coulomb barrier energy. Since the subsystem of ^{6}Li has a bound state as deuteron ($n + p = d$), a four-body channel (^{6}Li + T $\rightarrow n + p + \alpha +$ T) and a three-body channel (^{6}Li + T $\rightarrow d + \alpha +$ T) get entangled during scattering as a breakup channel. Both channels are precisely treated with the four-body version of the continuum-discretized coupled-channels method (four-body CDCC). Four-body CDCC well reproduces measured elastic cross sections with no adjustable parameter. We then estimate the channel coupling effects by dividing the breakup channel into four- and three-body channels. It is found that ^{6}Li breakup is mainly induced by a three-body channel.

1 Introduction

Projectile breakup is essential in reactions of weakly-bound nuclei and appears as a strong coupling effect between elastic and breakup channels. The continuum-discretized coupled-channels method (CDCC) was proposed for treating various kinds of channels including breakup (continuum) channels [1–3]. The coupling effect was first confirmed in deuteron scattering and later verified in halo nuclei near the drip-line in which a two-body projectile + target (T) three-body problem was assumed. Nowadays, CDCC is widely applied to describe three-body dynamics.

Our interest is now going to four-body dynamics in scattering of three-body projectiles. CDCC for three- and four-body scattering are now called three- and four-body CDCC, respectively. It is interesting to consider the difference of four-body scattering between ^{6}He and ^{6}Li as typical examples. Since ^{6}He is a Borromean nucleus in which no binary subsystem is bound, only a four-body channel (^{6}He + T $\rightarrow n + n + \alpha +$ T) exists as a breakup channel. This property makes four-body dynamics relatively simpler and this is the reason why four-body CDCC was first applied to ^{6}He scattering [4, 5]. On the other hand, ^{6}Li has a bound state in the $n + p$ subsystem. Therefore, not only a four-body channel (^{6}Li + T $\rightarrow n + p + \alpha +$ T) but also a three-body channel (^{6}Li + T $\rightarrow d + \alpha +$ T) exists in its breakup channel. This situation makes it more difficult to understand four-body dynamics of ^{6}Li scattering.

[a] e-mail: s-watanabe@phys.kyushu-u.ac.jp

^{6}Li + ^{209}Bi scattering near the Coulomb barrier energy ($E_{\rm b}^{\rm Coul} \approx 30$ MeV) was first analyzed with three-body CDCC based on the $d+\alpha$ + ^{209}Bi three-body model [6]. However, the calculation could not reproduce the measured elastic cross section without a normalization factor 0.8 to d-^{209}Bi and α-^{209}Bi optical potentials. This problem was solved by four-body CDCC based on the $n + p + \alpha$ + ^{209}Bi four-body model [7]. The calculation well describes the experimental data with no adjustable parameter. As an interesting result, it was reported that d breakup in ^{6}Li is strongly suppressed during the elastic scattering. In this work, we focus on four-body dynamics of ^{6}Li elastic scattering from the point of view of four- and three-body channel coupling effects.

2 Theoretical framework and Model Hamiltonian

We recapitulate four-body CDCC based on the $n + p + \alpha$ + T four-body model; see Ref. [3, 7] for the detail. The scattering state Ψ with the total energy E is governed by the four-body Schrödinger equation with the model Hamiltonian H_4:

$$(H_4 - E)\Psi = 0, \qquad H_4 = K_{\boldsymbol{R}} + U_n + U_p + U_\alpha + \frac{e^2 Z_{\rm Li} Z_{\rm T}}{R} + h_{np\alpha}, \tag{1}$$

where $K_{\boldsymbol{R}}$ stands for the kinetic energy operator with respect to the relative coordinate $\boldsymbol{R}$ between ^{6}Li and T, and U_x ($x = n, p, \alpha$) represents the optical potential between x and T. Since Coulomb-breakup effects are negligibly small for the present elastic scattering [6, 7], the Coulomb part of U_p and U_α is then approximated into $e^2 Z_{\rm Li} Z_{\rm T}/R$, where $Z_{\rm A}$ is the atomic number of nucleus A. $h_{np\alpha}$ denotes the internal Hamiltonian of ^{6}Li in the three-cluster model. In CDCC, Eq. (1) is solved in the model space P spanned by the ground and discretized continuum states of ^{6}Li:

$$P = \sum_{\gamma=0}^{N} |\Phi_\gamma\rangle\langle\Phi_\gamma|, \tag{2}$$

where Φ_γ represents the γ-th eigenstate with eigenenergy ε_γ; note that the $\gamma = 0$ and $\gamma = 1$–N correspond to the ground and excited states in P, respectively. The Φ_γ are obtained by diagonalizing $h_{np\alpha}$ with the Gaussian basis functions [8].

In order to disentangle breakup dynamics, we divide the CDCC model space P in the following way. P can be decomposed into the ground-state part P_0 and the breakup-state part P^* as $P = P_0 + P^*$ for

$$P_0 = |\Phi_0\rangle\langle\Phi_0|, \quad P^* = \sum_{\gamma=1}^{N} |\Phi_\gamma\rangle\langle\Phi_\gamma|. \tag{3}$$

For later discussion, P^* is further divided into a subspace $P_{np\alpha}$ dominated by $np\alpha$ configurations and a subspace $P_{d\alpha}$ by $d\alpha$ configurations. The subspaces are defined as follows. The probability of $d\alpha$ configurations in the breakup state Φ_γ is obtained by the overlap between Φ_γ and the d ground state $\phi^{(d)}$: $\Gamma_\gamma^{(d\alpha)} = \left|\langle\phi^{(d)}|\Phi_\gamma\rangle\right|^2$. We then define a breakup state with $\Gamma_\gamma^{(d\alpha)} > 0.5$ ($\Gamma_\gamma^{(d\alpha)} \leq 0.5$) as a $d\alpha$-dominant ($np\alpha$-dominant) state. The subspace $P_{d\alpha}$ ($P_{np\alpha}$) is a model space spanned by $d\alpha$-dominant ($np\alpha$-dominant) breakup states. Consequently, the model space P of CDCC calculations is expressed as

$$P = P_0 + P_{d\alpha} + P_{np\alpha}. \tag{4}$$

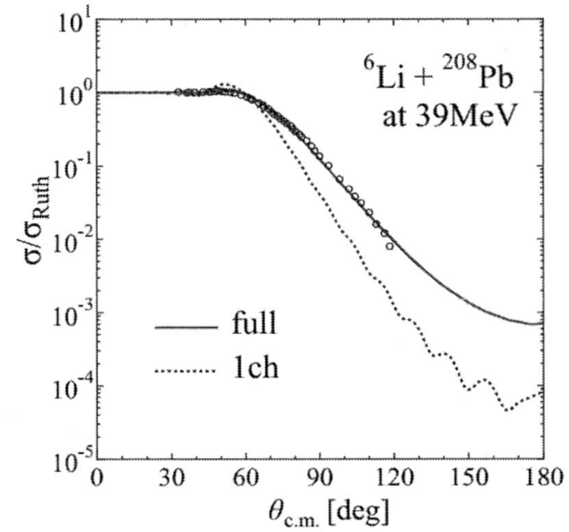

Figure 1. (Color online) Elastic cross sections (normalized by the Rutherford cross section) for ^{6}Li + ^{208}Pb scattering at 39 MeV. The solid line represents the result of four-body CDCC calculation with full channel coupling, whereas the dotted line shows the result of 1ch calculation with no channel coupling. The experimental data is taken from Ref. [11].

3 Results

We only show the results of ^{6}Li+^{208}Pb elastic scattering at 39 MeV; see Ref. [7] for the analysis of ^{6}Li+^{209}Bi scattering and the detail of model setting. As for U_n, we take the potential of Koning and Delaroche [9] determined from the measured elastic cross section of n + ^{209}Bi. For simplicity, the spin-orbit interaction is neglected and U_p is assumed to have the same geometry as U_n. The potential U_α is taken from Ref. [10] determined from measured differential cross sections of α + ^{209}Bi scattering at 19–22 MeV.

The angular distribution of elastic cross sections for ^{6}Li + ^{208}Pb scattering at 39 MeV are plotted in Fig 1. The one-channel (1ch) calculation with no breakup (dotted line) underestimates the experimental data, whereas the four-body CDCC calculation (solid line) reproduces the experimental data. The enhancement from the dotted to solid lines is induced by channel coupling effects, indicating ^{6}Li breakup effects are quite important. Thus, ^{6}Li scattering on heavy targets are well described by four-body CDCC with no adjustable parameter.

Next, we consider four- and three-body channel coupling effects on the present ^{6}Li scattering. For this purpose, the subspaces $P_{d\alpha}$ and $P_{np\alpha}$ are switched off from the full calculation, respectively. Figure 2 shows the angular distribution for ^{6}Li + ^{208}Pb scattering at 39 MeV, and the solid and dotted lines are the same as in Fig. 1. When the model space is limited to $P_0 + P_{np\alpha}$, the dot-dashed line is obtained [see Fig. 2(a)] and close to the result of 1ch calculation (dotted line). On the other hand, when the model space is limited to $P_0 + P_{d\alpha}$, we have the dashed line [see Fig. 2(b)] that well simulates the full calculation (solid line). Note that the number of $d\alpha$-dominant states in P is as few as about one-eighth of that of $np\alpha$-dominant states. The result indicates that the coupling between P_0 and $P_{d\alpha}$ is dominant. In other words, ^{6}Li breakup is mainly induced by the $d + \alpha$ breakup, and the d-breakup is suppressed in ^{6}Li scattering.

4 Summary

We have analyzed the four-body dynamics of ^{6}Li elastic scattering. The elastic scattering of ^{6}Li + ^{208}Pb at 39 MeV is well described by four-body CDCC based on the $n + p + \alpha +$ ^{208}Pb model. The breakup channels are approximately divided into four- and three-body channels and the coupling effects are estimated. ^{6}Li breakup is mainly caused by a strong transition between the elastic (P_0)

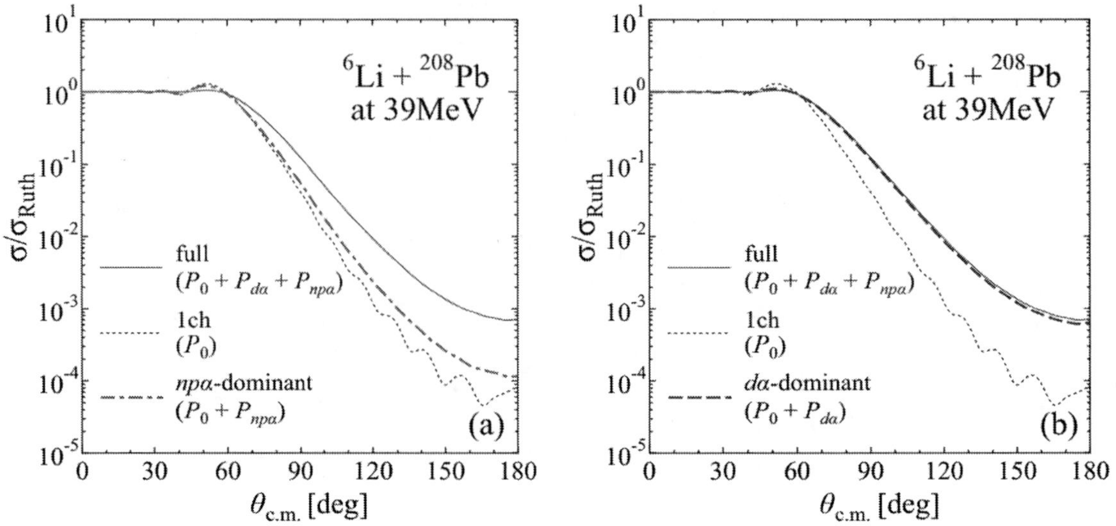

Figure 2. (Color online) Elastic cross sections (normalized by the Rutherford cross section) for ^{6}Li + ^{208}Pb scattering at 39 MeV. The solid (dotted) line represents the result of full (1ch) calculation. The dot-dashed line in the panel (a) shows the calculation in which the model space is limited to $P_0 + P_{np\alpha}$, whereas the dashed line in the panel (b) denotes the one with $P_0 + P_{d\alpha}$.

and three-body channel ($P_{d\alpha}$), suggesting d-breakups effect are suppressed in ^{6}Li scattering. We will investigate what causes the suppression, and the energy and target dependence in the forthcoming paper.

Acknowledgements

This work was supported in part by Grant-in-Aid for Scientific Research (KAKENHI) from Japan Society for the Promotion of Science 25·4319.

References

[1] M. Kamimura *et al.*, Prog. Theor. Phys. Suppl. **89**, 1 (1986)
[2] N. Austern *et al.*, Phys. Rep. **154**, 125 (1987)
[3] M. Yahiro *et al.*, Prog. Theor. Exp. Phys. **2012**, 01A206 (2012)
[4] T. Matsumoto *et al.*, Phys. Rev. C **70**, 061601(R) (2004)
[5] T. Matsumoto *et al.*, Phys. Rev. C **73**, 051602(R) (2006)
[6] N. Keeley *et al.*, Phys. Rev. C **68**, 054601 (2003)
[7] S. Watanabe *et al.*, Phys. Rev. C **86**, 031601(R) (2012)
[8] E. Hiyama, Y. Kino, and M. Kamimura, Prog. Part. Nucl. Phys. **51**, 223 (2003)
[9] A. J. Koning and J. P. Delaroche, Nucl. Phys. A **713**, 231 (2003)
[10] A. R. Barnett and J. S. Lilley, Phys. Rev. C **9**, 2010 (1974)
[11] S. Santra *et al.*, Phys. Rev. C **83**, 034616 (2011)

EPJ Web of Conferences **113**, 06008 (2016)
DOI: 10.1051/epjconf/201611306008

Pairing and quartetting in medium mass nuclei

M. Assié[1,a], B. Le Crom[1], L. Lefebvre[1], and J.-A. Scarpaci[2]

[1] *Institut de Physique Nucléaire, Université Paris-Sud-11 - CNRS/IN2P3, 91406 Orsay, France*
[2] *CSNSM, Université Paris-Sud-11 CNRS/IN2P3 91405 Orsay-Campus France.*

Abstract. Pairing and quartetting are competing phenomena in nuclei. In the self-conjugate nuclei from the *fp*-shell, the competition between T=0 pairing (deuteron-like) and alpha clusterisation is predicted to be affected by the spin-orbit interaction. In order to probe the α structure of *fp*-shell nuclei, the nuclear break-up of ^{40}Ca and ^{40}Ar has been measured at GANIL. The α spectroscopic factors obtained for the N=Z nucleus and the N≠Z nuclei do not reflect any difference. This might be an indication that the tail of the wave function in not sensitive to nuclear asymmetry. Perspectives for probing the deuteron-like pairing (T=0) are given.

1 Introduction

Pairing is a well know feature of the nuclear interaction. Beyond the mean field picture, including like-particle pairing describes nuclear structure effects such as the odd-even mass staggering. The Wigner term added to the mass formula could be a signature of the T=0 pairing which is specific to neutron-proton pairing. Like-particle pairing in the T=1 channel can be accounted for at normal density in the BCS approach by weakly coupled Cooper pairs.

Infinite nuclear matter BCS calculations in ref.[1] give the evolution of the size of the Cooper pair as a function of the density. For the low density region, the size of the Cooper pair is smaller than the inter-nucleon distance. This reduced size of the pair is the signature for condensation. The best candidates to observe such phenomenon in nuclei is the halo of light nuclei where the density can reach very low values. The size of neutron-proton pairs (see ref.[2]) and quartets (two neutrons and two protons) (see ref.[3]) can also be determined in such calculations. The condensation phenomenon can be observed at normal density for deuteron-like pairs and at lower density for alpha-like particles. So that in infinite matter, the deuteron condensation prevails over alpha condensation. In nuclei, the picture may change due to the effect of the spin-orbit interaction.

Indeed, the spin-orbit force is affecting the neutron-proton interaction, particularly in the T=0 channel. Shell model calculations [4] and quartets model [5] show that adding the spin-orbit interaction hinders neutron-proton pairing for nuclei in the *fp*-shell. So that the best candidates for the search for neutron-proton pairing are the self-conjugate *sd*-shell nuclei. On the other hand, α clusterization might be stronger in the *fp*-shell nuclei.

In order to investigate such phenomena, *fp*-shell nuclei have been studied through nuclear break-up. The emission of α particles from ^{40}Ca and ^{40}Ar has been measured in order to investigate the

[a] e-mail: assie@ipno.in2p3.fr

alpha structure of their ground state. Selected results from both studies will be presented in the two following sections.

2 Simultaneous study of alpha break-up from the N=Z nucleus ^{40}Ca and from the N≠Z nucleus ^{40}Ar

Alpha clusterisation is a well known phenomenon in excited states of nuclei close to the decay threshold as depicted in the Ikeda diagram. Very little is known about clusterisation in the ground state of nuclei. Nuclei which have a pronounced α-cluster structure in their ground states are for example ^{8}Be, ^{9}Be, ^{6}Li and ^{7}Li. ^{8}Be is formed out of two almost free α-particle, roughly 4 fm apart and barely touching one another. The two α particle are only held together via the Coulomb barrier. Because of the large distance between the two α particles, the 0^{+} ground state of ^{8}Be has a spherical density distribution close to one third of the saturation density. The α structure of the ground state of heavier nuclei would give an indication of quartetting at normal density.

The existence of α particles into the ground state of heavier nuclei such as ^{40}Ca has been investigated. As a comparison we have measured simultaneously the nuclear break-up of ^{40}Ar which is slightly isopsin asymmetric. We have performed the measurement of nuclear break-up of ^{40}Ar beam on a ^{40}Ca target at 35A MeV at GANIL. The set-up was using the SPEG spectrometer coupled with the MUST2 telescopes composed double sided Silicon detectors coupled to CsI detectors (see fig.1). The excitation energy resolution obtained with this set-up was of about 500 keV (FWHM).

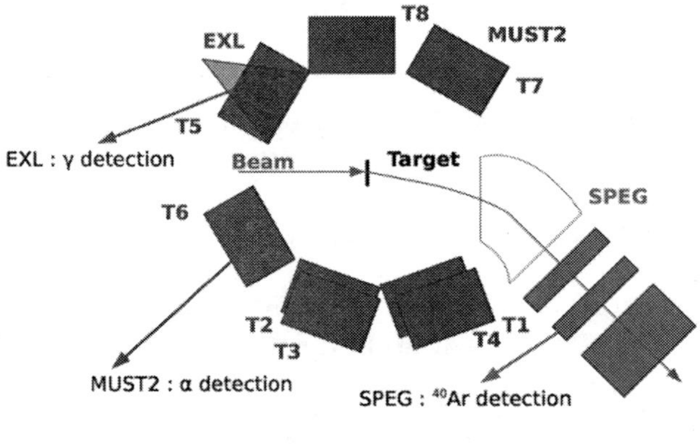

Figure 1. Experimental set-up on the SPEG line. The MUST2 Silicon detectors coupled to CsI were surrounding the ^{40}Ca target.

As for the nuclear break-up of ^{40}Ca, the excitation energy spectra of ^{36}Ar obtained by missing mass method is shown in fig.2. In the laboratory frame in backward direction, the statistical decay of the target gives the main contribution whereas in the forward direction, 3 peaks can clearly be seen. They can be attributed to the first 0^{+}, 2^{+} and 4^{+} states of ^{36}Ar although the assignment of the 4^{+} is more difficult (see spectrum of fig.2). The angular distribution extracted for the contributions of these three peaks were compared to a Time Dependent Schrödinger Equation (TDSE) calculation in order to extract the alpha spectroscopic factor of the ground state of ^{40}Ca (for more detail see ref.[6]). The TDSE calculations relies on the Buck potential [7] to reproduce the rotational bands in ^{40}Ca and the Pauli principle is accounted for by taking into account only the 6s, 5d, 4g waves.

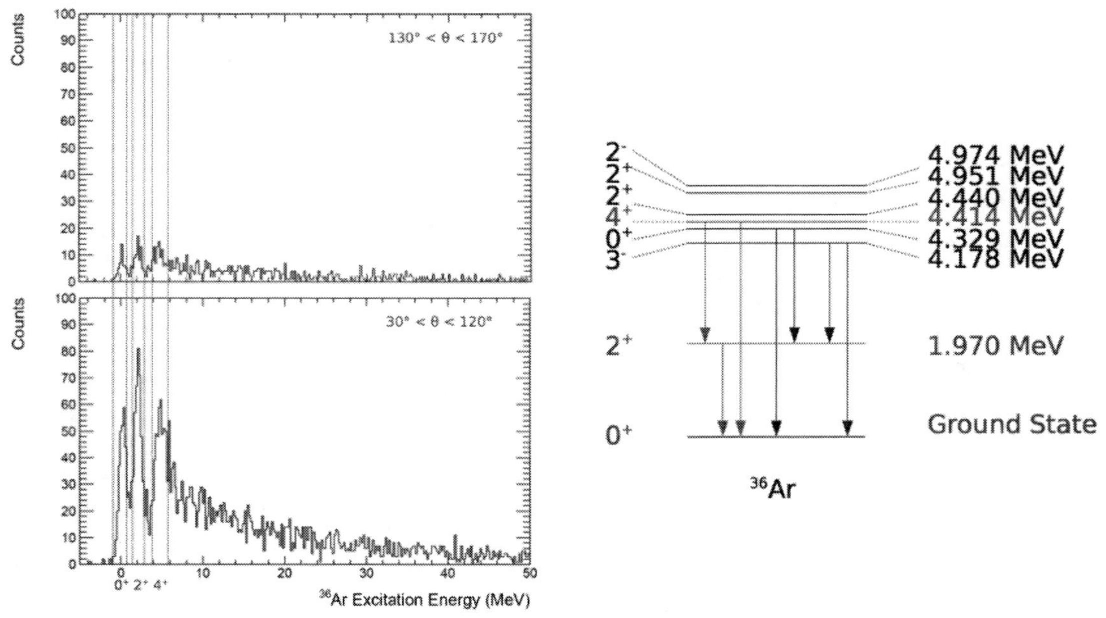

Figure 2. (Left) Missing energy spectrum of ^{36}Ar for an ^{40}Ca excitation energy range greater than 15 MeV. (Bottom) For α-particles detected between 30 deg. and 120 deg. in the laboratory system. (Top) For α-particles detected between 130 deg. and 170 deg. in the laboratory system. The blue gate represents the 0^+ ground state at 142 keV, the green one the 2^+ excited state at 2.1 MeV and the red one the 4^+ state around 4.75 MeV. (Right) Level scheme of ^{36}Ar.

The inclusive spectroscopic factor in the ground state of ^{40}Ca is found to be 3.9 ± 0.7. This is in good agreement with the shell model calculations from ref.[8] but much lower than the transfer reaction ^{40}Ca(d,^{6}Li) measurement from ref.[9].

The nuclear break-up of ^{40}Ar can be identified clearly in the spectra of parallel versus perpendicular momentum obtained in coincidence with ^{36}S in SPEG (see fig.3). A contribution from the statistical decay of the target can also be seen in all direction, whereas the breakup is located in the forward left side of the beam. The angular distribution obtained in this case are shown together with the theoretical angular distributions obtained from the TDSE calculations. Assuming the same contribution for the three waves (6s, 5d, 4g), an inclusive spectroscopic factor for α clusterisation in ^{40}Ar of 4.0±2.5 is obtained.

The α spectroscopic factors obtained in a single measurement for ^{40}Ca and ^{40}Ar are compatible within the error bars. Nuclear break-up is probing the peripheral part of the wave function which is probably not very sensitive to the slight asymmetry between these two nuclei.

3 Perspectives

In order to further investigate the relationship between deuteron-like correlations and α-like correlation, transfer reactions like (p,^{3}He) which allows for deuteron-like transfer (ΔT=0) but also ΔT=1 transfer are an efficient probe [10]. An experiment has been performed at GANIL to measure (p,^{3}He) and (d,α) reaction on the closed shell nucleus ^{56}Ni and on the open shell nucleus ^{52}Fe. By comparing

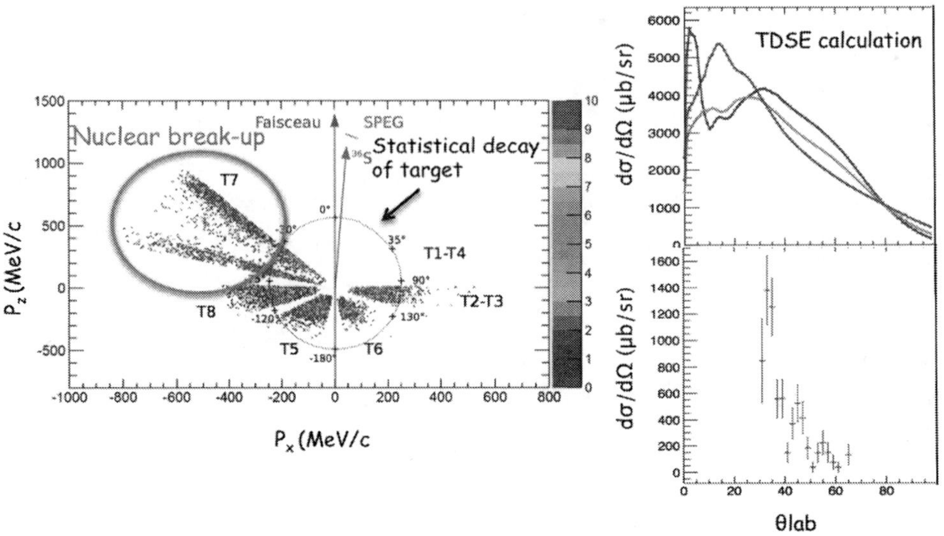

Figure 3. (Left) Parallel momentum distribution versus perpendicular momentum distribution. (Right) Angular distributions: (top) from TDSE calculations for the three contributing waves 6s, 5d, 4g (bottom) from the experiment.

the population of the T=0 and T=1 states in the residual nucleus, the relative strength of the two pairing channels can be inferred. The analysis of these data is underway. They will extend the systematics of deuteron transfer reactions to the *fp*-shell region where no data are available yet. The comparison between α and deuteron clusterisation will be given by the comparison of (d,α) and (d,^{6}Li) transfer reactions. It will open new perspectives to the understanding of self-conjugate nuclei.

References

[1] M. Matsuo, Phys. Rev. C **73**, 044309 (2006).
[2] U. Lombardo and P. Schuck, Phys. Rev. C **63** 038201 (2001).
[3] G. Röpke and A Schnell and P. Schuck, Phys. Rev. Lett. **80**, 3177 (1998).
[4] A. Gezerlis and G.F. Bertsch and Y.L. Luo, Phys. Rev. Lett. **106**, 252502 (2011).
[5] M. Sambataro and N. Sandulescu, Phys. Rev. C **91** 064318 (2015).
[6] L. Lefebvre, Proc. of Science (Bormio 2013) **069**.
[7] B. Buck *et al*, Phys. Rev. C **51**, 1840 (1995).
[8] W. Chung *et al*, Phys. Lett. B **79**, 381 (1978).
[9] K. Umeda *et al*, Nucl. Phys. A **429** 88 (1984).
[10] P. van Isäcker, D.D. Warner and A. Franck, Phys. Rev. Lett. **94**, 162502 (2005).

EPJ Web of Conferences **113**, 06009 (2016)
DOI: 10.1051/epjconf/201611306009

Contributions of chiral two- and three-nucleon interactions to closed shell nuclei

Yiharn Tzeng[1,a], Shwu-Yun Tsay Tzeng[2,b], and T. T. S. Kuo[3,c]

[1] *Institute of Physics, Academia Sinica, Taipei 11529, Taiwan, ROC*
[2] *Dept. of Electro-Optical Engineering, National Taipei University of Technology, Taipei 10608, Taiwan, ROC*
[3] *Department of Physics, State University New York at Stony Brook, Stony Brook , New York 11794, USA*

Abstract. Using two-nucleon force alone is usually inadequate to interpret nuclear systems' experimental data. We adopt a chiral N^3LO two-nucleon potential V_{2N} with the inclusion of an in-medium three-nucleon (NNN) force $\bar{V}_{3N}$ to calculate binding energies of closed-shell nuclei. The matrix elements of low momentum nucleon-nucleon potential V_{low-k} obtained from integrating the high momentum part of a realistic potentials is inputted in the particle-particle hole-hole ring diagram calculation to study nuclear properties. Nuclear binding energies are accurately reproduced. Without this three-nucleon force, the nuclear binding energy is too weak, as already known. The correction from ring diagrams of order higher than 1 can not be ignored.

The main purpose of this work is to investigate the contributions of a chiral N^3LO two-nucleon potential V_{2N} and an in-medium three-nucleon (3N) force V_{3N}[1] to binding energies of closed shell nuclei, such as ^{16}O, and ^{40}Ca. To achive this aim, we perform the calculation in the framework of a ring diagram expansion[2, 3], as shown in Fig. 1. The particle-particle hole-hole ring diagrams are summed to all orders in this expansion. This method has been applied to several calculations on nuclear matter and neutron star as well as on cold neutron matter at the unitary limit. Applications to finite nuclei[3] have been quite limited. Such ring-diagram calculations for closed-shell nuclei using realistic V_{2N} and V_{3N} have not been carried out. The low momentum V_{low-k} effective interaction matrix elements[4] of V_{2N} and V_{3N} are calculated for the uses in the ring diagram formulism. This V_{low-k} interaction has been successfully applied to finite nuclei in shell model calculations and studies in neutron matter.

The leading contribution to V_{3N} occurs at N^{2}LO in the chiral power counting and is composed of a long-range two-pion exchange component $V_{3N}^{(2\pi)} = \sum_{i\neq j\neq k} \frac{g_A^2}{8f_\pi^4} \frac{\vec{\sigma}_i\cdot\vec{q}_i\,\vec{\sigma}_j\cdot\vec{q}_j}{(\vec{q}_i^{\,2}+m_\pi^2)(\vec{q}_j^{\,2}+m_\pi^2)} F_{ijk}^{\alpha\beta}\tau_i^\alpha\tau_j^\beta$, a medium-range one-pion exchange term $V_{3N}^{(1\pi)} = -\sum_{i\neq j\neq k} \frac{g_A c_D}{8f_\pi^4\Lambda_\chi} \frac{\vec{\sigma}_j\cdot\vec{q}_j}{\vec{q}_j^{\,2}+m_\pi^2}\vec{\sigma}_i\cdot\vec{q}_j\,\vec{\tau}_i\cdot\vec{\tau}_j$, and a pure contact interaction $V_{3N}^{(ct)} = \sum_{i\neq j\neq k} \frac{c_E}{2f_\pi^4\Lambda_\chi}\vec{\tau}_i\cdot\vec{\tau}_j$, where $g_A = 1.29$, $f_\pi = 92.4$ MeV, $\Lambda_\chi = 700$ MeV, $m_\pi = 138.04$ MeV/c^2 is the average pion mass, $\vec{q}_i = \vec{p}_i^{\,\prime} - \vec{p}_i$ is the difference between the final and initial momentum of nucleon i and $F_{ijk}^{\alpha\beta} = \delta^{\alpha\beta}\left(-4c_1m_\pi^2 + 2c_3\vec{q}_i\cdot\vec{q}_j\right) + c_4\epsilon^{\alpha\beta\gamma}\tau_k^\gamma\vec{\sigma}_k\cdot\left(\vec{q}_i\times\vec{q}_j\right)$. The low-energy constants $c_1 = -0.76\,\text{GeV}^{-1}$, $c_3 = -4.78\,\text{GeV}^{-1}$, and $c_4 = 3.96\,\text{GeV}^{-1}$ appear already in the N^{2}LO two-nucleon

[a] e-mail: phtzengyh@sinica.edu.tw
[b] e-mail: sytsay@mail.ntut.edu.tw
[c] e-mail: kuo@tonic.physics.sunysb.edu

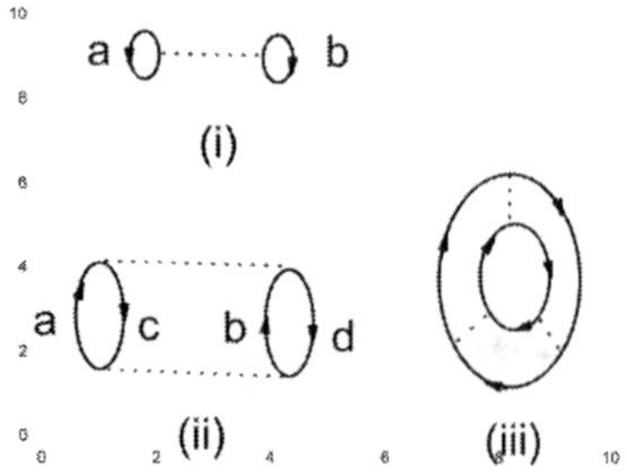

Figure 1. Some lowest order ring diagrams.

potential and are therefore constrained by low-energy NN phase shifts. The low-energy constants c_D and c_E are typically fit to reproduce the properties of light nuclei[5].

We then calculate effective low momentum V_{low-k} matrix elements from V_{2N} and V_{3N} for the use in the ring diagram calculations. Similar to the half-on-shell T-matrix in the usual Lippmann-Schwinger equation $T(k',k,k^2) = V_{2N}(k',k) + P\int_0^{\infty} q^2dqV_{2N}(k',q)\frac{1}{k^2-q^2} \times T(q,k,k^2)$, the effective V_{low-k} interaction is calculated from a T-matrix equivalence requirement.

$$T_{low-k}(p',p,p^2) = V_{low-k}(p',p) + P\int_0^{\Lambda} q^2dqV_{low-k}(p',q)\frac{1}{p^2-q^2} \times T_{low-k}(q,p,p^2),$$

with $T(p',p,p^2) = T_{low-k}(p',p,p^2)$; $(p',p) \leq \Lambda$,for V_{2N} and V_{3N}. In the above equations Λ denotes a momentum space cut-off (such as $\Lambda \simeq 2.0fm^{-1}$), and the symbol $P\int$ the principal-value integration. It was found that under this process, using different realistic potentials would lead to almost the same results in fitting NN phase shifts and deuteron properties. The Lee-Suzuki iteration method [6] has been employed in the calulation of V_{low-k}.

With these ring diagram summed to all orders, the ground-state energy shift from V is given as [2]

$\Delta E_0 = \int_0^1 d\lambda \sum_m \sum_{ijkl\in P} Y_m(ij,\lambda)Y_m^*(kl,\lambda) \times \langle ij|V|kl\rangle$, where (i,j,k,l) are each a shell-model s.p. wave function, and P denotes a chosen shell-model space composed of a set of hole (h) and particle (p) orbits. For example, for ^{16}O we take P as composed of the three h orbits $(0s_{1/2}, 0p_{3/2}, 0p_{1/2})$ and the seven p ones $(0d_{5/2}, ...1p_{1/2})$. The amplitudes Y above are calculated from an RPA-type equation,

$$\sum_{ef}[(\epsilon_i + \epsilon_j)\delta_{ij,ef} + \lambda(1-n_i-n_j)\langle ij|V|ef\rangle] \times Y_m(ef,\lambda) = \omega_m(\lambda)Y_m(ij,\lambda);$$

where $(i,j,e,f) \in P$,and λ a strength parameter, to be integrated from 0 to 1.

The occupation factors are $n_a = 1$ for $a = h$, and $= 0$ otherwise. Thus the amplitudes $Y_m(ij)$ has only either hh (i=h, j=h') or pp (i=p, j=p') components. The transition amplitudes Y in the equation can be classified into two types, one dominated by hh and the other by pp components. We include only the former, denoted by Y_m, for the calculation of the all-order sum of the $pphh$ ring diagrams.

We use HF s.p. spectrum ϵ_j in the above RPA equation, $\epsilon_j = \langle j|K_{sp}|j\rangle + \sum_h\langle jh|V|jh\rangle$where K_{sp} denotes the s.p. kinetic energy operator. Note that j and h are each oscillator s.p. wave function.

The parameters c_D and c_E in the one-pion exchange and contact terms of V_{3N} are taken from [5] where binding energies of $A = 3$ nuclei were fitted. To calculate the V_{low-k} matrix for V_{2N} and V_{3N} , we have used the Bertsch formula $\hbar\omega = 45.0A^{-1/3} - 25.0A^{-2/3}$ for the nuclear wave functions. The nucleon densities arising from the contact term of V_{3N} for the nuclei considered are chosen as those approximately at the RMS radii from the experimental charge density profiles[7] of the nuclei. The RMS radii for ^{16}O, and ^{40}Ca are 2.7013, and 3.4764 fm[8] respectively. In this way, we set the densities around the RMS radii to be $0.6\rho_0$ for both ^{16}O and ^{40}Ca, with ρ_0 the nucleon density in nuclear matter. We have also performed theoretical calculations using harmonic oscillator wave functions and obtained similar results.

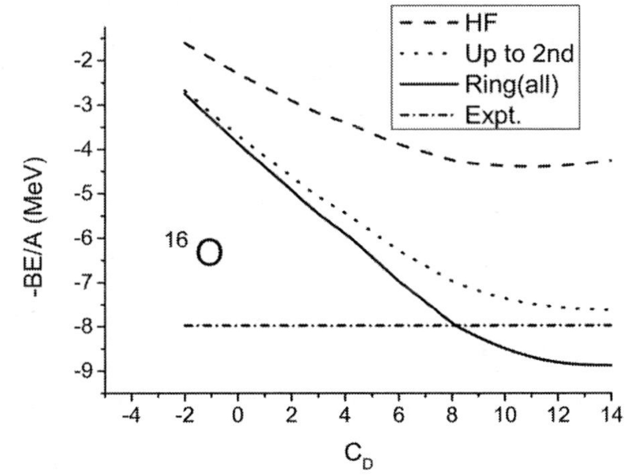

Figure 2. Dependence of $-BE/A$ of ^{16}O on the parameter c_D of V_{3N}.

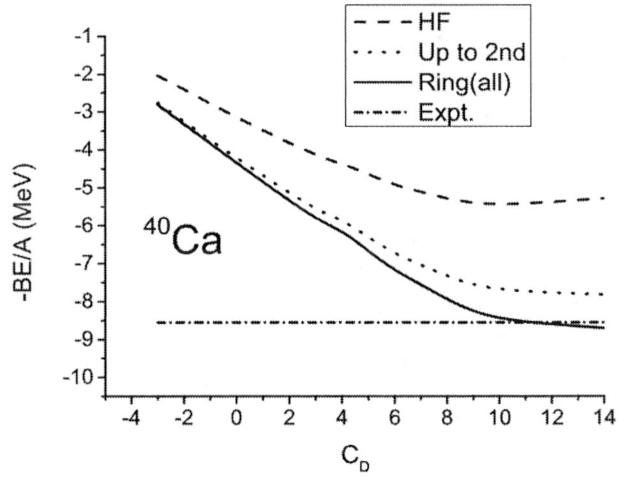

Figure 3. Dependence of $-BE/A$ of ^{40}Ca on the parameter c_D of V_{3N}.

Shown in Figs. 2-3 are the dependence of ground state energies per nucleon (or -BE/A) for ^{16}O and ^{40}Ca on the 3N force constant c_D appearing in V_{3N}. In these figures, results from the first order ring diagram (denoted as "HF"), up to the second order one(denoted as "Up to 2nd"), and all order ring diagrams (denoted as "Ring(all)") are all calculated with V_{3N} included. Experimental data[9] are displayed for comparison. We can see from the figures, although the contribution from diagrams up to the second order one improves quite significantly comparing to that from the first order one alone, contribution from higher order diagrams is needed to fill the discrepancy so that the experimental binding energy can be obtained. Another feature worth of mentioning is that for both nuclei, the ring diagram results all meet the experimental data at c_D=8.5 and its corresponding c_E from [5].

Also shown in Table 1, our results from V_{2N} + V_{3N} with parameter c_D= 8.5 and its corresponding c_E from [5] at densities ρ/ρ_0= 0.6 for both ^{16}O and ^{40}Ca fit the experimental data [9] quite well. In the Table we also examine the importance of V_{3N} to the nuclear binding energies for both nuclei that we considered. As expected, the binding energies obtained from V_{2N} alone are too weak. The deviation between results with and without V_{3N} gets wider when all ring diagrams are included.

Table 1. -BE/A (in MeV) of ^{16}O and ^{40}Ca, with the same notations as those in Figs.2-3, and c_D = 8.5 for both nuclei.

		ρ/ρ_0	HF	up to 2nd	Ring(all)	Expt
^{16}O	V_{2N}	–	-3.76	-5.17	-5.28	
	V_{2N}+V_{3N}	0.6	-4.30	-7.09	-8.09	-7.976
^{40}Ca	V_{2N}	–	-4.19	-5.89	-6.24	
	V_{2N}+V_{3N}	0.6	-5.37	-7.48	-8.14	-8.551

In summary, we have applied the ring diagram method to calculate binding energy of ^{16}O and ^{40}Ca by adding an in-medium three-nucleon potential V_{3N} to the chiral N^3LO two-nucleon potential V_{2N}. The ground state energy per nucleon so obtained fits the experimental data quite well for each nucleus considered when V_{3N} is added in and all orders of ring diagrams are included. Contributions from ring diagrams with orders higher than 2 can not be ignored. As expected, binding energies obtained with V_{2N} alone are too weak. Our study demonstrates the importance of three-body force in filling in the discrepancy of nuclear binding energies calculated by including only two-body force as compared to the experimental measurements. We believe that the three-nucleon force is also important in other properties of nuclear systems. We are working on this line and will report the results in other publications in the near future.

References

[1] E. Epelbaum, Prog. Part. Nucl. Phys. **57**, 654 (2006); J. W. Holt, N. Kaiser, W. Weise, PRC81, 024002 (2010).

[2] H. Q. Song, S. D. Yang and T. T. S. Kuo, Nucl. Phys. **A462**, 491 (1987).

[3] Yiharn Tzeng, T. T. S. Kuo, Chin. J. Phys. (Taipei) **25**, 326 (1987); Yiharn Tzeng, T. T. S. Kuo, Nucl. Phys. **A485**, 85 (1988).

[4] S. K. Bogner, T. T. S. Kuo and L. Coraggio, Nucl. Phys. **A684**, (2001) 432; S.K. Bogner, T. T. S. Kuo, L. Coraggio A. Covello, and N. Itaco, Phys. Rev. C **65**, 051301(R) (2002); S. K. Bogner, T. T. S. Kuo, and A. Schwenk, Phys. Rep. **386**, 1 (2003).

[5] P. Navrátil, V. G. Gueorguiev, J. P. Vary, W. E. Ormand, A. Nogga, PRL **99**, 042501 (2007).

[6] S. Y. Lee and K. Suzuki, Phys. Lett. **B91** , 173 (1980); K. Suzuki and S. Y. Lee, Prog. Theor. Phys. **64** , 2091(1980).

[7] H. de Vries, C.W. de Jager, and C. de Vries, At. Data Nucl. data Tables **36**, 495 (1987).

[8] I. Angeli, K. P. Marinova, Atomic and Nucl. Data Tables **99**, 69 (2013).

[9] B. Pfeiffer, K. Venkataramaniaih, U. Czok, C. Scheidenberger, Atomic and Nucl. Data Tables **100**, 403-535 (2014); also see Nuclear data from Brookhaven National Laboratory, http://www.nndc.bnl.gov/chart/

EPJ Web of Conferences **113**, 06010 (2016)
DOI: 10.1051/epjconf/201611306010

Spectroscopy of ^{17}C via one-neutron knockout reaction

Sunji Kim[1,a], Jongwon Hwang[1], Yoshiteru Satou[1], Nigel A. Orr[2], Takashi Nakamura[3], Yosuke Kondo[3], Julien Gibelin[2], N. Lynda Achouri[2], Thomas Aumann[4], Hidetada Baba[5], Franck Delaunay[2], Pieter Doornenbal[5], Naoki Fukuda[5], Naohito Inabe[5], Tadaaki Isobe[5], Daisuke Kameda[5], Daiki Kanno[3], Nobuyuki Kobayashi[3], Toshio Kobayashi[6], Toshiyuki Kubo[5], Sylvain Leblond[2], Jenny Lee[5], F. Miguel Marqués[2], Ryogo Minakata[3], Tohru Motobayashi[5], Daichi Murai[7], Tetsuya Murakami[8], Kotomi Muto[6], Tomohiro Nakashima[3], Noritsugu Nakatsuka[8], Alahari Navin[9], Seijiro Nishi[3], Shun Ogoshi[3], Hideaki Otsu[5], Hiromi Sato[5], Yohei Shimizu[5], Hiroshi Suzuki[5], Kento Takahashi[6], Hiroyuki Takeda[5], Satoshi Takeuchi[5], Ryuki Tanaka[3], Yasuhiro Togano[10], Adam G. Tuff[11], Marine Vandebrouck[12], and Ken-ichiro Yoneda[5]

[1]*Department of Physics and Astronomy, Seoul National University, 599 Gwanak, Seoul 151-742, Republic of Korea*
[2]*LPC-ENSICAEN, IN2P3-CNRS et Université de Caen, F-14050, Caen Cedex, France*
[3]*Department of Physics, Tokyo Institute of Technology, 2-12-1 O-Okayama, Meguro, Tokyo 152-8551, Japan*
[4]*Institut für Kernphysik, Technische Universität Darmstadt, D-64289 Darmstadt, Germany*
[5]*RIKEN Nishina Center, Hirosawa 2-1, Wako, Saitama 351-0198, Japan*
[6]*Department of Physics, Tohoku University, Aramaki Aoba 6-3, Aoba, Sendai, Miyagi 980-8578, Japan*
[7]*Department of Physics, Rikkyo University, Toshima, Tokyo 171-8501, Japan*
[8]*Department of Physics, Kyoto University, Kyoto 606-8502, Japan*
[9]*GANIL, CEA/DSM-CNRS/IN2P3, F-14076 Caen Cedex 5, France*
[10]*ExtreMe Matter Institute EMMI and Research Division, GSI Helmholtzzentrum für Schwerionenforschung GmbH, D-64291 Darmstadt, Germany*
[11]*Department of Physics, University of York, Heslington, York YO10 5DD, United Kingdom*
[12]*IPN Orsay, Université Paris Sud, IN2P3-CNRS, F-91406 Orsay Cedex, France*

Abstract. A spectroscopic study of ^{17}C was performed via the one-neutron knockout reaction of ^{18}C on a carbon target at RIKEN-RIBF. Three unbound states at excitation energies of 2.66(2), 3.16(5), and 3.97(3) MeV (preliminary) were observed. The energies are compared with shell-model calculations and existing measurements to deduce their spin-parities. From the comparison, the states at 2.66(2) and 3.97(3) MeV are suggested to be $1/2^-$ and $3/2^-$, respectively. From its decay property, the state at 3.16(5) MeV is indicated to be $9/2^+$.

1 Introduction

The unique features of neutron-rich nuclei, such as non-canonical shell structures differing from those near stability, have been a subject of intensive studies in both theory and experiment. Among the neutron-rich nuclei, neutron-rich carbon isotopes have attracted particular attention in recent years

[a]e-mail: sjkim06@snu.ac.kr

due to their anomalous level structures. For example, the ground state spin-parities (J^π) of ^{15,17,19}C nuclei, $1/2^+$, $3/2^+$, and $1/2^+$, do not follow naive shell-model expectations. The 2^+ excitation energy, $E_x(2_1^+) = 1588(20)$ keV, of $^{20}C_{14}$ turned out to be not much different from that of neighboring even-even $^{18}C_{12}$ of 1585(10) keV [1]. This migration of the 2_1^+ energy is distinct from that of oxygen isotopes, where $E_x(2_1^+)$ exhibits a sudden jump at $^{22}O_{14}$, signifying the emergence of the shell gap at the neutron number $N = 14$ [2]. The breakdown of the $N = 14$ gap in carbon isotopes has been ascribed to the near degeneracy of the $s_{1/2}$ and $d_{5/2}$ orbits and to the reduced neutron-neutron interaction when going from oxygen to carbon nuclei [1]. Detailed mechanisms and interplay of these effects, however, remain to be answered, requiring further studies. Here we present a spectroscopic study on a neutron-rich odd carbon isotope, ^{17}C, performed to furnish information on energy levels, especially above the neutron threshold. To populate neutron unbound states in ^{17}C, the one-neutron knockout reaction of ^{18}C on a carbon target was utilized.

2 Experiment

The experiment was performed at the Radioactive Isotope Beam Factory (RIBF) at RIKEN, operated by the RIKEN Nishina Center and the Center for Nuclear Study of the University of Tokyo, using the SAMURAI (Superconducting Analyser for MUlti particles from RAdio Isotope beams) spectrometer [3] during the first physics runs using the apparatus. For the spectroscopy of ^{17}C through the one-neutron knockout reaction from ^{18}C, beam particles were identified by the $B\rho$-TOF-ΔE method and the ^{18}C beam was selected. The ^{18}C beam with the mid-target energy of 245 MeV/nucleon and a momentum spread of ±3% impinged on a carbon reaction target having a thickness of 1.8 g/cm^2. The beam intensity was about 2300 pps and the mass resolution was $A/\Delta A = 770$ in σ.

The unbound states in ^{17}C produced by one-neutron knockout of ^{18}C immediately decayed into a ^{16}C fragment and a neutron. Downstream of the target, the charged particles were bent by a superconducting dipole magnet having a central magnetic field of 3 T. They were also identified by the $B\rho$-TOF-ΔE method and the ^{16}C fragment was identified in terms of both charge and mass. The fragment mass resolution was $A/\Delta A = 250$ in σ. The momentum vector of the neutron was obtained by the position and timing information from the neutron hodoscope, NEBULA, consisting of 120 modules of plastic scintillators for neutron detection (NEUT) and 24 modules of plastic scintillators for charged-particle veto (VETO). For NEUT, TOF resolution was 270 ps for a flight length of around 11 m. A γ-ray detector array, DALI2 [4], having 140 blocks of NaI(Tl) scintillators surrounded the target to measure γ rays emitted from the excited fragments.

3 Results and Discussion

The energies of the unbound states in ^{17}C were reconstructed by the momentum vectors of the ^{16}C fragment and neutron utilizing the invariant mass method. As preliminary results, three unbound states at the relative energies of 0.66(5), 1.92(1), and 3.23(3) MeV corresponding to the excitation energies (E_X) of 3.16(5), 2.66(2), and 3.97(3) MeV, respectively, were observed. The 3.16-MeV state was observed to decay into the 2^+ state in ^{16}C at 1.760(3) MeV [5]. The states at 2.66 and 3.97 MeV are associated with p-wave neutron knockout, as evidenced by their momentum distributions (both longitudinal and transverse ones) which are well described by the Glauber model calculations for $L = 1$ [6], and correspond well to the reported $1/2^-$ [2.71(2) MeV] and $3/2^-$ [3.93(2) MeV] states [7], respectively.

Fig. 1 compares the measured level energies with existing data [7–9] and shell-model calculations, separately for both negative (a) and positive (b) parity states. The J^π of the state at 3.16 MeV is not

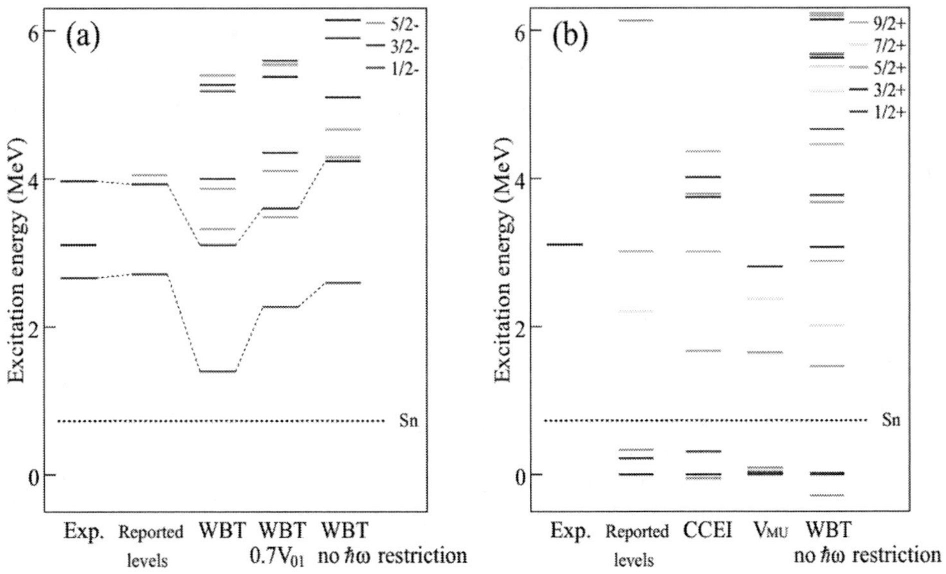

Figure 1. Negative (a) and positive (b) parity states in ^{17}C. "Exp." refers to the states observed in this study. For others see the text.

definitely known and this state is plotted in both panels of Fig. 1 in black. "WBT" and "WBT 0.7V_{01}" refer to the calculations using the WBT interaction [10] within the $1\hbar w$ basis and the *p-sd* model space and those with reduced two-body matrix elements (by multiplying a factor 0.7) for the $J = 0$ and $T = 1$ channels [7], respectively. "WBT no $\hbar w$ restriction" provides energies from the NuShellX code [11] which can perform $\hbar w$ unconstrained calculations. Calculations utilizing the coupled-cluster effective interaction [12] and the monopole based universal interaction in the *p-sd* model space [13] are denoted as "CCEI" and "V_{MU}", respectively.

Ueno et al. [7] reported that the $1/2^-$ energy is underpredicted by about 1 MeV with "WBT", while such an underprediction is significantly remedied with "WBT 0.7V_{01}". The present "WBT no $\hbar w$ restriction" reproduces the energy of the $1/2^-$ state well together with that of $3/2^-$. Concerning the "WBT no $\hbar w$ restriction" predictions for negative parity states, there is no candidate for the 3.16-MeV state. This state, instead, has two candidate states, $9/2^+$ and $3/2^+$, in the positive parity state diagram, Fig. 1 (b). In an attempt to clarify the nature of this state, its decay property was examined within the shell model. This state has been observed to populate predominantly the 2^+ state of ^{16}C [5]. For the $3/2^+$ assumption, there are two routes for its decay, namely via *s*-wave and *d*-wave neutron emissions. From a simple estimate for the partial width (Γ), assuming that it is proportional to the product of the spectroscopic factor (C^2S) for decay, the wave number ($k = \sqrt{2\mu E_{decay}}/\hbar c$), the radius of the decay daughter (R), and the transmission through the centrifugal barrier (v_l) [14] as

$$\Gamma \propto C^2S \cdot k \cdot R \cdot v_l, \tag{1}$$

we obtain that 97.5% of the decay is induced by *s*-wave. Here E_{decay} is the decay energy and μ is the reduced mass of the decay products. For this *s*-wave decay, the single-particle resonance width is predicted as 4.84 MeV [14]. A resonance having a width of this order will be hardly visible as a distinct peak structure. For d-wave decay, the single-particle width is calculated as 0.03 MeV [14]. The $9/2^+$ assignment for the 3.16-MeV state, thus, remains as a plausible solution (a $9/2^+$ state can decay to $^{16}C(2^+)$ only through the d-wave neutron emission). In one-neutron knockout from ^{18}C, the $9/2^+$ state is not populated via one-step processes; $9/2^+$ orbit occupancy of neutrons is expected to be zero. Multiple step processes, such as (a) inelastic excitation of ^{18}C followed by sequential decay into the $9/2^+$ state in ^{17}C and/or (b) initial population of $^{18}C(2^+)$ followed by $d_{5/2}$ neutron knockout, might be responsible for the population of the $9/2^+$ state.

4 Summary

Three unbound states at $E_x = 2.66(2)$, 3.16(5), and 3.97(3) MeV in ^{17}C were observed by the one-neutron knockout reaction of ^{18}C using the SAMURAI spectrometer in RIBF at RIKEN. While the reported shell-model calculation with the WBT interaction had to be carried out using the two-body matrix elements of the particles within the *sd* orbits to obtain a proper excitaion energy spectrum for the observed negative parity states in ^{17}C, the calculation utilizing the WBT interaction without $\hbar w$ restriction in the model space provided an appropriate description of them. The state at 3.16 MeV is suggested to be a $9/2^+$ state from an examination of its decay property.

Acknowledgements

This work was partly supported by the National Research Foundation of Korea(NRF) grant funded by the Korea government (MSIP) (Nos. 2R32-2008-000-10155-0 (WCU), 2010-0027136, and 014M2B2B1071110). Partial support via the French-Japanese LIA for Nuclear Structure Problems is also acknowledged.

References

[1] M. Stanoiu et al., Phys. Rev. C **78**, 034315 (2008)
[2] M. Stanoiu et al., Eur. Phys. J. A **20**, 95 (2004)
[3] T. Kobayashi et al., Nucl. Instrum. Methods Phys. Res., Sect. B **317**, 294 (2013)
[4] S. Takeuchi et al., Nucl. Instrum. Methods Phys. Res., Sect. A **763**, 596 (2014)
[5] S. Kim et al., JPS Conf. Proc. **6**, 030031 (2015)
[6] S. Kim, PhD dissertation, (2015)
[7] H. Ueno et al., Phys. Rev. C **87**, 034316 (2013)
[8] Z. Elekes et al., Phys. Lett. B **614**, 174 (2005)
[9] Y. Satou et al., Phys. Lett. B **660**, 320 (2008)
[10] E. K. Warburton and B. A. Brown, Phys. Rev. C **46**, 923 (1992)
[11] B. A. Brown and W. D. M. Rae, Nucl. Data Sheets **120**, 115 (2014)
[12] G. R. Jansen et al., Phys. Rev. Lett. **113**, 142502 (2014)
[13] C. Yuan et al., Phys. Rev. C **85**, 064324 (2012)
[14] A. Bohr and B. R. Mottelson, *Nuclear Structure* (World Scientific, Singapore, 1999) 439

EPJ Web of Conferences **113**, 06011 (2016)
DOI: 10.1051/epjconf/201611306011

Exploring the Structure of the Most Neutron-rich Boron and Carbon Isotopes

N.A.Orr[1,a]

[1] *LPC-Caen, ENSICAEN, Université de Caen Normandie, CNRS/IN2P3, 14050 Caen cedex, France*

Abstract. The most neutron-rich isotopes of boron and carbon are of considerable interest in terms of their few-body character - most notably through the possible formation of two-neutron haloes in the Borromean nuclei ^{19}B and ^{22}C. Here a series of recent investigations are briefly reviewed, including preliminary results from reaction studies of ^{18}B and ^{21}C. It is concluded that ^{22}C exhibits a prominent halo arising from a significant $\nu 2s_{1/2}^2$ valence neutron configuration. More generally, the structure of the dripline boron and carbon nuclei appears to be governed by the competition between the $\nu 1d_{5/2}$ and $\nu 2s_{1/2}$ single-particle orbitals, with a $5/2^+ - 1/2^+$ level inversion very probably occuring in ^{21}C.

1 Introduction

The investigation of the light neutron-rich dripline nuclei, including in particular those exhibiting a halo, is a central theme of nuclear structure physics. In recent years advances have been made through the development of increasingly more sophisticated models, such as ab intio approaches and those incorporating explicitly the effects of the continuum. Experimentally, with the advent of ever more powerful facilities – most recently, the RIKEN-RIBF – detailed studies have become possible for nuclei lying at and beyond the driplines for Z>4.

In this context, of particular interest are the heaviest candidate two-neutron halo systems, ^{19}B and ^{22}C and the associated unbound sub-systems ^{18}B and ^{21}C, the continuum states of which are critical to the defining the ^{17}B–n and ^{20}C–n interactions for three-body models. In addition, ^{18,19}B and ^{21,22}C are of considerable interest in terms of the evolution of shell-structure far from stability as they span the N=14 and 16 sub-shell closures below doubly-magic ^{22,24}O. Indeed, the level structures of ^{18}B and ^{21}C can shed light on the evolution of the $\nu 2s_{1/2}$ and $\nu 1d_{5/2}$ levels which may become degenerate in this region [1].

In terms of ascertaining the halo character of a nucleus, measurements of total reaction cross sections (ideally across a chain of isotopes) provide the first indications [2]. In practice, such measurements may be carried out at very low beam intensities, a considerable advantage at the dripline. In the case of ^{19}B measurements were undertaken some time ago at very high energy (800 MeV/nucleon) and interpreted in terms of a Glauber optical model analysis [3]. An RMS matter radius of 3.11±0.13 fm was deduced for ^{19}B as compared to 2.99±0.09 fm for ^{17}B. As such ^{19}B did not appear to exhibit a strongly developed two-neutron halo, but rather, through comparison with the less neutron rich boron isotopes, it was suggested that it may be better described as a ^{15}B core and 4 valence neutrons.

[a] e-mail: orr@lpccaen.in2p3.fr

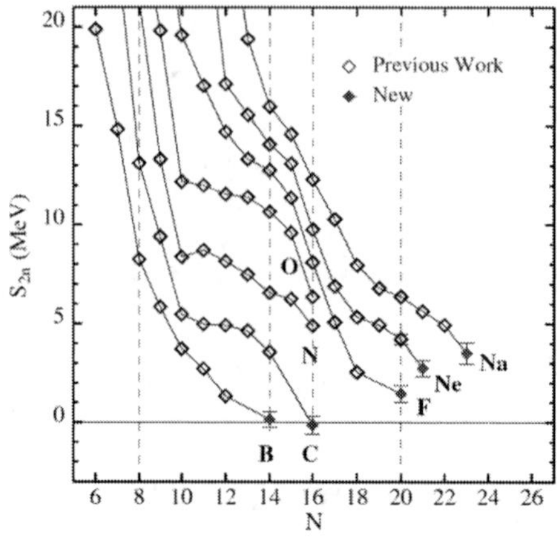

Figure 1: Two-neutron separation energies derived from experimentally determined masses. The N=8, 14, 16 and 20 (sub-) shell closures are indicated. Adapted from Ref. [6].

In the case of ^{22}C intermediate energy (40 MeV/nucleon) measurements undertaken more recently found a very large enhancement of the total reaction cross section with respect to the less neutron-rich neighbouring isotopes [4]. Using a similar Glauber optical limit approach an RMS matter radius of 5.4±0.9 fm was deduced – very much larger than systematics would suggest and very much larger than the well-established single-neutron halo ^{19}C. It was concluded that such a large matter radius may be attributed to a strong *s*-wave valence neutron configuration.

2 Masses and valence neutron binding energies

One of the key ingredients in characterising halo systems is the valence neutron binding energy. Determining experimentally the masses of boron and carbon isotopes at the neutron dripline is not possible at present, however, using high precision techniques such as those employing traps. Indeed, current ISOL based techniques are unable to provide even the weakest of beams of such nuclei. As such, in-flight fragmentation followed by a direct time-of-flight and rigidity measurement represents the only viable approach [5]. Such measurements have been made at GANIL with a ^{48}Ca beam and the SISSI+alpha spectrometer facility coupled to the SPEG spectrometer [6]. Using well established masses as references [7], the mass excesses of 16 light neutron-rich nuclei were determined, including those of ^{19}B and ^{22}C for the first time[1].

The two-neutron separation energies (S_{2n}) derived from the measured masses and previously established values [7] are reported in Fig. 1 [6]. It is clear that ^{19}B and ^{22}C lie at the very limits of binding – S_{2n}= 0.14±0.39 and –0.14±0.46 MeV respectively[2] – suggesting that prominent neutron haloes may occur. These binding energies may be employed, in conjunction with the radii deduced from the interaction cross section measurements, to provide a first insight into the valence neutron configurations.

In a simple picture [9], the matter radius, R_m, may be estimated as,

$$R_m^2 = \frac{A-k}{A}(R_c^2 + \frac{k}{A}r_v^2),$$

[1]The first mass measurements were also made for ^{29}F and ^{34}Na together with the most precise determinations for ^{23}N and the deformed, single-neutron halo system ^{31}Ne [8].

[2]Given that ^{22}C is bound, this may be taken as providing an upper limit (1σ) for the S_{2n} of 0.32 MeV.

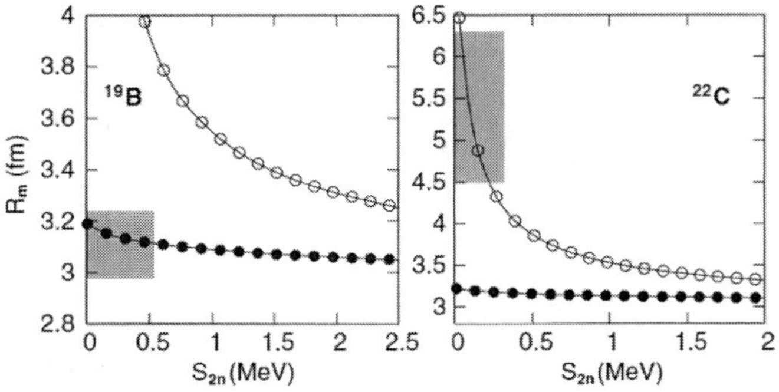

Figure 2: Matter radii (R_m) for ^{19}B and ^{22}C as a function of the two neutron binding energy (S_{2n}). Curves marked with symbols correspond to calculations assuming different valence neutron configurations: $\nu 2s^2_{1/2}$ open circles and $\nu 1d^2_{5/2}$ filled points. The experimentally deduced R_m [3, 4] and mass determinations [6] are shown (shaded rectangles). Adapted from Ref. [6].

where A is the mass number of the nucleus, k the number of valence neutrons, R_c the matter radius of the core nucleus, and r_ν the radius of the valence neutron distribution. The R_c adopted were: 2.99 ± 0.09 fm for ^{17}B [3] and 2.98 ± 0.05 fm for ^{20}C [10]. The valence neutron radius may be estimated using a Woods-Saxon potential (a = 0.65 fm and r_0 = 1.25 fm), where the depth is adjusted to ensure the correct binding of each valence neutron (n,ℓ) where it is assumed that the the binding energy (S_{2n}) is shared equally between them. The results for a range of valence neutron binding energies are displayed for both the s and d-wave neutron configurations in Fig. 2 along with the experimentally determined R_m and S_{2n}, where the matter radii for ^{19}B and ^{22}C noted in the preceding section were employed. As may be seen, the results for ^{19}B are well reproduced when the valence neutrons are assumed to be predominantly of d-wave character. This suggests that the radial extent of the valence neutron distribution will be significantly suppressed by the angular momentum barrier. On the contrary, the results for ^{22}C strongly support the devopment of a two-neutron halo with a predominately s-wave configuration in line with the conclusion of the total reaction cross section study [4] (Section 1).

3 Momentum Distributions

As was first demonstrated in some of the earliest studies of halo nuclei, the momentum distribution of the breakup prodcuts – in particular the core – can provide a clear signature of their halo character through their narrow widths [11, 12]. More recently, the momentum distributions from reactions on a low-Z target (such as Be and C) and associated cross sections have been developed as a spectroscopic tool [13, 14], as demonstrated, for example, by the "text book" case of the systematic study of Ref. [15]. In simple terms, the core-like reaction residue, following high-energy neutron removal (or "knockout") carries rather directly the imprint of the orbital angular momentum of the valence neutron. On the other hand, the cross section, when compared with reaction calculations (typically within the eikonal approach), provides a measurement of the spectroscopic strength between the initial (projectile ground-state) and final (core residue) state [14].

In the case of ^{22}C, the secondary beam intensities available during the early stage operations of the RIKEN-RIBF enabled inclusive measurements to be made of two-neutron removal on a carbon target at some 240 MeV/nucleon [16]. For comparison, measurements were also made on the neighbouring carbon isotopes, including ^{20}C. The momentum distributions were derived using a time-of-flight

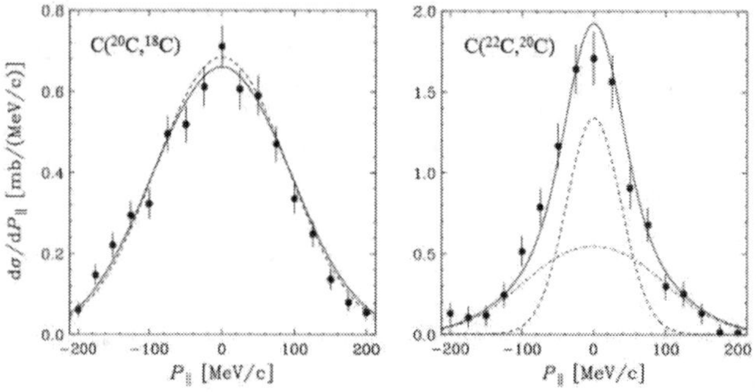

Figure 3: Inclusive longitudinal momentum distribution for two-neutron removal from ^{20}C and ^{22}C. The results of eikonal reaction theory calculations are shown (solid lines). In the case of ^{22}C, the components arising from *s*-wave (dashed) and *d*-wave (dot-dashed line) neutron removal are also shown. Adapted from Ref. [16].

technique and a spectrometer operating at zero-degrees. The results are shown in Fig. 3 whereby the narrowness of the momentum distribution for two-neutron removal from ^{22}C, as compared to ^{20}C, is clear.

In the case of a Borromen nucleus such as ^{22}C, essentially all the yield is produced via single-neutron removal followed by in-flight decay of the unbound ^{21}C [17]. As discussed in detail in Ref. [16], combining shell model predictions obtained using the WBP interaction [18], within the *psd* model space, with eikonal reaction calculations (including the recoil from the decay of the ^{21}C states) the ^{20}C core momentum distribution was very well reproduced, in both form and magnitude. As expected, the narrowness of the distribution arises from the *s*-wave removal component. This result not only confirms the importance of the $\nu 2s^2_{1/2}$ valence neutron configuration but also suggests that ^{21}C has low-lying $1/2^+$ and $5/2^+$ levels. In the case of ^{19}B, while no measurements are yet available of the core-residue momentum distribution, the shell model calculations suggest that closely spaced levels[3] arising from the $\nu 2s_{1/2}$ and $\nu 1d_{5/2}$ configurations dominate the low-lying level scheme of ^{18}B [19].

4 Invariant mass spectroscopy of ^{18}B and ^{21}C

In order to investigate directly the low-lying level structures of ^{18}B and ^{21}C, the technique of invariant mass spectroscopy following in-flight decay was employed. In addition, the complementary probes of neutron-removal from ^{19}B and ^{22}C, together with single-proton removal from ^{19}C and ^{22}N, were used to populate different final-states of ^{18}B and ^{21}C. In simple terms, proton removal is expected to populate final states with the same neutron configurations as the projectile ($\nu 2s_{1/2}$ in both cases), whilst neutron removal will probe directly the valence neutron configurations (here, most probably, $\nu 2s^2_{1/2}$ and $\nu 1d^2_{5/2}$). As for the momentum distribution study, the secondary beams were generated by the RIKEN-RIBF facility. The measurements were performed using the SAMURAI spectrometer [20] coupled to the large area neutron array NEBULA [21].

The preliminary results for the reconstructed fragment–neutron relative energy (E_{rel}) spectra – that is the energy above the single-neutron emission threshold – are displayed in Figs 4 and 5. In the case of proton removal from ^{19}C, the ^{17}B+n spectrum exhibits a very strong accumulation of strength at

[3]Multiplets, owing to the coupling to the $\pi 1p_{3/2}$ hole.

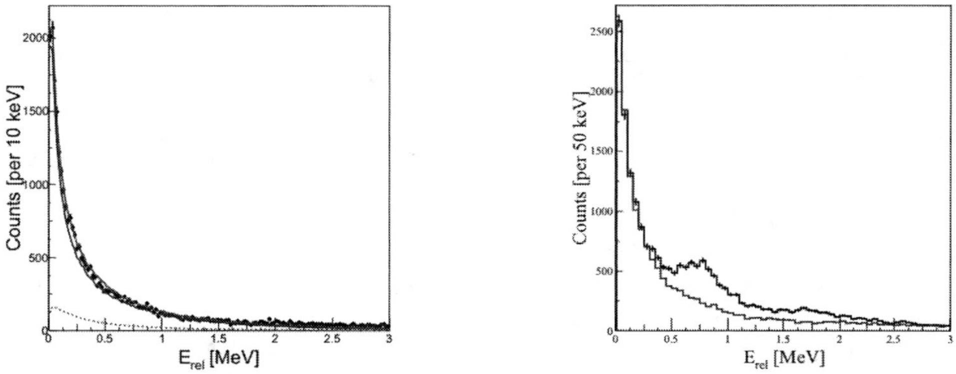

Figure 4: (Colour online) Relative energy spectra, as shown by the data points, for the C(^{19}C,^{17}B+n) (left) and C(^{19}B,^{17}B+n) (right panel) reactions. In the former a fit (red) is shown including a strongly interacting s-wave virtual state (blue) and a small non-resonant continuum contribution (dotted line). In the latter, comparison is shown with the spectrum, renormalised at low E_{rel}, for the C(^{19}C,^{17}B+n) reaction (blue histogramme).

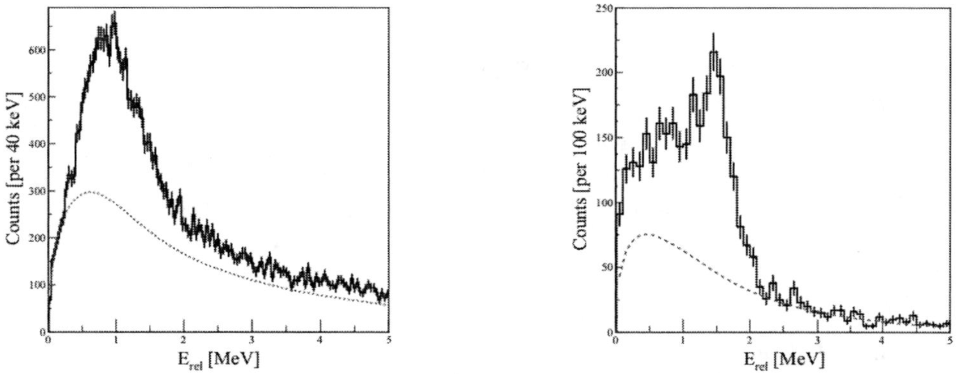

(a) C(^{22}N,^{20}C+n)

(b) C(^{22}C,^{20}C+n)

Figure 5: Relative energy spectra for the C(^{22}N,^{20}C+n) (left) and C(^{22}C,^{20}C+n) (right panel) reactions. Comparison is shown with non-resonant continuum distributions (dotted lines).

threshold. Such behaviour is indicative of a strongly interacting virtual s-wave state [22] and indeed the spectrum, as shown, may be very well described with such a state with a scattering length of less than -50 fm – in agreement with an earlier study [19] – and a small contribution from the non-resonant continuum [23]. In the case of neutron removal from ^{19}B, in addition to the strong s-wave strength a broad resonance-like structure is present at around 0.8 MeV with a possible second weaker structure at ~1.7 MeV. The former lies close to the predicted location of the 2^-– 4^- levels arising primarily from the $1d_{5/2}$ valence neutron [19].

Turning to the results for ^{21}C (Fig. 5), the relative energy spectrum for proton removal from ^{22}N is, in contrast to an earlier very low statistics study [24], dominated by a prominent resonance-like structure at around 1 MeV which may be identified, based on the selectivity of the reaction and width, as an s-wave resonance. In the case of the neutron removal from ^{22}C, a relatively narrow structure is observed close to 1.5 MeV above threshold, together with a broad distribution of lower-lying strength. A first examination of the reconstructed ^{21}C momentum distribution indicates that the narrow structure may be identified with d-wave neutron removal, whilst the low-lying strength is consistent with s-wave neutron removal. Further analysis will be required to confirm these propositions, however it may be noted that they are in line with the shell-model predictions for the ^{21}C level scheme, which predict an inversion of the lowest lying $5/2^+$ and $1/2^+$ states [16].

In addition to refining the analyses and interpretation, including examining ^{17}B and ^{20}C–γ–n coincidences, data acquired for dissociation of both the ^{19}B and ^{22}C beams on a lead target is also being analysed with the goal of extracting the E1 strength functions. The latter should provide further insight into the structure and halo character of both nuclei.

Acknowledgements

I am grateful to all my collaborators in the studies described here, including, in particular, Nobuyuki Kobayashi, Takashi Nakamura and Jeff Tostevin (momentum distributions), Laurent Gaudefroy and Wolfi Mittig (mass measurments), Sylvain Leblond, Miguel Marqués and Julien Gibelin (spectroscopy of ^{18}B and ^{21}C). Support from the Franco-Japanese LIA-International Associated Laboratory for Nuclear Structure Problems is also acknowledged.

References

[1] G. Jansen *et al.*, Phys. Rev. Lett. **113**, 142502 (2014)
[2] I. Tanihata *et al.*, Phys. Rev. Lett. **55**, 2676 (1985).
[3] T. Suzuki *et al.*, Nucl. Phys. A **658**, 313 (1999).
[4] K. Tanaka *et al.*, Phys. Rev. Lett. **104**, 062701 (2010).
[5] W. Mittig, A. Lépine-Szily, N.A. Orr, Ann. Rev. Nucl. Part. Sci. **47**, 27 (1997).
[6] L. Gaudefroy *et al.*, Phys. Rev. Lett. **109**, 20503 (2012).
[7] W. Meng *et al.*, Chin. Phys. C **36**, 1603 (2012).
[8] T. Nakamura *et al.*, Phys. Rev. Lett. **103**, 262501 (2009).
[9] P.G. Hansen, B. Jonson, Europhys. Lett. **4**, 409 (1987).
[10] A. Ozawa *et al.*, Nucl. Phys. A **691**, 599 (2001).
[11] T. Kobayashi *et al.*, Phys. Rev. Lett. **60**, 2599 (1988).
[12] N.A. Orr *et al.*, Phys. Rev. Lett. **69**, 2050 (1992).
[13] P.G. Hansen, J.A. Tostevin, Ann. Rev. Nucl. Part. Sci. **53**, 219 (2003).
[14] J.A. Tostevin, J. Phys. G **25**, 735 (1999).
[15] E. Sauvan *et al.*, Phys. Lett. B **491**, 1 (2000).
[16] N. Kobayashi *et al.*, Phys. Rev. **C83**, 054604 (2012).
[17] F. Barranco *et al.*, Phys. Lett. B **319**, 387 (1993).
[18] E.K. Warburton and B.A. Brown, Phys. Rev. C **46**, 923 (1992).
[19] A. Spyrou *et al.*, Phys. Lett. B **683**, 129 (2010).
[20] T. Kobayashi *et al.*, Nucl. Instr. Meth. B **317**, 294 (2013).
[21] Y. Kondo *et al.*, RIKEN Accel. Prog. Rep. **45**, 131 (2012);
[22] M. Thoennessen *et al.*, Phys. Rev. Lett. **59**, 111 (1999).
[23] G. Randisi, Phys. Rev. C **89**, 034320 (2014).
[24] S. Mosby *et al.*, Nucl. Phys. A **909**, 69 (2013).

EPJ Web of Conferences **113**, 06012 (2016)
DOI: 10.1051/epjconf/201611306012

Momentum distributions in light halo nuclei and structure constraints

L. A. Souza[1,a], F. F. Bellotti[2,3,b], T. Frederico[1,c], M. T. Yamashita[4,d], and Lauro Tomio[4,5e]

[1] *Instituto Tecnológico de Aeronáutica, DCTA, 12228-900, São José dos Campos, Brazil*
[2] *Department of Physics and Astronomy, Aarhus University, DK-8000 Aarhus C, Denmark*
[3] *Instituto de Fomento e Coordenação Industrial, 12228-901, São José dos Campos, SP, Brazil*
[4] *Instituto de Física Teórica, UNESP, 01156-970, São Paulo, Brazil*
[5] *CCNH, Universidade Federal do ABC, 09210-580, Santo André, Brazil.*

Abstract. The core recoil momentum distribution of neutron-rich isotopes of light exotic nuclei is studied within a three-body model, where the nuclei are described by a core and two neutrons, with interactions dominated by the s-wave channel. In our framework, the two-body subsystems should have large scattering lengths in comparison with the interaction range allowing to use a three-body model with a zero-range force. The ground-state halo wave functions in momentum space are obtained by using as inputs the two-neutron separation energy and the energies of the singlet neutron-neutron and neutron-core virtual states. Within our model, we obtain the momentum probability densities for the Borromean exotic nuclei ^{11}Li and ^{22}C. In the case of the core recoil momentum distribution of ^{11}Li, a fair reproduction of the experimental data was obtained, without free parameters, considering only the two-body low-energies. By analysing the obtained core momentum distribution in face of recent experimental data, we verify that such data are constraining the ^{22}C two-neutron separation energy to a value between 100 and 400 keV.

1 Introduction

The observation of a large enhancement of the reaction cross sections of ^{22}C on liquid hydrogen target at 40A MeV compared to the neighbour nuclei ^{19}C and ^{20}C by Tanaka and collaborators [1], analysed with a finite-range Glauber calculation under an optical-limit approximation suggested a matter root-mean-square (rms) radius of 5.4 ± 0.9 fm. In addition, their analysis showed that the two-valence neutrons occupy preferentially the $1s_{1/2}$ orbital. Such a large matter radius taken together with the ^{20}C one indicates a rms radius of the halo neutron orbit of 15 ± 4 fm estimated in [2]. The information on the rms matter radius was used to constrain the poorly known experimental value of the two-neutron separation energy of ^{22}C (S_{2n}) to be below 100 keV. This value was found consistently in calculations performed with a renormalized three-body model [2], with a shell model approach [3]

[a] e-mail: lucasufsj@gmail.com
[b] e-mail: ffbellotti@gmail.com
[c] e-mail: tobias@ita.br
[d] e-mail: yamashita@ift.unesp.br
[e] e-mail: tomio@ift.unesp.br

and with effective field theory with a contact interaction [4]. However, the comparison between the experimental data for the core recoil momentum distributions of ^{11}Li [5] and ^{22}C [6] indicates similar sizes of their halos composed by two neutrons, which suggest that the matter radius of this carbon isotope could be overestimated (for further discussions see [7]).

This motivates us to study the properties of low-momentum distributions in two-neutron halo light exotic nucleus, which are dominated by s-wave short-range two-body interactions in the neutron-neutron and neutron-core subsystems. Our aim is to investigate theoretically within a three-body model, the core recoil momentum distribution as obtained experimentally by the halo breakup in nuclear targets. Our interest is to provide a constraint to the parameters associated with the halo structure and two-neutron separation energy based on these data fitted to three-body model calculations.

The model calculations for the core recoil momentum distribution are performed within a zero-range three-body model applied to a core-neutron-neutron system [9, 10]. The detailed formulation of the momentum distributions was given in [8] and the inputs of the zero-range model are the scattering lengths, and one three-body scale, which is given by the two-neutron separation energy. Therefore, three low-energy observables are enough to determine another low-energy quantity through an appropriate universal scaling function computed within the zero-range model. For example, the width of the momentum distribution is universally correlated to these physical scales and once the two-body scattering lengths are given, it is possible to constraint S_{2n} if the width is known experimentally.

The data for ^{11}Li (see e.g. [5]) is used as a test of the reliability of our model analysis, as the low energy parameters used as input to our calculations are known. The recent experimental results for ^{20}C and ^{22}C [6] allow us, in principle, to constraint S_{2n} and the matter radius of ^{22}C. In the preliminary analysis presented in this contribution, we focus on the constrain to S_{2n} obtained by computing the core recoil momentum distribution of ^{22}C and fitting it to the experimental data from reference [6].

In sect. 2, we briefly sketch the zero-range three-body model and the core recoil momentum distribution formula, used in our analysis for the ^{11}Li and ^{22}C data. In sect. 3, we present our calculations for the momentum distributions compared to the data for these two cases, and provide a summary of our findings.

2 Model for the neutron-neutron-core system

The core recoil momentum distribution in the neutron-neutron-core (n-n-c) system is given by [8]:

$$n(q_c) = \int d^3 p_c |\langle \vec{q}_c \vec{p}_c | \Psi \rangle|^2, \tag{1}$$

where the relative momentum of the core to the nn subsystem is $\vec{q}_c$ and the relative momentum between the neutrons is $\vec{p}_c$, which is integrated above. The renormalized zero-range three-body model for the two-neutron halo nuclei is detailed in Ref.[9]. The s-wave three-body wave function for the model ($|\Psi\rangle$) can be written in terms of spectator functions, $\chi_{nn}(q_c)$ and $\chi_{nc}(|\vec{p}_c \pm \frac{\vec{q}_c}{2}|)$, as

$$\langle \vec{q}_c \vec{p}_c | \Psi \rangle = \frac{\chi_{nn}(q_c) + \chi_{nc}(|\vec{p}_c - \frac{\vec{q}_c}{2}|) + \chi_{nc}(|\vec{p}_c + \frac{\vec{q}_c}{2}|)}{E_3 + H_0}, \tag{2}$$

where the free Hamiltonian is

$$H_0 \equiv \frac{p_c^2}{2m_{nn}} + \frac{q_c^2}{2m_{nn,c}}. \tag{3}$$

Defining $\mathcal{A} \equiv \frac{m_c}{m_n}$, the reduced masses are given by $m_{nn} = \frac{m_n}{2}$ and $m_{nn,c} = \frac{2\mathcal{A}}{\mathcal{A}+2} m_n$, where m_n is neutron mass. The core is assumed structureless and the two neutrons are in a singlet spin state to supply the correct antisymmetrization of the neutrons state in the halo.

The spectator functions come from the solution of a coupled set of integral equations corresponds to the Faddeev equations subtracted at a given scale (see [9]). The inputs are the nn and nc scattering lengths, or in the Borromean case the virtual state energies. In addition to these low energy parameters the value of the subtraction energy scale $-\mu^2$ has to be provided. This is done by fixing S_{2n}. The value of μ can be moved towards infinity and a tower of Thomas-Efimov states appears, with the shallowest one having its infrared properties fixed only by the three low energy quantities, the virtual energies of the subsystems and S_{2n}.

The log-periodic behaviour appear in the ultraviolet region not affecting low momentum part of the wave function that indeed runs to a limit cycle. This behaviour is explored in details in [8], where it was also derived an universal ration between the magnitude of the log-periodic spectator functions in the limit of $\mu \rightarrow \infty$, which is a function of only the mass ratio A. In practice the first cycle is enough to get the proper ties of the wave function in the infrared momentum region, which is acceptable for our purposes and used in our computations of the wave function. In the actual case of ^{11}Li and ^{22}C we have $\mu^2 >> S_{2n}$, and therefore we are close to the universal limit cycle result in the low momentum region of the core recoil distribution of interest for the present study.

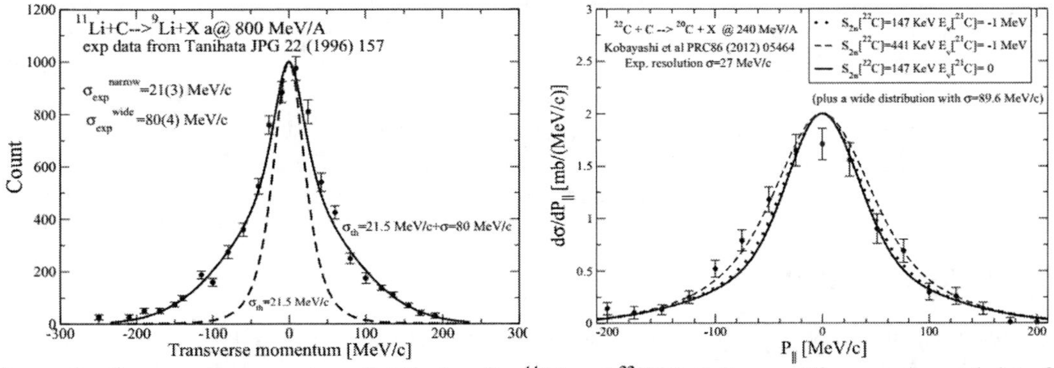

Figure 1. Core recoil momentum distribution for ^{11}Li and ^{22}C. *Left Frame:* The experimental data for ^{11}Li from [5] is compared to the zero-range model calculation with no free parameters (see the explanation in the text). Theory with a narrow distribution (dashed line) with σ_{th}=21.5 MeV/c. Theory with a narrow distribution added to a wide one (solid line) with σ_{th} (=21.5 MeV/c) + σ (=80 MeV/c). The resulting distributions are fitted to the peak. *Right Frame:* The experimental data for ^{22}C from [6] is compared to the zero-range model calculation with different input parameters. The results obtained for the virtual state energy of ^{21}C, $E_V\left[^{21}C\right]$=-1 MeV, are shown for two values of S_{2n} given by 441 keV (dashed line) and 147 keV (dotted line). The calculations for $E_V\left[^{21}C\right] = 0$ and S_{2n} = 147 keV are given by the solid line. The computed narrow distribution is folded with the experimental resolution of σ = 27 MeV/c. It is added to a wide one with width 89.6 MeV/c and fitted to the peak value of 2 mb/(MeV/c).

3 Results and Summary

We start the presentation of our results in figure 1 for ^{11}Li (left frame), by checking the reliability of the three-body approach to compute the core distribution density for a weakly bound two-neutron halo nuclei. In this case the s-wave virtual states of ^{10}Li and nn are fixed to $E_V\left[^{10}Li\right]$ =-50 keV and $E_V\left[nn\right]$ =-147 keV, respectively. The two-neutron separation energy of ^{11}Li is given by the experimental value of 369 keV [11]. We compare our results with the experimental data for the core transverse momentum distribution obtained from the ^{11}Li breakup on a carbon target at 800 MeV/A [5]. A wide distribution with σ =80 MeV/c is added to the computed narrow one with width σ = 21.5 MeV/c, which should be compared with the experimental one of 21(3) MeV/c. The wide momentum distribution is outside the description of our model. It should be associated with inner part of the halo neutron orbits close to the core region. We show in the figure the comparison with the experimental data, where the normalisations of the wide and narrow distributions are fitted to the data. After that we find a fair reproduction of the experimental momentum distribution as shown in the figure. This test indicates that our approach is a viable tool to extract information on the large two-neutron halo properties from the core momentum distribution.

We apply the neutron-neutron-core model to describe the halo of ^{22}C and the results are shown in figure 1 (right frame). We show a sample of results for the core recoil momentum distribution in ^{22}C compared with the data from [6] with given separation energies of 147 and 441 KeV. The virtual s-wave energy for the singlet nn state is fixed to $E_V\,[nn]$ =-147 keV. We allow a variation of the virtual state energy of ^{21}C with values 0 and -1 MeV [12] to spam different possibilities. We have added a wide momentum distribution to the narrow one as indicated in the figure. In addition the experimental resolution from [6] was folded with our theoretical results. We observe that the core recoil momentum distribution does not exhibit a strong sensitivity to the variation of $E_V\left[{}^{21}C\right]$ with S_{2n} = 147 keV. The comparison in the figure shows a difference between the distributions obtained with S_{2n} = 147 and 441 keV, which however are within the error of the experimental data. This suggests that the previous works based only the matter radius from [1] underestimate the two neutron separation energy.

In summary, our model was tested in the case of the core recoil momentum distribution of ^{11}Li, where we found a fair reproduction of the experimental data, using only the known low energy parameters. In view of that, we claim that the model can be useful to analyse experimental data on core momentum distribution of other halo nuclei. In particular, by analysing the recent experimental data on the core momentum distribution in ^{22}C, given by Kobaiashi et al. [6], we are able to conclude that the value of the ^{22}C two-neutron separation energy is within the interval of 100 and 400 keV.

Acknowledgements

This work was partially supported by the Brazilian agencies CAPES, FAPESP and CNPq.

References

[1] K. Tanaka et al. Phys. Rev. Lett. **104**, 062701 (2010).

[2] M. T. Yamashita, R. M. de Carvalho, T. Frederico, L. Tomio, Phys. Lett. B **697**, 90 (2011); Phys. Lett. B **715**, 282 (2012).

[3] H. T. Fortune, R. Sherr, Phys. Rev. C **85**, 027303 (2012).

[4] B. Acharya, C. Ji, D. R. Phillips, Phys. Lett. B **723**, 196 (2013).

[5] I. Tanihata, J. Phys. G **22**, 157 (1996).

[6] N. Kobayashi et al., Phys. Rev. C **86**, 054604 (2012).

[7] K. Riisager, Phys. Scr. **T152**, 014001 (2013).

[8] M. T. Yamashita, F. F. Bellotti, T. Frederico, D. V. Fedorov, A. S. Jensen, N. T. Zinner, Phys. Rev. A **87**, 062702 (2013).

[9] T. Frederico, A. Delfino, L. Tomio, M. T. Yamashita, Prog. Part. Nucl. Phys. **67**, 939 (2012).

[10] T. Frederico, L. Tomio, A. Delfino, M. T. Yamashita, M. R. Hadizadeh, Few-Body Syst. **51**, 87 (2011).

[11] M. Smith et al., Phys. Rev. Lett. **101**, 202501 (2008).

[12] S. Mosby et al., Nucl. Phys. A **909**, 69 (2013).

EPJ Web of Conferences **113**, 06013 (2016)
DOI: 10.1051/epjconf/201611306013

Properties of Lithium-11 and Carbon-22 at leading order in halo effective field theory

Bijaya Acharya[1,2,a] and Daniel R Phillips[1,b]

[1] *Department of Physics and Astronomy, University of Tennessee, Knoxville, TN 37996, USA*
[2] *Institute of Nuclear and Particle Physics and Department of Physics and Astronomy, Ohio University, Athens, OH 45701, USA*

Abstract.
We study the ^{11}Li and ^{22}C nuclei at leading order (LO) in halo effective field theory (Halo EFT). Using the value of the ^{22}C rms matter radius deduced in Ref. [1] as an input in a LO calculation, we simultaneously constrain the values of the two-neutron ($2n$) separation energy of ^{22}C and the virtual-state energy of the ^{20}C−neutron system (hereafter denoted ^{21}C). The $1-\sigma$ uncertainty of the input rms matter radius datum, along with the theory error estimated from the anticipated size of the higher-order terms in the Halo EFT expansion, gives an upper bound of about 100 keV for the $2n$ separation energy. We also study the electric dipole excitation of $2n$ halo nuclei to a continuum state of two neutrons and the core at LO in Halo EFT. We first compare our results with the ^{11}Li data from a Coulomb dissociation experiment and obtain good agreement within the theoretical uncertainty of a LO calculation. We then obtain the low-energy spectrum of $B(E1)$ of this transition at several different values of the $2n$ separation energy of ^{22}C and the virtual-state energy of ^{21}C. Our predictions can be compared to the outcome of an ongoing experiment on the Coulomb dissociation of ^{22}C to obtain tighter constraints on the two- and three-body energies in the ^{22}C system.

1 Introduction

The separation of scales between the size of the core and its distance from the halo nucleons allows the low-energy properites of halo nuclei to be studied using Halo EFT [2, 3], which is written in terms of the core and halo nucleons as degrees of freedom. Halo EFT yields relations for the low-energy observables as systematic expansions in the ratio of the short-distance scale set by the core size and excitation energies to the long-distance scale associated with the properties of the halo nucleons. At LO, the three-body wavefunction of the $2n$ halo nucleus is constructed with zero-range two-body interactions, which can be completely characterized by the neutron-neutron (nn) and the neutron-core (nc) scattering lengths [4]. However, a three-body coupling also enters at LO [5], necessitating the use of one piece of three-body data as input to render the theory predictive. It is convenient to fix the three-body force by requiring the three-body bound state to lie at $-E_B$, where E_B is the $2n$ separation

[a] e-mail: bacharya@utk.edu
[b] e-mail: phillips@phy.ohiou.edu

energy. The only inputs to the equations that describe a $2n$ halo are, therefore, E_B together with the energies of the nc virtual/real bound state, E_{nc}, and the nn virtual bound state, E_{nn}. The effects of interactions that are higher order in the Halo EFT power counting are estimated from the size of the ignored higher-order terms and then included as theory error bands.

In Ref. [1], Tanaka et al. measured the reaction cross-section of ^{22}C on a hydrogen target and, using Glauber calculations, deduced a ^{22}C rms matter radius of 5.4 ± 0.9 fm, implying that ^{22}C is an S-wave two-neturon halo nucleus. This conclusion is also supported by data on high-energy two-neutron removal from ^{22}C [6]. We used Halo EFT in Ref. [7], to calculate the rms matter radius of ^{22}C as a model-independent function of E_B and E_{nc}. Since the virtual-state energy of the unbound [8] ^{21}C is not well known [9], we used Halo EFT to find constraints in the (E_B, E_{nc}) plane using Tanaka et al.'s value of the rms matter radius.

We have also derived universal relations for the electric dipole excitation of two-neutron halo nuclei into the three-body continuum consisting of the core and the two neutrons in Halo EFT. Our LO calculation of the $B(E1)$ of this transition includes all possible rescatterings with S-wave nn and nc interactions, in both the initial and the final state. We compare our results with the ^{11}Li data from Ref. [10] and obtain a good agreement within the theoretical uncertainty. We predict the $B(E1)$ spectrum of ^{22}C for selected values of E_B and E_{nc}. These findings will be published in Ref. [11].

2 Matter radius constraints on the binding energy

In Fig. 1, we plot the sets of (E_B, E_{nc}) values that give a ^{22}C rms matter radius, $\sqrt{\langle R^2 \rangle}$, of 4.5 fm, 5.4 fm and 6.3 fm, along with the theoretical error bands. All sets of E_B and E_{nc} values in the plotted region that lie within the area bounded by the edges of these bands give an rms matter radius that is consistent with the value Tanaka et al. extracted within the combined $(1-\sigma)$ experimental and theoretical errors. The figure shows that, regardless of the value of the ^{21}C virtual energy, Tanaka et al.'s experimental result puts a model-independent upper limit of 100 keV on the $2n$ separation energy of ^{22}C. This is to be compared with another theoretical analysis of the matter radius datum of Ref. [1] in a three-body model by Ref. [12], which set an upper bound of 120 keV on E_B. Similarly, Ref. [13] used a correlation between the binding energy and the matter radius derived from a potential model to exclude $E_B > 220$ keV. Our constraint is stricter than the ones set by these studies. Although our conclusion is consistent with the experimental value of -140 (460) keV from a direct mass measurement [14], more studies are needed to further reduce the large uncertainty in the $2n$ separation energy. In this spirit, we study the $E1$ excitation of $2n$ halo nuclei to the three-body continuum.

3 The $B(E1)$ spectrum

We first present the result of our LO Halo EFT calculation of the $B(E1)$ for the break up of ^{11}Li into ^{9}Li and two neutrons at energy E in their center of mass frame. Only S-wave $^{9}\mathrm{Li}-n$ interactions are included. After folding with the detector resolution, we obtain the curve shown in Fig. 2 for $E_B = 369.15(65)$ keV [15] and $E_{nc} = 26$ keV [16]. The sensitivity to changes in E_{nc} is much smaller than the EFT error, represented by the purple band. Within the uncertainty of a LO calculation, a good agreement with the RIKEN data [10] is seen, despite the fact that ^{10}Li has a low-lying P-wave resonance which is not included in this calculation.

Figure 3 shows the dipole response spectrum for the break up of ^{22}C into ^{20}C and neutrons for three different combinations of E_B and E_{nc} which lie within the $1-\sigma$ confidence region shown in Fig. 1. These results agree qualitatively with those of a potential model calculation by Ref. [17]. A comparison of Fig. 3 with the forthcoming data [18] can provide further constraints on the (E_B, E_{nc})

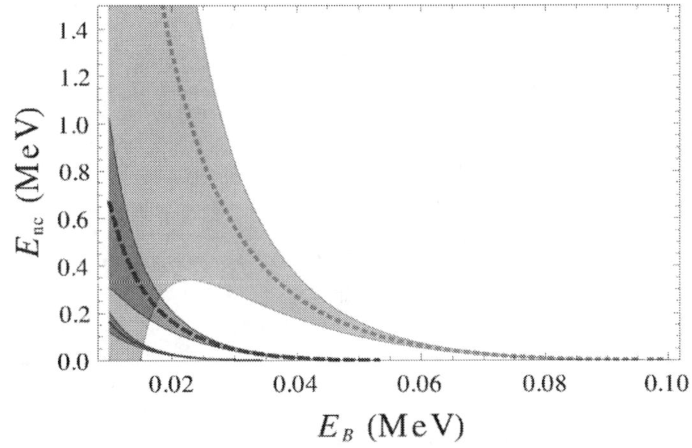

Figure 1. Plots of $\sqrt{\langle R^2 \rangle}$ = 5.4 fm (blue, dashed), 6.3 fm (red, solid), and 4.5 fm (green, dotted), with their theoretical error bands, in the (E_B, E_{nc}) plane. (Published in Ref. [7].)

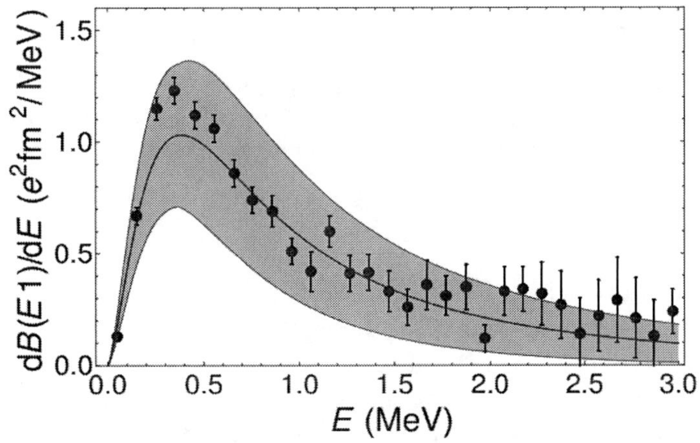

Figure 2. The dipole response spectrum for ^{11}Li after folding with the detector resolution (blue curve) with the theory error (purple band), and data from Ref. [10].

plane. However, the individual values of these energies thus extracted will have large error bars because different sets of (E_B, E_{nc}) values can give similar curves. This ambiguity can be removed by looking at the neutron-momentum distribution of the Coulomb dissociation cross section [11].

4 Conclusion

The matter radius and the $E1$ response of S-wave $2n$ halo nuclei were studied. We put constraints on the (E_B, E_{nc}) parameter space using the value of the ^{22}C matter radius. The calculated $B(E1)$ spectrum of ^{11}Li agrees with the experimental result within our theoretical uncertainty. Our ^{22}C result can be tested once the experimental data is available. Further improvements can be made by rigorously calculating the higher-order terms in the EFT expansion and by including higher partial waves.

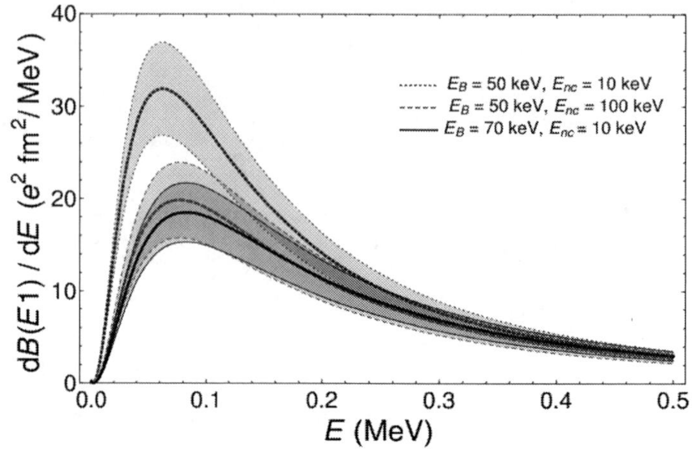

Figure 3. The dipole response spectrum for ^{22}C for E_B = 50 keV, E_{nc} = 10 keV (blue, dotted); E_B = 50 keV, E_{nc} = 100 keV (red, dashed) and E_B = 70 keV, E_{nc} = 10 keV (black), with their EFT error bands.

Acknowledgements

We thank our collaborators Chen Ji, Hans-Werner Hammer and Philipp Hagen. This work was supported by the US Department of Energy under grant DE-FG02-93ER40756. BA is grateful to the organizers of the conference for the opportunity to present this work and to UT for sponsoring his attendance.

References

[1] K. Tanaka et al., Phys. Rev. Lett. **104**, 062701 (2010).
[2] C. A. Bertulani, H.-W. Hammer and U. van Kolck, Nucl. Phys. A **712**, 37 (2002).
[3] P. F. Bedaque, H.-W. Hammer and U. van Kolck, Phys. Lett. B **569**, 159 (2003).
[4] D. B. Kaplan, M. J. Savage and M. B. Wise, Nucl. Phys. B **534**, 329 (1998).
[5] P. F. Bedaque, H.-W. Hammer and U. van Kolck, Phys. Rev. Lett. **82**, 463 (1999).
[6] N. Kobayashi et al., Phys. Rev. C **86**, 054604 (2012).
[7] B. Acharya, C. Ji and D. R. Phillips, Phys. Lett. B **723**, 196 (2013).
[8] M. Langevin et al., Phys. Lett. B, **150**, 71 (1985).
[9] S. Mosby et al., Nucl. Phys. A **909**, 69 (2013).
[10] T. Nakamura et al., Phys. Rev. Lett. **96**, 252502 (2006).
[11] B. Acharya , P. Hagen, H.-W. Hammer and D. R. Phillips, In preparation.
[12] M. T. Yamashita et al., Phys. Lett. B **697**, 90 (2011) [Erratum-ibid. B **715**, 282 (2012)].
[13] H. T. Fortune and R. Sherr, Phys. Rev. C **85**, 027303 (2012).
[14] L. Gaudefroy et al., Phys. Rev. Lett. **109**, 202503 (2012).
[15] M. Smith et al., Phys. Rev. Lett. **101**, 202501 (2008).
[16] National Nuclear Data Center, BNL, Chart of Nuclides (2013), *http://www.nndc.bnl.gov/chart/*.
[17] S. N. Ershov, J. S. Vaagen and M. V. Zhukov, Phys. Rev. C **86**, 034331 (2012).
[18] T. Nakamura, J. Phys. Conf. Ser. **445**, 012033 (2013).

EPJ Web of Conferences **113**, 06014 (2016)
DOI: 10.1051/epjconf/201611306014

Study of ^{19}C by One-Neutron Knockout

Jongwon Hwang[1,a], Sunji Kim[1], Yoshiteru Satou[1], Nigel A. Orr[2], Takashi Nakamura[3], Yosuke Kondo[3], Julien Gibelin[2], N. Lynda Achouri[2], Thomas Aumann[4], Hidetada Baba[5], Franck Delaunay[2], Pieter Doornenbal[5], Naoki Fukuda[5], Naohito Inabe[5], Tadaaki Isobe[5], Daisuke Kameda[5], Daiki Kanno[3], Nobuyuki Kobayashi[3], Toshio Kobayashi[6], Toshiyuki Kubo[5], Sylvain Leblond[2], Jenny Lee[5], F. Miguel Marqués[2], Ryogo Minakata[3], Tohru Motobayashi[5], Daichi Murai[7], Tetsuya Murakami[8], Kotomi Muto[6], Tomohiro Nakashima[3], Noritsugu Nakatsuka[8], Alahari Navin[9], Seijiro Nishi[3], Shun Ogoshi[3], Hideaki Otsu[5], Hiromi Sato[5], Yohei Shimizu[5], Hiroshi Suzuki[5], Kento Takahashi[6], Hiroyuki Takeda[5], Satoshi Takeuchi[5], Ryuki Tanaka[3], Yasuhiro Togano[10], Adam G. Tuff[11], Marine Vandebrouck[12], and Ken-ichiro Yoneda[5]

[1]*Department of Physics and Astronomy, Seoul National University, 599 Gwanak, Seoul 151-742, Republic of Korea*
[2]*LPC-ENSICAEN, IN2P3-CNRS et Université de Caen, F-14050, Caen Cedex, France*
[3]*Department of Physics, Tokyo Institute of Technology, 2-12-1 O-Okayama, Meguro, Tokyo 152-8551, Japan*
[4]*Institut für Kernphysik, Technische Universität, D-64289 Darmstadt, Germany*
[5]*RIKEN Nishina Center, Hirosawa 2-1, Wako, Saitama 351-0198, Japan*
[6]*Department of Physics, Tohoku University, Miyagi 980-8578, Japan*
[7]*Department of Physics, Rikkyo University, Toshima, Tokyo 171-8501, Japan*
[8]*Department of Physics, Kyoto University, Kyoto 606-8502, Japan*
[9]*GANIL, CEA/DSM-CNRS/IN2P3, F-14076 Caen Cedex 5, France*
[10]*ExtreMe Matter Institute EMMI and Research Division, GSI Helmholtzzentrum für Schwerionenforschung GmbH, D-64291 Darmstadt, Germany*
[11]*Department of Physics, University of York, Heslington, York YO10 5DD, United Kingdom*
[12]*Institut de Physique Nucléaire, Université Paris-Sud,IN2P3-CNRS, Université de Paris Sud, F-91406 Orsay, France*

Abstract. The spectroscopic structure of ^{19}C, a prominent one-neutron halo nucleus, has been studied with a ^{20}C secondary beam at 290 MeV/nucleon and a carbon target. Neutron-unbound states populated by the one-neutron knockout reaction were investigated by means of the invariant mass method. The preliminary relative energy spectrum and parallel momentum distribution of the knockout residue, ^{19}C*, were reconstructed from the measured four momenta of the ^{18}C fragment, neutron, and beam. Three resonances were observed in the spectrum, which correspond to the states at E_x = 0.62(9), 1.42(10), and 2.89(10) MeV. The parallel momentum distributions for the 0.62-MeV and 2.89-MeV states suggest spin-parity assignments of $5/2^+$ and $1/2^-$, respectively. The 1.42-MeV state is in line with the reported $5/2^+_2$ state.

[a]e-mail: hjw8707@snu.ac.kr

1 Introduction

Neutron-rich nuclei exhibit exotic features such as quenching of a shell gap or advent of a new magic number. Such a structure is caused by ascending or descending single-particle orbits from their original location in stable nuclei, and cannot be explained by the conventional shell model. Recently, three-body forces [1] and tensor forces [2] have been introduced to shell model calculations for neutron-rich nuclei. The lack of spectroscopic information of near-drip-line nuclei, however, is an obstacle to verifying those state-of-the-art theories. In this study, the level structure of ^{19}C has been investigated via the one-neutron knockout reaction.

^{19}C is the heaviest odd carbon isotope. Its $1/2^+$ ground state has a one-neutron halo structure [3]. Two bound states, $3/2^+_1$ and $5/2^+_1$, were reported from an in-beam γ-ray spectroscopy study [4], while the $5/2^+_2$ state at $E_x = 1.46(10)$ MeV in neutron-unbound continuum was observed in a (p, p') experiment [5]. Recent knockout measurements left room for a conjecture that the $5/2^+_1$ state is unbound [6, 7]. The present study addresses the issue whether or not the $5/2^+_1$ state is bound.

2 Experiment and Analysis

The experiment was carried out at the RI Beam Factory [8] (RIBF) at RIKEN Nishina Center for Accelerator-Based Science. A ^{20}C beam, separated by the BigRIPS separator [9, 10], produced using a ^{48}Ca primary beam at 345 MeV/nucleon, had an average intensity of 190 cps and the momentum acceptance of $\Delta P/P = \pm 3\%$. It impinged on a carbon target with a thickness of 1.8 g/cm^2 and produced ^{19}C isotopes by the one-neutron knockout reaction. The mid-target energy was 290 MeV/nucleon.

^{19}C populated in a neutron-unbound state decayed into a charged fragment, ^{18}C, and a neutron, which were measured by the SAMURAI spectrometer [11]. The charged fragment, separated by the dipole magnet, was detected by a plastic-scintillator hodoscope and two drift chambers (FDCs) placed before and after the magnet. The $B\rho$-TOF-ΔE method was used for the identification of the fragment. Its momentum was determined with the $B\rho$ and the direction reconstructed by the drift chamber. Neutron Detection System for Breakup of Unstable Nuclei with Large Acceptance (NEBULA) was used to determine the momentum vector of the decayed neutron using the TOF method. The detection efficiency of NEBULA was 31.6% for a threshold of 6 MeVee, measured by using the ^{7}Li$(p, n)^7$Be(g.s.+0.43 MeV) reaction at $E_p = 250$ MeV. DALI2 [12], the γ-ray detector array, surrounded the secondary target to observe the de-excitation γ rays from the charged fragment.

The relative energy ($E_{\rm rel}$) of the knockout residue (^{19}C*) was reconstructed from the four momenta of the ^{18}C fragment and decayed neutron. The background was subtracted by using data taken with an empty target. The geometrical acceptance was estimated with a Monte Carlo simulation taking into account the beam profile and geometry of the setup. Single Briet-Wigner shape functions were used to extract the parameters of the resonances with an empirical distribution for the non-resonant continuum in the fitting analysis. Response functions were generated by a simulation to take into account the experimental resolution, which was estimated to be $\Delta E_{\rm rel} \approx 0.40\sqrt{E_{\rm rel}}$ MeV in FWHM. The excitation energy (E_x) of the populated state corresponding to a resonance centered at $E_{\rm rel}$ is obtained by the following equation: $E_x = E_{\rm rel} + S_n + E^*$, where S_n is the one-neutron separation energy of ^{19}C (0.58(9) MeV [13]) and E^* is the excitation energy of the daughter nucleus. Note that no γ rays in coincidence with any of the observed resonances were identified for the ^{18}C + n channel.

The parallel momentum ($p_{\parallel}$) of ^{19}C* was reconstructed as well, which is a useful measure of the orbital angular momentum (l) of the knocked-out nucleon. By comparing the experimental distribution with the theoretical ones, the spin-parity of the populated state of ^{19}C was determined. The theoretical distributions for various l values were calculated by the code MOMDIS [14], which is based on the static density limit of the eikonal model. The core- and neutron-target S-matrices were obtained using the density-folding method using the NN profile function. The density of the carbon target was taken to be of a Gaussian form with a point-nucleon root-mean-square (rms) radius of 2.32 fm. The nucleon density distribution of the ^{19}C core was estimated from Hartree-Fock calculations using the SkX interaction [15]. For the nucleon-nucleon profile function, zero-range effective nucleon-nucleon interaction was used [16]. Neutron single-particle wave functions were calculated using the Woods-Saxon potential with a diffuseness $a_0 = 0.7$ fm and a reduced radius r_0 that was calculated to fulfill the relationship [17]: $r_{sp} = \sqrt{A/(A-1)}r_{HF}$ at the HF calculated binding energy of each orbit, where r_{sp} is the rms radius of the single-particle wave function and r_{HF} is the rms radius of the orbit deduced by the HF calculation for the beam nucleus. The calculated distributions were convoluted with an experimental resolution of 28 MeV/c in rms.

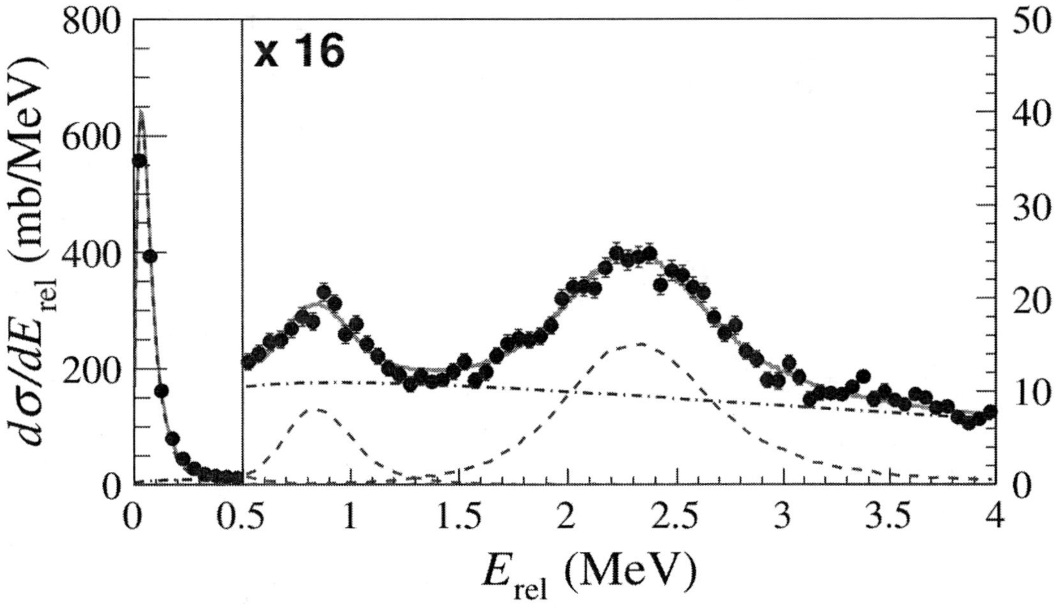

Figure 1. Preliminary relative energy spectrum of the ^{18}C + n system (*solid circle*) with statistical errors. The dashed and dot-dashed lines are the results of the fits for the resonances and a background component, respectively. The different scales for the y-axis are used below and above $E_{rel} = 0.5$ MeV.

3 Results and Discussion

Figure 1 shows the preliminary E_{rel} spectrum for the reaction of C(^{20}C,^{18}C + n), which was described using three resonances at $E_{rel} = 0.036(1)$, 0.84(3), and 2.31(2) MeV. The corresponding excitation energies are $E_x = 0.62(9)$, 1.42(10), and 2.89(10) MeV. While the first and second resonances are consistent with the $5/2_1^+$ and $5/2_2^+$ states reported by the knockout experiment [6] and the inelastic scattering measurement [5], respectively, the third one was observed for the first time in the present work.

The momentum distribution observed for the 0.62-MeV state was rather wide to be consistent with the $l = 2$ assignment, suggesting that the 0.62-MeV state was populated by the d-wave neutron knockout [18]. This is consistent with the suggested spin parity $5/2^+$ in Ref. [6]. For the 2.89-MeV state, the momentum distribution was found consistent with with the $l = 1$ distribution, which exhibits the p-wave knockout character. Considering the hierarchy of the orbits in the shell model, the 2.89-MeV state is likely to be the $1/2^-$ state [18]. This state is the firstly observed negative-parity state in ^{19}C.

4 Summary

The neutron-unbound states of ^{19}C have been investigated via a one-neutron knockout reaction with a carbon target. From the relative energy spectrum reconstructed by means of the invariant mass method, three states were identified at $E_{\rm rel}$ = 0.036(1), 0.84(3), and 2.31(2) MeV. They correspond to the states at E_x = 0.62(9), 1.42(10), and 2.89(10) MeV and the last one was observed for the first time. The spin-parity of the 0.62-MeV and 2.89-MeV states are determined to be $5/2^+$ and $1/2^-$, respectively, by comparing the parallel momentum distribution with the theoretical calculation. As a consequence, we provided direct evidence that the $5/2_1^+$ state is unbound. The analysis is still in progress, and more definitive results are expected to be obtained in the near future.

Acknowledgements

This work was partly supported by the NRF grant (R32-2008-000-10155-0 (WCU), 2010-0027136, 2014M2B2B1071110) of MSIP Korea. Partial support via the French-Japanese LIA for Nuclear Structure Problems is also acknowledged.

References

[1] T. Otsuka et al., Phys. Rev. Lett. **104**, 012501 (2010)
[2] T. Otsuka et al., Phys. Rev. Lett. **95**, 232502 (2005)
[3] T. Nakamura et al., Phys. Rev. Lett. **83**, 1112 (1999)
[4] Z. Elekes et al., Phys. Lett. B **614**, 174 (2005)
[5] Y. Satou et al., Phys. Lett. B **660**, 320 (2008)
[6] M. Thoennessen et al., Nucl. Phys. A **912**, 1 (2013)
[7] N. Kobayashi et al., Phys. Rev. C **86**, 054604 (2012)
[8] Y. Yano, Nucl. Instrum. Methods Phys. Res., Sect. B **261**, 1009 (2007)
[9] T. Kubo, Nucl. Instrum. Methods Phys. Res., Sect. B **204**, 97 (2003)
[10] T. Kubo et al., IEEE Trans. Appl. Supercond. **17**, 1069 (2007)
[11] T. Kobayashi et al., Nucl. Instrum. Methods Phys. Res., Sect. B **317**, 294 (2013)
[12] S. Takeuchi et al., Nucl. Instrum. Methods Phys. Res., Sect. A **763**, 596 (2014)
[13] M. Wang et al., Chin. Phys. C **36**, 1603 (2012)
[14] C. A. Bertulani and A. Gade, Com. Phys. Comm. **175**, 372 (2006)
[15] B. A. Brown, Phys. Rev. C **58**, 220 (1998)
[16] P. G. Hansen and J. A. Tostevin, Annu. Rev. Nucl. Part. Sci. **53**, 219 (2003)
[17] A. Gade et al., Phys. Rev. C **77**, 044306 (2008)
[18] J. W. Hwang, Ph.D. thesis, Seoul National University (2015)

EPJ Web of Conferences **113**, 06015 (2016)
DOI: 10.1051/epjconf/201611306015

Two neutron decay of ^{16}Be

A.E. Lovell[1,a], F.M. Nunes[1,b], and I.J. Thompson[2,c]

[1] *Department of Physics and Astronomy and National Superconducting Cyclotron Laboratory, Michigan State University, East Lansing, MI 48824, USA*
[2] *Lawrence Livermore National Laboratory, L-414, Livermore, California 94551, USA*

Abstract. Recently, the first example of two-neutron decay from the ground state of an unbound nucleus, ^{16}Be, was seen (A. Spyrou, *et. al.*, Phys. Rev. Lett. **108** 102501 (2012)). Three-body methods are ideal for exactly treating the degrees of freedom important for these decays. Using a basis expansion over hyperspherical harmonics and the hyperspherical R-matrix method, we construct a realistic model of ^{16}Be in order to investigate its decay mode and the role of the two-neutron interaction. The neutron-^{14}Be interaction is constrained using shell model predictions. We obtain a ground state for ^{16}Be that is under-bound by approximately 0.7 MeV with a width of approximately 0.17 MeV. For such a system, an attractive three-body force must be included to reproduce the experimental ground state energy.

1 Introduction

Exotic nuclei can be found across the nuclear chart, from light two-proton and two-neutron halo nuclei, such as ^{17}Ne and ^{6}He, to heavy two-proton emitters, such as ^{45}Fe, ^{48}Ni, and ^{54}Zn. These nuclei can be modeled as three-body systems, a core plus two neutrons or protons, and three-body wave functions can be used to investigate their properties as well as to give insights into their decay modes. (See for example [1].) The difference between dinucleon decay (where two correlated neutrons or protons decay from a nucleus) and three-body decay (where two uncorrelated nucleons decay from a nucleus) is of particular interest.

Two-proton decay was first theorized in 1960 [2] and was then first experimentally verified over 40 years later in experiments such as [3], [4], and [5]. These systems have been analyzed in terms of three-body models, where knowledge of their decay mechanism has led to a better understanding of their structure [1]. Although there has been much progress in the field of two-proton decay, much less is known about two-neutron radioactivity [6]. Recently, two-neutron decay from the continuum has been investigated theoretically [7].

An experiment performed in 2012 at the National Superconducting Cyclotron Laboratory first measured two-neutron radioactivity from the ground state of the unbound nucleus ^{16}Be [8]. Although ^{16}Be is an ideal candidate for two-neutron radioactivity (the lowest experimentally measured state in ^{15}Be is energetically inaccessible to one-neutron decay), there was controversy over this result when

[a] e-mail: lovell@nscl.msu.edu
[b] e-mail: nunes@nscl.msu.edu
[c] e-mail: IJT@ianthompson.org

it was first published [9]. In the original work [8], the authors used two extreme models to explain the ^{16}Be decay - one without a neutron-neutron interaction to model three-body decay [^{16}Be $\rightarrow$ ^{14}Be + n + n] and one with two strongly correlated neutrons to model dineutron decay [^{16}Be $\rightarrow$ ^{14}Be + (2n)]. However, in using a three-body formulation, we can exactly treat the degrees of freedom relevant for this decay - better understanding two-neutron decay and the neutron-neutron correlation in the nucleus.

2 Solutions to the three-body scattering problem

To solve the Schrödinger Equation in a three-body model, interactions between the three bodies must be defined. Typically, experimentally measured levels in the core + n system (here, ^{15}Be) would be used to constrain the interaction between these two bodies. However, only one level in ^{15}Be has been measured, at 1.8 MeV [10], and therefore, shell model calculations are used to constrain this interaction. Shell model calculations for ^{15}Be [11] give a $d_{5/2}$ state at 2.8 MeV, an $s_{1/2}$ state at 4.0 MeV, and a $d_{3/2}$ state at 7.0 MeV. In order to match the experimentally measured state in ^{15}Be, these shell model levels were lowered by 1 MeV in the current work. In the three-body model, the depth of three Woods-Saxon potentials (one for each of the s,p, and d partial waves) is adjusted to reproduce the shell model levels.

The neutron-neutron interaction is taken to be the GPT interaction [12]. Although this interaction is simpler than others such as AV18 [13] and Reid soft-core [14], it reproduces NN scattering observables up to 300 MeV, which is more than sufficient for the energy region considered in this work. The GPT interaction has often been used in the study of two-neutron halo nuclei (see for example, [15], [16], [17], and [18]). A three-body interaction can also be included in order to adjust the final resonance energy to match the experimental value.

For the three-body system, hyperspherical coordinates, ρ and θ, are used instead of the standard Jacobi coordinates [19]. This coordinate system allows the three-body wave function to be separated into hyperradial and hyperangular parts (where the hyperangular function can be constructed through a linear combination of known functions):

$$\Psi^{JM} = \frac{1}{\rho^{5/2}} \sum_{K\gamma}^{K_{max}} \chi^{J}_{K\gamma}(\rho) \mathcal{Y}^{JM}_{K\gamma}(\Omega_5, \sigma_1, \sigma_2, \xi). \quad (1)$$

This expansion introduces the hypermomentum, K, related to the relative angular momenta, l_x and l_y [19]. $K_{\max}$ determines the size of the model space.

Then, all that must be solved is the system of coupled hyperradial equations

$$\left(-\frac{\hbar^2}{2m}\left[\frac{d^2}{d\rho^2} - \frac{(K+3/2)(K+5/2)}{\rho^2}\right] - E\right)\chi^{J}_{K\gamma}(\rho) + \sum_{\gamma' K'} V_{KK'\gamma\gamma'}(\rho)\chi^{J}_{K'\gamma'}(\rho) = 0, \quad (2)$$

with scattering boundary conditions

$$\chi^{LKK_i}_{\gamma\gamma_i}(\kappa\rho) \longrightarrow \frac{i}{2}\left[\delta^{KK_i}_{\gamma\gamma_i} H^{-}_{K+3/2}(0,\kappa\rho) - \mathbf{S}^{LKK_i}_{\gamma\gamma_i} H^{+}_{K+3/2}(0,\kappa\rho)\right], \quad (3)$$

where $\kappa = \frac{\sqrt{2mE}}{\hbar}$. The index γ contains the quantum numbers l (total orbital angular momentum of the two neutrons relative to the core), S (total spin of the two neutrons), j (total angular momentum of the two neutrons relative to the core), I (the spin of the core), l_x (relative orbital angular momentum of the two neutrons), and l_y (relative orbital angular momentum of the core + (2n) system). The subscript i labels the incoming channel.

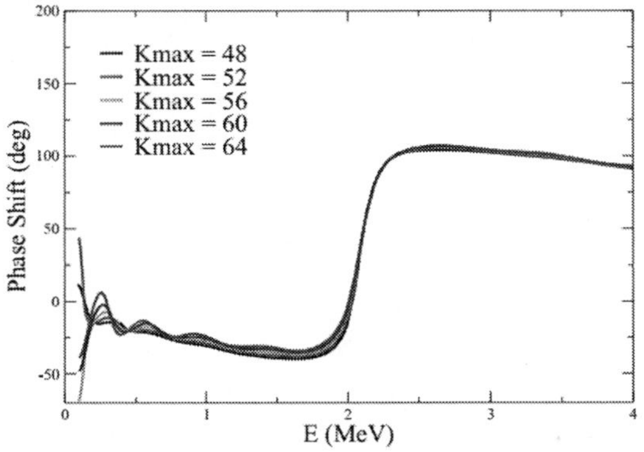

Figure 1. Phase shifts as a function of three-body energy of the $K = 0$ channel for the largest values of $K_{\max}$ used in the calculation.

Equation (2) is solved using the hyperspherical R-matrix method [19]. Through the R-matrix method, phase shifts can be calculated, allowing for the extraction of resonance energies and widths.

3 Preliminary Phase Shift Analysis and Future Work

In this work, energies are extracted from the phase shifts in the channel $K = 0$ at the points where the energy derivative of the phase shift has a maximum. In Figure 1, we show the convergence of the these phase shifts. This system required a large model space for convergence, including 80 hyperradial basis states, hypermomentum greater than 64, and over 80 hyperangular basis states. For values greater than $K_{\max} = 64$, numerical inaccuracies were induced due to the strong repulsive centrifugal barrier. By finding the peak of the derivative of the phase shifts in Figure 1, the resonance energy was calculated to be 2.07 MeV.

This ground state energy found above is under-bound by about 0.7 MeV when compared to the experimental ground state energy of 1.35 MeV [8]. Traditionally, three-body forces in three-body models account for the additional binding coming from degrees of freedom that are excluded from the model space. Here, an attractive three-body force must be added to the model in order to reproduce the experimental ground state of ^{16}Be.

A rough analysis of the $K = 0$ phase shift shows that the width is approximately 0.17 MeV. Due to the coupled-channel nature of these calculations, the interpretation of the phase shifts is not straightforward. Because of effects from the couplings between the channels, phase shifts do not rise sharply through 90° at the resonance energy. Therefore, it is not clear whether the ground state of the calculated system is given simply by the lowest resonance energy extracted from a single phase shift (as computed here) or if these phase shifts have to be weighted and combined, in the case of several resonance energies around the same region. One such method for this weighting would be to calculate the widths of the state through an integral relationship, such as that used in [20]. A study comparing the various approaches is underway.

Acknowledgements

The authors would like to acknowledge iCER and the High Performance Computing Center at Michigan State University for their computational resources. This work was supported by the National Science Foundation under Grant No. PHY-1403906. This work was performed under the auspices of the U.S. Department of Energy through NNSA contract DE-FG52-08NA28552 and by Lawrence Livermore National Laboratory under Contract DE-AC52-07NA27344.

References

[1] L.V. Grigorenko and M.V. Zhukov, Phys. Rev. C **68** 054005 (2003)
[2] V.I. Goldansky, Nucl. Phys. **19** 482 (1960)
[3] J. Giovinazzo, *et. al.*, Phys. Rev. Lett. **89** 102501 (2002)
[4] M. Pfützner, *et. al.*, Eur. Phys. Jour. A **14** 279 (2002)
[5] K. Miernik, *et. al.*, Phys. Rev. Lett. **99** 192501 (2007)
[6] L.V. Grigorenko, *et. al.*, Phys. Rev. C **84** 021303(R) (2011)
[7] L.V. Grigorenko and M.V. Zhukov, Phys. Rev. C **91** 064617 (2015)
[8] A. Spyrou, *et. al.*, Phys. Rev. Lett. **108** 102501 (2012)
[9] F.M. Marques, *et. al.*, Phys. Rev. Lett. **109** 239201 (2012)
[10] J. Snyder, *et. al.*, Phys. Rev. C **88** 031303(R) (2013)
[11] B.A. Brown, *Private Communications*, Nov. 21, 2013
[12] D. Gogny, P. Pires, and R. De Tourreil, Phys. Lett. **32B** 591 (1970)
[13] R.B. Wiringa, V.G.J. Stokes, and R. Schiavilla, Phys, Rev. C **51** 38 (1995)
[14] Roderick V. Reid, Jr. Ann. Phys. **50** 411 (1968)
[15] F.M. Nunes, J.A. Christley, I.J. Thompson, R.C. Johnson, and V.D. Efros, Nucl. Phys. A **609** 43 (1996)
[16] N.K. Timofeyuk, I.J. Thompson, and J.A. Tostevin, J. Phys. Conference Series **111** 012034 (2009)
[17] I. Brida, F.M. Nunes, and B.A. Brown, Nucl. Phys. A **775** 23 (2006)
[18] I. Brida and F.M. Nunes, Int. J. Mod. Phys. E **17** 2374 (2008)
[19] Ian J. Thompson and Filomena M. Nunes, *Nuclear Reactions for Astrophysics* (Cambridge University Press, United Kingdom, 2009) 221-225,279-286
[20] K.M. Nollett, Phys. Rev. C **86** 044330 (2012)

EPJ Web of Conferences **113**, 06016 (2016)
DOI: 10.1051/epjconf/201611306016

Is the isoscalar monopole resonance of the α-particle a collective mode?

Giuseppina Orlandini[1,2,a], Sonia Bacca[3,4,b], Nir Barnea[5,c], and Winfried Leidemann[1,2,d]

[1] *Dipartimento di Fisica, Università di Trento, Via Sommarive, 14 I-38123 Trento, Italy*
[2] *Istituto Nazionale di Fisica Nucleare, TIFPA, Via Sommarive, 14 I-38123 Trento, Italy*
[3] *TRIUMF, 4004 Wesbrook Mall, Vancouver, British Columbia V6J 2A3, Canada*
[4] *Department of Physics and Astronomy, University of Manitoba, Winnipeg, Manitoba R3T 2N2, Canada*
[5] *Racah Institute of Physics, Hebrew University, 91904, Jerusalem, Israel*

Abstract. In this contribution we review and clarify the arguments which might allow the interpretation of the isoscalar monopole resonance of ^{4}He as a collective breathing mode.

1 Introduction

The observable of interest in this contribution is the spectrum of ^{4}He when it responds to isoscalar monopole excitations. Such a spectrum can be measured by perturbative isoscalar probes, like for example a beam of α-particles.

At the end of the 60's inclusive electron scattering experiments [1, 2] put in evidence the existence of a narrow peak in the spectrum of ^{4}He. This was ascribed just to an isoscalar monopole resonant excitation (0^+ resonance) and the transition form factor to this resonant state was measured for different momentum transferred q up to 2 fm^{-1}. No measurement of the whole isoscalar monopole spectrum, however, is available for ^{4}He.

In the case of larger systems, such spectra have been the object of considerable activity, both from the experimental and theoretical points of view. The isoscalar monopole spectra of these nuclei exhibit visible "bumps", called isoscalar giant monopole resonance (GMR). The interest in these GMR's lies in the attempt to get an extrapolated value for the nuclear matter compressibility, a quantity of great astrophysical interest.

One has to notice that GMR's are visible in experiments at low momentum transfer and that the low energy part of the spectrum gives the main contribution to the compressibility. Of course, since for low q the wavelength that probes the target is large the spectrum will show mainly features that involve all constituents and therefore can reveal collective behaviors. Therefore the GMR's have been interpreted as signatures of "breathing modes".

[a] e-mail: giuseppina.orlandini@unitn.it
[b] e-mail: bacca@triumf.cau
[c] e-mail: nir@phys.huji.ac.il
[d] e-mail: winfried.leidemann@unitn.it

In an attempt to bridge few- and many-body physics we think it is interesting to ask the question whether also the measured 0^+ excited state in ^{4}He might be interpreted as a "collective" state. A similar question arose in Ref. [3], when the calculated dipole photonuclear cross section of ^{6}He surprisingly exhibited two prominent peaks. That result suggested a possible collective interpretation: the higher energy peak could have been ascribed to the displacement of the proton sphere against the neutron one (Gamow-Teller mode), that at lower energy to the displacement of the halo neutrons against a possible α-core.

The α-particle is a very compact system, with a similar binding energy per particle as heavier nuclei, and it is often considered the nucleus where few- and many-body methods can be benchmarked. Therefore in order to answer the question of the title, we apply to ^{4}He the same criteria as those, present in the many-body literature, used to judge the collectivity of "bumps" in the spectra. They consist in studying both the moments of the spectrum (sum rules) and the transition densities.

For an isoscalar monopole spectrum $S_M(q,\omega)$ the moments and relative sum rules are

$$m_k(q) = \int d\omega\, \omega^k S_M(q,\omega) = \langle 0|M(q)\, H^k\, M(q)|0\rangle\,, \tag{1}$$

where

$$M(q) = \frac{1}{2}\left(\sum_i j_0(qr_i) - \langle 0|\sum_i j_0(qr_i)|0\rangle\right). \tag{2}$$

For $k \geq 0$ these quantities can in principle be calculated avoiding the knowledge of scattering states for energies in the continuum.

Transition densities to specific states $|n\rangle$ are defined as

$$\rho_{tr}(\vec{r}) = \langle n|\hat{\rho}(\vec{r})|0\rangle\,. \tag{3}$$

If the 0^+ state in ^{4}He were an excitation of extremely collective character the strength would have a

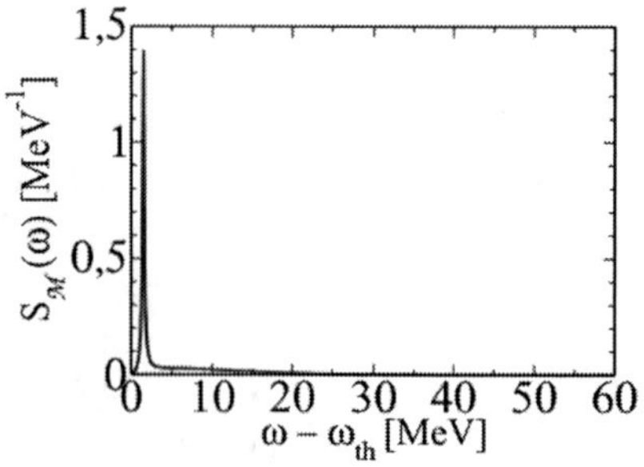

Figure 1. The inelastic isoscalar monopole strength of ^{4}He as a function of the excitation energy relative to threshold, for momentum transfer $q = 0.25$ fm^{-1}.

δ-function character. Therefore the strength of the resonance would exhaust all m_k. If this excitation were collective only to some degree, the ratio between the strength of the resonance and m_0 would give a measure of the *degree of collectivity*. Other sum rules for higher k would not be fit for the purpose, since they emphasize too much the spreading of the background strength to higher energy. Moreover, if the 0^+ state in ^{4}He were an excitation of extremely collective character, described as a breathing mode, the transition density would be zero at a value of r equal to the radius of the system.

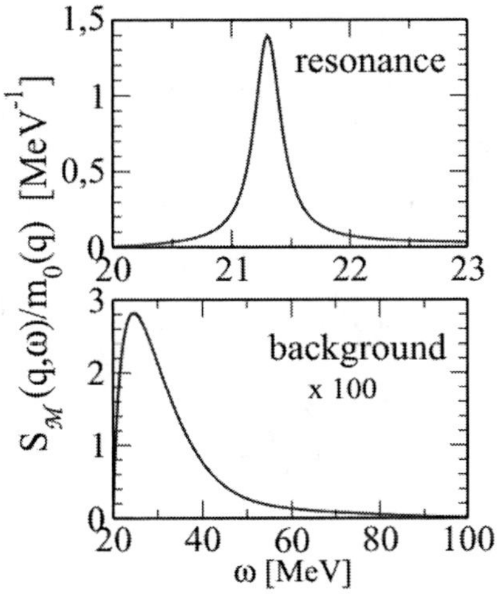

Figure 2. Upper panel: resonance contributions to the inelastic isoscalar monopole strength of ^{4}He as a function of the excitation energy; lower panel: background contribution magnified by a factor 100.

In [4] the transition form factor to the 0^+ excitation had been studied using the Lorentz Integral Transform (LIT) method [5, 6] using the Effective Interaction Hyperspherical Harmonics (EIHH) expansion [7]. In [8] we have focused on the isoscalar monopole strength distributions $S_M(q,\omega)$ for several fixed momentum transfers. There it has been shown that especially at low momentum transfer (e.g. q=0.25 fm^{-1}) the isoscalar monopole spectrum is indeed dominated by a single resonant "bump" close to threshold, as it is evident in Fig. 1. Since the LIT method has allowed to separate the resonance contribution from the background (see Ref. [4] for details) we show them separately here in Fig. 2. One can notice how far in energy the background extends, even if it remains very small. Figures 1 and 2 have been obtained with the chiral effective potential of Ref. [9], including both 2-body (at N3LO) and 3-body (at N2LO) contributions.

According to what was stated above, one can measure "the degree of collectivity" of the resonance state, by taking the ratio of the resonance area to the moment m_0. In this case one obtains 0.53. Values exceeding 50% are generally considered an indication of a rather strong collectivity (see e.g. [10]). Because of the extension of the background contribution at higher energy the ratio to m_1 is much smaller (34%), and a conclusion about the concentration of strength in the resonance, would be misleading if compared to $S_M(q,\omega)$ as in Fig. 1, which shows a single prominent narrow peak in the spectrum.

In Fig. 3 one can see the behavior of the transition density as a function of r. If one considers that the potential used in these calculations gives a root mean square radius of 1.46 fm one notices that this does indeed correspond almost to the point where the transition cross section crosses the zero axis, a typical signature of some kind of breathing mode. The results shown in Figs. 1-3 seem to point out that the two criteria that are generally accepted as indications of a collective behavior are fulfilled in the case of the isoscalar monopole excitation of ^{4}He. On the other hand it is clear that if one observes this system with such a large wavelength one is likely to observe the dynamics of all nucleons together. In this case it looks as if the resonance state $|R\rangle$ is obtained by a scaling transformation

$$|R\rangle = e^{i\alpha \sum_i p_i r_i}|0\rangle = e^{i\lambda[T,M^{LW}]}|0\rangle \tag{4}$$

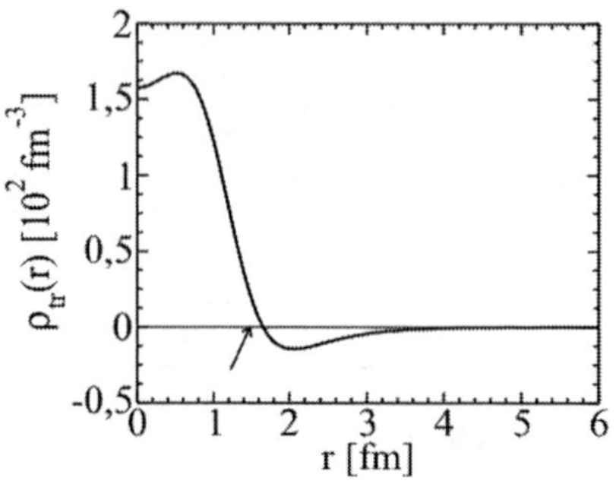

Figure 3. The transition form factor to the resonant state 0^+. The arrow indicates the position of the root mean square radius.

where $M^{LW} = \sum_i r_i^2 - \langle 0| \sum_i r_i^2 |0\rangle$ is proportional to the low-q (long wavelength) limit of M. The transformation in the previous equation acts scaling the positions r_i of all particles into αr_i, and therefore also of the hyperradius, namely one of the six *collective* coordinates defined by the group $GL^+(3, R)$ [11, 12] (breathing). In order to further prove that this is the case one could evaluate m_{-1} (connected to the compressibility) in terms of the moment m_3, in a similar case as it was done in Ref. [13], and compare it to the value obtained integrating the inverse energy weighted strength of Fig. 1. Work in this direction is in progress.

References

[1] R.F. Frosch, R.E. Rand, M.R. Yearian, H. Crannell, and L.R. Suelzle, Phys. Lett. **19**, 155 (1965); R.F. Frosch, R.E. Rand, H. Crannell, J.S. McCarthy, L.R. Suelzle, and M.R. Yearian Nucl. Phys. A **110**, 657 (1968)

[2] Th. Walcher, Phys. Lett. B, **31**, 442 (1970)

[3] S. Bacca, M.Marchisio, N. Barnea, W. Leidemann, G. Orlandini, Phys. Rev. Lett. **89** 052502 (2002)

[4] S. Bacca, N. Barnea, W. Leidemann, G. Orlandini, Phys. Rev. Lett. **110**, 042503 (2013).

[5] V.D. Efros, W. Leidemann, and G. Orlandini, Phys. Lett. B **338**, 130 (1994)

[6] V.D. Efros, W. Leidemann, G. Orlandini, and N. Barnea, J. Phys. G **34**, R459 (2007)

[7] N. Barnea, W. Leidemann, and G. Orlandini, Phys. Rev. C **61**, 054001 (2000); Nucl. Phys. A **693**, 565 (2001)

[8] S. Bacca, N. Barnea, W. Leidemann, G. Orlandini, Phys. Rev. C **91**, 024303 (2015)

[9] D.R. Entem and R. Machleidt, Phys. Rev. C **68**, 041001 (2003)

[10] D.J. Rowe, "Nuclear Collective Motion", Methuen & Co. LTD., London, 1970

[11] W. Zickendraht, J. Math. Phys. **12**, 1663 (1971)

[12] D. J. Rowe, and G. Rosensteel, Ann. Phys. **126**, 198 (1980)

[13] E. Lipparini, G. Orlandini, S. Stringari, and M. Traini, Nuovo Cim. **42 A** 296 (1977)

EPJ Web of Conferences **113**, 06017 (2016)
DOI: 10.1051/epjconf/201611306017

Scaling functions of two-neutron separation energies of ^{20}C with finite range potentials

M. A. Shalchi[1,a], M. R. Hadizadeh[2,b], M. T. Yamashita[1,c], Lauro Tomio[1,3,d], and T. Frederico[4,e]

[1] *Instituto de Física Teórica, UNESP, 01140-070, São Paulo, SP, Brazil*
[2] *Institute for Nuclear and Particle Physics and Department of Physics and Astronomy, Ohio University, Athens, OH, USA*
[3] *Centro de Ciências Naturais e Humanas, Universidade Federal do ABC, 09210-580, Santo André, SP, Brazil*
[4] *Instituto Tecnológico de Aeronáutica, DCTA, 12228-900, São José dos Campos, SP, Brazil.*

Abstract. The behaviour of an Efimov excited state is studied within a three-body Faddeev formalism for a general neutron-neutron-core system, where neutron-core is bound and neutron-neutron is unbound, by considering zero-ranged as well as finite-ranged two-body interactions. For the finite-ranged interactions we have considered a one-term separable Yamaguchi potential. The main objective is to study range corrections in a scaling approach, with focus in the exotic carbon halo nucleus ^{20}C.

1 Introduction and Model

In the present contribution our aim is to consider a general scaling model for three-body system with two non-identical ones near the unitary limit (when one or both two-body scattering lengths have absolute values very large), by considering bound neutron-core (nc) subsystem. The approach is exemplified for the exotic carbon halo nucleus ^{20}C, where an Efimov [1] excited energy state is studied within a three-body Faddeev formalism for $n - n -^{18} C$ (also named neutron-neutron-core, *nnc*, for more general neutron-rich exotic nuclei). More details on related studies, one can obtain from [2–7] and references therein. Therefore, the first task was to solve the coupled s−wave Faddeev integral equation using a zero-range potential to obtain the two-neutron (*nn*) separation energies for the ground and excited states of ^{20}C. Next, by fixing the energy of the virtual state of the *nn* subsystem to $E_{nn} = -143$ keV, we vary a three-body parameter in order to obtain different ^{19}C binding energies for the $n -^{18} C$ subsystem, such that we can also reproduce the ^{20}C ground-state energy, namely $E^{(0)}_{^{20}C}$. This procedure is followed by the calculation of the first excited-state energy $E^{(1)}_{^{20}C}$. In this way, we are able to obtain a scaling plot, which is conveniently given in dimensionless quantities, as $\sqrt{\left[E^{(1)}_{^{20}C} - E_{^{19}C}\right]/E^{(0)}_{^{20}C}}$ versus $\sqrt{E_{^{19}C}/E^{(0)}_{^{20}C}}$ [5, 8]. In order to study the range corrections, the same

[a] e-mail: shalchi@ift.unesp.br
[b] e-mail: hadizadm@ohio.edu
[c] e-mail: yamashita@ift.unesp.br
[d] e-mail: tomio@ift.unesp.br
[e] e-mail: tobias@ita.br

above steps are repeated, with the zero-range interaction being replaced by a separable Yamaguchi potential[9], with an appropriate range parameter β.

In the following we present the corresponding model-formalism, where for the two-body interactions we have used zero-ranged potentials (constant in momentum space), renormalized to give us the corresponding energies of the two-body subsystems (nn and nc), as well as finite-range interactions, given by one-term separable Yamaguchi form-factors. In the case of the finite-ranged interactions, the two-body potential, with strength λ and range-parameter β, is given by

$$v(p,p') = \lambda\, g(p)\, g(p') = \lambda \left(\frac{1}{p^2+\beta^2}\right)\left(\frac{1}{p'^2+\beta^2}\right). \quad (1)$$

The corresponding two-body t−matrix is also given analytically in a separable form, as follows:

$$t(p,p',E) = \tau(E)\, g(p)\, g(p'), \quad (2)$$

where

$$\tau(E) = \left[\lambda^{-1} - \int d^3p''\, g^2(p'')\, G_0(p'')\right]^{-1} = \left(\lambda^{-1} + \frac{2\pi^2\mu}{\beta(\beta \pm k)^2}\right)^{-1}. \quad (3)$$

The plus and minus signs are for bound and virtual states, respectively. μ is reduced mass and we are considering units such that $\hbar c = 1$. The parameters of the potential λ and β are given by the corresponding values of the subsystem energies, since the two-body t−matrix has poles for bound and virtual state energies. Next, the above definitions will be identified with the corresponding two-body subsystems by the respective labels nn and nc.

In our treatment for the two-neutron halo nuclei, we consider the two neutrons with masses given by $m_n \equiv m$ and the core with mass $m_A \equiv A\,m$. Such system can be studied in momentum space by the Faddeev formalism as a three-body problem with two different mass particles. With the assumption that the third particle is the core, with particles 1 and 2 being the neutrons, the coupled Faddeev equations for the $|\psi\rangle^{nn}$ and $|\psi\rangle^{nc}$ components of the subsystems nn and nc are given by

$$\begin{aligned} |\psi\rangle^{nn} &= G_{0nn}\, t_{nn}\, (|\psi\rangle^{cn} + |\psi\rangle^{nc}) = 2\, G_{0nn}\, t_{nn}\, |\psi\rangle^{nc}, \\ |\psi\rangle^{nc} &= G_{0nc}\, t_{nc}\, (|\psi\rangle^{nn} + |\psi\rangle^{nc}) = G_{0nc}\, t_{nc}\, (|\psi\rangle^{nn} + P_{12}|\psi\rangle^{nc}), \end{aligned} \quad (4)$$

where G_{0nn} (G_{0nc}) and t_{nn} (t_{nc}) are the three-body free propagator and two-body t−matrix for the nn (nc) subsystem, respectively. After partial-wave projection, in momentum space, for the s−wave, we have the following definitions for the wave-function components,

$$\begin{aligned} \psi_{nn}(p,q) &= 4\pi\, G_{0nn}\,(p,q)\, g_{nn}(p)\, F_1(q), \\ \psi_{nc}(p,q) &= 4\pi\, G_{0nc}\,(p,q)\, g_{nc}(p)\, F_2(q). \end{aligned} \quad (5)$$

with the corresponding Faddeev spectator functions, $F_1(q)$ and $F_2(q)$, given by

$$\begin{aligned} F_1(q) &= 4\pi\,\tau_{nn}\left(E - \frac{q^2(A+2)}{4mA}\right)\int dq'\, q'^2 \int dx\, \frac{2m}{2mE - (1+1/A)q^2 - 2q'^2 - 2qq'x} F_2(q'), \\ F_2(q) &= 2\pi\,\tau_{nc}\left(E - \frac{q^2(A+2)}{2m(A+1)}\right)\int dq'\, q'^2 \int dx\, \frac{2m}{2mE - 2q^2 + (1+1/A)q'^2 + 2qq'x} F_1(q') \\ &+ 2\pi\,\tau_{nc}\left(E - \frac{q^2(A+2)}{2m(A+1)}\right)\int dq'\, q'^2 \int dx\, \frac{2mA}{2mAE - (A+1)(q^2-q'^2) + 2qq'x} F_2(q'). \end{aligned} \quad (6)$$

One can easily verify, from Eqs. (1) and (2), that the results given by above formalism for separable one-term Yamaguchi interaction must converge to the zero-range results for very large values of β. In agreement with this expectation, next, we present some sample results for two cases where the nn virtual-state energy is fixed.

2 Results - Scaling plots and range corrections

(a) Three-body scaling plot for $E_{nn} = 0$

In Fig. 1 we have presented the scaling plots, which are correlating the binding energies of two successive ^{20}C states $E^{(N)}_{^{20}C}$ and $E^{(N+1)}_{^{20}C}$, for the case that the virtual-state energy of the nn subsystem is zero ($E_{nn} = 0$). In this case, $N = 0$ corresponds to the ground-state. The scaling plots are shown for zero-range and Yamaguchi potentials and for two cycles, with $N = 0$ and $N = 1$, implying that we need to obtain the first two excited states ($N = 1, 2$). As we can see, the first cycle, for zero-range and Yamaguchi potentials, have some deviation, which are showing clearly the range effect obtained by using the Yamaguchi finite-range interaction. However, for the second cycle ($N = 1$), the scaling plots for zero-range and Yamaguchi potentials are almost the same, which is not surprising since the ratio between the range and scattering length is very small and consequently the Yamaguchi scaling plot approaches the results obtained with the zero-range interaction.

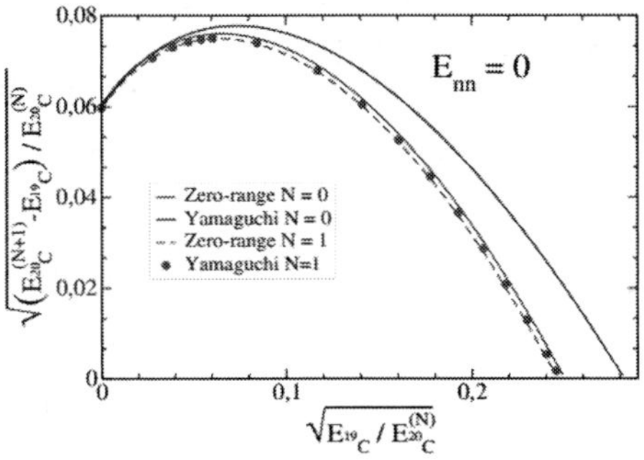

Figure 1. Scaling plots for the first and the second cycles, with $E_{nn} = 0$. The range effect in the scaling plot given by the Yamaguchi potential disappears in the second cycle, i.e. $N = 1$.

(b) Three-body scaling plot for $E_{nn} = -143$ keV

In this section we have presented our results for realistic case, when $E_{nn} = -143$ keV. As it can be verified from our formulation, for large values of parameter β, the coupled Faddeev equations for Yamaguchi potential goes to zero-range equations. As we can see in Fig. 2, the scaling plot for Yamaguchi potential for high values of β, or low values of range, is almost the same with zero-range results. This is very useful test to numerically verify the validity of our calculations.

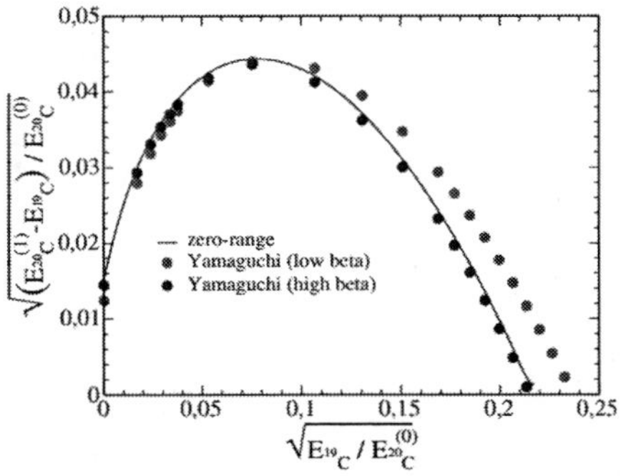

Figure 2. Scaling plots for zero-range and Yamaguchi potentials, with $E_{nn} = -143$ keV. The scaling plot for the Yamaguchi potential goes to zero-range results for large values of β (lower values of the effective range).

(c) Effective range correction

The correlation between the binding energy of two successive three-body states is given by a scaling function varying with the energies and ranges of the subsystems. By choosing a very small range for the nn subsystem, one can expand the scaling function in terms of the range of the nc subsystem, i. e. $x_{nc} = r_{nc}\sqrt{2\mu_{nc}E^{(N)}_{nnc}}$. For leading order expansion one can have

$$\frac{E^{(N+1)}_{nnc}}{E^{(N)}_{nnc}} = \epsilon\left(-\sqrt{\frac{E_{nn}}{E^{(N)}_{nnc}}}, \sqrt{\frac{E_{nc}}{E^{(N)}_{nnc}}}, r_{nn}\sqrt{2\,\mu_{nn}\,E^{(N)}_{nnc}}, r_{nc}\sqrt{2\,\mu_{nc}\,E^{(N)}_{nnc}}\right)$$
$$= \epsilon\left(-\sqrt{\frac{E_{nn}}{E^{(N)}_{nnc}}}, \sqrt{\frac{E_{nc}}{E^{(N)}_{nnc}}}\right) + \frac{\delta\epsilon}{\delta x_{nc}} r_{nc}\sqrt{2\,\mu_{nc}\,E^{(N)}_{nnc}} \quad (7)$$

In Fig 3 we show the results for the derivative $\frac{\delta\epsilon}{\delta x_{nc}} r_{nc}$ of this scaling function in terms of the energy of the nc subsystem. Our preliminary results were computed using large values of β for nn interaction, while for the nc potential we use for low β values. We compute the range effect in Fig 3, which comes essentially from the nc interaction, as the large values of β for the nn potential gives small effective range. Our calculation is done for the Yamaguchi potential, and should be viewed as a preliminary calculation, indicating the range correction pattern in the scaling plot of Fig. 2. We have only looked to the range effect in the nc potential, and with a particular form factor for the separable potential. We intend to study different form factors, in order to implement the range correction directly in the set of renormalised subtracted zero-range integral equation. This will allow us to obtain a model independent analysis of the range corrections in the energy scaling plots, namely the correlation between the energy of two close states. Other properties as sizes and momentum distribution should also be investigated in the future.

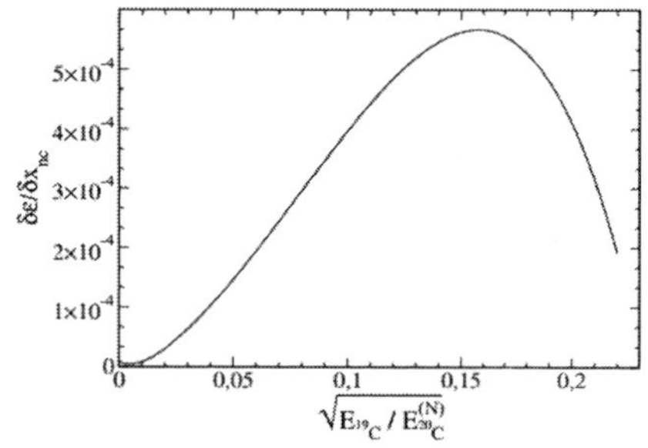

Figure 3. The derivation of ^{20}C scaling plot, given in Eq. (7), for E_{nn} =-143 keV as a function of the nc subsystem energy.

Acknowledgements

We acknowledge partial financial support from the Brazilian agencies FAPESP, CNPq and CAPES.

References

[1] V. Efimov, Phys. Lett. B **33**, 563 (1970); Nucl. Phys. A **210**, 157 (1973).
[2] A.E.A. Amorim, T. Frederico, L. Tomio, Phys. Rev. C **56**, R2378 (1997).
[3] M. T. Yamashita, T. Frederico, L. Tomio, Phys. Lett. B **660**, 339 (2008).
[4] D. L .Canham and H. Hammer, Eur. Phys. J. A **37**, 367 (2008)
[5] T. Frederico, A. Delfino, L. Tomio, M. T. Yamashita, Prog. Part. Nucl. Phys. **67**, 939(2012).
[6] D. R. Phillips, Few-Body Syst. **55**, 931 (2014).
[7] D. R. Phillips, Pramana **83**, 661 (2014).
[8] T. Frederico, L. Tomio, A. Delfino, M. R. Hadizadeh, M. T. Yamashita, Few-Body Syst. **51**, 87 (2011).
[9] Y. Yamaguchi, Phys. Rev. **95**, 1628 (1954).

EPJ Web of Conferences **113**, 07001 (2016)
DOI: 10.1051/epjconf/201611307001

Experimental investigations of the hypernucleus ${}^{4}_{\Lambda}$H

P. Achenbach[1,a], F. Schulz[1,b], S. Aulenbacher[1], J. Beričič[2], S. Bleser[1,3], R. Böhm[1], D. Bosnar[4], L. Correa[5], M. O. Distler[1], A. Esser[1], H. Fonvieille[5], I. Friščić[4], Y. Fujii[6], M. Fujita[6], T. Gogami[6,c], H. Kanda[6], M. Kaneta[6], S. Kegel[1], Y. Kohl[1], W. Kusaka[6], A. Margaryan[7], H. Merkel[1], M. Mihovilovič[1], U. Müller[1], S. Nagao[6], S. N. Nakamura[6], J. Pochodzalla[1], A. Sanchez Lorente[1,3], B. S. Schlimme[1], M. Schoth[1], C. Sfienti[1], S. Širca[2], M. Steinen[1,3], Y. Takahashi[6], L. Tang[8], M. Thiel[1], K. Tsukada[6,d], A. Tyukin[1], and A. Weber[1]
(A1 Collaboration)

[1] *Institut für Kernphysik, Johannes Gutenberg-Universität, D-55099 Mainz, Germany*
[2] *Department of Physics, University of Ljubljana, and Jožef Stefan Institute, SI-1000 Ljubljana, Slovenia*
[3] *Helmholtz Institute Mainz, D-55099 Mainz, Germany*
[4] *Department of Physics, University of Zagreb, HR-10002 Zagreb, Croatia*
[5] *LPC-Clermont, Université Blaise Pascal, CNRS/IN2P3, F-63177 Aubière Cedex, France*
[6] *Department of Physics, Tohoku University, Sendai, 980-8571, Japan*
[7] *Yerevan Physics Institute, 375036 Yerevan, Armenia*
[8] *Department of Physics, Hampton University, Hampton, Virginia 23668, USA*

Abstract. Negatively charged pions from two-body decays of stopped ${}^{4}_{\Lambda}$H hypernuclei were studied in 2012 at the Mainz Microtron MAMI, Germany. The momenta of the decay-pions were measured with unprecedented precision by using high-resolution magnetic spectrometers. A challenge of the experiment was the tagging of kaons from associated $K^{+}\Lambda$ production off a Be target at very forward angles. In the year 2014, this experiment was continued with a better control of the systematic uncertainties, with better suppression of coincident and random background, improved particle identification, and with higher luminosities. Another key point of the progress was the improvement in the absolute momentum calibration of the magnetic spectrometers.

1 Introduction

The structure of light Λ-hypernuclei and the precise determination of Λ binding (separation) energies has been the focus of recent experimental and theoretical programs. The Λ-hypernucleus ${}^{4}_{\Lambda}$H was investigated in 2012 by high-resolution spectroscopy at the Mainz Microtron MAMI, Germany [1]. In this experiment the binding energy of ${}^{4}_{\Lambda}$H was determined from the two-body charged decay mode with an unprecedented ±10 keV statistical uncertainty and ±90 keV systematic uncertainty to be B_{Λ} =

[a] e-mail: patrick@kph.uni-mainz.de
[b] Part of doctoral thesis.
[c] Present address: Graduate School of Science, Kyoto University, Kyoto 606-8502, Japan.
[d] Present address: Research Center for Electron Photon Science, Tohoku University, Sendai 982-0826, Japan.

2.12 MeV. This value is 80 keV different from emulsion data, the most complete compilation of which found $B_\Lambda = 2.04 \pm 0.04$ MeV using only three-body charged decay modes [2].

The new method of decay-pion spectroscopy has the potential to allow binding energy measurements of several light hypernuclei with a precision comparable or better than with the emulsion technique. The statistical uncertainty for the binding energies in light hypernuclei with the emulsion method ranged from a minimum of 20 keV for the most abundant hypernuclei $^5_\Lambda$He to more than 700 keV for $^8_\Lambda$He [2–4]. Systematic errors in the emulsion data are not shown, but D.H. Davis assumed errors of the order of 50 keV [5].

A major effort in hypernuclear physics is to understand the interaction between hyperons and nucleons. One motivation to study the $^4_\Lambda$H hypernucleus is the possible difference between the hyperon–proton and hyperon–neutron interaction which is manifest in the difference of Λ binding energies, ΔB^4_Λ, in the mirror pair of hypernuclei $^4_\Lambda$H and $^4_\Lambda$He. This difference of $\Delta B^4_\Lambda = 350 \pm 60$ (stat.) keV is one of the main sources of information about the charge symmetry breaking (CSB) in the ΛN interaction, see Ref. [6] for a recent review. The improvement of the uncertainties on the binding energy of $^4_\Lambda$He is outside the potentialities of decay-pion spectroscopy because this hypernucleus cannot decay in the two-body pionic channel. Combined with the emulsion data of $B_\Lambda(^4_\Lambda\text{He})$, the MAMI experiment indicates a large ΔB^4_Λ. This supports a CSB effect in the ΛN interaction being much larger than in the NN interaction.

In the year 2014 the MAMI experiment was continued. In this paper, the control of systematic uncertainties, the improvements in background suppression, and the extensive calibrations are discussed which reduced the systematic error for the binding energy extraction and yielded a more accurate absolute calibration of the magnetic spectrometers.

2 Improved experimental setup of the 2014 beam-time

The experimental setup is shown in Fig. 1. Kaons from the strangeness production reaction off the ^{9}Be target were tagged in forward direction by the KAOS spectrometer. Two different target thicknesses of 23 and 47 mg/cm^2 were used. The $^4_\Lambda$H hypernuclei were formed as hyperfragments and the π^- from the two-body decay $^4_\Lambda\text{H} \to {}^4\text{He} + \pi^-$ at rest were observed. The π^- momentum spectra were measured with two spectrometers SpekA and SpekC, described in Ref. [7]. In order to check systematic momentum uncertainties the acceptance of two spectrometers covered the $^4_\Lambda$H decay-momentum region simultaneously. The use of large solid angle spectrometers and good background rejection techniques are crucial ingredients for this experiment. Accidental background was reduced by heavier shielding of the photon beam-dump and stronger positron absorption in the KAOS spectrometer.

Energy-loss fluctuations in the scattering chamber windows could be eliminated in the 2014 beam-time by directly coupling the vacuum system of the spectrometers to the chamber. Two tungsten alloy collimators were placed behind the target to reduce the coincident background from quasi-free produced Σ^- decays in flight. The collimator blocks had thicknesses of 30 mm and their positions as shown in Fig. 1 (right) were optimized to stop low-momentum decay-pions originating from a decay-region a few cm behind the target.

The magnetic fields of the spectrometers were monitored continuously throughout the beam-time with NMR measurements every 5 minutes. Variations of the magnetic fields of the spectrometers corresponded to less than $\delta p_{stabil.} < 4$ keV/c as compared to 40 keV/c in 2012.

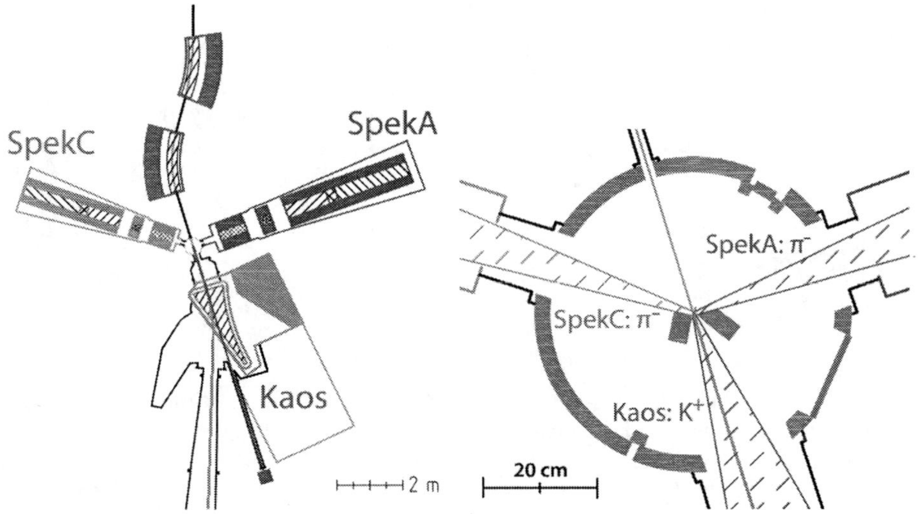

Figure 1. Left: Setup of electron beam-line and magnetic spectrometers in the experimental hall at the Mainz Microtron MAMI. The beam entered from top and was deflected by a magnetic chicane before hitting the target with an angle of 17° with respect to the incoming direction. About 1 m downstream of the target the beam was deflected by the KAOS spectrometer. Photons from Bremsstrahlung reactions in the target were stopped in a separate beam-dump. Right: Enlarged view of the target region showing the position of the tungsten alloy collimators. All spectrometers, the scattering chamber, and the beam-line formed a common vacuum system.

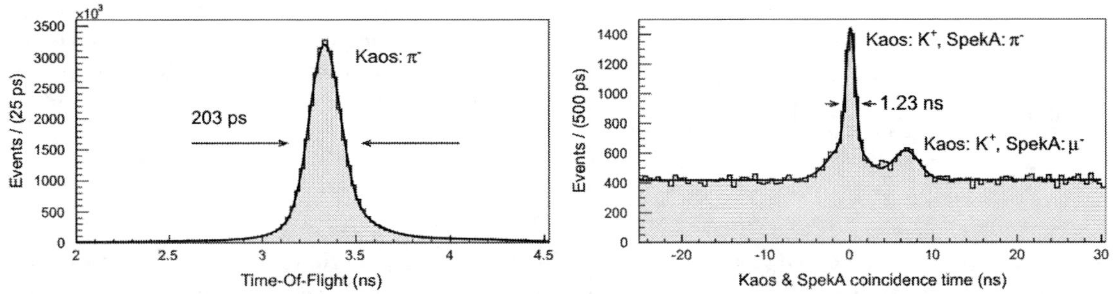

Figure 2. Left: Measured time-of-flight for pions inside the KAOS spectrometer along a normalized flight path length of 1 m. Right: Coincidence time distribution for kaons in the KAOS spectrometer and pions and muons in SpekA. The background includes mis-identified particles near to the central peak and a flat distribution of accidental coincidences.

3 Improved calibrations of the 2014 beam-time

3.1 TOF detector calibrations

For the identification of decay-pions it is crucial to have an efficient kaon tagging in combination with a sharp cut on the coincidence time due to the huge background of pions and protons at very forward angles. The coincidence requirement improves dramatically the signal-to-noise ratio but places stringent requirements on the calibration of the time-of-flight (TOF) detectors in the KAOS spectrometer. In particular, any changes of ambient parameters like temperature, beam intensity, or

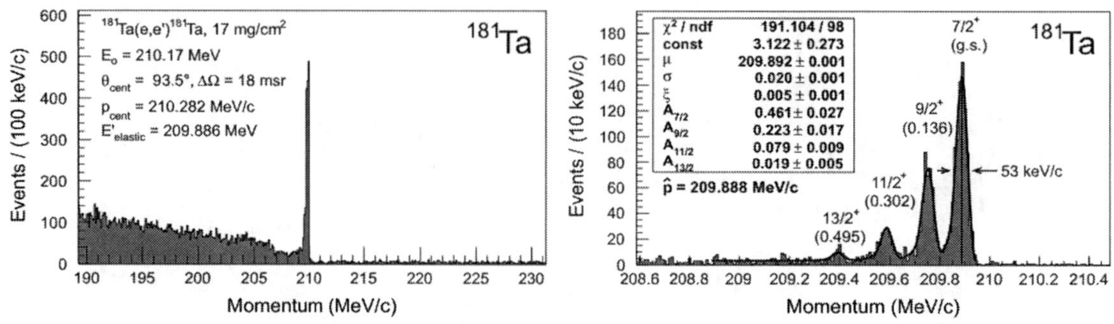

Figure 3. Left: Measured momentum spectrum for the ^{181}Ta$(e,e')^{181}$Ta reaction. Target thickness, reaction kinematics, spectrometer settings, and calculated energy for elastic scattering are given. Right: Fit of the peak region in the momentum spectrum with a sum of four Landau distributions convoluted with a Gaussian resolution function on top of a constant background. Three low-lying excited states were resolved and characterized by J^P quantum numbers and excitation energy. The ground-state FWHM of 52 keV/c and the peak maximum position $\hat{p}$ are indicated.

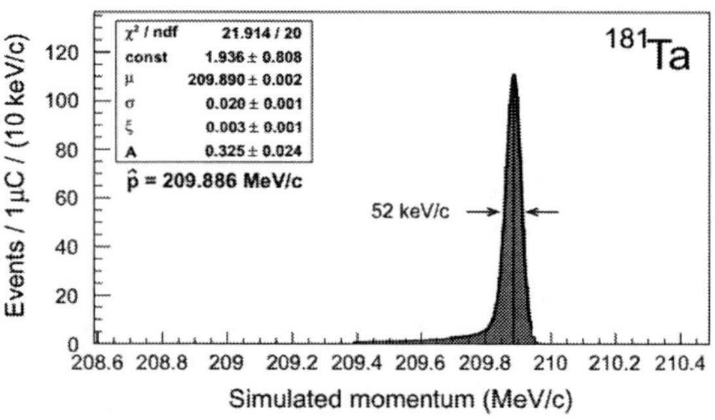

Figure 4. Simulated momentum spectrum for the elastic scattering reaction ^{181}Ta$(e,e')^{181}$Ta in the same kinematics as the measurements shown in Fig 3. The peak was fitted with a Landau distribution convoluted with a Gaussian resolution function. The FWHM of 52 keV/c and the peak maximum position $\hat{p}$ are indicated.

background particle flux could lead to a reduction in kaon tagging efficiency. For the 2014 beam-time an automatic correction was employed for every control setting of all 60 TOF detector segments for every one of the 1201 data runs. In total, more than 370 000 parameter values were calculated and used for the hypernuclei analysis. Fig. 2 shows the measured time-of-flight for pions inside the KAOS spectrometer after calibration. The FWHM $\approx$ 200 ps is a factor of 2 better than in the 2012 beam-time. This high resolution for the TOF measurement is also reflected in a better resolved coincidence time. Further improvements have been achieved for the measurement of the particle's energy-loss and Cerenkov light yield.

3.2 Absolute momentum calibration

Detailed measurements with electron beams of energies 195 and 210 MeV have been performed for an absolute momentum calibration and to study the properties of the two spectrometers SpekA and SpekC. As targets, a ^{181}Ta target of 6 μm foil thickness, corresponding to 10 mg/cm^2 mass thickness, and a ^{12}C target of 450 μm foil thickness, corresponding to 100 mg/cm^2 mass thickness, were used. Due to tantalum's large mass it experiences relatively little recoil compared to carbon, making it an excellent choice for calibration. The target foils were tilted by 54° with respect to the incoming beam.

The beam energy was measured in the third stage of the racetrack microtron cascade with an absolute accuracy of $\delta E_{beam} = \pm 160\,\text{keV}$ by exact determination of the beam position on the accelerator axis and in a higher return path.

As a result, Fig. 3 shows a momentum spectrum of electrons scattered on ^{181}Ta at a scattering angle of 93.5°, corrected for kinematic broadening. Three low-lying rotational states were clearly resolved. The spectrum was measured with SpekA using a solid angle of 18 msr. As compared to the full value of 28 msr the acceptance angle cuts reduced the aberration effects slightly. The elastic line for the reduced acceptance had a width of 53 keV/c (FWHM), corresponding to a momentum resolution of $\delta p/p \approx 2 \times 10^{-4}$. The width increased by 10 keV/c for the full solid angle.

The ground-state line was fitted with a Gaussian distribution corresponding to the spectrometer resolution $\mathcal{G}(p, \mu, \sigma)$ parameterized with the expectation value μ and variance σ^2. This distribution was folded with the energy-loss distribution represented by a numerical approximation of the Landau distribution $\mathcal{L}(p, p_0, \xi)$ with ξ quantifying the width of the distribution and a shift between the most probable value $p_0 = \mu + 0,22278298\,\xi$ and the position μ. The spectrometer response function used for the fitting of the calibration data is then $\mathcal{A}(p, \mu, \sigma, \xi, A) = A \cdot (\mathcal{G} * \mathcal{L})(p)$.

Several calibration measurements for different momentum settings were taken. For the computation of new spectrometer calibration values only tantalum elastic scattering data were employed. The calibration was fixed at the relative momentum closest to where the decay-pion line of $^4_\Lambda$H was observed. The data shown in the figure was taken at a 7 % shifted momentum setting. The difference between the calculated elastic line position $E'_{elastic}$ after energy-loss corrections and the observed line position was of the order of 2 keV/c.

For the determination of the energy-loss entering the calibration the spectra were simulated with a Monte Carlo code taking into account the energy-loss of the incoming beam in the target, the reaction vertex distribution inside the foil, the reaction kinematics, and the energy-loss of the outgoing electron in the target and the spectrometer exit foils. The simulation included ionization losses according to Landau's theory for thin absorbers and internal and external radiation losses. Fig. 4 shows the simulated elastic line for the same kinematics as the measurements shown in Fig. 3 using the design values for the momentum resolution. The shape and width of the line were reproduced.

The excitation spectrum of the $^{12}\text{C}(e, e')^{12}\text{C}$ reaction was used to check the linearity of the momentum scale. Fig. 5 (left) shows the excitation spectrum of carbon for one spectrometer setting. The expected positions of eight identified states, calculated from the known excitation energies, were indicated by arrows. Several of these states could be fitted for the same setting, in total 21 fits were performed. As an example the fit to the $J^P = 2^+$ state for this setting is shown in Fig. 5 (right). The carbon lines are wider than the tantalum lines because of the larger energy-loss straggling in the target. From the observed aberration effects along the full momentum acceptance differences of the order of 20 keV/c between the setting shown in Fig. 3 and the calibration point were observed.

In summary, the energy-loss calculation, fit procedure, remaining aberration effects, and the limited statistics can add up to a systematic error of 10 keV/c. Fluctuations in the magnetic field could lead to errors not larger than 5 keV/c. The whole calibration procedure was also applied to SpekC.

4 Discussion

A preliminary analysis of these data confirmed in both spectrometers the mono-energetic line from pionic $^4_\Lambda$H decays stopped in Be targets of two different thicknesses. In comparison to the 2012 beam-time the 2014 beam-time profited from improved calibrations which reduced the systematic error on the extraction of the $^4_\Lambda$H binding energy. The remaining systematic error of the binding energy originating from the calibration procedure is < 5 keV for SpekA and < 20 keV for SpekC while the

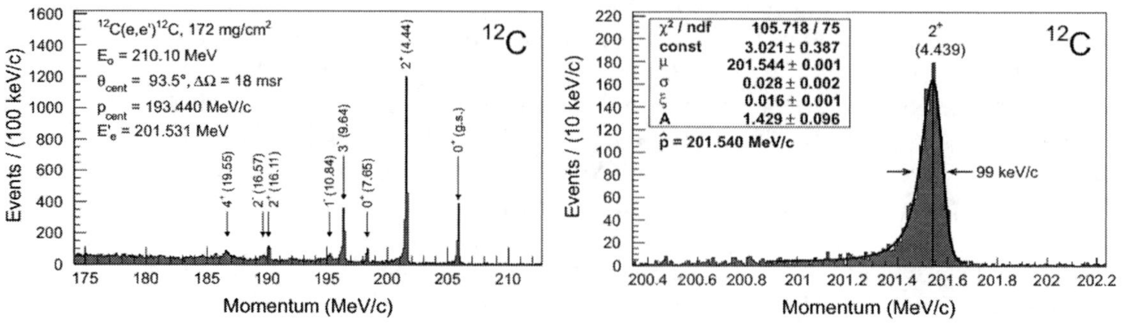

Figure 5. Left: Measured momentum spectrum for the $^{12}C(e, e')^{12}C$ reaction. Several excited states over the full momentum acceptance were resolved and characterized by J^P quantum numbers and excitation energy. Target thickness, reaction kinematics, spectrometer settings, and calculated scattering energy for excitation of the first $J^P = 2^+$ state are given. Right: Fit of the peak region for this state with a Landau distribution convoluted with a Gaussian resolution function on top of a constant background. The FWHM of 99 keV/c and the peak maximum position $\hat{p}$ are indicated.

dominant source of systematic uncertainty of 70 − 80 keV originates from the absolute accuracy of the beam energy measurement. The control and stability of the magnetic fields contributes < 5 keV compared to $\sim$ 30 keV in 2012. Measured values of the binding energy with the two spectrometers and the two different target thicknesses will be published in a forthcoming publication.

The calibrations were performed so that future improvements in the beam energy measurements could be used to correct a posteriori the uncertainties of the MAMI 2014 beam-time results. The correction will affect only the central momenta of the spectrometers. A typical energy spread of the extracted MAMI beam is 13 keV with the absolute energy known with an accuracy of $\delta E_{beam} \approx \pm 160$ keV. To establish a factor 10 improvement in precision a dipole magnet will be used as a beam-line spectrometer. This method requires the measurement of its field map with a precision in the range of μT and a stability of the setup on the level of 20 μm over an area of $5 \times 2\,m^2$.

Acknowledgements

This work was supported in part by Deutsche Forschungsgemeinschaft (SFB 1044), the Carl Zeiss Foundation, European Community Research Infrastructure Integrating Activity FP7, U.S.-DOE Contract No. DEFG02-97ER41047, the Strategic Young Researchers Overseas Visits Program for Accelerating Brain Circulation (R2201), and the Core-to-Core program (21002) of JSPS.

References

[1] A. Esser et al. (A1 Collaboration), Phys. Rev. Lett. **114**, 232501 (2015)
[2] M. Jurić et al., Nucl. Phys. B **52**, 1 (1973)
[3] G. Bohm et al., Nucl. Phys. B **4**, 511 (1968)
[4] W. Gajewski et al., Nucl. Phys. B **1**, 105 (1967)
[5] D.H. Davis, Nucl. Phys. A **547**, 369 (1992)
[6] A. Gal, Phys. Lett. B **744**, 352 (2015)
[7] K.I. Blomqvist et al. (A1 Collaboration), Nucl. Instrum. Meth. Phys. Res. A **403**, 263 (1998)

EPJ Web of Conferences **113**, 07002 (2016)
DOI: 10.1051/epjconf/201611307002

Study of the Hyperon-Nucleon Interaction in Exclusive Λ Photo-production off the Deuteron

Nicholas Zachariou[1,2,a], Yordanka Ilieva[2,b], and Tongtong Cao[2,c], for the CLAS Collaboration

[1] *University of Edinburgh, Edinburgh , UK*
[2] *University of South Carolina, Columbia SC, USA*

Abstract. The study of final-state interactions in exclusive hyperon photoproduction off the deuteron is a promising approach to extract information about the hyperon-nucleon (YN) interaction. First preliminary results on the azimuthal asymmetry Σ, as well as the polarization transfer coefficients O_x, O_z, C_x, and C_z for the reaction $\gamma d \to K^+\Lambda n$ initiated with linearly and circularly polarized photon beam are presented. The data were taken with the CLAS detector in Hall B of Jefferson Lab during the E06-103 experiment. The large kinematic coverage of the CLAS, combined with the exceptionally high quality of the experimental data, allows identifying and selecting final-state interaction events to extract single- and double-polarization observables and their kinematical dependencies.

1 Introduction

Understanding of the nucleon-nucleon (NN) and hyperon-nucleon (YN) interaction is necessary in order to obtain a comprehensive picture of the strong interaction. The NN interaction has been extensively studied for many decades, and a good understanding of this system at energies beyond the pion production threshold exists. On the other hand, direct studies of interactions between other members of the baryon octet remain scarce and inaccurate due to the experimental challenges imposed by short-lived hyperon beams and targets. Reliable data on the YN interaction will allow for realistic calculations of hypernuclear structure and hyperon matter, and will provide a more detailed understanding of the structure of neutron stars. In addition, the study of the YN interaction is important for understanding hadron dynamics involving strange quarks, and enables the extension of the baryon-baryon interaction to a more unified picture demanded by SU(3) symmetry.

YN potentials, constructed from NN potentials under the assumption of exact SU(3) symmetry, have several free parameters that must be constrained from experimental data. The best approach to constrain these is to use YN elastic scattering, but the challenges imposed by hyperon beams and targets result in data of poor quality and energy limitations. This has prompted interest in complimentary studies that allow indirect access to the dynamics of the YN interaction through hypernuclear studies and studies of final state interactions (FSI). Our work utilises the sufficient counting rates that are obtained with the use of modern accelerators to study FSI in exclusive hyperon photoproduction off a

[a]e-mail: nicholas@jlab.org
[b]e-mail: ilieva@sc.edu
[c]e-mail: caot@email.sc.edu

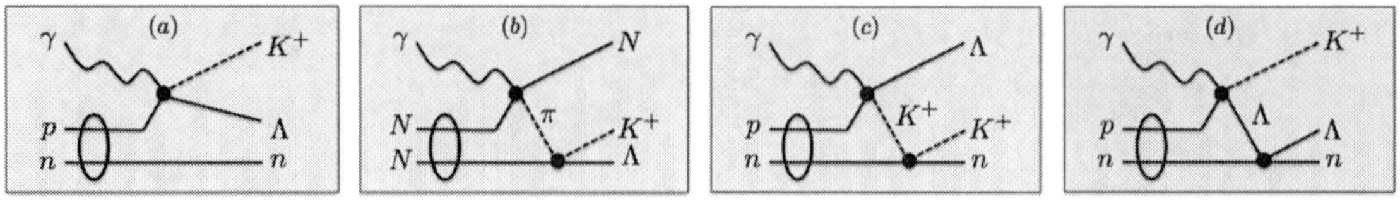

Figure 1. Four main mechanisms that contribute to the reaction $\gamma d \rightarrow K^+\Lambda n$ according to theoretical models [1, 2]: (a) quasi-free Λ photoproduction on the proton; (b) pion mediated production; (c) K^+ rescattering on spectator neutron; (d) Λ rescattering on spectator neutron.

deuterium target. The simplicity of the target, as well as the fact that the electromagnetic interaction is well understood, allows us to select kinematics where FSI between the hyperon and the spectator nucleon are greatly enhanced.

According to the available models [1, 2], several mechanisms contribute to the reaction $\gamma d \rightarrow K^+\Lambda n$; the main ones are shown in Fig. 1. The exclusivity of the reaction allow us to select kinematics that minimize contributions from the quasi-free (QF) mechanism, which dominates the cross section. This enhances contributions from FSI, giving us the ability to extract information of the dynamics of YN in a model-dependent way. Currently, the most comprehensive model for this reaction allows calculations of the unpolarized differential cross section, as well as single- and double- polarization observables [1, 3]. This model makes predictions using two YN potentials (Nijmegen NSC89 and NSC97f), both of which correctly predict the hypertrition binding energy [4] and take into account contributions from the QF reaction. The predictions clearly indicate sensitivities of polarization observables to the YN potentials at certain kinematics. A large sample of polarization observables is expected to place stringent constraints on the theoretical models and allow us to extract information of the dynamics of the YN interaction.

In addition, a new theoretical approach allows the extraction of the spin-averaged $\Lambda - n$ scattering lengths by studying the $\Lambda - n$ invariant mass distributions of observables [5]. This approach has been tested on inclusive K^+ hadron-production data and on pseudodata on deuteron photoproduction indicating that the current poorly-determined scattering lengths can be further constrained from FSI.

2 Experimental Procedure

Polarization observables for FSI in the reaction $\gamma d \rightarrow K^+\Lambda n$ were determined using data collected with the CEBAF Large Acceptance Spectrometer (CLAS) [6] housed in Hall B of Jefferson Lab in Newport News, Virginia. The experiment E06-103 utilised both linearly- and circularly-polarized photon beams and took place in 2006 and 2007 [7]. Photons with average linear polarization of 75%, produced via coherent bremsstrahlung, gave access to Σ, O_x, and O_z. Photons produced using a thin gold radiator with circular polarizations between 30% and 80% gave access to C_x and C_z. Three charged-track events were selected identifying the K^+ and the p-π^- pair from the Λ decay. The hyperon was reconstructed from the p-π^- invariant mass and the reaction was identified using the missing-mass technique. The upper panel of Fig. 2 shows the missing mass of the reaction $\gamma d \rightarrow K^+\Lambda X$, in which one can clearly identify missing-neutron events over a smooth background from Σ^0 and Σ^* photoproduction events. Background contributions were studied in detail utilizing a comprehensive event generator (developed by the authors) and a realistic detector simulation [8] and subtracted from the total yield to obtain a clean signal.

Contributions from the QF mechanism were significantly reduced by selecting events in which the final neutron had momentum higher than the typical Fermi momentum. The neutron momentum

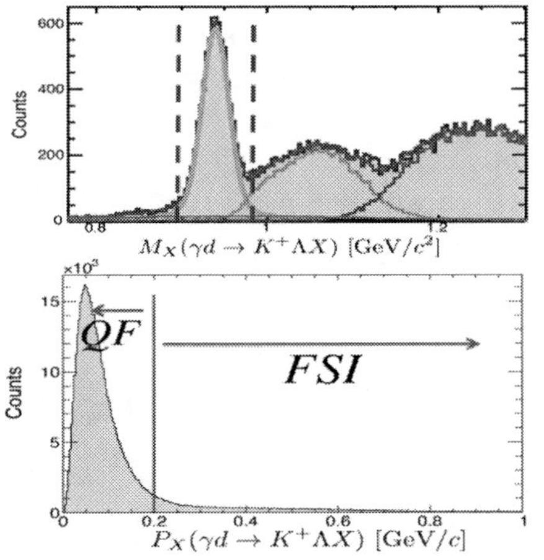

Figure 2. Top panel: Missing-mass distribution of the reaction $\gamma d \to K^+\Lambda X$ fitted with contributions from $\gamma d \to K^+\Sigma^0 X$ (green), $\gamma d \to K^+\Sigma^* X$ (purple), and accidental events (magenta) from simulated data. The vertical dotted lines indicate the cuts that select the events of interest and the fits are used to perform background subtraction. Bottom panel: Missing momentum of the reaction $\gamma d \to K^+\Lambda X$ indicating the cut applied to select FSI.

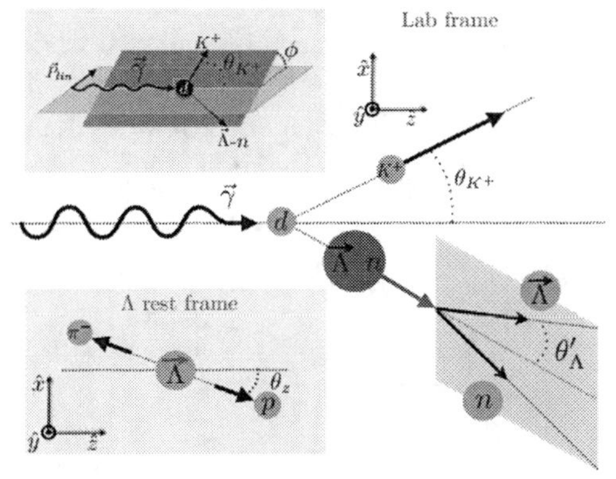

Figure 3. Reaction plane definition for $\gamma d \to K^+\Lambda n$. The upper left panel shows the definition of the kaon azimuthal angle ϕ in which Σ produces a modulation (see Eq. [1]). The lower left panel shows the definition of θ_z (and correspondingly θ_x) in which the double-polarization observables produce modulations. The main figure defines the kaon laboratory polar angle θ_{K^+}, and the hyperon angle θ'_Λ, which is in respect to an axis pointing along the momentum of the YN pair.

was reconstructed through four-momentum conservation, with the cut to select FSI indicated on the bottom panel of Fig. 2.

The polarization observables were determined using a maximum likelihood technique. A multi-dimensional minimization was performed on each of the linearly and circularly polarized data, and from this, each set of polarization observables were determined. The likelihood function was modelled according to the polarized cross section

$$\frac{d\sigma}{d\Omega} = \left.\frac{d\sigma}{d\Omega}\right|_0 \Big[1 - P_{lin}\Sigma\cos 2\phi + \alpha\cos\theta_x(-P_{lin}O_x\sin 2\phi - P_{circ}C_x) \\ -\alpha\cos\theta_y(-P_y + P_{lin}T\cos 2\phi) - \alpha\cos\theta_z(P_{lin}O_z\sin 2\phi + P_{circ}C_z)\Big], \qquad (1)$$

where $\left.\frac{d\sigma}{d\Omega}\right|_0$ is the unpolarized cross section, P_{lin} and P_{circ} are the degree of linear and circular polarization, respectively, and Σ, O_x, O_z, C_x, C_z, P_y, and T are the polarization observables. The angles ϕ, and $\theta_{x,y,z}$ are defined in Fig 3. The self-analyzing power of Λ, which allows the determination of the Λ polarization by studying the distribution of its decay products, is denoted by α.

3 Results and Discussion

Preliminary results for polarization observables of FSI in the reaction $\gamma d \to K^+\Lambda n$ were determined. Specifically, Σ, O_x, O_z, C_x, and C_z were determined for photon energies between 1.0 and 2.4 GeV, kaon momenta between 0.3 and 1.9 GeV/c, and θ'_Λ angles between 0 and 60 degrees as shown in Fig. 4. Invaluable information on kinematic regions where specific FSI are enhanced can be extracted from studying the differences between the observables determined for the QF reaction and FSI (see Fig. 4).

Our preliminary results clearly indicate that the data are of sufficient statistics to study kinematical dependencies of several polarization observables with adequate precision. Three-fold and four-fold differential results can also be extracted with acceptable statistical uncertainties. These are very important due to their larger sensitivity to details of the theoretical models. Such results were extracted for Σ, C_x, and C_z and a detailed study of these in collaboration with theorists is underway. The $\Lambda - n$

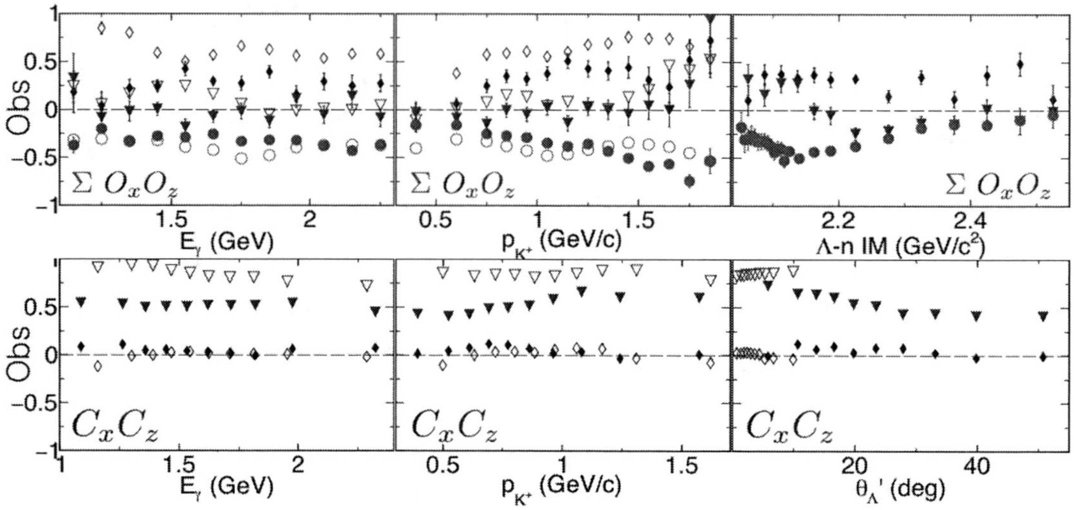

Figure 4. First preliminary results for FSI (solid markers) and QF (open markers) in the reaction $\gamma d \to K^+ \Lambda n$. The top row indicates polarization observables determined from data collected from the linearly polarized photon beam and the bottom row from the circularly polarized photon beam. The dependences of the observables on the incident photon energy (E_γ), kaon momentum (p_{K^+}), $\Lambda - n$ invariant mass, and the angle of Λ (θ'_Λ) are shown.

invariant mass dependence of these observables was also determined (see Fig. 4 upper right panel). These results will be used to extract the spin-averaged $\Lambda - n$ scattering length using the approach in Ref. [5]. It is expected that the result from this work will provide a scattering length that is better constrained than the one obtained from elastic scattering data.

Extraction of additional polarization observables, such us P_y, T, and especially beam-spin asymmetries defined using the azimuthal distribution of the Λ or the $\Lambda - n$ pair will give additional information on FSI and may allow us to identify kinematics where specific rescattering mechanism dominates. Our set of observables will be fitted to theoretical models in order to constrain the free parameters of YN potentials.

References

[1] A. Salam *et al.*, Phys. Rev. C **74**, 044004 (2006)
[2] J. M. Laget, Phys. Rev. C **75**, 014002 (2007)
[3] K. Miyagawa *et al.*, Phys. Rev. C **74**, 034002 (2006);
[4] K. Miyagawa *et al.*, Nucl. Phys. **A639**, 297c (1998)
[5] A. Gasparian *et al.*, Phys. Rev. C **69**, 034006 (2004)
[6] B. A. Mecking *et. al.*, Nucl. Instr. and Meth. A **503**, 513 (2003)
[7] P. Nadel-Turonski *et. al.*, Jefferson Lab PAC30 Proposal PR-06-103 (2006)
[8] E. Wolin, *GSIM Documentation,* http://www.jlab.org/Hall-B/document/gsim/userguide.html

EPJ Web of Conferences **113**, 07003 (2016)
DOI: 10.1051/epjconf/201611307003

Conjecture: A Possible $nn\Lambda$ Resonance

B. F. Gibson[1,a] and I. R. Afnan[2,b]

[1] *Theoretical Division, Los Alamos National Laboratory, Los Alamos, NM 87545, USA*
[2] *School of Chemical and Physical Sciences, Flinders University, GPO Box 2100, Adelaide 5001, Australia*

Abstract. We address the question of whether there might exist a resonance in the $nn\Lambda$ system, using a rank one separable potential formulation of the Hamiltonian. We explore the eigenvalues of the kernel of the Faddeev equation in the complex energy plane using contour rotation to allow us to analytically continue the kernel onto the second energy sheet. We follow the largest eigenvalue as the $n\Lambda$ potentials are scaled and the $nn\Lambda$ continuum is turned into a resonance and then into a bound state of the system.

1 Introduction

Recent experiments have suggested that a bound state of the $nn\Lambda$ system exists. [1] Several theoretical analyses demonstrate that such a bound state can not exist. [2–4] The question we wish to address is: Could we have a three-body resonance in the $nn\Lambda$ system even though all the interactions are predominantly s-waves? We consider the $nn\Lambda$ system with the pairwise interactions being rank one separable potentials that fit effective range parameters of the nn system, and those predicted by the Nijmegen model D [5] one-boson exchange potential for the $n\Lambda$ system. The use of rank one separable potentials allows us to easily analytically continue the Faddeev equations into the second complex energy plane in search of resonance poles by examining the eigenvalue spectrum of the kernel of the Faddeev equations as we did previously for Λ-d scattering [6].

2 Background

We have only limited data regarding $p\Lambda$ scattering, but we have no data regarding $n\Lambda$ scattering, because there are no free neutron targets. Λ hypernuclei provide only weak constraints. The hypertriton ${}^{3}_{\Lambda}$H is barely bound, being one of the largest halo nuclei known. [$B_\Lambda({}^{3}_{\Lambda}\mathrm{H}) = 0.13 \pm 0.05$ MeV]. The A=4 isodoublet ${}^{4}_{\Lambda}$H-${}^{4}_{\Lambda}$He exhibits significant Charge Symmetry Breaking, some 2-3 times that seen in the ^{3}H-^{3}He isodoublet. The uncertainty in the extracted parameters from the sparse $p\Lambda$ scattering data imply a wide range of variation is possible in the $n\Lambda$ interaction.

The HypHI collaboration reported evidence for a bound $nn\Lambda$ system ${}^{3}_{\Lambda}$n [1]. They observed both 2-body and 3-body decay modes. Such a bound state would provide a strong constraint on the $n\Lambda$ interaction, because the nn interaction is well known. Such a bound state could be observed directly

[a] e-mail: bfgibson@lanl.gov
[b] e-mail: iraj.afnan@flinders.edu.au

in a ^{3}H(e,e$'$K$^+$)$^3_\Lambda$n experiment at JLab, although a weakly bound system would imply a rather small cross section.

However, the existence of a $^3_\Lambda$n bound state has been strongly questioned. [2–4] Moreover, simple physics suggests that one would not expect a bound state. The hypertriton is barely bound and has as its core a deuteron. A $^3_\Lambda$n bound state would have as its core an unbound di-neutron pair. Could there instead exist a $nn\Lambda$ three-body resonance? If so, then one might still be able to utilize the electro-production reaction (or the HypHI heavy-ion collision) to constrain the $n\Lambda$ interaction.

3 Our Three-Body Model for $nn\Lambda$

For simplicity we utilize pairwise s-wave interactions of rank one separable form

$$V(k,k') = g(k)Cg(k') \qquad g(k) = 1/(k^2+\beta^2)\,,$$

where the nn potential strength C and range β are fitted to the effective range parameters: [7]

a_{nn} = -18.9 ± 0.4 fm and r_{nn} = 2.75 ± 0.11 fm

and the $n\Lambda$ strength and range are fitted to the Nijmegen model D $n\Lambda$ scattering lengths and effective ranges: [5]

a_s = -2.03 ± 0.32 fm and r_s = 3.66 ± 0.32 fm ,
a_t = -1.84 ± 0.10 fm and r_t = 3.32 ± 0.11 fm .

The use of rank one separable potentials allows us to simply analytically continue onto the second sheet of the energy plane in exploring for three-body resonances. We search for the resonance poles by examining the eigenvalue spectrum of the kernel of the Faddeev equations for the $nn\Lambda$ system. We used a similar technique to explore Λ-d scattering 20 years ago. [6].

As stated above, we analytically continue the Faddeev equations onto the second energy sheet. For a three-body system containing two identical Fermions interacting via Yamaguchi pairwise potentials, the homogeneous integral equation is of the form

$$\lambda_n(E)\,\phi_{n,k_\alpha}(q,E) = \sum_{k_\beta}\int_0^\infty dq'\; K^{JT}_{k_\alpha,k_\beta}(q,q';E)\,\phi_{n;k_\beta}(q',E)\,, \tag{1}$$

where the kernel of the integral equation is given by

$$K^{JT}_{k_\alpha,k_\beta}(q,q';E) = Z^{JT}_{k_\alpha,k_\beta}(q,q';E)\,\tau_{k_\beta}[E-\epsilon_\beta(q')]\,q'^2\,. \tag{2}$$

We analytically continue onto the second Riemann energy sheet by utilizing the transformation

$$q \to q\,e^{-i\theta} \qquad q' \to q'\,e^{-i\theta} \qquad \text{with} \qquad \theta > 0\,. \tag{3}$$

One limitation on the rotation angle θ is imposed by singularities of the kernel; the Born amplitude $Z^{JT}_{k_\alpha,k_\beta}$ requires that $\theta < \frac{\pi}{2}$, which gives us the region $\Im(E) < 0$ on the second Riemann sheet. The other source of singularity is the quasi-particle propagator $\tau_{k_\beta}[E-\epsilon_\beta(q')]$, but because there are no two-body bound states, this does not limit the rotation.

4 Results of the Eigenvalue Search

Let us consider the specific example in which we utilize the (^{1}S$_0$) nn and the ^{1}S$_0$ and ^{3}S$_1$ $n\Lambda$ potentials defined in the previous section. We searched in the complex energy plane for the largest eigenvalue of the kernel $\lambda(E) = 1$. We found a pole at:

$$E = -0.154 - 0.753\,i \text{ MeV} \quad \text{with eigenvalue} \quad \lambda(E) = 1.0000 + 0.0001\,i\,.$$

Because $\Re(E) < 0$, this pole does not correspond to a resonance.

This pole actually lies below the breakup threshold. Because the pole lies just below the threshold, we may ask how easy it might be to convert the pole into a resonance or even a bound state. We scale the strength of the $^1S_0$ and $^3S_1$ $n\Lambda$ potentials by the same factor s. We follow the path of the pole as it turns into a "resonance" and then into a bound state. See Figure 1. We observe that a change in strength of the order of 15% produces a resonance above the three-body breakup threshold. A change of about 40% produces a $nn\Lambda$ bound state.

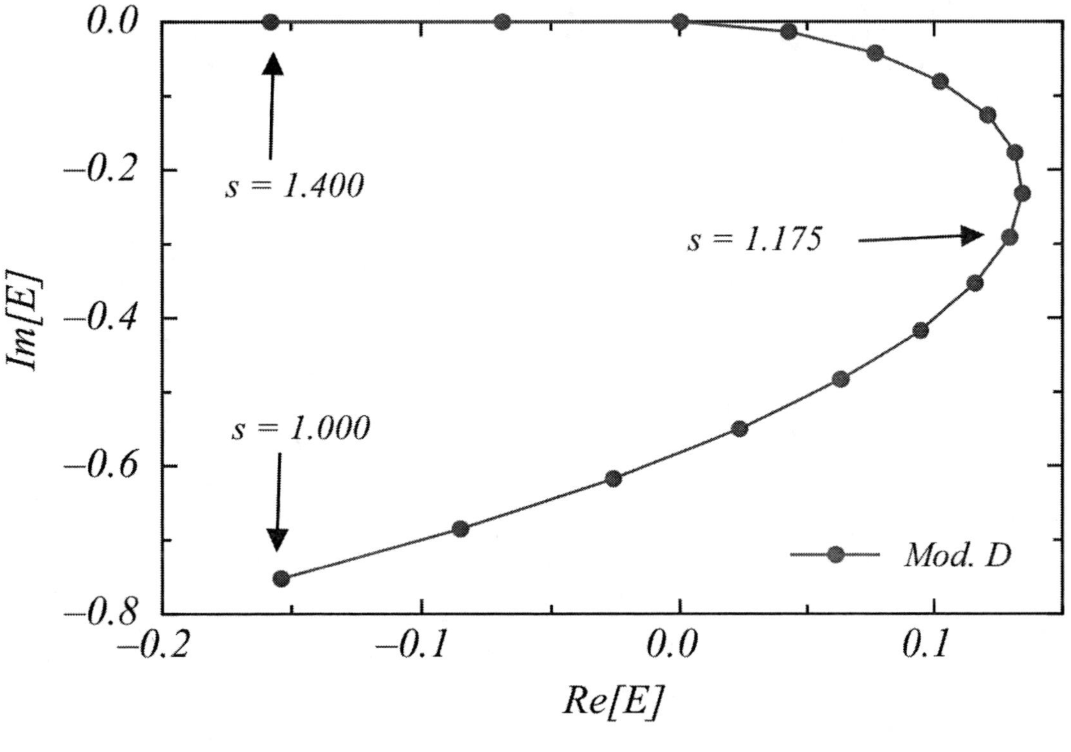

Figure 1. Trajectory of the pole as a function of s for the potentials defined in Section 3.

In other words we follow the trajectory of the "resonance" pole as a function of the strength s of the $n\Lambda$ interactions as s is increased from a value of 1.0 in increments (Δs) of 0.025. We find a sub-threshold resonance at values of $s = 1.000$ up to $s = 1.050$. For $s = 1.075$ up to $s = 1.350$ we obtain a resonance; in particular we obtain a resonance with $E = 0.1295 - 0.2915\,i$ MeV at $s = 1.175$. As s is further increased, we obtain a bound state with energy $E = -0.069$ MeV at $s = 1.375$ and $E = -0.158$ MeV for $s = 1.400$. Thus, one can see for this particular model that an $n\Lambda$ potential whose parameters lie within the uncertainty of the observed low energy $p\Lambda$ scattering parameters could easily produce a resonance in the $nn\Lambda$ system. Therefore, a $^3H(e,e'K^+)^3_\Lambda n$ experiment at JLab could possibly provide a significant constraint upon the $n\Lambda$ low energy scattering parameters.

Acknowledgements

The work of BFG was performed under the auspices of the National Nuclear Security Administration of the U.S. Department of Energy at Los Alamos National Laboratory under Contract No. DE-AC52-06NA25396.

References

[1] C. Rappold, *et al.* Phys. Rev. C **88**, 041001(R) (2012).
[2] H. Garcilazo, and A.Valcarce, Phys. Rev. C **89**, 057001 (2014).
[3] E. Hiyama, S. Ohnishi, B.F. Gibson, and Th. A. Rijken, Phys. Rev. C **89**, 061302(R) (2014).
[4] A. Gal, and H. Garcilazo, Phys. Lett. **B736**, 93 (2014).
[5] M. M. Nagels, T. A. Rijken, and J. J. de Swart, Phys, Rev. D **15**, 2547 (1977).
[6] I. R. Afnan and B. F. Gibson, Phys. Rev. C **47**, 1000 (1993).
[7] Q. Chen, *et al.* Phys. Rev. **77**, 054002 (2008).

EPJ Web of Conferences **113**, 07004 (2016)
DOI: 10.1051/epjconf/201611307004

A Λnn three-body resonance

H. Kamada[1,a], K. Miyagawa[2,b], and M. Yamaguchi[3,c]

[1] *Department of Physics, Faculty of Engineering, Kyushu Institute of Technology, 1-1 Sensuicho, Tobata, Kitakyushu 804-8550, Fukuoka, Japan*
[2] *Faculty of Applied Physics, Okayama University of Science, 1-1 Ridaicho, Okayama 700-0005 , Japan*
[3] *Research Center for Nuclear Physics, Osaka University, Ibaraki 567-0047, Osaka, Japan*

Abstract. We solved the Faddeev equations in the Λnn system ($J^\pi = \frac{1}{2}^+, T = 1$). There is no bound state but found a resonance state. Complex Energy Method helps to find the location of the resonance energy in the complex Riemann sheet. We obtained a resonance energy $E = 0.25 - 0.40i$ MeV using the realistic NN and YN Nijmegen potential. As a preliminary result the recent Nijmegen YN potential (NSC97f) gives 0.60± 0.05 - (0.25 ± 0.05)i MeV. We discuss importance of an irreducible three-body force to make it bound.

1 Introduction

The hypertriton of Λnp was observed and the Faddeev calcultions also achived to obtain a precise binding energy [1]. On the other hand, in the Λnn system, because the partial wave $^1S_0$ between two neutrons does not have sufficient attractive force, it has been expected that there is no bound state [1]. However, recently, a bound state was reported [2]. Hiyama *et al.*[3] addressed this issue and investigated it by the variational calculation with the Gauss expansion method. They insisted that there must be no bound state, however, a tiny modification of the tensor force between $\Lambda N - \Sigma N$ coupled channel might make it bound. The Faddeev calculations were confirmed [4] with a quark cluster model potential. From aspect of several hypernuclear systems it was examined [5]. They are consistent with the former work[1].

In this paper we revisit the Λnn system and look for the binding energy or the resonance one by solving the Faddeev equations using the realistic YN Nijmegen potential (NijmYN89) [6] and NN one (Nijm93)[7]. In our last work[1] the recent YN Nijmegen potential (NSC97f) [8] was not applied yet. In addition the Complex Energy Method (CEM) [9] is powerful and useful tool to solve the few-body system for the continuum energy regime. It has been used not only to the Faddeev calculation but to the four-body Yakubovsky calculation[10, 11].

The next section we briefly explain our Faddeev equations and the CEM. The results are demonstrated in sec. 3. Summary and outlook are given in sec. 4.

[a] e-mail: kamada@mns.kyutech.ac.jp
[b] e-mail: miyagawa@dap.ous.ac.jp
[c] e-mail: yamagu@rcnp.osaka-u.ac.jp

2 Faddeev equations and Complex Energy Method

The Faddeev equations of Λnn system consists of the two components $\psi^{(12)}$ and $\psi^{(13)}$ which links to the channel subsystems (*Nucleon*:1,2; *Hyperon*:3) described by the Jacobi coordinates.

$$\begin{aligned} \psi^{(12)} &= G_0\, T_{12}\, (\, 1 - P_{12}\,)\, \psi^{(13)}, \\ \psi^{(13)} &= G_0\, T_{13}\, (\, \psi^{(12)} - P_{12}\psi^{(13)}\,). \end{aligned} \tag{1}$$

where P_{ij} are permutation operators which exchanges particle i and j, and T_{ij} means the two-body transition matrix which couples Λn and Σn channels in case of YN interaction. Detail of the formalism is written in [1] as well as the case of Λpn system. After the partial-wave decompositions and the preparation of integral points, Eq. (1) becomes analogous to a large eigenvalue equation;

$$K(E)\vec{\psi} = \eta(E)\vec{\psi} \quad \text{with} \quad \vec{\psi} = \begin{pmatrix} \psi^{(12)}_{\Lambda nn} \\ \psi^{(12)}_{\Sigma nn} \\ \psi^{(13)}_{\Lambda nn} \\ \psi^{(13)}_{\Sigma nn} \end{pmatrix}. \tag{2}$$

The matrix $K(E)$ is regarded as the integral kernel in Eq.(1). The eigen value $\eta(E)$ must be 1 when the energy reaches to the resonance energy ($E = E_r - i\,\Gamma/2$). In order to solve the eigenvalue equation of the large rank there are 2 difficulties.

- The largest number of the absolute of the eigenvalue is not 1 because the NN and YN realistic potentials have a strong repulsive core.
- For the continuum states ($E > 0$) the equations have singularities of the integral kernel on the real axis of momenta.

Gauss-Seidel method works well to get the largest absolute value of the eigen value $|\eta|$, however, it becomes unphysical negative ($\eta^{neg} < 0$) because of the potential having a repulsive core. Using the negative eigenvalue η^{neg} we could converge the Gauss-Seidel iteration to the physical state as;

$$\vec{\psi}^{(n+1)} = \frac{K\vec{\psi}^{(n)} - \eta^{neg}\vec{\psi}^{(n)}}{1 - \eta^{neg}}, \tag{3}$$

where $\vec{\psi}^{(0)}$ is a nonzero arbitary vector which is of course a finite size. Therefore, the physical eigenvalue η^{phy} is obtained from

$$\eta^{phy} = (1 - \eta^{neg})\, r + \eta^{neg} \quad \text{with} \quad r \equiv \lim_{n\to\infty} \frac{\vec{\psi}^{(n+1)} \cdot \vec{\psi}^{(n)}}{\vec{\psi}^{(n)} \cdot \vec{\psi}^{(n)}}. \tag{4}$$

One expects $r = 1$ at $E = Er - i\,\Gamma/2$. The second difficulty can be avoided by the CEM on the physical Riemann sheet of the complex energy ($E = E_r + i\epsilon$). The typical $\epsilon > 0$ is about 1.0 MeV. Instead of the direct calculation at $E = E_r - i\,\Gamma/2$ some samples $\eta(E_r + i\epsilon)$ are calculated. Both the eigen energy of $E_r + i\epsilon$ and $E_r - i\,\Gamma/2$ are located on the same/close physical Riemann sheet of the complex energy plane. Using these samples $\eta(E_r + i\epsilon)$ we could extrapolate [9] $\eta^{phy}(E_r - i\,\Gamma/2) = 1$ of the resonance state.

3 Results

In case of the total spin ($J^{\pi} = \frac{1}{2}^{+}$) and the isospin ($T = 1$) for the Λnn system we expect a first resonance state (see Fig.4 of [1]). Table 1 shows the number of the coupled equations in Eq. (1). In

Table 1. Number of three-body couple channels and the binding energies E_b in MeV. j_{max} is the total spin in the subsystem. The last column show the Λnn binding energy with the strength factor s=1.20.

j_{max}	$\Lambda\ n$ channel	$\Sigma\ n$ channel	$n\ n$ channel	Sum	E_b
2	10	18	36	64	-0.224
3	12	26	52	90	-0.241
4	18	34	68	120	-0.243
5	20	42	84	146	-0.244

Table 2. The resonance energies $E_r - i\,\Gamma/2$. The factor s in the first column is multiplied to the original YN potential[6] in Eq.(5). Caution that in [3] NN potential was used Argonne V8 model and the factor s was multiplied only with the tensor part of $^3V_{N\Sigma-N\Lambda}$ in NSC97f potential.

s	NijmYN89 + Nijm93	NSC97f+ Nijm93	Hiyama *et al.* [3]
1.00	0.25 -0.40i	0.60± 0.05 -(0.25 ± 0.05)i	
1.05	0.15 -0.20i		
1.10	0.08 -0.15i		
1.20	-0.243 (bound)		-0.054 (bound)

order to obtain 3 digits accuracy for binding energy we need the partial waves up to j_{max} = 4 or 5. We take 34 integral points for two Jacobi momenta and 16 points for angle integral.

Before looking into the continuum state we can track the resonance pole from an artificial bound state which is forced by enlarging the strength of the YN potential. We introduced a factor s and define an artificial potential $\tilde{V}_{YN}$ as

$$\tilde{V}_{YN} = sV_{YN}. \tag{5}$$

Of course the potential $\tilde{V}_{YN}$ is switched into the original V_{YN} with $s = 1$. In fact the factor s is multiplied all partial waves of YN potentials except for the case of isospin 3/2.

We take the realistic YN Nijmegen potential (NijmYN89) [6] and NN one (Nijm93)[7]. The artificial bound state is found at E=-0.244 MeV under the strength factor s = 1.20 and j_{max} = 5. The convergency of the binding energy E_b is demonstrated in Table 1. Decreasing s and choosing j_{max} = 4 we calculate some eigenvalues η at $E = E_r + i\epsilon$. The number of samples η is about 20. We extrapolate the resonance energy out of these sampling $\eta(E_n)$ $(n = 1, 2, 3, \cdots, 20)$.The resonance energies are shown for each s in Table 2. Under the physical condition s = 1 we obtain the physical resonance energy E =0.25 -0.40i MeV. As the preliminary result we have E=0.60± 0.05 - (0.25 ± 0.05)i MeV using the recent NSC97f YN potential[8]. Comparing to NijmYN89 case the numerical stability is not good, we estimate the errors about ± 0.05MeV. Fig. 1 shows the position of the resonance poles in the complex energy plane.

4 Summary and Outlook

We solved the Faddeev equations in the Λnn system $(J^\pi = \frac{1}{2}^+, T = 1)$. There is no bound state but found a resonance state. The Complex Energy Method helps to find the location of the resonance energy in the complex Riemann sheet. We obtained $E = 0.25 - 0.40i$ MeV using Nijm93[7] and NjimYN89[6] models. Though preliminary, we got a result of E=0.60± 0.05 - (0.25 ± 0.05)i MeV using the recent NSC97f YN potential[8] instead of NijmYN89 model.

Our results suggest that there is no bound state which does not agree with the experimental report [2]. However, the remaining possibility of the exsistance of the bound state may be supported by

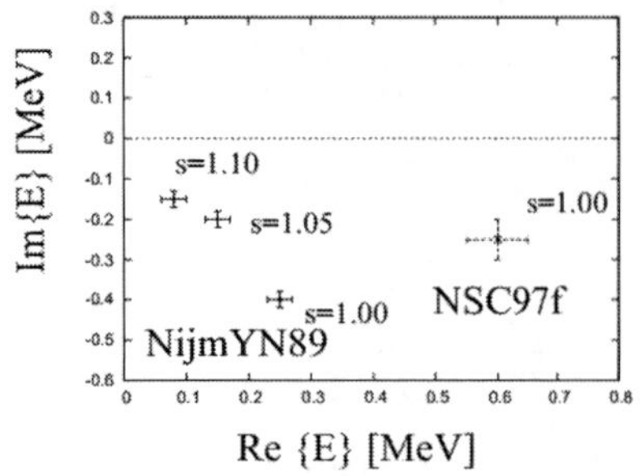

Figure 1. The resonence pole positions in the complex energy plane. The resonance energies from NijmYN89 model are marked with solid errorbars as well as NSC97f's does with dashed one.

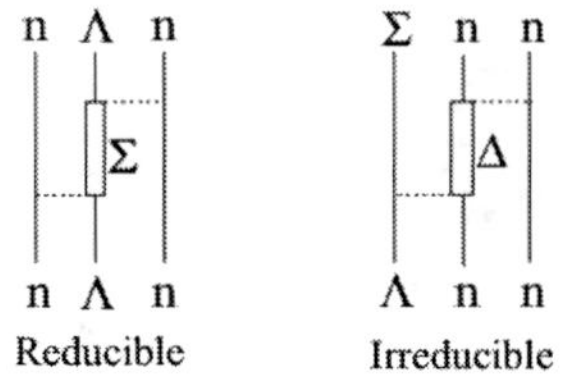

Figure 2. Diagrams for a reducible three-body force and a irreducible three-body force. The process of the irreducible three-body force contains with the Δ isobar particle transition.

the other YN potentials [12–14] or a three-body force. In the sense of the meson theoretical picture Fig. 2 shows diagrams of typical three-body forces. The left diagram of the pictures is a reducible one which is already taken into account in the YN two-body potential. On the other hand the right diagram involved with a Δ isobar excitation process is an irreducible one which might be seriously missing. There is a tiny lack of energy (~ 0.5 MeV) for binding. For example, the case of triton (^{3}H) needs more than 0.5 MeV from a three-body force.

Acknowledgements

The numerical calculations were performed on the interactive server at RCNP, Osaka University, Japan, and on the supercomputer cluster of the JSC, Jülich, Germany.

References

[1] K. Miyagawa, H. Kamada, W. Glöckle, and V. Stoks, Phys. Rev. C **51**, 2905 (1995).
[2] C. Rappold *et al.*,Phys. Rev. C **88**, 041001(R) (2013).
[3] E. Hiyama, S. Ohnishi, B. F. Gibson and Th. A. Rijken, Phys. Rev. C **89**, 061302(R) (2014).
[4] H. Garcilazo *et al.*, Phys. Rev. C **89**, 057001 (2014).
[5] A. Gal *et al.*, Phys. Lett. **B736**, 93 (2014).
[6] P.M.M. Maessen, Th.A. Rijken, and J.J. de Swart, Phys. Rev. C **40**, 2226 (1989).
[7] V. G. J. Stoks, R. A.M. Klomp, C. P. F. Terheggen, and J. J. de Swart, Phys. Rev. C **49**, 2950 (1994).
[8] Th. A. Rijken, V. G. J. Stokes, and Y. Yamamoto, Phys. Rev. C **59**, 21 (1999).
[9] H. Kamada, Y. Koike, W. Glöckle, Prog. Theor. Phys. **109**, 869 (2003).
[10] E. Uzu, H. Kamada, Y. Koike, Phys. Rev. C **68**, 061001(R) (2003).
[11] A. Deltuva, A. C. Fonseca, Phys. Rev. C **90** , 044002 (2014).
[12] A. G. Reuber, K. Holinde, and J. Speth, Czech. J. Phys. **42**, 1115 (1992); J. Haidenbauer *et al.*, Phys. Rev. **72**, 044005 (2005).
[13] J. Haidenbauer *et al.*, Nucl. Phys. A **915**, 14 (2013).
[14] K. Tominaga *et al*, Nucl. Phys. A**642**, 483 (1998).

EPJ Web of Conferences **113**, 07005 (2016)
DOI: 10.1051/epjconf/201611307005

$\bar{K}pp$ and $\bar{K}\bar{K}p$ Clusters

Roman Ya. Kezerashvili[1,a], Shalva M. Tsiklauri[2], Igor N. Filikhin[3], Vladimir M. Suslov[3], and Branislav Vlahovic[3]

[1] *New York City College of Technology, The City University of New York, Brooklyn, NY 11201, USA*
[2] *Borough of Manhattan Community College, The City University of New York, New York, NY 10016, USA*
[3] *North Carolina Central University, Durham, NC 27707, USA*

Abstract. The three-body kaonic $\bar{K}pp$ and $\bar{K}\bar{K}p$ clusters are studied using the method of hyperspherical harmonics (HH) in momentum representation and differential Faddeev equations (DFE). We use different NN and $\bar{K}N$ interactions. Results obtained by the methods of HH and DFE are in reasonable agreement. The binding energy and the width show very strong dependence on the $\bar{K}N$ potential. We have two different classes of results: the results based on phenomenological strong $\bar{K}N$ potential, and the results obtained with much weaker chiral SU(3)-based $\bar{K}N$ potential.

1 Introduction

The light kaonic $\bar{K}pp$ and $\bar{K}\bar{K}p$ clusters represent three-body systems and have been treated in the framework of various theoretical approaches such as variational methods [1–5], the method of Faddeev equations [6–9], and the method of hyperspherical harmonics (HH) [10], [11]. All the aforementioned approaches predict the existence of a bound state for the $\bar{K}NN$ system. The different predictions for the binding energy and the width are in considerable disagreement, varying within the ranges 9–95 MeV and 34–110 MeV, respectively [12]. To resolve this controversial situation that perhaps stems from the ambiguity of the $\bar{K}N$ interaction, the importance of the NN interaction or methods of three-body calculations, the further theoretical investigations are apparently needed. We present three-body nonrelativistic calculations using a potential model for the kaonic $\bar{K}pp$ and $\bar{K}\bar{K}p$ systems and applying two methods: the method of HH in momentum representation and the Faddeev equations in configuration space. Calculations for a binding energy and width of the kaonic $\bar{K}pp$ system are performed using three different potentials for the NN interaction, as well as two different potentials for the description of the kaon-nucleon interaction. The latter we also use for study of the $\bar{K}\bar{K}p$ system. Such approach allows one to understand the dependence of the bound state and the width of the kaonic three-body system on the method of calculations, the importance of nucleon-nucleon interaction, and the key role of the kaon-nucleon interaction.

2 Theoretical framework

The binding energy and the wave function of the three nonrelativistic particle can be obtained by solving the Schrödinger equation

$$(\widehat{T} + V_{123})\Psi = E\Psi, \tag{1}$$

[a] email: rkezerashvili@citytech.cuny.edu

Table 1. The binding energy B and the width for the $\bar{K}pp$ system for different NN and $\bar{K}N$ interactions.

	MeV	AV14+AY	MT+AY	T+AY	AV14+HW	MT+HW	T+HW
HH	B	46.2	46.5	46.3	–	20.5	20.6
	Γ	66.7	84.3	74.5	–	48.1	49.5
DFE	B	47.3	46.0	46.3	21.6	20.4	20.6

where $\widehat{T}$ is the operator of the kinetic energy and the potential energy $V_{123} = V_{NN} + V_{\bar{K}N_1} + V_{\bar{K}N_2}$ for the $\bar{K}NN$ system and $V_{123} = V_{\bar{K}\bar{K}} + V_{\bar{K}p} + V_{\bar{K}p}$ for the $\bar{K}\bar{K}p$ system. Here V_{NN} is the nucleon-nucleon potential, $V_{\bar{K}N_1} + V_{\bar{K}N_2}$ is the sum of a pairwise effective antikaon interaction with the first and second nucleon, respectively, and $V_{\bar{K}\bar{K}}$ is the $\bar{K}\bar{K}$ interaction. The effective interactions of the $\bar{K}N$ and $\bar{K}\bar{K}$ two-body subsystems are discussed in detail in Refs. [1], [2], [5], [13]. Below, we use two effective $\bar{K}N$ interactions. One is the energy-independent phenomenological $\bar{K}N$ potential from Refs. [1] (AY potential), and the other is the energy-dependent effective $\bar{K}N$ interaction [13] derived within the chiral SU(3) effective field theory that we refer to as the HW potential. We use three NN potentials [14]: the realistic Argonne V14 (AV14), the semi-realistic Malfliet and Tjon MT-I-III (MT), and the Tamagaki G3RS that we refer to as the T potential. We solve Eq. (1) using two different methods: the method of Faddeev equations in configuration space and the method of hyperspherical functions in momentum representation. In the framework of both methods we introduce the trees of the Jacobi coordinates for three particles in configuration space and the conjugated set of momenta in the momentum space. In the Faddeev method the appropriate division of the 3-body wave function in the configuration space into the Faddeev components leads to the system of differential equations in configuration space known as the differential Faddeev equations (DFE) [16]. For the $\bar{K}pp$ cluster the total wave function is decomposed into the sum of the Faddeev components U and W corresponding to the $(pp)\bar{K}$ and $(\bar{K}p)p$ types of rearrangements: $\Psi = U + W - PW$, where P is the permutation operator for two identical fermions. For the $\bar{K}\bar{K}p$ system the decomposition has the form $\Psi = U + W + PW$, where P is the permutation operator for two identical bosons and the Faddeev components U and W correspond to the $(\bar{K}\bar{K})p$ and $(\bar{K}p)\bar{K}$ types of rearrangements. For the both kaonic systems the set of the Faddeev equations can be reduced to the system of two equations for the components U and W [17], which allows one to find the total wave functions and binding energies for the $\bar{K}pp$ and $\bar{K}\bar{K}p$ systems.

To use the method of HH in momentum representation one can write the Schrödinger equation (1) in an integral form $\Psi(\mathbf{p},\mathbf{q}) = -\frac{1}{(2\pi)^6}\int G(\mathbf{p},\mathbf{q}) < \mathbf{p}'\mathbf{q}'|V_{123}|\mathbf{pq} > \Psi(\mathbf{p}',\mathbf{q}')d\mathbf{p}'d\mathbf{q}'$, where $< \mathbf{p}'\mathbf{q}'|V_{123}|\mathbf{pq} > = \frac{1}{(2\pi)^6}\int V_{123}\exp\left[i(\mathbf{q}-\mathbf{q}')\mathbf{x}+i(\mathbf{p}-\mathbf{p}')\mathbf{y}\right]d\mathbf{x}d\mathbf{y}$ is the Fourier transformation of the V_{123} interactions, which are defined above. At the next step we introduce the hyperradii ϱ and $\varkappa$, and two sets of five angles denoted by Ω_ρ and $\Omega_\varkappa$ which define the direction of the vector ρ and vector $\varkappa$, in the six dimensional configuration and momentum spaces, respectively [15]. Following Refs. [15] and [11] we expand the wave function of three bound particles in terms of the symmetrized hyperspherical harmonics $\Phi_\mu^{l_pl_q}(\Omega_\varkappa,\sigma,\tau)$ in momentum representation: $\Psi(\mathbf{p},\mathbf{q}) = \sum_{\mu l_p l_q} u_\mu^{l_pl_qL}(\varkappa)\Phi_\mu^{l_pl_qL}(\Omega_\varkappa,\sigma,\tau)$, where μ is the grand angular momentum, L is the total orbital momentum, and l_p and l_q are the angular momenta corresponding to the Jacobi momenta p and q that are conjugated to the standard Jacobi coordinates for three particles. Here $\Phi_\mu^{l_pl_q}(\Omega_\varkappa,\sigma,\tau)$ are written as a sum of products of spin and isospin functions and HH, using the Raynal-Revai coefficiants. For the system $\bar{K}pp$ the wave function is antisymmetrized with respect to two protons, while for the $\bar{K}\bar{K}p$ system it is symmetrized with respect to two kaons. For the hyperradial functions $u_\mu^{l_ql_pL}(\varkappa)$ we obtain the coupled integral equations. By

Table 2. The bound state energies of $\bar{K}p$ (E_2) and $\bar{K}\bar{K}p$ (B) systems, and the root-mean-square radius of the $\bar{K}$ distribution. ΔE is the binding energy measured from the two-body threshold.

	$\bar{K}\bar{K}$	$\bar{K}p$	$< r^2 >^{1/2}$, fm	E_2, MeV	B, MeV	ΔE, MeV
DFE	AY	AY	1.35	30.3	31.7	1.4
	$V_{\bar{K}\bar{K}} = 0$	AY			35.2	4.9
HH	AY	AY	1.36	30.0	31.7	1.7
	$V_{\bar{K}\bar{K}} = 0$	AY			35.3	5.3
DFE	HW	HW	1.94	11.16	unbound	
	$V_{\bar{K}\bar{K}} = 0$	HW			12.17	1.01
HH	HW	HW	1.96	11.42	11.43	0.01
	$V_{\bar{K}\bar{K}} = 0$	HW			12.21	0.79

solving the coupled integral equations one can find the hyperradial functions $u_{\mu}^{l_q l_p L}(\varkappa)$ for a given L and the binding energies for the $\bar{K}pp$ and $\bar{K}\bar{K}p$ systems.

3 Results and discussion

$\bar{K}pp$ system. Results of our calculations for the $\bar{K}pp$ cluster are presented in table 1. For the calculations of the binding energy and the width with the methods of HH and DFE we use MT, T, and AV14 potentials for the NN interaction, while for the $\bar{K}N$ interaction we use the energy-dependent effective HW and the phenomenological AY potentials with the range parameter b=0.47 fm and b=0.66 fm, respectively. Such an approach allowed us to examine how the $\bar{K}pp$ cluster's structure depends on different choices of the $\bar{K}N$ interactions for the same NN potential, as well as its dependence on different choices of the NN interaction for the same $\bar{K}N$ interaction, and to understand the sensitivity of the system to the input interactions. The analysis of the results presented in table 1 shows that the values for the binding energy for the $\bar{K}pp$ system obtained by both methods are in reasonable agreement, and that the ground state energy is not sensitive to the NN interaction. However, there is a very strong dependence on the kaon-nucleon interaction. The energy of the bound state, as well as the width calculated for the AY potential [1] are more than twice as big as for the energy-dependent chiral $\bar{K}N$ potential [13]. Therefore, the highest binding energies are those that are obtained based on the phenomenological AY potential. Discrepancies obtained for the binding energy using the same potentials but different methods are mostly related to a problem of an equivalent representation of the potentials in momentum and configuration spaces.

$\bar{K}\bar{K}p$ system. We study a possible bound state of the $\bar{K}\bar{K}p$ cluster with $S = -2$, $I = 1/2$, $J^{+} = 1/2^{+}$ using the effective s–wave AY and HW potentials assuming that this state is formed due to the strong $\bar{K}p$ attraction. The strength of the s–wave $\bar{K}\bar{K}$ interaction for the isospin $I = 0$ is zero due to Bose statistics, and we consider a weak repulsion for the isospin $I = 1$ that reproduces the scattering lengths $a_{\bar{K}\bar{K}}$=-0.14 fm for the range parameter b=0.66 fm (AY potential) and b=0.47 fm (HW potential). The results of calculations for the binding energies for the $\bar{K}p$ and $\bar{K}\bar{K}p$, the bound $\bar{K}\bar{K}p$ state without $\bar{K}\bar{K}$ interaction, and the root-mean-square radius of the $\bar{K}$ distribution are presented in table 2. For the AY potential, the $\bar{K}\bar{K}p$ system is still bound even with a much stronger $\bar{K}\bar{K}$ repulsion, while for the HW potential there is no bound state found within the DFE calculation and there is a bound state with energy 0.01 MeV relative to the $\bar{K}p$ +$\bar{K}$ threshold for the HH calculation. Thus, although the $\bar{K}p$ with $I = 1$ is attractive, the attraction is not enough to overcome the $\bar{K}\bar{K}$ repulsion. For the width within the method of HH we obtain 58.6 MeV and 20.8 MeV with the AY and HW potentials, respectively.

The results for the binding energy of the $\bar{K}\bar{K}p$ system obtained by the methods of HH and DFE are in good agreement, and there is reasonable agreement between our calculations and results obtained using a variational method [5] and the Faddeev calculations [9].

4 Conclusion

The calculations show that we have two different classes of results: the results based on the phenomenological strong $\bar{K}N$ potential predict a strong bound state of the $\bar{K}pp$ system, and the results obtained with much weaker HW potential indicate a shallow bound state for this cluster. Therefore, the binding energy of the $\bar{K}pp$ system depends entirely on the ansatz for the $\bar{K}N$ interaction and substantially changes when we use the AY and HW potentials. On the other hand, the binding energy of the $\bar{K}pp$ and $\bar{K}\bar{K}p$ systems obtained by the methods of HH and DFE are in reasonable agreement. Thus, results are not sensitive to the method of calculations. The sensitivity of the bound state of the $\bar{K}pp$ system to details of the NN interaction is marginal for the s-wave MT and T potentials and is slightly more pronounced for the l-dependent AV14 potential. Our calculations, as well as calculations within other theoretical approaches [1–3], [6–11], show that the width is more than twice as much as the binding energy. Thus, we are facing a situation in which the states have a much larger width than the binding energy. That makes the experimental observation of such states challenging and it may be hard to identify the resonances. However, the continuation of the experimental search for the quasi-bound kaonic clusters still remains important.

Acknowledgements

This work is supported by the NSF Grant No. HRD-1345219.

References

[1] Y. Akaishi and T. Yamazaki, Phys. Rev. C **65**, 044005 (2002); *ibid* C **76**, 045201 (2007).

[2] A. Doté, T. Hyodo, and W. Weise, Nucl. Phys. A **804**, 197 (2008); Phys. Rev. C **79**, 014003 (2009).

[3] S. Wycech and A. M. Green, Phys. Rev. C **79**, 014001 (2009).

[4] A. Doté, T. Inoue, and T. Myo, Prog. Theor. Exp. Phys. 043D02 (2015).

[5] Y. Kanada En'yo and D. Jido, Phys. Rev. C **78**, 025212 (2008).

[6] N.V. Shevchenko, A. Gal, and J. Mares, and J. Révai, Phys. Rev. C **76**, 044004 (2007).

[7] Y. Ikeda and T. Sato, Phys. Rev. C **76**, 035203 (2007); *ibid* C **79**, 035201 (2009).

[8] M. Bayar and E. Oset, Nucl. Phys. A **883**, 57 (2012); *ibid* A **914**, 349 (2013).

[9] S. Maeda, Y. Akaishi, and T. Yamazaki, Proc. Jpn. Acad., Ser. B **89**, 418 (2013).

[10] N. Barnea, A. Gal, and E. Z. Liverts, Phys. Lett. B **712**, 132 (2012).

[11] R. Ya. Kezerashvili and Sh. M. Tsiklauri, EPJ Web Conf. **81**, 02022 (2014).

[12] A.O.Tokiyasu, et al., LEPS Collaboration, Phys. Lett. B **728**, 616 (2014).

[13] T. Hyodo and W. Weise, Phys. Rev. C **77**, 035204 (2008).

[14] R.B. Wiringa, R.A. Smith, and T.L. Ainsworth, Phys. Rev. C **29**, 1207 (1984); R. A. Malfliet and J. A. Tjon, Nucl. Phys. A **127**, 161 (1969); R. Tamagaki, Prog. Theor. Phys. **44**, 905 (1970).

[15] R.I Jibuti, N.B. Krupennikova, and V.Yu. Tomchinsky, Nucl. Phys. A **276**, 421 (1977).

[16] L. D. Faddeev and S. P. Merkuriev, *Quantum Scattering Theory for Several Particle Systems*, (Kluwer Academic, Dordrecht, 1993).

[17] I. N. Filikhin, A. Gal, and V. M. Suslov, Phys. Rev. C **68**, 024002 (2003); Nucl. Phys. A **743**, 194 (2004).

EPJ Web of Conferences **113**, 07006 (2016)
DOI: 10.1051/epjconf/201611307006

Strangeness in nuclei and neutron stars: a challenging puzzle

Diego Lonardoni[1], Alessandro Lovato[1], Stefano Gandolfi[2], and Francesco Pederiva[3,4]

[1] *Physics Division, Argonne National Laboratory, Lemont, Illinois 60439, USA*
[2] *Theoretical Division, Los Alamos National Laboratory, Los Alamos, New Mexico 87545, USA*
[3] *Department of Physics, University of Trento, Via Sommarive 14, I-38123 Trento, Italy*
[4] *INFN-Trento Institute for Fundamental Physics and Application, Trento, Italy*

Abstract. The prediction of neutron stars properties is strictly connected to the employed nuclear interactions. The appearance of hyperons in the inner core of the star is strongly dependent on the details of the underlying hypernuclear force. We summarize our recent quantum Monte Carlo results on the development of realistic two- and three-body hyperon-nucleon interactions based on the available experimental data for light- and medium-heavy hypernuclei.

1 Introduction

Neutron stars are among the most compact and dense objects in the Universe, with typical masses up to $\sim 2\,M_\odot$ and radii $R \sim 10$ km. Their central densities can be several times larger than nuclear saturation density, $\rho_0 = 0.16$ fm^{-3}. At densities larger than $\sim 2\rho_0$, when the nucleon chemical potential is large enough, the conversion of nucleons into hyperons might become energetically favorable, and Pauli blocking would prevent hyperons from decaying by limiting the phase space available to nucleons. This would lead to a reduction of the Fermi pressure exerted by the baryons and to a softening of the equation of state (EOS). As a consequence, the maximum mass determined by the equilibrium condition between gravitational and nuclear forces would be reduced.

Currently there is no general agreement (even qualitative) among the predicted results for the EOS of strange baryonic matter and the maximum mass of a neutron star (NS), that violates the constraints given by the recently measured value of $\sim 2\,M_\odot$ of the millisecond pulsars PSR J1614-2230 [1] and PSR J0348+0432 [2], see Ref. [3]. This apparent inconsistency between NS mass observations and theoretical calculations is a long standing problem known as the *hyperon puzzle*. It has to be ascribed to a combination of an incomplete knowledge of the forces governing the system and to the concurrent use of approximate theoretical many-body techniques. Its solution requires a more accurate theoretical analysis and a thorough experimental investigation of the hyperon-nucleon force in a variety of systems, ranging from light to medium and heavy hypernuclei.

We employ a quantum Monte Carlo computational technique to tackle the general problem of the hyperon-nucleon interaction. The algorithm, auxiliary field diffusion Monte Carlo (AFDMC), has been introduced by Schmidt and Fantoni [4] as an extension of the usual diffusion Monte Carlo method to deal in an efficient way with spin/isospin-dependent Hamiltonians. It has been successfully applied to solve the many-body problem in a non perturbative fashion for strongly correlated systems, both in the standard [5–8] and strange nuclear sector [3, 9–11].

2 Hypernuclei: from few- to many-body systems

As reported in Refs. [9–11], an accurate analysis of the Λ separation energy B_Λ of light- and medium-heavy hypernuclei has been recently carried out using a phenomenological hypernuclear interaction analogous to the Argonne-Illinois nucleon-nucleon force [12–19]. The ΛN two-body component of this interaction has been fit to the existing Λp scattering data. The ΛNN three-body sector instead has been re-fit to the available Λ separation energies of single-Λ hypernuclei over a wide mass range by means of AFDMC calculations. It has been shown that the repulsive nature of the three-body hyperon-nucleon interaction is the key to satisfactorily reproduce the ground state properties of light- to medium-heavy hypernuclei within a unique theoretical framework. Fig. 1 shows the AFDMC results for B_Λ when using the complete $\Lambda N + \Lambda NN$ force. The agreement with experimental data is good in the mass range investigated and for the Λ particle in different single particle states. For more details see Refs. [9–11].

In the light sector, a two-body ΛN charge symmetry breaking interaction has been also employed [10, 13]. As shown in Tab. 1, AFDMC results reproduce the hyperon separation energies and the energy splitting of the mirror hypernuclei ${}^4_\Lambda$H and ${}^4_\Lambda$He, according to the old emulsion data [20]. ${}^3_\Lambda$H is weakly bound compared to deuterium and the neutral hypernucleus ${}^3_\Lambda$n is predicted to be unbound, in agreement with other theoretical results [21–23], contradicting the possible experimental evidence reported in Ref. [24].

In order to validate the phenomenological approach adopted in quantum Monte Carlo calculations for hypernuclei, a benchmark among different few-body methods has been recently proposed.

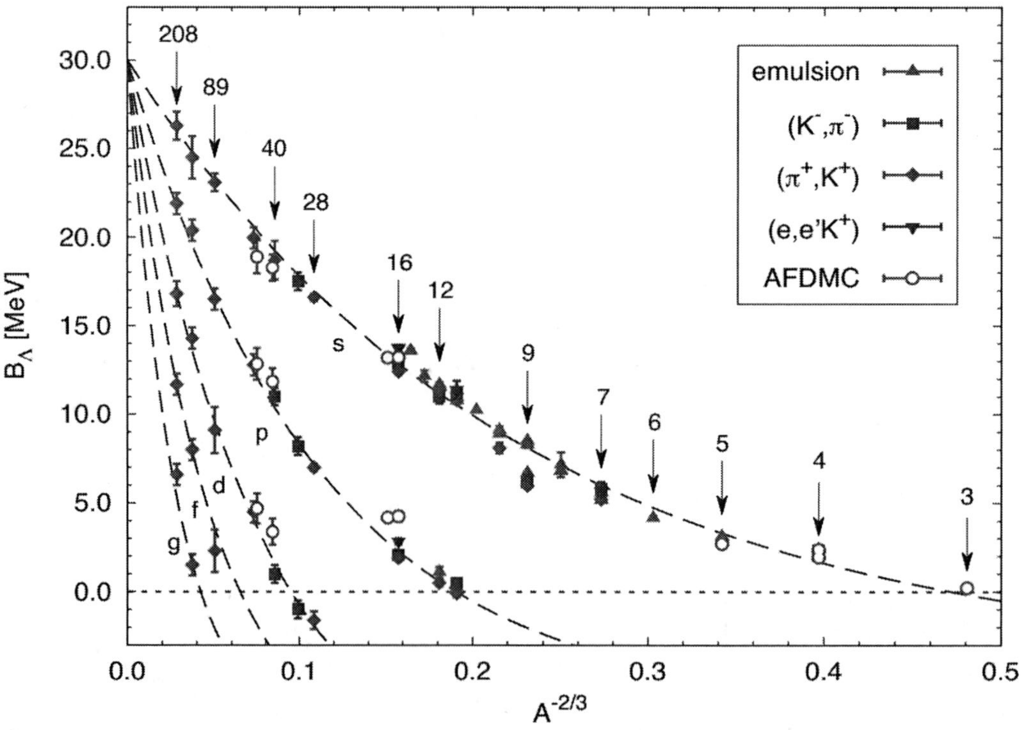

Figure 1. Measured (see Ref. [11] for the complete list of references) and computed Λ separation energies as a function of $A^{-2/3}$. Results for the Λ particle in different single particle states are shown.

Table 1. AFDMC and experimental [20] Λ separation energies for the lightest hypernuclei. In the case of $A = 4$ the separation energy difference ΔB_Λ is also shown. Evidence for the neutral hypernucleus ${}^3_\Lambda$n has been reported in Ref. [24], but no B_Λ value has been reported. All energies are in MeV.

System	B_Λ	B_Λ^{exp}	ΔB_Λ	ΔB_Λ^{exp}
${}^3_\Lambda$n $\left(\frac{1}{2}^+\right)$	unbound	unknown	—	—
${}^3_\Lambda$H $\left(\frac{1}{2}^+\right)$	0.23(9)	0.13(5)		
${}^4_\Lambda$H (0^+)	1.95(9)	2.04(4)	0.42(11)	0.35(6)
${}^4_\Lambda$He (0^+)	2.37(9)	2.39(3)		

In table 2 we report preliminary results for ${}^5_\Lambda$He for two different exact methods, AFDMC and Non-symmetrized Hyperspherical Harmonics (HSHH) [25]. In both strange and non-strange sectors the potentials employed are at the two-body level only. Results of these two methods agree for both the soft (Minnesota [26]) and hard (Argonne V4′ [27]) potentials.

Table 2. Binding energies and Λ separation energies for ${}^5_\Lambda$He for two different combinations of nucleon-nucleon and hyperon-nucleon potentials and two different methods. All energies are in MeV.

${}^5_\Lambda$He	AFDMC		NSHH	
NN Minn [26] $+ \Lambda N$	E	-37.69(8)	E	-37.77(10)
	B_Λ	6.95(9)	B_Λ	6.99(10)
NN AV4′ [27] $+ \Lambda N$	E	-39.46(12)	E	-39.54(10)
	B_Λ	6.70(16)	B_Λ	6.84(10)

3 Neutron stars: from hypernuclei to the infinite medium

Precise AFDMC calculations have been performed to first assess the parameters of the interaction from measured hyperon separation energies, and then to extrapolate the results to the case of neutron matter with strangeness. However, parametrizations of the potential predicting relatively small differences in the Λ separation energies of hypernuclei give very different results for the properties of the infinite medium [3]. The resulting EOS spans the whole regime extending from the appearance of a substantial fraction of hyperons at $\sim 2\rho_0$ to the absence of Λ particles in the entire density range of the star. This yields a sizable effect on the predicted NS structure. Therefore, the derivation of realistic hypernuclear potential models is of paramount importance to properly assess the role of hyperons in NSs and reconcile theoretical predictions with astrophysical observations. This demands a precise and systematic experimental investigation of hypernuclear properties over a wide range of masses. For instance, a recent study of the isospin dependence of the present three-body hyperon-nucleon force shows the difficulty in extracting the information on the Hamiltonian from currently available experimental information on hypernuclei [11].

Acknowledgements

This work was supported by the U.S. Department of Energy, Office of Science, Office of Nuclear Physics, under the NUCLEI SciDAC grant (D.L., A.L., S.G.), by the Department of Energy, Office of Science, Office of Nuclear Physics, under Contract No. DE-AC02-06CH11357 (A.L.), and by DOE under Contract No. DE-AC52-06NA25396 and Los Alamos LDRD grant (S.G.). This research used resources of the National Energy Research Scientific Computing Center (NERSC), which is supported by the Office of Science of the U.S. Department of Energy under Contract No. DE-AC02-05CH11231, and computing time provided by Institutional Computing (IC) at LANL.

References

[1] P. B. Demorest, T. Pennucci, S. M. Ransom, M. S. E. Roberts, and J. W. T. Hessels, Nature 467, 1081 (2010).
[2] J. Antoniadis et al., Science 340, 1233232 (2013).
[3] D. Lonardoni, A. Lovato, S. Gandolfi, and F. Pederiva, Phys. Rev. Lett. **114**, 092301 (2015).
[4] K. E. Schmidt and S. Fantoni, Phys. Lett. B 446 (1999) 99.
[5] S. Gandolfi, J. Carlson, and S. C. Pieper, Phys. Rev. Lett. **106**, 012501 (2011).
[6] S. Gandolfi, J. Carlson, and S. Reddy, Phys. Rev. C **85**, 032801 (2012).
[7] S. Gandolfi, A. Lovato, J. Carlson, and K. E. Schmidt, Phys. Rev. C **90**, 061306 (2014).
[8] J. Carlson, S. Gandolfi, F. Pederiva, S. C. Pieper, R. Schiavilla, K. E. Schmidt, and R. B. Wiringa arXiv:1412.3081 (2015).
[9] D. Lonardoni, S. Gandolfi, and F. Pederiva, Phys. Rev. C **87**, 041303(R) (2013).
[10] D. Lonardoni, F. Pederiva, and S. Gandolfi, Phys. Rev. C **89**, 014314 (2014).
[11] F. Pederiva, F. Catalano, D. Lonardoni, A. Lovato, and S. Gandolfi, arXiv:1506.04042 (2015).
[12] A. R. Bodmer, Q. N. Usmani, and J. Carlson, Phys. Rev. C **29**, 684 (1984).
[13] A. R. Bodmer and Q. N. Usmani, Phys. Rev. C **31**, 1400 (1985).
[14] A. R. Bodmer and Q. N. Usmani, Nucl. Phys. A 477, 621 (1988).
[15] A. A. Usmani, Phys. Rev. C **52**, 1773 (1995).
[16] A. A. Usmani, S. C. Pieper, and Q. N. Usmani, Phys. Rev. C **51**, 2347 (1995).
[17] Q. N. Usmani and A. R. Bodmer, Phys. Rev. C **60**, 055215 (1999).
[18] A. A. Usmani and F. C. Khanna, J. Phys. G 35, 025105 (2008).
[19] M. Imran, A. A. Usmani, M. Ikram, Z. Hasan, and F. C. Khanna, J. Phys. G 41, 065101 (2014).
[20] M. Jurič et al., Nucl. Phys. B 52, 1-30 (1973).
[21] E. Hiyama, S. Ohnishi, B. F. Gibson, and Th. A. Rijken, Phys. Rev. C **89**, 061302(R) (2014).
[22] A. Gal and H. Garcilazo, Physics Letters B 736 (2014) 93–97.
[23] J.-M. Richard, Q. Wang, and Q. Zhao, Phys. Rev. C **91**, 014003 (2015).
[24] C. Rappold et al., Phys. Rev. C **88**, 041001 (2013).
[25] S. Deflorian, N. Barnea, W. Leidemann, and G. Orlandini, Few-Body Syst. 54, 1879 (2013).
[26] D. R. Thompson, M. Lemere, and Y. C. Tang, Nucl. Phys. A 286 (1977) 53 .
[27] R. B. Wiringa and S. C. Pieper, Phys. Rev. Lett. **89**,18 (2002).

EPJ Web of Conferences **113**, 07007 (2016)
DOI: 10.1051/epjconf/201611307007

${}^{6}_{\Lambda\Lambda}$He in cluster effective field theory

S.-I. Ando[1,a] and Y. Oh[2,b]

[1] *Department of Information Display, Sunmoon University, Asan, Chungnam 31460, Korea*
[2] *Department of Physics, Kyungpook National University, Daegu 41566, Korea*

Abstract. The bound state of ${}^{6}_{\Lambda\Lambda}$He is studied as a three-body ($\Lambda\Lambda\alpha$) system in a cluster effective field theory at leading order. We find that the system exhibits the limit cycle that is associated with the formation of bound states called the Efimov states. This implies that the three-body contact interaction should be promoted to the leading order. The relationship of the binding energy and the $\Lambda\Lambda$ scattering length is discussed as well as the role of the scale of this system.

1 Introduction

Although the first observation of ${}^{6}_{\Lambda\Lambda}$He was reported in 1960s [1], its clear track was only recently caught in the emulsion experiment of KEK-E373 Collaboration [2], which is known as the "NAGARA" event. When combined with the recent information, the two-Λ separation energy $B_{\Lambda\Lambda}$ of ${}^{6}_{\Lambda\Lambda}$He is estimated to be $B_{\Lambda\Lambda} = 6.93 \pm 0.16$ MeV [3]. Theoretical studies on double-Λ hypernuclei are expected to shed light on the understanding of baryon-baryon interactions in the strangeness sector, which may help search for new exotic systems [4, 5].

In this contribution, we present our recent work [6] in which the effective field theory (EFT) method at leading order (LO) is applied to explore the structure of ${}^{6}_{\Lambda\Lambda}$He as a three-body ($\Lambda\Lambda\alpha$) cluster system.[1] One advantage of EFTs at very low energies is that it can provide a model-independent and systematic perturbative method to analyze a system, where one can introduce a high momentum separation scale Λ_H between relevant degrees of freedom of low energy and irrelevant degrees of freedom of high energy of the system. For a review of this approach, we refer to Refs. [8, 9] and references therein. In the study of three-body systems, it was suggested by Bedaque *et al.* to modify the counting rules so that the three-body contact interaction should be promoted to LO to renormalize the limit cycle [10]. The limit cycle, a cyclic singularity, in a renormalization group analysis was suggested by Wilson [11]. In the unitary limit, the limit cycle is known to be associated with the formation of bound states called the Efimov states [12]. Whether the system exhibits the limit cycle or not is easily examined by using the homogeneous part of the integral equation in the asymptotic limit, which was obtained earlier by Danilov [13]. This allows one to investigate the nature of bound states in three-body systems by using the properties of the limit cycle.

[a] e-mail: sando@sunmoon.ac.kr
[b] e-mail: yohphy@knu.ac.kr

[1] This approach was also applied to relate the formation of the bound state ${}^{4}_{\Lambda\Lambda}$H and the the S-wave scattering of Λ + ${}^{3}_{\Lambda}$H below the hypertriton breakup threshold by treating ${}^{4}_{\Lambda\Lambda}$H as a three-body ($\Lambda\Lambda d$) system in an EFT at LO [7].

In the present study, we treat the α particle as an elementary field. The binding energy of the α particle is known to be $B_4 \simeq 28.3$ MeV and its excitation energy is $E_1 \simeq 20.0$ MeV. Thus the large momentum scale of the α-cluster theory is estimated to be $\Lambda_H \sim \sqrt{2\mu E_1} \simeq 170$ MeV with μ being the reduced mass of the (^{3}H, p) or (^{3}He, n) system so that $\mu \simeq \frac{3}{4}m_N$ with m_N being the nucleon mass. On the other hand, we choose the binding momentum of $^5_\Lambda$He as the typical momentum scale Q of the theory. Since the Λ separation energy of $^5_\Lambda$He is $B_\Lambda \simeq 3.12$ MeV, the binding momentum of $^5_\Lambda$He as the $\Lambda\alpha$ cluster system is then $\gamma_{\Lambda\alpha} = \sqrt{2\mu_{\Lambda\alpha}B_\Lambda}$, where $\mu_{\Lambda\alpha}$ is the reduced mass of the $\Lambda\alpha$ system. This leads to $\gamma_{\Lambda\alpha} \simeq 73.2$ MeV and our expansion parameter becomes $Q/\Lambda_H \sim \gamma_{\Lambda\alpha}/\Lambda_H \simeq 0.43$.

2 $\Lambda\Lambda\alpha$ bound state

The effective Lagrangian at LO for this system can be found in Ref. [6]. With the given Lagrangian the bound state of the $\Lambda\Lambda\alpha$ system is examined through the homogeneous part of the coupled integral equations that are written in terms of the scattering amplitude of the S-wave $\Lambda + {}^5_\Lambda$He scattering. The obtained LO coupled integral equations are determined by three parameters: $\gamma_{\Lambda\alpha}$, $a_{\Lambda\Lambda}$, and $g(\Lambda_c)$. Here, $\gamma_{\Lambda\alpha} = 73.2$ MeV is the binding momentum of $^5_\Lambda$He, and $a_{\Lambda\Lambda}$ is the scattering length of the S-wave $\Lambda\Lambda$ scattering and its value is deduced from the data on the reaction of ^{12}C$(K^-, K^+\Lambda\Lambda X)$ in Ref. [14] as $a_{\Lambda\Lambda} = -1.2 \pm 0.6$ fm. The coupling of the three-body contact interaction $g(\Lambda_c)$ is fixed by $B_{\Lambda\Lambda} = 6.93$ MeV as a function of the sharp cutoff Λ_c introduced in the coupled integral equations. Further details of the integral equations can be found in Ref. [6].

As mentioned above, the limit cycle behavior of the system can be explored by studying the homogeneous part of the integral equation in the asymptotic limit. As shown in Eq. (9) of Ref. [6], when we exclude the three-body interaction in the asymptotic limit $p \sim l \gg k, E, 1/a_{\Lambda\Lambda}, \gamma_{\Lambda\alpha}$, we have

$$1 = C_1\, I_1(s) + C_2\, I_2(s)\, I_3(s), \tag{1}$$

where

$$C_1 = \frac{1}{2\pi}\frac{m_\alpha}{\mu_{\Lambda\alpha}}\sqrt{\frac{\mu_{\Lambda(\Lambda\alpha)}}{\mu_{\Lambda\alpha}}}, \quad C_2 = \frac{\sqrt{2m_\Lambda \mu_{\Lambda(\Lambda\alpha)}\mu_{\alpha(\Lambda\Lambda)}}}{\pi^2 \mu^{3/2}_{\Lambda\alpha}}, \tag{2}$$

and $\mu_{\alpha(\Lambda\Lambda)} = 2m_\Lambda m_\alpha/(2m_\Lambda + m_\alpha)$, etc. The functions $I_{1,2,3}(s)$ are obtained by the Mellin transformation and their explicit expressions can be found in Ref. [6]. The imaginary solution $s = \pm i s_0$ indicates the limit cycle solution and we found the solution as

$$s_0 = 1.0496\cdots. \tag{3}$$

This shows that the $\Lambda\Lambda\alpha$ system exhibits the limit cycle and thus the three-body contact interaction should be included at LO for renormalization.

In Fig. 1, we plot $g(\Lambda_c)$ as a function of Λ_c with $a_{\Lambda\Lambda} = -0.6, -1.2, -1.8$ fm. One can verify that the curves clearly exhibit the limit cycle and the first divergence appears at $\Lambda_c \sim 1$ GeV. In addition, a larger value of $|a_{\Lambda\Lambda}|$ behaves as giving a larger attractive force and shifts the curves of $g(\Lambda_c)$ to the left as shown in Fig. 1. Furthermore, the value of s_0 can be estimated from the curves in Fig. 1. The $(n+1)$-th values of Λ_n at which $g(\Lambda_c)$ vanishes can be parameterized as $\Lambda_n = \Lambda_0 \exp(n\pi/s_0)$ in the discrete scaling symmetry. By using the second and third values of Λ_n for the three values of $a_{\Lambda\Lambda}$, we find $s_0 = \pi/\ln(\Lambda_2/\Lambda_1) \simeq 1.05$, which is in a very good agreement with the value of Eq. (3).

Figure 2 shows $B_{\Lambda\Lambda}$ as a function of $1/a_{\Lambda\Lambda}$, while $g(\Lambda_c)$ is renormalized at the point marked by a filled square, i.e., $B_{\Lambda\Lambda} = 6.93$ MeV and $1/a_{\Lambda\Lambda} = -2.0$ fm^{-1}. This leads to $g(\Lambda_c) \simeq -0.715, -0.447, -0.254$ for $\Lambda_c = 170, 300, 430$ MeV, respectively. Open squares are the estimated values from the

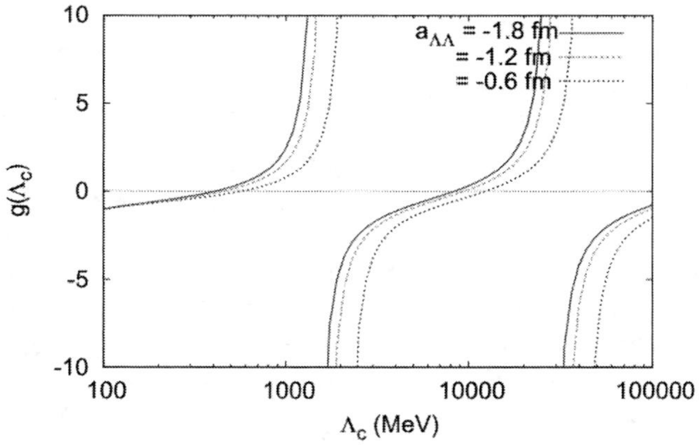

Figure 1. The coupling $g(\Lambda_c)$ as a function of Λ_c for $a_{\Lambda\Lambda} = -0.6, -1.2, -1.8$ fm, where the values of $g(\Lambda_c)$ are fitted by $B_{\Lambda\Lambda} = 6.93$ MeV of ${}^{6}_{\Lambda\Lambda}$He.

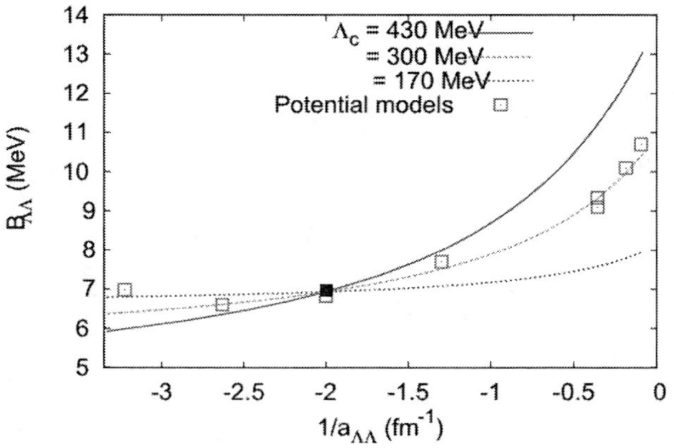

Figure 2. The two-Λ separation energy $B_{\Lambda\Lambda}$ as a function of $1/a_{\Lambda\Lambda}$ for $\Lambda_c = 170$, 300, 430 MeV, where $g(\Lambda_c)$ is renormalized at the point of $B_{\Lambda\Lambda} = 6.93$ MeV and $1/a_{\Lambda\Lambda} = -2.0$ fm^{-1} that is marked by a filled square. Open squares are the results from the potential models given in Table 5 of Ref. [15].

potential models given in Table 5 of Ref. [15]. The results show that the curves are sensitive to the cutoff value and the outputs from the potential models are remarkably well reproduced by the curve with $\Lambda_c = 300$ MeV.

3 Discussion and conclusions

To summarize, the hypernucleus ${}^{6}_{\Lambda\Lambda}$He was studied in cluster EFT at LO as a three-body ($\Lambda\Lambda\alpha$) system. The three-body contact interaction of the system exhibits the limit cycle, so it is needed to be promoted to LO so that the result is independent of the cutoff. The equation to examine the limit cycle for ${}^{6}_{\Lambda\Lambda}$He is developed, which reproduces the numerical results for the limit cycle. The two-Λ separation energy $B_{\Lambda\Lambda}$ of ${}^{6}_{\Lambda\Lambda}$He can be well obtained even without introducing the three-body contact interaction, but it requires $\Lambda_c = 570 \sim 410$ MeV for $a_{\Lambda\Lambda} = -0.6 \sim -1.8$ fm. The cutoff values are comparable to the large scale of the theory, $\Lambda_H \simeq 170$ MeV. This pattern is similar to that discovered already for the formation of three-body bound state without the three-body contact interaction such as the triton in pionless EFT [10, 16]. We also found that the dependence on $a_{\Lambda\Lambda}$ is significant but the range of $a_{\Lambda\Lambda}$ cannot be narrowed until we have more precise and ample experimental data for various physical quantities of the system.

It should also be noted that the equation we developed in Eq. (1) depends on the masses and the spin-isospin quantum numbers of the state but not on the details of dynamics. It is well known that the imaginary solution of that equation implies the existence of the Efimov states in the unitary limit [10]. Even though our system is not close to the unitary limit, the presence of imaginary solution could imply the existence of a bound state. Therefore, the integral equations in the asymptotic limit for three-body cluster systems would be useful to search for exotic states.[2]

In order to study the correlation between $B_{\Lambda\Lambda}$ and $a_{\Lambda\Lambda}$ we introduce the three-body contact interaction. By varying the value of Λ_c, we found that the results of potential models can be remarkably well reproduced with $\Lambda_c = 300$ MeV, which would demonstrate a nontrivial role of the two-pion exchange as a long range mechanism of the $\Lambda\Lambda$ interaction. However, including two-pion exchanges in the present approach is highly nontrivial as it requires to have $\Lambda_c > \Lambda_H \simeq 170$ MeV and such a large cutoff demands the inclusion of various short range (or high momentum) degrees of freedom such as the first excitation state of the α. Investigating such mechanisms is beyond the scope of the present work but deserves further studies.

Acknowledgements

S.-I.A. is grateful to the organizers of the FB21 conference for hospitality and support. He also acknowledges the support from the Sunmoon University Research Grant of 2015. Y.O. was supported by the National Research Foundation (Grant No. NRF-2013R1A1A2A10007294) funded by the Ministry of Education of Korea. This work was also supported in part by the Ministry of Science, ICT, and Future Planning (MSIP) and the National Research Foundation of Korea under Grant No. NRF-2013K1A3A7A06056592 (Center for Korean J-PARC Users).

References

[1] D. J. Prowse, Phys. Rev. Lett. **17**, 782 (1966).
[2] H. Takahashi *et al.*, Phys. Rev. Lett. **87**, 212502 (2001).
[3] K. Nakazawa for KEK-E176, E373 and J-PARC E07 Collaborators, Nucl. Phys. A **835**, 207 (2010).
[4] E. Hiyama, T. Motoba, T. A. Rijken, and Y. Yamamoto, Prog. Theor. Phys. Suppl. **185**, 1 (2010).
[5] A. Gal, Prog. Theor. Phys. Suppl. **186**, 270 (2010).
[6] S.-I. Ando and Y. Oh, Phys. Rev. C **90**, 037301 (2014).
[7] S.-I. Ando, G.-S. Yang, and Y. Oh, Phys. Rev. C **89**, 014318 (2014).
[8] P. F. Bedaque and U. van Kolck, Ann. Rev. Nucl. Part. Sci. **52**, 339 (2002).
[9] E. Braaten and H.-W. Hammer, Phys. Rep. **428**, 259 (2006).
[10] P. F. Bedaque, H. W. Hammer, and U. van Kolck, Nucl. Phys. A **676**, 357 (2000).
[11] K. G. Wilson, Phys. Rev. D **3**, 1818 (1971).
[12] V. N. Efimov, Yad. Fiz. **12**, 1080 (1970), [Sov. J. Nucl. Phys. **12**, 589 (1971)].
[13] G. S. Danilov, Zh. Eksp. Teor. Fiz. **40**, 498 (1961), [Sov. Phys. JETP **13**, 349 (1961)].
[14] A. Gasparyan, J. Haidenbauer, and C. Hanhart, Phys. Rev. C **85**, 015204 (2012).
[15] I. Filikhin and A. Gal, Nucl. Phys. A **707**, 491 (2002).
[16] S.-I. Ando and M. C. Birse, J. Phys. G **37**, 105108 (2010).
[17] S.-I. Ando, U. Raha, and Y. Oh, Phys. Rev. C **92**, 024325 (2015).

[2] This method is recently applied to study the existence of a putative $nn\Lambda$ bound system in Ref. [17].

EPJ Web of Conferences **113**, 07008 (2016)
DOI: 10.1051/epjconf/201611307008

Hyperon binding energy in $^{6}_{\Lambda}$He and $^{7}_{\Lambda}$He

Igor Filikhin[1,a], Vladimir Suslov[1], and Branislav Vlahovic[1]

[1] *North Carolina Central University, 1801 Fayetteville St., Durham, NC, USA*

Abstract. The three-body approach based on the configuration space Faddeev equations for systems of non-identical particles is proposed to describe light hypernuclei (A=6,7, S=-1) with α particle clustering. We focus on the model $(\alpha+\Lambda)+n+n$ for $^{7}_{\Lambda}$He hypernucleus for which the first experimental data have been recently reported. New evaluation for hyperon binding energy in $^{7}_{\Lambda}$He is done by using a relation between energies of the spin doublet (1^-,2^-) of $^{6}_{\Lambda}$He and the $^{7}_{\Lambda}$He ground state. Energies of low-lying levels of $^{6}_{\Lambda}$He hypernucleus are calculated within the cluster $\alpha+\Lambda+n$ model.

1 Introduction

P-shell hypernuclei, like the $^{7}_{\Lambda}$He and the $^{6}_{\Lambda}$He are being considered as appropriate systems to study the ΛN interaction and in particular their spin dependence [1]. The $^{7}_{\Lambda}$He hypernucleus is of some interest for studying the CSB effect of ΛN interaction [2]. The first reliable experimental $^{7}_{\Lambda}$He data obtained at the Jlab experiments (E01-011, E05-115) have been reported in Refs. [2, 3]. Theoretical investigations for $^{7}_{\Lambda}$He within the frameworks of three and four-body cluster models were undertaken in Refs. [4–7]. Up to now, the experimental data and predictions for the ground state energy are in disagreement. The experimental values for hyperon energy B_Λ are larger than the theoretical ones [8].

The $^{6}_{\Lambda}$He hypernucleus has been theoretically studied in Ref. [6, 9] within the cluster $\alpha+\Lambda+n$ model. Experimental data for $^{6}_{\Lambda}$He are restricted by the value of 1^- ground state energy. The recent theoretical result [6] for the energy is within experimental errors, but slightly overebound the system. The spin doublet (1^-, 2^-) of $^{6}_{\Lambda}$He is interesting due to study of spin dependence of the ΛN interaction [9].

In the present work, three-body cluster approach is applied for studying the $^{7}_{\Lambda}$He and the $^{6}_{\Lambda}$He nucleus. The configuration space Faddeev equations [10] are used for calculation of low-lying level energies of these hypernuclei. The system $\alpha+\Lambda+n+n$ can be consider as the four-body model of $^{7}_{\Lambda}$He hypernucleus that includes the bound subsystems $\alpha+\Lambda$ as $^{5}_{\Lambda}$He nucleus, $\alpha+\Lambda+n$ as $^{6}_{\Lambda}$He nucleus and $\alpha+n+n$ as ^{6}He nucleus. Thus, the consideration will be completed if all subsystems have been taken into account. It is clear that some correlations can be found between energies of the systems.

2 Theoretical assumptions

$^{7}_{\Lambda}$He hypernucleus is modeled as the cluster system $(\alpha+\Lambda)+n+n$ [5, 11]. Here, the $\alpha+\Lambda$ pair is considered as $^{5}_{\Lambda}$He hypernucleus in the ground state with the hyperon energy of 3.12 MeV. The

[a] e-mail: ifilikhin@nccu.edu

folding procedure is applied to construct the ${}^{5}_{\Lambda}$He-n potential [5]. In the $\alpha + \Lambda + n$ system, the Λn interaction is averaged by coordinates of a hyperon in ${}^{5}_{\Lambda}$He. One can assume that the two-body ${}^{5}_{\Lambda}$He-n model is a good approximation for ${}^{6}_{\Lambda}$He hypernicleus. However, an adjustment for the coefficients of spin-dependent part of effective ${}^{5}_{\Lambda}$He-n potential in p-wave is required. The adjustment can be done by reproducing the binding energy of the 1^- and the excitation energy of the 2^- states of ${}^{6}_{\Lambda}$He nucleus.

The s- and p-wave components of the folding potential are written in the form $V^{l=0,t(s)}_{\text{Core}-n} = \gamma V^{l=0}_{\alpha n} + V^{f,t(s)}_{\Lambda n}$, $V^{l=1,t(s)}_{\text{Core}-n} = V^{l=1}_{\alpha n} + \eta^{t(s)} V^{f,t(s)}_{\Lambda n}$. Here are three free parameters γ, η^t, η^s and Core means ${}^{5}_{\Lambda}$He. In this expression, the upper indicies s and t mean singlet and triplet spin states of the pair Core + n. The coefficient γ is adjusted using the experimental value for the ${}^{6}_{\Lambda}$He bound state energy within the $\alpha + n + n$ model [11]. We choose the spin-dependent coefficient $\eta^{t(s)}$ to reproduce the experimental and theoretical data for the 2^- and 1^- states of the ${}^{6}_{\Lambda}$He hypernucleus, respectively. $V^{f,t(s)}_{\Lambda N}$ is the triplet(singlet) component of the folding ΛN potential: $V^{f,t(s)}_{\Lambda N}(|\vec{r}|) = \int d\vec{\xi}\rho(|\vec{\xi}|)V^{s(t)}_{\Lambda N}(|\vec{r} - \vec{\xi}|)$, where $\rho(|\vec{r}|)$ is the Λ-particle density function for the ${}^{5}_{\Lambda}$He hypernucleus ($\int d\vec{r}\rho(|\vec{r}|) = 1$). $V^{s(t)}_{\Lambda N}$ is the s-wave component of ΛN potentials. $V^{l}_{\alpha N}$ is the partial component (s or p-wave) of the αN potential (spin-orbit coupling is not taken into account). Finally, we present the effective Core − n potential in the form averaged over spins: $V^{l,\text{eff.}}_{\text{Core}-n} = \frac{3}{4} V^{l,t}_{\text{Core}-n} + \frac{1}{4} V^{l,s}_{\text{Core}-n}$. Because the experimental value for the 2^- state is still unknown, such three-body model encounters a difficulty. We have obtained the energy of 2^- as a result of our calculations for ${}^{6}_{\Lambda}$He within the cluster $\alpha + \Lambda + n$ model.

Numerical modeling for Core + $n + n$ and $\alpha + \Lambda + n$ systems is based on the differential Faddeev equations. The set of potentials for the models includes the simulated NSC97f potential [12] for ΛN interaction, the TH $\alpha\Lambda$ potential taken from [13] and αn potential proposed in Ref. [14]. For nucleon-nucleon interaction we used the Malfliet-Tjon potential (MT-I-III) corrected by Friar et al. [15]. Calculations with this semi-realistic potential reproduce well the experimental data for three- and four-nucleon systems.

3 Numerical Results

The ${}^{7}_{\Lambda}$He binding energy is evaluated by calculating the correlation between the ${}^{7}_{\Lambda}$He binding energy and the excitation energy of the 2^- state of ${}^{6}_{\Lambda}$He nucleus. The strengths of the singlet and triplet components of the ${}^{5}_{\Lambda}$He-n potential consider as variable parameters of the model. These parameters were chosen to reproduce the ${}^{6}_{\Lambda}$He(2^-) excitation energy obtained by direct three-body $\alpha + \Lambda + n$ calculation. The calculated value for the binding energy of the ${}^{6}_{\Lambda}$He(1^-) is -0.65 MeV. The energy is measured from the ${}^{5}_{\Lambda}$He+n threshold. The energy of the ${}^{6}_{\Lambda}$He(2^-) resonance is about 0.11 MeV. Because the experimental value for the ${}^{6}_{\Lambda}$He ground state was reported to be E_Λ=-0.17± 0.10 MeV, our cluster model underestimates the binding energy of the hypernucleus. It is not surprising that the similar cluster system $\alpha + n + n$ is also weakly bound when calculation is performed with pair potentials only. To reproduce the ground state energy of ^{6}He, a three-body potential is used. One can assume that a weak three-body force is also needed to reproduce the experimental ${}^{6}_{\Lambda}$He data. The correlation between the Λ-hyperon energy B_Λ in ${}^{7}_{\Lambda}$He and the excitation energy E_x of the 2^- state of ${}^{6}_{\Lambda}$He hypernucleus is presented in figure 1a). The 1^- and 2^- energies are defined by the adjusting $\eta^{t(s)}$ parameters of the ${}^{5}_{\Lambda}$He−n folding potential. The correlation has linear dependence. We calculate its dependence for the different values $E_{g.s.}$ of the ${}^{6}_{\Lambda}$He(1^-) ground state energy. The obtained value is 0.18 MeV for the 2^- excitation energy of ${}^{6}_{\Lambda}$He within three-body model. This result is presented by the vertical dashed line. The corresponding ${}^{6}_{\Lambda}$He(1^-) binding energy $E_{g.s.}$ is -0.16 MeV as is following from figure 1. The correlated value of the ${}^{7}_{\Lambda}$He hyperon binding energy is 5.69 MeV. The result is close to recently reported experimental value of 5.68±0.03±0.25 MeV [3]. This value is noted by the horizontal line Jlab(E01-111) with the experimental error given by the square left bracket.

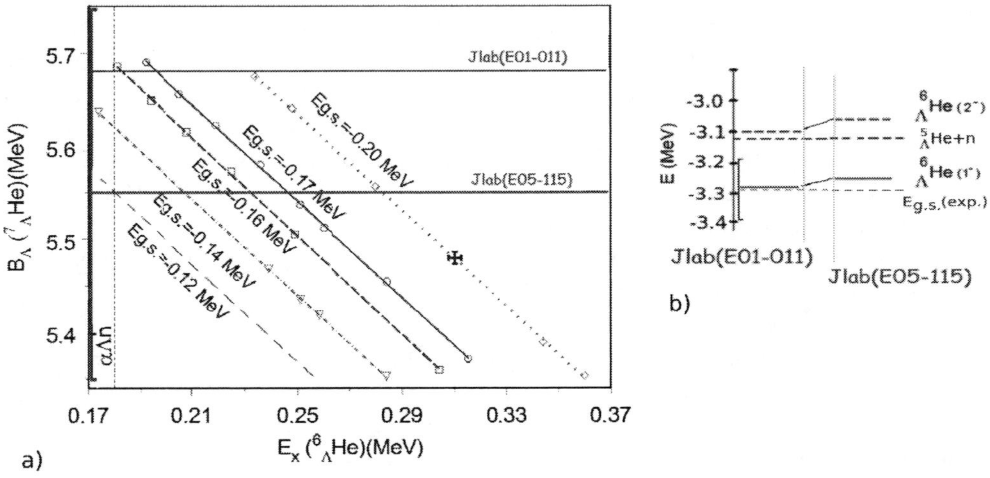

Figure 1. a) The correlation between the binding energy B_Λ of Λ-hyperon in ${}^7_\Lambda$He and the excitation energy E_x of the 2^- state of ${}^6_\Lambda$He hypernucleus for different values $E_{g.s.}$ of the ${}^6_\Lambda$He(1^-) ground state binding energy. The energies of 1^- and 2^- states are defined by the $\eta^{t(s)}$ parameters of the ${}^5_\Lambda$He–n folding potential. The three-body $\alpha\Lambda n$ result for ${}^6_\Lambda$He is presented by vertical dashed line. The experimental data are shown by horizontal line with the experimental error given by the square bracket. Result of calculations [6] is shown by a cross. b) The 1^- ground state energy of ${}^6_\Lambda$He obtained by the correlation of 1a) for two experimental data (Jlab E01-011 and E05-115) for ${}^7_\Lambda$He. The ${}^6_\Lambda$He data are shown by horizontal line with the experimental error given by the square bracket. The level 2^- is presented for the excitation energy E_x of 0.18 MeV.

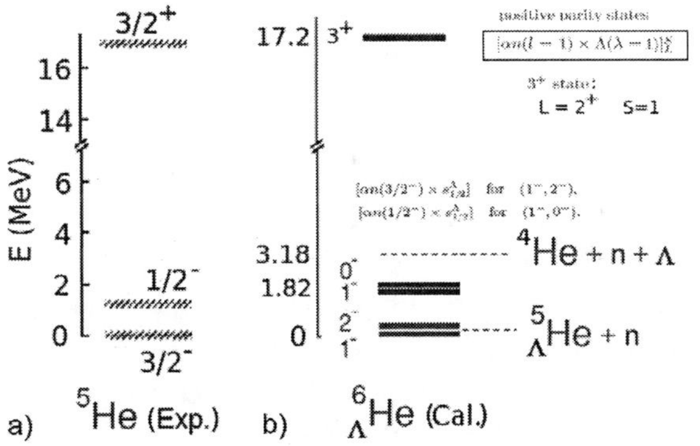

Figure 2. Excitation energies of low-lying levels of a) ^{5}He (experimental data) and b) ${}^6_\Lambda$He (calculated). The total spin J^π of each level is shown. Description of the spin-orbital configurations for the ${}^6_\Lambda$He levels is given from Ref.[9]. Λ + Core and $\alpha + \Lambda + n$ thresholds are shown by horizontal dashed line.

Result of calculations [6] is shown by a cross. Note that this result lies on the correlation dependence with $E_{g.s.}$ about -0.2 MeV. The difference between our calculation and the calculation [6] may be related with different sets of pair potential used. The result of our evaluation based on the E05-115 experimental data (5.55±0.10±0.11 MeV [2]) predicting the value of -0.12 MeV for the ${}^6_\Lambda$He(1^-) binding energy disagrees with existing experimental ${}^6_\Lambda$He(1^-) data. This situation is illustrated in figure 1b). To estimate the energies and widths of low-lying resonance states, we used the method of analytical continuation in the coupling constant [16]. A variant of this method with an additional non-physical three-body potential is employed. The strength parameter of this potential is considered as a variational parameter (coupling constant) for the analytical continuation of the bound state energy into the complex plane [14]. The structure of the ${}^6_\Lambda$He spectrum is studied within the three-body

$\alpha + \Lambda + n$ model. We present results of the Faddeev calculations for several low-lying levels in figure 2. Comparison with ^{5}He experimental data shows that direct relation can be establish between the spectra of ^{5}He and $^6_\Lambda$He. The spectrum of $^6_\Lambda$He is more complicated due to existence additional spin variable which forms the spin-doublets with small spacings reflecting the weak spin dependence of ΛN interaction. Description of the spin-orbital configurations for the $^6_\Lambda$He spectrum has been done in Ref. [9]. However, the excitation energies of the levels in [9] are quite different from ones shown in figure 2 due to strong ΛN potential used in [9] which overestimates the $^6_\Lambda$He binding energy. It has been previously shown in Ref. [11] that the low-lying states of the $^7_\Lambda$He nucleus can be qualitatively classified by analogy for corresponding states of the ^{6}He nucleus, within the ^{6}He(J^π) + $\Lambda(s)$ form. This analogy is generally supported in Ref. [7] where the calculations for the low-lying $^7_\Lambda$He states have been performed by using the four-body cluster model. The recent $^7_\Lambda$He experimental data [2] show the ground state and the excited state peaks of ^{7}Li$(e, eK^+)^7_\Lambda$He reaction (Jlab E05-115). The second peak is assumed to be related with the first 2^+ level of ^{6}He. According to the analysis in Ref. [7], the second 2^+ level of ^{6}He has to generate third excitation peak corresponding to energy $3/2_2$ ($5/2_2$) level of the $^7_\Lambda$He.

4 Conclusion

In the presented work we focused on the correlation between energies of the spin doublet (1^-,2^-) of $^6_\Lambda$He and the $^7_\Lambda$He ground state. The correlation allowed us to evaluate both $^6_\Lambda$He and $^7_\Lambda$He hyperon energies and to show relation to the existing experimental data. Another established correlation is between spectra of core nucleus and hypernucleus like ^{5}He and $^6_\Lambda$He, ^{6}He and $^7_\Lambda$He. These correlations are possible due to known "glue-like role" of hyperon in nuclear matter.

This work is supported by NSF (HRD-1345219). The authors thank Prof. R. Kezerashvili and Prof. E. Hiyama for useful discussion.

References

[1] O. Hashimoto, H. Tamura, Progress in Particle and Nuclear Physics **57** 564-653 (2006)

[2] S.N. Nakamura, Hypernuclear Workshop May 27-29, 2014, Jefferson Lab, Newport News, VA

[3] S.N. Nakamura et al. Phys. Rev. Lett. **110** 012502 (2013)

[4] E. Hiyama, M. Kamimura, T. Motoba, T. Yamada, Y. Yamamoto, Phys. Rev. C **53** 2075 (1996)

[5] I. Filikhin, V.M. Suslov and B. Vlahovic, J. Phys. G **31** 389-400 (2005)

[6] E. Hiyama, S. Ohnishi, M. Kamimura, Y. H. Yamamoto, Nucl. Phys. A **908** 29-39 (2013)

[7] E. Hiyama, M. Isaka, M. Kamimura, T. Myo, T. Motoba, arXiv:1504.07735

[8] A. Gal, J. Millener, Physics Letters B **725** 445-450 (2013)

[9] T. Motoba, H. Bando and K. Ikeda, Prog. Theor. Phys. **70** 189-221 (1983)

[10] L. D. Faddeev and S. P. Merkuriev, *Quantum Scattering Theory for Several Particle Systems* (Kluwer Academic, Dordrecht, 1993) pp. 398

[11] I. Filikhin, V.M. Suslov and B. Vlahovic, Physics of Atomic Nuclei, **82** 580-587 (2009)

[12] I. Filikhin and A. Gal, Phys. Rev. Lett. **89** 172502 (2002)

[13] Y.C. Tang, R.C. Herndon, Phys. Rev. B **138** 637 (1965)

[14] I. Filikhin, V.M. Suslov, B. Vlahovic, Few-Body Systems **50** 255-257 (2011)

[15] J.L. Friar et al. Phys. Rev. C **42** 1838 (1990)

[16] V.I. Kukulin, V. M. Krasnopolsky and J. Horacek, *Theory of Resonances* (Kluwer Academic, Dordrecht, 1989) pp. 354

EPJ Web of Conferences **113**, 08001 (2016)
DOI: 10.1051/epjconf/201611308001

van der Waals and Casimir-Polder interactions between neutrons

James F. Babb[1,a] and Mahir S. Hussein[1,2,3,4,b]

[1] *ITAMP, Harvard-Smithsonian Center for Astrophysics, MS 14, 60 Garden St., Cambridge, MA 02138*
[2] *Instituto de Estudos Avançados, Universidade de São Paulo C. P. 72012, 05508-970 São Paulo-SP, Brazil*
[3] *Instituto de Física, Universidade de São Paulo, C. P. 66318, 05314-970 São Paulo, Brazil*
[4] *Departamento de Física, Instituto Tecnológico de Aeronáutica, CTA, São José dos Campos, S.P., Brazil*

Abstract. We investigate the van der Waals interaction between neutrons using the theory of Casimir and Polder, wherein the potential for asymptotically large separations falls off as the inverse seventh power, and compare it to the similar interaction between a neutron and a proton, for which the asymptotic interaction falls off as the inverse fourth power. Modifications of the formalism to extend the validity to smaller separations using dynamic electric and magnetic dipole polarizability data are discussed.

1 Introduction

The long-range interaction between a hydrogen atom and an electron is $-\alpha_H/2R^4$, where R is the separation distance and α_H is the electric dipole polarizability of the atom. We use atomic units throughout. The potential arises from the charge-induced dipole interaction. The long-range interaction between a neutron and a nucleus arises in the same way. It is $-\alpha_n Z^2/2R^4$, where Z is the nuclear charge and α_n is the neutron electric dipole polarizability. A physical picture of the atomic polarizability is the response of the electron in the presence of an applied static electric field; a similar picture for the neutron might be the response of the pion cloud to a static electric field [1, 2].

In brief, Thaler [3] and others considered the effect of the potential on neutron scattering from heavy nuclei. Experiments on the transmission of neutrons through lead used the potential above to obtain a value for α_n [4]. A number of experiments were carried out, essentially dependent in some way on using this potential to measure α_n, and these are reviewed in Refs. [5, 6].

As expected with the above physical picture, the effects of external photons can also be related to the neutron dipole polarizabilities and much recent work is focused on using Compton scattering from the neutron bound in the deuteron to extract α_n and the magnetic dipole polarizability β_n [2, 6]. The Compton scattering amplitudes are related to the dynamic polarizabilities $\alpha_n(\omega)$ and $\beta_n(\omega)$, which describe the response of the neutron to the photon of frequency ω, and are identical to the static polarizability values at zero frequency [6], *viz.* $\alpha_n(0) = \alpha_n$, $\beta_n(0) = \beta_n$. Theoretical methods using chiral effective field theory (χEFT) yielded values of the dynamic polarizabilities [2, 7, 8].

We are using dynamic polarizabilities from χEFT to investigate the long-range interactions between two neutrons and between a neutron and a proton [9]. In this short report, we describe the direction of the project.

[a] e-mail: jbabb@cfa.harvard.edu
[b] e-mail: hussein@if.usp.br

2 Casimir-Polder potential

The long-range interaction between two neutrons is given formally for asymptotically large separations by the Casimir-Polder potential [10], where $c = 137$ is the speed of light,

$$V_{nn}^{\infty}(r) = -(c/4\pi)[23(\alpha_n\alpha_n + \beta_n\beta_n) - 14(\alpha_n\beta_n)]R^{-7} + O(R^{-9}), \quad R \sim \infty. \tag{1}$$

For neutrons, we take $\alpha_n = (11.8 \pm 1.1) \times 10^{-4}$ fm^3 and β_n is $(3.7 \pm 1.2) \times 10^{-4}$ fm^3 [11]. [Note that Eq. (1) is the asymptotic limit of a more general equation involving dynamic polarizabilities (see, for example, Ref. [12]) that describes the neutron-neutron interaction for all distances at separations r sufficiently large that the interaction is purely electromagnetic.]

An early analysis considering parts of Eq. (1) and the general equation and its effect on the neutron scattering length was carried out in 1973 by Arnold [13]. At the time of his analysis, β_n was unknown and the accepted value of α_n was twice today's value. In particular, the χEFT predicts a dynamic polarizability, $\alpha_n(\omega)$, which contains a reference to the excited state of the neutron, and this will have an influence on the nn interaction. This hitherto not considered effect is also being investigated.

The long-range interaction between a proton and a neutron is notably different from Eq. (1),

$$V_{pn}^{\infty}(R) = -(1/2)\alpha_n R^{-4} + (1/4\pi c M_p)(11\alpha_n + 5\beta_n)R^{-5} + O(R^{-7}), \quad R \sim \infty, \tag{2}$$

where M_p is the proton mass [14]. [Similarly to Eq. (1), Eq. (2) is the asymptotic limit of a general equation involving dynamic polarizabilities [12] that describes the proton-neutron interaction at all distances sufficiently large that the interaction is purely electromagnetic.]

A detailed exploration of Eqs. (1) and (2) and their more general forms using dynamic polarizabilities for applications to neutron-neutron and neutron-proton scattering and for neutron-wall interactions, is in progress [9].

Acknowledgements

The Institute for Theoretical Atomic, Molecular, and Optical Physics (ITAMP) is supported in part by a grant from the NSF to the Smithsonian Institution and Harvard University. MSH is also partly supported by the Brazilian funding agencies CAPES/ITA, CNPq, FAPESP, INCT-IQ/MCT, and CEPID-CEPOF.

References

[1] D. Drechsel, T. Walcher, Rev. Mod. Phys. **80**, 731 (2008)
[2] H.W. Grießhammer, J.A. McGovern, D.R. Phillips, G. Feldman, Prog. Part. Nucl. Phys. **67**, 841 (2012), arXiv: 1203.6834
[3] R.M. Thaler, Phys. Rev. **114**, 827 (1959)
[4] J. Schmiedmayer, P. Riehs, J.A. Harvey, N.W. Hill, Phys. Rev. Lett. **66**, 1015 (1991)
[5] F. Wissmann, M.I. Levchuk, M. Schumacher, Eur. Phys. J. A **1**, 193 (1998)
[6] M. Schumacher, Prog. Part. Nucl. Phys. **55**, 567 (2005)
[7] H.W. Grießhammer, T.R. Hemmert, Phys. Rev. C **65**, 045207 (2002)
[8] R.P. Hildebrandt, H.W. Grießhammer, T.R. Hemmert, B. Pasquini, Eur. Phys. J. A **20**, 293 (2004)
[9] J.F. Babb, M.S. Hussein, in preparation
[10] G. Feinberg, J. Sucher, Phys. Rev. A **2**, 2395 (1970)
[11] K.A. Olive et al., Chin. Phys. C **38** (2014)
[12] J.F. Babb, *Adv. At. Molec. Opt. Phys.* (Academic Press, San Diego, 2010), Vol. 59, p. 1
[13] L.G. Arnold, Phys. Lett. B **44**, 401 (1973)
[14] J. Bernabéu, R. Tarrach, Ann. Phys. (N.Y.) **102**, 323 (1976)

EPJ Web of Conferences **113**, 08002 (2016)
DOI: 10.1051/epjconf/201611308002

Alpha clustering in Ti isotopes: ^{40,44,48}Ca + α resonant scattering

Sam Bailey[1,a], Martin Freer[1,b], Tzany Kokalova[1,c], Carl Wheldon[1], Robin Smith[1], Joseph Walshe[1], Lovro Prepolec[2], Neven Soić[2], Vedrana Tokić[2], Miguel Marqués[3], Lynda Achouri[3], Franck Delaunay[3], Marian Parlog[3], Quentin Deshayes[3], Beatriz Fernández-Dominguez[4], and Bertrand Jacquot[5]

[1] *School of Physics and Astronomy, University of Birmingham, Birmingham, B15 2TT, UK*
[2] *Institut Ruder Bošković, Bijenička cesta 54, 10000 Zagreb, Croatia*
[3] *Laboratoire de Physique Corpusculaire de Caen, 6 bd Maréchal Juin, 14050 Caen Cedex, France*
[4] *Universidade de Santiago de Compostela, Praza do Obradoiro, 15782 Santiago de Compostela, Spain*
[5] *Grand Accélérateur National d'Ions Lourds, Bd Henri Becquerel, BP 55027 - 14076 Caen Cedex 05, France*

Abstract. Measurements were made of the ^{4}He(^{40,44,48}Ca,α) resonant scattering reactions at 180° and up to $E_{c.m.}$ ~11.5 MeV, using the Thick Target Inverse Kinematics technique. These measurements are discussed, with a focus on assessing their usefulness for investigating α-clustering in medium mass ^{44,48,52}Ti nuclei.

1 Introduction

Clustering in nuclei allows the techniques of few-body physics to be applied to the many body nuclear problem, yet identifying nuclei which exhibit cluster structures is far from trivial. Alpha-clustering has been shown to play an important role in the structure of many α-conjugate and neutron rich light nuclei [1]. However, it is equally apparent that mean field approaches to nuclear structure describe the properties of heavy nuclei very well, suggesting that clusters do not contribute significantly to their structure. This leads to the question of to what extent does α-clustering persist into medium and heavy mass nuclei.

The core+Xn+α molecular cluster structure is an essential component in the structure of ^{20,21,22}Ne [2, 3]. Here ^{16}O acts as a core, and any additional neutrons act as valence particles providing additional binding to the α + ^{16}O cluster structure. This structure is thought to be exceptionally pronounced due to the inert nature of the doubly magic ^{16}O core. It has been hypothesized that ^{44}Ti may exhibit a similar cluster structure to ^{20}Ne, due to its doubly magic ^{40}Ca core. If confirmed this would provide evidence for α-clustering in medium mass nuclei.

In the present work ^{44,48,52}Ti are investigated, allowing the interaction of valence neutrons with the ^{40}Ca+α structure to be examined and compared with ^{20}Ne. Additionally since ^{48}Ca is also a doubly magic nucleus, this allows the evolution of this cluster structure across an entire shell to be explored.

[a] e-mail: s.c.bailey@pgr.bham.ac.uk
[b] e-mail: m.freer@bham.ac.uk
[c] e-mail: t.kokalova@bham.ac.uk

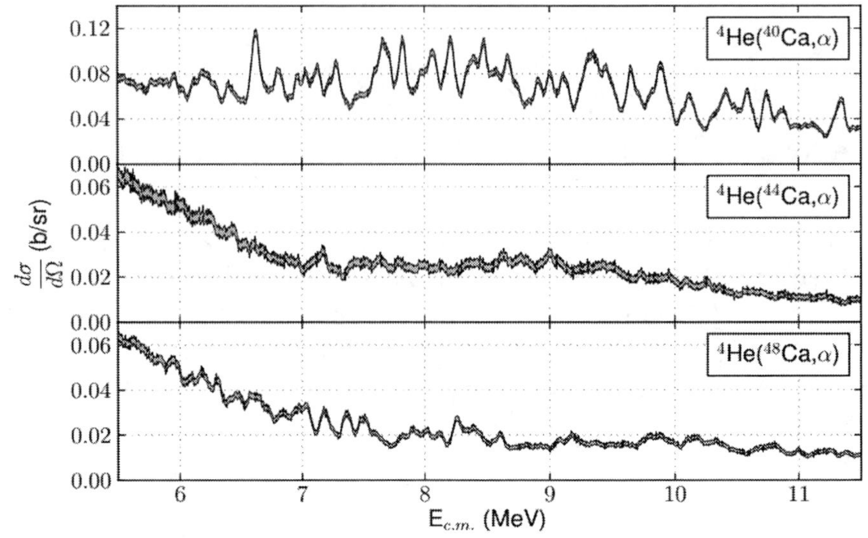

Figure 1. Measured spectra. All measurements taken at a scattering angle of 180° and normalized to the Rutherford scattering cross section at low $E_{c.m.}$. Grey region represents a 90% confidence interval.

2 Experimental Work

Measurements were made of the ^{4}He(^{40,44,48}Ca,α) resonant scattering reactions, at a scattering angle of 180° using the Thick Target Inverse Kinematics technique [4]. The experiment was performed at GANIL, France using ^{40,44,48}Ca beams incident on a thick ^{4}He gas target, with beam energies of 180, 207 and 234 MeV respectively. The scattered α-particles were measured using Double Sided Silicon Strip Detectors, which were placed at the end of the reaction chamber at 0° to the beam line, corresponding to a scattering angle of 180° in inverse kinematics. The beam loses energy as it traverses the gas, tracing out the continuous excitation spectra shown in Figure 1.

The resonant structure of these spectra is dependent entirely on the properties of the states in the compound nuclei: ^{44,48,52}Ti. This dependence is expressed using R-Matrix theory [5], which formulates explicitly the differential cross-section in terms of the energies, spins, parities and reduced widths of the compound nuclear states. The analysis will focus on developing techniques based on R-Matrix theory to probe the structure of ^{44,48,52}Ti, using these measurements of resonant structure. Since the states in these measurements are being populated through the α+^{40,44,48}Ca channels it is expected the α-clustered states will be preferentially populated, providing an excellent insight into the degree of α-clustering in ^{44,48,52}Ti.

References

[1] W. von Oertzen, M. Freer, and Yoshiko Kanada-En'yo, Physics Reports **432**, 43–113 (2006)

[2] W. von Oertzen, Eur. Phys. J. **A11**, 403 (2001)

[3] G. V. Rogachev et al., Phys. Rev. C **64**, 051320 (2001)

[4] K. P. Artemov et al., Sov. J. Nucl. Phys. **52**, 408–411 (1990)

[5] A. M. Lane and R. G. Thomas, Rev. Mod. Phys. **30**, 257 (1958)

EPJ Web of Conferences **113**, 08003 (2016)
DOI: 10.1051/epjconf/201611308003

Low-energy calculations for nuclear photodisintegration

S. Deflorian[1,2,a], V.D. Efros[3], and W. Leidemann[1,2]

[1] *Dipartimento di Fisica, Università di Trento, I-38123 Trento, Italy*
[2] *Istituto Nazionale di Fisica Nucleare, TIFPA, I-38123 Trento, Italy*
[3] *Russian Research Centre "Kurchatov" Institute, Kurchatov Square, 123182 Moscow, Russia*

Abstract. In the Standard Solar Model a central role in the nucleosynthesis is played by reactions of the kind ${}^{A_1}_{Z_1}X_1 +^{A_2}_{Z_2} X_2 \to^{A_1+A_2}_{Z_1+Z_2} Y + \gamma$, which enter the proton-proton chains. These reactions can also be studied through the inverse photodisintegration reaction. One option is to use the Lorentz Integral Transform approach, which transforms the continuum problem into a bound state-like one. A way to check the reliability of such methods is a direct calculation, for example using the Kohn Variational Principle to obtain the scattering wave function and then directly calculate the response function of the reaction.

Our goal is to study nuclear reactions of the kind

$$ {}^{A}_{Z}N + \gamma \to X_1 + X_2 \tag{1}$$

close to the threshold with charged fragments X_1 and X_2. Such reactions are relevant in astrophysics: the inverse reaction of the photodisintegration of a nucleus is the hadronic capture, which plays an important role in the stellar nucleosynthesis, and is a central part of the proton-proton chain.
This study is carried out with *ab initio* methods.

The Lorentz integral transform (LIT) method is an ab initio method that can also be used to study inclusive reactions. In fact that it allows to solve a continuum problem with bound state methods.
In general, we are interested in the response function $R(E)$ of a given reaction. Instead of directly determining the response functions, its Lorentz integral transform can be considered:

$$L(\sigma) = \int dE\mathcal{L}(E,\sigma)R(E) \ , \ \mathcal{L}(E,\sigma) = \frac{1}{(E-\sigma_R)^2+\sigma_I^2} \tag{2}$$

where the kernel is a Lorentzian with $\sigma = \sigma_R + i\sigma_I$. This is a transform with a *controlled resolution*: in fact the parameters σ_R and σ_I control the position and width of the Lorentzian, respectively. For inclusive reactions, the transform $L(\sigma)$ can be calculated as the squared norm of a *localized* state $\tilde{\Psi}$ depending on the operator inducting the reaction. Thanks to these properties, the LIT can be calculated using bound state methods, such as expansions on a complete basis and subsequent diagonalization.
Due to the Coulomb barrier the cross section becomes very small close to the threshold. We want to check if the resolution of the LIT calculation can be made sufficiently high in order to allow a stable inversion of the transform at very low energies: near the reaction threshold, the precision of the LIT calculation may be not sufficient because a high density of LIT states is needed, since the width

[a]e-mail: s.deflorian.2@unitn.it

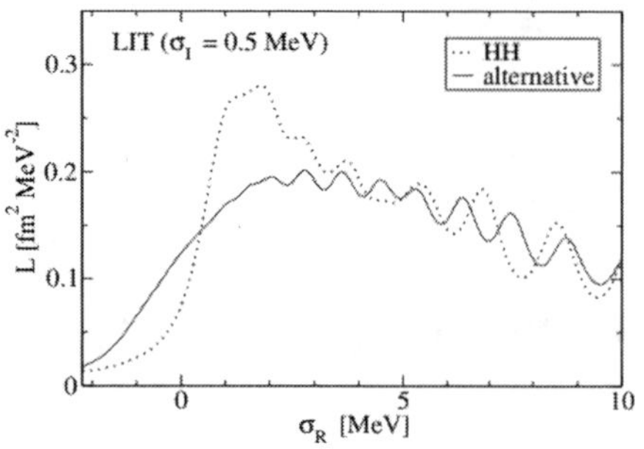

Figure 1. Preliminary results for the LIT of the ^{3}He photodisintegration response function in unretarded dipole approximation, with the Malfliet-Tjon potential. In comparison to the results for the alternative basis (solid line), for the HH basis (dots), with $K_{MAX} = 20$, $N_{MAX} = 10$ (as defined in [2]), the strength is shifted to energies above the three-body breakup threshold (here at $\sigma_R = 0$), due to a lack of LIT states in the low-energy region.

σ_I of each Lorentzian must be sufficiently small to ensure that the low-energy part of the transform doesn't get too strong contributions from other parts of the cross section. Extending the basis leads to a larger number of Hamiltonian eigenstates in general increases the density of such states. It has been observed in recent calculations using a HH (hyperspherical harmonic) basis for the ^{4}He isoscalar monopole resonance [1] that the density of LIT states is not sufficiently high below the three-body breakup threshold to allow to determine the very small width of the resonance (Γ_{exp} = 270 keV).

A possible solution to the lack of LIT states near the reaction threshold may be in the use of an alternative basis that is more appropriate to the type of problem studied. Instead of using a HH basis, we directly use the relative coordinate $\mathbf{r}_{rel}$ for the fragments. Such a coordinate does not enter explicitly the usual HH formalism. Such a method has been used with success in the description of the above mentioned isoscalar monopole resonance $M(q,\omega)$ of the ^{4}He nucleus [2] (see also the contribution of W. Leidemann to this volume). A similar calculation is presently carried out for the ^{3}He photodisintegration (see Fig. 1). After a completion of the LIT calculations, inversions will be performed. One possibility to check the reliability of the inversion of the LIT is to perform a direct calculation of the cross section: such a comparison between the two methods is the aim of the present work. This method in the framework of a HH approach has been already introduced in [3] and has been further completed using modern realistic potential models (see e.g. [4] and [5]). To perform such a calculation, one has to find the scattering wave function of the product fragments of the reaction. Such a wave function can be written as the sum of an asymptotic and an inner part; the asymptotic part is (almost) known, if the wave functions of each fragment are known, and consists of a linear combination of regular and irregular Coulomb functions:

$$|\Psi_{sc}\rangle = |\Omega_{asym}\rangle + |\Psi_{in}\rangle \ , \ |\Omega_{asym}\rangle = \left[|\phi_A\rangle|\phi_B\rangle\left(Y_L^M(\theta,\varphi)R_L^R(r) + T^J_{LM,L'M'}Y_L^M(\theta,\varphi)R_L^I(r)\right)\right]_{J,J_z} . \quad (3)$$

Using the Kohn Variational Principle, one finds that the solution satisfies a set of linear equations that have to be solved in order to find the correct wave function which satisfies the Schrödinger equation $(H-E)|\Psi_{sc}\rangle = 0$. Such calculations are currently being performed, and results will be presented in a forthcoming work.

References

[1] S. Bacca, N. Barnea, W. Leidemann, G. Orlandini, PRL **110**, 042503 (2013)

[2] W. Leidemann, PRC **91**, 054001 (2015)

[3] B.N. Zakhariev, V.V. Pustovalov, V.D. Efros, Yad. Fiz. **8**, 406 (1968)

[4] M. Viviani, A. Kievsky, L.E. Marcucci, S. Rosati, R. Schiavilla, PRC **61**, 064001 (2000)

[5] M. Viviani, A. Deltuva, R. Lazauskas, J. Carbonell, A.C. Fonseca, A. Kievsky, L.E. Marcucci, S. Rosati, PRC **84**, 054010 (2011)

EPJ Web of Conferences **113**, 08004 (2016)
DOI: 10.1051/epjconf/201611308004

Helium atom under pressure

Jérémy Dohet-Eraly[1,a] and Daniel Baye[2,b]

[1] *TRIUMF, 4004 Wesbrook Mall, Vancouver BC V6T 2A3, Canada*
[2] *Physique Quantique, and Physique Nucléaire Théorique et Physique Mathématique, C.P. 229, Université libre de Bruxelles (ULB), 1050 Brussels, Belgium*

Abstract. Hard-sphere confinement is used to study helium atoms under pressure. The confined-helium Schrödinger equation is solved with a high accuracy by a Lagrange-mesh method.

The effects of high pressure on a helium gas can be estimated by studying the helium atom in a hard confinement, i.e. confined at the centre of an impenetrable spherical cavity, for different cavity radii. Contrary to the free helium atom, the confined helium has not been described with a very high accuracy until recently, when we have developed a Lagrange-mesh method to study this system [1]. This method improves by several order of magnitudes the accuracy of previous approaches [2–4].

The outlines of the model are the following. The assumed-infinite-mass nucleus is fixed and the electrons are characterized by coordinates $\boldsymbol{r}_1$ and $\boldsymbol{r}_2$ with respect to this nucleus. In atomic units, the Hamiltonian of the helium atom reads

$$H = -\frac{1}{2}\Delta_1 - \frac{1}{2}\Delta_2 - \frac{2}{r_1} - \frac{2}{r_2} + \frac{1}{r_{12}}, \tag{1}$$

where $\boldsymbol{r}_{12} = \boldsymbol{r}_1 - \boldsymbol{r}_2$ and Δ_1 and Δ_2 are the Laplacians with respect to $\boldsymbol{r}_1$ and $\boldsymbol{r}_2$. The confinement is introduced by forcing the wave function into some spherical cavity of radius R ($r_1, r_2 \leq R$). The wave function $\psi(r_1, r_2, r_{12})$ of an S state must thus verify the Schrödinger equation

$$H\psi(r_1, r_2, r_{12}) = E\psi(r_1, r_2, r_{12}) \tag{2}$$

and vanishes at $r_1 = R$ and $r_2 = R$. The coordinates (r_1, r_2, r_{12}) are advantageously replaced by the coordinates (u, v, w) defined over [0,1] by

$$u = \frac{r_1 - r_2 + r_{12}}{2R - r_1 - r_2 + r_{12}}, \quad v = \frac{-r_1 + r_2 + r_{12}}{2R - r_1 - r_2 + r_{12}}, \quad w = \frac{r_1 + r_2 - r_{12}}{2R}. \tag{3}$$

The confinement implies that the wave function $\psi(u, v, w)$ vanishes at $u = 1$, $v = 1$, and $w = 1$.

The Schrödinger equation is solved by the Lagrange-mesh method [5–7], an approximate variational approach taking the form of a system of mesh equations by computing the Hamiltonian and overlap matrix elements with a Gauss quadrature. Using the coordinates (u, v, w) is essential for an easy treatment of the confinement and an high accuracy of the Gauss quadrature. The ground-state energy and wave function are obtained by diagonalizing a rather large (matrix dimension$\approx 10^3 \sim 2\times10^4$) but sparse matrix. Ground-state energies and mean interparticle distances for several confinement radii R are given in table 1. The pressure acting on the confined helium atom is also given in table 1. It is

[a] e-mail: jdoheter@triumf.ca
[b] e-mail: dbaye@ulb.ac.be

Table 1. Ground-state energy and mean interparticle distances of a helium atom confined in a sphere of radius R and pressure acting on such a confined helium atom [1]. Atomic units are used. The powers of ten are indicated between brackets.

R	E	$\langle r_{12} \rangle$	$\langle r_1 \rangle = \langle r_2 \rangle$	P
0.1	906.562 422 919 888	0.069 580 382 884 2	0.049 501 246 340 1	1.507 426 738 64[5]
1.0	1.015 754 976 048 4	0.643 664 253 878	0.441 796 632 103 3	9.510 085 662 1 [-1]
10.0	−2.903 724 375 687	1.422 070 172 936	0.929 472 251 212	2.711 [-12]

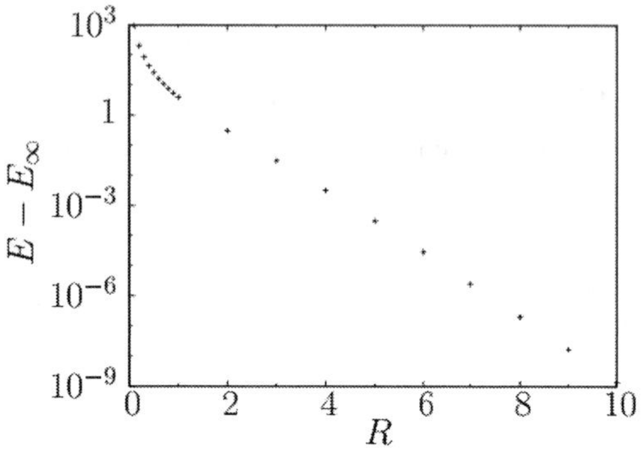

Figure 1. Difference between the ground-state energies of confined helium (E) and of free helium (E_∞) as a function of the confinement radius R. Atomic units are used.

calculated from the radius dependence of the energies $E(R)$ by the formula [2–4],

$$P = -\frac{1}{4\pi R^2} \frac{dE}{dR}, \tag{4}$$

where the derivative is evaluated by a finite-difference formula. A similar accuracy is obtained for the energy and interparticle distances of the first excited singlet level and the lowest triplet level [1]. Although the Lagrange-mesh method can be applied easily for small and large radii, it is particularly efficient for small radii (smaller variational basis size and better accuracy). The dependence of the ground-state energy on the confinement radius R is shown graphically in Fig. 1. For large confinement radii ($R \gtrsim 2$ fm), the confined helium energy tends nearly exponentially to the free helium energy. A parametrization of this curve could allow an easy evaluation of the pressure at very large radii, difficult to reach by direct calculations.

Acknowledgements

This text presents research results of the interuniversity attraction pole programme P7/12 initiated by the Belgian-state Federal Services for Scientific, Technical and Cultural Affairs. TRIUMF receives funding via a contribution through the National Research Council Canada.

References

[1] D. Baye, J. Dohet-Eraly, Phys. Chem. Chem. Phys. (2015), DOI:10.1039/C5CP00110B

[2] N. Aquino, A. Flores-Riveros, J. Rivas-Silva, Phys. Lett. A **307**, 326 (2003)

[3] C. Laughlin, S.I. Chu, J. Phys. A **42**, 265004 (2009)

[4] S. Bhattacharyya, J.K. Saha, P.K. Mukherjee, T.K. Mukherjee, Phys. Scripta **87**, 065305 (2013)

[5] D. Baye, P.-H. Heenen, J. Phys. A **19**, 2041 (1986)

[6] M. Hesse, D. Baye, J. Phys. B **32**, 5605 (1999)

[7] D. Baye, Phys. Rep. **565**, 1 (2015)

EPJ Web of Conferences **113**, 08005 (2016)
DOI: 10.1051/epjconf/201611308005

Non-Symmetrized Hyperspherical Harmonics Method Applied to Light Hypernuclei

F. Ferrari Ruffino[1,2,a], N. Barnea[3], S. Deflorian[1,2], W. Leidemann[1,2], and G. Orlandini[1,2]

[1] *Dipartimento di Fisica, Università di Trento, Via Sommarive, 14 I-38123 Trento, Italy*
[2] *Istituto Nazionale di Fisica Nucleare, TIFPA, Via Sommarive, 14 I-38123 Trento, Italy*
[3] *Racah Institute of Physics, Hebrew University, 91904, Jerusalem, Israel*

Abstract. We have adapted the non-symmetrized hyperspherical harmonics method (NSHH) in order to treat light hypernuclei. In the past the method has been applied in the atomic and nuclear context dealing with identical particle systems exclusively. We have generalized and optimized the formalism in presence of two different species of particles, namely nucleons and hyperons. Preliminary benchmark results with a modern realistic 2-body nucleon-hyperon interaction are provided.

1 Introduction

Despite the recent development of hypernuclear physics both in theory and experiment, the actual knowledge of the hyperon-nucleon (YN) interaction is limited and the amount of data available to define a realistic YN potential is far from being comparable to the NN case: the standard set employed in the modern hyperon-nucleon interactions comprises 35 selected Λp low energy scattering data [2] and a few YN data at higher energies, against the over 4000 NN data in the range $0 \div 350$ MeV. Quite a variety of potential models involving hyperons exist, ranging from purely phenomenological ones [3] to models based on chiral effective field theory [4]. It is evident the necessity of testing the quality of these interactions by comparing experimental and theoretical results, like Λ separation energies of hypernuclei. For such a check ab initio calculations are fundamental, because their results only depend on the chosen interaction, leading to a clearcut conclusion about the validity of the chosen model.
Our recent progress lies in this direction, providing a new technique to perform ab-initio calculations in the hypernuclear sector, which presents a few new features compared to the ordinary nuclear case like, for example, the presence of different particles. As interaction model we have chosen the potential defined in [1], which includes $\Lambda - \Sigma$ baryon mixing.

2 NSHH method for hypernuclei

The NSHH method is based on the formalism developed by M. Gattobigio et al. in [5] and later extended by S. Deflorian et al. in [6]. The wave functions are expressed in hyperspherical Jacobi coordinates and the chosen basis of the Hilbert space is composed by a hyperradial part (we use Laguerre

[a] e-mail: f.ferrariruffino@unitn.it

Table 1. Separation energies B_Λ in MeV for different hypernuclei and interactions. S stands for the NSC97f YN interaction as in [1]. S' and S" differ from S by a factor in the mixing part of the potential.

$\mathbf{V_{NN} + V_{YN}}$	**Hypernucleus**	**NSHH**	**GEM**
AV8'+S	$^3_\Lambda$H	0.19	0.19
AV8'+S'	$^3_\Lambda$H	0.35	0.36
AV8'+S"	$^3_\Lambda$H	0.69	0.72
AV8'+S	$^4_\Lambda$H	2.29	2.33

polynomials), a hyperspherical part (the hyperspherical harmonics functions) and the spin-isospin part. The basis has no good behaviour under particle permutations, so the real problem is the selection of the physical states we are looking for (fermionic in our case). This is done by using one specific property of the Casimir operator $\hat{C}(A)$ of the permutation group S_A, defined as: $\hat{C}(A) = \sum_{j>i=1}^{A} \hat{P}_{ij}$, where $\hat{P}_{ij}$ is the transposition operator for particles i and j. In fact the operator $\hat{C}(A)$ commutes with the Hamiltonian and its largest and smallest eigenvalues, λ_s and λ_a respectively, correspond to the symmetric and antisymmetric eigenfunctions. Therefore we consider a pseudo-hamiltonian defined as: $\tilde{H} = H + \gamma\hat{C}(A)$. Any eigenvalue $\tilde{E}_n^x$ of $\tilde{H}$ is given by $E_n^x + \gamma\lambda_x$ with $x = s, m, a$ (m stands for mixed symmetry). Since we are mainly interested in the ground state of a fermionic system we just have to put γ large enough to make $\tilde{E}_0^a$ the lowest eigenvalue of $\tilde{H}$ and obtain E_0^a by subtracting the quantity $\gamma\lambda_a$. In this way we do not need the exact diagonalization of $\tilde{H}$, but we can use much faster procedures to solve the new eigenvalue problem only for the lowest values (e.g. the Lanczos algorithm).
When we deal with hypernuclei, our system is composed by two species of particles, so the Casimir operator must act independently on the two subsets. It splits then into two terms: $\hat{C}(A) = \hat{C}_{NN}(A - n_Y) + \hat{C}_{YY}(n_Y)$, where n_Y is the number of hyperons. If $n_Y = 1$ obviously $\hat{C}_{YY} = 0$.
We then add an effective 2-body interaction procedure [7] based on Lee-Suzuki theory in order to consider a wide range of potentials which cannot be treated properly with a bare interaction approach.

3 Results

In Table 1 we show the results obtained with the AV8' NN potential and the NSC97f simulated YN interaction [1]. Since the latter model uses the Σ baryon degree of freedom explicitly, it was necessary to extend part of the NSHH formalism. We show a comparison between our results and the ones obtained with the Gaussian Expansion Method (GEM) by E. Hiyama et al. in [1].
A more complete and extensive benchmark calculation is in progress, comparing five different ab initio methods: the present NSHH, the GEM (with E. Hiyama, Riken) the No Core Shell Model (with D. Gazda, ECT*), the Auxiliary Field Diffusion Monte Carlo (with D. Lonardoni, Argonne) and the Faddeev-Yakubowsky method (with A. Nogga, Jülich).

References

[1] E. Hiyama, S. Ohnishi, B.F. Gibson, Th.A. Rijken, Phys. Rev. C **89**, 061302 (2014).
[2] H-J. Schulze, T. Rijken, Phys. Rev. C **88**, 024322 (2013).
[3] A.A. Usmani, F.C. Khanna, J.Phys.G **35**, 025105 (2008).
[4] J. Haidenbauer et al., Nucl. Phys. A **915**, 24 (2013).
[5] M. Gattobigio, A. Kievsky, M. Viviani, Phys. Rev. C **83**, 024001 (2011).
[6] S.Deflorian, N. Barnea, W. Leidemann, G. Orlandini, Few-Body Syst. **54**, 1879 (2013).
[7] N. Barnea, W. Leidemann, G. Orlandini, Nucl. Phys. A **693**, 565 (2001).

EPJ Web of Conferences **113**, 08006 (2016)
DOI: 10.1051/epjconf/201611308006

Λnn bound state with three-body potential

Igor Filikhin[1,a], Vladimir Suslov[1], and Branislav Vlahovic[1]

[1]North Carolina Central University, 1801 Fayetteville St., Durham, NC, USA

Abstract. Formalism of the configuration-space Faddeev equations is applied to analyze the bound state problem of ΛNN system. The simulated NSC97f potential for ΛN interaction and semi-realistic NN potentials are used for calculations. We predict a broad Λnn resonance near threshold. A spin dependent three-body ΛNN potential has been proposed to explain experimental evidence for ${}^3_\Lambda n$ bound state. The potential is defined by a Gaussian function of a hyper-radius. It was found that such potential with appropriate spin dependence does not support a bound state of ${}^3_\Lambda n$. A short range form of the potential generates non-physical solutions of the Faddeev equations.

The HypHI Collaboration has recently reported evidence for a bound state of $\Lambda + n + n$ system [1]. However, the bound ${}^3_\Lambda n$ state has not been found under theoretical analysis (see for instance [2]). In the present work we construct a phenomenological three-body ΛNN potential with spin dependence that forms a bound state in the channel $\Lambda + n + n$ (T=1, S=1/2). Such potential does not affect the ground state energy of ${}^3_\Lambda$H hypernucleus fixed by the experimental value of 0.13(5) MeV for the hyperon separation (or binding) energy

Formalism of the configuration-space Faddeev equations was applied for $\Lambda + n + n$ and $\Lambda + n + p$ systems within an s-wave approach. The simulated NSC97f potential [3] for ΛN interaction and MT-I-III semi-realistic NN potential [4] were used for calculations. This model reproduces well the hyperon binding energy for ${}^3_\Lambda$H, with the value of 0.14 MeV (see also Ref. [5]). We have found a broad resonance in the channel $\Lambda + n + n$. The results of the calculations are presented in figure 1a). The method of analytical continuation in the coupling constant is used to calculate the parameters of the resonance [6]. The coupling constant is the coefficient γ which scales triplet component of the ΛN potential. The complex value of the Pade approximant calculated by set of negative energies (circles in figure 1a) for γ=1 gives the energy and width of resonance: $E(\gamma = 1) = E_r - i\frac{\Gamma}{2}$. Our result for the energy of the resonance is 0.2 MeV, that agrees with [7]. Within the presented model, the three-body ΛNN potential is treated as a perturbation of the Hamiltonian. It is defined as one range Gaussian: $V_{3bf}(\rho) = -\delta \exp(-\alpha\rho^2) S(s_\Lambda, s_{NN})$, where ρ is the hyper-radius: $\rho^2 = x^2 + y^2$, with x, y the mass scaled Jacobi coordinates. The function $S(s_\Lambda, s_{NN})$ depends on the spin variables of a hyperon (s_Λ) and NN pair (s_{NN}). Two free parameters δ and α have to be adjusted. We assume the ratio $S(1/2, 1)/S(1/2, 0)$ to be equal 1 or 1/3 [2]. The calculations for α=0.1 fm^{-2} are shown in figure 1b). The three-body ΛNN potential resulted in a bound state of ${}^3_\Lambda n$ generates Λnp over-bound states for S=1/2 and 3/2. The spin dependence, which provides a bound state of ${}^3_\Lambda n$ and keeps the ${}^3_\Lambda$H ground state energy, cannot be considered as realistic. It was found that the short range form (when α>0.1 fm^{-2}) of the potential generates non-physical solutions of the Faddeev equations (see figure 2).

[a]e-mail: ifilikhin@nccu.edu

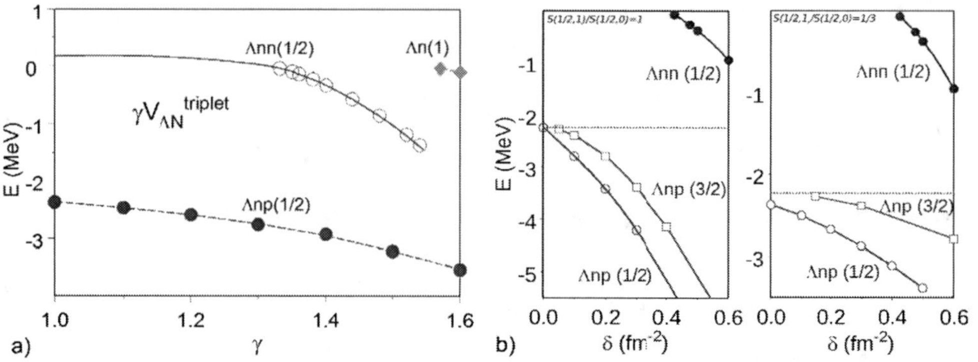

Figure 1. a) Bound state $\Lambda + n + n$ and $\Lambda + n + p$ energies (circles) calculated for different values of γ. The line for Λnn represents the real part of the Pade approximant. The resonance energy corresponds to the value of the approximant for γ=1. The three-body potential is not taken into account. The energy is measured from the $\Lambda + n + n$ threshold. b) $\Lambda + n + n$ and $\Lambda + n + p$ energies (solid and open circles) calculated for different values of δ (α=0.1 fm^{-2}). The δ parameter is scaled by the value $\hbar^2/m$, m is mass of nucleon. (Left) $S(1/2, 1)/S(1/2, 0)$=1, (Right) $S(1/2, 1)/S(1/2, 0)$=1/3.

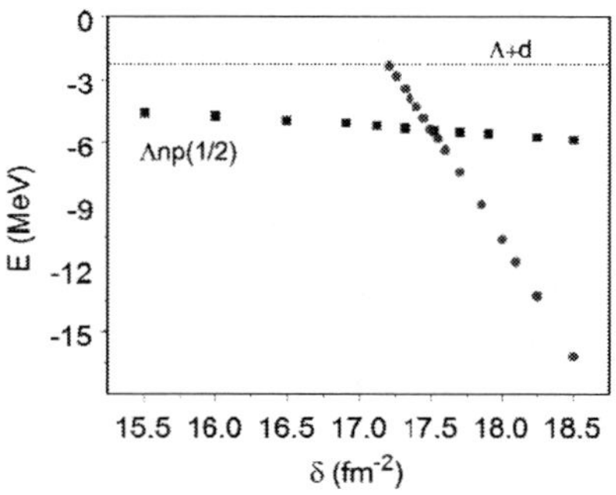

Figure 2. Bound state $\Lambda + n + p$ energies calculated for different values of the δ parameter of the three body potential (α=0.9 fm^{-2}). The squares (circles) correspond to physical (non-physical) solution of the Faddeev equations. $\Lambda + d$ threshold is shown by horizontal dashed line.

We have found a broad Λnn resonance near threshold with the energy of 0.2 MeV and the width of 2 MeV. It is not possible to construct a three-body potential for both Λnn and Λnp systems with reasonable spin dependence under the condition of existence of a ${}^{3}_{\Lambda}n$ bound state. The fact of the appearance of non-physical solutions reflects the known problem of implementation of three-body force into the Faddeev equation formalism.

The authors thank Prof. Satoshi N. Nakamura for useful discussion. This work is supported by NSF (HRD-1345219) and NASA (NNX09AV07A).

References

[1] C. Rappold et al. [HypHI Collaboration] Phys. Rev. C **88** 041001(R) (2013)
[2] A. Gal, H. Garcilazo, Phys. Lett. B **736** 93 (2014)
[3] I. Filikhin and A. Gal, Phys. Rev. Lett. **89** 172502 (2002)
[4] R.A. Malfiet and J.A. Tjon, Nucl. Phys. A **127** 161 (1969)
[5] I. Filikhin, V.M. Suslov and B. Vlahovic, J. Phys. G **31** 389 (2005)
[6] I. Filikhin, V.M. Suslov, B. Vlahovic, Few-Body Systems **50** 255 (2011)
[7] V.B. Belyaev, S.A. Rakityansky, W. Sandhas, Nucl. Phys. A **803** 210 (2008)

EPJ Web of Conferences **113**, 08007 (2016)
DOI: 10.1051/epjconf/201611308007

Design of an Experiment to Measure a_{nn} Using $^3H(\gamma, pn)n$ at HIγS*

F.Q.L. Friesen[1,a], M.W. Ahmed[2], B.J. Crowe[2], A.S. Crowell[1], L.C. Cumberbatch[1], B. Fallin[1], Z. Han[1], C.R. Howell[1], R.M. Malone[1], D. Markoff[2], W. Tornow[1], and H. Witała[3]

[1]*Department of Physics, Duke University and Triangle Universities Nuclear Laboratory, Durham, NC, USA*
[2]*Department of Mathematics and Physics, North Carolina Central University and Triangle Universities Nuclear Laboratory, Durham, NC, USA*
[3]*M. Smoluchowski Institute of Physics, Jagiellonian University, Cracow, Poland*

Abstract. We provide an update on the development of an experiment at TUNL for determining the $^1S_0$ neutron-neutron (nn) scattering length (a_{nn}) from differential cross-section measurements of three-body photodisintegration of the triton. The experiment will be conducted using a linearly polarized gamma-ray beam at the High Intensity Gamma-ray Source (HIγS) and tritium gas contained in thin-walled cells. The main components of the planned experiment are a 230 Ci gas target system, a set of wire chambers and silicon strip detectors on each side of the beam axis, and an array of neutron detectors on each side beyond the silicon detectors. The protons emitted in the reaction are tracked in the wire chambers and their energy and position are measured in silicon strip detectors. The first iteration of the experiment will be simplified, making use of a collimator system, and silicon detectors to interrogate the main region of interest near 90° in the polar angle. Monte-Carlo simulations based on rigorous 3N calculations have been conducted to validate the sensitivity of the experimental setup to a_{nn}.

1 Introduction

The goal of the experiment is to determine a_{nn} to an uncertainty less than ±1 fm. This will be the first determination of this quantity using a photodisintegration reaction. The initial measurements will be carried out with a 15-MeV gamma-ray beam where the total cross section is at its maximum [1]. The measurements will be performed with sufficient accuracy to resolve discrepancies in existing three-body photodisintegration cross-section data. Witała and Glöckle showed that increasing the strength of the nn S-wave force in the CD-Bonn potential by about 10% could resolve the standing discrepancy observed in nn quasi-free scattering in nd breakup [2]. Our data for the $^3H(\gamma, pn)$ reaction will provide another determination of the strength of the nn S-wave interaction. Also, data for the $^3H(\gamma, p)$, $^3H(\gamma, pn)$, $^3He(\gamma, p)$, and $^3He(\gamma, pp)$ reactions provide assessments of ab initio 3N calculations, including tests for long-range 3N force effects.

*This research supported in part by the DOE Office of Nuclear Physics Grant Number DE-FG02-97ER41033
[a]e-mail: fqf@tunl.duke.edu

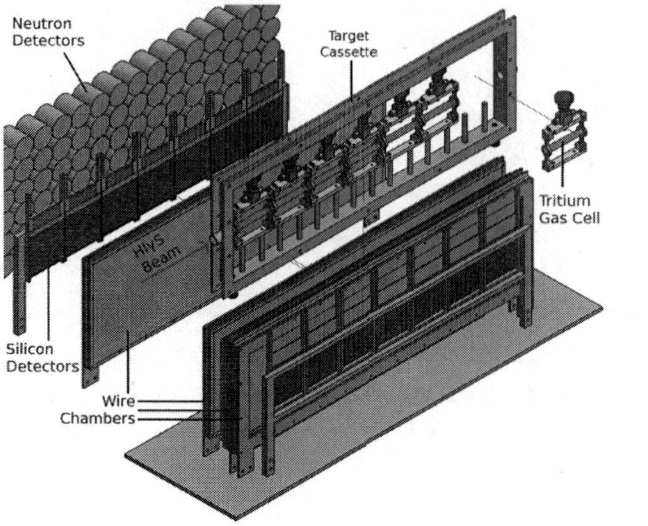

Figure 1. Diagram of the experimental setup for tracking-based measurements of ^{3}H and ^{3}He photodisintegration differential cross-sections. The gas targets are mounted to a central open-sided structure (target cassette) that lies along the beam axis. Two planes of wire chambers followed by a plane of large-area silicon strip detectors are on each side of the targets. The setup is about 80 cm long x 20 cm tall x 30 cm wide. A leak-tight stainless steel box will enclose the entire apparatus and provide additional containment of the tritium gas, as well as electrical and optical shielding for the detectors. A ^{3}H scrubbing system (not shown) will connect to the internal volume of the target cassette through the bottom plate.

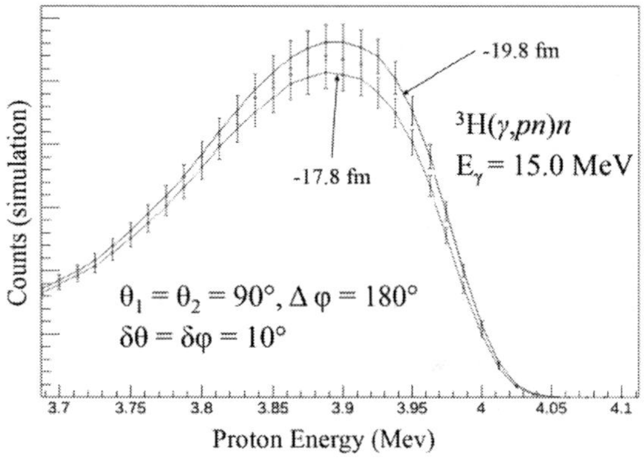

Figure 2. The predicted scattering length sensitivity for coincident detection of a proton and a neutron assuming a single gamma energy. The difference in azimuthal angle between the particles has been constrained to 175° ± 5°. The polar (θ) angles have been constrained to 90° ± 5°. This figure includes propagation of particles through the tracking apparatus. Error bars are 3% for reference.

2 Experimental Setup

The collimator system to be used in the first iteration of the experiment will constrain the trajectories of the emitted charged particles. Neutrons detected in coincidence will be used to identify 2-body events kinematically. Cross sections measured with the tracking system will be determined with uncertainties better than ±5% in the angle range from 45° to 135° relative to the beam axis and for an azimuthal angular acceptance of approximately ±60°.

The main kinematic configuration of interest for a_{nn} measurements is the nn final state interaction, in which the proton and unbound pair of neutrons are emitted colinearly in opposite directions, such that the proton energy is maximal. This configuration will be isolated using measured charged particle energy, time differences between detection of particles, and relative locations of detected particles.

References

[1] R. Skibinski et al. Phys. Rev. C 67, 054002 (2003)
[2] H. Witała and W. Glöckle, Phys. Rev. C 83, 034004 (2011)

EPJ Web of Conferences **113**, 08008 (2016)
DOI: 10.1051/epjconf/201611308008

Three Dimensional SRG Evolution of the NN Interactions Using Picard Iteration

M. R. Hadizadeh[1,a], K. A. Wendt[2,3], and Ch. Elster[1]

[1] *Institute of Nuclear and Particle Physics and Department of Physics and Astronomy, Ohio University, Athens, OH 45701, USA*
[2] *Department of Physics and Astronomy, University of Tennessee, Knoxville, TN 37996, USA*
[3] *Physics Division, Oak Ridge National Laboratory, Oak Ridge, TN 37831, USA*

Abstract. The Similarity Renormalization Group (SRG) evolution of nucleon-nucleon (NN) interactions is calculated directly as function of momentum vectors for realistic potentials. To overcome the stiffness of the SRG flow equations in differential form for far off diagonal matrix elements, the differential equation is transformed to an integral form without employing a partial wave decomposition.

The SRG evolution of NN interaction in a three-dimensional representation, without using a partial wave decomposition, can be obtained by solving the differential flow equation [1]

$$\begin{aligned} \frac{dV^{\pi ST}_{s\Lambda\Lambda'}(\mathbf{p},\mathbf{p}')}{ds} &= -\left(\frac{p^2}{m}-\frac{p'^2}{m}\right)^2 V^{\pi ST}_{s\Lambda\Lambda'}(\mathbf{p},\mathbf{p}') \\ &+ \frac{1}{4}\sum_{\Lambda''=-1}^{\Lambda''=+1}\int d^3p''\left(\frac{p^2}{m}+\frac{p'^2}{m}-\frac{2p''^2}{m}\right)V^{\pi ST}_{s\Lambda\Lambda''}(\mathbf{p},\mathbf{p}'')\,V^{\pi ST}_{s\Lambda''\Lambda'}(\mathbf{p}'',\mathbf{p}'). \end{aligned} \tag{1}$$

The input to this formulation is an operator form of the NN interaction, where the total spin of the system, S, is treated in a helicity representation [2]. Dependent on the value of S, one needs to solve a single equation for $S=0$ and four coupled equations for $S=1$. The non-linear SRG differential equations can become extremely stiff and take a prohibitively long time to evolve to large values of the flow parameter s. To overcome this problem, we transform the differential equation to an integral form via a Picard ansatz and solve it iteratively. By integrating both sides of the differential equation (1) over the flow parameter s from zero to a specific value s, the SRG flow evolution can be written schematically as

$$V_s = V_{NN} + A\int_0^s ds\, V_s + B\int_0^s ds\, V_s\, V_s, \tag{2}$$

which can be solved iteratively as

$$V_s^{(n)} = V_{NN} + A\int_0^s ds\, V_s^{(n-1)} + B\int_0^s ds\, V_s^{(n-1)}\, V_s^{(n-1)}; \quad n=1,2,... \tag{3}$$

The iteration is started with the bare NN interaction V_{NN}. The SRG integral equations are successfully solved for the Bonn-B and a Chiral N3LO potential and evolved to a large flow parameter s, or small

[a] e-mail: hadizadm@ohio.edu

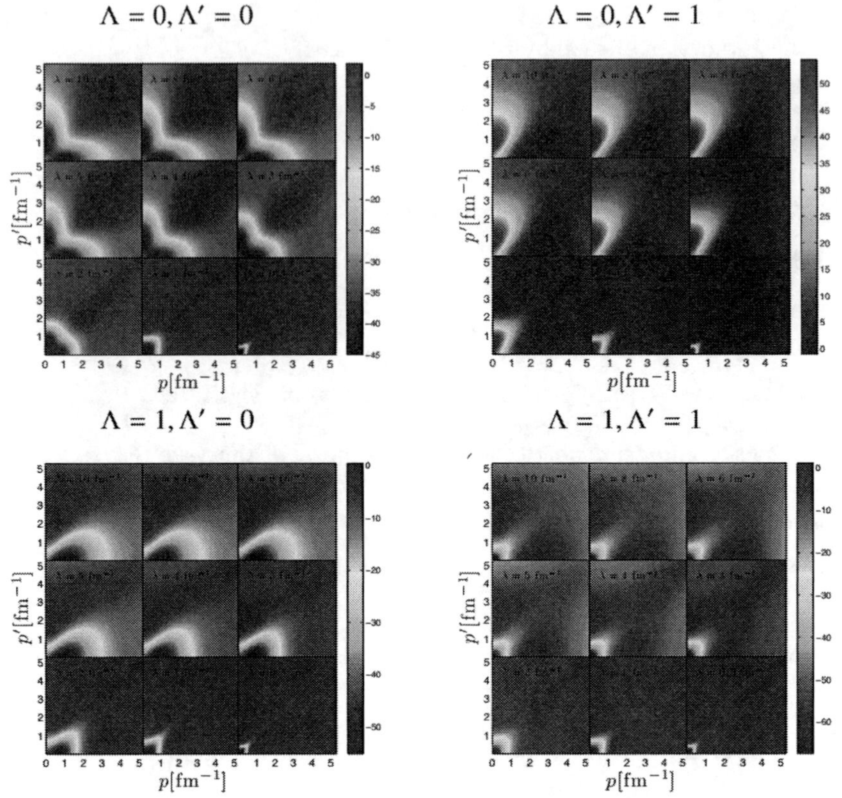

Figure 1. The matrix elements of evolved NN interaction $V^{\pi ST}_{s\Lambda\Lambda'}(p, p', x = \frac{1}{2})$ (in units MeV fm^3) calculated from Bonn-B potential by solving 3D form of SRG integral equation using Picard iteration method.

$\lambda = s^{-1/4}/\sqrt{\frac{\hbar^2}{m}} = 0.5\ \text{fm}^{-1}$. In Fig. 1 we show the matrix elements of the evolved Bonn-B interaction, $V^{\pi ST}_{\lambda\Lambda'\Lambda}(p, p', x = \frac{1}{2})$, for total spin $S = 1$, isospin $T = 0$, parity $\pi = 1$ and different helicity eigenvalues $\Lambda', \Lambda = 0, 1$. The successful implementation of the 3D evolution of NN interactions paves the path to consider a 3D evolution of three-nucleon forces.

Acknowledgements

This work was performed under the auspices of the National Science Foundation under Contract No. NSF-PHY-1005587 with Ohio University, the U. S. Department of Energy, Office of Science, Office of Nuclear Physics, under Contracts No. DE-FG02-93ER40756 with Ohio University, DEFG02-96ER40963 with the University of Tennessee and DE-AC05- 00OR22725 with Oak Ridge National Laboratory. Partial support by the Institute of Nuclear and Particle Physics at Ohio University is acknowledged. We thank the Ohio Supercomputer Center (OSC) for the use of their facilities under Grant No. PHS0206.

References

[1] M. R. Hadizadeh, K. A. Wendt, and Ch. Elster, in preparation.

[2] I. Fachruddin, Ch. Elster, W. Glöckle, Phys. Rev. C **62**, 044002 (2000).

EPJ Web of Conferences **113**, 08009 (2016)
DOI: 10.1051/epjconf/201611308009

Monte-Carlo Simulation of ^{3}H(γ, pn)n and ^{3}He(γ, pp)n Experiments at HIγS*

Z. Han[1,a], F.Q.L. Friesen[1], C.R. Howell[1], M.W. Ahmed[2], B.J. Crowe[2], A.S. Crowell[1], L.C. Cumberbatch[1], B. Fallin[1], D. Ticehurst[1], W. Tornow[1], and H. Witała[3]

[1] *Department of Physics, Duke University and Triangle Universities Nuclear Laboratory, Durham, NC, USA*
[2] *Department of Physics, North Carolina Central University and Triangle Universities Nuclear Laboratory(TUNL), Durham, NC, USA*
[3] *M. Smoluchowski Institute of Physics, Jagiellonian University, Cracow, Poland*

Abstract. We are developing an experiment to measure the two and three-body (γ, p) differential cross sections (DCS) for ^{3}H and ^{3}He. These data will be used to determine the $^1S_0$ nn scattering length (a_{nn}) and np scattering length (a_{np}) respectively. This paper describes features of the Monte-Carlo (MC) simulation that will aid in the optimization of the experimental design and the data analysis approach.

1 Introduction

Experiments to measure the differential cross sections for photodisintegration of ^{3}H and ^{3}He are being developed at the Triangle University Nuclear Laboratory (TUNL). The goal of this project is to provide data for assessing the theoretical treatment of meson-exchange currents in photodisintegration of nuclei and for investigating long-range features of three-nucleon interactions. Measurements will be performed using a linearly polarized gamma-ray beam at the High Intensity Gamma-ray Source (HIγS) at TUNL. The proton energy spectrum will be measured at several polar angles in the range from 45° to 135°. The first measurements will be performed with an incident γ-ray beam energy of 15 MeV. Higher energies may be investigated in future studies. In the conceptual experiment design, trajectories and energies of the emitted protons will be determined with wire chambers and silicon detectors located on either side of thin-walled gas targets.

2 Monte-Carlo Simulation Overview and Preliminary Simulation Results

The simulations are important for guiding the experiment design, the data acquisition strategy and for comparing theory to data for interpretation of results. The main focus of this work is to determine nucleon-nucleon scattering lengths from the photodisintegration data. The simulation is based on the GEANT4 toolkit. Using theoretical calculations by Witała and Skibinski [1,2], photodisintegration events are generated inside the gas targets, and the emitted protons are propagated from the

*This work is supported in part by the U.S. Department of Energy under grant Nos. DE-FG02-97ER41033 and DE-SC0005367 and by the Polish National Science Center under Grant No.DEC-2013/10/M/ST2/00420.
[a] e-mail: zh27@duke.edu

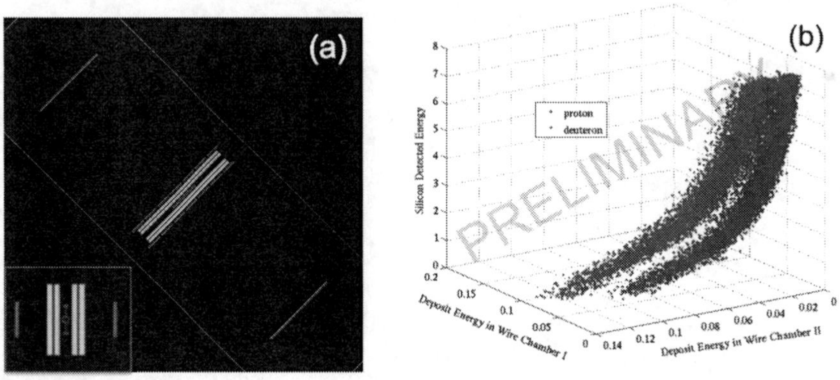

Figure 1. (a) Experiment geometry used in the simulation, includes two wire chambers (white), seven silicon strip detectors (green), and a neutron detector (blue) on each side. (b) Expected signal in silicon detector as a function of energy deposited in wire chambers without including finite detector resolution for protons (red) and deuterons (blue).

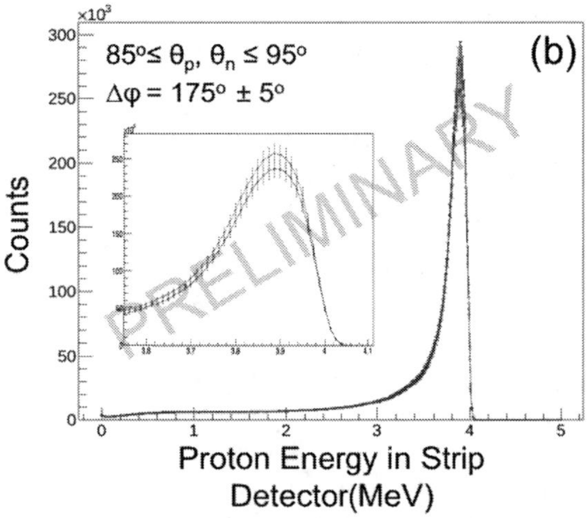

Figure 2. Simulated exclusive proton energy spectra for ^{3}H(γ, pn) of two different scattering length a_{nn}= -17.8fm (red) and a_{nn}= -19.8fm (blue) at the silicon detector. Events shown are constrained such that 85° $< \theta_p, \theta_n <$ 95°, and $\Delta\phi$ = 175° ± 5°. The error bars are 3% for reference.

reaction site through the detection apparatus. The simulations of the proton energy spectra will allow direct comparison between measurements and theory. Count rate and background estimations will be performed with simulations, which are used in experiment planning and data analysis.

The goals of our simulation are as following: (a) to aid in comparing the measurements to theory; (b) to analyze systematic uncertainties; (c) to estimate count rate of the reactions of interest and background for projecting statistical accuracy for determining a_{nn}; (d) to evaluate the use of $\Delta E/E$ for charged particle identification.

References

[1] H. Witała and W. Glöckle, Phys. Rev. C 83, 034004 (2011)
[2] R. Skibinski et al. Phys. Rev. C 67, 054002 (2003)

EPJ Web of Conferences **113**, 08010 (2016)
DOI: 10.1051/epjconf/201611308010

Separable Forces for (d, p) Reactions in Momentum Space

L. Hlophe[1], V. Eremenko[1,5,], Ch. Elster[1,a], F. M. Nunes[2,], I. J. Thompson[3,], G. Arbanas[4,], and J. E. Escher[3]

TORUS Collaboration[†] (http://reactiontheory.org)

[1] *Institute of Nuclear & Particle Physics and Department of Physics & Astronomy, Ohio University, Athens, OH, 45710, USA*
[2] *NSCL, Michigan State University, East Lansing, MI 48824, USA*
[3] *Lawrence Livermore National Laboratory L-414, Livermore, CA 94551, USA*
[4] *Nuclear Science and Technology Division, Oak Ridge National Laboratory, Oak Ridge, TN 37831, USA*
[5] *Institute of Nuclear Physics, Moscow State University, Moscow 119991, Russia*

Abstract. Treating (d, p) reactions in a Faddeev-AGS framework requires the interactions in the sub-systems as input. We derived separable representations for the neutron- and proton-nucleus interactions from phenomenological global optical potentials. In order to take into account excitations of the nucleus, excitations need to be included explicitly, leading to a coupled-channel separable representation of the optical potential.

The (d, p) scattering problem can be viewed as a three-body problem and thus be described exactly by the Faddeev equations, which are more readily solved in momentum space, specifically when transfer or break up are considered. However, when considering (d, p) reactions involving heavier nuclei, currently employed screening techniques for solving Faddeev equations with charged particles break down. At present, methods are developed to solve the Faddeev equations in the Coulomb basis, however, those rely on the short range forces being separable. For the neutron-proton subsystem separable potentials are readily available in literature. This is not true for nucleon-nucleus subsystems, for which interactions are often described by Woods-Saxon type optical potentials.

The method of deriving a separable representation of any arbitrary real potential proposed by Ernst, Shakin, and Thaler (EST) [1] is well suited. EST separable potentials have the property that at specific chosen energies the wavefunctions corresponding to the original potential and its separable representation are identical. In order to apply the EST method to optical potentials, it must be generalized to non-Hermitian potentials. Based on this generalized EST scheme neutron-nucleus optical potentials for ^{48}Ca, ^{132}Sn, and ^{208}Pb [2] were derived. Here a rank-5 separable interaction was sufficient to provide a good description of the neutron-nucleus scattering observables from 0 to 50 MeV.

For proton-nucleus optical potentials one has to consider the point Coulomb interaction, which is seen at large distances, and a short range Coulomb potential describing the charged nuclear sphere. In order to extend the EST scheme to include the point Coulomb interaction, one has to recast it in a Coulomb instead of a plane wave basis. This extended EST scheme was used to construct a separable representation of the proton-nucleus optical potential for ^{48}Ca and ^{208}Pb [3]. The cross-sections

[a] e-mail: elster@ohio.edu

corresponding to the CH89 proton-nucleus phenomenological potential [4] and its rank-5 separable representation are shown in Fig. 1. The good agreement between the two results illustrates the success of the EST scheme in reproducing on-shell properties of the proton-nucleus optical potentials.

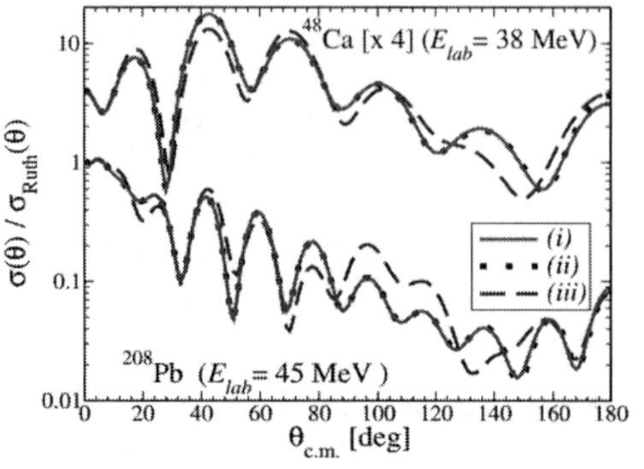

Figure 1. (color online) Unpolarized differential cross section for elastic scattering of protons from ^{48}Ca (upper) at 38 MeV laboratory kinetic energy and ^{208}Pb (lower) at 45 MeV divided by the Rutherford cross section as a function of the c.m. angle. The cross section for ^{48}Ca is multiplied by 4. The solid lines (*i*) depict the cross section calculated in momentum space based on the rank-5 separable representation of the CH89 [4] optical potential,while the dotted lines (*ii*) represent the corresponding coordinate space calculations. The dash-dotted lines (*iii*) show calculations in which the short-ranged Coulomb potential is omitted.

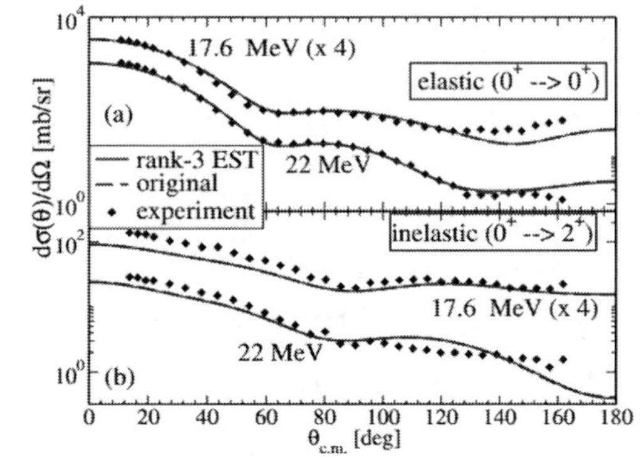

Figure 2. (color online) Unpolarized differential cross section for scattering of neutrons off ^{12}C. Panel (a) shows the elastic scattering cross section while the inelastic $0^+ \rightarrow 2^+$ cross section is shown in panel (b). The (blue) dashed line shows results obtained using the Olsson optical potential [5]. The (red) solid line indicates results obtained using a separable representation of the Olsson optical potential with EST points at 5, 16.5, and 45 MeV. Experimental data from Ref. [5] is depicted by black diamonds. The cross sections at 17.6 MeV are scaled up by a factor of four.

In order to develop a potential for a deformed nucleus, non-spherical contributions (excitations) need to be added. A separable rank-3 representation of the n+^{12}C Olsson [5] coupled-channels optical potential including the 2^+ and 4^+ states of ^{12}C was constructed. The EST support points in the 0 to 50 MeV energy regime were determined to be 5, 16.5, and 45 MeV. In Fig. 2 elastic and inelastic scattering cross sections computed using the Olsson potential (dashes) and its separable representation (solid line) are shown. We observe that the quality of the coupled-channels separable representation matches that of the single-channel case [2, 3].

Summarizing, we constructed separable representations of nuetron- and proton-nucleus phenomenological optical potentials, including non-spherical contributions, which can serve as input to Faddeev type (d, p) or (d, n) reaction calculations.

Acknowledgements

This work was performed in part under the auspices of the US Department of Energy, Office of Science of Nuclear Physics, under the topical collaborations in nuclear theory program No. DE-SC0004087 (TORUS Collaboration), and under Contract DE-FG02-93ER40756 with Ohio University.

References

[1] D.J. Ernst, C.M. Shakin, R.M. Thaler, Phys. Rev. **C9** 1780 (1974).
[2] L. Hlophe *et. al.* (TORUS Collab.), Phys. Rev. C **88**, 064608 (2013).
[3] L. Hlophe *et. al.* (TORUS Collab.), Phys. Rev. C**90**, 061602(R) (2014).
[4] R.L. Varner *et. al.*, Phys. Rept. **201** 57 (1991).
[5] B. Olsson *et. al.*, Nuclear Physics A**496**, 505 (1989).

EPJ Web of Conferences **113**, 08011 (2016)
DOI: 10.1051/epjconf/201611308011

Green's Function Monte Carlo Calculations with Two- and Three-Nucleon Interactions from Chiral Effective Field Theory

J. E. Lynn[1,a]

[1] *Theoretical Division, Los Alamos National Laboratory, Los Alamos, New Mexico 87545, USA*

Abstract. I discuss our recent work on Green's function Monte Carlo (GFMC) calculations of light nuclei using local nucleon-nucleon interactions derived from chiral effective field theory (EFT) up to next-to-next-to-leading order (N^2LO). I present the natural extension of this work to include the consistent three-nucleon (3N) forces at the same order in the chiral expansion. I discuss our choice of observables to fit the two low-energy constants which enter in the 3N sector at N^2LO and present some results for light nuclei.

1 Introduction

Accurate predictions of properties of light nuclei require at least two ingredients: (1) validated nuclear interactions and (2) accurate many-body computational methods. While quantum Chromodynamics is known to be the ultimate source of the first, arguably the best current path to validated nuclear interactions is through chiral effective field theory (EFT) [1, 2]. The second may come in many forms – hyperspherical harmonics, the no-core shell model, and Green's function Monte Carlo (GFMC) to name a few – but among the most accurate of these is the GFMC method. Recently we have demonstrated the powerful effect of combining these two ingredients (which was not possible before) [3] by studying properties of the lightest nuclei $A = 3, 4$ with chiral interactions up to N^2LO, neglecting however, the corresponding three-nucleon (3N) interaction which enters at N^2LO. Now we include the consistent 3N interaction arising at N^2LO in the chiral expansion, fitting the two low-energy constants which enter at this order to the ^{4}He binding energy and to n-α P-wave elastic scattering phase shifts [4].

2 The Fits

We fit the two low-energy constants c_D and c_E to ^{4}He binding energy Fig. 1a and to P-wave n-α scattering phase shifts Fig. 1b. Details about the various 3N interactions $V_C + V_D^{(i)} + V_E^{(j)}$, with $i = 1, 2, 3$ and $j = \tau, \mathbb{1}, P$, which amount to different operator choices for the terms proportional to c_D and c_E are available in [4]. It is significant that chiral interactions at N^2LO have sufficient freedom to fit both properties of light nuclei (Fig. 1a and Table 1) and the splitting of the P-wave n-α scattering phase shifts (Fig. 1b). See [4] for more details.

[a] e-mail: joel.lynn@gmail.com

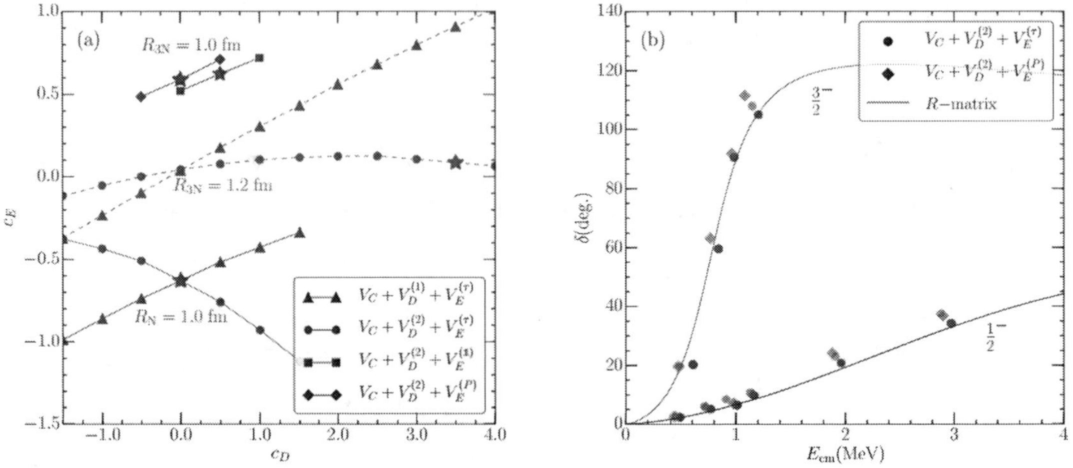

Figure 1. Fitting c_D and c_E. The GFMC statistical uncertainties are the size of the symbols or smaller. Blue (Red) symbols correspond to $R_{3N} = 1.0$ fm ($R_{3N} = 1.2$ fm), with R_{3N} a 3N cutoff parameter. (a) Curves of c_E vs. c_D obtained by fitting the ^{4}He binding energy. The fits were obtained at the points: The lines are a guide to the eye. The stars correspond to the values of c_D and c_E which fit both the ^{4}He binding energy and the n-α P-wave phase shifts. No fit to both observables can be obtained for the case with $R_{3N} = 1.2$ fm and $V_D^{(1)}$. See Ref. [4] for details. (b) P-wave n-α elastic scattering phase shifts compared with an R-matrix analysis of the data.

Table 1. Comparison of results for $A = 3$ in this work with experimental values. The values in parenthesis are the GFMC statistical errors.

	R_0 [fm]	E_B [MeV]		$\sqrt{\langle r_{pt}^2 \rangle}$ [fm]	
		GFMC	Exp.	GFMC	Exp.
^{3}H	1.0	-8.33(1)	-8.48	1.55(3)	1.59
	1.2	-8.35(4)		1.55(4)	
^{3}He	1.0	-7.66(1)	-7.72	1.77(2)	1.76
	1.2	-7.63(4)		1.77(1)	

Acknowledgements

This work was supported the U.S. Department of Energy, Office of Nuclear Physics, and by the NUCLEI SciDAC program.

References

[1] E. Epelbaum, H.-W Hammer, and Ulf-G. Meißner, Rev. Mod. Phys. **81**, 1773–1825 (2009)
[2] R. Machleidt, and D. R. Entem, Phys. Rep. **503**, 1–75 (2011)
[3] J. E. Lynn, J. Carlson, E. Epelbaum, S. Gandolfi, A. Gezerlis, and A. Schwenk, Phys. Rev. Lett. **113**, 192501 (2014)
[4] J. E. Lynn, I. Tews, J. Carlson, S. Gandolfi, A. Gezerlis, and A. Schwenk, in preparation (2015)

EPJ Web of Conferences **113**, 08012 (2016)
DOI: 10.1051/epjconf/201611308012

Non-relativistic Neutron Deuteron Scattering

A. Margaryan[1,a], J. Vanasse[1,2,b], and R.P. Springer[1,c]

[1] *Duke University, Durham, NC*
[2] *Ohio University, Athens, OH*

Abstract. We discuss the calculation of polarization observables, including A_y, in *nd* scattering to next-to-next-to-next-to-leading order in pionless effective field theory.

We present preliminary results of the first next-to-next-to-next-to-leading order (N^{3}LO) calculation of the *nd* scattering amplitude in the framework of nonrelativistic pionless effective field theory (EFT$_{\not\pi}$). In this theory, the typical momentum exchange in the scattering must be much smaller then the mass of the pion. The power counting parameter for EFT$_{\not\pi}$ is the ratio $\frac{Q}{\Lambda_{\not\pi}}$, where Q is the typical momentum exchange in the scattering and $\Lambda_{\not\pi}$ is the EFT$_{\not\pi}$ breakdown scale, $\Lambda_{\not\pi} \sim m_\pi$. The EFT$_{\not\pi}$ interaction terms in the two-body sector up to N^{3}LO are the two two-nucleon-to-dibaryon vertices (for the $^3S_1$ and $^1S_0$ channels) at LO, the effective range term at NLO, the SD-mixing interaction at N^{2}LO, and the shape parameter and two-body P-wave contact interaction terms at N^{3}LO. Three-body interaction terms enter first at LO and a new energy dependent term appears at N^{2}LO. The two-body interaction coefficients are matched onto *NN* scattering data. At LO the three-body interaction coefficient is matched onto the doublet S-wave *nd* scattering length and the N^{2}LO energy dependent three-body force to the triton binding energy.

The calculation of the amplitude for *nd* scattering requires summing an infinite set of diagrams. This sum does not factorize as it does in the two-body case; instead an integral equation must be solved numerically [1]. The nth order correction to the *nd* scattering amplitude is given by the integral equation shown in Fig. 1 [2–4]. An important part of this calculation is the two-body P-wave contact interaction diagram for *nd* scattering shown in Fig. 2. To solve the equation shown in Fig. 1 to a given order we project it onto different partial waves and then solve the projected equations numerically [5, 6]. After obtaining the numerical solution for the amplitude we are able to calculate any parity conserving observable of interest. One of the most important observables is A_y, which measures the asymmetry between the cross sections induced by nucleons of opposite transverse spin polarization on an unpolarized deuteron target [7]:

$$A_y = \frac{\frac{d\sigma}{d\Omega}|_\uparrow - \frac{d\sigma}{d\Omega}|_\downarrow}{\frac{d\sigma}{d\Omega}|_\uparrow + \frac{d\sigma}{d\Omega}|_\downarrow} \ . \tag{1}$$

Varying the N^{3}LO coefficients within the EFT$_{\not\pi}$ uncertainty gives the results in Fig 3, which shows good agreement with the differential cross section and reasonable agreement with A_y for a variety of energies. We can also pursue the calculation to higher orders to attempt to improve the agreement.

[a] e-mail: am343@duke.edu
[b] e-mail: jjv9@phy.duke.edu
[c] e-mail: rps@phy.duke.edu

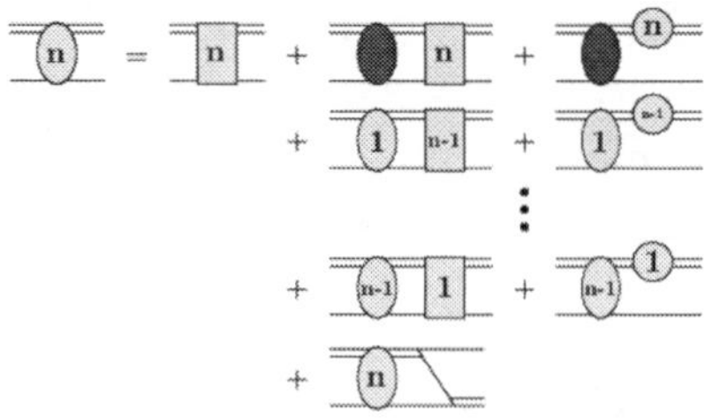

Figure 1. Thin line is nucleon propagator, double line is dibaryon propagator. For $m = 1, \ldots, n$ oval with an m is the mth order correction to the nd scattering amplitude, rectangle with an m is N^{m}LO corrections that involve all three-nucleons, and this includes three-body forces, circle with m is the mth order correction to the dibaryon propagator. The solid red oval is the LO nd scattering amplitude.

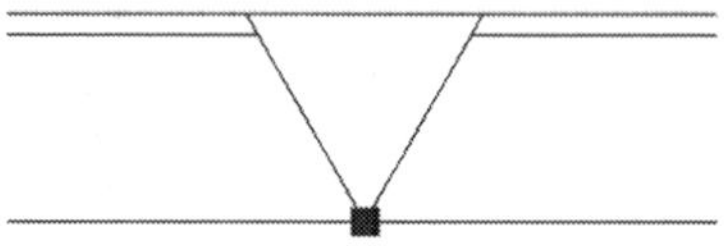

Figure 2. Diagram contributing at N^{3}LO. The square is a two-body P-wave interaction vertex. This diagram is included in Fig. 1 in the rectangle with $n = 3$.

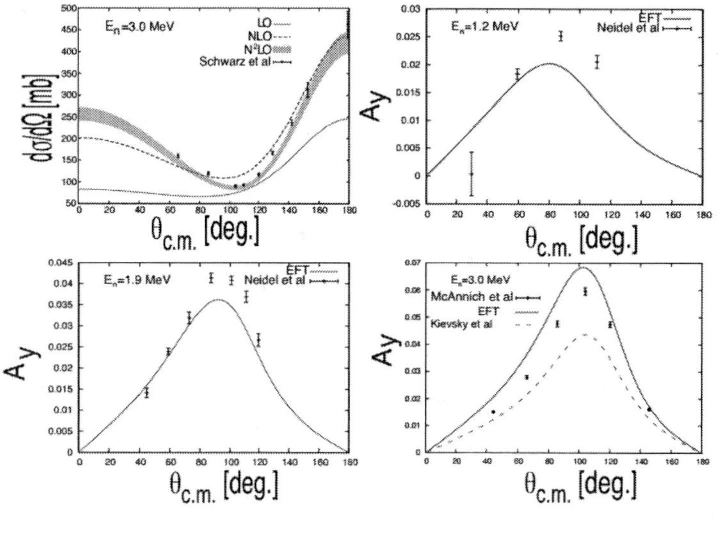

Figure 3. Top left is the EFT$_{\not\pi}$ results up to N^{2}LO and data [8] for the cross-section at a neutron lab energy of $E_n = 3.0$ MeV. The band on N^{2}LO result represents the estimated error of about 6%. On the other three figures the solid red line represents our preliminary N^{3}LO results for A_y at $E_n = 1.2$ and 1.9 MeV with data from [9] and at $E_n = 3.0$ MeV with data from [10] . On the bottom right figure the dashed green line comes from potential model calculations [11].

Acknowledgements

This material is based upon work supported by the U.S. Department of Energy Office of Science, Office of Nuclear Physics, under Award Number DE-FG02-05ER41368.

References

[1] G.V. Skornyakov, K.A.Ter-Martirosian, Sov. Phys. JETP **4**, 648 (1957)

[2] P.F. Bedaque, H. Hammer, U. van Kolck, Phys. Rev. C **58**, 641 (1998), nucl-th/9802057

[3] P.F. Bedaque, U. van Kolck, Phys. Lett. **B428**, 221 (1998), nucl-th/9710073

[4] P.F. Bedaque, H.W. Grießhammer, Nucl. Phys. A **671**, 357 (2000), nucl-th/9907077

[5] J. Vanasse, Phys. Rev. C **88**, 044001 (2013), 1305.0283

[6] J. Vanasse, "Perturbative Techniques in Three-Body Nuclear Systems," Proceedings of Few Body 21

[7] D. Huber, J.L. Friar, Phys. Rev. C **58**, 674 (1998), nucl-th/9803038

[8] P. Schwarz, H. Klages, P. Doll, B. Haesner, J. Wilczynski, B. Zeitnitz, J. Kecskemeti, Nuclear Physics A **398**, 1 (1983)
[9] E. Neidel, W. Tornow, D.G. Trotter, C. Howell, A. Crowell, R. Macri, R. Walter, G. Weisel, J. Esterline, H. Witała et al., Physics Letters B **552**, 29 (2003)
[10] J.E. McAninch, L.O. Lamm, W. Haeberli, Phys. Rev. C **50**, 589 (1994)
[11] A. Kievsky, M. Viviani, L.E. Marcucci, Few Body Syst. **54**, 2395 (2013)

EPJ Web of Conferences **113**, 08013 (2016)
DOI: 10.1051/epjconf/201611308013

The ^{3}H–^{3}He Charge Radii Difference

L. S. Myers[1,a], J. R. Arrington[2], and D. W. Higinbotham[1]

[1] *Thomas Jefferson National Accelerator Facility, Newport News, VA 23606, USA*
[2] *Physics Division, Argonne National Laboratory, Argonne, IL 60439, USA*

Abstract. The upcoming E12-14-009 [1] experiment at Jefferson Lab will determine the ratio of the electric form factors for the A=3 mirror nuclei ^{3}He and ^{3}H. The measurement will use a 1.1 GeV electron beam, a special collimator plate to allow for simultaneous optics measurements, and the low-activity tritium target being prepared for Jefferson Lab. By observing the dependence of the form factor ratio as a function of Q^2 over 0.05–0.09 GeV^2, the dependence of the radii extraction on the shape of the form factors is minimized. As a result, we anticipate the uncertainty of the extracted charge radii difference to be 0.03 fm, a reduction of 70% from the current measurement. Using precise measurements of the ^{3}He charge radius from isotopic shift or μHe measurements [2–4], we can deduce the absolute ^{3}H charge radius. The results will provide a direct comparison to recent calculations of the charge radii.

1 Introduction

The E14009 experiment [1] proposes to measure the ratio of the electric form factors of ^{3}He and ^{3}H over a range of Q^2 from 0.05 to 0.09 GeV^2. From this measurement, the relative charge radii for the A=3 nuclei can be determined. From this relationship, the experiment plans to extract the charge radius difference with an uncertainty of ~15% – a significant reduction from the current 50% uncertainty [5]. The JLab PAC42 approved the experiment for the requested 1.5 days of beam time.

2 Experimental Layout

This experiment will be a single-arm measurement of elastic electron scattering from ^{3}He and ^{3}H. It will utilize one of the high-resolution spectrometers in Hall A positioned at 12.5° and 15.0°. The setup is very similar to previous form factor measurements in Hall A, and the systematics are well-understood.

There are two unique features of the setup for E14009. First, the electron beam will be set to 1.1 GeV (the lowest achievable energy at the upgraded Jefferson Lab accelerator) and the current will be limited to only 5 μA. The energy is necessary to get down to Q^2 of 0.05–0.09 GeV^2. At these momentum transfers, the cross sections and scattering rates are large. The reduced current significantly reduces the effects of deadtimes in the data acquisition.

[a] e-mail: lmyers@jlab.org

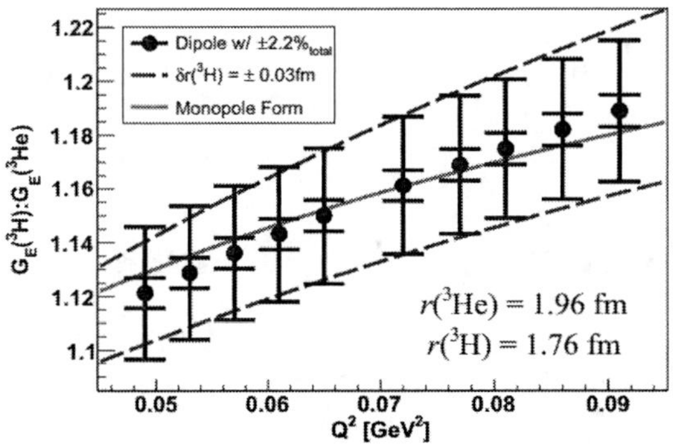

Figure 1. The expected results of the measurement of the charge form factor ratio of ^{3}He and ^{3}H. The ^{3}He radius is fixed to 1.96 fm [5]. The solid line assumes a ^{3}H radius of 1.76 fm with ±0.03 fm used to calculate the dashed lines.

The other unique feature of E14009 is the custom collimator plate to be placed on the spectrometer. This plate consists of a tapered center slot and rows of small holes above and below the slot. This design serves three purposes: (1) the center slot further reduces the event rate in the spectrometer, (2) the tapered design reduces the rate on the low-angle side thereby balancing the rates in each of the five Q^2 bins, and (3) the rows of holes above and below the center slot allow for in situ optics measurements.

3 Expected Sensitivity

The count rates for the elastic electrons are expected to be $\sim 10^5$ counts/hr per Q^2 bin. Data taking for each target will take 1.5 hrs at each HRS angle. As a result, the systematic uncertainties (~ 2%) will dominate the statistical (<0.5%). Within the systematic uncertainty, the largest contribution is expected to come from the relative thickness of the ^{3}H and ^{3}He targets. This uncertainty (~1.5-2.0%) will ultimately be determined using deep inelastic scattering from both targets.

The anticipated results for the experiment are shown in Fig. 1. Assuming a conservative estimate of the uncertainties, the ^{3}He–^{3}H can be determined to ~0.03 fm, a reduction of 70% from the current uncertainty. The model dependence of the radius extraction is constrained by examining the evolution of the form factor ratio over the full range of Q^2.

This measurement will also be able to extract the most precise value of the ^{3}H charge radius by using ^{3}He radii available from isotopic shift measurements [2, 3] and soon from muonic ^{3}He [4]. The proton and neutron matter radii in the A=3 nuclei can be determined from these results (assuming isospin symmetry) which can then be compared to ab initio calculations.

References

[1] L. S. Myers, D. W. Higinbotham, and J. R. Arrington, arXiv:1408.5283 [nucl-ex] (2014).
[2] D. Shiner, R. Dixson, and V. Vedantham, Phys. Rev. Lett. **74**, 3553 (1995).
[3] D. C. Morton, Q. Wu, and G. W. F. Drake, Phys. Rev. A **73**, 034502 (2006).
[4] T. Nebel *et al.*, Hyperfine Interactions **212**, 195 (2012).
[5] A. Amroun *et al.*, Nucl. Phys. A **579**, 596 (1994).

EPJ Web of Conferences **113**, 08014 (2016)
DOI: 10.1051/epjconf/201611308014

A New Treatment Below the Three-Body Break up Threshold in the NNπ System

Shinsho Oryu[1,a], Yasuhisa Hiratsuka[1,b] and Takashi Watanabe[1,c]

[1] *Department of Physics, Faculty of Science and Technology, Tokyo University of Science*

Abstract. The two-body threshold behavior at NN′ and πD are investigated by using the multi-channel Lippmann-Schwinger equations with an energy dependent two-body quasi potential, which are analytically continued from the three-body Faddeev equations at the three-body break up threshold. Our calculated NN′ and πD scattering lengths show better agreement with the experimental data for NN and πD systems than those from the original NNπ three-body Faddeev equations.

Lovelace's idea in the early 1960s that the two-body N+N→ N+N reaction should be described by the three-body ($N_2\pi$)+N_1 → N_2 + (πN_1) reaction [1], can be accomplished by the introduction of an energy dependent two-body quasi potential (E2Q) below the three-body break up threshold [2]. The three-body Faddeev equations for the NNπ system are analytically continued to the multi-channel two-body Lippmann-Schwinger (MLS) equations by the E2Q [2] where the P_{11} bound state of (Nπ)≡ N′ is normalized to a nucleon N in this approach. The denominator of the Born term in the three-body Faddeev formalism is given by $\omega_j(k_j) = \sqrt{k_j^2 + m_j^2}$ $(j = 1,\ 2,\ 3)$ with the pion mass $m_3 = m_\pi \equiv m$ and the nucleon mass $m_1 = m_2 \equiv M$, and the three-body total energy is $\sqrt{S}$,

$$D_{\rm Fadd} = \sqrt{S} - \omega_1(k_1) - \omega_2(k_2) - \omega_3(k_3) = \sqrt{S} + m - \omega_1(\bar{k}_1) - \omega_2(\bar{k}_2) - \omega_3(\bar{k}_3) \equiv D_{\rm E2Q}, \tag{1}$$

where $D_{\rm E2Q}$ is defined as the denominator of the E2Q. The non relativistic approximation for Eq.(1) gives,

$$D_{\rm E2Q} \approx (E + m) - \bar{k}_1^2/2m_1 - \bar{k}_2^2/2m_2 - (\bar{\mathbf{k}}_1 - \bar{\mathbf{k}}_2)^2/2m_3 \equiv E_{\rm cm} - \bar{k}_{1,2}^2/2\mu_{1,2} - \bar{z}_{1,2} \tag{2}$$

$$D_{\rm Fadd} \approx E - k_1^2/2m_1 - k_2^2/2m_2 - (\mathbf{k}_1 - \mathbf{k}_2)^2/2m_3 \equiv E - k_{1,2}^2/2\mu_{1,2} - z_{1,2} \tag{3}$$

where $\bar{k}_{1,2}$ indicates $\bar{k}_1$ or $\bar{k}_2$, and $E_{\rm cm} = E + m$ is the c.m. energy in the E2Q with $E = \sqrt{S} - 2M - m$, while $\mu_{1,2}^{-1} = M^{-1} + (M+m)^{-1}$ and $\mu_3^{-1} = m^{-1} + (2M)^{-1}$ are the reduced masses, respectively. Because the inter-nucleon $\bar{k}_{1,2}$ is obtained from $k_{1,2}$ by absorbing energy m, therefore we can define

$$\bar{k}_{1,2}^2/2\mu_{1,2} \equiv k_{1,2}^2/2\mu_{1,2} + m. \tag{4}$$

Substituting Eq.(4) to Eq.(2), and compairing with Eq.(3), we obtain $\bar{z}_{1,2} = z_{1,2}$ which are the two-body sub-energies $\bar{z}_1$ or $\bar{z}_2$ in the virtual three-body system. Summing up k_i^2 with respect to $i = 1, 2, 3$,

[a] e-mail: oryu@rs.noda.tus.ac.jp
[b] e-mail: yasuhisa1059@gmail.com
[c] e-mail: watanabe@ph.noda.tus.ac.jp

then we obtain $\bar{k}_3^2/2\mu_3 = k_3^2/2\mu_3 - m^2/2M$ where $\bar{k}_3$ is a virtual momentum for $\bar{k}_3^2 < 0$. Let us call the new MLS equations with the E2Q the *E2Q* equation [2].

In solving the E2Q and the Faddeev equations, the first difference occurs "at the NN′ threshold" where $E_{cm} = 0$ and the integral variable $0 \le \bar{k}_{1,2} < \infty$ in the E2Q equations, therefore $\{D_{E2Q}\}^{-1}$ has a singular cut, while in the original Faddeev equations, $E = -m$ and $0 \le k_i < \infty$, so that $\{D_{Fadd}\}^{-1}$ is a regular function. The second appears in Eq.(4), $0 \le \bar{k}_{1,2}^2 \le 2m\mu_{1,2}$ gives $-2m\mu_{1,2} \le k_{1,2}^2 \le 0$, therefore the integral range of $\bar{k}_{1,2}$ is larger than $k_{1,2}$. In addition, a phenomenon at the three-body break up threshold $E = 0$ gives rise to the Efimov effect [3], with an infinite two-body scattering length, where $E = 0$ brings about a singular cut which is the same as our $E_{cm} = 0$ case just mentioned above [2]. In a fourth difference, a phenomenon below the three-body threshold emerges as a long range NN′ interaction from an energy average of the E2Q [2]. We find that the calculated "NN′" and πD scattering lengths for the E2Q equations seem to be in better agreement with the experimental "NN" and πD data than the original three-body Faddeev calculations (Table 1).

Table 1. The "NN′" and πD scattering lengths are calculated using the original three-body Faddeev equation (Org-Fadd) and the E2Q calculation (E2Q). The potential-A is given by Thomas [4] and the potential-B is proposed by Fuda [5]. The E2Q results show good agreement with the experimental data both in the "NN" ($^3S_1$) and the πD cases [6, 7].

Method	Scattering length [fm]	System/State
Org-Fadd (potential-A[4])	0.280	NN′ $^3S_1$
Org-Fadd (potential-B[5])	2.85	NN′ $^3S_1$
E2Q (potential-B[5])	4.66	NN′ $^3S_1$
EXP [6]	5.424±0.004	NN $^3S_1$
Org-Fadd (potential-A[4])	0.033	πD
Org-Fadd (potential-B[5])	-0.019+0.019i	πD
E2Q (potential-B[5])	-0.023+0.019i	πD
EXP [7]	-0.038+0.009i	πD

Finally, it is stressed that the Hamiltonian *below* the three-body break up threshold is different from that *above* the threshold where the calculation should be carried out not with the "original" Faddeev equations but with the E2Q or modified Faddeev equations. This difference is emphasized in the NNπ system, because the two-body binding energy (*i.e.* the pion mass in the P_{11} state) is much larger than that in the nuclear three-body system, where such a discrepancy may be neglected.

References

[1] C. Lovelace, Phys. Rev. **135**, B1225-B1249 (1964).

[2] S. Oryu, Phys. Rev. **C86**, 044001-1-10 (2012); *ibid.*, Few-Body Syst, **54**,1-4, 283-286 (2013); Y. Hiratsuka, S. Oryu and T. Watanabe, Proc. of the 6th APFB Conf. (Adelaide 2014).

[3] V. Efimov, Phys. Lett. B**33** 563-564 (1970); *ibid.*, Nucl. Phys. **A210**, 157-188 (1973).

[4] A. W. Thomas, Nucl. Phys. **A258**, 417-446 (1976).

[5] M. G. Fuda, Phys. Rev. **C52**, 2875-2884 (1995).

[6] O. Dumbrajs *et al.*, Nucl. Phys. **B216**, 277-335 (1983).

[7] P. Hauser *et al.*, Phys. Rev. **C58**, R1869-R1872 (1998).

EPJ Web of Conferences **113**, 08015 (2016)
DOI: 10.1051/epjconf/201611308015

Accurate wave functions in the presence of long-ranged potentials

George Rawitscher

[1] *Physics Department, University of Connecticut, Storrs CT 06269, USA*

Abstract. For attractive potentials of the form $1/r^3$ or $1/r$ obtaining accurate wave functions at long distances can be numerically challenging because of the many oscillations involved. In this case high accuracy can be achieved using the phase-amplitude representation of the one-dimensional wave function.

For the cases where the attractive potentials are of the $1/r^3$ or $1/r$ forms (dipole dipole interaction between atomic molecules, or, the Coulomb potential, respectively), and also for the solution of three-body clusters with halo conditions, accurate wave functions at large distances are required. The usual procedure is to solve the Schrödinger equation, and follow the wave function $\psi(r)$, with all its many oscillations, from small distances to large ones, accumulating large errors in the process.

A way to bypass this difficulty is to resort to the phase-amplitude (Ph-A) representation of the one-dimensional wave function. This representation has the form $\psi(r) = y(r)\, sin[\phi(r)]$, where y and ϕ are the amplitude and phase, respectively. The main advantage is that both quantities are nearly monotonic functions of the distance r, hence their calculation requires far fewer mesh points than the wave function itself. One difficulty is that the differential equation for y, given by W. E. Milne in 1930 [1], has a non-linear term of the form $1/y^3$. That equation was subsequently solved iteratively by Seaton and Peach in 1963 [2] in the form

$$\frac{k}{y_{n+1}^2} = [w + D_n]^{1/2}, \quad n = 0, 1, 2, ... \tag{1}$$

where $D(r) = (d^2y/dr^2)/y$, and $w(r) = k^2 - V$. Here V is the central potential assumed to be attractive, and k is the wave number. The factor $\hbar^2/2m$ has already been taken into account for the potential and the energy $E = k^2$, so that both are now given in units of inverse length squared. Once y is known, the phase can be obtained according to $\phi(r) = k \int_0^r [y(r')]^{-2}\, dr'$. For $n = 0$ one sets $D_0 = 0$, and y_1 is the WKB approximation.

The present approach consists in expanding the amplitude y in terms of Chebyshev polynomials, $y(x) = \sum_{s=0}^{N} a_s T_s(x), \quad -1 \le x \le 1$. In order to avoid having to calculate the derivatives of y, the quantities D_n are obtained recursively by means of

$$D_{n+1} = \frac{5}{16}\frac{(V')^2}{(w + D_n)^2} + \frac{1}{4}\frac{V''}{(w + D_n)}, \quad n = 0, 1, 2, ... \tag{2}$$

This equation can be solved non-iteratively in terms of the roots of a cubic equation [3] which is very fast (by a factor 10-20) compared to the conventional methods of solving the Schrödinger

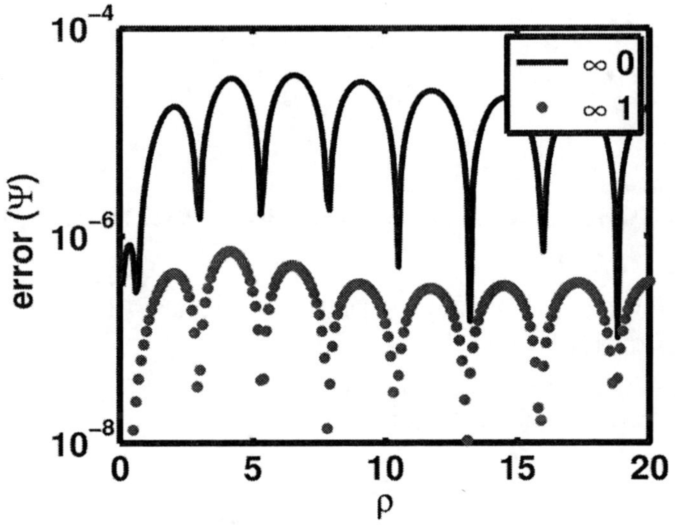

Figure 1. The accuracy of the wave function for a Coulomb potential with $\eta = -2$, calculated in the radial interval $[0, 200]$.The result labeled ∞ 0 is based on Eq. (2), while for ∞ 1 the first order derivative of D_n is included in that equation.

equation. If the derivatives of D are included in Eq. (2), then the accuracy of the Ph-A result is increased considerably, as is illustrated in Fig. 1 for the case of a Coulomb potential "rounded" [3] near $r = 0$, with $\eta = -2$ and $N = 100$, calculated in the radial $\rho = kr$ interval 0 to 200.

In summary, the computational time for the Ph-A wave function is much faster than for conventional methods, the number of support points is also much smaller, the asymptotic value of the amplitude is automatically equal to unity, and the Ph-A calculation can be carried out for long range potentials to large distances. A generalization to repulsive potentials is underway.

References

[1] W. E. Milne, Phys. Rev. 35 (1930) 863;

[2] M. J. Seaton and G. Peach Proc. Phys. Soc. 79 (1962) 1296 doi:10.1088/0370-1328/79/6/127;

[3] G. Rawitscher, *A Spectral Phase-Amplitude method for propagating a wave function to large distances,* Comp. Phys. Comm. **191** (2015) 33

EPJ Web of Conferences **113**, 08016 (2016)
DOI: 10.1051/epjconf/201611308016

The Breakup Cross Section of the D+D Reaction at 6.94 MeV

A.L. Richard[1,a], C.R. Brune[1], D.C. Ingram[1], S. Dhakal[1], A. Karki[2],
T.N. Massey[1], J.E. O'Donnell[1,b], and C.E. Parker[1]

[1] *Ohio University, Athens, OH 45701, USA*
[2] *University of Vermont, Burlington, VT 05405, USA*

Abstract. The D+D reactions are well known and widely used for a variety of purposes, mainly because of the use of the $D(d,n)^3He$ reaction as a mono-energetic neutron source. The least studied of the D+D reactions is the $D(d,n)$pD reaction known as the deuteron breakup reaction, which produces a continuum of neutrons at energies below the mono-energetic peak. The neutron energy distribution as a function of angle for the cross section, $\frac{d^2\sigma}{d\Omega dE}$, of the D(*d,n*)pD reaction has been measured using a 6.94-MeV pulsed deuteron beam incident upon a D_2 gas target. The time-of-flight technique was used to determine the energy of the neutrons detected in an array of two lithium glass scintillators and one NE-213 scintillator. The breakup cross section was determined as low as 225-keV neutron energy in the lithium glass detectors.

1 Introduction and Experiment

The D+D reactions yield mono-energetic neutrons from the $D(d,n)^3He$ reaction and a continuum of low-energy neutrons from the $D(d,n)$pD reaction for 7-MeV incident deuterons. The $D(d,n)$pD reaction occurs when the incident deuteron energy is above the threshold of 4.45 MeV, causing one of the deuterons to break up into its constituent particles, the proton and the neutron. This reaction produces a continuum of neutrons at energies lower than the mono-energetic neutrons from the $D(d,n)^3He$ reaction.

In this work, the D(*d,n*)pD reaction has been studied for the purpose of use as a neutron source for the active interrogation of hidden fissile materials. The D+D reaction yields mono-energetic neutrons up to approximately 8 MeV based upon the incident deuteron energy and yields breakup neutrons at lower energies. The D+D reaction, as a neutron source, is attractive because unlike mono-energetic neutron sources that use radioactive materials, such as tritium, special regulations are not required. Neutrons from the $D(d,n)^3He$ reaction have been fully characterized, but neutrons resulting from the breakup reaction have not received this level of attention. Prior to this work, the D(*d,n*)pD cross section was not measured below a neutron energy of 1.5 MeV, where a substantial contribution to the cross section occurs.

[a] e-mail: ar286106@ohio.edu
[b] deceased

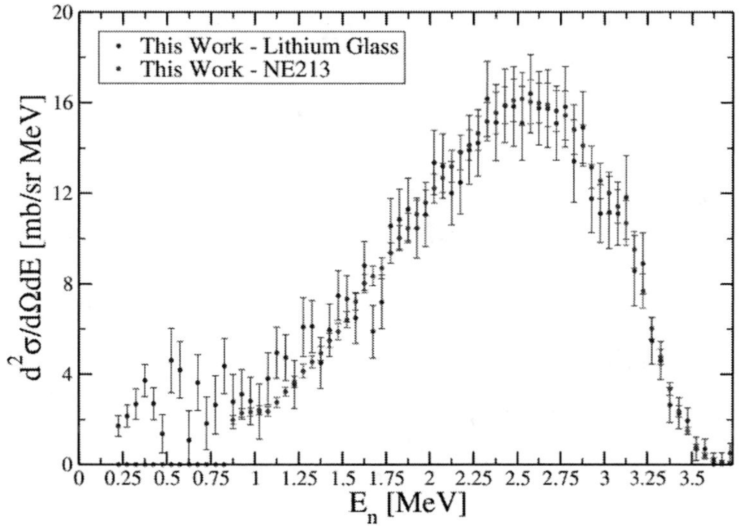

Figure 1. The measured neutron energy distribution of the cross section of the D*(d,n)*pD reaction at 0° for an incident deuteron energy of 6.94 MeV. The data were measured using NE-213 and lithium glass detectors.

2 Results and Discussion

The neutron energy distribution as a function of angle for the cross section, $\frac{d^2\sigma}{d\Omega dE}$, of the D*(d,n)*pD reaction has been measured at the Edwards Accelerator Laboratory of Ohio University, using a 6.94-MeV pulsed deuteron beam incident upon a D_2 gas target. Data for the cross section range between laboratory angles of 0° to 60°, in 5° increments. The 0° distribution is shown in Fig. 1. The time-of-flight technique was used to determine the energy of the neutrons detected in an array consisting of two lithium glass scintillators and one NE-213 scintillator. The NE-213 scintillator has a good neutron detection efficiency over a wide energy range and is practical for use down down to about 1 MeV. Lithium glass scintillators have a non-zero efficiency below 1 MeV, and were utilized to detect continuum neutrons from the breakup reaction down to 225 keV.

The breakup cross section was determined for neutron energies as low as 225 keV in the lithium glass detectors, well below the previous limits. Preliminary results from this work can be found in [1]. Simultaneous fits to the $\frac{d^2\sigma}{d\Omega dE}$ to test the self-consistency of the data are underway, which will allow for extrapolating cross section results to other laboratory angles and neutron energies as well as interpolating between existing results.

Acknowledgements

This project was funded in part by the Defense Threat Reduction Agency (DTRA) through grant number HDTRA1-09-1-0059 and U.S. D.O.E. (NNSA) through grant number DE-NA0001837. The authors would like to thank D.K. Jacobs and D.E. Carter for their assistance with the setup and operation of the accelerator and electronics.

References

[1] A.L. Richard, Master's thesis, Ohio University, Athens, Ohio, 2014.

EPJ Web of Conferences **113**, 08017 (2016)
DOI: 10.1051/epjconf/201611308017

Electron-molecule scattering with analytic static potential approach

Dipti[1] Lalita Sharma[1] Rajesh Srivastava[1,a] and Al Stauffer[2]

[1] *Department of Physics, Indian Institute of Technology Roorkee, Roorkee 247667 India*
[2] *Department of Physics and Astronomy, York University, Toronto, Canada M3J 1P3*

Abstract. A newly proposed analytic method to calculate static potential has been applied to study the elastic scattering of electron with CO_2 molecule. The molecule has been represented as a test case by using $STO-3G$ Gaussian wave functions initially. Results obtained for differential cross-section show good agreement when compared with experimental reported results.

1 Introduction

Optical potential model calculations have proved to be a very successful approach to study elastic electron-atom scattering. However, studying electron scattering from molecules using optical potential model calculations is complicated by the fact that molecules are a multi-centered target with the nuclei of the constituent atoms being a center of charge. Thus one of the most important parts of this scattering calculation is to obtain the static potential, which represents the interaction of the incident electronwith the unperturbed charge distribution of the molecule. This job is very difficult and has been a challenge if multi-center wavefunctions are to be used for the molecular target.

Most of the calculations, therefore, used independent atom model where the electron-molecule scattering results are obtained by first considering the scattering of electron from the individual constituent atoms of the molecule and then getting the results for the molecule in approximate way whose reliability is questionable. In the present work, we adopt an approach where an analytic static potential is first generated from the multi-center molecular wavefunctions which can then be directly utilized in the optical potential method for studying the general case of electron scattering from molecules. Recently we have applied this analytical static potential approach to study electron scattering from water molecules represented by Gaussian wavefunctions [1]. Our method has given a very good description of electron-water molecule elastic scattering.

2 Results

In the present work we have extended our method in studying the elastic electron scattering from the linear molecule CO_2 using such an analytical static potential. We have obtained the spherically

[a] e-mail: rajsrfph@gmail.com

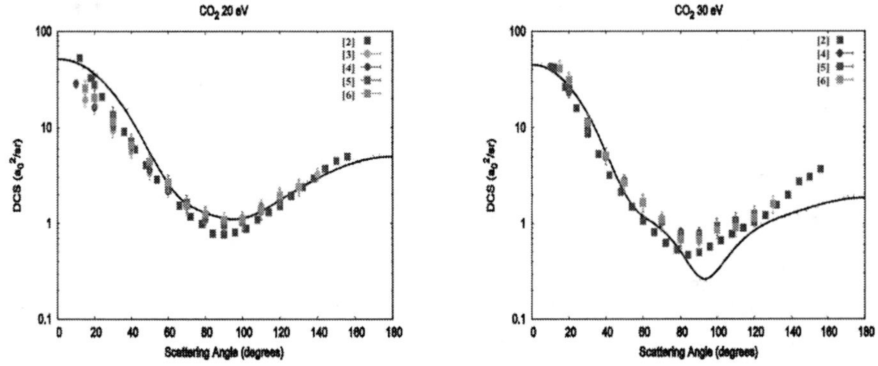

Figure 1. Differential cross-section of the elastic scattering of electron-CO_2 molecule at incident electron energy 20 and 30 eV.

symmetric part of the static interaction between an electron and CO_2 molecule represented by Gaussian wave functions. Since this molecule has no permanent dipole moment, we expect our method to produce accurate results for elastic scattering. As an illustration, we have presented our results for differential cross-section (DCS) of the elastic electron scattering of CO_2 molecule at electron impact energy 20 and 30 eV in the Figure 1. Our results are compared with existing experimental data [2–6] for this process to assess the accuracy of the method and we observed good agreement. Details of the method to obtain the static potential along with the description of full optical potential model can be followed from our earlier paper [1]. With the present work we show that it is possible to carry out the integrations analytically in order to find the spherically symmetric static potential due to the electronic charge distribution of an arbitrary molecule which is represented by Gaussian wavefunctions. Thus the accuracy of this potential is limited only by the accuracy of the molecular wavefunctions. Since our method is analytic , provided the molecule is represented by the well-known Gaussian wavefunctions, it can be applied to arbitrary molecular systems and requires much less in the way of computer resources than methods which depend on direct numerical integration of the wave function. There is no error in our method of evaluating of the static potential of the wave function and thus accurate Gaussian wavefunctions for a molecule will lead to correspondingly accurate potentials.

Acknowledgements

Work is supported by DAE-BRNS Mumbai and CSIR, New Delhi.

References

[1] T. Das, A. D. Stauffer and R. Srivastava, Eur. Phys. J. D **68**, 4 (2014).

[2] T. W. Shyn, W. E. Sharp and G. R. Carignan, Phys. Rev. A **17**, 1855 (1978).

[3] D. F. Register, H. Nishimura and S.Trajmar, J. Phys. B **13**, 1651 (1980).

[4] J. C. Gibson, M. A. Green, K. W. Trantham, S. J. Buckman, P. J. O. Teubner and M. J. Brunger, J. Phys. B **32**, 213 (1999).

[5] I. Kanik, D. C. McCollum and J. C. Nickel, J. Phys. B **22**, 1225 (1989).

[6] H. Tanaka, T. Ishikawa, T. Masai, T. Sagara, L. Boesten, M. Takekawa, Y.Itikawa and M. Kimura, Phys. Rev. A **57**, 1798 (1998).

EPJ Web of Conferences **113**, 08018 (2016)
DOI: 10.1051/epjconf/201611308018

Halo Effective Field Theory of ^{6}He

Arbin Thapaliya[1,a], Chen Ji[2,b], and Daniel Phillips[1,c]

[1] *INPP and Deparment of Physics and Astronomy, Ohio University, Athens, Ohio 45701 USA*
[2] *TRIUMF, 4004 Wesbrook Mall, Vancouver, BC V6T 2A3, Canada*

Abstract. ^{6}He has a cluster structure with a tight ^{4}He (α) core surrounded by two loosely bound neutrons (n) making it a halo nucleus. The leading-order (LO) Halo Effective Field Theory (EFT) [1, 2] calculations using momentum-space Faddeev equations pertinent to a bound ^{6}He were carried out in [3]. In this work, we investigate ^{6}He up to next-to-leading order (NLO) within Halo EFT.

1 Introduction

In halo nuclei, the valence nucleon distribution extends much further out relative to the size of the tightly bound core. The cluster structure of ^{6}He exhibits separation of scales and is suitable for a Halo EFT treatment [1, 2]. ^{6}He constitutes a Borromean three-body system, i.e., no two of the three bodies bind independently, but all three together can form a bound state.

2 Three-body formalism

The valence neutrons of ^{6}He interact with the α-core predominantly through a p-wave ($^2P_{3/2}$) resonance while the interaction between the two neutrons is in the relative $^1S_0$ partial wave. The three-body t-matrix for the scattering (unphysical) of a neutron off the $n\alpha$ pair can be represented as a set of Feynman diagrams shown in figure 1. In [3], the authors find that a three-body input is required for

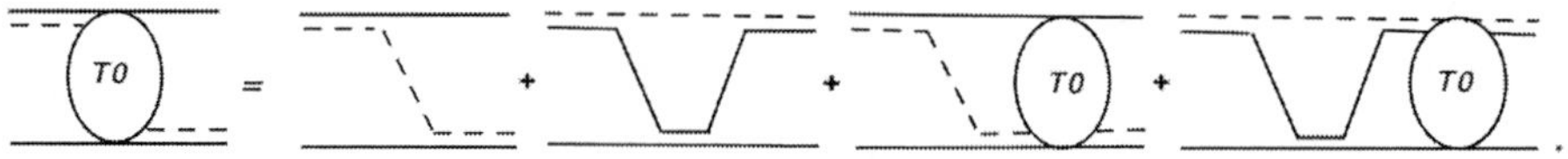

Figure 1. Integral equation for the LO three-body t-matrix. The binding of ^{6}He is represented as the pole of this t-matrix at negative energy. Solid (Dashed) line represents neutron (α). Double solid line (Solid and dashed) line represents LO nn ($n\alpha$) dimer propagator.

a consistent renormalization of the LO integral equation. (Cf. the calculation in a different Halo EFT power counting of Ref. [4].) After introducing a three-body force, they show that the renormalized

[a] e-mail: at311509@ohio.edu
[b] e-mail: jichen@triumf.ca
[c] e-mail: phillid1@ohio.edu

LO Faddeev components F_n and F_α are cutoff independent for large cutoffs. The Faddeev component F_n (or equivalently $T0$) will serve as an input to the NLO three-body t-matrix as shown in figure 2. Therefore, having an approximate analytic form of $F_n(q)$ will be useful for the NLO calculations. We provide an analytic fit that closely resembles the actual F_n at large q and can be represented as

$$F_n(q) \cong 945\frac{1}{q^{0.49}} \times \cos(0.925q^{0.088}\ln(q) + 0.15). \tag{1}$$

3 NLO calculations

The NLO piece of the ${}^1S_0$ nn dimer propagator, the NLO piece of the ${}^2P_{3/2}$ $n\alpha$ dimer propagator and the contact $n\alpha$ vertex in the ${}^2S_{1/2}$ channel enter the NLO three-body amplitude, $T1$, for the $nn\alpha$ system. $T1$ can be represented diagrammatically as shown in figure 2. We have investigated

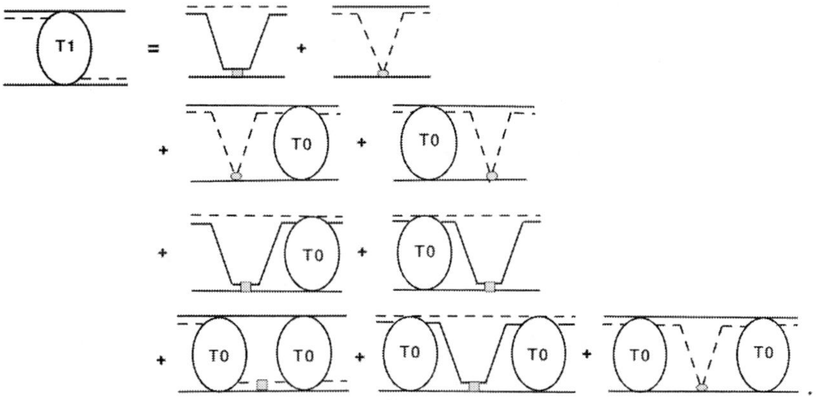

Figure 2. Diagrams that contribute to the NLO three-body t-matrix. Solid (Dashed) line represents neutron (α). Double solid line (Solid and dashed) line represents nn ($n\alpha$) dimer. Propagators with gray squares represent the NLO pieces of the nn and $n\alpha$ dimers respectively. The gray ovals represent the ${}^2S_{1/2}$ $n\alpha$ vertex.

the divergences associated with each of the above NLO diagrams. The last three diagrams have the highest degree of divergence, and are the only ones relevant when considering NLO contributions in the bound-state problem. We will renormalize these diagrams using an NLO three-body $nn\alpha$-counterterm. With the renormalized result in hand, we will study the NLO Faddeev components and investigate how the NLO corrections affect the ^{6}He radii. Finally, we will also explore the role of the ${}^2P_{1/2}$ channel of the $n\alpha$ interaction in the binding of ^{6}He.

Acknowledgements

This work was supported in parts by the U.S. Department of Energy under grant DE-FG02-93ER40756, the Natural Sciences and Engineering Research Council (NSERC), and the National Research Council of Canada.

References

[1] C. A. Bertulani, H. W. Hammer, and U. Van Kolck, Nucl. Phys. A **712**, 37 (2002).
[2] P. F. Bedaque, H. W. Hammer, and U. van Kolck, Phys. Lett. B **569**, 159 (2003).
[3] C. Ji, Ch. Elster, and D. R. Phillips, Phys. Rev. C **90**, 044004 (2014).
[4] J. Rotureau and U. van Kolck, Few Body Syst. **54**, 725 (2013).